SEVEN-MEMBERED HETEROCYCLIC COMPOUNDS CONTAINING OXYGEN AND SULFUR

This is the twenty-sixth volume in the series

THE CHEMISTRY OF HETEROCYCLIC COMPOUNDS

THE CHEMISTRY OF HETEROCYCLIC COMPOUNDS

A SERIES OF MONOGRAPHS

ARNOLD WEISSBERGER and EDWARD C. TAYLOR

Editors

SEVEN-MEMBERED HETEROCYCLIC COMPOUNDS CONTAINING OXYGEN AND SULFUR

Edited by

Andre Rosowsky

CHILDREN'S CANCER RESEARCH FOUNDATION, INC.
BOSTON, MASSACHUSETTS

WILEY-INTERSCIENCE
A division of

JOHN WILEY & SONS, INC.
NEW YORK · LONDON · SYDNEY · TORONTO

Library of Congress Cataloging in Publication Data:

Rosowsky, Andre.
Seven-membered heterocyclic compounds containing oxygen and sulfur.

(The Chemistry of heterocyclic compounds, v. 26)
Includes bibliographical references.
1. Oxepins. 2. Thiepins. I. Title.

QD405.R68 547′.595 70-39735

ISBN 0-471-38210-8

Printed in the United States of America.

10 9 8 7 6 5 4 3 2 1

The Chemistry of Heterocyclic Compounds

The chemistry of heterocyclic compounds is one of the most complex branches of organic chemistry. It is equally interesting for its theoretical implications, for the diversity of its synthetic procedures, and for the physiological and industrial significance of heterocyclic compounds.

A field of such importance and intrinsic difficulty should be made as readily accessible as possible, and the lack of a modern detailed and comprehensive presentation of heterocyclic chemistry is therefore keenly felt. It is the intention of the present series to fill this gap by expert presentations of the various branches of heterocyclic chemistry. The subdivisions have been designed to cover the field in its entirety by monographs which reflect the importance and the interrelations of the various compounds, and accommodate the specific interests of the authors.

In order to continue to make heterocyclic chemistry as readily accessible as possible new editions are planned for those areas where the respective volumes in the first edition have become obsolete by overwhelming progress. If, however, the changes are not too great so that the first editions can be brought up-to-date by supplementary volumes, supplements to the respective volumes will be published in the first edition.

ARNOLD WEISSBERGER

Research Laboratories
Eastman Kodak Company
Rochester, New York

EDWARD C. TAYLOR

Princeton University
Princeton, New Jersey

Preface

The field of heterocyclic chemistry has undergone many significant developments during recent times, both in theory and at the more practical levels of laboratory synthesis and commercial application. Not the least among these developments has been a rapidly growing interest on the part of chemists throughout the world in the chemistry of heterocyclic systems other than the classically popular five- and six-membered varieties.

In recognition of the impact made upon the chemical literature by this trend, there appeared in 1964 in the Heterocyclic Compounds Series a two-part volume which contained upward of 4000 references and dealt entirely with three- and four-membered rings. After the publication of this volume, it became apparent that heterocyclic systems containing *more than six members* had suffered the same neglect as their small-ring counterparts, and that a survey of such systems was very clearly justified. In the present volume a start in this direction will be made by giving an account of the current state of knowledge concerning *seven-membered* oxygen-containing rings (oxepins) and sulfur-containing rings (thiepins). Nitrogen-containing seven-membered rings (azepines) and mixed heteroatomic seven-membered rings (oxathiepins, oxazepines, thiazepines, etc.) will form the subject of subsequent volumes to be published at a later date.

The unique physical and chemical properties conferred upon the oxepin ring by its valence tautomeric character, which is reminiscent of cyclooctatetraene, have engendered considerable theoretical excitement. Particularly elegant contributions have been made in this area by Vogel and his collaborators, as indicated in Chapter I. Other noteworthy recent studies—only a few of the many that have been published during the past decade—are those of van Tamelen and Carty, and of Paquette and co-workers, on the photochemical synthesis and photochemically induced rearrangement of oxepins; of Rhoads and Cockroft on the synthesis of 2,5-dihydrooxepin by thermolytic rearrangement of *cis*-2-vinylcyclopropanecarboxaldehyde; and of Schweizer and Parham, as well as Ando and co-workers, on the thermal isomerization of 2-oxanorcaranes to 2,3-dihydrooxepins. The important investigations of these workers, and others too numerous to mention here, attest to the fact that heterocyclic chemistry has come of age.

Like the monocyclic oxepins, condensed oxepins have received extensive attention during recent years. Of the many known systems composed of two rings (Chapter II), the isomeric 1-benzoxepins, 2-benzoxepins, and 3-benzoxepins have been studied most thoroughly. Again, only a few outstanding investigations can be cited here. They include the work of Vogel and co-workers, as well as Sondheimer and Shani, on the synthesis of 1-benzoxepin by valence tautomerization of 9,10-oxidonaphthalene; of Schweizer and co-workers on the synthesis of 1-benzoxepin and 2,3-dihydro-1-benzoxepin via several ingenious applications of the Wittig reaction; and of Dimroth and co-workers, as well as Jorgenson and more recently Ziegler and Hammond, on the preparation and chemical transformations of 3-benzoxepins.

Condensed oxepin systems composed of three rings (Chapter III) and more than three rings (Chapter IV) bear witness to the seemingly limitless variety of chemical structures that can be generated in the laboratory through the creative efforts of imaginative and perseverant synthetic organic chemists. No fewer than sixty different systems containing three or more rings are reviewed (excluding those derived from, or related to, complex natural products, which are treated in separate chapters), and their number is multiplying at an astonishing rate. Work of fundamental theoretical importance which has been done with polycyclic oxepins includes, for example, the classic preparation and characterization by Linstead and Doering of the six possible stereoisomeric perhydrodiphenic anhydrides (dodecahydrodibenz[*c*,*e*]oxepin-5,7-diones), and elegant spectroscopic studies by Mislow and co-workers with optically active, sterically-hindered 5,7-dihydrodibenz[*c*,*e*]oxepins of the bridged biphenyl type. Far from being merely of routine interest, these investigations have played a significant role in the formulation of modern conformational theory.

Oxepin chemistry is not devoid of important practical aspects, as evidenced by the eminent position enjoyed in the polymer field by two members of the oxepin family, the 1,3-dioxepins (Chapter V) and ε-caprolactones. The latter class of compounds, whose commercial importance is reflected in a voluminous patent literature, was to have been reviewed in a separate chapter, together with the chemistry of adipic anhydrides. Regrettably, illness has prevented the author of this chapter from completing his task in time for publication in the present volume. If possible, this material will appear at a later date as part of the Heterocyclic Compounds Series.

Seven-membered oxygen heterocycles also occupy a prominent place in the chemistry of natural products. In the terpene field (Chapter VI), a number of polycyclic ε-lactones of plant origin have been described, one well-known example being the bitter principle limonin. In addition, many oxepins are encountered in the terpene literature, as well as among steroids (Chapter VII), in the form of ε-lactones derived from six-membered ketone rings via Baeyer-Villiger oxidation. Many of these oxidative degradation products have performed key roles in the structural elucidation of the parent terpene or steroid. In the domain of sugar chemistry (Chapter VIII), numerous 1,6-anhydrohexoses and 2,7-anhydroheptuloses are known which can be viewed as bridged oxepin derivatives, the naturally occurring compound sedoheptulosan being a familar example of the latter category. Finally, oxepins are encountered even among alkaloids (Chapter IX), the best-known being strychnine and cularine.

Less widely studied but nonetheless of considerable theoretical and practical interest are the monocyclic seven-membered sulfur heterocycles (Chapter X) and their condensed systems (Chapter XI). Although thiepin itself apparently lacks the relative stability of oxepin and has thus far successfully eluded synthesis, the sulfone of thiepin has recently been prepared and subjected to thorough physicochemical investigation by Mock and co-workers. Sulfoxides and sulfones of reduced thiepins and dithiepins have been examined in some detail by a number of workers because of the insight which these compounds might afford into the stereochemical and conformational aspects of sulfur *d*-orbitals. Among the condensed thiepins, 1-, 2-, and 3-benzothiepins, as well as their sulfoxides and sulfones, have been accorded much attention in the past few years. Sulfur extrusion reactions have been of particular interest as evidenced by the continuing studies in this area by Traynelis and co-workers, among others. Also of great interest are the recently described syntheses by Schlessinger and Ponticello of thieno-[3,4-*d*]thiepin and furo[3,4-*d*]thiepin, and of the corresponding sulfoxides and sulfones. These studies, and other which are undoubtedly in progress as this is being written, are unfolding many of the most exciting new horizons in sulfur chemistry.

I acknowledge with admiration and gratitude the efforts of the eight expert authors who collaborated so patiently with me in the preparation of this volume. I also wish to give heartfelt thanks to Dr. Arnold Weissberger and Dr. Edward C. Taylor for their stimulating encouragement throughout the planning and execution of this work, and to the publishers and their staff for their efficient handling of the project at every stage. Finally, special thanks are due to my own family for their devoted support from the outset of this long and very arduous endeavor.

ANDRE ROSOWSKY

Laboratories of Organic Chemistry
Children's Cancer Research Foundation
Boston, Massachusetts
October 1971

Contents

Tables

SEVEN-MEMBERED HETEROCYCLIC COMPOUNDS CONTAINING OXYGEN AND SULFUR

CHAPTER I

Oxepins and Reduced Oxepins

ANDRE ROSOWSKY

Children's Cancer Research Foundation and Department of Biological Chemistry, Harvard Medical School, Boston, Massachusetts

A. Oxepins

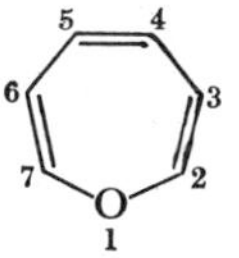

R.I. 356

1. Theoretical Interest

The preparation of oxepin (**1**) has represented a considerable synthetic challenge for a number of years, and has stimulated the ingenious efforts of many outstanding chemists. The theoretical impetus for these efforts was first supplied by the qualitative observations (36) that **1** can be regarded as an analog of cyclooctatetraene in the same sense that furan is an analog of benzene. The possibility of such an electronic relationship was supported by molecular orbital calculations suggesting that **1** might possess a certain amount of aromatic character, despite the fact that it appears to violate the $(4n + 2)$ requirement for aromaticity (55). By analogy with the closely related case of cycloheptatriene and norcaradiene, it was also postulated (36) that **1** represents a "valence tautomer" of hypothetical epoxy diene **1A**. This

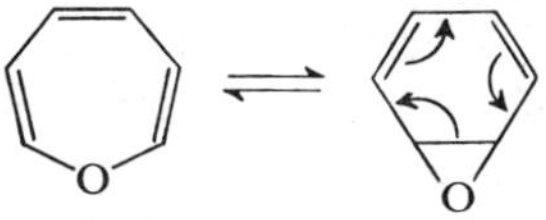

1 **1A**

far-reaching concept must be credited for the remarkably rapid development of the chemistry of oxepins and other heterocyclic tropilidene analogs which has occurred during the last several years (65, 69).

2. Preparation

The first successful synthesis of the unsubstituted parent compound oxepin (**1**) was reported in 1964 by Vogel, Schubart, and Boell (66). 1,2-Dibromo-4,5-epoxycyclohexane (**2**) was subjected to the action of 1,5-diazabicyclo-[3.4.0]non-5-ene, a compound which has gained recognition recently as an extraordinarily mild reagent for dehydrohalogenation. The formation of **1** was postulated to occur via thermal reorganization of transient epoxydiene **1A**, the previously predicted (36) "valence tautomer" of **1**. In a subsequent paper, Vogel and co-workers (64) once again reported the conversion of **2** into **1**, this time with sodium methoxide in refluxing ether. Physical evidence was cited, however, which indicated that **1** and **1A** were both present in solution and underwent rapid interconversion at ordinary temperatures.

Br Br N N or NaOMe O O O

2 **1A** **1**

The synthetic approach eventually brought to fruition by Vogel and co-workers had actually been explored several years earlier by Meinwald and Nozaki (36), with γ-collidine or the sodium salt of ethylene glycol as dehydrohalogenating agents. Unfortunately, the reaction conditions employed by the latter authors were too strenuous, so that phenol was the only product which could be isolated and identified. This outcome can be understood readily now that it is known (66) that **1** undergoes facile thermal rearrangement to phenol at temperatures above 70°.

Several other unsuccessful approaches to the synthesis of **1** were described be Meinwald and Nozaki (36). A common feature of all these efforts is that they were directed, in each case, toward the preparation of epoxydiene **1A**, which was expected to undergo thermal isomerization to **1**. In one attempted route, for example, it was hoped that "α-benzene tetrachloride" (**3**) could be converted into **1** by way of epoxide **4**, which was to be dehalogenated. This approach had to be abandoned because **3** proved unexpectedly resistant to epoxidation with either peracetic or trifluoroperacetic acid.

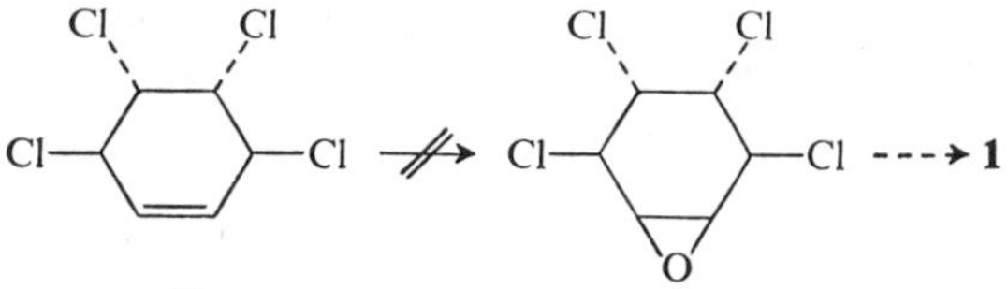

Cl Cl Cl Cl
Cl—Cl ↛ Cl—Cl ---→ 1
O
3 4

In a second approach, 1,4-cyclohexadiene was converted into *trans*-4,5-diacetoxycyclohexene (**5**) by successive treatment with peracetic acid and acetic anhydride. Oxidation of **5** with perbenzoic acid then yielded epoxide **6**. Pyrolysis of **6** at 505° afforded phenol, in about 30% yield, as the only characterizable product.

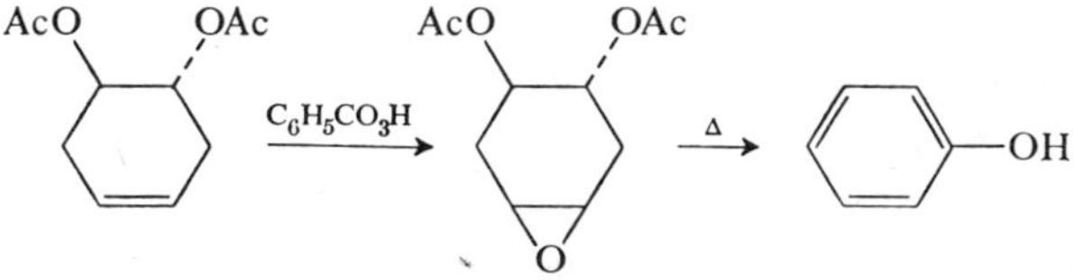

AcO OAc AcO OAc
$C_6H_5CO_3H$ → Δ → —OH
O
5 6

In still another variation, 1,4-cyclohexadiene was converted into 4,5-epoxycyclohexene (**7**) by oxidation with monoperphthalic acid. Allylic bromination of **7** with *N*-bromosuccinimide afforded a monobromo compound which was arbitrarily assigned structure **8**. Efforts to dehydrobrominate **8** by treatment with silver oxide in ether once again yielded only phenol.

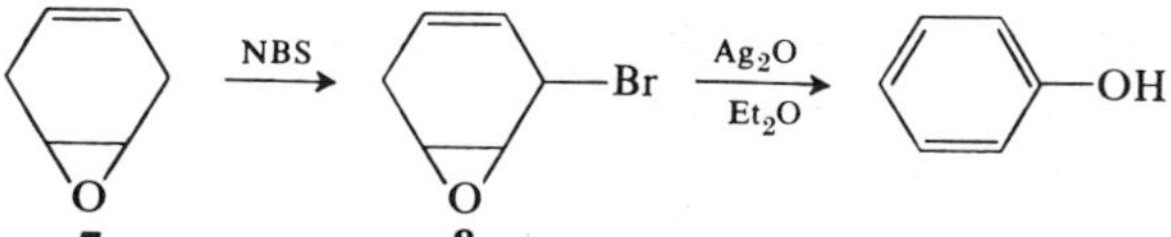

NBS → —Br Ag_2O / Et_2O → —OH
O O
7 8

Mention may also be made here of an ingenious plan to generate **1** by epoxidation of 7-ketonorbornadiene (**9**), followed by thermal decarbonylation of the resulting epoxy ketone, **10**. This attractive scheme was thwarted by the unexpected failure of **9** to undergo epoxidation under a variety of conditions (30).

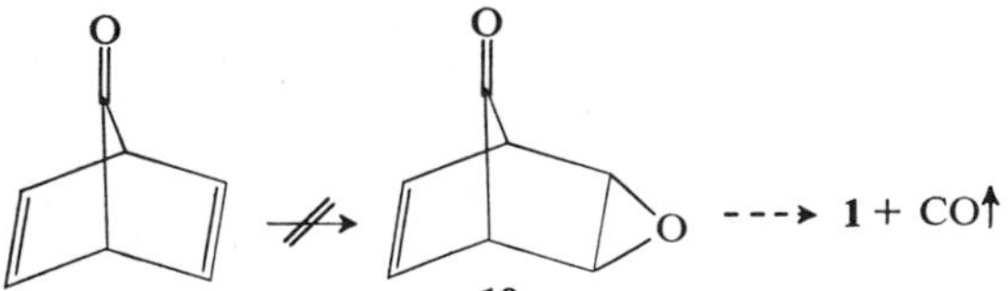

O O
↛ O ---→ 1 + CO↑
9 10

The basic route developed by Vogel and co-workers for the synthesis of **1** has been extended to several substituted oxepin derivatives. Thus, 4,5 epoxy-4-methylcyclohexene (**11**) was brominated, and the resulting dibromide, **12**, was converted into 2-methyloxepin (**13**) upon treatment with sodium methoxide (65, 70). Alternatively, allylic bromination of **11** afforded 6-bromo-4,5-epoxy-4-methylcyclohexene (**14**), which was transformed into **13** under the influence of potassium *tert*-butoxide (65, 70). In a similar fashion, bromination of 4,5-epoxy-4,5-dimethylcyclohexene (**15**) afforded dibromide **16**, and dehydrobromination of the latter with sodium methoxide led to 2,7-dimethyloxepin (**17**) (65, 66). The isomeric 4,5-dimethyloxepin (**18**) has also been prepared via epoxides **19** and **20** (65). Lack of stability except when in solution has precluded the isolation of **18** in the solid state.

11; $R_1 = R_2 = R_3 = H, R_4 = Me$
15; $R_1 = R_2 = H, R_3 = R_4 = Me$
19; $R_1 = R_2 = Me, R_3 = R_4 = H$

Br_2; NaOMe, Et_2O

12; $R_1 = R_2 = R_3 = H, R_4 = Me$
16; $R_1 = R_2 = H, R_3 = R_4 = Me$
20; $R_1 = R_2 = Me, R_3 = R_4 = H$

NBS; KOBu-*tert*, Et_2O

14

13; $R_1 = R_2 = R_3 = H, R_4 =$
17; $R_1 = R_2 = H, R_3 = R_4 =$
18; $R_1 = R_2 = Me, R_3 = R_4$

According to Vogel and Guenther (65), substitution at the 2-position by a group capable of entering into conjugation with the oxepin ring produces a stabilizing effect. This was demonstrated by the synthesis of 2-acetyloxepin (**21**) from 1-acetyl-1,2-epoxy-4,5-dibromocyclohexane (**22**). Dehydrobromination was effected smoothly in the presence of 1,5-diazabicyclo[4.3.0]non-5-ene. It is noteworthy that attempts to dehydrobrominate this particular epoxide with sodium methoxide or potassium *tert*-butoxide led only to polymeric products. Unlike 2-methyloxepin (**13**), which is in equilibrium with a significant proportion of its "valence tautomer" at ordinary temperatures, **21** shows no evidence of being in equilibrium with the corresponding epoxide form.

22 → **21**

Several interesting attempts to generate the oxepin ring system by expansion of the six-membered pyran ring in pyrylium perchlorate (**23**) and 2,6-diphenylpyrylium perchlorate (**24**) were reported by Whitlock and Carlson (68).

Both **23** and **24** gave rise to complex mixtures of products upon treatment with diazomethane. In the case of **24**, the only identifiable product was found to by 2,6-diphenyl-4*H*-pyran (**25**). An authentic specimen of **25** was prepared by reduction of **24** with sodium borohydride. No evidence could be found for the presence of 2,7-diphenyloxepin (**26**) under the conditions studied. When **24** was treated with ethyl diazoacetate in acetonitrile at 0°, a complex mixture was formed once again, from which a 9% yield of **25** could be isolated by thin-layer chromatography. The yield of **25** was raised to 14% upon addition of triethylamine. If the reaction was conducted in the presence of one equivalent of 2,6-lutidine as a proton acceptor, the major product (18%) recoverable from the reaction mixture proved to be 2,6-diphenyl-4-carbethoxymethylene-4*H*-pyran (**27**), rather than the hoped-for 4-carbethoxy-2,7-diphenyloxepin (**28**). Several plausible reasons were advanced to explain the failure of this seemingly rational synthetic approach to oxepin synthesis.

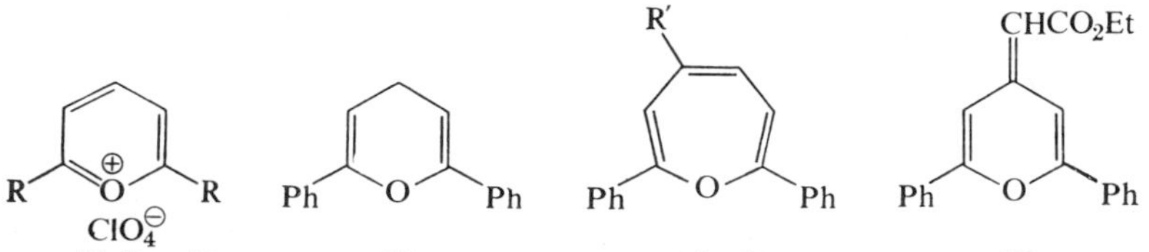

R R $ClO_4^{\ominus}$ Ph Ph Ph Ph R′ Ph Ph $CHCO_2Et$

23; R=H **24**; R=Ph **25** **26**; R′=H **28**; R′=CO_2Et **27**

Still another approach to the problem of oxepin synthesis was devised by Masamune and Castellucci (33). Addition of dichlorocarbene to 3-acetoxy-2,3-dihydropyran (**29**) afforded a mixture of acetates **30** and **31**. The mixture was saponified readily by the action of sodium methoxide at room temperature, and the resulting mixture of diols **32** and **33** was converted into a single ketone, **34**, by gentle oxidation with chromic acid (Jones reagent). Upon heating, **34** underwent dehydrochlorination and rearrangement to an acid- and base-sensitive compound, to which was assigned structure **35**. Although **35** could, in principle, be written as the tautomeric species 3-chloro-6-hydroxyoxepin (**35A**), the authors concluded on the basis of uv

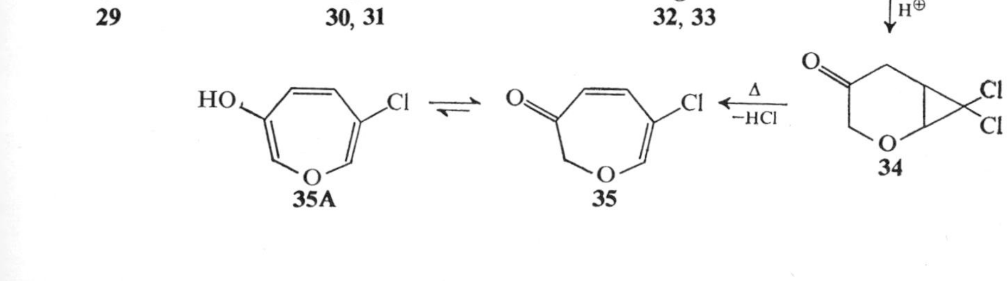

AcO **29** $:CCl_2$ AcO Cl Cl **30, 31** NaOMe HO Cl Cl **32, 33** CrO_3 $H^{\oplus}$ O Cl Cl **34** Δ −HCl O Cl **35** ⇌ HO Cl **35A**

and nmr evidence that the product actually exists almost exclusively in keto form **35**.

Westoo reported in 1959 that the condensation of 2,5-hexanedione with α-cyanoacetamide gave a product which was formulated tentatively as 2-amino-3-carboxamido-4,7-dimethyloxepin (**36**) (67). Hydrolysis of this compound with concentrated hydrochloric acid gave a product assumed to be 2-amino-4,7-dimethyloxepin-3-carboxylic acid (**37**). Buchanan and co-workers (6) reinvestigated the structure of the alleged **36** via X-ray crystallographic analysis on the hydrobromide salt. The results of this study demonstrated that the presumed oxepins were actually cyclopentadiene derivatives **38** and **39**, respectively.

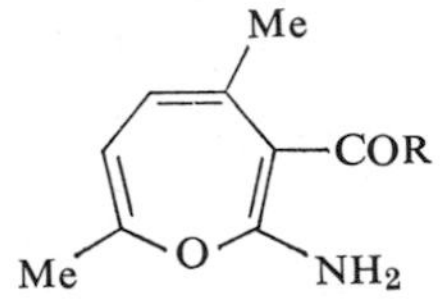

Me, COR, NH_2, O, Me

RCO, NH_2, Me, COMe

36; R = NH_2
37; R = OH

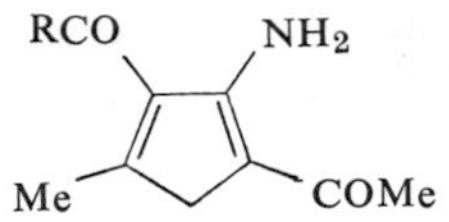

38; R = NH_2
39; R = OH

A very interesting method of synthesis of substituted oxepins which promises to be of considerable preparative value is that developed recently by Prinzbach and co-workers (49). Irradiation of the Diels-Alder adduct of furan and dimethyl acetylenedicarboxylate afforded a tetracyclic isomer, which was assigned structure **40**. Above 100°, this compound was found to undergo thermal rearrangement to a product formulated as dimethyl oxepin-4,5-dicarboxylate (**41**). Catalytic reduction of **41** afforded dimethyl 2,3,4,5,6,7-hexahydrooxepin-4,5-dicarboxylate, which was identified by comparison

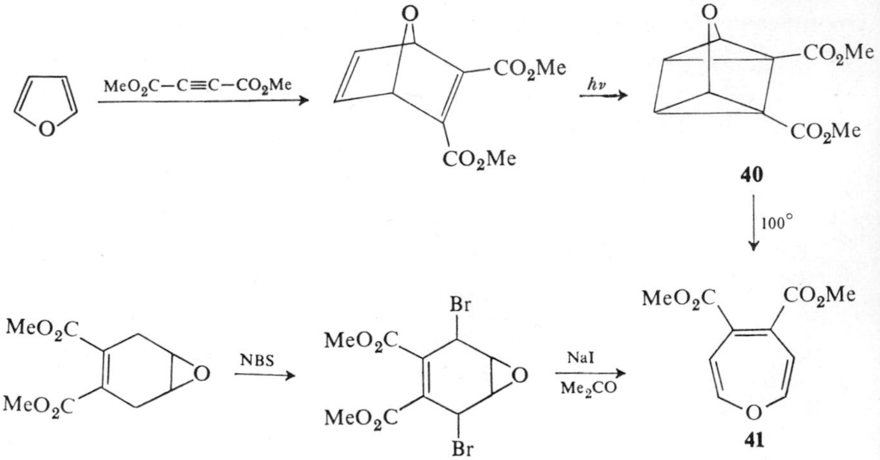

$MeO_2C{-}C{\equiv}C{-}CO_2Me$ → ; $h\nu$ → **40**; 100° ↓ **41**; NBS → ; NaI, Me_2CO →

O, CO_2Me, CO_2Me; O, CO_2Me, CO_2Me

MeO_2C, MeO_2C, O; Br, MeO_2C, MeO_2C, O, Br; MeO_2C, CO_2Me, O with an authentic specimen. An alternate route to **41**, reported more recently by Vogel and co-workers (65), involved allylic bromination of dimethyl 4,5-epoxy-3,6-dihydrophthalate with *N*-bromosuccinimide, followed by dehydrobromination with sodium iodide in acetone.

Perhaps the most elegant and direct method devised to date for the synthesis of oxepins is that reported in 1967 by van Tamelen and Carty (63). Epoxidation of bicyclo[2.2.0]hexa-2,5-diene ("Dewar benzene") with *m*-chloroperbenzoic acid afforded a 75% yield of 2,3-epoxybicyclo[2.2.0]hex-5-ene (**42**), which was transformed into oxepin (**1**) upon irradiation or heating. Although substituted oxepins have not been prepared by this approach as yet, it would seem reasonable to suppose that a great variety of such compounds could be obtained from the corresponding benzene derivatives, and that these reactions would be of profound theoretical interest.

$m\text{-}ClC_6H_4CO_3H$ → **42** $\xrightarrow{h\nu \text{ or } \Delta}$ **1**

3. Physical Properties

a. *Infrared Absorption*

As of the time that this review was being written, no systematic ir spectrophotometric studies had been reported with oxepin and its simple substituted derivatives. It may be assumed, in the absence of other published data, that the value of 1660 cm^{-1} (6.02 μ) given by Vogel and co-workers for the —CH=CH—O— (enol ether) double bond stretching frequency in 2,7-dimethyloxepin (**17**) (66) is typical for this type of compound.

b. *Ultraviolet Absorption*

In their initial account of the synthesis of oxepin (**1**), Vogel and co-workers (66) stated that this compound absorbs at λ_{max} 271 mμ (log ε 3.15) in cyclohexane. In the same paper, the absorption maximum of 2,7-dimethyloxepin (**17**) was said to occur at 297 mμ (log ε 3.25), a bathochromic displacement which appeared to be consistent with alkyl substitution at each of the terminal sites of the conjugated triene system. However, in an extension of the original work, Vogel and co-workers (64) reported that, in isooctane solution, **1** exhibits a maximum at 271 mμ and also a somewhat weaker shoulder at λ_{max} 305 mμ (log ε 2.95). They concluded that the 271 mμ peak previously assigned to **1** should, in fact, be ascribed to the "valence tautomeric" epoxydiene form, whose concentration is solvent-dependent, and that **1** is responsible only for the shoulder at 305 mμ. From their analysis of the uv spectral data, they estimated that, in isooctane solution, **1** and **1A** are present in a ratio of approximately 7:3. In 85% water/15% methanol solution, on the other hand, the ratio was estimated to be about 9:1. It is also of interest that 2,7-dimethyloxepin (**17**), in contrast to **1**, did not exhibit this type of solvent effect, presumably because tautomerization to an epoxydiene form does not occur to an appreciable extent at room temperature. 2-Methyloxepin (**13**) was found to resemble **1**, rather than **17**, in that the uv spectrum showed the presence of two "valence tautomeric" components in varying proportions, depending on the solvent used (65, 70). On the other hand, 2-acetyloxepin (**21**) exhibited uv spectral features indicating the virtual absence of an epoxydiene form. In cyclohexane, this compound absorbed at λ_{max} (log ε) 222 (4.04), 315 (3.40), and 360 (3.08) mμ, and the spectrum was essentially unaffected by changes of solvent (65). Another oxepin derivative on which some uv data is available is dimethyl 4,5-oxepindicarboxylate (**41**), which Prinzbach and co-workers (49) reported as having λ_{max} 304 mμ. According to Vogel and Guenther (65), this compound, like **1** and **13**, is in rapid equilibrium with its "valence tautomer," as evidenced by the strong solvent-dependent character of its uv absorption bands.

c. *Nuclear Magnetic Resonance*

Vogel and co-workers (66) noted that the nmr spectrum of oxepin (**1**) bears a strong resemblance to that of cycloheptatriene. The spectrum of **1** was observed to consist of three multiplets of equal area, centered at τ 3.9, τ 4.4, and τ 4.7, respectively. The spectrum of 2,7-dimethyloxepin (**17**) was found to exhibit only two multiplets of equal area, centered at τ 4.1 and τ 4.7, and also a singlet at τ 8.1, corresponding to the two methyl groups.

In an extension of their original nmr analysis, Vogel and co-workers (64) determined that the "multiplet" at τ 4.7 was actually a doublet with superimposed fine structure. This signal was assigned to the α-protons, while the τ 4.4 and τ 3.9 multiplets were assigned to the γ-protons and β-protons, respectively. Furthermore, the nmr spectrum of **1** was studied by Vogel and co-workers (64) in relation to the possible effect of temperature upon the rate of interconversion between **1** and its "valence tautomer," **1A**. When the nmr spectrum was determined at 105°, the α-proton doublet at τ 4.7 coalesced into a broad singlet and was displaced *upfield*. At the same time, the γ-proton multipliet at τ 4.4 was broadened too, but was displaced *downfield*. This effect was not observed at 105° with **17**. The strong temperature dependence of the nmr spectrum of **1** was ascribed by Vogel and co-workers to the existence of a rapid tautomeric equilibrium between **1** and **1A**.

Additional studies concerning the nmr spectral characteristics of **1** have been reported by Guenther and Hinrichs (20–24). When a solution of **1** in a 2:1 mixture of bromotrifluoromethane and *n*-pentane was cooled below −113°, there was observed a general broadening of the spectrum, splitting of the α-proton signal, and finally superposition of a second spectral pattern consistent with the presence of **1A**.

The nmr spectrum of 2-methyloxepin (**13**) was examined at various temperatures in bromotrifluoromethane by Guenther, Schubart, and Vogel (70). At −119°, the signal initially observed at τ 8.20 was reportedly split into two signals of unequal intensity at τ 8.09 and τ 8.45, respectively. The lower field signal was assigned to the 2-methyl group in **13**, while the higher field signal was assigned to the bridgehead methyl group in the epoxy diene tautomer. The complete coalescence of these peaks at 37° was attributed to the fact that, at this temperature, the rate of interconversion of the two tautomeric forms is very rapid on the nmr time scale.

The nmr spectrum of 2,7-dimethyloxepin (**17**) was analyzed by Vogel and co-workers (24, 65, 66). In carbon disulfide solution at ordinary temperatures, signals were observed at τ 4.19, τ 4.75, and τ 8.19. The τ 4.19 resonance was assigned to the two γ-protons, and the τ 4.75 signal was ascribed to the two β-protons. Neither these signals, nor the τ 8.19 signal produced by the methyl groups, were broadened significantly upon lowering of the temperature, even to −110°. Moreover, whereas the nmr patterns of **1** and **13** were manifestly solvent-dependent, the spectrum of **17** remained unaffected by changes of solvent. These characteristics are consistent with the relatively high degree of stability ascribed to **17**.

According to Vogel and co-workers (65), the two methyl groups in 4,5-dimethyloxepin (**18**) produce a resonance signal at τ 8.16, while the two α-protons and two β-protons give rise to absorption at τ 5.68 and τ 4.2, respectively. The spectrum of **18** is both temperature-dependent and solvent-dependent, and the extent of this effect indicates that the epoxy diene–oxepin equilibrium lies preponderantly on the side of the epoxy diene in this case. The nmr spectrum of dimethyl 4,5-oxepindicarboxylate (**41**) was likewise examined by Vogel and co-workers (65) and found to exhibit the features of a "valence tautomeric" equilibrium favoring the epoxide form. The two α-protons and two β-protons in **41** show peaks at τ 5.10 and τ 3.44, respectively, at room temperature. The shift to lower field relative to **18** is presumably the result of deshielding by the two electron-withdrawing carbomethoxy groups.

It is interesting that although two carbomethoxy groups at the γ,γ′-positions of the oxepin ring do not provide sufficient resonance energy to favor the oxepin form over the epoxy diene form, the presence of a single acetyl group at the α-position in 2-acetyloxepin (**21**) suffices to make the epoxide component undetectable. This is indicated by the nmr spectrum of **21**, which is reported to be independent of temperature and solvent (65).

In summary, nmr spectrometry must be credited for providing an enormous amount of useful information concerning the physical and chemical character of the oxepin ring. Modern refinements in nmr technique, such as double resonance, computer simulation of complex spectra, and the use of high field magnets, will undoubtedly lead to a still greater understanding of the geometry and electronic fine structure of this extremely interesting seven-membered heterocyclic system.

d. *Thermodynamic Properties*

Guenther (65, 70) has performed a number of important calculations concerning the "valence tautomeric" equilibrium between oxepins and epoxy dienes on the basis of nmr data. From the changes in line shape of the spectrum of oxepin (**1**) itself at various temperatures, the enthalpy for the isomerization of **1A** to **1** was determined to be 1.7 ± 0.4 kcal/mole, and the entropy difference between the two tautomers was found to be 10.5 ± 8.3 cal/(deg)(mole). Although the energy of **1A** is lower than that of **1** by 1.7 kcal/mole, the equilibrium concentration of **1** is considerably greater than that of **1A**, apparently as a result of the significant entropy gain associated with the rearrangement. The free energy term for this equilibrium was found to be −1.3 kcal/mole. By comparison of the experimentally observed enthalpies with the theoretically derived values, it has been estimated that the resonance stabilization of **1** is at best only of the order of magnitude of tropilidene, that is, 7 kcal/mole.

Similar calculations have been carried out for 2-methyloxepin (**13**) (65). In this case the enthalpy term for the epoxy diene–oxepin equilibrium turns out to be only 0.4 ± 0.2 kcal/mole, while the entropy difference is reduced to 5.0 cal/(deg)(mole). The activation energy for the conversion of **13** into its valence tautomer is approximately the same as that for the **1**–**1A** equilibrium, but the presence of the methyl group causes a significant decrease in the energy difference between **13** and the epoxy diene form. According to Vogel and Guenther (65), this effect can result either from hyperconjugation or from σ-bond stabilization.

4. Chemical Reactions

Although relatively little is known at present concerning the chemistry of the oxepin ring, it appears likely that rapid advances will be made in this area now that the fundamental synthetic problem has been solved.

Attention has already been drawn to the facile rearrangement of oxepin (**1**) itself to phenol (36, 64, 66). The ease with which this important transformation occurs is perhaps unfortunate, since it limits the number of reactions in which the seven-membered ring can be expected to remain intact. The only such reactions thus far reported, in fact, have been the catalytic reduction of **1** to hexamethylene oxide (**44**) (66) and the catalytic reduction of dimethyl 4,5-oxepindicarboxylate (**41**) to its hexahydro derivative, **42** (49). So facile is the rearrangement of **1** to phenol, that even under mild hydrogenating conditions a small yield of cyclohexanol can be isolated as a side product (64).

When **1** is reduced with lithium aluminum hydride, the product is 5-hydroxycyclohexa-1,3-diene, an unstable compound which readily loses water to give benzene (65). Likewise, condensation of **1** with methyl lithium reportedly gives a 9:1 mixture of *cis*- and *trans*-1-methyl-6-hydroxycyclohexa-2,4-diene (65).

The strong tendency of **1** to undergo transformation via its valence tautomer, **1A**, imposes a further limitation upon the number of possible reactions during which the seven-membered ring can survive. In this respect the behavior of **1** is singularly reminiscent of cyclooctatetraene. Thus in the condensation of **1** with maleic anhydride the product is **45**, the Diels-Alder adduct derived from **1A** (64, 66). Similarly, **17** and **21** afford **46** and **47**, respectively, upon treatment with maleic anhydride (64, 66), while **1** yields **48** upon treatment with dimethyl acetylenedicarboxylate (65).

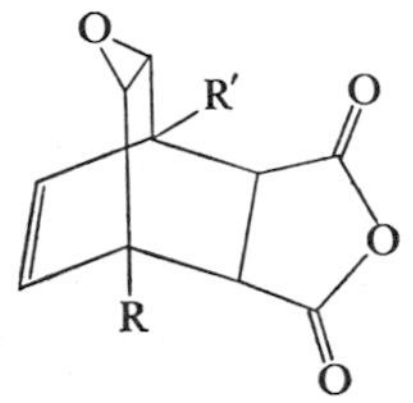

45; R = R′ = H
46; R = R′ = Me
47; R = COMe, R′ = H

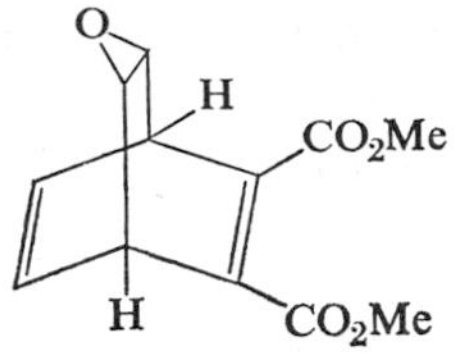

48

Interesting photochemically induced rearrangements have been reported recently with **1** (65) and **17** (44). The products have been identified as *cis*-2-oxabicyclo[3.2.0]hepta-3,6-diene (**49**) and *cis*-1,3-dimethyl-2-oxabicyclo-[3.2.0]hepta-3,6-diene (**50**), respectively. It is noteworthy that the photoisomerization of **17** was reported (43) to be fairly slow in comparison with the corresponding process in the 2,3-dihydrooxepin series (45). Conceivably, this difference in reactivity may be viewed as a reflection of aromatic character in the oxepin ring.

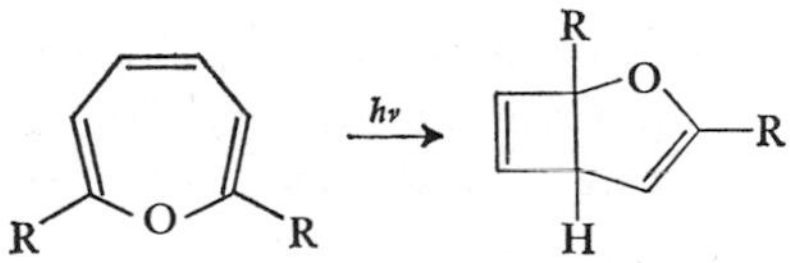

1; R = H
17; R = Me

49; R = H
50; R = Me

a. *Addendum*

A diketone described in 1966 by Kubota and co-workers (71) can be formally written as a fully aromatic oxepin derivative. This compound was obtained from the natural pigment derivative "flavonol red" by acid hydrolysis and was assigned structure **51**. A diacetate and a dimethoxy derivative were prepared, and the latter was degraded by ozonolysis into anhydride **52**.

Fischer and co-workers (72) reported in 1967 that 2,7-dimethyloxepin reacted with iron pentacarbonyl under uv irradiation, giving two distinct products separable by column chromatography. The microanalytical and spectral properties of these adducts corresponded to the structures $(C_8H_{10}O)Fe_2(CO)_6$ and $(C_8H_{10}O)Fe(CO)_3$, respectively.

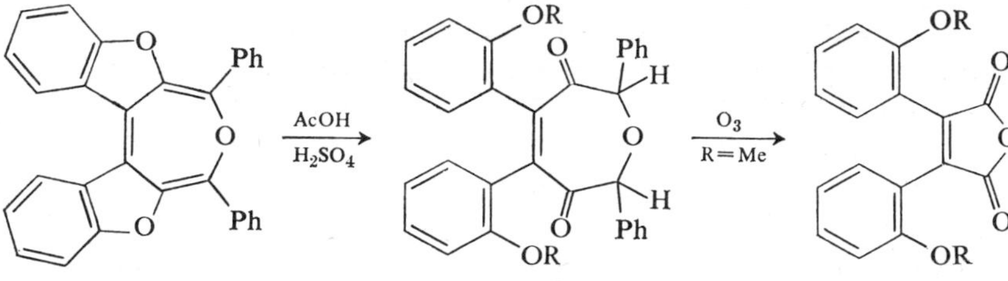

AcOH, H_2SO_4 → **51**; R=H; O_3, R=Me → **52**; R=Me

Further light on the benzene oxide–oxepin equilibrium was shed in 1968 by Jerina and co-workers (73), who obtained a mixture of 4-methyloxepin (**53**) and its epoxide tautomer, **53A**, upon treatment of dibromo epoxide **54** with potassium *tert*-butoxide in ether or tetrahydrofuran at −20°. For the preparation of **54**, toluene was reduced to diene **55**, the trisubstituted double bond in **55** was brominated selectively, and the resultant dibromide, **56**, was epoxidized to **54** with trifluoroperacetic acid. The uv absorption maximum of **53**–**53A** was observed at 273 mμ in cyclohexane solution and at 267 mμ in aqueous methanol. Reaction with maleic anhydride afforded the Diels-Alder adduct corresponding to the epoxide tautomer, as in the case of other unstable oxepins. Conversion of **53** to *p*-cresol is extremely facile, the half-life of **53** being only 24 hr even in scrupulously base-cleaned glassware.

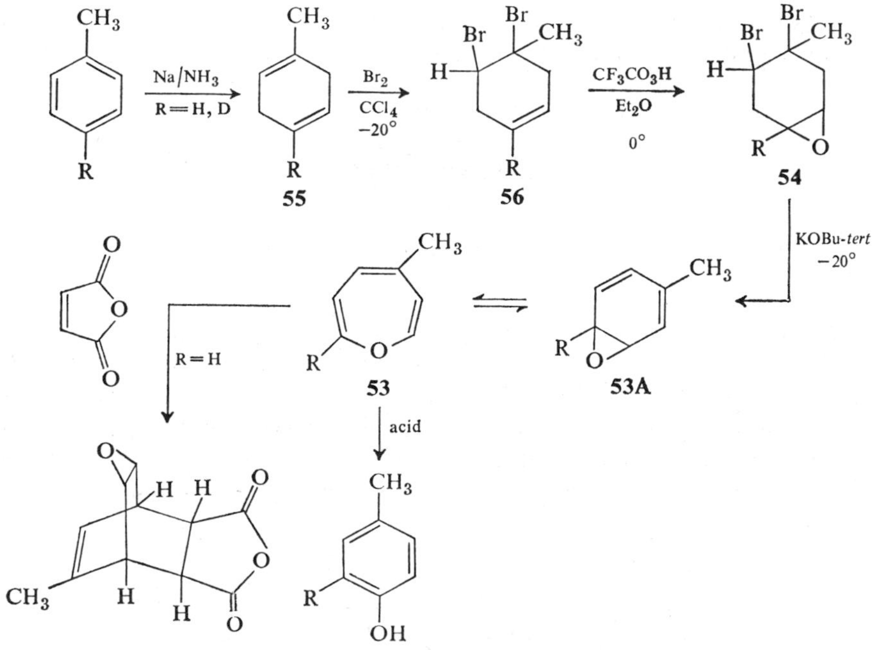

Na/NH_3, R=H, D → **55**; Br_2, CCl_4, −20° → **56**; CF_3CO_3H, Et_2O, 0° → **54**; KOBu-*tert*, −20° → **53A** ⇌ **53**; R=H (maleic anhydride); acid

Starting with *p*-deuteriotoluene, Jerina and co-workers were able to determine the position of the deuterium atom in **53** and also in the *p*-cresol arising from **53**. The movement of deuterium from the *para* to the *meta* position in this reaction sequence follows the pattern observed during the microsomal or direct chemical *para*-hydroxylation of toluene, which has come to be termed the "NIH shift." According to Jerina and co-workers, **53–53A** may in fact participate in this remarkable process as a highly transient intermediate.

An interesting nonenzymatic nucleophilic ring opening reaction of oxepin (**1**) has been reported by DeMarinis and Berchtold (78). Although **1** was inert to sodium amide, treatment with sodium nitrite at room temperature resulted

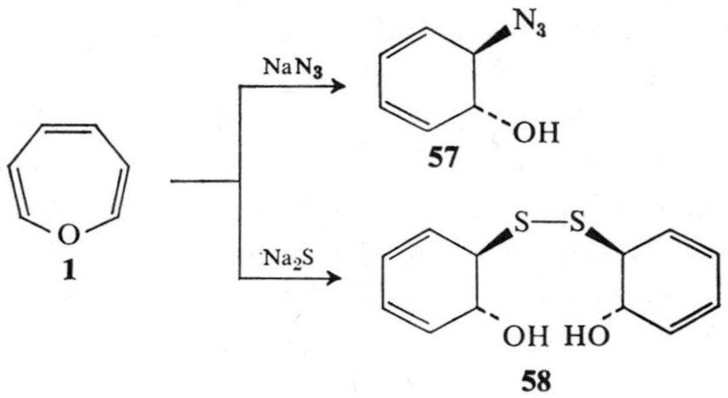

NaN_3 → **57**; Na_2S → **58**

in a 55% yield of a diene formulated as *trans*-5-azido-6-hydroxy-1,3-cyclohexadiene (**57**). Reaction of **1** with sodium sulfide led to disulfide **58** in 37% yield.

B. Dihydrooxepins

1. Introduction

Of the four theoretically possible isomers of dihydrooxepin, only 2,3-dihydrooxepin (**59**), 4,5-dihydrooxepin (**60**), and 2,5-dihydrooxepin (**62**) have been prepared so far. Attempts to prepare the symmetrical 2,7-dihydrooxepin (**61**) have led only to the isolation of **59**. It has been suggested (35) that **59** may be more stable than **61** because conjugation between the oxygen atom and the diene system can be achieved in **59**, but not in **61**.

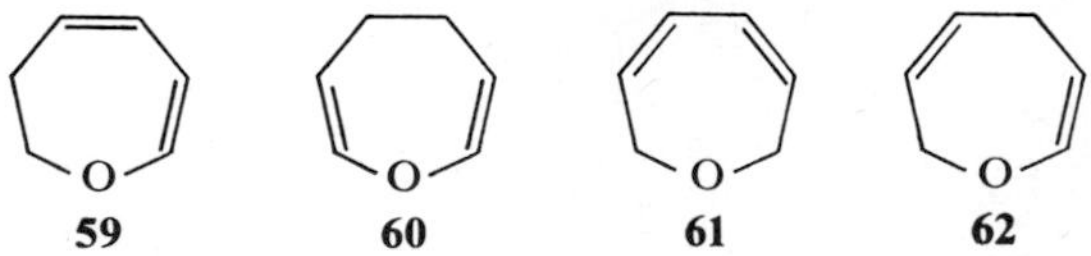

59 **60** **61** **62**

2. Preparation

Successful syntheses of **59** were reported simultaneously in 1960 by two independent groups. In one approach, Meinwald and co-workers (35) converted 2,3,6,7-tetrahydrooxepin (**63**), in stepwise fashion, into *trans*-4,5-dibromo-2,3,4,5,6,7-hexahydrooxepin (**64**), 4-(*N*,*N*-dimethylamino)-2,3,4,7-tetrahydrooxepin (**65**), and the *N*-oxide **66**. Pyrolysis of crude **66** at 80–100° (5 mm) gave a 55% yield of **59**, based upon the amount of **65** actually consumed. The formation of **59** from **66** is considered to involve direct 1,4-elimination during the pyrolysis step, rather than the alternative possibility of initial *cis*-elimination to 2,7-dihydrooxepin (**61**) and thermal isomerization of **61** to **59** in a separate step.

63 $\xrightarrow[\text{CCl}_4,\ 0°]{Br_2}$ **64** (Br, Br) $\xrightarrow[\text{C}_6\text{H}_6,\ 95-100°]{Me_2NH}$ **65** (NMe$_2$) $\xrightarrow[\text{MeOH},\ -10°]{H_2O_2}$ **66** (O←NMe$_2$) $\xrightarrow{80-100°}$ **59**

A variation of this basic approach involved conversion of **63** into 4,5-epoxy-2,3,4,5,6,7-hexahydrooxepin (**67**) by epoxidation with perbenzoic acid, acetolysis of **67**, and finally pyrolysis of the resulting diacetate, **68**, at 525°. Although Meinwald and co-workers (35) hoped to achieve the synthesis of **61** via this route, the product actually turned out to be **59**. From a preparative standpoint, however, this approach was less satisfactory than the pyrolysis of **66**.

63 $\xrightarrow[\text{CHCl}_3]{C_6H_5CO_3H}$ **67** $\xrightarrow[p\text{-TsOH}]{AcOH-Ac_2O}$ **68** (AcO, OAc) $\xrightarrow{525°}$ [OAc intermediate] $\longrightarrow$ **59**

A more direct synthesis of **59** was devised by Schweizer and Parham (50). These workers found that addition of chlorocarbene to dihydropyran afforded a 28% yield of a 40:60 mixture of two isomeric adducts which could be separated by fractional distillation. The more abundant isomer was found to undergo facile conversion into **59** upon being heated above 120°, and was formulated as *exo*-2-oxa-7-chloronorcarane (**69**). The other isomer (bp 176.5°/1 atm) was not affected by distillation at atmospheric pressure and was assigned the *endo* structure, **70**. The formulation of the more abundant, more easily rearranged isomer as **69** was based, in part, upon the argument that the transition state leading to the *exo* adduct would be expected to involve somewhat less steric hindrance than the transition state leading to the *endo* isomer. Furthermore, it was proposed that, in the formation of **59**, the expulsion of the chlorine atom must be assisted by anchimeric (neighboring group) participation of the oxygen atom, and that this process can occur effectively with isomer **69** but not with isomer **70**.

The structural assignments made by Schweizer and Parham were subsequently reversed by DePuy and co-workers (11), and the later assignments supported experimentally by Ando and co-workers (2). In an extension of the Woodward-Hoffman rules, DePuy and co-workers found that, in the solvolytic cleavage of cyclopropyl compounds, groups *cis* to the leaving group must rotate "outward." On this basis these investigators concluded that only an adduct possessing the *endo* configuration can rearrange to **59**, the isomerization of the *exo* structure being sterically quite unfavorable (see below). They proposed, therefore, that the more abundant isomer of 2-oxa-7-chloronorcarane is actually **70** and not **69** as postulated earlier by Schweizer and Parham.

:CHCl

Cl H H H O

70

175° quinoline

H O H

→ **59**

H Cl H H O

69

H H O

The stereochemical assignments postulated by DePuy and co-workers were later verified experimentally by Ando and co-workers (2) on the basis of an analysis of the nmr coupling constants of the cyclopropane protons in the two isomers in question. According to the Japanese workers, when a mixture containing **69** and **70** in a ratio of approximately 1.1–1.2:1.0 is heated at 120° for 3 hr in quinoline solution, **59** is obtained in 70% yield (based on **70** actually consumed), and the *exo* isomer, **69**, is recovered unchanged.

Schweizer and Parham also investigated the thermal elimination of hydrogen chloride from 2-oxa-7,7-dichloronorcarane (**71**), which was prepared by addition of dichlorocarbene to dihydropyran. When **71** was heated to 140–150° under reduced pressure in quinoline solution, 6-chloro-2,3-dihydrooxepin (**72**) was obtained in 83% yield. The formation of **72** from **71** was assumed to occur according to the process depicted. The conversion of **71** into **72** was also investigated recently by Ando and co-workers (2). These authors performed the reaction at 120° and reported a yield of 80%. As in the case of **70**, it is the *endo*-chloro substituent which is being expelled during isomerization.

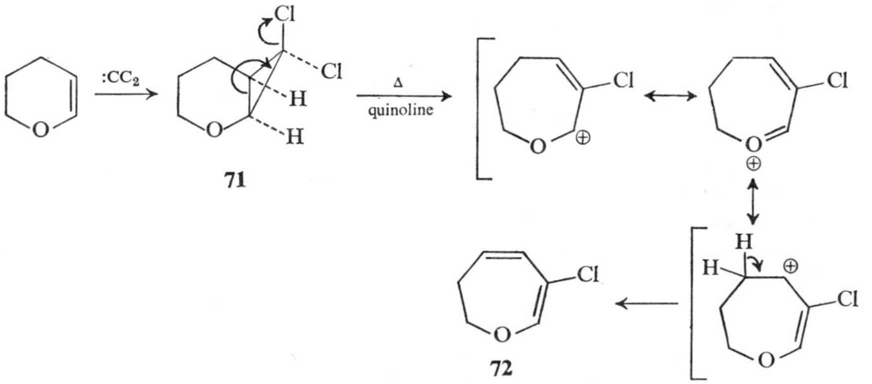

:CC$_2$ → **71** —Δ, quinoline→ [cation resonance forms] → **72**

Two other 2-oxanorcarane derivatives were examined by Ando and co-workers (2) as 2,3-dihydrooxepin precursors. When a 1.5–1.6:1.0 mixture of isomeric 7-chloro-7-fluoro-2-oxanorcaranes **73** and **74** was heated at 120° for 3 hr in quinoline, 6-fluoro-2,3-dihydrooxepin (**75**) was obtained in 77% yield (based on the amount of **73** consumed), and **74** was recovered unchanged. Similarly, when a 1.6–1.8:1.0 mixture of *exo*-7-fluoro-2-oxanorcarane (**76**) and *endo*-7-fluoro-2-oxanorcarane (**77**) was heated at 120° in quinoline for 10 hr and then at 140° for an additional 5 hr, **59** was obtained, together with some unchanged **76**. The concerted disrotatory mechanism proposed by DePuy and co-workers for the isomerization of **70** to **59** was invoked by the Japanese authors for the rearrangement of **73** and **77** as well.

Somewhat different results were reported recently by Nerdel and co-workers (40) in connection with a study of the reaction of dichlorocarbene with dihydropyran in a sealed vessel at an elevated temperature. The dichlorocarbene was generated *in situ* by dehydrohalogenation of chloroform under the influence of ethylene oxide. Under these conditions there was formed a 57% yield of a substance identified as 6-chloro-7-chloroethoxy-2,3,4,7-tetrahydrooxepin (**78**), together with a small amount of the expected 7,7-dichloro-2-oxanorcarane (**71**). Apparently the initial dichlorocarbene adduct, **71**, underwent further reaction with excess ethylene oxide to form **78**, perhaps by a mechanism similar to that which Schweizer and Parham (50) had proposed earlier for the transformation of **71** into **72**. It is of interest that Nerdel and co-workers did not observe the formation of any **72** and were apparently unaware of the earlier work when they published their own findings.

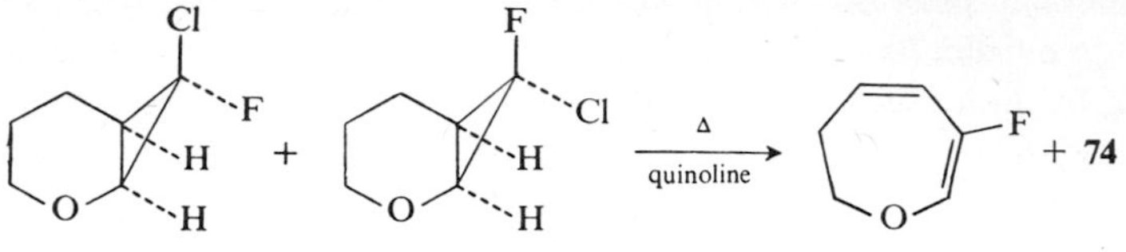

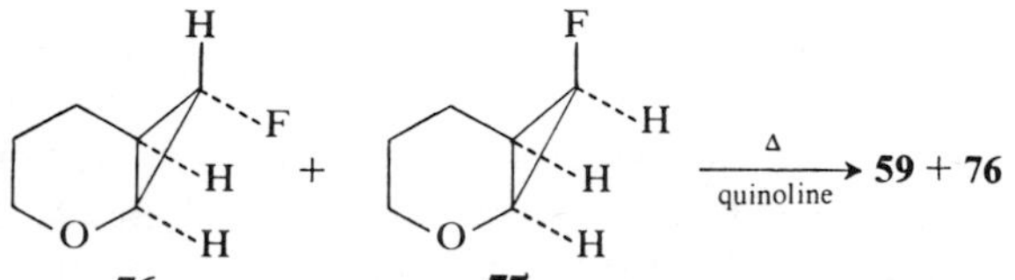

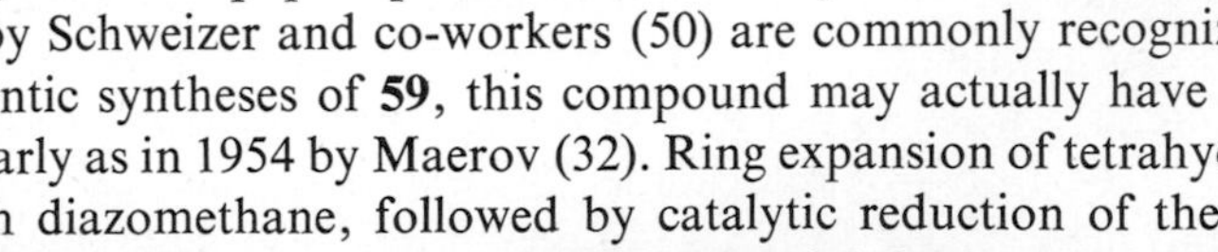

Although the two papers published in 1960 by Meinwald and co-workers (35) and by Schweizer and co-workers (50) are commonly recognized as the first authentic syntheses of **59**, this compound may actually have been prepared as early as in 1954 by Maerov (32). Ring expansion of tetrahydropyran-4-one with diazomethane, followed by catalytic reduction of the resultant 2,3,4,5,6,7-hexahydrooxepin-4-one (**79**) to 2,3,4,5,6,7-hexahydrooxepin-4-ol (**80**) and boric acid-catalyzed dehydration of **80**, afforded a crude mixture which was believed to contain 2,3,4,7-tetrahydrooxepin (**81**) and 2,3,6,7-tetrahydrooxepin (**82**). Treatment of this mixture with *N*-bromosuccinimide afforded a monoallylic bromide. This intermediate was then dehydrobrominated directly to a dihydrooxepin which may well have been **59**, although a firm structure was never actually assigned.

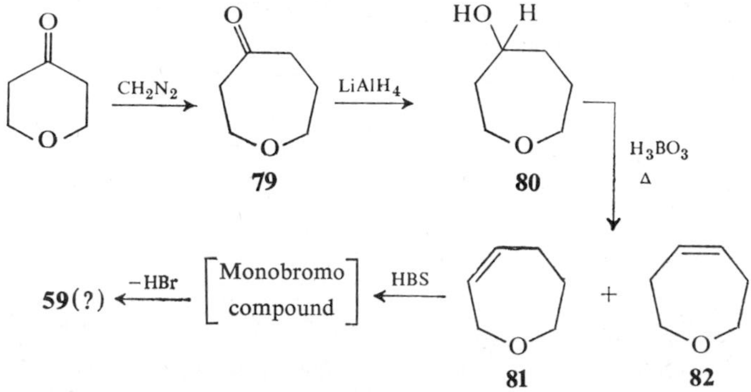

The synthesis of 4,5-dihydrooxepin (**60**) was first reported in 1961 by Pohl (48), who observed the formation of this compound during the pyrolysis of 2-acetoxy-2,3,4,5-tetrahydrooxepin (**83**) at 470°.

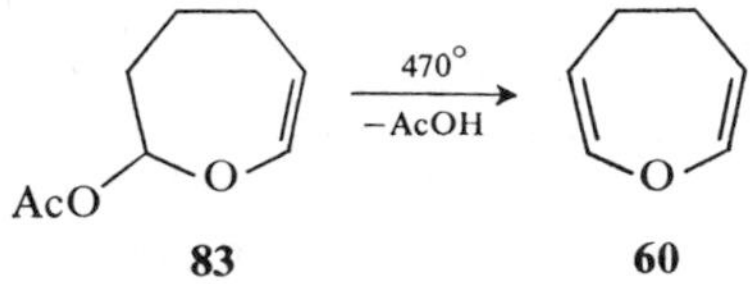

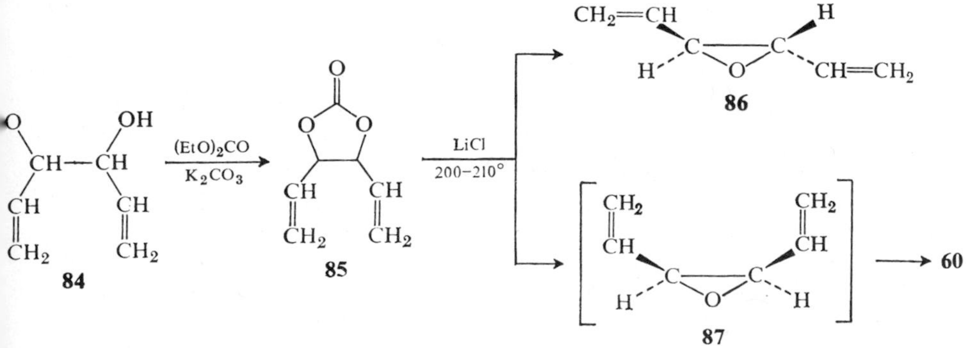

A second method of preparation of **60** was published in 1963 by Braun (5). Acrolein was converted into 1,5-hexadiene-3,4-diol (**84**) in 90% yield under the influence of a zinc–copper couple. Condensation of **84** with diethyl carbonate then gave *sym*-divinylethylene carbonate (**85**) in 74% yield. Finally, pyrolysis of **85** at 200–210° in the presence of lithium chloride yielded **60** together with a second product which was identified as *trans*-1,2-divinylethylene oxide (**86**). Separation of these two compounds required a combination of fractional distillation and preparative gas–liquid chromatography (glc). The formation of **60** was postulated to take place by way of *cis*-1,2-divinylethylene oxide (**87**), a fugitive species which can be considered a "valence tautomer" of **60**. Although the origin of **86** was not discussed, this product was presumably formed because neither **84** nor **85** was stereochemically homogeneous in this particular experiment.

The findings of Braun were later confirmed and extended by Stogryn and co-workers (54, 62). Diene **84** was converted into 3-chloro-4-acetoxy-1,5-hexadiene (**88**) by reaction with acetyl chloride in the presence of calcium chloride. Treatment of **88** with sodium hydroxide and potassium hydroxide in aqueous ethylene glycol at 40–50° (10–15 mm) produced a mixture of **86** and **87**. Although the two isomeric epoxides could not be separated by fractional distillation, glc analysis showed that **86** and **87** were formed in a ratio of 2:1 under these conditions. After being heated overnight on a steam bath, a mixture originally containing 64.4% of **86**, 27.8% of **87**, and 7.8% of **60** was found to contain 61.8% of **86**, 38.2% of **60**, and no trace of **87**. When **88** was added slowly to a reactor containing sodium hydroxide, potassium hydroxide, and water at 170°, a mixture of **60** and **86** was formed. The ratio of **60** to **86** in this mixture was determined by glc analysis to be 1:2. Fractional distillation failed to effect separation of these compounds, but a pure specimen of **60** was obtained via preparative glc. After being heated at 230° for 17 hr, **86** which was initially 93% pure was converted into a mixture containing 42.9% of **60** and 48.3% of **86**. It appeared, therefore, that even the *trans*-epoxide, **86**, was capable of isomerization into **60**, provided that sufficient thermal energy was supplied to the system.

$$\mathbf{84} \xrightarrow[\text{CaCl}_2]{\text{AcCl}} \underset{\mathbf{88}}{CH_2{=}CH{-}CH(Cl){-}CH(OAc){-}CH{=}CH_2} \xrightarrow{\text{base}} \begin{cases} \xrightarrow{<50^\circ} \mathbf{86} + \mathbf{87} \\ \xrightarrow{>170^\circ} \mathbf{86} + \mathbf{60} \end{cases}$$

(**86** + **87** → **86** + **60** on steam bath; **86** → **60** at 230°)

Although the exact nature of the processes discovered by Braun and by Stogryn and co-workers was not unraveled fully by these investigators, additional light was shed recently by Vogel and Guenther (65) as a result of a careful kinetic study of the isomerization of **86** and **87**. In each case, the kinetics of the gas-phase reaction was found to be first-order. However, some significant differences were observed between the enthalpy and entropy values for the two processes. The enthalpy of reaction for the rearrangement of **86**, the *trans*-epoxide, was 36.0 kcal/mole, and the entropy change was −0.4 cal/(deg)(mole). In contrast, the isomerization of **87**, the *cis*-epoxide, proceeded with an enthalpy of reaction of 24.6 kcal/mole and an entropy change of −11.3 cal/(deg)(mole). Thus **87** has a relatively low activation energy and a highly negative activation entropy, in keeping with a synchronous reorganization of bonding electrons, while **86** has a fairly high activation energy but only a slightly negative activation entropy, indicating a nonconcerted process. According to Vogel and Guenther, the kinetic data for **86** can be accommodated by a mechanism involving a diradical intermediate and is supported by their additional finding, not reported by the earlier workers, that a side product accompanying **60** during the rearrangement of **86** is 2-vinyl-2,3-dihydrofuran.

Two interesting reactions representing possible routes to **60** were investigated by Maerov (32) but, unfortunately, neither of them afforded the desired product. In one of these approaches 2-hydroxymethylene-2,3-dihydropyran (**89**) was heated in the presence of acid alumina to give, unexpectedly, 3,8-dioxabicyclo[3.2.1]octane (**90**). In the second approach 2-aminomethyl-2,3-dihydropyran (**91**) was subjected to nitrous acid deamination with formation, in this case, of adipaldehyde as the product. It was proposed that **91** undergoes a Demjanow-type rearrangement to some unspecified labile oxepin intermediate, which is then cleaved to adipaldehyde in the acidic medium used for the reaction.

O CHOH $\xrightarrow[\Delta]{Al_2O_3}$ O O

89 **90**

O CH_2NH_2 $\xrightarrow{HNO_2}$ $OHC(CH_2)_4CHO$

91

Condensation of the Grignard reagent derived from 2-acetoxy-4-bromo-2-butene (**92**) with a second molecule of **92**, followed by direct hydrolysis of the coupling product with acid, was reported by Matschinskaya and coworkers (34) to give a 68% yield of a product which they formulated as 2,7-dimethyl-4,5-dihydrooxepin (**93**). Upon being subjected to similar treatment the Grignard reagents derived from 2-acetoxy-4-bromo-2-pentene (**94**) and 2-acetoxy-4-bromo-4-phenyl-2-butene (**95**) allegedly afforded 2,4,5,7-tetramethyl-4,5-dihydrooxepin (**96**) and 2,7-dimethyl-4,5-diphenyl-4,5-dihydrooxepin (**97**), respectively. However, this report seems to warrant further investigation in view of Pohl's subsequent observation (48) that in **60** itself, at any rate, the seven-membered ring is quite labile in the presence of acid.

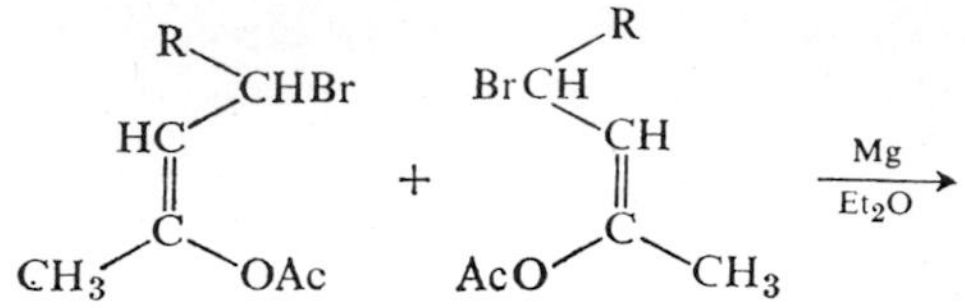

92; R = H
94; R = Me
95; R = Ph

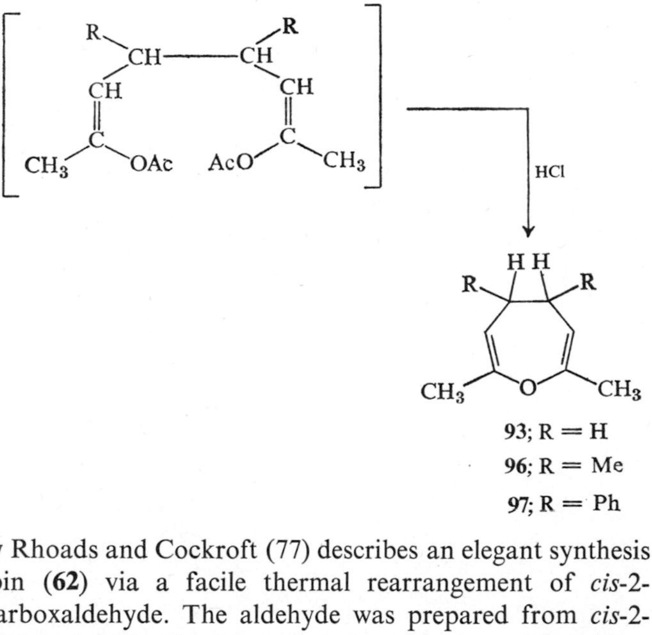

HCl

93; R = H
96; R = Me
97; R = Ph

A recent paper by Rhoads and Cockroft (77) describes an elegant synthesis of 2,5-dihydrooxepin (**62**) via a facile thermal rearrangement of *cis*-2-vinylcyclopropanecarboxaldehyde. The aldehyde was prepared from *cis*-2-vinylcyclopropanecarbonyl chloride by conversion to the acylaziridine followed by lithium aluminum hydride reduction. The kinetics of the reaction showed a striking similarity to those observed by Vogel and Guenther (65) for the formation of 4,5-dihydrooxepin (**60**) from *cis*-1,2-divinylethylene oxide. In the present example, the enthalpy of reaction was determined to be 23 kcal/mole, with an entropy change of −6 cal/(deg)(mole). These values are consistent with a synchronous reorganization of bonding electrons.

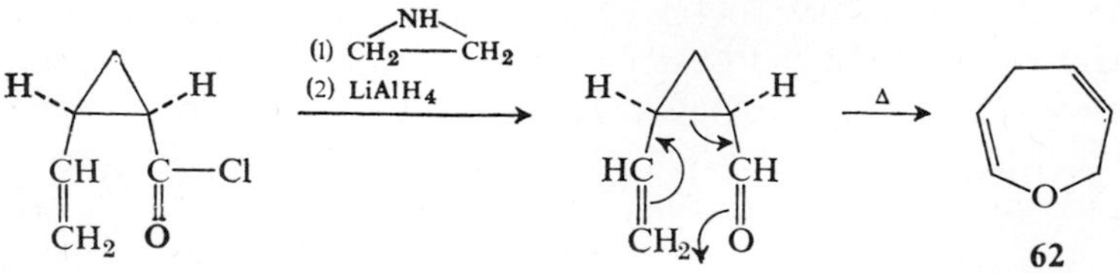

(1) CH_2—NH—CH_2
(2) $LiAlH_4$

Δ

62

However, it is of interest to note that **62**, unlike **60**, is unstable at room temperature, reverting to the starting aldehyde with a half-life of approximately 1 day.

3. Physical Properties

a. *Infrared Absorption*

Infrared absorption maxima of 2,3-dihydrooxepin (**59**) have been reported in detail by Meinwald and co-workers (35) and also by Schweizer and Parham (50). With minor exceptions the values found by these two groups of investigators are in close agreement. The double bonds in **59** can be expected to give rise to different —CH=CH— stretching frequencies since they are not alike. This was, indeed, found to be the case, strong peaks being observed at 6.05 and 6.17 μ in one specimen (35) and at 6.09 and 6.21 μ in the other (50).

The principal features of the ir absorption spectrum of 4,5-dihydrooxepin (**60**) were reported first by Braun (5) and later by Stogryn and co-workers (54). According to Braun, the following peak assignments can be made in the spectrum of **60**: 3055 cm^{-1} (3.27 μ), olefinic C—H stretch; 2940 cm^{-1} (3.41 μ) and 2860 cm^{-1} (3.49 μ), methylene C—H stretch; 1650 cm^{-1} (6.06 μ), nonconjugated olefin C=C stretch; 1448 cm^{-1} (6.90 μ), allylic CH_2 deformation; and 1238 cm^{-1} (8.09 μ), enol ether. It is perhaps of interest to note that while Braun reported only a single value for the C=C stretching frequency, which is in accord with the fact that **60** contains one type of double bond, the specimen synthesized by Stogryn and co-workers seemed to exhibit a pair of peaks at 6.05 and 6.10 μ. No explanation was given for this anomaly.

2,5-Dihydrooxepin (**62**) has been reported (77) to show the following characteristic ir peaks: 1655 cm^{-1} (6.05 μ), vinyl ether; 1270 and 1105 cm^{-1} (7.88 and 9.05 μ), C—O stretching modes.

b. *Ultraviolet Absorption*

2,3-Dihydrooxepin (**59**) has been reported by Meinwald and co-workers (35) to absorb at λ_{max} 260 mμ (log ε = 3.95) in *iso*-octane solution. Ultraviolet absorption values for 4,5-dihydrooxepin (**60**) have not been published.

c. *Nuclear Magnetic Resonance*

Nmr spectrometry has played an important role in permitting unambiguous structures to be assigned to the dihydrooxepins. Thus in the case of **59**, the following peak assignments were made by Schweizer and Parham (50):

τ 3.69, doublet (C-7 proton); τ 4.32, "pseudo doublet" (C-4 and C-5 protons); τ 5.23, triplet (C-6 proton); τ 5.89, triplet (C-2 proton); τ 7.47, "pseudo quartet" (C-3 proton). The same interpretation was made independently by Meinwald and co-workers (35) with the exception that the C-4, C-5, and C-6 protons were not resolved.

The nmr spectrum of 6-chloro-2,3-dihydrooxepin (**72**) was found by Schweizer and Parham (50) to exhibit the following signals: τ 3.36, singlet (C-7 proton); τ 4.23, multiplet (C-4 and C-5 protons); τ 5.89, triplet (C-2 proton); τ 7.44, "pseudo quartet" (C-3 proton).

The symmetrical character of the 4,5-dihydrooxepin (**60**) molecule was revealed most clearly by its nmr spectrum, which consisted of the following peaks, according to Braun (5): τ 3.42, doublet (C-2 and C-7 protons); τ 5.3, multiplet (C-3 and C-6 protons); τ 7.3, multiplet (C-4 and C-5 protons).

The structure of 2,5-dihydrooxepin (**62**) was evident from its nmr spectrum, which contained signals at τ 3.87, doublet of triplets (C-7 proton); and τ 3.95–4.20, τ 5.50–5.85, and τ 7.00–7.30, complex multiplets (C-2 and C-6 protons, C-3 and C-4 protons, and C-5 protons, respectively) (77).

4. *Chemical Reactions*

2,3-Dihydrooxepin (**59**) was found by Meinwald and co-workers (35) to behave like a typical acid-labile enol ether, being cleaved readily upon treatment with dilute hydrochloric acid at room temperature. A mixture of carbonyl-containing products was obtained, but the composition of the mixture was not reported. The facility with which **59** underwent acid-catalyzed ring opening was cited as chemical evidence excluding 2,7-dihydrooxepin (**61**) as a possible alternate structure for this compound.

Meinwald and co-workers also found that **59** acts as a typical diene in giving a Diels-Alder adduct, **98**, upon reaction with 1,1,2,2-tetracyanoethylene (TCNE).

The catalytic reduction of **59** and also of **72** was investigated by Schweizer and Parham (50). Whereas **59** was converted readily into hexamethylene oxide (**99**), ring cleavage occurred in the case of **72** with formation of 1-hexanol as the sole identifiable product.

Paquette and co-workers (45) have carried out an interesting study of the

59; R = H
72; R = Cl

H_2/Pt; R = H → 99 (89%); R = Cl → $CH_3(CH_2)_5OH$ (63%)

photoisomerization of **59** and **72**. Upon irradiation in ether solution at room temperatures, these compounds underwent rearrangement to the previously unknown *cis*-2-oxabicyclo[3.2.0]hept-6-ene (**100**) and *cis*-7-chloro-2-oxabicyclo[3.2.0]hept-6-ene (**101**), respectively. According to the authors, dihydrooxepins undergo this type of isomerization more easily than their fully unsaturated counterparts (43, 44). Whether this difference is to be ascribed to any degree of aromaticity in the oxepin system is a matter of speculation.

59; (R=H) $\xrightarrow{h\nu}$ 100; (R=H)
72; (R=Cl) $\xrightarrow{h\nu}$ 101; (R=Cl)

Hydrogenation of 4,5-dihydrooxepin (**60**) has been found to proceed smoothly with formation of **99** in high yield (5, 48). Both platinum oxide and Raney nickel can be used as catalysts.

60 $\xrightarrow[\text{catalyst}]{H_2}$ **99**

As in the case of **59**, acid treatment of **60** as been found to effect rapid isomerization. According to Pohl (48), exposure of **60** to warm aqueous acetic acid for 10 min results in the formation of cyclopentene-1-carboxaldehyde (**102**).

60 $\xrightarrow{AcOH}$ CHO **102**

C. Tetrahydrooxepins

1. Introduction

Three isomeric tetrahydrooxepins are possible in principle, the only difference among them being the location of the double bond. All three unsubstituted parent compounds, 2,3,4,5-tetrahydrooxepin (**103**), 2,3,4,7-tetrahydrooxepin (**104**), and 2,3,6,7-tetrahydrooxepin (**105**), have been reported, as well as several substituted derivatives of **103** and **104**.

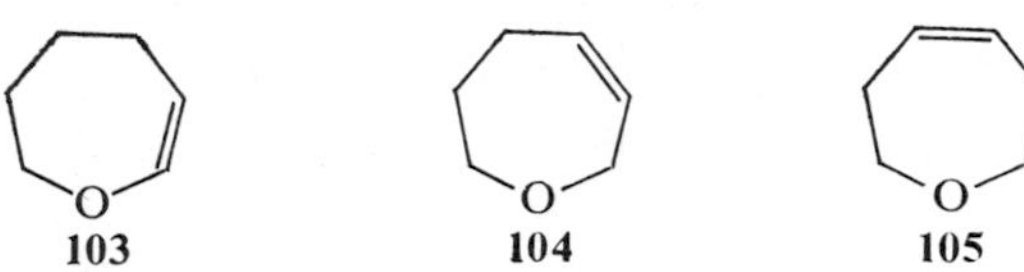

2. Preparation

Although certain of its substituted derivatives had been described earlier, **103** itself remained unknown until 1965 when Larkin (31) identified this compound as the principal product of the catalytic dehydrogenation of 1,6-hexanediol at 250°. A catalyst consisting of 15% copper and 0.85% chromium on kieselguhr was used. A number of other products were also identified, including ε-caprolactone, 2-methylcyclopentanone, cyclopentane carboxaldehyde, and cyclopentanemethanol. A variety of experimental conditions were investigated, and it was found that the composition of the product mixture could be altered considerably, depending especially upon the nature of the catalyst. The formation of **103** from 1,6-hexanediol is explained most reasonably by assuming that the primary dehydrogenation product is 6-hydroxyhexanal, which undergoes internal hemiketalization to 2,3,4,5,6,7-hexahydro-2-hydroxyoxepin (**106**). Elimination of a molecule of water from **106** then leads to **103**. A more recent synthesis is that reported by Nerdel and co-workers (40) who obtained **103** in 70% yield upon heating 2-(2′-chloroethoxy)-2,3,4,5,6,7-hexahydrooxepin (**107**) at 190° in quinoline.

The conversion of straight-chain 6-hydroxycarbonyl compounds into 2,3,4,5-tetrahydrooxepins under the influence of dehydrating agents may be a fairly general method, but the scope of the reaction has never been explored in a systematic fashion. The only example of such a transformation, in fact, was that reported in 1963 by Teisseire and Corbier (56), involving treatment of 8-hydroxy-2,6-dimethyl-3-octanone with phosphoric acid to yield 2,3,4,5-tetrahydro-7-isopropyl-4-methyloxepin (**108**).

$HO(CH_2)_6OH \xrightarrow[\text{Cu-Cr on kieselguhr}]{250°} [HO(CH_2)_5CHO \rightleftharpoons$ **106** $]$ $\xrightarrow{-H_2O}$ **103**

107 (OCH$_2$CH$_2$Cl) $\xrightarrow[\text{190° 80–100 min}]{\text{quinoline}}$ **103**

$\xrightarrow[\Delta]{H_3PO_4}$ **108**

An interesting method of synthesis of 2-acyloxy-2,3,4,5-tetrahydrooxepins, which was first reported in 1958 by Shine and Snyder (51, 53, 61), involved air oxidation of cyclohexene in the presence of an acyl anhydride and a suitable radical initiator, such as azobisisobutyronitrile. Thus in the presence of acetic and propionic anhydrides, cyclohexene afforded moderate yields of 2-acetoxy-2,3,4,5-tetrahydrooxepin (**109**) and 2,3,4,5-tetrahydro-2-propionyloxyoxepin (**110**), respectively. Some 2-cyclohexen-1-one was also isolated. As an extension of their original investigation, Snyder and co-workers (53) likewise carried out the oxidation of 3-methyl-1-cyclohexene in acetic anhydride, using uv light to initiate the reaction. In this case only a low yield of the oxepin derivative was formed together with larger amounts of a ketone which was formulated as either 3-methyl-2-cyclohexe-1-ene or 4-methyl-2-cyclohexen-1-ene. A key step in this reaction is probably the formation of a cyclohexene hydroperoxide ester which undergoes subsequent ring enlargement via a stepwise or concerted process.

$\xrightarrow[(RCO)_2O + \text{initiator}]{O_2}$ O—C(=O)—R + (cyclohexenone)

109; R = Me
110; R = Et

The synthesis of **104** was reported for the first time in 1958 by Olsen and Bredoch (42). The four-step sequence developed by these authors began with the addition of diazomethane to tetrahydropyran-4-one, which resulted in the formation of a mixture of spiroepoxide **111** and ring-expanded ketone **112**.

Reduction of **112** and acetylation of the resulting alcohol, **113**, afforded 4-acetoxy-2,3,4,5,6,7-hexahydrooxepin (**114**). Deacetylation of **114** was achieved by treatment with *p*-toluenesulfonic acid at 300–325°. Repetition of this work by Meinwald and co-workers (35) in 1960 revealed that the deacetylation of **114** did not proceed unidirectionally, as claimed by Olsen and Bredoch, but actually produced a 30:70 mixture of **104** and **105**. Similar experiments had been performed as early as 1954 by Maerov (32), who converted tetrahydropyran-4-one into **113** and observed that treatment of **113** with hot boric acid afforded a mixture of **104** and **105**.

CH$_2$N$_2$; 23% → **111**; 60% → **112**; LiAlH$_4$ → **113**; Ac$_2$O → **114**; **113**: H$_3$BO$_3$, Δ; **114**: *p*-TsOH, Δ → **104** + **105**

2,3,4,7-Tetrahydrooxepins carrying additional substituents have also been reported by Meinwald and co-workers (35). Thus 4-*N*,*N*-dimethylamino-2,3,4,7-tetrahydrooxepin (**115**) was prepared by reaction of *trans*-4,5-dibromo-2,3,4,5,6,7-hexahydrooxepin (**116**) with excess dimethylamine at elevated temperature and pressure, and *N*-oxide **117** was prepared by treatment of **115** with hydrogen peroxide.

116 (Br, Br) → Me$_2$NH, C$_6$H$_6$, 95-100° → **115** (Me$_2$N) → H$_2$O$_2$, MeOH, −10° → **117** (Me$_2$N→O)

Pure **105** was synthesized unambiguously in 1958 by Meinwald and Nozaki (36), using the multistep sequence shown. Ring closure to **105** was achieved

$$HC\equiv C(CH_2)_4OH \xrightarrow[HCl]{\text{dihydropyran}} HC\equiv C(CH_2)_4O\text{-THP} \xrightarrow[LiNH_2/NH_3]{\text{ethylene oxide}} HO(CH_2)_2C\equiv C(CH_2)_4O\text{-THP}$$

$$\xrightarrow{H_2/Pd} HO(CH_2)_2CH{=}CH(CH_2)_4O\text{-THP} \quad (cis\text{-})$$

$$\xrightarrow[(2)\ HCl]{(1)\ TsCl/C_5H_5N} TsO(CH_2)_2CH{=}CH(CH_2)_4OH \quad (cis\text{-}\mathbf{118})$$

$$cis\text{-}\mathbf{118} \xrightarrow[Me_2CO]{NaI} I(CH_2)_2CH{=}CH(CH_2)_4OH \quad (cis\text{-}\mathbf{119})$$

$$cis\text{-}\mathbf{118} \xrightarrow[t\text{-}BuOH]{KOBu\text{-}t} \mathbf{105} + CH_2{=}CHCH{=}CH(CH_2)_2OH \quad (\mathbf{120})$$

$$cis\text{-}\mathbf{119} \xrightarrow[Et_2O]{Ag_2O} \mathbf{105} + \mathbf{120}$$

either by treatment of tosylate **118** with potassium *tert*-butoxide in diglyme or by the action of silver oxide upon iodo compound **119**. The latter route gave a better yield of product, even though it comprised an extra step. Cyclization to **105** was accompanied by some elimination to open-chain by-products, one of which was identified tentatively as 3,5-hexadien-1-ol (**120**). Certain improvements in the synthesis were reported in 1960 by Meinwald and co-workers (35), but the essential aspects of their approach remained unchanged.

3. Physical Properties

a. *Infrared Absorption*

The ir absorption spectrum of 2,3,4,5-tetrahydrooxepin (**103**) has been found by Larkin (31) to exhibit three strong olefinic bands, located at 1655, 1650, and 1641 cm^{-1} (6.04, 6.06, and 6.09 μ), respectively. This is in contrast to 2,3,4,7-tetrahydrooxepin (**104**), which was reported by Olsen and Bredoch (42) to absorb strongly at 1655 cm^{-1} (6.04 μ) only. Single —CH=CH— stretching frequencies have likewise been observed by Meinwald and co-workers (35) at 1660 cm^{-1} (6.02 μ) in 2,3,6,7-tetrahydrooxepin (**105**) and at 1655 cm^{-1} (6.04 μ) in 4-*N*,*N*-dimethylamino-2,3,4,7-tetrahydrooxepin (**115**).

b. *Nuclear Magnetic Resonance*

As in the case of the fully unsaturated oxepins and dihydrooxepins, nmr spectrometry has been a vital tool in the assignment of unambiguous structures for **103** and **105**.

The nmr spectrum of **103** has been reported (31) to consist of the following peaks: τ 3.68, doublet (C-7 proton); τ 5.25, quartet (C-6 proton); τ 6.05, triplet (C-2 protons); τ 8.9, multiplet (C-5 protons); τ 9.2, multiplet (C-3 and C-4 protons).

The symmetrical nature of **105** can be deduced readily from its simple nmr spectrum, which consists of only a triplet (C-4 and C-5 protons) and two multiplets (C-2 and C-7 protons, and C-3 and C-6 protons, respectively) (35).

The nmr spectrum of **104** has not been published to date.

4. Chemical Reactions

2,3,4,5-Tetrahydrooxepins can be regarded as cyclic enol ethers and, as such, would be expected to react rapidly with water under acidic conditions.

Thus treatment of **103** with boiling aqueous ethanolic hydrochloric acid results in the formation of 6-hydroxyhexanal which exists partly in the form of its hemiacetal, 2-hydroxy-2,3,4,5,6,7-hexahydrooxepin (**106**) (31). It may be noted that this process is the reverse of the reaction taking place during the catalytic dehydrogenation of 1,6-hexanediol (31).

$$\mathbf{103} \xrightarrow[H_2O-EtOH]{HCl} \left[HO(CH_2)_5CHO \rightleftharpoons \mathbf{106} \right]$$

The chemical properties of 2-acyloxy-2,3,4,5-tetrahydrooxepins have been explored in considerable detail by Shine and Snyder (51). Reduction of 2-acetoxy-2,3,4,5-tetrahydrooxepin (**109**) in the presence of palladium–charcoal gave 2-acetoxy-2,3,4,5,6,7-hexahydrooxepin (**121**), the seven-membered ring remaining intact under these conditions. When platinum was used as the catalyst, on the other hand, **109** was reduced to 1,6-hexanediol and acetic acid. Oxidation of **109** with neutral potassium permanganate yielded glutaric acid, the product corresponding to fission of the 2,3-double bond. The facile acid-catalyzed ring-opening of **109** and **110** was also demonstrated in the reaction of these enol ether esters with 2,4-dinitrophenylhydrazine, which afforded the 2,4-DNP derivatives of adipaldehyde and cyclopentene-1-carboxaldehyde. Similarly, **109** was converted into adipaldehyde dioxime and cyclopentene-1-carboxaldehyde oxime upon treatment with hydroxylamine hydrochloride. Cyclopentene-1-carboxaldehyde was presumed to form via an acid-catalyzed intramolecular aldol condensation reaction of adipaldehyde. An alternative explanation might be that **109**

109 (AcO, O):

- $\xrightarrow{H_2/Pd\text{-}C}$ **121** (AcO, O)
- $\xrightarrow[\text{or } LiAlH_4]{H_2/Pt}$ $HO(CH_2)_6OH$
- $\xrightarrow[\text{Ni or Pt}]{H_2/NH_3}$ $H_2N(CH_2)_6NH_2$
- $\xrightarrow[\text{neutral}]{KMnO_4}$ $HO_2C(CH_2)_4CO_2H$
- $\xrightarrow[R\,=\,OH \text{ or } 2,4\text{-}(NO_2)_2C_6H_3NH]{RNH_2}$ $RN{=}CH(CH_2)_4CH{=}NR$ + cyclopentenyl–$CH{=}NR$

undergoes partial deacetylation to 4,5-dihydrooxepin, which then rearranges to the aldehyde as demonstrated by Pohl (48).

In addition to the aforementioned reactions with **109**, potassium permanganate oxidation and condensation with hydroxylamine hydrochloride have been used for the characterization of the product obtained upon air oxidation of 3-methyl-1-cyclohexene in the presence of acetic anhydride (53).

Another oxidative transformation in the 2,3,4,5-tetrahydro series is the ozonolysis of **108**, which was reported in 1964 by Corbier and Teisseire (9). The primary cleavage product was formulated as aldehyde ester **122**. Further oxidation of this substance with moist silver oxide afforded the corresponding acid ester which gave, upon alkaline hydrolysis and oxidation with potassium permanganate, a mixture of acids. One of these was identified as 3-methylglutaric acid. The ozonolysis of the 7-isopropylidene-4-methyl compound, **123**, was also investigated as part of this work. Presumably the primary oxidation product in this case was ε-caprolactone derivative **124**.

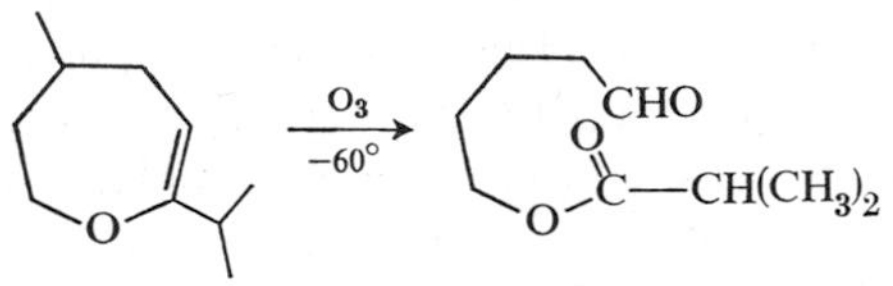

O_3, $-60°$: **108** → **122** (CHO, O–C(=O)–$CH(CH_3)_2$)

O_3, $-60°$: **123** → **124**

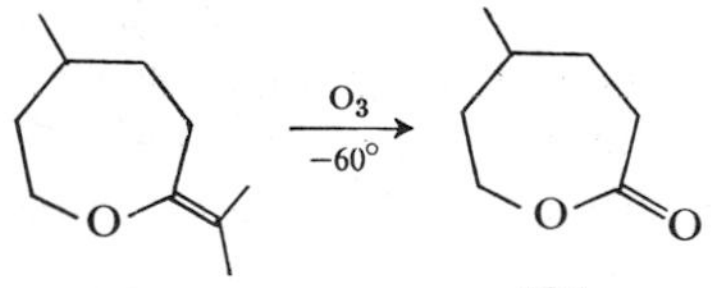

Few chemical reactions involving 2,3,4,7-tetrahydrooxepins have been reported. Olsen and Bredoch (42) stated that the compound which they believed to be **104** was degraded to oxalic acid and succinic acid upon oxidation with alkaline potassium permanganate. The absence of malonic acid among the oxidation products was cited as evidence that **105** was not present. The validity of this observation was cast into doubt, however, by the more recent work of Meinwald and co-workers (35) which suggests that the material formulated by Olsen and Bredoch as pure **104** was actually a mixture of **104** and **105**, with the latter isomer being preponderant. Catalytic reduction of this mixture yielded hexamethylene oxide (**125**) (35, 42). Additional reactions which may be cited here include (*1*) the oxidation of **115** with hydrogen peroxide, (*2*) the pyrolysis of the resulting amine oxide, **117**, to 2,3-dihydrooxepin (**126**) (35), and (*3*) the recently reported reduction of

6-chloro-7-(2-chloroethoxy)-2,3,4,7-tetrahydrooxepin (**127**) to 7-(2-chloroethoxy)-2,3,4,5,6,7-hexahydrooxepin (**128**) (40)

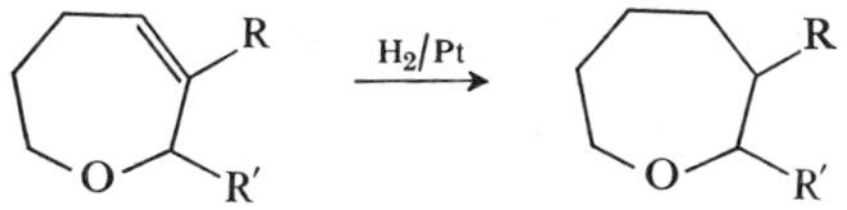

104; R = R′ = H
127; R = Cl, R′ = $ClCH_2CH_2O$
H_2/Pt
125; R = R′ = H
128; R = Cl, R′ = $ClCH_2CH_2O$

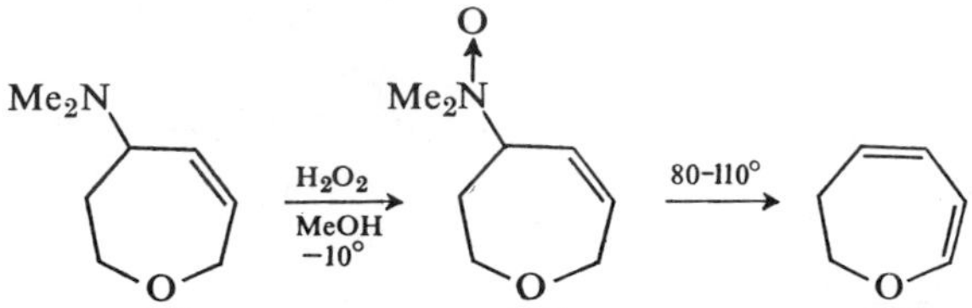

Me_2N
115
H_2O_2, MeOH, −10°
117
80–110°
126

Unlike isomers **103** and **104**, which might be expected to behave more like vinylic and allylic ethers, respectively, 2,3,6,7-tetrahydrooxepin (**105**) exhibits typical olefinic ether properties. Among these may be mentioned

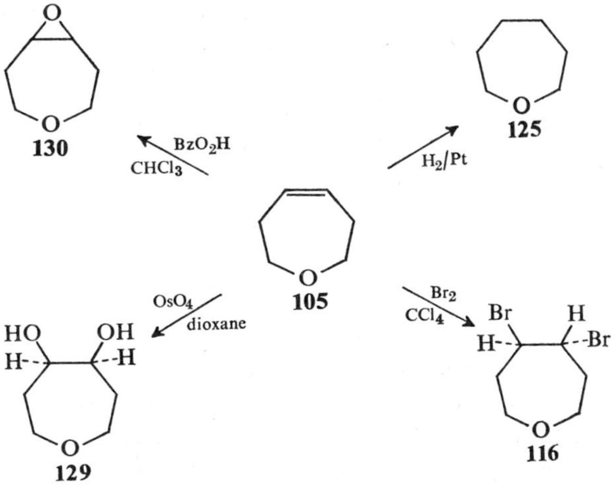

130 ← BzO_2H, $CHCl_3$ — **105** — H_2/Pt → **125**
129 ← OsO_4, dioxane — **105** — Br_2, CCl_4 → **116**

catalytic reduction to **125** (35), bromination to *trans*-4,5-dibromo-2,3,4,5,6,7-hexahydrooxepin (**116**) (36), osmium tetroxide oxidation to *cis*-4,5-dihydroxy-2,3,4,5,6,7-hexahydrooxepin (**129**) (36), and perbenzoic acid oxidation to 4,5-epoxy-2,3,4,5,6,7-hexahydrooxepin (**130**) (36).

D. Hexahydrooxepins (Oxepans)

1. Preparation

a. *Hexamethylene Oxide*

The earliest reported synthesis of hexamethylene oxide (**131**) appeared in 1914 in a paper by Franke and Lieben (14) pertaining to the reaction of sulfuric acid with 1,6-hexanediol. The steam-volatile fraction isolated from this reaction proved to be a mixture of cyclic ethers. Oxidation of the mixture with potassium permanganate afforded formic, acetic, propionic, glutaric, and succinic acid. On this basis, Franke and Lieben concluded that the cyclic ethers formed as a result of dehydration were **131**, 2-methyltetrahydropyran (**132**), and 2-ethyltetrahydrofuran (**133**). Additional evidence supporting this assignment was obtained subsequently in the same laboratory (13). Oxidation of the mixture of cyclic ethers with chromic acid yielded 4-ketocaproic acid and 5-ketocaproic acid. Cleavage of the cyclic ethers with hydrobromic acid, conversion of the resulting mixture of dibromo compounds into dinitriles, and hydrolysis of the latter led to a mixture of suberic, 2-methylpimelic, and 2-ethyladipic acids. The formation of significant quantities of **131** during the sulfuric acid dehydration of 1,6-hexanediol was also confirmed recently by Traynelis and co-workers (57).

$$HO(CH_2)_6OH \xrightarrow[\text{steam distn}]{H_2SO_4} \textbf{131} + \textbf{132} + \textbf{133}$$

In 1944 Mueller and Vanc (39) published a study on the reaction of a series of terminal diols with anhydrous hydrogen bromide. The diols investigated included 1,4-butanediol, 1,6-hexanediol, 1,7-heptanediol, and 1,9-nonanediol. Interesting differences were observed, depending upon the chain length of the diol, the reaction temperature, the length of heating, and the rate of removal of volatile products from the reaction zone. Although conditions were found which allowed tetrahydrofuran to be formed in nearly 69% yield from 1,4-butanediol, the highest yield of **131** that could be achieved from 1,6-hexanediol was only 7.2%. The main products, in this case, were dibromohexane and bis(6-bromo-1-hexyl) ether, and it was shown that these compounds probably arise by further reaction of **131** with hydrogen bromide.

Another method for the preparation of **131** from 1,6-hexanediol was reported more recently by Freidlin and co-workers (15, 16). Passage of the diol over a calcium phosphate catalyst at 320–420° afforded a complex mixture

$$HO(CH_2)_nOH \xrightarrow{HBr} \begin{cases} \xrightarrow[135°,\ 1.75\ hr]{n=4} \text{tetrahydrofuran } (68.7\%) + Br(CH_2)_4Br\ (24.4\%) + Br(CH_2)_4O(CH_2)_4Br\ (1.3\%) \\ \xrightarrow[195°,\ 2.5\ hr]{n=6} \mathbf{131}\ (7.2\%) + Br(CH_2)_6Br\ (57.1\%) + Br(CH_2)_6O(CH_2)_6Br\ (24.4\%) \end{cases}$$

of dehydration products. The mixture was fractionated by vacuum distillation and gas–liquid chromatography. In addition to **131**, other products identified included **132**, **133**, 5-hexen-1-ol, and several isomeric hexadienes. Similar results were also observed when 1,6-hexanediol was heated at 200–300° in the presence of alumina (19). The maximum yield of **131** realized by this route was 30%. A detailed report concerning the application of gas–liquid chromatography to this problem has appeared (52).

Two recent French patents (75, 76) describe the catalytic dehydration of 1,6-hexanediol to **131**, 2-methyltetrahydropyran, and 5-hexen-1-ol. Varying proportions of these three products were formed, depending on the combination of catalyst and temperature. In each case, however, some water must be passed over the catalyst bed at the same time as the diol. Hexamethylene oxide was the sole product when dehydration was conducted at 265° in the presence of a catalyst consisting of 20% aluminum chromate and 10% boric acid on alumina. When the temperature was lowered to 240°, the yield of hexamethylene oxide fell to 63% and a small amount of 2-methyltetrahydropyran was observed.

A practical laboratory method for the conversion of 1,6-hexanediol into **131** was discovered recently by Traynelis and co-workers (57), who noted the efficacy of hot dimethyl sulfoxide as a dehydrating agent for a wide assortment of alcohols. Under optimum conditions, **131** was obtained in yields of nearly 70%. At the present time this procedure appears to be the method of choice for the preparation of **131**.

$$HO(CH_2)_6OH \xrightarrow[190°,14hr]{DMSO(1:2)} \mathbf{131} \quad (70\%)$$

Another open-chain precursor from which **131** can be prepared is 6-bromo-1-hexanol. Treatment with base causes ring closure via intramolecular nucleophilic displacement of bromide ion (18).

$$Br(CH_2)_6OH \xrightarrow{KOH} \left[Br\text{—}CH_2 \quad CH_2\text{—}O^{\ominus} \right] \longrightarrow \mathbf{131}$$

Kirrmann and Hamaide (17, 28) reported in 1957 that methyl ω-bromoalkyl ethers, upon being heated in the presence of ferric chloride, afforded excellent yields of cyclic ethers with concomitant formation of methyl bromide. The volatile products were removed by vacuum distillation as soon as they were formed. Although nearly quantitative yields of tetrahydrofuran and tetrahydropyran were obtained, the yield of **131** was 70%. The starting materials were prepared by partial methanolysis of the appropriate α,ω-dibromides. The formation of cyclic ethers was postulated to take place via an oxonium salt.

$$MeO(CH_2)_6Br \xrightarrow[\Delta]{FeCl_3} \left[\text{cyclic } \overset{\oplus}{O}\text{–Me} \; Br^{\ominus} \right] \longrightarrow \mathbf{131} + MeBr$$

Reference has already been made to the catalytic reduction of unsaturated oxepins. Compounds that have been converted into **131** in this manner include 2,3,4,7-tetrahydrooxepin (42), 2,3,6,7-tetrahydrooxepin (36), 2,3-dihydrooxepin (50), 4,5-dihydrooxepin (5), and oxepin itself (64, 66). Although they obviously do not constitute practical methods for the preparation of **131**, these reactions have been of considerable importance in establishing the correct structure of the compounds being reduced.

b. *Substituted Oxepans*

(1) Ring Closure Reactions. Only one example of the synthesis of a substituted hexamethylene oxide derivative by intramolecular cyclization of an open-chain precursor exists in the literature. This consists of the long-known report by Michiels (37) that treatment of 2,7-dimethyl-2,7-octanediol with warm dilute sulfuric acid results in the formation of a product formulated as 2,2,7,7-tetramethyloxepan (**134**). It seems likely that this reaction is far more general in scope than one isolated example would indicate.

$$(Me)_2C(OH)(CH_2)_4C(OH)(Me)_2 \xrightarrow[\Delta]{\text{dil } H_2SO_4} \underset{\mathbf{134}}{\text{2,2,7,7-tetramethyloxepan (Me, Me, O, Me, Me)}}$$

(2) Ring Expansion Reactions. The reaction of diazomethane with tetrahydropyran-4-one (**135**) has been utilized by Maerov (32) and more recently by Olsen and Bredoch (42) for the synthesis of oxepan-4-one (**136**).

In the same manner Pohl (47) has reported the preparation of 2,7-dimethyloxepan-4-one (**137**) by ring expansion of 2,6-dimethyltetrahydropyran-4-one (**138**).

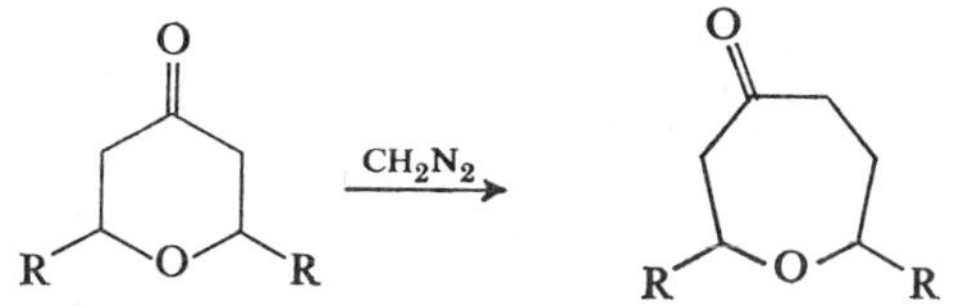

$$\xrightarrow{CH_2N_2}$$

135; R = H
138; R = Me

136; R = H
137; R = Me

An interesting extension of the same reaction was described in 1960 by Korobitsyna and Pivnitskii (29), who found that treatment of 2,2,5,5-tetramethyltetrahydrofuran-3-one (**139**) with diazomethane resulted in the formation of a mixture of 2,2,6,6-tetramethyltetrahydropyran-3-one (**140**), 2,2,6,6-tetramethyltetrahydropyran-4-one (**141**), 2,2,7,7-tetramethyloxepan-3-one (**142**), and 2,2,7,7-tetramethyloxepan-4-one (**143**).

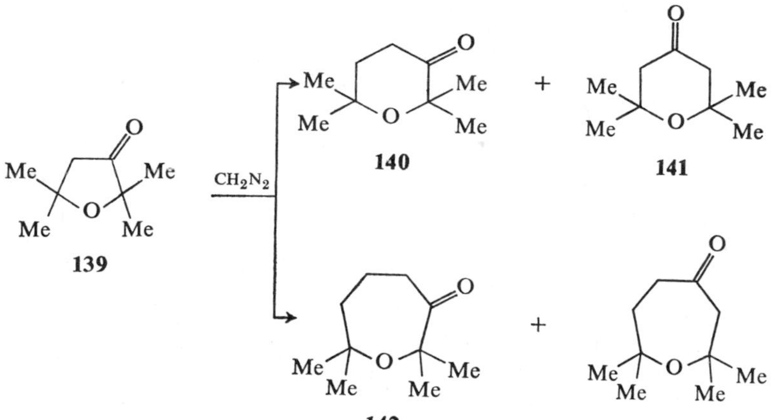

139 $\xrightarrow{CH_2N_2}$ **140** + **141**; **142** + **143**

(3) Two-Component Reactions. Another general approach to the synthesis of substituted oxepans involves construction of the seven-membered ring via a two-component reaction in which one of the building blocks already contains an ether linkage. Thus the disodium salt of 1,1,2,2-tetracarbethoxyethane undergoes condensation with bis(2-bromoethyl) ether with formation of 4,4,5,5-tetracarbethoxyoxepan (**144**) (1).

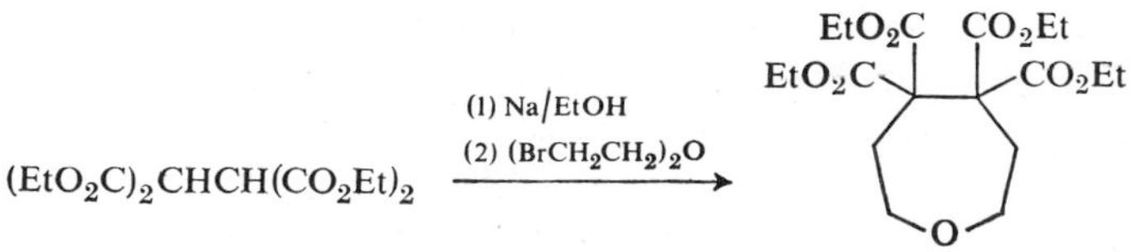

$$(EtO_2C)_2CHCH(CO_2Et)_2 \xrightarrow[(2)\ (BrCH_2CH_2)_2O]{(1)\ Na/EtOH}$$

144

In a similar manner, it was reported by Ali-Zade and Arbuzov (1) and more recently by Matschinskaya and co-workers (34) that treatment of bis-(2,2-dicarbethoxyethyl) ether with magnesium, followed by addition of 1,2-dibromoethane to the resulting magnesium salt, resulted in the formation of 3,3,6,6-tetracarbethoxyoxepan (**145**).

$(EtO_2C)_2CH$ $CH(CO_2Et)_2$ O $\xrightarrow[(2)\ BrCH_2CH_2Br]{(1)\ Mg/EtOH}$ EtO_2C CO_2Et EtO_2C CO_2Et O

145

2. Physical Properties of Hexamethylene Oxide

Arnett and Wu (3) have investigated the basicity of **131** in relation to other cyclic ethers. Their method involved partitioning of the ether between an inert organic solvent and a series of standard acid solutions of known H_0 strength, followed by gas–liquid chromatographic analysis of the organic layer. The pK_a values determined in this manner for **131**, tetrahydropyran, and tetrahydrofuran were −2.02, −2.79, and −2.08, respectively. It is interesting that the five- and seven-membered cyclic ethers appear to be significantly more basic than the six-membered homolog. According to Arnett and Wu, the observed order of pK_a values for the cyclic ethers can be explained more satisfactorily in terms of reverse *I*-strain than in terms of the usual polar or steric arguments. It is assumed that the unshared electron pairs on the oxygen atom of a cyclic ether such as **131** are sterically equivalent to medium-sized groups. Repulsion between these electrons and the electrons in neighboring C—H bonds is substantial but can be relieved by coordination of the oxygen electrons with a proton. Because of the unique conformational properties of six-membered rings, protonation is less effective in relieving nonbonded repulsion in tetrahydropyran than in tetrahydrofuran or **131**.

In a more recent investigation Barakat and co-workers (4) determined the association constants for the hydrogen-bonded complexes between thiocyanic acid and various cyclic ethers. Their method involved measurement of the intensity of the nonbonded N—H stretching bands of the complexes in carbon tetrachloride solution at several temperatures. The enthalpy, entropy, and free energy of formation of the complexes were obtained by standard thermodynamic calculations. For trimethylene oxide, tetrahydrofuran, tetrahydropyran, and **131**, the association constants were found to be 86.1, 72.0, 53.7, and 38.6 l/mole, respectively. Thus using complex formation with thiocyanic acid as a measure, one would rank the basicities (i.e., electron-sharing capacities) of cyclic ethers as $4 < 5 < 6 < 7$, which is somewhat in conflict with the earlier results of Arnett and Wu.

Two ir spectroscopic studies involving **131** have been published. The first, reported in 1952 by Tschamler and Leutner (59), was an analysis of the C—O—C stretching bands in cyclic ethers in terms of ring size. For **131**, tetrahydropyran, and tetrahydrofuran, these bands were found to occur at 1130 cm^{-1} (8.85 μ), 1090 cm^{-1} (9.16 μ), and 1070 cm^{-1} (9.35 μ), respectively. The second study, published in 1957 by Kirrmann and Hamaide (28), extended this correlation to include the C—H stretching, CH_2 deformation, and ring vibration frequencies as well.

3. Chemical Reactions of Hexamethylene Oxide

Like other cyclic ethers, **131** undergoes ring cleavage upon vigorous acid treatment. Thus heating with concentrated hydrobromic acid at 100–110° affords a high yield of 1,6-dibromohexane (39, 42) together with a small amount of bis(6-bromo-1-hexyl) ether (39).

$$\mathbf{131} \xrightarrow[\substack{100\text{–}110^\circ \\ 18\text{hr}}]{\text{HBr}} Br(CH_2)_6Br + Br(CH_2)_6O(CH_2)_6Br$$

In the presence of either strong mineral acid, a metal halide such as phosphorus pentafluoride or boron trifluoride, or a variety of trialkyloxonium salts, **131** can also undergo polymerization (41). When other cyclic ethers or cyclic acetals are included, **131** can also function as a copolymer (41). Similarly, copolymerization has been observed between **131** and maleic anhydride at 120–125° in the presence of phosphorus pentafluoride (26).

Habermeier and co-workers (74) reported in 1967 that hexamethylene oxide and other cyclic ethers, such as tetrahydrofuran and tetrahydropyran, can be polymerized in the presence of an anhydride and a Lewis-acid catalyst. Anhydrides used for this purpose included succinic anhydride and phthalic anhydride, and catalysts reported were antimony pentachloride–acetyl chloride complex and triethyloxonium fluoroborate.

Ring cleavage has also been brought about by heating with methyl iodide in a sealed tube. The resulting 1,6-diiodohexane was not isolated, being converted directly into 1,6-bis(2-naphthoxy)hexane by reaction with sodium 2-naphthoxide (38).

$$\mathbf{131} \xrightarrow[160^\circ,\ 48\text{ hr}]{\text{MeI}} \left[\text{(seven-membered ring } O^{\oplus}\text{–Me)}\ \ I^{\ominus}\right] \longrightarrow I(CH_2)_6I$$

Reference has been made previously to a number of reactions of substituted oxepans which do not involve rupture of the seven-membered ring. These include the conversion of **136** into oxepan-4-ol, 4-acetoxyoxepan, and 2,3,6,7-tetrahydrooxepin and 2,3,4,7-tetrahydrooxepin (35, 42), the reaction of *trans*-4,5-dibromooxepan with dimethylamine (35), and the acetylation of *cis*-4,5-dihydroxyoxepan (35). These transformations all attest to the relative stability of the fully saturated seven-membered ether function.

4. Preparation and Reactions of Cyclic Acetals, Ketals, and Related Compounds

The ability of ε-hydroxyaldehydes to exist in the form of seven-membered hemiacetals was first recognized by Helferich and Sparmberg (25), who reported in 1931 that alkaline hydrolysis of 6-acetoxy-4-methylhexanal, which was derived from citronellol via ozonolysis and acetylation, afforded a mixture of 6-hydroxy-4-methylhexanal and 2-hydroxy-5-methyloxepan (**146**). A positive Fehling's test indicated that hemiacetal formation had not proceeded to completion. When the equilibrium mixture was treated with methanol in the presence of acid, a single compound giving a negative Fehling's test was formed. This product was formulated as 2-methoxy-5-methyl-oxepan (**147**). Acid hydrolysis of **147** regenerated the mixture of 6-hydroxy-4-methylhexanal and **146**.

The work of Helferich and Sparmberg was confirmed and extended by Eschinazi (12) in 1961. Ozonolysis of citronellol in acetic acid gave, upon distillation, a product whose refractive index, optical rotation, and ir absorption characteristics underwent a systematic change as a function of

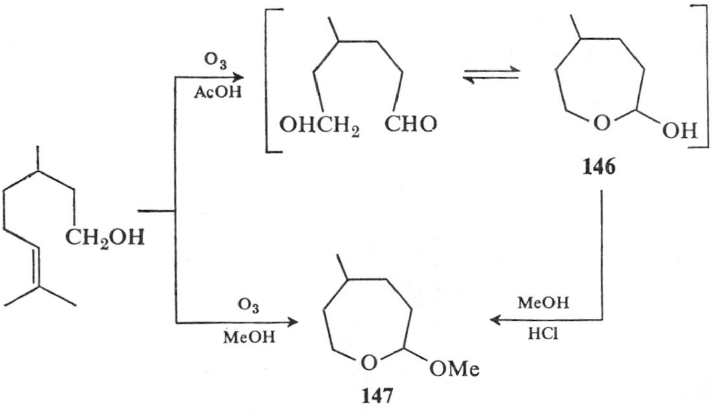

O_3, AcOH; OHCH$_2$ CHO ⇌ O OH **146**

CH_2OH; O_3, MeOH; MeOH, HCl; O OMe **147**

time. The gradual disappearance of OH and C═O peaks and concomitant emergence of new C—O bands after about 24 hr was consistent with gradual ring closure of the initially formed ε-hydroxyaldehyde. When ozonolysis was conducted in methanol, moreover, the previously reported stable compound, **147**, was isolated directly together with other, unidentified higher-boiling products.

An excellent study of the influence of chain length upon ease of formation of cyclic acetals from ε-hydroxyaldehydes was carried out in 1952 by Hurd and Saunders (27). An estimate of the position of equilibrium was made by comparing the uv extinction coefficients of ε-hydroxyaldehydes with those of the corresponding ε-methoxyaldehydes. At 25° in 75% dioxane, 6-hydroxyhexanal was calculated to exist in the open-chain form, **148A**, to the extent of 85%. At 35° the value increased to 89%. The seeming reluctance of 6-hydroxyhexanal to undergo ring closure to a hemiacetal contrasted sharply with the ease of cyclization of 5-hydroxypentanal and 4-hydroxybutyraldehyde, which were found to exist in the open-chain form only to the extent of 6.1 and 11.4%, respectively, at 25°. This difference was attributed to a combination of ring strain effects and probability factors.

For the preparation of 6-hydroxyhexanal, 2-hydroxycyclohexanone (adipoin) was oxidized to methyl 6,6-dimethoxyhexanoate with lead tetraacetate in methanol. Lithium aluminum hydride reduction then gave 6,6-dimethoxy-1-hexanol, and hydrolysis of the acetal function with dilute acid yielded the desired product.

O, OH; $Pb(OAc)_4$ / MeOH → MeO_2C $CH(OMe)_2$ → $LiAlH_4$

$OHCH_2$ $CH(OMe)_2$ → HCl

$[HO(CH_2)_5CHO \rightleftharpoons$ O, OH$]$

148A **148**

An alternate synthesis of 6-hydroxyhexanal was devised more recently by Perrine (46), starting from 1-benzyloxy-6,6-diethoxyethane. Hydrogenolysis over palladium–charcoal afforded a compound formulated tentatively as 2-ethoxyoxepan (**149**). Acid hydrolysis of **149** gave 6-hydroxyhexanal, which was isolated in the form of its 2,4-DNP derivative.

$CH(OEt)_2$, O, $C_6H_5CH_2$ $\xrightarrow{H_2/Pd-C}$ O, OEt **149** $\xrightarrow{HCl}$ $[HO(CH_2)_5CHO \rightleftharpoons$ **148**]

Still another method of preparation of 6-hydroxyhexanal, reported in 1958 by Colonge and Corbet (7, 8), involved nitrous acid deamination of 2-aminomethyltetrahydropyran (**150**). In addition to 6-hydroxyhexanal the reaction yielded a second product identified as 2-hydroxymethyltetrahydropyran (**151**).

O, CH_2NH_2 **150** $\xrightarrow{HNO_2}$ $[HO(CH_2)_5CHO \rightleftharpoons$ **148**] + O, CH_2OH **151** (26%)

48%

An interesting recent paper describing the synthesis of a 2-alkoxyoxepan was published by Nerdel and co-workers (40), who obtained 2-(2′ chloroethoxy)oxepan (**152**) in 75% yield upon hydrogenation of 6-chloro-7-(2-chloroethoxy)-2,3,4,7-tetrahydrooxepin (**153**) in alkaline methanol in the presence of nickel catalyst.

Cl, O, OCH_2CH_2Cl **153** $\xrightarrow[\text{alk MeOH}]{H_2/Ni}$ O, OCH_2CH_2Cl **152**

Like ε-hydroxyaldehydes, ε-hydroxyketones can exist, in principle, in the form of seven-membered hemiketals. Thus treatment of 7-hydroxy-2-heptanone with dilute methanolic hydrochloric acid has been reported to give 2-methoxy-2-methyloxepan (**154**) (10). Similarly, the reaction of 1-hydroperoxy-1-methylcyclohexane (**155**) with acetic acid has been found to yield 2-acetoxy-2-methyloxepan (**156**) (60). Acid hydrolysis of **156** affords 7-hydroxy-2-heptanone.

A related type of oxepan derivative was described in 1961 by Treibs and Schoellner (58) in a paper dealing with the photocatalyzed oxidation of cyclohexene. Among the numerous products formed when the reaction was carried out in methanol containing a small amount of sulfuric acid was a compound formulated as 2,7-dimethoxyoxepan (157). This substance can

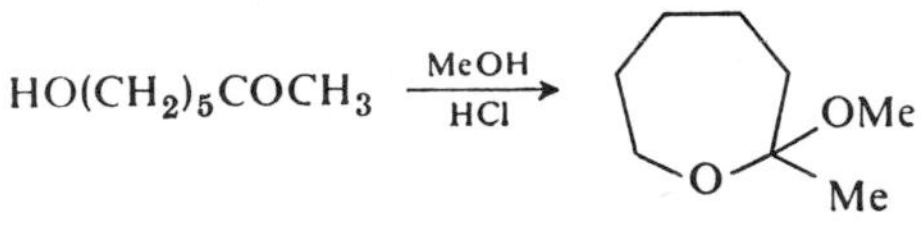

$HO(CH_2)_5COCH_3$ —MeOH/HCl→ OMe, Me

154 (44%)

OOH, Me —AcOH→ OAc, Me —HCl→ $HO(CH_2)_5COCH_3$

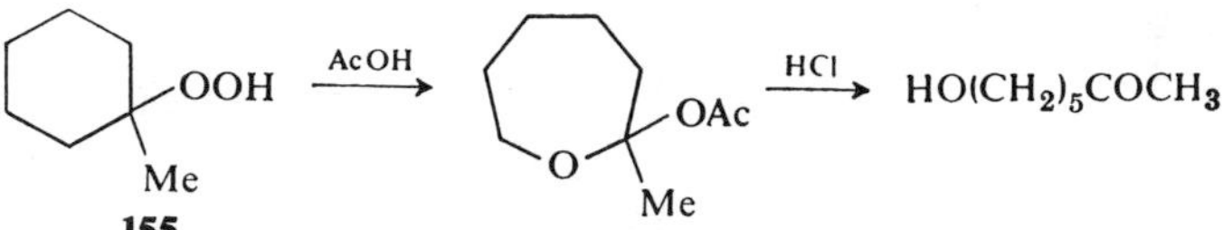

155 **156**

be derived formally from 1,6-hexanedialdehyde. Treatment of **157** with 2,4-dinitrophenylhydrazine in the presence of acid did, in fact, afford the 2,4-DNP derivative of 1,6-hexanedialdehyde. The formation of **157** was postulated to occur via the pathway shown.

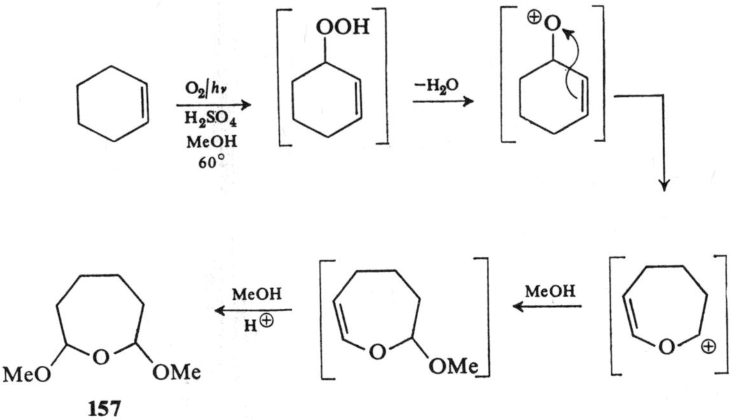

$O_2/h\nu$, H_2SO_4, MeOH, 60° → [OOH] —$-H_2O$→ [⊕O] → [O ⊕] —MeOH→ [O OMe] —MeOH, H⊕→ MeO O OMe

157

E. Tables

The physical constants for the various oxepins and reduced oxepins discussed in this chapter are tabulated in Table I-1.

TABLE I-1. Oxepins and Reduced Oxepins

Empirical formula	Name of compound	Physical constants	Derivatives	Refs.
$C_6H_5ClO_2$	6-Chloro-2,3-dihydrooxepin-3-one			33
C_6H_6O	Oxepin	b_{30} 38°C n_D^{20} 1.5162		66
C_6H_7ClO	6-Chloro-2,3-dihydrooxepin	b_{25} 75°C b_{78} 103–107°C n_D^{25}1.5167–1.5178		2, 50
C_6H_7FO	6-Fluoro-2,3-dihydrooxepin			2
C_6H_8O	2,3-Dihydrooxepin	bp 118–120°C $n_D^{25.5}$ 1.4949–1.4950		35
		b_{100} 61°C n_D^{25} 1.4965		50
	4,5-Dihydrooxepin	bp 108°, 110–112°, 113°C n_D^{25} 1.4632		5, 54, 48
$C_6H_{10}Br_2O$	*trans*-4,5-Dibromo-2,3,4,5,6,7-hexahydrooxepin	mp 29°C (pentane)		35
$C_6H_{10}O$	2,3,4,5-Tetrahydrooxepin	bp 113–114°, 114.5–115°C n_D^{25} 1.5445 d_{20}^{25} 0.9331		31, 40
	2,3,4,7-Tetrahydrooxepin	bp 98–100°C n_D^{20} 1.4562 d_{20} 0.9266		42
	2,3,6,7-Tetrahydrooxepin	bp 118–119°C $n_D^{25.5}$ 1.4548		36, 42

$C_6H_{10}O_2$	2,3,4,5,6,7-Hexahydrooxepin-4-one	b_8 68°C n_D^{20} 1.4611 d_{20} 1.0682	Phenylsemicarbazone, m. 168–169° (EtOH) 2,4-DNP, m. 173–174° (MeOH)	
	4,5-Epoxy-2,3,4,5,6,7-hexahydrooxepin	b_{30} 77.5–78°C $n_D^{25.5}$ 1.4600		35
$C_6H_{12}O$	2,3,4,5,6,7-Hexahydrooxepin (Oxepan, Hexamethylene oxide)	b_{760} 119–120°C b_{749} 104.5–106°C b_{741} 121°C $n_D^{19.5}$ 1.4358 n_D^{20} 1.4393, 1.4355 n_D^{25} 1.4365 $d_{19.5}$ 0.8875 d_4^{20} 0.9036 $d_4^{21.1}$ 0.8537		14–17, 36, 39, 42, 50, 57
$C_6H_{12}O_2$	2,3,4,5,6,7-Hexahydrooxepin-4-ol	b_3 84–86°C n_D^{22} 1.471 d_{22} 1.0595	Phenylurethan, mp 84–86°C (ligroin) α-Naththylurethan, mp 119–121°C (ligroin)	42
$C_6H_{12}O_3$	2,3,4,5,6,7-Hexahydrooxepin-*cis*-4,5-diol	mp 42–45°C subl 60°C (0.05 mm)		35
C_7H_8O	2-Methyloxepin	b_{11} 33.5–34°C		22, 65
$C_8H_8O_2$	2-Acetyloxepin	$b_{0.4}$ 51°C		65
$C_8H_{10}O$	2,7-Dimethyloxepin	b_{14} 51°C n_D^{20} 1.5045		65, 66
	4,5-Dimethyloxepin			65
$C_8H_{12}Cl_2O_2$	6-Chloro-7-(2-chloroethoxy)-2,3,4,7-tetrahydrooxepin	$b_{0.3}$ 83.5–84°C n_D^{20} 1.4996		40
$C_8H_{12}O$	4,5-Dihydro-2,7-dimethyloxopin	b_{10} 52–54°C, b_{14} 51°C n_D^{20} 1.4298, 1.5045 d_4^{20} 0.891		34, 65, 66

TABLE I-1.—(*Contd.*)

Empirical formula	Name of compound	Physical constants	Derivatives	Refs.
$C_8H_{12}O_3$	2-Acetoxy-2,3,4,5-tetrahydrooxepin	b_3 52°C n_D^{20} 1.4580		51
$C_8H_{14}O_2$	2,3,4,5,6,7-Hexahydro-2,7-dimethyl-oxepin-4-one	b_{11} 81–82°C		47
$C_8H_{14}O_3$	2-Acetoxy-2,3,4,5,6,7-hexahydrooxepin	b_3 61–62°C n_D^{20} 1.4472		51
	4-Acetoxy-2,3,4,5,6,7-hexahydrooxepin	b_9 86°C n_D^{20} 1.450 d_{20} 1.0595		42
$C_8H_{15}ClO_2$	2-(2-Chloroethoxy)-2,3,4,5,6,7-hexahydrooxepin	b_{12} 100–102°C n_D^{20} 1.4638		40
$C_8H_{15}NO$	4-(*N*,*N*-Dimethylamino)-2,3,4,7-tetrahydrooxepin	$b_{1.4}$ 42–43°C $n_D^{25.5}$ 1.4754–1.4756	Picrate, mp 136–138°C (EtOH)	35
$C_8H_{15}NO_2$	4-(*N*,*N*-Dimethylamino)-2,3,4,7-tetrahydrooxepin *N*-oxide	Oil	Picrate, mp 164–166°C (EtOH)	35
$C_8H_{16}O_2$	2-Ethoxy-2,3,4,5,6,7-hexahydrooxepin			46
	2,3,4,5,6,7-Hexahydro-2-methoxy-2-methyloxepin	b_{740} 135–136°C		10
	2,3,4,5,6,7-Hexahydro-2-methoxy-5-methyloxepin	b_{28} 65–67°C n_D^{20} 1.4350 $[\alpha]^{25}$ +97°		12
$C_8H_{16}O_3$	2,3,4,5,6,7-Hexahydro-2,7-dimethyloxepin	b_{15} 82–84°C n_D^{20} 1.432		58
$C_9H_{14}O_2$	2,3,4,5-Tetrahydro-2-propionyloxyoxepin	$b_{0.5}$ 44–48°C n_D^{20} 1.4586		51
$C_9H_{14}O_3$	2-Acetoxy-2,3,4,5-tetrahydro-2-methyloxepin	b_2 61–62°C n_D^{20} 1.4583		53
$C_9H_{16}O_3$	2-Acetoxy-2,3,4,5,6,7-hexahydro-2-methyloxepin	b_4 98–104°C		60

$C_{10}H_{10}O_5$	Dimethyl oxepin-4,5-dicarboxylate			49
$C_{10}H_{16}O$	4,5-Dihydro-2,4,5,7-tetramethyloxepin	b_7 75–77°C n_D^{20} 1.4487 d_4^{20} 0.879		34
$C_{10}H_{16}O_5$	*cis*-4,5-Diacetoxy-2,3,4,5,6,7-hexahydrooxepin	$b_{0.2}$ 80–83°C $n_D^{25.5}$ 1.4535		35
	trans-4,5-Diacetoxy-2,3,4,5,6,7-hexahydrooxepin	$b_{0.4}$ 91–91.5°C n_D^{26} 1.4521		35
	Dimethyl 2,3,4,5,6,7-hexahydrooxepin-4,5-dicarboxylate			49
$C_{10}H_{18}O$	2,3,4,5,6,7-Hexahydro-2-isopropylidene-5-methyloxepin			9
	2,3,4,5-Tetrahydro-7-isopropyl-4-methyloxepin	b_{10} 63–63.5°C n_D^{15} 1.4529 d_{15} 0.8923		56
$C_{10}H_{18}O_2$	2,3,4,5,6,7-Hexahydro-2,2,7,7-tetramethyloxepin-3-one			29
	2,3,4,5,6,7-Hexahydro-2,2,7,7-tetramethyloxepin-4-one			29
$C_{10}H_{20}O$	2,3,4,5,6,7-Hexahydro-2,2,7,7-tetramethyloxepin	b_{756} 156–157°C		37
$C_{18}H_{28}O_9$	3,3,6,6-Tetracarbethoxy-2,3,4,5,6,7-hexahydrooxepin	$b_{1.5}$ 158–160°C n_D^{21} 1.4385 d_{21}^{21} 1.1110		1, 34
	4,4,5,5-Tetracarbethoxy-2,3,4,5,6,7-hexahydrooxepin	b_5 208–212°C n_D^{21} 1.4561 d_{21}^{21} 1.1572		1
$C_{20}H_{20}O$	4,5-Dihydro-2,7-dimethyl-4,5-diphenyloxepin	b_6 104–106°C n_D^{20} 1.5066 d_4^{20} 0.955		34
$C_{30}H_{22}O_5$	2,7-Dihydro-4,5-di(*o*-hydroxyphenyl)-2,7-diphenyloxepin-3,6-dione	mp 211–212°C dec	Diacetate, mp 162.5–163°C	71
$C_{32}H_{26}O_5$	2,7-Dihydro-4,5-di(*o*-methoxyphenyl)-2,7-diphenyloxepin-3,6-dione	mp 194–195°C		71

F. References

1. I. Ali-Zade and B. A. Arbuzov, *J. Gen. Chem. USSR*, **13,** 113 (1943); *Chem. Abstr.*, **38,** 352 (1944).
2. T. Ando, H. Yamanaka, and W. Funasaka, *Tetrahedron Lett.*, **1967,** 2587.
3. E. M. Arnett and C. Y. Wu, *J. Amer. Chem. Soc.*, **84,** 1684 (1962).
4. T. M. Barakat, M. J. Nelson, S. M. Nelson, and S. D. E. Pullin, *Trans. Faraday Soc.*, **62,** 2674 (1966).
5. R. A. Braun, *J. Org. Chem.*, **28,** 1383 (1963).
6. G. L. Buchanan, J. A. Hamilton, T. A. Hamor, and G. A. Sims, *Acta Chem. Scand.*, **16,** 776 (1962).
7. J. Cologne and P. Corbet, *C. R. Acad. Sci., Paris*, **247,** 2144 (1958).
8. J. Cologne and P. Corbet, *Bull. Soc. Chim. Fr.*, **1960,** 287.
9. B. Corbier and P. Teisseire, *Recherches* (Paris), **14,** 93 (1964); *Chem. Abstr.*, **63,** 4334 (1965).
10. R. Criegee and W. Schnorrenberg, *Justus Liebigs Ann. Chem.*, **560,** 141 (1948).
11. C. H. DePuy, L. G. Schnack, and J. W. Hausser, *J. Amer. Chem. Soc.*, **88,** 3343 (1966).
12. H. E. Eschinazi, *J. Org. Chem.*, **26,** 3072 (1961).
13. A. Franke, A. Kroupa, F. Schweitzer, M. Winischhofer, H. Klein-Lohr, M. Just, M. Hackl, I. v. Reyher, and R. Bader, *Monatsh. Chem.*, **69,** 167 (1936).
14. A. Franke and F. Lieben, *Monatsh. Chem.*, **35,** 1431 (1914).
15. L. Kh. Friedlin, V. Z. Sharf, and M. A. Abidov, *Neftekhimiya*, **5,** 558 (1965); *Chem. Abstr.*, **63,** 16303 (1965).
16. L. Kh. Freidlin, V. Z. Sharf, and N. S. Andreev, *Izv. Akad. Nauk SSSR, Otd. Khim. Nauk*, **1961,** 373; *Chem. Abstr.*, **55,** 19774 (1961).
17. French Patent 1,165,673; *Chem. Abstr.*, **54,** 19731 (1960).
18. G. Gailer, Dissertation, Vienna University, 1937, cited by W. Luerken and E. Mueller, in *Methoden der Organischen Chemie* (Heuben-Weyl), 4th ed., E. Mueller, Ed., Georg Thieme Verlag, Stuttgart, 1966, Vol. 6/4, p. 450.
19. German Patent 840,844; *Chem. Abstr.*, **52,** 16388 (1958).
20. H. Guenther, *Angew. Chem.*, **77,** 1022 (1965).
21. H. Guenther, *Angew. Chem.*, **77,** 1083 (1965).
22. H. Guenther, *Z. Naturforsch.*, **20b,** 948 (1965).
23. H Guenther, *Tetrahedron Lett.*, **1965,** 4085.
24. H. Guenther and H. H. Hinrichs, *Tetrahedron Lett.*, **1966,** 787.
25. B. Helferich and G. Sparmberg, *Chem. Ber.*, **64B,** 104 (1931).
26. A. Hilt, K. H. Reichert, and K. Hamann, *Makromol. Chem.*, **101,** 246 (1967).
27. C. D. Hurd and W. H. Saunders, Jr., *J. Amer. Chem. Soc.*, **74,** 5324 (1952).
28. A. Kirrmann and N. Hamaide, *Bull. Soc. Chim. Fr.*, **1957,** 789.
29. I. K. Korobitsyna and K. K. Pivnitskii, *Zh. Obshch. Khim.*, **30,** 4008 (1960); *Chem. Abstr.*, **55,** 22277 (1961).
30. S. S. Labana, *Dissertation Abstr.*, **24,** 3544 (1964).
31. D. R. Larkin, *J. Org. Chem.*, **30,** 335 (1965).
32. S. B. Maerov, *Dissertation Abstr.*, **14,** 765 (1954).
33. S. Masamune and N. T. Castellucci, *Chem. Ind.* (London), **1965,** 184.
34. I. W. Matschinskaya, W. A. Barchasch, and A. T. Pruschenko, *Zh. Obshch. Khim.*, **30,** 2362 (1960); *Chem. Abstr.*, **55,** 8327 (1961).
35. J. Meinwald, D. W. Dicker, and N. Danieli, *J. Amer. Chem. Soc.*, **82,** 4087 (1960).

36. J. Meinwald and H. Nozaki, *J. Amer. Chem. Soc.*, **80,** 3132 (1958).
37. L. Michiels, *Bull. Soc. Chim. Belges*, **27,** 25 (1913); *Chem. Abstr.*, **7,** 3602 (1913).
38. A. Mueller, E. Funder-Fritzsche, W. Konar, and E. Rintersbacher-Wlasak, *Monatsh. Chem.*, **84,** 1206 (1953).
39. A. Mueller and W. Vanc, *Chem. Ber.*, **77B,** 669 (1944).
40. F. Nerdel, J. Buddrus, W. Brodowski, and P. Weyerstahl, *Tetrahedron Lett.*, **1966,** 5385.
41. Netherlands Patent Appl. 6,514,413; *Chem. Abstr.*, **65,** 10687 (1966).
42. S. Olsen and R. Bredoch, *Chem. Ber.*, **91,** 1589 (1958).
43. L. A. Paquette, *Trans. N. Y. Acad. Sci.*, **28,** 397 (1966).
44. L. A. Paquette and J. H. Barrett, *J. Amer. Chem. Soc.*, **88,** 1718 (1966).
45. L. A. Paquette, J. H. Barrett, R. P. Spitz, and R. Pitcher, *J. Amer. Chem. Soc.*, **87,** 3417 (1965).
46. T. D. Perrine, *J. Org. Chem.*, **18,** 1356 (1953).
47. G. Pohl, Dissertation, Marburg University, 1961, cited by W. Luerken and E. Mueller, in *Methoden der Organischen Chemie* (Houben-Weyl), 4th ed., E. Mueller, Ed., Georg Thieme Verlag, Stuttgart, 1966, Vol. 6/4, p. 459.
48. G. Pohl, Dissertation, Marburg University, 1961, cited by W. Luerken and E. Mueller, in *Methoden der Organischen Chemie* (Houben-Weyl), 4th ed., E. Mueller, Ed., Georg Thieme Verlag, Stuttgart, 1966, Vol. 6/4, p. 466.
49. H. Prinzbach, M. Arguelles, and E. Druckrey, *Angew. Chem.*, **78,** 1057 (1966).
50. E. E. Schweizer and W. E. Parham, *J. Amer. Chem. Soc.*, **82,** 4085 (1960).
51. H. J. Shine and R. H. Snyder, *J. Amer. Chem. Soc.*, **80,** 3064 (1958).
52. N. I. Shuikin, B. L. Lebedev, and V. V. An, *Izv. Akad. Nauk SSSR, Otd. Khim. Nauk*, **1962,** 1868; *Chem. Abstr.*, **58,** 3889 (1963).
53. R. H. Snyder, H. J. Shine, K. A. Leibbrand, and P. O. Tawney, *J. Amer. Chem. Soc.*, **81,** 4299 (1959).
54. E. L. Stogryn, M. H. Gianni, and A. J. Passanante, *J. Org. Chem.*, **29,** 1275 (1964).
55. A. Streitwieser, Jr., *Molecular Orbital Theory for Organic Chemists*, Wiley, New York, 1961, p. 280.
56. P. Teisseire and B. Corbier, *Recherches* (Paris), **13,** 78 (1963); *Chem. Abstr.*, **61,** 7049 (1964).
57. V. J. Traynelis, W. L. Hergenrother, H. T. Hanson, and J. A. Valicenti, *J. Org. Chem.*, **29,** 123 (1964).
58. W. Treibs and R. Schoellner, *Chem. Ber.*, **94,** 2983 (1961).
59. H. Tschamler and R. Leutner, *Monatsh. Chem.*, **83,** 1502 (1952).
60. U.S. Patent 2,594,322; *Chem. Abstr.*, **47,** 1196 (1953).
61. U.S. Patent 2,881,185; *Chem. Abstr.*, **53,** 13180 (1959).
62. U.S. Patent 3,261,848; *Chem. Abstr.*, **65,** 13842 (1966).
63. E. E. van Tamelen and D. Carty, *J. Amer. Chem. Soc.*, **89,** 3922 (1967).
64. E. Vogel, W. A. Boell, and H. Guenther, *Tetrahedron Lett.*, **1965,** 609.
65. E. Vogel and H. Guenther, *Angew. Chem.*, **79,** 429 (1967).
66. E. Vogel, R. Schubart, and W. A. Boell, *Angew. Chem.*, **76,** 535 (1964).
67. G. Westoo, *Acta Chem. Scand.*, **13,** 604 (1959).
68. H. W. Whitlock, Jr., and N. A. Carlson, *Tetrahedron*, **20,** 2101 (1964).
69. J. Wolters, *Chem. Weekbl.*, **62,** 588 (1966).
70. H. Guenther, R. Schubart, and E. Vogel, *Z. Naturforsch.*, **22b,** 25 (1967).
71. T. Kubota, N. Ichikawa, K. Matsuo, and K. Shibata, *Tetrahedron Lett.*, **1966,** 4671.
72. E. O. Fischer, C. G. Kreiter, H. Ruehle, and K. Schwarzhans, *Chem. Ber.*, **100,** 1905 (1967).

73. D. M. Jerina, J. W. Daly, and B. Witkop, *J. Amer. Chem. Soc.*, **90**, 6523 (1968).
74. J. Habermeier, K. H. Reichert, and K. Hamann, *J. Polym. Sci., Part C*, **16**, 2131 (1967).
75. French Patent 1,499,784; *Chem. Abstr.*, **69**, 9029 (1968).
76. French Patent 1,499,783; *Chem. Abstr.*, **69**, 6279 (1968).
77. S. J. Rhoads and R. D. Cockroft, *J. Amer. Chem. Soc.*, **91**, 2815 (1969).
78. R. M. DeMarinis and G. A. Berchtold, *J. Amer. Chem. Soc.*, **91**, 6525 (1969).

CHAPTER II

Oxepin Ring Systems Containing Two Rings

ANDRE ROSOWSKY

Children's Cancer Research Foundation and Department of Biological Chemistry, Harvard Medical School, Boston, Massachusetts

A. Fused Ring Systems

1. 2-Oxabicyclo[5.1.0]octanes

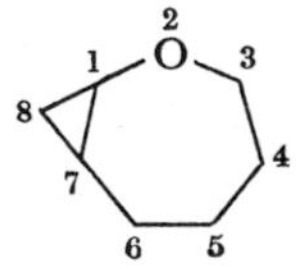

A derivative of this ring system was reported in 1966 by Nederl and co-workers (106) who obtained 8,8-dichloro-2-oxabicyclo[5.1.0]octane (**1**) in 41% yield upon addition of dichlorocarbene to 2,3,4,5-tetrahydrooxepin (**2**) in the presence of ethylene oxide. Accompanying **1** was a 39% yield of 3-chloro-2-(β-chloroethoxy)-8*H*-4,5,6,7-tetrahydrooxecin (**3**).

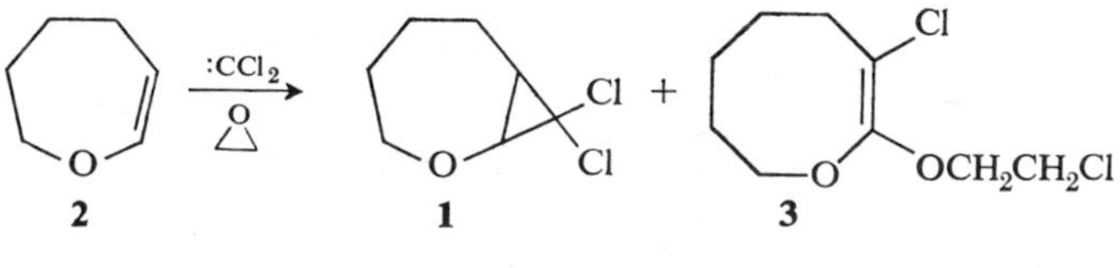

2. 2-Oxabicyclo[5.2.0]nonanes

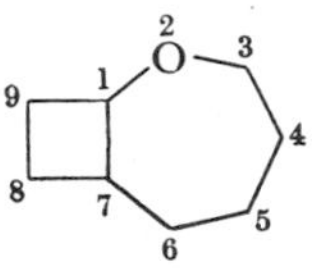

The 2-oxabicyclo[5.2.0]nonane ring system was generated first in 1964 by Corey and co-workers via the oxidation of 6,8,8-trimethylbicyclo[4.2.0]-octan-1-one (**4**) with peracetic acid (35). The starting ketone was obtained as one of the products in the photocondensation of 3-methyl-2-cyclohexenone and isobutylene. The peracetic acid oxidation product was identified as 7,9,9-trimethyl-2-oxabicyclo[5.2.0]nonan-3-one (**5**) on the basis of a combination of spectral and chemical evidence. The seven-membered lactone carbonyl group gave rise to an ir absorption band at 1755 cm^{-1} (5.70 μ). The nmr spectrum contained a singlet at τ 6.2 which was consistent with the bridgehead methinyl proton. Alkaline hydrolysis of **5** gave a hydroxy acid, and chromic acid oxidation of the latter afforded a keto acid exhibiting an ir peak at 1770 cm^{-1} (5.65 μ), as would be expected in cyclobutanone structure **6**. This evidence ruled out the alternative structure, 7,9,9-trimethyl-3-oxabicyclo-[5.2.0]nonan-2-one, for **5**.

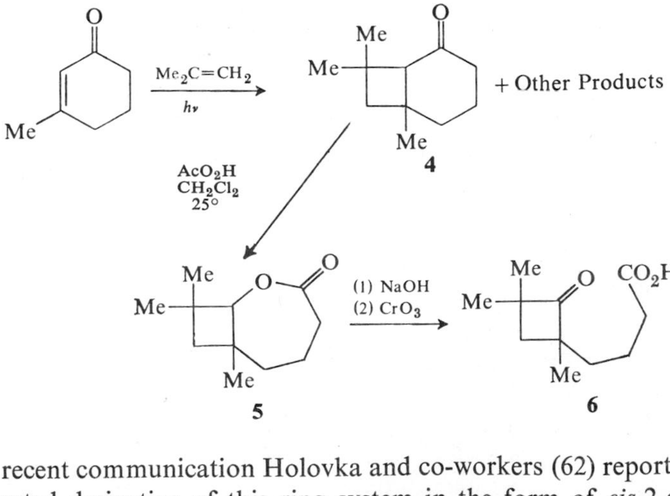

In a recent communication Holovka and co-workers (62) reported a highly unsaturated derivative of this ring system in the form of *cis*-2-oxabicyclo-[5.2.0]nona-3,5,8-triene (**7**). This compound was formed in 6% yield during the uv irradiation of the epoxide derived from cyclooctatetraene. Three other photoisomerization products were identified by means of gas–liquid chromatography, nmr analysis, and classical chemical evidence. The structures of these products, together with their relative yields, are presented here. According to the authors, a common intermediate in the conversion of cyclooctatetraene epoxide to its photoisomers is the elusive species 1-oxacyclononatetraene (**8**), which can be regarded as a vinylog of oxepin obeying the $(4n + 2)$ criterion for aromaticity.

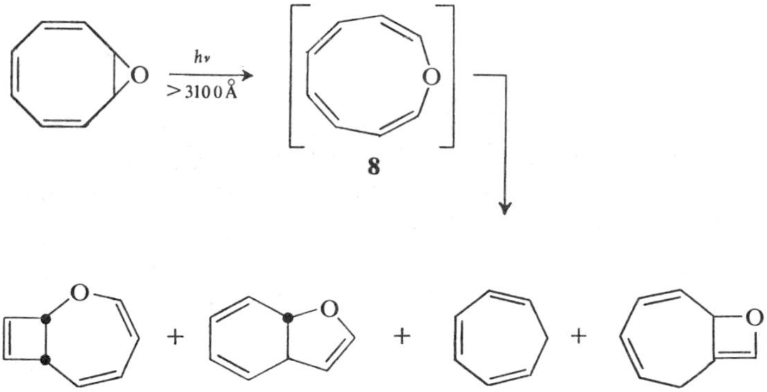

hν
>3100Å
8
7 (6%) + 45% + 12% + 10%

3. 2*H*-Cyclopent[*b*]oxepins

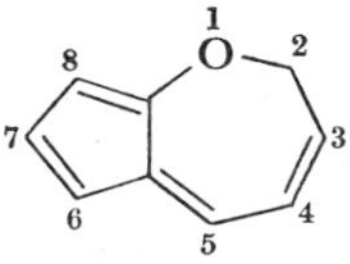

R.I. 11,924

The cyclopent[*b*]oxepin ring system was reported in 1961 by Obara (72). Treatment of ethyl 4-(2-hydroxycyclopentyl)butyrate with alkali caused ring closure to the lactone 3,4,5,5a,6,7,8,8a-octahydro-2*H*-cyclopent[*b*]oxepin-2-one (**9**). Reduction of **9** with lithium aluminum hydride afforded 4-(2-hydrocyclopentyl)butyraldehyde, while heating with polyphosphoric acid resulted in the interesting ring contraction shown.

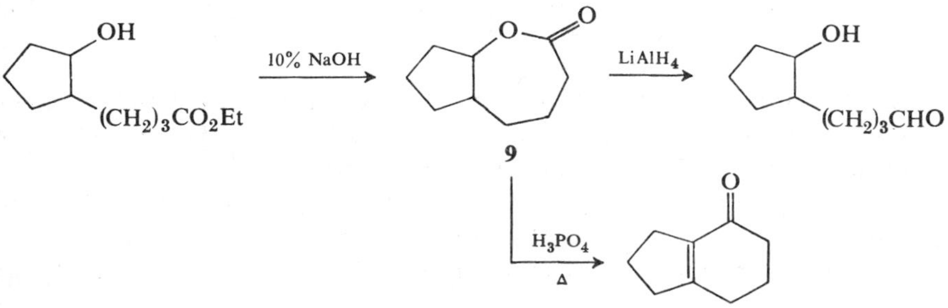

OH
$(CH_2)_3CO_2Et$
10% NaOH
9
$LiAlH_4$
OH
$(CH_2)_3CHO$
H_3PO_4
Δ

4. 2*H*-Cyclopent[*d*]oxepins

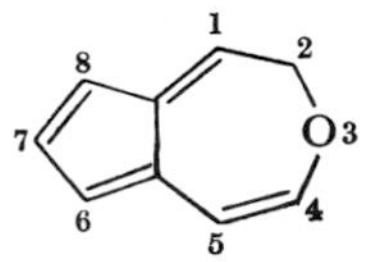

The isomeric 2*H*-cyclopent[*d*]oxepin ring system was produced in 1958 by Granger and co-workers (55) by reaction of *trans*-1,2-cyclopentanediacetic acid with a hot mixture of acetic anhydride and acetyl chloride. This gave a 72% yield of the anhydride, *trans*-1,4,5,5a,6,7,8,8a-octahydro-2*H*-cyclopent[*d*]oxepin-2,4-dione (**10**), with some contamination by ketone **11**. Treatment of **10** with sodium methoxide yielded **11**, along with the monomethyl ester of the starting diacid.

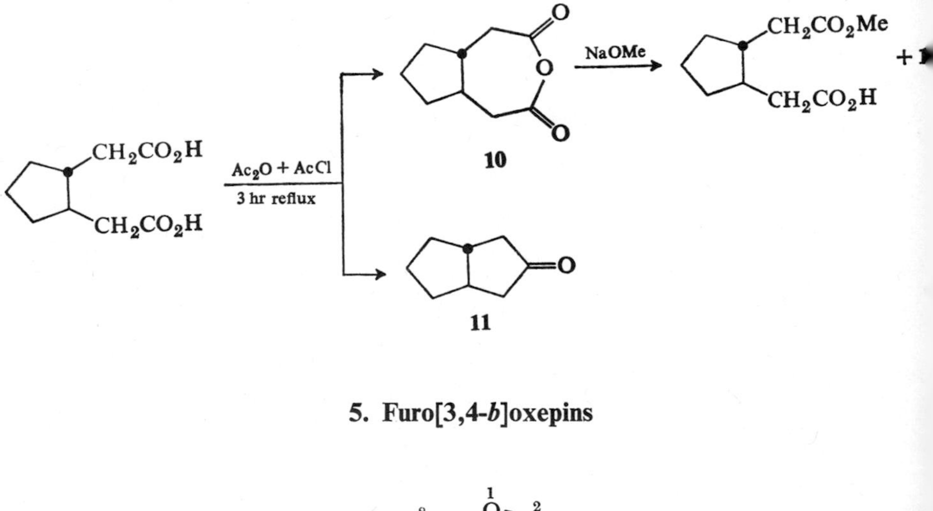

CH_2CO_2H CH_2CO_2H $Ac_2O + AcCl$ 3 hr reflux **10** NaOMe CH_2CO_2Me CH_2CO_2H + **11**

5. Furo[3,4-*b*]oxepins

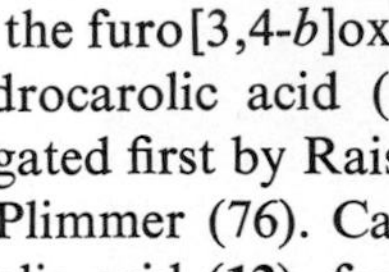

R. I. 1429

An interesting derivative of the furo[3,4-*b*]oxepin ring system was found in nature in the form of dehydrocarolic acid (**12**), a metabolic product of *Penicillium cinerascens* investigated first by Raistrick and co-workers (20, 30) and studied recently also by Plimmer (76). Catalytic reduction of **12** in the presence of alkali affords carolic acid (**13**), from which is obtained dihydro derivative **14** upon vacuum distillation. Simple hydrolysis of **12** results in

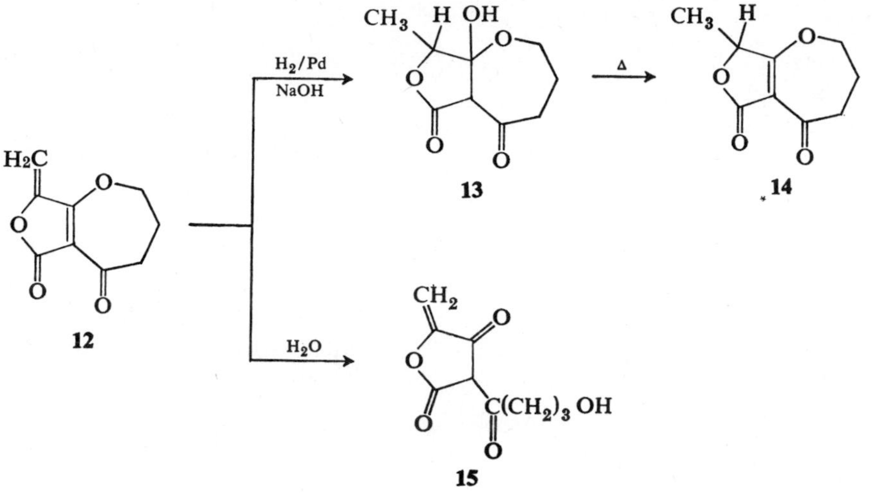

CH_3 H OH

H_2/Pd

NaOH

Δ

CH_3 H

O

O

O

O

O

O

O

O

13

14

H_2C

O

O

O

O

O

12

CH_2

O

H_2O

O

O

$C(CH_2)_3OH$

O

15

OAc

$CH_3CHCOCl$

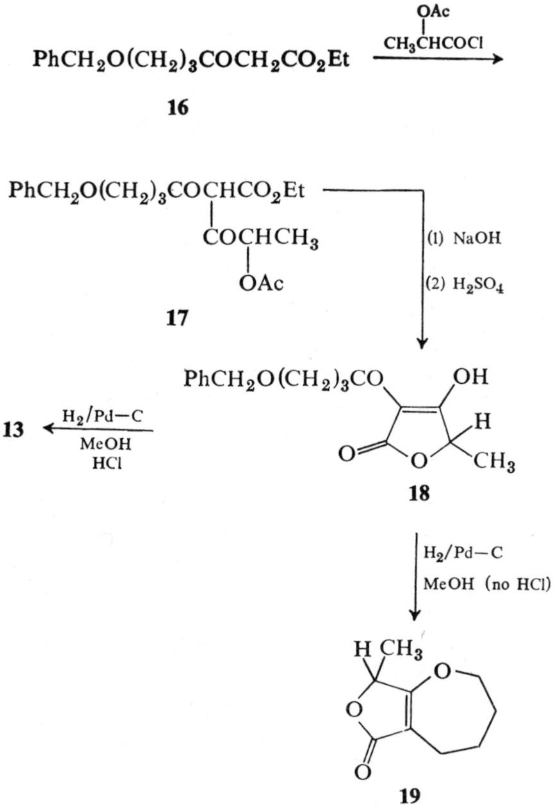

$PhCH_2O(CH_2)_3COCH_2CO_2Et$

16

$PhCH_2O(CH_2)_3COCHCO_2Et$

$COCHCH_3$

OAc

(1) NaOH

(2) H_2SO_4

17

$PhCH_2O(CH_2)_3CO$ OH

H

13 $\leftarrow$ $H_2/Pd{-}C$ MeOH HCl

O O CH_3

18

$H_2/Pd{-}C$

MeOH (no HCl)

H CH_3

O

O

O

19

cleavage of the seven-membered enol ether with formation of the unusual lactone, **15**.

A total synthesis of carolic acid was reported in 1967 by Sudo and co-workers (105). Condensation of ethyl 6-benzyloxy-3-oxohexanoate **(16)** with 2-acetoxypropionyl chloride gave ester **17**. Alkaline hydrolysis of the acetate blocking group, followed by lactonization with sulfuric acid, transformed **17** into **18**. Catalytic hydrogenation of **18** in the presence of palladium–charcoal afforded carolic acid (**13**), provided that a trace of hydrochloric acid was added to the methanol solvent. When the acid was omitted, over-reduction occurred, with formation of desoxy derivative **19**.

6. Furo[3,4-*d*]oxepins

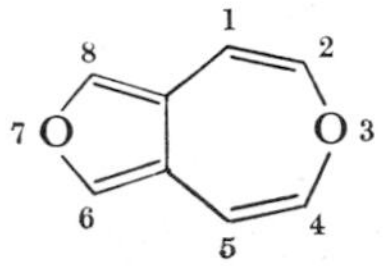

A representative of this ring system was synthesized in 1965 by Krapcho and Mundy (67) via treatment of diol **20** with *p*-toluenesulfonyl chloride in pyridine under reflux. The chemical and spectral properties of the product were consistent with structure **21**. The ir C—O stretching frequencies for the five- and seven-membered cyclic ether moieties in **21** were observed at 1088 cm^{-1} (9.18 μ) and 1057 cm^{-1} (9.46 μ), respectively. The O—CH_2 protons in the two rings could be differentiated on the basis of their nmr chemical shifts, which had values of τ 6.6 and τ 6.1, respectively. If chemical shift is taken as a measure of electron density, the O—CH_2 protons which are most highly shielded in **21** are those on the five-membered ring. This would indicate that there is a higher electron density on the carbon atom bearing these protons, and a correspondingly lower electron density on the adjacent oxygen, in the five-membered ring than in the seven-membered ring. The recent findings of Barakat and co-workers (7) concerning the relative basicities of tetrahydrofuran and hexamethylene oxide support this conclusion.

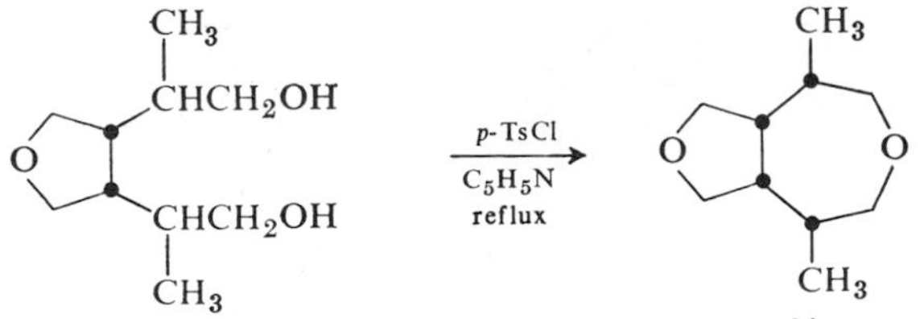

CH_3 / $CHCH_2OH$ / $CHCH_2OH$ / CH_3 — *p*-TsCl, C_5H_5N, reflux → CH_3 / CH_3

20 **21**

7. 2*H*-Oxepino[2,3-*b*]pyrroles

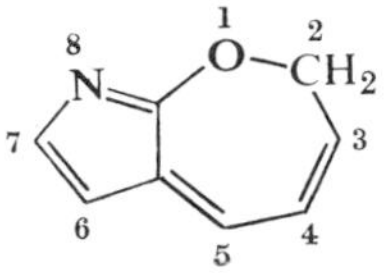

R.I. 8039

Fiesselmann and Ehmann (128) reported a member of this ring system in 1958 in the form of lactone **22**, which they obtained in 48% yield upon treatment of **23** with acetyl chloride and pyridine. Methanolysis of **22** in the presence of sulfuric acid afforded ester **24**.

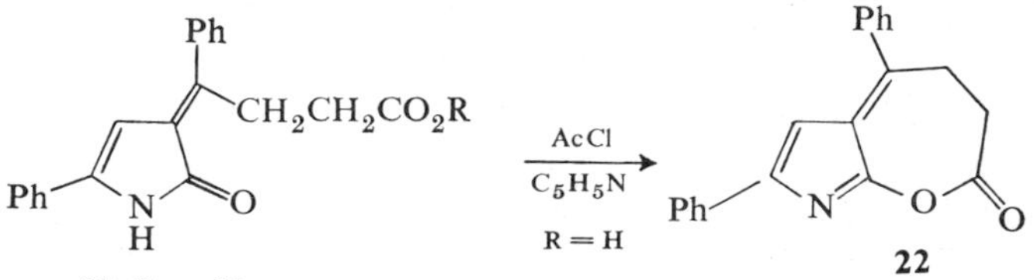

23; R = H
24; R = Me

AcCl, C_5H_5N, R = H

22

8. Thieno[3,4-*d*]oxepins

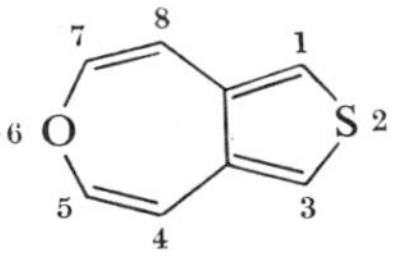

The first synthesis of a member of this interesting ring system was reported in 1966 by Dimroth and co-workers (123). Bischloromethylation of 2,5-dimethylthiophene, followed by reaction with sodium methoxide and oxidation with dinitrogen tetroxide, led to 2,5-dimethylthiophene-3,4-dialdehyde (**25**). Condensation of **25** with the bifunctional Wittig reagent from bis(2-bromoethyl) ether, triphenylphosphine, and sodium methoxide

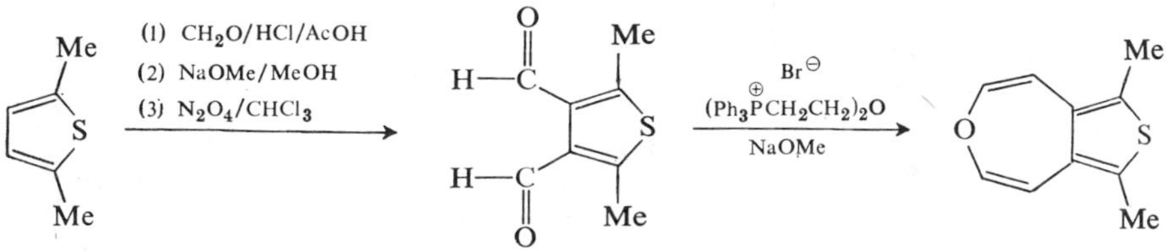

(1) $CH_2O/HCl/AcOH$
(2) NaOMe/MeOH
(3) $N_2O_4/CHCl_3$

25

$Br^{\ominus}$ $(Ph_3\overset{\oplus}{P}CH_2CH_2)_2O$, NaOMe

26

gave 1,3-dimethylthieno[3,4-*d*]oxepin (**26**) in low yield. Compound **26** proved to be extremely unstable, decomposing after a few weeks upon storage at 0°.

9. 1-Benzoxepins

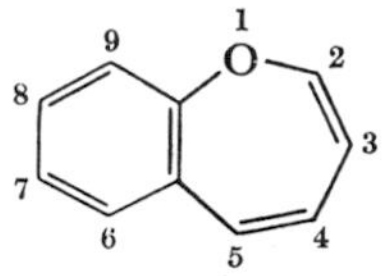

R.I. 1847

a. *Synthesis, Chemical Reactions, and Physical Properties of the Parent Compound*

The fully unsaturated compound 1-benzoxepin (**28**) has been known only since 1964 when Vogel and co-workers (95) and Sondheimer and Shani (86) almost simultaneously disclosed its synthesis by the same approach. This involved bromination of 9,10-oxido-1,4,5,8,9,10-hexahydronaphthalene (**29**) in chloroform solution, dehydrobromination of the resultant tetrabromide, **30**, by treatment with base, and finally chromatography on silica gel (95) or alumina (86). Dehydrobromination was carried out with potassium *tert*-butoxide in ether at −10° in one case (95) and with potassium hydroxide in ethanol at 50–55° in the other (86). The main dehydrobromination product was found, in both instances, to be 1,6-oxido[10]annulene (**31**). According to Vogel and co-workers, the primary dehydrobromination product was actually **31**, which then underwent partial isomerization to **28** during purification

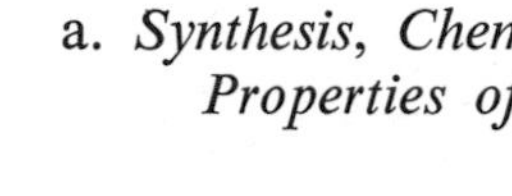

$\xrightarrow[CHCl_3]{Br_2}$

29

base

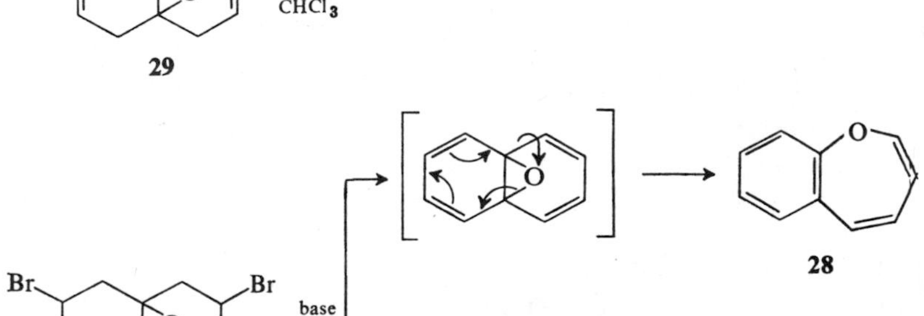

30

28

31

by silica gel chromatography. Sondheimer and Shani, on the other hand, postulated that dehydrobromination proceeded via two independent and simultaneous pathways, one leading to **31** and the other to **28**, as shown. In support of this view, Sondheimer and Shani pointed out that **28** and **31** were both recovered unchanged upon treatment with ethanolic potassium hydroxide at 50–55°, the reagent used for the initial dehydrobromination.

The structure of **28** was established chemically by catalytic hydrogenation to 2,3,4,5-tetrahydro-1-benzoxepin (homochroman) (**32**) and comparison with an authentic specimen prepared by an independent route.

28 $\xrightarrow{H_2/Pd\text{-}C}$ **32**

An interesting alternate route to **28**, which was reported in 1968 by Schweizer and co-workers (110), involves condensation of epoxide **33** with triphenylphosphonium hydrobromide and treatment of the resultant phosphonium salt, **34**, with sodium ethoxide. An intramolecular Wittig reaction with concomitant dehydration affords **28** in 5% yield. While the yield is low, the shortness of the synthetic sequence and ease of preparation of the starting epoxide from salicyaldehyde and glycidol offer an advantage over the earlier approach. On the other hand, formation of **28** is accompanied by at least one major side reaction which yields salicylaldehyde and its *O*-allyl derivative, **35**, via the hypothetical intermediates shown.

33 (CHO; $OCH_2CH{-}CH_2$ epoxide) $\xrightarrow{Ph_3P\cdot HBr}$

34 (CHO; $OCH_2CHCH_2\overset{\oplus}{P}Ph_3Br^{\ominus}$, OH) $\xrightarrow{NaOEt}$ [CHO; $OCH_2CHCH{=}PPh_3$, OH] $\longrightarrow$ **28** + Ph_3PO

34 $\xrightarrow{NaOEt}$ [CHO; $OCH_2CHCH_2\overset{\oplus}{P}Ph_3$, $O^{\ominus}$] $\longrightarrow$ **35** (CHO; $OCH_2CH{=}CH_2$) + salicylaldehyde (CHO; OH)

The spectral properties of **28** are of interest in relation to the possible quasi-aromatic character of oxepins and their benzo derivatives. According to Sondheimer and Shani (86), **28** is a yellow-green liquid with uv absorption bands at 211, 231, and 288 mμ (ε 14,700, 10,700, and 2900) in ethanol solution. The ir absorption spectrum shows peaks at 6.08 μ (olefinic C=C) and 6.26 μ (aromatic C=C). The nmr spectrum shows a complex band at τ 2.50–3.20 (four benzenoid protons), a doublet at τ 3.37 (C-2 proton), a doublet at τ 3.79 (C-5 proton), a double doublet centered at τ 4.03 (C-3 proton), and a double doublet centered at τ 4.60 (C-4 proton).

The synthesis of substituted analogs of **28** by the route above has not been reported to date. It would undoubtedly be of interest to determine whether such compounds could be made, starting from 1- or 2-substituted derivatives of **29**, and also to determine whether the substituents would ultimately appear in the benzene or oxepin moiety of the product.

b. *Synthesis and Chemical Reactions of Dihydro-*1*-benzoxepins*

There are, in principle, three possible dihydro derivatives of 1-benzoxepin in which the benzene ring retains its aromaticity. They are 2,3-dihydro-1-benzoxepin (**36**), 2,5-dihydro-1-benzoxepin (**37**), and 4,5-dihydro-1-benzoxepin (**38**). Neither **37** nor any of its derivatives has been reported to date, although an unsuccessful attempt to prepare 2,5-dihydro-1-benzoxepin-5-one was described in 1964 by Fontaine and Maitte (49).

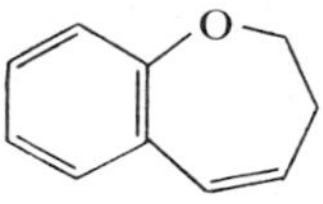

O O O

36 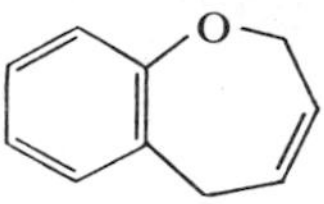 **37** 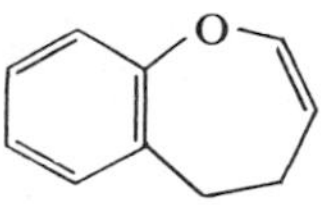 **38**

The earliest reported synthesis of 2,3-dihydro-1-benzoxepin derivatives was the work of Sen and Roy (83), who condensed levulinic acid with salicylaldehyde in the presence of dry hydrogen chloride and obtained a reddish-violet compound which they formulated as 4-acetyl-2,3-dihydro-1-benzoxepin-2-one (**39**). In similar fashion the reaction of levulinic acid with

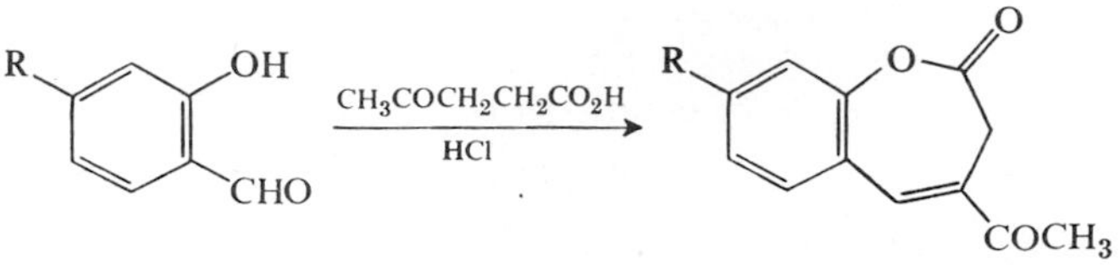

R OH CHO $\xrightarrow[\text{HCl}]{CH_3COCH_2CH_2CO_2H}$ R O O $COCH_3$

39; R = H
40; R = OH

2,4-dihydroxybenzaldehyde gave a bluish-violet product considered to be 4-acetyl-2,3-dihydro-8-hydroxy-1-benzoxepin-2-one (**40**).

The only other known example of a 2,3-dihydro-1-benzoxepin-2-one derivative was reported in 1962 by Smith and Bealor (85) who investigated the reaction of 2,4-dimethylphenol with diethyl 2-ketoglutarate. Treatment with sulfuric acid at 0–5°, followed by reaction with potassium hydroxide in aqueous alcohol under reflux, afforded two products. The first of these, isolated in 15% yield, was formulated as 2,3-dihydro-7,9-dimethyl-1-benzoxepin-2-one-5-carboxylic acid (**41**). The second, which was almost equally abundant, was assigned structure **42**. Alternative structures for both products were rejected on the basis of physical and chemical evidence. Catalytic hydrogenation of **41** furnished tetrahydro derivative **43**.

Me OH + CO_2Et CH_2 CH_2 C=O CO_2Et (1) H_2SO_4 (2) KOH

41 (15%) H_2/Pt AcOH **43**

42 (22.5%)

2,3-Dihydro-1-benzoxepins not containing a carbonyl group in the oxepin moiety have appeared in the literature on four occasions. In the earliest instance Dann and Arndt (41) reported in 1954 that dehydration of 2,3,4,5-tetrahydro-5-hydroxy-7,8-dimethyl-1-benzoxepin (**44**) with hydrochloric or acetic acid, and of 2,3,4,5-tetrahydro-5-hydroxy-5,7,8-trimethyl-1-benzoxepin (**45**) with phosphorus pentoxide or acetic acid, afforded 2,3-dihydro-7,8-dimethyl-1-benzoxepin (**46**) and 2,3-dihydro-5,7,8-trimethyl-1-benzoxepin (**47**), respectively. Compound **46** was characterized, in part, by conversion into dibromide **48**. Compounds **44** and **45** were prepared from 2,3,4,5-tetrahydro-7,8-dimethyl-1-benzoxepin-5-one (**49**) by reduction with aluminum isopropoxide and reaction with methyl magnesium iodide, respectively.

The approach of Dann and Arndt was utilized again by Fontaine and Maitte (49, 111) for the synthesis of 2,3-dihydro-1-benzoxepin (**36**) and 5-ethyl-2,3-dihydro-1-benzoxepin (**50**). Reduction of 2,3,4,5-tetrahydro-1-benzoxepin-5-one (**51**) with aluminum isopropoxide and dehydration of the resulting alcohol, **52**, with *p*-toluenesulfonic acid gave **36** in 82% yield. Addition of ethyl magnesium bromide to **51** afforded alcohol **53**, and treatment of the latter with *p*-toluenesulfonic acid gave a mixture of **50** and *cis/trans*-isomeric exocyclic olefins **54** and **55**. Analysis of the dehydration mixture by gas chromatography indicated **50** to be present in the mixture to the extent of approximately 50%. In similar fashion, condensation of **51** with methyl and phenyl Grignard reagents led to alcohols **56** and **57**, respectively. Dehydration of **56** afforded a mixture of 2,3-dihydro-5-methyl-1-benzoxepin (**58**) and 2,3,4,5-tetrahydro-5-methylene-1-benzoxepin (**59**), while dehydration of **57** gave 2,3-dihydro-5-phenyl-1-benzoxepin (**60**) as the sole product. Catalytic hydrogenation of **36** over palladium–charcoal catalyst afforded 2,3,4,5-tetrahydro-1-benzoxepin (**32**). Similar reduction of the other olefins gave the corresponding tetrahydro derivatives.

Oxidation of **36** and **60** has recently been reported by Fontaine (129) to give *cis*-2,3,4,5-tetrahydro-1-benzoxepin-4,5-diol (**61**) and *cis*-2,3,4,5-tetrahydro-5-phenyl-1-benzoxepin-4,5-diol (**62**), respectively. Under the influence of copper sulfate, **61** yields the pinacolic rearrangement product 2,3,4,5-tetrahydro-1-benzoxepin-4-one (**63**). Similarly, upon treatment with *p*-toluenesulfonic acid, **62** undergoes rearrangement to 2,3,4,5-tetrahydro-5-phenyl-1-benzoxepin-4-one (**64**).

$KMnO_4$; $CuSO_4$ (R = H) or *p*-TsOH (R = Ph)

36; R = H
60; R = Ph

61; R = H
62; R = Ph

63; R = H
64; R = Ph

An alternate, though less effective, procedure for the synthesis of **36** was reported in 1963 by Schweizer and Schepers (82). The seven-membered ring was generated in this case via an intramolecular Wittig condensation of ylide **65**. For the preparation of **65**, salicylaldehyde was alkylated with 1,3-dibromopropane, the resulting **66** was converted into phosphonium salt **67** by treatment with triphenylphosphine, and **67** was transformed into **65** under the influence of base. Unfortunately, the final ring closure furnished only a small yield of the desired product, the major portion being comprised of cyclic ether **68**. Chemical proof for the structure of **36** was secured by

OH, CHO → ($Br(CH_2)_3Br$, NaOH) → $O(CH_2)_3Br$, CHO **66** → (Ph_3P, EtOAc or C_6H_5Cl) → $O(CH_2)_3\overset{\oplus}{P}Ph_3Br^{\ominus}$, CHO **67** → (NaOR) → [$OCH_2CH_2CH{=}PPh_3$, CHO] **65** → **68** (H, CH_3) + **36**

catalytic hydrogenation to 2,3,4,5-tetrahydro-1-benzoxepin (**32**), which was also prepared via an alternate, unambiguous route.

The origin of **68** in the above cyclization reaction has not yet been fully elucidated. While **36** undergoes rearrangement to **68** in 94% yield upon treatment with sodium ethoxide in refluxing ethanol after 14 days, this process does not appear to be more than partly responsible for the high yield of **68** obtained after a much shorter time under the conditions of the Wittig reaction. According to Schweizer and Schepers, ylide **65** probably rearranges under the influence of base, generating a new ylide which can then give rise to **68**. The direct base-induced isomerization of **36** to **68** is nonetheless quite interesting, since it is the first reported example of a ring contraction of a seven-membered oxygen heterocycle wherein the oxygen atom remains part of the new hetero ring.

Another method for the preparation of **36**, introduced by Schweizer and co-workers (81) in 1968, consists of heating a mixture of cyclopropyl triphenyl phosphonium bromide and the sodium salt of salicylaldehyde at 150–160°. Once again a major by-product in this reaction is cyclic ether **68**. When the reaction is performed under dry fusion conditions, **36** and **68** are formed in the ratio of 60:40, the combined yield of the two products being 15%. When it is carried out in the presence of mineral oil, the reaction affords **36** and **68** in the ratio of 65:35, with an over-all yield of 18%.

In a follow-up of their earlier work, Schweizer and co-workers (108, 109) recently considered several possible mechanisms for the formation of **36** and **68**. The origin of **68** has been especially problematic, and no mechanism could be devised that explained all the experimental findings completely. Optimal conditions for the synthesis of **36** were found to involve treatment of **67** with 0.95 equivalent of sodium methoxide in very dilute dimethylformamide; the yield of **36** under these conditions was about 70%. On the other hand, treatment of **67** with excess sodium methoxide in refluxing methanol led to an 88% yield of **68**. When **67** was treated with sodium methoxide in deuteriomethanol, $\alpha,3\text{-}d_2$-**68** was obtained. A number of different solvents were examined in order to determine the possible influence of solvent polarity upon the course of the reaction. It was concluded that solvent polarity alone did not play a dramatic role, but that a proton-donating solvent was critical for the formation of **68**. The ratio of **36** to **68** in dioxane was 98:2; in pyrrole it was 15:85. Both dioxane and pyrrole are solvents of fairly low dielectric constant (in contrast to DMF and methanol), but only pyrrole can serve as a proton source.

CHO
OH
$+ Ph_3P^{\oplus}$—cyclopropyl $Br^{\ominus}$ $\xrightarrow{\Delta}$ **36 + 68**

An interesting method of synthesis of substituted 2,3-dihydro-1-benzoxepin derivatives was reported recently by Marcaillou and co-workers (130). This involves condensation of dibromocarbene with **69** and treatment of the resulting adduct with silver ion to give 3-bromo-2,3-dihydro-4-methoxy-1-benzoxepin (**70**) in 90% yield.

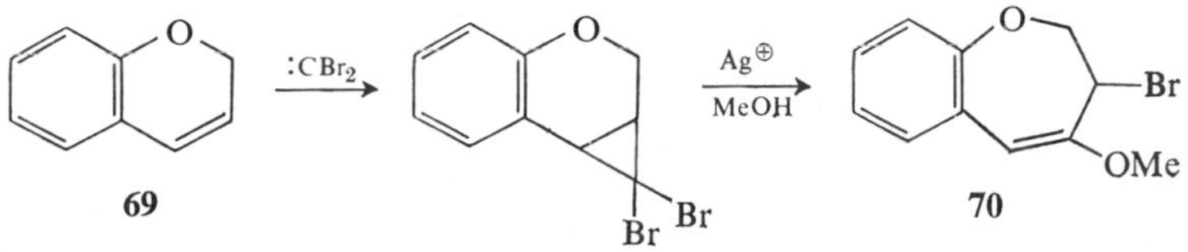

:CBr_2 ; $Ag^{\oplus}$ / MeOH

69 **70**

4,5-Dihydro-1-benzoxepin (**38**) was unknown until Fontaine (129) reported in 1968 that this compound can be prepared from tetralin by photooxidation, followed by heating of the resultant 2-acetoxy-2,3,4,5-tetrahydro-1-benzoxepin (**71**) to eliminate acetic acid.

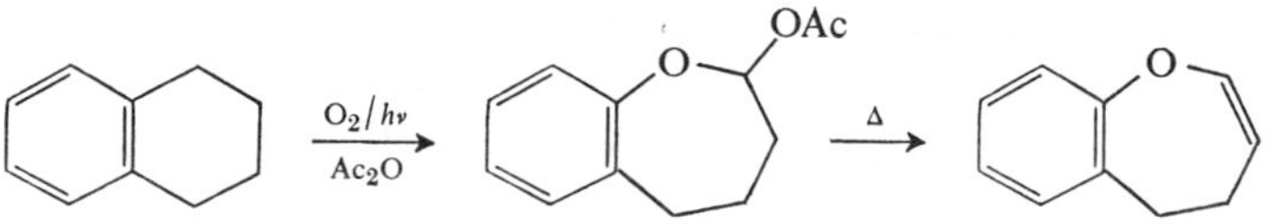

$O_2/h\nu$ / Ac_2O ; Δ

71 **38**

c. *Synthesis and Chemical Reactions of Tetrahydro-*1*-benzoxepins*

2,3,4,5-Tetrahydro-1-benzoxepin-5-one (**51**) was first synthesized in 1931 by Powell and Anderson (77) starting from 4-phenoxybutyric acid. Treatment with phosphorus pentoxide in benzene gave only a poor yield of **51**, however, and the use of phosphorus pentachloride, thionyl chloride, or aluminum chloride/thionyl chloride was completely unsuccessful. Dann and Arndt (41) reexamined this approach in 1954 and found that **51** could be prepared quite satisfactorily by using polyphosphoric acid. In similar fashion, 4-(3′,4′-dimethylphenoxy)butyric acid was readily transformed into 2,3,4,5-tetrahydro-7,8-dimethyl-1-benzoxepin-5-one (**49**). The effectiveness of added xylene in the polyphosphoric acid cyclization of 4-phenoxybutyric acid, and also the conversion of 4-phenoxybutyryl chloride into **51** under the influence of stannic chloride, were reported by Fontaine and Maitte (49, 111).

The use of liquid hydrogen fluoride as a reaction medium for the preparation of **49**, **51**, and related compounds was investigated in some detail by Dann and Arndt (41). Although **49**, 2,3,4,5-tetrahydro-7-methyl-1-benzoxepin-5-one (**72**), and 2,3,4,5-tetrahydro-8-methoxy-1-benzoxepin-5-one (**73**) were obtained successfully by allowing 4-(3′,4′-dimethylphenoxy)butyric

$O(CH_2)_3COCl$ → $SnCl_4$ / CCl_4

R, R, $O(CH_2)_3CO_2H$ → PPA or PPA/xylene, R = H, Me

49; R = Me
51; R = H

acid, 4-(4′-methylphenoxy)butyric acid, and 4-(3′-methoxyphenoxy)butyric acid, respectively, to stand in hydrogen fluoride at room temperature for 10 days, only polymers of the type shown were formed when either 4-phenoxybutyric acid itself or 4-(2′,5′-dimethylphenoxy)butyric acid was subjected to these reaction conditions. It appears, from these observations, that the *para* position in the phenoxybutyric acid must be blocked in order to prevent the formation of *para*-substituted polymers, except when a suitably located activating substituent is present. In the case of **73**, cyclization occurs despite the absence of a *para*-blocking group because the 3′-methoxy substituent promotes cyclization sufficiently to overcome the tendency for *para*-polymerization. Dann and Arndt (41) also reported an unsuccessful effort to prepare **49** by condensation of 3,4-dimethylphenol with γ-butyrolactone in the presence of anhydrous hydrogen fluoride.

R, R′, $O(CH_2)_3CO_2H$ → HF, room temp, 10 days

49; R = R″ = Me
72; R = H,R′ = Me
73; R = MeO,R′ = H

R = R′ = H → $O(CH_2)_3C$

A 1963 patent (52) reported the use of 17 different 2,3,4,5-tetrahydro-1-benzoxepin-5-ones containing methyl and/or halogen substituents at the 6-, 7-, 8-, and/or 9-positions as starting materials for the preparation of a series of spirohydantoins with anticonvulsant properties. Neither the method of preparation nor the physical constants of the products are given in *Chemical Abstracts.*

A wide assortment of typical ketone reactions have been carried out with 2,3,4,5-tetrahydro-1-benzoxepin-5-ones. Clemmensen reduction with metallic zinc in hydrochloric acid has been used to transform **49** into 2,3,4,5-tetrahydro-7,8-dimethyl-1-benzoxepin (**74**) (41), as well as 7-chloro-2,3,4,5-tetrahydro-1-benzoxepin-5-one (**75**), 7,9-dichloro-2,3,4,5-tetrahydro-1-benzoxepin-5-one (**76**), and 7-chloro-2,3,4,5-tetrahydro-9-methyl-1-benzoxepin-5-one (**77**) into the corresponding reduced compounds, **78–80**.

49; R = R′ = Me, R″ = H
75; R = Cl, R′ = R″ = H
76; R = R″ = Cl, R′ = H
77; R = Cl, R′ = H, R″ = Me

74; R = R′ = Me R″ = H
78; R = Cl, R′ = R″ = H
79; R = R″ = Cl, R′ = H
80; R = Cl, R′= H, R″ = Me

Reduction with aluminum isopropoxide has been used to convert **51** into 2,3,4,5-tetrahydro-5-hydroxy-1-benzoxepin (**52**) (49, 111), and **49** into 2,3,4,5-tetrahydro-5-hydroxy-7,8-dimethyl-1-benzoxepin (**44**) (41). Addition of ethyl magnesium bromide to **51** (49, 111) and of methyl magnesium iodide to **49** (41) gives 5-ethyl-2,3,4,5-tetrahydro-5-hydroxy-1-benzoxepin (**53**) and 2,3,4,5-tetrahydro-5-hydroxy-5,7,8-trimethyl-1-benzoxepin (**45**), respectively. The dehydration of alcohols **44**, **45**, **52**, and **53** has been cited earlier in connection with the synthesis of 2,3-dihydro-1-benzoxepins. The reaction of **51** with methyl and phenyl Grignard reagents was also described recently by Fontaine (111).

Treatment of **49** with phosphorus pentachloride in benzene gives 5,5-dichloro-2,3,4,5-tetrahydro-7,8-dimethyl-1-benzoxepin (**81**) (41). Bromination of **51** in ether (49, 111) and of **49** in carbon disulfide (41) affords 4-bromo-2,3,4,5-tetrahydro-1-benzoxepin-5-one (**82**) and 4-bromo-2,3,4,5-tetrahydro-7,8-dimethyl-1-benzoxepin-5-one (**83**), respectively. Bromination of **49** in chloroform, on the other hand, yields 4,4-dibromo-2,3,4,5-tetrahydro-7,8-dimethyl-1-benzoxepin-5-one (**84**) (41). Attempted dehydrobromination of **82** with piperidine fails to give the expected α, β-unsaturated ketone (49).

Treatment of α-bromoketone **83** with sodium acetate affords 4-acetoxy-2,3,4,5-tetrahydro-7,8-dimethyl-1-benzoxepin-5-one (**85**), which is likewise obtained upon oxidation of **49** with lead tetraacetate (41). Alkaline hydrolysis

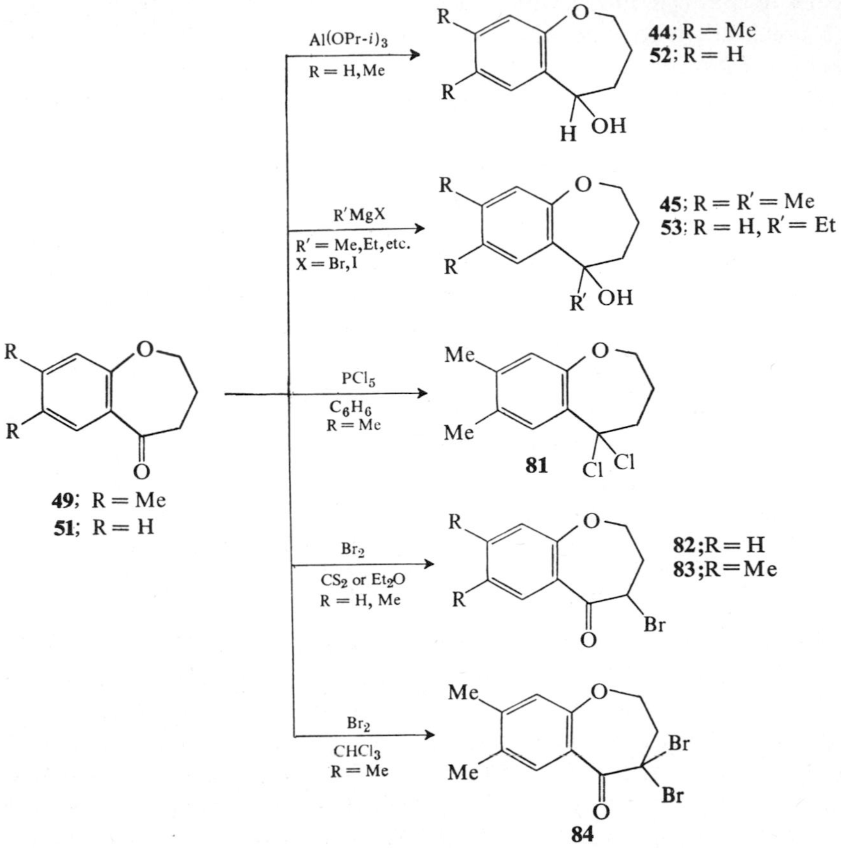

$Al(OPr\text{-}i)_3$, R = H, Me → 44; R = Me / 52; R = H

R'MgX, R' = Me, Et, etc., X = Br, I → 45; R = R' = Me / 53; R = H, R' = Et

PCl_5, C_6H_6, R = Me → 81

Br_2, CS_2 or Et_2O, R = H, Me → 82; R = H / 83; R = Me

Br_2, $CHCl_3$, R = Me → 84

49; R = Me
51; R = H

of **83**, and also acid hydrolysis of **85**, affords 2,3,4,5-tetrahydro-4-hydroxy-1-benzoxepin-5-one (**86**) (41). Several reactions of α-bromoketone **82** were also described recently by Fontaine (111). These included the formation of a ketal and the replacement of bromine by an acetoxy or methoxy group. In the last instance the action of sodium methoxide upon **82** gave not only the simple displacement product but also the Favorski rearrangement product, chroman-1-carboxylic acid.

Condensation of **49** with hydroxylamine and reduction of the oxime with sodium amalgam affords 5-amino-2,3,4,5-tetrahydro-7,8-dimethyl-1-benzoxepin (**87**) (41). Mannich condensation with formaldehyde and dimethylamine has been used to convert **51** and **49** into amino ketones **88** and **89**, respectively (41, 49). Further reaction of **88** with phenylmagnesium bromide gives amino alcohol **90**, which affords propionate ester **91** upon treatment with propionyl chloride (11, 93). Compounds such as **91** are claimed to dossess hypotensive and analgesic properties (11, 93).

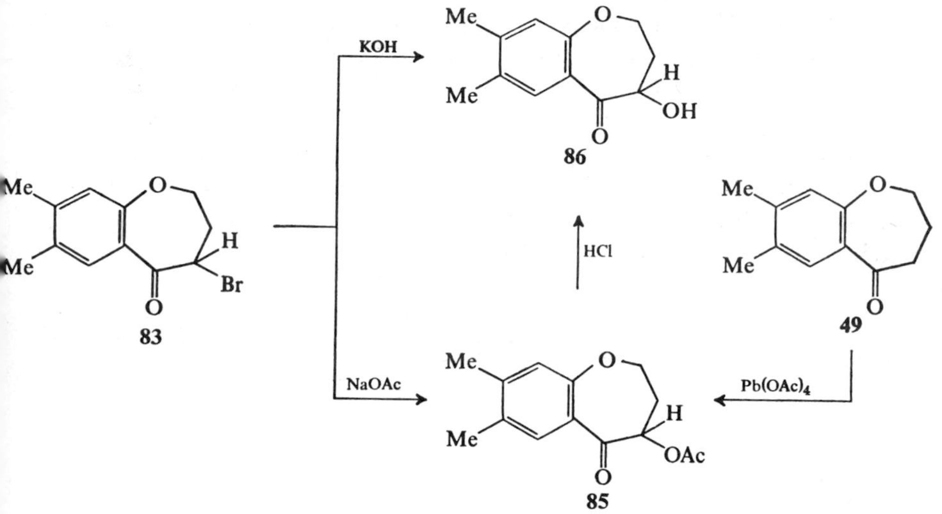

KOH

86

HCl

83

49

NaOAc

$Pb(OAc)_4$

85

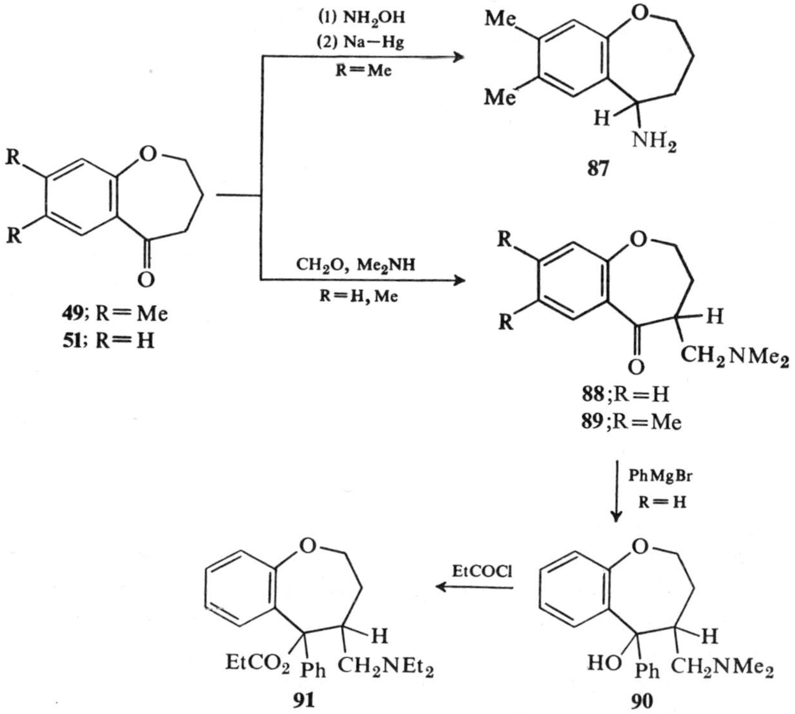

(1) NH_2OH
(2) Na—Hg
R=Me

87

49; R=Me
51; R=H

CH_2O, Me_2NH
R=H, Me

88; R=H
89; R=Me

PhMgBr
R=H

EtCOCl

91

90

Other reactions of **49** which have been reported (41) include condensation with benzaldehyde, oxidation with selenium dioxide, and oximation with isoamyl nitrite. The products obtained from these reactions are 4-benzylidene-2,3,4,5-tetrahydro-7,8-dimethyl-1-benzoxepin-5-one (**92**), 2,3,4,5-tetrahydro-7,8-dimethyl-1-benzoxepin-4,5-dione (**93**), and 2,3,4,5-tetrahydro-7,8-dimethyl-4-oximino-1-benzoxepin-5-one (**94**), respectively. The reaction of **49** with selenium dioxide appears to be capricious and difficult to reproduce. Catalytic reduction has been used to transform oxime **94** into 4-amino-2,3,4,5-tetrahydro-7,8-dimethyl-1-benzoxepin-5-one (**95**)

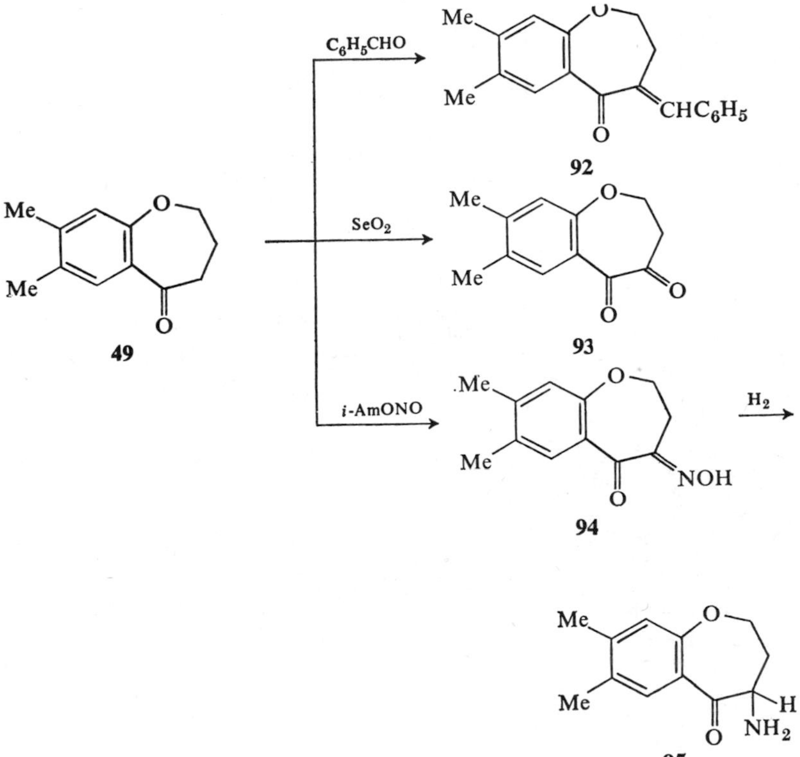

Me Me O O **49** — C_6H_5CHO → Me Me O O CHC_6H_5 **92**; — SeO_2 → Me Me O O O **93**; — *i*-AmONO → Me Me O O NOH **94** — H_2 → Me Me O O H NH_2 **95**

Additional reactions of **51** reported recently include the formation of ketal **96** by condensation with ethylene glycol (111), the preparation of 2,3,4,5-tetrahydro-5-methylene-1-benzoxepin (**97**) via the Wittig reaction (111), the formation of 5-carbethoxymethylene-2,3,4,5-tetrahydro-1-benzoxepin (**98**) via the Reformatsky reaction (111), and the formation of 5-chloro-2,3-dihydro-1-benzoxepin-4-carboxaldehyde (**99**) upon treatment with the Vielsmeier reagent (125).

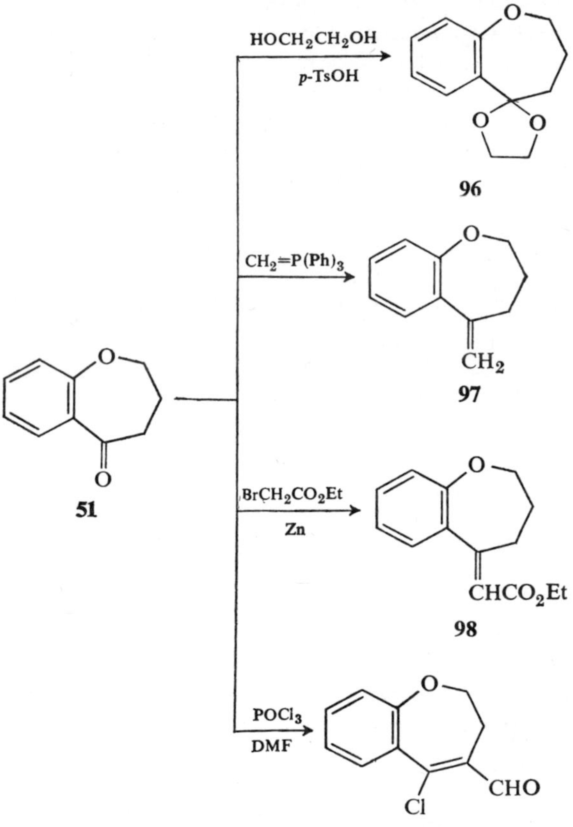

HOCH$_2$CH$_2$OH
p-TsOH
96
CH$_2$=P(Ph)$_3$
CH$_2$
97
51
BrCH$_2$CO$_2$Et
Zn
CHCO$_2$Et
98
POCl$_3$
DMF
CHO
Cl
99

Mention was made earlier in this chapter of the recently reported synthesis of 2,3,4,5-tetrahydro-1-benzoxepin-4-one (**63**) and the 5-phenyl analog (**64**) via pinacolic rearrangements of the corresponding 1,2,3,4-tetrahydro-1-benzoxopin-4,5-diols (129). This constitutes the only available method of synthesis of this particular type of 1-benzoxepin derivative to date.

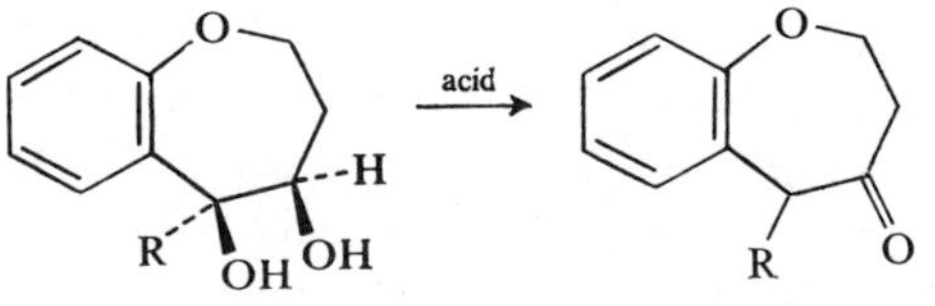

acid

63;R=H
64;R=Ph

The earliest reported synthesis of a 2,3,4,5-tetrahydro-1-benzoxepin-2-one was contained in a 1933 patent (50) dealing with the persulfate oxidation of 1-tetralone in methanol solution. Saponification of the resulting methyl 4-(2′-hydroxyphenyl)butyrate gave the corresponding acid, which was then cyclized to 2,3,4,5-tetrahydro-1-benzoxepin-2-one by heating. A similar reaction sequence transformed 6-methoxy-1-tetralone into 2,3,4,5-tetrahydro-7-methoxy-1-benzoxepin-2-one (**97**). Chlorination of **96** gave 3-chloro-2,3,4,5-tetrahydro-1-benzoxepin-2-one (**98**).

$K_2S_2O_8$, H_2SO_4, MeOH, R=H, MeO

OH, R, $(CH_2)_3CO_2Me$

(1) NaOH (2) Δ R=H, MeO

H, Cl, Cl_2, R=H

98

96;R=H
97;R=MeO

A closely related type of reaction is the Baeyer-Villiger oxidation of keto diester **99**, reported in 1954 by Jeger and co-workers (64) to give seven-membered lactone **100** upon prolonged treatment with perbenzoic acid in chloroform solution.

$PhCO_3H$, $CHCl_3$, room temp 10 days

CO_2Me, Me, $(CH_2)_3CO_2Me$

99 **100**

The sole example of a 1-benzoxepin-2,5-dione reported to date is 2,3,4,5-tetrahydro-4-(4′-methoxyphenyl)-6,8-dimethoxy-1-benzoxepin-2,5-dione (**101**), allegedly formed in low yield when 3,5-dimethoxy-2-(4′-methoxyphenyl)-acetylphenol (**102**) is allowed to react with ethyl bromoacetate (99). The expected simple *O*-alkylation product (**103**) was not observed. The formation of **102** may be supposed to proceed as shown.

1-Benzoxepin-3,5-diones were not known prior to 1965 when Hofmann (60, 107) reported that acid-catalyzed isomerization of spiroepoxide **104** produced 2,3,4,5-tetrahydro-4-phenyl-1-benzoxepin-3,5-dione (**105**) in 75% yield. Suitable catalysts for this rearrangement included sulfuric acid, boron

MeO OH C=O MeO CH2 OMe **102** $BrCH_2CO_2Et$ K_2CO_3 Me_2CO MeO OCH_2CO_2Et C=O MeO CH_2 MeO **103** MeO OH CH_2CO_2Et C—CH MeO O OMe MeO O O MeO O **101** OMe

trifluoride, and alumina. Infrared and nmr evidence indicated that **105** exists in the diketone form both in the solid state and in nonpolar solvents. This finding suggests that there is insufficient resonance energy available through quasi-aromaticity of the 1-benzoxepin ring system to stabilize the dienol tautomer. Acetylation of **105** under mild conditions gave a product formulated as 5-acetoxy-2,3-dihydro-4-phenyl-1-benzoxepin-3-one (**106**). Further acetylation of **106** at a higher temperature afforded 3,5-diacetoxy-4-phenyl-1-benzoxepin (**107**). Under the influence of boiling acetic anhydride containing sodium acetate, **105**–**107** all underwent isomerization to a common product which was identified as 1,3,4-triacetoxy-2-phenylnaphthalene. A mechanism involving valence bond tautomerism was postulated by Hofmann for this interesting reaction.

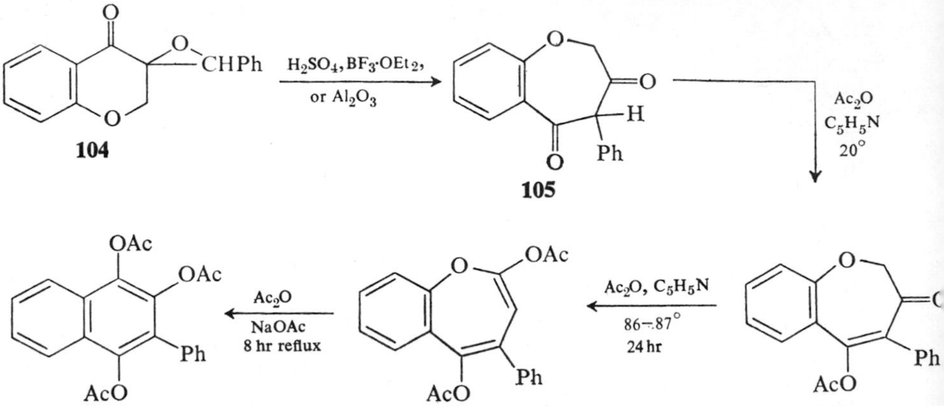

104 → (H_2SO_4, $BF_3{\cdot}OEt_2$, or Al_2O_3) → 105 → (Ac_2O, C_5H_5N, 20°) → 106 → (Ac_2O, C_5H_5N, 86–87°, 24 hr) → 107 → (Ac_2O, NaOAc, 8 hr reflux)

In an extension of their earlier work, Hofmann and Westermacher (126) recently prepared several other 1-benzoxepin-3,5-diones via acid-catalyzed rearrangement of the appropriate spiroepoxides. Diketones **108–111**, monoenol acetate **112**, and dienol acetates **113** and **114** were obtained.

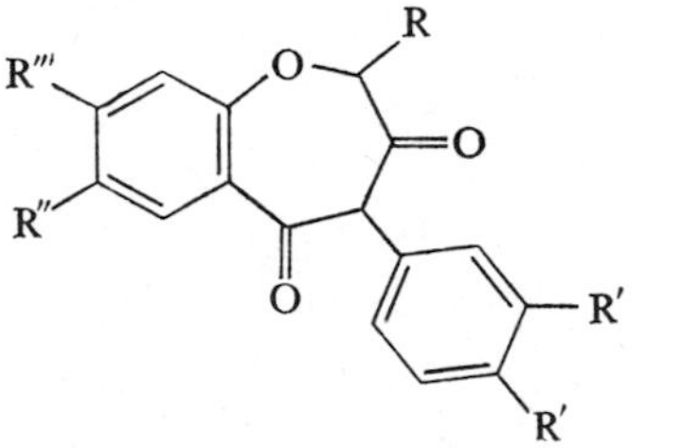

108;R=R′=R″=R‴=H
109;R=R″=H, R′=R‴=MeO
110;R=R″=Me, R′=R‴=H
111;R=Ph, R′=R″=R‴=H

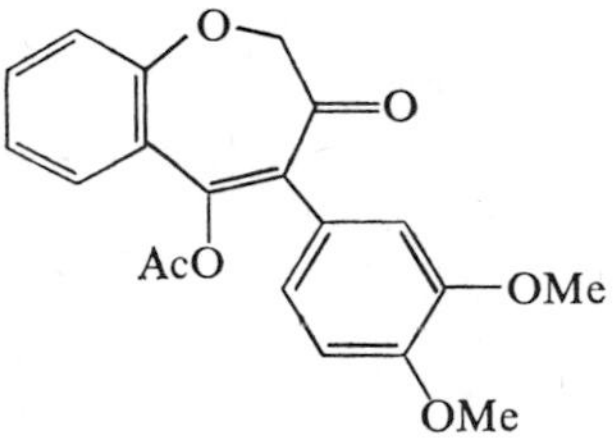

112

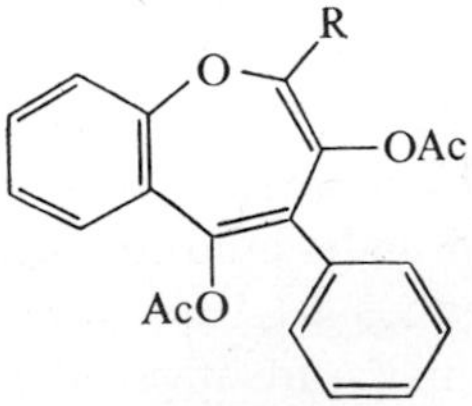

113;R=He
114;R=M

1-Benzoxepin-3,5-diones were also reported recently by Tyman and Pickles (89) in connection with a study of the chemistry of certain *o*-acetophenoxyacetates. Reaction of ethyl 2-(2′-acetyl-4′-bromo-5′-methylphenoxy)-propionate with phosphorus pentachloride in benzene, or with sodium ethoxide, led to the formation of a compound identified as 7-bromo-2,3,4,5-tetrahydro-2,8-dimethyl-1-benzoxepin-3,5-dione (**115**). The same product was likewise obtained, together with other compounds, when 5-bromo-2-hydroxy-4-methylacetophenone was allowed to react with ethyl 2-bromopropionate in acetone which contained potassium carbonate. Similarly, cyclization of ethyl 2-(2′-acetylphenoxy)propionate afforded 2,3,4,5-tetrahydro-2-methyl-1-benzoxepin-3,5-dione (**116**). The identification of **115** was based upon its nmr spectral features and upon its alkaline hydrolysis to 2-(2′-acetyl-4′-bromo-5′-methylphenoxy)propionic acid. In accord with the findings of Hofmann (60) with **105**, the physical properties of **115** were consistent with a diketone structure, indicating a very low degree of aromatic character in the 1-benzoxepin ring system.

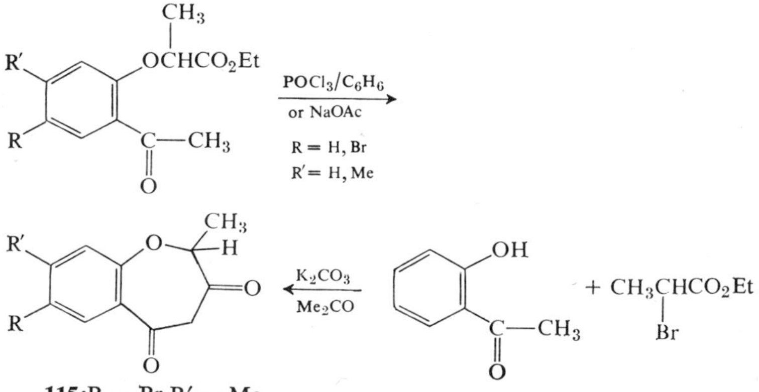

CH_3 OCHCO$_2$Et R′ R C—CH$_3$ O

$POCl_3/C_6H_6$ or NaOAc

R = H, Br
R′ = H, Me

R′ O CH$_3$ H =O R O

K_2CO_3 / Me_2CO

OH C—CH$_3$ O + CH$_3$CHCO$_2$Et Br

115;R = Br,R′ = Me
116;R = R = H′

When ethyl 2-acetyl-4-bromo-5-methylphenoxyacetate was treated with sodium ethoxide, the predominant products were the expected furan derivatives shown. However, a small quantity of a substance which appears to be 7-bromo-2,3,4,5-tetrahydro-8-methyl-1-benzoxepin-3,5-dione (**117**) was also isolated. According to Tyman and Pickles (89), compounds such as **117** are

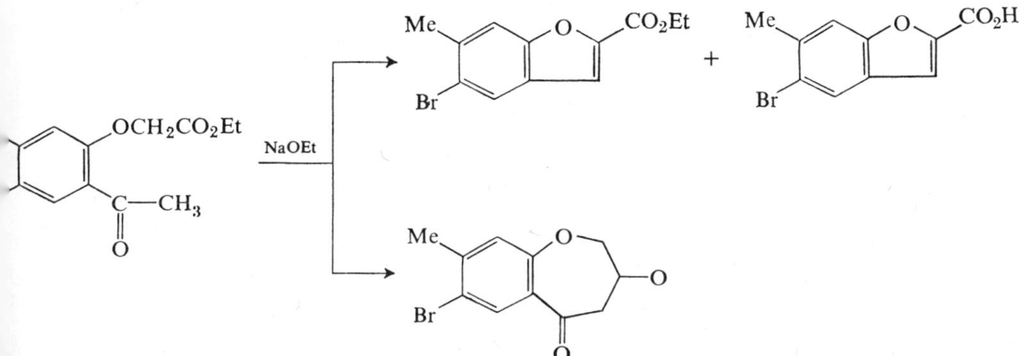

OCH$_2$CO$_2$Et C—CH$_3$ O

NaOEt

Me O CO$_2$Et Br + Me O CO$_2$H Br

Me O O Br O

117

probably formed quite generally under these conditions and were overlooked earlier because of low yields or rapid hydrolysis during workup.

In an extension of this work, Tyman and Pickles (89) also reported that condensation of diethyl bromomalonate and *o*-hydroxyacetophenone, followed by ring closure of the resulting intermediate, apparently gave some ethyl 2,3,4,5-tetrahydro-1-benzoxepin-3,5-dione-2-carboxylate (**118**) along with other products of undetermined structure.

OH; C—CH_3; O; $BrCH(CO_2Et)_2$ → $OCH(CO_2Et)_2$; C—CH_3; O; NaOEt → CO_2Et; O; H; O; O

118

The first synthesis of 2,3,4,5-tetrahydro-1-benzoxepin (homochroman) (**32**) was reported in 1949 by Cagniant (27). Condensation of *o*-methoxyphenethyl magnesium bromide with ethylene oxide and treatment of the resulting 4-(*o*-methoxyphenyl)-1-butanol with hydrobromic acid and acetic acid gave *o*-(4-bromobutyl)phenol, which underwent cyclization to **32** under the influence of base. The same series of transformations was also used by Cagniant for the synthesis of 2,3,4,5-tetrahydro-7-methyl-1-benzoxepin (**119**). The yield of **32** (35%) was initially claimed to be lower than the yield of chroman obtainable via the same approach. However, Baddeley and co-workers (5) developed an improved method for the cyclization, which enabled them to prepare **32** in 64% yield. Their modification involved heating with potassium carbonate in methyl ethyl ketone at high dilution. Other, more recent methods of synthesis of **32** have been cited earlier. These include the catalytic reduction of 2,3-dihydro-1-benzoxepin (**36**) (82) and the catalytic reduction of 1-benzoxepin (**28**) itself (95).

An interesting extension of Cagniant's work was published in 1961 by Newman and Meker (71), who obtained **119** in yields of 81 and 86%, respectively, upon treatment of 2-(4′-bromobutyl)-4-methylphenol with sodium methoxide in refluxing methanol, and with lithium in liquid ammonia. The

OMe; R; CH_2CH_2MgBr → (O, R = H, Me) → OMe; R; $(CH_2)_4OH$ → (HBr, AcOH)

OH; R; $(CH_2)_4Br$ → (base) → O; R

32; R = H
119; R = Me

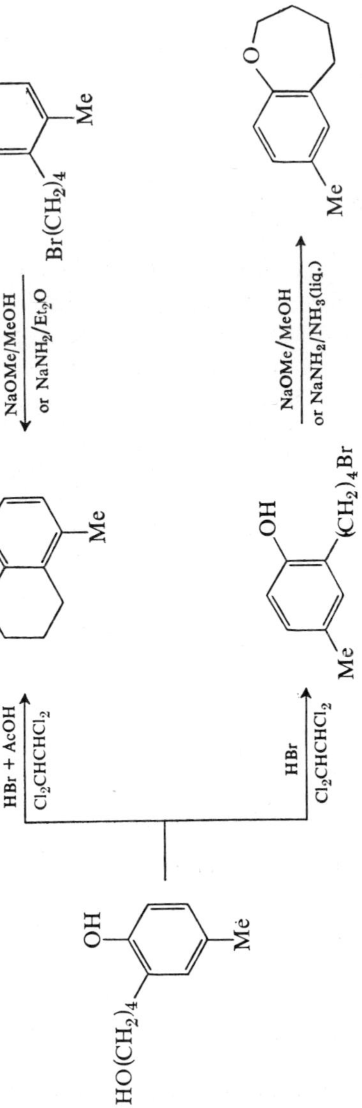

HO$(CH_2)_4$ OH Me

HBr + AcOH, $Cl_2CHCHCl_2$ → OH Me

NaOMe/MeOH or $NaNH_2/Et_2O$ ← Br$(CH_2)_4$ OH Me

HBr, $Cl_2CHCHCl_2$ → Me OH $(CH_2)_4$Br

NaOMe/MeOH or $NaNH_2/NH_3$(liq.) → Me O

119

bromo compound was produced in high yield by treatment of 2-(4′-hydroxybutyl)-4-methylphenol with hydrogen bromide in 1,1,2,2-tetrachloroethane in the absence of added acetic acid. In the presence of acetic acid an intramolecular Friedel-Crafts type alkylation occurred unexpectedly, which gave 5,6,7,8-tetrahydro-4-methyl-1-naphthol as the sole product. The latter was identical with the product formed upon treatment of 3-(4′-bromobutyl)-4-methylphenol with sodium methoxide in methanol or sodium amide in ether.

An unusually direct method for the preparation of a 2,3,4,5-tetrahydro-1 benzoxepin derivative was discovered by Hart and co-workers (34, 58) in the course of an investigation of the reaction of phenol with 2-methyl-5-chloro-2-pentene. When the reactants were heated at 150° a 28% yield of a product which was formulated as 2,3,4,5-tetrahydro-5,5-dimethyl-1-benzoxepin (**120**) was obtained. Simultaneously formed were two isomeric phenolic compounds which were identified by unambiguous synthesis as 5,6,7,8-tetrahydro-5,5-dimethyl-1-naphthol and 5,6,7,8-tetrahydro-8,8-dimethyl-2-naphthol. When the reactants were treated with potassium carbonate in boiling acetone, simple *O*-alkylation occurred with formation of 2-methyl-5-phenoxy-2-pentene. In the presence of one drop of acid, this intermediate underwent quantitative conversion into **120**. Acid cleavage of **120** produced a mixture of the aforementioned isomeric naphthols.

HBr + AcOH, 3hr reflux

150°, 12 hr

120

K_2CO_3, Me_2CO, reflux

$O(CH_2)_2CH{=}CMe_2$

2,3,4,5-Tetrahydro-5,5-diphenyl-1-benzoxepin (**121**), an interesting and exceedingly hindered homochroman derivative, was reported in 1968 by

$Ph_3C(CH_2)_3OH \xrightarrow[\text{70\%, no } O_2]{Pb(OAc)_4,\ C_6H_6}$ **121** + (1,1-diphenylindane) + $Ph_3C(CH_2)_3OAc$

121

Starnes (113) as one of the products of the reaction of 4,4,4-triphenyl-1-butanol with lead tetraacetate at 70° in the absence of atmospheric oxygen. Other products obtained in this reaction were 1,1-diphenylindane and 4,4,4-triphenyl-1-butyl acetate.

Friedel-Crafts acylation of 2,3,4,5-tetrahydro-1-benzoxepin (**32**) with acetyl chloride in the presence of aluminum chloride occurs at the *O*-activated 7-position, giving 7-acetyl-2,3,4,5-tetrahydro-1-benzoxepin (**122**) (5, 27). Acylation can be effected similarly with propionyl chloride, giving 2,3,4,5-tetrahydro-7-propionyl-1-benzoxepin (**123**) (27) Oxidation with sodium hypochlorite affords 2,3,4,5-tetrahydro-1-benzoxepin-7-carboxylic acid (**124**), which can be esterified to **125**. Lithium aluminum hydride reduction of **125** yields 2,3,4,5-tetrahydro-1-benzoxepin-7-methanol (**126**), from which 7-chloromethyl-2,3,4,5-tetrahydro-1-benzoxepin (**127**) is obtained upon treatment with dry hydrogen chloride gas (5).

32 → ($RCOCl$, $AlCl_3$, R = Me,Et) → **122**; R=Me, **123**; R=Et → (NaOCl, R = Me) → **124** (92%) → (EtOH, HO) → **125** → ($LiAlH_4$) → **126** → (HCl) → **127**

An interesting method of synthesis was utilized by Baddeley and Cooke (3) for the preparation of 2,3,4,5-tetrahydro-1-benzoxepin-2-carboxylic acid (**128**). Baeyer-Villiger oxidation of 2-bromo-1-benzosuberone with perbenzoic acid in chloroform, followed successively by oxidation with cold alkaline hydrogen peroxide and warming on the steam bath in the presence of base, gave **128**. According to a recent paper by Clark and Williams (112), further treatment of **128** with oxalyl chloride leads to acid chloride **129**, from which amide **130** is obtained upon reaction with dimethylamine. Lithium aluminum hydride reduction converts **130** into amine 2,3,4,5-tetrahydro-2-(*N*,*N*-dimethylamino)methyl-1-benzoxepin (**131**), and reaction of **131** with methyl bromide yields a quaternary salt, **132** that can be considered a bridged analog of choline phenyl ether.

2,3,4,5-Tetrahydro-5-phenyl-1-benzoxepin-5-carboxylic acid (**133**), 5-amino-2,3,4,5-tetrahydro-5-phenyl-1-benzoxepin (**134**), and a number of derivatives of **133** and **134** were prepared by Zaugg and co-workers (12, 13, 91, 100) via a novel ring expansion reaction involving the direct enlargement of a five-membered ring to a seven-membered ring. Five-membered lactone **135** was gradually transformed into **133** upon standing in base at room temperature. Similarly, treatment of **135** with sodium methoxide in methanol at room temperature afforded a 97% yield of methyl ester **136**. When the chloro analog, **137**, was treated with sodium methoxide or sodium ethoxide, esters **136** and **138** were isolated, but the yields were low. Alkaline hydrolysis of **136** afforded **133**. Acid chloride **139** was obtained upon treatment of **133** with thionyl chloride, and amide **140** was formed by amination of **139**. An alternate synthesis of **140** consisted of allowing **135** to react with ammonia (91). Sodium hypobromite treatment gave urethan **141**, from which **134** was produced on alkaline hydrolysis. Ring contraction and regeneration of lactone **135** was effected by the action of hydrobromic acid on **133**. Cleavage of the seven-membered ether ring in **133** was found to proceed more readily than in the five- and six-membered homologs. This finding supports the view that the seven-membered ring in these compounds is relatively strained.

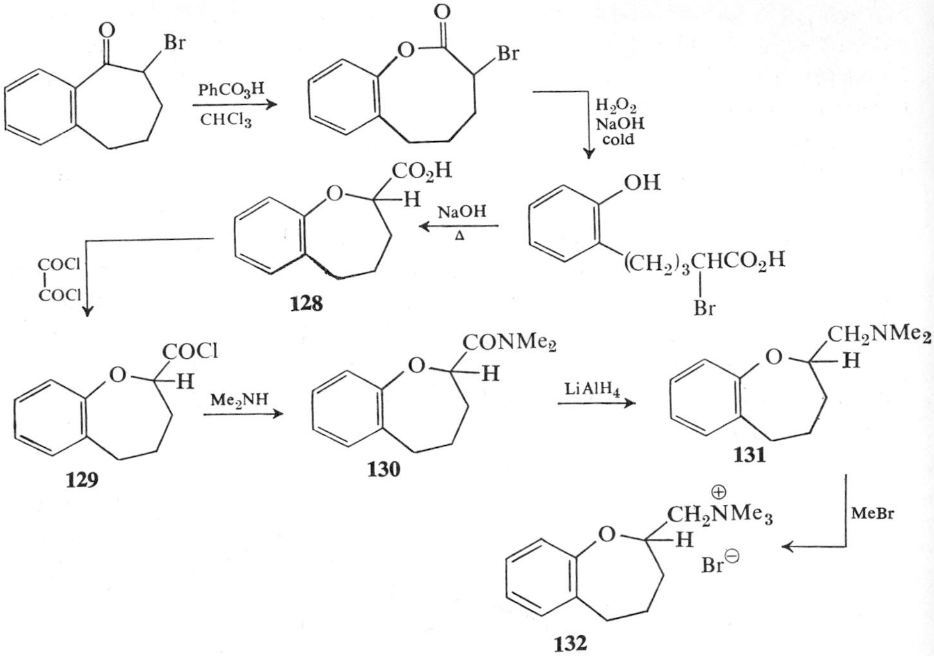

In an extension of the above work, Zaugg and co-workers (14, 91, 101) synthesized a large series of ester derivatives as analgesic and hypotensive agents. Acid chloride **139** was subjected to the action of various basic alcohols, including 2-(*N*,*N*-diethylamino)ethanol, 2-(1-piperidino)ethanol, and similar compounds. Alternately, lactone **135** was allowed to react directly with the sodium salts of the same alcohols. Acid chloride **139** was also found to react readily with 2-(*N*,*N*-diethylamino)ethyl mercaptan.

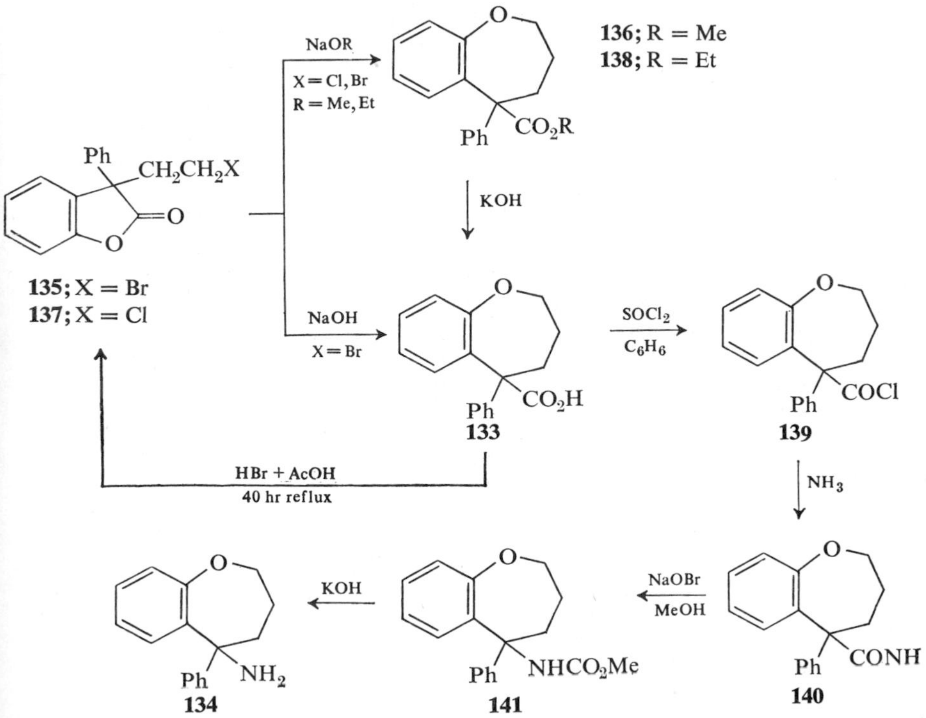

In a further extension of this work, the reduction of **140** with lithium aluminum hydride led to 5-aminomethyl-2,3,4,5-tetrahydro-5-phenyl-1-benzoxepin (**142**) (94). Other amides derived from **133** were reduced similarly. The products were claimed to possess analgesic, antispasmodic, local anesthetic, and hypotensive activities.

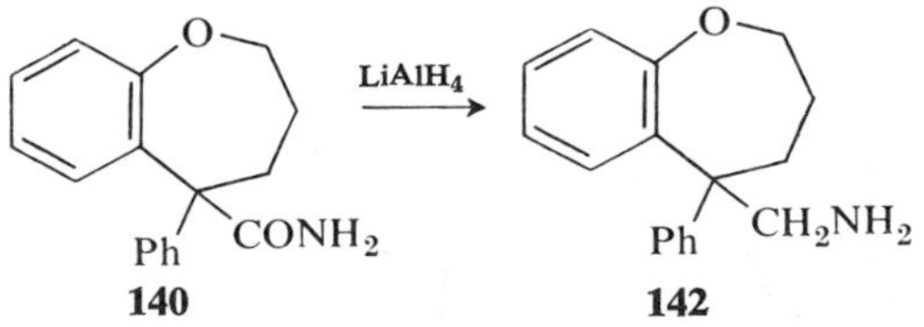

An interesting route to 2,3,4,5-tetrahydro-1-benzoxepins containing substituents at the 3-position was discovered by Zaugg and co-workers (121, 122). Treatment of γ-butyrolactone derivatives **143** and **144** with sodium methoxide resulted in ring closure to bridged lactone **145**, and further reaction of **145** with ammonia gave 2,3,4,5-tetrahydro-5-phenyl-1-benzoxepin-5-carboxamide (**146**). Similarly, γ-butyrolactone derivative **147** was

143; X = Cl
144; X = Br

NaOMe
$MeOCH_2CH_2OMe$
X = Cl, Br

145

HCl/AcOH

148

NaOMe
$MeOCH_2CH_2OMe$

147

HCl

NH_3

146

HX

150; X = Cl
151; X = Br

X = Cl

149

$LiAlH_4$
$MeOCH_2CH_2OMe$

NaOR, NaH,
or $NaNH_2$
X = Cl

$LiAlH_4$
$MeOCH_2CH_2OMe$

157

152

158

transformed into bridged lactone **148** under the influence of sodium methoxide, and condensation of **148** with morpholine hydrochloride afforded *N*-cyclopropyl 2,3,4,5-tetrahydro-3-morpholino-5-phenyl-1-benzoxepin-5-carboxamide (**149**). Reaction of **148** with dry hydrogen chloride or bromide in 2-butanone solution led to the formation of 3-chloro and 3-bromo derivatives **150** and **151**, respectively. Condensation of **150** with morpholine yielded **149**, while dehydrohalogenation with 1,4-diazabicyclo[2.2.2]octane regenerated iminolactone **148**. Interestingly, treatment of **150** with sodium alkoxides, sodium hydride, or sodamide yielded, not **148**, but an isomeric compound formulated as a bridged lactam, **152**.

A remarkable skeletal rearrangement was also described by Zaugg and Michaels (124) upon reduction of imino lactone **148** with lithium aluminum hydride in 1,2-dimethoxyethane. The expected product, 2,3,4,5-tetrahydro-5-phenyl-5-(*N*-propylaminomethyl)-1-benzoxepin-3-ol (**153**), was obtained in only 24% yield. The remainder of the product was found to consist of chroman derivative **154** and methyl *n*-propyl amine. Similar results were obtained upon reduction of imino lactone **155**, but lactone **145** gave only the expected diol, 2,3,4,5-tetrahydro-3-hydroxy-5-phenyl-1-benzoxepin-5-methanol (**156**). Interestingly, when lactam **152** was reduced under the same conditions the product was not the fully reduced amine, **157**, but instead amino carbinol **158**. However, **157** was formed smoothly upon reduction of chloro amide **150**.

145; R = O
148; R = N–◁
155; R = NPr-*n*

$LiAlH_4$, $MeOCH_2CH_2OMe$

R=N–◁ or NPr-*n* → **153** (CH_2NHPr-*n*, OH, Ph) + **154** (Ph) + MeNHPr-*n*

R=O → **156** (CH_2OH, OH, Ph)

d. *Synthesis and Chemical Reactions of Other Reduced 1-Benzoxepins*

An almost fully reduced derivative of 1-benzoxepin was reported in 1965 by Borowitz and Williams (18) in the form of 2,3,4,5,6,7,8,9-octahydro derivative **159**. To date, **159** remains the only example of its class in the literature. For the synthesis of **159**, cyclohexanone pyrrolidine enamine was alkylated with 1-acetoxybutyl bromide, the resulting acetate was hydrolyzed to 2-(4-hydroxybutyl)cyclohexanone, and the latter was cyclized by dehydration and azeotropic removal of water in boiling benzene containing a catalytic amount of *p*-toluenesulfonic acid. Oxidation of **159** with *m*-chloroperbenzoic acid afforded 10-hydroxy-6-ketodecanoic acid lactone in good yield.

N
$Br(CH_2)_4OAc$, PhMe, reflux
$(CH_2)_4OAc$
KOH, alc
$(CH_2)_4OH$
p-TsOH, C_6H_6, reflux
159
m-$ClC_6H_4CO_3H$, CH_2Cl_2, reflux

e. *Physicochemical Properties of 2,3,4,5-Tetrahydro-1-benzoxepins*

A number of useful kinetic and spectroscopic studies comparing 2,3,4,5-tetrahydro-1-benzoxepins with their five- and six-membered homologs have been reported during the past 12 years.

Baddeley and co-workers (5) observed in 1956 that bromination of the aromatic ring in 2,3,4,5-tetrahydro-1-benzoxepin (**32**) itself occurred much more slowly than in chroman or 2,3-dihydrobenzofuran. Thus 2,3-dihydrobenzofuran and chroman underwent bromination 76 times and 28 times faster, respectively, than anisole. Under the same conditions **32** was only 1.4

times more reactive than anisole. This gradation in reactivity was found to parallel quite closely the variation in uv absorption characteristics of the three compounds (Table II-1). The clear-cut bathochromic and hypsochromic effects observed in the series were taken to mean that mesomeric interaction between the oxygen atom and the benzene ring, and hence activation toward electrophilic attack by bromine, become subject to considerable steric hindrance as the number of atoms in the cyclic ether increases from five to seven. The uv spectrum of **32** has also been discussed recently by Cagniant and Cagniant (28).

As an extension of this work, Baddeley and co-workers (5) also carried out a kinetic study of the solvolysis of 7-chloromethyl-2,3,4,5-tetrahydro-1-benzoxepin (**127**), its five- and six-membered homologs, and a number of related compounds. As shown in Table II-2, there is a substantial variation in solvolysis rate with increasing ring size, the rate constant being 150–200 times larger for the furan derivative than for the oxepin derivative. Once again this difference was ascribed to the decreasing ability of the oxygen atom to interact mesomerically with the developing benzyl carbonium ion as the number of atoms in the ether ring increases from five to seven.

Similar results were obtained in another study by Baddeley and Cooke (3) who compared the uv spectral properties, pK_a values, and rates of aromatic bromination of 2,3,4,5-tetrahydro-1-benzoxepin-2-carboxylic acid (**128**) and its five- and six-membered homologs (Table II-3). There is a clear-cut bathochromic shift in both the low and high wavelength bands as the size of the ether ring increases, and also a decrease in the extinction coefficient. Furthermore there appears to be a decrease in the acidity of the compounds as the size of the ether ring becomes larger and a decrease in the rate of bromination relative to phenoxyacetic acid. As in the previous investigation (5), these findings can be accommodated with the assumption that there is a greater degree of steric inhibition of resonance in **128** than in its lower ring homologs. The authors also noted that the rates of bromination were uniformly lower for the carboxylic acids than for the unsubstituted parent compounds. This was attributed to the inductive effect of the carboxyl group, which tends to lessen the resonance capability of the oxygen *p*-orbitals and thereby deactivates the benzene ring toward electrophilic substitution.

In a follow-up study published in 1961, Baddeley and Smith (4) compared the rates of bromination of **32** and 3,4-dihydro-2*H*-1,5-benzodioxepin, and also the rates of solvolysis of **127** and its dioxepin analog. In each case the dioxepin derivative proved to be less reactive than its oxepin counterpart. These observations were rationalized satisfactorily on the basis of the assumption that, in the dioxepins, one oxygen atom exerts an inductive effect which opposes the mesomeric effect of the other.

2,3,4,5-Tetrahydro-5,5-dimethyl-1-benzoxepin (**120**) has been a compound of considerable physicochemical interest, particularly with regard to its structural geometry. In 1958 Hart and Wagner (59) examined the uv absorption spectrum of this compound and observed a significant bathochromic displacement with respect to analogs containing a five- or six-membered ether ring. They estimated that the O—CH_2 bond in the seven-membered ether deviates from planarity by approximately 70°, and predicted that this substance should be resolvable into *d*- and *l*-isomers because of the *gem*-dimethyl groups, which can prohibit rapid inversion of the ring to which they are attached. The nonplanar geometry of **120**, and also of 2,3,4,5-tetrahydro-5,5-dimethyl-1-benzoxepin-9-carboxylic acid, was confirmed in 1962 by the X-ray crystallographic analysis of Sundaralingam and Jeffrey (88).

An interesting consequence of the nonplanar geometry of **120** is the rather remarkable difference in pK_a between this compound and open-chain phenyl ethers. Using the H_0 method, Arnett and Wu (1, 2) determined the pK_a of **120** to be -1.94. This value may be compared, for example, with that of anisole, whose pK_a was found to be -6.54 by the same method.

Detailed ir, uv, and nmr spectral values were also reported for **120** by Hart and co-workers (58) in 1963. The C—O stretching frequencies for the seven-membered ether moiety were found at 1286 cm^{-1} (7.77 μ) and 1226 cm^{-1} (8.15 μ). The uv absorption spectrum in cyclohexane solution showed maxima at 271.5 mμ (ε 725) and 266 mμ (ε 700). The nmr spectrum exhibit signals at τ 8.67 (singlet, *gem*-Me_2), τ 8.2–8.5 (multiplet, 4 CH_2 protons, τ 6.14 (triplet, OCH_2 protons), and τ 2.8–3.2 (complex aromatic pattern).

An interesting comparison of the uv absorption maxima of 2,3,4,5-tetrahydro-2-(*N*,*N*-dimethylamino)methyl-1-benzoxepin methobromide (**132**) with those of its five- and six-membered homologs was reported in 1967 by Clark and Williams (112). Calculations were made of the angle of deformation (θ) which demonstrated clearly the deviation from coplanarity of the seven-membered ring in **132** (Table II-4).

10. 2-Benzoxepins

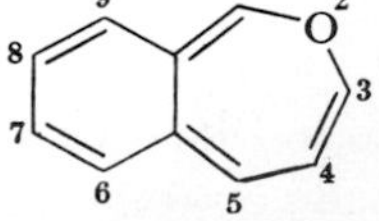

R.I. 1848

The earliest synthesis of a 2-benzoxepin derivative not containing a carbonyl group at the 1- or 3-positions was reported in 1958 by Bersch and co-workers (15) who obtained 4,5-dihydro-3-phenyl-1*H*,3*H*-2-benzoxepin (**160**) and 4,5-dihydro-3,3-diphenyl-1*H*,3*H*-2-benzoxepin (**161**) by the ring closure reaction shown. As in other reactions of the Hofmann type, a molecule of trimethylamine is produced.

$$\text{o-}(\overset{\oplus}{C}H_2NMe_3I^{\ominus})C_6H_4CH_2CH_2C(OH)(R)C_6H_5 \xrightarrow[R=H,\,Ph]{Ag_2O,\ 60^\circ} \mathbf{160};\ R=H,\ \mathbf{161};\ R=Ph + Me_3N\uparrow$$

The unsubstituted parent compound, 4,5-dihydro-1*H*,3*H*-2-benzoxepin (**162**), was first synthesized in 1962 by Riechè and Gross (79) via a straightforward two-step sequence starting from 3-phenyl-1-propanol. Chloromethylation with formaldehyde and concentrated hydrochloric acid gave a 95% yield of chloromethyl 3-phenylpropyl ether, which underwent intramolecular Friedel-Crafts reaction in 68% yield under the influence of aluminum chloride in carbon disulfide. The same approach was also used by Rieche and Gross (79) for the preparation of 4,5-dihydro-3-methyl-1*H*,3H-2-benzoxepin (**163**) from 1-methyl-3-phenyl-1-propanol. The cyclic ethers were purified most conveniently via their oxonium complexes with potassium ferricyanide.

$$C_6H_5CH_2CH_2CH(R)OH \xrightarrow[R=H,\,Me]{CH_2O/HCl} C_6H_5CH_2CH_2CH(R)OCH_2Cl \xrightarrow[CS_2]{AlCl_3} \mathbf{162};\ R=H,\ \mathbf{163};\ R=Me$$

The seven-membered ether ring in **162** has been subjected to several unexceptional chemical transformations (46). Oxidation with alkaline potassium permanganate results in complete degradation, giving phthalic acid. Heating with zinc chloride in acetic anhydride affords diacetate **164**, and treatment with fuming hydrobromic acid yields dibromide **165**. Photocatalyzed oxidation at 100° converts **162** into 4,5-dihydro-1*H*,3*H*-2-benzoxepin-1-one (**166**). When the reaction is conducted at room temperature, a hydroperoxide presumed to have structure **167** is formed, but this intermediate cannot be obtained pure.

Additional transformations of **162** were reported in 1967 by Normant (119). Oxidation with chromic acid yielded 3-(*o*-carboxyphenyl)-1-propanol (**168**), presumably via intermediate cyclic hemiacetal **169**. On the other hand, oxidation with selenium dioxide in boiling xylene afforded a mixture of 3-(*o*-hydroxymethyl)phenylpropionaldehyde (**170**) and 3-(*o*-formylphenyl) propionaldehyde (**171**). In this case, attack by the oxidizing agent occurred at the 3-position, giving intermediate cyclic hemiacetal **172**.

$KMnO_4/OH^{\ominus}$

CO_2H

CO_2H

$ZnCl_2/Ac_2O$

165–170°

CH_2OAc

$(CH_2)_3OAc$

164

O

162

HBr

100°

CH_2Br

$(CH_2)_3Br$

165

OOH

O

O

O

O_2

$h\nu$, 100°

O

167

166

HO

O

CrO_3

AcOH

CO_2H

$(CH_2)_3OH$

169

168

O

162

SeO_2

xylene

Δ

O

OH

R

$(CH_2)_3OH$

172

170; $R{=}CH_2OH$

171; $R{=}CHO$

Treatment of **162** with *N*-bromosuccinimide resulted in the formation of 1-bromo-4,5-dihydro-1*H*,3*H*-2-benzoxepin (**173**) (119). Further reaction of this interesting α-haloether with sodium ethoxide afforded 1-ethoxy-4,5-dihydro-1*H*,3*H*-2-benzoxepin (**174**), and acid hydrolysis of **174** gave 3-(*o*-formylphenyl)-1-propanol (**175**). Condensation of **173** with diethylamine and ethyl magnesium bromide yielded 1-(*N*,*N*-diethylamino)-4,5-dihydro-1*H*,3*H*-2-benzoxepin (**176**) and 1-ethyl-4,5-dihydro-1*H*,3*H*-2-benzoxepin (**177**), respectively.

NEt_2 O Et_2NH Br **176** O O NBS Et O **162** **173** EtMgBr NaOEt **177** OEt CHO O $H^{\oplus}$ $(CH_2)_3OH$ **174** **175**

Oxidation of **162** with nitric acid led to 3-(*o*-carboxyphenyl)propionic acid (**178**); catalytic hydrogenation in the presence of Raney nickel gave 3-(*o*-tolyl)-1-propanol (**179**).

CO_2H HNO_3 $(CH_2)_3CO_2H$ O **178** CH_3 **162** H_2/Ni $(CH_2)_3OH$ **179**

An interesting method of preparation of lactone **166** was devised in 1935 by Krollpfeiffer and Müller (68). As shown, 2-bromo-1-tetralone was allowed to react with pyridine, and the resulting pyridinium bromide salt, **180**, was converted to a second, ring-opened pyridinium salt, **181**, by successive treatment with base and acid. Vacuum distillation transformed the latter into **166** in 35% yield with simultaneous elimination of a molecule of pyridine hydrochloride. This method is strongly reminiscent of the more recently described procedure of Bersch and co-workers (15) for the preparation of **160** and **161**. The seven-membered lactone ring in **166** was cleaved readily by dilute alkaline hydrolysis and was regenerated from the resulting hydroxy acid, **182**, by distillation at atmospheric pressure (63).

O Br $\xrightarrow{C_5H_5N}$

O $\overset{\oplus}{N}$ $Br^{\ominus}$ (1) NaOH (2) HCl $\longrightarrow$ CO_2H $(CH_2)_3\overset{\oplus}{N}$ $Cl^{\ominus}$

180 **181**

Δ

CO_2H $(CH_2)_3OH$ dil $OH^{\ominus}$ Δ O O

182 **166**

A saturated derivative of 1*H*,3*H*-2-benzoxepin was reported in 1962 by Paquette and Nelson (74) in connection with a projected study of the Baeyer-Villiger oxidation of *trans*-2-decalone. This compound, *trans*-4,5,5a,6,7,8,9,-9a-octahydro-1*H*,3*H*-2-benzoxepin-3-one (or *trans*-3-oxa-4-oxobicyclo[5.4.0]-undecane) (**183**), was synthesized by the series of reactions shown. The C=O stretching frequency of the seven-membered lactone ring in **183** was found to be at 1740 cm^{-1} (5.75 μ).

1*H*,3*H*-2-Benzoxepin-1,3-dione (**184**) has been reported on two occasions. Boeseken and Slooff (17) observed the formation of this anhydride in 1930 during a study of the peracetic acid oxidation of 1,2-naphthoquinone. The diacid obtained from this reaction underwent dehydration to **184** upon being heated at its melting point. In an extension of this work, Karrer and Schneider (66) reported in 1947 that perbenzoic acid oxidation of 1,2-naphthoquinone afforded **184** directly, together with a peroxy compound believed to have the structure shown. Anhydride **184** was also prepared by the reaction of peracetic acid with naphthalene. Ring opening of thc seven-membered anhydride was achieved with either acid or base, giving the same diacid as had been obtained originally by Boeseken and Slooff.

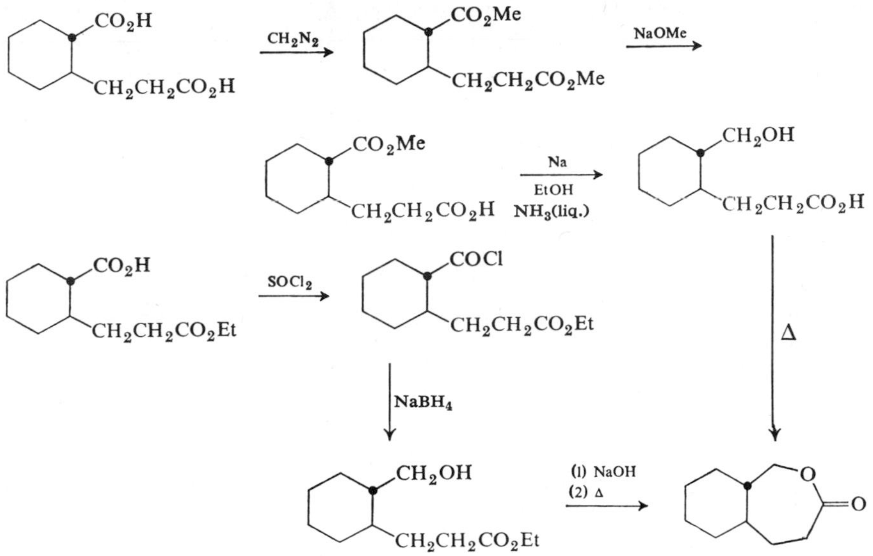

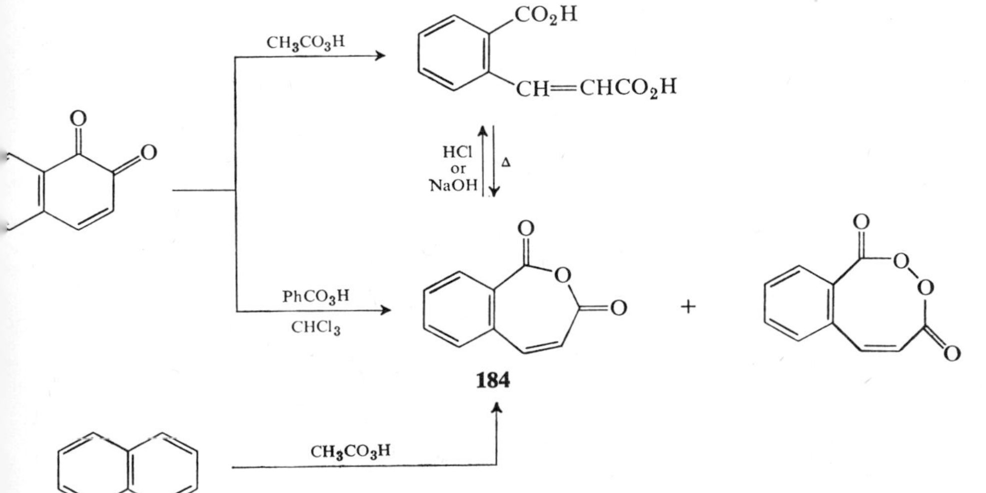

A derivative of **184** was obtained in 1960 by Chatterjea and Mukherjee (29) from the condensation of phthalaldehydic acid and ethyl phenylacetate in the presence of sodium methoxide as a catalyst. After separation by standard extraction procedures, two products were isolated. These were identified as 4-phenyl-1*H*,3*H*-2-benzoxepin-1,3-dione (**185**) and the corresponding dicarboxylic acid (**186**). Cleavage of the seven-membered ring in anhydride **185** with hydrochloric acid gave **186**.

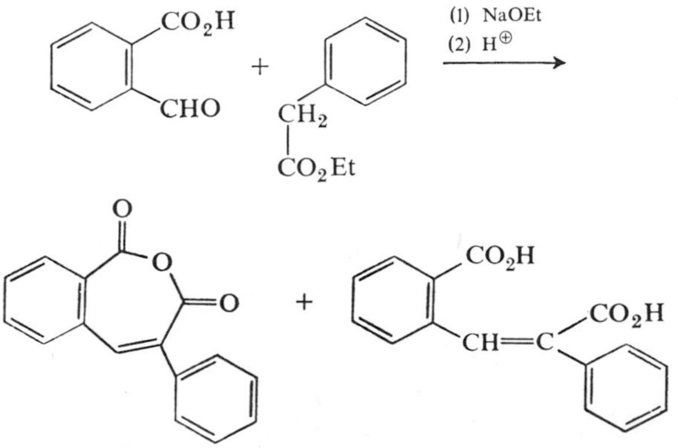

(1) NaOEt
(2) $H^{\oplus}$

185 **186**

Two dihydro derivatives of **168** were reported in 1962 by Broquet-Borgel and Quelet (24), namely 4,5-dihydro-4,4-dimethyl-1*H*,3*H*-2-benzoxepin-1,3-dione (**187**) and 4,5-dihydro-4,4,8-trimethyl-1*H*,3*H*-2-benzoxepin-1,3-dione (**188**). For the preparation of these seven-membered anhydrides, the butyrolactone derivative shown was allowed to undergo Friedel-Crafts condensation

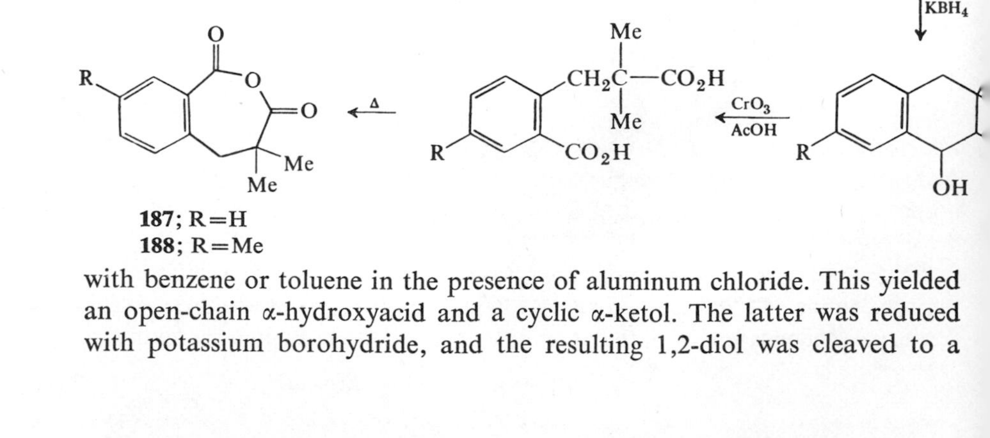

$AlCl_3$

R = H, Me

KBH_4

CrO_3 AcOH

Δ

187; R=H
188; R=Me

with benzene or toluene in the presence of aluminum chloride. This yielded an open-chain α-hydroxyacid and a cyclic α-ketol. The latter was reduced with potassium borohydride, and the resulting 1,2-diol was cleaved to a diacid with chromium trioxide in acetic acid. Sublimation of the diacid afforded the anhydride.

A novel synthesis of a 2-benzoxepin derivative involving the reaction of the natural product variotin (**189**) with maleic anhydride was reported in 1966 by Takeuchi and Yonehara (120). Ozonolysis of the adduct, **190**, afforded α-hydroxycaproic acid.

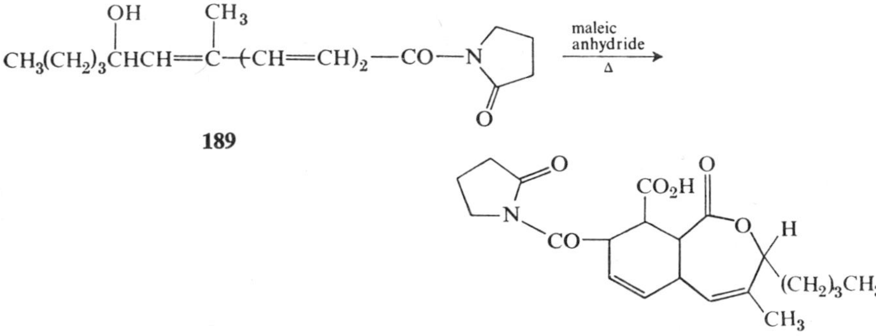

189

190

11. 3-Benzoxepins

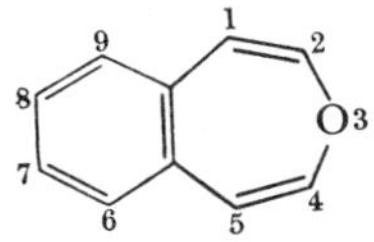

R.I. 1849

Fully unsaturated 3-benzoxepin derivatives were first described in 1957 by Dimroth and Freyschlag (45, 46), who obtained dimethyl 3-benzoxepin-2,4-dicarboxylate (**191**) from the reaction of phthalaldehyde and dimethyl 2,2′-oxydiacetate in the presence of sodium methoxide or potassium *tert*-butoxide as the condensing agent. Hydrolysis of **191** furnished 3-benzoxepin-2,4-dicarboxylic acid (**192**) Treatment of **192** with thionyl chloride and pyridine afforded an acid chloride, which was converted directly to diethyl 3-benzoxepin-2,4-dicarboxylate (**193**) by reaction with ethanol in pyridine solution. Dimethyl ester **191** was regenerated from **192** by reaction with diazomethane. All of these transformations were subsequently confirmed by other workers (63, 65). According to Jorgenson (65), diacid **192** exists in three crystalline modifications possessing different ir spectra in potassium bromide, but with identical uv absorption spectra and melting points.

Compounds **191** and **192** were observed to possess a rather unexpected degree of chemical stability, particularly toward acid hydrolysis. Dimroth

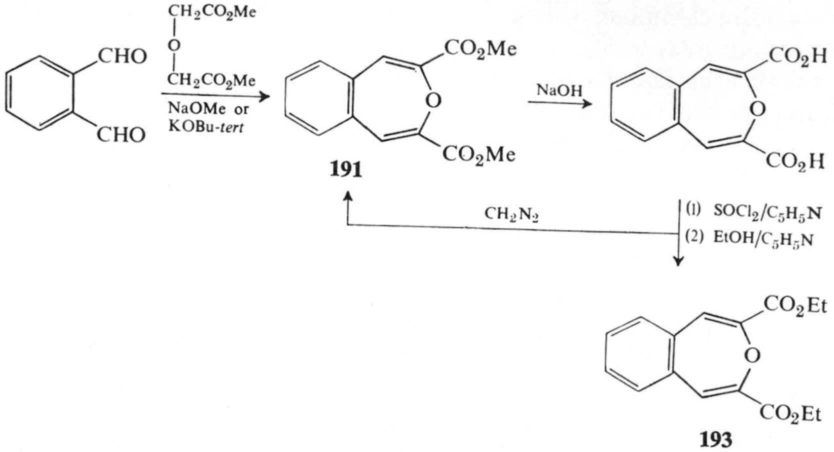

and Freyschlag (45) reported that bromination and potassium permanganate oxidation were both unsuccessful. Huisgen and co-workers (63), however, did succeed in converting ester **191** into a dibromide under forcing conditions. This dibromide, which presumably possessed structure **194**, was reconverted into **191** under the influence of zinc metal; pyridine, on the other hand, left the dibromo compound intact. Pyrolysis of **192** at 300° under reduced pressure afforded a 44% yield of 2-hydroxy-3-naphthoic acid, while distillation in the presence of copper powder produced naphthalene-2,3-dicarboxylic acid. The detailed mechanisms of these thermal reactions are not readily evident from the structures of the starting materials. Huisgen and co-workers reported that diester **191** failed to react with several common reagents,

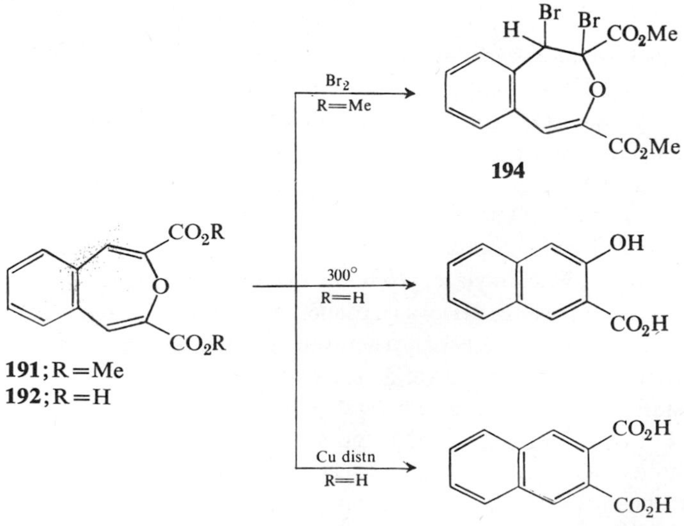

including maleic anhydride, diethyl diazodicarboxylate, tetracyanoethylene, benzoyl chloride, and acetic anhydride.

The most striking illustration of the stability of **191** and **192**, however, was provided by the behavior of these compounds in the presence of strong acids. Thus **192** was found to be unaffected by prolonged storage in 80% sulfuric acid at room temperature or by treatment with 50% sulfuric acid under reflux (45, 46), while diester **191** was recovered unchanged after exposure to 48% hydrobromic acid (63) or 75% sulfuric acid (65). Jorgenson also reported that **191** could be dissolved in concentrated deuteriosulfuric acid and recovered without undergoing any deuterium incorporation, as evidenced by ir or nmr spectra.

Several other fully unsaturated 3-benzoxepins were prepared by Jorgenson during the course of a study confirming and extending the earlier work of Dimroth and Freyschlag, and of Huisgen and co-workers. Reduction of diester **191** with hydrogen over palladium–carbon afforded dimethyl 1,2-dihydro-3-benzoxepin-2,4-dicarboxylate (**195**). Catalytic reduction of diacid **192**, on the other hand, led to the uptake of two molecules of hydrogen, giving 1,2,4,5-tetrahydro-3-benzoxepin-2,4-dicarboxylic acid (**196**). Reduction of diester **191** with lithium aluminum hydride, or with a mixture of lithium aluminum hydride and aluminum chloride, furnished 3-benzoxepin-2,4-dimethanol (**197**). Oxidation of **197** with manganese dioxide yielded monoaldehyde **198**, but attempts to prepare the dialdehyde by further oxidation of **198** were unsuccessful. Diol **197**, unlike diacid **192** and diester **191**, proved to be quite acid-sensitive, as evidenced by its rapid decomposition upon being heated in ethanol containing a trace of mineral acid. Diacetate **199**, likewise a rather unstable compound, was prepared from **198** by treatment with acetic anhydride in pyridine.

Several explanations have been advanced for the acid-stable character of **191** and **192**, and a detailed discussion has been given by Jorgenson. The initial view of Dimroth and Freyschlag that this property is the result of aromaticity in the 3-benzoxepin ring system appears to conflict with Jorgenson's finding that compounds **197–199** exhibit the typical acid sensitivity of vinyl ethers. According to the latter author, compounds **191** and **192** undergo protonation on oxygen, rather than on the olefinic carbons or on the carbonyl oxygen of the acid or ester function. This conclusion is in accord with the lack of deuterium incorporation of **191**, the reluctance of **192** to undergo acid-catalyzed decarboxylation, and the pK_a values for the oxonium ions derived from **191** and **192**, which were both found to be -5.30 ± 0.10 from spectral measurements in 95% sulfuric acid–5% dioxane (65).

The parent compound itself, 3-benzoxepin (**200**), was first reported in 1961 by Dimroth and Pohl (47, 48), who prepared it from phthalaldehyde by means of a double Wittig condensation employing the ylide derived from bis(bromomethyl) ether. The structure of **200** was proved by its reduction with hydrogen over platinum, which afforded 1,2,4,5-tetrahydro-3-benzoxepin (**201**). The identity of **201** was established by comparison with an authentic sample prepared by sulfuric acid treatment of 1,2-bis(2-hydroxyethyl)-benzene. Like other 3-benzoxepins not substituted with stabilizing groups, **200** was found to be acid-labile, giving a product formulated as indene-3-carboxaldehyde upon treatment with refluxing concentrated hydrochloric acid (47). Bromination or chlorination of **200** in carbon tetrachloride afforded oily dihalides of uncertain structure, from which **200** could be regenerated by zinc dust distillation (48).

CHO, CHO → ($Br^{\ominus}$ $Br^{\ominus}$ $Ph_3\overset{\oplus}{P}CH_2OCH_2\overset{\oplus}{P}Ph_3$, NaOMe) → O **200** → (HCl, reflux) → CHO

200 → (H_2/Pt) → **201**

CH_2CH_2OH, CH_2CH_2OH → (H_2SO_4) → O **201**

A theoretically significant, although preparatively less useful, photochemical method of synthesis of 3-benzoxepins was reported recently by Ziegler and Hammond (114, 115). Irradiation of dilute solutions of 1,4-epoxy-1,4-dihydro(or deuterio)naphthalene in ether, cyclohexane, or absolute ethanol in degassed tubes at 2537 Å gave **200** (or its 2,4-d_2 derivative) in 6% yield. Irradiation of 1-methyl-1,4-epoxy-1,4-dihydronaphthalene, which was prepared from benzyne and 2-methylfuran, afforded a 2% yield of 2-methyl-3-benzoxepin (**202**); similarly, photoisomerization of 1,4-epoxy-1,4-dihydro-1,4-diacetoxymethylnaphthalene produced 1,3-diacetoxymethyl-3-benzoxepin (**199**) in 5% yield. The mechanism of photoisomerization is presumed to involve a quadricyclic intermediate as shown; however, attempts to isolate or trap this species have not been successful.

The apparent photostability of the 3-benzoxepins resulting from this synthesis is noteworthy in view of the reported photocyclization reactions of oxepin (97) and 2,7-dimethyloxepin (73).

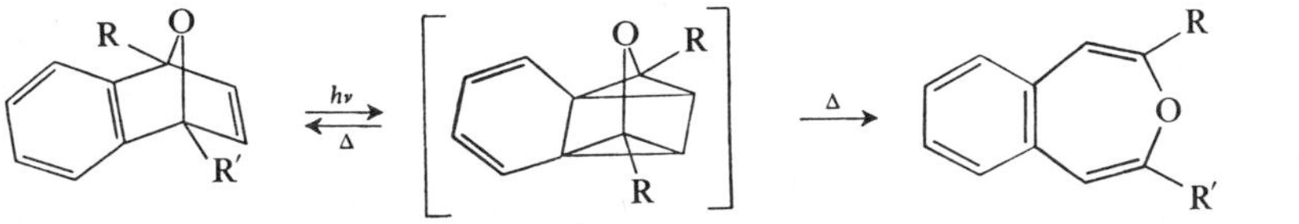

200; R = R′ = H or D
202; R = Me, R′ = H
199; R = R′ = CH_2OAc

The earliest reported preparation of a fully saturated 3-benzoxepin derivative was the work of Borsche and Lange (19) who obtained 1,2,3,5,5a,-6,7,8,9,9a-decahydro-3-benzoxepin-2,4-dione (**203**) upon treatment of cyclohexane-1,2-diacetic acid with acetic anhydride. The stereochemistry of the ring junction was not fully defined. In all probability, both the starting diacid and the resulting anhydride consisted of mixtures of *cis* and *trans* isomers. Cleavage of the seven-membered ring in **203** was achieved readily by reaction with hydrochloric acid.

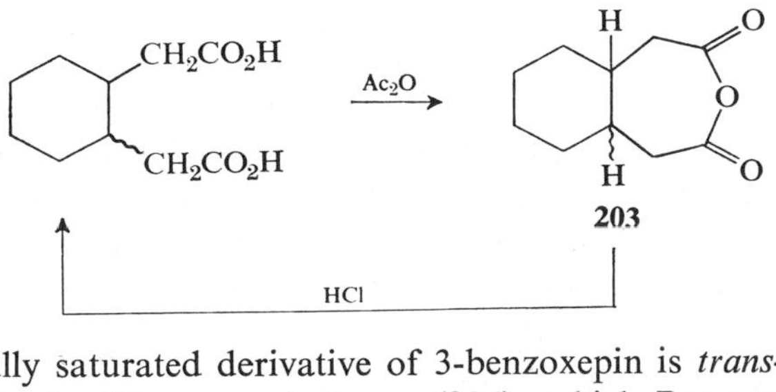

203

Another fully saturated derivative of 3-benzoxepin is *trans*-1,2,4,5,5a,6,-7,8,9,9a-decahydro-3-benzoxepin-2-one (**204**), which Paquette and Nelson prepared in 1962 in connection with a study of the Baeyer-Villiger oxidation of *trans*-2-decalone (74). Reaction of *trans*-cyclohexane-1,2-diacetic acid with acetic anhydride and a trace of acetyl chloride afforded a crude anhydride, which was treated directly with methanol to give methyl hydrogen *trans*-cyclohexane-1,2-diacetate. Reduction of this monoester with sodium in liquid ammonia and ethanol, followed by direct vacuum distillation of the resulting hydroxy acid, yielded the desired seven-membered lactone.

CH_2CO_2H / CH_2CO_2H —Ac_2O/AcCl→ (anhydride) —MeOH→ CH_2CO_2Me / CH_2CO_2H —Na, NH_3/EtOH→ CH_2CH_2OH / CH_2CO_2H —vacuum distn→ **204**

204

A number of interesting derivatives of the 3-benzoxepin ring system were synthesized by Burgstahler and Abdel-Rahman (26) using 1,1,4,4-tetramethyl-2-tetralone (25) as the key starting material. Oxidation of this compound with selenium dioxide gave 1,1,4,4-tetramethyl-2,3-tetralindione, which was reduced to the corresponding diol with lithium aluminum hydride. Reduction of the diketone with hydrogen over platinum afforded 3-hydroxy-1,1,4,4-tetramethyl-2-tetralone. Lead tetraacetate oxidation of this α-ketol yielded 1,2,4,5-tetrahydro-1,1,5,5-tetramethyl-3-benzoxepin-2,4-dione (**205**). The same anhydride was also produced when the diketone was subjected to the action of sodium ethoxide in an oxidative atmosphere. Reaction of the diol with lead tetraacetate under carefully defined conditions yielded *o*-phenylenediisobutyraldehyde, which was converted readily into its cyclic bishemiacetal, 1,2,4,5-tetrahydro-2,4-dihydroxy-1,1,5,5-tetramethyl-3-benzoxepin (**206**), on attempted crystallization from moist solvents. Alkaline hydrolysis of **205** resulted in rupture of the seven-membered anhydride ring and formation of *o*-phenylenediisobutyric acid.

In addition to its extraordinarily facile conversion to bishemiacetal **206**, *o*-phenylenediisobutyraldehyde served as the precursor for several other previously unknown derivatives of 3-benzoxepin (26). Thus, under the influence of base, the dialdehyde readily underwent an internal Cannizzaro reaction, giving 1,2,4,5-tetrahydro-1,1,5,5-tetramethyl-3-benzoxepin-2-one (**207**) in 77% yield. The dialdehyde likewise gave **207** on attempted bisthioketalization in the presence of either hydrochloric acid or boron trifluoride etherate. Lactone **207** was also obtained by reaction of anhydride **205** with sodium amalgam. Reduction of *o*-phenylenediisobutyraldehyde with lithium aluminum hydride at −40° afforded 1,2,4,5-tetrahydro-2-hydroxy-1,1,5,5-tetramethyl-3-benzoxepin (**208**). Hemiacetal **208** was likewise produced if lactone **207** was reduced under these conditions. Reaction of **208** with methanol in the presence of catalytic amounts of acid furnished 1,2,4,5-tetrahydro-2-methoxy-1,1,5,5-tetramethyl-3-benzoxepin (**209**).

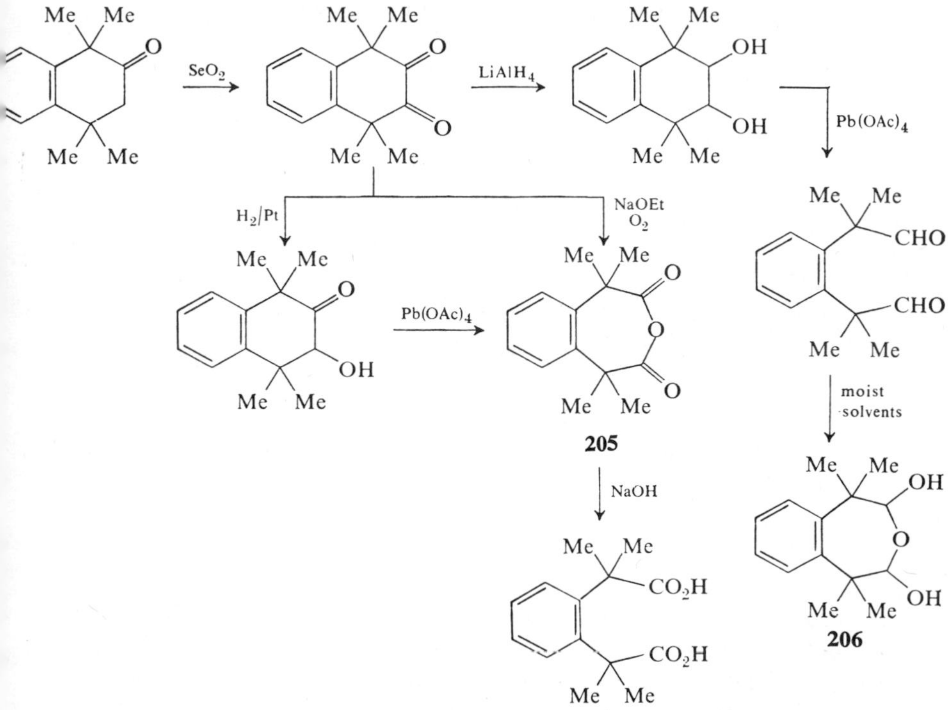

When the reduction of *o*-phenylenediisobutyraldehyde with lithium aluminum hydride was carried out in refluxing ether instead of at −40°, the product formed was *o*-phenylenediisobutyl alcohol. Reaction of this diol with *p*-toluenesulfonyl chloride in pyridine gave the ditosylate ester. Treatment of the latter with lithium aluminum hydride in refluxing tetrahydrofuran afforded a mixture of 1,2,4,5-tetrahydro-1,1,5,5-tetramethyl-3-benzoxepin (**210**) and *o*-di-*tert*-butylbenzene. The cyclic ether, which was the preponderant product, was separated readily from the less polar hydrocarbon by chromatography on acid-washed alumina.

o-Phenylenediisobutyric acid was prepared independently by Barclay and co-workers (8) in connection with a program aimed at the synthesis of *o*-di-*tert*-butylbenzene by classical chemical methods. The above authors reported in 1962 that treatment of the diacid with acetic anhydride under reflux effected ring closure to anhydride **205** in 92% yield. The same anhydride was also obtained in two other reactions, namely the oxidation of *o*-phenylenediisobutyric acid with periodic acid in glacial acetic acid at 100° and with lead tetraacetate in dioxane at 75°. In each case, anhydride **205** was accompanied by an approximately equal amount of a six-membered lactone, to which

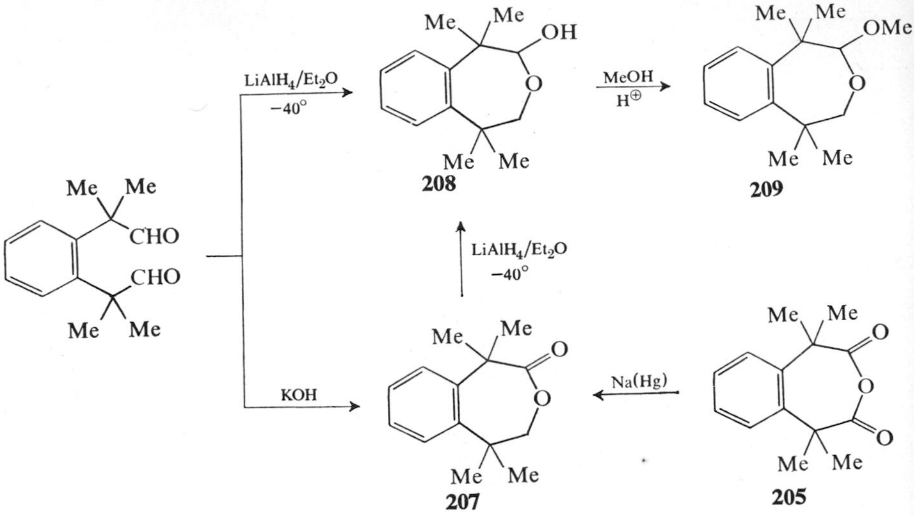

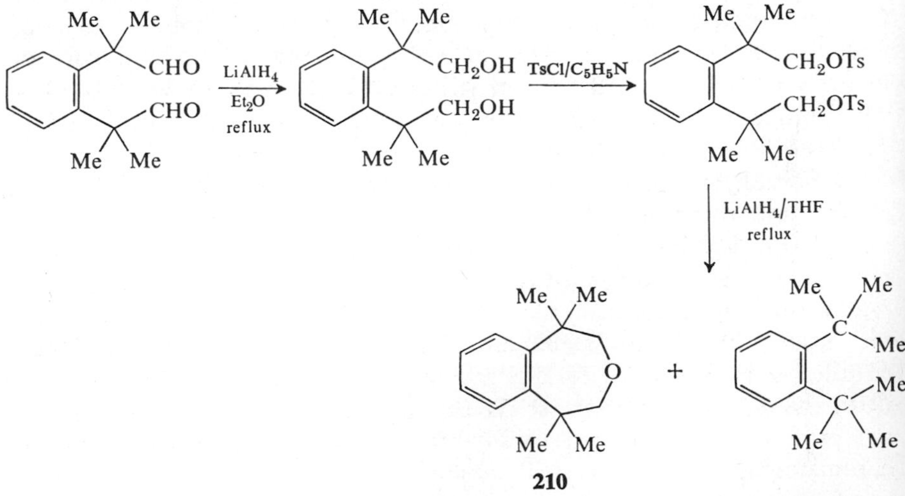

structure **211** was assigned. The dimethyl ester of *o*-phenylenediisobutyric acid was reduced with lithium aluminum hydride, the resulting diol was converted to a ditosylate, and the latter was transformed into *o*-di-*tert*-butylbenzene by hydride reduction in ether. Interestingly, there was no indication of ring closure to cyclic ether **210** when ether was used for the reduction of the ditosylate.

Me Me CO_2H CO_2H Me Me; Ac_2O reflux → **205**; HIO_4/AcOH 100° or $Pb(OAc)_4$/dioxane 75° → **205**, **211**; (1) CH_2N_2 (2) $LiAlH_4$ (3) $TsCl/C_5H_5N$ (4) $LiAlH_4$

Barclay and co-workers found that the anhydride derived from *o*-phenylenediisobutyric acid was different from a substance previously assigned the same structure by Colonge and Lagier (31, 32). The French authors claimed to have obtained anhydride **205** on chromic acid oxidation of a hydrocarbon which they believed to be 1,1,5,5-tetramethyl-6,7-benzosuberane. This hydrocarbon, which was formed by condensation of benzene with 2-chloro-2,6-dimethyl-5-heptene in the presence of aluminum chloride, was shown by Barclay and co-workers to be, in fact, 1-isopropyl-4,4-dimethyltetralin. Thus the earlier assignment of structure **205** to the chromic acid oxidation product appeared to be erroneous.

Another anhydride derived from the 3-benzoxepin ring system was reported in 1963 by Vulfson and Iodko (98). This compound, 1,2,4,5-tetrahydro-1-methyl-3-benzoxepin-2,4-dione **(212)**, allegedly undergoes an interesting thermal rearrangement, giving 2-tetralone and carbon dioxide upon being heated at 190–230° under reduced pressure.

212 $\xrightarrow[\textit{in vacuo}]{190-230^\circ}$ 2-tetralone $+ CO_2\uparrow$

B. Spirans

1. 1,7-Dioxaspiro[5.6]dodecanes

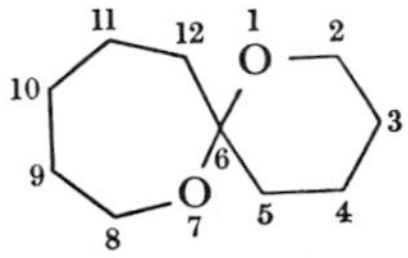

R.I. 8137

The spiroketal 1,7-dioxaspiro[5.6]dodecane (**213**) was prepared in 1958 by Stetter and Rauhut (87). The ethylene ketal of 5-ketosebacid acid was reduced with lithium aluminum hydride, and the resulting dihydroxy ketal **214**, was transketalized under the influence of *p*-toluenesulfonic acid.

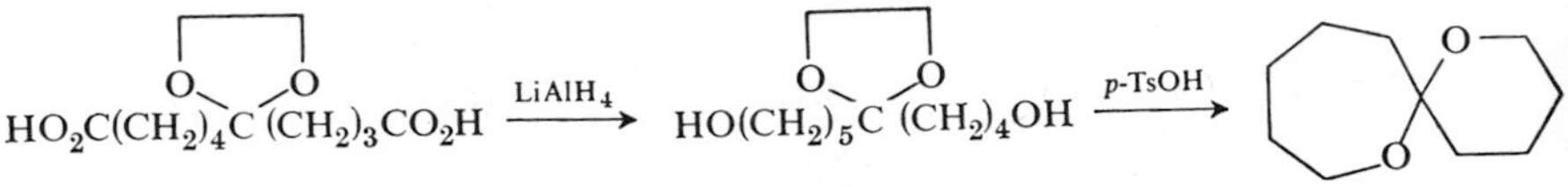

$$HO_2C(CH_2)_4C(OCH_2CH_2O)(CH_2)_3CO_2H \xrightarrow{LiAlH_4} HO(CH_2)_5C(OCH_2CH_2O)(CH_2)_4OH \xrightarrow{p\text{-TsOH}} \mathbf{213}$$

214 **213**

2. 1,4,6-Trioxaspiro[4.6]undecanes

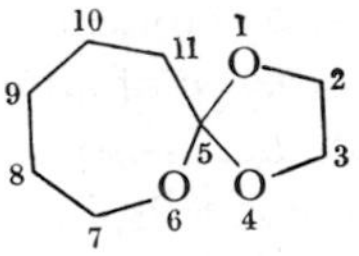

1,4,6-Trioxaspiro[4.6] undecane (**215**) was prepared in 1967 by Inokawa and co-workers (102) by condensation of ethylene oxide with ε-caprolactone. Further reaction of **215** with ethanol in the presence of boron trifluoride etherate afforded mainly ethyl ε-hydroxycaproate, together with other products. Spiran **215** represents an interesting type of intramolecular ortho-ester derivative which is quite reactive, as evidenced by the report (103) that **215**, and also the 10-chloro derivative of **215**, can be copolymerized with *s*-trioxane, the cyclic trimer of formaldehyde.

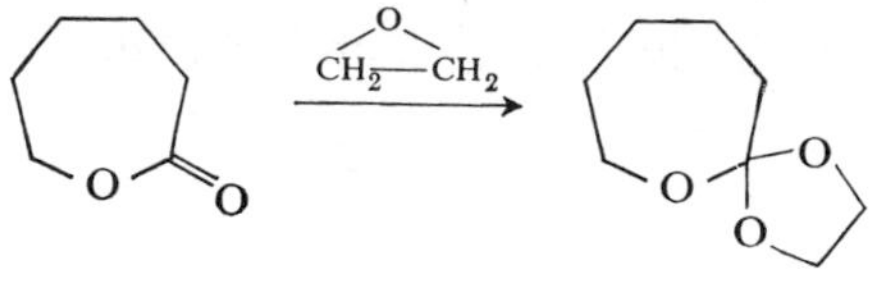

CH_2—CH_2 (O)

215

C. Bridged Systems

1. 2-Oxabicyclo[3.2.2]nonanes

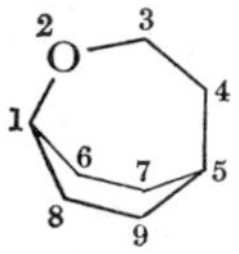

R.I. 10,087

The 2-oxabicyclo[3.2.2]nonane ring system entered the chemical literature via the work of Plieninger and Keilich (75), who prepared the bridged lactone 2-oxabicyclo[3.2.2]nonan-3-one-5-carboxylic acid (**216**) according to

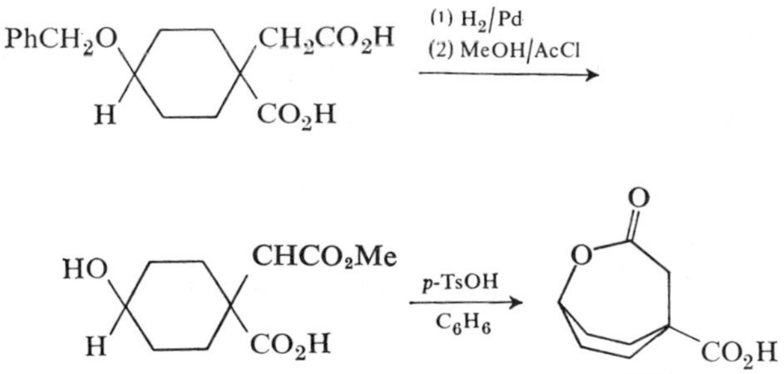

$PhCH_2O$, H, CH_2CO_2H, CO_2H —(1) H_2/Pd (2) MeOH/AcCl→

HO, H, $CHCO_2Me$, CO_2H —p-TsOH, C_6H_6→ CO_2H

216

the reaction sequence shown. The intramolecular transesterification which resulted in the formation of the seven-membered lactone in the last step was brought about with *p*-toluenesulfonic acid. It is interesting to note that transesterification occurred in preference to simple lactonization, apparently because the latter process would involve the formation of a more strained ring system.

Another member of the 2-oxabicyclo[3.2.2]nonane ring system was reported in 1960 by Meinwald and co-workers (70), who isolated from the condensation of tetracyanoethylene with 2,3-dihydrooxepin an adduct which could have structure **217**. However, alternative nonbridged structures for this adduct are also possible.

$(NC)_2C{=}C(CN)_2$, C_6H_6

217

2. 3-Oxabicyclo[3.2.2]nonanes

R.I. 1850

The isomeric 3-oxabicyclo[3.2.2]nonane ring system was reported first in 1934 by Malachowski and Jankiewiczowna (69) in connection with an interesting study of polymeric anhydrides. Treatment of *cis*-1,4-cyclohexanedicarboxylic acid with acetic anhydride at steam bath temperature was claimed to give "poly-*cis*-anhydride," which was allegedly isomerized to "poly-*trans*-anhydride" upon being heated above 230°. When either of these polymeric anhydrides was heated carefully under reduced pressure with a free flame, care being taken to cool the distillate immediately and to prevent its contact with moisture, the monomeric anhydride 3-oxabicyclo[3.2.2]-nonane-2,4-dione (**218**) was isolated. This monomer was also reported briefly in 1958 by Hall (57), and more recently in a German patent (104).

HO_2C–C₆H₁₀–CO_2H →(Ac_2O, steam bath) "poly-*cis*-anhydride" →(>230°) "poly-*trans*-anhydride" →(Δ) **218**; "poly-*cis*-anhydride" →(Δ) **218**; **218** →(H_2O) diacid

218

3-Oxabicyclo[3.2.2]nonane (**219**) itself was obtained recently by Brown and Seaton (118) as one of the products of the vapor phase dehydration of *cis*-1,4-cyclohexanedimethanol. Other products formed in this reaction were the unsaturated alcohols shown.

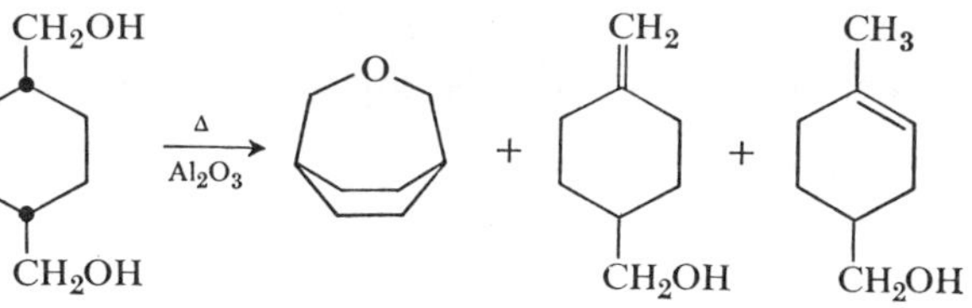

CH$_2$OH, CH$_2$OH → (Δ, Al$_2$O$_3$) **219** + CH$_2$ / CH$_2$OH + CH$_3$ / CH$_2$OH

3. 9-Oxabicyclo[3.3.2]decanes

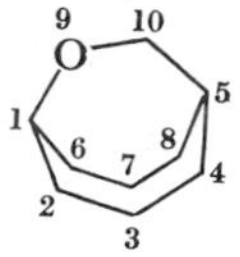

Another related ring system is exemplified by the bridged seven-membered lactone, 9-oxabicyclo[3.3.2]decan-10-one (**220**), which Cope and Gale (33) synthesized in 1963 from bicyclo[3.3.1]nonan-9-one by oxidation with peracetic acid. Reduction of **220** with lithium aluminum hydride in refluxing tetrahydrofuran afforded 5-hydroxymethylcyclooctanol (**221**).

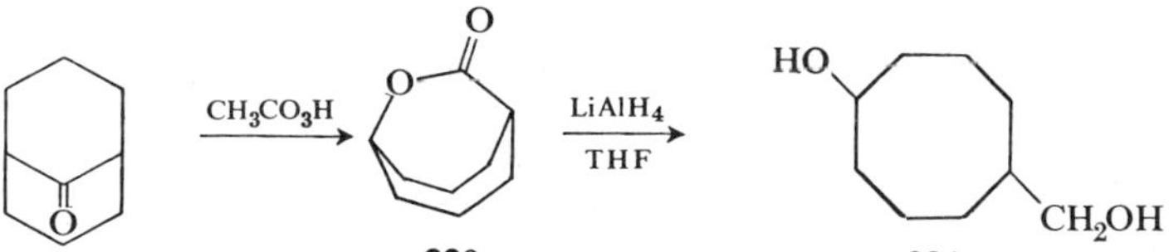

CH_3CO_3H → **220** → (LiAlH$_4$, THF) **221** (HO, CH$_2$OH)

4. 11-Oxabicyclo[4.4.1]undecanes

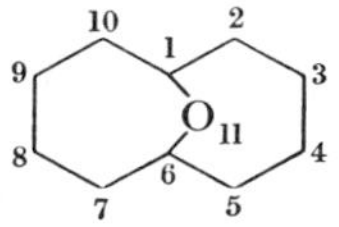

R.I. 2011

Still another bicyclic system containing a seven-membered ether structure is the 11-oxabicyclo[4.4.1]undecane ring system, which has received considerable attention ever since its initial discovery in 1944 by Criegee (36). Other designations commonly employed in the literature are the terms "1,6-epoxycyclodecane" and "1,6-oxidocyclodecane," which refer to reduced derivatives, and "1,6-oxido[10]annulene," which refers to the fully unsaturated, pseudoaromatic system. In the present review, the more formal bicyclo[4.4.1]undecane system of nomenclature will be retained for the partly or fully reduced derivatives, and the term "1,6-oxido[10]annulene" will be used in the discussion on fully unsaturated members of this family.

a. *Partly or Fully Saturated Derivatives*

Entry into the 11-oxabicyclo[4.4.1]undecane ring system was achieved first (36) via the key intermediate *trans*-9-decalyl hydroperoxide, which was obtained from decalin by reaction with oxygen at 110°. Mild treatment of the hydroperoxide with acetic anhydride gave *trans*-9-decalyl peroxyacetate. This rather labile compound tended to rearrange to a product which Criegee formulated as 1-acetoxy-11-oxabicyclo[4.4.1]undecane (**222**). The assignment of structure **222** to the rearranged product was supported by the finding that alkaline hydrolysis gave 6-hydroxycyclodecanone. Under the influence of acetic anhydride at 110°, *trans*-9-decalyl hydroperoxide afforded a 40% yield of **222** directly. Similarly, *trans*-9-decalyl peroxybenzoate could be prepared from the hydroperoxide under mild conditions, and then rearranged to 1-benzoyloxy-11-oxabicyclo[4.4.1]undecane (**223**) by simple heating, or

OOH; 110°, O_2; $BzCl/C_5H_5N$, 100°; OOBz; Ac_2O, 110°; OBz; O; OAc; O; OOAc; **222**; **223**; NaOH; O; OH

by treatment with pyridine or a mixture of pyridine and pyridine hydrochloride. Direct conversion of the hydroperoxide into **223** could be effected by reaction with benzoyl chloride in pyridine at 100°.

The oxidation of decalin was also investigated in some detail in 1956 by Holmquist and co-workers (61), who noted the formation of only minute amounts (2.6%) of **222** when the reaction was performed in acetic anhydride at 115–120° in the presence of azodicyclohexanecarbonitrile. Copious amounts of decalin were recovered unchanged, as might be expected, and a substantial proportion of the oxidized product (13%) appeared to consist of isomeric decalols and decalones, which were not separated. In addition, a trace of 6-ketodecanoic acid was also isolated from the reaction.

A related procedure described by Schenk and Schulte-Elte (80) in 1958 is worthy of mention because it provides a novel avenue for the conversion of octahydronaphthalene into an octahydroazulene. In this instance, treatment of $\Delta^{4(10)}$-octalin-9-hydroperoxide with 3,5-dinitrobenzoyl chloride in pyridine yielded the rearranged compound, **224**, directly. Under the influence of alcoholic base, **224** underwent further rearrangement to $\Delta^{9(10)}$-octahydroazulen-4-one.

OOH
3,5-dinitrobenzoyl chloride
C_5H_5N
$OOCC_6H_4(NO_2)_2$-3,5
$OCC_6H_4(NO_2)_2$-3,5
NaOH
MeOH
224

In an extension of the initial work on this ring system, Criegee and Kaspar (38) reported in 1948 the synthesis of 1-(*p*-nitrobenzoyloxy)-11-oxabicyclo-[4.4.1]undecane (**225**) from *trans*-9-decalyl *p*-nitroperbenzoate, and noted the ease with which this ester underwent rearrangement in relation to the analog lacking the *p*-nitro substituent (38, 39). Also in 1948, Criegee and Schnorrenberg (40) described the synthesis of 1-methoxy-11-oxabicyclo-[4.4.1]undecane (**226**), a compound which earlier workers had tried to prepare without success (37). 6-Hydroxycyclodecanone was found to give a relatively modest yield of **226** when kept at 0° in methanol containing dilute

MeOH/HCl
reflux
OMe
OH
MeOH/HCl
0°
OMe
226
HCl
reflux

hydrochloric acid. The hemiketal was isolated by steam distillation, and 80% of the starting material was recovered unchanged. However, when the reaction was carried out under reflux, as previously reported (37), the product was not **226**, but 6-methoxycyclodecanone. Hydrolysis of **226** with aqueous acid regenerated 6-hydroxycyclodecanone in high yield.

The pioneering work of Criegee and his students aroused widespread theoretical speculation concerning possible mechanistic pathways for the rearrangement of various types of peroxy esters. A perspective on the early thinking in this area can be found in review articles by Criegee (37) and Bartlett (9), both of which appeared in 1950.

In the decade that followed, several groups of workers carried out detailed kinetic and isotope-labeling studies which are now considered textbook classics. In the course of these studies, a number of new 11-oxabicyclo[4.4.1]-undecane derivatives were prepared.

In 1953, Goering and Olson (54) investigated the rearrangement of *trans*-9-decalyl perbenzoate to **223**, and of *trans*-9-decalyl *p*-nitroperbenzoate to **225.** Decompositions were carried out at various temperatures in anhydrous as well as aqueous methanol mixtures, and in the presence of added lithium *p*-nitrobenzoate and lithium benzoate. The composition of each product mixture was analyzed in detail, and the reaction kinetics were measured. The findings of Criegee and co-workers (36–38) were confirmed and extended, especially with respect to the dependence of the reaction rate upon temperature and solvent polarity.

The rearrangement of *trans*-9-decalyl perbenzoate and *trans*-9-decalyl *p*-nitroperbenzoate to **223** and **225**, respectively, was also studied kinetically by Bartlett and Kice (10). In addition, these authors prepared the *p*-bromobenzoate, *p*-methylbenzoate, and *p*-methoxybenzoate esters of *trans*-9-decalyl hydroperoxide, and rearranged them to 1-(*p*-bromobenzoyloxy)-11-oxabicyclo[4.4.1]undecane (**227**), 1-(*p*-methylbenzoyloxy)-11-oxabicyclo-[4.4.1]undecane (**228**), and 1-(*p*-methoxybenzoyloxy)-11-oxabicyclo[4.4.1]-undecane (**229**), respectively. The rate of reaction was enhanced by the

R R
OO—C=O OO—C=O
MeOH →
O

225; R = NO_2
227; R = Br
228; R = Me
229; R = MeO

electron-donating *p*-methyl and *p*-methoxy substituents, and retarded by the electron-withdrawing *p*-bromo and *p*-nitro substituents.

Although **223** is by far the predominant product formed during the rearrangement of *trans*-9-decalyl perbenzoate, Bartlett and Kice (10) also isolated a small amount of an olefinic substance which they identified tentatively as 11-oxabicyclo[4.4.1]undec-1-ene (**230**). Catalytic reduction of this material resulted in the uptake of 0.75 mole of hydrogen and the disappearance of the olefinic ir absorption band at 5.99 μ. Treatment of **230** with 50% acetic acid on the steam bath afforded 6-hydroxycyclodecanone. A completely homogeneous sample of **230** could not be obtained at the time, the principal contaminants probably being **226** and 6-methoxycyclodecanone. The product formed upon hydrogenation of **230** was presumably the parent member of the series, 11-oxabicyclo[4.4.1]undecane (**231**), itself. This substance had, in fact, been obtained previously (90), together with a smaller amount of cyclodecane, upon catalytic hydrogenation of 6-hydroxycyclodecanone at high temperature.

Two recent patents are likewise of interest in connection with the unusual internal enol ether, **230**. The first of these (23) describes a method of preparation of **230** which starts from *trans*-9-decalyl hydroperoxide instead of *trans*-9-decalyl perbenzoate. It is claimed that a 72% yield of **204** can be realized via this modification. Treatment with hot acetic acid is said to convert **230** into 6-hydroxycyclodecanone in 95% yield, in accord with the earlier findings of Bartlett and Kice (10).

The second patent (92) describes the oxidation of $\Delta^{5(10)}$-octalin with

OOH; AcOH, room temp; **230**; H_2/Pt; **231**; AcOH, reflux; H_2O_2, H_2SO_4, <10°; OH; O; **232**; O; OH; H_2/Pt, Δ; **231** +; H_2SO4; 0

hydrogen peroxide in sulfuric acid at low temperature. In this case, apparently, the first-formed hydroperoxide rearranges to **230**, but the latter is hydrated extremely rapidly under the influence of strong acid. The product isolated ultimately is 6-hydroxycyclodecanone. Further treatment of this product with strong acid leads to the formation of cyclodec-5-enone. It is very likely (116) that 6-hydroxycyclodecanone exists in equilibrium with its hemiketal tautomer, 1-hydroxy-11-oxabicyclo[4.4.1]undecane (**232**), much as in the case, for example, of 6-hydroxyhexanal and its seven-membered cyclic hemiacetal tautomer.

In connection with the above discussion, it is of interest that Prelog and Küng (78) have reported 6-hydroxy-6-methylcyclodecanone, which was prepared by chromic acid oxidation of 1-methyl-1,6-cyclodecanediol, to be in equilibrium with the hemiketal tautomer, 1-hydroxy-6-methyl-11-oxabicyclo[4.4.1]undecane (**233**). Reduction of the mixture of tautomers with lithium aluminum deuteride gave the starting diol with specific deuterium incorporation at the 6-position. The extent to which the bridgehead methyl group may stabilize or destabilize the hemiketal form in this case is not known.

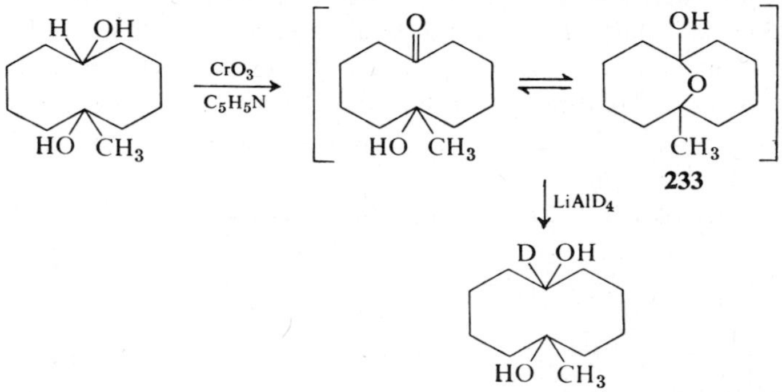

H OH; HO CH₃ → (CrO_3, C_5H_5N) [O; HO CH₃ ⇌ OH, O, CH₃] **233** ↓ $LiAlD_4$ D OH; HO CH₃

Of great importance in elucidating the mechanism of the rearrangement of *trans*-9-decalyl perbenzoate to **223** has been the classic study of Denney and Denney (42, 43) involving O^{18}-labeling techniques. When *trans*-9-decalyl perbenzoate labeled with O^{18} in the carbonyl group was treated with methanol or acetic acid under the usual rearrangement conditions, virtually all of the O^{18} in the rearranged product, **223**, was still in the carbonyl group. This

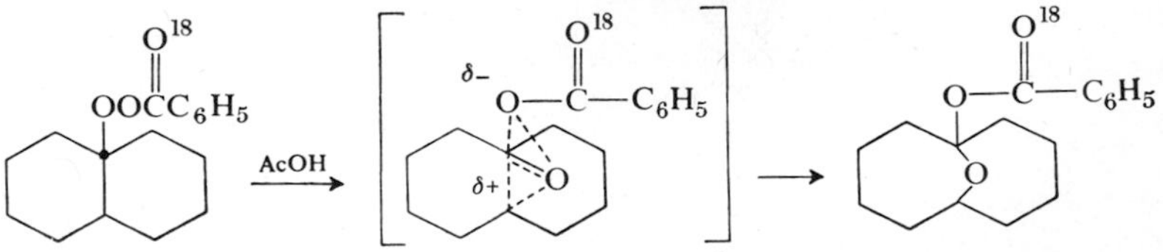

O^{18}; $OOCC_6H_5$ → (AcOH) [O^{18}; δ− O—C—C_6H_5; δ+ O] → O^{18}; O—C—C_6H_5; O

223

observation militated against any mechanism invoking complete departure of the benzoate group from the decalyl moiety, since O^{18} scrambling would be expected in such a process. The mechanism best supported by the data appears to be one involving a delocalized oxonium ion intermediate of the type shown.

As an interesting outgrowth of their studies on the chemistry of *trans*-9-decalyl perbenzoate, Denney and co-workers (44) also investigated the reaction of this peroxyester with tri-*n*-butyl phosphine. The major product was found to be *trans*-9-decalyl benzoate, but a modest quantity of **223** was also isolated.

b. *1,6-Oxido[10]annulenes*

Perhaps the most interesting members of the 11-oxabicyclo[4.4.1]undecane ring system are those which are known collectively as 1,6-oxido[10]annulenes. These novel, fully unsaturated compounds were selected for investigation on the basis of their obvious analogy to the 1,6-methano[10]annulene ring system, which is known to exhibit a certain degree of aromatic character. Like their carbocyclic counterparts, the 1,6-oxido[10]annulenes can be considered to possess a peripheral 10π-electron system, thereby satisfying the requirements of the Hückel $4n + 2$ rule for aromaticity. Essentially all that is known today of the chemistry of this fascinating series is the result of elegant work performed independently in the laboratories of Sondheimer (86) and Vogel (95), respectively. For an excellent historical perspective on this work within the broader content of oxepin chemistry, the reader should consult the recent review article by Vogel and Guenther (97).

The synthesis of the parent compound, 1,6-oxido[10]annulene (**234**), can be accomplished in four straightforward steps, starting from naphthalene (84, 86, 95). Reduction with sodium in liquid ammonia, selective monoepoxidation of the tetrasubstituted 9,10-double bond, and bromination of the two remaining double bonds gives the key tetrabromoepoxide shown. Alternatively, bromination can be carried out prior to epoxidation, but this approach is much less satisfactory because bromination of the triene gives a complex mixture, only a small part of which is the required unsaturated tetrabromide. Dehydrobromination of the tetrabromoepoxide with excess potassium hydroxide in ethanol at approximately 50° gives a 50% yield of **234**, together with a 20% yield of 1-benzoxepin (84, 86). The two products can be separated by chromatography on alumina. Sodium methoxide may be used in place of potassium hydroxide in the dehydrobromination step. Another procedure, which is claimed to give **234** as the sole product in 60% yield, involved dehydrobromination of the tetrabromo compound with potassium *tert*-butoxide in ether at −10° (95).

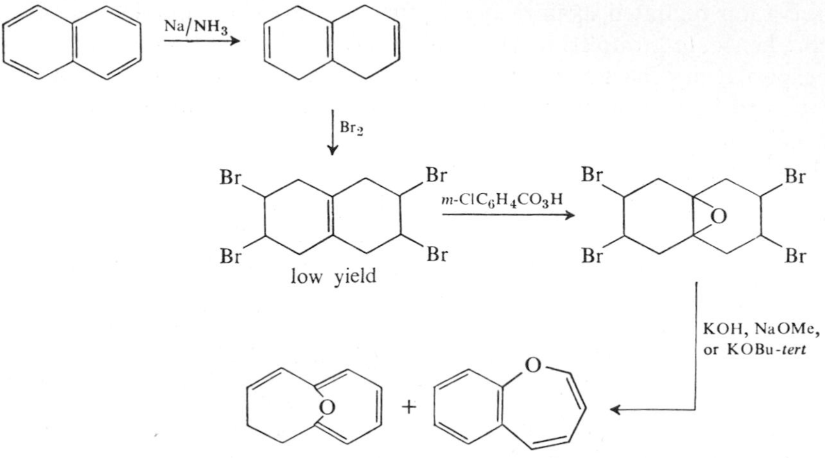

Na/NH$_3$

Br$_2$

Br Br Br Br

m-$ClC_6H_4CO_3H$

Br Br Br Br

low yield

KOH, NaOMe, or KOBu-*tert*

O + O

234

According to Sondheimer and Shani (86), the formation of **234** and 1-benzoxepin can be rationalized on the basis of a common unsaturated epoxide intermediate, which can undergo bond reorganization in the manner shown. Path A involves direct valence isomerization to **234**; Path B proceeds via a second postulated epoxide intermediate which has not actually been demonstrated to have a finite lifetime. Two independent routes had to be proposed on the grounds that **234** did not undergo further rearrangement upon treatment with base or alumina.

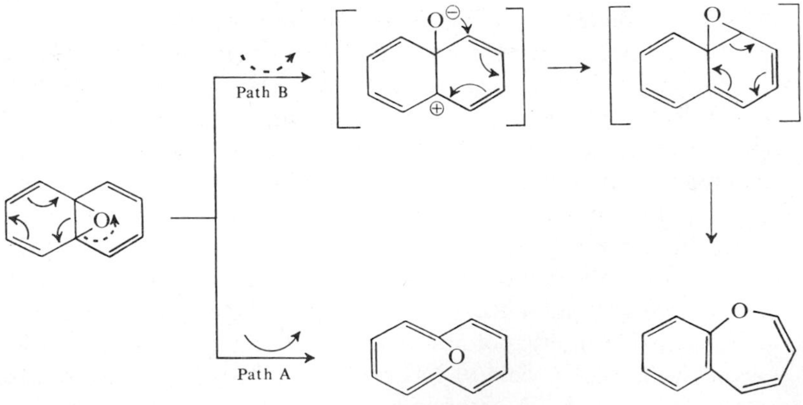

Path B

Path A

234

In contrast to its relative inertness toward basic reagents, and also toward heat, oxygen, and light, **234** undergoes facile isomerization under the influence of acidic reagents (84, 86, 95). Treatment with acid-washed alumina or silica gel produces extensive rearrangement to 1-benzoxepin, with simultaneous formation of small amounts of 1-naphthol. When boron trifluoride etherate is used as the Lewis acid, the major product becomes 1-naphthol. However, it appears unlikely that 1-benzoxepin is an intermediate in the isomerization of **234** to 1-naphthol, since only a small yield of 1-naphthol can be isolated upon treatment of 1-benzoxepin with boron trifluoride etherate or concentrated sulfuric acid. According to one report (95), in fact, treatment of 1-benzoxepin with acids yields, not 1-naphthol, but carbonyl compounds arising from cleavage of the oxepin ring. It is also of interest that, in the presence of aqueous acetic acid at 50°, **234** can be transformed into a mixture of approximately equal parts of 1-naphthol, 2-naphthol, and 1-benzoxepin (84). The exact course of these skeletal rearrangements remains unelucidated at present.

234

SiO_2 → 50% + 2.5%

$BF_3 \cdot OEt_2$ → " (2%) + " (70%)

AcOH, 50° → " (25%) + " (25%) + 25%

Removal of the oxygen atom in **234** can be achieved by catalytic hydrogenation over palladium–charcoal (84). The product obtained, when reduction is stopped after 5 min and ethanol is used as the solvent, is naphthalene. If pentane is used as the solvent and hydrogenation is allowed to proceed for several hours, the product is tetralin. When reduction is effected by means of lithium aluminum hydride, the sole product in high yield is naphthalene. Other reagents that have been reported to bring about the conversion of **234** into naphthalene are triphenylphosphine and $Cr(CO)_3(NH_3)_3$ (97).

An indication of the degree of aromaticity in **234** is the fact that this compound undergoes certain aromatic electrophilic substitution reactions without skeletal disruption. For example, treatment with cupric nitrate in acetic

$H_2/Pd-C$, $LiAlH_4$, Ph_3P, or $Cr(CO)_3(NH_3)_3$

234

$H_2/Pd-C$
n-pentane

anhydride affords a mixture of 2-nitro-1,6-oxido[10]annulene (**235**) and 3-nitro-1,6-oxido[10]annulene (**236**) in approximately equal proportions (84, 86, 95). The isomeric mononitro derivatives can be separated by alumina chromatography, and their structures can be deduced on the basis of nmr spectra. Isomerization of **236** to a nitro derivative of 1-benzoxepin under the influence of acid has been reported, but the position of the nitro group in the product has not been determined with certainty. Attempted reduction of **235** and **236** to the corresponding amines has not been successful thus far.

NO_2 NO_2

$Cu(NO_3)_2$ / Ac_2O

234 **235** **236**

Other electrophilic substitution reactions attempted with **234** have been less successful than nitration, mainly because of the ease of isomerization of the 1,6-oxido[10]annulene ring system. Thus mild treatment of **234** with acetic anhydride and boron trifluoride etherate in chloroform affords 1-acetoxynaphthalene in 70% yield (84). Furthermore, reaction of **234** with oleum in dioxane and addition of methyl iodide to the silver salt of the resulting sulfonation product apparently gives only naphthalenoid derivatives, as evidenced by uv spectral data. Finally, treatment of **234** with excess bromine in chloroform at room temperature gives two isomeric tetrabromo compounds in yields of 65 and 10%, respectively. The major isomer has been identified as 1,4,5,8-tetrabromo-9,10-epoxy-1,4,5,8-tetrahydronaphthalene. When the bromination is performed in dichloromethane at −78°, it is possible to isolate a dibromo compound which has been formulated as 1,4-dibromo-9,10-epoxy-1,4-dihydronaphthalene (96). Upon treatment with sodium iodide in acetone, both the dibromo and tetrabromo adducts revert quantitatively to **234**. Dehydrobromination of the dibromo and tetrabromo adducts has been achieved smoothly with potassium *tert*-butoxide, and the resulting products have been identified as 2-bromo-1,6-oxido[10]annulene

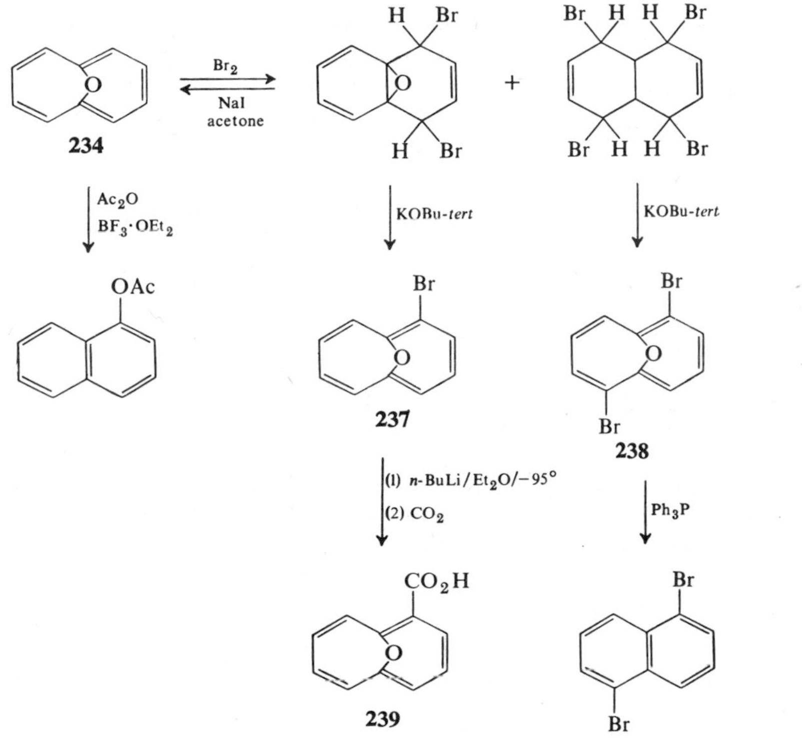

(**237**) and 2,7-dibromo-1,6-oxido[10]annulene (**238**), respectively (97). Treatment of **237** with *n*-butyllithium in ether at $-95°$, followed by reaction with carbon dioxide, affords 1,6-oxido[10]annulene-2-carboxylic acid (**239**). Deoxygenation of **238** with triphenylphosphine leads to the formation of 1,5-dibromonaphthalene, thereby proving the structure of **238** unequivocally with respect to bromine substitution.

The unusual structure of the 1,6-oxido[10]annulenes has led to a number of detailed studies of their physical properties. These studies include a theoretical analysis of the uv absorption spectrum (16), a report on the magnetic circular dichroism spectrum (22), investigations of the nmr (56, 84, 117), and esr (53) spectra, an X-ray crystallographic analysis (6), and determinations of the dipole moment (21). All of these precise measurements support the assigned structure of **234** and confirm its aromatic character.

The ir spectrum of **234** is characterized by the appearance of the C=C stretching band at the unusually high wavelength of 6.5 μ, which is more reminiscent of benzenoid absorption than olefinic absorption. The uv absorption spectrum in ethanol solution exhibits maxima at 255, 299, and 393 mμ, and is quite similar to that of 1,6-methano[10]annulene, thereby indicating the absence of significant interaction between the oxygen atom and the peripheral 10 π-electron system. The nmr spectrum consists of an AA′BB′

system with $\tau_A = 2.54$ and $\tau_B = 2.74$, and with the coupling constants $J_{AB} = 8.8, J_{BB'} = 9.3, J_{AB'} = 0.3$, and $J_{AA'} = 1.1$ Hz. The low-field absorption and type of coupling pattern shown by **234** is consistent with the presence of a ring current, and militates against a system of localized π-bonds.

5. 3-Oxa-6,7-dithiabicyclo[3.2.2]nonanes

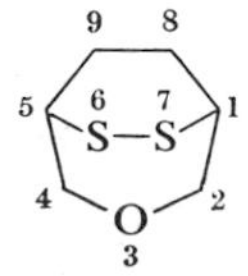

R.I. 10070

A patent disclosure has been made (127) concerning the synthesis of anhydride **240** by reaction of *meso*-α,α′-dithioadipic acid (**241**) with acetyl chloride under reflux. Cleavage of the seven-membered anhydride ring in **240** with sodium methoxide affords the monomethyl ester (**242**). Treatment of **240** with ammonia results in the formation of ammonium salt **243**, while reaction with sodium in liquid ammonia leads to reduced ammonium salt **244**. Anhydride **240** finds use as a hair-waving agent.

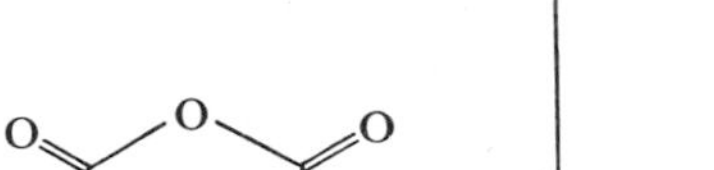

241 $\xrightarrow{AcCl}$

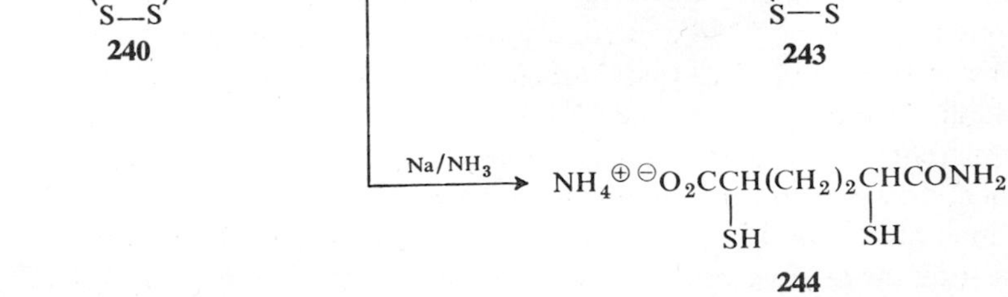

240 $\xrightarrow{NaOMe}$ $Na^{\oplus}\,{}^{\ominus}O_2C$–(S–S ring)–$CO_2Me$ **242**

240 $\xrightarrow{NH_4OH}$ $NH_4^{\oplus}\,{}^{\ominus}O_2C$–(S–S ring)–$CONH_2$ **243**

240 $\xrightarrow{Na/NH_3}$ $NH_4^{\oplus}\,{}^{\ominus}O_2CCH(SH)(CH_2)_2CH(SH)CONH_2$ **244**

D. Tables

The physical constants for the 1-benzoxepins (Section A-9), 2-benzoxepins (Section A-10), and 3-benzoxepins (Section A-11) are included in Tables II-5, II-6, and II-7, respectively.

TABLE II-1 Uv Data and Bromination Rate of 2,3,4,5-Tetrahydro-1-benzoxepin and Lower Ring Homologs

Compound	Relative bromination rate (anisole = 1.0)	λ_{max} (mμ)	ε
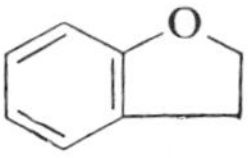 [2,3-dihydrobenzofuran structure]	76	283	3095
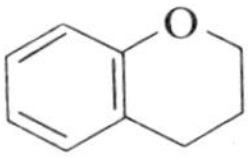 [chroman structure]	28	276	2125
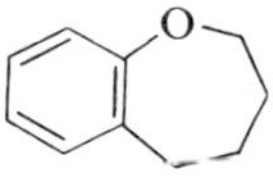 [2,3,4,5-tetrahydro-1-benzoxepin structure]	1.4	267	678

TABLE II-2 Solvolysis Kinetics of 7-Chloromethyl-2,3,4,5-tetrahydro-1-benzoxepin and Lower Ring Homologs

	Rates of solvolysis in 90% EtOH at 0 and 25°			
Compound	$k \times 10^5$ min^{-1} at 0°	$k \times 10^5$ min^{-1} at 25°	E_a (kcal/mole)	$A \times 10^{-11}$ min^{-1}
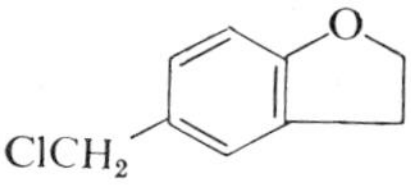 $ClCH_2$ [dihydrobenzofuran structure]	2250	3630	17.3	15
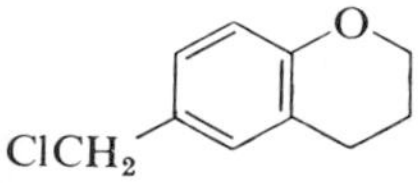 $ClCH_2$ [chroman structure]	585	1060	18.8	55
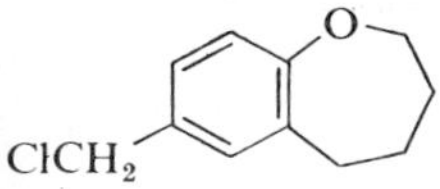 $ClCH_2$ [tetrahydrobenzoxepin structure]	14.0	18.5	21.0	39

TABLE II-3 Uv Data and Bromination Rate of 2,3,4,5-Tetrahydro-1-benzoxepin-2-carboxylic Acid and Lower Ring Homologs.

Compound	λ_{max} (mμ)	ε	pK_a	Relative bromination rate in AcOH at 20° (phenoxyacetic acid = 1.0)
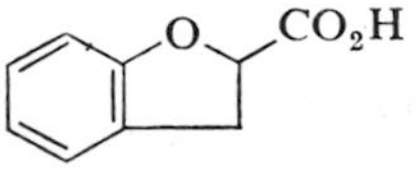 2,3-dihydrobenzofuran-2-carboxylic acid (CO_2H)	279 285	2670 2375	3.25	30
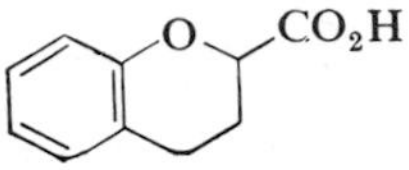 chroman-2-carboxylic acid (CO_2H)	273.5 281	1800 1860	3.28	16
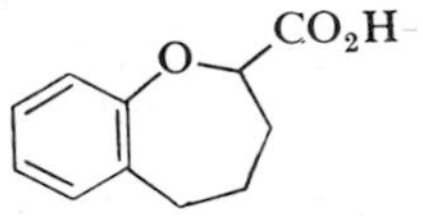 2,3,4,5-tetrahydro-1-benzoxepin-2-carboxylic acid (CO_2H)	265 270	570 555	3.63	0.22

TABLE II-4 Uv Data and Molecular Geometry of 2,3,4,5-Tetrahydro-2-(N,N-dimethylamino)methyl-1-benzoxepin Methobromide and Lower Ring Homologs.

Compound	λ_{max} (mμ)	ε	θ (°)
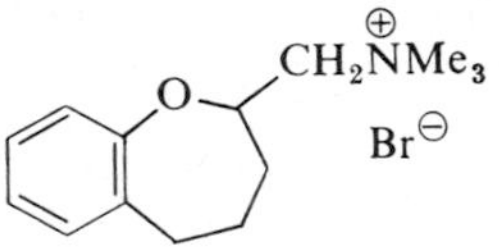 benzoxepin, $CH_2\overset{\oplus}{N}Me_3$ $Br^{\ominus}$	264.5 270	528 520	63.5
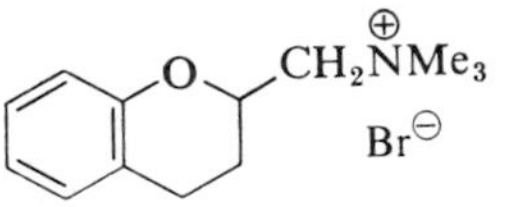 chroman, $CH_2\overset{\oplus}{N}Me_3$ $Br^{\ominus}$	273.5 279	1850 1720	34
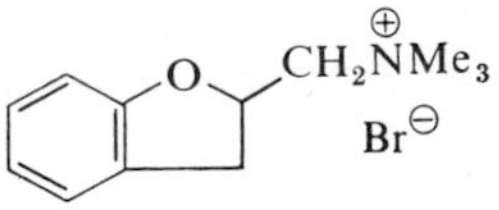 dihydrobenzofuran, $CH_2\overset{\oplus}{N}Me_3$ $Br^{\ominus}$	277.5 282.5	2680 2350 (sh)	0

TABLE II-5. 1-Benzoxepins

Empirical formula	Name of compound	Physical constants	Derivatives	Refs.
$C_{10}H_8O$	1-Benzoxepin	b_{14} 101–102°C $b_{0.5}$ 50°C		86, 95
$C_{10}H_8Cl_2O_2$	7,8-Dichloro-2,3,4,5-tetrahydro-1-benzoxepin-5-one		Hydantoin, mp 237–241°C	52
	7,9-Dichloro-2,3,4,5-tetrahydro-1-benzoxepin-5-one			51
	8,9-Dichloro-2,3,4,5-tetrahydro-1-benzoxepin-5-one		Hydantoin, mp 273–279°C	52
$C_{10}H_9BrO_2$	7-Bromo-2,3,4,5-tetrahydro-1-benzoxepin-5-one		Hydantoin, mp 249–259°C	52
	4-Bromo-2,3,4,5-tetrahydro-1-benzoxepin-5-one	$b_{0.3}$ 135–136°C		111
$C_{10}H_9ClO_2$	3-Chloro-2,3,4,5-tetrahydro-1-benzoxepin-2-one	mp 63°C		50
	7-Chloro-2,3,4,5-tetrahydro-1-benzoxepin-5-one		Hydantoin, mp 240–245°	51, 52
	8-Chloro-2,3,4,5-tetrahydro-1-benzoxepin-5-one		Hydantoin, mp 218–222°C	52
	9-Chloro-2,3,4,5-tetrahydro-1-benzoxepin-5-one		Hydantoin, mp 251–256°C	52
$C_{10}H_9FO_2$	7-Fluoro-2,3,4,5-tetrahydro-1-benzoxepin-5-one		Hydantoin, mp 253–257°C	52
$C_{10}H_{10}O$	2,3-Dihydro-1-benzoxepin	b_{18} 113–114°C $b_{0.07}$ 35°C $n_D^{23.5}$ 1.5877, 1.5926 $d_4^{23.5}$ 1.076		49, 82

TABLE II-5.—(*contd.*)

Empirical formula	Name of compound	Physical constants	Derivatives	Refs.
$C_{10}H_{10}O_2$	2,3,4,5-Tetrahydro-1-benzoxepin-2-one	mp 130°C		50
	2,3,4,5-Tetrahydro-1-benzoxepin-5-one	b_2 106°C, b_4 100–130°C n_D^{21} 1.565, n_D^{20} 1.564 $d_4^{21.5}$ 1.155	2,4-DNP, mp 242°C Semicarbazone, mp 228–229°C Oxime, mp 99°C Hydantoin, mp 245–248°C	41, 49, 52, 77, 111
$C_{10}H_{10}Cl_2O$	7,9-Dichloro-2,3,4,5-tetrahydro-1-benzoxepin	b_{10} 250–260°C		51
$C_{10}H_{11}ClO$	7-Chloro-2,3,4,5-tetrahydro-1-benzoxepin	b_{10} 136–142°C		51
$C_{10}H_{12}O$	2,3,4,5-Tetrahydro-1-benzoxepin	b_7 86–88°C, b_{13} 102°C b_{18} 102°C mp 28–29°, 30°C $n_D^{20.5}$ 1.5412 n_D^{21} 1.5398 $d_4^{20.5}$ 1.049		5, 27, 49, 82, 95
$C_{10}H_{12}O_2$	2,3,4,5-Tetrahydro-5-hydroxy-1-benzoxepin	mp 75–76°C		49
$C_{10}H_{12}O_3$	2,3,4,5-Tetrahydro-4,5-dihydroxy-1-benzoxepin	mp 138°C		49
$C_{10}H_{13}ClO$	7-Chloromethyl-2,3,4,5-tetrahydro-1-benzoxepin	$b_{0.2}$ 113°C		5
$C_{10}H_{16}O$	2,3,4,5,6,7,8,9-Octahydro-1-benzoxepin	b_{11} 91–92°C		18
$C_{11}H_9BrO_3$	7-Bromo-2,3,4,5-tetrahydro-8-methyl-1-benzoxepin-3,5-dione	mp 162–164°C		89

$C_{11}H_9ClO_2$	5-Chloro-2,3-dihydro-1-benzoxepin-4-carboxaldehyde	mp 65–66°C		125
$C_{11}H_{10}O_3$	2,3,4,5-Tetrahydro-2-methyl-1-benzoxepin-3,5-dione	mp 33–34°C		89
$C_{11}H_{11}ClO_2$	7-Chloro-2,3,4,5-tetrahydro-8-methyl-1-benzoxepin-5-one		Hydantoin, mp 260–263°C	52
	7-Chloro-2,3,4,5-tetrahydro-9-methyl-1-benzoxepin-5-one		Hydantoin, mp 257–263°C	51, 52
	2,3,4,5-Tetrahydro-1-benzoxepin-2-carboxylic acid chloride	b_{13} 149°C		112
$C_{11}H_{12}O$	2,3,4,5-Tetrahydro-5-methylene-1-benzoxepin	b_{14} 127–134°C		111
$C_{11}H_{12}O_2$	2,3,4,5-Tetrahydro-7-methyl-1-benzoxepin-5-one	$b_{0.9}$ 114–116°C n_D^{18} 1.560	Oxime, mp 71.5–72.5°C (C_6H_6-petroleum ether) Semicarbazone, mp 188.5–189.5°C (MeOH) Hydantoin, mp 231–238°C	41, 52
	2,3,4,5-Tetrahydro-8-methyl-1-benzoxepin-5-one		Hydantoin, mp 213–217°C	52
	2,3,4,5-Tetrahydro-9-methyl-1-benzoxepin-5-one		Hydantoin, mp 225–229°C	52
$C_{11}H_{12}O_3$	2,3,4,5-Tetrahydro-7-methoxy-1-benzoxepin-2-one	b_{14} 180°C		50
	2,3,4,5-Tetrahydro-8-methoxy-1-benzoxepin-5-one	$b_{0.6}$ 140°C n_D^{19} 1.576	Oxime, mp 99–100°C Semicarbazone, mp 193.5–194.5°C	41
	2,3,4,5-Tetrahydro-1-benzoxepin-2-carboxylic acid	mp 100–102.5°C		3, 112
	2,3,4,5-Tetrahydro-1-benzoxepin-7-carboxylic acid	mp 165–165.5°C (aq EtOH)		5
	2,3,4,5-Tetrahydro-4-methoxy-1-benzoxepin-5-one		2,4-DNP, mp 196°C	111

TABLE II-5.—(*contd.*)

Empirical formula	Name of compound	Physical constants	Derivatives	Refs.
$C_{11}H_{13}ClO$	7-Chloro-2,3,4,5-tetrahydro-9-methyl-1-benzoxepin	b_{14} 144–148°C		51
$C_{11}H_{14}O$	2,3,4,5-Tetrahydro-7-methyl-1-benzoxepin	b_{2-3} 88–90°C b_{11} 109°C $n_D^{14.5}$ 1.5379		27
$C_{11}H_{14}O_2$	2,3,4,5-Tetrahydro-1-benzoxepin-7-methanol	$b_{0.19}$ 128–129°C	Phenylurethan, mp 83–83.5°C (ligroin)	5
$C_{12}H_{10}O_3$	4-Acetyl-2,3-dihydro-1-benzoxepin-2-one	mp 151° (alcohol)	Phenylhydrazone, mp 132°C Dibromide, mp 128°C	83
$C_{12}H_{10}O_4$	4-Acetyl-2,3-dihydro-8-hydroxy-1-benzoxepin-2-one	mp > 200°C		83
$C_{12}H_{11}BrO_3$	7-Bromo-2,3,4,5-tetrahydro-2,8-dimethyl-1-benzoxepin-3,5-dione	mp 111–112°C	2,4-DNP, mp 259–260°C Semicarbazone, mp 184–186°C	89
$C_{12}H_{12}Br_2O_2$	4,4-Dibromo-2,3,4,5-tetrahydro-7,8-dimethyl-1-benzoxepin-5-one	mp 76–78°C (MeOH)		41
$C_{12}H_{12}O_3$	2,3,4,5-Tetrahydro-7,8-dimethyl-1-benzoxepin-4,5-dione	$b_{0.6}$ 120–160°C mp 118–121° (ligroin)	4-Oxime, mp 180–182°C dec (C_6H_6) 4,5-Dioxime, mp 188–189°C (aq MeOH)	41
$C_{12}H_{12}O_4$	4-Acetoxy-2,3,4,5-tetrahydro-1-benzoxepin	$b_{0.04}$ 111–113°C		111
$C_{12}H_{13}BrO_2$	4-Bromo-2,3,4,5-tetrahydro-7,8-dimethyl 1-benzoxepin-5-one	mp 83–85°C (MeOH)		41

$C_{12}H_{13}ClO_2$	7-Chloro-2,3,4,5-tetrahydro-6,8-dimethyl-1-benzoxepin-5-one		Hydantoin, mp 218–223°C	52
$C_{12}H_{14}O$	2,3-Dihydro-7,8-dimethyl-1-benzoxepin	mp 32–34°C (MeOH)		41
	5-Ethylidene-2,3,4,5-tetrahydro-1-benzoxepin (*cis* + *trans*)	b_{19} 134–135°C (impure)		49
	5-Ethyl-2,3-dihydro-1-benzoxepin	b_{19} 134–135°C (impure)		49
$C_{12}H_{14}O_2$	2,3,4,5-Tetrahydro-6,8-dimethyl-1-benzoxepin-5-one		Hydantoin, mp 148–152°C	52
	2,3,4,5-Tetrahydro-7,8-dimethyl-1-benzoxepin-5-one	$b_{0.1}$ 130–140°C mp 61.5–62°C (petroleum ether)	Oxime, mp 109–110°C (C_6H_6-petroleum ether) Semicarbazone, mp 199–200°C (*n*-BuOH)	41
	7-Ethyl-2,3,4,5-tetrahydro-1-benzoxepin-5-one		Hydantoin, mp 225–229°C	52
	7-Acetyl-2,3,4,5-tetrahydro-1-benzoxepin	b_5 155–157°C, b_{11} 171–172°C $n_D^{14.5}$ 1.5628 n_D^{15} 1.5642	Semicarbazone, mp 205–205.5°C (EtOH) 212.5°C dec (alcohol)	5, 27
$C_{12}H_{14}O_3$	2,3,4,5-Tetrahydro-4-hydroxy-1-benzoxepin-5-one	mp 99–100°C (aq MeOH)	O-Acetate, $b_{0.01}$ 100–120°C, mp 92–93°C (MeOH)	41
$C_{12}H_{14}Br_2O$	4,5-Dibromo-2,3,4,5-tetrahydro-7,8-dimethyl-1-benzoxepin	mp 127–130°C (aq MeOH or petroleum ether)		41
$C_{12}H_{14}Cl_2O$	5,5-Dichloro-2,3,4,5-tetrahydro-7,8-dimethyl-1-benzoxepin	b_1 120°C mp 55–58°C (petroleum ether)		41

TABLE II-5.—(*contd.*)

Empirical formula	Name of compound	Physical constants	Derivatives	Refs.
$C_{12}H_{15}NO_2$	4-Amino-2,3,4,5-tetrahydro-7,8-dimethyl-1-benzoxepin		Hydrochloride, mp 225°C dec ($MeOH-Et_2O$)	41
$C_{12}H_{16}O$	2,3,4,5-Tetrahydro-5,5-dimethyl-1-benzoxepin	$b_{1.9}$ 85–95°C mp 47–48°C (petroleum ether)		34, 58, 88
	5-Ethyl-2,3,4,5-tetrahydro-1-benzoxepin	b_{12} 112°C n_D^{22} 1.5287		49
	2,3,4,5-Tetrahydro-7,8-dimethyl-1-benzoxepin	$b_{0.4}$ 70–90°C n_D^{19} 1.535		41
$C_{12}H_{16}O_2$	5-Ethyl-2,3,4,5-tetrahydro-5-hydroxy-1-benzoxepin	mp 69°C		49
	2,3,4,5-Tetrahydro-5-hydroxy-7,8-dimethyl-1-benzoxepin	mp 96–96.5°C (petroleum ether)		41
$C_{12}H_{17}NO$	5-Amino-2,3,4,5-tetrahydro-7,8-dimethyl-1-benzoxepin	$b_{0.6}$ 130°C	Hydrochloride, mp 243–246°C	41
$C_{13}H_{12}O_4$	2,3-Dihydro-7,9-dimethyl-1-benzoxepin-2-one-5-carboxylic acid	mp 245–246.5°C ($MeNO_2$)		85
$C_{13}H_{16}O$	2,3-Dihydro-5,7,8-trimethyl-1-benzoxepin	mp 60–61°C(aq MeOH)		41
$C_{13}H_{16}O_2$	8-Ethyl-2,3,4,5-tetrahydro-6-methyl-1-benzoxepin-5-one		Hydantoin, mp 203–207°C	52
	2,3,4,5-Tetrahydro-7-propionyl-1-benzoxepin	b_{11} 178–179°C $n_D^{14.5}$ 1.5530	Semicarbazone, mp 180°C dec (alcohol)	27

$C_{13}H_{16}O_3$	2,3,4,5-Tetrahydro-5,5-dimethyl-1-benzoxepin-9-carboxylic acid			88
$C_{13}H_{17}NO_2$	4-(N,N-Dimethylamino)methyl-2,3,4,5-tetrahydro-1-benzoxepin-5-one		Hydrochloride, mp 160–162°C	11
	N,N-Dimethyl-2,3,4,5-tetrahydro-1-benzoxepin-2-carboxamide	mp 71–72°C		112
$C_{13}H_{18}O_2$	2,3,4,5-Tetrahydro-5-hydroxy-5,7,8-trimethyl-1-benzoxepin	mp 109–110°C (petroleum ether)		41
$C_{13}H_{19}NO$	2,3,4,5-Tetrahydro-2-(*N*,*N*-dimethylamino)-methyl-1-benzoxepin	b_{10} 135°C	Methobromide, mp 256°C	112
$C_{15}H_{21}NO_2$	4-(*N*,*N*-Dimethylamino)methyl-2,3,4,5-tetrahydro-7,8-dimethyl-1-benzoxepin-5-one	n_D^{20} 1.547	Picrate, mp 166–167°C (*n*-BuOH)	41
$C_{16}H_{12}O_3$	2,3,4,5-Tetrahydro-4-phenyl-1-benzoxepin-3,5-dione	mp 99–101°C (MeOH)		60, 126
$C_{16}H_{17}NO$	5-Amino-2,3,4,5-tetrahydro-5-phenyl-1-benzoxepin		Hydrochloride, mp 207–208°C (alcohol-Et_2O)	13, 100
$C_{17}H_{15}ClO_2$	2,3,4,5-Tetrahydro-5-phenyl-1-benzoxepin-5-carboxylic acid chloride	mp 100–101°C		100
$C_{17}H_{16}O_3$	2,3,4,5-Tetrahydro-5-phenyl-1-benzoxepin-5-carboxylic acid	mp 181–183, 183–184°C (C_6H_6)		12, 100
$C_{17}H_{17}NO_2$	2,3,4,5-Tetrahydro-5-phenyl-1-benzoxepin-5-carboxamide	mp 154–155°C (C_6H_6)		91, 100
$C_{17}H_{17}NO_3$	2,3,4,5-Tetrahydro-3-hydroxy-5-phenyl-1-benzoxepin-5-carboxamide	mp 188–189°C (MeCOEt-EtOH)		122
$C_{17}H_{19}NO$	5-Aminomethyl-2,3,4,5-tetrahydro-5-phenyl-1-benzoxepin		Hydrochloride, mp 249°C	94
$C_{17}H_{20}O_6$	4-Carbomethoxy-5-(2-carbomethoxyethyl)-5-methyl-1-benzoxepin-2-one	mp 132–133°C (Me_2CO-Et_2O)		64

TABLE II-5.—(*contd.*)

Empirical formula	Name of compound	Physical constants	Derivatives	Refs.
$C_{18}H_{14}O_4$	5-Acetoxy-2,3-dihydro-4-phenyl-1-benzoxepin-3-one	mp 125–127°C		60
$C_{18}H_{16}O_3$	2,3,4,5-Tetrahydro-2,7-dimethyl-4-phenyl-1-benzoxepin-3,5-dione			126
$C_{18}H_{18}O_3$	5-Carbomethoxy-2,3,4,5-tetrahydro-5-phenyl-1-benzoxepin	mp 110–111°C (MeOH)		100
$C_{18}H_{19}NO_3$	Methyl N-(2,3,4,5-tetrahydro-5-phenyl-1-benzoxepin-5-yl) urethan	mp 110–111°C (cyclohexane)		100
$C_{19}H_{18}O_2$	4-Benzylidene-2,3,4,5-tetrahydro-7,8-dimethyl-1-benzoxepin-5-one	mp 92–93.5°C (MeOH)		41
$C_{19}H_{18}O_6$	2,3,4,5-Tetrahydro-6,8-dimethoxy-4-(4-methoxyphenyl)-1-benzoxepin-2,5-dione	mp 227°C dec (Me_2CO)		99
	2,3,4,5-Tetrahydro-8-methoxy-4-(3′,4′-dimethoxyphenyl)-1-benzoxepin-3,5-dione			126
$C_{19}H_{20}O_3$	5-Carbethoxy-2,3,4,5-tetrahydro-5-phenyl-1-benzoxepin	mp 110–111°C (MeOH or EtOH)		12, 100
$C_{19}H_{23}NO$	5-(*N*,*N*-Dimethylamino)methyl-2,3,4,5-tetrahydro-5-phenyl-1-benzoxepin		Hydrochloride, mp 250°C	94
$C_{19}H_{23}NO_2$	4-(*N*,*N*-Dimethylamino)methyl-2,3,4,5-tetrahydro-5-hydroxy-5-phenyl-1-benzoxepin	mp 116–117°C	Propionate, mp 172–174°C	11
$C_{20}H_{16}O_5$	3,5-Diacetoxy-4-phenyl-1-benzoxepin	mp 133.5–135°C		60, 126
$C_{20}H_{20}BrNO_2$	*trans*-3-Bromo-*N*-cyclopropyl-5-phenyl-2,3,4,5-tetrahydro-1-benzoxepin-5-carboxamide	mp 165–166°C (MeCOEt-Et_2O)		122

$C_{20}H_{20}ClNO_2$	*trans*-3-Chloro-*N*-cyclopropyl-5-phenyl-2,3,4,5-tetrahydro-1-benzoxepin-5-carboxamide	mp 176–177°C (MeCOEt)		122
$C_{21}H_{18}O_5$	3,5-Diacetoxy-2-methyl-4-phenyl-1-benzoxepin			126
$C_{21}H_{20}O_7$	5-Acetoxy-2,3-dihydro-8-methoxy-4-(3′,4′-dimethoxyphenyl)-1-benzoxepin			126
$C_{21}H_{25}NO_3$	2-(*N*,*N*-Dimethylamino)ethyl-2,3,4,5-tetrahydro-5-phenyl-1-benzoxepin-5-carboxylate		Hydrochloride, mp 214–215°C dec	101
$C_{22}H_{16}O_3$	2,3,4,5-Tetrahydro-2,4-diphenyl-1-benzoxepin-3,5-dione			126
$C_{22}H_{20}O$	2,3,4,5-Tetrahydro-5,5-diphenyl-1-benzoxepin			113
$C_{22}H_{26}N_2O_2$	2,3,4,5-Tetrahydro-5-(4-methyl-1-piperazinyl)carbonyl-5-phenyl-1-benzoxepin	mp 166°C		91
$C_{22}H_{27}NO_3$	2-(*N*,*N*-Dimethylamino)-1-methylethyl 2,3,4,5-tetrahydro-5-phenyl-1-benzoxepin-5-carboxylate		Hydrochloride, mp 200–201°C	14, 101
$C_{22}H_{28}N_2O$	2,3,4,5-Tetrahydro-5-(4-methyl-1-piperazinyl)methyl-5-phenyl-1-benzoxepin		Dihydrochloride, mp 253°C	94
$C_{23}H_{27}NO_3$	2-(1-Pyrrolidinyl)ethyl 2,3,4,5-tetrahydro-5-phenyl-1-benzoxepin-5-carboxylate		Hydrochloride, mp 205–206°C	14, 101
	4-Hexahydroazepinyl 2,3,4,5-tetrahydro-5-phenyl-1-benzoxepin-5-carboxylate		Hydrochloride, mp 211–212°C	101
$C_{23}H_{29}NO_3$	2-(*N*,*N*-Diethylamino)ethyl 2,3,4,5-tetrahydro-5-phenyl-1-benzoxepin-5-carboxylate		Hydrochloride, mp 194–195°C	14, 101
$C_{23}H_{29}NO_2S$	2-(*N*,*N*-Diethylamino)ethyl 2,3,4,5-tetrahydro-5-phenyl-1-benzoxepin-5-carbothioate		Hydrochloride, mp 165–166°C	14, 101

TABLE 11.5.—(*contd.*)

Empirical formula	Name of compound	Physical constants	Derivatives	Refs.
$C_{24}H_{27}NO_3$	3α-Tropanyl 2,3,4,5-tetrahydro-5-phenyl-1-benzoxepin-5-carboxylate		Hydrochloride, mp 131–132°C	14, 101
$C_{24}H_{28}N_2O_3$	*trans-N*-Cyclopropyl-2,3,4,5-tetrahydro-3-morpholino-5-phenyl-1-benzoxepin-5-carboxamide		Hydrochloride, mp 168–169°C	122
$C_{24}H_{29}NO_3$	1-Methyl-2-(1-pyrrolidinyl)ethyl 2,3,4,5-tetrahydro-1-benzoxepin-5-carboxylate		Hydrochloride, mp 199–200°C	14, 101
	2-(1-Piperidinyl)ethyl 2,3,4,5-tetrahydro-5-phenyl-1-benzoxepin-5-carboxylate		Hydrochloride, mp 203–204°C	14, 101
	2-(3-Hexahydroazepinyl)methyl 2,3,4,5-tetrahydro-5-phenyl-1-benzoxepin-5-carboxylate		Hydrochloride, mp 233–234°C	101
$C_{24}H_{30}N_2O_3$	2-(4-Methyl-1-piperazinyl)ethyl 2,3,4,5-tetrahydro-5-phenyl-1-benzoxepin-5-carboxylate		Dihydrochloride, mp 242–244°C, dec	14, 101
$C_{25}H_{31}NO_3$	1-Methyl-2-(1-piperidinyl)ethyl 2,3,4,5-tetrahydro-5-phenyl-1-benzoxepin-5-carboxylate		Hydrochloride, mp 218–220°C	14, 101
	2-(2-Hexahydroazepinyl)ethyl 2,3,4,5-tetrahydro-5-phenyl-1-benzoxepin-5-carboxylate		Hydrochloride, mp 222–226°C	101
$C_{25}H_{33}NO_3$	2-(*N*,*N*-Diisopropylamino)ethyl 2,3,4,5-tetrahydro-5-phenyl-1-benzoxepin-5-carboxylate		Hydrochloride, mp 135–136°C	14, 101
$C_{26}H_{33}NO_3$	1-Methyl-2-(1-hexahydroazepinyl)ethyl 2,3,4,5-tetrahydro-5-phenyl-1-benzoxepin-5-carboxylate		Hydrochloride, mp 223–224°C	14, 101

Empirical formula	Name of compound	Physical constants	Derivatives	Refs.
$C_{10}H_6O_3$	1*H*,3*H*-2-Benzoxepin-1,3-dione	mp 152°C		17, 66
$C_{10}H_{10}O_2$	4,5-Dihydro-1*H*,3*H*-2-benzoxepin-1-one	$b_{0.01}$ 95–99°C mp 54–56°C (petroleum ether)		68, 79
$C_{10}H_{11}BrO$	1-Bromo-4,5-dihydro-1*H*,3*H*-2-benzoxepin			119
$C_{10}H_{12}O$	4,5-Dihydro-1*H*,3*H*-2-benzoxepin	$b_{0.13}$ 55°C mp 23.5–24°C n_D^{20} 1.5464		79, 119
$C_{10}H_{16}O_2$	*trans*-4,5,5a,6,7,8,9,9a-Octahydro-1*H*,3*H*-2-benzoxepin-3-one	mp 77.5–78° (C_6H_6-petroleum ether)		74
$C_{11}H_{14}O$	4,5-Dihydro-3-methyl-1*H*,3*H*-2-benzoxepin	$b_{11.5}$ 115°C $n_D^{19.5}$ 1.5347		79
$C_{12}H_{13}O_3$	4,5-Dihydro-4,4-dimethyl-1*H*,3*H*-2-benzoxepin-1,3-dione	mp 168°C (petroleum ether)		24
$C_{12}H_{16}O$	1-Ethyl-4,5-dihydro-1*H*,3*H*-2-benzoxepin	b_7 105°C n_D^{25} 1.5325		119
$C_{12}H_{16}O_2$	1-Ethoxy-4,5-dihydro-1*H*,3*H*-2-benzoxepin			119
$C_{13}H_{14}O_3$	4,5-Dihydro-4,4,8-trimethyl-1*H*,3*H*-2-benzoxepin-1,3-dione	mp 146°C (MeOH)		24
$C_{14}H_{21}NO$	4,5-Dihydro-1-(*N*,*N*-dimethylamino)methyl-1*H*,3*H*-2-benzoxepin	$b_{0.1}$ 95°C n_D^{25} 1.5240	Picrate, mp 91°C	119
$C_{16}H_{10}O_3$	4-Phenyl-1*H*,3*H*-2-benzoxepin-1,3-dione	mp 255°C (AcOH)		29
$C_{16}H_{16}O$	4,5-Dihydro-3-phenyl-1*H*,3*H*-2-benzoxepin	$b_{0.05}$ 162°C mp 72°C		15
$C_{21}H_{27}NO_6$	3-*n*-Butyl-5a,8,9,9a-tetrahydro-4-methyl-8-(2-oxopyrrolidino)carbonyl-1*H*,3*H*-2-benzoxepin-1-one-9-carboxylic acid	mp 137–137.5°C (*i*-PrOH)		120
$C_{22}H_{20}O$	4,5-Dihydro-3,3-diphenyl-1*H*,3*H*-2-benzoxepin	mp 190°C (alcohol)		15

TABLE II-7. 3-Benzoxepins

Empirical formula	Name of compound	Physical constants	Derivatives	Refs.
$C_{10}H_8O$	3-Benzoxepin	mp 83–84°C	Picrate, mp 125–127°C (MeOH)	47, 48, 123
$C_{10}H_{12}O$	1,2,4,5-Tetrahydro-3-benzoxepin	b_{12} 103°C n_D 1.5450		47, 48
$C_{10}H_{14}O_3$	1,2,4,5,5a,6,7,8,9,9a-Decahydro-3-benzoxepin-2,4-dione	b_{10} 180°C		19
$C_{10}H_{16}O_2$	*trans*-1,2,4,5,5a,6,7,8,9,9a-Decahydro-3-benzoxepin-2-one	$b_{0.2}$ 113°C mp 31–32°C n_D^{25} 1.4892		74
$C_{11}H_{10}O_3$	1,2,4,5-Tetrahydro-1-methyl-3-benzoxepin-2,4-dione			98
$C_{12}H_8O_5$	3-Benzoxepin-2,4-dicarboxylic acid			45, 46, 63, 65
$C_{12}H_{10}O_3$	2-Formyl-3-benzoxepin-4-methanol	mp 122–124°C (C_6H_6-petroleum ether)		65
$C_{12}H_{12}O_3$	3-Benzoxepin-2,4-dimethanol	mp 89.5–90.5°C (C_6H_6)	Diacetate, mp 45–46°C (*n*-pentane)	65
$C_{12}H_{12}O_5$	1,2,4,5-Tetrahydro-3-benzoxepin-2,4-dicarboxylic acid	mp 196–198°C (MeOH-$CHCl_3$)		65

$C_{14}H_{12}O_5$	Dimethyl 3-benzoxepin-2,4-dicarboxylate	mp 108–109, 110.5–111°C	45, 46, 63, 65
$C_{14}H_{12}Br_2O_5$	Dimethyl 1,2-dibromo-1,2-dihydro-3-benzoxepin-2,4-dicarboxylate		63
$C_{14}H_{14}O_5$	Dimethyl 1,2-dihydro-3-benzoxepin-2,4-dicarboxylate	mp 76–77.5°C (MeOH)	65
$C_{14}H_{16}O_3$	1,2,4,5-Tetrahydro-1,1,5,5-tetramethyl-3-benzoxepin	mp 97–98°C (petroleum ether)	8, 26
$C_{14}H_{18}O_2$	1,2,4,5-Tetrahydro-1,1,5,5-tetramethyl-3-benzoxepin-2-one	mp 59–60°C (petroleum ether) n_D^{30} 1.5419 (supercooled)	26
$C_{14}H_{20}O$	1,2,4,5-Tetrahydro-1,1,5,5-tetramethyl-3-benzoxepin	$b_{0.4}$ 58–60°C mp 45.5–47°C (petroleum ether) n_D^{25} 1.5219	26
$C_{14}H_{20}O_2$	1,2,4,5-Tetrahydro-1,1,5,5-tetramethyl-3-benzoxepin-2-ol	mp 142–143°C(C_6H_6-petroleum ether)	26
$C_{14}H_{20}O_3$	1,2,4,5-Tetrahydro-1,1,5,5-tetramethyl-3-benzoxepin-2,4-diol	mp 131–132°C (C_6H_6-petroleum ether)	26
$C_{15}H_{22}O_2$	1,2,4,5-Tetrahydro-2-methoxy-1,1,5,5-tetramethyl-3-benzoxepin	mp 47–48°C (petroleum ether)	26
$C_{16}H_{16}O_5$	Diethyl 3-benzoxepin-2,4-dicarboxylate		46

E. References

1. E. M. Arnett and C. Y. Wu, *J. Amer. Chem. Soc.*, **82,** 5660 (1960).
2. E. M. Arnett and C. Y. Wu, *J. Amer. Chem. Soc.*, **84,** 1684 (1962).
3. G. Baddeley and J. R. Cooke, *J. Chem. Soc.*, **1958,** 2797.
4. G. Baddeley and N. H. P. Smith, *J. Chem. Soc.*, **1961,** 2516.
5. G. Baddeley, N. H. P. Smith, and M. A. Vickars, *J. Chem. Soc.*, **1956,** 2455.
6. N. A. Bailey and R. Mason, *Chem. Commun.*, **1967,** 1039.
7. T. M. Barakat, M. J. Nelson, S. M. Nelson, and A. D. E. Pullin, *Trans. Faraday Soc.*, **62,** 2674 (1966).
8. L. R. C. Barclay, C. E. Milligan, and N. D. Hall, *Can. J. Chem.*, **40,** 1664 (1962).
9. P. D. Bartlett, *Rec. Chem. Progr.*, **11,** 47 (1950).
10. P. D. Bartlett and J. L. Kice, *J. Amer. Chem. Soc.*, **75,** 5591 (1953).
11. Belgian Patent 612,970; *Chem. Abstr.*, **58,** 3406 (1963).
12. Belgian Patent 618,528; *Chem. Abstr.*, **59,** 6372 (1963).
13. Belgian Patent 618,529; *Chem. Abstr.*, **59,** 7500 (1963).
14. Belgian Patent 618,672; *Chem. Abstr.*, **59,** 3897 (1963).
15. H. W. Bersch, R. Meyer, A. v. Mletzko, and K. H. Fischer, *Arch. Pharm.*, **291,** 82 (1958); *Chem. Abstr.*, **52,** 14628 (1958).
16. H. R. Blattmann, W. A. Boell, E. Heilbronner, G. Hohlneicher, E. Vogel, and J. P. Weber, *Helv. Chim. Acta*, **49,** 2017 (1966).
17. J. Boeseken and G. Slooff, *Rec. Trav. Chim. Pays-Bas*, **49,** 100 (1930).
18. I. J. Borowitz and G. J. Williams, *Tetrahedron Lett.*, **1965,** 3813.
19. I. W. Borsche and E. Lange, *Justus Liebigs Ann. Chem.*, **434,** 219 (1923).
20. A. Bracken and H. Raistrick, *Biochem. J.*, **41,** 569 (1947).
21. W. Bremser, H. T. Grunder, E. Heilbronner, and E. Vogel, *Helv. Chim. Acta*, **50,** 84 (1967).
22. B. Briat, D. A. Schooley, R. Records, E. Bunnenberg, C. Djerassi, and E. Vogel, *J. Amer. Chem. Soc.*, **90,** 4691 (1968).
23. British Patent 963,945; *Chem. Abstr.*, **61,** 9414 (1964).
24. C. Broquet-Borgel and R. Quelet, *Bull. Soc. Chim. Fr.*, **1962,** 1882.
25. H. A. Bruson, F. W. Grant, and E. Bobko, *J. Amer. Chem. Soc.*, **80,** 3633 (1958).
26. A. W. Burgstahler and M. O. Abdel-Rahman, *J. Amer. Chem. Soc.*, **85,** 173 (1963).
27. P. Cagniant, *C. R. Acad. Sci., Paris* **229,** 889 (1949).
28. D. Cagniant and P. Cagniant, *Bull. Soc. Chim. Fr.*, **1966,** 228.
29. J. N. Chatterjea and H. Mukherjee, *J. Indian Chem. Soc.*, **37,** 443 (1960).
30. P. W. Clutterbuck, H. Raistrick, and F. Reuter, *Biochem. J.*, **29,** 300 (1935).
31. J. Colonge and A. Lagier, *C. R. Acad. Sci., Paris,* **225,** 1160 (1947).
32. J. Colonge and A. Lagier, *Bull. Soc. Chim. Fr.*, **1949,** 27.
33. A. C. Cope and D. M. Gale, *J. Amer. Chem. Soc.*, **85,** 3743 (1963).
34. J. L. Corbin, H. Hart, and C. R. Wagner, *J. Amer. Chem. Soc.*, **84,** 1740 (1962).
35. E. J. Corey, J. D. Bass, R. LeMahieu, and R. B. Mitra, *J. Amer. Chem. Soc.*, **86,** 5570 (1964).
36. R. Criegee, *Chem. Ber.*, **77B,** 722 (1944).
37. R. Criegee, *Fortschr. Chem. Forsch.*, **1,** 508 (1950).
38. R. Criegee and R. Kaspar, *Justus Liebigs Ann. Chem.*, **560,** 127 (1948).
39. R. Criegee, R. Kaspar, and W. Dietrich, *Justus Liebigs Ann. Chem.*, **560,** 135 (1948).
40. R. Criegee and W. Schnorrenberg, *Justus Liebigs Ann. Chem.*, **560,** 141 (1948).

41. O. Dann and W. D. Arndt, *Justus Liebigs Ann. Chem.*, **587,** 38 (1954).
42. D. B. Denney and D. G. Denney, *J. Amer. Chem. Soc.*, **77,** 1706 (1955).
43. D. B. Denney and D. G. Denney, *J. Amer. Chem. Soc.*, **79,** 4806 (1957).
44. D. B. Denney, W. F. Goodyear, and B. Goldstein, *J. Amer. Chem. Soc.*, **83,** 1726 (1961).
45. K. Dimroth and H. Freyschlag, *Angew. Chem.*, **69,** 95 (1957).
46. K. Dimroth and H. Freyschlag, *Chem. Ber.*, **90,** 1623 (1957).
47. K. Dimroth and G. Pohl, *Angew. Chem.*, **73,** 436 (1961).
48. K. Dimroth, G. Pohl, and H. Follmann, *Chem. Ber.*, **99,** 634 (1966).
49. G. Fontaine and P. Maitte, *C. R. Acad. Sci., Paris*, **258,** 4583 (1964).
50. German Patent 562,827; *Chem. Abstr.*, **27,** 1224 (1933).
51. German Patent 1,088,981; *Chem. Abstr.*, **55,** 27383 (1961).
52. German Patent 1,135,915; *Chem. Abstr.*, **58,** 3440 (1963).
53. F. Gerson, E. Heilbronner, W. A. Boell, and E. Vogel, *Helv. Chim. Acta*, **48,** 1494 (1965).
54. H. L. Goering and A. C. Olson, *J. Amer. Chem. Soc.*, **75,** 5853 (1953).
55. R. Granger, P. Nau, and J. Nau, *Bull. Soc. Chim. Fr.*, **1958,** 531.
56. H. Guenther, *Z. Naturforsch.*, **20b,** 948 (1965).
57. H. K. Hall, Jr., *J. Amer. Chem. Soc.*, **80,** 6412 (1958).
58. H. Hart, J. L. Corbin, C. R. Wagner, and C.-Y. Wu, *J. Amer. Chem. Soc.*, **85,** 3269 (1963).
59. H. Hart and C. R. Wagner, *Proc. Chem. Soc.*, **1958,** 284.
60. H. Hofmann, *Angew. Chem.*, **77,** 864 (1965).
61. H. E. Holmquist, H. S. Rothrock, C. W. Theobald, and B. E. Englund, *J. Amer. Chem, Soc.*, **78,** 5339 (1956).
62. J. M. Holovka, P. D. Gardner, C. B. Strow, M. L. Hill, and T. V. Van Auken, *J. Amer. Chem. Soc.*, **90,** 5041 (1968).
63. R. Huisgen, E. Laschtuvka, I. Ugi, and A. Kammermeier, *Justus Liebigs Ann. Chem.*, **630,** 128 (1960).
64. O. Jeger, R. Mirza, V. Prelog, Ch. Vogel, and R. B. Woodward, *Helv. Chim. Acta*, **37,** 2295 (1954).
65. M. J. Jorgenson, *J. Org. Chem.*, **27,** 3224 (1962).
66. P. Karrer and L. Schneider, *Helv. Chim. Acta*, **30,** 859 (1947).
67. A. P. Krapcho and B. P. Mundy, *J. Heterocycl. Chem.*, **2,** 355 (1965).
68. F. Krollpfeiffer and A. Müller, *Chem. Ber.*, **68B,** 1169 (1935).
69. R. Malachowski and J. Jankiewiczowna, *Chem. Ber.*, **67B,** 1786 (1934).
70. J. Meinwald, D. W. Dicker, and N. Danieli, *J. Amer. Chem. Soc.*, **82,** 4087 (1960).
71. M. S. Newman and A. B. Meker, *J. Org. Chem.*, **26,** 336 (1961).
72. H. Obara, *Nippon Kagaku Zasshi*, **82,** 65 (1961); *Chem. Abstr.*, **57,** 16427 (1962).
73. L. A. Paquette, and J. H. Barrett, *J. Amer. Chem. Soc.*, **88,** 1718 (1966).
74. L. A. Paquette and N. A. Nelson, *J. Org. Chem.*, **27,** 2272 (1962).
75. H. Plieninger and G. Keilich, *Chem. Ber.*, **92,** 2897 (1959).
76. J. R. Plimmer, *J. Org. Chem.*, **29,** 511 (1964).
77. S. G. Powell and L. Anderson, *J. Amer. Chem. Soc.*, **53,** 811 (1931).
78. V. Prelog and W. Küng, *Helv. Chim. Acta*, **39,** 1394 (1956).
79. A. Rieche and H. Gross, *Chem. Ber.*, **95,** 91 (1962).
80. G. O. Schenk and K. H. Schulte-Elte, *Justus Liebigs Ann. Chem.*, **618,** 185 (1958).
81. E. E. Schweizer, C. J. Berninger, and J. G. Thompson, *J. Org. Chem.*, **33,** 336 (1968).
82. E. E. Schweizer and R. Schepers, *Tetrahedron Lett.*, **1963,** 979.
83. R. N. Sen and B. C. Roy, *J. Indian Chem. Soc.*, **7,** 401 (1930).

84. A. Shani and F. Sondheimer, *J. Amer. Chem. Soc.*, **89,** 6310 (1967).
85. R. V. Smith and M. D. Bealor, *J. Org. Chem.*, **27,** 3092 (1962).
86. F. Sondheimer and A. Shani, *J. Amer. Chem. Soc.*, **86,** 3168 (1964).
87. H. Stetter and H. Rauhut, *Chem. Ber.*, **91,** 2543 (1958).
88. M. Sundaralingam and G. A. Jeffrey, *Acta Cryst.*, **15,** 1058 (1962).
89. J. H. P. Tyman and R. Pickles, *Tetrahedron Lett.*, **1966,** 4993.
90. U.S. Patent 2,544,737; *Chem. Abstr.*, **45,** 8553 (1951).
91. U.S. Patent 3,122,551; *Chem. Abstr.*, **61,** 1844 (1964).
92. U.S. Patent 3,123,619; *Chem. Abstr.*, **60,** 14406 (1964).
93. U.S. Patent 3,142,682; *Chem. Abstr.*, **61,** 9479 (1964).
94. U.S. Patent 3,156,688; *Chem. Abstr.*, **62,** 2764 (1965).
95. E. Vogel, M. Biskup, W. Pretzer, and W. A. Boell, *Angew. Chem.*, **76,** 785 (1964).
96. E. Vogel, W. A. Boell, and M. Biskup, *Tetrahedron Lett.*, **1966,** 1569.
97. E. Vogel and H. Guenther, *Angew. Chem.*, **79,** 429 (1967).
98. N. S. Vulfson and M. O. Iodko, *Sb. Statei, Nauchn. Issled. Inst. Organ. Poluprod. Krasitelei*, **2,** 143 (1961); *Chem. Abstr.*, **58,** 487 (1963).
99. W. B. Whalley and G. Lloyd, *J. Chem. Soc.*, **1956,** 3213.
100. H. E. Zaugg, R. W. DeNet, and R. J. Michaels, Jr., *J. Org., Chem.*, **26,** 4821 (1961).
101. H. E. Zaugg, R. W. DeNet, and R. J. Michaels, Jr., *J. Org. Chem.*, **26,** 4828 (1961).
102. S. Inokawa, T. Ogata, H. Yoshida, S. Suzuki, H. Susumura, and Y. Yamase, *Nippon Kagaku Zasshi*, **88,** 1100 (1967); *Chem. Abstr.*, **68,** 10978 (1968).
103. French Patent 1,409,957; *Chem. Abstr.*, **65,** 4052 (1966).
104. German Patent 1,232,684; *Chem. Abstr.*, **66,** 6269 (1967).
105. R. Sudo, A. Kaneda, and N. Itoh, *J. Org. Chem.*, **32,** 1844 (1967).
106. F. Nederl, J. Buddrus, W. Brodowski, and P. Weyerstahl, *Tetrahedron Lett.*, **1966,** 5385.
107. H. Hofmann and H. Westernacher, *Angew. Chem., Int. Ed. Eng.*, **5,** 958 (1966).
108. E. E. Schweizer, E. T. Schaffer, C. T. Hughes, and C. J. Berninger, *J. Org. Chem.*, **31,** 2907 (1966).
109. E. E. Schweizer, C. J. Berninger, D. M. Crouse, R. A. Davis, and R. Schepers Logothetis, *J. Org. Chem.*, **34,** 207 (1969).
110. E. E. Schweizer, M. S. El-Bakoush, K. K. Light, and K. H. Oberle, *J. Org. Chem.*, **33,** 2591 (1968).
111. G. Fontaine, *Ann. Chim.* (Paris), [14]**3,** 179 (1968).
112. E. R. Clark and S. G. Williams, *J. Chem. Soc., B*, **1967,** 859.
113. W. H. Starnes, Jr., *J. Org. Chem.*, **33,** 2767 (1968).
114. G. R. Ziegler and G. S. Hammond, *J. Amer. Chem. Soc.*, **90,** 513 (1968).
115. G. R. Ziegler, *J. Amer. Chem. Soc.*, **91,** 446 (1969).
116. W. J. Mijs, K. S. De Vries, J. G. Westra, H. A. A. Gaur, J. Smidt, and J. Vriend, *Rec. Trav. Chim. Pays-Bas*, **87,** 580 (1968).
117. W. A. Boell, *Tetrahedron Lett.*, **1968,** 2595.
118. V. L. Brown and W. H. Seaton, *J. Heterocycl. Chem.*, **5,** 575 (1968).
119. C. Normant, *C. R. Acad. Sci., Paris*, **264,** 1665 (1967).
120. S. Takeuchi and H. Yonehara, *Tetrahedron Lett.*, **1966,** 5197.
121. H. E. Zaugg and R. J. Michaels, *J. Org. Chem.*, **28,** 1801 (1963).
122. H. E. Zaugg, R. J. Michaels, A. D. Schaeffer, A. M. Wenthe, and W. H. Washburn, *Tetrahedron*, **22,** 1257 (1966).
123. K. Dimroth, G. Pohl, and H. Follmann, *Chem. Ber.*, **99,** 634 (1966).
124. H. E. Zaugg and R. J. Michaels, *J. Org. Chem.*, **31,** 1322 (1966).
125. M. Weissenfels, H. Schurig, and G. Huehsam, *Z. Chem.*, **6,** 471 (1966).

126. H. Hofmann and H. Westermacher, *Chem. Ber.*, **102,** 205 (1969).
127. U.S. Patent 2,930,799; *Chem. Abstr.*, **54,** 20881 (1960).
128. H. Fiesselmann and W. Ehmann, *Chem. Ber.*, **91,** 1713 (1958).
129. G. Fontaine, *Ann. Chim.* (Paris), **3,** 469 (1968).
130. C. Marcaillou, G. Fontaine, and P. Maitte, *C. R. Acad. Sci., Paris*, **267,** 846 (1968).

CHAPTER III

Oxepin Ring Systems Containing Three Rings

ANDRE ROSOWSKY

Children's Cancer Research Foundation and Department of Biological Chemistry, Harvard Medical School, Boston, Massachusetts

A. Fused Ring Systems

1. Cyclopenta[5.6]pyrano[4,3-*b*]oxepins

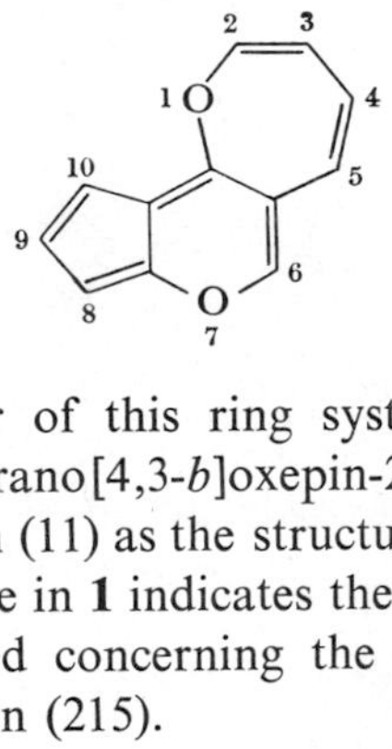

The sole known member of this ring system to date is 3,4,5,8,9,10-hexahydrocyclopenta[5.6]pyrano[4,3-*b*]oxepin-2,6-diene (**1**), which was proposed in 1963 by Baldwin (11) as the structure for the dimer of ethylenebisketene (**2**). The dotted line in **1** indicates the sense of this novel dimerization. A report has appeared concerning the use of **1** as a promoter in ε-caprolactam polymerization (215).

$$2\ O{=}C{=}CH(CH_2)_2CH{=}C{=}O \longrightarrow$$

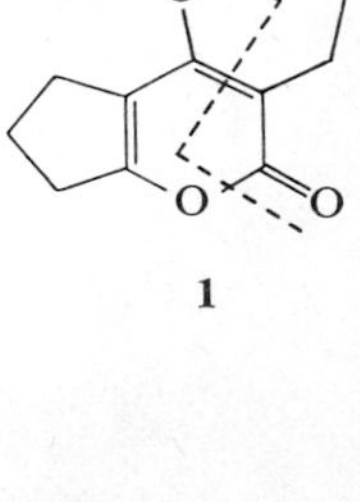

2 **1**

2. 8*H*-1,3-Dioxolo[4,5-*h*][3]benzoxepins

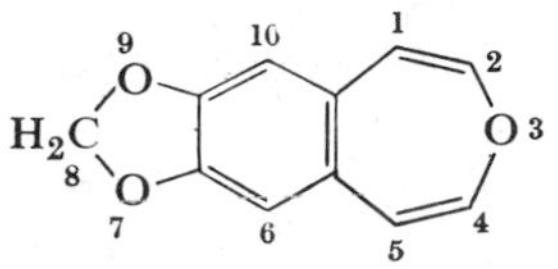

R.I. 10398

The only reference to this ring system at present is the paper of Dallacker and co-workers (51), who reported the synthesis of 8*H*-dioxolo[4,5-*h*][3]-benzoxepin-2,4-dicarboxylic acid (**3**) via the reaction shown.

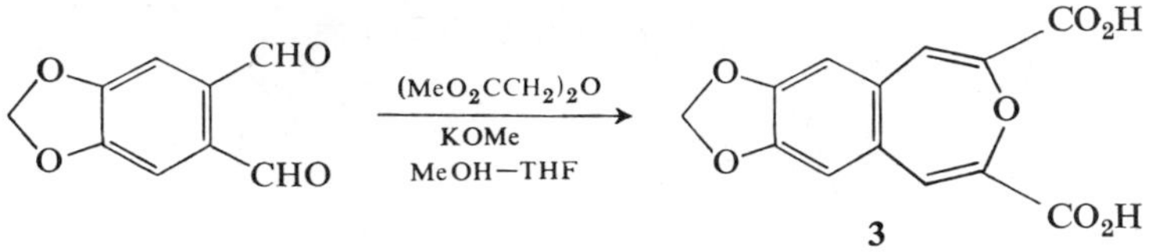
CHO
CHO
$(MeO_2CCH_2)_2O$
KOMe
MeOH—THF
CO_2H
CO_2H

3

3. Dibenz[*b*,*d*]oxepins

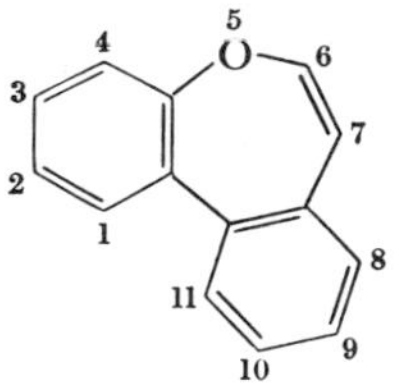

R.I. 3700

The chemistry of the dibenz[*b*,*d*]oxepin ring system has been virtually ignored. A rare example is 6,7-dihydrodibenz[*b*,*d*]oxepin (**4**), which was reported in 1925 by Sieglitz and Koch (148). For the preparation of this compound, the potassium salt of *o*-nitrophenol was condensed with phenethyl bromide to give 2-(*o*-nitrophenoxy)ethylbenzene, which was then reduced to the corresponding amine. Diazotization and hydrolysis gave, in addition to the expected phenol, **5**, a product which was formulated as **4** on the basis of its

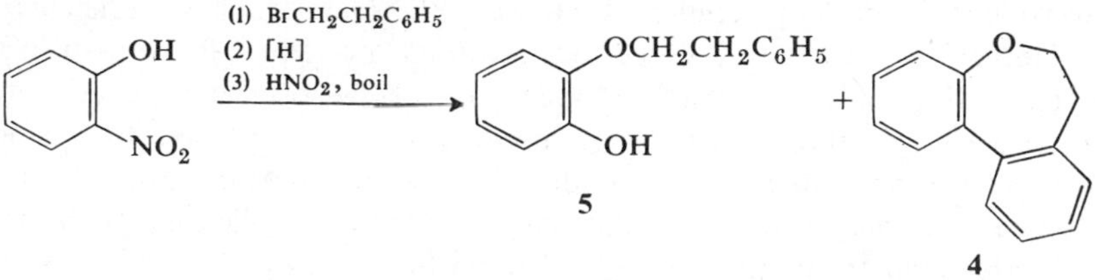
OH
NO_2
(1) $BrCH_2CH_2C_6H_5$
(2) [H]
(3) HNO_2, boil
$OCH_2CH_2C_6H_5$
OH
+

5

4

typical ether like properties. The formation of **4** apparently resulted from a ring closure of the Pschorr type, with elimination of nitrogen from the diazonium ion. Treatment of **4** with phosphorus and hydriodic acid at high temperature produced a mixture of phenanthrene and partly reduced phenanthrenes, which could be converted into 9,10-phenanthraquinone upon oxidation with chromic acid.

The dibenz[*b*,*d*]oxepin ring system has also been encountered on rare occasion in the field of natural products. In this connection the work of Grove and Riley (72), and more recently of Wenkert and Strike (173), in the area of abietic and podocarpic acid chemistry may be cited.

4. Dibenz[*b*,*e*]oxepins

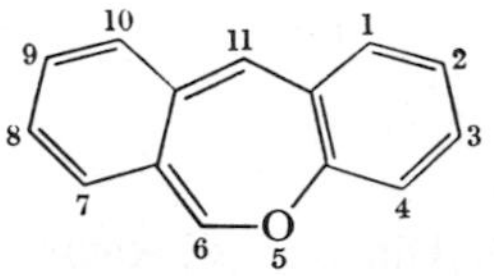

R.I. 3697

The earliest report describing the synthesis of a dibenz[*b*,*e*]oxepin was published in 1924 by Orndorff and Kline (132). These workers observed the formation of a neutral lactone upon treatment of *o*-(2,4-dihydroxybenzoyl)-benzoic acid with acetic anhydride under reflux. The product was formulated as 3-acetoxydibenz[*b*,*e*]oxepin-6,11-dione (**6**). Similarly, when the above acid was brominated in acetic acid and then subjected to the action of boiling acetic anhydride, there was obtained a dibromo analog of **6** which was presumably 3-acetoxy-2,4-dibromodibenz[*b*,*e*]oxepin-6,11-dione (**7**). Reaction of **6** with methanol or ethanol gave rise to the corresponding esters of the starting acid, indicating the relatively labile character of the seven-membered lactone ring in this compound.

No further mention of the dibenz[*b*,*e*]oxepin ring system appeared in the literature until 1951, when Berti (29) reported the interesting pyrolysis of 2-chloro-3-(2,4-dihydroxyphenyl)-2-methyl-3-phenyl-1-indanone (**8**), which afforded a chlorine-free product formulated as 3-hydroxy-11-α-methyl-benzylidenedibenz[*b*,*e*]oxepin-6(11*H*)-one (**9**). The probable existence of two isomeric products was suggested, although no actual fractionation was described. Saponification resulted in cleavage of the seven-membered lactone and formation of the corresponding dihydroxy acid. Treatment of this acid, **10**, or of the original hydroxy lactone, **9**, with hot acetic anhydride in the presence of a trace of sodium acetate gave rise to acetoxy lactone **11**, in which the stereochemistry of the exocyclic double bond was not specified. Similar

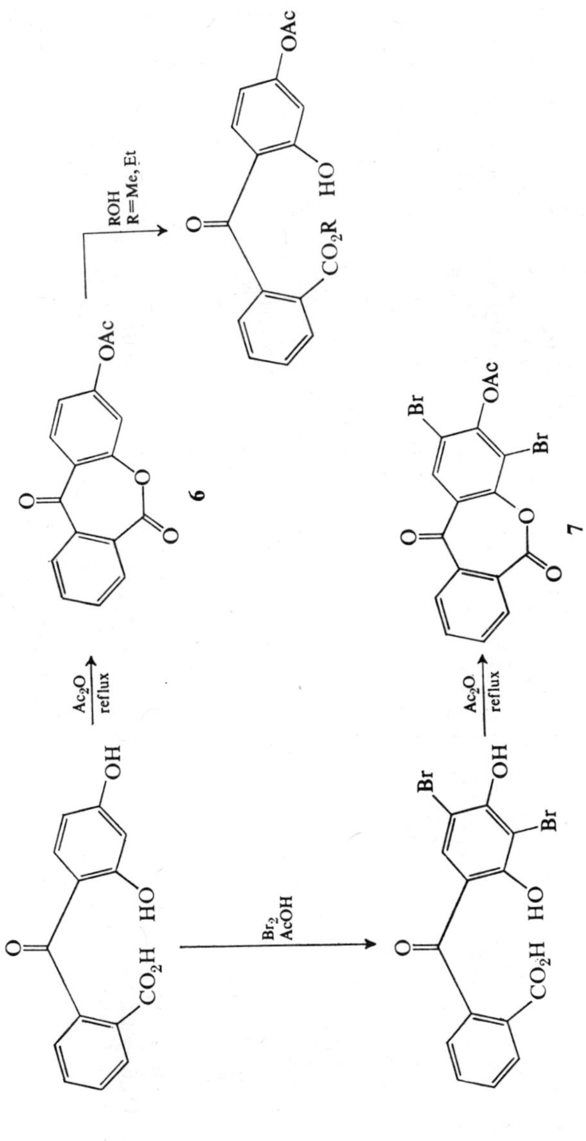

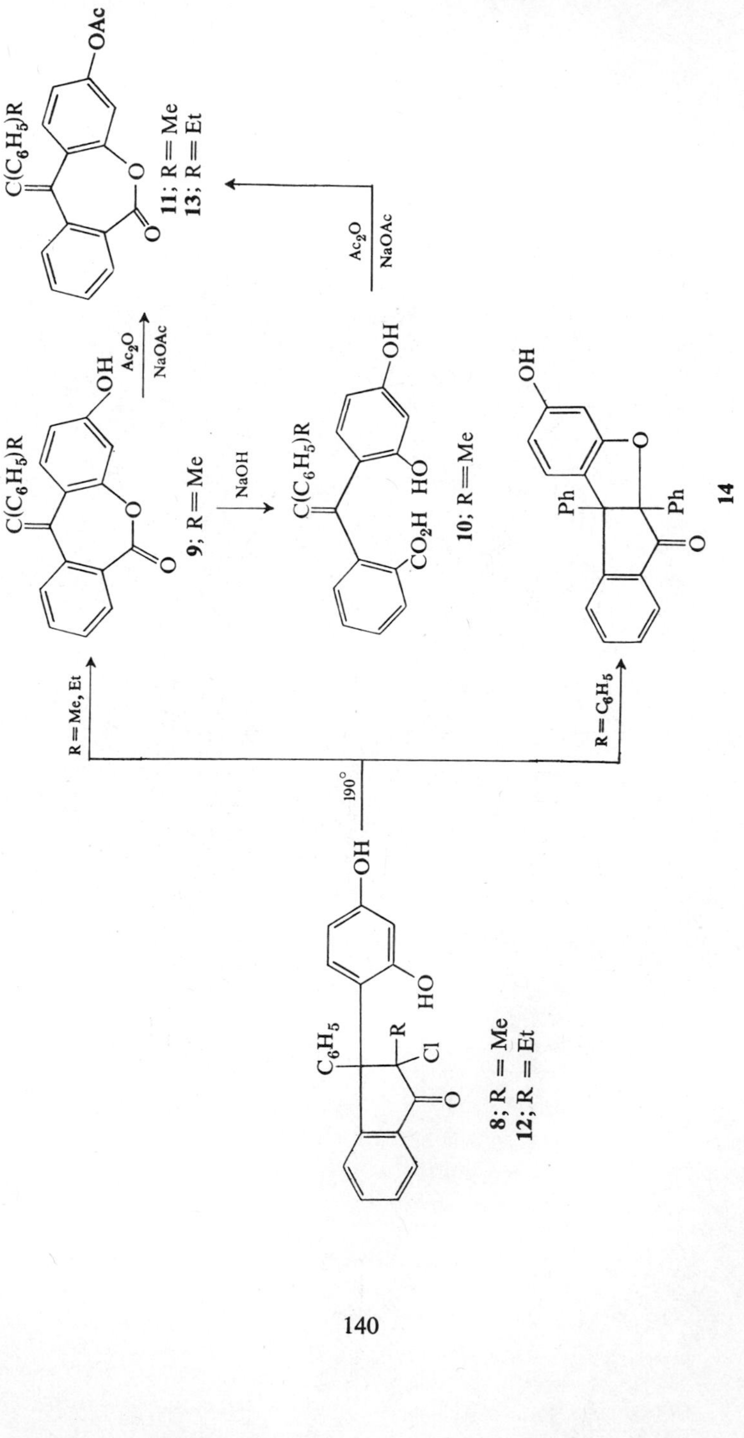

8; R = Me
12; R = Et

190°

R = Me, Et

R = C_6H_5

9; R = Me

Ac_2O / NaOAc

NaOH

10; R = Me

11; R = Me
13; R = Et

14

results were observed with 2-chloro-2-ethyl-3-(2,4-dihydroxyphenyl)-3-phenyl-1-indanone (**12**), which yielded 3-acetoxy-11-α-ethylbenzylidene-dibenz[*b*,*e*]oxepin-6(11*H*)-one (**13**). A different course was followed by 2-chloro-3-(2,4-dihydroxyphenyl)-2,3-diphenyl-1-indanone, however. Under the usual pyrolysis conditions, this compound afforded a *nonlactonic* product, to which structure **14** was assigned.

As part of an active interest in rigid middle-sized and large-sized rings, Baker and co-workers (10) investigated the effect of a series of commonly used cyclodehydrating agents upon *o*-(2-hydroxybenzoyl)benzoic acid and *o*-(2-hydroxy-5-methylbenzoyl)benzoic acid. In agreement with the earlier observation of Orndorff and Kline (132), the English authors found that, under the influence of acetic anhydride in refluxing benzene, the hydroxy acids **15** and **16** underwent lactonization in nearly quantitative yield, giving dibenz[*b*,*e*]oxepin-6,11-dione (**17**) and 2-methyldibenz[*b*,*e*]oxepin-6,11-dione (**18**), respectively. On occasion there was obtained, instead of **17**, a different product for which structure **19** was proposed. When other reagents were used in place of acetic anhydride, the products obtained were not **17** and **18**, but instead were fourteen-membered dimeric lactides **20** and **21**. The reagents found to produce these lactides included trifluoroacetic anhydride, thionyl chloride in the presence of *N*,*N*-diethylaniline, phosphorus pentoxide, and phosphorus oxychloride. Alkaline hydrolysis of **17** and **18** regenerated thc

Ac_2O/C_6H_6 reflux

17; R = H
18; R = Me

or

AcO O OAc O R

19; R = Me

O R CO_2H HO

15; R = H
16; R = Me

$(CF_3CO)_2O$ or other reagents

20; R = H
21; R = Me

open-chain hydroxy acids with great ease. The reaction of **18**, for example, was 96% complete in only 30 sec when carried out in 2*N* sodium hydroxide under reflux.

According to Baker and co-workers (10), the cyclization brought about by acetic anhydride involves preliminary acetylation of the phenolic hydroxyl group, as shown. In the view advanced by these authors, the starting hydroxy acid can exist in two configurations, one of which involves internal hydrogen bond formation between the phenolic hydroxyl and the ketonic carbonyl groups. A second configuration, in which this hydrogen bond is geometrically impossible to achieve, is in equilibrium with the hydrogen-bonded form. Acetic anhydride apparently reacts rapidly with the phenolic hydroxyl of the nonbonded form to give an acetylated intermediate which cyclizes readily to the seven-membered lactone with elimination of a molecule of acetic acid. The equilibrium between the two configurations of the starting material is thus shifted continuously in favor of the nonbonded form until lactonization is complete. With trifluoroacetic anhydride, and with other reagents that lead to the fourteen-membered lactide product, it appears that the reaction proceeds by way of mixed anhydride intermediates in which the molecular geometry is unfavorable for the formation of a seven-membered ring.

R, OH, CO_2H $\xrightarrow{Ac_2O}$ [R, OAc, CO_2H] $\xrightarrow{-AcOH}$ Seven-membered Lactone

⇅

R, O--H, CO_2H $\xrightarrow[\text{and other reagents}]{(CF_3CO)_2O}$ [Mixed Anhydride] ⟶ Fourteen- membered lactide

Strong support for the above mechanism was provided when Baker and co-workers (10) studied the action of the same series of cyclodehydrating agents upon *o*-(2-hydroxybenzyl)benzoic acid (**22**) and *o*-(2-hydroxy-5-methylbenzyl)benzoic acid (**23**), in which the type of internal hydrogen bonding displayed by the corresponding ketones, **15** and **16**, is precluded.

In this case only the seven-membered lactones dibenz[*b*,*e*]oxepin-6(11*H*)-one (**24**) and 2-methyldibenz[*b*,*e*]oxepin-6(11*H*)-one (**25**) were obtained, with no evidence for the formation of dimeric lactides. The yields of product realized with each reagent are shown. Alkaline hydrolysis of **24** and **25** readily afforded the starting hydroxy acids, **22** and **23**, in nearly quantitative recovery.

An interesting reaction with some possible bearing upon the mechanistic arguments advanced by Baker is the one described recently by Sarngadharan and Seshadri (209), wherein 3-phenylphthalan is transformed into **17** upon treatment with boiling acetic anhydride.

The use of acetic anhydride for the formation of seven-membered lactone rings in the dibenz[*b*,*e*]oxepin series was also investigated by Lamchen (105), who reported the conversion of *o*-(2-hydroxybenzoyl)benzoic acid (**15**) and

R
CO_2H HO
$\xrightarrow{-H_2O}$
R
O
O

22; R = H
23; R = Me

24; R = H
25; R = Me

Reagent	% Yield (R = H)	% Yield (R = Me)
$Ac_2O/NaOAc$	80	70
$(CF_3CO)_2O$	75	99
$SOCl_2/PhNEt_2$	49	65
P_4O_{10}	59	54
$POCl_3$	51	50

o-(2-hydroxy-4-methoxybenzoyl)benzoic acid (**26**) into **17** and 3-methoxydibenz[*b*,*e*]oxepin-6,11-dione (**27**), respectively. It is of interest to note, however, that the melting point of **17** as reported by Lamchen differs markedly

O
O
Ph H
$\xrightarrow[\Delta]{Ac_2O}$
O
O
O **17**

from the value found by Baker and co-workers (10). The compound formulated as **27** was subjected to the action of a number of reagents, all of which cleaved the seven-membered lactone ring. Thus treatment with acetic acid, ethanol, ethylamine, and diethylamine led to the formation of the starting hydroxy acid (**26**), its ethyl ester (**28**), its *N*-ethyl amide (**29**), and its *N*,*N*-diethylamide (**30**) respectively. Reaction with ammonia gave isoindole **31**, while condensation with hydrazine yielded phthalazine **32**.

A further example of the lactonization of an *o*-(2-hydroxybenzyl)benzoic acid derivative was added to the literature in 1958 by Hubacher (78), who assigned structure **33** to the neutral compound obtained on heating hydroxy acid **34** at 220°. Acetylation and methylation of **33** afforded 3-acetoxy-11-phenyldibenz[*b*,*e*]oxepin-6(11*H*)-one (**35**) and 3-methoxy-11-phenyldibenz-[*b*,*e*]oxepin-6(11*H*)-one (**36**), respectively. Acetate ester **35** was also obtained

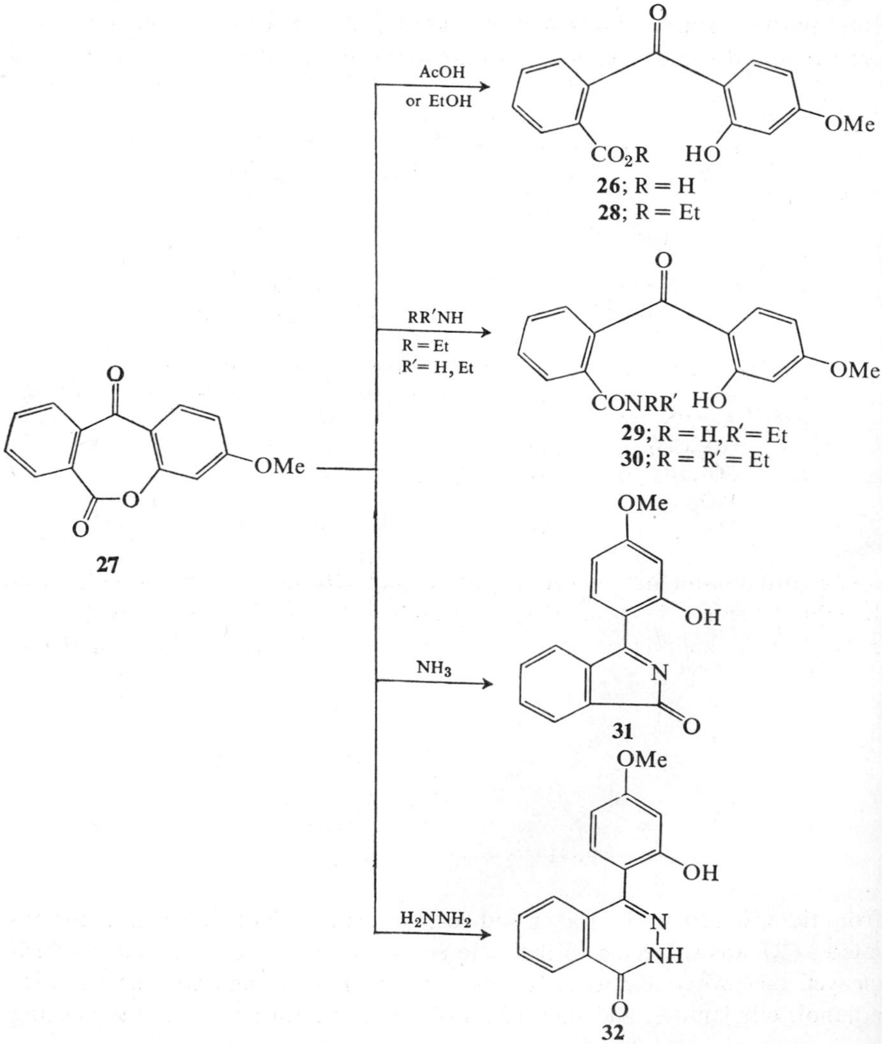

27

31

32

directly from **34** by reaction with acetic anhydride in the presence of sodium acetate. Hydroxy acid **34** was prepared from *o*-benzoylbenzoic acid by fusion with resorcinol at 120° in the presence of zinc chloride, followed by reductive saponification with sodium hydroxide in the presence of Raney nickel.

(1) resorcinol/$ZnCl_2$
(2) NaOH/Ni

CO_2H **34** HO OH

220° Ac_2O NaOAc

36 OMe **33** OH Ac_2O H_2SO_4 120° **35** OAc

Dibenz[*b*,*e*]oxepins were seldom encountered in the chemical literature until 1962, when it was first reported (150, 151, 176) that certain compounds belonging to this ring system exhibit significant psychopharmacological properties. Following this discovery, the number of dibenz[*b*,*e*]oxepins appearing in the literature underwent a sharp increase, as might be expected.

Pharmacologically active dibenz[*b*,*e*]oxepin derivatives are most commonly obtained via dibenz[*b*,*e*]oxepin-11(6*H*)-ones. The latter can be prepared in good yield by cyclodehydration of suitably substituted *o*-(phenoxymethyl)-benzoic acids. Among the reagents that have been utilized satisfactorily for this purpose may be mentioned ethyl pyrophosphate (18, 151, 152), phosphorus pentoxide (176), thionyl chloride (152), polyphosphoric acid (149), liquid hydrogen fluoride (176), and trifluoroacetic anhydride (20). In addition, it has been reported that *o*-(phenoxymethyl)benzoyl chlorides can likewise be used to prepare dibenz[*b*,*e*]oxepin-11(6*H*)-ones, cyclization being effected either thermally or under the influence of aluminum chloride in carbon disulfide or nitrobenzene (18, 151).

Dibenz[*b*,*e*]oxepin-11(6*H*)-one (**37**) itself has been reported on several occasions, and it is of interest to compare the efficacy of various reagents that have been used to bring about ring closure. Stach and Spingler (151–153) reported that, upon being heated at 100–110° in ethyl orthophosphate (a mixture of phosphorus pentoxide and ethanol), *o*-phenoxymethylbenzoic acid afforded **37** in 85.5% yield. Stack and Spingler (152) also found that

$\xrightarrow{-H_2O}$

37; R = R′ = H

reaction of the acid with thionyl chloride under reflux likewise gave **37**, but in only 71% yield. At about the same time Winthrop and co-workers (176) observed that cyclization occurred in 68% yield when carried out in boiling xylene in the presence of phosphorus pentoxide and Celite, but gave only an anhydride and polymeric by-products when carried out in polyphosphoric acid at 95°, in concentrated sulfuric acid at room temperature, or in liquid hydrogen fluoride. A conflicting claim regarding the use of polyphosphoric acid was made by Stach and Bickelhaupt (149), who stated that this reagent brought about ring closure in 80–90% yield. Most recently it has been disclosed in a patent (20) that trifluoroacetic anhydride at room temperature produces cyclization in 76% yield.

Substituted dibenz[*b*,*e*]oxepin-11(6*H*)-ones which have been prepared by cyclization of the corresponding *o*-phenoxymethylbenzoic acids include the 2-bromo (151, 152), 2-chloro (123, 151, 152, 176), 2-trifluoromethyl (20), 2-methyl (151, 152), 3-methyl (152), 2-methoxy (151, 152), and 2-(*N*,*N*-dimethyl)sulfamyl (20, 123) derivatives. The required *o*-phenoxymethylbenzoic acid precursors have been obtained by several related approaches. These include (*1*) reaction of the sodium salt of a *m*- or *p*-substituted phenol with *o*-bromomethylbenzoyl bromide in ethanol, and saponification of the resulting ethyl *o*-phenoxymethylbenzoate (152); (*2*) condensation between *o*-cyanobenzyl bromide and phenol in the presence of sodium methoxide, followed by alkaline hydrolysis of the resulting phenyl *o*-cyanobenzyl ether (176); and (*3*) condensation between ethyl *o*-bromomethylbenzoate and phenol in the presence of sodium hydroxide, and saponification of the resulting ethyl *o*-phenoxymethylbenzoate (20).

An ingenious method of ring closure used recently (38) for the preparation of dibenz[*b*,*e*]oxepins involves treatment of 1-bromo-2-(2-bromobenzyloxy)-benzene with *n*-butyllithium, and condensation of the resulting organolithium intermediate with ethyl γ-(*N*,*N*-dimethylamino)butyrate or ethyl β-(*N*,*N*-dimethylamino)propionate to give *N*,*N*-dimethyldibenz[*b*,*e*]oxepin-$\Delta^{11(6H),\gamma}$-propylamine (**38**) and *N*,*N*-dimethyldibenz[*b*,*e*]oxepin-$\Delta^{11(6H),\beta}$-ethylamine (**39**), respectively.

11-Alkylidene derivatives of the dibenz[*b*,*e*]oxepin ring system such as **38** and **39** have attracted considerable attention because of their potential value as psychopharmacodynamic agents. Initially these compounds were synthesized by reaction of various 3-(*N*,*N*-dialkylamino)-1-propyl magnesium

$CH(CH_2)_nNMe_2$

Br Br

n-BuLi

Et_2O

Li Li

38; n = 2
39; n = 1

chlorides with dibenz[*b*,*e*]oxepin-11(6*H*)-ones (18, 150, 151, 176). The tertiary alcohols obtained in this fashion were then dehydrated to the corresponding exocyclic olefins by treatment with hot hydrochloric acid. Substituted dibenz[*b*,*e*]oxepin-11(6*H*)-ones used in this connection included the 2-methyl, 2-methoxy, and 2-chloro derivatives. More recently the 2-(*N*,*N*-dimethyl)sulfamyl and 2-trifluoromethyl derivatives were also used (20). Dialkylamine moieties which have been incorporated into the Grignard reagent include the dimethylamino, piperidino, *N*-methylpiperazino, and benzylmethylamino groups.

It is of interest to note that when dibenz[*b*,*e*]oxepin-11(6*H*)-one (**37**) is allowed to react with three, rather than two, equivalents of Grignard reagent, and the temperature and reaction time are increased likewise, the product

$X(CH_2)_3MgCl$ (2 equiv)

THF, 40°, 3hr

37; R=H

HO $(CH_2)_3X$ R

40; R = H, X = Me_2N

HCl

$CH(CH_2)_2X$ R

38; R = H, X = Me_2N

obtained is not **40** as initially thought (176), but actually 9,10-dihydro-9-[3-(*N*,*N*-dimethylamino)propyl]anthracene-9,10-diol (**41**) (175). The structure of **41** was confirmed by independent synthesis involving (*1*) condensation of 9,10-anthraquinone with one equivalent of 3-(*N*,*N*-dimethylamino)-1-propyl magnesium chloride, and (*2*) lithium aluminum hydride reduction of the resulting hydroxy ketone, **42**. Furthermore, treatment of **41** with acetic anhydride in pyridine at room temperature resulted in the formation of

$(CH_2)_3NMe_2$

OAc
43

Ac_2O
C_5H_5N

O

HO $(CH_2)_3NMe_2$

$Me_2N(CH_2)_3MgCl$(3 equiv)
THF, 16hr reflux

O

HO H

37

41

$LiAlH_4$

O

HO $(CH_2)_3NMe_2$

$Me_2N(CH_2)_3MgCl$(1 equiv)
THF-C_6H_6, reflux

O

O

42

9-acetoxy-10-[3-(*N*,*N*-dimethylamino)propyl]anthracene (**43**), identical with an authentic specimen prepared by another route.

Some insight into the possible mechanism of formation of **41** from **37** was provided by the finding that **40**, upon being exposed to the action of three equivalents of Grignard reagent in boiling tetrahydrofuran, likewise gave **41** in high yield. Two possible pathways, A and B, were proposed (175) on the basis of this observation, as shown on facing page.

The stereochemistry of the exocyclic double bond in 11(6*H*)-alkylidenedibenz[*b*,*e*]oxepins was not investigated in detail until recently. It has now been demonstrated that a mixture of isomeric *cis*- and *trans*-olefins is produced during the acid-catalyzed dehydration of alcohol **40** (20). Interestingly, one geometrical isomer proved to be a much more potent antidepressant drug than the other. The individual isomers were obtained in pure form by the following complex series of operations. The mixed olefin isomers arising from acid treatment of alcohol **40** were converted into their maleate salts. The pure *trans*-maleate salt was isolated from the mixture by repeated crystallization from alcohol, and was thence converted, via the free base,

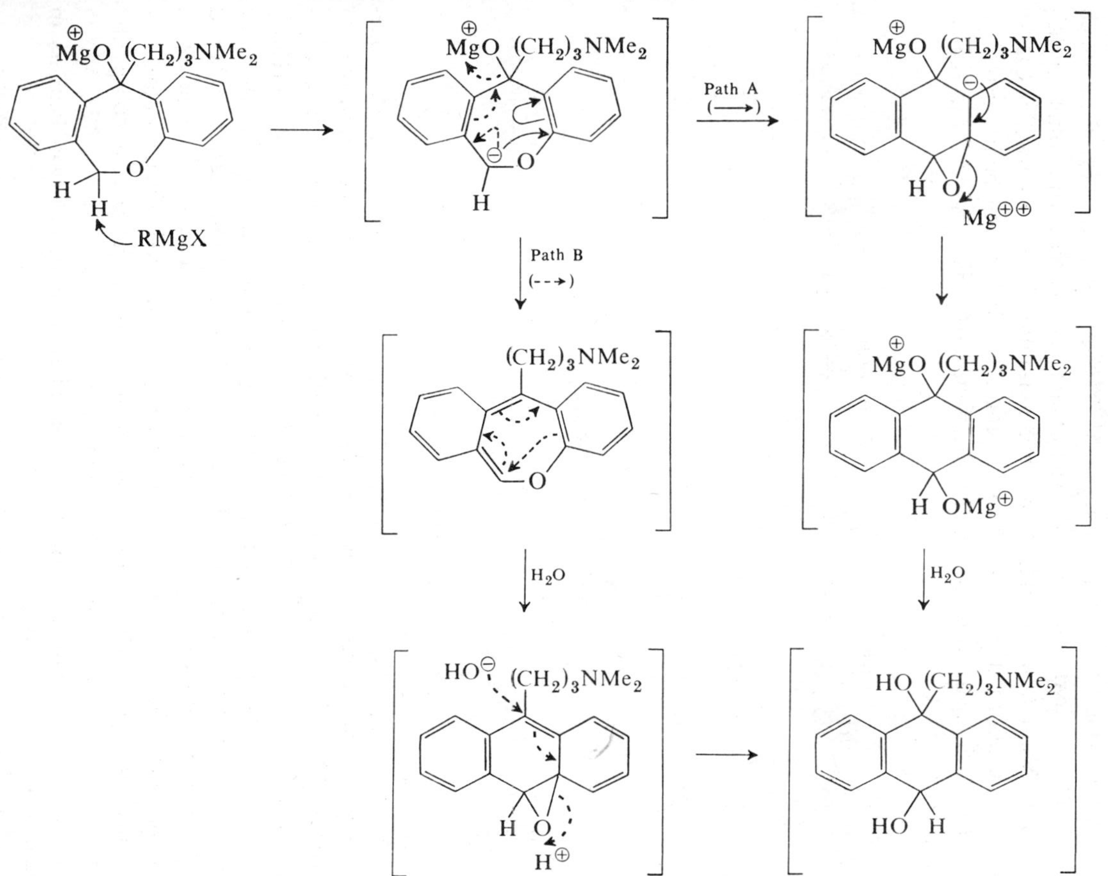
RMgX
Path A
Path B
H2O
H2O

into the pure hydrochloride salt, mp 192–193°. Treatment of alcohol **40** with cyanogen bromide, followed by alkaline hydrolysis, gave a mixture of *cis*- and *trans*-*N*-methyldibenz[*b*,*e*]oxepin-$\Delta^{11(6H),\gamma}$-propylamine (**44**), from which a single, pure hydrochloride salt was obtained by careful fractional crystallization. Methylation of this desmethyl compound by the Eschweiler-Clarke procedure afforded pure *cis*-**38** · HCl, mp 209–210°. Upon being heated in hydrochloric acid, *trans*-**38** · HCl underwent partial isomerization to the *cis* form.

(1) HCl reflux
(2) fract crystn of maleate salts → *trans*-**38**·maleate —(1) NaOH (2) HCl→ *trans*-**38**·HCl, mp 192–193°

HCl, hot (partial isomerization) ↓

HO (CH$_2$)$_3$NMe$_2$

40

(1) CNBr
(2) NaOH →

CH(CH$_2$)$_2$NHMe

cis- and *trans*-**44**

(1) fract crystn of HCl salts
(2) CH$_2$O/HCO$_2$H → *cis*-**38**·HCl, mp 209–210°

Another approach to the synthesis of 11(6*H*)-alkylidenedibenz[*b*,*e*]oxepins from the corresponding 11(6*H*)-ketones was described recently in a patent (123). This method involves a Wittig reaction between a triphenylphosphine *N*,*N*-dialkylaminoalkylene ylide and a dibenz[*b*,*e*]oxepin-11(6*H*)-one. The ylides can be prepared by reaction of triphenylphosphine with one equivalent of an α,ω-dibromoalkane (e.g., 1,3-dibromopropane), followed by reaction with one equivalent of a primary or secondary amine (e.g., ethylamine, dimethylamine, or piperazine). Substituted dibenz[*b*,*e*]oxepin-11(6*H*)-ones which have been used in addition to the parent compound include the 2-chloro and 2-*N*,*N*-dimethylsulfamyl derivatives. The Wittig condensation step is said to produce a mixture of geometrical isomers in which the *cis*/*trans* ratio is 4.

11-Alkylidenedibenz[*b*,*e*]oxepin-6(11)*H*-ones were not known until very recently when it was reported in a patent disclosure (39) that dibenz[*b*,*e*] oxepin-6,11-diones yield ring-opened adducts on treatment with 3-(*N*,*N*-dimethylamino)-1-propyl magnesium chloride, and that these adducts can be recyclized to dibenz[*b*,*e*]oxepin derivatives under the influence of acetic anhydride containing a trace of pyridine. Other Grignard reagents studied in

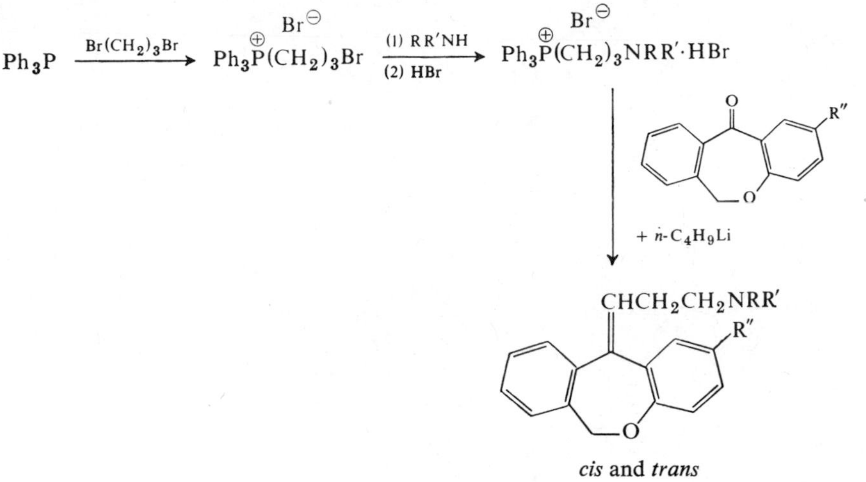

$$Ph_3P \xrightarrow{Br(CH_2)_3Br} Ph_3\overset{\oplus}{P}(CH_2)_3Br\ Br^{\ominus} \xrightarrow[(2)\ HBr]{(1)\ RR'NH} Ph_3\overset{\oplus}{P}(CH_2)_3NRR'\cdot HBr\ Br^{\ominus}$$

+ n-C_4H_9Li

$CHCH_2CH_2NRR'$

R″

cis and *trans*

this connection included 3-(piperidino)-1-propyl, 2-methyl-3-(*N*,*N*-dimethylamino)-1-propyl, 3-(4-morpholino)-1-propyl, and 3-(4-methyl-1-piperazino)-1-propyl magnesium chloride. In addition to dibenz[*b*,*e*]oxepin-6,11-dione

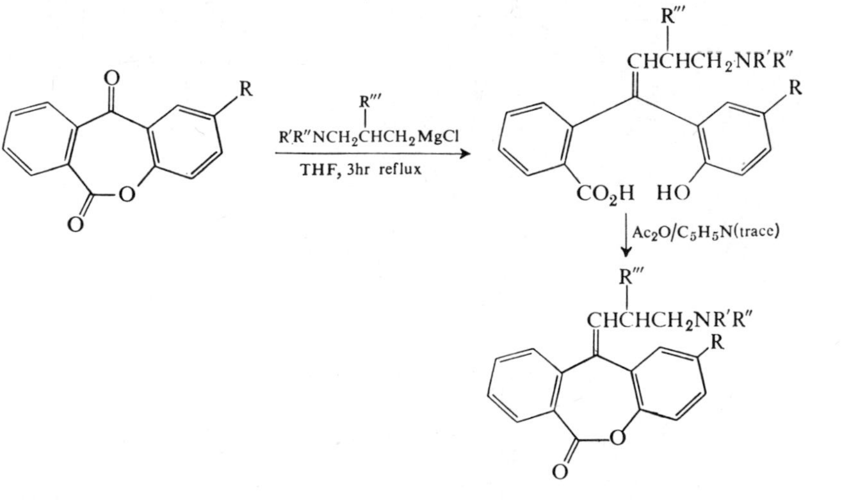

$R'R''NCH_2CH(R''')CH_2MgCl$

THF, 3hr reflux

$CHCH(R''')CH_2NR'R''$

CO_2H HO

Ac_2O/C_5H_5N(trace)

$CHCH(R''')CH_2NR'R''$

itself, the reaction was also carried out with the 2-methyl and 2-chloro derivatives.

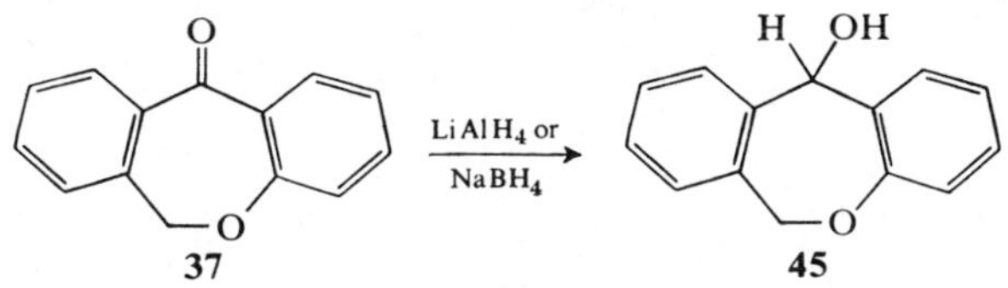

$LiAlH_4$ or $NaBH_4$

37 → **45**

CH(CH$_2$)$_2$N(Me)—CH$_2$Ph

46

$\xrightarrow{ClCO_2Et}$

CH(CH$_2$)$_2$N(Me)—CO$_2$Et

47

LiAlH$_4$/THF or H$_2$/Pd

H (CH$_2$)$_3$NMe$_2$ CH$_3$ HO

49

LiAlH$_4$/THF

$\underset{LiAlH_4/Et_2O}{\overset{ClCO_2Et}{\rightleftharpoons}}$

CH(CH$_2$)$_2$NMe$_2$

38

KOH

CH(CH$_2$)$_2$NHMe

44

$\xrightarrow[MeOH]{CH_2O}$

CH(CH$_2$)$_2$N(CHO)Me

48

NaBH$_4$

A number of chemical reactions have been carried out with dibenz[*b*,*e*]-oxepins in which the ring system itself remains intact. Most of these involve only the side chain of certain 11(6*H*)-alkylidene derivatives, rather than functional groups attached directly to the dibenz[*b*,*e*]oxepin ring system. A noteworthy exception is the reduction of dibenz[*b*,*e*]oxepin-11(6*H*)-one (**37**) to 6,11-dihydrodibenz[*b*,*e*]oxepin-11-ol (**45**) with sodium borohydride or lithium aluminum hydride (169).

Side chain transformations which may be cited here include (*1*) the reaction of *N*-benzyl-*N*-methyl (150) and *N*,*N*-dimethyl (37) compounds **46** and **38**, respectively, with ethyl chloroformate; (*2*) the hydrolysis and decarboxylation of the resulting *N*-carbethoxy-*N*-methyl derivative, **47**, with potassium hydroxide (19, 30); (*3*) the *N*-formylation of the resulting *N*-methyl compound,

Me
$CH(CH_2)_2NCO_2Et$
O
47

HBr/HOAc

$(CH_2)_2NHMe$
HO
50

CH_2O/HCO_2H

$CH(CH_2)_2NMe_2$
O
38

HBr/HOAc

$(CH_2)_2NMe_2$
HO
51

44, to **48**, and the subsequent regeneration of **38** from **48** by reduction with sodium borohydride (30); and, finally, (*4*) reduction of **47** back to **38** with lithium aluminum hydride, and further hydrogenolysis of **38** to an open-chain compound, **49**, by catalytic reduction or more vigorous treatment with lithium aluminum hydride (30).

An interesting rearrangement was discovered recently by Stach and co-workers (30) during a careful study of the chemistry of **47**. Treatment with hydrobromic acid in acetic acid was found to give not the previously reported simple hydrolysis and decarboxylation product (37, 150), but instead a compound which was formulated as 1-(*o*-hydroxyphenyl)-2-(2-*N*-methyl-aminoethyl)-3*H*-indene (**50**). Eschweiler-Clarke methylation of **50** yielded the corresponding *N,N*-dimethyl derivative, **51**, which was likewise obtained upon reaction of **38** with hydrobromic acid in acetic acid solution.

Upon treatment with hydrogen chloride and calcium chloride in benzene (210) or hydrogen chloride alone in chloroform (211), 6,11-dihydrodibenz-[*b,e*]oxepin-11-ol (**45**) is converted into 11-chloro-6,11-dihydrodibenz[*b,e*]-oxepin (**52**). Condensation of **52** with *N*-nitrosopiperazine, followed by

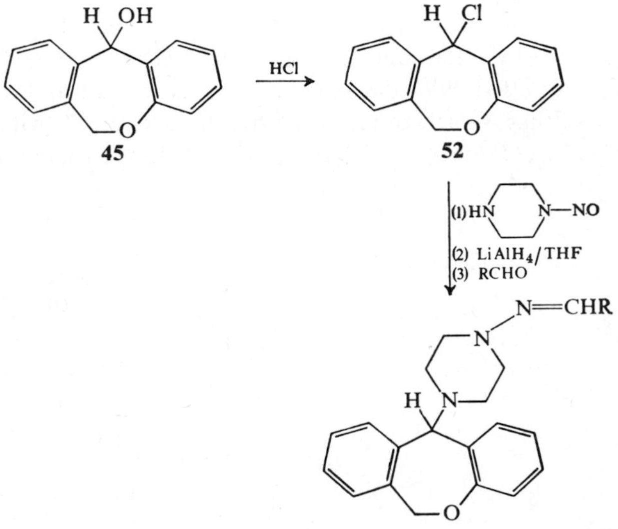

H OH
45
HCl
H Cl
52
(1) HN N—NO
(2) $LiAlH_4$/THF
(3) RCHO
N=CHR

reduction of the nitroso function with lithium aluminum hydride and reaction with various aldehydes, has led to a series of Schiff bases with anticonvulsant properties (211).

5. Dibenz[*b,f*]oxepins

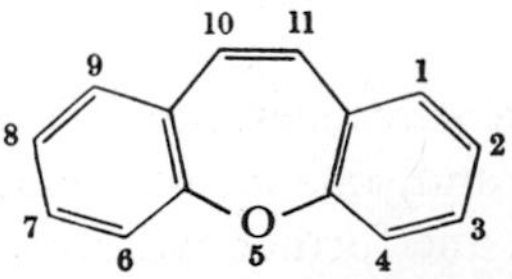

R.I. 3696

The first example of a dibenz[*b,f*]oxepin was discovered in 1911 by Pschorr and Knöffler (137), during a study of the action of nitrous acid on α-(2,4-dimethoxyphenyl)-β-(2-amino-3,4-dimethoxyphenyl)acrylic acid (**53**). In addition to the expected Pschorr cyclization product, **54**, there was formed an anomalous by-product in which loss of one methoxy group had occurred. The structure assigned to this compound was 3,4,7-trimethoxydibenz[*b,f*]-oxepin-10-carboxylic acid (**55**).

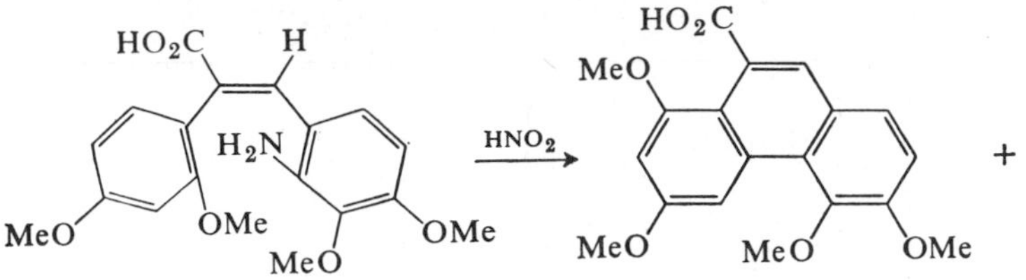

It was almost 40 years before another dibenz[*b,f*]oxepin was reported in the chemical literature. Interest in the synthesis of this ring system was kindled by the structural elucidation of the alkaloid cularine by Manske (112). As part of a program of synthesis of various model compounds, Manske and Ledingham (113) devised a straightforward synthesis of dibenz-[*b,f*]oxepin-10(11*H*)-one (**56**) starting from *o*-phenoxyphenylacetic acid (**57**). Treatment with thionyl chloride gave the acid chloride, which was cyclized readily under the influence of aluminum chloride in nitrobenzene. Using the same approach, Manske and Ledingham also prepared 6-methoxydibenz[*b,f*]oxepin-10(11*H*)-one (**58**). The yield of **58** was significantly lower than that of **56.** This can be ascribed to a deactivating inductive effect by the methoxyl group or to a steric effect (the latter being more likely). Compound **56** was likewise prepared in 1954 by Kimoto and co-workers (97); the same approach was employed as before, except that the intermediate acid chloride was obtained by treatment of the acid with oxalyl chloride in benzene. In an extension of the earlier work, Kulka and Manske (102) found liquid hydrogen fluoride to be an effective reagent for the direct cyclization of a series of *o*-phenoxyphenylacetic acids. Using this reagent, they prepared 2,3,6-trimethoxydibenz[*b,f*]oxepin-10(11*H*)-one (**59**) in 15% yield, 2,3,7-trimethoxydibenz[*b,f*]oxepin-10(11*H*)-one (**60**) in 60% yield, 2,3,8-trimethoxydibenz[*b,f*]oxepin-10(11*H*)-one (**61**) in 53% yield, and 2,3,6-trimethoxy-9-methyldibenz[*b,f*]oxepin-10(11*H*)-one (**62**) in 3% yield. Structure **60** was selected in preference to the alternate 2,3,9-trimethoxy structure on the premise that ring closure should give the least sterically hindered product. This assumption found support in the very low yield obtained in the synthesis of **62**. The relatively poor yield of **59** was likewise ascribed to a steric effect. The fact that **60** and **61** were isolated in roughly equal yields was

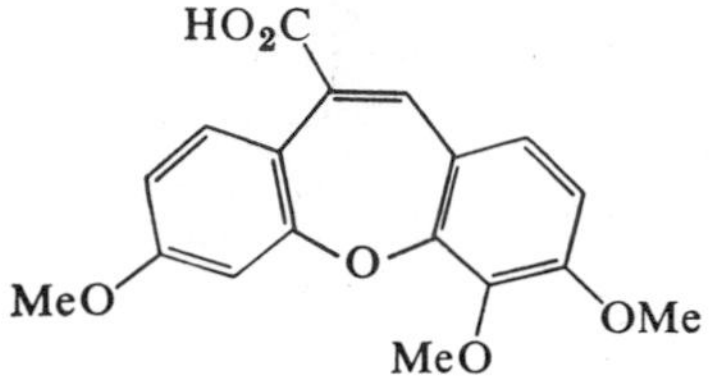

(1) $SOCl_2$ or $(COCl)_2$
(2) $AlCl_3/CS_2$ or $PhNO_2$

liquid HF

57; R=R′=R″=H

	R	R′	R″
56	H	H	H
58	H	6-MeO	H
59	MeO	6-MeO	H
60	MeO	7-MeO	H
61	MeO	8-MeO	H
62	MeO	7-MeO	Me

taken to mean that electronic effects play a rather minor role in determining the rate of the cyclization process.

Another effective reagent for the cyclodehydration of *o*-phenoxyphenylacetic acids was reported in 1957 by Loundon and Summers (109), who obtained 2-nitrodibenz[*b*,*f*]oxepin-10(11*H*)-one (**63**) in 80% yield by heating 5-nitro-2-phenoxyphenylacetic acid in polyphosphoric acid at 100° for 2 hr. In similar fashion these workers prepared compounds **56**, **58**, and also 6-methoxy-2-nitrodibenz[*b*,*f*]oxepin-10(11*H*)-one (**64**) in high yields. More

PPA
Δ

	R	R′	R″
63	NO_2	H	H
64	NO_2	H	6-MeO
65	MeO	MeO	H
66	H	H	8-MeO
59	MeO	MeO	6-MeO
66A	H	H	7,8-MeO

recent applications of the same reagent have been the syntheses of 2,3-dimethoxydibenz[*b,f*]oxepin-10(11*H*)-one (**65**) from 2-phenoxy-4,5-dimethoxyphenylacetic acid (90), of 8-methoxydibenz[*b,f*]oxepin-10(11*H*)-one (**66**) from *o*-(4-methoxyphenoxy)phenylacetic acid (124), of 2,3,6-trimethoxydibenz[*b,f*]oxepin-10(11*H*)-one (**59**) from 4,5-dimethoxy-2-(*o*-methoxyphenoxy)phenylacetic acid (89), and of 7,8-dimethoxydibenz[*b,f*]oxepin-10(11*H*)-one (**66A**) from *o*-(3,4-dimethoxyphenoxy)phenylacetic acid (201).

Polyphosphoric acid has likewise been a useful reagent for the synthesis of dibenz[*b,f*]oxepin-10-carboxylic acids (109). For example, treatment of 3-nitro-6-phenoxyphenylpyruvic acid with hot polyphosphoric acid has been reported to give a 75% yield of 2-nitrodibenz[*b,f*]oxepin-10-carboxylic acid (**67**). Reduction of **67** with hydrogen over a palladium–strontium carbonate catalyst yielded 2-amino-10,11-dihydrodibenz[*b,f*]oxepin-10-carboxylic acid (**68**). Nitrous acid deamination transformed the latter into 10,11-dihydrodibenz[*b,f*]oxepin-10-carboxylic acid (**69**). Decarboxylation of **67** in refluxing quinoline in the presence of copper bronze afforded 2-nitrodibenz[*b,f*]-oxepin (**70**), and oxidation of **70** with osmium tetroxide in pyridine gave 10,11-dihydro-2-nitrodibenz[*b,f*]oxepin-10,11-diol (**71**). Polyphosphoric acid cyclodehydration likewise effected the ring closure of 2-(*o*-anisyloxy)-5-nitrophenylpyruvic acid with formation of 6-methoxy-2-nitrodibenz[*b,f*]-oxcpin-10-carboxylic acid (**72**) in 70% yield.

Dibenz[*b,f*]oxepin (**73**) itself was prepared first by Manske and Ledingham (113) via Meerwein-Pondorff reduction of **56** and direct dehydration of the crude alcohol, **74**, with *p*-toluenesulfonic acid. Potassium permanganate oxidation of **73** in acetone afforded a mixture of 2,2′-dicarboxydiphenyl ether (**75**) and xanthone (**76**) (113). A similar sequence was used more recently by Alam and MacLean (4) to prepare **73** from **56**, except that the reduction of **56** to **74** was carried out with sodium borohydride. 10,11-Dihydrodibenz-[*b,f*]oxepin-10-ol (**74**) was likewise isolated by Alam and MacLean as the major product in the nitrous acid deamination of 10-aminomethylxanthene (**77**). The other products formed were olefins of undetermined structure and a compound identified as 10-hydroxymethylxanthene (**78**).

Another method of synthesis of **73** was reported in 1957 by Anet and Bavin (5). This consisted of reducing xanthene-10-carboxylic acid with lithium aluminum hydride and heating the resulting alcohol, **78**, in refluxing xylene in the presence of phosphorus pentoxide. The seven-membered ring is probably generated via a concerted Wagner-Meerwein process. The same approach was also utilized by Anet and Bavin to prepare 10-methyldibenz-[*b,f*]oxepin (**79**). Methyl xanthene-10-carboxylate (**80**) was methylated using sodium amide in liquid ammonia as the base and methyl iodide as the alkylating agent, and the resulting ester, **81**, was reduced to the corresponding alcohol, **82**, with lithium aluminum hydride. The alcohol was converted

HO_2CCH_2
NO_2
R
PPA
Δ
HO_2C
NO_2
R **67**; R=H
72; R=MeO
Cu
quinoline
Δ
(R=H)
NO_2
70
OsO_4
C_5H_5N
$H_2/Pd\text{-}SrCO_3$
(R = H)
HO_2C H
NH_2
68
(1) HNO_2
(2) H_3PO_2
HO_2C H
69
HO H H OH
NO_2
71

H CH_2NH_2 77 HNO_2 H CH_2OH 78 Olefins HO H 74 56 $Al(OPr\text{-}i)_3$ or $NaBH_4$ *p*-TsOH 120–140° 10 min 73 $KMnO_4$ acetone CO_2H HO_2C 75 + 76

into tosylate ester **83** in the conventional manner, and the latter was allowed to react with formic acid to give **79**. Oxidation of **79** with potassium permanganate in acetone afforded **75**. Interestingly, no **79** was obtained when 10-ethylxanthol was dehydrated by passage over a column of activated alumina; the only product formed in this case was 10-ethylidenexanthene.

A closely related example which may be cited here is the synthesis of 10-phenyldibenz[*b*,*f*]oxepin (**84**) described in 1964 by Bergmann and Rabinowitz (27). Treatment of xanthene with *n*-butyllithium and condensation of the resultant anion with benzaldehyde afforded alcohol **85**. Dehydration of **85** in boiling formic acid containing some *p*-toluenesulfonic acid and then gave **84**.

An instructive comparison of the uv spectra of **73** and **79** was made by Anet and Bavin (5). Absorption maxima in methanol were observed at 235 and 288 mμ with **73**, and at 235 and 280 mμ with **79**. The 8 mμ bathochromic shift in the long-wavelength band of the dibenz[*b*,*f*]oxepin chromophore when a methyl substituent is placed at the 10-position indicates that the seven-membered ring is capable of being distorted rather easily in this tricyclic system. That **79** is less planar in geometry than **73** is also apparent from the fact that **79** melts at only 57–58.5°, whereas **73** melts at 109–110°.

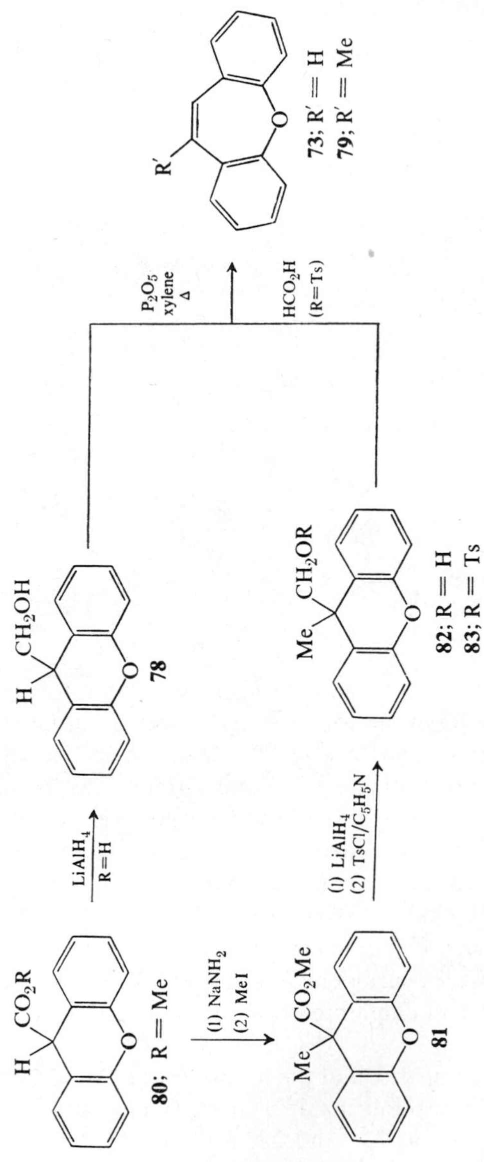
80; R = Me
H
CO$_2$R
LiAlH$_4$
R = H
78
H
CH$_2$OH
(1) NaNH$_2$
(2) MeI
81
Me
CO$_2$Me
(1) LiAlH$_4$
(2) TsCl/C$_5$H$_5$N
82; R = H
83; R = Ts
Me
CH$_2$OR
P$_2$O$_5$
xylene
Δ
HCO$_2$H
(R = Ts)
R′
73; R′ = H
79; R′ = Me

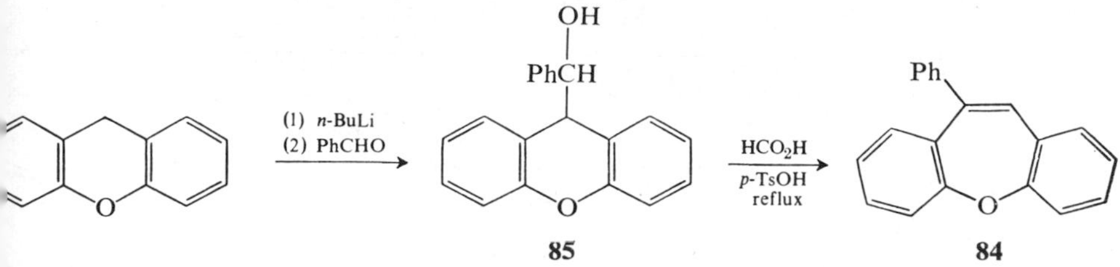

OH PhCH (1) n-BuLi (2) PhCHO HCO$_2$H p-TsOH reflux Ph

85 **84**

Still another synthesis of **73** was reported in 1957 by Loundon and Summers (109). Starting from **56**, oxime **86** was prepared and reduced to 10-acetamido-10,11-dihydrodibenz[*b,f*]oxepin (**87**) by catalytic hydrogenation in acetic anhydride solution. Treatment of **87** with phosphorus pentoxide in boiling xylene produced **73**.

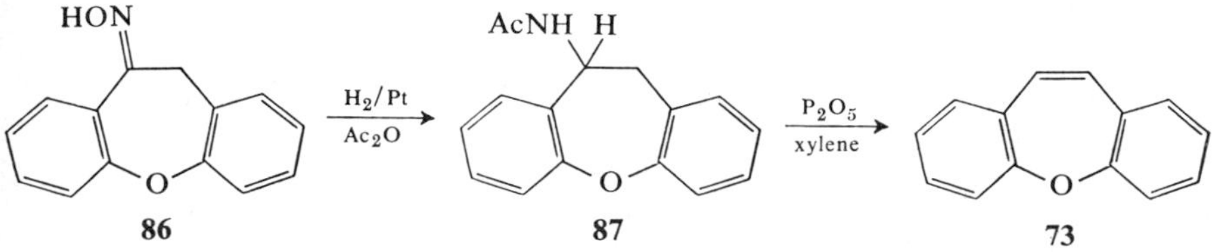

HON H_2/Pt Ac_2O AcNH H P_2O_5 xylene

86 **87** **73**

The most recently disclosed method of preparation of **73** is that of Whitlock (174), involving the interesting reaction of xanthylium perchlorate (**88**) with diazomethane.

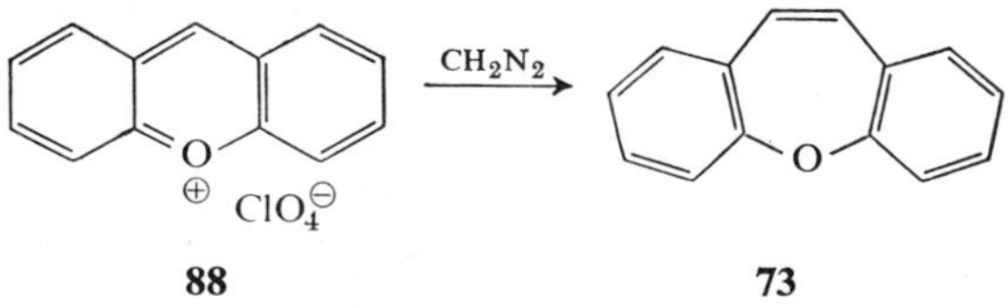

CH_2N_2 $\oplus$ $ClO_4^{\ominus}$

88 **73**

A number of chemical transformations of dibenz[*b,f*]oxepin (**73**) were described recently by Bavin and co-workers (193), tending to support the view that this compound possesses at least some degree of aromaticity. Upon being heated in deuteriotrifluoroacetic acid at 100°, **73** underwent exchange of the C-10 proton to the extent of about 35%, as shown by the decrease in intensity of the τ 3.39 singlet in the nmr spectrum. Nitration of **73** afforded a 22% yield of 10-nitrodibenz[*b,f*]oxepin (**89**), and reduction of the latter with hydrazine and palladium–charcoal gave oxime **86**. Hydrolysis of **86** with hydrochloric acid yielded **56**. Bromination of **73** led to dibromide **90** in 61% yield, and dehydrobromination of **90** with potassium *tert*-butoxide gave 10-bromodibenz[*b,f*]oxepin (**91**) in 64% yield. Treatment of **91** with cuprous cyanide in dimethylformamide and pyridine gave a 71% yield of 10-cyanodibenz[*b,f*]oxepin (**92**). Further dehydrobromination of **91** was achieved by Tochtermann and co-workers (194) with sodamide. The resultant

73 → (HNO_3) → 89 → (H_2NNH_2, Pd/C) → 86

73 → (Br_2) → 90 → (KOBu-*tert*) → 91 → (CuCN, DMF-C_5H_5N) → 92

91 → ($NaNH_2$) → [93] → (PhCH=C(Ph)—C(Ph)=CHPh) → tetraphenyl Diels-Alder adduct

benzynelike intermediate, **93**, was trapped in the form of a Diels-Alder adduct with 1,2,3,4-tetraphenylbutadiene.

An unusual approach to the synthesis of the dibenz[*b*,*f*]oxepin ring system was developed by Bergmann and co-workers (199, 222). Bromination of 4,4′-dibromo-2,2′-dimethyldiphenyl ether with *N*-bromosuccinimide in the presence of a catalytic amount of benzoyl peroxide afforded a tetrabromide. Treatment of this intermediate with six equivalents of phenyllithium gave a 66.5% yield of a compound initially considered to be 2,8-dibromo-10,11-dihydrodibenz[*b*,*f*]oxepin (**94**), together with 2.3% of a product believed to be 10,11-dihydrodibenz[*b*,*f*]oxepin (**95**) (199). The correctness of these assignments was challenged (221), and the dimeric nature of the products was subsequently recognized. In a reexamination of the original work, 2,2′-dimethyldiphenyl ether was brominated and the resulting dibromide treated with one equivalent of phenyllithium, rather than six. Under these conditions the main product, dimer **96**, was accompanied by a 20% yield of authentic **95** (222).

CH_3 CH_3 R NBS Bz_2O_2 R=H, Br R CH_2Br $BrCH_2$ R O PhLi

R R O **94**; R = Br **95**; R = H

R R O O R R **96**; R = H

10,11-Dihydrodibenz[*b,f*]oxepin (**95**) was also prepared recently by Hess and co-workers (202) by catalytic hydrogenation of dibenz[*b,f*]oxepin (**73**). Metallation of **94** with *n*-butyllithium, followed by carbonation, gave 10,11-dihydrodibenz[*b,f*]oxepin-4,6-dicarboxylic acid (**97**). The dimethyl

95 (1) *n*-Buli (2) CO_2 (3) CH_2N_2 CO_2R CO_2R **97**; R = H **98**; R = Me (1) $LiAlH_4$ (2) PBr_3

101 PhLi X=Br CH_2X CH_2X **99**; X = OH **100**; X = Br

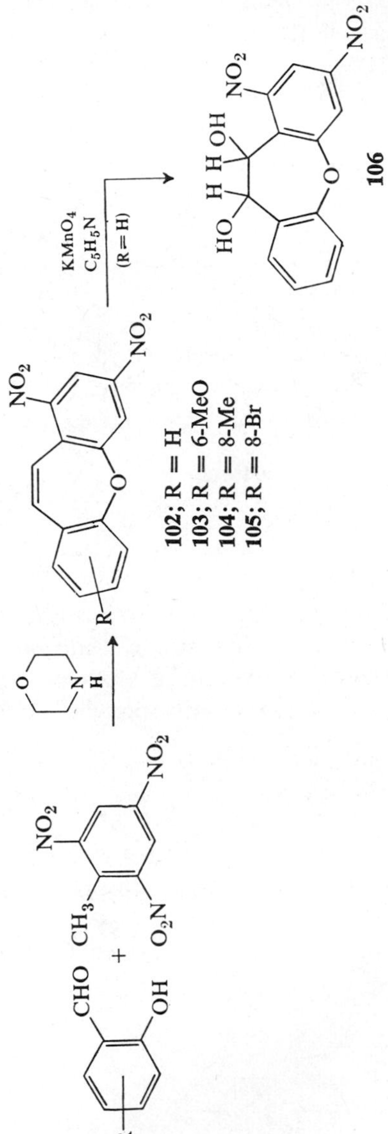

CHO
OH
R
+
CH_3
NO_2
O_2N
NO_2

NO_2
NO_2
R
O

102; R = H
103; R = 6-MeO
104; R = 8-Me
105; R = 8-Br

$KMnO_4$
C_5H_5N
(R = H)

HO H H OH
O
NO_2
NO_2

106

ester, **98**, obtained from **97** by reaction with diazomethane, was reduced to 10,11-dihydrodibenz[*b,f*]oxepin-4,6-dimethanol (**99**) with lithium aluminum hydride. Treatment of this diol with phosphorus tribromide led to 4,6-bis(bromomethyl) derivative **100**, and further reaction of the dibromide with phenyllithium afforded a 75% yield of the interesting O-bridged *meta*-cyclophane **101**.

A novel single-step synthesis of certain members of the dibenz[*b,f*]oxepin ring system from easily accessible precursors was reported in 1962 by Hoyer and Vogel (76). This consisted of heating 2,4,6-trinitrotoluene with salicylaldehyde in the presence of morpholine. Using salicylaldehyde and its 3-methoxy, 5-methyl, and 5-bromo derivatives, they prepared 1,3-dinitrodibenz[*b,f*]oxepin (**102**), 6-methoxy-1,3-dinitrodibenz[*b,f*]oxepin (**103**), 8-methyl-1, 3-dinitrodibenz[*b,f*]oxepin (**104**), and 8-bromo-1,3-dinitrodibenz-[*b,f*]oxepin (**105**), respectively. Oxidation of **102** with potassium permanganate in pyridine afforded 10,11-dihydro-1,3-dinitrodibenz[*b,f*]oxepin-10,11-diol (**106**). When the reaction medium was acetone, permanganate oxidation effected cleavage of the seven-membered ring, with formation of salicylaldehyde.

Dibenz[*b,f*]oxepins containing aminoalkyl side chains have recently aroused some interest as possible central nervous system stimulants, antihistamines, and antispasmodics. A recent patent disclosure (168) deals with the synthesis of such compounds via a Wagner-Meerwein process. Thus reaction of 10-[3-(*N,N*-dimethylamino)propyl]xanthene-10-methanol (**107**) with a dehydrating agent yields 10-[3-(*N,N*-dimethylamino)propyl]dibenz-[*b,f*]oxepin (**108**). Other side chains which have been used in this work include

HOCH$_2$ (CH$_2$)$_n$X ... R′ ... O ... R $\xrightarrow{-H_2O}$ X(CH$_2$)$_n$ ... R′ ... O ... R

107; R = R′ = H, n = 3, X = Me$_2$N **108**; R=R′ = H, n = 3, X = Me$_2$N

the 2-(pyrrolidino)ethyl, 2-(*N,N*-dimethylamino)ethyl, and 3-(piperidino)-propyl groups. Nuclear substituents which have been used include the trifluoromethyl, methyl, chloro, bromo, and methylthio groups. With xanthene-10-methanol derivatives containing a single aromatic substituent, two possible Wagner-Meerwein products can be obtained, in principle, depending upon the relative migratory aptitudes of the substituted and unsubstituted phenyl moieties. Detailed product analyses and structure proofs have not been published to date, but it is not unreasonable to predict that the direction of rearrangement would obey the usual substitution rules for this type of carbonium ion-mediated reaction.

Other aminoalkyl-substituted dibenz[*b,f*]oxepins with useful chemotherapeutic properties have been reported recently. Starting from 10-chloro-10,11-dihydrodibenz[*b,f*]oxepin (**109**), for example, Protiva and co-workers (136) prepared 10,11-dihydro-10-[2-(*N,N*-dimethylamino)ethoxy]dibenz[*b,f*]oxepin (**110**) by reaction with 2-(*N,N*-dimethylamino)ethanol, and *N*-methyl-*N*′-(10,11-dihydrodibenz[*b,f*]oxepin-10-yl)piperazine (**111**) by reaction with *N*-methylpiperazine. Compounds **110** and **111** are both claimed to be effective neuroleptic agents (136).

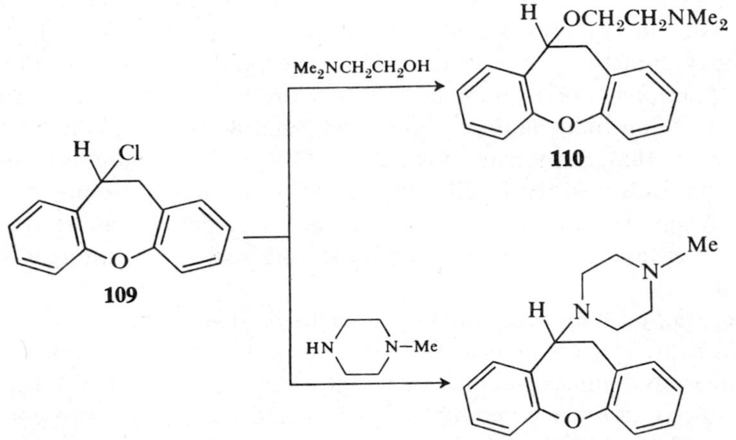

H Cl

109

$Me_2NCH_2CH_2OH$

H $OCH_2CH_2NMe_2$

110

HN N—Me

H N N Me

111

Central depressive agents with sedative and adrenolytic activity were prepared from 10-bromomethyldibenz[*b,f*]oxepin (**112**) and its 2-methoxy and 2-chloro analogs, **113** and **114** (124). The bromides were synthesized by reaction of the corresponding 10-methyl compounds with *N*-bromosuccinimide, and were condensed readily with such amines as dimethylamine, monomethylamine, pyrrolidine, *N*-(2-hydroxyethyl)piperazine, and *N*-methylpiperazine. Amines were likewise condensed with 10-(1-bromoethyl)dibenz[*b,f*]oxepin (**115**).

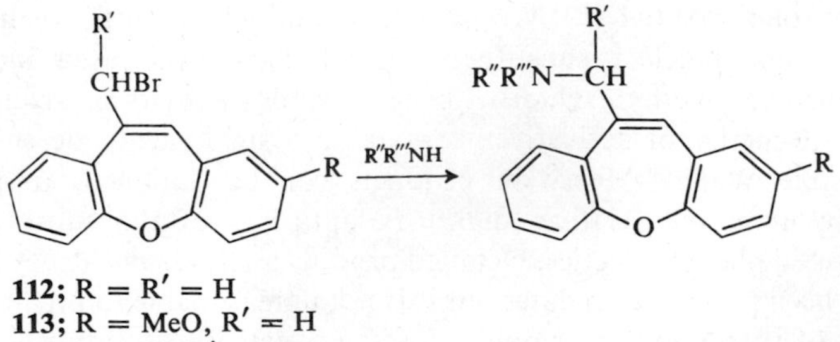

R′ CHBr R

R″R‴NH

R′ R″R‴N—CH R

112; R = R′ = H
113; R = MeO, R′ = H
114; R = Cl, R′ = H
115; R = H, R′ = Me

2-Methoxy-10-methyldibenz[*b*,*f*]oxepin (**116**), which was required for the preparation of bromide **113**, was synthesized in several steps, starting from 8-methoxydibenz[*b*,*f*]oxepin-10(11*H*)-one (**66**) (124). Addition of methyl iodide to the anion of **66**, with sodium amide as the base, yielded 8-methoxy-11-methyldibenz[*b*,*f*]oxepin-10(11*H*)-one (**117**). Lithium aluminum hydride reduction of **117** furnished 10,11-dihydro-8-methoxydibenz[*b*,*f*]oxepin-10-ol (**118**), which was converted into the corresponding 10-chloro derivative under the influence of a suitable chlorinating agent. Dehydrohalogenation with potassium *tert*-butoxide then gave **116**.

MeO O O **66** (1) $NaNH_2$/PhMe (2) MeI MeO O Me O **117** $LiAlH_4$ HO H Me H MeO O **118** (1) $POCl_3$ (2) KOBu-*tert* MeO Me O **116**

Dibenz[*b*,*f*]oxepin-10(11*H*)-one (**56**) has served as the point of departure for a number of interesting synthetic schemes in addition to those already mentioned. Thus Mathys and co-workers (116) in 1956 reported the conversion of **56** into the hitherto unknown dibenz[*b*,*f*]oxepin-10,11-dione (**119**) by oxidation with selenium dioxide in aqueous dioxane. The same reaction was carried out at about the same time by Rees (141), except that acetic acid was used as the solvent. Diketone **119** could be reduced to 10,11-diacetoxydibenz-[*b*,*f*]oxepin (**120**) upon treatment with zinc in acetic anhydride buffered with sodium acetate (116). A facile ring contraction of the benzilic rearrangement type was brought about under the influence of base. Whether the α-hydroxy or α-methoxy acid, **121** or **122**, was formed depended upon the use of alcoholic (116) or aqueous (141) base. A novel extension of this ring contraction was also observed by Erdtman and Spetz (58) when **119** was allowed to react with sodium sulfite. The product in this case was formulated as sodium xanthene-10-sulfonate (**123**). The dipole moment of **119** was determined with the view of assessing the possible contribution of canonical form **119A** to the ground state of the α-diketone system. The results of this investigation led to the conclusion that **119** is relatively nonpolar (116), that is, that tropolonoid properties should not be observed (141).

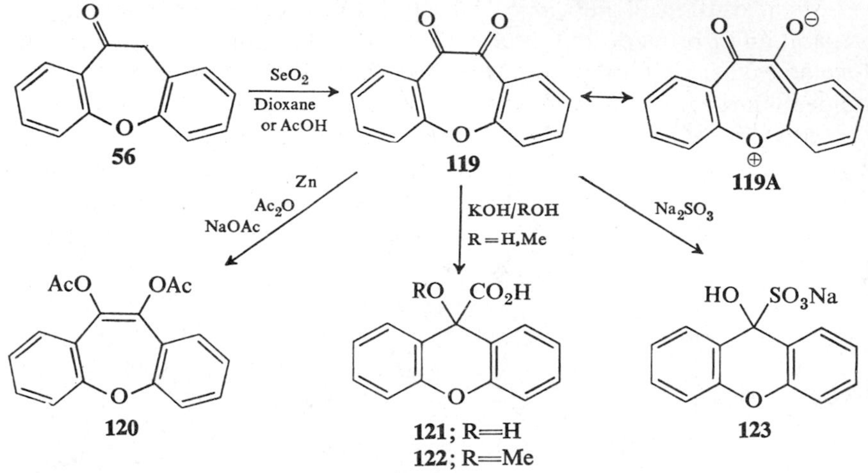

Dibenz[*b,f*]oxepin-10(11*H*)-one (**56**) has also been used as the starting point for the synthesis of a number of amino-substituted dibenz[*b,f*]oxepin derivatives of potential pharmacological interest (93). Reduction of the oxime derivative of **56** has been effected with sodium amalgam or palladium–charcoal in hydrochloric acid. Under these conditions, 10-amino-10,11-dihydrodibenz[*b,f*]oxepin (**124**) is produced directly. If the oxime is reduced catalytically in the presence of acetic anhydride, on the other hand, the *N*-acetyl derivative, **87**, of **124** is obtained. Acid hydrolysis of this amide then gives **124**. Another method of preparation of **124** is to subject **56** to the Leuckart reaction with ammonium formate. The optical antipodes of **124** have been separated by resolution with D- and L-mandelic acid. Methylation of **124** by the Eschweiler-Clarke procedure affords 10,11-dihydro-10-(*N,N*-dimethylamino)dibenz[*b,f*]oxepin (**125**). This compound has likewise been resolved, using optically active camphorsulfonic and tartaric acids to prepare the required diastereomeric salts. Treatment of **56** with dimethylformamide under reflux afforded 10-formamido-10,11-dihydrodibenz[*b,f*]oxepin (**126**), which was then reduced to 10,11-dihydro-10-(*N*-methylamino)dibenz[*b,f*]-oxepin (**127**) with lithium aluminum hydride. Bromination of **56** with *N*-bromosuccinimide, followed directly by condensation with dimethylamine in a sealed tube, led to the formation of 11-(*N,N*-dimethylamino)dibenz-[*b,f*]oxepin-10(11*H*)-one (**128**). Reduction of the latter with aluminum isopropoxide or lithium aluminum hydride furnished 10,11-dihydro-11-(*N,N*-dimethylamino)dibenz[*b,f*]oxepin-10-ol (**129**), for which analgesic properties have been claimed (84). Finally, reaction of **56** with *n*-butyl

131 $\xleftarrow{LiAlH_4}$ 130 $\xleftarrow[HCl,\ Et_2O]{n\text{-}BuONO}$ 56 $\xrightarrow[170-180^\circ,\ 5\ hr]{HCONH_2}$ 126 $\xrightarrow{LiAlH_4}$ 127

HO H NH$_2$ H (131); O NOH (130); HCNH(=O) H (126); MeNH H (127)

56 $\xrightarrow[(2)\ Me_2NH/C_6H_6]{(1)\ NBS/CCl_4/Bz_2O_2}$ 128

56 $\xrightarrow[(2)\ H_2/Pd\text{-}C/HCl\ or\ Na\text{-}Hg]{(1)\ H_2NOH}$ 124

O NMe$_2$ H (128); O (56); H$_2$N H (124)

128 $\xrightarrow{LiAlH_4}$ 129

56 $\xrightarrow[(2)\ H_2/Pt/Ac_2O]{(1)\ H_2NOH}$ 87 $\xrightarrow{HCl}$ 124

124 $\xrightarrow{CH_2O/HCO_2H}$ 125

HO H NMe$_2$ H (129); AcNH H (87); Me$_2$N H (125)

nitrite in ether containing hydrogen chloride yielded 11-oximinodibenz[*b,f*]-oxepin-11-one (**130**), from which 11-amino-10,11-dihydrodibenz[*b,f*]oxepin-10-ol (**131**) was prepared by lithium aluminum hydride reduction. This amino alcohol is likewise claimed to exhibit useful analgesic activity (85).

According to a recent patent (195), bromination of dibenz[*b,f*]oxepin-10(11*H*)-one (**56**) in carbon disulfide yields α-bromoketone **132**; similarly, 2-chlorodibenz[*b,f*]oxepin-10(11*H*)-one (**133**) affords **134**. Condensation of **132** and **134** with *N*-methylpiperazine gives *N*-methyl-*N'*-(dibenz[*b,f*]-oxepin-10(11*H*)-on-11-yl)piperazine (**135**) and its 2-chloro derivative, **136**, respectively. Use of *N*-(2-hydroxyethyl)piperazine with **134** led to **137.** These products are claimed to possess nerve-blocking and anticonvulsive properties.

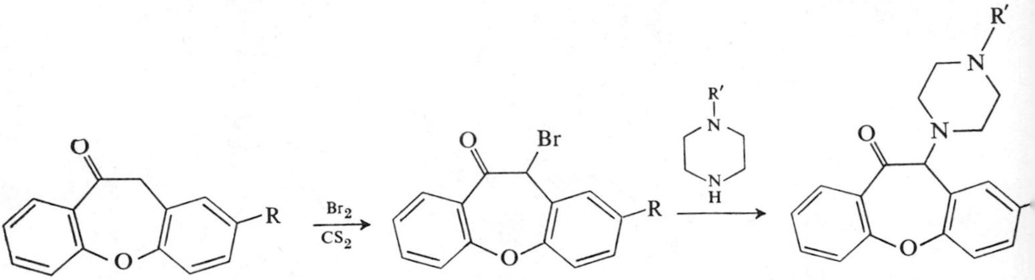

56; R = H
133; R = Cl

$\xrightarrow[CS_2]{Br_2}$

132; R = H
134; R = Cl

135; R = H, R' = Me
136; R = Cl, R' = Me
137; R = Cl, R' = HOCH$_2$

Another group of psychopharmacologically active compounds was obtained from dibenz[*b,f*]oxepin-10(11*H*)-one (**56**) by reaction with methyl magnesium iodide and immediate acid-catalyzed dehydration of the resulting alcohol to 10-methylene-10,11-dihydrodibenz[*b,f*]oxepin (**138**), bromination of **138** to **139**, and finally treatment of **139** with methylamine,

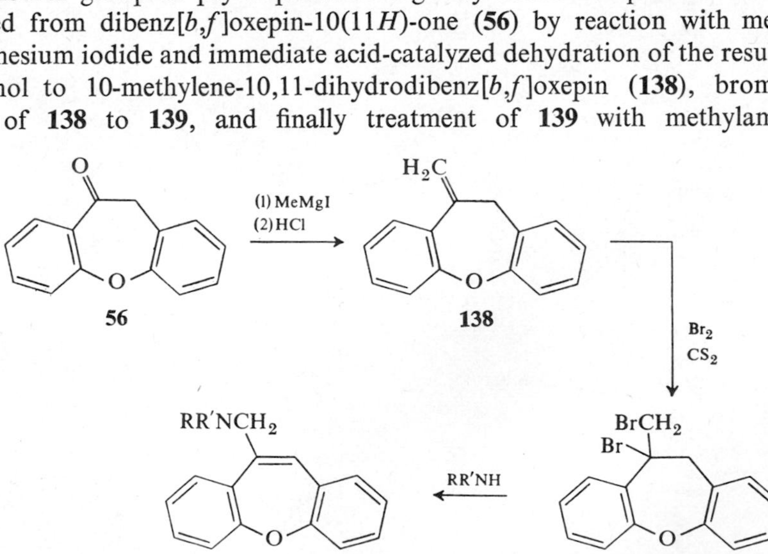

56

(1) MeMgI
(2) HCl

138

Br_2
CS_2

139

RR'NH

dimethylamine, pyrrolidine, *N*-methylpiperazine, or *N*-(2-hydroxyethyl) piperazine (196). The products are alleged to exhibit adrenolytic, sedative, and narcotic properties.

Still another series of psychopharmacologic agents was synthesized (197) from **56** upon treatment with lithium amide and alkylation with 1-chloro-3-(*N*,*N*-dimethylamino)propane. The product, 11-[3-(*N*,*N*-dimethylamino)-propyl]dibenz[*b*,*f*]oxepin-10(11*H*)-one (**140**), gave *N*-cyano derivative **141** upon reaction with cyanogen bromide. Acid hydrolysis (von Braun degradation) transformed **141** into 11-[3-(*N*-methylamino)propyl]dibenz-[*b*,*f*]oxepin-10(11*H*)-one (**142**). 8-Methoxy analogs were prepared via the same approach, starting from 8-methoxydibenz[*b*,*f*]oxepin-10(11*H*)-one (**66**) (197). The pharmacology of **140** and **142** has been discussed by Theobald and co-workers (198).

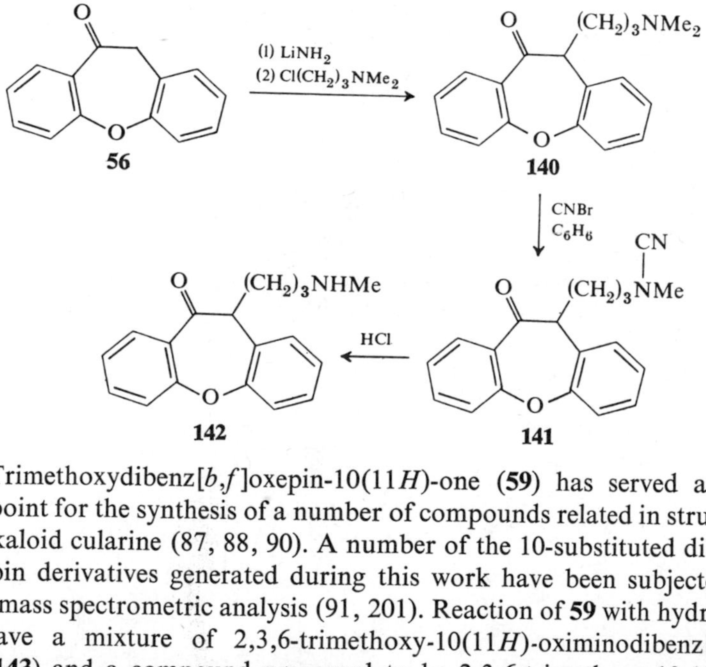

2,3,6-Trimethoxydibenz[*b*,*f*]oxepin-10(11*H*)-one (**59**) has served as the starting point for the synthesis of a number of compounds related in structure to the alkaloid cularine (87, 88, 90). A number of the 10-substituted dibenz-[*b*,*f*]oxepin derivatives generated during this work have been subjected to nmr and mass spectrometric analysis (91, 201). Reaction of **59** with hydroxyl-amine gave a mixture of 2,3,6-trimethoxy-10(11*H*)-oximinodibenz[*b*,*f*]-oxepin (**143**) and a compound presumed to be 2,3,6-trimethoxy-10,11-bis-(oximino)dibenz[*b*,*f*]oxepin (**144**). Catalytic reduction of **143** afforded 10-amino-10,11-dihydro-2,3,6-trimethoxydibenz[*b*,*f*]oxepin (**145**), while reduction of **144** with lithium aluminum hydride (or by catalytic hydrogenation) yielded 10,11-diamino-10,11-dihydro-2,3,6-trimethoxydibenz[*b*,*f*]oxepin (**146**). Direct catalytic reduction of **59** itself gave 10,11-dihydro-2,3,6-trimethoxydibenz[*b*,*f*]oxepin-10-ol (**147**), from which 2,3,6-trimethoxy-dibenz[*b*,*f*]oxepin (**148**) was produced by heating in the presence of

59 $\xrightarrow{H_2NOH}$ 143 + 144

143 $\xrightarrow{H_2/Pt}$ 145

144 $\xrightarrow{H_2/\text{catalyst or } LiAlH_4}$ 146

59 $\xrightarrow{H_2/Pt/AcOH \text{ or } LiAlH_4}$ 147

145 $\xrightarrow{HNO_2}$ 147

147 $\xrightarrow{p\text{-TsOH}}$ 148

p-toluenesulfonic acid. A recently reported alternative preparation of **147** involves nitrous acid deamination of **145** (201).

Reaction of **59** with formamide and formic acid (Leuckart reaction) gave amide **149**, and acid hydrolysis converted the latter into **145** (87). When the Leuckart reaction was performed with formamide alone, a side product was obtained, to which structure **150** was assigned.

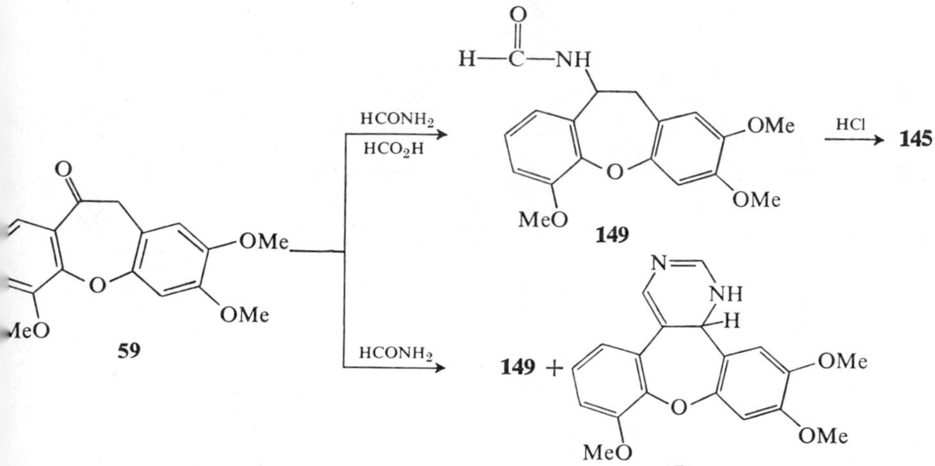

Condensation of **59** with aminoacetal, followed by cyclodehydration of the initially formed Schiff base, **151**, with sulfuric acid (87, 182) or polyphosphoric acid (183), afforded 6,9,10-trimethoxy-12*H*-benz[6,7]oxepino[2,3,4-*ij*]isoquinoline (**152**), a key intermediate in the synthesis of cularine.

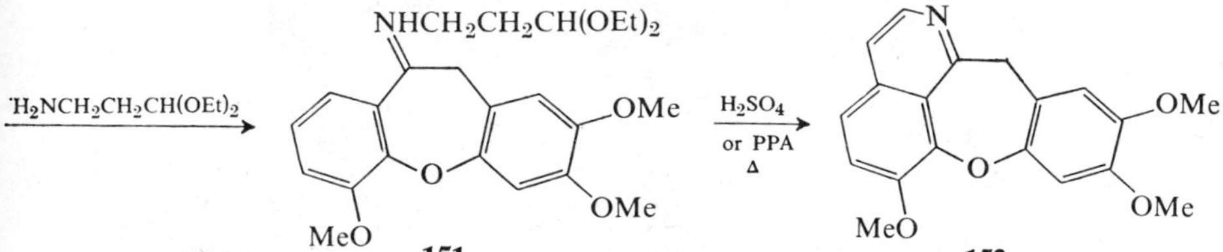

Upon reaction with phenylhydrazine and subsequent cyclization according to the Fischer indole synthesis, **59** gave another condensed, nitrogen-containing derivative, 6,7,10-trimethoxy-14*H*-dibenz[2,3:6,7]oxepino[4,5-*b*]-indole (**153**) (184).

An interesting synthesis reported by Kametani and Ogasawara (185) in 1964 involved vigorous alkaline hydrolysis of azlactone **154** with barium hydroxide. The product, 9-(2-aminoethyl)-10,11-dihydro-2,3,6-trimethoxy-dibenz[*b,f*]oxepin-10-ol (**155**), was converted to 10,11-dihydro-2,3,6-trimethoxy-9-vinyldibenz[*b,f*]oxepin-10-ol (**156**) by exhaustive alkylation and Hofmann degradation. Oxidation of **156** yielded acid **157**.

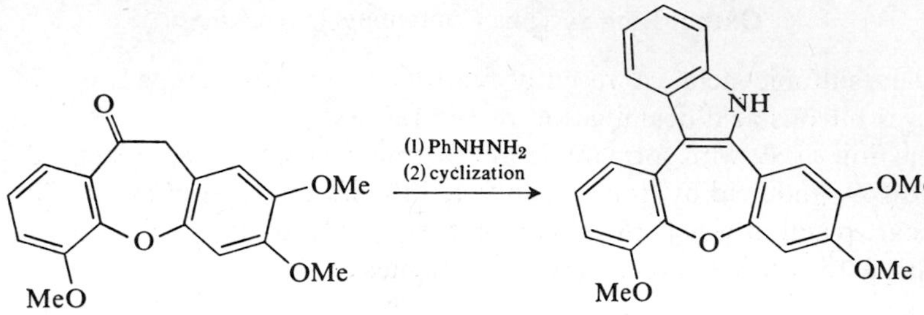

(1) $PhNHNH_2$
(2) cyclization

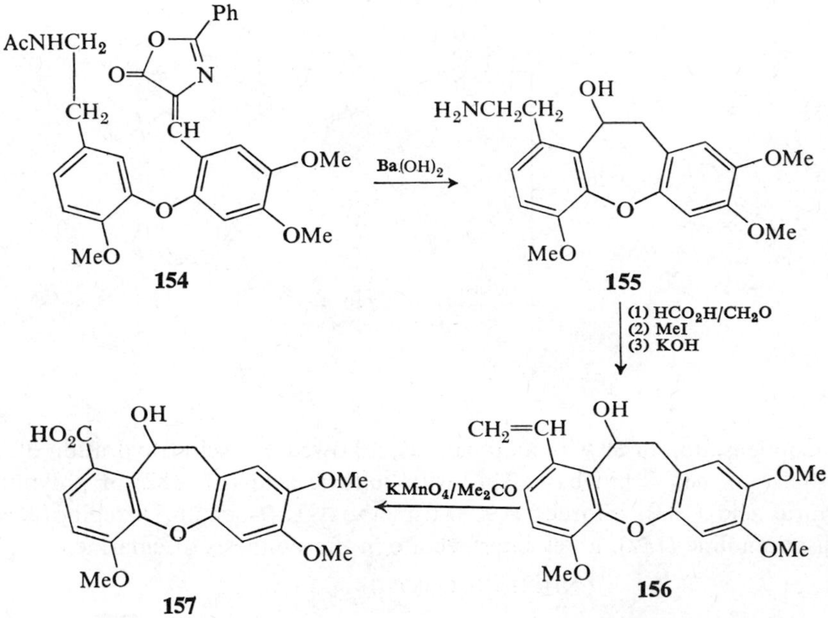

153

154

$Ba(OH)_2$

155

(1) HCO_2H/CH_2O
(2) MeI
(3) KOH

156

$KMnO_4/Me_2CO$

157

65

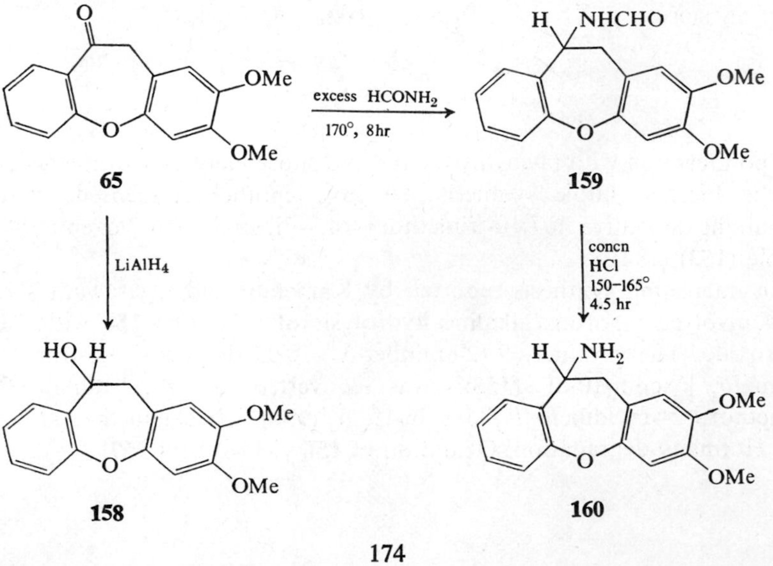

excess $HCONH_2$
170°, 8 hr

159

concn HCl
150–165°
4.5 hr

160

$LiAlH_4$

158

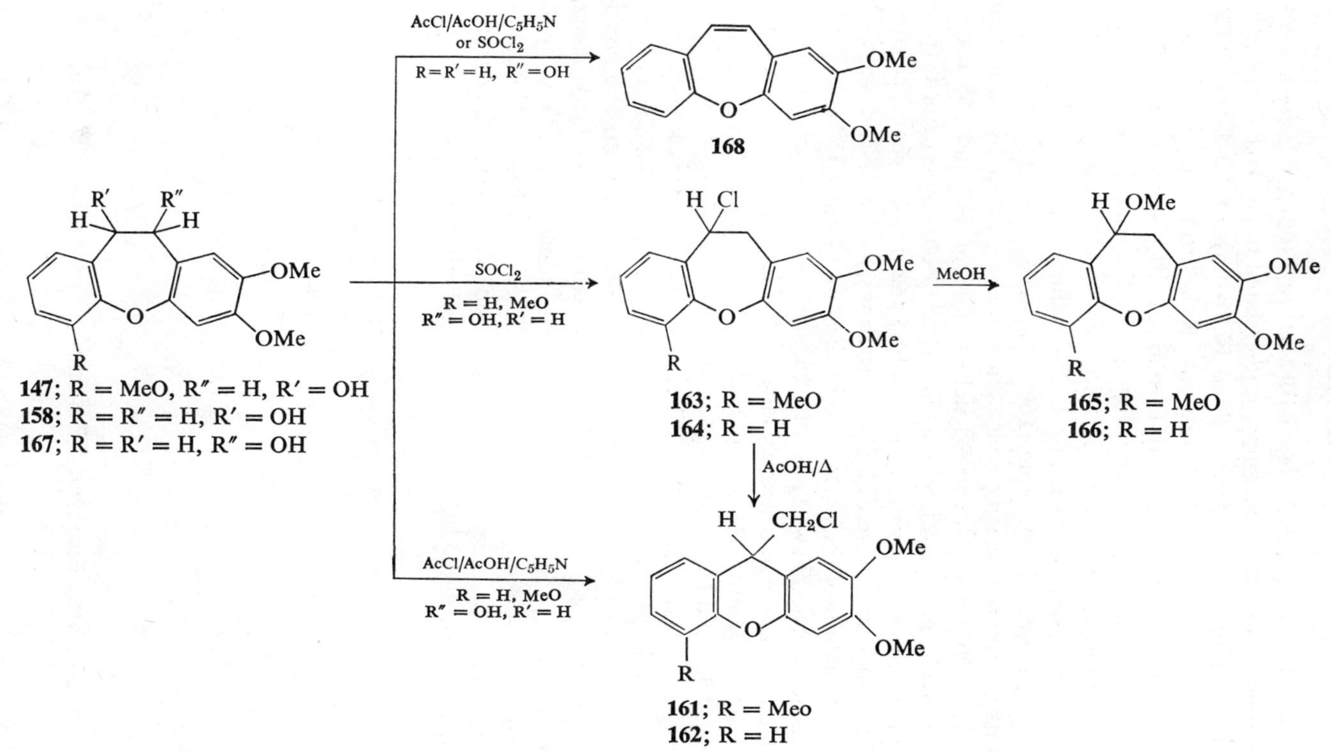

147; R = MeO, R″ = H, R′ = OH
158; R = R″ = H, R′ = OH
167; R = R′ = H, R″ = OH
AcCl/AcOH/C_5H_5N
or $SOCl_2$
R = R′ = H, R″ = OH
168
$SOCl_2$
R = H, MeO
R″ = OH, R′ = H
163; R = MeO
164; R = H
MeOH
165; R = MeO
166; R = H
AcOH/Δ
CH_2Cl
AcCl/AcOH/C_5H_5N
R = H, MeO
R″ = OH, R′ = H
161; R = Meo
162; R = H

2,3-Dimethoxydibenz[*b*,*f*]oxepin-10(11*H*)-one (**65**) has likewise been used for the preparation of model compounds related to cularine. Thus reduction with lithium aluminum hydride yielded 10,11-dihydro-2,3-dimethoxydibenz-[*b*,*f*]oxepin-10-ol (**158**), while reaction with formamide in an autoclave (Leuckart reaction) furnished 10-formamido-10,11-dihydro-2,3-dimethoxy-dibenz[*b*,*f*]oxepin (**159**). Interesting polycyclic oxepin derivatives were obtained as by-products in this reaction unless a large excess of formamide was used. Hydrolysis of **159** led to the formation of 10-amino-10,11-dihydro-2,3-dimethoxydibenz[*b*,*f*]oxepin (**160**).

A rather remarkable ring contraction was observed by Kametani and co-workers (201) upon attempted dehydration of **147** and **158** with acetyl chloride–acetic acid–pyridine reagent. The products obtained from this reaction unexpectedly proved to be **161** and **162**, respectively. The structures of these compounds were established rigorously by nmr and mass spectral analysis. Reaction of the acetate derivatives of **147** and **158** with hot acetyl chloride–acetic acid likewise afforded **161** and **162**. Treatment of **147** and **158** with thionyl chloride in benzene yielded chlorides **163** and **164**, which underwent rearrangement to **161** and **162** upon being heated in acetic acid. Methanolysis of **163** and **164** afforded 10,11-dihydro-2,3,6,10-tetramethoxydibenz-[*b*,*f*]oxepin (**165**) and 10,11-dihydro-2,3,6-trimethoxydibenz[*b*,*f*]oxepin (**166**), respectively. The disposition of methoxy groups appears to play a determinant role with respect to the course of rearrangement. This conclusion was prompted by the finding that 10,11-dihydro-7,8-dimethoxydibenz[*b*,*f*]-oxepin-10-ol (**167**), derived from the corresponding ketone, **66A**, by sodium borohydride reduction, gave only 2,3-dimethoxydibenz[*b*,*f*]oxepin (**168**) upon treatment with acetyl chloride–acetic acid–pyridine reagent or attempted chlorination with thionyl chloride in benzene–ether solution.

6. Dibenz[*c*,*e*]oxepins

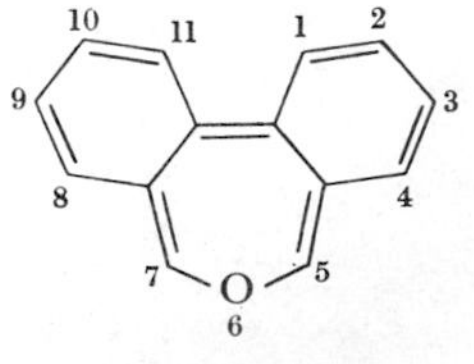

R.I. 3701

a. *Dibenz*[c,e]*oxepin-5,7-diones* (*Diphenic Anhydrides*)

(1) Synthesis. The earliest mention of dibenz[*c*,*e*]oxepin-5,7-dione (**169**) (or diphenic anhydride, as it is known more commonly) was made in 1877

by the German chemist Anschütz, who observed the formation of this seven-membered anhydride when diphenic acid was treated with acetyl chloride (6) or acetic anhydride (7). Shortly after this work, Graebe and Aubin (69) likewise reported the facile ring closure of diphenic acid under the influence of hot acetic anhydride. Numerous subsequent reports have confirmed the observations of these early workers, and have extended them to substituted diphenic acids. Of all the dehydrating agents which have been investigated over the years for the preparation of diphenic anhydrides from diphenic acids, acetic anhydride has endured as the reagent of choice. Graebe and Aubin (69) reported that the formation of **169** from diphenic acid was complete after 1 hr at 120°, and this has been corroborated in a modern patent (167). Shorter reaction times can be used with equal success if a suitable increase in temperature is made. Thus Hunn (79) has reported complete conversion of diphenic acid into **169** upon heating in acetic anhydride at 170° for 15 min. On occasion the addition of equal parts of acetic acid to the acetic anhydride has also been found satisfactory (147, 165). Several other dehydrating agents have been investigated, but none of these has proved particularly effective. Phosphorus pentachloride, for example, was found by Graebe and Mensching (70) and later by Oyster and Adkins (133) to give a mixture of **169** and a second product which was formulated as fluorenone-4-carboxylic acid (**170**). Phosphorus oxychloride likewise afforded a mixture of **169** and **170** (133). According to Graebe and Aubin (69), phosphorus

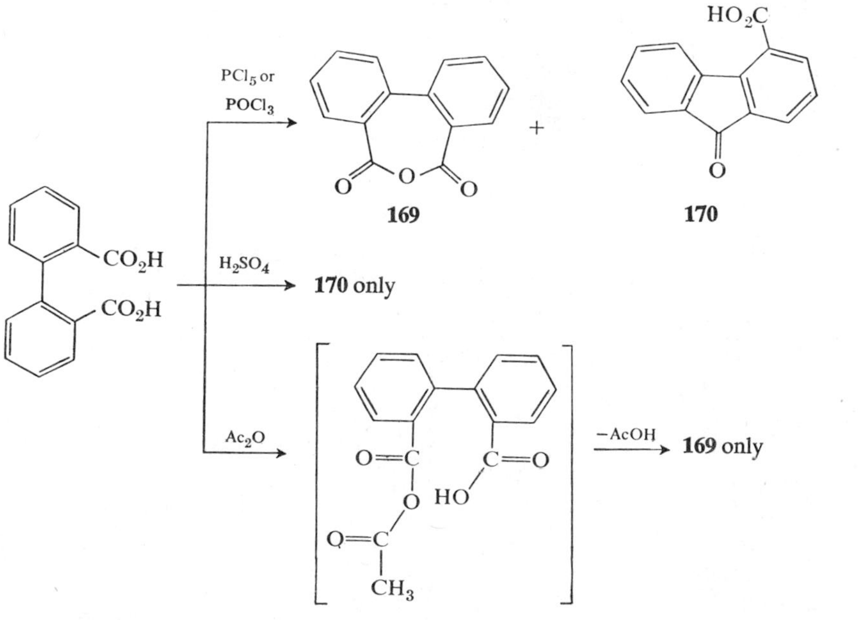

fructose
C_5H_5N
12 hr reflux

$C_5H_5N{\cdot}HO_2C$

reflux

$+ \; C_5H_5N$

Ac_2O
reflux

169

$H^{\oplus}$

Free Acid

Ac_2O
reflux

Δ

trichloride, and also stannic chloride, give rise to **169** exclusively, but this work has never been confirmed. Sulfuric acid, which was initially believed to produce **169** (70), was later found to give, in fact, only **170** (71, 165). Thionyl chloride was reported by Bell and Robinson (23) to convert diphenic acid mainly into its acid chloride with only a low yield of **169** being produced. It has also been reported recently (101, 170) that **169** can be obtained from diphenic acid in unspecified yield upon treatment with dry chlorine gas at 140–220°. Finally it may be mentioned that merely heating diphenic acid above its melting point for 30 min has been claimed to give no anhydride (165), and that treatment of **170** with a mixture of acetic anhydride and acetic acid for 6 hr at 140–150° reportedly leads to the slow formation of **169** in low yield (165). It is reasonable to suppose that the seemingly unique ability of acetic anhydride to cause the formation of the seven-membered ring in **169** is not the result of dehydrating action alone. Rather it seems probable that a mixed anhydride intermediate is involved as indicated.

Although they are of no particular synthetic usefulness, several reactions from precursors other than diphenic acid have been reported to give rise to **169**. A rather novel example, described in 1940 by Diels and Thiele (54), involves the reaction of 9,10-phenanthraquinone with fructose in the presence of pyridine. The initial product is an ester derived from diphenic acid and 9,10-dihydroxyphenanthrene, the latter arising from the quinone under the reducing influence of fructose. The observations of Diels and Thiele are summarized in the scheme shown.

Another example of the conversion of 9,10-phenanthraquinone into **169** was provided in 1961 by Ramirez and co-workers (140). Reaction with trimethyl phosphite in methylene chloride at room temperature, followed by ozonolysis at −70°, gave upon work-up a 50% yield of diphenoyl peroxide (violently explosive when heated to *ca.* 70° or under impact, but relatively stable at low temperatures). The peroxide, **171**, was transformed into **169** in better than 80% yield by oxygen abstraction with triphenyl phosphine or trimethyl phosphite. A 1:1 adduct between the quinone and trimethyl phosphite was isolated. This compound, which was formulated by Ramirez and co-workers as pentavalent phosphorus ester **172**, can be ozonized in a separate step to give the peroxide.

Another preparation of diphenic anhydride from a starting material other than diphenic acid is the catalytic oxidation of phenanthrene, which was described in 1955 by Brooks (40). The reaction is carried out at 300° over a vanadium pentoxide catalyst. Other products formed besides diphenic anhydride are 2-carboxy-2′-hydroxybiphenyl lactone (**173**) and 1,2-naphthalenedicarboxylic acid anhydride (**174**).

Reduced diphenic anhydrides made their first appearance in the literature in 1933, when Vocke (171) investigated the catalytic hydrogenation of

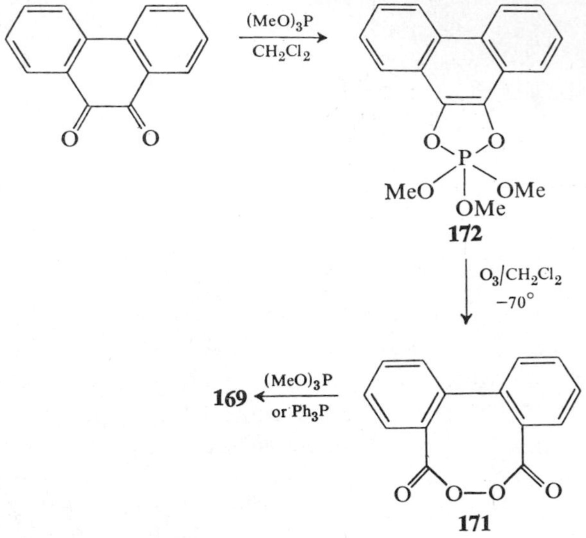

(MeO)$_3$P, CH$_2$Cl$_2$

MeO OMe OMe

172

O$_3$/CH$_2$Cl$_2$ −70°

169 ← (MeO)$_3$P or Ph$_3$P

171

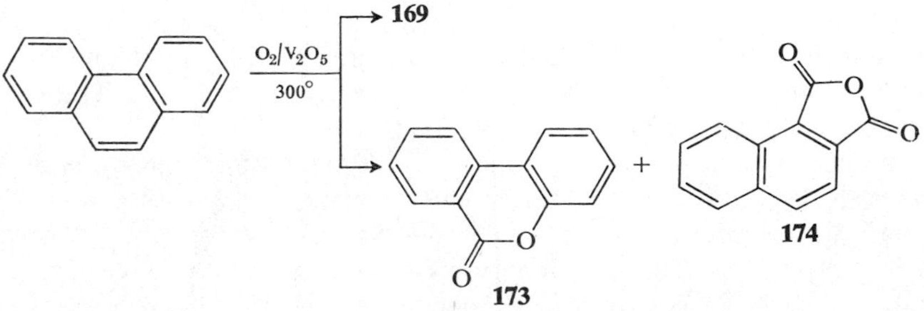

O$_2$/V$_2$O$_5$ 300° → **169**

173 + **174**

diphenic acid and reported the isolation of two hexahydrodiphenic acids melting at 242° and 220°, respectively. Separation of these acids was achieved by fractional crystallization of their barium salts, followed by recrystallization from acetic acid. Treatment of the isomer melting at 220° with hot acetic anhydride afforded a hexahydrodiphenic anhydride melting at 120°. Reexamination of this work in 1942 by Linstead and Davis (106) led to the formulation of the diacid melting at 242° as the *cis* isomer and of the diacid melting at 220° as the *trans* isomer. Heating of the *cis* diacid for 5 hr in refluxing acetic anhydride afforded the *cis* anhydride, **175**, which did not solidify. This compound had, for some reason, escaped identification in Vocke's laboratory. Heating the *trans* diacid for 4 hr in refluxing acetic anhydride produced the *trans* anhydride, **176**, which melted at 115–116° and was apparently the same substance as the material melting at 120° reported earlier by Vocke.

CO_2H CO_2H — H_2/Pt → CO_2H CO_2H, mp 242° — Ac_2O, reflux, 5hr → **175**; oil

→ CO_2H CO_2H, mp 220° — Ac_2O, reflux, 4hr → **176**; mp 115–116°

By carrying out catalytic hydrogenation until both aromatic rings are reduced, and by then performing a series of steps involving stereochemical inversions at various asymmetric centers, it is possible to isolate and identify six separate and distinct dodecahydrodiphenic acids. This difficult work was begun by Vocke (171), then pursued by Linstead and Walpole (108) and by Marvel and White (114), and finally brought to an elegant conclusion in 1942 by Linstead and Doering (107). Only the findings drawn from the definitive paper of Linstead and Doering need be presented here. Six dodecahydrodiphenic anhydrides were prepared and characterized. All of them were obtained by treatment of the corresponding diacids with boiling acetic anhydride. The reported heating times varied from 1 to 20 hr, depending upon the particular diacid being cyclized. In some cases the anhydrides were characterized chemically be reconversion to the starting diacid or by methanolysis to the corresponding monomethyl ester. Stereochemical formulas for the six anhydrides, **177–182**, and for the diacids from which they were derived are shown.

Anti Series:

mp 247° — Ac_2O, reflux →

177; *trans-trans*, mp 242°

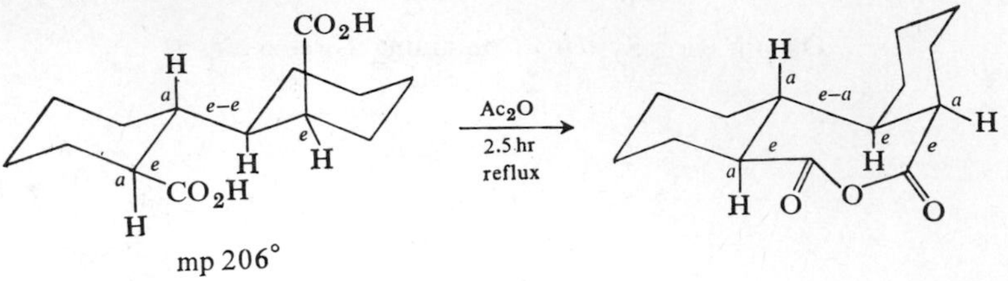

CO_2H H a e–e e H H e a CO_2H H

mp 206°

Ac_2O 2.5 hr reflux

H a e–a a H e H e a H O O O

178; *trans-cis*, mp 91.5–93°

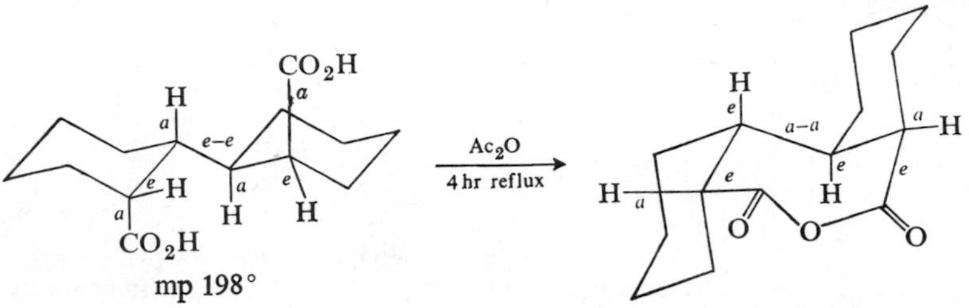

CO_2H a H a e–e e H a e H H a CO_2H

mp 198°

Ac_2O 4 hr reflux

H e a–a a H e H e H a O O O

179; *cis-cis*, mp 95–95.6°(99–100°)

Syn Series:

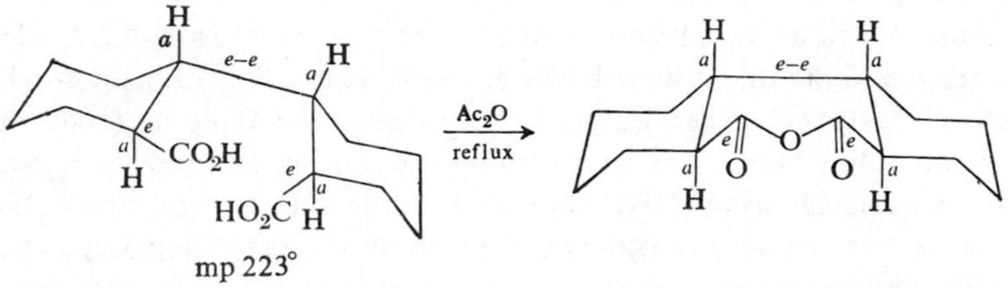

H a e–e H a e a CO_2H H e a HO_2C H

mp 223°

Ac_2O reflux

H H a e–e a e O e a O O a H H

180; *trans-trans*, mp 105–106.5°

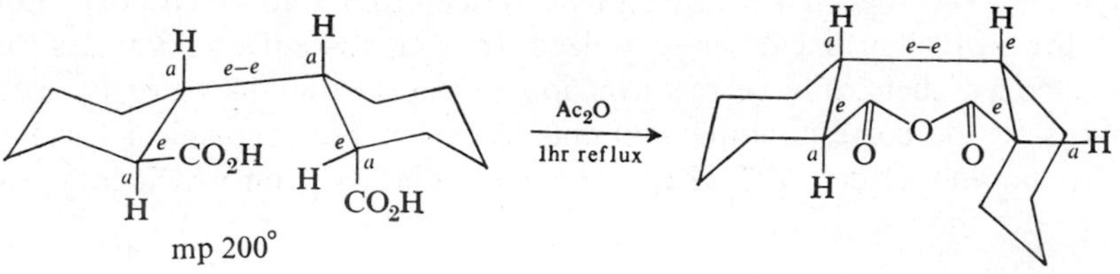

H a e–e H a e CO_2H a H e a H CO_2H

mp 200°

Ac_2O 1hr reflux

H H a e–e e e O e a O O a H H

181; *trans-cis*, mp 104–104.5°

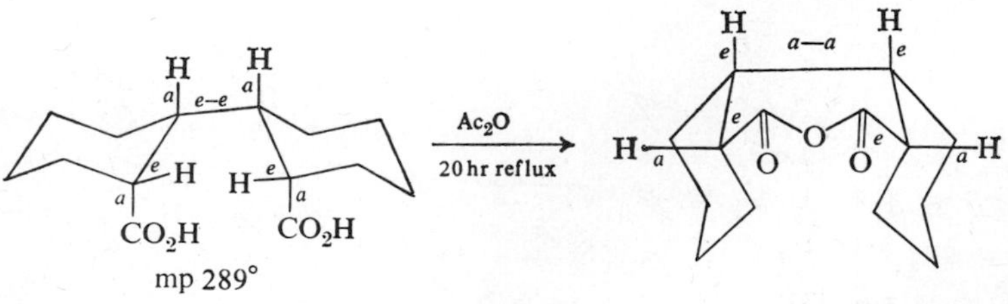

H H a e–e a e H a CO_2H H e a CO_2H

mp 289°

Ac_2O 20 hr reflux

H H e a—a e H e O O e O a a H

182; *cis-cis*, mp 146–147°

Alkyl-substituted diphenic anhydrides have been reported principally in connection with the chemistry of the hydrocarbon retene, which is isolated from the tar of certain resinous woods and fossil resins. Specifically, retenequinone yields 9-isopropyl-4-methyldiphenic anhydride (**183**) upon oxidation with hydrogen peroxide and treatment of the resulting diacid with acetyl chloride or acetic anhydride (2, 64, 92, 111). Under conditions favoring aromatic acylation the diacid appears to give, instead of **183**, a mixture of 1-acetyl-9-isopropyl-4-methyldiphenic anhydride (**184**) and 3-acetyl-9-isopropyl-4-methyldiphenic anhydride (**185**) (126, 127). Reduction of the 1-acetyl group in **184** has also been carried out with formation of 1-ethyl-9-isopropyl-4-methyldiphenic anhydride (**186**) (126).

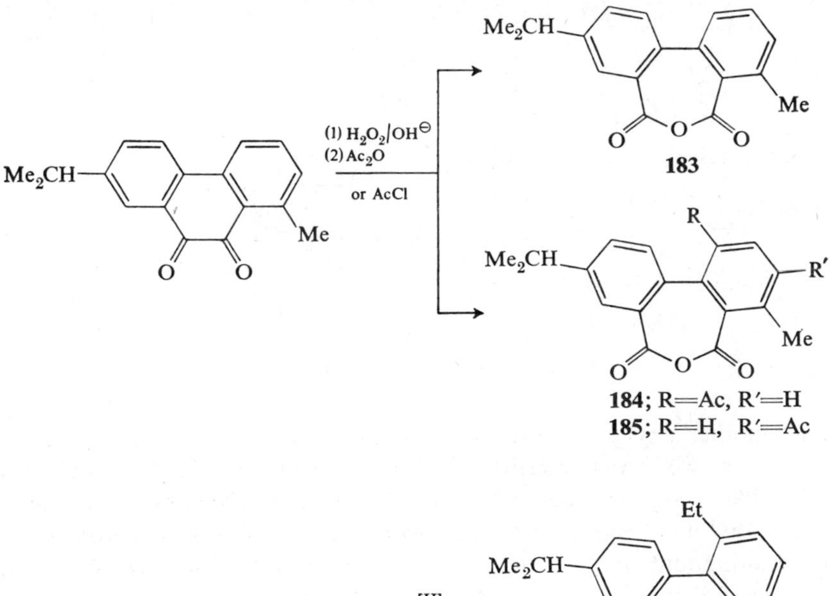

(1) $H_2O_2/OH^{\ominus}$ (2) Ac_2O or AcCl

183

184; R=Ac, R′=H
185; R=H, R′=Ac

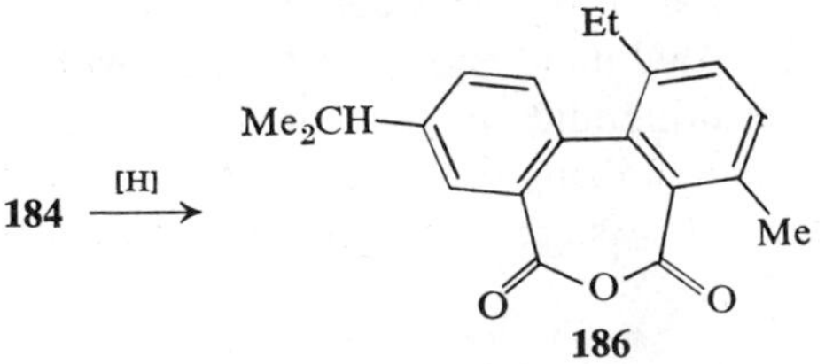

184 $\xrightarrow{[H]}$ **186**

The thermal decarboxylation of 2,3-dimethylphenyl-2′,4′,5′-tricarboxylic acid with copper powder in refluxing quinoline was reported in 1937 by Ghigi (68). In addition to the desired fully decarboxylated hydrocarbon, there was isolated an anhydride which lost only a single molecule of carbon dioxide. A suggested structure for this product was 2,3-dimethyldiphenic anhydride (**187**), but the alternative possibility of 3,4-dimethyl-6-phenylphthalic anhydride (**188**) was not ruled out. In this connection it is of interest that Kenner (94) obtained no seven-membered anhydride upon heating biphenyl-2,2′,3,3′-tetracarboxylic acid but instead isolated the double phthalic anhydride derivative **189**.

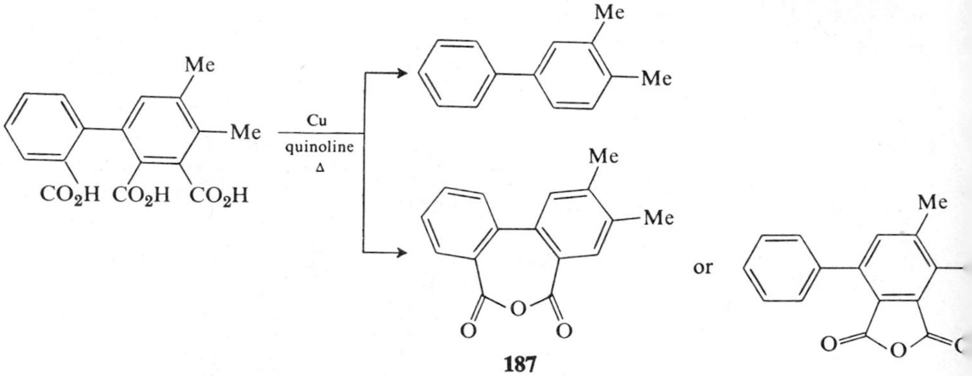

Cu quinoline Δ

187

or

188

Δ

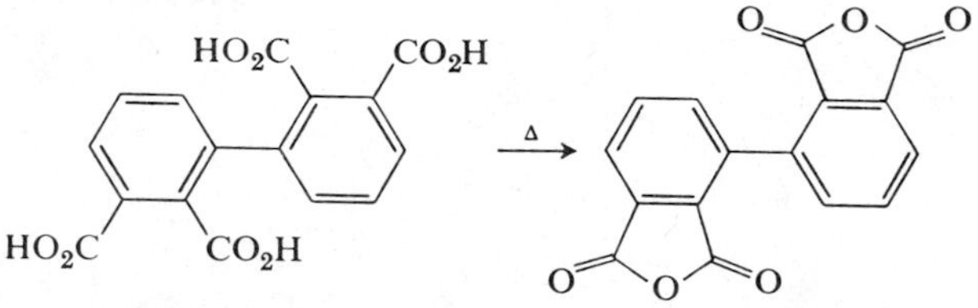

189

Still another alkyl-substituted diphenic anhydride, this one a partly reduced derivative as well, was described by Fieser and Dunn (62). The Diels-Alder adduct **190**, which was derived from 3,7-dimethyl-1,2-naphthoquinone and 2,3-dimethylbutadiene, was treated with alkaline hydrogen peroxide, and the resulting dicarboxylic acid was converted into 1,4-dihydro-2,3,4a,9-tetramethyldibenz[*c*,*e*]oxepin-5,7-dione (**191**) in 53% yield upon simple heating. The stereochemistry of the system was not specified.

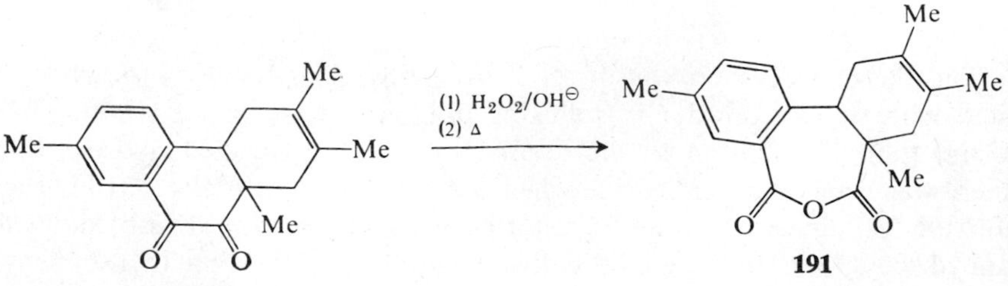

(1) $H_2O_2/OH^{\ominus}$
(2) Δ

190

191

Halogen-substituted diphenic anhydrides which have been reported in the literature to date include the 2,10-dichloro (214), 3,9-dichloro (81), 4,8-dichloro (80), 2,10-dibromo (49), 3,9-dibromo (163), and 2,10-diiodo (203) derivatives **192–197**. Compounds **192**, **194**, and **196** were made from the corresponding diphenic acids by treatment with acetic anhydride under reflux for 15–30 min. Ring closure to **196** was reportedly achieved also by simple heating of the diacid in the solid state. For the synthesis of **195**, the substituted diphenic acid was heated at 145° for 11 hr in a mixture of acetic anhydride and acetic acid.

Nitro-substituted diphenic anhydrides can likewise be obtained from the corresponding diphenic acid derivatives by heating in acetic anhydride. The 2-nitro (24, 59), 3-nitro (23, 122), 2,10-dichloro-3,9-dinitro (214), 2,10-dinitro (138), and 3,9-dinitro (23) compounds, **198–202**, have all been prepared by this method. Compound **199** has been prepared also by addition of thionyl chloride to the substituted diphenic acid (23). The order of addition has been shown to be important in the latter example, the bisacid chloride being favored greatly if the diphenic acid is added to the thionyl chloride instead of vice versa. It is rather curious that **202** allegedly forms a molecular complex with acetic anhydride (23), a peculiarity which might account for three early reports (122, 142, 166) of failures to prepare this compound. Also of interest is the claim (23) that 1-nitrodiphenic anhydride could not be prepared by the usual reaction of the appropriate diphenic acid derivative with acetic anhydride under reflux.

Diphenic anhydrides bearing amide functions have appeared in the literature on two occasions. In one instance (3), treatment of 2,2′-diaminodiphenic acid dihydrochloride and 4,4′-diaminodiphenic acid dihydrochloride

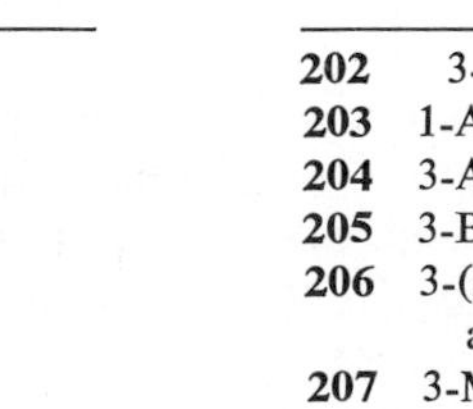

	R	R′
192	3-Cl	9-Cl
193	2-Cl	10-Cl
194	4-Cl	8-Cl
195	2-Br	10-Br
196	3-Br	9-Br
197	2-I	10-I
198	2-NO_2	H
199	3-NO_2	H
200	2-Cl-3-NO_2	10-Cl-9-NO_2
201	2-NO_2	10-NO_2
202	3-NO_2	9-NO_2
203	1-AcNH	11-AcNH
204	3-AcNH	9-AcNH
205	3-BzNH	9-BzNH
206	3-(*m*-nitrobenzamido)	9-(*m*-nitrobenzamido)
207	3-MeO	9-MeO
208	3-MeO-4-Me	9-MeO-8-Me
209	1,2-$(MeO)_2$	10,11-$(MeO)_2$
210	3,4-$(MeO)_2$	8,9-$(MeO)_2$

with acetic anhydride led to simultaneous ring closure and *N*-acetylation with formation of 1,11-diacetamidodiphenic anhydride (**203**) and 3,9-diacetamidodiphenic anhydride (**204**), respectively. In another case (138), 3,3′-diaminodiphenic acid was benzoylated and *m*-nitrobenzoylated, and the resulting diamides were subjected to the action of boiling acetic anhydride to effect ring closure to anhydrides **205** and **206**, respectively.

Methoxy-substituted diphenic anhydrides which have been reported include the 3,9-dimethoxy- and 3,9-dimethoxy-4,8-dimethyl compounds, **207** and **208** (35), and also the 1,2,10,11-tetramethoxy and 3,4,8,9-tetramethoxy compounds, **209** and **210** (14), all of which were produced from the corresponding diphenic acids under the influence of acetic anhydride.

Anhydride **211** has been obtained from piloquinone, a naturally-occurring red phenanthraquinone pigment found in certain *Streptomyces* species, on alkaline hydrogen peroxide oxidation followed by treatment with acetic anhydride (66,135).

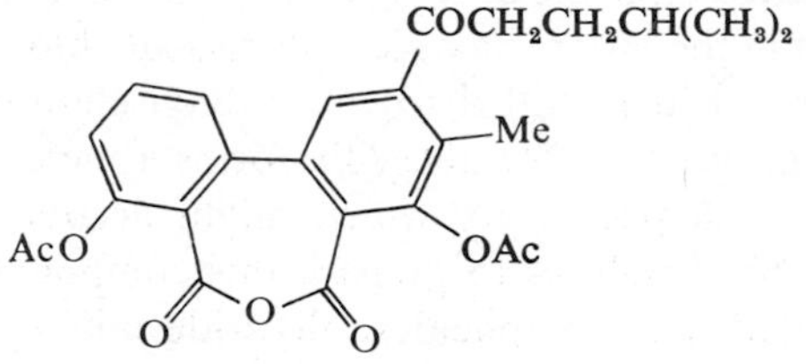

$COCH_2CH_2CH(CH_3)_2$

Me

AcO OAc

O O O

211

(2) Reactions. It was reported as early as the 1880s by Graebe and coworkers (69, 70) that diphenic anhydride itself undergoes a thermally-induced decarboxylation reaction with formation of fluorenone. This observation led to revision of an earlier claim (71) that diphenic anhydride can be produced by treatment of diphenic acid with concentrated sulfuric acid at 100–120°. The product of this reaction was now formulated as fluorenone-4-carboxylic acid (69). In 1923 Underwood and Kochmann (165) investigated the action of Lewis acids upon diphenic anhydride. Treatment with stannic chloride at 120–130°, or with zinc chloride at 225°, resulted in the formation of fluorenone-4-carboxylic acid as the sole product, no condensation with phenol occurring unless higher temperatures and longer reaction times were employed. In a later study of the condensation of diphenic anhydride with phenol under various conditions, Bell and Briggs (21) observed that, in the presence of aluminum chloride in tetrachloroethylene, the anhydride afforded a mixture of fluorenone-4-carboxylic acid and diphenic acid. Extending this work, Bell and Briggs (22) then reported that attempted condensation of diphenic anhydride with *o*- or *p*-aminophenol, or with salicylic acid, in the presence of stannic chloride led to fluorenone-4-carboxylic acid exclusively, no reaction being observed with these particular phenols.

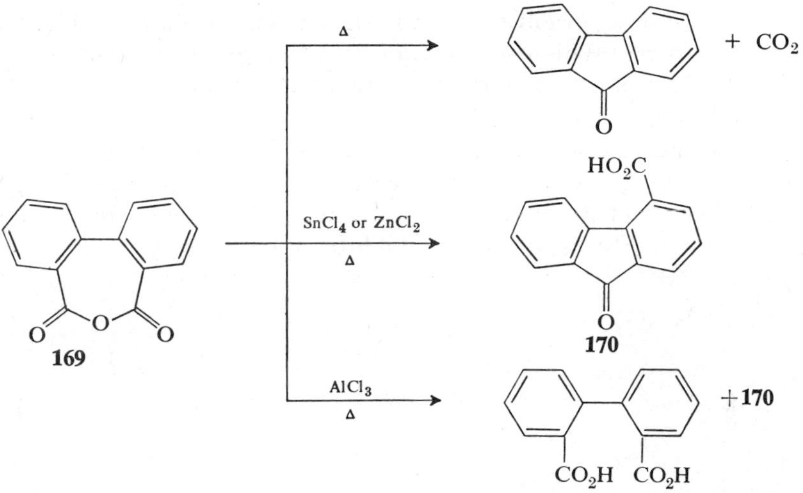

Examples of thermal ring contractions involving substituted diphenic anhydrides can also be found in the literature. Thus Huntress and coworkers (80, 81) reported in 1933 that heating 2,10-, 3,9-, and 4,8-dichlorodiphenic anhydrides, at temperatures ranging from 400° for the first two to 310° for the last, gave 3,6-, 2,7-, and 1,8-dichlorofluorenone, respectively. These findings provided an explanation for the fact that thermal cyclization of 3,3′-dichlorodiphenic acid afforded a mixture of 1,8- and 1,6-dichlorofluorenone, **212** and **213**. The 1,8-dichloro compound was assumed to originate from a rearrangement of the initially formed anhydride, while the 1,6-dichloro isomer had to be produced via decarboxylation of 3,8-dichlorofluorenone-4-carboxylic as shown.

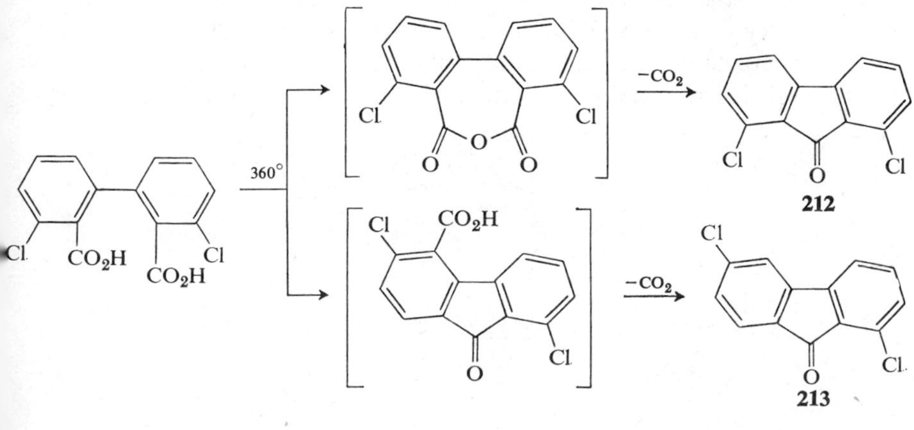

The facile thermal decarboxylation of diphenic anhydride has not been investigated in detail from a mechanistic standpoint. Quite possibly the geometry of the seven-membered ring in this compound plays a significant role in the process. With the exception of some dipole moment studies (105a, 105b, 128), however, there is a great lack of data in the literature concerning this important point.

Alkaline hydrolysis of diphenic anhydride itself readily affords diphenic acid (79). Reported examples of the regeneration of diphenic acids upon alkaline hydrolysis of substituted diphenic anhydrides include the reactions of the 9-isopropyl-4-methyl (64), 3,9-dibromo (163), and 3-nitro (122) derivatives shown.

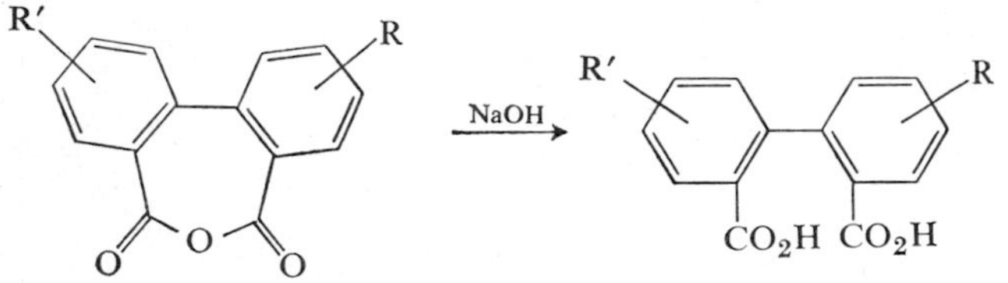

R′ R NaOH R′ R CO_2H CO_2H

O O O

	R	R′
169	H	H
183	4-Me	9-isopropyl
196	3-Br	9-Br
199	3-NO_2	H

Cleavage of the seven-membered ring in diphenic anhydride is likewise achieved readily by reaction with various alcohols. Under mild neutral conditions monoesters of diphenic acid are produced (69). At high temperatures in a sealed tube (69), or under the influence of sulfuric acid (147), diphenic anhydride can be converted directly into diesters of diphenic acid. Symmetrically disubstituted diphenic anhydrides that have been cleaved to monoesters of the corresponding diphenic acids include the 2,10-dibromo (49), 2,10-dibenzamido (138), 3,9-dimethoxy (134), and 3,9-diacetoxy (134) compounds. Single products are formed in these cases because of the structural symmetry of the molecules. In an unsymmetrical molecule like 9-isopropyl-4-methyldiphenic anhydride (**183**) (2), reaction with an alcohol should lead, in principle, to a mixture of isomeric monoesters as indicated.

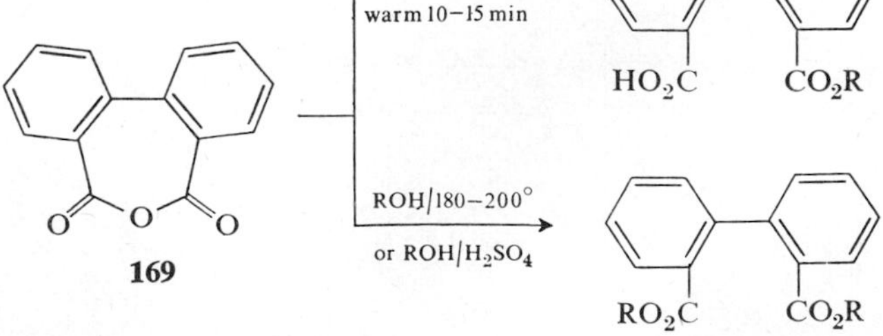

ROH warm 10–15 min HO_2C CO_2R

O O O **169**

ROH/180–200° or ROH/H_2SO_4 RO_2C CO_2R

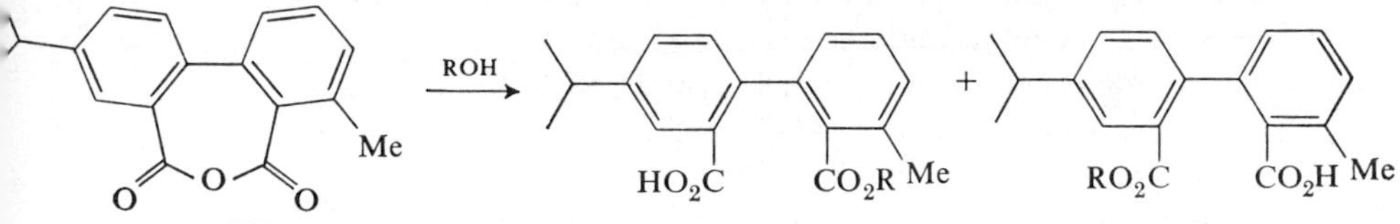

ROH

HO_2C CO_2R Me + RO_2C CO_2H Me

Me

183

A substantial number of reactions have been reported that involve nucleophilic attack by various types of amines upon diphenic anhydride and its substituted derivatives. The reaction of diphenic anhydride with ammonia itself, which gives diphenamic acid, was first described in 1888 by Graebe and Aubin (69) and has been the subject of a recent patent as well (167). Other amines which have been condensed with diphenic anhydride to give the corresponding monoamide derivatives include allylamine (56), aniline (22), *N*-methylaniline (22), phenylhydrazine (69), *o*-phenylenediamine (34), *p,p'*-diaminobiphenyl (146), and *N*-phenylsulfonylhydrazine (48). Urea and thiourea have also been condensed with diphenic anhydride (22). With urea, a mixture of diphenamic acid and diphenimide is reportedly obtained. With thiourea, on the other hand, instead of diphenamic acid, there is

R	R′
H	H
Ph	H
Ph	Me
PhNH	H
$PhNHSO_2$	H

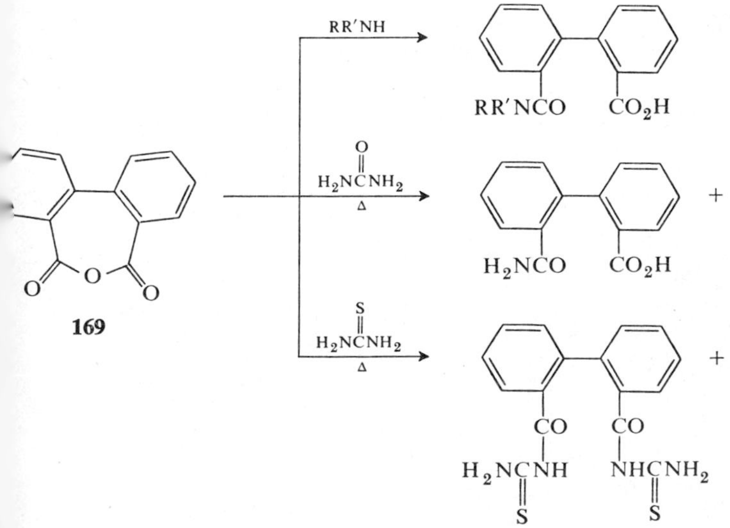

169

RR′NH → RR′NCO CO_2H

$H_2NC(=O)NH_2$, Δ → H_2NCO CO_2H + (diphenimide, N-H)

$H_2NC(=S)NH_2$, Δ → CO–NHC(=S)NH_2 / CO–NHC(=S)NH_2 +

isolated a product stated to contain two molecules of thiourea. Upon being heated, this material undergoes elimination of thiourea to give diphenimide. Semicarbazide has been reported to react with diphenic anhydride to give a compound with the empirical formula $C_{29}H_{17}N_3O_5$, but no structure has been assigned to this product.

The reaction of diphenic anhydride with aqueous allylamine (56) is of interest because the product first isolated proved to be *N*-allyldiphenimide (**214**) rather than the expected *N*-allyldiphenamic acid (**215**). The latter was formed extremely fast, however, upon subsequent treatment of **214** with alkali.

169 $\xrightarrow{CH_2{=}CHCH_2NH_2}$ **214** (N–$CH_2CH{=}CH_2$) $\xrightarrow{NaOH}$ **215** ($CH_2{=}CHCH_2NHCO$, CO_2H)

Reactions of substituted diphenic anhydrides with amines can likewise be found. These include the condensation of 2,10-dibromodiphenic anhydride with ammonia (49), of diphenic anhydride with ammonia or thiourea (22), and of 2,10-dibenzamidodiphenic anhydride with aniline (138). A single product is generated in each of these reactions since the starting materials are all symmetrical in structure. With unsymmetrical diphenic anhydrides, such as the 9-isopropyl-4-methyl derivative (64), the direction of ring opening is not unique. In a carefully analyzed case, 3-nitrodiphenic anhydride (**199**) was found to give both possible monoamide products, the yields being 32 and 23%, respectively (122). Structures **216** and **217** were assigned to the amides on the basis of their further transformation into known phenanthridone derivatives **218** and **219** upon alkaline hypobromite oxidation.

199 (NO_2) $\xrightarrow{NH_4OH}$ **216** (32%) (NO_2; CO_2H, $CONH_2$) $\xrightarrow{KOBr}$ **218** (NH)

199 $\xrightarrow{NH_4OH}$ **217** (23%) (NO_2; H_2NCO, CO_2H) $\xrightarrow{KOBr}$ **219** (NH)

Phenols have been known for many years to react with diphenic anhydrides. As early as 1880 Graebe and Mensching (70) observed that, upon being heated in the presence of zinc chloride, diphenic anhydride and phenol underwent condensation to give a phenolphthalein-like indicator known as "phenoldiphenein." The same compound was prepared almost half a century later by Dutt (57) and also by Underwood and Kochmann (165). The later workers used stannic chloride as the catalyst. Chemical reactions that were carried out in order to characterize this material included the formation of tetrabromo, tetraiodo, and dinitro derivatives (162), fusion with potassium hydroxide to give phenol and biphenyl-2-carboxylic acid (161), and alkylation with methyl iodide (161). With *o*-cresol in the presence of stannic chloride at 125–135°, diphenic anhydride gave an indicator known as "*o*-cresoldiphenein", which could be transformed into dibromo, diiodo, and dinitro derivatives (162). The same reaction has been carried out also with resorcinol in the presence of zinc chloride (33, 57) or stannic chloride (165), and with pyrogallol (57, 165) and *m*-(*N*,*N*-dimethylamino)phenol (57) as well. The product derived from resorcinol is called "resorcinoldiphenein". Ultraviolet spectral studies were carried out with phenoldiphenein, *o*-cresoldiphenein, and resorcinoldiphenein by Underwood and co-workers (164).

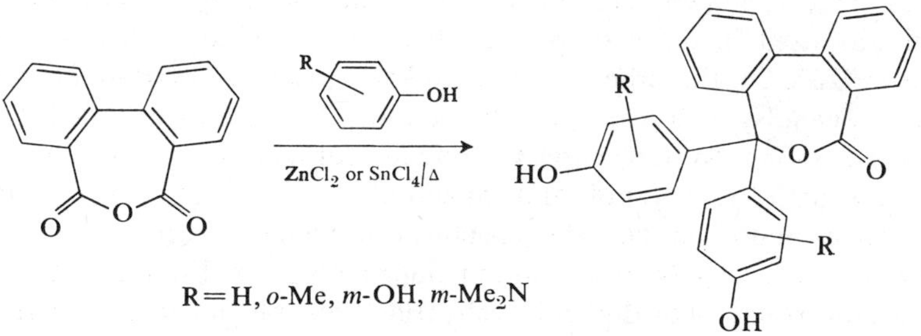

$ZnCl_2$ or $SnCl_4/\Delta$

R=H, *o*-Me, *m*-OH, *m*-Me_2N

Two substituted diphenic anhydrides have been reported to undergo condensation with resorcinol (3). Upon being heated for 25–30 hr at 110° in the presence of zinc chloride, 3,9-diacetamidodiphenic anhydride and resorcinol yielded the expected oxepin derivative. Treatment of 1,11-diacetamidodiphenic anhydride with resorcinol and zinc chloride at 110° likewise afforded an oxepin derivative. Above 110°, on the other hand, the reaction allegedly gave a fluorene derivative as shown at top of page 192.

The reaction of phenols with diphenic anhydride was also investigated in great detail by Bell and Briggs (21). Upon being heated for only 5 hr at 125° in the presence of phenol the anhydride was converted exclusively into the monophenyl ester, **220**, of diphenic acid, as might be expected. In the presence of added stannic chloride, however, there was obtained a mixture of the diphenyl ester, **221**, of diphenic acid and the Fries rearrangement product, **222**, as shown. The formation of **222** was suppressed if *p*-chlorophenol was used in place of phenol, the sole product in this case being monoester **223**. On the other hand, only Fries products were isolated when alkylated phenols, such as *o*-, *m*-, or *p*-cresol, 2,4-, 2,5-, or 3,4-xylenol, or thymol, were used. In these cases the initially formed esters of diphenic acid apparently underwent rapid and quantitative rearrangement to various monoketones and diketones, the structures of which were not always elucidated fully. Interestingly, if aluminum chloride was used in place of stannic chloride, the reaction of diphenic anhydride with phenol in the absence of solvent afforded only the nonrearranged monophenyl ester, **220**. With tetrachloroethane added, the products formed were fluorenone-4-carboxylic acid (**170**) and diphenic acid. The findings of Bell and Briggs are of interest as an extension of the earlier work on phenoldiphenein and its analogs.

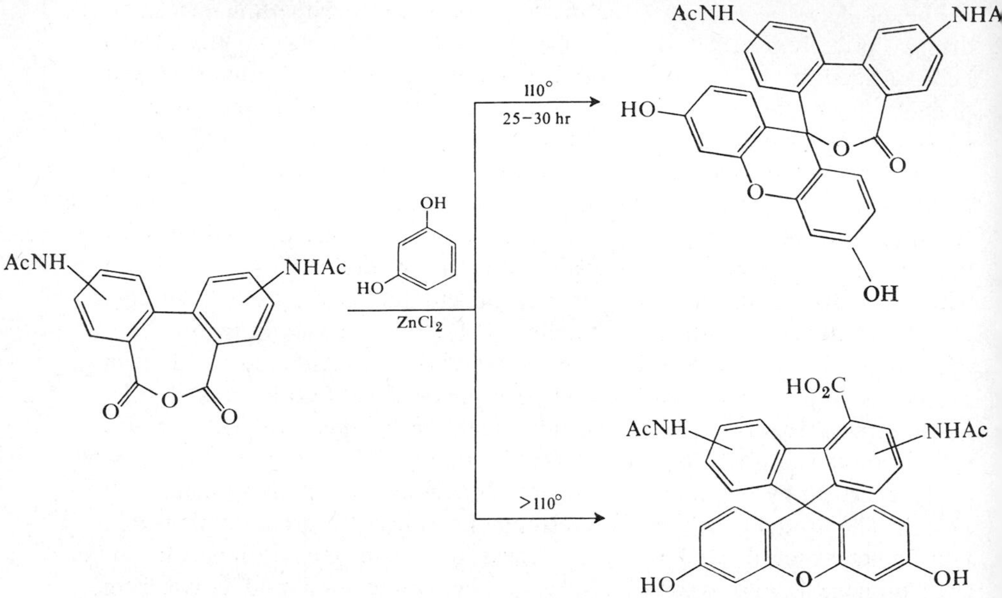

According to a recent report by Dabrowski and Okon (204), diphenic anhydride also reacts with hydroquinone and 1,4-dimethoxybenzene in the presence of zinc chloride at 200–250°. A black amorphous high-melting product of uncertain structure was isolated. Spectral evidence suggested this material to be a phthalein rather than a polyester.

Another group of compounds which has been found to react with diphenic anhydrides with rupture of the seven-membered ring is comprised of the organometallic reagents. The earliest example of this type of reaction was reported in 1930 by Sergeev (145), who noted that treatment of diphenic

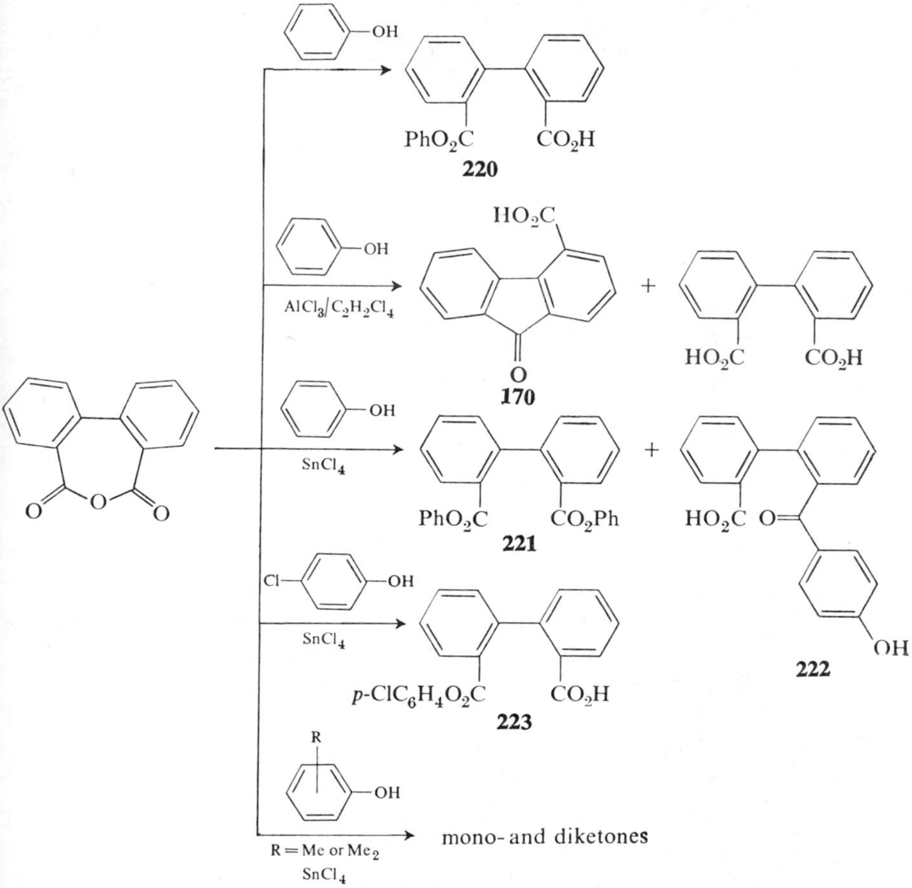

OH

PhO_2C CO_2H

220

HO_2C

OH

$AlCl_3/C_2H_2Cl_4$

+

HO_2C CO_2H

O

170

OH

$SnCl_4$

+

O O O

PhO_2C CO_2Ph

221

HO_2C O=

Cl OH

$SnCl_4$

OH

222

p-$ClC_6H_4O_2C$ CO_2H

223

R

OH

mono- and diketones

R = Me or Me_2

$SnCl_4$

anhydride itself with phenylmagnesium bromide in ether led to the formation of a mixture of hydroxy acid **224** and hydroxy ketone **225**. In a more recent study Nightingale and co-workers (125) converted diphenic anhydride into 1-benzoylbiphenyl-1′-carboxylic acid (**226**) by reaction with diphenyl cadmium in toluene.

An important reaction of diphenic anhydrides involves Friedel-Crafts acylation of aromatic hydrocarbons in the presence of a Lewis acid catalyst. The first published report of the use of diphenic anhydride in this connection appeared in 1938 in a paper by Bell and Briggs (21). Condensation was reported to occur smoothly with *o* and *m*-xylene, and also with mesitylene and naphthalene, with formation of monoketones as the sole products. With *p*-xylene, for some unknown reason, a mixture of monoketone and diketone was produced. The reactions were conducted in benzene solution at room temperature in the presence of aluminum chloride, no reaction being observed

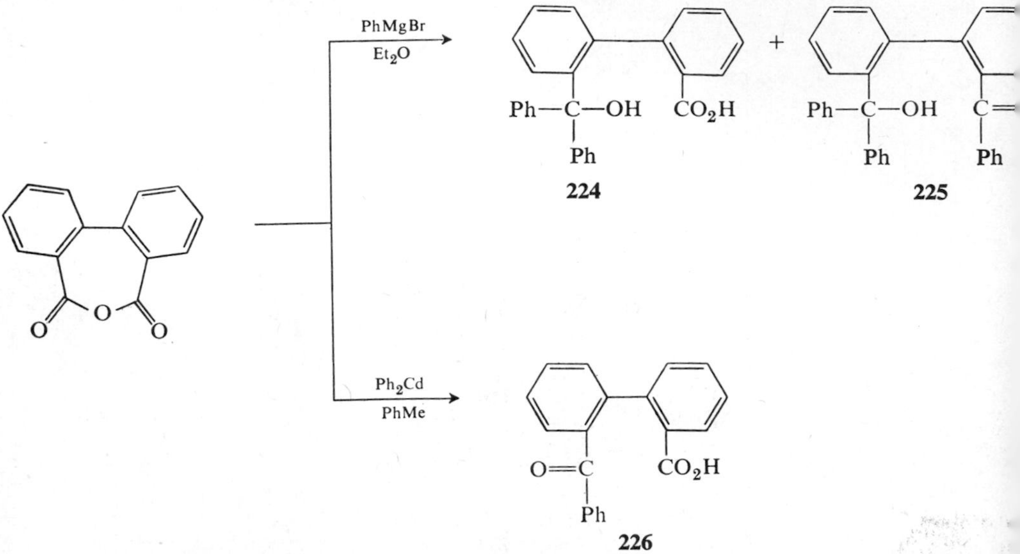

PhMgBr / Et_2O → 224 + 225

Ph_2Cd / PhMe → 226

Ph—C—OH (Ph) CO_2H **224**

Ph—C—OH (Ph) C= (Ph) **225**

O=C (Ph) CO_2H **226**

between diphenic anhydride and this solvent under these conditions. Toluene and ethylbenzene were likewise unreactive. With biphenyl, tar formation was predominant, some diphenic acid being recovered as well. In a later study Nightingale and co-workers (125) likewise used diphenic anhydride to acylate *o*-xylene, mesitylene, and naphthalene. With aluminum chloride as the catalyst and tetrachloroethane as the solvent instead of benzene, monoketones were obtained in yields of 62.5, 80%, and 18%, respectively.

The rate-enhancing influence of electron-releasing substituents in the compound being acylated was demonstrated with phenetole, anisole, and resorcinol dimethyl ether (21). These activated molecules afforded, respectively, monoketone **227**, a mixture of monoketone **228** and diketone **229**, and diketone **230**.

In a further extension of their initial investigation, Bell and Briggs (22) also acylated *m*-xylene and mesitylene with 3,9-dinitrodiphenic anhydride (**200**), aluminum chloride being used as the catalyst once again. As in the case of diphenic anhydride itself, the only products isolated were monoketones **231** and **232**, respectively.

An interesting reaction of diphenic anhydride which may be regarded as involving ring cleavage and recyclization was discovered in 1939 by Taurins (160). Upon treatment with quinaldine and zinc chloride at 170–180° for 2.5 hr, diphenic anhydride was found to give "quinodiphenone," to which structure **233** was assigned. This substance was characterized as a typical ketone by formation of phenylhydrazone and 2,4-dinitrophenylhydrazone derivatives, and was degraded to diphenic acid by oxidation with peracetic

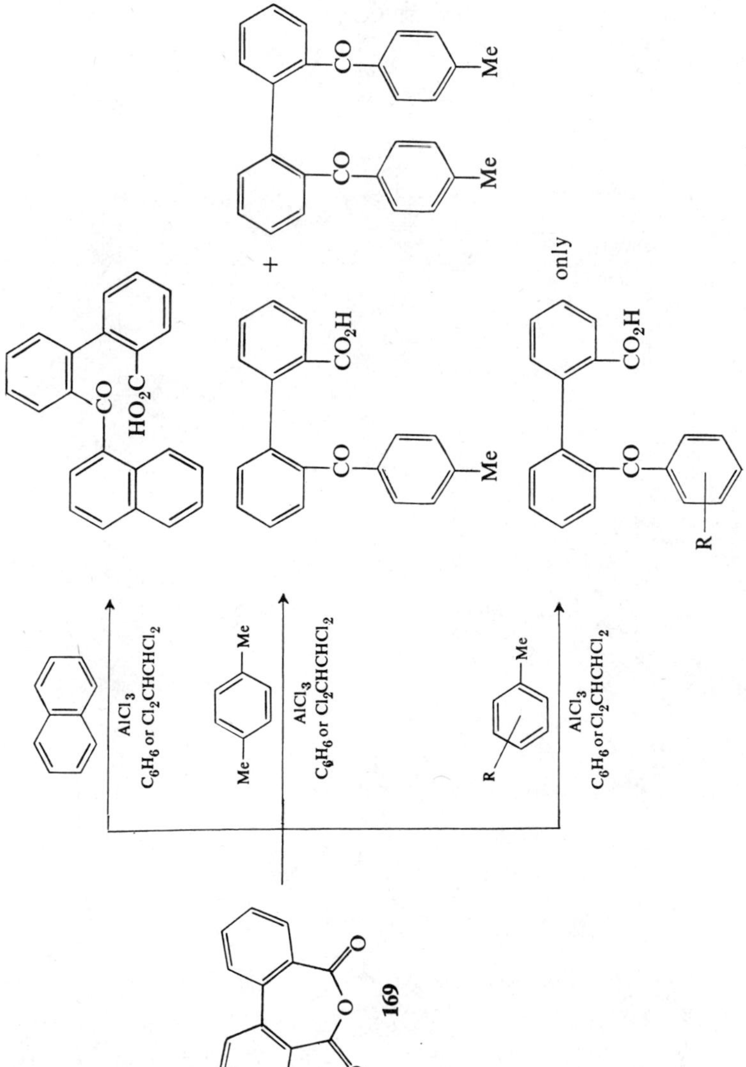

169

$AlCl_3$

C_6H_6 or $Cl_2CHCHCl_2$

CO

HO_2C

Me

CO_2H

+

Me

only

R

169

OEt
$AlCl_3$
CO_2H CO
OEt
227

OMe
$AlCl_3$
CO_2H CO
OMe
228
+
CO CO
MeO OMe
229

MeO OMe
$AlCl_3$
MeO CO CO OMe
MeO OMe
230

O_2N NO_2
O O O
200

R
Me Me
$AlCl_3$
R = H, Me

O_2N NO_2
R CO CO_2H
Me
Me

231; R = H
232; R = Me

acid. Other nitrogenous components used successfully in this condensation were 6-methylquinaldine, α-picoline, 2,4- and 2,6-lutidine, and 2,4,6-collidine, which gave compounds **234–238**, respectively, by analogy with the reaction of quinaldine. Because of the varying reactivity of the α-methyl group in the substituted pyridines and quinolines, it was necessary to employ temperatures as high as 240° and to extend the reaction up to 30 hr in certain cases. Shortly after the work of Taurins appeared, Bell and Briggs (22) likewise reported the reaction of diphenic anhydride with quinaldine.

169 → ($ZnCl_2$, Δ)

	R	R′	R″	% Yield
233	—(CH=CH)$_2$–		H	49
234	—CH=C(CH$_3$)—CH=CH—		H	11
235	H	H	H	25
236	Me	H	H	16
237	H	H	Me	16
238	Me	H	Me	11

Somewhat related to the above-mentioned reaction is the condensation of diethyl malonate with diphenic anhydride, which was reported in 1952 by Lucien and Taurins (110). In this case the anhydride effectively acylated an active methylene group instead of an activated methyl. Condensation with diethyl malonate in the presence of zinc chloride, followed by acid hydrolysis and decarboxylation, let to the formation of a product identified as dibenzo-[1,2;3,4]cycloheptadiene-5,7-dione (**239**). On the other hand, if the anhydride was allowed to react with diethyl sodiomalonate in the absence of solvent, subsequent hydrolysis and decarboxylation gave 2′-acetylbiphenyl-2-carboxylic acid (**240**). If benzene was used as the solvent for the diethyl sodiomalonate reaction, hydrolysis and decarboxylation were found to give a mixture of 2,2′-diacetylbiphenyl (**241**) and **239**. The intermediates presumed to take part in these transformations are depicted here.

Likewise of related interest is the reaction described in 1967 by Aleksiev and Milosev (206) wherein diphenic anhydride is heated with phenylacetic acid and sodium acetate to give a mixture of 6-phenyldibenzo[1,2:3,4]-cycloheptadiene-5,7-dione (**242**) and diphenic acid. When the reactants are heated at a lower temperature in the presence of acetic anhydride, with or without added triethylamine, the product is enol ester **243**.

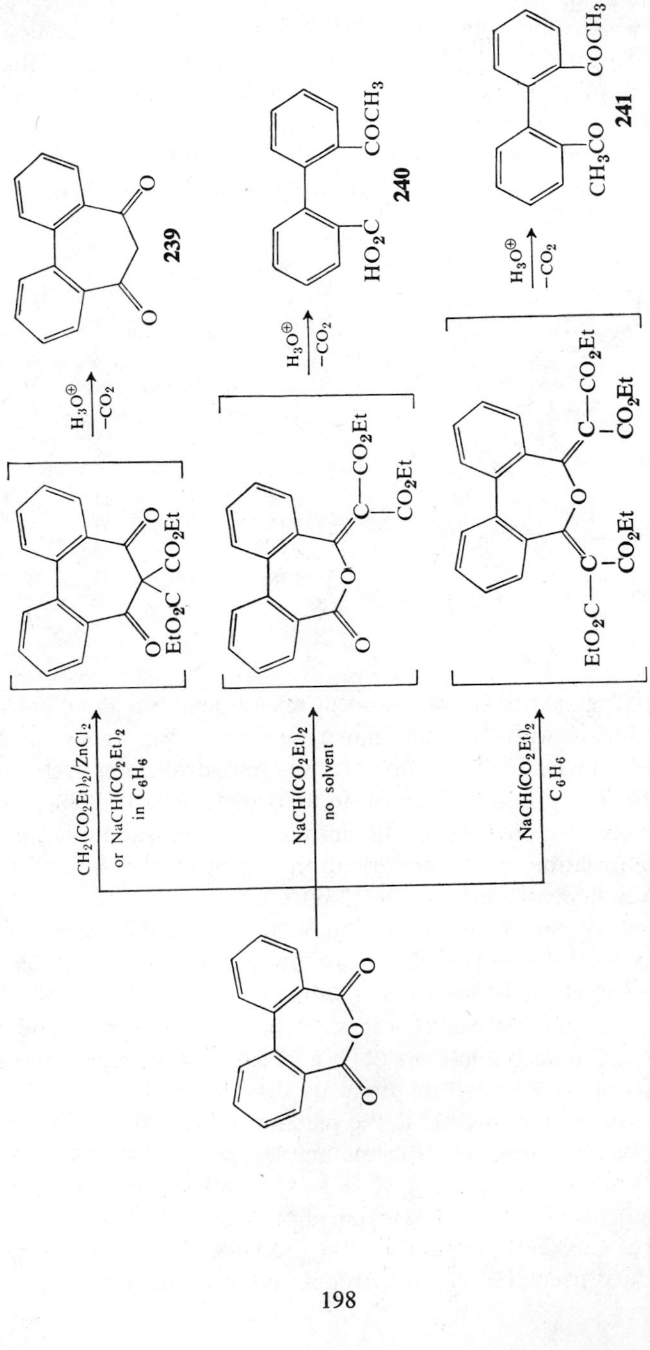

$CH_2(CO_2Et)_2/ZnCl_2$ or $NaCH(CO_2Et)_2$ in C_6H_6

EtO_2C CO_2Et

$H_3O^{\oplus}$, $-CO_2$

239

$NaCH(CO_2Et)_2$ no solvent

C—CO_2Et, CO_2Et

$H_3O^{\oplus}$, $-CO_2$

HO_2C $COCH_3$

240

$NaCH(CO_2Et)_2$ C_6H_6

EtO_2C—C—CO_2Et C—CO_2Et, CO_2Et

$H_3O^{\oplus}$, $-CO_2$

CH_3CO $COCH_3$

241

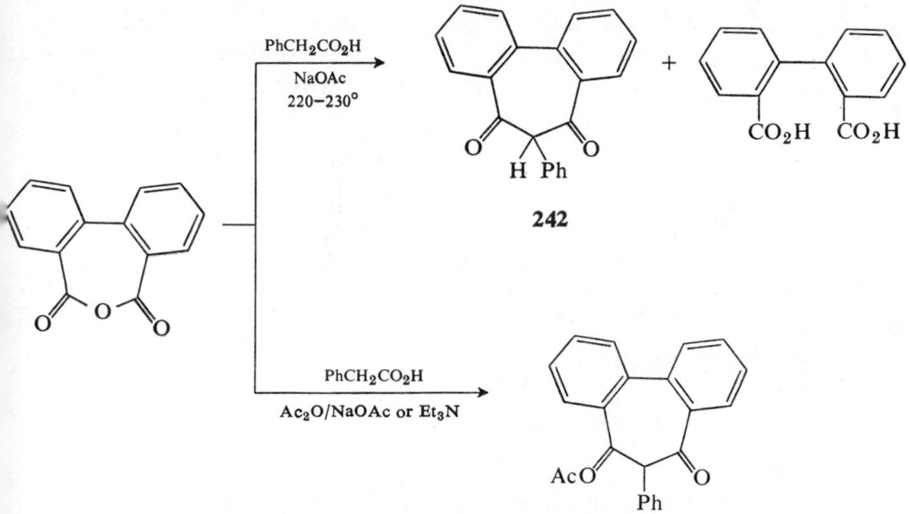

PhCH$_2$CO$_2$H
NaOAc
220–230°

H Ph

242

CO$_2$H CO$_2$H

PhCH$_2$CO$_2$H
Ac$_2$O/NaOAc or Et$_3$N

AcO Ph

243

Rupture of the seven-membered ring in diphenic anhydride is also observed under the influence of certain reducing agents. Thus Chatterjee (44) reported in 1935 that treatment of the anhydride with 5% sodium amalgam in alcohol resulted in the formation of a mixture of diphenic acid, a monoalkyl ester of diphenic acid corresponding to the alcohol used, and dibenz[*c*,*e*]oxepin-5(7*H*)-one (**244**). This product was isolated in only 5% yield, however. More recently Madeja (181) has studied the reduction of diphenic anhydride with lithium aluminum hydride. Once again diphenic acid and **244** were isolated although the principal product was 2,2′-bis(hydroxymethyl)-biphenyl **245**. The procedure of choice for the synthesis of **244** was discovered

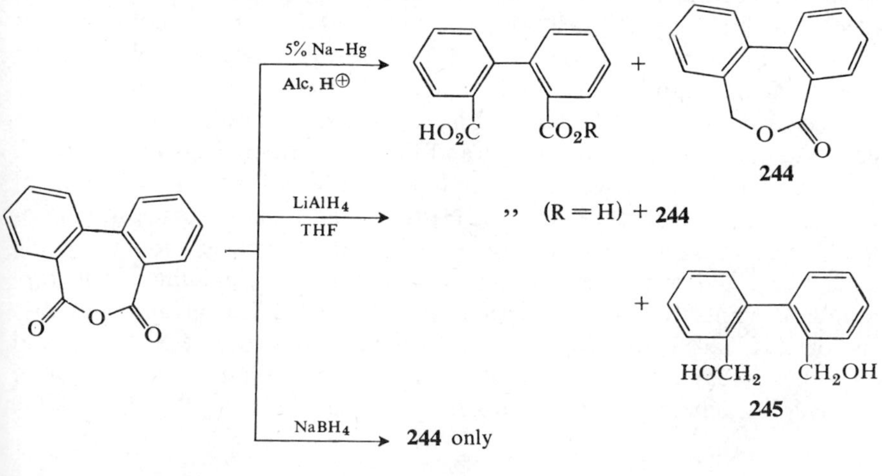

5% Na–Hg
Alc, H$^{\oplus}$

HO$_2$C CO$_2$R

244

LiAlH$_4$
THF

" (R = H) + **244**

HOCH$_2$ CH$_2$OH

245

NaBH$_4$ → **244** only

in 1968 by Belleau and Chevalier (205); this consists of reduction of diphenic anhydride with sodium borohydride. By this method the yield of **244** is 87%.

Additional reagents which lead to ring opening in diphenic anhydride are phosphorus pentachloride (69, 70, 205) and hydrazoic acid (43). The latter is of interest as a method of synthesis of phenanthridone (**246**).

$\xrightarrow{HN_3}$

246

b. *Dibenz*[c,e]*oxepin-5*(7H)-*ones*

Reduction of one of the carbonyl groups in diphenic anhydrides leads, formally, to dibenz[*c*,*e*]oxepin-5(7*H*)-ones. A number of ε-lactones of this type have been reported in the chemical literature. In some cases they can be obtained by treatment of anhydrides with appropriate reagents (see preceding section); in others they are formed by cyclization of suitable open-chain precursors, as will be seen below.

The earliest reported dibenz[*c*,*e*]oxepin-5(7*H*)-one derivative was "phenol-diphenein," which Graebe and Mensching (70) synthesized in 1880 from diphenic anhydride and phenol. The subsequently reported syntheses and reactions of this compound (57, 161, 162, 164), and also of analogs corresponding to alteration of either the phenolic (57, 165) or anhydride (3) component, have already been discussed.

The unsubstituted parent compound itself, dibenz[*c*,*e*]oxepin-5(7*H*)-one (**244**), was first prepared in 1911 by Kenner and Turner (95), and again in 1954 by Stephenson (154), from 2,2′-bis(formyl)biphenyl (**246**) via an internal Cannizzaro reaction and spontaneous lactonization of the initially formed hydroxy acid, **247**. The striking ease of formation of the lactone has been attested to recently by Bacon and Bankhead (8) and also by Sturrock and co-workers (155). According to the English authors (8), oxidation of 2,2′-bis(hydroxymethyl) biphenyl (**245**) with dinitrogen tetroxide in chloroform solution affords mainly **244** together with a smaller amount of the dialdehyde, **246**, which can be separated from **244** by column chromatography. Attempts to obtain the open-chain hydroxy acid by alkaline hydrolysis of **244** failed to give entirely lactone-free material. Sturrock and coworkers (155) claimed to have effected nearly quantitative lactonization of the hydroxy acid merely by heating in ethanol under reflux. In this case the hydroxy acid was prepared from the corresponding dialdehyde via a Cannizzaro reaction, and the dialdehyde was obtained conveniently by ozonolysis of phenanthrene in 1:1 *tert*-butyl alcohol/water.

NaOH/EtOH room temp
CHO CHO
246
N_2O_4 / $CHCl_3$
KOH/H_2O reflux
CH_2OH CH_2OH
245
CH_2OH CO_2H
247
EtOH reflux
O O
244

7-Substituted derivatives of **244** have been prepared by several general approaches, all of which utilize an organometallic reagent at some stage of the synthetic sequence. The earliest example of this type was provided in 1929 by Sergeev (145) who condensed diphenic anhydride with phenylmagnesium bromide and obtained a hydroxy acid, **224**, and a hydroxy ketone (or cyclic hemiketal), **225**. The acid was separated from the neutral product by extraction with alkali and was lactonized to 7,7-diphenyldibenz[*c*,*e*]oxepin-5(7*H*)-one (**248**) by simple heating. Treatment of **248** with phosphorus and iodine in acetic acid afforded 2-benzhydryl-2′-biphenylcarboxylic acid (**249**).

Ph_2COH CO_2H
224
Δ
Ph Ph O O
248
P/I_2 AcOH
Ph_2CH CO_2H
249
PhMgBr
Ph_2COH COPh
225
Ph Ph O OH Ph
225A

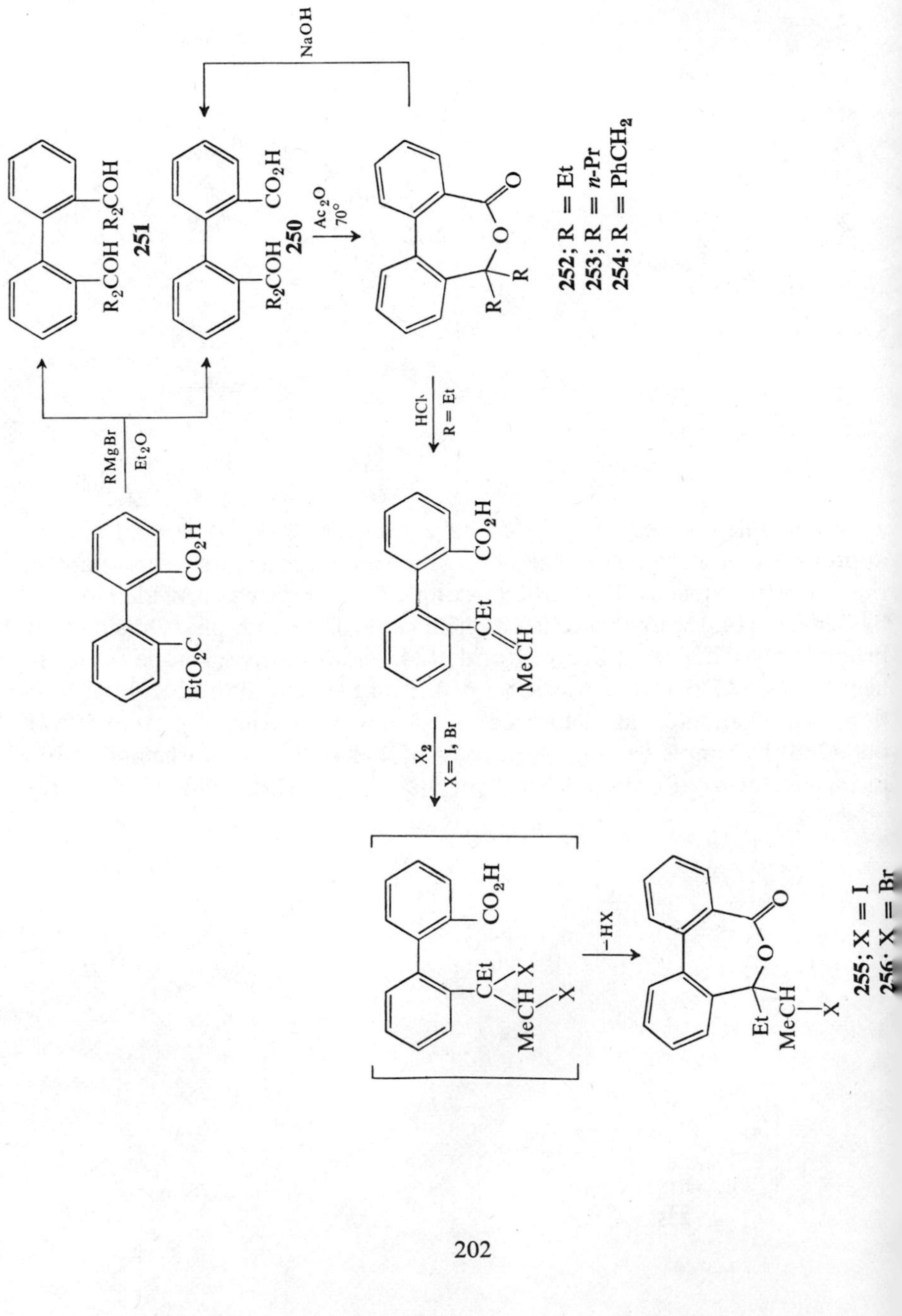

EtO_2C CO_2H

$RMgBr$, Et_2O

R_2COH R_2COH
251

R_2COH CO_2H
250

Ac_2O, 70°

NaOH

R R O O

252; R = Et
253; R = *n*-Pr
254; R = $PhCH_2$

HCl, R = Et

CEt CO_2H
MeCH

X_2, X = I, Br

CEt CO_2H
MeCH X
X

−HX

Et MeCH O O
X

255; X = I
256; X = Br

Reaction with additional phenylmagnesium bromide transformed **248** into the previously obtained hydroxy ketone, **225**.

7,7-Disubstituted dibenz[*c*,*e*]oxepin-5(7*H*)-ones were also prepared by Corbellini and Vigano (47), starting from the monoethyl ester of diphenic acid. Reaction with ethylmagnesium bromide gave a hydroxy acid, **250**, and a diol, **251**. The acid was separated from the neutral diol by extraction with alkali, and was cyclized readily to 7,7-diethyldibenz[*c*,*e*]oxepin-5(7*H*)-one (**252**) upon treatment with acetic anhydride at 70°. Replacement of ethylmagnesium bromide in this sequence with *n*-propylmagnesium bromide and benzylmagnesium bromide led to 7,7-di-*n*-propyl and 7,7-dibenzyl analogs **253** and **254**, respectively. Reaction of the diphenic ester with *o*- and *p*-tolylmagnesium iodide, and likewise with 1- and 2-naphthylmagnesium bromide, afforded low yields of products which could not be purified satisfactorily. Upon treatment with alkali, **252** was cleaved to the open-chain hydroxy acid, **250**. Under the influence of acid, on the other hand, **252** gave a mixture of *cis* and *trans* olefin acids, as shown. Reaction of one of the olefin isomers with iodine or bromine resulted in the formation of lactones **255** and **256**, respectively.

Another illustration of the use of organometallic reagents for the synthesis of 7,7-disubstituted dibenz[*c*,*e*]oxepin-5(7*H*)-ones was provided in 1953 by Nightingale and co-workers (125). Aluminum chloride-catalyzed acylation of *o*-xylene with diphenic anhydride afforded the expected keto acid, **257**. The acid chloride, **258**, was then prepared and converted into lactones **259** and **260** by condensation with diethyl and di-*n*-butylcadmium, respectively. Condensation of **258** with di-*p*-tolylcadmium afforded only an open-chain diketone, **261**.

Recent studies have also been reported with 7,7-dimethyldibenz[*c*,*e*]-oxepin-5(7*H*)-one (**262**) which can be prepared in the same manner as other 7,7-dialkyl analogs. Interesting rearrangements of **262** have been observed by Everitt and Turner (60). Upon being heated in the presence of alumina, **262** is converted into the open-chain olefin acid **263**. Under the influence of hot formic acid or 50% sulfuric acid, on the other hand, **262** undergoes isomerization to 9,9-dimethylfluorene-4-carboxylic acid (**264**). The nmr spectrum of **262** has been described by Sutherland and Ramsay (156) in connection with a study of molecular geometry and ring inversion in several members of the dibenz[*c*,*e*]oxepin ring system.

Dibenz[*c*,*e*]oxepin-5(7*H*)-one derivatives containing substitution on the aromatic rings have not been the objects of any systematic study. One incompletely characterized example in the literature is the lactone that was isolated upon ozonolysis of 4,5,5′-trimethoxy-2-(β-styryl)-2′-vinylbiphenyl (**265**) in chloroform, followed by direct reduction of the resulting dialdehyde, **266**, with zinc in acetic acid. Dialdehyde **266** was not isolated, and the choice of structures **267** and **268** for the final lactone was left open.

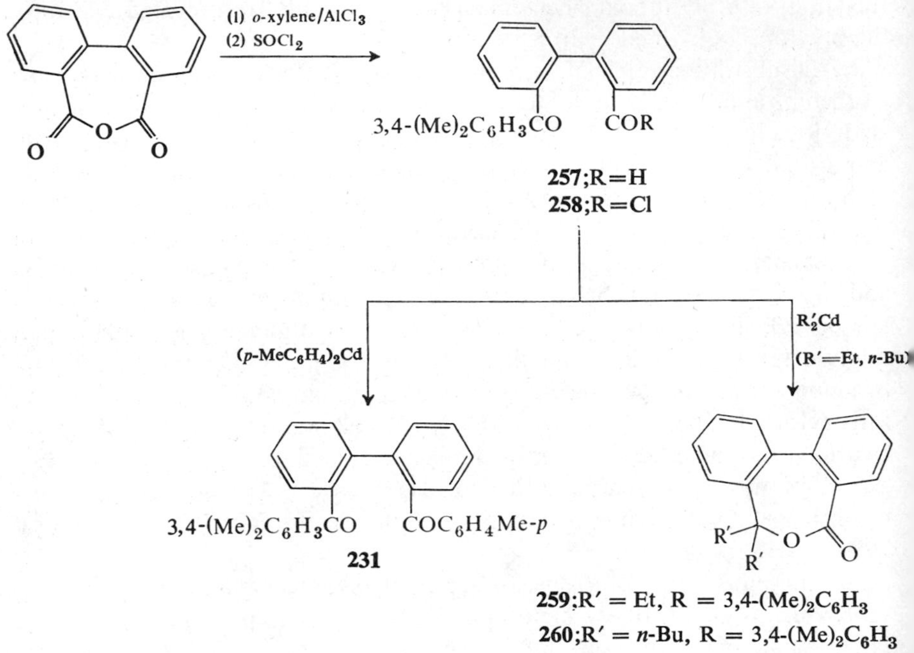

(1) o-xylene/$AlCl_3$
(2) $SOCl_2$

3,4-$(Me)_2C_6H_3CO$ COR

257;R=H
258;R=Cl

$(p\text{-}MeC_6H_4)_2Cd$

R'_2Cd
(R'=Et, n-Bu)

3,4-$(Me)_2C_6H_3CO$ $COC_6H_4Me\text{-}p$
231

R' O O
R'

259;R' = Et, R = 3,4-$(Me)_2C_6H_3$
260;R' = n-Bu, R = 3,4-$(Me)_2C_6H_3$

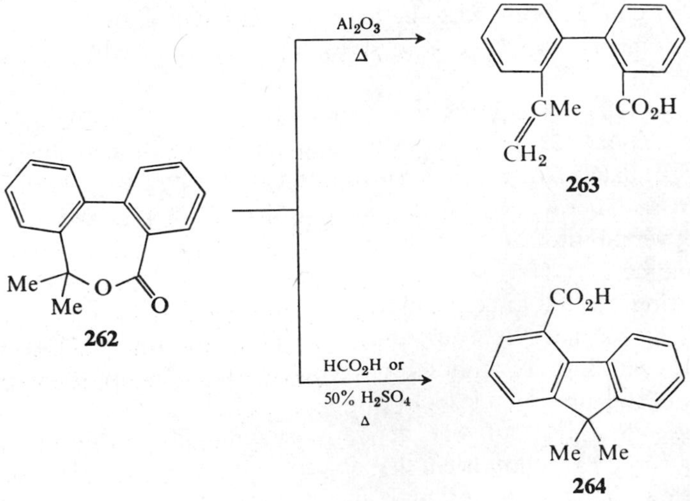

Al_2O_3
Δ

CMe CO_2H
CH_2
263

Me O
Me O
262

HCO_2H or
50% H_2SO_4
Δ

CO_2H
Me Me
264

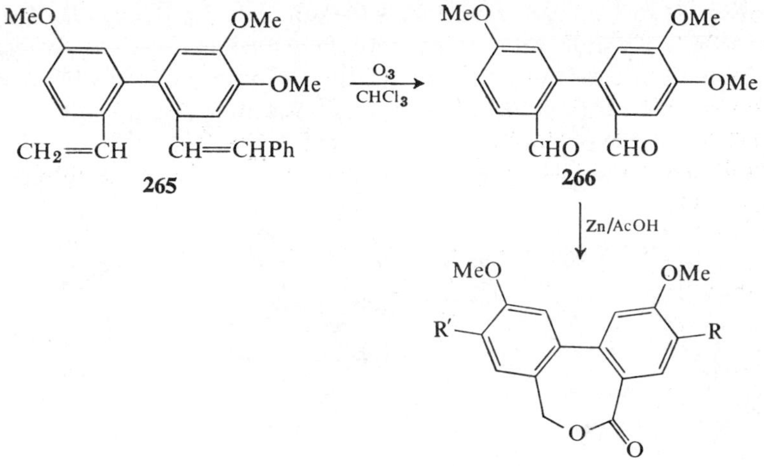

MeO OMe OMe $CH_2{=}CH$ $CH{=}CHPh$ **265** $\xrightarrow[CHCl_3]{O_3}$ MeO OMe OMe CHO CHO **266**

Zn/AcOH

MeO OMe R′ R O O

267;R = H, R′ = MeO

268;R = MeO, R′ = H

Two other examples of dibenz[*c*,*e*]oxepin-5(7*H*)-one derivatives with multiple aromatic substitution were reported in 1966 by Pecherer and Brossi (207) in the form of lactones **269** and **270**. As indicated, the first of these was obtained by a conventional acid-catalyzed lactonization while the latter was formed via an intramolecular tosylate displacement.

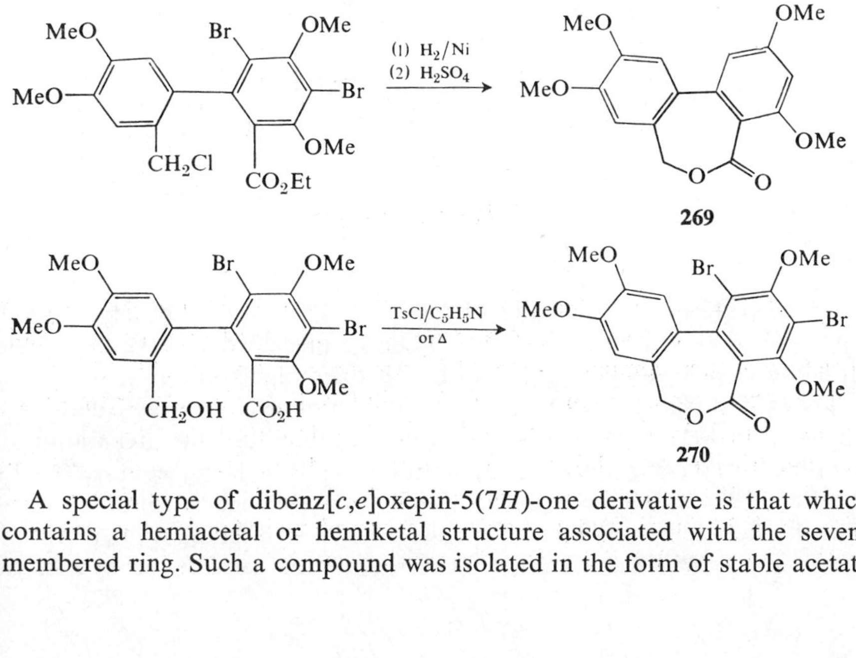

MeO MeO Br OMe Br OMe CH_2Cl CO_2Et $\xrightarrow{(1)\ H_2/Ni\quad (2)\ H_2SO_4}$ MeO MeO OMe OMe O O **269**

MeO MeO Br OMe Br OMe CH_2OH CO_2H $\xrightarrow[\text{or } \Delta]{TsCl/C_5H_5N}$ MeO MeO Br OMe Br OMe O O **270**

A special type of dibenz[*c*,*e*]oxepin-5(7*H*)-one derivative is that which contains a hemiacetal or hemiketal structure associated with the seven-membered ring. Such a compound was isolated in the form of stable acetate derivative **271** by Cook and co-workers (48) and later by Bailey (9). Reaction of diphenic anhydride with phenylsulfonylhydrazide, followed by heating with sodium acetate in ethylene glycol, afforded 2-formylbiphenyl-2′-carboxylic acid (**272**). Reaction of the latter with acetic anhydride in the presence of a trace of sulfuric acid generated **271** readily. Ring opening was accomplished with equal ease by treatment of **271** with dilute sulfuric acid or acetic acid.

(1) $PhSO_2NHNH_2$
(2) $NaOAc/HOCH_2CH_2OH$

CHO CO_2H

272

HO H O O

dil H_2SO_4 + AcOH

Ac_2O + H_2SO_4 (trace)

AcO H O O

271

c. *Dibenz*[c,e]*oxepins*

(1) Synthesis. No examples of the fully unsaturated dibenz[*c*,*e*]oxepin ring system are known to date, and it can be presumed that the quinonoid structure of such compounds would render them highly unstable.

The earliest reported synthesis of a 5,7-dihydrodibenz[*c*,*e*]oxepin derivative appeared in 1931 when Wittig and Leo (178) described the preparation of 5,7-dihydro-1,11-dimethoxy-5,5,7,7-tetraphenyldibenz[*c*,*e*]oxepin (**273**) by reaction of **274** with hot acetic acid. The diol was obtained by condensation of dimethyl 2,2′-dimethoxydiphenate with excess phenyllithium. As an extension of this study, Wittig and Petri (179) carried out the same sequence with the added refinement that diol **274** was resolved into optically active forms prior to cyclization. This pioneering investigation was the forerunner of a great deal of subsequent work on the stereochemistry of bridged biphenyls, much of which is still actively in progress today.

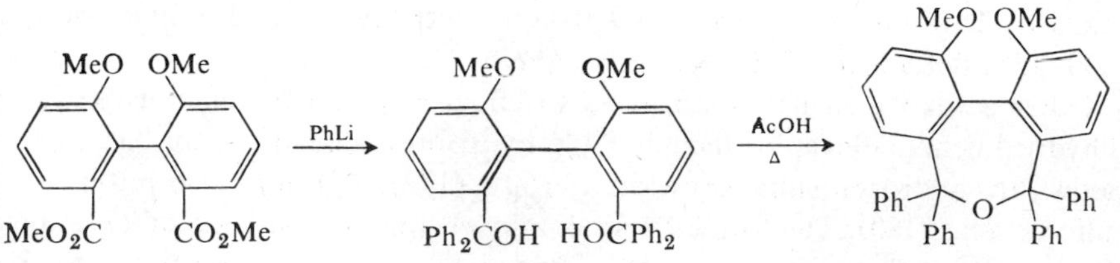

The two papers of Wittig and co-workers were followed closely by a report by Bennett and Wain (25) concerning the synthesis of 5,7-dihydro-5,5,7,7-tetramethyldibenz[*c*,*e*]oxepin (**275**) from the corresponding open-chain diol. Ring closure was effected in 72% yield by treatment of the diol with 70% sulfuric acid at room temperature. Following this work a number of other authors reported successful syntheses of 5,7-dihydrodibenz[*c*,*e*]oxepins by treatment of diols with sulfuric acid under various conditions of temperature and acid strength. The most dilute acid reported thus far has been 10% sulfuric acid, which Beaven and co-workers (16) used to prepare 5,7-dihydro-2,10-bis(hydroxymethyl)dibenz[*c*,*e*]oxepin (**276**). Somewhat stronger conditions (refluxing 20% sulfuric acid) have been used for the synthesis of

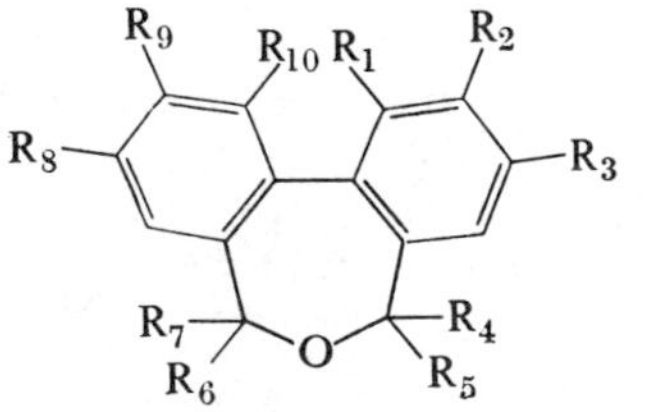

	R_1	R_2	R_3	R_4	R_5	R_6	R_7	R_8	R_9	R_{10}
275	H	H	H	Me	Me	Me	Me	H	H	H
276	H	CH_2OH	H	H	H	H	H	H	CH_2OH	H
277	H	H	H	Me	H	Me	H	H	H	H
278	H	H	H	Me	Me	H	H	H	H	H
279	H	H	H	H	H	H	H	H	H	H
280	Cl	H	Cl	H	H	H	H	Cl	H	Cl
281	MeO	H	H	H	H	H	H	H	H	MeO
282	H	MeO	H	H	H	H	H	H	MeO	H
283	H	MeO	MeO	H	H	H	H	MeO	MeO	H
284	NO_2	H	H	H	H	H	H	H	H	H
285	Me	H	H	H	H	H	H	H	H	Me

5,7-dihydro-5,6-dimethyldibenz[*c*,*e*]oxepin (**277**) and, more recently, 5,7-dihydro-5,5-dimethyldibenz[*c*,*e*]oxepin (**278**). Ring closure of diols has been achieved most commonly by treatment with 50% sulfuric acid. Compounds which have been made in this fashion include 5,7-dihydrodibenz[*c*,*e*]oxepin (**279**) itself (74), and also the 1,3,9,11-tetrachloro (74), 1,11-dimethoxy (15, 75), 2,10-dimethoxy (16), 2,3,9,10-tetramethoxy (115, 139), and 1-nitro (130, 131) derivatives, **280**–**284**, respectively.

Other acids which have been used to effect ring closure of diols to 5,7-dihydrodibenz[*c*,*e*]oxepins include 24% hydrobromic acid (steam bath) and potassium hydrogen sulfate (fusion at 160°) (172, 177), and also *p*-toluenesulfonic acid (180). The former were employed for the synthesis of **279**, and the latter for the synthesis of 5,7-dihydro-1,11-dimethyldibenz[*c*,*e*]oxepin (**285**) by Wittig and Zimmerman (180). Compound **285** is of interest because the starting diol was resolved into two diastereomeric forms prior to cyclization. Each diastereomeric diol gave an optically active product upon ring closure, the levorotatory diol yielding a dextrorotatory ether, and vice versa.

Two additional syntheses of 5,7-dihydrodibenz[*c*,*e*]oxepins merit mention here because diols are undoubtedly involved as intermediates. Reduction of dimethyl 3,4′,5,5′-tetramethoxydiphenate with lithium aluminum hydride in refluxing tetrahydrofuran has been reported (100) to yield 5,7-dihydro-2,3,8,-10-tetramethoxydibenz[*c*,*e*]oxepin (**286**) directly. This finding is in contrast to the reductions of the closely analogous dimethyl 4,4′,5,5′-tetramethoxy (115, 139) and 4,4′,5,5′,6,6′-hexamethoxy (98, 99) compounds with lithium aluminium hydride, which are reported to give diols as products.

MeO OMe OMe MeO MeO_2C CO_2Me $\xrightarrow[\text{THF, reflux}]{LiAlH_4}$ MeO OMe OMe MeO O

286

A second synthesis of 5,7-dihydrodibenz[*c*,*e*]oxepins which probably proceeds via a diol intermediate is the reaction of *o*-toluidine with formaldehyde in 98% sulfuric acid. The product of this condensation has been identified as 3,9-diamino-5,7-dihydro-2,10-dimethyldibenz[*c*,*e*]oxepin (**287**) (61). In similar fashion, *o*-anisidine has been converted into 3,9-diamino-5,7-dihydro-2,10-dimethoxydibenz[*c*,*e*]oxepin (**288**) (61).

If it is desired to convert a diol into a cyclic ether under nonacidic conditions, a commonly used approach is to prepare a monotosylate ester and then subject this intermediate to the action of base. Ring closure, in this instance, involves an intramolecular nucleophilic displacement reaction with

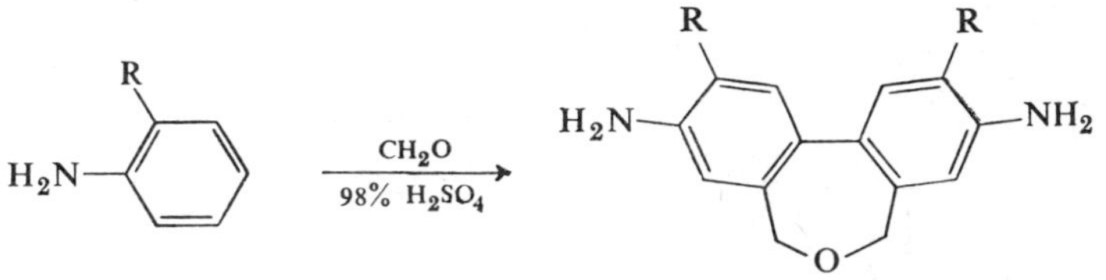

$$\text{H}_2\text{N-C}_6\text{H}_4\text{-R} \xrightarrow[98\%\ H_2SO_4]{CH_2O}$$

287; R = Me
288; R = OMe

tosylate as the leaving species. A synthesis illustrating this general route is the recently described preparation of 5,7-dihydro-1,2,3,9,10,11-hexamethoxy-dibenz[*c*,*e*]oxepin (**289**) (98, 99). This compound was obtained directly upon treatment of the corresponding diol with *p*-toluenesulfonyl chloride in pyridine, without isolation of the intermediate monotosylate ester. The open-chain diol was prepared by lithium aluminum hydride reduction of the appropriate diphenic ester derivative. Another recent example of this type of ring closure was provided in the synthesis of 5,7-dihydro-2,3,8,10-tetramethoxydibenz[*c*,*e*]oxepin (**290**) by Pecherer and Brossi (207).

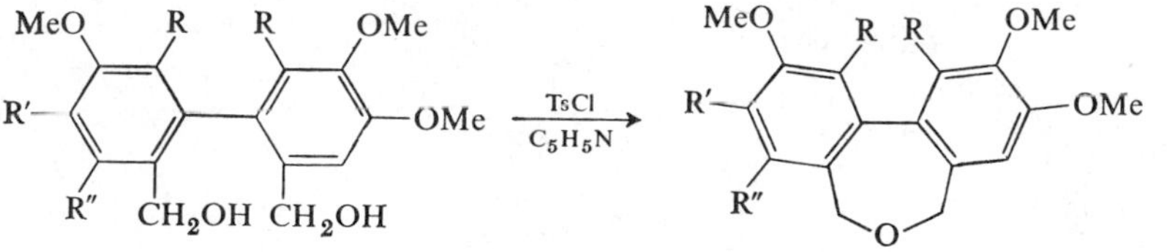

$$\xrightarrow[C_5H_5N]{TsCl}$$

289; R = R′ = MeO, R″ = H
290; R = R′ = H, R″ = MeO

Other intermediates for the preparation of 5,7-dihydrodibenz[*c*,*e*]oxepins are 2,2′-bis(halomethyl)biphenyls. Thus 2,2′-bis(chloromethyl)-4,4′5,5′-tetra-methoxybiphenyl (**291**) has been converted into 5,7-dihydro-2,3,9,10-tetramethoxydibenz[*c*,*e*]oxepin (**292**) under the influence of various reagents, including refluxing aqueous acetic acid, alkaline potassium permanganate, and lead or copper nitrate in acetic acid (115, 139). The bis(bromomethyl) analog, **293**, undergoes conversion into **292** under the influence of sodium hydroxide, and into a mixture of **292** and open-chain dinitrile **294** upon treatment with potassium cyanide (50).

A further example of the same type involves the action of silver oxide upon dimethyl 6,6′-bis(bromomethyl)diphenate (**295**) in aqueous acetone solution, which has been found to give 1,11-dicarbomethoxy-5,7-dihydrodibenz[*c*,*e*]-oxepin (**296**) (118, 119).

An occasionally used synthesis of cyclic ethers involves heating quaternary ammonium iodide compounds containing a suitably situated hydroxy substituent under the conditions of the Hofmann elimination with silver oxide as the base. An illustration of this approach is the synthesis of 5,7-dihydrodibenz[*c*,*e*]oxepin (**279**) and 5,7-dihydro-5-methyldibenz[*c*,*e*]oxepin (**297**) which was reported in 1958 by Bersch and co-workers (28).

A special type of 5,7-dihydrodibenz[*c*,*e*]oxepin is that in which a hemiacetal or hemiketal structure is present as part of the seven-membered ring. The possibility that such a compound could exist was recognized as early as 1929 by Sergeev (145), who suggested that condensation of phenylmagnesium bromide with 7,7-diphenyldibenz[*c*,*e*]oxepin-5(7*H*)-one (**248**) leads to an equilibrium mixture consisting of 5,7-dihydro-5,7,7-triphenyldibenz[*c*,*e*]-oxepin-5-ol (**225A**) and its open-chain hydroxy ketone form, **225**. No other instances of this type have been discovered to date in the dibenz[*c*,*e*]oxepin series. However, there is no immediately apparent reason why many more such compounds could not be prepared, especially if the hydroxy group in the hemiacetal or hemiketal form were stabilized as an ester or other derivative.

PhMgBr

Ph Ph O O **248**

Ph O OH Ph Ph **225A**

Ph_2COH COPh **225**

(2) Physical Properties. Credit belongs to Wittig and Leo (178) for being the first to recognize that 5,7-dihydrodibenz[*c*,*e*]oxepins can exhibit the phenomenon of molecular asymmetry. These workers prepared optically active 5,7-dihydro-1,11-dimethoxy-5,5,7,7-tetraphenyldibenz[*c*,*e*]oxepin (**273**) and reported that rapid racemization of this compound occurred at 315°. The high temperature required for the racemization of **273** is a reflection of the energy barrier opposing free rotation of the two phenyl rings about their common axis. In this particular example there are two important contributions to this inversion barrier, one being that of the methoxy groups and the other being that of the four phenyl groups.

With less highly substituted compounds inversion is much more rapid, and optically active diastereomers cannot be isolated under normal laboratory conditions. At low temperatures, however, it is possible to slow the rate of inversion sufficiently to allow detection of discrete species by nmr spectrometry and to permit estimation of the energy barrier in quantitative terms. Thus Oki and co-workers (130) reported in 1963 that inversion in 5,6-dihydrodibenz[*c*,*e*]oxepin (**279**) itself was too rapid to permit detection even at −90°, and estimated the energy barrier for the racemization of this compound to be somewhat less than 9 kcal/mole. Working independently, Kurland and co-workers (103) calculated that the inversion barrier for **279** at −84° is approximately 9 kcal/mole, whereas the value for 5,7-dihydro-dibenz[*c*,*e*]thiepin is 17 kcal/mole at +42°. More recently, Sutherland and Ramsey (156) compared the inversion barrier of **279** with the inversion barriers of 5,7-dihydro-5,5-dimethyldibenz[*c*,*e*]oxepin (**278**) and 5,7-dihydro-5,5,7,7-tetramethyldibenz[*c*,*e*]oxepin (**275**). Values of 9.2 and 13.2 kcal/mole were obtained for compounds **279** and **275**, respectively. No broadening of the nmr signal was observed with **278** even at −60°.

Substituents placed at the 1- and/or 11-positions of the 5,7-dihydrodibenz-[*c*,*e*]oxepin ring system are very effective in hindering ring inversion, and warming is necessary in order to detect rapid ring inversion by nmr spectrometry. Examples that may mentioned in this connection include 5,7-dihydro-1-nitrodibenz[*c*,*e*]oxepin (**284**) (130, 131), 5,7-dihydro-1,11-dimethyldibenz-[*c*,*e*]oxepin (**285**) (118, 119, 130, 131), 5,7-dihydro-1,11-bis(hydroxymethyl)-dibenz[*c*,*e*]oxepin (**276**) (118, 119), 1,11-dicarbomethoxy-5,7-dihydrodibenz-[*c*,*e*]oxepin (**296**) (118, 119), and 5,7-dihydrodibenz[*c*,*e*]oxepin-1-ol (208, 212).

An important physical tool for establishing the geometry of 1,11-disubstituted derivatives of 5,7-dihydrodibenz[*c*,*e*]oxepin (**279**) has been uv absorption spectroscopy. Among the first to employ this approach in the study of this particular ring system were Beaven and co-workers (15, 16), who noted significant differences in the position and intensity of uv absorption maxima in **279** (15), 5,7-dihydro-1,11-dimethoxydibenz[*c*,*e*]oxepin (**281**) (15), and 5,7-dihydro-2,10-dimethoxydibenz[*c*,*e*]oxepin (**282**) (16). The bathochromic and hypsochromic shifts displayed by **281** relative to **282** are presumably attributable to an unfavorable interaction between the methoxy groups at the 1- and 11-positions, which tends to make the energy difference between the ground state and the planar excited state larger in **281** than in **282**.

In a slightly later paper, Beaven and Johnson (17) compared the absorption characteristics of **279** with those of 5,7-dihydro-5,7-dimethyldibenz[*c*,*e*]-oxepin (**277**) and 5,7-dihydro-5,5,7,7-tetramethyldibenz[*c*,*e*]oxepin (**275**). An interesting effect was observed as the number of substituents on the seven-membered ring increased. In going from **279** to **277**, there appeared to be a small hypsochromic and hypochromic displacement of the benzenoid band in the 250 mμ region. On the other hand, in going from **277** to **275**, although the intensity of the benzenoid band continued to decrease, its position underwent a *bathochromic* shift to a wavelength even longer than in **279**.

Following the work of Beaven and co-workers, a number of other authors commented upon the uv absorption features of **279**, including Suzuki (157, 158), Braude and Forbes (36), and Birkeland and co-workers (31). Braude and Forbes likewise discussed the spectrum of 5,7-dihydro-1,11-dimethyldibenz[*c*,*e*]oxepin (**285**), and Mislow and co-workers (63) included both **279** and **285** in their classic studies on the optical rotatory dispersion of compounds possessing molecular asymmetry. In addition the spectrum of 1,3,9,11-tetrachloro-5,7-dihydrodibenz[*c*,*e*]oxepin (**280**) was reported by Hall and Minhaj (74), and that of 5,7-dihydro-2,3,9,10-tetramethoxydibenz-[*c*,*e*]oxepin (**283**) was reported by Quelet and Matarasso-Tchiroukhine (115, 139). It is of interest to note that the benzenoid absorption in **280** was found to occur at a longer wavelength than in **279**, and that the intensity of the peak was greater in **280** than in **279**. This bathochromic shift is somewhat surprising when viewed in terms of the simple steric argument presented earlier for 1,11-disubstituted 5,7-dihydrodibenz[*c*,*e*]oxepin derivatives.

(3) Chemical Reactions. 5,7-Dihydrodibenz[*c*,*e*]oxepins have been found to undergo a varied assortment of chemical reactions involving the seven-membered ether ring. In the present review these may be grouped conveniently into the categories of ring contraction reactions and ring cleavage reactions, respectively. In addition passing mention may be made of (*1*) the observation by Wittig and Petri (179) that 5,7-dihydro-5,5,7,7-tetramethyldibenz[*c*,*e*]oxepin (**275**) forms a molecular complex with 1-naphthol, catechol, and other phenolic compounds; and (*2*) the report by Huang and Lee (77) that, under appropriate conditions, 5,7-dihydrodibenz-[*c*,*e*]oxepin (**279**) can be made to yield free radical species.

An interesting ring contraction reaction discovered in 1951 by Wittig and co-workers (177) involves the action of phenyllithium upon **279**. Under the influence of this strongly basic reagent a smooth rearrangement of the seven-membered ring took place with formation of 9,10-dihydro-9-hydroxyphenanthrene in 77% yield. The mechanism proposed for this ring contraction by the German workers is shown. In 1953 the same transformation was reported by Weinheimer and co-workers (172) except that potassium amide in liquid ammonia was used as the base. In this work a trace of phenanthrene was also found to accompany the main rearrangement product.

279 —PhLi, Et_2O, 7 days→ [carbanion, $Li^{\oplus}$] → [alkoxide, $O^{\ominus}$] → 9,10-dihydro-9-hydroxyphenanthrene (OH, H)

Another "ring contraction" reaction observed in this series is that of 5,7-dihydro-5,7,7-triphenyldibenz[*c*,*e*]oxepin-5-ol (**225A**), from which 4-benzoyl-9,9-diphenylfluorene (**298**) was obtained by Sergeev (145) upon treatment with hydrobromic acid, acetyl chloride, or hydrochloric acid in glacial acetic acid. Since **225A** is the cyclic hemiketal form of an open-chain hydroxy ketone that can lead to the observed product by simple cyclodehydration, the use of the term "ring contraction" to describe this reaction should probably be viewed with reservation.

Ring cleavage reactions that have been reported in the 5,7-dihydrodibenz-[*c*,*e*]oxepin series can be classified as oxidative or hydrolytic. An example of the oxidative type was provided by Quelet and Matarasso-Tchiroukhine (115, 139), who prepared 4,4′,5,5′-tetramethoxydiphenic aldehyde (**299**) in 90% yield by reaction of 5,7-dihydro-2,3,9,10-tetramethoxydibenz[*c*,*e*]-oxepin (**283**) with potassium dichromate in acetic acid under reflux. When sodium dichromate was used as the oxidant and the reaction time was shortened from 3 to 1 hr, the product was 2,3,8,9-tetramethoxyphenanthraquinone (**300**).

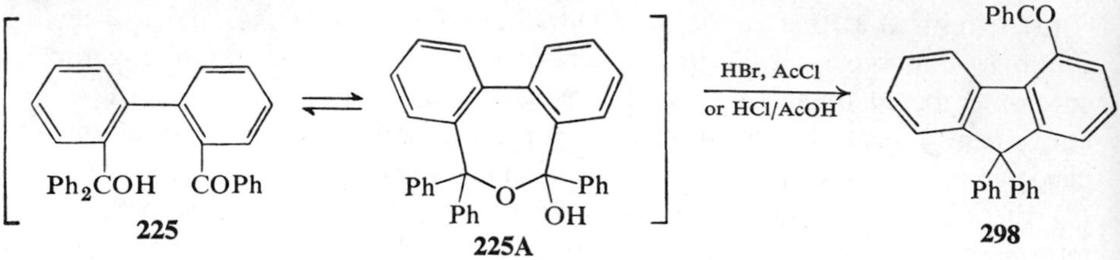

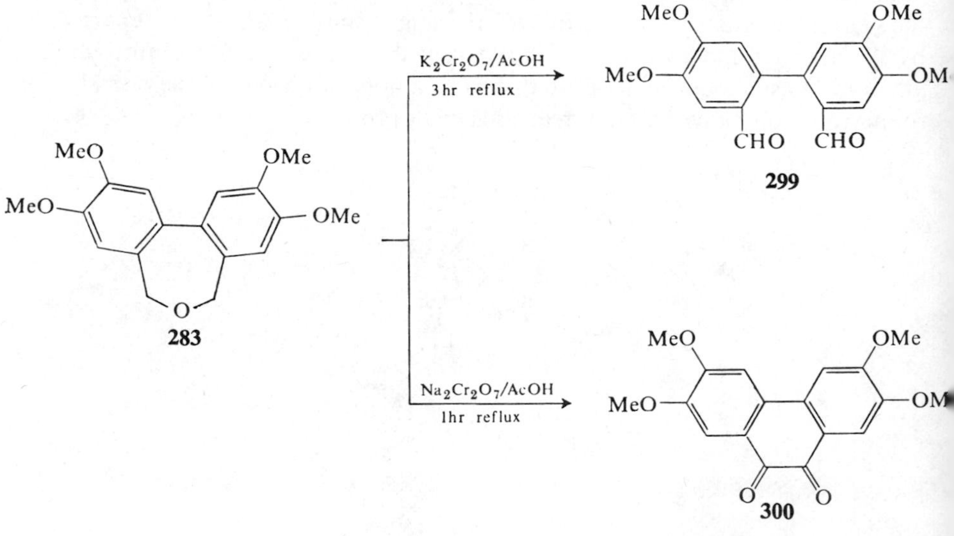

Hydrolytic cleavage of the seven-membered ether ring is exemplified by the reaction of 5,7-dihydro-5,5-dimethyldibenz[*c*,*e*]oxepin (**278**) with lithium aluminum hydride, which has been found by Sutherland and Ramsay (156) to give an 85% yield of **301**.

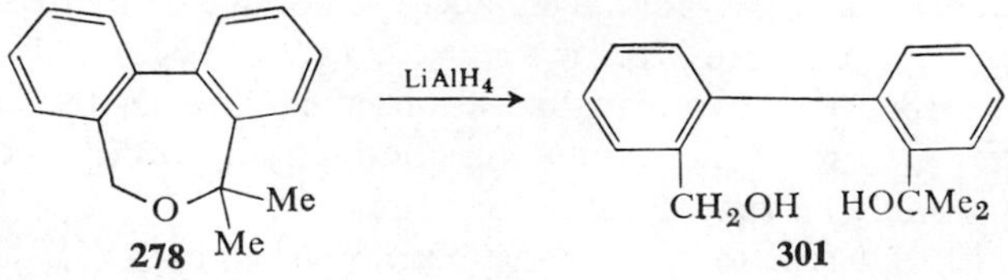

A further illustration of hydrolytic cleavage in this series is the reaction of 3,9-diamino-5,7-dihydro-2,10-dimethyldibenz[*c*,*e*]oxepin (**287**) with acetic anhydride under reflux, which has been reported by Farrar (61) to give diacetate **302**. If the reaction is carried out at room temperature in the presence of pyridine, the ether ring remains intact, the product isolated being simply the bis-*N*-acetyl derivative, **303**.

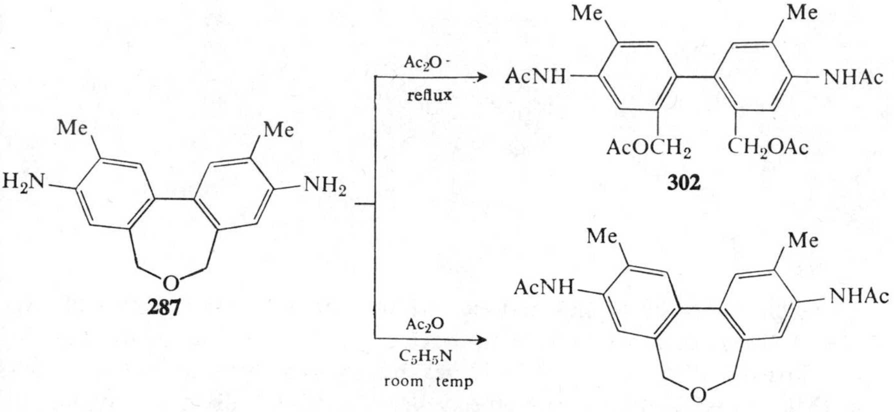

Me Me Ac_2O reflux AcNH NHAc $AcOCH_2$ CH_2OAc **302**

Me Me H_2N NH_2 O **287**

Me Me AcNH NHAc Ac_2O C_5H_5N room temp O **303**

7. Naphth[1,2-*b*]oxepins

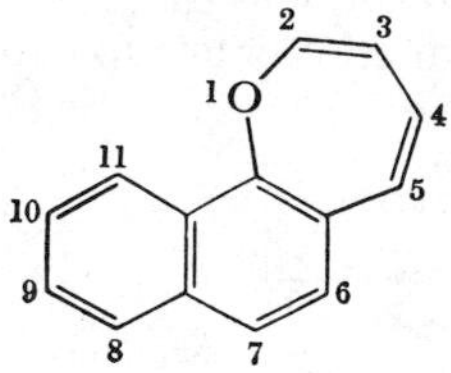

2 3 1O 4 11 10 5 9 6 8 7

This ring system was unknown until 1966 when Cagniant and co-workers reported the synthesis (41) and uv absorption characteristics (42) of 2,3,4,5-tetrahydronaphth[1,2-*b*]oxepin (**304**). The sodium salt of 1-naphthol was condensed with ethyl 4-bromobutyrate, and the resultant ester was saponified to 4-(1-naphthoxy)butyric acid (**305**). Treatment of **305** with hot polyphosphoric acid afforded 2,3-dihydro-4*H*-naphth[1,2-*b*]oxepin-5-one (**306**) in 80% yield, and reduction of **306** by the Huang-Minlon modification of the Wolff-Kishner method gave **304** in 90% yield. Reaction of **304** with acetyl chloride in the presence of aluminum chloride led to the formation of the expected 7-acetyl derivative, **307**. Wolff-Kishner reduction transformed the latter into 7-ethyl-1,2,3,4-tetrahydronaphth[1,2-*b*]oxepin (**308**).

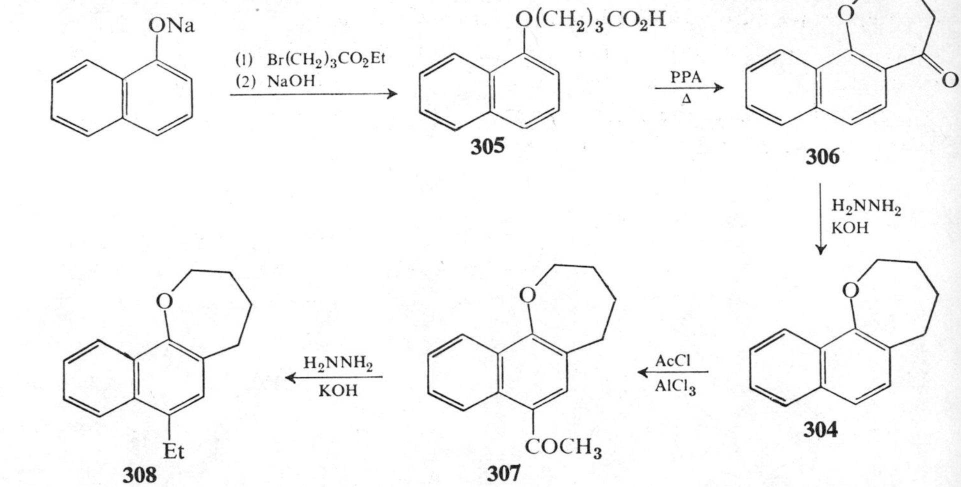

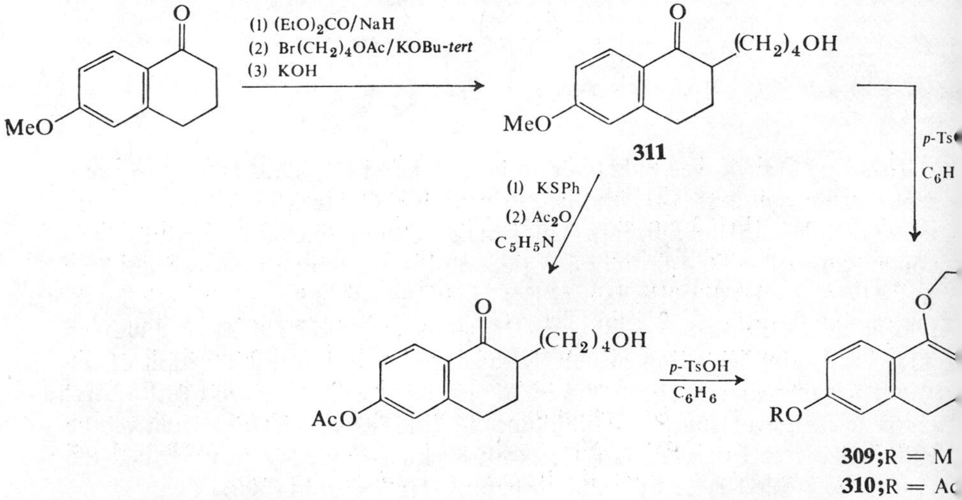

A recent paper by Immer and Bagli (200) contains the synthesis of two other members of this ring system, 9-acetoxy-2,3,4,5,6,7-hexahydronaphth-[1,2-*b*]oxepin (**309**) and 2,3,4,5,6,7-hexahydro-9-methoxynaphth[1,2-*b*]oxepin (**310**). Condensation of 6-methoxy-1-tetralone with diethyl carbonate in the presence of sodium hydride gave the expected 2-carbethoxy derivative, which was alkylated with 4-bromobutyl acetate in the presence of potassium *tert*-butoxide. Alkaline hydrolysis and decarboxylation then gave 2-(4-hydroxybutyl)-6-methoxy-1-tetralone (**311**), and treatment of the latter with *p*-toluenesulfonic acid in a Dean-Stark apparatus afforded **309** in 74% yield. For the synthesis of **310** the methoxy group in **311** was converted into acetoxy by reaction with potassium thiophenoxide in DMSO and acetylation with acetic anhydride. Cyclization of the resultant product was effected, once again, under the influence of *p*-toluenesulfonic acid.

8. Naphth[1,2-*c*]oxepins

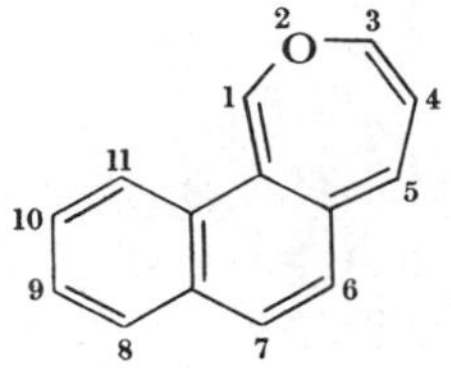

The sole example of this type of compound in the literature at the present time is hemiketal **312**, which Fuson and Shachat (65) isolated upon ozonolysis of olefin **313**. As might be expected, ir spectral evidence indicates that **312** exists in equilibrium with an open-chain hydroxy aldehyde form, **312A**. This

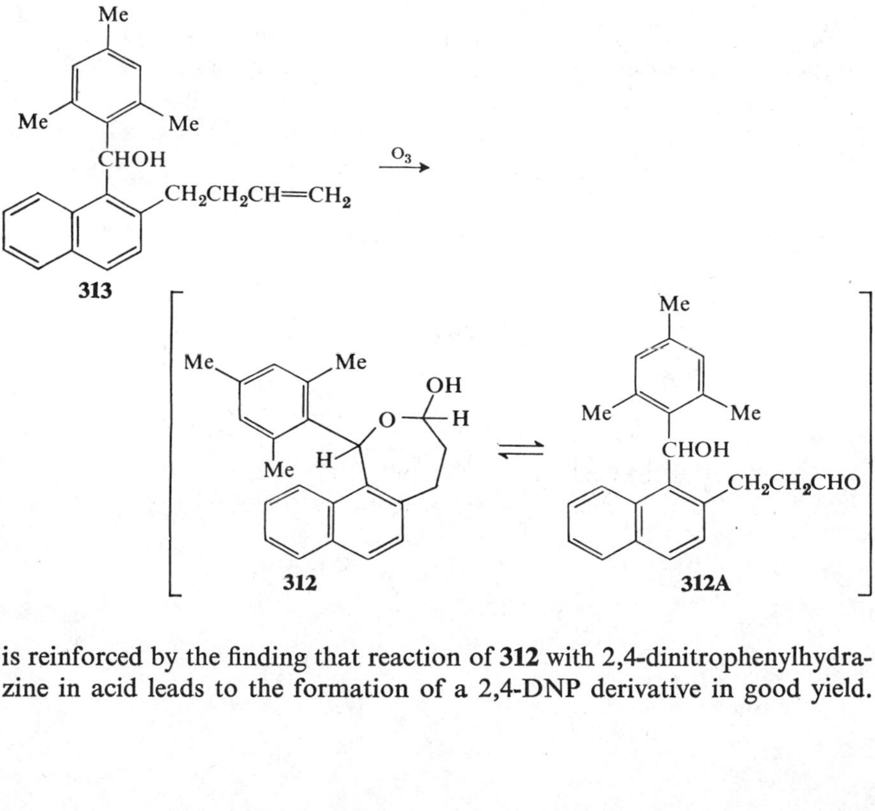

313 $\xrightarrow{O_3}$ [**312** $\rightleftharpoons$ **312A**]

is reinforced by the finding that reaction of **312** with 2,4-dinitrophenylhydrazine in acid leads to the formation of a 2,4-DNP derivative in good yield.

9. Naphth[2,1-*b*]oxepins

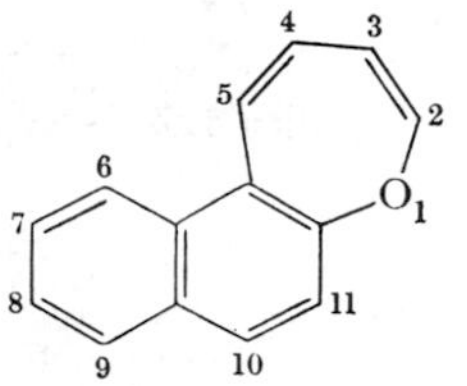

R.I. 3703

The earliest reported member of this series was 4,5-dihydro-3*H*-naphth-[2,1-*b*]oxepin-2-one (**314**), which Schroeter and co-workers (143) obtained in 1930 upon vacuum distillation of 4-(2-hydroxy-1-naphthyl)butyric acid (**315**).

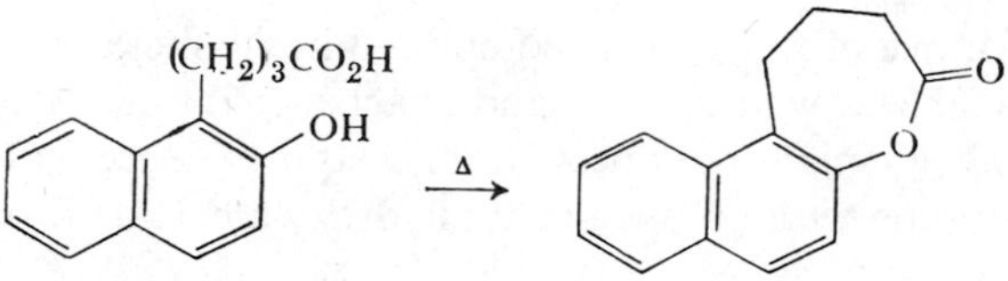

$(CH_2)_3CO_2H$ OH Δ O O

315 **314**

Also first described in 1930 was the synthesis of a compound which was formulated tentatively as 4-acetyl-3*H*-naphth[2,1-*b*]oxepin-2-one (**316**). In this work Sen and Roy (144) reported that the acid-catalyzed reaction between 2-hydroxy-1-naphthaldehyde and levulinic acid afforded a 40% yield of **316** together with 20% of another compound which was not identified.

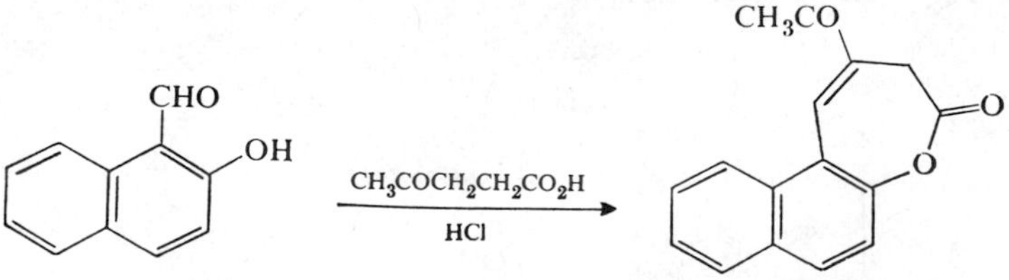

CHO OH $CH_3COCH_2CH_2CO_2H$ HCl CH_3CO O O

316

2,3,4,5-Tetrahydronaphth[2,1-*b*]oxepin (**317**) was prepared in 1955 by Chatterjea (45), starting from 3-(2-methoxy-1-naphthoyl)propionic acid

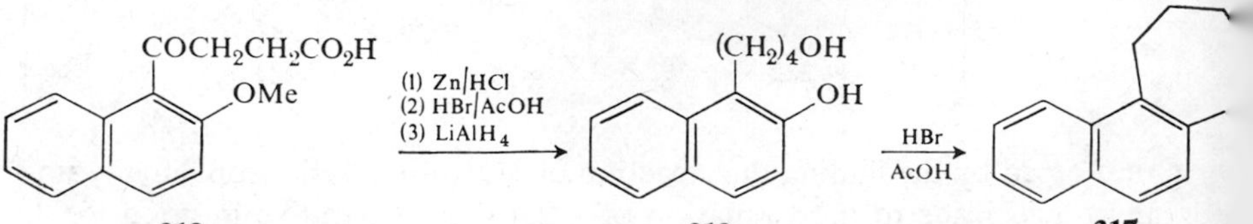

$COCH_2CH_2CO_2H$ OMe (1) Zn/HCl (2) HBr/AcOH (3) $LiAlH_4$ $(CH_2)_4OH$ OH HBr AcOH

318 **319** **317**

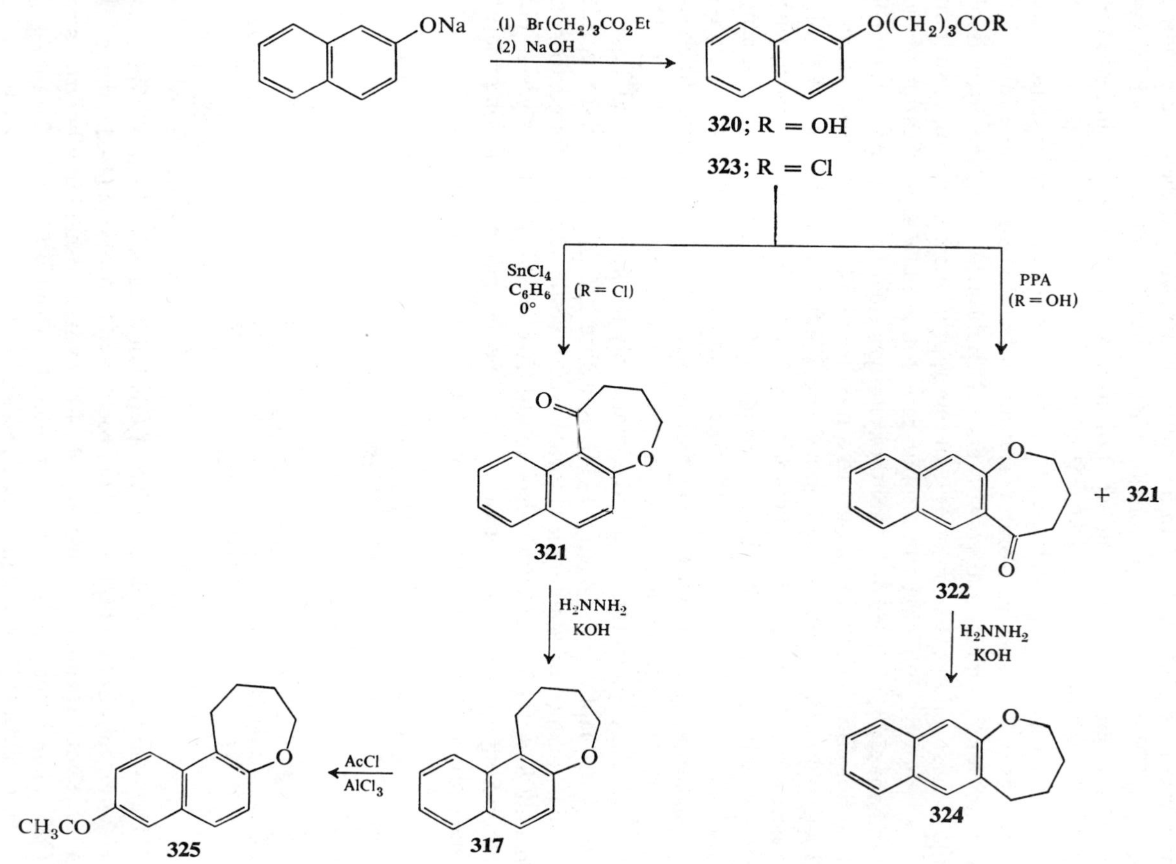

ONa
(1) $Br(CH_2)_3CO_2Et$
(2) NaOH
$O(CH_2)_3COR$
320; R = OH
323; R = Cl
$SnCl_4$
C_6H_6
0°
(R = Cl)
PPA
(R = OH)
321
322
+ 321
H_2NNH_2
KOH
H_2NNH_2
KOH
317
324
AcCl
$AlCl_3$
CH_3CO
325

(**318**). Clemmensen reduction gave 4-(2-methoxy-1-naphthyl)butyric acid, and treatment of the latter with hydrobromic acid yielded the corresponding naphthol. Reduction with lithium aluminum hydride then gave 4-(2-hydroxy-1-naphthyl)butanol (**319**), and cyclization of this diol under the influence of dry hydrogen bromide in acetic acid afforded **317** in modest yield.

In 1966 Cagniant and Charaux (41) reported a more satisfactory alternative synthesis of **317**, starting from the sodium salt of 2-naphthol. In a follow-up of this work Cagniant and co-workers (42) published a detailed analysis of the uv absorption spectrum of **317**. In this approach condensation of sodium 2-naphthoxide with ethyl 4-bromobutyrate, followed by saponification of the resultant ester, afforded 4-(2-naphthoxy)butyric acid (**320**). Treatment of **320** with hot polyphosphoric acid led to a mixture of **321** and the linear isomer, 3,4-dihydro-2*H*-naphth[2,3-*b*]oxepin-5-one (**322**). However, when **320** was was converted first into its acid chloride, **323**, and the latter was treated with stannic chloride in benzene at 0°, a 65% yield of **321** was obtained with no evidence for the formation of **322**. Reduction of **321** by the Huang-Minlon modification of the Wolff-Kishner reaction gave **317** in 72% yield. Interestingly, when this procedure was carried out with the mixture of **321** and **322**, the reduced product turned out to contain only the linear compound 2,3,4,5-tetrahydronaphth[2,3-*b*]oxepin (**324**). Reaction of **317** with acetyl chloride in the presence of aluminum chloride was investigated and found to give mainly 8-acetyl-2,3,4,5-tetrahydronaphth[2,1-*b*]oxepin (**325**), but gas–liquid chromatographic analysis demonstrated the concomitant formation of seven other products of undetermined structure.

10. Naphth[2,3-*b*]oxepins

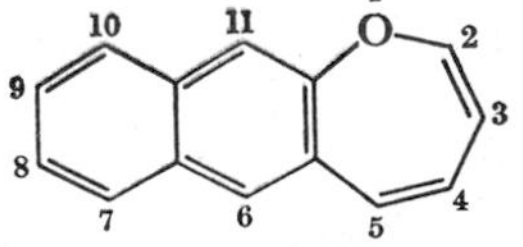

R.I. 3698

The oldest known representative of this family is 4,5-dihydro-3*H*-naphth-[2,3-*b*]oxepin-2-one (**322**), which Schroeter and co-workers (143) obtained in 1930 upon vacuum distillation of 4-(2-hydroxy-3-naphthyl)butyric acid (**326**). The latter was prepared conveniently from 4-(2-amino-3-naphthyl)-butyric acid (**327**) via a Sandmeyer reaction.

Another entry into this ring system was reported in 1966 by Cagniant and Charaux (41), consisting of the reaction of 4-(2-naphthoxy)butyric acid with hot polyphosphoric acid. Although the main product is the expected angular isomer, 3,4-dihydro-2*H*-naphth[2,1-*b*]oxepin-5-one (**321**), a small

$\xrightarrow[\text{H}_2\text{SO}_4,\ \text{alcohol}]{\text{NaNO}_2}$

327 (NH_2, $(CH_2)_3CO_2H$)

326 (OH, $(CH_2)_3CO_2H$) $\xrightarrow{\Delta}$ 322

amount of **322** can also be isolated. When the mixture of **322** and **321** was subjected to the Huang-Minlon modification of the Wolff-Kishner reduction, the product proved, unexpectedly, to contain only the linear isomer 2,3,4,5-tetrahydronaphth[2,3-*b*]oxepin (**324**). The uv absorption spectrum of **324** has been analyzed in detail by Cagniant and co-workers (42).

$O(CH_2)_3CO_2H$ $\xrightarrow[\Delta]{\text{PPA}}$ 321 + 32

$\downarrow$ H_2NNH_2, KOH

324

A polyacetylated derivative of **322** has been reported by Kuroda and Akajima (104) in connection with the structural elucidation of spinochrome M, a pigment isolated from the spines of the sea urchin. Reductive (zinc dust) acetylation of **328** afforded a product which was formulated as 6,7,8(or 9),10,11-pentaacetyl-5*H*-naphth[2,3-*b*]oxepin-2-one (**329**).

A quinonoid derivative of this ring system was reported in 1967 by Machatzke and co-workers (191). Treatment of 2-(4-chlorobutyl)-3-hydroxy-1,4-naphthoquinone (**330**) with sodium iodide in acetone containing sodium carbonate did not give the expected iodo compound, but instead a product identified as 2,3,4,5-tetrahydro-6*H*,11*H*-naphth[2,3-*d*]oxepin-6,11-dione (**331**).

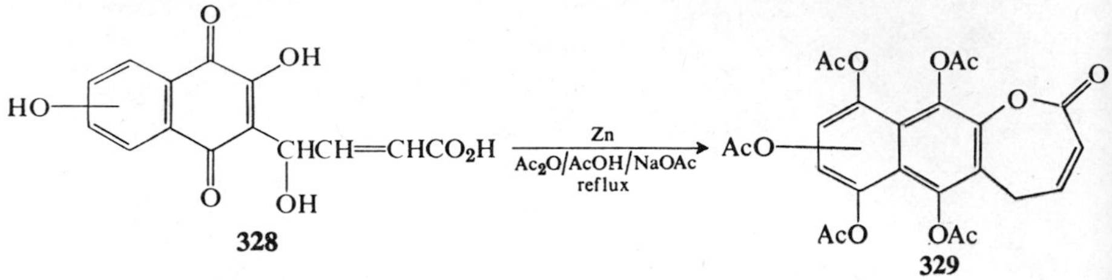

O
OH
HO
CHCH=CHCO2H
O
OH
Zn
Ac2O/AcOH/NaOAc
reflux
AcO
AcO
OAc
O
O
AcO
OAc

328 329

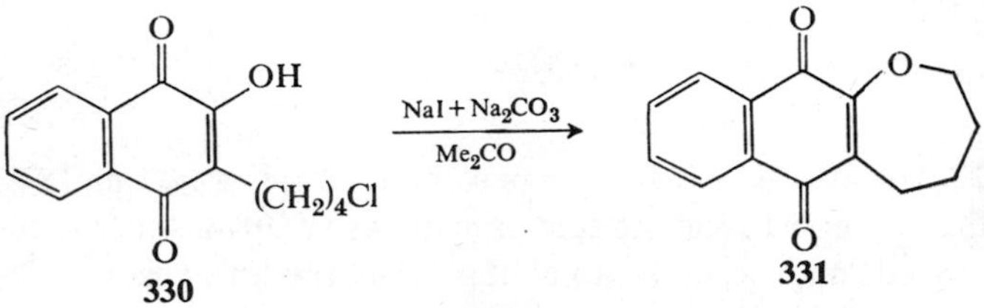

O
OH
(CH2)4Cl
O
NaI + Na2CO3
Me2CO
O
O
O

330 331

11. Naphth[2,3-*d*]oxepins

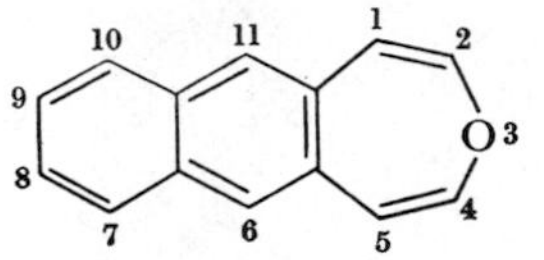

10
11
1
2
9
O 3
8
7
6
5
4

R.I. 3699

Only one reference to this ring system exists in the literature to date. This is a report by Tanaka (159) in 1935 that alkaline ferricyanide oxidation of the natural product hystazarin affords the interesting quinone anhydride **332.**

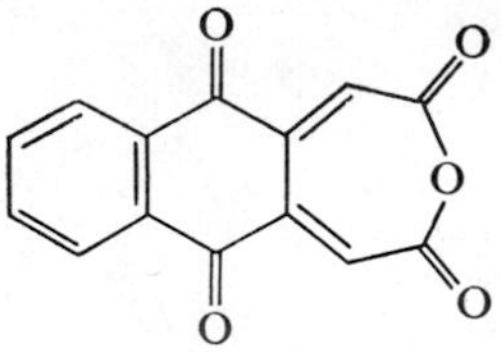

O
O
O
O
O

332

12. 1*H*-Naphth[1,8-*cd*]oxepins

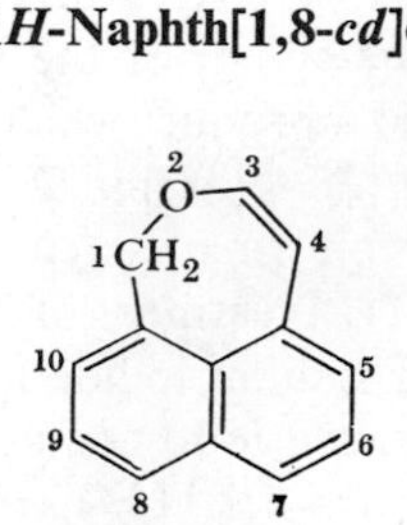

2
3
O
1 CH2
4
10
5
9
6
8
7

A single representative of this ring system was described in 1966 by Ghighi and co-workers (192) in the form of anhydride **333**, which was obtained, along with several other products, upon alkaline hydrolysis of 8-chloroacenaphthylene-7-carboxylic acid (**334**) and treatment with sulfuric acid.

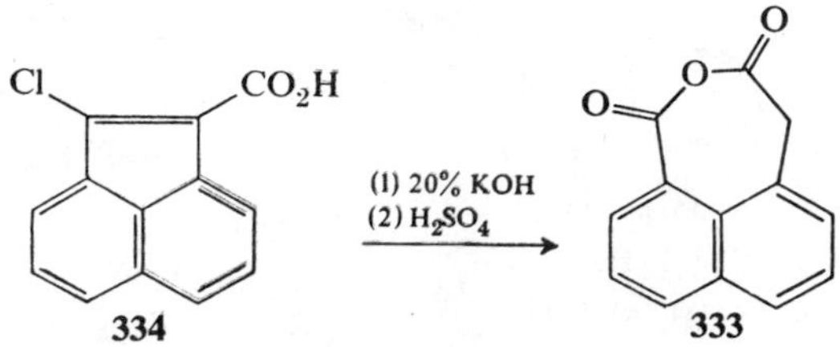

(1) 20% KOH
(2) H_2SO_4

334 **333**

13. [2]Benzoxepino[4,5-*b*]pyridine

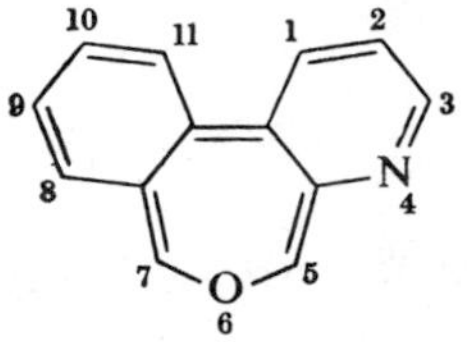

R.I. 3652

This ring system made its first and, thus far, only entry into the chemical literature in 1902 when Marckwald and Dettmer (190) reported that treatment of 3-azadiphenic acid (**335**) with hot acetic anhydride furnished 5*H*,7*H*-[2]benzoxepino[4,5-*b*]pyridine-5,7-dione (**336**). Reaction of **336** with ammonia led to a cleavage of the oxepin ring with formation of amide **337**.

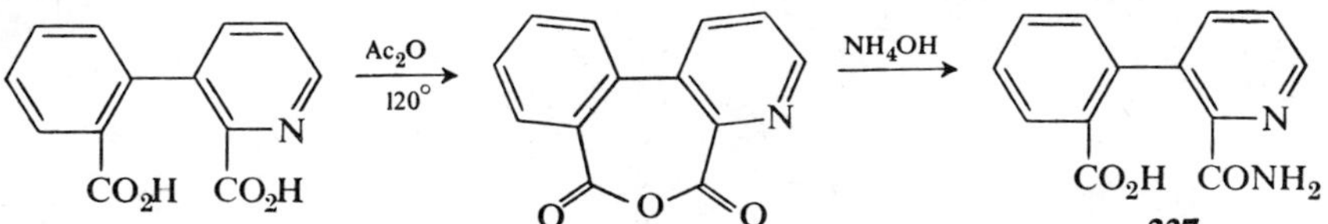

Ac_2O, 120° ; NH_4OH

335 **336** **337**

14. 4*H*-Pyrano[3,2-*h*][1]benzoxepins

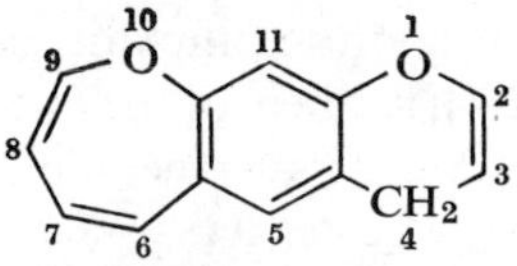

This ring system came to light only recently as a result of investigations by Dean and co-workers (52, 53), as well as McCabe and co-workers (117), concerning the structures of ptaeroxylin (**338**) and its derivatives, ptaeroxylinol (**339**), karenin (**340**), dehydroptaeroxylin (**341**), dihydroptaeroxylin (**342**), ptaeroglycol (**343**), and ptaeroxylone (**344**).

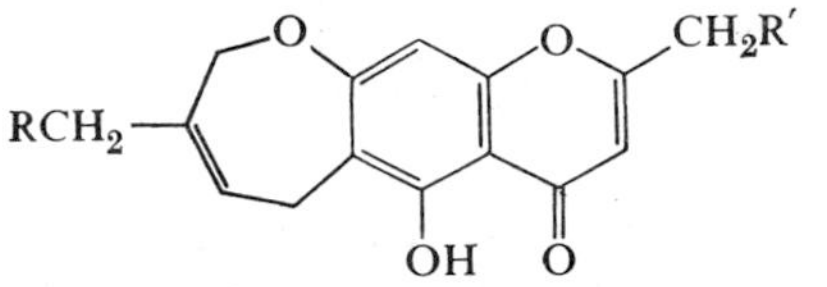

338; R = R′ = H
339; R = OH, R′ = H
340; R = H, R′ = OH

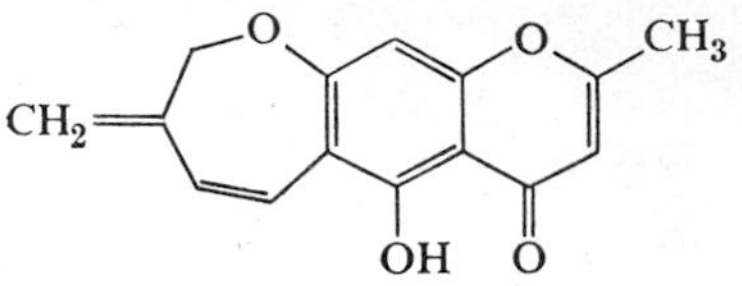

341

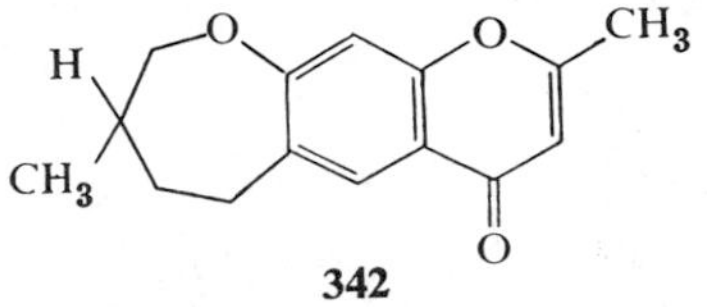

342

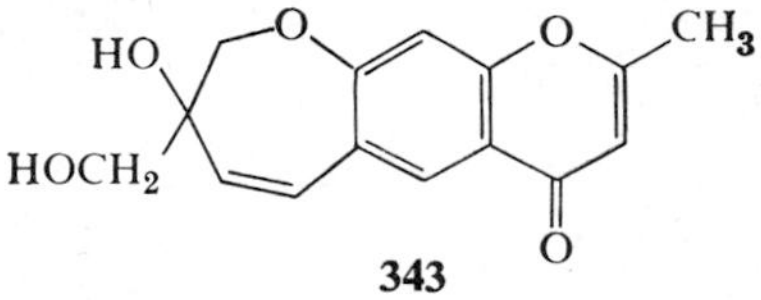

343

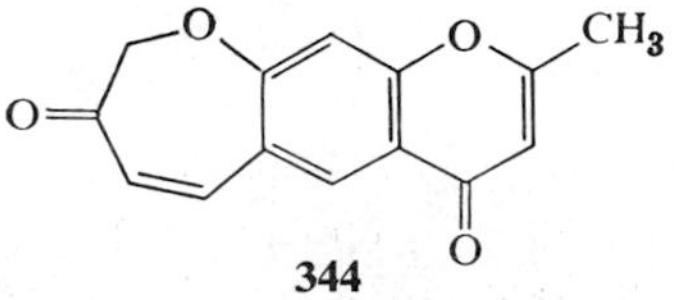

344

15. 6*H*-Oxepino[3,2-*c*][1,2]benzothiazines

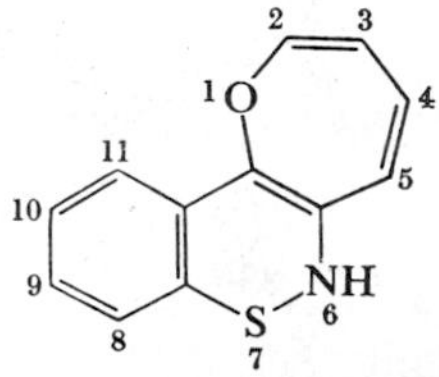

A member of this tricyclic system was synthesized in 1967 by Zinnes and co-workers (188) starting from sodium saccharin (**345**). Condensation with 1,5-dichloro-2-pentanone afforded *N*-alkyl derivative **346**, and treatment of **346** with two equivalents of sodium ethoxide gave an 8% yield of 2,3,4,5-tetrahydro-6*H*-oxepino[3,2-*c*][1,2]benzothiazin-5-one 7,7-dioxide (**347**), together with a larger amount of an incompletely characterized product probably consisting, in part, of ester **348**. When **346** was treated with three, rather than two, equivalents of sodium ethoxide, an 8% yield of **347** was obtained once again, but the main product (40% yield) was cyclopropane

NNa SO_2 **345**

$ClCH_2CO(CH_2)_3Cl$

N—$CH_2CO(CH_2)_3Cl$ SO_2 **346**

NaOEt

CO_2Et $SO_2NHCH_2CO(CH_2)_3Cl$ **348**

NaOEt

CO_2Et $SO_2N\overset{\ominus}{H}CHCO(CH_2)_3Cl$

H —$CO(CH_2)_3Cl$ NH SO_2

NaOEt

O NH SO_2 **347**

NaOEt

H —COCH CH_2 CH_2 NH SO_2 **349**

HBr AcOH

HBr AcOH

H —$CO(CH_2)_3Br$ NH SO_2 **350**

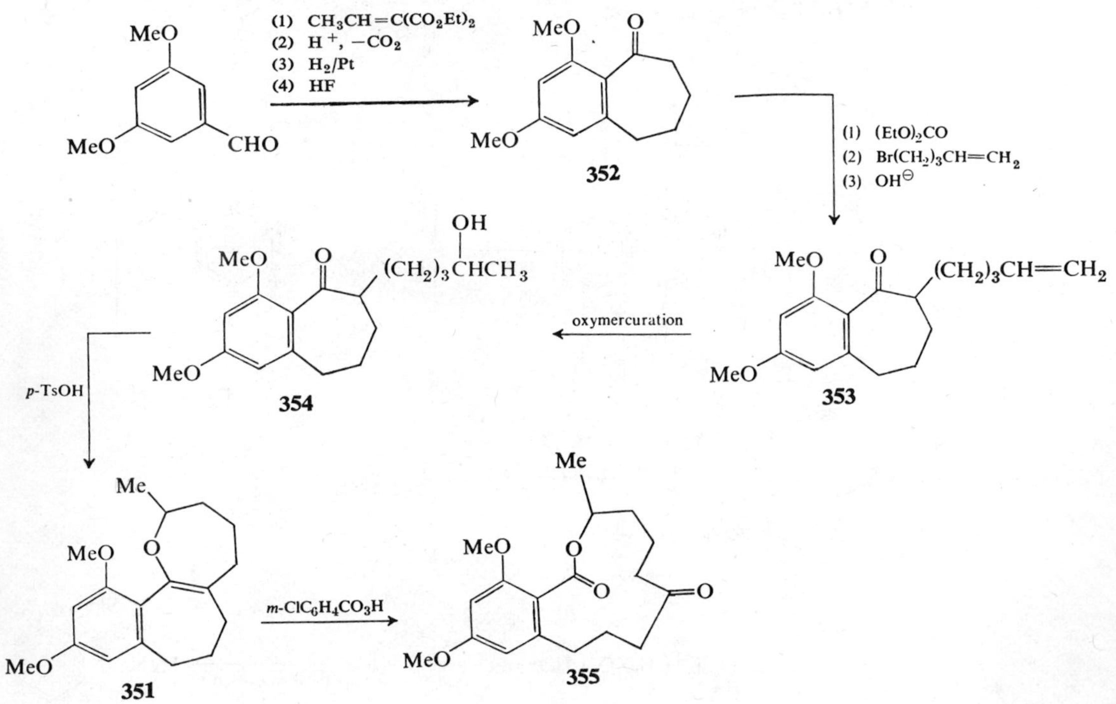

MeO

MeO CHO

(1) $CH_3CH{=}C(CO_2Et)_2$
(2) H^+, $-CO_2$
(3) H_2/Pt
(4) HF

MeO O

MeO

352

(1) $(EtO)_2CO$
(2) $Br(CH_2)_3CH{=}CH_2$
(3) $OH^{\ominus}$

MeO O $(CH_2)_3CH{=}CH_2$

MeO

353

oxymercuration

OH

MeO O $(CH_2)_3CHCH_3$

MeO

354

p-TsOH

Me

MeO O

MeO

351

m-$ClC_6H_4CO_3H$

Me

MeO O O

O

MeO

355

349. According to the authors, the transformation of **346** into **347** and **349** proceeds via the steps indicated, and is therefore analogous to the Gabriel-Colman rearrangement of structurally related *N*-alkylphthalimide derivatives. Support for this mechanism comes from the fact that treatment of either **347** or **349** with hydrogen bromide affords bromide **350**, and that **347** can be regenerated from **350** upon treatment with base.

16. 6*H*-Benzo[3,4]cyclohept[2,1-*b*]oxepins

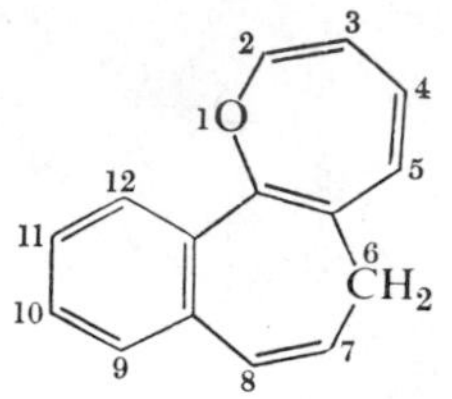

The first report of this ring system appeared in 1968 when Bagli and Immer (213) described the synthesis of 2,3,4,5,7,8-hexahydro-10,12-dimethoxy-2-methyl-6*H*-benzo[3,4]cyclohept[2,1-*b*]oxepin (**351**) starting from 3,5-dimethoxybenzaldehyde. This compound was transformed into 7,9-dimethoxy-1-benzosuberone (**352**) and thence into 7,9-dimethoxy-2-(4-penten-1-yl)-1-benzosuberone (**353**) via the reaction sequence outlined. Oxymercuration and demercuration of **353** yielded 2-(4-hydroxy-1-pentyl)-7,9-dimethoxy-1-benzosuberone (**354**), and cyclodehydration of **354** under the influence of acid gave **351**. Oxidation of **351** with *m*-chloroperbenzoic acid provided a convenient route to twelve-membered lactone **355**.

B. Bridged and Spiran Ring Systems

1. Spiro[naphthalene-1(2*H*),4′-oxepanes]

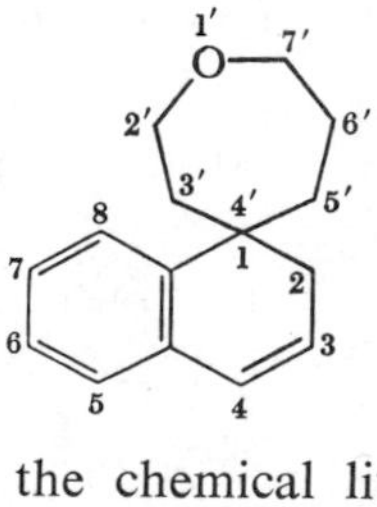

R.I. 3704

This ring system entered the chemical literature through the work of Barltrop and Saxton (13) who reported in 1952 that, under the influence of hydrobromic acid in acetic acid solution, methoxy nitrile **356** underwent hydrolysis and spontaneous lactonization to **357**. For the synthesis of **356**, acrylonitrile was condensed with 1-(2-methoxyethyl)-2-tetralone (**358**) in the presence of base.

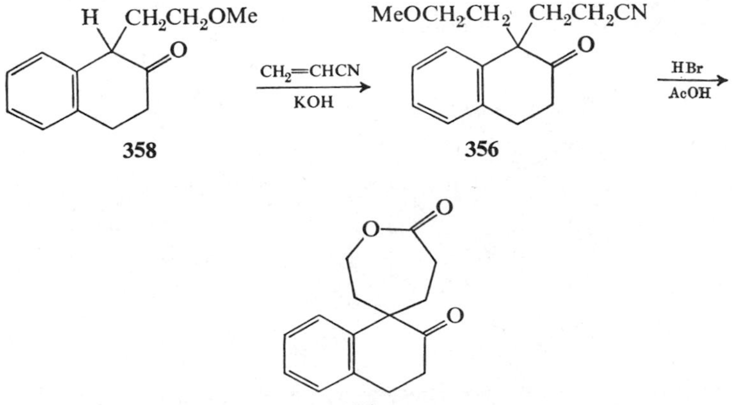

H CH_2CH_2OMe — $CH_2{=}CHCN$ / KOH → $MeOCH_2CH_2$ CH_2CH_2CN — HBr / AcOH →

358 **356**

357

2. 1,5-Ethano-3-benzoxepins

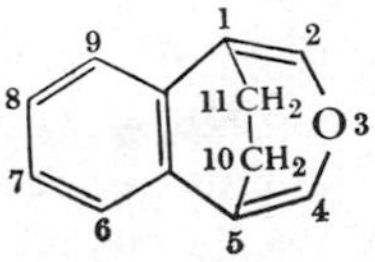

R.I. 8636

Doering and Goldstein (55) first synthesized this bridged 3-benzoxepin derivative in 1959, starting from *cis*-1,2,3,4-tetrahydronaphthalene-1,4-dicarboxylic acid (**359**). Upon being heated at 100° in acetic anhydride, this diacid underwent facile ring closure to **360** in a manner reminiscent of the ready formation of diphenic anhydrides from diphenic acid. Reaction of **360** with ammonia afforded the corresponding imide, **361**. The same transformations were reported in 1962 by Cioranescu and co-workers (189).

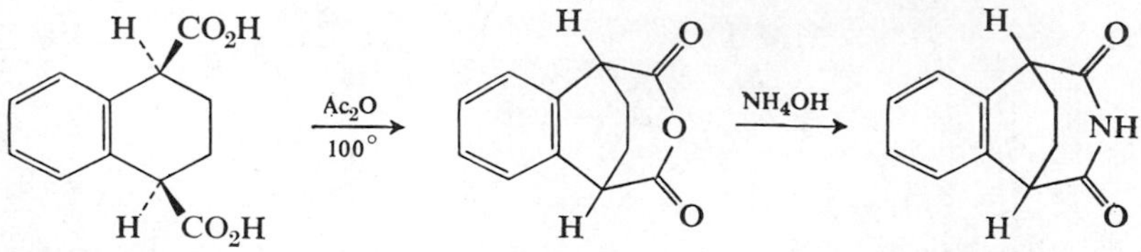

H CO_2H — Ac_2O / 100° → — NH_4OH → NH

H CO_2H

359 **360** **361**

3. 5a,9a-[2]Buteno-3-benzoxepins

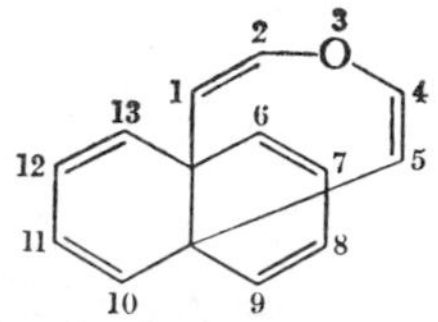

A member of this novel "oxapropellane" system was described in 1966 by Altman and co-workers (216) in the form of anhydride **362**, which was obtained in 78% yield upon treatment of *cis*-4a,8a-bis(carboxymethyl)decalin-2,6-dione (**363**) with boiling acetic anhydride.

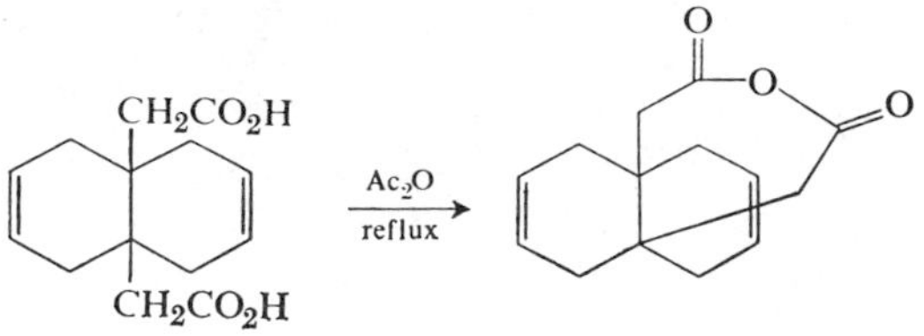

4. 15,16-Dioxatricyclo[8.4.1.1^{3,8}]hexadecanes

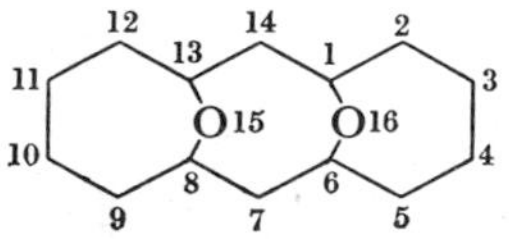

In a fascinating extension of their work on 1,6-oxido[10]annulenes, Vogel and co-workers (186, 187) synthesized the fully unsaturated derivative, 1,6:8,13-bisoxido[14]annulene (**364**), starting from 1,4,5,6,9,10-hexahydroanthracene (**365**). Epoxidation of **365** with perbenzoic acid led to a mixture of *syn*-epoxide **366** and *anti*-epoxide **367**. The *syn*/*trans* ratio in the mixture was estimated to be 60:40 by nmr spectral analysis. Further reaction of each isomer with *N*-bromosuccinimide afforded a crude mixture of tetrabromides, which was dehydrobrominated directly by treatment with potassium *tert*-butoxide in tetrahydrofuran at 0°. Interestingly, the dehydrobromination of the tetrabromide(s) derived from **366** gave a 5% yield of a red compound showing all the properties expected for structure **364**, whereas the same reaction with the tetrabromide(s) derived from **367** afforded only a colorless product still containing intact epoxide rings. The latter product, to which structure **368** was assigned, could not be isomerized either to **364** or to the isomeric *anti* form, **369**. The failure of **367** to undergo transformation via its tetrabromide(s) into **364** was consistent with molecular orbital considerations, according to which resonance stabilization of the annulene system should be sterically possible only in the *syn* form.

$PhCO_3H$

365 **366** **367**

(1) Br_2/CCl_4
(2) KOBu-*tert*

(1) Br_2/CCl_4
(2) KOBu-*tert*

368

−2H

364 **369**

The structure of **364** is supported by a number of chemical and physical lines of evidence. On the physical side, the uv spectrum displays absorption maxima at 306 mμ (ε 169,000), 345 mμ (ε 14,400) 382 mμ (ε 8,500), and 555 mμ (ε 775). These spectral features are consistent with a 14 π-electron annulene system and point to a significant degree of aromaticity. Aromaticity is likewise indicated by the nmr spectrum of **364**, which exhibits signals at low field strengths consistent with a ring current. The C-7 and C-14 protons give rise to a singlet at τ 2.06, while the remaining protons generate a typical AA′BB′ pattern at somewhat higher field strength (τ_A 2.25 and τ_B 2.40). The fairly high dipole moment of **364** (3.25 Debye units) bears out the fact that the oxygen atoms are situated on the same side of the molecule. Chemical evidence for the aromaticity of **364** includes (*1*) its considerable thermal stability, (*2*) its insensitivity to oxygen, and (*3*) its reactivity toward various electrophilic reagents.

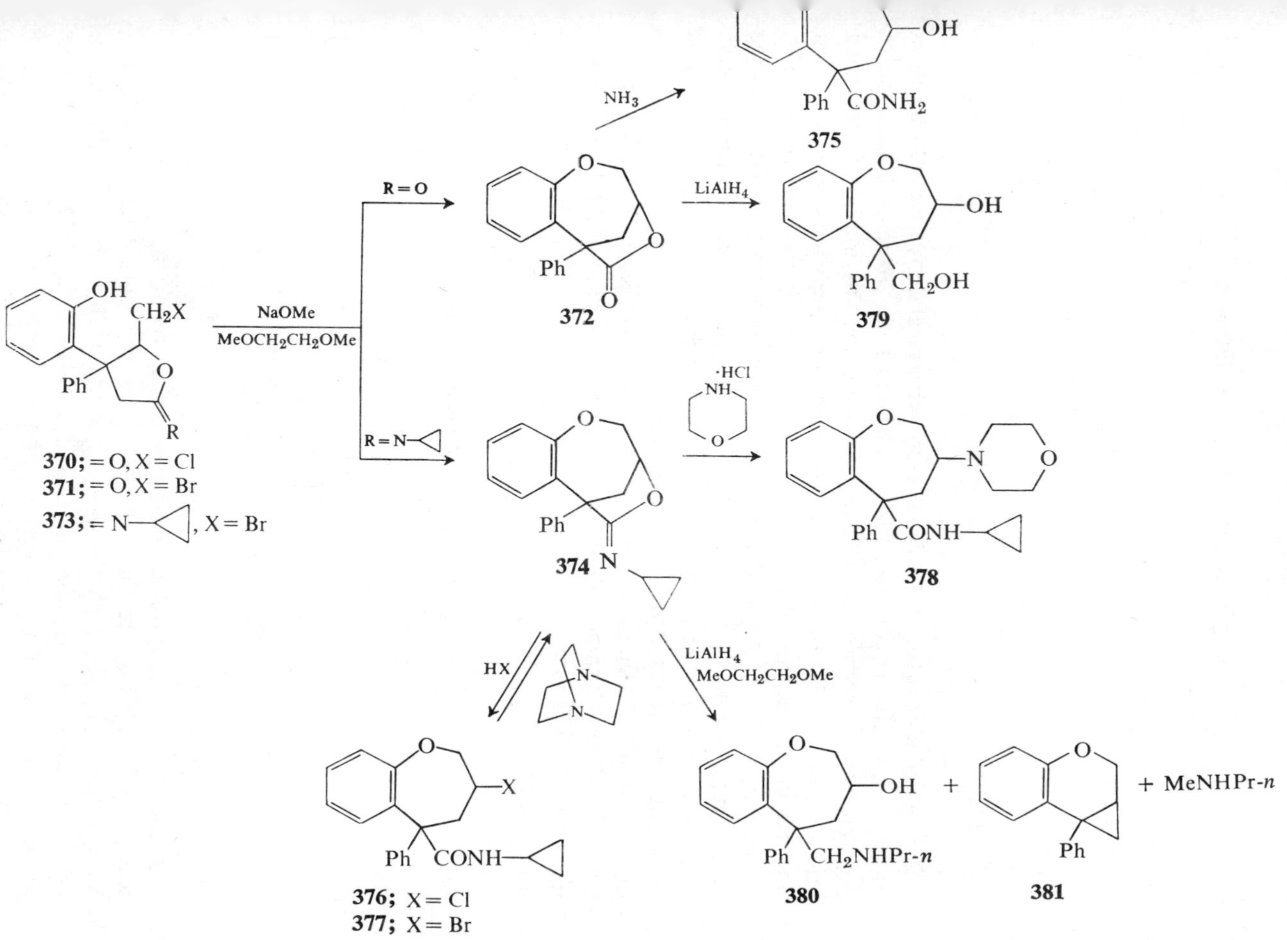

OH

CH_2X

Ph

O

R

370; = O, X = Cl
371; = O, X = Br
373; = N–cyclopropyl, X = Br

NaOMe

$MeOCH_2CH_2OMe$

R = O

R = N–cyclopropyl

NH_3

Ph $CONH_2$ OH

375

Ph O O

372

$LiAlH_4$

Ph CH_2OH OH

379

Ph N

374

·HCl morpholine

Ph CONH–cyclopropyl

378

HX

$LiAlH_4$
$MeOCH_2CH_2OMe$

Ph CONH–cyclopropyl X

376; X = Cl
377; X = Br

Ph CH_2NHPr-*n* OH

380

\+

Ph

381

\+ MeNHPr-*n*

5. 3,6-Methano-1,4-benzodioxocins

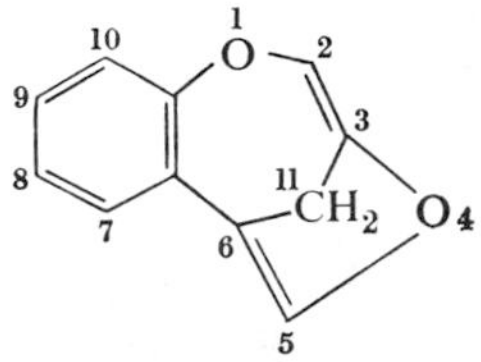

Zaugg and co-workers (217–219) described several derivatives of this novel ring system in connection with their studies on 1-benzoxepins Treatment. of γ-butyrolactone derivatives **370** or **371** with sodium methoxide afforded bridged lactone **372**, while similar treatment of imino lactone **373** yielded bridged imino lactone **374**. Acid hydrolysis of **374** likewise led to **372**. Ammonolysis of bridged lactone **372** gave ring-opened product **375**. Cleavage of the imino lactone function in **374** with hydrogen chloride or bromide gave 1-benzoxepins **376** and **377**, and condensation of **374** with morpholine hydrochloride led to **378**. Dehydrohalogenation of **376** or **377** with 1,4-diazabicyclo[2.2.2]octane was accompanied by recyclization to **374**. Reduction of **372** with lithium aluminum hydride yielded diol **379**, but similar treatment of **373** unexpectedly led to a mixture of **380**, **381**, and methyl *n*-propyl amine.

6. 3,6-Metheno-1,4-benzoxazocins

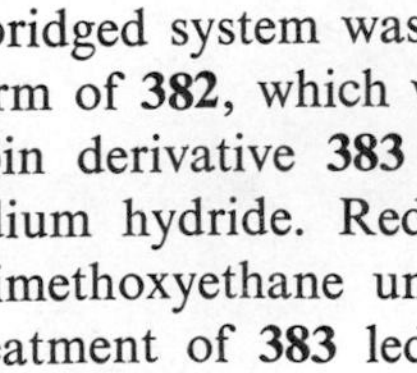

A member of this novel bridged system was reported by Zaugg and coworkers (217–219) in the form of **382**, which was obtained upon dehydrohalogenation of 1-benzoxepin derivative **383** with sodium methoxide or ethoxide, sodamide, or sodium hydride. Reduction of **382** with lithium aluminum hydride in 1,2-dimethoxyethane unexpectedly gave amino carbinol **384**, while similar treatment of **383** led to the completely reduced product, **385**.

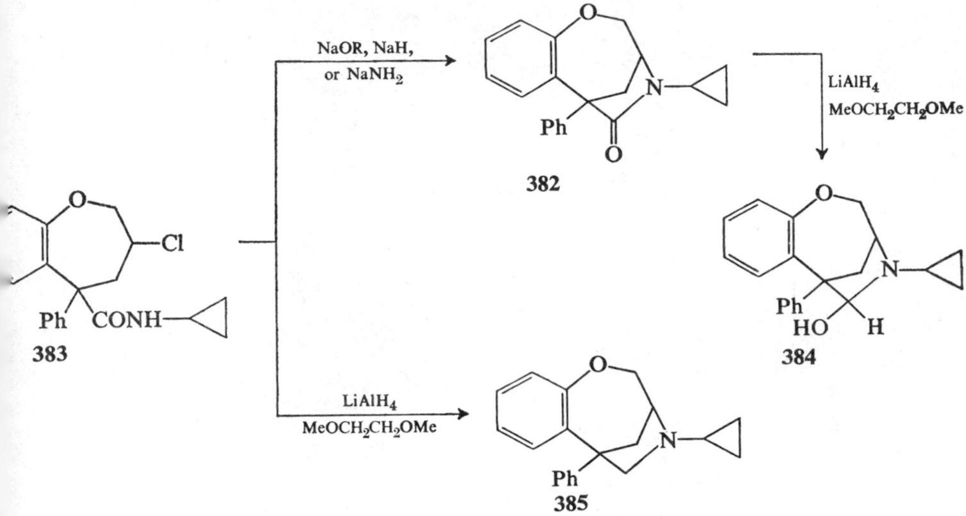

NaOR, NaH, or $NaNH_2$

382

$LiAlH_4$ $MeOCH_2CH_2OMe$

383

384

$LiAlH_4$ $MeOCH_2CH_2OMe$

385

7. 2*H*-Spiro[1-benzoxepin-5,2′-dioxoles]

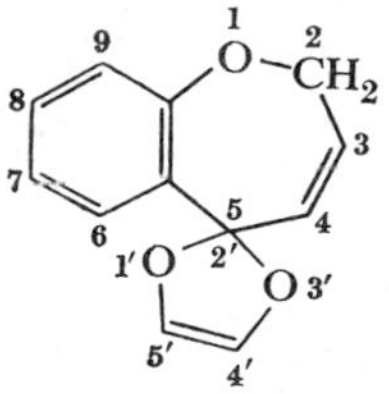

Two members of this ring system were reported in 1968 by Fontaine (220) in the form of spiroketals **386** and **387**, which were obtained from 2,3,4,5-tetrahydro-1-benzoxepin-5-one (**388**) and 4-bromo-2,3,4,5-tetrahydro-1-benzoxepin-5-one (**389**), respectively, by condensation with ethylene glycol in the presence of acid.

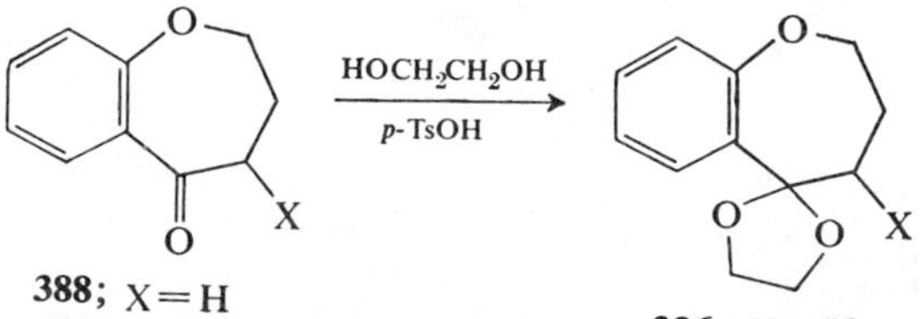

$HOCH_2CH_2OH$ *p*-TsOH

388; X=H
389; X=Br

386; X=H
387; X=Br

C. Tables

The physical constants of known dibenz[*b*,*e*]oxepins, dibenz[*b*,*f*]oxepins, and dibenz[*c*,*e*]oxepins are listed in Tables III-1, III-2, and III-3, respectively.

TABLE III-1. Dibenz[*b*,*e*]oxepins

Empirical formula	Name of compound	Physical constants	Derivatives	Refs.
$C_{14}H_6Br_2O_4$	2,4-Dibromo-3-hydroxydibenz[*b*,*e*]oxepin-6,11-dione		*O*-Acetate, mp 200–201°C dec (Ac_2O)	132
$C_{14}H_8O_3$	Dibenz[*b*,*e*]oxepin-6,11-dione	mp 179–180°C (?) mp 84–85°C		10,105
$C_{14}H_8O_4$	3-Hydroxydibenz[*b*,*e*]oxepin-6,11-dione		*O*-Acetate, mp 135–136°C (AcOH)	132
$C_{14}H_9BrO_2$	2-Bromodibenz[*b*,*e*]oxepin-11(6*H*)-one	$b_{0.05}$ 165–168°C mp 135–137°C		18, 151, 152
$C_{14}H_9ClO_2$	2-Chlorodibenz[*b*,*e*]oxepin-11(6*H*)-one	$b_{0.5}$ 162–166°C mp 126–127°C		18, 151, 152, 176
$C_{14}H_{10}O_2$	Dibenz[*b*,*e*]oxepin-6(11*H*)-one	mp 98–99°C		10
	Dibenz[*b*,*e*]oxepin-11(6*H*)-one	mp 71–72°C $b_{0.2}$ 142–145°C		18, 20, 151, 152, 176
$C_{14}H_{11}ClO$	11-Chloro-6,11-dihydrodibenz[*b*,*e*]oxepin	mp 75–77, 80–82°C		210, 211
$C_{14}H_{11}ClO_2$	2-Chloro-6,11-dihydrodibenz[*b*,*e*]-oxepin-11-ol	mp 140°C (*i*-PrOH)		176
$C_{14}H_{12}O_2$	6,11-Dihydrodibenz[*b*,*e*]oxepin-11-ol	mp 88–89°C (*n*-hexane)		169, 176, 211
$C_{15}H_9F_3O_2$	2-Trifluoromethyldibenz[*b*,*e*]-oxepin-11(6*H*)-one	mp 108.5–109.5°C		20
$C_{15}H_{10}O_3$	2-Methyldibenz[*b*,*e*]oxepin-6,11-dione	mp 145°C		10
$C_{15}H_{10}O_4$	3-Methoxydibenz[*b*,*e*]oxepin-6,11-dione	mp 167°C (Ac_2O)		105
$C_{15}H_{12}O_2$	2-Methyldibenz[*b*,*e*]oxepin-11(6*H*)-one	$b_{0.1}$ 147–150°C mp 108–109°C		18, 151, 152
	2-Methyldibenz[*b*,*e*]oxepin-6(11*H*)-one	mp 121°C		10
	3-Methyldibenz[*b*,*e*]oxepin-11(6*H*)-one	$b_{0.1}$ 140–147°C mp 71–72°C (ligroin–Et_2O)		18, 152

$C_{15}H_{12}O_3$	2-Methoxydibenz[*b*,*e*]oxepin-11(6*H*)-one	mp 93–94°C $b_{0.05}$ 162–166°C		151, 152
$C_{16}H_{15}NO_4S$	*N*,*N*-Dimethyldibenz[*b*,*e*]oxepin-11(6*H*)-one-2-sulfonamide			20
$C_{17}H_{16}O_2$	11-Allyl-6,11-dihydrodibenz[*b*,*e*]oxepin-11-ol			20
$C_{18}H_{19}NO$	*N*,*N*-Dimethyldibenz[*b*,*e*]-oxepin-$\Delta^{11(6H),\beta}$-ethylamine	$b_{0.3}$ 175–178°C		38
	N-Methyldibenz[*b*,*e*]oxepin-$\Delta^{11(6H),\gamma}$-propylamine	mp 63–65, 65–67°C (ligroin) $b_{0.05}$ 158–164°C $b_{0.01}$ 168–172°C	Hydrochloride, mp 235–236°C (*i*-PrOH or dioxane–Et_2O) mp 241–242°C (*cis* and *trans*) mp 225–226.5°C (*cis* only) Hydrobromide, mp 223–225°C	19, 20, 30, 37, 123
$C_{18}H_{19}N_3O_2$	6,11-Dihydro-11-(4-nitroso-1-piperazinyl)-dibenz[*b*,*e*]oxepin	mp 190–191°C		211
$C_{18}H_{21}N_3O$	11-(4-Amino-1-piperazinyl)-6,11-dihydrodibenz[*b*,*e*]oxepin	mp 146–147°C		211
$C_{19}H_{18}ClNO_2$	2-Chloro-11-[3-(*N*,*N*-dimethylamino)-propylidene]dibenz[*b*,*e*]oxepin-6(11*H*)-one	mp 96–97°C	Hydrochloride, mp 261–262°C	39
$C_{19}H_{19}NO_2$	11-[3-(*N*,*N*-Dimethylamino)propylidene]-dibenz[*b*,*e*]oxepin-6(11*H*)-one		Hydrochloride, mp 254–255°C	39
$C_{19}H_{20}ClNO$	2-Chloro-*N*,*N*-dimethyldibenz-[*b*,*e*]oxepin-$\Delta^{11(6H),\gamma}$-propylamine	mp 140–144°C $b_{0.3}$ 175–181°C	Hydrochloride, mp 216–218, 220–222°C	105, 123, 150
$C_{19}H_{21}NO$	*N*,*N*-Dimethyldibenz[*b*,*e*]oxepin-Δ-$^{11(6H),\gamma}$-propylamine	mp 118–119°C $b_{0.03}$ 154–157°C $b_{0.1}$ 166–170°C	Hydrochloride, mp 184°C (*i*-PrOH), 184–186°C (MeCOEt),	20, 30, 38, 105, 123, 150

TABLE III-1.—(*contd.*)

Empirical formula	Name of compound	Physical constants	Derivatives	Refs.
$C_{19}H_{21}NO$ (*cont.*)		$b_{0.2}$ 260–270°C	228–230°C (4:1 *cis*/*trans*), 192–193°C (*trans*), 209–210.5°C (*cis*) Maleate, mp 161–164°C (dioxan), 168–169°C (*cis* + *trans*), 172–173°C (EtOH) (pure *trans*)	
$C_{19}H_{22}ClNO_2$	2-Chloro-11-[3-(*N*,*N*-dimethylamino)-propyl]-6,11-dihydrodibenz[*b*,*e*]oxepin-11-ol	mp 144–145°C (*i*-PrOH)		150
$C_{19}H_{22}N_2O$	6,11-Dihydro-11-(4-methyl-1-piperazinyl)dibenz[*b*,*e*]oxepin		Maleate, mp 175–177°C (EtOH)	210
$C_{19}H_{23}NO_2$	11-[3-(*N*,*N*-Dimethylaminopropyl]-6,11-dihydrodibenz[*b*,*e*]oxepin-11-ol	mp 119–120°C (*i*-PrOH)		20, 150, 175
$C_{20}H_{14}O_3$	3-Hydroxy-11-phenyldibenz[*b*,*e*]oxepin-6(11*H*)-one	mp 242–242.5°C subl 200°C (0.01 mm)	*O*-Acetate, mp 174–175°C (EtOAc)	78
$C_{20}H_{20}ClNO_2$	2-Chloro-11-[2-methyl-3-(*N*,*N*-dimethylamino)propylidene]dibenz-[*b*,*e*]oxepin-6(11*H*)-one		Hydrochloride, mp 263–264°C dec	39
$C_{20}H_{21}NO_2$	11-[2-Methyl-3-(*N*,*N*-dimethyl-amino)propylidene]dibenz[*b*,*e*]oxepin-6(11*H*)-one		Hydrochloride, mp 200–201°C	39
	2-Methyl-11[3-(*N*,*N*-dimethylamino)-propylidene]dibenz[*b*,*e*]oxepin-6(11*H*)-one	mp 94–96°C	Hydrochloride, mp 247–249°C dec	39
$C_{20}H_{23}NO$	*N*,*N*,2-Trimethyldibenz[*b*,*e*]-oxepin-$\Delta^{11(6H),\gamma}$-propylamine	mp 125–127°C $b_{0.3}$ 164–167°C	Hydrochloride, mp 176–178°C	105, 150

$C_{20}H_{23}NO_2$	2-Methoxy-*N*,*N*-dimethyldibenz-[*b*,*e*]oxepin-$\Delta^{11(6H),\gamma}$-propylamine	mp 107–111°C $b_{0.3}$ 183–185°C	Hydrochloride, mp 183–185°C	105, 150
$C_{20}H_{24}N_2O_3S$	6,11-Dihydro-*N*,*N*-dimethyl-11-[3-(N-methylamino)propylidene]-dibenz[*b*,*e*]oxepin-2-sulfonamide		Hydrochloride, mp 199–201°C	20
$C_{20}H_{25}NO_2$	11-[3-(*N*,*N*-Dimethylamino)propyl]-6,11-dihydro-2-methyldibenz[*b*,*e*]-oxepin-11-ol	mp 126–128°C (*i*-PrOH)		150
$C_{20}H_{25}NO_3$	11-[3-(*N*,*N*-Dimethylamino)propyl]-6,11-dihydro-2-methoxydibenz[*b*,*e*]-oxepin-11-ol	mp 113–117°C (*i*-PrOH)		150
$C_{21}H_{16}O_3$	3-Methoxy-11-phenyldibenz[*b*,*e*]-oxepin-6(11*H*)-one	mp 191.9–192.7°C (EtOH)		78
$C_{21}H_{20}ClNO_3$	3-[3-(2-Chlorodibenz[*b*,*e*]oxepin-6-one-11-ylidene)propyl]morpholine	mp 141–142°C	Hydrochloride, 265°C dec	39
$C_{21}H_{21}NO_3$	3-[3-(Dibenz[*b*,*e*]oxepin-6-one-11-ylidene)propyl]morpholine		Hydrochloride, mp 220–222°C	39
$C_{21}H_{23}NO_2$	2-Methyl-11-[2-methyl-3-(*N*,*N*-dimethylamino)propylidene]dibenz-[*b*,*e*]oxepin-6(11*H*)-one		Hydrochloride, mp 251–252°C dec	39
$C_{21}H_{23}NO_3$	Ethyl *N*-3-(dibenz[*b*,*e*]oxepin-11(6*H*)-ylidenepropyl)-*N*-methylcarbamate	$b_{0.1}$ 181–183°C		30, 37, 150
$C_{21}H_{24}N_2O$	1-(Dibenz[*b*,*e*]oxepin-11(6*H*)-ylidene)propylpiperazine		Dihydrochloride, mp 189–193°C	123
$C_{21}H_{26}N_2O_3S$	6,11-Dihydro-*N*,*N*-dimethyl-11-[3-(*N*,*N*-dimethylamino)propylidene]-dibenz[*b*,*e*]oxepin-2-sulfonamide		Hydrochlorides, mp 225–227°C (*cis*), 214–217°C (*trans*)	123
$C_{22}H_{16}O_3$	3-Hydroxy-11-α-methylbenzylidene-dibenz[*b*,*e*]oxepin-6(11*H*)-one (mixture of *cis* and *trans*)	mp 200–207°C	*O*-Acetate, mp 190–191°C (1:1 C_6H_6–petroleum ether)	29
$C_{22}H_{23}NO_2$	1-(3-Dibenz[*b*,*e*]oxepin-6-one-11-ylidene)propylpiperidine		Hydrochloride, mp 211–212°C dec	39

TABLE III-1.—(*contd.*)

Empirical formula	Name of compound	Physical constants	Derivatives	Refs.
$C_{22}H_{23}NO_3$	3-[3-(2-Methyldibenz[*b*,*e*]oxepin-6-one-11-ylidene)propyl]morpholine		Hydrochloride, mp 255–260°C dec	39
$C_{22}H_{24}N_2O_2$	1-Methyl-3-[3-(dibenz[*b*,*e*]oxepin-6-one-11-ylidene)propyl]piperazine		Dihydrochloride, mp 248–250°C dec	39
$C_{22}H_{25}NO$	1-(3-Dibenz[*b*,*e*]oxepin-11(6*H*)-ylidenepropyl)piperidine	mp 140–143°C $b_{0.2}$ 190–195°C	Succinate, mp 136–138°C (*i*-PrOH)	105, 150
$C_{22}H_{26}N_2O$	1-(3-Dibenz[*b*,*e*]oxepin-11(6*H*)-ylidenepropyl)-4-methylpiperazine	mp 151–155°C dec $b_{0.1}$ 200–205°C	Hydrochloride, mp 256–258°C (*i*-PrOH)	105, 150
$C_{22}H_{27}NO_2$	6,11-Dihydro-11-(3-piperidinopropyl)-dibenz[*b*,*e*]oxepin-11-ol	mp 142–143°C (*i*-PrOH)		150
$C_{22}H_{28}N_2O_2$	6,11-Dihydro-11-[3-(4-methyl-1-piperazinyl)propyl]dibenz-[*b*,*e*]oxepin-11-ol	mp 153–155°C (*i*-PrOH)		150
$C_{23}H_{18}O_3$	11-Ethylbenzylidene-3-hydroxy-dibenz[*b*,*e*]oxepin-6(11*H*)-one		*O*-Acetate, mp 152–153°C (petroleum ether)	29
$C_{23}H_{25}NO_2$	1-[3-(2-Methyldibenz[*b*,*e*]oxepin-6-one-11-ylidene)propyl]piperidine		Hydrochloride, mp 245–247°C dec	39
$C_{23}H_{26}N_2O_2$	1-Methyl-3-[3-(2-methyldibenz[*b*,*e*]-oxepin-6-one-11-ylidene)propyl]-piperazine	$b_{0.1}$ 232–240°C	Dihydrochloride, mp 269–271°C dec	39
$C_{23}H_{27}ClN_2O_2$	4-[3-(2-Chlorodibenz[*b*,*e*]oxepin-11(6*H*)-ylidene)propyl]-1-piperazine-ethanol		Dihydrochloride, mp 257–259°C	123
$C_{24}H_{24}N_4O$	6,11-Dihydro-11-[4-(3-pyridylazinyl)-1-piperazinyl]dibenz[*b*,*e*]oxepin	mp 222–224°C		211
	6,11-Dihydro-11-[4-(4-pyridylazinyl)-1-piperazinyl]dibenz[*b*,*e*]oxepin	mp 205–206°C		211

$C_{24}H_{29}ClN_2O_3$	3-[4-[3-(2-Chlorodibenz[*b*,*e*]oxepin-11(6*H*)-ylidene)propyl]-1-piperazinyl]-propane-1,2-diol		Dihydrochloride, mp 253–255°C	123
$C_{25}H_{25}NO$	*N*-Benzyl-*N*-methyldibenz[*b*,*e*]-oxepin-$\Delta^{11(6H),\gamma}$-propylamine	mp 104–107°C $b_{0.15}$ 210–215°C $b_{0.1}$ 220–230°C	Hydrochloride, mp 197–199°C (*i*-PrOH)	105, 150
$C_{25}H_{25}N_3O_2$	6,11-Dihydro-11-[4-(*p*-hydroxyphenyl-azinyl)-1-piperazinyl]dibenz[*b*,*e*]oxepin	mp 199–200°C		211
$C_{25}H_{27}NO_2$	11-[3-(*N*-Benzyl-*N*-methylamino)propyl]-6,11-dihydrodibenz[*b*,*e*]oxepin-11-ol	mp 106–107°C (*i*-PrOH)		150
$C_{26}H_{25}N_3O_2$	6,11-Dihydro-11-[4-(3,4-methylene-dioxyphenylazinyl)-1-piperazinyl]-dibenz[*b*,*e*]oxepin	mp 171–172°C		211
$C_{27}H_{36}N_4O_4S$	4-[3-[2-(*N*,*N*-Dimethylsulfamoyl)dibenz-[*b*,*e*]oxepin-11(6*H*)-ylidene]propyl]-*N*-methyl-1-piperazinepropionamide		Dihydrochloride, mp 222–224°C	123

TABLE III-2. Dibenz[*b,f*]oxepins

Empirical formula	Name of compound	Physical constants	Derivatives	Refs.
$C_{14}H_7BrN_2O_4$	8-Bromo-1,3-dinitrodibenz[*b,f*]oxepin	mp 174°C		76
$C_{14}H_8N_2O_5$	1,3-Dinitrodibenz[*b,f*]oxepin	mp 154°C (AcOH)		76
$C_{14}H_8O_3$	Dibenz[*b,f*]oxepin-10,11-dione	mp 118°C (petroleum ether), 119°C (C_6H_6)	Mono-2,4-DNP, mp 267°C (AcOH), 270°C ($CHCl_3$)	116, 141
$C_{14}H_9BrO$	10-Bromodibenz[*b,f*]oxepin	b_{10} 205–210°C		193
$C_{14}H_9NO_3$	10-Oximinodibenz[*b,f*]oxepin-11-one	mp 180–181°C		85, 96
	2-Nitrodibenz[*b,f*]oxepin	mp 130°C (MeOH)		109
	10-Nitrodibenz[*b,f*]oxepin	mp 102–103°C		193
$C_{14}H_{10}Br_2O$	2,8-Dibromo-10,11-dihydrodibenz[*b,f*]-oxepin	mp 210°C		199
	10,11-Dibromo-10,11-dihydrodibenz[*b,f*]-oxepin	mp 152.5–153.5°C		193
$C_{14}H_{10}N_2O_6$	10,11-Dihydro-1,3-dinitrodibenz[*b,f*]-oxepin-10,11-diol	mp 188–189°C dec (C_6H_6–petroleum ether)		76
$C_{14}H_{10}O$	Dibenz[*b,f*]oxepin	mp 111°C (MeOH–Me_2CO)		4, 5, 109, 113, 174, 193
$C_{14}H_{10}O_2$	Dibenz[*b,f*]oxepin-10(11*H*)-one	mp 56°C (*n*-hexane) b_3 162–165°C	Oxime, mp 135°C (C_6H_6–ligroin), 137°C (aq MeOH) 2,4-DNP, mp 237–238°C	4, 97, 109, 113, 193
$C_{14}H_{11}ClO$	10-Chloro-10,11-dihydrodibenz[*b,f*]oxepin			136
$C_{14}H_{11}NO_2$	2-Aminodibenz[*b,f*]oxepin-10(11*H*)-one	mp 158°C	Oxime, mp 184° (C_6H_6)	109
	10(11*H*)-Oximinodibenz[*b,f*]oxepin	mp 136–137°C		193
$C_{14}H_{11}NO_5$	10,11-Dihydro-2-nitrodibenz[*b,f*]oxepin-10,11-diol	mp 196°C (C_6H_6–MeOH)		109

$C_{14}H_{12}O$	10,11-Dihydrodibenz[*b*,*f*]oxepin	mp 154°C (C_6H_6–hexane)		199, 202
$C_{14}H_{12}O_2$	10,11-Dihydrodibenz[*b*,*f*]oxepin-10-ol	$b_{0.2}$ 100–103°C, oil		4
$C_{14}H_{13}NO$	10-Amino-10,11-dihydrodibenz[*b*,*f*]oxepin		*d*,*l*-Hydrochloride, mp 265°C (HCl), 277°C (H_2O or EtOH) *d*-Hydrochloride, mp 258–260°, $[\alpha]_D^{23}$ +104.2° *l*-Hydrochloride, mp 260–262°, $[\alpha]_D^{23}$ −105° *l*-(*D*-Mandalate), mp 178–181°C *N*-Acetyl, mp 139°C (C_6H_6–ligroin)	82, 83, 96, 109
$C_{14}H_{13}NO_2$	10-Amino-10,11-dihydrodibenz[*b*,*f*]-oxepin-11-ol	mp 183–185°C (EtOH)	Hydrochloride, mp 233–234°C dec	85, 96
$C_{15}H_9NO$	10-Cyanodibenz[*b*,*f*]oxepin	mp 159–160°C		193
$C_{15}H_9NO_5$	2-Nitrodibenz[*b*,*f*]oxepin-10-carboxylic acid	mp 224°C (C_6H_6–ligroin)		109
$C_{15}H_{10}N_2O_4$	8-Methyl-1,3-dinitrodibenz[*b*,*f*]oxepin	mp 148°C		76
$C_{15}H_{10}N_2O_5$	6-Methoxy-1,3-dinitrodibenz[*b*,*f*]oxepin	mp 187.5°C		76
$C_{15}H_{11}BrO$	10-Bromomethyldibenz[*b*,*f*]oxepin	mp 88–90°C		124
$C_{15}H_{11}NO_5$	6-Methoxy-2-nitrodibenz[*b*,*f*]oxepin-10(11*H*)-one	mp 195°C (C_6H_6–ligroin)	Oxime, mp 211°C (C_6H_6–ligroin)	109
$C_{15}H_{12}O$	10-Methyldibenz[*b*,*f*]oxepin	mp 57–58.5°C ($CHCl_3$)		5
$C_{15}H_{12}O_3$	2-Methoxydibenz[*b*,*f*]oxepin-11(10*H*)-one	$b_{0.015}$ 163–165°C		124
	8-Methoxydibenz[*b*,*f*]oxepin-11(10*H*)-one	mp 85°C (Et_2O–*n*-hexane), 93°C (ligroin)	Oxime, mp 196°C (MeOH)	109, 113

TABLE III-2.—(*contd.*)

Empirical formula	Name of compound	Physical constants	Derivatives	Refs.
$C_{15}H_{12}O_3$ (*cont.*)	10,11-Dihydrodibenz[*b*,*f*]oxepin-10-carboxylic acid	mp 186°C (C_6H_6)		109
$C_{15}H_{13}NO_2$	*N*-(10,11-Dihydrodibenz[*b*,*f*]oxepin-10-yl)-formamide	mp 108–109°C		86, 96
$C_{15}H_{13}NO_3$	2-Amino-10,11-dihydrodibenz[*b*,*f*]oxepin-10-carboxylic acid	mp 210°C dec (aq alcohol)	Hydrochloride, mp 262°C dec	109
$C_{15}H_{15}NO$	10,11-Dihydro-10-(*N*-methylamino)dibenz-[*b*,*f*]oxepin	b_3 160–163°C	Hydrochloride, mp 195–196°C	86, 96
$C_{16}H_{11}NO_6$	6-Methoxy-2-nitrodibenz[*b*,*f*]oxepin-10-carboxylic acid	mp 250°C dec		109
$C_{16}H_{12}O_5$	10,11-Dihydrodibenz[*b*,*f*]oxepin-4,6-dicarboxylic acid	mp 234–235°C (AcOH)		202, 221
$C_{16}H_{14}Br_2O$	4,6-Bis(bromomethyl)-10,11-dihydrodibenz-[*b*,*f*]oxepin	mp 124–125°C (hexane)		202
$C_{16}H_{14}NO_2$	10-(*N*,*N*-Dimethylamino)dibenz[*b*,*f*]oxepin-11(10*H*)-one	mp 119–121°C	Hydrochloride, mp 185–186°C dec	84, 96
$C_{16}H_{14}O_2$	2-Methoxy-10-methyldibenz[*b*,*f*]oxepin	$b_{0.001}$ 144–146.5°C		124
$C_{16}H_{14}O_3$	2-Methoxy-10-methyldibenz[*b*,*f*]oxepin-11(10*H*)-one	$b_{0.015}$ 163–165°C		124
	2,3-Dimethoxydibenz[*b*,*f*]oxepin	mp 114–116°C (MeOH)		201
$C_{16}H_{14}O_4$	2,3-Dimethoxydibenz[*b*,*f*]oxepin-11(10*H*)-one	mp 136°C (MeOH or C_6H_6–hexane)		201
$C_{16}H_{14}O_4$	2,3-Dimethoxydibenz[*b*,*f*]oxepin-10(11*H*)-one	mp 115–116°C (MeOH)		90, 102
$C_{16}H_{15}ClO_2$	11-Chloro-10,11-dihydro-10-methyldibenz-[*b*,*f*]oxepin			124
$C_{16}H_{15}ClO_3$	10-Chloro-10,11-dihydro-2,3-dimethoxy-dibenz[*b*,*f*]oxepin	mp 126–127°C (Et_2O–$CHCl_3$)		201

$C_{16}H_{15}NO$	10-(*N*-Methylamino)methyldibenz[*b*,*f*]-oxepin	$b_{0.004}$ 145°C	Hydrochloride, mp 185–188°C	124, 196
$C_{16}H_{16}O_3$	10,11-Dihydro-2-methoxy-10-methyldibenz-[*b*,*f*]oxepin-11-ol	mp 132–135.5°C		124
	10,11-Dihydrodibenz[*b*,*f*]oxepin-4,6-dimethanol	mp 123–124°C		202
$C_{16}H_{16}O_4$	10,11-Dihydro-2,3-dimethoxydibenz[*b*,*f*]-oxepin-10-ol	mp 167.5–169°C (EtOH)	*O*-Acetyl, mp 114–115.5°C	90
	10,11-Dihydro-2,3-dimethoxydibenz[*b*,*f*]-oxepin-11-ol	mp 147–148°C (C_6H_6–hexane)		201
$C_{16}H_{17}NO$	10,11-Dihydro-10-(*N*,*N*-dimethylamino)-dibenz[*b*,*f*]oxepin	b_2 148–152°C	*d*,*l*-Hydrochloride, mp 205–206°C *d*-Hydrochloride, mp 217°C, $[\alpha]_D^{16}$ +148.2° *l*-Hydrochloride, mp 217°C, $[\alpha]_D^{24}$ −148.7° Oxalate, mp 204–205°C *l*-(*d*-Camphorsulfonate), mp 184–187°C *l*-(*D*-Tartrate), mp 143–146°C	96
$C_{16}H_{17}NO_2$	11-Hydroxy-10-(*N*,*N*-dimethylamino)-10,11-dihydrodibenz[*b*,*f*]oxepin	mp 101–103°C	Hydrochloride, mp 225°C dec	84, 96
$C_{16}H_{17}NO_3$	10-Amino-10,11-dihydro-2,3-dimethoxydibenz[*b*,*f*]oxepin		Hydrochloride, mp 250–251°C dec (EtOH–Et_2O) *N*-Formyl, mp 146–147°C (EtOH)	90

TABLE III-2.—(*contd.*)

Empirical formula	Name of compound	Physical constants	Derivatives	Refs.
$C_{17}H_{16}ClNO$	2-Chloro-10-(*N*,*N*-dimethylamino)methyldibenz[*b*,*f*]oxepin	mp 68–69°C	Hydrochloride, mp 242–244°C	124
$C_{17}H_{16}N_2O$	10,11-Dioximino-2,3,6-trimethoxydibenz[*b*,*f*]oxepin	mp 100°C		88
$C_{17}H_{16}O_4$	2,3,6-Trimethoxydibenz[*b*,*f*]oxepin	mp 115–116.5°C (MeOH)		88
$C_{17}H_{16}O_5$	2,3,6-Trimethoxydibenz[*b*,*f*]oxepin-10(11*H*)-one	mp 133–134°C		102
	2,3,7-Trimethoxydibenz[*b*,*f*]oxepin-10(11*H*)-one	mp 138–139°C		102
	2,3,8-Trimethoxydibenz[*b*,*f*]oxepin-10(11*H*)-one	mp 119–120°C		102
$C_{17}H_{17}ClO_4$	10-Chloro-10,11-dihydro-2,3,6-trimethoxydibenz[*b*,*f*]oxepin	mp 116–117°C (Et_2O–$CHCl_3$)		201
$C_{17}H_{17}NO$	10-(*N*,*N*-Dimethylamino)methyldibenz[*b*,*f*]oxepin	$b_{0.004}$ 128–130°C	Hydrochloride, mp 234–236°C	124, 196
	10-[1(*N*-Methylamino)ethyl]dibenz[*b*,*f*]oxepin	$b_{0.005}$ 140–145°C	Hydrochloride, mp 235–237°C	124
$C_{17}H_{17}NO_2$	2-Methoxy-10-(*N*-methylamino)methyldibenz[*b*,*f*]oxepin		Hydrochloride, mp 151.5–154°C	124
$C_{17}H_{17}NO_4$	10(11*H*)-Oximino-2,3,6-trimethoxydibenz[*b*,*f*]oxepin	mp 147°C (EtOH)		88
$C_{17}H_{18}O_5$	10,11-Dihydro-2,3,6-trimethoxydibenz[*b*,*f*]oxepin-10-ol	mp 167°C, 169–170°C (EtOH)	*O*-Acetyl, mp 111.5–113°C	88, 90, 201
$C_{17}H_{19}NO_4$	10-Amino-10,11-dihydro-2,3,6-trimethoxydibenz[*b*,*f*]oxepin	mp 160–161 C	*N*-Chloroacetyl, mp 182–183°C	87, 88

			N-Trichloroacetyl, mp 139°C Hydrochloride, mp >285°C *N*-Formyl, mp 192°C	
$C_{17}H_{20}N_2O_4$	10,11-Diamino-10,11-dihydro-2,3,6-trimethoxydibenz[*b*,*f*]oxepin	mp 117°C (C_6H_6–petroleum ether)		88
$C_{18}H_{14}O_4$	10,11-Diacetoxydibenz[*b*,*f*]oxepin	mp 138°C (C_6H_6–*n*-hexane) $b_{0.005}$ 130–135°C subl		116
$C_{18}H_{16}O_6$	3,4,7-Trimethoxydibenz[*b*,*f*]oxepin-10-carboxylic acid	mp 253°C (MeOH)	Methyl ester, mp 182°C	137
	Dimethyl 10,11-dihydrodibenz[*b*,*f*]oxepin-4,6-dicarboxylate	mp 94.5–95°C		202
$C_{18}H_{18}O_5$	2,3,6,-Trimethoxy-9-methyldibenz[*b*,*f*]-oxepin-10(11*H*)-one	mp 106°C		102
	10-Acetoxy-10,11-dihydro-2,3-dimethoxydibenz[*b*,*f*]oxepin	mp 104–105°C (EtOH)		201
	11-Acetoxy-10,11-dihydro-2,3-dimethoxydibenz[*b*,*f*]oxepin	mp 94–95°C (C_6H_6–hexane)		201
$C_{18}H_{18}O_7$	10,11-Dihydro-10-hydroxy-2,3,6-trimethoxydibenz[*b*,*f*]oxepin-9-carboxylic acid	mp 178°C (EtOH-hexane)		185
$C_{18}H_{19}NO$	10-[1-(*N*,*N*-Dimethylamino)ethyl]dibenz-[*b*,*f*]oxepin	$b_{0.005}$ 135–140°C	Hydrochloride, mp 130–132°C	124
$C_{18}H_{19}NO_2$	11-[3-(*N*-Methylamino)propyl]dibenz-[*b*,*f*]oxepin-10(11*H*)-one		Hydrochloride, mp 141–143°C (*i*-PrOH)	197
$C_{18}H_{19}NO_2$	2-Methoxy-10-(*N*,*N*-dimethylamino)methyl-dibenz[*b*,*f*]oxepin		Hydrochloride, mp 204–208.5°C	124
$C_{18}H_{21}NO_2$	10,11-Dihydro-10-[2-(*N*,*N*-dimethyl-amino)ethoxy]dibenz[*b*,*f*]oxepin		Hydrogen maleate, mp 110–113°C ($EtOH–Et_2O$)	136

TABLE III-2.—(*contd.*)

Empirical formula	Name of compound	Physical constants	Derivatives	Refs.
$C_{19}H_{18}N_2O_2$	11-[3-(*N*-Cyano-*N*-methylamino)propyl]-dibenz[*b*,*f*]oxepin-10(11*H*)-one	$b_{0.08}$ 220–225°C		197
$C_{19}H_{19}Cl_2NO$	*x*,*y*-Dichloro-10-[3-(*N*,*N*-dimethylamino)-propyl]dibenz[*b*,*f*]oxepin		Maleate, —	168
$C_{19}H_{19}NO$	10-Pyrrolidinomethyldibenz[*b*,*f*]oxepin	$b_{0.01}$150–155°C	Hydrochloride, mp 193–196°C	124, 196
$C_{19}H_{20}ClNO$	2-Chloro-10-(*N*,*N*-diethylamino)methyl-dibenz[*b*,*f*]oxepin	$b_{0.94}$ 155–157°C	Hydrochloride, mp 182–184°C	124
$C_{19}H_{20}O_5$	10,11-Dihydro-2,3,6-trimethoxy-9-vinyldibenz[*b*,*f*]oxepin-10-ol	mp 120–125°C		185
$C_{19}H_{20}O_7$	Methyl 10,11-dihydro-10-hydroxy-2,3,6-trimethoxydibenz[*b*,*f*]oxepin-9-carboxylate	mp 66–67°C (EtOH–hexane)		185
$C_{19}H_{21}NO$	10-[3-(*N*,*N*-Dimethylamino)propyl]-dibenz[*b*,*f*]oxepin		Maleate, —	168
$C_{19}H_{21}NO_2$	11-[3-(*N*,*N*-Dimethylamino)propyl]-dibenz[*b*,*f*]oxepin-10(11*H*)-one	$b_{0.01}$ 166–169°C	Hydrochloride, mp 191–195°C (*i*-PrOH)	197
$C_{19}H_{21}NO_3$	2-Methoxy-11-[3-(*N*-methylamino)propyl]-dibenz[*b*,*f*]oxepin-10(11*H*)-one		Hydrochloride, mp 186–187°C (*i*-PrOH)	197
$C_{19}H_{22}N_2O$	1-(10,11-Dihydrodibenz[*b*,*f*]oxepin-10-yl)-4-methylpiperazine		Maleate, mp 128–130°C ($EtOH–Et_2O$)	136
$C_{19}H_{23}NO_5$	9-(2-Aminoethyl)-10,11-dihydro-2,3,6-trimethoxydibenz[*b*,*f*]oxepin-10-ol		Picrate, mp202°C(MeOH)	185
$C_{20}H_{14}O$	10-Phenyldibenz[*b*,*f*]oxepin	mp 103°C (*n*-BuOH)		27
$C_{20}H_{20}F_3NO$	*x*-Trifluoromethyl-10-[3-(*N*,*N*-dimethyl-amino)propyl]dibenz[*b*,*f*]oxepin		Citrate, —	168
$C_{20}H_{21}NO$	10-(2-Pyrrolidinoethyl)dibenz[*b*,*f*]oxepin		Citrate, —	168

$C_{20}H_{22}N_2O$	1-Methyl-4-(dibenz[*b,f*]oxepin-10-yl)-methylpiperazine	mp 82–83°C	Dihydrochloride, mp 210–215°C	124, 196
$C_{20}H_{23}NOS$	10-[3-(*N,N*-Dimethylamino)propy l]-*x*-methylthiodibenz[*b,f*]oxepin		Hydrochloride, —	168
$C_{20}H_{23}NO_3$	2-Methoxy-11-[3-(*N,N*-dimethylamino)-propyl]dibenz[*b,f*]oxepin-10(11*H*)-one	$b_{0.005}$ 168–172°C		197
$C_{21}H_{24}N_2O_2$	1-Methyl-4-(2-methoxydibenz[*b,f*] oxepin-10-yl)piperazine		Dihydrochloride, mp 232.3–236°C	124
	1-(2-Hydroxyethyl)-4-(dibenz[*b,f*]-oxepin-10-yl)methylpiperazine		Dihydrochloride, mp 220–225°C	124, 196
$C_{21}H_{25}NO$	10-[2-(*N,N*-Diethylamino)ethyl]-*x*-methyldibenz[*b,f*]oxepin		Citrate, —	168
$C_{21}H_{27}NO_5$	10,11-Dihydro-9-(2-*N,N*-dimethyl-aminoethyl)-2,3,6-trimethoxydibenz[*b,f*]-oxepin-10-ol	$b_{0.1}$180–182°C	Methiodide, mp 231.5–232°C(MeOH) Picrate, mp 150–152°C (Me_2CO–EtOH–Et_2O)	185
$C_{22}H_{24}BrNO$	*x*-Bromo-10-piperidinopropyldibenz[*b,f*]-oxepin		Citrate, —	168
$C_{22}H_{26}N_2O_3$	1-(2-Hydroxyethyl)-4-(2-methoxydibenz-[*b,f*]oxepin-10-yl)methylpiperazine			124

TABLE III-3. Dibenz[*c,e*]oxepins

Empirical formula	Name of compound	Physical constants	Derivatives	Refs.
$C_{14}H_4Cl_2N_2O_7$	2,10-Dichloro-3,9-dinitrodibenz[*c,e*]-oxepin-5,7-dione	mp 240°C (C_6H_6)		214
$C_{14}H_6Br_2O_3$	2,10-Dibromodibenz[*c,e*]oxepin-5,7-dione	mp 217–218°C		49
	3,9-Dibromodibenz[*c,e*]oxepin-5,7-dione	mp 304–305°C		163
$C_{14}H_6Cl_2O_3$	2,10-Dichlorodibenz[*c,e*]oxepin-5,7-dione	mp 199°C (MeCN), 201°C, 206°C (C_6H_6–ligroin)		80, 214
	3,9-Dichlorodibenz[*c,e*]oxepin-5,7-dione	mp 308–310°C (Ac_2O–AcOH)		81
	4,8-Dichlorodibenz[*c,e*]oxepin-5,7-dione	mp 257–258°C (C_6H_6–ligroin)		80
$C_{14}H_6N_2O_7$	2,10-Dinitrodibenz[*c,e*]oxepin-5,7-dione	mp 265°C		138
	3,9-Dinitrodibenz[*c,e*]oxepin-5,7-dione (·Ac_2O)	mp 232–234°C (sintered at 125°)		23
$C_{14}H_7NO_5$	2-Nitrodibenz[*c,e*]oxepin-5,7-dione	mp 194–195°C		24, 59
	3-Nitrodibenz[*c,e*]oxepin-5,7-dione	mp 205–207°C (Ac_2O), 207–207.5°C		23, 122
$C_{14}H_8Cl_4O$	1,3,9,11-Tetrachloro-5,7-dihydrodibenz[*c,e*]-oxepin	mp 151–152.5°C (MeOH)		73
$C_{14}H_8O_3$	Diphenic anhydride	mp 206° (C_6H_6–ligroin), 210–213°C, 212°C, 216–217°C ($C_2H_4Cl_2$), 220°C, 222–226°C, 224.5–225.5°C (PhCl)		6, 69, 70, 79, 101, 133, 147, 167
$C_{14}H_{10}O_2$	Dibenz[*c,e*]oxepin-5(7*H*)-one	mp 132°C (alcohol), 136–136.5°C (C_6H_6)		8, 44, 95, 154, 155, 181, 205
$C_{14}H_{10}O_3$	1,2,3,4,4a,11b-Hexahydrodibenz[*c,e*]oxepin-5,7-dione	mp 220°C, 242°C		106, 171

$C_{14}H_{11}NO_3$	5,7-Dihydro-1-nitrodibenz[*c*,*e*]oxepin	mp 115°C	130, 131
$C_{14}H_{12}O$	5,7-Dihydrodibenz[*c*,*e*]oxepin	mp 71.5–72°C, 72.5–73°C, 73°C (alcohol)	28, 63, 74, 171, 172, 177
$C_{14}H_{18}O_3$	1,2,3,4,4a,7a,8,9,10,11,11a,11b-Dodecahydrodibenz[*c*,*e*]oxepin-5,7-diones (10 isomers, A–J)	A, mp 86°C B, mp 265°C C, mp 115°C D, mp 135°C E, mp 242°C F, mp 143°C G, mp 105–106°C H, mp 103–104°C I, mp 146–147°C J, mp 104–104.5°C	107, 108, 114, 171
$C_{16}H_{12}O_3$	2,3-Dimethyldibenz[*c*,*e*]oxepin-5,7-dione (?)	mp 190–191°C	68
$C_{16}H_{12}O_4$	7-Acetoxydibenz[*c*,*e*]oxepin-5(7*H*)-one	mp 125°C, 138–139°C (EtOH, MeOH)	9, 48
$C_{16}H_{12}O_5$	3,9-Dimethoxydibenz[*c*,*e*]oxepin-5,7-dione		14, 134
$C_{16}H_{14}O_2$	7,7-Dimethyldibenz[*c*,*e*]oxepin-5(7*H*)-one		60, 156
$C_{16}H_{16}O$	5,7-Dihydro-1,11-dimethyldibenz[*c*,*e*]oxepin	(±)-, mp 130–131°C (+)-, mp 114.3°C (C_6H_6), $[\alpha]_D^{20}$ +100.2° (MeOH) (−)-, mp 92.3°C (C_6H_6), $[\alpha]_D^{20}$ −77.8° (MeOH)	119, 121, 131, 180
	5,7-Dihydro-5,5-dimethyldibenz[*c*,*e*]oxepin	mp 68–69° (*n*-pentane)	156
	5,7-Dihydro-5,7-dimethyldimethylbenz-[*c*,*e*]oxepin	mp 77–80°C (MeOH)	17, 73

TABLE III-3.—(*contd.*)

Empirical formula	Name of compound	Physical constants	Derivatives	Refs.
$C_{16}H_{16}O_3$	5,7-Dihydro-1,11-dimethoxydibenz[*c*,*e*]-oxepin	mp 136°C		75
	5,7-Dihydro-2,10-dimethoxydibenz[*c*,*e*]-oxepin	mp 159–160°C		16
	5,7-Dihydro-1,11-bis(hydroxymethyl)-dibenz[*c*,*e*]oxepin	(+)-, mp 139–140.5°C $[\alpha]^{23}_{D}$ + 32° ($CHCl_3$)		118, 119
	5,7-Dihydro-2,10-bis(hydroxymethyl)-dibenz[*c*,*e*]oxepin	mp 192°C dec	Bis-phthalate, mp 179–179.5°C (Aq. AcOH)	16
$C_{17}H_{16}O_5$	1,2,10(or 2,10,11)-Trimethoxydibenz[*c*,*e*]-oxepin-5(7*H*)-one	mp 147°C		26
$C_{18}H_{12}O_7$	3,9-Diacetoxydibenz[*c*,*e*]oxepin-5,7-dione	mp 216–219°C (toluene)		134
$C_{18}H_{14}N_2O_5$	1,11-Diacetamidodibenz[*c*,*e*]oxepin-5,7-dione	Unstable		3
	3,9-Diacetamidodibenz[*c*,*e*]oxepin-5,7-dione	No mp		3
$C_{18}H_{16}O_3$	3-Isopropyl-8-methyldibenz[*c*,*e*]oxepin-5,7-dione	mp 112–112.5°C (Ac_2O), 155°C		1, 2, 64, 92, 111
$C_{18}H_{16}O_5$	3,9-Dimethoxy-4,8-dimethyldibenz[*c*,*e*]-oxepin-5,7-dione	mp 241–242°C		14
	1,11-Bis(carbomethoxy)-5,7-dihydrodibenz-[*c*,*e*]oxepin	(±)-, mp 185–186°C (MeOH) (−)-, $[\alpha]^{24}_{D}$–377° (C_6H_6)		119
$C_{18}H_{16}O_7$	1,2,10,11-Tetramethoxydibenz[*c*,*e*]-oxepin-5,7-dione	mp 195.5–196°C		35
	3,4,8,9-Tetramethoxydibenz[*c*,*e*]oxepin-5,7-dione	mp 172–173°C		35

$C_{18}H_{17}BrO_2$	7-(α-Bromomethyl)-7-ethyldibenz[*c*,*e*]-oxepin-5(7*H*)-one			47
$C_{18}H_{17}IO_2$	7-Ethyl-7-(α-iodoethyl)dibenz[*c*,*e*]oxepin-5(7*H*)-one	mp 111°C dec		47
$C_{18}H_{18}O_2$	7,7-Diethyldibenz[*c*,*e*]oxepin-5(7*H*)-one	mp 104°C (EtOH)		47
$C_{18}H_{20}O$	5,7-Dihydro-5,5,7,7-tetramethyldibenz-[*c*,*e*]oxepin	mp 90–91.8°C (MeOH)		25, 73, 156
$C_{18}H_{20}O_3$	1,4,4a,11b-Tetrahydro-2,3,4a,9-tetra-methyldibenz[*c*,*e*]oxepin-5,7-dione	mp 97–98°C		62
$C_{18}H_{20}O_5$	5,7-Dihydro-2,3,8,10-tetramethoxydibenz-[*c*,*e*]oxepin	mp 167°C (MeOH–petroleum ether)		100
	5,7-Dihydro-2,3,9,10-tetramethoxydibenz-[*c*,*e*]oxepin	mp 249°C (C_6H_6), 256°C (C_6H_6), 258°C (C_6H_6)		50, 115, 139
$C_{20}H_{16}O$	5,7-Dihydro-3-phenyldibenz[*c*,*e*]oxepin	mp 137–138°C (cyclohexane)		32
$C_{20}H_{18}O_4$	Phenoldiphenein	mp 250–251°C (AcOH)	Tetrabromo, mp 213–214°C (AcOH) Diidodo, mp 216–217°C (AcOH) Dinitro, mp 212.5–213°C (absolute alcohol)	57, 162, 165
	1-Acetyl-9-isopropyl-4-methyldibenz-[*c*,*e*]oxepin-5,7-dione	mp 167–168°C		126
	3-Acetyl-9-isopropyl-4-methyldibenz[*c*,*e*]-oxepin-5,7-dione	mp 152–153°C		127
$C_{20}H_{22}O_2$	7,7-Di-*n*-propyldibenz[*c*,*e*]oxepin-5(7*H*)-one	mp 92°C (EtOH)		47
$C_{20}H_{24}O$	5,7-Dihydro-1,2,3,9,10,11-hexa-methoxydibenz[*c*,*e*]oxepin	mp 146.5–147°C		98, 99
$C_{21}H_{18}O_2$	5-(*p*-Anisyl)-5,7-dihydrodibenz[*c*,*e*]oxepin	mp 157°C (EtOH)		28

TABLE III-3.—(*contd.*)

Empirical formula	Name of compound	Physical constants	Derivatives	Refs.
$C_{21}H_{18}O_6$	5-Dicarbomethoxymethylenedibenz[*c*,*e*]-oxepin-7-one	mp 95°C (C_6H_6)		110
$C_{22}H_{22}O_4$	*o*-Cresoldiphenein	mp 218.5–219.5°C (AcOH)	Dibromo, mp 203–204°C (alcohol) Diiodo, mp 206–207°C (alcohol) Dinitro, mp 212.5–213°C (alcohol)	162
$C_{24}H_{22}O_2$	7-Ethyl-7-(3,4-xylyl)dibenz[*c*,*e*]oxepin-5(7*H*)-one	Glass		125
$C_{25}H_{23}O_8$	4,8-Diacetoxy-2-(4-methylpentanoyl)-3-methyldibenz[*c*,*e*]oxepin-5,7-dione	mp 162–164°C		135
$C_{26}H_{18}O_2$	7,7-Diphenyldibenz[*c*,*e*]oxepin-5(7*H*)-one	mp 190°C (alcohol)		145
$C_{26}H_{20}O$	5,7-Dihydro-3-(*p*-biphenylyl)dibenz[*c*,*e*]-oxepin	mp 217–219°C (EtOAc)		32
$C_{26}H_{26}O_2$	7-*n*-Butyl-7-(3,4-xylyl)dibenz[*c*,*e*]oxepin	Glass		125
$C_{28}H_{16}N_4O_9$	2,10-Bis(*m*-nitrobenzamido)dibenz[*c*,*e*]-oxepin-5,7-dione	mp 296–297°C		138
$C_{28}H_{18}N_2O_5$	2,10-Bis(benzamido)dibenz[*c*,*e*]oxepin-5,7-dione	mp 283–284°C dec		138
$C_{28}H_{22}O_2$	7,7-Dibenzyldibenz[*c*,*e*]oxepin-5(7*H*)-one	mp 146°C (EtOH)		47
$C_{28}H_{22}O_4$	7,7-Di-*p*-anisyldibenz[*c*,*e*]oxepin-5(7*H*)-one	mp 150–151°C (MeOH)		161
$C_{32}H_{24}O_2$	5,7-Dihydro-5,7,7-triphenyldibenz[*c*,*e*]-oxepin-5-ol	mp 189–191°C		145
$C_{40}H_{32}O_3$	1,11-Dimethoxy-5,5,7,7-tetraphenyl-dibenz[*c*,*e*]oxepin	mp 314–316°C		178
$C_{56}H_{48}O_7$	1,2,3,9,10,11-Hexa(benzyloxy)-5,7-dihydro-dibenz[*c*,*e*]oxepin	(±)-, mp 116.5–117°C (C_6H_6-hexane) (+)-, oil, $[\alpha]_D^{25}$ +65° (C_6H_6)		121

D. References

1. D. E. Adelson and M. T. Bogert, *J. Amer. Chem. Soc.*, **58,** 2236 (1936).
2. D. E. Adelson, T. Hasselstrom, and M. T. Bogert, *J. Amer. Chem. Soc.*, **58,** 871 (1936).
3. H. Adkins, E. F. Steinberg, and E. Pickering, *J. Amer. Chem. Soc.*, **46,** 1917 (1924).
4. S. N. Alam and D. B. MacLean, *Can. J. Chem.*, **43,** 3433 (1965).
5. F. A. L. Anet and P. M. G. Bavin, *Can. J. Chem.*, **35,** 1084 (1957).
6. R. Anschütz, *Chem. Ber.*, **10,** 326 (1877).
7. R. Anschütz, *Chem. Ber.*, **10,** 1884 (1877).
8. R. G. R. Bacon and R. Bankhead, *J. Chem. Soc.*, **1963,** 839.
9. P. S. Bailey, *J. Amer. Chem. Soc.*, **78,** 3811 (1956).
10. W. Baker, D. Clark, W. D. Ollis, and T. S. Zealley, *J. Chem. Soc.*, **1952,** 1452.
11. J. E. Baldwin, *J. Org. Chem.*, **28,** 3112 (1963).
12. B. K. Banerjee, *J. Indian Chem. Soc.*, **12,** 4 (1935).
13. J. A. Barltrop and J. E. Saxton, *J. Chem. Soc.*, **1952,** 1038.
14. R. A. Barnes and R. W. Faessinger, *J. Org. Chem.*, **26,** 4544 (1961).
15. G. H. Beaven, D. M. Hall, M. S. Lesslie, and E. E. Turner, *J. Chem. Soc.*, **1952,** 854.
16. G. H. Beaven, D. M. Hall, M. S. Lesslie, E. E. Turner, and G. R. Bird, *J. Chem. Soc.*, **1954,** 131.
17. G. H. Beaven and E. A. Johnson, *J. Chem. Soc.*, **1957,** 651.
18. Belgian Patent 623,259; *Chem. Abstr.*, **60,** 10659 (1964).
19. Belgian Patent 631,009; *Chem. Abstr.*, **61,** 1843 (1964).
20. Belgian Patent 641,498; *Chem. Abstr.*, **64,** 719 (1966).
21. F. Bell and F. Briggs, *J. Chem. Soc.*, **1938,** 1561.
22. F. Bell and F. Briggs, *J. Chem. Soc.*, **1941,** 282.
23. F. Bell and P. H. Robinson, *J. Chem. Soc.*, **1927,** 1695.
24. F. Bell and P. H. Robinson, *J. Chem. Soc.*, **1927,** 2234.
25. G. M. Bennett and R. L. Wain, *J. Chem. Soc.*, **1936,** 114.
26. K. W. Bentley and R. Robinson, *J. Chem. Soc.*, **1952,** 947.
27. E. D. Bergmann and M. Rabinowitz, *Israel J. Chem.*, **1,** 125 (1963); *Chem. Abstr.*, **60,** 7997 (1964).
28. H. W. Bersch, R. Meyer, A. v. Mletzko, and K. H. Fischer, *Arch. Pharm.*, **291,** 82 (1958); *Chem. Abstr.*, **52,** 14628 (1958).
29. G. Berti, *Gazz. Chim. Ital.*, **81,** 570 (1951).
30. F. Bickelhaupt, K. Stach, and M. Thiel, *Monatsh. Chem.*, **95,** 485 (1964).
31. S. P. Birkeland, G. H. Daub, F. N. Hayes, and D. G. Ott, *Z. Phys.*, **159,** 516 (1960).
32. S. P. Birkeland, G. H. Daub, F. N. Hayes, and D. G. Ott, *J. Org. Chem.*, **26,** 2662 (1961).
33. F. Bischoff and H. Adkins, *J. Amer. Chem. Soc.*, **45,** 1030 (1923).
34. A. Bistrzycki and K. Fassler, *Helv. Chim. Acta*, **6,** 519 (1923).
35. B. M. Bogoslovskii and V. S. Krasnova, *J. Gen. Chem. USSR*, **7,** 1543 (1937); *Chem. Abstr.*, **31,** 8527 (1937).
36. E. A. Braude and W. F. Forbes, *J. Chem. Soc.*, **1955,** 3776.
37. British Patent 996,255; *Chem. Abstr.*, **63,** 8330 (1965).
38. British Patent 1,004,683; *Chem. Abstr.*, **64,** 8158 (1966).
39. British Patent 1,059,898; *Chem. Abstr.*, **66,** 10741 (1967).
40. J. D. Brooks, *J. Appl. Chem.* (London), **5,** 250 (1955); *Chem. Abstr.*, **50,** 4886 (1956).
41. P. Cagniant and C. Charaux, *Bull. Soc. Chim. Fr.*, **1966,** 3249.

42. D. Cagniant, C. Charaux, and P. Cagniant, *Bull. Soc. Chim. Fr.*, **1966,** 3644.
43. G. Caronna, *Gazz. Chim. Ital.*, **71,** 475 (1941); *Chem. Abstr.*, **37,** 118 (1943).
44. N. Chatterjea, *J. Indian Chem. Soc.*, **12,** 418 (1935).
45. J. N. Chatterjee, *J. Indian Chem. Soc.*, **32,** 203 (1955).
46. A. Corbellini and M. Angeletti, *Atti Accad. Naz. Lincei, Rend. Cl. Sci. Fis. Mat. Nat.* **15,** 968 (1932); *Chem. Abstr.*, **27,** 715 (1933).
47. A. Corbellini and C. Vigano, *Gazz. Chim. Ital.*, **65,** 735 (1935); *Chem. Abstr.*, **30,** 2187 (1936).
48. J. W. Cook, G. T. Dickson, J. Jack, J. D. Loudon, J. McKeown, J. MacMillan, and W. F. Williamson, *J. Chem. Soc.*, **1950,** 139.
49. Ch. Courtot and J. Kronstein, *Chim. Ind.* (Paris), **45,** 66 (1941); *Chem. Abstr.*, **37,** 2369 (1943).
50. R. I. T. Cromartie, J. Harley-Mason, and D. G. P. Wanningama, *J. Chem. Soc.*, **1958,** 1982.
51. F. Dallacker, K. W. Glombitza, and M. Lipp, *Justus Liebigs Ann. Chem.*, **643,** 82 (1961).
52. F. M. Dean, B. Parton, N. Somvichien, and D. A. H. Taylor, *Tetrahedron Lett.*, **1967,** 3459.
53. F. M. Dean and D. A. H. Taylor, *J. Chem. Soc., C*, **1966,** 114.
54. O. Diels and W. E. Thiele, *J. Prakt. Chem.*, **156,** 186 (1940).
55. W. E. Doering and M. J. Goldstein, *Tetrahedron*, **5,** 53 (1959).
56. M. Dominikiewicz, *Arch. Chem. Farm.*, **3,** 141 (1937); *Chem. Abstr.*, **32,** 2916 (1938).
57. S. Dutt, *J. Chem. Soc.*, **1923,** 225.
58. H. Erdtman and A. Spetz, *Acta Chem. Scand.*, **10,** 1427 (1956).
59. P. M. Everitt, S. M. Loh, and E. E. Turner, *J. Chem. Soc.*, **1960,** 4587.
60. P. M. Everitt and E. E. Turner, *J. Chem. Soc.*, **1957,** 3477.
61. W. V. Farrar, *J. Appl. Chem.* (London), **14,** 389 (1964); *Chem. Abstr.*, **62,** 7753 (1965).
62. L. F. Fieser and J. T. Dunn, *J. Amer. Chem. Soc.*, **59,** 1021 (1937).
63. D. D. Fitts, M. Siegel, and K. Mislow, *J. Amer. Chem. Soc.*, **80,** 480 (1958).
64. H. P. Fogelberg, *Ann. Acad. Sci. Fennicae*, **29A**(4), 7 pp. (1927); *Chem. Abstr.*, **22,** 1153 (1928).
65. R. C. Fuson and N. Shachat, *J. Org. Chem.*, **22,** 1394 (1957).
66. A. Gaudemer and J. Polonsky, *Bull. Soc. Chim. Fr.*, **1963,** 1918.
67. German Patent 562,827; *Chem. Abstr.*, **27,** 1224 (1933).
68. L. Ghigi, *Chem. Ber.*, **70B,** 2469 (1937).
69. C. Graebe and C. Aubin, *Justus Liebigs Ann. Chem.*, **247,** 260 (1888).
70. C. Graebe and C. Mensching, *Chem. Ber.*, **13,** 1302 (1880).
71. C. Graebe and C. Mensching, *Chem. Ber.*, **20,** 846 (1887).
72. J. F. Grove and B. J. Riley, *J. Chem. Soc.*, **1961,** 1105.
73. D. M. Hall, J. E. Ladbury, M. S. Lesslie, and E. E. Turner, *J. Chem. Soc.*, **1956,** 3475.
74. D. M. Hall and F. Minhaj, *J. Chem. Soc.*, **1957,** 4584.
75. D. M. Hall and E. E. Turner, *J. Chem. Soc.*, **1951,** 3072.
76. H. Hoyer and M. Vogel, *Monatsh. Chem.*, **93,** 766 (1962).
77. R. L. Huang and H. H. Lee, *J. Chem. Soc.*, **1964,** 2500.
78. M. Hubacher, *J. Org. Chem.*, **23,** 1400 (1958).
79. E. B. Hunn, *J. Amer. Chem. Soc.*, **45,** 1024 (1923).
80. E. H. Huntress and I. S. Cliff, *J. Amer. Chem. Soc.*, **55,** 2559 (1933).
81. E. H. Huntress, I. S. Cliff, and E. R. Atkinson, *J. Amer. Chem. Soc.*, **55,** 4262 (1933).
82. Japanese Patent 8784 ('58); *Chem. Abstr.*, **54,** 4636 (1960).

83. Japanese Patent 8785 ('58); *Chem. Abstr.*, **54,** 4636 (1960).
84. Japanese Patent 5125 ('59); *Chem. Abstr.*, **54,** 14285 (1960).
85. Japanese Patent 6287 ('59); *Chem. Abstr.*, **54,** 15413 (1960).
86. Japanese Patent 14,723 ('61); *Chem. Abstr.*, **56,** 10114 (1962).
87. T. Kametani and K. Fukumoto, *J. Chem. Soc.*, **1963,** 4289.
88. T. Kametani and K. Fukumoto, *Chem. Pharm. Bull.* (Tokyo), **11,** 1322 (1963).
89. T. Kametani, K. Fukumoto, and T. Nakano, *Yakugaku Zasshi*, **82,** 1307 (1962); *Chem. Abstr.*, **58,** 13913 (1963).
90. T. Kametani and C. Kibayashi, *Yakugaku Zasshi*, **84,** 642 (1964); *Chem. Abstr.*, **61,** 13278 (1964).
91. T. Kametani, S. Shibuya, and C. Kibayashi, *Chem. Pharm. Bull.* (Tokyo), **16,** 34 (1968).
92. K. J. Karrman and P. V. Laakso, *Acta Chem. Scand.*, **1,** 449 (1947).
93. K. Kawai, *Nippon Yakurigaku Zasshi*, **56,** 971 (1960); *Chem. Abstr.*, **55,** 21374 (1961).
94. J. Kenner, *J. Chem. Soc.*, **103,** 613 (1913).
95. J. Kenner and E. G. Turner, *J. Chem. Soc.*, **99,** 2101 (1911).
96. S. Kimoto, K. Asaki, and S. Saito, *Yakugaku Zasshi*, **77,** 652 (1957); *Chem. Abstr.*, **51,** 16497 (1957).
97. S. Kimoto, K. Kimura, and S. Muramatsu, *J. Pharm. Soc. Jap.*, **74,** 358 (1954); *Chem. Abstr.*, **49,** 5373 (1955).
98. N. Kochetkov, A. Ya. Khorlin, and O. S. Chizhov, *Izv. Akad. Nauk SSSR, Otd. Khim. Nauk*, **1962,** 856; *Chem. Abstr.*, **57,** 13704 (1962).
99. N. K. Kochetkov, A. Ya. Khorlin, O. S. Chizhov, and V. I. Scheichen, *Tetrahedron Lett.*, **1961,** 730.
100. H. Kondo and K. Takeda, *Itsuu Kenkyusho Nempo*, **9,** 33 (1958); *Chem. Abstr.* **54,** 1580 (1960).
101. L. P. Kulev, V. A. Salskii, and G. V. Shiskin, *Izv. Tomsk. Politekhn. Inst.*, **112,** 23 (1963); *Chem. Abstr.*, **61,** 13230 (1964).
102. M. Kulka and R. H. F. Manske, *J. Amer. Chem. Soc.*, **75,** 1322 (1953).
103. R. J. Kurland, M. B. Rubin, and W. B. Wise, *J. Chem. Phys.*, **40,** 2426 (1964).
104. C. Kuroda and M. Akajima, *Proc. Jap. Acad.*, **27,** 343 (1951); *Chem. Abstr.*, **47,** 6386 (1953).
105. M. Lamchen, *J. Chem. Soc.*, **1962,** 4695.
105a. R. J. W. Le Fevre and A. Sundaram, *J. Chem. Soc.*, **1962,** 4009.
105b. R. J. W. Le Fevre and H. Vine, *J. Chem. Soc.*, **1938,** 967.
106. R. P. Linstead and S. B. Davis, *J. Amer. Chem. Soc.*, **64,** 2006 (1942).
107. R. P. Linstead and W. E. Doering, *J. Amer. Chem. Soc.*, **64,** 1991 (1942).
108. R. P. Linstead and A. L. Walpole, *J. Chem. Soc.*, **1939,** 850.
109. J. D. Loundon and L. A. Summers, *J. Chem. Soc.*, **1957,** 3807.
110. H. W. Lucien and A. Taurins, *Can. J. Chem.*, **30,** 208 (1952).
111. H. Lund, *Acta Chem. Scand.*, **3,** 748 (1949).
112. R. H. F. Manske, *J. Amer. Chem. Soc.*, **72,** 55 (1950).
113. R. H. F. Manske and A. E. Ledingham, *J. Amer. Chem. Soc.*, **72,** 4797 (1950).
114. C. S. Marvel and R. V. White, *J. Amer. Chem. Soc.*, **62,** 2739 (1940).
115. E. Matarasso-Tchiroukhine, *Ann. Chim.* (Paris), [13] **3,** 405 (1958).
116. F. Mathys, V. Prelog, and R. B. Woodward, *Helv. Chim. Acta*, **39,** 1095 (1956).
117. P. H. McCabe, R. McCrindle, and R. D. Murray, *J. Chem. Soc., C*, **1967,** 145.
118. K. Mislow and M. A. W. Glass, *J. Amer. Chem. Soc.*, **83,** 2780 (1961).
119. K. Mislow, M. A. W. Glass, H. B. Hopps, E. Simon, and G. H. Wahl, Jr., *J. Amer. Chem. Soc.*, **86,** 1710 (1964).

120. K. Mislow, M. A. W. Glass, R. E. O'Brien, P. Rutkin, D. H. Steinberg, and C. Djerassi, *J. Amer. Chem. Soc.*, **82,** 4740 (1960).
121. K. Mislow, M. A. W. Glass, R. E. O'Brien, P. Rutkin, D. H. Steinberg, J. Weiss, and C. Djerassi, *J. Amer. Chem. Soc.*, **84,** 1455 (1962).
122. F. J. Moore and E. H. Huntress, *J. Amer. Chem. Soc.*, **49,** 1324 (1927).
123. Netherlands Patent Appl. 6,411,861; *Chem. Abstr.*, **63,** 16366 (1965).
124. Netherlands Patent Appl. 6,508,284; *Chem. Abstr.*, **64,** 19574 (1966).
125. D. V. Nightingale, W. S. Wagner, and R. H. Wise, *J. Amer. Chem. Soc.*, **75,** 4701 (1953).
126. G. A. Nyman, *Ann. Acad. Sci. Fennicae, Ser. A*, **48**(6), 32 pp. (1937); *Chem. Abstr.* **33,** 8192 (1939).
127. G. A. Nyman, *Ann. Acad. Sci. Fennicae. Ser. A*, **55**(11), 9 pp. (1940); *Chem. Abstr.*, **37,** 5056 (1943).
128. P. F. Oesper and C. P. Smyth, *J. Amer. Chem. Soc.*, **64,** 768 (1942).
129. M. Okajima, *Sci. Papers Inst. Phys. Chem. Res.* (Tokyo), **53,** 356 (1959); *Chem. Abstr.*, **54,** 1742 (1960).
130. M. Oki, H. Iwamura, and N. Hayakawa, *Bull. Chem. Soc. Jap.*, **36,** 1542 (1963); *Chem. Abstr.*, **60,** 6368 (1964).
131. M. Oki, H. Iwamura, and N. Hayakawa, *Bull. Chem. Soc. Jap.*, **37,** 1865 (1964); *Chem. Abstr.*, **62,** 7612 (1965).
132. W. R. Orndorff and E. Kline, *J. Amer. Chem. Soc.*, **46,** 2276 (1924).
133. L. Oyster and H. Adkins, *J. Amer. Chem. Soc.*, **43,** 208 (1921).
134. H. R. Patel, D. W. Blackburn, and G. L. Jenkins, *J. Amer. Pharm. Assoc.*, **46,** 51 (1957).
135. J. Polonsky, B. C. Johnson, P. Cohen, and E. Lederer, *Bull. Soc. Chim. Fr.*, **1963,** 1909.
136. M. Protiva, J. O. Jilek, J. Metysova, V. Seidlova, I. Jarkovsky, J. Metys, E. Adlerova, I. Ernest, K. Pelz, and J. Pomykacek, *Farmaco, Ed. Sci.*, **20,** 721 (1965); *Chem. Abstr.*, **64,** 5090 (1966).
137. R. Pschorr and G. Knöffler, *Justus Liebigs Ann. Chem.*, **382,** 50 (1911).
138. F. Pufahl, *Chem. Ber.*, **62B,** 2817 (1929).
139. R. Quelet and E. Matarasso-Tchiroukhine, *C.R. Acad. Sci., Paris*, **244,** 467 (1957).
140. F. Ramirez, N. B. Desai, and R. B. Mitra, *J. Amer. Chem. Soc.*, **83,** 492 (1961).
141. A. H. Rees, *Chem. Ind.* (London), **1957,** 76.
142. J. Schmidt and A. Kämpf, *Chem. Ber.*, **36,** 3734 (1903).
143. G. Schroeter, A. Gluschke, S. Götzky, J. Huang, G. Irmisch, E. Laves, O. Schrader, and G. Stier, *Chem. Ber.*, **63B,** 1308 (1930).
144. R. N. Sen and B. C. Roy, *J. Indian Chem. Soc.*, **7,** 401 (1930).
145. P. G. Sergeev, *J. Russ. Phys. Chem. Soc.*, **61,** 1421 (1929); *Chem. Abstr.*, **24,** 1365 (1930).
146. A. Shimomura, *Mem. Coll. Sci. Univ. Kyoto Ser. A*, **8,** 19 (1925); *Chem. Abstr.*, **19,** 2196 (1925).
147. H. Shioda and S. Kato, *Yuki Gosei Kagaku Kyokai Shi*, **17,** 91 (1959); *Chem. Abstr.*, **53,** 8058 (1959).
148. A. Sieglitz and H. Koch, *Chem. Ber.*, **58B,** 78 (1925).
149. K. Stach and F. Bickelhaupt, *Angew. Chem.*, **74,** 752 (1962).
150. K. Stach and F. Bickelhaupt, *Monatsh. Chem.*, **93,** 896 (1962).
151. K. Stach and H. Spingler, *Angew. Chem.*, **74,** 31 (1962).
152. K. Stach and H. Spingler, *Monatsh. Chem.*, **93,** 889 (1962).
153. K. Stach and F. Bickelhaupt, *Conf. Hung. Therap. Invest. Pharmacol.*, **2** (Budapest, 1962), 80 (Pub. 1964); *Chem. Abstr.*, **62,** 7726 (1965).

154. E. F. M. Stephenson, *J. Chem. Soc.*, **1954,** 2354.
155. M. G. Sturrock, E. L. Cline, and K. R. Robinson, *J. Org. Chem.*, **28,** 2340 (1963).
156. I. O. Sutherland and M. V. J. Ramsay, *Tetrahedron*, **21,** 3401 (1965).
157. H. Suzuki, *Bull. Chem. Soc. Jap.*, **27,** 597 (1954); *Chem. Abstr.*, **49,** 10739 (1955).
158. H. Suzuki, *Bull. Chem. Soc. Jap.*, **32,** 1357 (1959); *Chem. Abstr.*, **54,** 18063 (1960).
159. M. Tanaka, *J. Chem. Soc. Jap.*, **56,** 198 (1935); *Chem. Abstr.*, **29,** 4353 (1935).
160. A. Taurins, *J. Prakt. Chem.*, **153,** 189 (1939).
161. H. W. Underwood, Jr., and G. E. Barker, *J. Amer. Chem. Soc.*, **52,** 4082 (1930).
162. H. W. Underwood, Jr., and G. E. Barker, *J. Amer. Chem. Soc.*, **58,** 642 (1936).
163. H. W. Underwood, Jr., and L. A. Clough, *J. Amer. Chem. Soc.*, **51,** 583 (1929).
164. H. W. Underwood, Jr., L. Harris, and G. Barker, *J. Amer. Chem. Soc.*, **58,** 643 (1936).
165. H. W. Underwood, Jr., and E. L. Kochmann, *J. Amer. Chem. Soc.*, **45,** 3071 (1923).
166. H. W. Underwood, Jr., and E. L. Kochmann, *J. Amer. Chem. Soc.*, **46,** 2073 (1924).
167. U.S. Patent 2,693,465; *Chem. Abstr.*, **49,** 12546 (1955).
168. U.S. Patent 3,100,207; *Chem. Abstr.*, **60,** 1719 (1964).
169. U.S. Patent 3,324,235; *Chem. Abstr.*, **64,** 12654 (1966).
170. USSR Patent 127,656; *Chem. Abstr.*, **54,** 22504 (1960).
171. F. Vocke, *Justus Liebigs Ann. Chem.*, **508,** 1 (1933).
172. A. J. Weinheimer, S. W. Kantor, and C. R. Hauser, *J. Org. Chem.*, **18,** 801 (1953).
173. E. Wenkert and D. P. Strike, *J. Amer. Chem. Soc.*, **86,** 2044 (1964).
174. H. W. Whitlock, Jr., *Tetrahedron Lett.*, **1961,** 593.
175. S. O. Winthrop, *Tetrahedron Lett.*, **1963,** 1113.
176. S. O. Winthrop, M. A. Davis, F. Herr, J. Stewart, and R. Gaudry, *J. Med. Pharm. Chem.*, **5,** 1207 (1962).
177. G. Wittig, P. Davis, and G. Koenig, *Chem. Ber.*, **84,** 627 (1951).
178. G. Wittig and M. Leo, *Chem. Ber.*, **64B,** 2395 (1931).
179. G. Wittig and H. Petri, *Justus Liebigs Ann. Chem.*, **505,** 17 (1933).
180. G. Wittig and H. Zimmermann, *Chem. Ber.*, **86,** 629 (1953).
181. R. Madeja, *Poznan Towarz. Przyjaciol. Nauk, Wydzial Mat. Przyrod., Prace Komisiji Mat. Przyrod.*, **10,** 17 (1965); *Chem. Abstr.*, **64,** 3403 (1966).
182. T. Kametani and I. Fukumoto, *Chem. Ind.* (London), **1963,** 291.
183. T. Kametani, S. Shibuya, and I. Noguchi, *Yakugaku Zasshi*, **85,** 667 (1965); *Chem. Abstr.*, **63,** 13228 (1965).
184. T. Kametani, K. Fukumoto, and K. Masuko, *Yakugaku Zasshi*, **83,** 1052 (1963); *Chem. Abstr.*, **60,** 10732 (1964).
185. T. Kametani and K. Ogasawara, *J. Chem. Soc.*, **1964,** 4142.
186. E. Vogel, M. Biskup, A. Vogel, and H. Guenther, *Angew. Chem.*, **78,** 755 (1966).
187. E. Vogel and H. Guenther, *Angew. Chem.*, **79,** 429 (1967).
188. H. Zinnes, R. A. Comes, and J. Shavel, Jr., *J. Med. Chem.*, **10,** 223 (1967).
189. E. Cioranescu, A. Bucur, A. Mihai, G. Matcescu, and C. D. Nenitzescu, *Bull. Soc. Chim. Fr.*, **1962,** 471.
190. W. Marckwald and H. Dettmer, *Chem. Ber.*, **35,** 296 (1902).
191. H. Machatzke, W. R. Vaughan, C. L. Warren, and G. R. White, *J. Pharm. Sci.*, **56,** 86 (1967).
192. E. Ghighi, A. Drusiani, and G. Giovanninetti, *Ann. Chim.* (Rome) **56,** 786 (1966); *Chem. Abstr.*, **66,** 1036 (1967).
193. P. M. G. Bavin, K. D. Bartle, and D. W. Jones, *J. Heterocycl. Chem.*, **5,** 327 (1968).
194. W. Tochtermann, K. Oppelaender, and M. N.-D. Hoang, *Justus Liebigs Ann. Chem.*, **701,** 117 (1967).
195. Netherlands Patent Appl. 6,605,741; *Chem. Abstr.*, **66,** 6164 (1967).

196. Netherlands Patent Appl. 6,514,187; *Chem. Abstr.*, **65,** 12183 (1966).
197. Netherlands Patent Appl. 6,515,702; *Chem. Abstr.*, **65,** 13672 (1966).
198. W. Theobald, O. Buech, C. Morpurgo, and W. Schindler, *Antidepressant Drugs, Proc. Int. Symp., 1st, Milan,* 205 (1966); *Chem. Abstr.*, **67,** 115430 (1967).
199. E. D. Bergmann, I. Shahak, and Z. Aizenshtat, *Tetrahedron Lett.*, **1968,** 3469.
200. H. Imer and J. H. Bagli, *J. Org. Chem.*, **33,** 2457 (1968).
201. T. Kametani, S. Shibuya, and W. D. Ollis, *J. Chem. Soc., C,* **1968,** 2877.
202. B. A. Hess, Jr., A. S. Bailey, and V. Boekelheide, *J. Amer. Chem. Soc.*, **89,** 2746 (1967).
203. H. Braeuking, F. Binnig, and H. A. Staab, *Chem. Ber.*, **100,** 880 (1967).
204. R. Dabrowski and K. Okon, *Biul. Wojskow. Akad. Tech.*, **14,** 163 (1965); *Chem. Abstr.*, **65,** 3179 (1966).
205. B. Belleau and R. Chevalier, *J. Amer. Chem. Soc.*, **90,** 6865 (1968).
206. B. Aleksiev and M. Milosev, *Chem. Ber.*, **100,** 701 (1967).
207. B. Pecherer and A. Brossi, *Helv. Chim. Acta,* **49,** 2261 (1966).
208. M. Oki and H. Iwamura, *J. Amer. Chem. Soc.*, **89,** 576 (1967).
209. M. G. Sarngadharan and T. R. Seshadri, *Indian J. Chem.*, **4,** 503 (1966).
210. Czechoslovakian Patent 123,846; *Chem. Abstr.*, **69,** 4120 (1968).
211. U.S. Patent 3,401,165; *Chem. Abstr.*, **70,** 394 (1969).
212. M. Oki, H. Iwamura, and T. Nishida, *Bull. Chem. Soc. Jap.*, **41,** 656 (1968).
213. J. F. Bagli and H. Immer, *Can. J. Chem.*, **46,** 3115 (1968).
214. C. D. Weis, *Helv. Chim. Acta,* **51,** 1582 (1968).
215. Netherlands Patent Appl. 6,406,750; *Chem. Abstr.*, **64,** 16016 (1966).
216. J. Altman, E. Babad, J. Itzchaki, and D. Ginsburg, *Tetrahedron Suppl.*, **8**(1), 279 (1966).
217. H. E. Zaugg and R. J. Michaels, *J. Org. Chem.*, **28,** 1801 (1963).
218. H. E. Zaugg, R. J. Michaels, A. D. Schaeffer, A. M. Wenthe, and W. H. Washburn, *Tetrahedron,* **22,** 1257 (1966).
219. H. E. Zaugg and R. J. Michaels, *J. Org. Chem.*, **31,** 1322 (1966).
220. G. Fontaine, *Ann. Chim.* (Paris), [14]**3,** 179 (1968).
221. B. A. Hess, Jr., A. S. Bailey, B. Bartusek, and V. Boekelheide, *J. Amer. Chem. Soc.*, **91,** 1665 (1969).
222. E. D. Bergmann, I. Shahak, and Z. Aizenshtat, *Tetrahedron Lett.*, **1969,** 2007.

CHAPTER IV

Oxepin Ring Systems Containing More Than Three Rings

ANDRE ROSOWSKY

Children's Cancer Research Foundation and Department of Biological Chemistry, Harvard Medical School, Boston, Massachusetts

A. Tetracyclic Systems

1. Fused Ring Systems

a. *1*H-*Dibenzo*[b, f]*cyclopentoxepins*

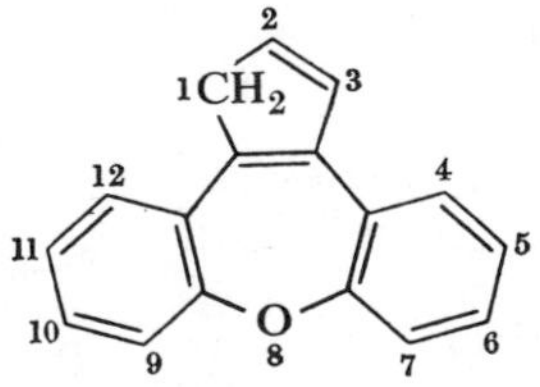

R.I. 4849

The sole reported member of this tetracyclic ring system is **1**, which Lippmann and Fritsch (36) claimed to be the principal product in the condensation of acetone and salicylaldehyde in the presence of zinc chloride at 140°. Accompanying **1** was the open-chain diphenol **2**, from which **1** could be obtained upon heating.

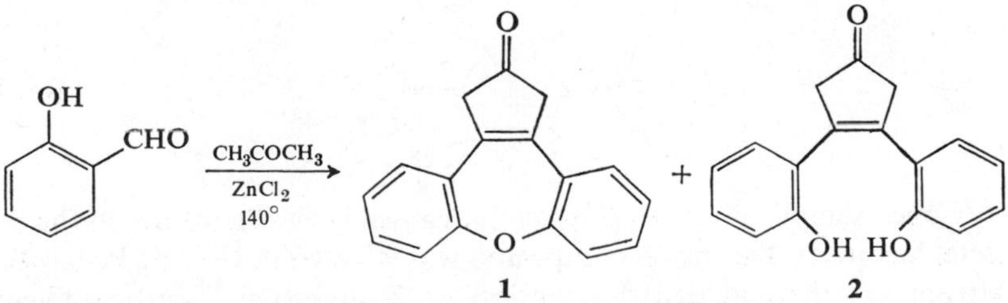

b. 4H-*Fluoreno*[4,5-cde]*oxepins*

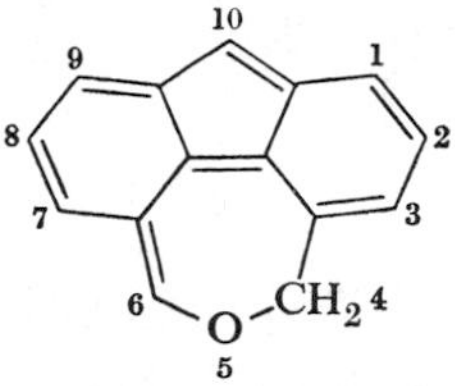

R.I. 4859

This ring system was first synthesized in 1934 by Kruber (34), who obtained anhydride **3** upon thermal cyclization of fluorenone-4,5-dicarboxylic acid (**4**).

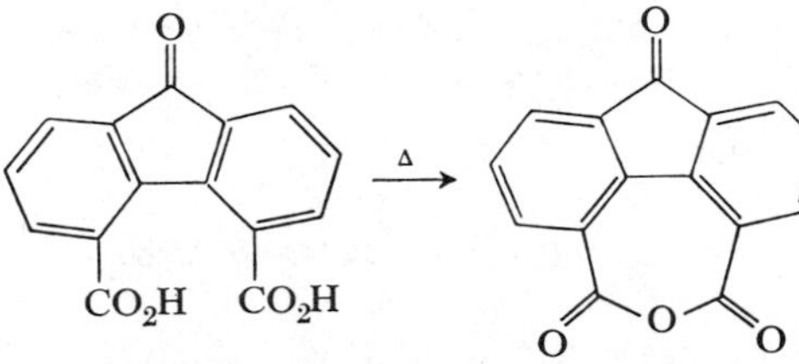

More recently, Takahashi (66) reported that oxidation of 5-methylfluorene-4-carboxylic acid (**5**) and 5-ethylfluorene-4-carboxylic acid (**6**) with sodium dichromate in refluxing acetic acid afforded lactones **7** and **8**, respectively.

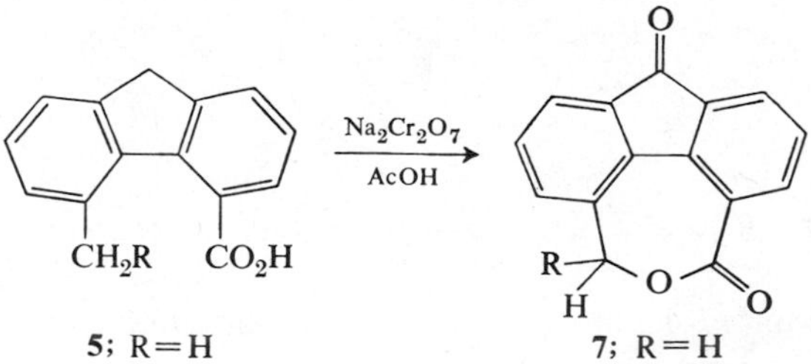

5; R=H
6; R=Me

7; R=H
8; R=Me

c. 4H-*Isobenzofuro*[1,7-cd][2]*benzoxepins*

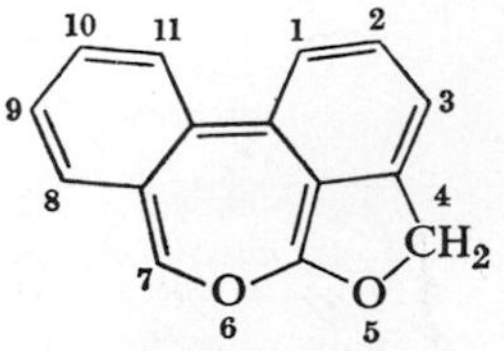

R.I. 4838

Only one example of this ring system is known in the literature, in the form of acetal lactone **9**. This novel compound was isolated in 1951 by Pozzo-Balbi (53) from the thermal decarboxylation of 2-carboxy-6-(2-carboxyphenyl)-phenylglyoxylic acid (**10**). Treatment of **9** with base led to ring opening and formation of 2-carboxy-6-(2-carboxyphenyl)benzaldehyde (**11**). Reduction of **9** with zinc in acid afforded 4-(2-carboxyphenyl)-3*H*-isobenzofuran-1-one (**12**).

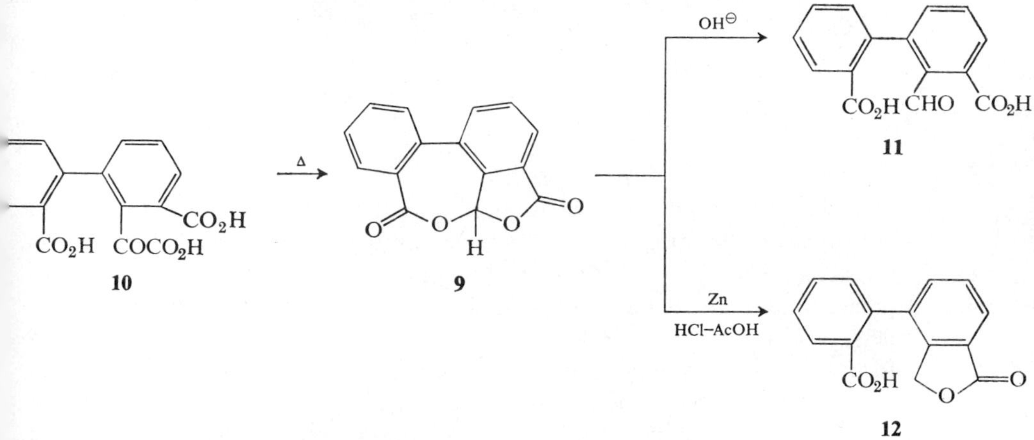

d. *Benzofuro*[*3,2*-c][*1*]*benzoxepins*

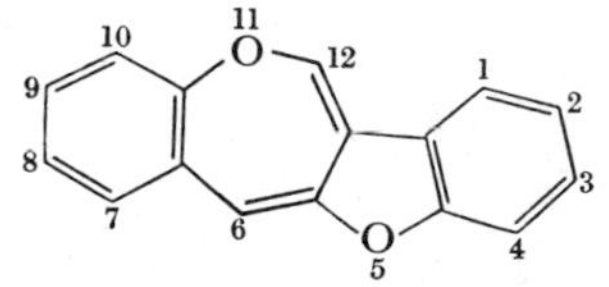

R.I. 9002

A single compound is known in this ring system to date. This was obtained in 1957 by Chatterjea (93) in the form of 6*H*-benzofuro[3,2-*c*][1]benzoxepin-12-one (**13**), the lactone arising from 2-(*o*-hydroxybenzyl)benzofuran-3-carboxylic acid (**14**) upon treatment with thionyl chloride. Acid **14** was prepared by isomerization of 3-(*o*-hydroxyphenacyl)isocoumaranone (**15**) in the presence of acid.

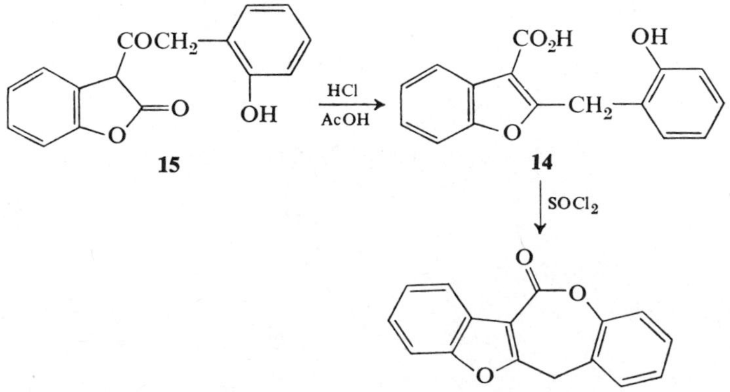

e. *Benzo*[d]*dioxolo*[*4,5*-h][*2*]*benzoxepins*

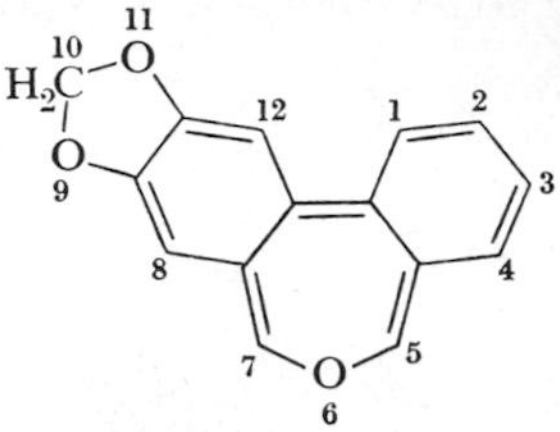

R.I. 8987

Ikeda and co-workers (94) carried out the first chemical synthesis of derivatives of this ring system in connection with their structural studies on the alkaloid tazzetine. A crossed Ullmann condensation between 6-bromopiperonal and methyl *o*-iodobenzoate afforded a complex mixture of products, one of which was aldehyde ester **16**. This compound was separated from the other components of the mixture by column chromatography, and then reduced with sodium borohydride to diol **17** and lactone **18**. Upon treatment with sulfuric acid in alcohol, diol **17** underwent ring closure to 5,7-dihydrobenzo[*d*]dioxolo[4,5-*h*][2] benzoxepin (**19**).

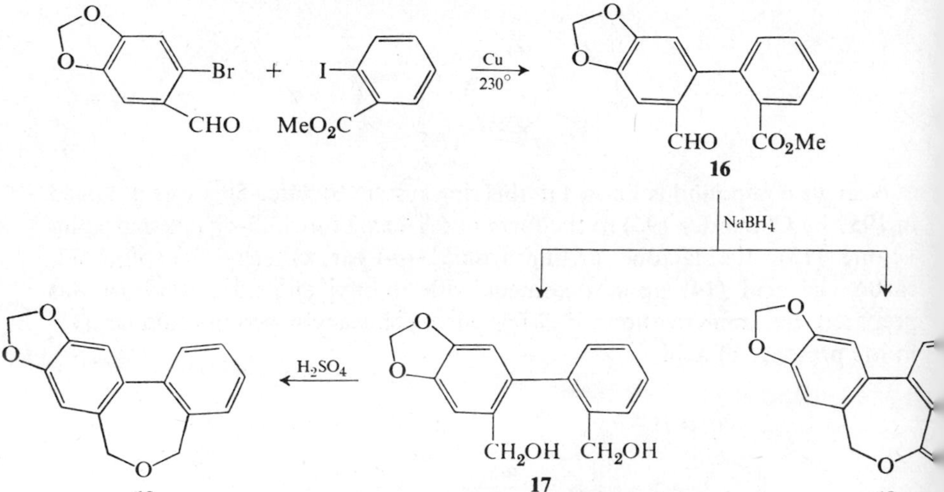

f. [*1*]*Benzothieno*[*2,3*-c][*1*]*benzoxepins*

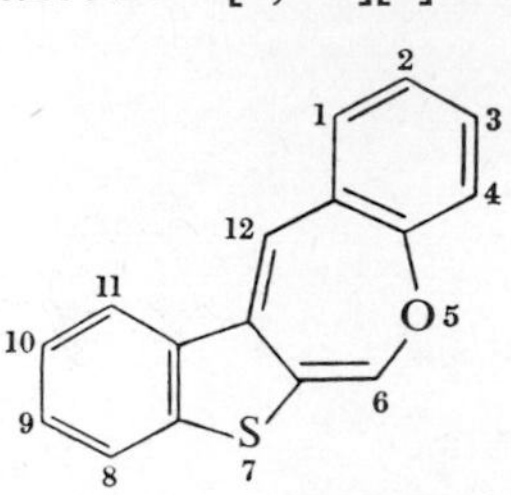

R.I. 4841

According to Peters and Walker (95), treatment of 3-(2′,4′-dihydroxybenzoyl)thianaphthene-2-carboxylic acid (**20**) with boiling acetic anhydride alone, or with acetyl chloride in refluxing xylene, afforded 3-acetoxy[1]-benzothieno[2,3-*c*][1]benzoxepin-6,12-dione (**21**). Similarly, dibromide **22** was cyclized to 3-acetoxy-2,4-dibromo[1]benzothieno[2,3-*c*][1]benzoxepin-6,12-dione (**23**) under the influence of acetyl chloride in xylene, and **20** was converted into benzoate ester **24** upon reaction with benzoyl chloride instead of acetyl chloride. Gentle alkaline hydrolysis of **21** gave [1]benzothieno-[2,3-*c*][1]benzoxepin-3-ol (**25**). However, cleavage of the seven-membered lactone ring could also be effected readily.

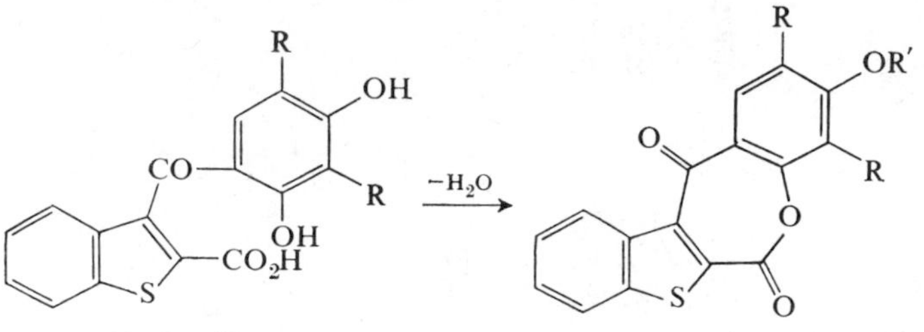

$\xrightarrow{-H_2O}$

20; R = H
22; R = Br

21; R = H, R′ = Ac
23; R = Br, R′ = Ac
24; R = H, R′ = Bz
25; R = R′ = H

g. 5H-*Pyrido*[2′,3′:5,6]*oxepino*[2,3-b]*indoles*

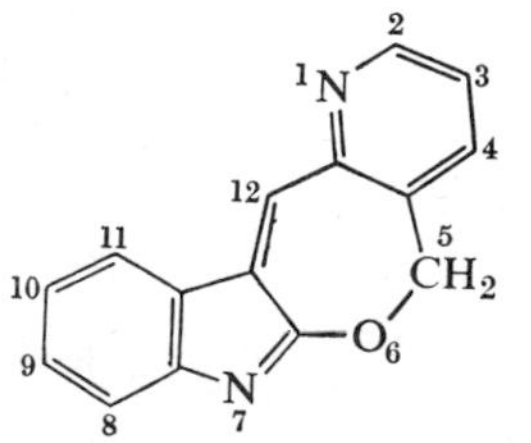

R.I. 8990

Entry into this interesting tetracyclic system was achieved in 1958 by Plieninger and co-workers (96). Base-catalyzed condensation of isatin or *N*-methylisatin with dimethyl 2-methylpyridinedicarboxylate gave unsaturated diacids **26** and **27**, respectively. Catalytic hydrogenation of these compounds over Raney nickel in methanol containing some triethylamine yielded the corresponding reduced monoesters **28** and **29**, and treatment of the latter compounds with boiling acetic anhydride led to lactone esters **30** and **31**. If monoester **29** was hydrolyzed with base prior to lactonization, the product was lactone acid **32**. Methanolysis of **30** or **31** afforded diesters **33** and **34**, respectively. An interesting property of **32** was the fact that it could

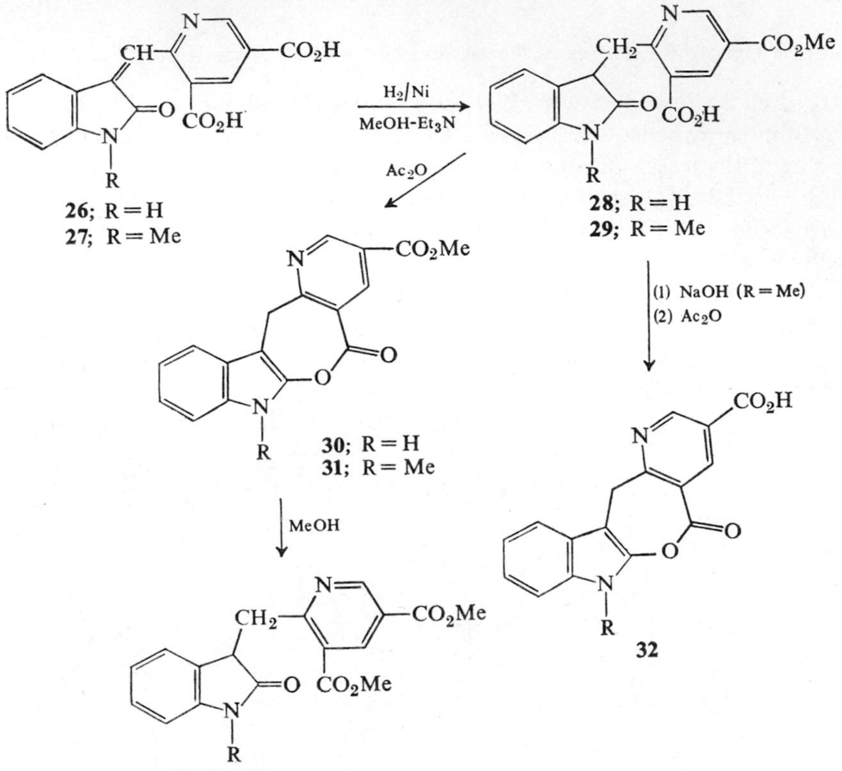

26; R = H
27; R = Me

H_2/Ni
MeOH-Et_3N

28; R = H
29; R = Me

Ac_2O

30; R = H
31; R = Me

(1) NaOH (R = Me)
(2) Ac_2O

32

MeOH

33; R = H
34; R = Me

H_2NNH_2

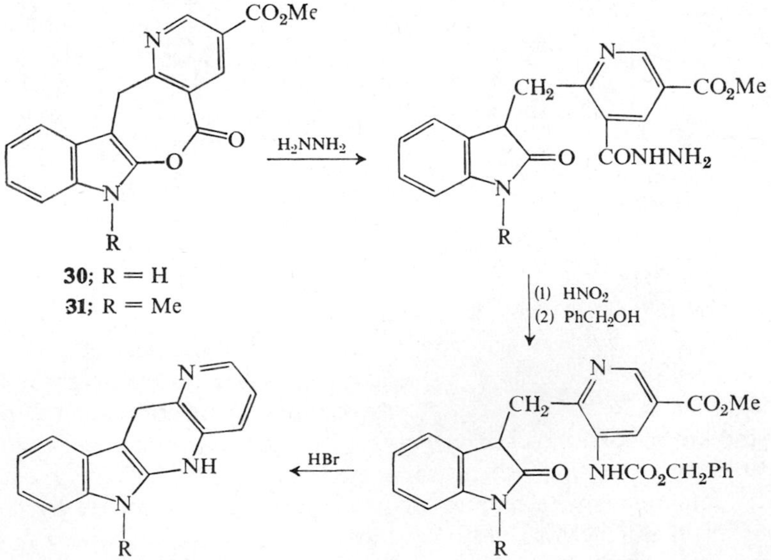

30; R = H
31; R = Me

(1) HNO_2
(2) $PhCH_2OH$

HBr

be resolved into optical isomers through diastereomeric quinidine salts. The ability of **32** to exist in two optically active forms is a result of the distorted geometry of the seven-membered ring, which leads to molecular asymmetry.

A novel transformation of the oxepin ring in **30** or **31** into a pyridine ring was performed via the reaction sequence depicted.

h. *4,9-Dioxa-1,7-diazacyclohepta*[def]*fluorenes*

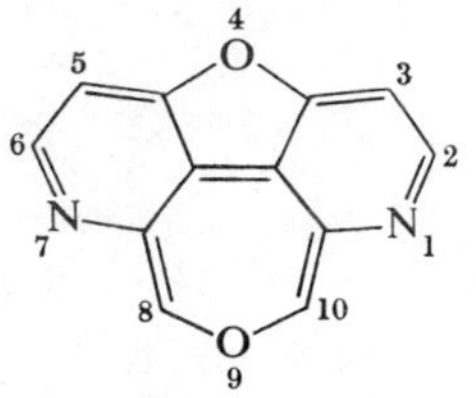

R.I. 4835

The lone known example of this ring system is anhydride **35**, which was prepared in 1939 by Jacini and Salini (27) by treatment of the corresponding diacid **36** with hot concentrated sulfuric acid. Reaction of **35** with ammonia and phenylhydrazine resulted in cleavage of the oxepin ring and formation of the monoamide and monophenylhydrazide, **37** and **38**, respectively.

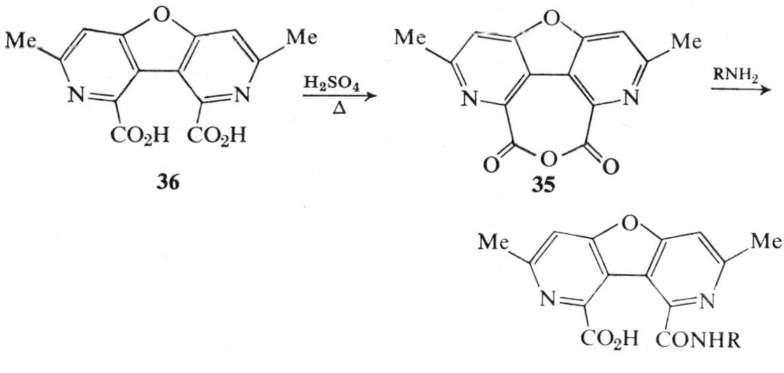

$\xrightarrow[\Delta]{H_2SO_4}$ $\xrightarrow{RNH_2}$

36 **35**

37; R = H
38; R = PhNH

i. *Tribenz*[b,d,f]*oxepins*

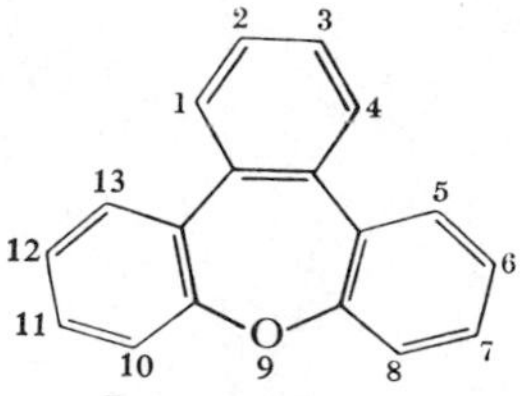

A synthesis of tribenz[*b*,*d*,*f*]oxepin (**39**), disclosed recently in a British patent (87), consists of heating sulfonic acid chloride **40** in octachloronaphthalene at 250°, with a stream of nitrogen through the system to remove evolved sulfur dioxide and hydrogen chloride.

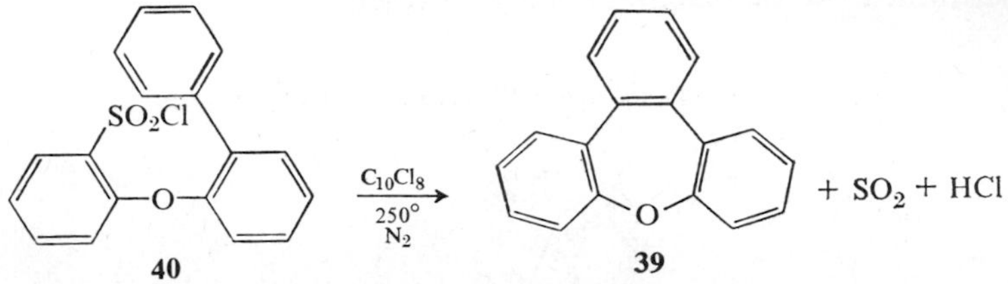

SO_2Cl — $C_{10}Cl_8$, 250°, N_2 → + SO_2 + HCl

40 **39**

A second, ingenious synthesis of a derivative of tribenz[*b*,*d*,*f*]oxepin was reported in 1967 by Tochtermann and co-workers (88). This involved dehydrobromination of 10-bromodibenz[*b*,*f*]oxepin (**41**) with sodamide followed by addition of 1,2,3,4-tetraphenylbutadiene to the resultant benzynelike intermediate to give 1,2,3,4-tetraphenyltribenz[*b*,*d*,*f*]oxepin (**42**).

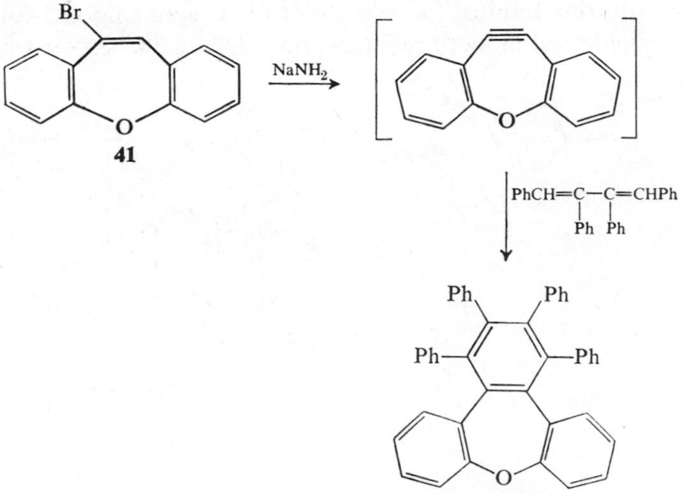

Br — $NaNH_2$ →

41

PhCH=C(Ph)—C(Ph)=CHPh

42

j. *Benzo*[c]*naphth*[*2,1*-e]*oxepins*

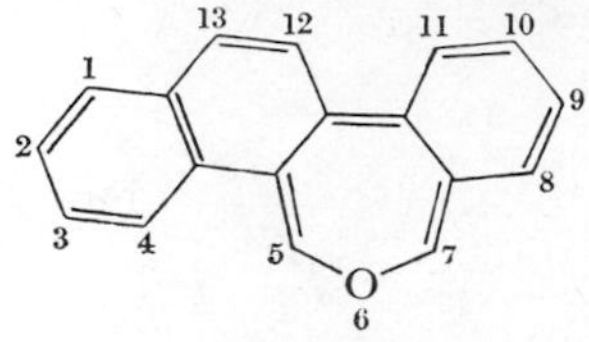

R.I. 5302

An example of this ring system was obtained in 1904 by Graebe and Gnehm (24) in the form of anhydride **43**. This compound was formed in 94% yield when the corresponding dicarboxylic acid, **44**, was heated in acetic anhydride. Hydrolysis of **43** regenerated the diacid. Reaction with methanol afforded a monoester, **45**, which was arbitrarily formulated as being derived by methanolysis of the benzoic, rather than naphthoic, half of the anhydride moiety. Treatment of **43** with dilute ammonia gave a monoamide, **46**, that was formulated similarly. Thermal decarboxylation of **43** led to the expected benzofluorenone, **47**.

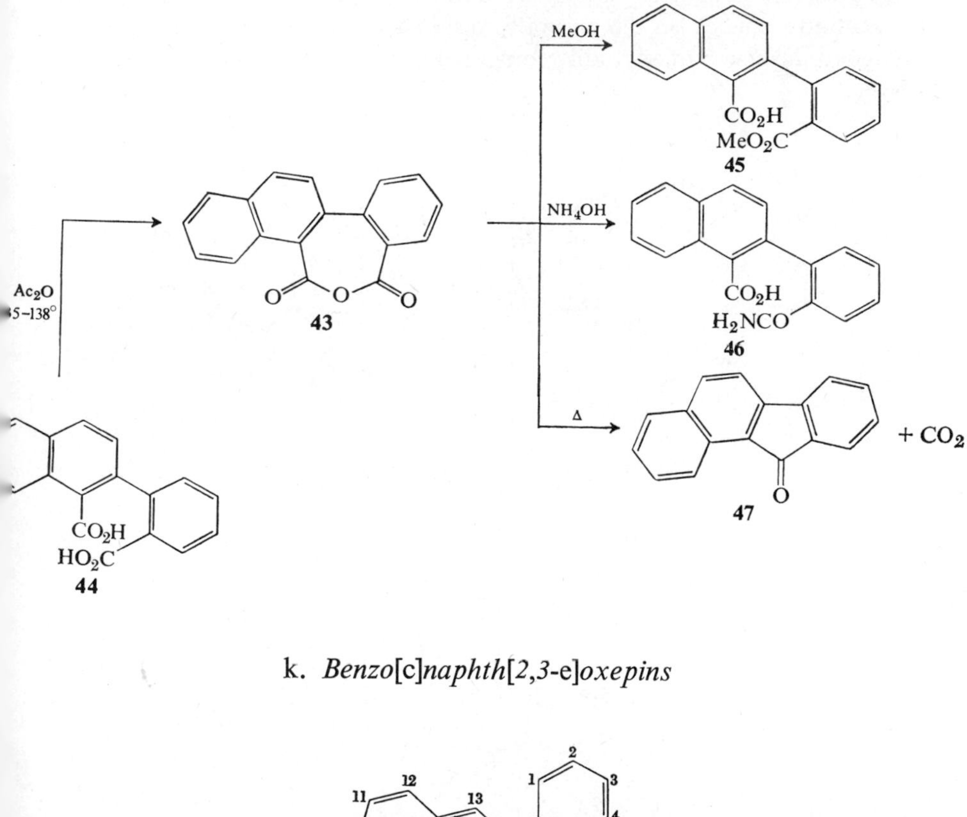

MeOH

CO_2H
MeO_2C
45

NH_4OH

Ac$_2$O
5–138°

43

CO_2H
H_2NCO
46

Δ

$+ CO_2$

47

CO_2H
HO_2C
44

k. *Benzo*[c]*naphth*[*2,3*-e]*oxepins*

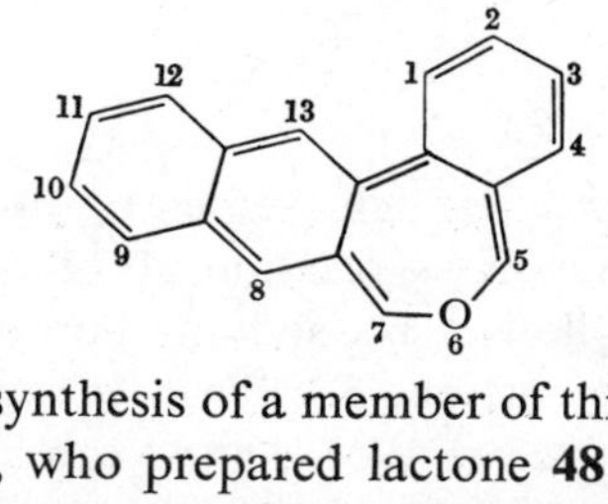

R.I. 10923

The first report of the synthesis of a member of this ring system was that of Hadler and Kryger (25), who prepared lactone **48** from the corresponding hydroxy acid, **49**, upon treatment with *p*-toluenesulfonic acid.

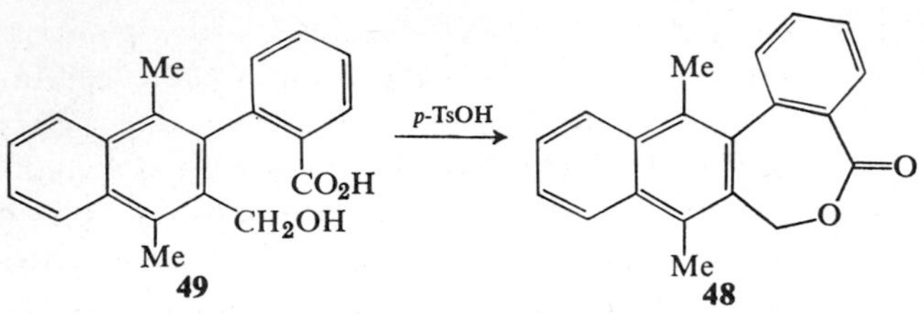

A second representative of the benzo[*c*]naphth[2,3-*e*]oxepin ring system was described in 1960 by Simonitsch and co-workers (63) in connection with degradative studies on the naturally occurring compound chartreusin. The dicarboxylic acid **50** was readily converted into its anhydride, **51**, upon being heated.

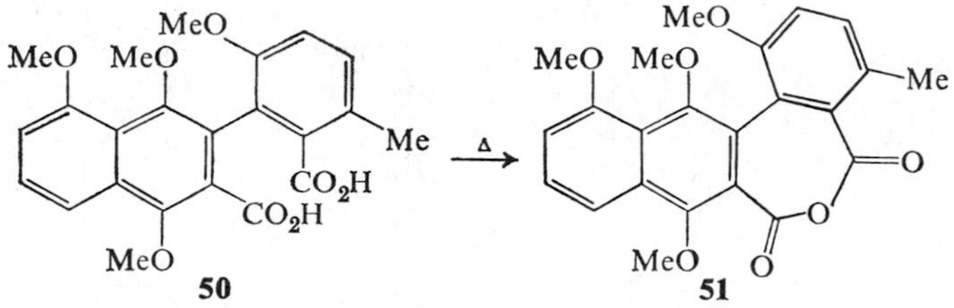

1. *Benzo*[e]*naphth*[*1,2*-b]*oxepins*

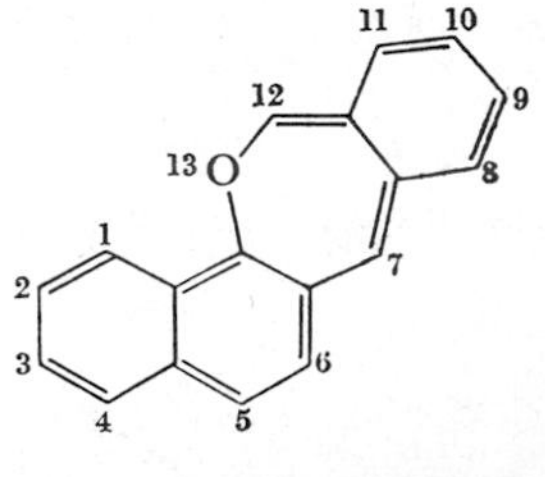

R.I 9102.

Pandit and Kulkarni (52) first achieved entry into this ring system in 1958 via a two-step synthesis involving reaction of 5,8-dimethoxy-1-tetralone and phthalaldehydic acid, followed by sodium borohydride reduction of the resultant keto acid, **52**, to lactone **53**. In an extension of this work, Shroff and co-workers (61) reported in 1961 that **53** can be reduced catalytically to **54**, which is apparently a mixture of *cis* and *trans* isomers.

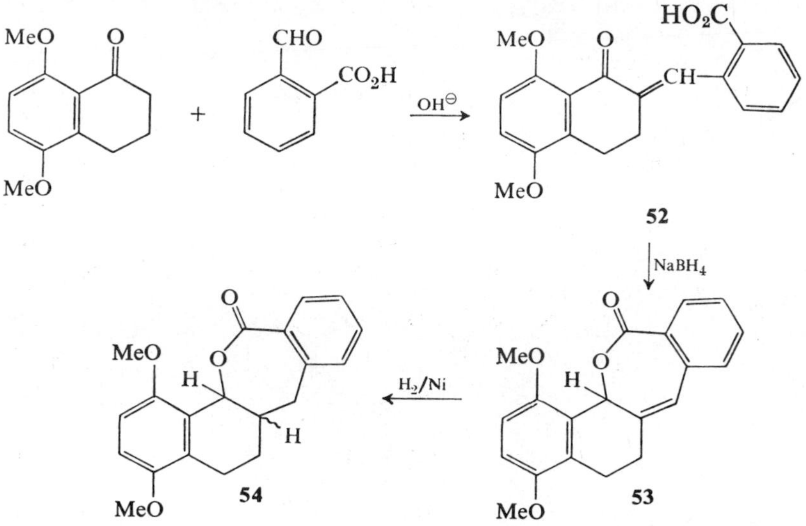

MeO O CHO CO_2H + $OH^{\ominus}$ MeO O HO_2C CH MeO MeO 52

$NaBH_4$

O MeO O H H 54 MeO H_2/Ni O MeO O H 53 MeO

m. *Benzo*[e]*naphth*[*2,1*-b]*oxepins*

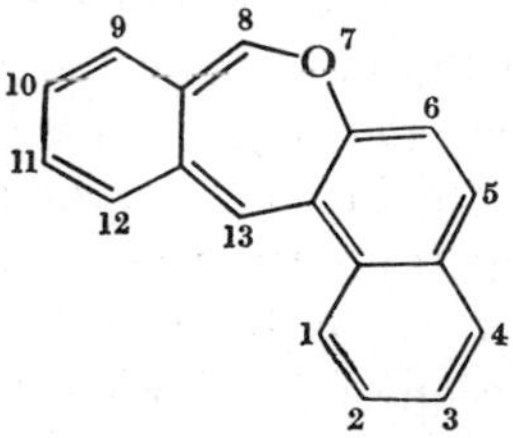

9 8 7 10 11 12 13 6 5 1 2 3 4

R.I. 5301

The first reported syntheses of members of this ring system were those of Szeki and co-workers (65) in 1930 and of Fieser (23) in 1931. Phthalic anhydride and 2-methoxynaphthalene, upon being heated in tetrachloroethylene at 100° in the presence of aluminum chloride, gave hydroxy acid **55**. Lactonization of **55** occurred readily and gave a product melting at 195°, to which structure **56** was assigned. Reduction of **55** with metallic zinc in alkali afforded hydroxy acid **57**, and lactonization of the latter yielded **58**. Similarly, ring closure of lactone **60** was achieved by heating acid **59** with a mixture of acetic and sulfuric acids (65).

According to Fieser, treatment of **56** with aluminum chloride in benzene results in the formation of an isomeric product of unidentified structure melting at 196°. The same isomeric product was supposedly also formed when phthalic anhydride was condensed with 2-naphthol, rather than 2-methoxynaphthalene, in the presence of aluminum chloride at 250°. Although several degradative transformations were carried out, the structure of the isomer of **56** could not be elucidated at the time.

In 1947 Badger (5) reinvestigated the reaction of phthalic anhydride and 2-methoxynaphthalene, but the mystery of Fieser's unidentified compound remained unsolved. When the reaction of 2-methoxynaphthalene and phthalic anhydride was carried out in nitrobenzene at 0°, the product was the expected 1-(*o*-carboxybenzoyl)-2-methoxynaphthalene, while in refluxing benzene the product proved to be the compound isolated by Fieser and formulated as **55**. Lactonization of **55** afforded **56**, and hydrolysis of **56** with concentrated sulfuric acid at 60–70° regenerated the open-chain hydroxy acid in good yield.

n. *Phenanthro*[*4,5*-bcd]*oxepins*

This ring system is represented in the literature only by the natural product thebenol (**61**) which was described in 1925 by Sieglitz and Koch (62).

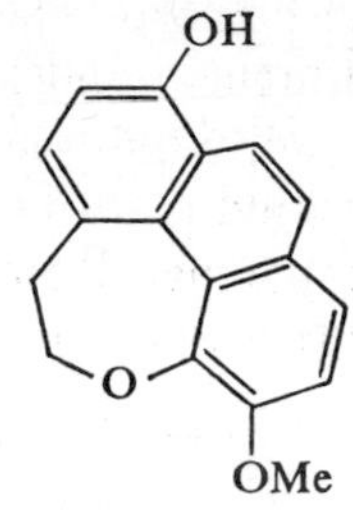

OH

OMe

61

o. *Phenanthro*[*4,5*-cde]*oxepins*

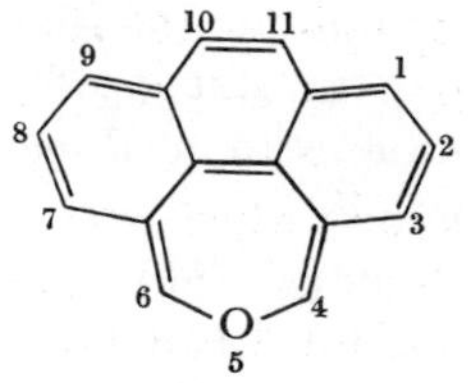

R.I. 5306

The earliest paper describing the phenanthro[4,5-*cde*]oxepin ring system was that of Newman and Whitehouse (46) who reported in 1949 that dehydration of phenanthrene-4,5-dimethanol under various conditions led to excellent yields of 4,6-dihydrophenanthro[4,5-*cde*]oxepin (**62**). Reagents used for this purpose included hydrogen chloride in benzene, thionyl chloride in a mixture of benzene and pyridine, phosphorus pentasulfide in benzene, hydrogen bromide in benzene, and *p*-toluenesulfonyl chloride in pyridine. Simple heating of the diol at the melting point was likewise sufficient to effect ring closure. An alternate method of synthesis reported by Badger and coworkers (6) in 1950 involves successive treatment of the diol with phosphorus tribromide and lithium aluminum hydride. Since the direct dehydration of the diol to **62** is so facile, the latter two-step procedure offers no particular advantage. Reductive cleavage of the oxepin ring in **62** was effected by Newman and Whitehouse (46) via the action of phosphorus and 47% hydriodic acid at 165°, the product being 4,5-dimethylphenanthrene.

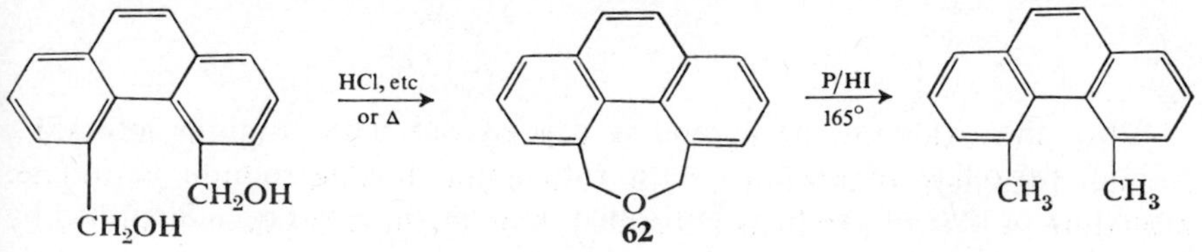

CH_2OH CH_2OH $\xrightarrow[\text{or }\Delta]{\text{HCl, etc}}$ **62** $\xrightarrow[165^\circ]{\text{P/HI}}$ CH_3 CH_3

Phenanthro[4,5-*cde*]oxepins in higher oxidation states than **62** have also been described. Thus phenanthrene-4,5-dicarboxylic acid is reportedly converted into phenanthro[4,5-*cde*]oxepin-4,6-dione (**63**) in 70–75% yield upon being heated above the melting point (17, 40). The diacid is prepared conveniently via ozonolysis of pyrene. Reaction of **63** with ammonium hydroxide occurs readily with formation of the expected monoamide, **64**.

CO_2H CO_2H —260–290°→ **63** —NH_4OH→ CO_2H $CONH_2$ **64**

4*H*-Phenanthro[4,5-*cde*]oxepin-6-one (**65**) has been prepared by Newman and Whitehouse (46) in two steps starting from 4-formylphenanthrene-5-carboxylic acid. Hydrogenation with a Raney nickel–aluminum catalyst afforded a 34% yield of a lactone which was formulated as 10,11-dihydro-4*H*-phenanthro[4,5-*cde*]oxepin-6-one (**66**). Dehydrogenation of **66** with sulfur at elevated temperatures furnished **65** in 66% yield. Alternatively, **65** could be prepared directly from the aldehyde by reduction with aluminum isopropoxide in toluene, the formation of the reduced hydroxy acid being accompanied by spontaneous lactonization.

CO_2H CHO —H_2/Ni–Al, 10% NaOH, 90°→ **66** —S, 230–235°→ **65**

CO_2H CHO —Al(OPr-*i*)$_3$, PhMe→ **65**

When the oxidation of pyrene is carried out with osmium tetroxide-sodium periodate reagent, or with ruthenium dioxide–sodium periodate, a mixture of several products is formed. One of these has been identified by

Oberender and Dixon (47) as 4-hydroxy-4*H*-phenanthro[4,5-*cde*]oxepin-6-one (**67**), on the basis of its ir spectral features, and also its reaction with diazomethane which gave a methyl ester .

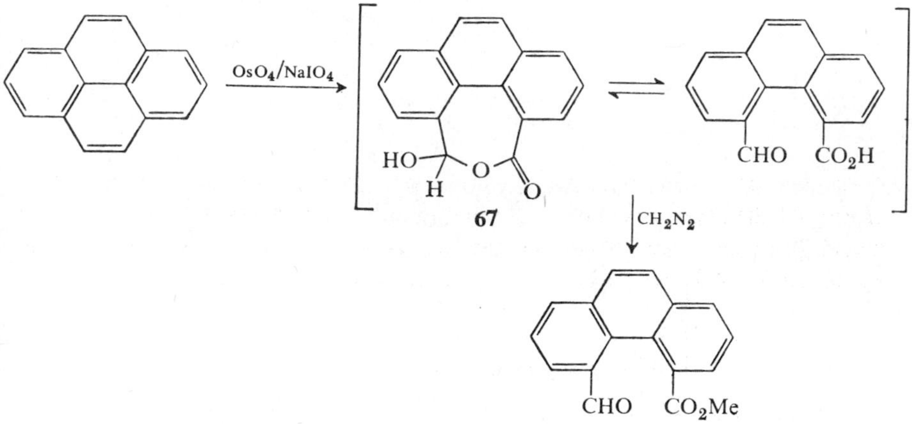
$OsO_4/NaIO_4$
HO
H
O
O
CHO
CO_2H
CH_2N_2
CHO
CO_2Me

67

p. *Phenanthro*[*9,10*-d]*oxepins*

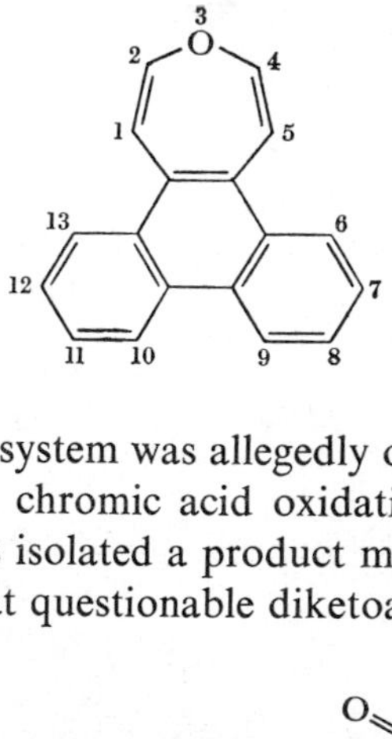
3
O
2
4
1
5
13
6
12
7
11
10
9
8

R.I. 5305

A derivative of this ring system was allegedly obtained in 1911 by Willgerodt and Albert (69) upon chromic acid oxidation of 9,10-diacetylphenanthrene (**68**). These workers isolated a product melting above 360° to which they assigned the somewhat questionable diketoanhydride structure **69**.

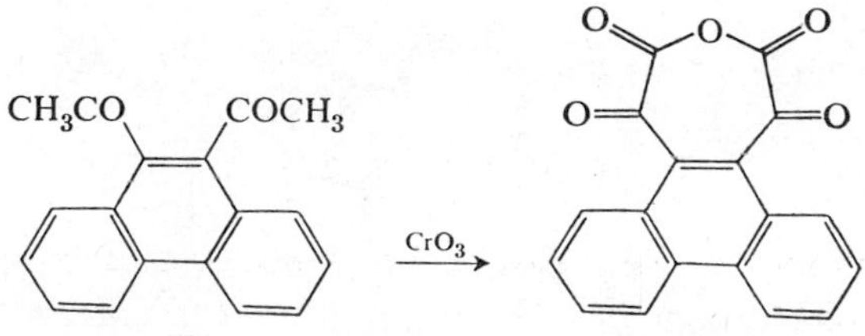
CH_3CO
$COCH_3$
CrO_3
O
O
O
O
O

68 **69**

q. [2]*Benzopyrano*[8,1-bc][1]*benzoxepins*

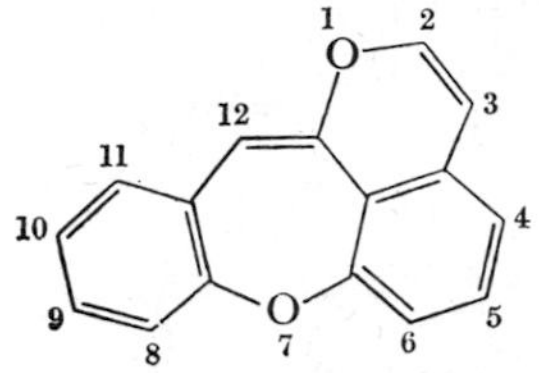

As part of a very extensive synthetic program directed toward the preparation of the alkaloid cularine, Kametani and co-workers (31, 32) subjected diacid **70** to the action of polyphosphoric acid and thereby obtained lactone **71**. Reaction of **71** with ammonia afforded lactam **72**.

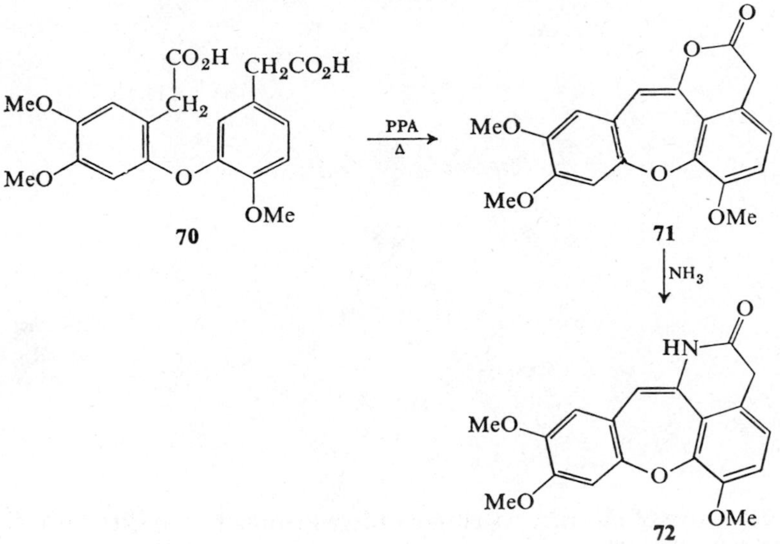

r. 1H-*Benz*[6,7]*oxepino*[2,3,4-ij]*isoquinolines*

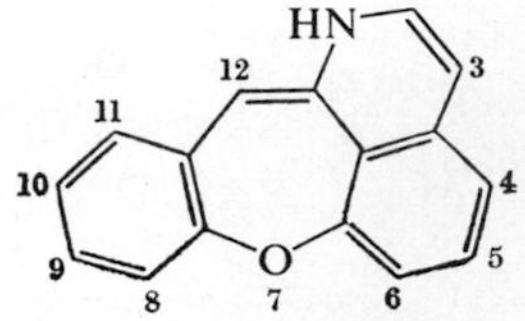

The chemistry of this ring system has been studied in great detail by Kametani and co-workers (28, 29, 31, 32, 81–84) in connection with their work on cularine (**73**), cularidine (**74**), cularimine (**75**), and related alkaloids.

Condensation of 10,11-dihydro-2,3,6-trimethoxydibenz[*b,f*]oxepin-10-one (**76**) with 2-aminoacetal followed by treatment with sulfuric (28, 81) or polyphosphoric acid (82) (the Pomerantz-Frisch reaction) gave 6,9,10-trimethoxy-12*H*-benz[6,7]oxepino[2,3,4-*ij*]isoquinoline (**77**). Catalytic hydrogenation of **77** in the presence of platinum oxide afforded **75**, and sodium borohydride reduction of the methiodide of **77** led to **73** (28, 81). Oxidation of **77** with chromium trioxide in pyridine yielded 6,9,10-trimethoxy-12*H*-benz[6,7]oxepino[2,3,4-*ij*]isoquinolin-12-one (**78**) (82).

MeO MeO O OMe O **76**

(1) $H_2NCH_2CH_2CH(OEt)_2$
(2) H_2SO_4

MeO MeO N O OMe **77**

CrO_3

H_2/PtO_2

(1) MeI
(2) $NaBH_4$

MeO MeO O N O OMe **78**

MeO MeO HN O OMe **75**

MeO MeO Me N O OR
73; R=Me
74; R=H

In another approach to the synthesis of this ring system, amino acid **79** was cyclized to eleven-membered ring amide **80**, and the latter was transformed into 2,3-dihydro-6,9,10-trimethoxy-12*H*-benz[6,7]oxepino[2,3,4-*ij*]-isoquinoline (**81**) upon treatment with phosphorus oxychloride (29, 83).

A third entry into the benz[6,7]oxepino[2,3,4-*ij*]isoquinoline ring system was via lactone **82** which was obtained from diacid **83** upon treatment with polyphosphoric acid. Ammonolysis of **82** furnished 2,3-dihydro-6,9,10-trimethoxy-1*H*-benz[6,7]oxepino[2,3,4-*ij*]isoquinoline-2-one (**84**), and catalytic hydrogenation of **84** over platinum oxide gave 2-oxocularimine (**85**) (31, 32).

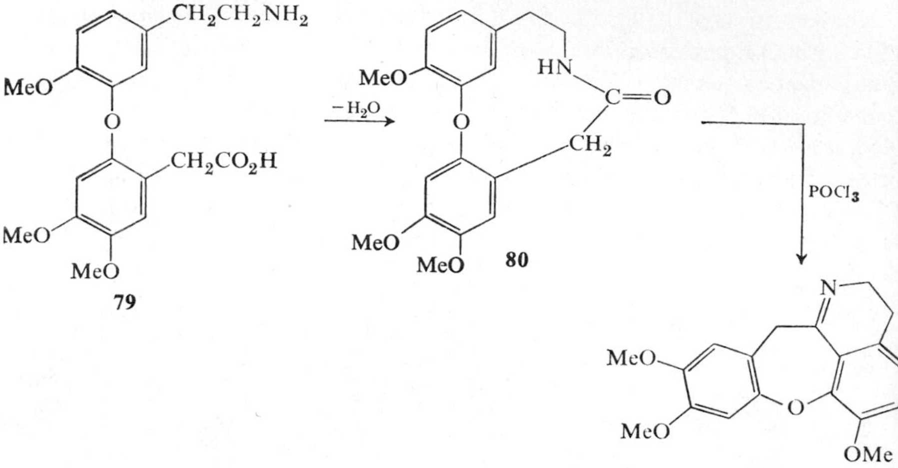

79

80

81

Reduction of **85** with lithium aluminum hydride in tetrahydrofuran gave a mixture of **75** (31, 32) and a compound identified as 2,3,12,12a-tetrahydro-6,9,10-trimethoxy-1*H*-benz[6,7]oxepino[2,3,4-*ij*]isoquinolin-2-ol (**86**) (32). Acetylation of **86** with acetic anhydride in pyridine gave **87**, and methylation of **86** by the Eschweiler-Clarke procedure yielded **88**. Dehydration of **86** with phosphorus pentoxide gave 2,3-dihydro-6,9,10-trimethoxy-1*H*-benz[6,7]-oxepino[2,3,4-*ij*]isoquinoline (**89**), and chlorination of **86** with phosphorus

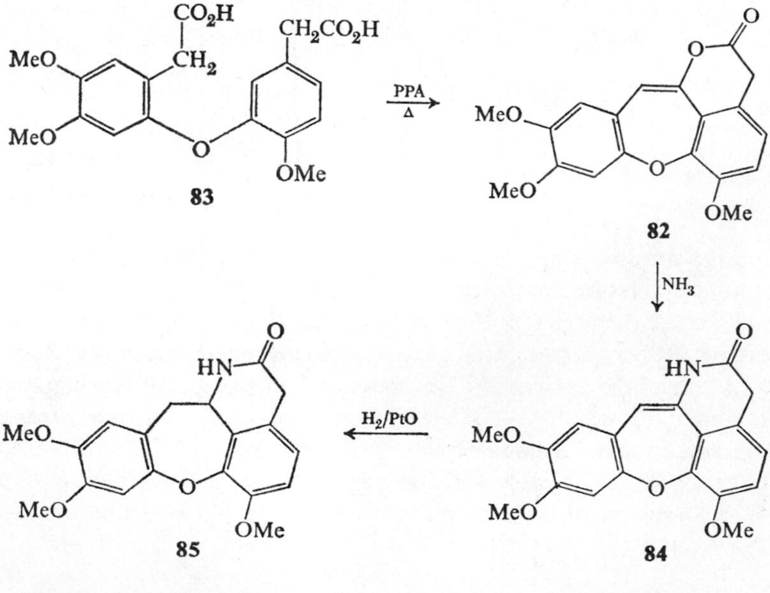

83

82

85

84

oxychloride gave 2-chloro derivative **90**. Treatment of either **89** or **90** with zinc metal in acetic and hydrochloric acid yielded **75**. An alternative method for converting **85** into **75** consisted of thiation with phosphorus pentasulfide followed by electrolytic reduction of resultant thioamide; the yield of **75** via this approach was very low, however. Eschweiler-Clarke methylation of **75** led to **73**.

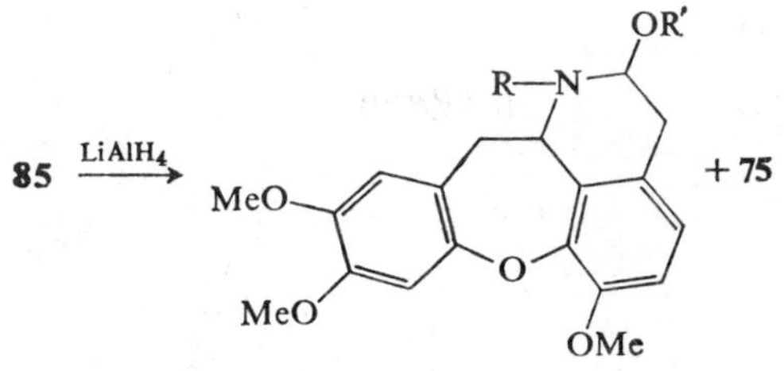
85
LiAlH4
OR'
R—N
MeO
MeO
O
OMe
+ 75

86; R = R' = H
87; R = R' = Ac
88; R = Me, R' = H

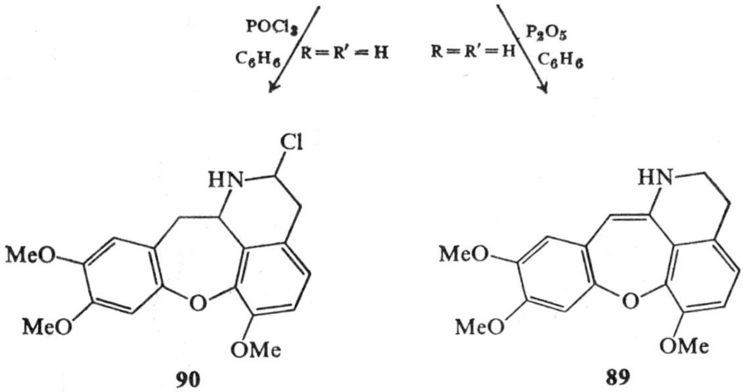
POCl3
C6H6
R = R' = H
P2O5
C6H6
R = R' = H
Cl
HN
MeO
MeO
O
OMe
HN
MeO
MeO
O
OMe

90 **89**

Still another synthetic route leading to the benz[6,7]oxepino[2,3,4-*ij*]-isoquinoline ring system involved cyclization of 9-[2-(*N*,*N*-dimethylamino)-ethyl]-10,11-dihydro-2,3,6-trimethoxydibenz[*b*,*f*]oxepin-10-ol (**91**) to cularine methochloride (**92**) by treatment with thionyl chloride (30). Upon reaction with potassium iodide, **92** was transformed into the corresponding methiodide derivative, **93**. For the synthesis of **91**, azlactone **94** was subjected to vigorous alkaline hydrolysis with concentrated barium hydroxide, and the resulting amine, **95**, was methylated.

The structure of cularidine (**74**) was recently elucidated by chemical means (84), and also on the basis of nmr and mass spectral evidence (85).

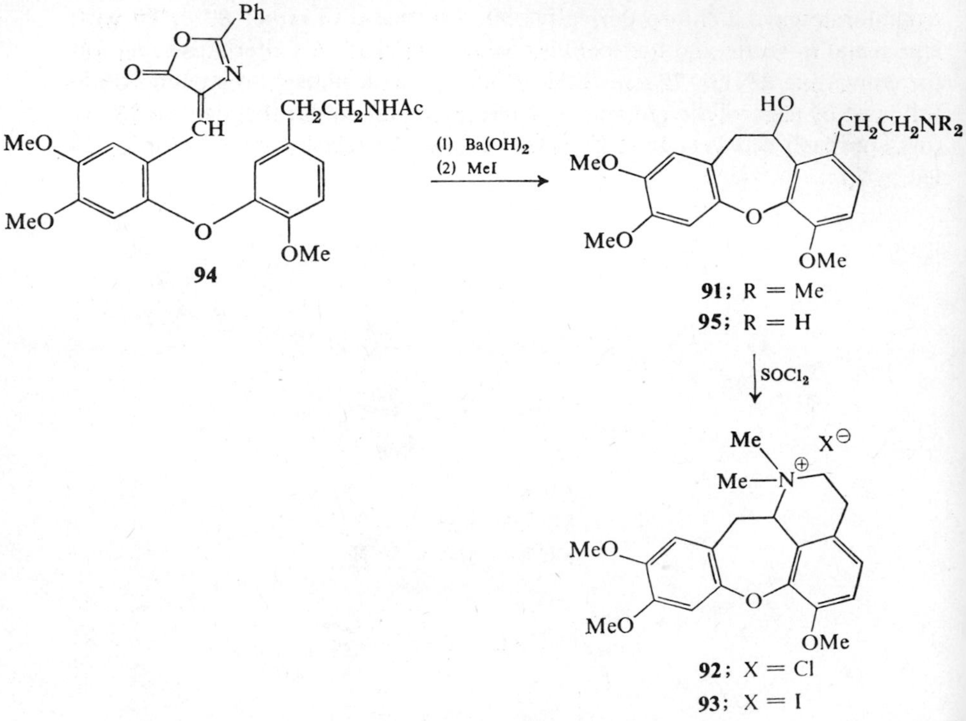

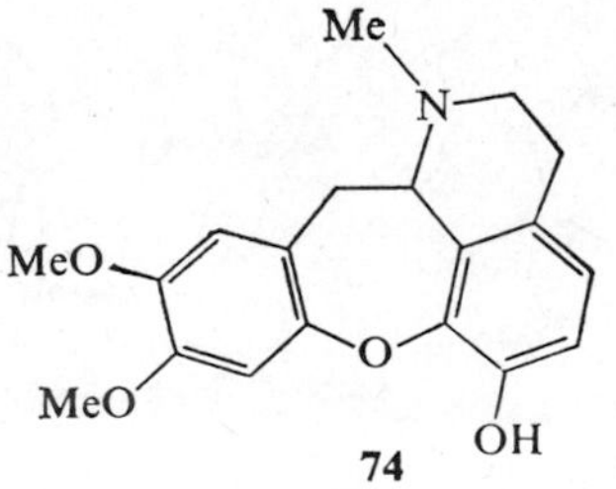

Ph
O
O
N
CH
CH_2CH_2NHAc
MeO
MeO
O
OMe
94

(1) $Ba(OH)_2$
(2) MeI

HO
$CH_2CH_2NR_2$
MeO
MeO
O
OMe
91; R = Me
95; R = H

$SOCl_2$

Me
$X^{\ominus}$
Me—$N^{\oplus}$
MeO
MeO
O
OMe
92; X = Cl
93; X = I

Me
N
MeO
MeO
O
OH
74

s. *Dibenz*[*2,3*:*6,7*]*oxepino*[*4,5*-d]*pyrimidines*

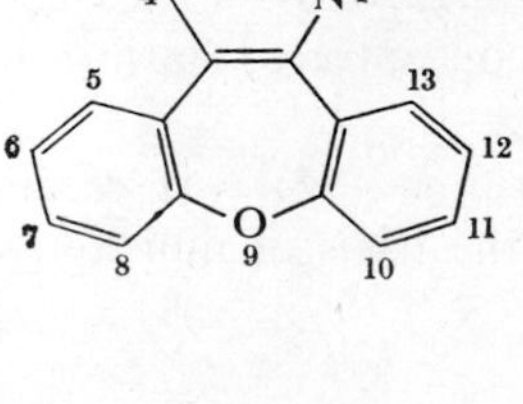

3 N 2 N 1 4 5 13 6 12 7 11 8 O 9 10

Kametani and co-workers (28, 86) reported two members of this ring system in the form of dihydro derivatives **96** and **97** which were isolated as by-products in the Leuckart reaction of 10,11-dihydro-2,3-dimethoxydibenz-[*b,f*]oxepin-10-one (**98**) and 10,11-dihydro-2,3,6-trimethoxydibenz[*b,f*]-oxepin-10-one (**99**) with formamide. The principal products were amides **100** and **101**. Large amounts of formamide or added formic acid tended to minimize the formation of **96** and **97**.

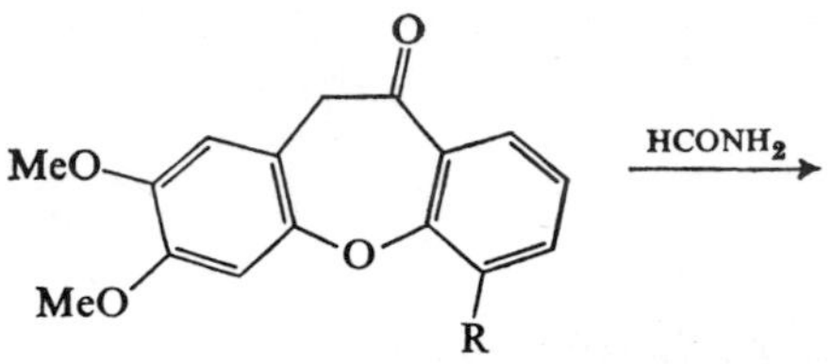

MeO MeO O O R $\xrightarrow{HCONH_2}$

98; R = H
99; R = MeO

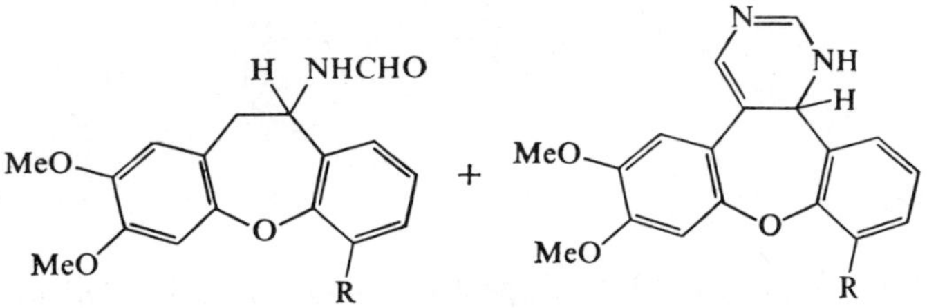

H NHCHO MeO MeO O R + N NH H MeO MeO O R

100; R = H
101; R = MeO

96; R = H
97; R = MeO

t. [*1*]*Benzoxepino*[*4,5*-b]*quinoxalines*

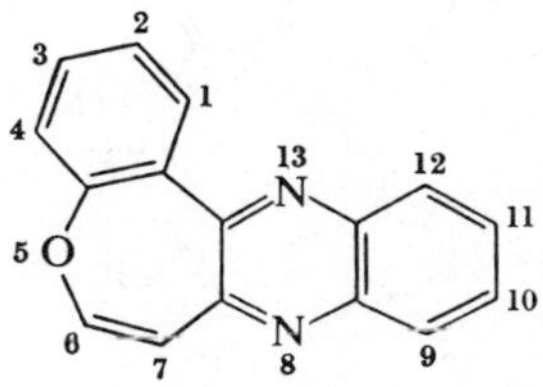

2 3 1 4 13 N 12 11 5 O 10 N 6 7 8 9

R.I. 5277

Only one representative of this ring system has been reported to date in the form of 6,7-dihydro-2,3-dimethyl[1]benzoxepino[4,5-*b*]quinoxaline (**102**) which Dann and Arndt (18) prepared in 1954 via condensation of 2,3-dihydro-7,8-dimethyl-1-benzoxepin-4,5-dione (**103**) with *o*-phenylenediamine.

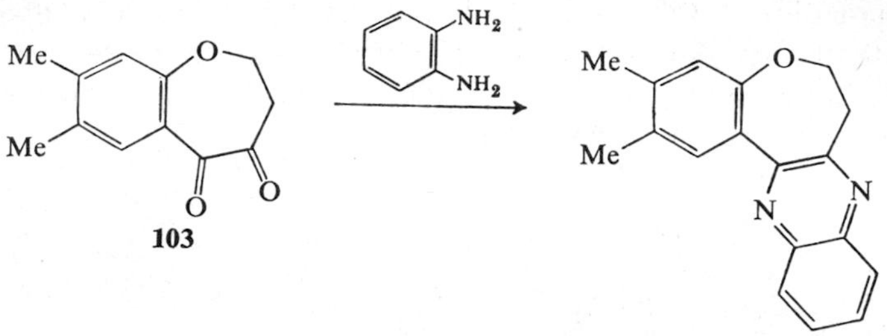

Me Me O O O **103** NH_2 NH_2 Me Me O N N **102**

u. [2]*Benzoxepino*[6,5,4-cde]*cinnolines*

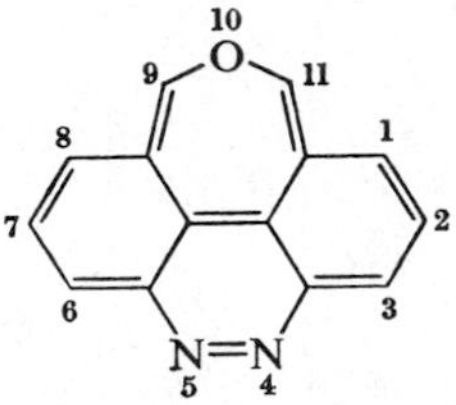

10 O 9 11 8 1 7 2 6 3 N=N 5 4

R.I. 10,919

Holt and Hughes (92) unexpectedly obtained 9,11-dihydro[2]benzoxepino-[6,5,4-*cde*]cinnoline (**104**) in place of the desired dibromide upon reaction of diol **105** with phosphorus tribromide or hydrogen bromide. The diol was prepared from 6,6′-dinitrodiphenic acid by reduction with lithium aluminum hydride.

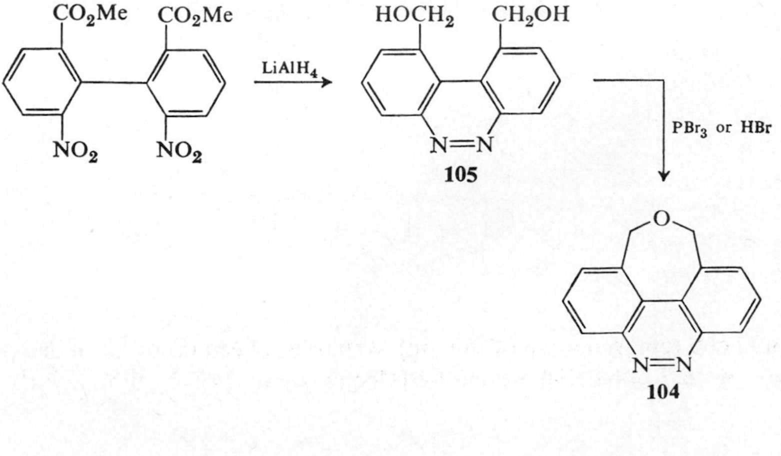

CO_2Me CO_2Me NO_2 NO_2 $LiAlH_4$ $HOCH_2$ CH_2OH N=N **105** PBr_3 or HBr O N=N **104**

v. *4*H,*6*H-*[2]Benzoxepino[6,5,4-*def*][2]benzoxepins*

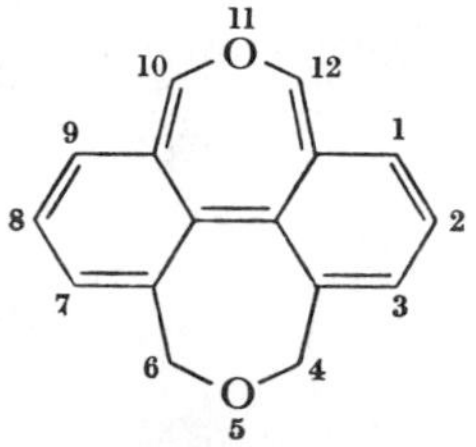

One of the two members of this ring system known to date is 4*H*,6*H*-[2]benzoxepino[6,5,4-*def*][2]benzoxepin (**106**) which Mislow and co-workers (42, 43), and also Oki and co-workers (48–50), have synthesized and studied in great detail. This compound is a most interesting type of doubly-bridged biphenyl derivative in that it is capable of existing in two enantiomeric twisted forms and therefore belongs to the unusual symmetry class D_2. Molecules that belong to this point group possess three mutually perpendicular twofold axes of symmetry, but lack reflection symmetry. The Cahn-Ingold-Prelog designation (14) assigned to the two enantiomeric forms of **106** is shown.

(*R*)- (*S*)-

Several synthetic routes to **106** have been developed. The most direct method consists of treating 2,2′,6,6′-tetrakis(bromomethyl)biphenyl with moist silver oxide in refluxing aqueous acetone. This procedure is satisfactory for the preparation of racemic **106**, but cannot be used to obtain optically active forms. An alternative stepwise sequence, which affords optically active material, starts with (−)-(*R*)-6,6′-dimethyl-2,2′-diphenic acid dimethyl ester (**107**). Bromination with *N*-bromosuccinimide followed by ring closure with silver oxide yields (−)-(*R*)-5,7-dihydrodibenz[*c*,*e*]-oxepin-1,11-dicarboxylic acid dimethyl ester (**108**) and reduction of this diester with lithium aluminum hydride gives (+)-(*R*)-5,7-dihydrodibenz[*c*,*e*]oxepin-1,11-dimethanol (**109**). Dehydration of optically active **109** with concentrated sulfuric acid at −20° affords (+)-(*R*)-**106**, but reaction of the diol with *p*-toluenesulfonyl chloride in pyridine at room temperature, or with acid under more strenuous conditions, leads to racemic **106**. Optically active **106** must be purified with great caution inasmuch as its half-life is only about 10 min at room temperature. Furthermore, **106** appears to be rather unstable when heated at or near its melting temperature in the presence of air. It has been suggested by Mislow and co-workers (43) that the high number of sensitive benzyl carbons in **106** is responsible for the facile autoxidative decomposition of this compound.

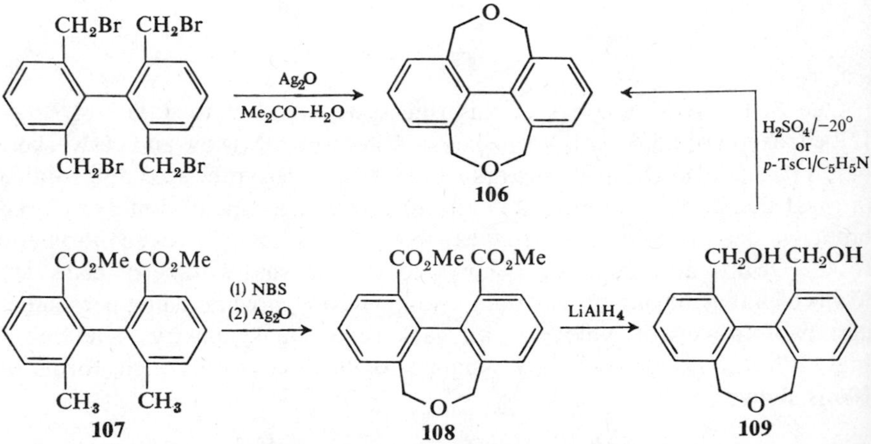

CH_2Br CH_2Br / CH_2Br CH_2Br — Ag_2O, $Me_2CO{-}H_2O$ → **106** ← $H_2SO_4/-20°$ or p-$TsCl/C_5H_5N$

CO_2Me CO_2Me / CH_3 CH_3 **107** — (1) NBS (2) Ag_2O → CO_2Me CO_2Me **108** — $LiAlH_4$ → CH_2OH CH_2OH **109**

The nmr spectrum of **106** has been studied in considerable detail by Mislow and co-workers (43) and also by Oki and co-workers (48–50), especially with reference to the problem of racemization. It is of interest that the geminal methylene protons in **106** are stereochemically, and hence magnetically, nonequivalent as evidenced by the presence of two sharp singlets at τ 5.45 and τ 5.82. The aromatic protons give rise to a multiplet centered at τ 2.48. Inspection of molecular models reveals clearly that two types of methylene protons are present, one type being much more shielded than the other by the π orbitals of the phenyl rings. In agreement with the general rule that this type of magnetic nonequivalence must be solvent-dependent, the chemical shifts of the methylene protons were found to be as follows (solvent and τ values are given): CCl_4, 5.53 and 5.89; C_6H_6, 5.51 and 5.81; CS_2, 5.59 and 5.95; $CDCl_3$, 5.45 and 5.82; C_5H_5N, 5.30 and 5.73; $C_5H_5NO_2$, 5.38 and 5.81. With the exception of benzene the peak separation in this series appears to increase with the dielectric constant of the solvent.

The uv absorption spectrum of **106** has been measured in isooctane (43). The conjugated biphenyl band at 256 mμ can be compared with corresponding band in 5,7-dihydrodibenz[*c*,*e*]oxepin (250 mμ) and 5,7-dihydro-1,11-dimethyldibenz[*c*,*e*]oxepin (243 mμ). The position of this band can be correlated with the angle of torsion between the phenyl rings, shorter λ_{max} values being associated with decreasing coplanarity.

The rate of racemization of optically active **106** has been determined and compared with that of other types of doubly-bridged biphenyls (42, 43). At 10.1° the half-life of optically active **106** is 54 min, and the activation energy for racemization is 20.4 kcal/mole. At 23.3° the half-life is reduced to 11 min. This may be compared with the value for 5,7-dihydrodibenz[*c*,*e*]oxepin, which is 9.3 kcal/mole. Theoretical calculations have been made of the strain energy E_s or **106**, E_s being defined as the difference in enthalpy between the ground state and transition state conformations. The calculated E_s value of 22 kcal/mole is in remarkable good agreement with the experimentally determined activation energy for racemization.

The only other representative of the [2]benzoxepino[6,5,4-*def*][2]benzoxepin ring system reported thus far has been dianhydride **110**, which was isolated upon oxidation of pyrene at 450° in the presence of a vanadium oxide–silver oxide catalyst (16). An isomeric anhydride is also formed which has been identified as **111**.

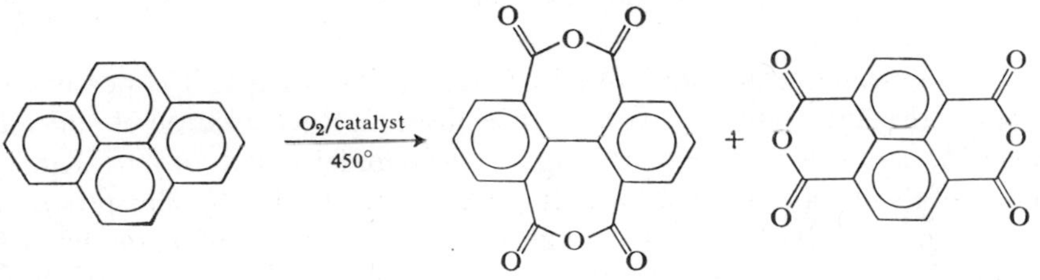

O_2/catalyst
450°

110 **111**

w. *4*H,*6*H-*[2]Benzothiepino[6,5,4*-def*][2]benzoxepins*

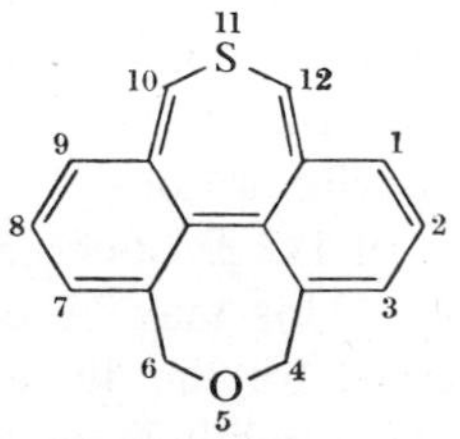

This ring system has been synthesized by Mislow and co-workers (42, 43) in the form of the compound 4*H*,6*H*-[2]benzothiepino[6,5,4-*def*][2]benzoxepin (**112**). Starting with 2,2′-bis(bromomethyl)-6,6′diphenic acid dimethyl ester (**113**), reaction with sodium disulfide afforded 5,7-dihydrodibenz[*c*,*e*]-thiepin-1,11-dicarboxylic acid (**114**). Treatment with diazomethane, followed by reduction with lithium aluminum hydride, gave 5,7-dihydrodibenzo[*c*,*e*]-thiepin-1,11-dimethanol (**115**). Finally, cyclization of this diol under the influence of *p*-toluenesulfonyl chloride in pyridine afforded the desired **112**. Alternatively, diol **115** could be cyclized under the influence of *p*-toluene-sulfonic acid.

CO_2Me CO_2Me / CH_2Br CH_2Br — Na_2S → R R / S — (1) CH_2N_2 (2) $LiAlH_4$ (3) $TsCl/C_5H_5N$ → O / S

113 **114**; R = CO_2H **112**

115; R = CH_2OH

When optically active **113** is used in the above synthesis, it is possible to carry out the entire sequence without racemization. Racemization of optically active **112** can be achieved by heating in xylene under reflux for 2 hr. Kinetic studies of the racemization show that the activation energy for this process is 30.6 kcal/mole at 104.9°, and the half-life is 181 min. At 84.6° the half-life is extended to 31.7 hr, and at 125.3° it is reduced to 23 min. It is of interest that **112**, in contrast to the dioxa analog (see preceding section), appears to be stable at the melting point.

The nmr spectrum of **112** conforms entirely to expectations. The aromatic protons appear as a multiplet centered at τ 2.62, and the benzylic protons as two pairs of singlets. The benzylic protons adjacent to oxygen give rise to resonance signals at τ 5.53 and τ 5.90, while those next to sulfur give peaks at τ 6.47 and τ 6.66 because of the poorer deshielding capacity of sulfur. Magnetic nonequivalence is indicated by the fact that four, rather than two, peaks are observed for the benzylic protons.

The uv absorption spectrum of **112** in isooctane solution shows a biphenyl conjugation band at 253.5 mμ. This may be compared with the value of 256 mμ which has been observed for the dioxa analog. The hypsochromic shift is presumably the result of a slightly greater torsional angle between the aromatic rings, which is in turn due to the difference in bond length between C—O and C—S and the difference in bond angle between —CH_2OCH_2— and —CH_2SCH_2—.

x. *6*H-[1]*Benzoxepino*[*5,4*-b][*1,5*]*benzodiazepines*

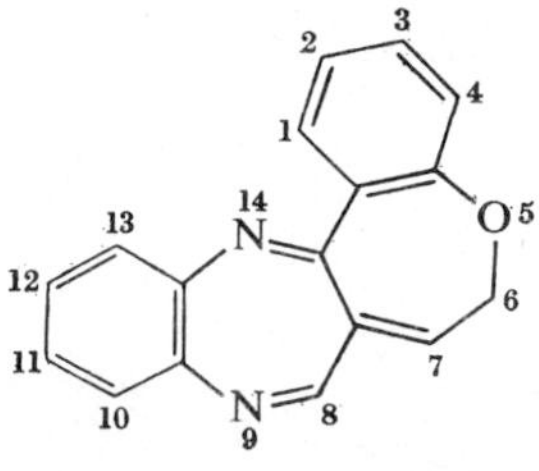

Weisenfels and co-workers (89) prepared a member of this ring system in the form of 7,14-dihydro-6*H*-[1]benzoxepino[5,4-*b*][1,5]benzodiazepine (**116**). This compound was obtained by condensation of *o*-phenylenediamine and 5-chloro-2,3-dihydro-1-benzoxepin-4-carboxaldehyde (**117**).

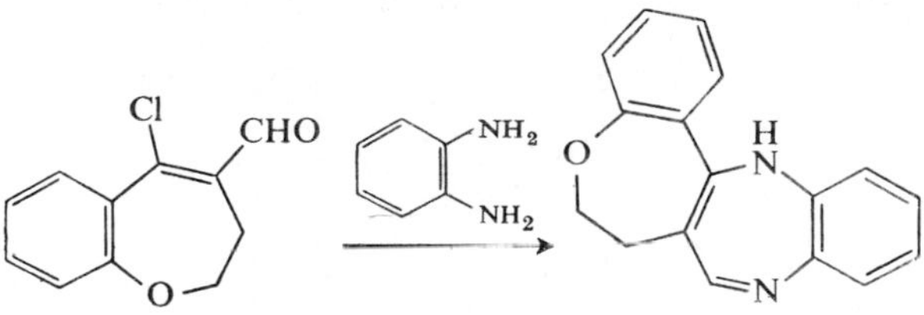

117 **116**

2. Bridged and Spiran Ring Systems

a. *4,4'*(5H,5'H)-*Spirobi*[*1-benzoxepins*]

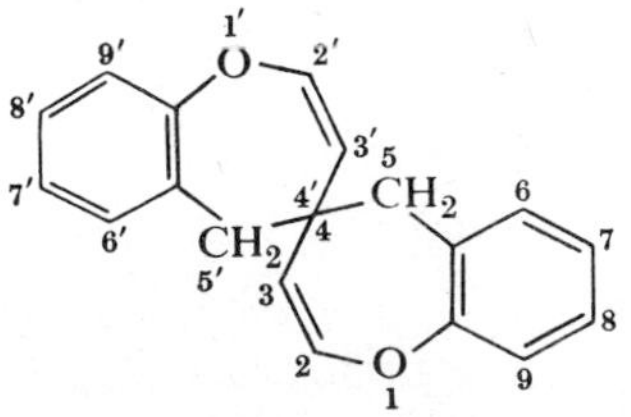

R.I. 5355

The sole example of this ring system in the literature was reported in 1924 by Leuchs and Reinhart (35) who obtained a product which they believed to be **118** upon treatment of 2,2-bis(2-phenoxyethyl)malonyl chloride (**119**) with ferric chloride at 100°. A second product was also isolated which was formulated as diphenyl 2,2-bis(2-chloroethyl)malonate (**120**).

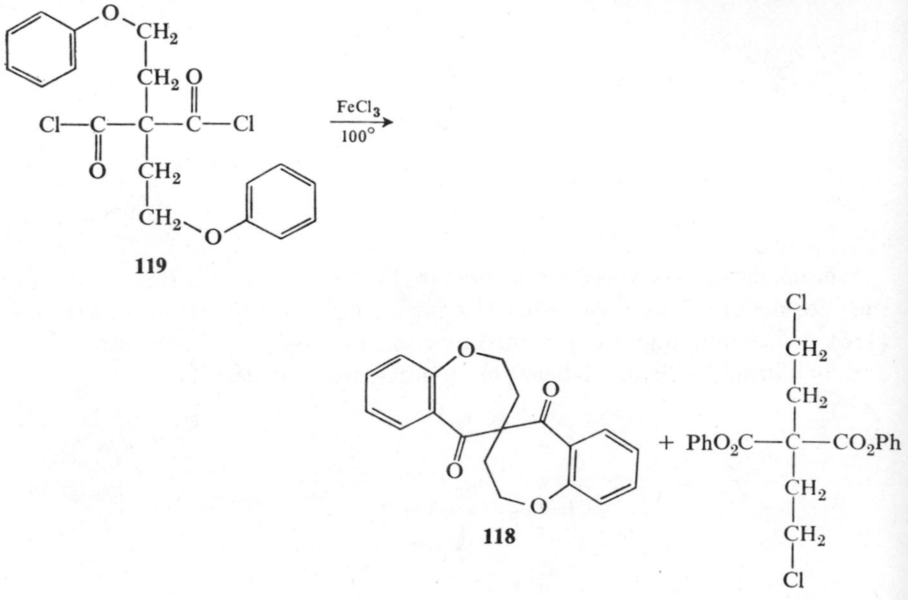

119 $\xrightarrow[100^\circ]{FeCl_3}$ **118** + **120**

b. *9,10-Methanoxymethanoanthracenes*

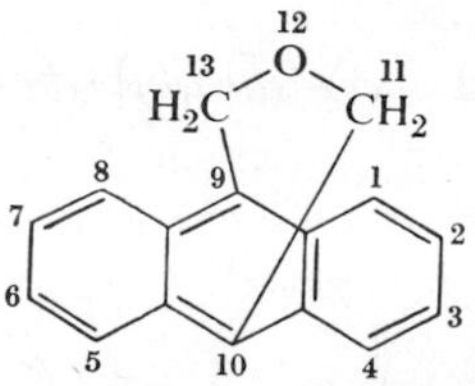

R.I. 5307

The earliest entry into this bridged ring system was made by Mathieu (38, 39) via the reaction of *cis*-9,10-dihydroanthracene-9,10-dicarboxylic acid (**121**) with acetic anhydride under reflux. A 90% yield of a compound having the expected properties for anhydride **122** was obtained. Interestingly, when *trans*-9,10-dihydroanthracene-9,10-dicarboxylic acid was treated similarly, except for a longer reflux time, the same anhydride was produced. Alkaline hydrolysis of **122** regenerated the *cis*-diacid, **121** (39). Reaction with methanol and morpholine transformed **122** into the monomethyl ester and monomorpholide of **121**, respectively.

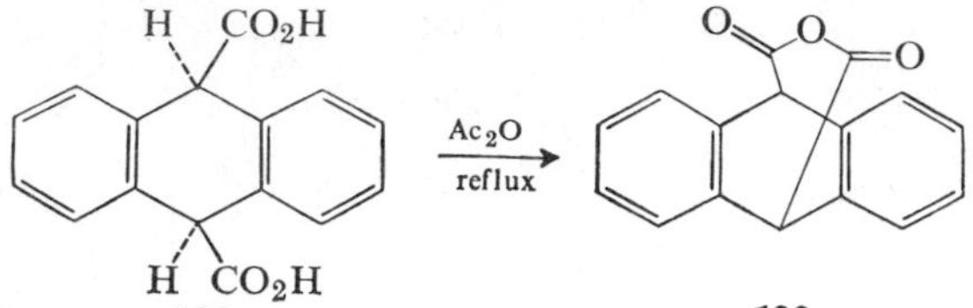

$\xrightarrow[\text{reflux}]{Ac_2O}$

121 **122**

A convenient method of preparation of **122** from anthracene involves reaction of the hydrocarbon with metallic sodium in ether, followed by addition of carbon dioxide (Dry Ice). The resulting diacid is then cyclized to **121** under the influence of acetic anhydride. With 9-methyl- and 9,10-dimethylanthracene this sequence affords modest yields of anhydrides **123** and **124**, respectively (8, 9).

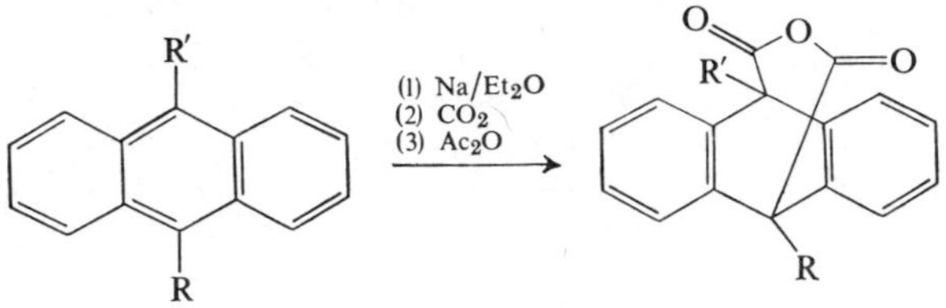

(1) Na/Et_2O
(2) CO_2
(3) Ac_2O

123; R = H, R′ = Me
124; R = R′ = Me

The reaction of **122** with aluminum chloride in carbon disulfide was investigated by Dufraisse and Rigaudy (72). Anthracene, 5-anthracenecarboxylic acid, carbon monoxide, and carbon dioxide were among the products formed under these conditions. Reactions were also carried out with benzene and toluene in the presence of aluminum chloride, but once again the readiness with which **122** underwent decarboxylation and decarbonylation limited its usefulness as a Friedel-Crafts acylating agent (54).

Reactions of **122** with Grignard reagents were studied by Rigaudy and co-workers (55–58). With methyl magnesium bromide, and anisole as the solvent, a 70% yield of tertiary alcohol **125** was obtained. The use of phenyl magnesium bromide, on the other hand, led to only a 15–20% yield of the expected alcohol, **126**, together with a 75% yield of ketone **127**. *p*-Tolyl and *p*-anisyl magnesium bromide behaved similarly. The difference between the alkyl and aryl Grignard reagents was accentuated by dilution of the anisole solvent with several parts of ether. Heating of **125** to 200° produced bridged seven-membered lactone **128**. Upon acid hydrolysis, **128** underwent cleavage to 10-isopropylidene-9,10-dihydro-9-anthracenecarboxylic acid (**129**). Treatment of **125** with sulfuric acid also afforded a trace of lactone **128** but gave mainly **129**. Similarly, heating **126** to 270° generated lactone **130** in 70% yield, and hydrolysis of the latter with 20% sulfuric acid led to the formation of 10-benzhydrylidene-9,10-dihydro-9-anthracenecarboxylic acid (**131**).

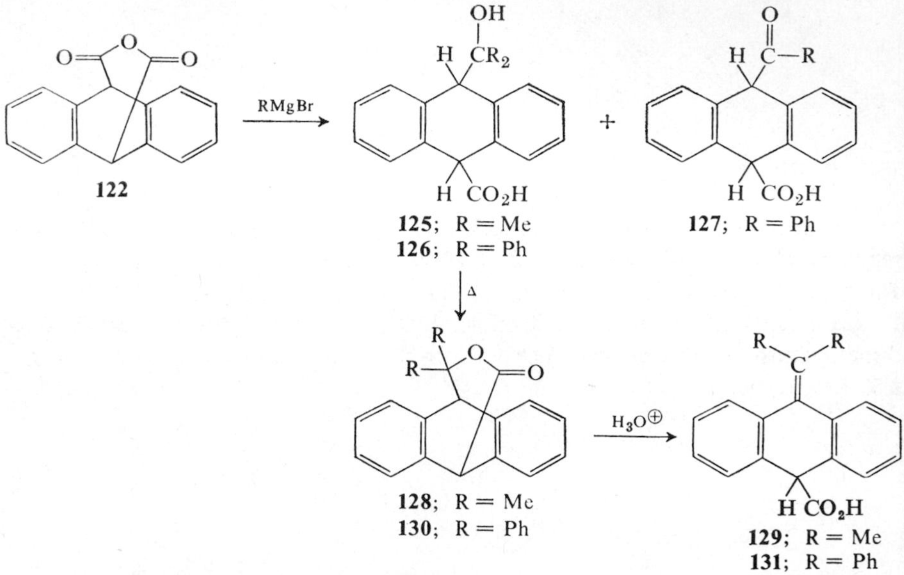

122 → (RMgBr) **125**; R = Me, **126**; R = Ph + **127**; R = Ph

125/**126** → (Δ) **128**; R = Me, **130**; R = Ph → ($H_3O^{\oplus}$) **129**; R = Me, **131**; R = Ph

By carrying out stepwise addition of phenyl magnesium bromide and methyl magnesium bromide it has been possible to prepare a mixed tertiary carbinol, **132**, from which lactone **133** was obtained readily upon heating (58).

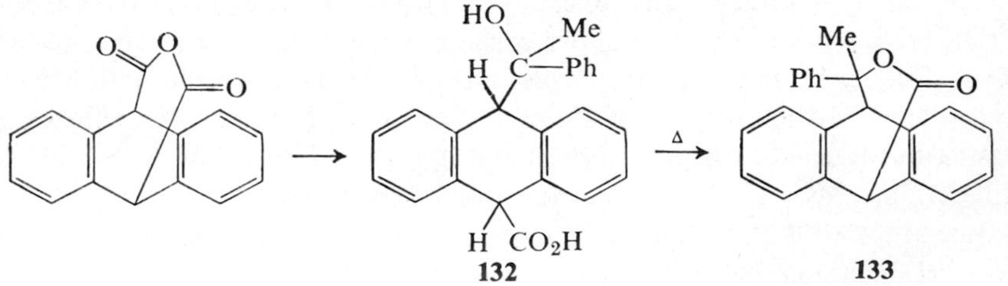

→ **132** → (Δ) **133**

c. *10,9-(Epoxyethano)anthracenes*

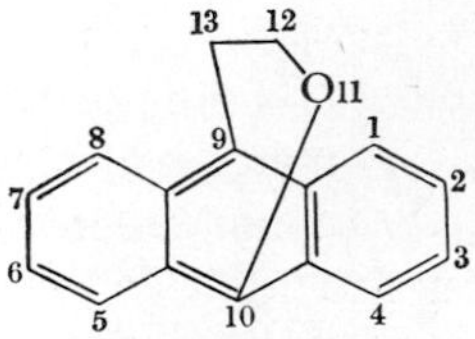

This tetracyclic bridged oxepin ring system was generated for the first time in 1964 by Brown and Cookson (13) in the form of 1,3-dipolar addition

product **134** which they obtained upon allowing tetracyanoethylene oxide to add across the 9,10-positions of anthracene.

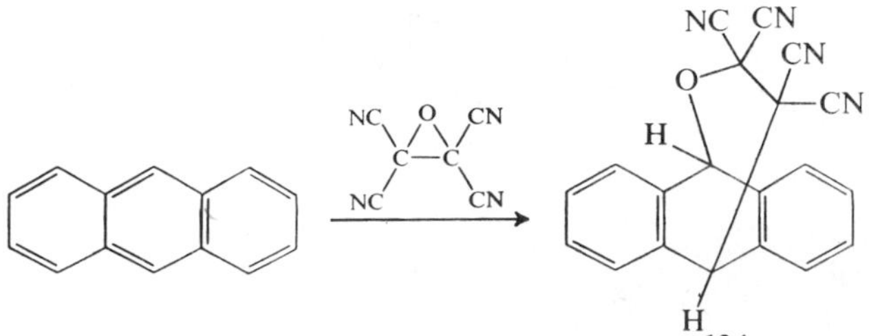

134

d. *4,6-Ethano(and etheno)dibenz*[b,f]*oxepins*

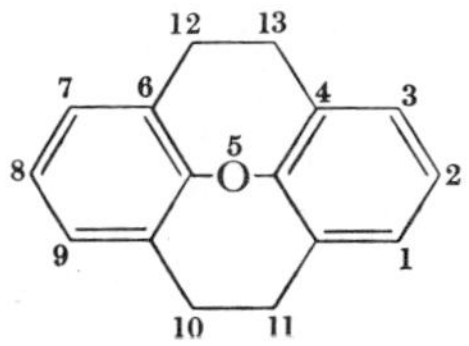

Two members of this novel O-bridged *meta*-cyclophane ring system were synthesized recently by Hess and co-workers (90), starting from 10,11-dihydrodibenz[*b,f*]oxepin (**135**). Successive reaction of **135** with *n*-butyllithium, carbon dioxide, and diazomethane afforded dimethyl 10,11-dihydrodibenz[*b,f*]oxepin-4,6-dicarboxylate (**136**). Reduction of **136** with lithium aluminum hydride gave 10,11-dihydrodibenz[*b,f*]oxepin-4,6-dimethanol (**137**) and treatment of the latter with phosphorus tribromide yielded the corresponding dibromo derivative, **138**. Upon cyclization under the influence of phenyllithium, **138** was converted into the 4,6-ethano derivative, **139**, in 75% yield. Bromination of **139** with *N*-bromosuccinimide followed by dehydrobromination with potassium *tert*-butoxide led to the 4,6-etheno derivative, **140**. This fully unsaturated compound exhibits a number of interesting physical and chemical properties, among which may be mentioned its gradual conversion to pyrene, 1,6-pyrenequinone, and 1,8-pyrenequinone in trifluoroacetic acid solution. According to Hess and co-workers (90), protonation of **140** generates valence tautomeric oxonium ion **141**, in which a pseudoaromatic 14 π-electron system is present. This species may serve as an oxidizing agent to convert trifluoroacetic into peroxytrifluoroacetic acid. Oxidation of the pyrene arising from this transformation then gives the observed 1,6- and 1,8-quinones.

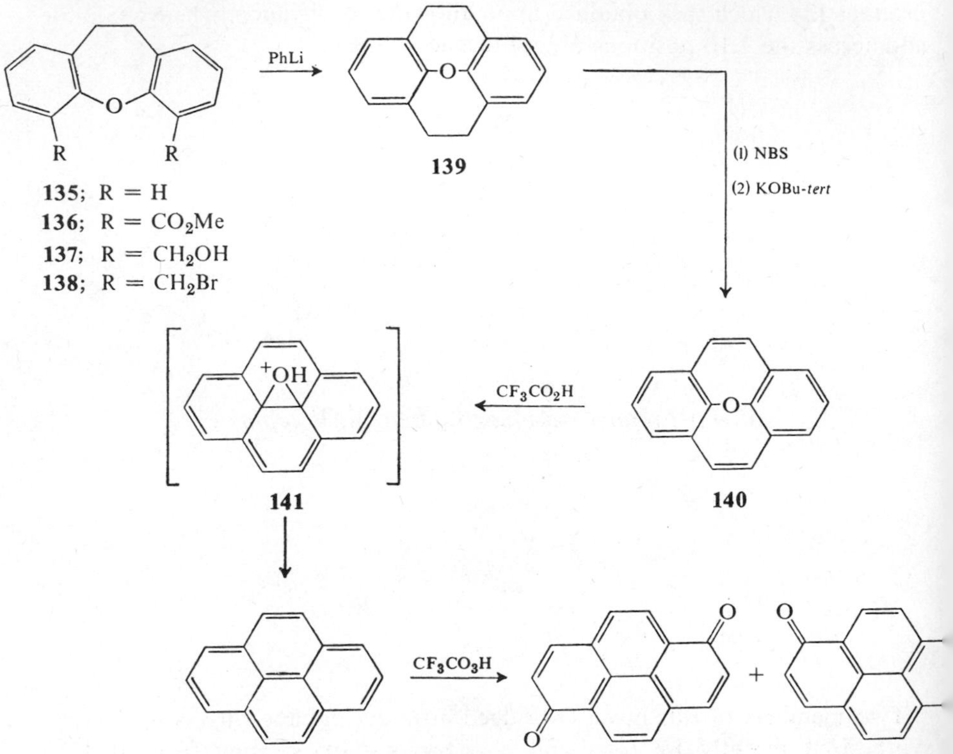

R R

PhLi

139

(1) NBS
(2) KOBu-*tert*

135; R = H
136; R = CO_2Me
137; R = CH_2OH
138; R = CH_2Br

^{+}OH

CF_3CO_2H

141

140

CF_3CO_3H

B. Pentacyclic Systems

1. Fused Ring Systems

a. *14*H-*Dibenz*[*2,3*:*6,7*]*oxepino*[*4,5*-b]*indoles*

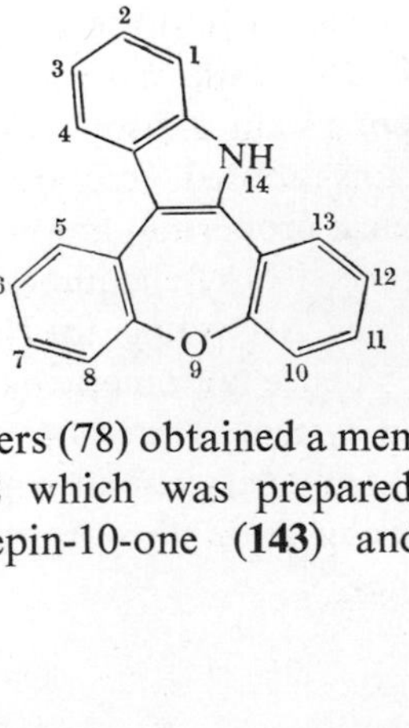

Kametani and co-workers (78) obtained a member of this ring system in the form of compound **142** which was prepared from 10,11-dihydro-2,3,6-trimethoxydibenz[*b*,*f*]oxepin-10-one (**143**) and phenylhydrazine via the Fischer indole synthesis.

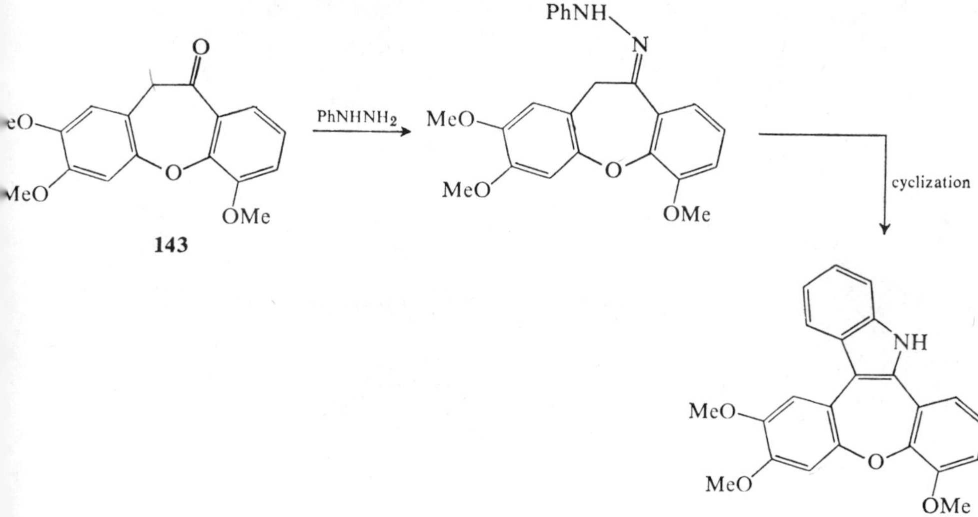

PhNH N O MeO MeO OMe 143 PhNHNH2 MeO MeO OMe cyclization NH MeO MeO O OMe 142

b. *Oxepino[3,4-*b*:6,5-*b*′]bisbenzofurans*

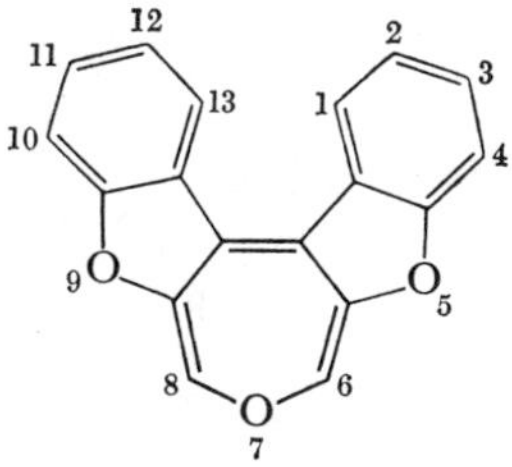

12 11 10 13 1 2 3 4 9 O O 5 8 O 7 6

Kubota and co-workers (79) reported in 1966 that the natural pigment "flavonol red," which can be obtained from 3-hydroxyflavanone (**144**) upon treatment with acetic anhydride and sodium acetate, possesses structure **145**. Ultraviolet irradiation transforms **145** into an isomeric compound called "photoflavonol red." Ozonolysis of **145** followed by reductive decomposition of the ozonide affords benzyl alcohol and salicylic acid. Acid hydrolysis of **145** yields a compound which has been formulated as **146**. Reduction of **145** with sodium in liquid ammonia produces a dihydroxy compound, and catalytic reduction leads to a hexahydro derivative. A number of other transformations and interconversions have been carried out, but the structures of some of the reaction products are still uncertain.

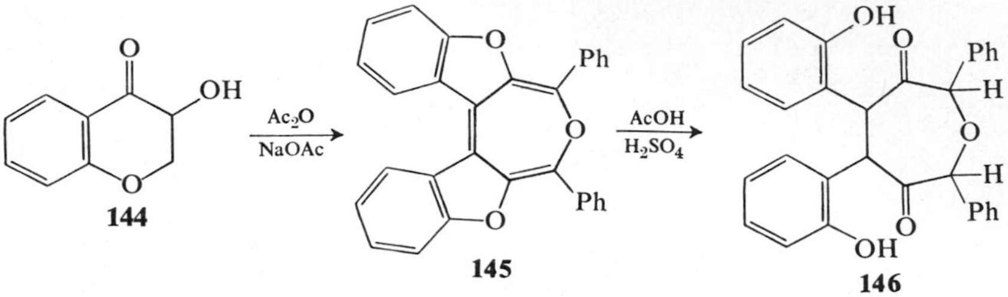

144 $\xrightarrow[\text{NaOAc}]{\text{Ac}_2\text{O}}$ 145 $\xrightarrow[\text{H}_2\text{SO}_4]{\text{AcOH}}$ 146

c. *Oxepino[4,5-*b*:2,3-*e*′]bisbenzofurans*

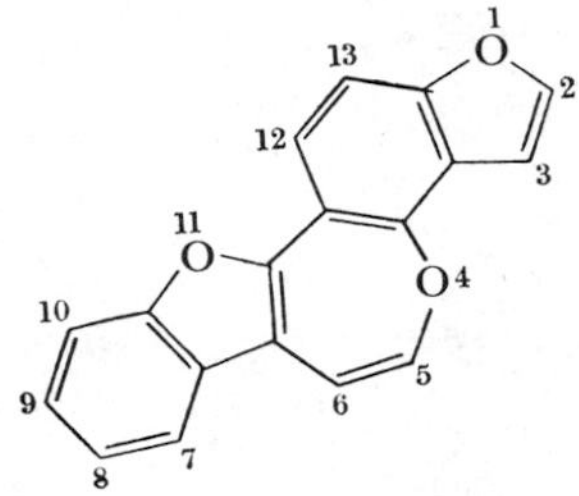

A representative of this pentacyclic ring system was obtained in 1966 by Adam and co-workers (1) in the form of hemiacetal **147**, a derivative of the natural product rotenol.

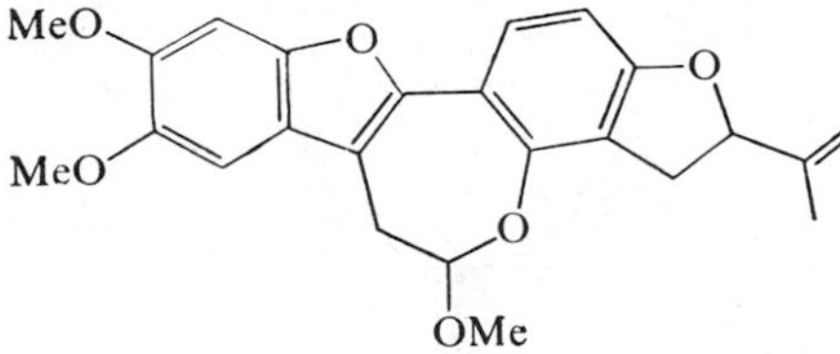

147

d. *Naphth[2′,1′:6,7]oxepino[3,4-*b*]indolizines*

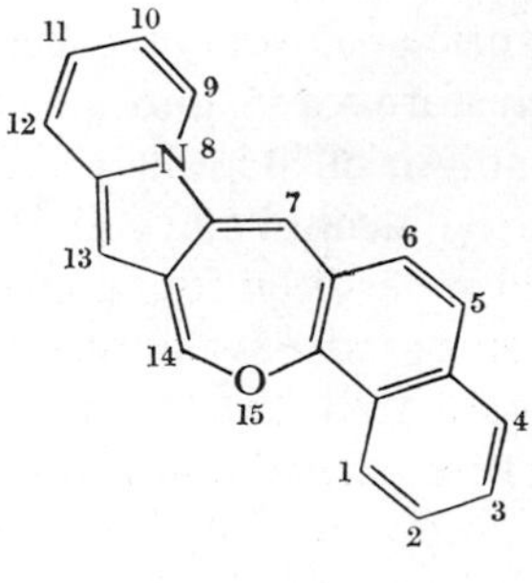

R.I. 13,783

An example of this novel ring system was synthesized in 1962 by Stepanov and Grineva (64), starting from indolizine-2-carboxylic acid and phenyl 1-hydroxy-2-naphthoate. The reactants were heated at 120–180° under reduced pressure, and when all the liberated phenol had distilled off, the residue was treated with base to give a 54% yield of a product formulated as **148**. Saponification of the seven-membered lactone ring in **148** was achieved by refluxing with 2*N* sodium hydroxide.

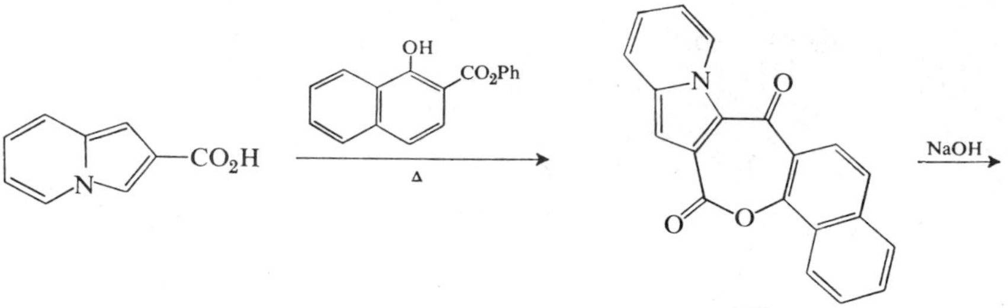

148

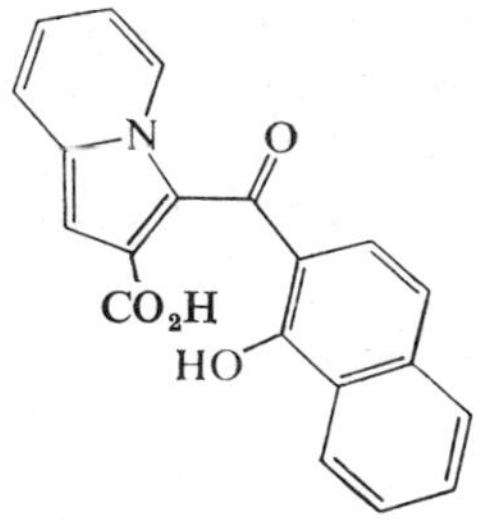

e. *Dinaphth*[2,3-c:2′,3′-e]*oxepins*

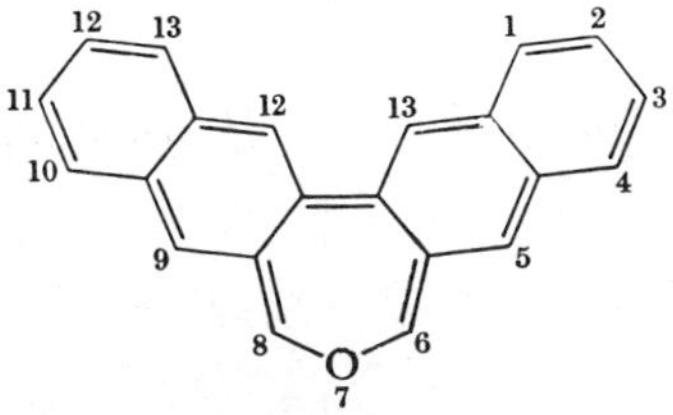

R.I. 9394

This pentacyclic ring system was first synthesized in 1957 by Badger and co-workers (7), starting from diol **149**. Ring closure of **149** to **150** was achieved by passing dry hydrogen chloride gas through a solution of the diol in diphenyl ether at 100°. The same cyclic ether, **150**, was also obtained adventitiously when the corresponding dibromide, **151**, was treated with phenyllithium with the intent of preparing hydrocarbon **152.**

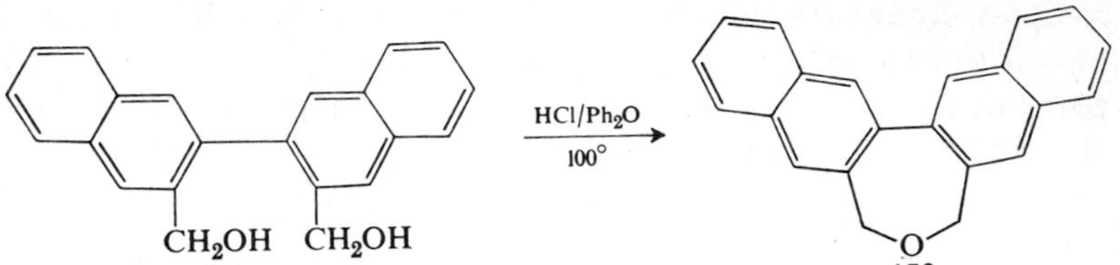
CH₂OH CH₂OH
HCl/Ph₂O
100°
O

149 **150**

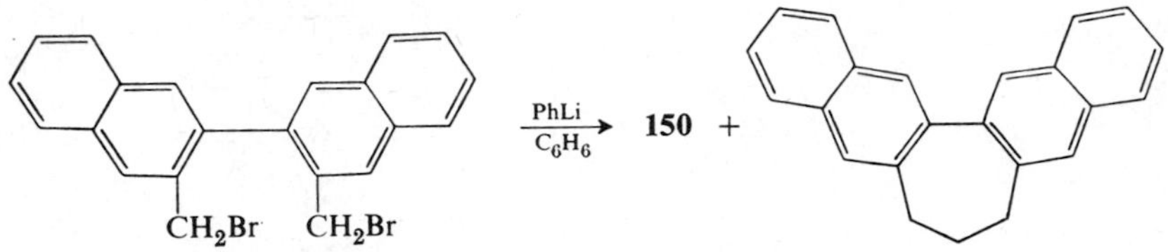
CH₂Br CH₂Br
PhLi
C₆H₆
150 +

151 **152**

More recently a derivative of dinaphth[2,3-*c*:2′,3′-*e*]oxepin ring system was prepared by Bacon and Bankhead (3) in the form of anhydride **153**. As with other anhydrides of the diphenic type, this compound was obtained from the corresponding diacid, **154**, upon heating.

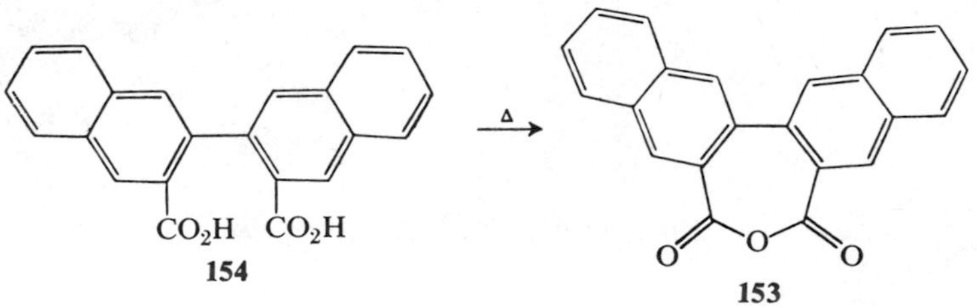
CO₂H CO₂H
Δ
O
O
O

154 **153**

f. *Dinaphth*[2,1-c:*1′*,2′-e]*oxepins*

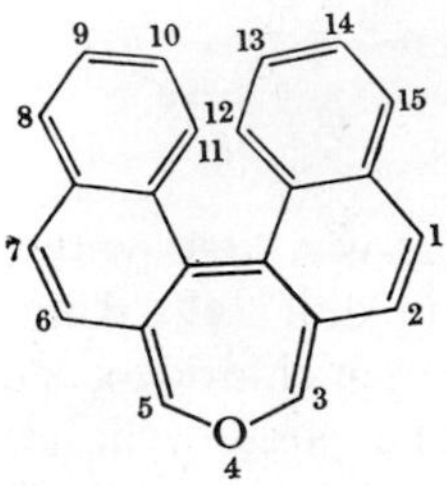
9 10 13 14
8 12 15
11
7 1
6 2
5 3
O
4

R.I. 11,212

No examples of this ring system were known until the recent work of Mislow and co-workers (41–44) who synthesized and studied the optical rotatory dispersion characteristics of 2,7-dihydrodinaphth[2,1-*c*:1′,2′-*e*]-oxepin (**155**) as one of a large number of bridged, optically active biaryl derivatives. For the synthesis of racemic **155**, (±)-1,1-′-binaphthalene-2,2′-dicarboxylic acid (**156**) was reduced to (±)-2,2′-bis(hydroxymethyl)-1,1′-binaphthyl (**157**) and the latter was heated in boiling benzene in the presence of a trace of *p*-toluenesulfonic acid. By similar treatment of (−)-diacid the (+)-oxepin was also prepared. According to the Cahn-Ingold-Prelog convention, (+)-**155** has the (*S*)-configuration.

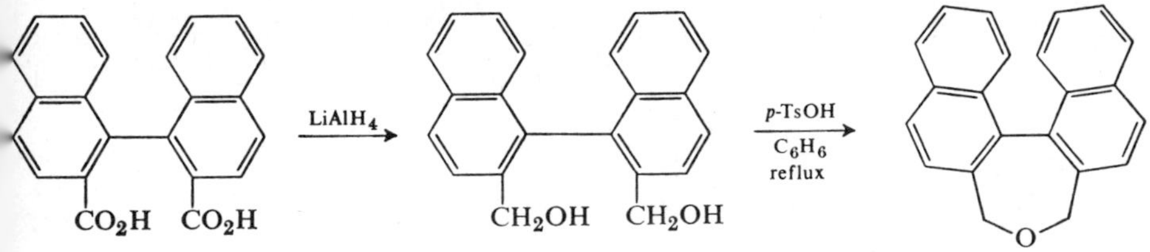

CO_2H CO_2H

156

$\xrightarrow{LiAlH_4}$

CH_2OH CH_2OH

157

$\xrightarrow[\text{reflux}]{p\text{-TsOH},\ C_6H_6}$

O

155

g. *Benzo*[c]*phenanthro*[2,3-e]*oxepins*

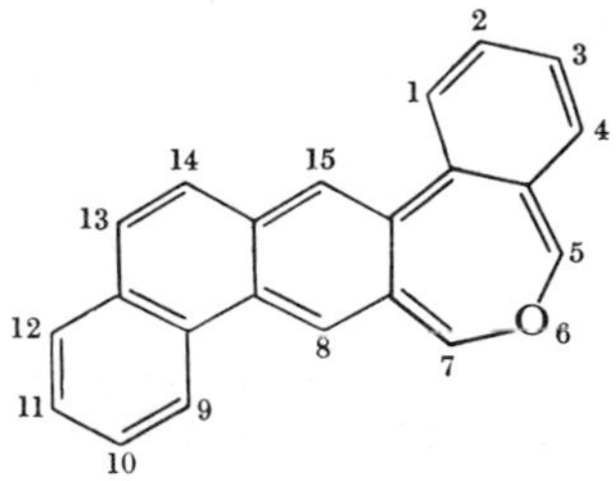

1 2 3 4 5 6 7 8 9 10 11 12 13 14 15

R.I. 9395

Bhargava and Heidelberger (10) first described a member of this ring system in 1956 in the form of anhydride **158** which they obtained in 70–80% yield upon heating diacid **159** in refluxing acetic anhydride. Reaction of **158** with ethanolic ammonia afforded a mixture of isomeric amides **160** and **161**. Anhydride **158** has also been synthesized recently by Moriconi and co-workers (45). Diacid **159** can be obtained conveniently via ozonolysis of dibenz[*a*,*h*]anthracene.

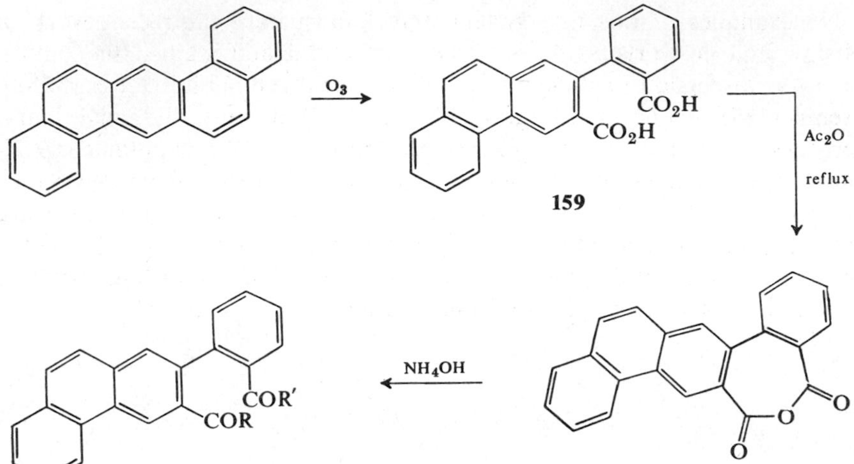

O_3 → **159** (CO_2H, CO_2H) — Ac_2O, reflux → **150** — NH_4OH →

160; R = OH, R′ = NH_2
161; R = NH_2, R′ = OH

h. *Oxepino*[*3,4*-b:*5,6*-b′]*diquinolines*

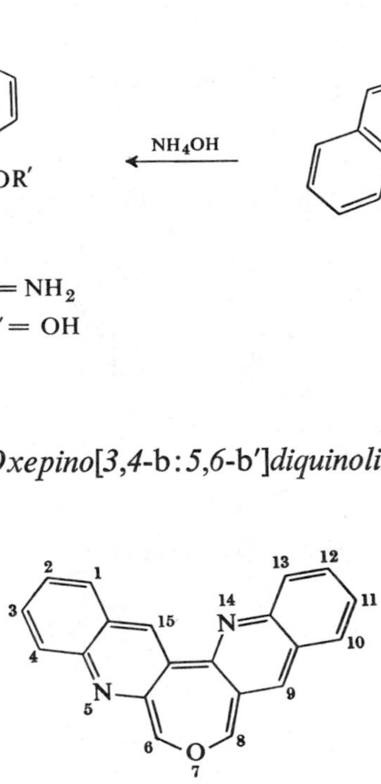

R.I. 6421

According to Baczynski and Niementowski (4), quinacridonic acid (**162**) is converted into anhydride **163** upon being heated. Their report constitutes the sole reference to this ring system to date.

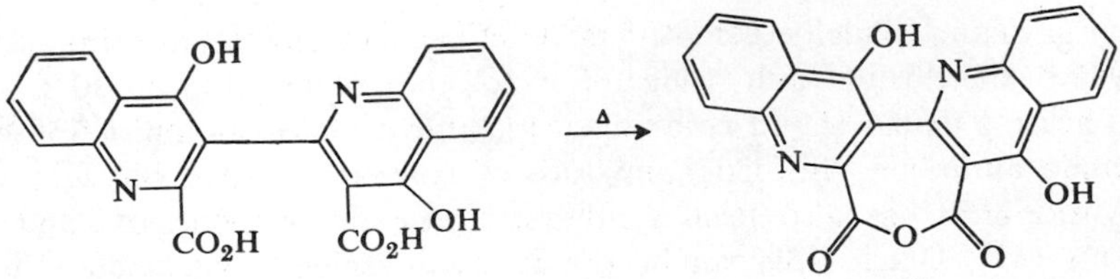

162 — Δ → **163**

i. *2*H-*Benzo[8,9]phenanthro[4,5*-bcd]*oxepins*

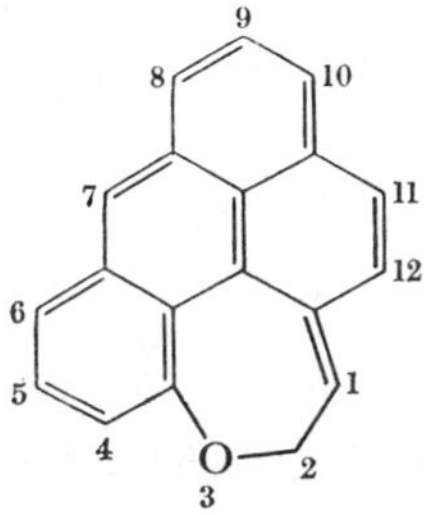

A member of this pentacyclic ring system was reported in 1968 by Falshaw and co-workers (73) in the form of lactone **164** which was obtained by a Baeyer-Villiger oxidation as shown.

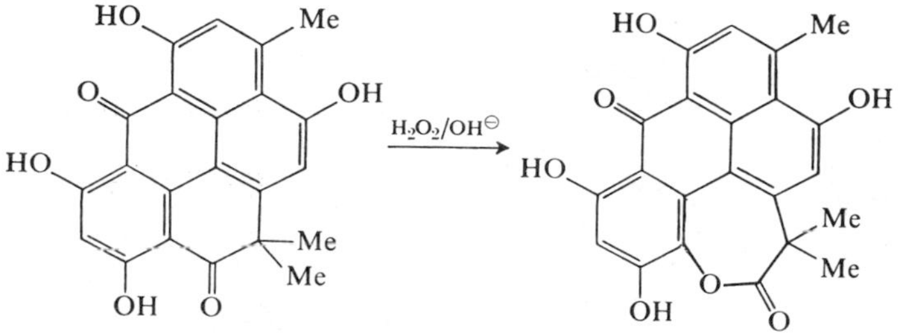

$H_2O_2/OH^{\ominus}$

164

j. *7*H,*15*H-*Bisoxepino[3',4':4,5]pyrrolo[1,2*-a:*1',2'*-d]*pyrazines*

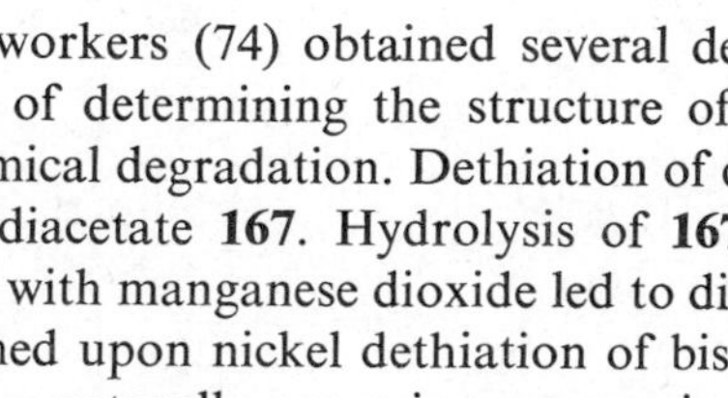

Nagarajan and co-workers (74) obtained several derivatives of this ring system in the course of determining the structure of the natural product aranotin (**165**) by chemical degradation. Dethiation of diacetylaranotin (**166**) with nickel afforded diacetate **167**. Hydrolysis of **167** gave diol **168**, and oxidation of the latter with manganese dioxide led to diketone **169**. Diacetate **167** was likewise formed upon nickel dethiation of bisdethiodi(methylthio)-acetylaranotin (**170**), a naturally-occurring companion of aranotin. Treatment of **170** with aluminum amalgam gave partly dethiated product **171**, while catalytic hydrogenation resulted in formation of the fully dethiated tetrahydro derivative, **172**. Hydrolysis of **172** afforded diol **173**, and oxidation of **173** with dimethylsulfoxide–acetic anhydride led to diketone **174**.

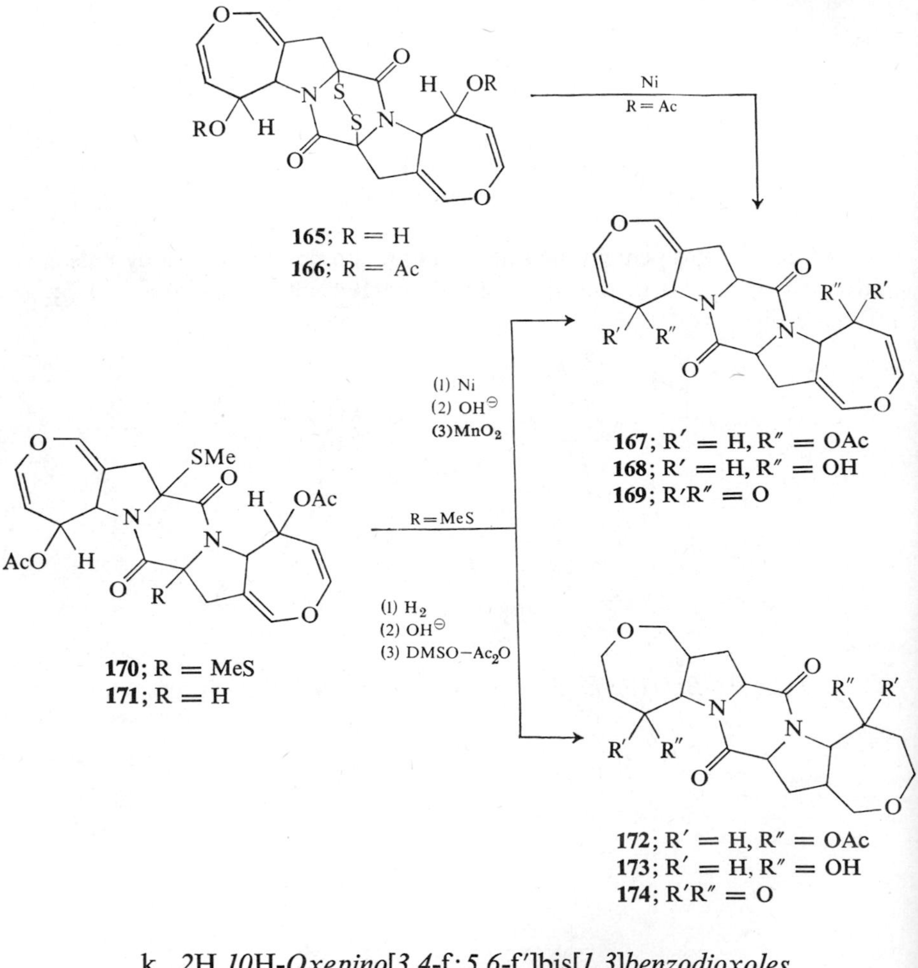

165; R = H
166; R = Ac

Ni, R = Ac

167; R′ = H, R″ = OAc
168; R′ = H, R″ = OH
169; R′R″ = O

(1) Ni
(2) $OH^{\ominus}$
(3) MnO_2

R = MeS

(1) H_2
(2) $OH^{\ominus}$
(3) DMSO–Ac_2O

170; R = MeS
171; R = H

172; R′ = H, R″ = OAc
173; R′ = H, R″ = OH
174; R′R″ = O

k. *2*H,*10*H-*Oxepino*[*3*,*4*-f:*5*,*6*-f′]bis[*1*,*3*]*benzodioxoles*

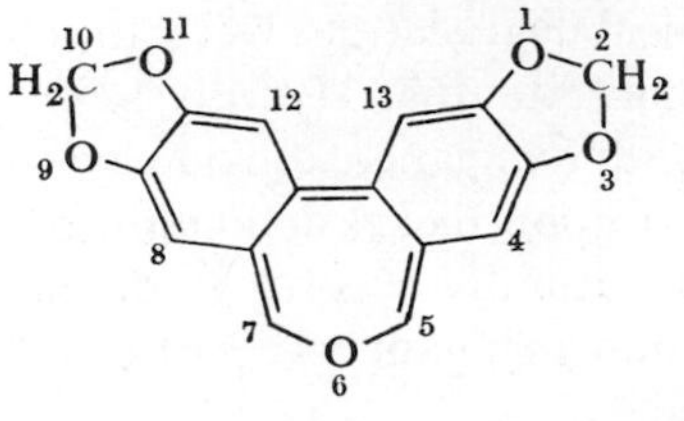

A single member of this ring system is known in the form of the 5,7-dihydro derivative, **175**. This compound was prepared in 1966 by Dallacker and Adolphen (97) from 3,4,3′,4′-bis(methylenedioxy)diphenyl (**176**). Chloromethylation with formalin and hydrogen chloride in acetic acid at 90–110° and alkaline treatment of the resultant **177** gave **175**. The oxepin ring was generated in a single step when **176** was treated with formalin and hydrogen bromide in acetic acid at 40–45°.

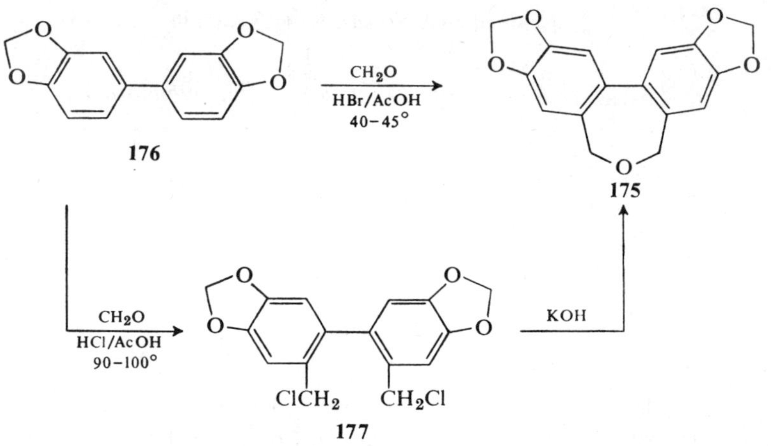

176

CH_2O
HBr/AcOH
40–45°

175

CH_2O
HCl/AcOH
90–100°

$ClCH_2$ CH_2Cl

177

KOH

1. *Dibenz*[*2,3*:*6,7*]*oxepino*[*4,5*-b]*quinoxalines*

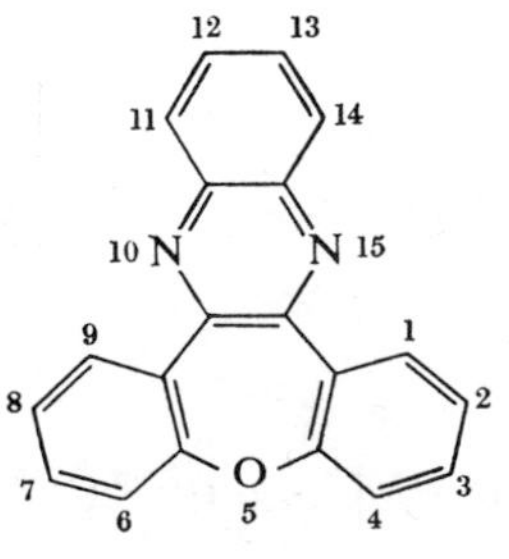

R.I. 9386

The sole known member of this ring system is the parent compound, **178**, which Mathys and co-workers (91) obtained in 1956 by condensation of dibenz[*b,f*]oxepin-10,11-dione (**179**) with *o*-phenylenediamine.

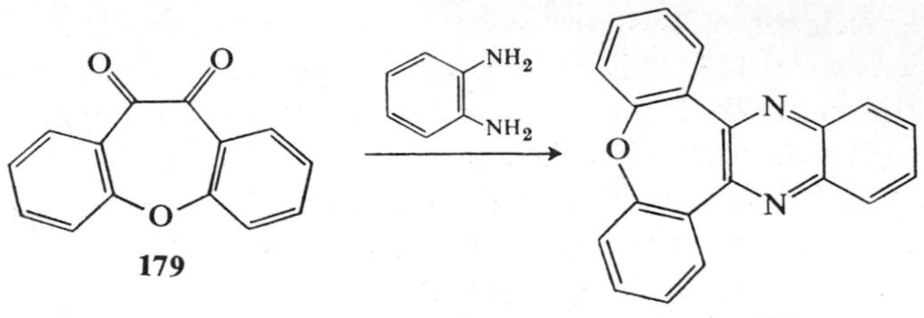

179

178

2. Bridged and Spiran Ring Systems

a. *4,6-(Methanoxymethoxy)cycloprop*[f]*isobenzofurans*

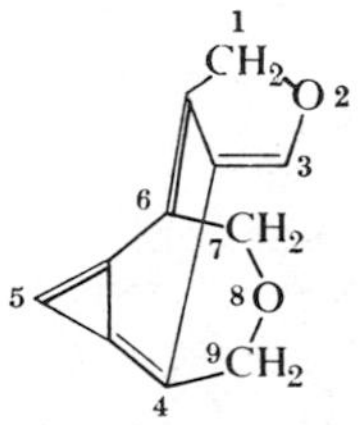

R.I. 3989

Only a single compound has been described in this series, consisting of dianhydride **180**, which Alder and Jacobs (2) prepared from tropylidene by Diels-Alder reaction with maleic anhydride followed by oxidation of the adduct with sodium permanganate in dilute alkali.

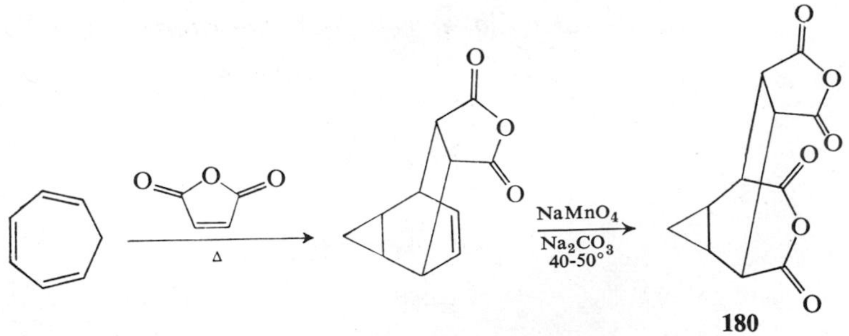

$NaMnO_4$

Na_2CO_3 40-50°

180

b. *5-Oxapentacyclo*[$5.4.0.0^{2,10}.0^{3,9}.0^{8,11}$]*undecanes*

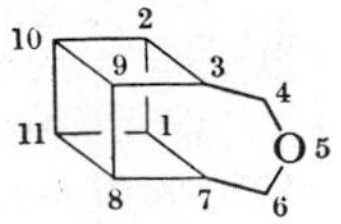

A member of this cagelike ring system was synthesized in 1966 by Chin and co-workers (77) in the form of anhydride **181** which was obtained via the reaction sequence depicted.

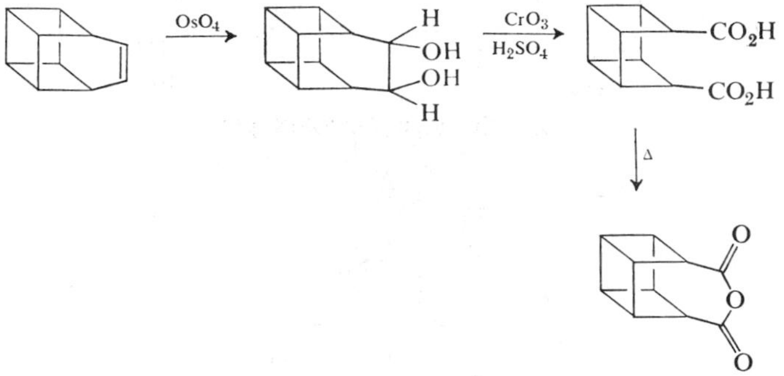

OsO_4 — CrO_3, H_2SO_4 — CO_2H — CO_2H — Δ

181

C. Hexacyclic Systems

1. Acenaphtho[1,2-*d*]dibenz[*b*,*f*]oxepins

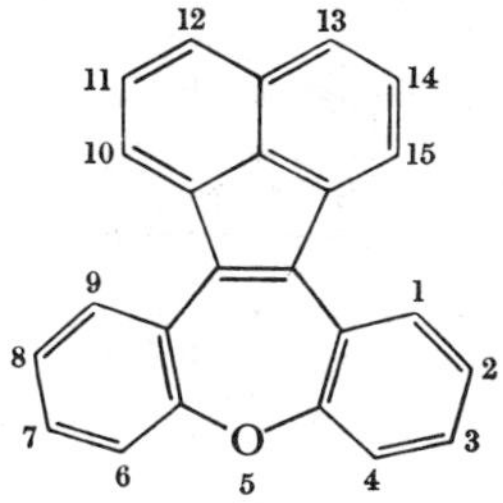

R.I. 6880

The first example of this type of hexacyclic oxepin derivative was prepared in 1934 by Matei and Bogdan (37) from acenaphthene-1,2-dione and *p*-cresol in the presence of sulfuric acid. The structure assigned to the product was **182**. In an extension of the earlier work, Bogdan (12) reported 4 years later that, under the influence of a larger amount of sulfuric acid, the above reactants gave a different product to which structure **183** was assigned. The same ketone was obtained upon treatment of **182** with additional acid. The formation of **183** from **182** was assumed to involve a pinacolic rearrangement. When 2,4-dimethylphenol was treated with concentrated sulfuric acid and glacial acetic acid, there was produced a mixture of diol and ketone. On the other hand, treatment of *o*- and *m*-cresol with alcoholic hydrogen chloride allegedly gave only rearranged ketones.

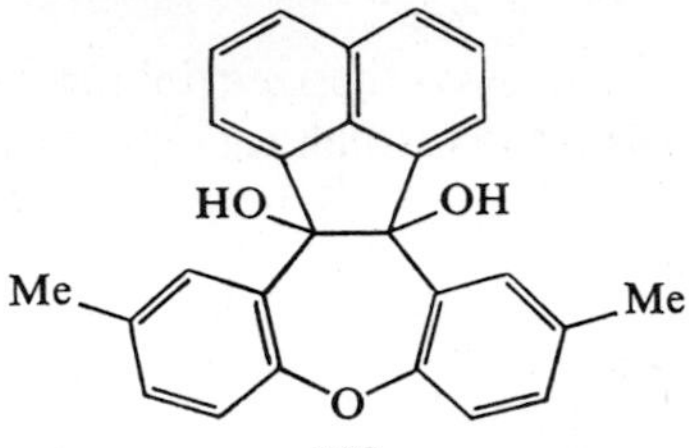

HO OH Me Me O

182

Me O Me O

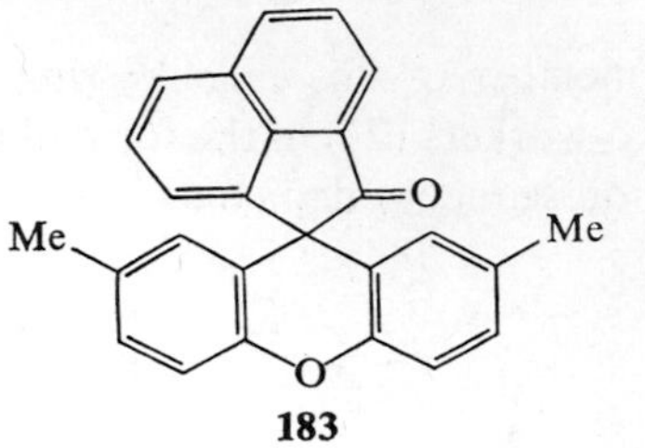

183

2. Piceno[6,7-*cde*]oxepins

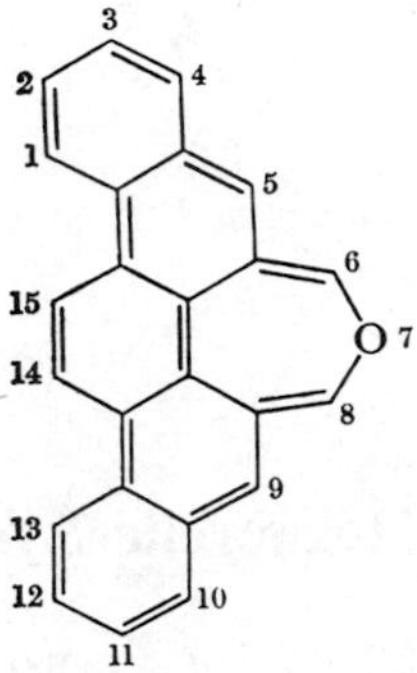

R.I. 7048

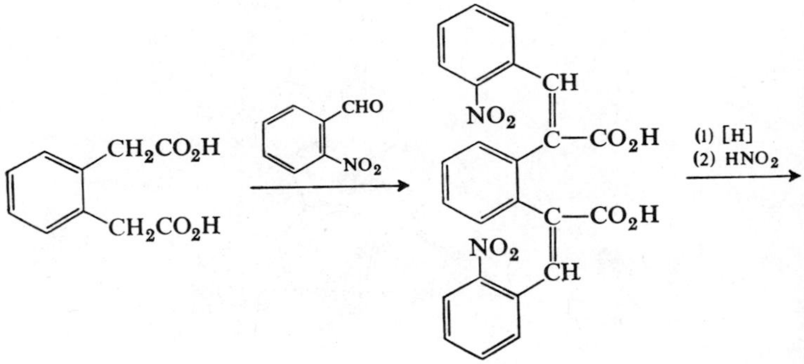

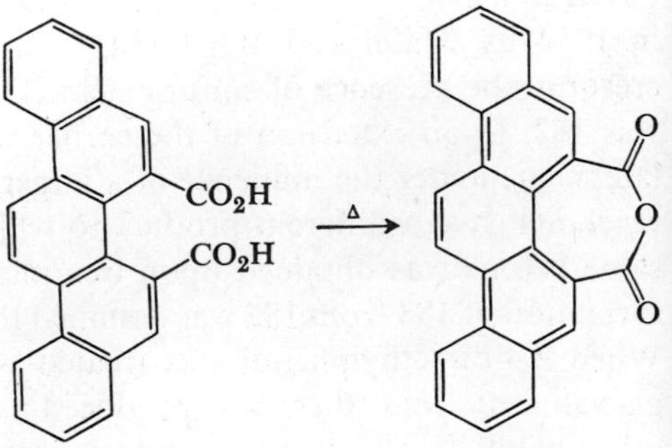

CHO NO_2 → CH NO_2 C—CO_2H C—CO_2H NO_2 CH (1) [H] (2) HNO_2 →

CH_2CO_2H CH_2CO_2H

CO_2H CO_2H Δ → O O O

185 **184**

The lone example of this ring system was reported in 1937 by Waldmann and Pitschak (68) in the form of anhydride **184** which was obtained readily from acid **185**. The synthesis of **185** is noteworthy in that it involves a double Pschorr cyclization as shown.

3. 16*H*-Dibenzo[*b*,*h*][2]benzoxepino[3,4,5-*de*][1,6]naphthyridines

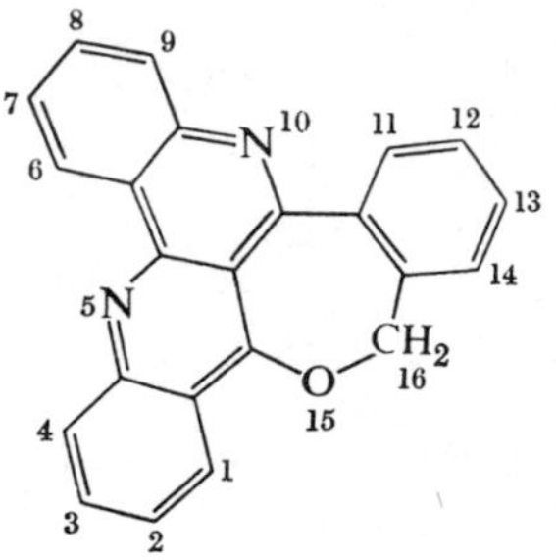

R.I. 7044

An example of this hexacyclic ring system was prepared by Hope and Anderson (26) by heating hydroxy acid **186** in diphenyl ether for 10 min. with formation of lactone **187**.

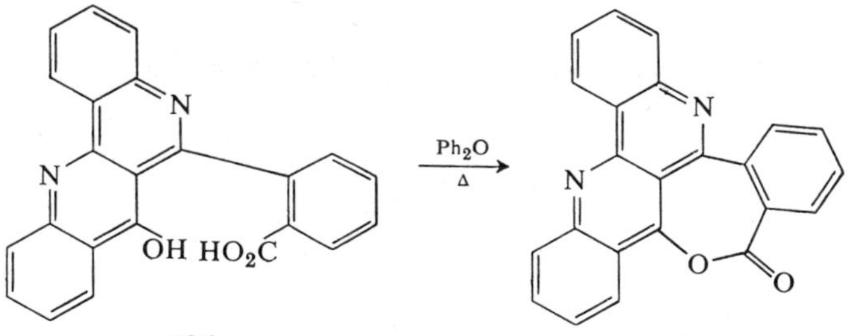

186 $\xrightarrow[\Delta]{Ph_2O}$ **187**

4. Oxepino[3′,4′,5′,6′:4,5]phenanthro[9,10-*b*]quinoxalines

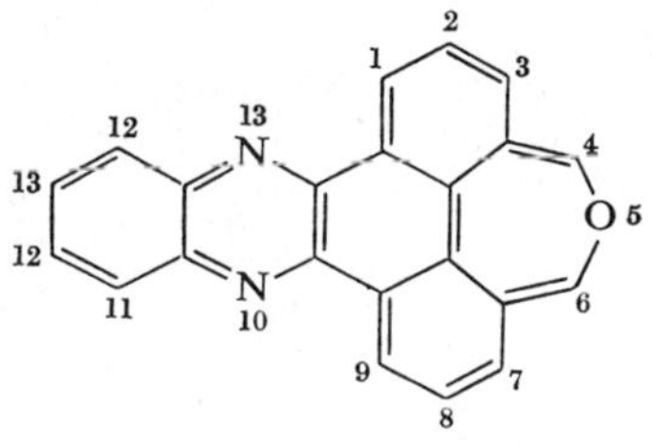

R.I. 7043

Vollmann and co-workers (67) obtained a derivative of this ring system by condensation of *o*-phenylenediamine with 9,10-phenanthrenequinone-4,5-dicarboxylic acid (**188**). Instead of the expected diacid, **189**, crystallization afforded only anhydride **190**.

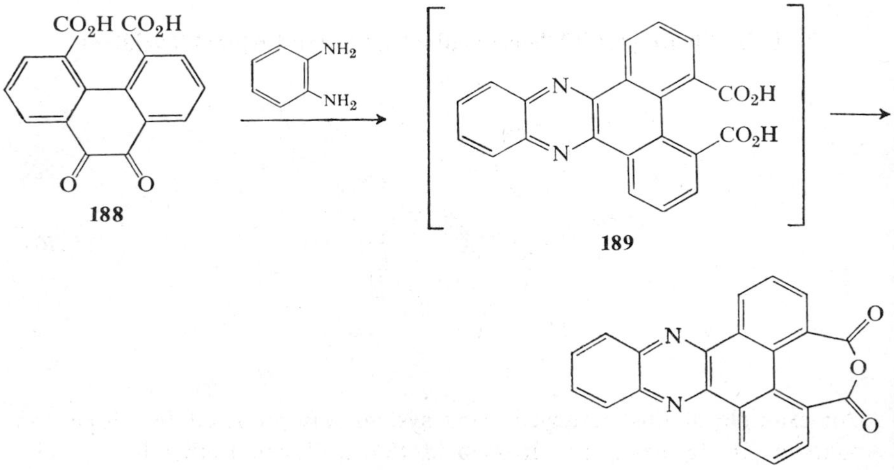

CO_2H CO_2H NH_2 NH_2 N N CO_2H CO_2H O O

188 **189**

N N O O O

190

5. Spiro[dibenz[*c*,*e*]exepin-5,9′-fluorenes]

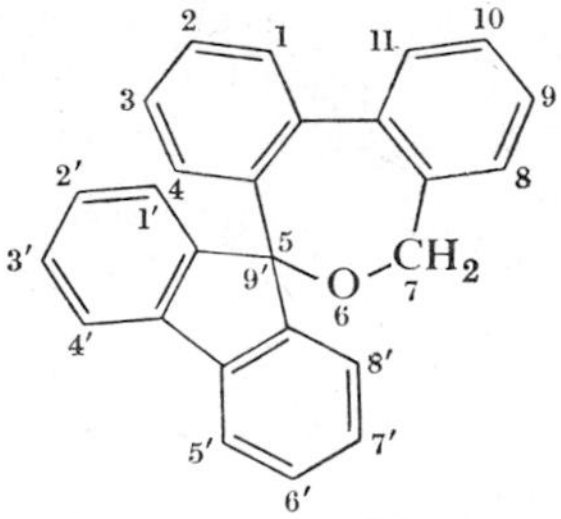

R.I. 6881

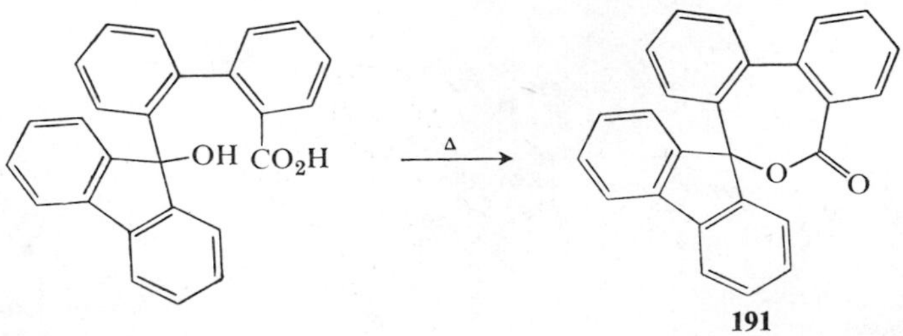

OH CO_2H Δ O O

191

This ring system is represented in the literature only by lactone **191** which was reported by Klinger and Lonnes (33) some 70 years ago.

6. Spiro[dibenz[*c*,*e*]oxepin-5,9′-xanthenes]

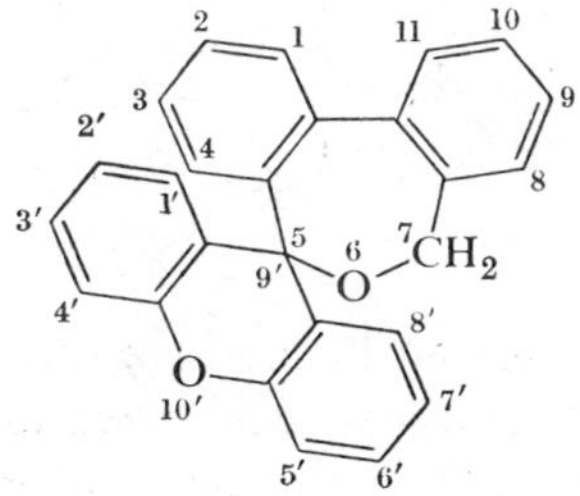

R.I. 7047

The reaction of diphenic anhydride with resorcinol at elevated temperatures in the presence of zinc chloride was investigated independently by Dutt (22) and by Bischoff and Adkins (11). The product, known by the trivial name "resorcindiphenein," exhibits indicator properties and is considered to have structure **192**. Consistent with this formulation is the fact that this substance yields a diacetyl derivative, presumed to be **193**, upon reaction with acetic anhydride, and a tetrabromo derivative, presumably **194**, upon bromination in glacial acetic acid. Reaction of **194** with acetic anhydride affords another diacetyl derivative which must have structure **195**.

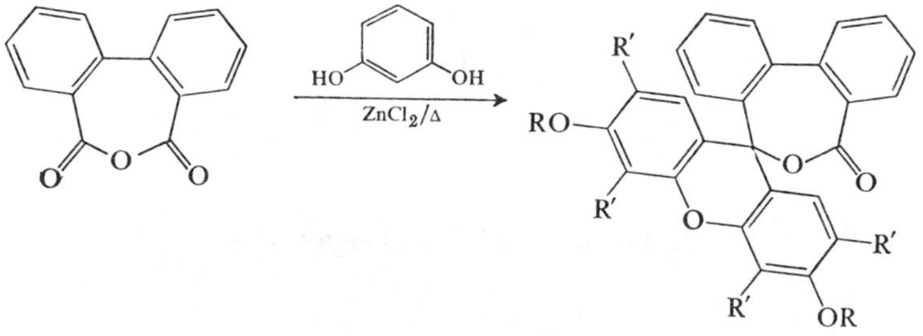

$ZnCl_2/\Delta$

192; R = R′ = H

193; R = Ac, R′ = H

194; R = H, R′ = Br

195; R = Ac, R′ = Br

D. Heptacyclic Systems

1. Difluoreno[1,2-*c*:2′,1′-*e*]oxepins

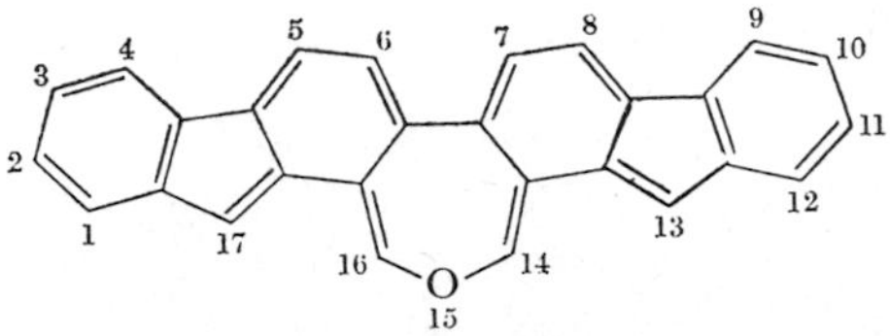

R.I. 7173

This ring system is represented in the literature by anhydride **196**, which Zinke and Ammerer (70) obtained upon treatment of 2,2′-bis(fluoren-9-one-1,1′-dicarboxylic acid) **(197)** with acetic anhydride. Upon being heated **196** underwent the usual type of decarboxylative ring contraction giving triketone **198**.

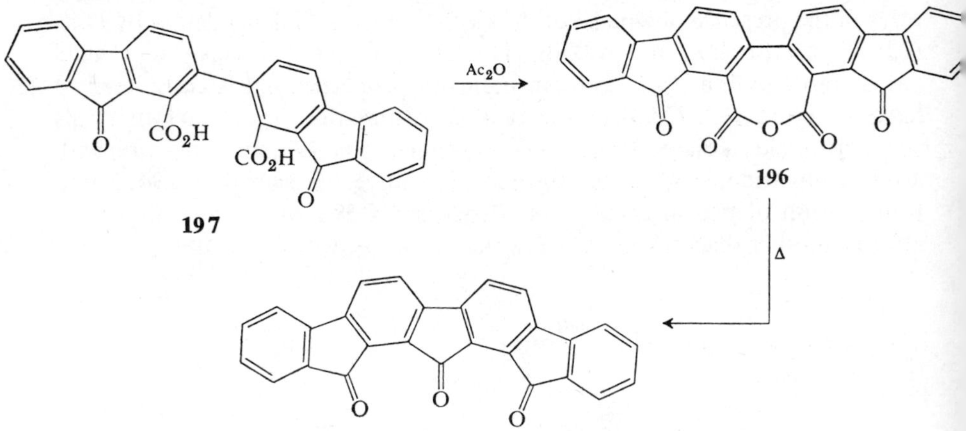

CO_2H CO_2H **197** Ac_2O → **196** Δ **198**

2. Dianthra[1,2-*c*:2′,1′-*e*]oxepins

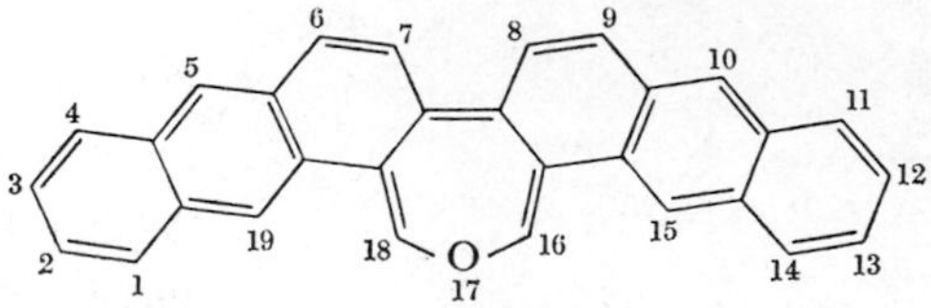

R.I. 7399

This ring system was synthesized in 1935 by Scholl and co-workers (60) starting from violanthrone. Oxidation with chromic acid afforded 2,2′-bis(anthraquinone-1,1′-dicarboxylic acid) (**199**), and treatment of the latter with benzoyl chloride in nitrobenzene under reflux gave a product to which anhydride structure **200** was assigned.

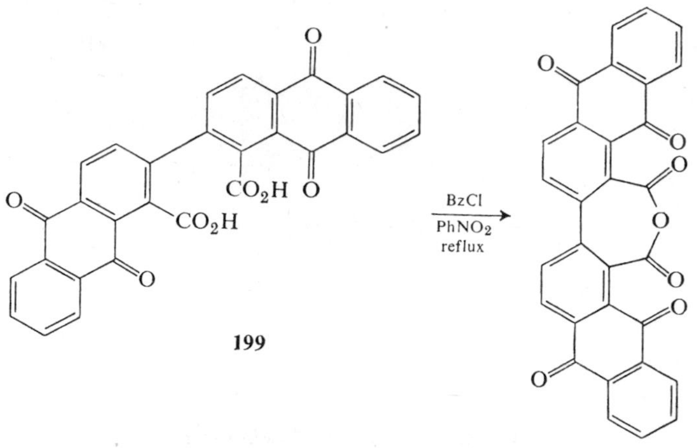

199

BzCl / $PhNO_2$ / reflux

200

3. Dianthra[9,1-*bc*:1′,9′-*ef*]oxepins

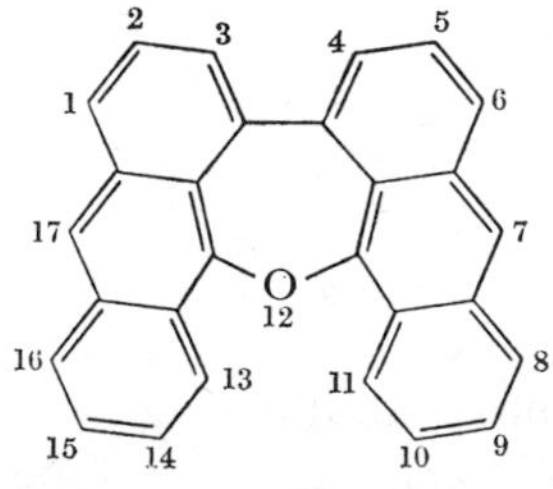

R.I. 7401

Dufraisse and Sauvage (21, 59) claimed to have obtained a member of this ring system in the form of cyclic ether **201**. This extraordinarily high-melting compound (mp 516–518°) was formed when 1,1′-bis(anthraquinone) (**202**) was condensed with four moles of phenyllithium, and the resulting tetrahydroxy compound, **203**, was treated with warm acetic acid.

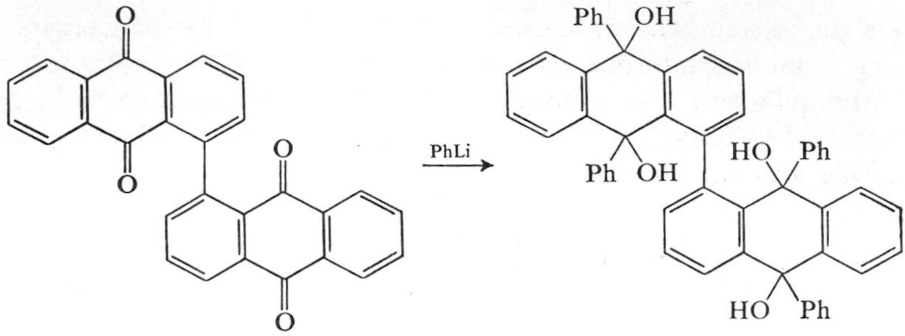

202

PhLi →

203

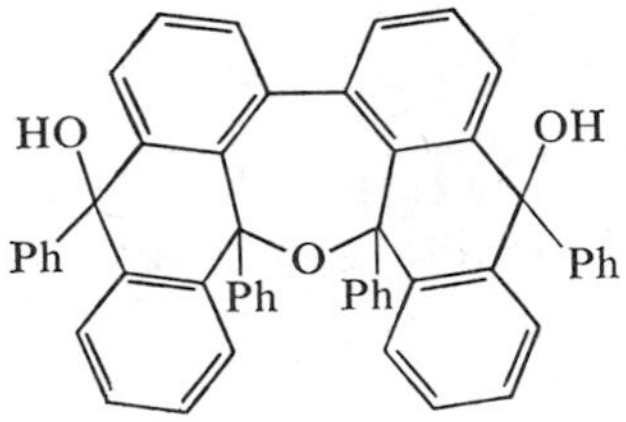

201

4. Dianthra[2,1-*c*:1′,2′-*e*]oxepins

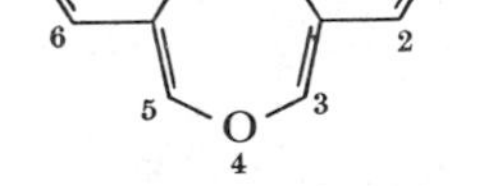

R.I. 9680

The synthesis of a member of this ring system was reported in 1957 by Badger and co-workers (7) and again the following year by Fitts and co-workers (80). 1,1′-Dianthryl-2,2′-dicarboxylic acid was esterified with diazomethane, the diester was reduced with lithium aluminum hydride, and the resultant diol was cyclized to **204** with *p*-toluenesulfonic acid in refluxing benzene. This sequence can be carried with racemic or diastereomeric compounds. The slow racemization rate of optically active **204** is reminiscent of other bridged biaryl systems such as the dinaphth[2,1-*c*:1′,2′-*e*]oxepins (41, 43, 44).

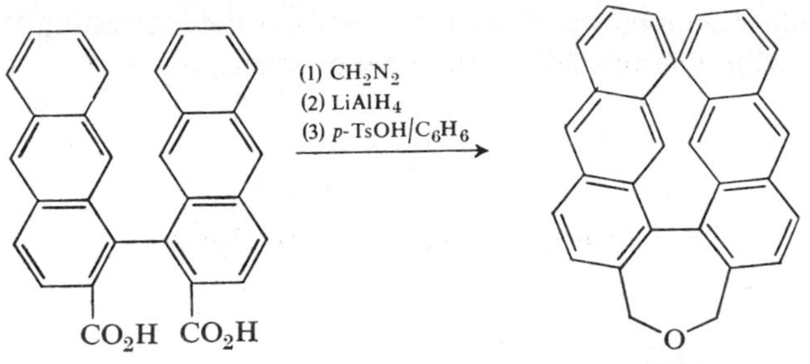
(1) CH2N2
(2) LiAlH4
(3) p-TsOH/C6H6
CO2H CO2H
O

204

5. Diphenanthro[9,10-*b*:9′,10′-*f*]oxepins

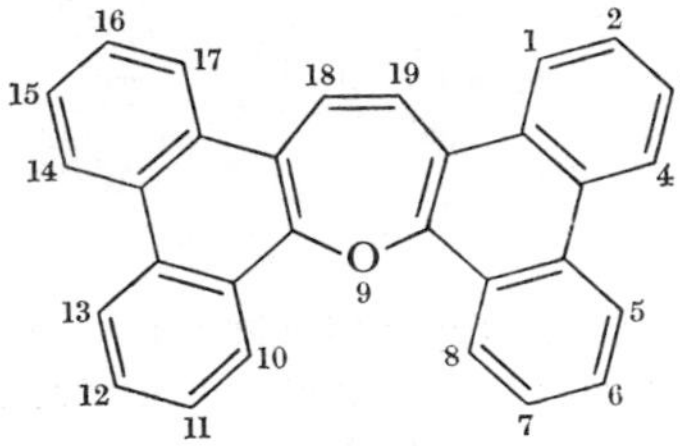
16
17
15
14
18
19
1
2
3
4
O
9
13
12
10
11
8
7
5
6

R.I. 7400

An example of this heptacyclic oxepin derivative was described in 1938 by Diels and Kassebart (20). Upon reaction with acetic anhydride in pyridine,

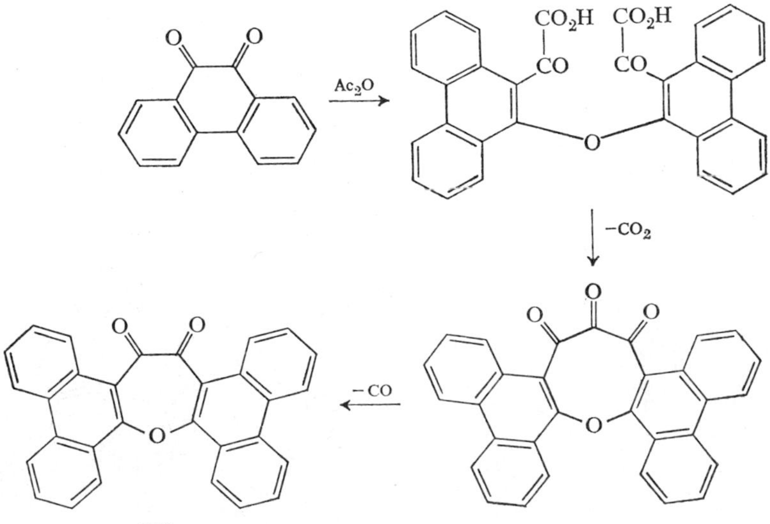
O O
Ac2O
CO2H CO2H
CO CO
O
−CO2
O
O O
−CO
O O
O
O

205

9,10-phenanthraquinone was alleged to undergo the interesting transformation shown, with diketone **205** as the final product.

6. Diphenanthro[9,10-*c*:9′,10′-*e*]oxepins

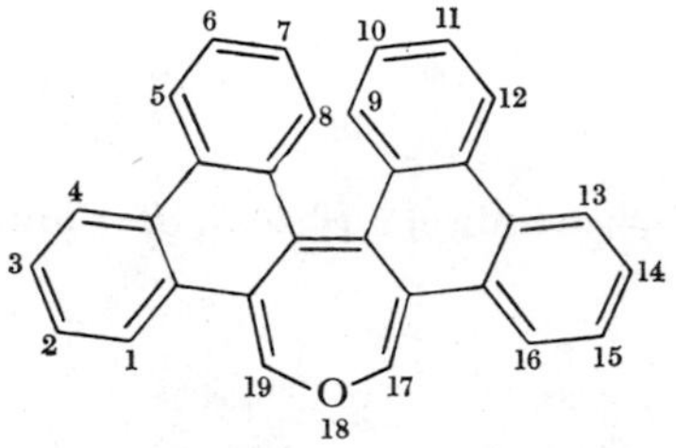

R.I. 11,460

De Ridder and Martin (19) prepared a member of this ring system in the form of anhydride **206** which was formed readily from dicarboxylic acid **207** under the influence of acetic anhydride. Upon being heated **206** underwent decarboxylation to give the ring-contracted ketone **208**.

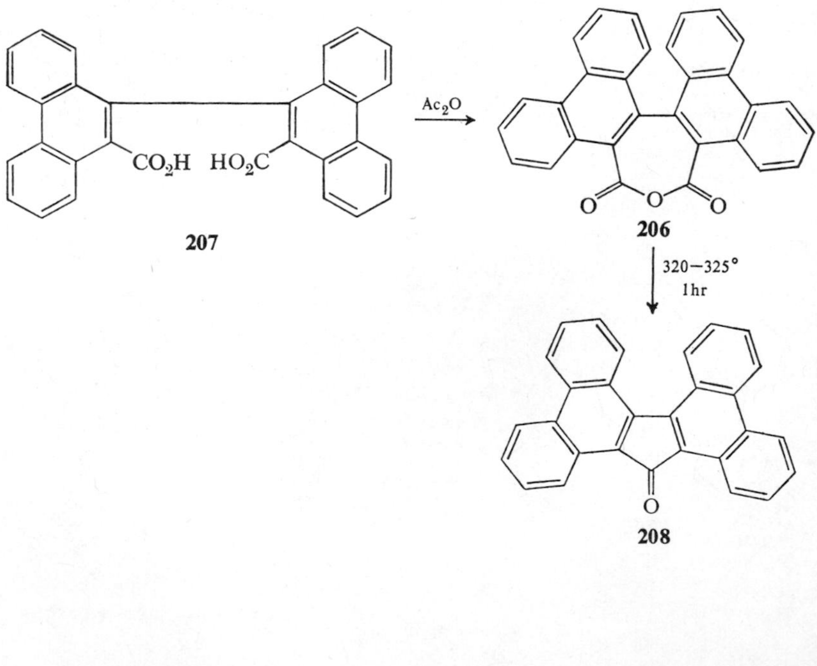

CO_2H HO_2C

207

Ac_2O

206

320–325°
1 hr

208

7. Benzo[6,7]perylo[1,12-*cde*]oxepins

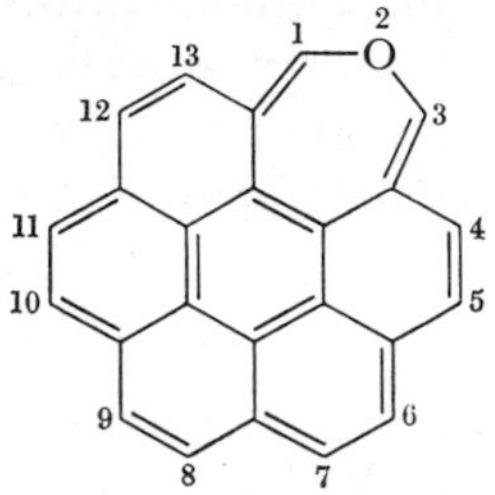

R.I. 7402

Ott and Zinke (51) obtained an example of this highly condensed system starting from the 1,2-quinone, **209**, derived from coronene. Alkaline hydrogen peroxide treatment afforded diacid **210** which underwent cyclization to anhydride **211** upon being heated in nitrobenzene. Reaction of **211** with aniline under reflux afforded the *N*-phenylimide analog **212**.

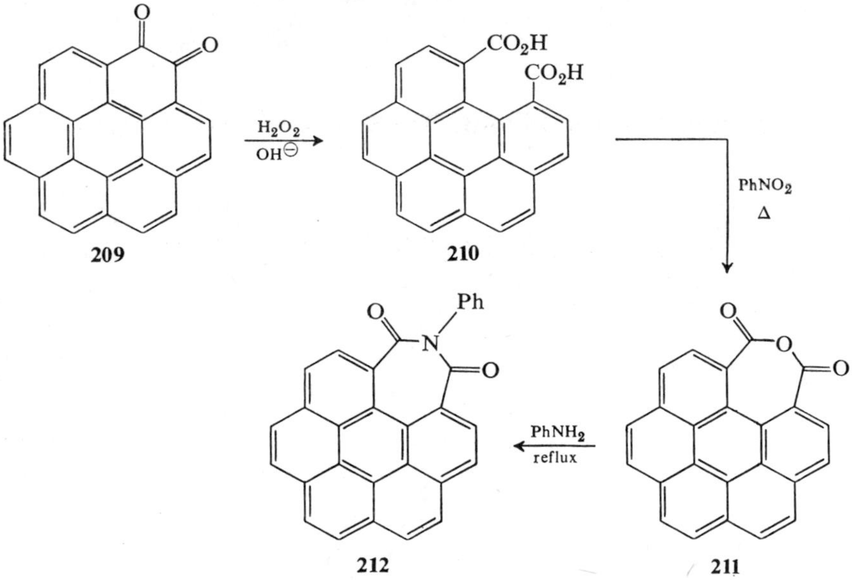

H_2O_2, $OH^{\ominus}$ → ; $PhNO_2$, Δ → ; $PhNH_2$, reflux →

209 **210** **212** **211**

8. 8*H*,16*H*-7*a*,15*a*-Epidithio-7*H*,15*H*-Bisoxepino[3′,4′:4,5]-pyrrolo[1,2-*a*:1′,2′-*d*]pyrazines

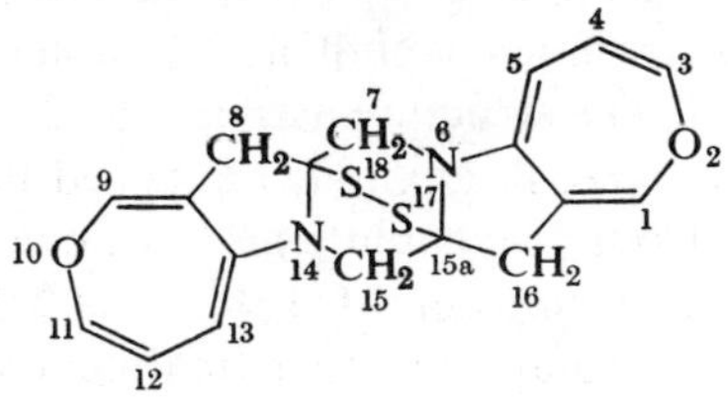

A member of this unusual ring system recently appeared in the literature in the form of the natural product aranotin (**213**) which is a fungal metabolite exhibiting interesting antiviral properties. The structure of **213** was determined on the basis of chemical degradation (74), nmr and circular dichroism spectra (75), and X-ray crystallography (76). Acetylation of **213** affords diacetylaranotin (**214**).

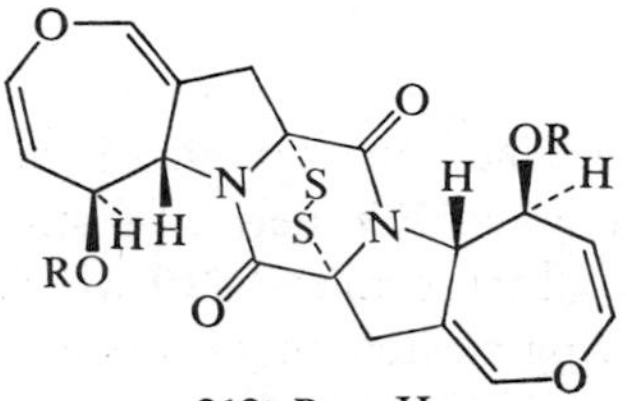

213; R = H
214; R = Ac

E. Systems Containing More Than Seven Rings

1. Acenaphtho[1,2-*d*]dinaphth[2,1-*b*:1′,2′-*f*]oxepins

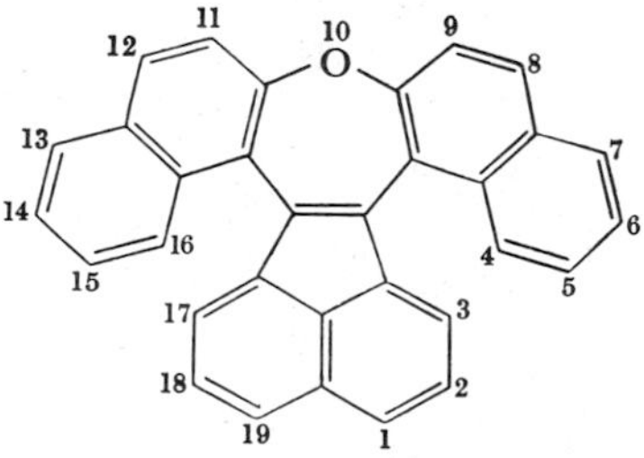

R.I. 7473

An example of this octacyclic oxepin derivative was prepared in 1934 by Matei and Bogdan (37) from acenaphthene-1,2-dione and 2-naphthol in the presence of sulfuric acid. The structure assigned to the product was **215**. In a paper published 2 years later Bogdan (12) assigned the same structure to a compound obtained from acenaphthene-1,2-dione and 2-naphthol in alcoholic hydrochloric acid, and reported that diol **215** underwent a pinacol rearrangement to ketone **216** upon further treatment with sulfuric acid.

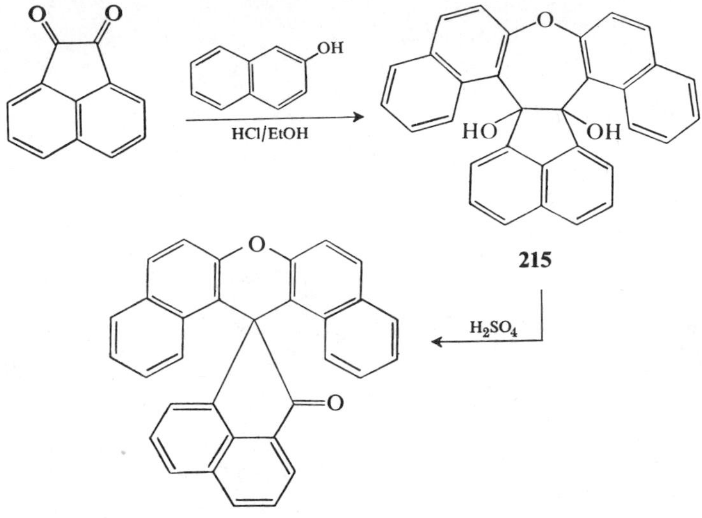

OH

HCl/EtOH

HO OH

215

H_2SO_4

216

2. Dinaphtho[2,1-*e*:2′,1′-*e*′]phenanthro[10,1-*bc*:9,8-*b*′,c′]bisoxepins

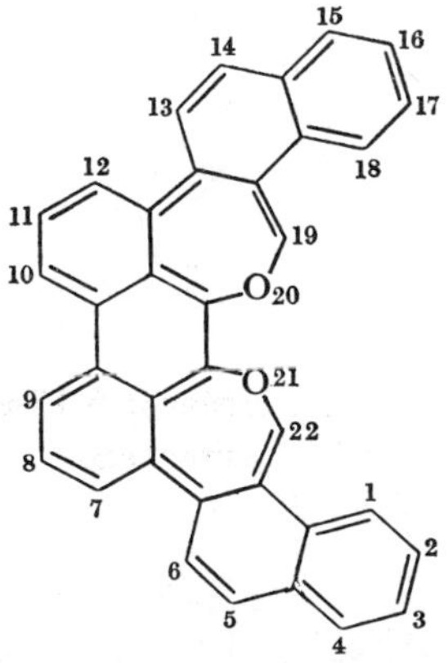

R.I. 7631

Starting from hydrocarbon **217**, Zinke and co-workers (71) gained access into this nine-ring system via compounds **218–221**, which were obtained as shown.

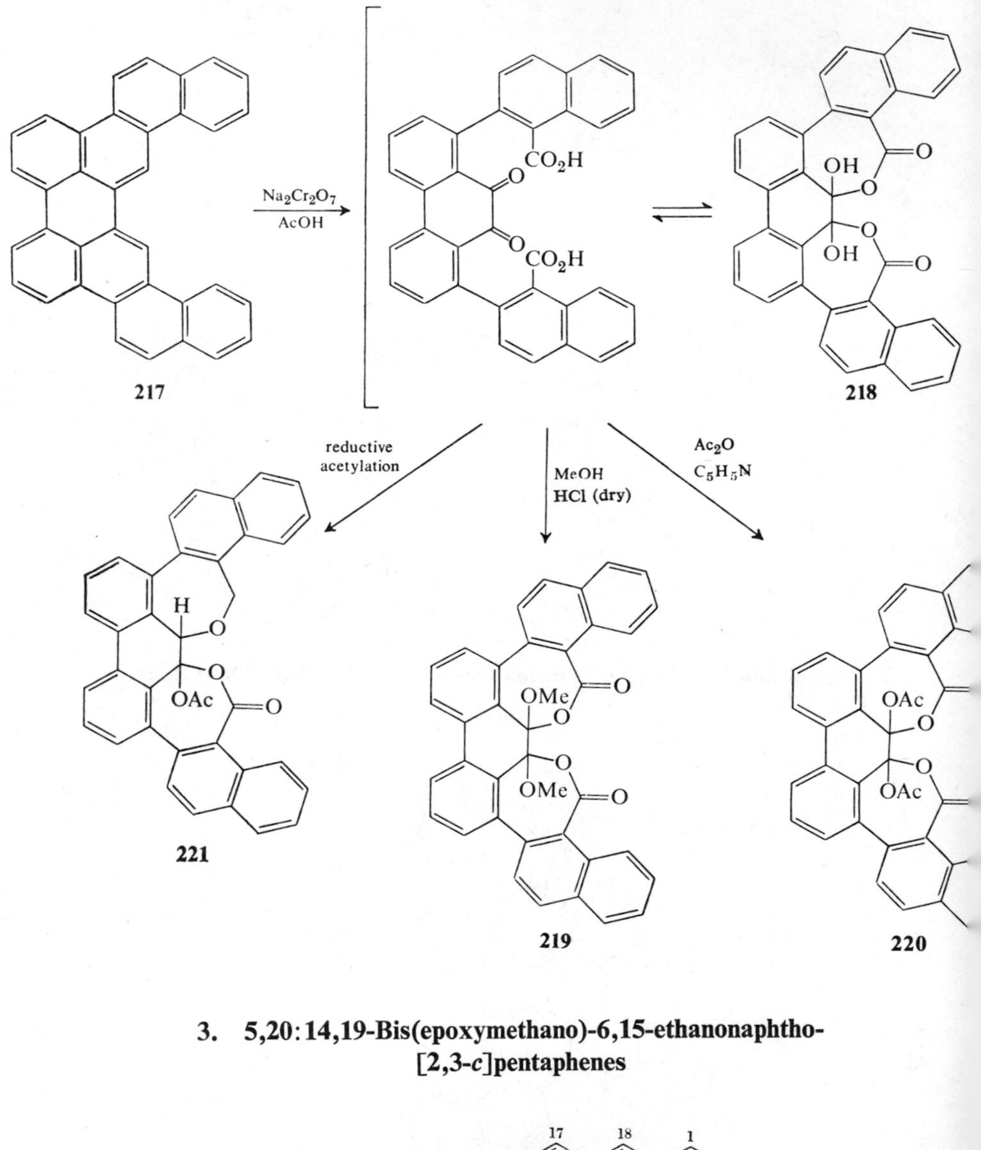

217

$Na_2Cr_2O_7$
AcOH

CO_2H
CO_2H

OH
OH

218

reductive acetylation

MeOH
HCl (dry)

Ac_2O
C_5H_5N

H
OAc

221

OMe
OMe

219

OAc
OAc

220

3. 5,20:14,19-Bis(epoxymethano)-6,15-ethanonaphtho-[2,3-*c*]pentaphenes

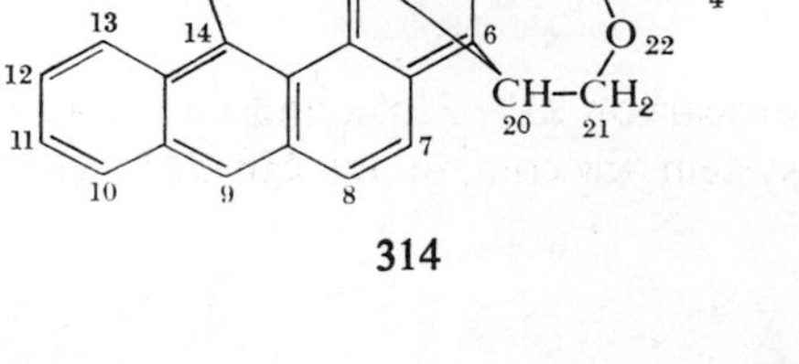

314

R.I. 7659

This novel bridged ring system, which is the largest known ring system containing an oxepin ring, was reported by Clar (15) in 1948. Condensation of pentaphenebisquinone with maleic anhydride followed successively by reduction with zinc and treatment with sodium hydroxide afforded a compound to which structure **222** was assigned. This substance underwent facile decarboxylation to pentaphene upon attempted crystallization from xylene, or upon being heated in acetic anhydride or nitrobenzene.

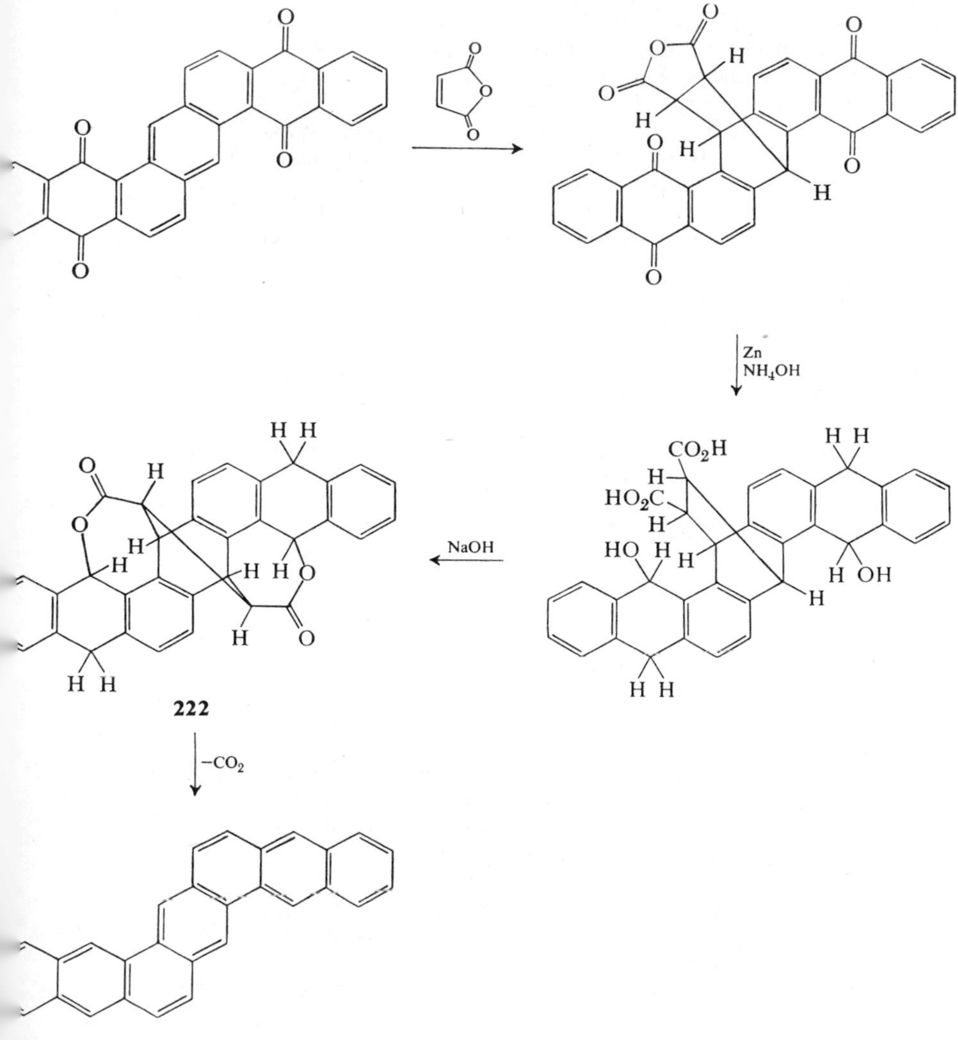

Zn
NH_4OH

NaOH

222

$-CO_2$

F. References

1. D. J. Adam, L. Crombie, and D. A. Whiting, *J. Chem. Soc., C*, **1966,** 550.
2. K. Alder and G. Jacobs, *Chem. Ber.*, **86,** 1528 (1953).
3. R. G. R. Bacon and R. Bankhead, *J. Chem. Soc.*, **1963,** 839.
4. W. Baczynski and S. Niementowski, *Chem. Ber.*, **52B,** 479 (1919).
5. G. M. Badger, *J. Chem. Soc.*, **1947,** 940.
6. G. M. Badger, J. E. Campbell, J. W. Cook, R. A. Raphael, and A. I. Scott, *J. Chem. Soc.*, **1950,** 2326.
7. G. M. Badger, P. R. Jeffries, and R. W. L. Kimber, *J. Chem. Soc.*, **1957,** 1837.
8. A. H. Beckett and R. G. Lingard, *J. Chem. Soc.*, **1961,** 588.
9. A. H. Beckett, R. G. Lingard, and B. A. Mulley, *J. Chem. Soc.*, **1953,** 3328.
10. P. M. Bhargava and C. Heidelberger, *J. Amer. Chem. Soc.*, **78,** 3671 (1956).
11. F. Bischoff and H. Adkins, *J. Amer. Chem. Soc.*, **45,** 1030 (1923).
12. H. Bogdan, *Ann. Sci. Univ. Jassy, I*, **25,** 645 (1939); *Chem. Abstr.*, **34,** 414 (1940).
13. P. Brown and R. C. Cookson, *Proc. Chem. Soc.*, **1952,** 1038.
14. R. S. Cahn, C. K. Ingold, and V. Prelog, *Experientia*, **12,** 81 (1956).
15. E. Clar, *Chem. Ber.*, **81,** 169 (1948).
16. Czechoslovakian Patent 111,551; *Chem. Abstr.*, **62,** 6447 (1965).
17. C. Danheux, L. Hanoteau, R. H. Martin, and G. van Binst, *Bull. Soc. Chim. Belges*, **72,** 289 (1963).
18. O. Dann and W. D. Arndt, *Justus Liebigs Ann. Chem.*, **587,** 38 (1954).
19. R. De Ridder and R. H. Martin, *Bull. Soc. Chim. Belges*, **69,** 534 (1960).
20. O. Diels and R. Kassebart, *Justus Liebigs Ann. Chem.*, **536,** 78 (1938).
21. C. Dufraisse and G. Sauvage, *C. R. Acad. Sci., Paris*, **221,** 665 (1945).
22. S. Dutt, *J. Chem. Soc.*, **123,** 224 (1923).
23. L. F. Feiser, *J. Amer. Chem. Soc.*, **53,** 3546 (1931).
24. C. Graebe and R. Gnehm, Jr., *Justus Liebigs Ann. Chem.*, **335,** 118 (1904).
25. H. I. Hadler and A. C. Kryger, *J. Org. Chem.*, **25,** 1896 (1960).
26. E. Hope and J. S. Anderson, *J. Chem. Soc.*, **1936,** 1474.
27. G. Jacini and A. Salini, *Gazz. Chim. Ital.*, **69,** 717 (1939).
28. T. Kametani and K. Fukumoto, *J. Chem. Soc.*, **1963,** 4289.
29. T. Kametani, K. Fukumoto, and T. Nakano, *Chem. Pharm. Bull.* (Tokyo), **11,** 1299 (1963); *Chem. Abstr.*, **60,** 10730 (1964).
30. T. Kametani and K. Ogasawara, *J. Chem. Soc.*, **1964,** 4142.
31. T. Kametani, S. Shibuya, S. Seino, and K. Fukumoto, *Tetrahedron Lett.*, **1964,** 25.
32. T. Kametani, S. Shibuya, S. Seino, and K. Fukumoto, *J. Chem. Soc.*, **1964,** 4146.
33. H. Klinger and C. Lonnes, *Chem. Ber.*, **29,** 2152 (1896).
34. O. Kruber, *Chem. Ber.*, **67B,** 111 (1934).
35. H. Leuchs and F. Reinhart, *Chem. Ber.*, **57B,** 1208 (1924).
36. E. Lippmann and R. Fritsch, *Chem. Ber.*, **38,** 1626 (1905).
37. I. Matei and E. Bogdan, *Chem. Ber.*, **67B,** 1834 (1934).
38. J. Mathieu, *Ann. Chim.* (Paris), [11] **20,** 215 (1945).
39. J. Mathieu, *C. R. Acad. Sci., Paris*, **219,** 620 (1944).
40. H. Medenwald, *Chem. Ber.*, **86,** 287 (1953).
41. K. Mislow, E. Bunnenberg, R. Records, K. Wellman, and C. Djerassi, *J. Amer. Chem. Soc.*, **85,** 1342 (1963).
42. K. Mislow and M. A. W. Glass, *J. Amer. Chem. Soc.*, **83,** 2780 (1961).

43. K. Mislow, M. A. W. Glass, H. B. Hopps, E. Simon, and G. H. Wahl, Jr., *J. Amer. Chem. Soc.*, **86,** 1710 (1964).
44. K. Mislow, M. A. W. Glass, R. E. O'Brien, P. Rutkin, D. H. Steinberg, J. Weiss, and C. Djerassi, *J. Amer. Chem. Soc.*, **84,** 1455 (1962).
45. E. T. Moriconi, W. F. O'Connor, W. J. Schmitt, G. W. Cogswell, and B. P. Fürer, *J. Amer. Chem. Soc.*, **82,** 3441 (1960).
46. M. S. Newman and H. S. Whitehouse, *J. Amer. Chem. Soc.*, **71,** 3664 (1949).
47. F. G. Oberender and J. A. Dixon, *J. Org. Chem.*, **24,** 1226 (1959).
48. M. Oki and H. Iwamura, *Tetrahedron*, **24,** 2377 (1968).
49. M. Oki, H. Iwamura, and N. Hayakawa, *Bull. Chem. Soc. Jap.*, **36,** 1542 (1963); *Chem. Abstr.*, **60,** 6368 (1964).
50. M. Oki, H. Iwamura, and N. Hayakawa, *Bull. Chem. Soc. Jap.*, **37,** 1865 (1964); *Chem. Abstr.*, **62,** 7612 (1965).
51. R. Ott and A. Zinke, *Monatsh. Chem.*, **84,** 1132 (1953).
52. A. L. Pandit and A. B. Kulkarni, *Curr. Sci.*, **27,** 254 (1958).
53. T. Pozzo-Balbi, *Gazz. Chim. Ital.*, **81,** 125 (1951).
54. J. Rigaudy, *Ann. Chim. (Paris)*, [12]**5,** 398 (1950); *Chem. Abstr.*, **45,** 596 (1951).
55. J. Rigaudy and L. H. Danh, *C. R. Acad. Sci., Paris*, **256,** 5370 (1963).
56. J. Rigaudy and J. M. Farthouat, *C. R. Acad. Sci., Paris*, **236,** 1173 (1953).
57. J. Rigaudy and J. M. Farthouat, *C. R. Acad. Sci., Paris*, **238,** 2431 (1954).
58. J. Rigaudy and K.-V. Thang, *Bull. Soc. Chim. Fr.*, **1959,** 1618.
59. G. Sauvage, *Ann. Chim.* (Paris), [12]**2,** 844 (1947); *Chem. Abstr.*, **42,** 4999 (1948).
60. R. Scholl, E. J. Müller, and O. Böttger, *Chem. Ber.*, **68B,** 45 (1935).
61. H. D. Shroff, A. L. Pandit, and A. B. Kulkarni, *J. Sci. Ind. Res.*, **20D,** 599 (1961); *Chem. Abstr.*, **57,** 13700 (1962).
62. A. Sieglitz and H. Koch, *Chem. Ber.*, **58B,** 78 (1925).
63. E. Simonitsch, W. Eisenhuth, O. A. Stamm, and H. Schmid, *Helv. Chim. Acta*, **43,** 58 (1960).
64. F. N. Stepanov and N. I. Grineva, *Zh. Obshch. Khim.*, **32,** 1529 (1962); *Chem. Abstr.*, **58,** 4499 (1963).
65. T. Szeki, E. Izsakovics, G. Mohalyi, and H. Simonfai, *Acta Lit. Ac. Sci. Univ. Hung. Francisco-Josephinae, Sect. Chem., Mineral. Phys.*, **2**(1), 1 (1930); *Chem. Abstr.*, **25,** 4519 (1931).
66. S. Takahashi, *Agr. Biol. Chem.* (Tokyo), **26,** 401 (1962); *Chem. Abstr.*, **59,** 8670 (1963).
67. H. Vollman, H. Becker, M. Corell, H. Streeck, and G. Langbein, *Justus Liebigs Ann. Chem.*, **531,** 1 (1937).
68. H. Waldmann and G. Pitschak, *Justus Liebigs Ann. Chem.*, **531,** 183 (1937).
69. C. Willgerodt and B. Albert, *J. Prakt. Chem.*, [2]**84,** 383 (1911).
70. A. Zinke and L. Ammerer, *Monatsh. Chem.*, **84,** 423 (1953).
71. A. Zinke, F. Bossert, and E. Ziegler, *Monatsh. Chem.*, **80,** 204 (1949).
72. C. Dufraisse and J. Rigaudy, *C. R. Acad. Sci., Paris*, **221,** 625 (1945).
73. C. P. Falshaw, W. D. Ollis, M. Watanabe, M. M. Dhar, A. W. Khan, and V. C. Vora, *Chem. Commun.*, **1968,** 374.
74. R. Nagarajan, L. L. Huckstep, D. H. Lively, D. C. DeLong, M. M. Marsh, and N. Neuss, *J. Amer. Chem. Soc.*, **90,** 2980 (1968).
75. N. Neuss, R. Nagarajan, B. B. Molloy, and L. L. Huckstep, *Tetrahedron Lett.*, **1968,** 4467.
76. D. B. Cosulich, N. R. Nelson, and J. H. van den Henden, *J. Amer. Chem. Soc.*, **90,** 6519 (1968).

77. C. G. Chin, H. W. Cuts, and S. Masamune, *Chem. Commun.*, **1966,** 880.
78. T. Kametani, K. Fukumoto, and K. Mesuko, *Yakugaku Zasshi*, **83,** 1052 (1963); *Chem. Abstr.*, **60,** 10732 (1964).
79. T. Kubota, N. Ichikawa, W. Brodowski, and P. Weyerstahl, *Tetrahedron Lett.*, **1966,** 5385.
80. D. D. Fitts, M. Siegel, and K. Mislow, *J. Amer. Chem. Soc.*, **80,** 480 (1958).
81. T. Kametani and K. Fukumoto, *Chem. Ind.* (London), **1963,** 291.
82. T. Kametani, S. Shibuya, and I. Noguchi, *Yakugaku Zasshi*, **85,** 667 (1965); *Chem. Abstr.*, **63,** 13228 (1965).
83. T. Kametani, K. Fukumoto, S. Shibuya, and T. Nakano, *Chem. Pharm. Bull.* (Tokyo), **11,** 1299 (1963); *Chem. Abstr.*, **60,** 10730 (1964).
84. R. H. F. Manske, *Can. J. Chem.*, **44,** 561 (1966).
85. T. Kametani, S. Shibuya, C. Kibayashi, and S. Sasaki, *Tetrahedron Lett.*, **1966,** 3215.
86. T. Kametani and C. Kibayashi, *Yakugaku Zasshi*, **84,** 642 (1964); *Chem. Abstr.*, **61,**
87. 13278 (1964).
British Patent 1,016,373; *Chem. Abstr.*, **64,** 9697 (1966).
88. W. Tochtermann, K. Oppelaender, and M. N.-D. Hoang, *Justus Liebigs Ann. Chem.*, **701,** 117 (1967).
89. M. Weisenfels, H. Schurig, and G. Huehsam, *Chem. Ber.*, **100,** 584 (1967).
90. B. A. Hess, Jr., A. S. Bailey, and V. Boekelheide, *J. Amer. Chem. Soc.*, **89,** 2746 (1967)
91. F. Mathys, V. Prelog, and R. B. Woodward, *Helv. Chim. Acta*, **39,** 1095 (1956).
92. P. F. Holt and A. N. Hughes, *J. Chem. Soc.*, **1960,** 3216.
93. J. N. Chatterjea, *J. Indian Chem. Soc.*, **34,** 299 (1957).
94. T. Ikeda, W. Taylor, Y. Tsuda, S. Uyeo, and H. Yajima, *J. Chem. Soc.*, **1956,** 4749.
95. A. T. Peters and D. Walker, *J. Chem. Soc.*, **1956,** 1429.
96. H. Plieninger, M. S. v. Wittenau, and B. Kiefer, *Chem. Ber.*, **91,** 2095 (1958).
97. F. Dallacker and G. Adolphen, *Justus Liebigs Ann. Chem.*, **694,** 110 (1966).

CHAPTER V

Dioxepins and Trioxepins

CHESTER E. PAWLOSKI

Ag-Organics Dept. Dow Chemical Company, Midland, Michigan

A. Monocyclic Dioxepins

1. History, Structures, and Nomenclature

Three monounsaturated seven-membered ring systems containing two oxygens are possible: 1,2-dioxepin (**1** and **2**), 1,3-dioxepin (R.I. 350) (**3** and **4**), and 1,4-dioxepin (R.I. 351) (**5** and **6**). There are presently no known derivatives of 1,2-dioxepin, but a large number of 1,3-dioxepins are known. In 1949 Covenhaver and Bigelow (1) reported that *cis*-2-butene-1,4-diol reacted with aldehydes by the usual methods to give cyclic acetals that would polymerize. The most elementary 1,3-dioxepin was prepared by Reppe (2) when he reacted *cis*-2-butene-1,4-diol with formaldehyde to give a cyclic acetal. Brannock and Lappin (3) prepared the first 1,3-dioxepins of the type shown in **3**.

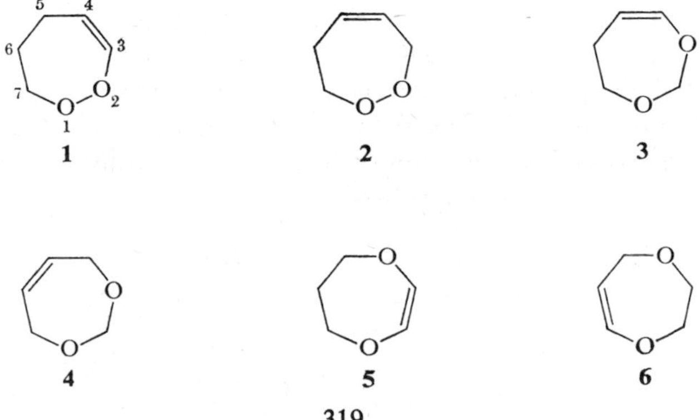

Very little is known of the 1,4-dioxepins. Only derivatives of structure **5** are known. Rio (5) prepared one of these when he treated 9,10-dihydroxy-9,10-diethynylphenyl-9,10-dihydroanthracene with concentrated sulfuric acid and isolated 2-phenyl-3-(10-phenylethynyl-9-anthryl)-1,4-dioxepin. Other mention of 1,4-dioxepins is made by Forrester (6).

The numbering of the rings is shown in **1**. These compounds are sometimes referred to as dioxepenes or dioxacycloheptenes in the literature. Spiro derivatives are named according to the length of the chain; for example, structure **7** is named 7,12-dioxaspiro[5.6]dodec-9-ene (R.I. 8146). The numbering of this ring is shown.

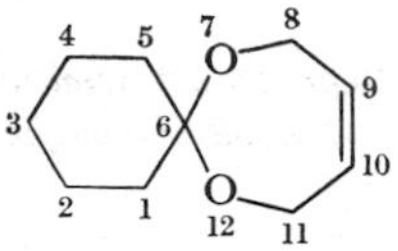

7

2. Methods of Preparation of 1,3-Dioxepins

a. *Reaction of* cis-*2-Butene-1,4-diol with Aldehydes*

Aldehydes react with *cis*-2-butene-1,4-diol as shown in Eq. 1. The reaction is catalyzed by the use of a strong acid such as *p*-toluenesulfonic or concentrated sulfuric acid. Best yields are obtained by using a solvent, benzene or toluene, as a reaction medium. The solvent acts as an azeotroping agent for removal of the water formed during the reaction (3). When a solvent is not employed, the product distills from the reaction mixture with the water, and separates from the water phase on cooling. Distillation usually requires reaction temperatures of 180–200°. In most instances the acid is neutralized before distillation. An exception to this is the preparation of 4,7-dihydro-1,3-dioxepin from formaldehyde. Best yields of this product are obtained when the dioxepin is distilled from the acidic reaction medium. If the mixture is neutralized before distillation, only a 25% yield is obtained along with a viscous nonvolatile polymer. Apparently a linear polymer is formed during the reaction and under acid conditions is converted to a volatile acetal (3). Pattison (4) also found the reaction of formaldehyde without a solvent proceeded violently to form tarry residues.

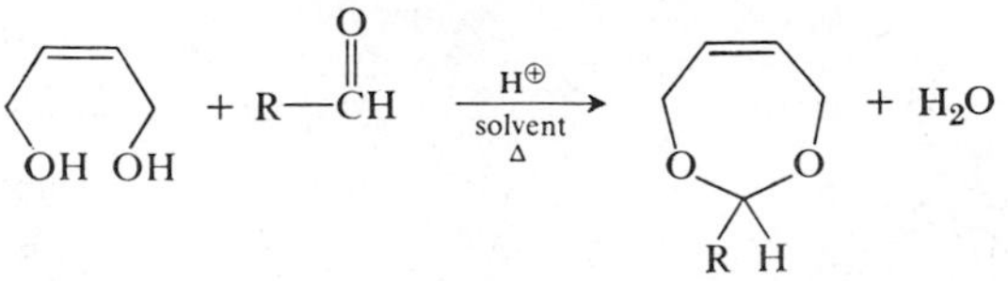

$$\text{OH OH (cis-2-butene-1,4-diol)} + \text{R—CHO} \xrightarrow[\text{solvent, }\Delta]{\text{H}^{\oplus}} \text{2-R,2-H-4,7-dihydro-1,3-dioxepin} + H_2O \quad (1)$$

Generally, aldehydes such as paraformaldehyde, isobutyraldehyde, 2-methylbutyraldehyde, benzaldehyde, and crotonaldehyde give high yields of the desired 1,3-dioxepin derivative. Some of the unsaturated aldehydes, such as acrolein, yield largely polymer (7).

b. *Reaction of* cis-*2-Butene-1,4-diol with Ketones*

In general alkyl ketones such as acetone, 4-methyl-2-pentanone, or 3-pentanone (3,7) do not yield a 1,3-dioxepin when allowed to react with *cis*-2-butene-1,4-diol in the same manner as the aldehydes. The products obtained are usually unreacted ketones and a dehydration product of the diol, 2,5-dihydrofuran. However, Kimel and Leimgruber (8) found that by using a large excess of the ketone, a fair yield of the 1,3-dioxepin derivative is obtained. They treated the diol with a large excess of acetone, with sodium sulfate and concentrated sulfuric acid as catalysts, and obtained 2,2-dimethyl-4,7-dihydro-1,3-dioxepin. Phenyl ketones such as acetophenone, benzophenone, and *p*-chloroacetophenone give low yields of the desired 1,3-dioxepin derivative (7).

Cyclic ketones react quite readily with *cis*-2-butene-1,4-diol to give high yields of spiro derivatives of 1,3-dioxepin (Eq. 2). Brannock and Lappin (3) were the first to show that this reaction took place in benzene, with *p*-toluenesulfonic acid as a catalyst, to yield 7,12-dioxaspiro[5.6]dodec-9-ene (**8**) from cyclohexanone.

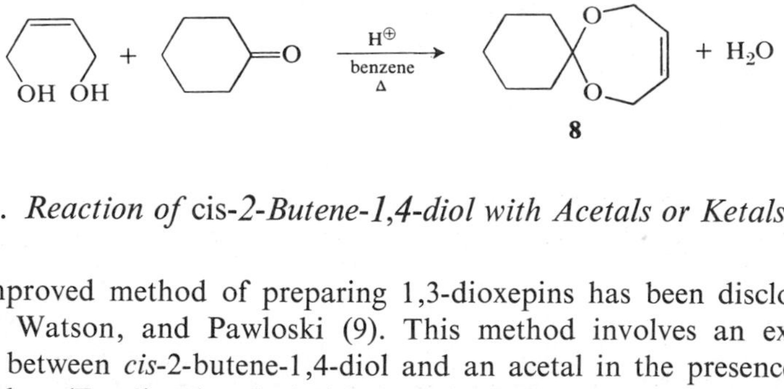

OH OH + cyclohexanone(=O) $\xrightarrow[\text{benzene, }\Delta]{H^{\oplus}}$ **8** + H_2O (2)

c. *Reaction of* cis-*2-Butene-1,4-diol with Acetals or Ketals*

An improved method of preparing 1,3-dioxepins has been disclosed by Sterling, Watson, and Pawloski (9). This method involves an exchange reaction between *cis*-2-butene-1,4-diol and an acetal in the presence of an acid catalyst (Eq. 3). The alcohol formed in the exchange must be distilled from the mixture since the reaction is reversible. The initial reaction is usually endothermic.

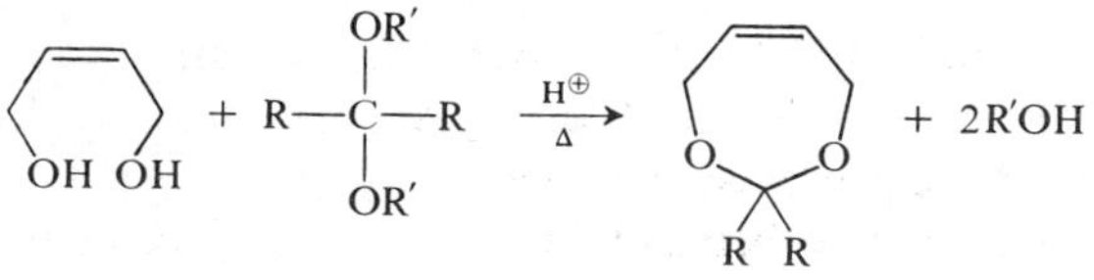

$$\text{OH OH} + R\!-\!C(OR')_2\!-\!R \xrightarrow[\Delta]{H^{\oplus}} \text{(2,2-R,R-4,7-dihydro-1,3-dioxepin)} + 2R'OH \quad (3)$$

High yields of substituted 1,3-dioxepins can be obtained by using a molar excess of acetal; however, increased yields are also obtained by using solvents such as benzene or ethyl acetate in place of excess acetal. Any nonoxidizing acidic catalyst such as sulfuric, dichloroacetic, *p*-toluenesulfonic, or phosphoric acid can be used. The reaction is usually carried out in a reactor equipped with a distillation column. After the alcohol formed by the reaction is removed, the mixture is neutralized before further purification of the product.

Acetals such as acetaldehyde diethyl acetal and benzaldehyde dimethyl acetal, or ketals such as acetone dimethyl ketal, acetone butyl methyl ketal, acetone dibutyl ketal, 2-butanone dimethyl ketal, 3-pentanone dimethyl ketal, cyclohexanone dimethyl ketal, and bromo-2-propanone dimethyl ketal, can be condensed with *cis*-2-butene-1,4-diol to yield substituted 1,3-dioxepins in yields of 50% or better. An exception is dimethoxymethane which yields only 3–5% of the desired 4,7-dihydro-1,3-dioxepin. Acetals of aldehydes and ketals of cyclic ketones give higher yields than ketals of alkyl ketones. Mixed acetals may also be employed.

This method is superior to previous methods because dialkyl derivatives can be obtained in high yields. For example, reaction of *cis*-2-butene-1,4-diol with acetone dimethyl ketal affords a 50% or better yield of 2,2-dimethyl-4,7-dihydro-1,3-dioxepin (**9**) (Eq. 4). The reaction of the diol with acetone itself gives poor yields.

$$\text{(cis-HOCH}_2\text{CH=CHCH}_2\text{OH)} + CH_3C(OCH_3)_2CH_3 \xrightarrow[\Delta]{H^{\oplus}} \mathbf{9}\ (\text{2,2-dimethyl-4,7-dihydro-1,3-dioxepin; O, O; } CH_3\ CH_3) + 2CH_3OH \qquad (4)$$

The acetals used in the preparation of 1,3-dioxepins can be prepared in a number of ways. The well-known reaction of alcohols with aldehydes is shown in Eq. 5.

$$R{-}\overset{O}{\overset{\|}{C}}H + ROH \underset{}{\overset{H^{\oplus}}{\rightleftarrows}} \underset{\text{hemi-acetal}}{RH\overset{OH}{\overset{|}{C}}{-}OR} \underset{H^{\oplus}}{\overset{ROH}{\rightleftarrows}} \underset{\text{acetal}}{R\overset{OR}{\overset{|}{\underset{OR}{\underset{|}{C}}}}H} \qquad (5)$$

Acetals and ketals can also be prepared by condensing another acetal or ketal with a ketone or aldehyde (Eq. 6). Acetone dimethyl ketal reacts with cyclohexanone in the presence of an acid catalyst to yield 80% cyclohexanone dimethyl ketal. Similarly benzaldehyde yields 75% benzaldehyde dimethyl acetal; acrolein, 60% acrolein dimethyl acetal; and 1,2,3,6-tetrahydrobenzaldehyde, 82% 1,2,3,6-tetrahydrobenzaldehyde dimethyl acetal (7).

$$R\text{—}\underset{\mathrm{OR'}}{\overset{\mathrm{OR'}}{\mathrm{C}}}\text{—}R \xrightarrow[\Delta]{\mathrm{H}^{\oplus}} \begin{cases} \xrightarrow{R''\text{—}\overset{\mathrm{O}}{\overset{\|}{\mathrm{C}}}\mathrm{H}} R''\text{—}\underset{\mathrm{OR'}}{\overset{\mathrm{OR'}}{\mathrm{CH}}} + R\overset{\mathrm{O}}{\overset{\|}{\mathrm{C}}}R \\ \xrightarrow{R''\text{—}\overset{\mathrm{O}}{\overset{\|}{\mathrm{C}}}\text{—}R''} R''\text{—}\underset{\mathrm{OR'}}{\overset{\mathrm{OR'}}{\mathrm{C}}}\text{—}R'' + R\overset{\mathrm{O}}{\overset{\|}{\mathrm{C}}}R \end{cases} \tag{6}$$

d. *Reaction of* cis-*2-Butene-1,4-diol with an Aldehyde and an Acetal or Ketal*

The reaction of an acetal or ketal with an aldehyde followed by treatment of the new acetal or ketal with *cis*-2-butene-1,4-diol, has been described in Section A-2-c. Sterling, Watson, and Pawloski (10) have found that these reactions can be combined into a double exchange reaction to yield a 1,3-dioxepin formed from the aldehyde used in the reaction (Eq. 7). The acetal of the aldehyde is thought to form first, reacting subsequently with the diol to yield the dioxepin. The lower-boiling by-products are distilled from the reaction mixture. Equimolar portions of diol, aldehyde, and acetal give good yields, but the best results are obtained when an excess of the aldehyde and acetal is used, or when a solvent such as benzene or ethyl acetate is employed in place of excess reactants.

$$\text{cis-HOCH}_2\text{CH=CHCH}_2\text{OH} + R\text{—}\overset{\mathrm{O}}{\overset{\|}{\mathrm{C}}}\mathrm{H} + R'\text{—}\underset{\mathrm{OR'}}{\overset{\mathrm{OR'}}{\mathrm{C}}}\text{—}R' \xrightarrow[\Delta]{\mathrm{H}^{\oplus}} \text{(1,3-dioxepin, C-2 bearing R, H)} + R'\overset{\mathrm{O}}{\overset{\|}{\mathrm{C}}}R' + 2R'\mathrm{OH} \tag{7}$$

One should be careful in selecting the acetals or ketals used in this double exchange reaction. Compounds such as acetone dimethyl ketal or acetone diethyl ketal give the best yields of 1,3-dioxepins from aldehydes. The by-product is acetone which reacts very slowly with *cis*-2-butene-1,4-diol under the conditions used. Use of an aldehyde acetal results in a mixture of products, as does the use of ketones in the double exchange reaction.

Inasmuch as short-chain aldehydes give an exothermic reaction when first mixed with the acetal and diol, care should be taken to cool the reactants initially. Longer-chain aldehydes such as butyraldehyde give an endothermic reaction. Yields as high as 97% can be obtained, as in the reaction of the diol with acetone dimethyl ketal and *n*-butyraldehyde, which leads to 2-propyl-4,7-dihydro-1,3-dioxepin (**10**) (Eq. 8). Usually yields of 80% or better are realized.

$$\text{cis-HOCH}_2\text{CH=CHCH}_2\text{OH} + C_3H_7\text{—}\overset{\overset{O}{\|}}{C}H + CH_3\underset{OCH_3}{\overset{OCH_3}{|}}{C}CH_3 \xrightarrow[\Delta]{H_2SO_4} \mathbf{10}\ (\text{2-H, 2-}C_3H_7\text{-4,7-dihydro-1,3-dioxepin}) + CH_3\overset{\overset{O}{\|}}{C}CH_3 + 2CH_3OH \quad (8)$$

e. *Reaction of* cis-*2-Butene-1,4-diol with a Ketone and an Acetal or Ketal*

Sterling, Watson, and Pawloski (10) also disclosed that substituted 1,3-dioxepins can be prepared by the double exchange reaction of *cis*-2-butene-1,4-diol with an acetal and a ketone (Eq. 9). The reaction is carried out in the same manner as described in Section A-2-c.

$$\text{cis-HOCH}_2\text{CH=CHCH}_2\text{OH} + R'\text{—}\overset{\overset{O}{\|}}{C}\text{—}R' + R\text{—}\underset{OR}{\overset{OR}{|}}{C}\text{—}R \xrightarrow[\Delta]{H^{\oplus}} (\text{2-}R'\text{,2-}R'\text{-4,7-dihydro-1,3-dioxepin}) + R\overset{\overset{O}{\|}}{C}R + 2ROH \quad (9)$$

This reaction is not so clean-cut as the double exchange with aldehydes. For example, acetone dimethyl ketal, 2-butanone, and *cis*-2-butene-1,4-diol give a 47% yield of 2,2-dimethyl-4,7-dihydro-1,3-dioxepin derived from the ketal, and only a 36% yield of the desired 2-ethyl-2-methyl-4,7-dihydro-1,3-dioxepin from the ketone. For this reason care should be used in selecting reactants. Ketals of alkyl ketones give higher yields of the desired 1,3-dioxepins than do acetals of aldehydes or ketals of cyclic ketones. For example, acetaldehyde diethyl acetal, 2-butanone, and *cis*-2-butene-1,4-diol yield only 3% of the desired 2-ethyl-2-methyl-4,7-dihydro-1,3-dioxepin together with a 78% yield of 2-methyl-4,7-dihydro-1,3-dioxepin from the acetal.

Cyclic ketones such as cyclohexanone give high yields of the spiro dioxepin. Cyclohexanone, acetone dimethyl ketal, and *cis*-2-butene-1,4-diol give an 88% yield of 7,12-dioxaspiro[5.6]dodec-9-ene (**8**). Yields of 80% or better can be obtained from other cyclic ketones.

Haloketones such as bromoacetone readily give high yields of the desired halo derivative of 1,3-dioxepin. Bromoacetone gives an 88% yield of 2-bromomethyl-2-methyl-4,7-dihydro-1,3-dioxepin when allowed to react with a ketal and *cis*-2-butene-1,4-diol.

The double exchange reaction with phenyl ketones gives low conversions. Acetophenone, acetone dimethyl ketal, and *cis*-2-butene-1,4-diol give a 25% yield of the desired 2-phenyl-2-methyl-4,7-dihydro-1,3-dioxepin, along with a 45% yield of 2,2-dimethyl-4,7-dihydro-1,3-dioxepin. Even with these lower yields the double exchange method is superior to direct reaction of the diol with ketones.

f. *Reaction of* cis-*2-Butene-1,4-diol with Trialkyl Orthoformates*

It was stated earlier that *cis*-2-butene-1,4-diol reacts with an acetal or ketal in the presence of an acidic catalyst to form a 1,3-dioxepin (see Section A-2-c). It was also found (7) that this reaction takes place when a trialkyl orthoformate such as trimethyl orthoformate is allowed to react with the diol in a similar manner (Eq. 10). The method differs somewhat from others in that good yields are obtained at ambient temperatures. High reaction temperatures produce decomposition products.

$$\text{(cis-diol: OH OH)} + HC(OR)_3 \xrightarrow{H^{\oplus}} \text{(1,3-dioxepin with H, OR at C-2)} + 2ROH \qquad (10)$$

In a typical experiment trimethyl orthoformate, *cis*-2-butene-1,4-diol, and a trace of concentrated sulfuric acid are heated in a distillation flask. The low boilers are removed by distillation until the reaction temperature reaches 114°. After cooling, sodium carbonate is added to neutralize the acidic catalyst. Distillation is resumed under reduced pressure. Only a 10% yield of the desired 1,3-dioxepin is obtained. On the other hand, when the reaction mixture is stirred for 2 hr at ambient temperatures, treated with sodium carbonate, and distilled under reduced pressures, a 65% yield of the desired 2-methoxy-4,7-dihydro-1,3-dioxepin is produced.

Other trialkyl orthoesters such as trimethyl orthoacetate or triethyl orthopropionate react in a similar manner.

g. *Dehydration of an Aldehyde 4-Hydroxy-2-buten-1-yl Hemiacetal*

Sturzenegger and Zelauskas (32) found that *cis*-2-butene-1,4-diol reacts with an aldehyde at or below 50° without a catalyst to form the hemiacetal of the aldehyde. For example, formaldehyde yields formaldehyde 4-hydroxy-2-buten-1-yl hemiacetal; benzaldehyde yields benzaldehyde

4-hydroxy-2-buten-1-yl hemiacetal; and crotonaldehyde yields crotonaldehyde 4-hydroxy-2-buten-1-yl hemiacetal. Despite the suggestion that ketones react in a similar manner (32), it has been our experience that ketones react slowly with the diol even in the presence of an acid catalyst, especially alkyl ketones such as acetone.

In a typical reaction isobutyraldehyde is added dropwise under a nitrogen blanket to *cis*-2-butene-1,4-diol at reaction temperatures of 30–33°. The product is a colorless liquid consisting of substantially pure isobutyraldehyde 4-hydroxy-2-buten-1-yl hemiacetal.

The dioxepin is prepared by adding the above hemiacetal dropwise to refluxing benzene containing a trace of *p*-toluenesulfonic acid. The rate of addition is such that no more than 10% of the added hemiacetal remains unreacted at any time. The reaction mixture is cooled to 25° after all of the liberated water has been collected. The acidic catalyst is neutralized, and the product is distilled to give a high yield of pure 2-isopropyl-4,7-dihydro-1,3-dioxepin.

The reactions involved in this process are summarized in Eq. 11; any nonoxidizing acidic catalyst can be used.

$$\mathrm{RCH{=}O} + \text{HOCH}_2\text{CH{=}CHCH}_2\text{OH} \longrightarrow \text{HOCH}_2\text{CH{=}CHCH}_2\text{OCH(OH)R} \xrightarrow[\Delta]{H^{\oplus}} \text{2-R-4,7-dihydro-1,3-dioxepin} + H_2O \quad (11)$$

h. *Reaction of* cis-*2-Butene-1,4-diol with Acetylenes*

Another method developed by Sterling, Watson, and Pawloski (11) for the preparation of 1,3-dioxepins involves the exothermic addition of *cis*-2 butene-1,4-diol to an acetylene in the presence of red mercuric oxide, boron trifluoride etherate, and a trace of an acid (Eq. 12). The reaction temperature is controlled by metering the addition of the acetylenic compound to the reaction mixture at a desired rate. Reaction temperatures are 50–100°. The reaction mixtures are neutralized before purification by distillation.

$$\text{HOCH}_2\text{CH{=}CHCH}_2\text{OH} + \mathrm{RC{\equiv}CH} \xrightarrow{\text{Catalyst}} \text{2-R-2-CH}_3\text{-4,7-dihydro-1,3-dioxepin} \quad (12)$$

This is a poor method for preparing 1,3-dioxepins, as can be seen from Table V-1, in which acetylene compounds condensed with *cis*-2-butene-1,4-diol and products thereby obtained are listed. The main products in most instances are 2,5-divinyl-1,4-dioxane and a ketone. The 2,5-divinyl-1,4-dioxane is formed by the reaction of the diol with itself (12). The water formed from this reaction then reacts with the acetylene to yield a ketone (13). Apparently ether formation is more rapid than the addition of diol to the acetylene.

i. *Reaction of* cis-*2-Butene-1,4-diol with Vinyl Ethers*

A patent issued to Seib (14) discloses the reaction of *cis*-2-butene-1,4-diol with methyl vinyl ether at 140–180° to give 2-methyl-4,7-dihydro-1,3-dioxepin (**11**) in 78% yield (Eq. 13). Although no mention is made of other derivatives, this would be a good method for the synthesis of other 1,3-dioxepins in high yields.

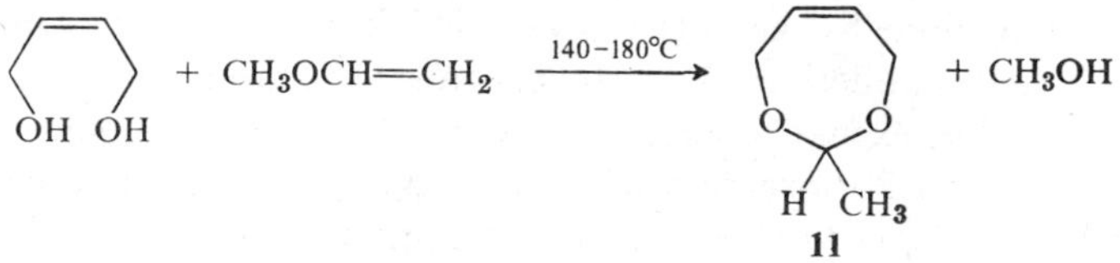

$$\text{(HOCH}_2\text{CH=CHCH}_2\text{OH)} + CH_3OCH{=}CH_2 \xrightarrow{140-180°C} \mathbf{11} + CH_3OH \qquad (13)$$

OH OH; H CH₃; **11**

j. *Dehydrohalogenation of Halogenated 1,3-Dioxepanes*

Brannock and Lappin (3) found that bromine adds to 4,7-dihydro-1,3-dioxepin to yield 5,6-dibromo-4,7-dihydro-1,3-dioxepane (**12**). They further discovered that dehydrobromination of this dihalo dioxepane in methanolic sodium methoxide affords two types of substituted 1,3-dioxepins (Eq. 14). 5-Bromo-4,7-dihydro-1,3-dioxepin (**13**) is obtained as one product since it has an inactive bromine. 5-Bromo-4,5-dihydro-1,3-dioxepin (**14**) is also

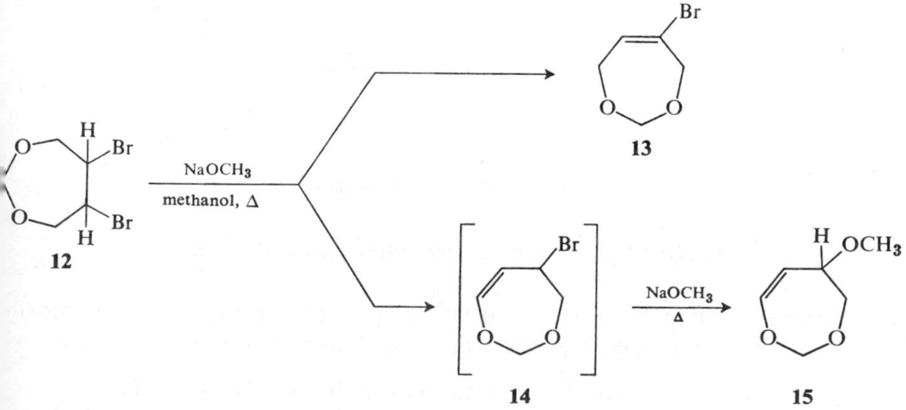

$$\mathbf{12} \xrightarrow[\text{methanol, }\Delta]{NaOCH_3} \mathbf{13} + [\mathbf{14}] \xrightarrow[\Delta]{NaOCH_3} \mathbf{15} \qquad (14)$$

generated, but only as an intermediate which reacts with sodium methoxide (since it is an allylic bromide) to yield 5-methoxy-4,5-dihydro-1,3-dioxepin (**15**) as the other product. Similarly a large number of substituted 1,3-dioxepins can be prepared by this method.

k. *Miscellaneous Reactions*

Attempts to prepare 2-alkoxy-4,7-dihydro-1,3-dioxepins as shown in Eqs. 15–17 were unsuccessful. Both ethyl acetate and methyl acetate were condensed with *cis*-2-butene-1,4-diol and acetone dimethyl ketal as described in Section A-2-d. Apparently the only effect these esters had was to provide an excellent medium for the acetal exchange reaction, 82% yields of 2,2-dimethyl-4,7-dihydro-1,3-dioxepin being obtained (7).

Likewise unsuccessful was the attempted reaction of a formate with the diol and acetone dimethyl ketal (Eq. 16). Ethyl formate, like the esters, gave only the 1,3-dioxepin derived from the exchange reaction with the ketal (7). Attempted double exchange reactions with an alkyl carbonate (Eq. 17) failed as well; the 1,3-dioxepin arising from the acetal was obtained as the sole product in 80% yield when ethylene carbonate was used (7).

OH OH + $CH_3C(OCH_3)_2CH_3$ $\xrightarrow{H^{\oplus},\ \Delta}$

with ROC(=O)R → 2-R-2-OR-4,7-dihydro-1,3-dioxepin (15)

with ROC(=O)H → 2-H-2-OR-4,7-dihydro-1,3-dioxepin (16)

with ROC(=O)OR → 2,2-di-RO-4,7-dihydro-1,3-dioxepin (17)

3. Types of 1,3-Dioxepins

a. *Alkyl- and Aryl-Substituted 1,3-Dioxepins*

2-Alkyl-4,7-dihydro-1,3-dioxepins can be prepared in high yields by most of the following methods, starting from *cis*-2-butene-1,4-diol:

(*1*) Reaction with aldehydes (Section A-2-a; for yields see Table V-2).
(*2*) Reaction with acetals of alkyl aldehydes (Section A-2-c; Table V-3).

(*3*) Double exchange reaction involving an alkyl aldehyde and a ketal (Section A-2-d; Table V-4).

(*4*) Addition to vinyl ether (Section A-2-i).

(*6*) Addition to acetylene (Section A-2-h; Table V-1).

2-Phenyl-1,3-dioxepin can be prepared by all of the above methods except the last.

The synthesis of 2,2-dialkyl-4,7-dihydro-1,3-dioxepins is similar to that of 2-alkyl derivatives, except that ketones and ketals are employed. The results of the reaction of *cis*-2-butene-1,4-diol with ketals are shown in Table V-3. Table V-5 lists the results of the double exchange reaction of the diol with a ketone and acetal. Yields are not as high as those obtained from aldehydes. Inasmuch as some dioxepin derivative is usually obtained from the acetal employed, care should be taken to select an acetal that gives a product with a boiling point above that of the desired 1,3-dioxepin. Table V-2 shows that very little of the desired product is obtained on reaction of *cis*-2-butene-1,4-diol with alkyl ketones. However, some success can be experienced in the addition of the diol to lower alkyl acetylenes as described in Section A-2-h.

Diphenyl derivatives and 2-alkyl-2-phenyl-1,3-dioxepins are obtained in low yields by the condensation of *cis*-2-butene-1,4-diol with phenyl ketones (Table V-6). Somewhat higher yields are experienced from the double exchange reaction of a phenyl ketone with the diol and a ketal (Table V-5). Low yields of 2-phenyl-2-alkyl-4,7-dihydro-1,3-dioxepins are obtained by the addition of the diol to phenyl acetylenes (Section A-2-h).

Alkyl-substituted 4,7-dihydro-1,3-dioxepins can be copolymerized (19–23, 28, 29, 34) and are useful as intermediates for the preparation of insecticides (25–27). Monroe found them to be useful in breaking emulsions of crude petroleum and aqueous liquids (31).

b. *Bis(1,3-Dioxepins)*

The reaction of a diacetal with *cis*-2-butene-1,4-diol yields a bis(1,3-dioxepin) derivative. For example, condensation with 1,1,3,3-tetramethoxypropane (Eq. 18) yields 2,2′-methylenebis(4,7-dihydro-1,3-dioxepin) (**16**) (16). Yields of 72–92% are obtained when 1 mole of diol is employed per mole of diacetal. The reaction of 2 moles of diol per mole of 1,1,3,3-tetraethoxypropane gives only a 17% yield of the desired bis(1,3-dioxepin) plus a considerable amount of polymeric ethers. Other by-products such as 4,7-dihydro-1,3-dioxepin-2-acetaldehyde dimethyl acetal or 4,7-dihydro-1,3-dioxepin-2-acetaldehyde diethyl acetal are obtained in low yields.

Another method for the preparation of bis(1,3-dioxepins) was disclosed by Reinhardt (17), who found that treatment of 2,6-diethoxytetrahydropyran

$$\text{cis-HOCH}_2\text{CH=CHCH}_2\text{OH} + \text{HC(OCH}_3)_2\text{CH}_2\text{CH(OCH}_3)_2 \xrightarrow[\Delta]{H^+} \mathbf{16} + 4CH_3OH \quad (18)$$

with *cis*-2-butene-1,4-diol and gaseous HCl as a catalyst yielded 2,2′-trimethylenebis(4,7-dihydro-1,3-dioxepin) (**17**) (Eq. 19). The ethanol and water formed during the reaction were removed by distillation. Other alkoxypyrans such as 2-alkoxy-3,4-dihydropyrans and ring-substituted 2-alkoxy-3,4-dihydropyrans and 2,6-dialkoxytetrahydropyrans can also be used.

$$2\ \text{cis-HOCH}_2\text{CH=CHCH}_2\text{OH} + \text{2,6-(OC}_2\text{H}_5)_2\text{-tetrahydropyran} \xrightarrow[\Delta]{HCl} \mathbf{17}\ (\text{–CH}_2\text{CH}_2\text{CH}_2\text{–}) + 2C_2H_5OH + H_2O \quad (19)$$

The reaction of *cis*-2-butene-1,4-diol with a dialdehyde also yields the desired bis(1,3-dioxepins). As indicated in Eq. 20, condensation with terephthaldehyde in benzene yields 2,2′-*p*-phenylenebis(4,7-dihydro-1,3-dioxepin) (**18**) (16). Other methods that could be employed are (*1*) reaction with a dialdehyde or diketone and an acetal or ketal in a double exchange reaction, and (*2*) reaction with the acetal of a dialdehyde or ketal of a diketone.

$$2\ \text{cis-HOCH}_2\text{CH=CHCH}_2\text{OH} + \text{OHC–C}_6\text{H}_4\text{–CHO} \xrightarrow[\text{benzene}]{H^{\oplus},\Delta} \mathbf{18} + 2H_2O \quad (20)$$

These compounds are useful as chain transfer agents during the polymerization of vinyl monomers (16).

c. *2-Alkenyl- and 2-Alkadienyl-4,7-dihydro-1,3-dioxepins*

Brannock and Lappin (3) prepared 2-propenyl-4,7-dihydro-1,3-dioexpin by allowing *cis*-2-butene-1,4-diol to react with crotonaldehyde. Sterling, Watson, and Pawloski (10) found these derivatives could be prepared by condensation of the diol with an unsaturated ketone or aldehyde and ketal. Table V-7 lists some of the results from various preparations of these derivatives. In many instances only polymeric ethers are obtained.

The alkadienyl-1,3-dioxepins are also listed on Table V-7. These compounds were found to be chain transfer agents by Sterling and Pawloski (18).

d. *2-(Cyclohexen-1-yl)-4,7-dihydro-1,3-dioxepins*

Cyclohexen-1-yl-1,3-dioxepin derivatives can be prepared in high yields by the reaction of *cis*-2-butene-1,4-diol with a cyclohexene-1-carboxaldehyde, or with a ketal and the previously mentioned carboxaldehyde. For example (Eq. 21), condensation of the diol with 2,4,6-trimethyl-3-cyclohexene-1-carboxaldehyde in benzene with an acid catalyst gives an 84% yield of 2-(2,4,6-trimethyl-3-cyclohexen-1-yl)-4,7-dihydro-1,3-dioxepin (**19**). Reaction with acetone dimethyl ketal and 1,2,3,6-tetrahydrobenzaldehyde (Eq. 22) affords 2-(cyclohexen-1-yl)-4,7-dihydro-1,3-dioxepin (**20**) in 87% yield. Similarly, various other derivatives can be prepared by these methods. These compounds are useful as parasiticides and herbicides. They can be copolymerized with butadiene to give latex polymers and vinyl rubber products (18).

OH OH + CH_3–(H, CH_3, H, H, CH_3)–CHO $\xrightarrow[\text{benzene}]{H^{\oplus},\Delta}$ **19** + H_2O (21)

OH OH + $CH_3C(OCH_3)_2CH_3$ + (H)–CHO $\xrightarrow[\Delta]{H^{\oplus}}$ **20** + CH_3COCH_3 + $2CH_3OH$ (22)

e. *Spiro Derivatives*

Brannock and Lappin (3) found that cyclohexanone reacts with *cis*-2-butene-1,4-diol in the same manner as aldehydes to give a 73% yield of 7,12-dioxaspiro[5.6]dodec-9-ene. Sterling, Watson, and Pawloski (10) found these spiro derivatives can also be prepared by condensation of the diol with cyclohexanone dimethyl ketal, or a cyclic ketone and acetal. Derivatives made by the above methods are shown in Table V-8. These compounds are useful as parasiticides, bactericides, fungicides, and ascaricides. Their copolymers with butadiene have improved tensile, soft, lubricant, and elastic properties (20).

f. *Halogenated 1,3-Dioxepins*

The earliest reported halogeno-substituted 1,3-dioxepin derivatives were prepared by Brannock and Lappin (3) by dehydrobromination of 5,6-dibromo-4,7-dihydro-1,3-dioxepane (**12**) to yield 5-bromo-4,7-dihydro-1,3-dioxepin (**13**) and an ether derivative as a by-product (Eq. 14). Other types can be prepared by reaction of *cis*-2-butene-1,4-diol with halo benzaldehydes (7), halo acetals (9), halo aldehydes or ketones and ketals (10), and addition of the diol to an acetylenic halide (11). Table V-9 lists the known halo substituted 1,3-dioxepins. Best yields were produced by condensation of the diol with halo acetals, halo benzaldehydes, and halo aldehydes or ketones and acetals; poor yields from the addition to an acetylenic halide.

g. *x-(4,7-Dihydro-1,3-dioxepin-2-yl)phenols*

Sterling and Pawloski (21) found that *o*-hydroxybenzaldehyde, acetone dimethyl ketal, and *cis*-2-butene-1,4-diol would react in benzene with an acid catalyst to give a 57% yield of *o*-(4,7-dihydro-1,3-dioxepin-2-yl) phenol (**21**) (Eq. 23). Similarly, *p*-anisaldehyde afforded a 44% yield of 2-(*p*-methoxyphenyl)-4,7-dihydro-1,3-dioxepin. When *p*-hydroxybenzaldehyde was allowed to react with the diol in toluene in the presence of an acid catalyst, only a 10% yield of *p*-(4,7-dihydro-1,3-dioxepin-2-yl)phenol was obtained. *o*-Hydroxybenzaldehyde gave the same results. Generally polymers were obtained as products. Table V-10 lists a number of phenyl-substituted 1,3-dioxepin derivatives.

$$\text{(cis-2-butene-1,4-diol)} + \text{(o-hydroxybenzaldehyde)} + CH_3C(OCH_3)_2CH_3 \xrightarrow[\text{benzene}]{H^{\oplus},\Delta} \mathbf{21} + CH_3COCH_3 + 2CH_3OH \qquad (23)$$

These compounds are active as parastiticides for the control of insects, worms, bacterial or fungal organisms, and protozoan organisms. They also copolymerize with dienes to yield improved latex polymers and vinyl rubbers (21).

h. *Ether Derivatives*

Ether derivatives of 1,3-dioxepin can be prepared in good yields by the reaction of a trialkyl orthoester with *cis*-2-butene-1,4-diol in the presence of an acidic catalyst (see Section A-2-f), or by the reaction of an alkoxy acetal with *cis*-2-butene-1,4-diol as shown in Eq. 24 (see also Section A-2-c). Some care must be taken in selecting the alkoxy acetal for reaction because secondary alkoxy acetals such as 3-methoxybutyraldehyde dimethyl acetal give a mixture of the desired ether derivative and an unsaturated derivative which cannot be easily separated from the product.

$$\text{HOCH}_2\text{CH=CHCH}_2\text{OH} + \text{R'OCH}_2\text{CH(OR)}_2 \xrightarrow[\Delta]{\text{H}^{\oplus}} \text{2-H-2-(CH}_2\text{OR')-4,7-dihydro-1,3-dioxepin} + 2\,\text{ROH} \qquad (24)$$

Tables V-11 and V-12 list the products and yields from the various preparations. These compounds have been found to be selective herbicides (7).

Attempts to prepare ether derivatives by condensation of *cis*-2-butene-1,4-diol with an alkyl formate, or an alkyl formate and an acetal were unsuccessful (see Section A-2-k).

i. *Keto-Substituted 1,3-Dioxepins*

Sterling and Pawloski (22) found that condensation of *cis*-2-butene-1,4-diol with 4,4-dimethoxy-2-butanone (Eq. 25) gave a 73% yield of (4,7-dihydro-1,3-dioxepin-2-yl)-2-propanone (**22**). They also found that the double exchange reaction involving 2,3-butanedione and acetone dimethyl ketal affords a 75% yield of (4,7-dihydro-2-methyl-1,3-dioxepin-2-yl)methyl ketone. Similarly a 17% yield of 4-(4,7-dihydro-2-methyl-1,3-dioxepin-2-yl)-2-butanone was obtained from 2,5-hexanedione. Other derivatives could be prepared by using similar methods. These compounds are useful as parasiticides and as constituents in latexes and vinyl rubber products (22).

$$\text{HOCH}_2\text{CH=CHCH}_2\text{OH} + \text{CH}_3\text{COCH}_2\text{CH(OCH}_3)_2 \xrightarrow[\Delta]{\text{H}^{\oplus}} \underset{\mathbf{22}}{\text{2-H-2-(CH}_2\text{COCH}_3\text{)-4,7-dihydro-1,3-dioxepin}} + 2\,\text{CH}_3\text{OH} \qquad (25)$$

j. *2-Furyl and 2-Pyranyl-4,7-dihydro-1,3-dioxepins*

The reaction of *cis*-2-butene-1,4-diol with furaldehyde and acetone dimethyl ketal (Eq. 26) gives a 53% yield of 2-furyl-4,7-dihydro-1,3-dioxepin (**23**). Similarly, 2-(3,4-dihydro-2*H*-pyran-2-yl)-4,7-dihydro-1,3-dioxepin is prepared from 3,4-dihydro-2*H*-pyran-2-carboxaldehyde in a 17% yield. Other derivatives can be prepared in a similar manner, or by reaction of the diol with a derivative of the desired aldehyde. Sterling and Pawloski (23) found these compounds to be parasiticides and useful in the manufacture of improved latex polymers and vinyl rubber products.

$$\text{(HOCH}_2\text{CH=CHCH}_2\text{OH)} + CH_3C(OCH_3)_2CH_3 + \text{2-furaldehyde (CHO)} \xrightarrow[\Delta]{H^{\oplus}} \textbf{23} + 2CH_3OH + CH_3\overset{O}{\overset{\|}{C}}CH_3 \quad (26)$$

OH OH; **23**

k. *Esters of 4,7-Dihydro-1,3-dioxepin-2-alkylcarboxylic Acids*

Esters of 4,7-dihydro-1,3-dioxepin-2-alkylcarboxylic acids are not easily prepared in high yields. The reaction of *cis*-2-butene-1,4-diol with butyl levulinate or allyl levulinate in benzene in the presence of an acid catalyst gives none of the desired ester (7). Levulinic acid butyl ester dimethyl ketal gives only a 2% yield of butyl 2-methyl-4,7-dihydro-1,3-dioxepin-2-propionate. This product can also be obtained in 7% yield from acetone dimethyl ketal, butyl levulinate, and the diol. Higher yields are obtained from alkyl acetoacetates. The reaction of ethyl acetoacetate, acetone dimethyl ketal, and the diol gives a 66% yield of ethyl 4,7-dihydro-2-methyl-1,3-dioxepin-2-acetate (**24**) (Eq. 27). Similarly the methyl ester is obtained from methyl acetoacetate. Pawloski and Sterling (24) found these ester derivatives to be parasiticides, being particularly useful in the control of ticks.

$$\text{(HOCH}_2\text{CH=CHCH}_2\text{OH)} + CH_3\overset{O}{\overset{\|}{C}}CH_2\overset{O}{\overset{\|}{C}}OC_2H_5 + CH_3C(OCH_3)_2CH_3 \xrightarrow[\text{benzene}]{H^{\oplus},\Delta} \text{2-}CH_3\text{, 2-}CH_2\overset{O}{\overset{\|}{C}}OC_2H_5\text{-4,7-dihydro-1,3-dioxepin } (\textbf{24}) + 2CH_3OH + CH_3\overset{O}{\overset{\|}{C}}C\ldots$$

OH OH; CH₃ CH₂COC₂H₅; **24**

l. *Miscellaneous Derivatives*

Attempts to prepare an alcohol derivative of 1,3-dioxepin by allowing *cis*-2-butene-1,4-diol to react with aldol in benzene in the presence of an acid catalyst resulted only in the formation of crotonaldehyde (7). Low yields (3–5%) of 2-methyl-4,7-dihydro-1,3-dioxepin-2-methylol were obtained from the reaction of acetol, acetone dimethyl ketal, and the diol, or by the addition of the diol to propargyl alcohol (see Section A-2-h).

Odaira and co-workers (33) found that by using a 450-W high-pressure mercury arc as an irradiation source, ethyl cyanoformate could be condensed with furan to form unique bridged dioxepins (Eq. 28). A 16.2% yield of the bridged dioxepin was obtained and a 8.1% yield of the bridged 1,4-dioxepin.

$NC\overset{O}{\overset{\|}{C}}OC_2H_5$ + [furan] $\xrightarrow{h\nu}$ [bridged 1,3-dioxepin, CN, OC_2H_5] + [bridged 1,4-dioxepin, CN, OC_2H_5] (28)

4. Identification of 1,3-Dioxepins

Brannock and Lappin (3) confirmed the structure of 4,7-dihydro-1,3-dioxepin by hydrogenating it and comparing the spectrum of the product with a known sample of 1,3-dioxepane. Further proof was obtained by hydrolysis of the 1,3-dioxepane with 33% sulfuric acid and distillation of a volatile product whose ir spectrum was identical to that of an authentic sample of tetrahydrofuran.

5. Reactions of 1,3-Dioxepins

a. *Halogen Addition to the Double Bond*

The double bond in 1,3-dioxepin will undergo halogen addition in the same manner as in other unsaturated hydrocarbons. In the addition of bromine to 4,7-dihydro-1,3-dioxepin (Eq. 29), first described by Brannock and Lappin (3), reaction in carbon tetrachloride at ice-bath temperatures gave a 76% yield of 5,6-dibromo-4,7-dihydro-1,3-dioxepane (**25**). Pattison (4) obtained the same product by using chloroform as the solvent and a temperature of −55°. He also added chlorine to 4,7-dihydro-1,3-dioxepin, isolating 5,6-dichloro-4,7-dihydro-1,3-dioxepane (**26**) in 75% yield.

$$\text{4,7-dihydro-1,3-dioxepin} + X_2 \xrightarrow{X=Cl, Br} \text{5,6-dihalo-1,3-dioxepane} \quad (29)$$

25; X = Br
26; X = Cl

b. *Hydrogenation of the Double Bond*

Addition of hydrogen to the double bond in 1,3-dioxepins proceeds readily in yields of 85% or better. For example (Eq. 30), Brannock and Lappin (3) found that 4,7-dihydro-1,3-dioxepin would undergo hydrogenation over Raney nickel at 1000 psi and 100° to 4,7-dihydro-1,3-dioxepane (**27**). Other 1,3-dioxepin derivatives could be hydrogenated similarly.

$$\text{1,3-dioxepin} + H_2 \xrightarrow[100°, \text{ Raney Ni}]{1000 \text{ psi}} \text{1,3-dioxepane } (\mathbf{27}) \quad (30)$$

c. *Reactions with Acids*

1,3-Dioxepins should react with aqueous sulfuric acid to yield dihydrofuran and a carbonyl compound (Eq. 31). Although no mention of this reaction has been made in the literature, 4,7-dihydro-1,3-dioxepane does undergo this fragmentation in 33% sulfuric acid to tetrahydrofuran and formaldehyde (3). Weaker acid solutions should yield *cis*-2-butene-1,4-diol and a carbonyl compound (Eq. 32).

$$\text{1,3-dioxepin} \xrightarrow[\Delta]{33\% \ H_2SO_4} \text{dihydrofuran} + HCH{=}O \quad (31)$$

$$\text{1,3-dioxepin} \xrightarrow[\Delta]{\text{dilute } H^{\oplus}} \text{HOCH}_2\text{CH=CHCH}_2\text{OH} + HCH{=}O \quad (32)$$

The reaction of a 1,3-dioxepin with 1*N* methanolic hydrogen chloride solution should produce *cis*-2-butene-1,4-diol and an acetal (Eq. 33) in the same manner as does 5,6-dibromo-4,7-dihydro-1,3-dioxepane (3).

$$\text{1,3-dioxepin} + 2CH_3OH \xrightarrow[\Delta]{HCl} \text{HOCH}_2\text{CH=CHCH}_2\text{OH} + (CH_3O)_2CH_2 \quad (33)$$

1,3-Dioxepins will react with a carboxylic acid to form a half ester and carbonyl compound. For example (Eq. 34), treatment of 2,2-dimethyl-4,7-dihydro-1,3-dioxepin with acetic acid produces 4-hydroxy-2-butenyl acetate and acetone (7)

$$\text{(2,2-dimethyl-4,7-dihydro-1,3-dioxepin)} + \text{AcOH} \xrightarrow{\Delta} \text{HO–CH}_2\text{CH=CHCH}_2\text{–OAc} + CH_3\overset{O}{\overset{\|}{C}}CH_3 \qquad (34)$$

d. *Diels-Alder Reactions*

(1) Condensation with Cyclopentadienes. Hausweiler and co-workers (25) found that 4,7-dihydro-1,3-dioxepin would undergo a Diels-Alder reaction with hexachlorocyclopentadiene in xylene at 140° under nitrogen (Eq. 35) giving 6,7,8,9,10,10-hexachloro-1,5,5a,6,9,9a-hexahydro-6,9-methano-2,4-benzodioxepin (**28**) in 90% yield.

$$\text{(hexachlorocyclopentadiene)} + \text{(4,7-dihydro-1,3-dioxepin)} \xrightarrow[\text{140°, 24 hr.}]{\text{xylene}} \mathbf{28} \qquad (35)$$

Further chlorination of this product yields an insecticide commercially known as Bayer 38920 (26) which is very active against caterpillars, browntail moths, gypsy moths, and ants (25).

(2) 6,9-Methano-2,4-benzodioxepin Derivatives. A number of 6,9-methano-2,4-benzodioxepin (R.I. 10286) derivatives were prepared by Hausweiler and co-workers (27) from substituted 1,3-dioxepins and cyclopentadienes. These compounds are listed in Table V-13 together with the reactants used for their preparation. Physical properties are also tabulated.

e. *Polymerization*

The copolymerization of trioxane and 4,7-dihydro-1,3-dioxepin in the presence of a Lewis acid, such as boron trifluoride etherate or a diazonium salt, to yield high molecular weight products has been described (28). Similarly copolyacetals are made from trioxane, a cyclic ether, and 4,7-dihydro-1,3-dioxepin. Crystalline products are obtained by using less than

10% by weight of 1,3-dioxepin (or cyclic formal or cyclic ether) and trioxane with the previously mentioned catalysts. With higher proportions the products are amorphous and elastic, and may cross-link. Cyclic ethers, such as ethylene oxide and epichlorohydrin, work equally well. The terpolymers are elastic, stable masses.

Jennings and Rose (34) also made copolymers of 1,3-dioxepin and trioxane in the presence of an electrophilic catalyst. A film cast from the copolymer at 180–185° was tough, highly crystalline, and had a melting range of 160–166°. Other copolymers were prepared with 2-isopropyl-4,7-dihydro-1,3-dioxepin, 4,4,7,7-tetramethyl-1,3-dioxepin, and 4,7-dimethyl-4,7-diethyl-1,3-dioxepin.

Pawloski and Sterling found that many substituted 1,3-dioxepins copolymerize with such dienes as butadiene to form latexes and vinyl rubber products with desirable and improved tensile, lubricant, soft, and elastic properties (19–24,30).

6. Physical Properties of 1,3-Dioxepins

1,3-Dioxepins are stable to alkaline conditions. Acids cause the formation of 2,5-dihydrofuran, a carbonyl compound, and some polymer. Rondestvedt and Mantell (29) found these compounds quite stable to heat. 2-Isopropyl-4,7-dihydro-1,3-dioxepin remained unchanged when heated to 390–403° over pumice. Table V-14 lists the physical properties of a number of substituted 1,3-dioxepins. Physical properties of a different type are listed in Table V-15.

B. Condensed Dioxepins

1. Benzodioxepins

a. *Nomenclature and Structures*

Six basic ring benzodioxepin systems will be considered within the scope of this review: 5*H*-1,2-benzodioxepin (**29**), 2*H*-1,3-benzodioxepin (R.I. 12071) (**30**), 5*H*-1,4-benzodioxepin (R.I. 1835) (**31**), 2*H*-1,5-benzodioxepin (R.I. 1836) (**32**), 1*H*-2,3-benzodioxepin (R.I. 1837) (**33**), and 3*H*-2,4-benzodioxepin (R.I. 10084) (**34**). Of these, only derivatives of 5*H*-1,2-benzodioxepin are still unknown to date. The numbering of the ring system is shown in **29**. All other rings are numbered similarly. These compounds are sometimes referred to in the literature as benzodioxacycloheptanes.

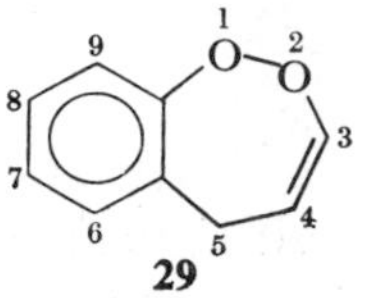

29 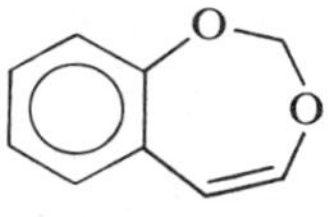 **30** 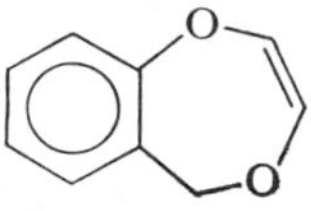 **31**

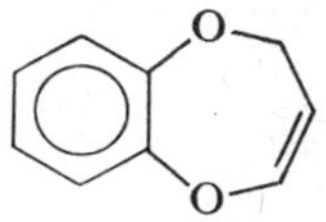

32 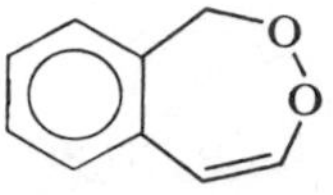 **33** 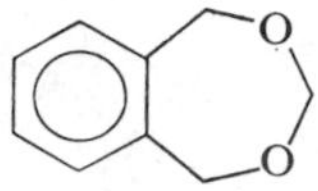 **34**

b. 2H-*1,3-Benzodioxepins*

A derivative of 2*H*-1,3-benzodioxepin was prepared by Zaugg and Michaels (44) on treatment of methyl *trans*-bromomethyl-4-phenyl-4-chromancarboxylate with 10% aqueous potassium hydroxide at reflux temperatures for 48 hr (Eq. 36). The bridged product, 4,5-dihydro-2,5-methano-2-methyl-5-phenyl-1,3-benzodioxepin-4(2*H*)-one (**35**) (R.I. 12521), was obtained as a glass. Recrystallization from ethanol gave a solid product (mp 132–133°C). The proton magnetic resonance spectrum was determined.

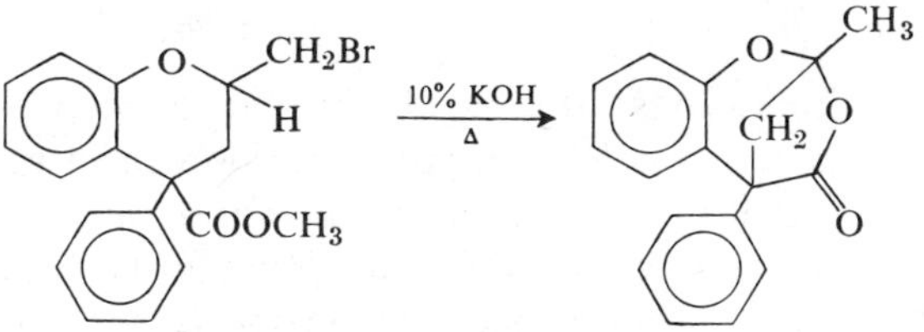

CH_2Br H $COOCH_3$ —10% KOH, Δ→ CH_3 CH_2 **35** (36)

c. 5H-*1,4-Benzodioxepins*

(1) Derivatives Found in Nature. Alertsen (45) established the structur of ageratochromene by degradation and synthesis of a yellow oil obtained on steam distillation of the whole plant *A. houstonianum*. The product, 2,3-dihydro-3,5,7,8-tetramethoxy-2,2-dimethyl-5*H*-1,4-benzodioxepin (**36**) (mp 120.5–123.5°), is the only known example of a 5*H*-1,4-benzodioxepin other than ketone derivatives.

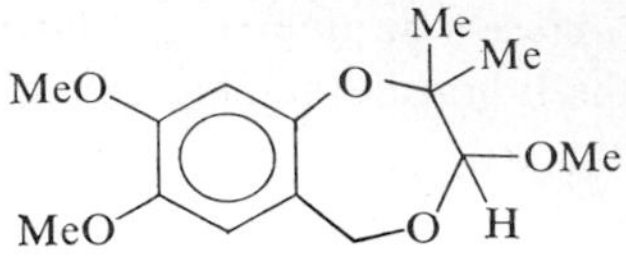

36

(2) METHODS OF PREPARATION. It may be possible to prepare 2,3-dihydro-5*H*-1,4-benzodioxepin (**37**) by allowing 2-hydroxymethylphenol to react with 1,2-dibromoethane under basic conditions (Eq. 37).

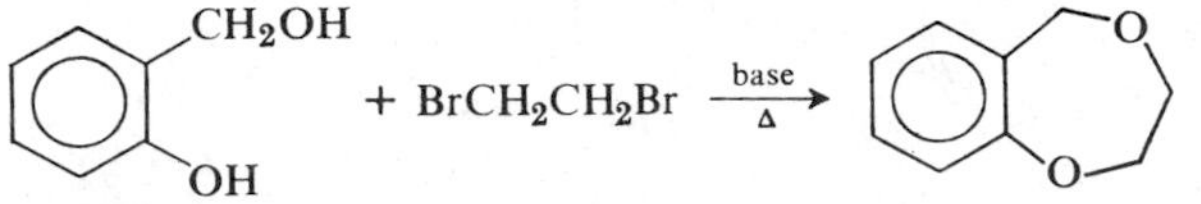

$$+ BrCH_2CH_2Br \xrightarrow[\Delta]{\text{base}} \quad (37)$$

37

Another type of 1,4-benzodioxepin (Eq. 38) was obtained by Yale and co-workers (46) on treatment of 3-(*o*-formylphenoxy)-1,2-propanediol with ethanol and ammonium chloride at reflux temperatures which yielded 3,6-epoxy-2,3,4,6-tetrahydro-1,5-benzodioxocin (**38**) (bp 113–114°C/1.0 mm, mp 63–64°C) (R.I. 3149).

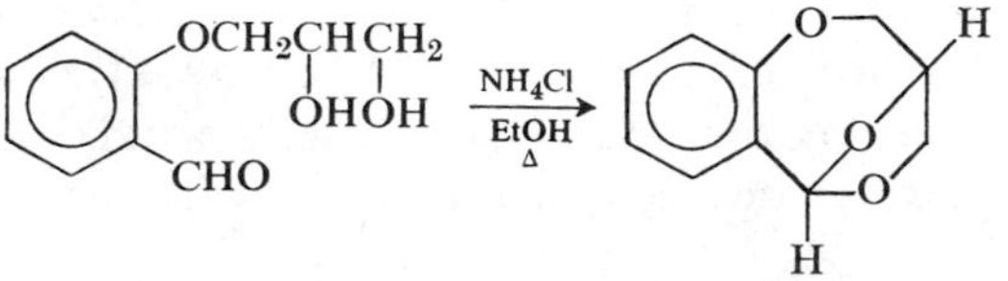

$$\xrightarrow[\text{EtOH, } \Delta]{NH_4Cl} \quad (38)$$

38

This compound was also isolated as a by-product by Stephenson (47) from the condensation of salicylaldehyde with epichlorohydrin at 100° for 6 hr in the presence of a small amount of pyridine as a catalyst (Eq. 39). The desired product was *o*-(3-chloro-2-hydroxypropyl)benzaldehyde. The by-product reacted with 2,4-dinitrophenylhydrazine to yield the 2,4-dinitrophenylhydrazone derivative of *o*-(2,3-dihydroxypropyl)benzaldehyde.

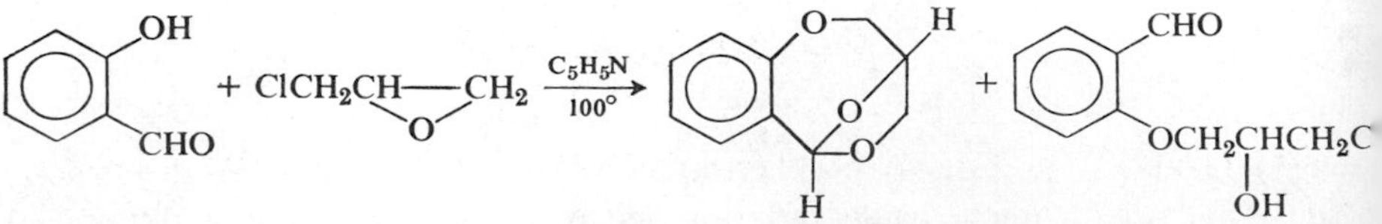

$$+ ClCH_2CH\!-\!CH_2 \xrightarrow[100^\circ]{C_5H_5N} \textbf{38} + \quad (39)$$

(3) 5*H*-1,4-BENZODIOXEPIN-(*x*)-ONES Weizmann and co-workers (48) found a cyclic ether (mp 82°C) as a by-product from the reaction of the sodium salt of salicyclic acid with ethylene chlorohydrin at 140°C in the

presence of Cu–bronze as a catalyst. This could be 2,3-dihydro-5*H*-1,4-benzodioxepin-5-one (**39**). As indicated in Eq. 40, the main product was 2-hydroxyethyl salicylate.

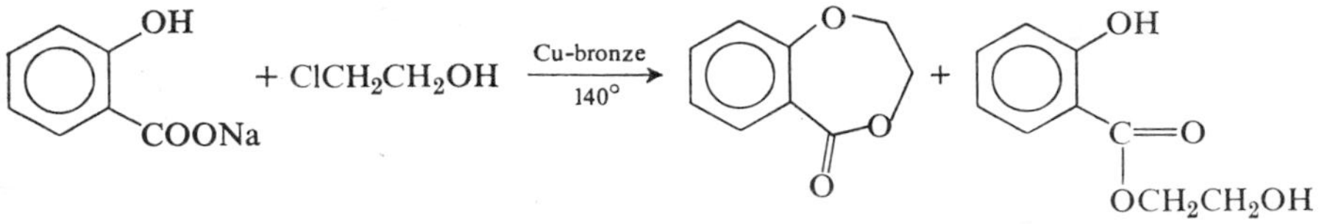

OH
COONa
+ ClCH2CH2OH
Cu-bronze
140°
O
O
O
+
OH
C=O
OCH2CH2OH

39

(40)

Dawkins and Mulholland (49) synthesized another type of ketone derivative by reaction of the methyl ester of 2-acetyl-6-chloro-3,5-dimethoxyphenoxyacetic acid with 3*N* HCl at reflux temperatures. The product was identified as 9-chloro-2,3-dihydro-5-hydroxy-6,8-dimethoxy-5-methyl-5*H*-1,4-benzodioxepin-3-one (**40**) (mp 143–144°C). This compound, when treated with diazomethane in ether (Eq. 41), yielded 9-chloro-2,3-dihydro-5,6,8-trimethoxy-5-methyl-5*H*-1,4-benzodioxepin-3-one (mp 97°C). The structures of **40** and **41** were determined on the basis of ir and uv spectra.

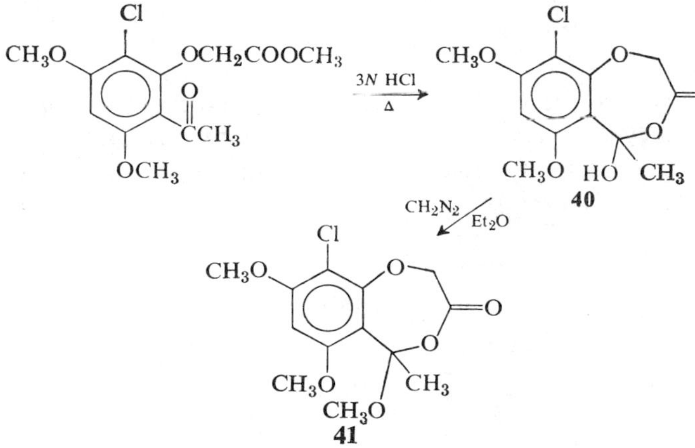

Cl
CH3O
OCH2COOCH3
O
CCH3
OCH3
3N HCl
Δ
Cl
CH3O
O
O
O
CH3O
HO
CH3
40
CH2N2
Et2O
Cl
CH3O
O
O
O
CH3O
CH3
CH3O
41

(41)

Bokhari and Whalley (50) found that 2-carboxymethoxy-4,6,2′,4′,6′-pentamethoxydesoxybenzoin reacted with phosphorus pentoxide in benzene at reflux temperatures (Eq. 42) to yield 2,3-dihydro-6,8-dimethoxy-5-(2,4,6-trimethoxybenzylidene)-5*H*-1,4-benzodioxepin-3-one (**42**) (mp 240°C). Hydrolysis of **42** with boiling aqueous–methanolic 20% potassium hydroxide solution for 15 min regenerated the starting acid.

(4) 5*H*-1,4-BENZODIOXEPIN-(*x*,*y*)-DIONES. An early preparation of a dione derivative of 5*H*-1,4-benzodioxepin discovered by Perkins and Yates (51) involved heating 2-carboxy-5,6-dimethoxyphenoxy acetic acid in acetic

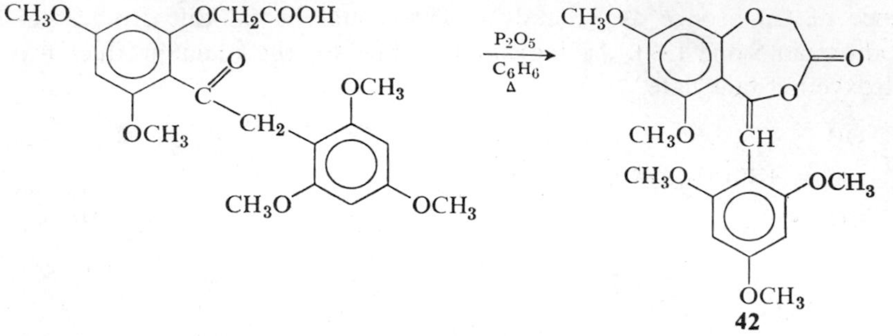

(42)

42

anhydride (Eq. 43) to yield 2,3-dihydro-8,9-dimethoxy-5*H*-1,4-benzodioxepin-3,5-dione (**43**) (mp 175°C), the anhydride of the starting acid.

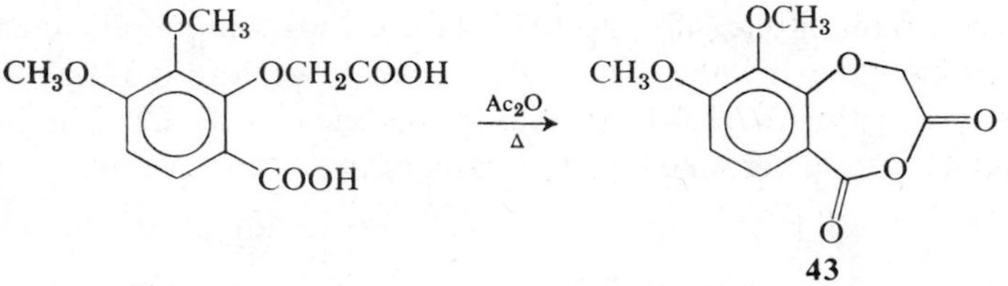

(43)

43

Another dione was prepared by Kagan and Birkenmeyer (52). They treated chloroacetylsalicylic acid with sodium iodide in anhydrous acetone at reflux temperatures and treated the resulting iodide with trimethyl amine to obtain 2,3-dihydro-5*H*-1,4-benzodioxepin-2,5-dione (**44**) (mp 115–115.5°C) (Eq. 44).

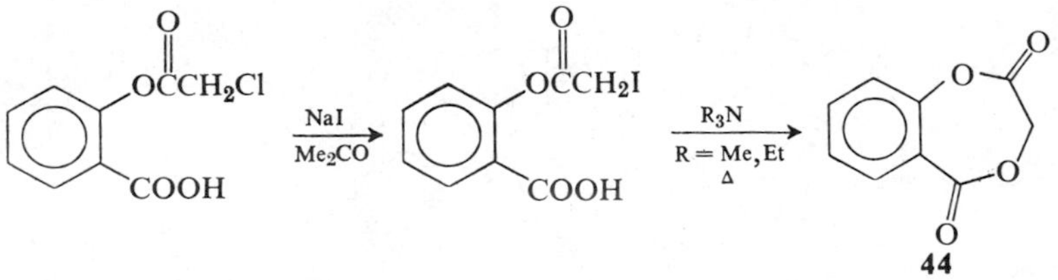

(44)

44

A patent issued to Kagan and Birkenmeyer (53) discloses that other amines such as *N*,*N*-dialkylanilines, pyridines, quinolines, and alkylpiperidines can also be used as dehydrohalogenation agents. The reaction must be carried out under substantially anhydrous conditions, otherwise the desired product will be converted into salicyloylglycolic acid. The uv absorption spectrum of **44** (in ethanol solution) exhibits maxima at 239 and 307 mμ. The ir absorption spectrum (mineral oil mull) exhibits maxima at 1795, 1730, 1613, 1586, 1320, 1223, 1205, 797, 785, 765, 702, and 685 cm^{-1}. The product is an anti-inflammatory agent.

2,3-Dihydro-5*H*-1,4-benzodioxepin-2,5-dione reacts with water at 20° to yield salicyloylglycolic acid (Eq. 45). It also reacts with primary and secondary amines to form 1-*R*-carbonylmethyl salicylates. For example (Eq. 45), pyrrolidine reacts with the dione in acetone under anhydrous conditions to yield 1-pyrrolidinylcarbonylmethyl salicylate. Similarly other amines such as methyl amine, diethyl amine, cyclohexyl amine, aniline, morpholine, piperidine, piperazine, and ethylenediamine should react to yield 1-*R*-carbonylmethyl salicylates. These compounds exhibit keratolytic and antiviral activity (53).

44 —H_2O→ OH / $COOCH_2COOH$; —pyrrolidine→ OH / $COOCH_2\overset{O}{\overset{\|}{C}}N$ (45)

(5) PHYSICAL PROPERTIES. The physical properties of 5*H*-1,4-benzodioxepins are listed in Table V-16.

d. 2H-*1,5-Benzodioxepins*

(1) ALKYL DERIVATIVES. One of the earliest preparations of a 2*H*-1,5-benzodioxepin was accomplished by Ziegler and co-workers (54) by reaction of 3-bromopropyl-*o*-hydroxyphenyl ether with potassium carbonate in boiling amyl alcohol (Eq. 46). They also found that condensation of 1,3-dibromopropane with 1,2-dihydroxybenzene in the presence of sodium methoxide in boiling methanol (Eq. 47) yielded the same product (**45**) (55). It was similarly prepared by Thompson (56) for use as an intermediate in preparing other 1,5-benzodioxepin derivatives. Cass and co-workers (57) used a similar

OH / $OCH_2CH_2CH_2Br$ —K_2CO_3, *n*-AmOH→ **45** (46)

OH / OH + $BrCH_2$–CH_2–$BrCH_2$ —K_2CO_3 or NaOMe, MeOH→ **45** (47)

procedure but substituted potassium carbonate for the sodium methoxide, They measured heats of combustion and derived the resonance energies. 2*H*-1,5-Benzodioxepin is useful as an intermediate for making perfumes.

Hartmut and co-workers (58) found that a similar reaction using 1,3-dichloro-2-methylenepropane gave an unsaturated 1,5-benzodioxepin as shown in Eq. 48. Refluxing 1,2-dihydroxybenzene and 1,3-dichloro-2-methylenepropane in ethanol for 24 hr gave a 67% yield of 3-methylene-2,4-dihydro-3*H*-1,5-benzodioxepin (**46**) (bp 113–113.5°C/10 mm). Hydrogenation of this product gave 3-methyl-2,4-dihydro-3*H*-1,5-benzodioxepin (**47**) (bp 111–112°C/12 mm).

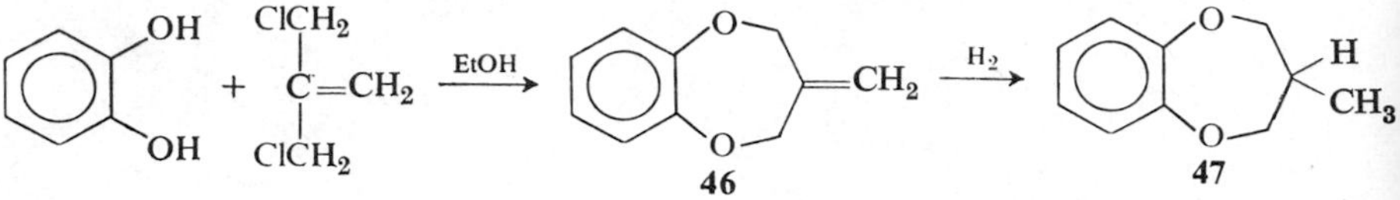

(48)

(2) Aldehydes. Tomita and Takahashi (59) found that 3,4-dihydroxybenzaldehyde reacted with 1,3-dibromopropane and sodium methoxide in methanol at 120° in an autoclave (Eq. 49) to yield 3,4-dihydro-2*H*-1,5-benzodioxepin-7-carboxaldehyde (**48**) as a yellow oil (semicarbazone, mp 188–190°C).

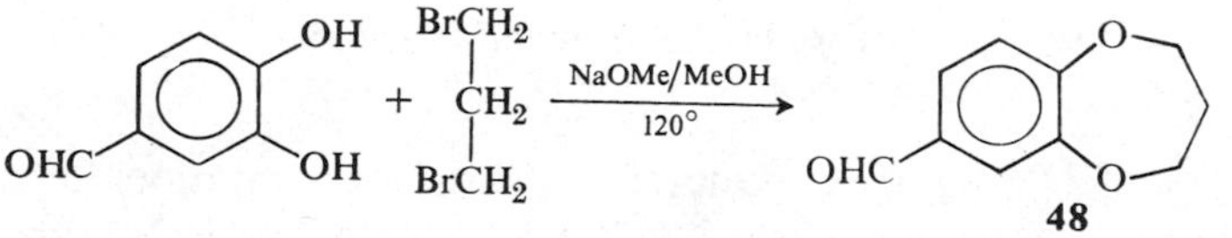

(49)

Baddeley and Smith (61) also prepared this aldehyde from 3,4-dihydro-2*H*-1,5-benzodioxepin by the Gattermann method (60) (bp 127–129°C/0.25 torr) (semicarbazone, mp 183–184°C).

(3) Ketones. In a Belgian patent (62) the reaction of bromine with 7-acetyl-3,4-dihydro-2*H*-1,5-benzodioxepin (**49**) in carbon disulfide to yield

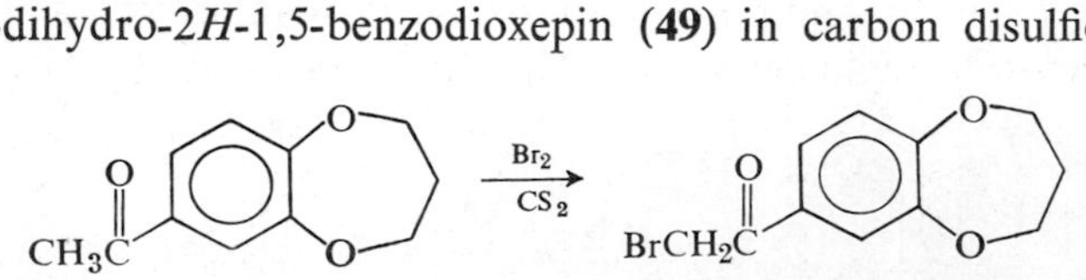

(50)

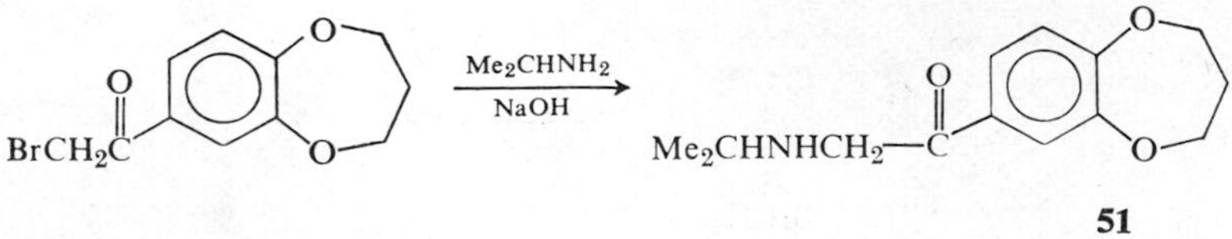

(51)

7-(bromoacetyl)-3,4-dihydro-2*H*-1,5-benzodioxepin (**50**) was disclosed (Eq. 50). No mention of the preparation of the starting acetyl derivative was made. The bromoacetyl derivative, when treated with isopropylamine in methanol (Eq. 51), yielded 3,4-dihydro-7-(isopropylaminoacetyl)-2*H*-1,5-benzodioxepin (**51**), isolated as a hydrochloride salt (mp 204–205°C).

(4) ALCOHOLS. An alcohol derivative of 1,5-benzodioxepin was prepared by Rosnati and De Marchi (63) who condensed 3,4-dihydro-2*H*-1,5-benzodioxepin-3-one (**52**) with methylmagnesium iodide in ether (Eq. 52) to form 3-hydroxy-3-methyl-3,4-dihydro-2*H*-1,5-benzodioxepin (**53**) (mp 77–79°C).

$$\mathbf{52} \xrightarrow[\text{Et}_2\text{O}]{\text{MeMgI}} \mathbf{53} \quad (52)$$

Another type of alcohol was prepared by Baddeley and Smith (61) by reduction of 3,4-dihydro-1,5-benzodioxepin-7-carboxaldehyde with lithium aluminum hydride. The product, 3,4-dihydro-2*H*-1,5-benzodioxepin-7-methanol, was isolated in the form of a phenylurethane derivative (mp 91–91.5°C).

An amino alcohol was prepared by reduction of 3,4-dihydro-7-isopropylaminoacetyl-2*H*-1,5-benzodioxepin hydrochloride (**51** · HCl) with sodium borohydride in methanol (Eq. 53). The product, 1-(3,4-dihydro-1,5-benzodioxepin-7-yl)-2-isopropylaminoethanol hydrochloride (**54** · HCl) (mp 163–164°C) (62), is an adrenergic agent and is of interest in the treatment and prevention of coronary afflictions.

$$\mathbf{51}\cdot\text{HCl} \xrightarrow[\text{MeOH}]{\text{NaBH}_4} \mathbf{54}\cdot\text{HCl} \quad (53)$$

HOCH—$CH_2NHCHMe_2\cdot HCl$

A recent patent (64) discloses the preparation of other amino alcohols by reaction of 6-(3-chloro-2-hydroxypropoxy)-3,4-dihydro-2*H*-1,5-benzodioxepin with primary amines in a closed vessel at 100° for 10 hr. When isopropylamine is used as a reactant, 6-(2-hydroxy-3-isopropylaminopropoxy)-3,4-dihydro-2*H*-1,5-benzodioxepin (mp 66–68°C) is obtained.

(5) HALIDE DERIVATIVES. Few halide derivatives of 1,5-benzodioxepin are known. Previously mentioned in this chapter are a 7-(bromoacetyl)

derivative (62) and a 6-(3-chloro-2-hydroxypropoxy) derivative (64). 7-Chloromethyl-3,4-dihydro-2*H*-1,5-benzodioxepin was prepared from 3,4-dihydro-2*H*-1,5-benzodioxepin-7-methanol by Baddeley and Smith (61). Leonard and Koo (65) used 3,4-dihydro-2*H*-1,5-benzodioxepin-2-carbonyl chloride as an intermediate for the preparation of the 2-carboxamide derivatives.

Other halo 1,5-benzodioxepins can be prepared by the method shown in Eq. 54. Since 1,3-dihydroxybenzene and 1,3-dibromopropane react in the presence of a base to form a 1,5-benzodioxepin, halo-substituted 1,2-dihydroxybenzenes should react in a similar manner.

Hal–C₆H₃(OH)₂ + $BrCH_2CH_2CH_2Br$ $\xrightarrow{K_2CO_3}$ Hal–(1,5-benzodioxepin) (54)

Baddeley and co-workers (72) have reported the bromination of 1,5-benzodioxepin without specifying the structure of the products.

(6) Amines. In 1954 Thompson (56) found that 7-amino-3,4-dihydro-2H-1,5-benzodioxepin is obtained on treatment of 7-nitro-3,4-dihydro-2H-1,5-benzodioxepin with hydrogen in methanol with nickel–diatomaceous earth as a catalyst. A number of secondary amine derivatives were prepared by Leonard and Koo (65) by allowing 3,4-dihydro-2*H*-1,5-benzodioxepin-2-carbonyl chloride (**55**) to react with a primary amine. The resultant carboxamide derivative (**56**), when treated with lithium aluminum hydride in ether at reflux temperatures for 48 hr, produced 50% yields of 2-aminoalkyl-3,4-dihydro-2*H*-1,5-benzodioxepins (**57**) (Eq. 55). Amines such as methylamine, isopropylamine, butylamine, allylamine, aminopyridines, morpholine, pyrrolidine, hydroxyalkylamines, piperazine, and piperidine were used. Table V-17 lists the products and their physical properties. These derivatives have pharmacological value.

55 (H, COCl) $\xrightarrow{RNH_2}$ **56** (H, CONHR) $\xrightarrow[Et_2O]{LiAlH_4}$ **57** (H, CH_2NHR) (55)

In patent disclosures (66, 67) Augstein and co-workers described the synthesis of compounds of the type shown in structure **58** (Table V-18). These compounds were prepared by reacting a 2-aminoalkyl-substituted

3,4-dihydro-2*H*-1,5-benzodioxepin with 1-amidino-3,5-dimethylpyrazole sulfate in water at reflux temperatures for 3 hr. For example, condensation of 2-aminomethyl-3,4-dihydro-2*H*-1,5-benzodioxepin with 1-amidino-3,5-dimethylpyrazole sulfate yielded bis(2-guanidinomethyl-3,4-dihydro-2*H*-1,5-benzodioxepin) sulfate (mp 255–258°C).

In another method disclosed by Augstein and co-workers (66, 67), treatment of a substituted 2-aminoalkyl-3,4-dihydro-2*H*-1,5-benzodioxepin with a suitable *S*-alkyl isothiouronium salt in an inert polar solvent at temperatures ranging from 20° to 120° for periods of 1 to 72 hr likewise yielded a guanidinium salt. Preferred reaction temperatures were 60–100° with reaction times of 1 to 20 hr. Approximately equimolar amounts of starting materials were employed. Isothiouronium salts with low-molecular weight *S*-alkyl group were preferred. Solvents such as water, lower alcohols, *N*,*N*-dialkylamides, such as *N*,*N*-dimethylformamide, dialkyl sulfoxides, and sulfones, and mixtures of these, were used as reaction media. For example, equal portions of 2-(*N*-methylaminomethyl)-3,4-dihydro-2*H*-1,5-benzodioxepin and 2-methyl-2-thiopseudourea sulfate were allowed to react in 30% aqueous ethanol at reflux temperatures for 48 hr to yield bis[2-(3-methylguandino)methyl-3,4-dihydro-2*H*-1,5-benzodioxepin] sulfate. These compounds (Table V-18) are used as antihypertensive agents.

Augstein and co-workers (66, 67) further found that other salts could be prepared by the aforementioned method. For example, reaction of 2-(*N*-methylaminomethyl)-3,4-dihydro-2*H*-1,5-benzodioxepin with 1,2,3-trimethyl-2-thiopseudourea hydroiodide in aqueous ethanol at reflux temperatures yielded *N*-methyl-2-(3-methylguanidino)methyl-3,4-dihydro-2*H*-1,5-benzodioxepin hydroiodide, which on treatment with 10*N* aqueous sodium hydroxide solutions yielded the free amine. A number of other such amines were obtained in the same manner (Table V-19). The hydrohalide salts were prepared by treatment of the free base with hydrogen halide gas in ether–acetone solution. Similarly the nitrate, sulfate or bisulfate, phosphate or acid phosphate, acetate, lactate, citrate, tartrate or bitartrate, oxalate, succinate, maleate, gluconate, saccharate, methanesulfonate, and benzenesulfonate salts of each of the 2-guanidinoalkyl-3,4-dihydro-2*H*-1,5-benzodioxepins can be prepared. These salts are antihypertensive agents.

(7) Amides. Leonard and Koo (68) found that 1,2-dihydroxybenzene, potassium carbonate, and an appropriate dibromoalkylcarboxamide react in acetone at reflux temperatures to yield amide derivatives of 1,5-benzodioxepin (Eq. 56). They also found that these amides can be prepared by condensing 3,4-dihydro-2*H*-1,5-benzodioxepin-2-carbonyl chloride with an amine (65) (see Section B-1-d-(6), or by allowing an amine derivative such as 3,4-dihydro-2*H*-1,5-benzodioxepin-2-ethylamine to react with a carbonyl

chloride such as 3,4,5-trimethoxybenzoyl chloride (see, for example, Eq. 57) (69). Table V-20 lists a number of 1,5-benzodioxepin amide derivatives. These compounds are useful as tranquilizers and sedatives.

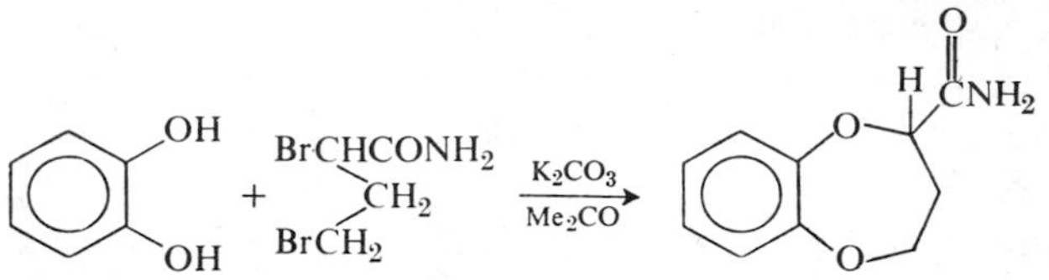

(56)

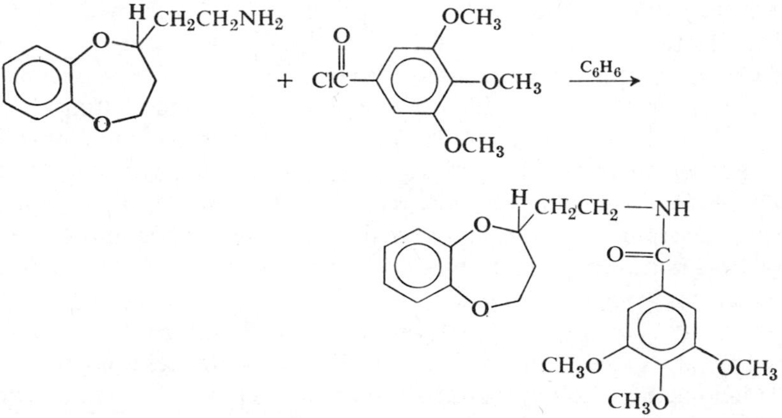

(57)

(8) ESTERS. Although no carboxylic acids of 1,5-benzodioxepin are known, a few ester derivatives have been prepared. Leonard and Koo (68) found that these could be prepared by reaction of 1,2-dihydroxybenzene, potassium carbonate, and an appropriate dibromoalkyl ester in acetone. Ethyl 2,4-dibromobutyrate was converted in this manner (Eq. 58) into ethyl 3,4-dihydro-2*H*-1,5-benzodioxepin-2-carboxylate (**59**) (bp 112–121°C/0.5 torr, n_D^{25} 1.5203) in 30% yield. Other derivatives could be prepared similarly.

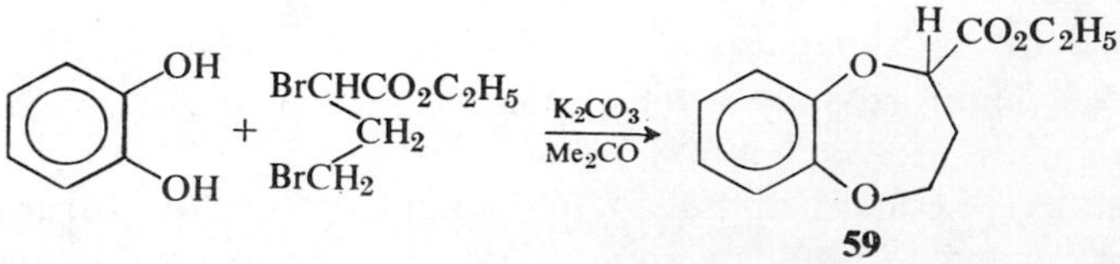

(58)

59

(9) PHENOLS. Thompson (56) has shown that treatment of 7-amino-3,4-dihydro-2*H*-1,5-benzodioxepin (**60**) with 10% sulfuric acid and sodium nitrite in water at 0–5°C, followed by reaction of the diazonium salt solution with boiling copper sulfate solution, yields 7-hydroxy-3,4-dihydro-2*H*-1,5-benzodioxepin (**61**) (mp 72–75°C) (Eq. 59). Alkylation of this phenol derivative

with *tert*-butyl alcohol in phosphoric acid–acetic anhydride–glacial acetic acid solution at 70–80° afforded 8-*tert*-butyl-7-hydroxy-3,4-dihydro-2*H*-1,5-benzodioxepin (**62**) (mp 133–135°C) (Eq. 60). These compounds are useful as intermediates for the preparation of medicinals and as oxidation inhibitors (70).

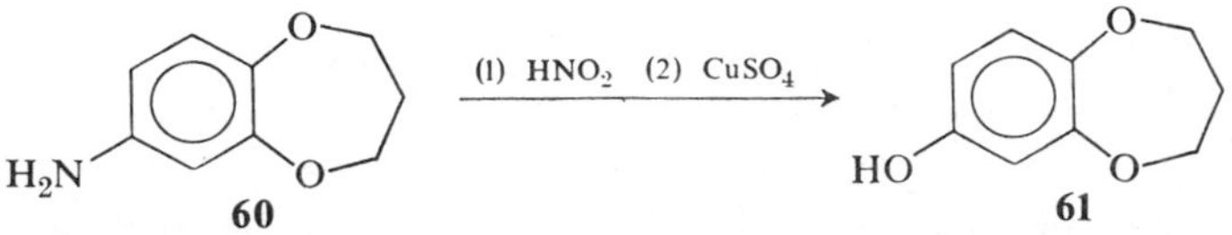

$$\mathbf{60}\ \xrightarrow{(1)\ HNO_2\quad(2)\ CuSO_4}\ \mathbf{61} \tag{59}$$

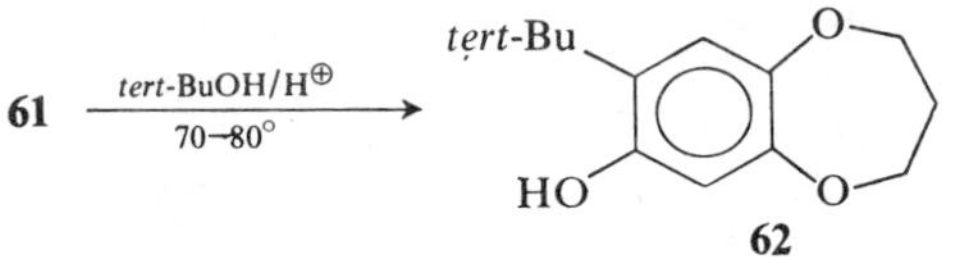

$$\mathbf{61}\ \xrightarrow[70\text{–}80^\circ]{tert\text{-}BuOH/H^{\oplus}}\ \mathbf{62} \tag{60}$$

(10) Miscellaneous Derivatives. Thompson (56) found that 7-nitro-3,4-dihydro-2*H*-1,5-benzodioxepin (**63**) (mp 106–107°C) can be prepared by allowing 3,4-dihydro-2*H*-1,5-benzodioxepin (**45**) to react with nitric acid in glacial acetic acid (Eq. 61). Treatment of 3,4-dihydro-2*H*-1,5-benzodioxepin-7-carboxaldehyde (**48**) with nitromethane in ethanol (Eq. 62) yields 7-nitrovinyl-3,4-dihydro-2*H*-1,5-benzodioxepin (**64**) as a yellow solid (mp 92–93°C). Tomita and Takahashi (59) used this compound as an intermediate for the preparation of other dioxepino derivatives.

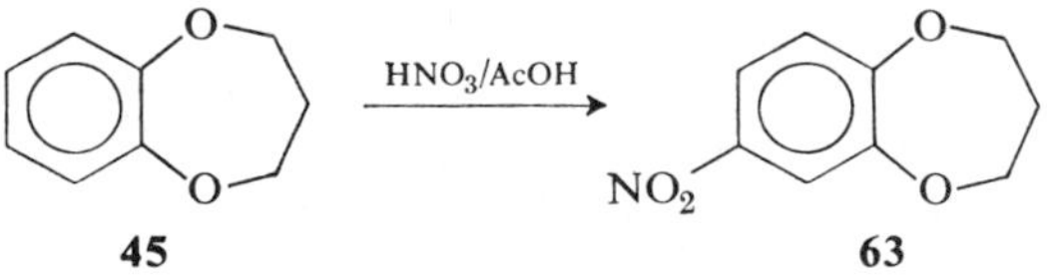

$$\mathbf{45}\ \xrightarrow{HNO_3/AcOH}\ \mathbf{63} \tag{61}$$

$$\mathbf{48}\ \xrightarrow[0\text{–}3^\circ]{MeNO_2/EtOH}\ \mathbf{64} \tag{62}$$

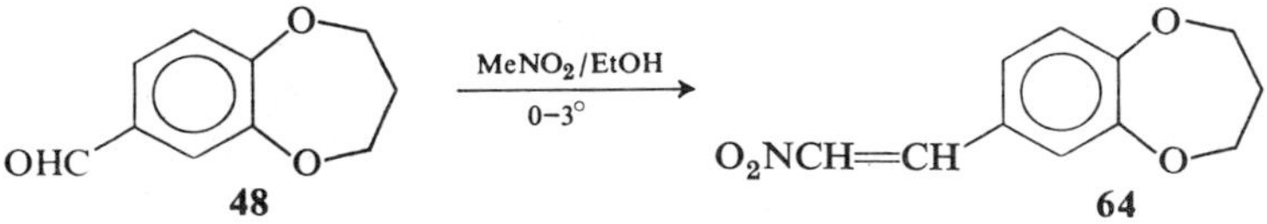

(11) 1,5-Benzodioxepin-3-ones. Rosnati and De Marchi (63) isolated a small amount of a crystalline product other than the desired 1,4-benzodioxane-2-carboxylic acid from the oxidation of 2-hydroxymethyl-1,4-benzodioxane with alkaline permanganate. On further investigation they found that this compound could also be formed from a by-product of the preparation of 2-hydroxymethyl-1,4-benzodioxane from 1,2-dihydroxybenzene and 1,3-dichloro-2-hydroxypropane. On the basis of the nmr spectrum the structure of the unknown compound was established as 2,4-dihydro-3*H*-1,5-benzodioxepin-3-one (mp 40–42°C). The *p*-nitrophenylhydrazone (mp

210–211°C dec), 2,4-dinitrophenylhydrazone (mp 164° dec), and semicarbazone (mp 230–233°C dec) derivatives were characterized. As further proof of the structure, 2,4-dihydro-3*H*-1,5-benzodioxepin was prepared by reduction of the semicarbazone derivative with potassium hydroxide, hydrazine hydrate, and diethylene glycol at 160–170°. The spectrum of this product was compared with that of an authentic sample prepared via an alternate route.

Hartmut and co-workers (58) found that ozonization of 3-methylene-2,4-dihydro-3*H*-1,5-benzodioxepin also produced 2,4-dihydro-3*H*-1,5-benzodioxepin-3-one. The 2,4-diphenylhydrazone derivative (mp 164–165°C) and semicarbazone derivative (mp 230–233°C) had the same melting points as those found by Rosnati and De Marchi (63).

(12) Derivatives of 2*H*-1,5-Benzodioxepin-4-ones. One of the earliest discoveries of seven-membered unsaturated ring systems was made by Engles and co-workers (71) when they treated trimethyl brazilein with hydrogen peroxide in acetic acid (Eq. 63). They considered the product to be the lactone of 2′,2,5-trihydroxy-α,4,5′-trimethoxy-β′-phenoxy-β-phenylisobutyric acid (mp 218°C), which is more precisely called 3,8-dimethoxy-3-(2,5-dihydroxy-4-methoxybenzyl)-3,4-dihydro-2*H*-1,5-benzodioxepin-4-one (**65**). Similar reaction of tetramethoxy dihydro brazileinol yielded 3,8-dimethoxy-3-(2-hydroxy-4,5-dimethoxybenzyl)-3,4-dihydro-2*H*-1,5-benzodioxepin-4-one (mp 160°C). The reaction of either of these compounds with dimethyl sulfate in strong alkaline solution gave a methyl ester product (Eq. 64).

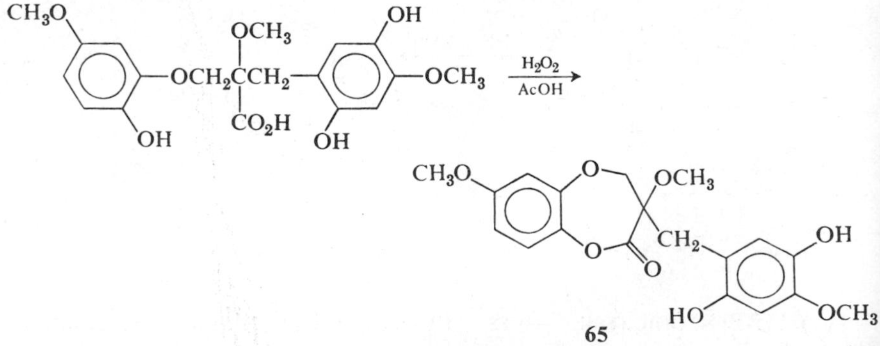

$$\xrightarrow[\text{AcOH}]{H_2O_2}$$

65

(63)

$$\textbf{65} \xrightarrow[\text{NaOH}]{Me_2SO_4}$$

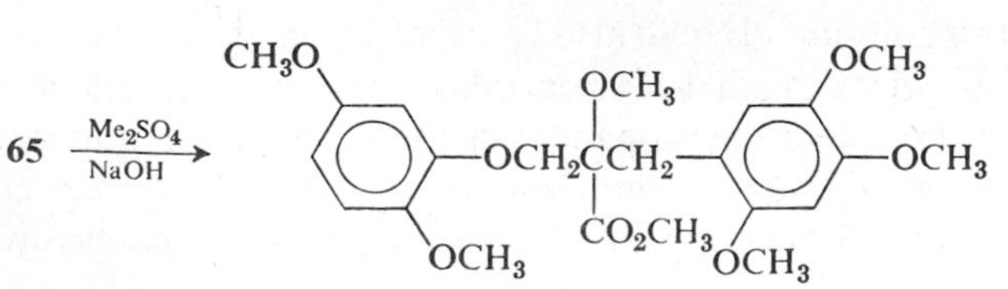

(64)

(13) 1,5-BENZODIOXEPIN-2,4-DIONES. A dione derivative of 1,5-benzodioxepin was prepared by Ziegler and co-workers (74) by reaction of 1,2-dihydroxybenzene with diethylbenzylbis(2,4-dichlorophenyl) malonate at 270–300° (Eq. 65). The products were identified as 3-benzyl-2,3-dihydro-2*H*-1,5-benzodioxepin-2,4-dione (**66**) (mp 253–255°C) and 3-benzyl-4,8-dihydroxycoumarin. A third product, 2,4-dichlorophenol, could be removed during the reaction because of its volatility.

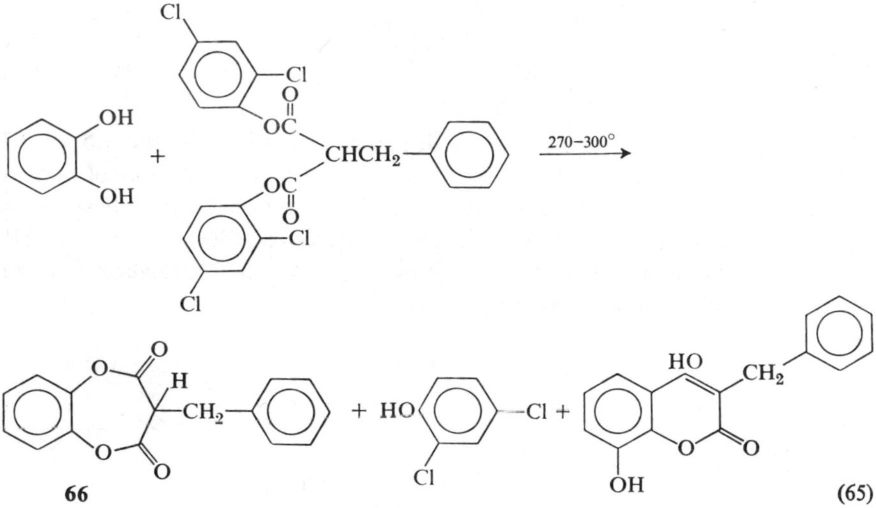

(65)

(14) PHYSICAL PROPERTIES OF 3,4-DIHYDRO-2*H*-1,5-BENZODIOXEPINS. These are listed in Table V-21.

e. *1*H-*2,3-Benzodioxepins*

(1) 1*H*-2,3-BENZODIOXEPIN DERIVATIVES. Warnell and Shriner (75) found that oxidation of indene at 0–5° in ethanol with ozone (Eq. 66) yielded 99% 4-ethoxy-4,5-dihydro-1*H*-2,3-dioxepin-1-ol (**67**) as a white nonexplosive solid (mp 105–108°C). The product was not crystalline but was similar in appearance to paraffin wax. Its structure was established by conversion to known acid or aldehyde compounds. Further reaction with ethanol and a few drops of concentrated sulfuric acid as a catalyst afforded a 66% yield of 1,4-diethoxy-4,5-dihydro-1*H*-2,3-benzodioxepin (**68**) as white needles (mp 80–81°C).

(2) 1*H*-2,3-BENZODIOXEPIN-1-ol DERIVATIVES. 4-Ethoxy-4,5-dihydro-1*H*-2,3-dioxepin-1-ol (**67**) gave homophthalic acid on oxidation with hydrogen

peroxide (Eq. 67). The diethoxy derivative also yielded this acid. Reduction of **67** with zinc in ethanol–water–acetic acid (Eq. 67) gave a 98% yield of homophthaldehyde as a yellow oil. This aldehyde was very unstable and was

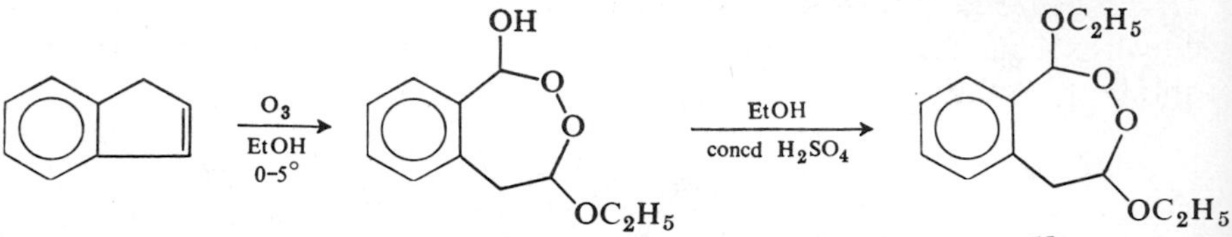

OH

O_3 EtOH 0–5°

EtOH concd H_2SO_4

OC_2H_5

67 **68** (66)

identified via its bis-*p*-nitrophenylhydrazone derivative. Reduction of **67** with lithium aluminum hydride in ether yielded homophthalyl alcohol as a colorless oil. Reaction of **67** with ethanol and 5% sodium hydroxide solution occurred exothermally to give *o*-carboxyphenylacetaldehyde as unstable yellow crystals which liquefied on standing and had to be characterized in the form of a stable semicarbazone derivative.

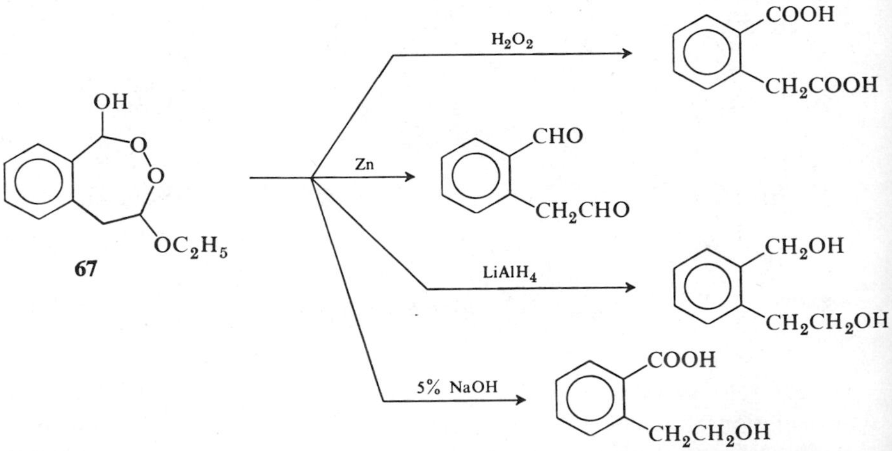

H_2O_2 COOH CH_2COOH

Zn CHO CH_2CHO

$LiAlH_4$ CH_2OH CH_2CH_2OH

5% NaOH COOH CH_2CH_2OH

OH OC_2H_5 **67**

(67)

(3) 1,4-Epoxy-1*H*-2,3-Benzodioxepins. A number of 1,4-epoxy-1*H*-2,3-benzodioxepin derivatives (R.I. 2658) have been prepared by oxidation of substituted indenes with ozone in various solvents. Bailey (76) oxidized 2,3-diphenylindene with 6% ozone in acetic acid (Eq. 68), obtaining 1,4-epoxy-1,4-diphenyl-4,5-dihydro-1*H*-2,3-benzodioxepin (**69**) (mp 122–123°C) in 80% yield. With 2% ozone in cyclohexane solution, a 64% yield was obtained. Criegee and co-workers (77) oxidized 1-phenylindene with ozone in

hexane at $-10°$ to $+10°$, isolating a 19% yield of 1,4-epoxy-5-phenyl-4,5-dihydro-1*H*-2,3-benzodioxepin (mp 89°C). Similarly a 15% yield of 1,4-epoxy-4-phenyl-4,5-dihydro-1*H*-2,3-benzodioxepin (mp 98°C) was obtained from 2-phenylindene.

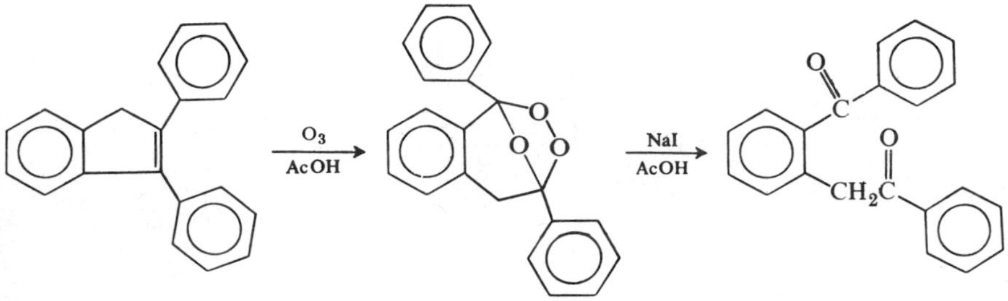

O_3 / AcOH → **69** → NaI / AcOH

(68)

Kuhn and Schulz (78) found that a novel epoxy-2,3-benzodioxepin was obtained on oxidation of 1-(*tert*-butylphenylvinylidiene)-3-*tert*-butylindene with ozone at $-85°$ in pentane (Eq. 69). A 79% yield of 1,4-epoxy-1-*tert*-butyl-5-(*tert*-butylphenylvinylidiene)-4,5-dihydro-1*H*-2,3-benzodioxepin (**70**) (mp 140°C) was obtained. This product, when reduced with lithium aluminum hydride, yielded a diol.

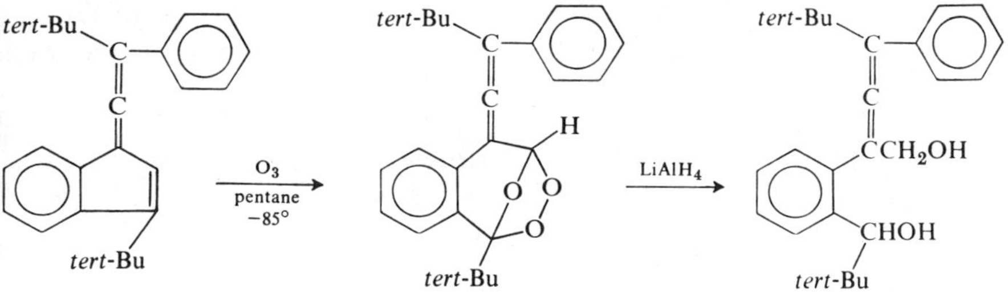

O_3 / pentane / $-85°$ → **70** → $LiAlH_4$

(69)

McMorris and Anchel (79) oxidized 7-methoxy-5-(2-methoxyethyl)-2,4,6-trimethylindene with ozone in ethyl acetate at Dry Ice/acetone bath temperatures to prepare 1,4-epoxy-6-methoxy-8-(2-methoxyethyl)-4,7,9-trimethyl-4,5-dihydro-1*H*-2,3-benzodioxepin (**71**) (mp 95–96°C). This compound, on treatment with chromium trioxide in aqueous acetic acid and subsequent reaction with ethyl acetate, yielded 3-acetyl-5,7-dimethyl-6-(2-methoxyethyl)-phthalide (Eq. 70).

Treatment of epoxybenzodioxepins with sodium iodide in acetic acid leads to reduction. 1,4-Epoxy-1,4-diphenyl-4,5-dihydro-1*H*-2,3-benzodioxepin (**69**)

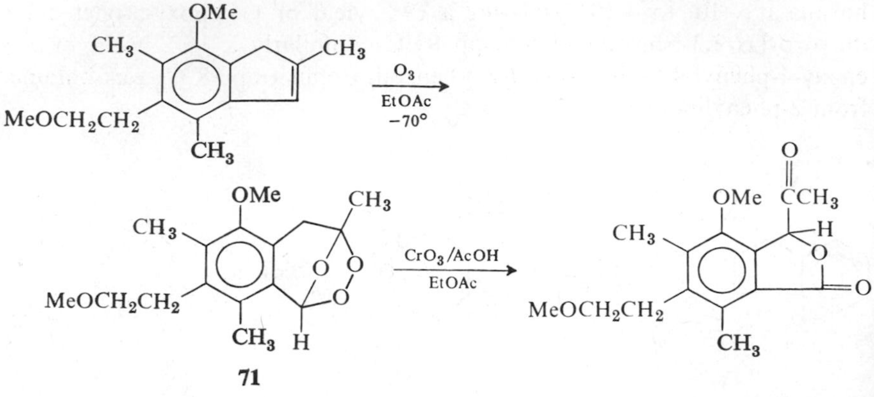

71

(70)

can be transformed into *o*-benzoyldesoxybenzoin in 66% yield (Eq. 68). A 77% yield of the same product was also obtained on reduction of **69** with hydrogen in dioxane and palladium–calcium carbonate as a catalyst (76).

Epoxybenzodioxepins also react with hydrazines to yield hydrazone derivatives. Thus **69** reacted with phenylhydrazine in ethanol and acetic acid to yield the phenylhydrazone of *o*-benzoyldesoxybenzoin (76).

(4) 1,4-Epoxy-1*H*-2,3-Benzodioxepin-5-one Derivatives. Criegee and co-workers (80) prepared a number of 1,4-epoxy-4,5-dihydro-1*H*-2,3-benzodioxepin-5-one derivatives by oxidizing a substituted indone with ozone at 0°. Thus oxidation of 2-phenyl-3-ethylindone (Eq. 71) yielded 1,4-epoxy-1-ethyl-4,5-dihydro-4-phenyl-1*H*-2,3-benzodioxepin-5-one (**72**) (mp 50°C dec). In similar fashion 2,3-diphenylindone yielded 1,4-epoxy-4,5-dihydro-1,4-diphenyl-1*H*-2,3-benzodioxepin-5-one (mp 105°C), which decomposed

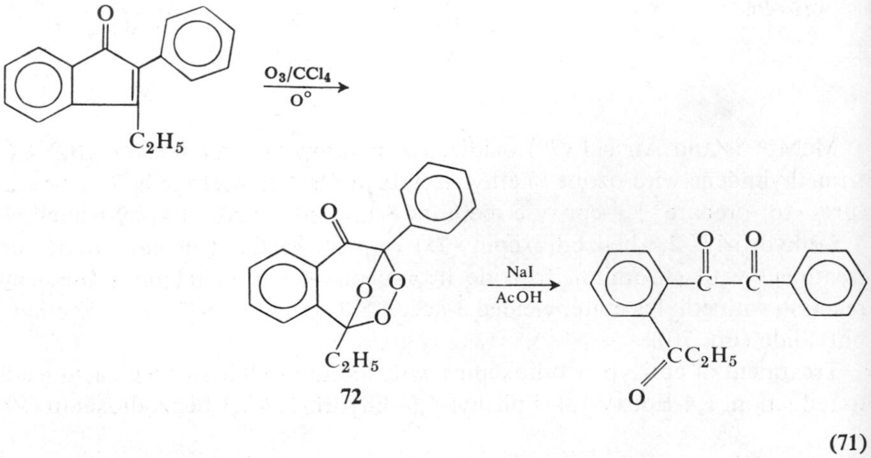

72

(71)

on standing; and 2-methyl-3-phenylindone yielded 1,4-epoxy-4,5-dihydro-1-phenyl-4-methyl-1*H*-2,3-benzodioxepin-5-one (mp 90°C), which also decomposed on standing (oxime derivative, mp 140–141°C). These products formed explosive compounds when recrystallized from methanol. The decomposition products formed on standing may be anhydrides and esters.

Reduction of **72** with sodium iodide in acetic acid yielded the triketone shown in Eq. 71. Other ozonide derivatives behaved similarly.

Another ketone derivative was prepared by Rigaudy and Auburn (81) who oxidized 3-phenyl-2-(*o*-methoxycarbonylphenyl)indole with ozone (Eq. 72) to obtain 1,4-epoxy-4,5-dihydro-1-phenyl-4-(*o*-carbomethoxyphenyl)-1*H*-2,3-benzodioxepin-5-one (**73**) (mp 122–128°C, dec). Treatment of **73** with potassium iodide led to reduction and ring cleavage to a triketone.

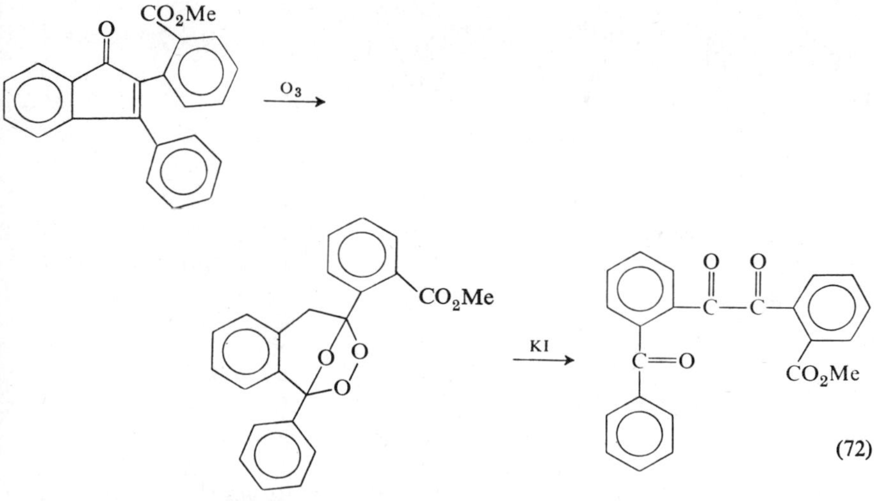

CO_2Me O $\xrightarrow{O_3}$

CO_2Me O O O $\xrightarrow{KI}$ O O C—C C=O CO_2Me

73

(72)

Similar reaction of 3-phenyl-2-(carboxyphenyl)indole (Eq. 73) failed to yield the expected ketone, giving instead an alcohol derivative (**74**) (mp 160–170°C, explodes).

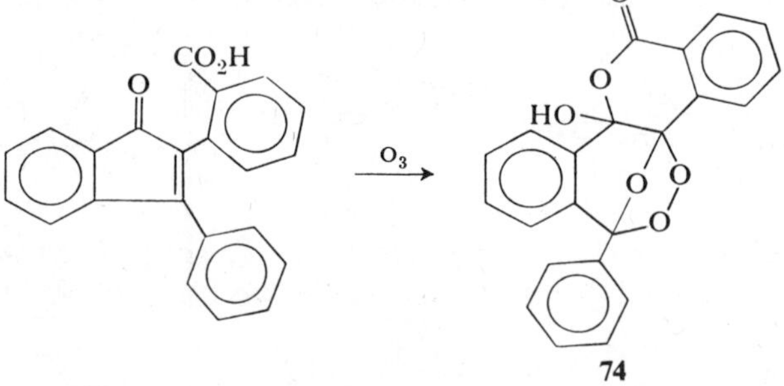

CO_2H O $\xrightarrow{O_3}$ O O HO O O O

74

(73)

(5) Physical Properties of 1*H*-2,3-Benzodioxepins. These are listed in Table V-22.

f. 3H-*2,4-Benzodioxepins*

(1) Methods of Preparation. Derivatives of 2,4-benzodioxepin were prepared by Grewe and Struve (82) by reaction of bromoacetaldehyde diethyl acetal with 1,2-benzenedimethanol in the presence of *p*-toluenesulfonic acid with simultaneous removal of ethanol (Eq. 74). The product was 3-(bromomethyl)-1,5-dihydro-2,4-benzodioxepin (**75**) (mp 98°C). Other 2,4-benzodioxepins could be prepared by condensation of 1,2-benzenedimethanol with acetals in the same manner.

$$+ BrCH(OC_2H_5)CH(OC_2H_5) \xrightarrow[\Delta]{H^{\oplus}} CH_2Br + 2C_2H_5OH \quad (74)$$

75

Small (83) prepared 1,5-dihydro-3*H*-2,4-benzodioxepin (**76**) (mp 37–38°C) by heating 1,2-benzenedimethanol with paraformaldehyde and an acid catalyst in benzene (Eq. 75). The water liberated during the reaction was removed by azeotropic distillation with the solvent. He further found that 1,5-dihydro-3*H*-2,4-benzodioxepin and trioxane would rapidly copolymerize to form high molecular weight oxymethylene polymers.

$$\xrightarrow[\Delta]{CH_2O/H^{\oplus}} + H_2O \quad (75)$$

76

Many of the methods employed to prepare 1,3-dioxepins could be used to prepare 2,4-benzodioxepin derivatives. One such compound, 3-ethyl-1,5-dihydro-3*H*-2,4-benzodioxepin, was made by Friebolin and co-workers (84) in connection with an nmr investigation.

A ring-substituted 2,4-benzodioxepin was prepared by Dallacker and co-workers (85) by heating 4,5-methylenedioxy-1,2-benzenedimethanol with

$$\xrightarrow[\Delta]{CH_2O/HCl} + H_2O \quad (76)$$

77

paraformaldehyde and HCl as a catalyst (Eq. 76). The product was identified as 1,3-dioxolo[4,5-*h*][2,4]benzodioxepin (**77**) (mp 148.5–49.5°C) (R.I. 10396).

Grewe and Struve treated 3-(3-bromomethyl)-1,5-dihydro-3*H*-2,4-benzodioxepin (**78**) with potassium metal in benzene and *tert*-butyl alcohol (Eq. 77) for 5 hr and obtained 3-methylene-1,5-dihydro-2,4-benzodioxepin (**79**) (mp 44°C). This product, on reduction with activated zinc and methylene iodide, yielded 3,3-spiroethano-1,5-dihydro-3*H*-2,4-benzodioxepin (**80**) (mp 78°C). Hydrogenation of the spiro derivative over brown PdO gave cyclopropanone hydrate.

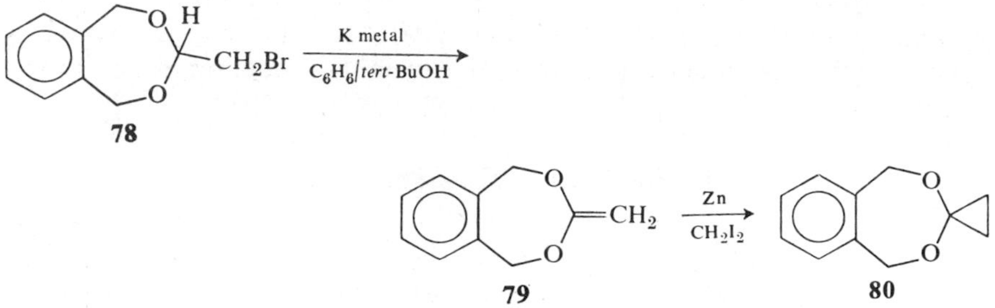

CH_2Br — K metal, C_6H_6/*tert*-BuOH →

78

$=CH_2$ — Zn, CH_2I_2 →

79 **80**

(77)

(2) Physical Properties. These are listed in Table V-23.

2. Dibenzodioxepins

a. *History, Structures, and Nomenclature*

There are two known ring types of dibenzodioxepins: 11*H*-dibenzo[*b*,*e*]-[1,4]dioxepin (**81**) (R.I. 3683) and dibenzo[*d*,*f*][1,3]dioxepin (**82**) (R.I. 12755). The numbering of these rings is shown.

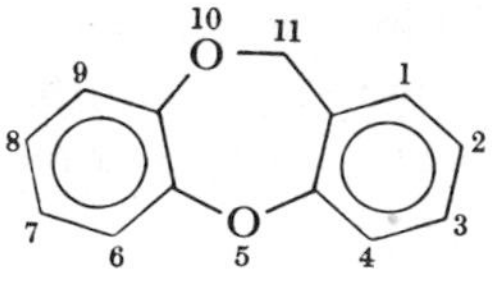

81 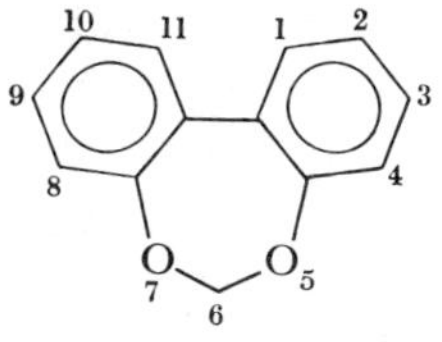 **82**

Some of the earliest derivatives of 11*H*-dibenzo[*b*,*e*][1,4]dioxepin, usually lactone substances, were found by extracting lichens. Davidson and coworkers (86) found in 1943 that extraction of the lichen *Lecanora gangaleoides* yielded gangaleoidin (**83**). The chemical name for this substance is 2,4-dichloro-3 -hydroxy-8-methoxy-1, 9-dimethyl-11-oxo-11*H*-dibenzo[*b*,*e*][1, 4]-dioxepin-7-carboxylic acid.

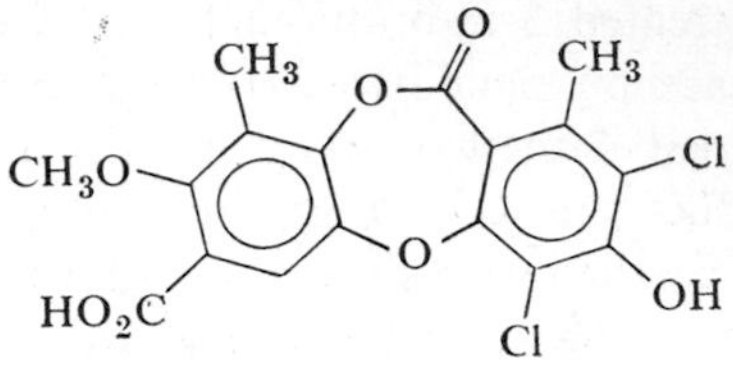

83

Derivatives of dibenzo[*d*,*f*][1,3]dioxepin were not known until 1962. Hewgill (87) was one of the first to characterize a member of this type ring system when he obtained a spiro derivative from the oxidation of 4-methoxy-3-*tert*-butylphenol.

b. *Derivatives Found in Nature*

A number of dibenzodioxepin derivatives occur in nature as lactones, usually in lichens and molds. One of the earliest dibenzodioxenpins obtained from lichens was psoromic acid, 6-formyl-7-hydroxy-2-methoxy-1,9-dimethyl-11-oxo-11*H*-dibenzo[*b*,*e*][1,4]dioxepin-4-carboxylic acid (**84**). It was characterized by Asahina and others (88–91).

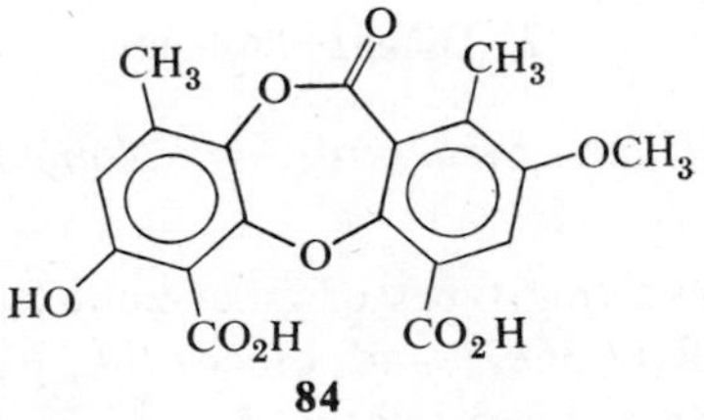

84

Murphy and co-workers (92) obtained variolaric acid by extracting Irish-grown *Leconora parella* with acetone for 20 hr. The product, 7-hydroxy-9-methyl-1-isobenzofuro[5,6-*b*][1,4]benzodioxepin-3,10-dione (**85**) (R.I. 4836), was purified by washing with hot ethanol and acetone and recrystallization from hot 80% aqueous acetone.

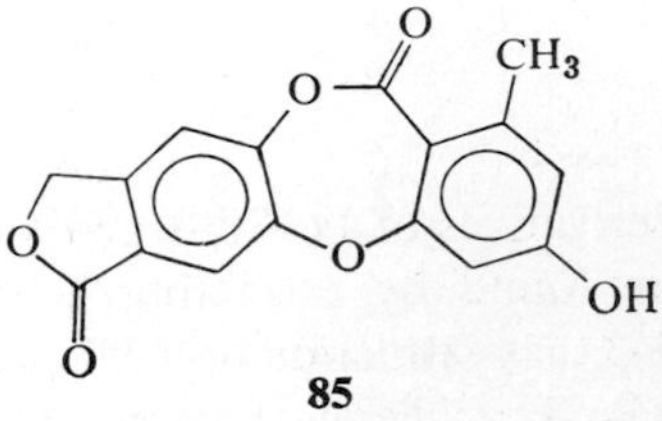

85

Neelakantan and co-workers (93) isolated from the lichen *Teloschistes flavicans* a chlorodepsidone which they called vicanicin. On the basis of analytical evidence and spectral studies, as well as degradation reactions, structure **86** was proposed for vicanicin. Its chemical name is 2,4-dichloro-3-hydroxy-8-methoxy-1,6,9-trimethyl-11*H*-dibenzo[*b*,*e*][1,4]dioxepin-11-one (mp 248–250°C, needles).

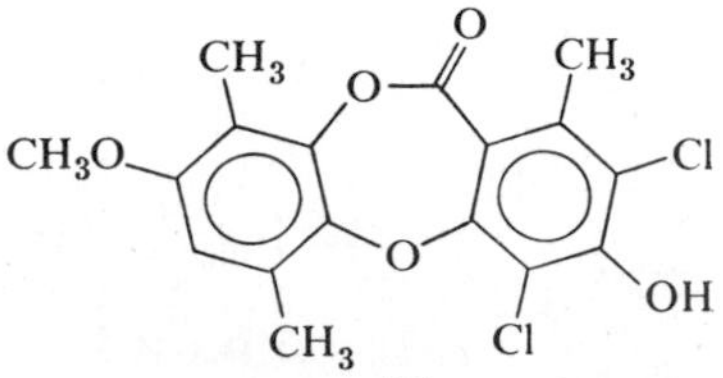

86

Schatz (94) used a number of depsidones from lichens for soil studies: salacinic acid, 1,3-dihydro-1,4,10-trihydroxy-5-(hydroxymethyl)-3,7-dioxo-7*H*-isobenzofuro[4,5-*b*][1,4]benzodioxepin-11-carboxaldehyde (R.I. 13259) (**87**); fumarprotocetrartic acid, the fumarate ester of 4-formyl-3,8-dihydroxy-9-(hydroxymethyl)-1,6-dimethyl-11-oxo-11*H*-dibenzo[*b*,*e*][1,4]dioxepin-7-carboxylic acid (**88**); lobaric acid, 3-methoxy-8-hydroxy-6-pentyl-1-valeroyl-11-oxo-11*H*-dibenzo[*b*,*e*][1,4]dioxepin-7-carboxylic acid (**89**); and physodic acid, 3,8-dihydroxy-6-pentyl-1-(valeroylmethyl)-11-oxo-11*H*-dibenzo[*b*,*e*]-[1,4]dioxepin-7-carboxylic acid (**90**). The last two mentioned have powerful chelating capacities (95, 96).

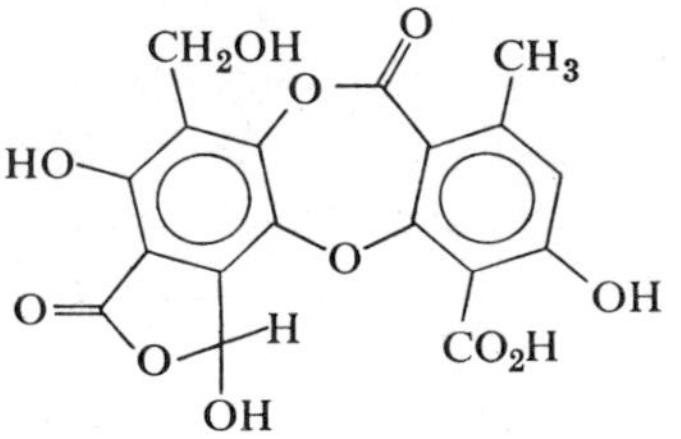

87

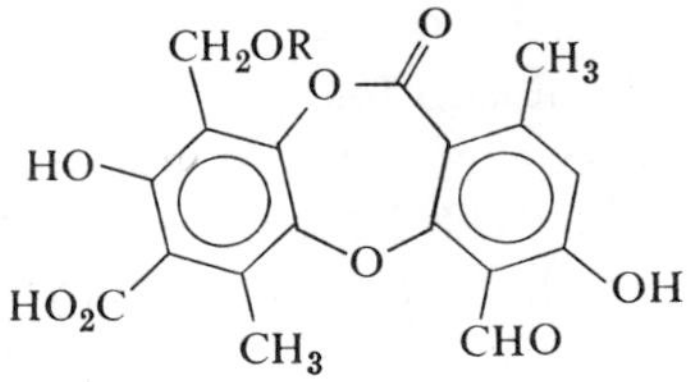

88 (R = $HO_2CCH{=}CHCO$)

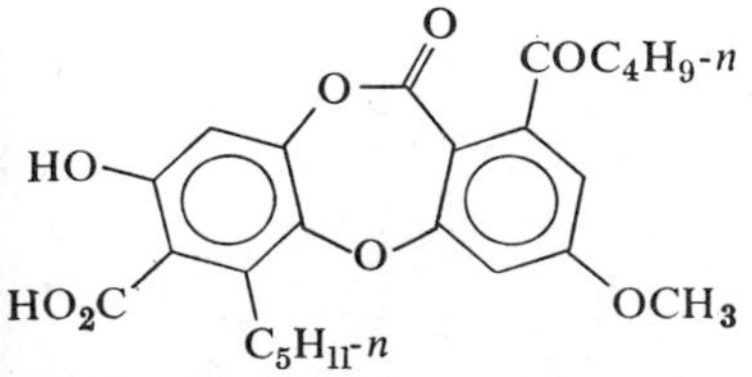

89

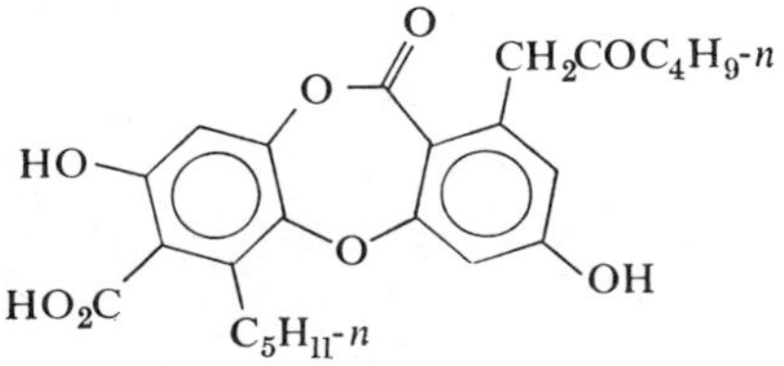

90

Dean and co-workers (97), after extensive chemical investigation, proposed structure **91** for nidulin, the principal metabolic product of the mold *Aspergillus nidulans*. Detailed analysis of the nmr spectra of nidulin, *O*-methylisonidulin, and dihydronidulin led Beach and Richards (98, 99) to conclude that a methyl group is present on the "B" ring as well as the "A" ring, as determined earlier by Dean and co-workers. Hence the previously undetermined five carbon residue must contain a methyl and a butenyl fragment (see structure **91A**). The exact positions of ring constituents in the "B" ring have not been agreed upon (100).

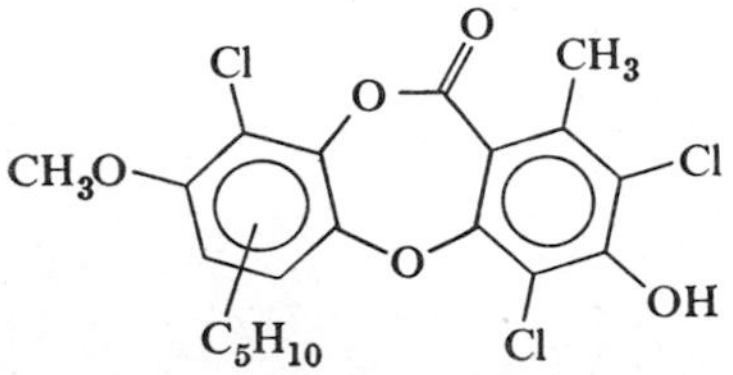

Nidulin (**91**) is chemically known as 2,4,*w*-trichloro-3-hydroxy-*x*-methoxy-1,*y*-dimethyl-*z*-(1-methylpropenyl)-11*H*-dibenzo[*b*,*e*][1,4]dioxepin-11-one (mp 180–181°C). On hydrogenation it gives 2,4,*w*-trichloro-3-hydroxy-*x*-methoxy-1,*y*-dimethyl-*z*-isobutyl-11*H*-dibenzo[*b*,*e*][1,4]dioxepin-11-one (mp 147–150°C).

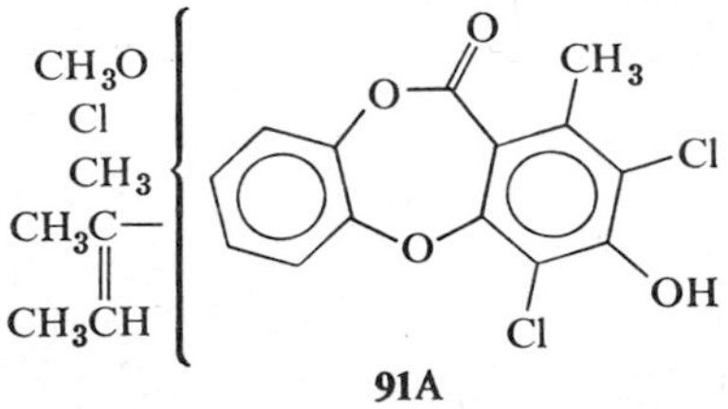

Siegfried and Pierre (101) tabulated the nmr spectral data for many of the aforementioned naturally-occurring dibenzodioxepins.

c. *Preparation of 11*H-*Dibenzo*[b,e][*1,4*]*dioxepin*

11*H*-Dibenzo[*b*,*e*][1,4]dioxepin (**92**) was prepared by Inubushi (102). Treatment of 2-hydroxy-2′-hydroxymethyldiphenyl ether with hydrogen bromide yielded 2-hydroxy-2′-bromomethyldiphenyl ether, which underwent cyclization to the desired dibenzodioxepin in the presence of alkali (Eq. 78). This "depsidan," as **92** is sometime called, had bp 112–114°C/0.2 torr. A dimer was also obtained as a product. The desired dioxepin has been prepared similarly by Tomita and co-workers (103).

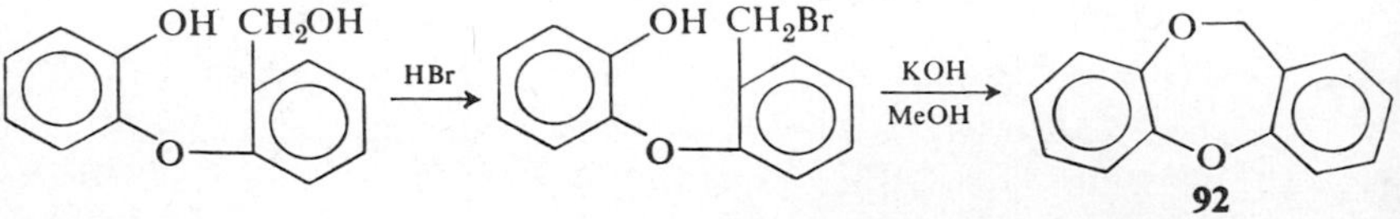

(78)

d. *11*H-*Dibenzo*[b,e][*1,4*]*dioxepin-11-ones*

(1) Methods of Preparation. Tomita and co-workers (104) have found that treatment of 2-(2-hydroxyphenoxy)benzoic acid with phosphorus

pentoxide in chloroform (Eq. 79) affords 11*H*-dibenzo[*b*,*e*][1,4]dioxepin-11-one (**93**) (mp 58–62°C, needles).

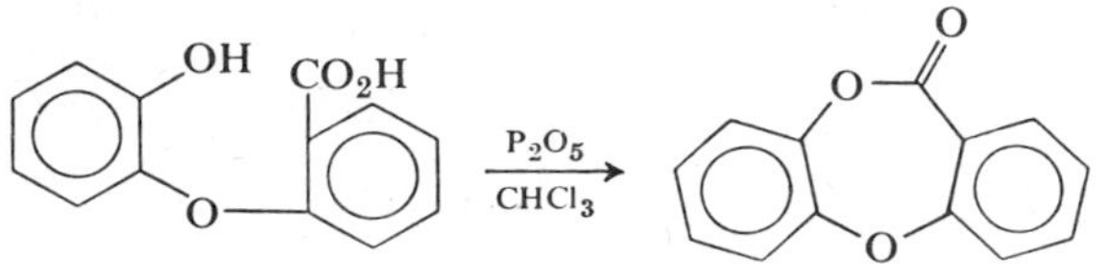

$$\xrightarrow[\text{CHCl}_3]{\text{P}_2\text{O}_5} \quad (79)$$

93

The lactonization of 2-(2-hydroxyphenoxy)benzoic acid was also accomplished in 70% yield by reaction with acetic anhydride, and by treatment of the acid with *β*-naphthalenesulfonic acid, although Noyce and Weldon (105) found the latter method less satisfactory. The best method of synthesis of these lactones, according to Noyce and Weldon, involves the use of thionyl chloride and pyridine in dry ether (Eq. 80). An essentially quantitative yield of 11*H*-dibenzo[*b*,*e*][1,4]dioxepin-11-one (**93**) (mp 65.5–66°C) was obtained. This method was also used to prepare a number of derivatives which are listed in Table V-24.

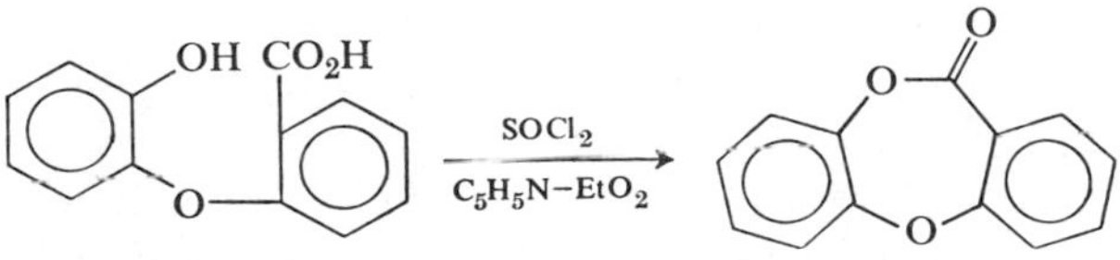

$$\xrightarrow[\text{C}_5\text{H}_5\text{N}-\text{EtO}_2]{\text{SOCl}_2} \quad (80)$$

93

Another method of synthesis of these "depsidones," as they are sometimes called, was devised by Ungnade and Rubin (108). Demethylation of 2-(2-methoxyphenoxy)benzoic acid with potassium hydroxide in ethylene glycol and acidification gave a 70% yield of 11*H*-dibenzo[*b*,*e*][1,4]dioxepin-11-one (**93**). This lactone underwent hydrolysis in 5% aqueous sodium hydroxide and recyclization on acidification. Acidification of the sodium salt of 2-(2-hydroxyphenoxy)benzoic acid likewise yielded lactone **93**.

(2) Reactions. Ring opening is a typical reaction of these lactone derivatives. Noyce and Weldon (105) found that heating 11*H*-dibenzo-[*b*,*e*][1,4]dioxepin-11-one in methanol and concentrated HCl at reflux temperatures, followed by a careful neutralization, afforded a mixture of 2-(2-hydroxyphenoxy)benzoic acid and its methyl ester (Eq. 81). Similarly Barry and Twomey (109) have found that hydrolysis of 2,4,7,9-tetrachloro-3-hydroxy-8-methoxy-1,6-dimethyl-11*H*-dibenzo[*b*,*e*][1,4]dioxepin-11-one with aqueous sodium hydroxide at boiling temperatures yields a mixture of 4,6-dichloro-5-hydroxy-3-methyl-2-(3,5-dichloro-2-hydroxy-4-methoxyphenoxy)benzoic acid and a decarboxylation product. Dean and co-workers

(97) found that *O*-methylnidulin reacted with $2N$ aqueous sodium hydroxide to yield *O*-methylnidulinic acid.

93 —(concd HCl / MeOH)→ [OH, CO_2H product] + [OH, CO_2CH_3 product] (81)

Noyce and Weldon (106) have shown that these lactone derivatives, when treated with sodium methoxide in methanol, are converted quantitatively into the methyl ester of the benzoic acid (Eq. 82). Dean and others (97) have used a similar reaction to transform nidulin into methyl nidulinate.

93 —(NaOMe / MeOH)→ [OH, CO_2CH_3 product] (82)

Cleavage of 11*H*-dibenzo[*b*,*e*][1,4]dioxepin (**93**) with sodium in ether and liquid ammonia yields a mixture of 2-(2-hydroxyphenoxy)toluene, *o*-cresol, and phenol (103).

When a hydroxy group is present on the ring, as in nidulin, dibenzo[*b*,*e*]-[1,4]dioxepin-11-ones undergo reactions typical of phenols. Dean and co-workers (97) allowed nidulin to react with methyl iodide and potassium carbonate in acetone or with diazomethane, isolating 2,4,*w*-trichloro-3,*x*-dimethoxy-1,*y*-dimethyl-*z*-(1-methylpropenyl)-11*H*-dibenzo[*b*,*e*][1,4]dioxepin-11-one (mp 143–145°C). Neelakantan and co-workers (93) reported that a similar alkylation with methyl iodide and potassium carbonate in acetone (Eq. 83) converted vicanicin (**86**) into 2,4-dichloro-3,8-dimethoxy-1,6,9-trimethyl-11*H*-dibenzo[*b*,*e*][1,4]dioxepin-11-one (**94**) (mp 193–194°C). Reaction of **86** with ethyl iodide gave 2,4-dichloro-3-ethoxy-8-methoxy-1,6,9-trimethyl-11*H*-dibenzo[*b*,*e*][1,4]dioxepin-11-one (colorless needles, mp 185–186°C). Methylation of diploicin (**95**) gave 2,4,7,9-tetrachloro-3,8-dimethoxy-1,6-dimethyl-11*H*-dibenzo[*b*,*e*][1,4]dioxepin-11-one (**96**) (109).

86 —(MeI / K_2CO_3/Me_2CO)→ **94** (83)

Barry and Twomey (109) reported the reaction of diploicin with ethyl bromoacetate (Eq. 84) which yields 3-(carbethoxymethoxy)-2,4,7,9-tetrachloro-8-methoxy-1,6-dimethyl-11*H*-dibenzo[*b*,*e*][1,4]dioxepin-11-one (**97**) (mp 179–180°C).

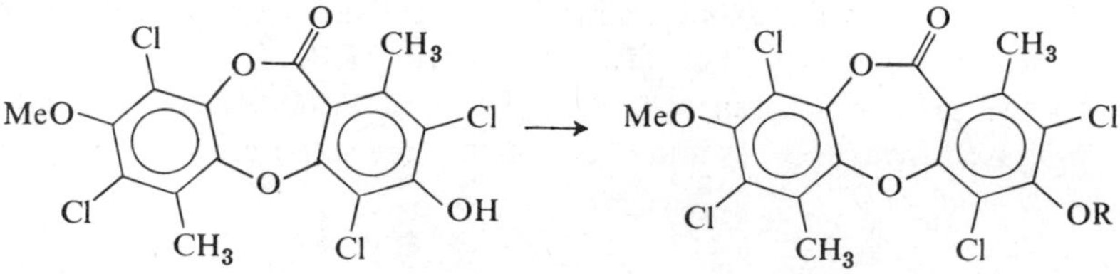

95

(84)

96; R=Me
97; $R{=}CH_2CO_2Et$

Neelakantan and co-workers (93) found that vicanicin (**86**) also reacts with benzoyl chloride in pyridine (Eq. 85) to yield 2,4-dichloro-3-benzoyloxy-8-methoxy-1,6,9-trimethyl-11*H*-dibenzo[*b*,*e*][1,4]dioxepin-11-one (**98**) (colorless tablets, mp 190–191°C). They also obtained esters by condensation with an acid anhydride. Vicanicin reacts with acetic anhydride alone or in the presence of concentrated sulfuric acid to yield 2,4-dichloro-3-acetoxy-8-methoxy-1,6,9-trimethyl-11*H*-dibenzo[*b*,*e*][1,4]dioxepin-11-one (**99**) (thick rectangular prisms, mp 213–214°C). Dean and co-workers (97) found that nidulin reacted in a similar manner with acetic anhydride to yield 2,4,*w*-trichloro-3-acetoxy-*x*-methoxy-1,*y*-dimethyl-*z*-(1-methylpropenyl)-11*H*-dibenzo[*b*,*e*][1,4]dioxepin-11-one (plates or slender rods, mp 167–168°C).

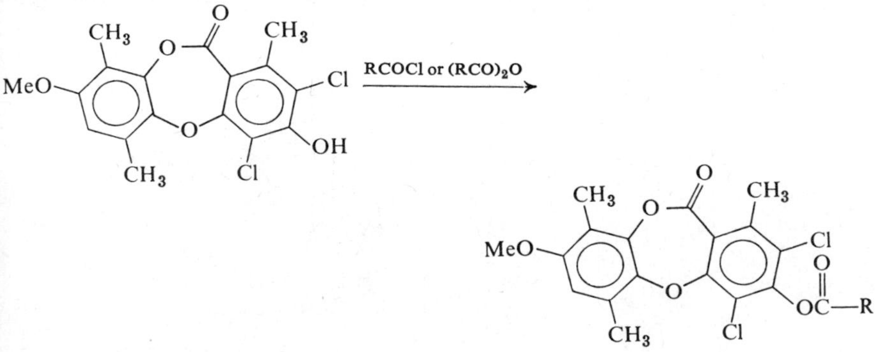

$RCOCl$ or $(RCO)_2O$

98; R= Ph
99; R= Me

(85)

e. *Preparation of 6*H-*Dibenzo*[d,f][*1,3*]*dioxepin Derivatives*

Hewgill and Hewitt (110) found that 2,2′-dihydroxybiphenyl reacts with acetone in the presence of an acid catalyst (Eq. 86), yielding 6,6-dimethyl-6*H*-dibenzo[*d,f*][1,3]dioxepin (**100**). Similarly cyclohexanone affords spirocyclohexane-1,6′-dibenzo[*d,f*][1,3]dioxepin (**101**) (R.I. 13378), and paraformaldehyde gives 6*H*-dibenzo[*d,f*][1,3]dioxepin (**102**). Other derivatives can be prepared from ketones and aldehydes in the same manner.

$$\text{(2,2'-dihydroxybiphenyl)} + RC(=O)R \xrightarrow[\Delta]{H^{\oplus}} \text{(6-R,6-R-dibenzodioxepin)} + H_2O \qquad (86)$$

100; R = Me
101; R = $(CH_2)_5$
102; R = H

Yoder and Zuckerman (114) condensed 2,2′-dihydroxybiphenyl with ketones in the presence of phosphorus pentoxide at 80° to form 6*H*-dibenzo-[*d,f*][1,3]dioxepins. For example, treatment of an equimolar mixture of 2,2′-dihydroxybiphenyl and 5-nonanone with excess phosphorus pentoxide at 80°C (Eq. 87) furnished 6,6-dibutyl-6*H*-dibenzo[*d,f*][1,3]dioxepin (**103**) (mp 56.8–57°C), with prominent ir absorption at 1240s, 1200m, 1040m, 960m, 755m, 715m, and 508m cm^{-1}. Similarly acetone yielded 6,6-dimethyl-6*H*-dibenzo[*d,f*][1,3]dioxepin (**101**) (mp 72.2–76°C) with prominent ir absorption at 1240s, 1190s, 1195sh, 1120m, 970m, 890m, 810m, 760s, 765s, 715m, and 550m cm^{-1}. Acetophenone yielded 6-methyl-6-phenyl-6*H*-dibenzo[*d,f*][1,3]-dioxepin (**104**) (plates, mp 100.3–100.5°C) with prominent ir absorption at 1240s, 1190m, 1125s, 1010s, 930m, 945m, 905m, 753s, 765m, 720m, and 700s cm^{-1}.

$$\text{(2,2'-dihydroxybiphenyl)} + O{=}C(R)(R') \xrightarrow{P_2O_5} \text{(6-R,6-R'-dibenzodioxepin)} + H_2O \qquad (87)$$

103; R = R′ = *n*-Bu
104; R = Ph, R′ = Me

Yoder and Zukerman (114) also prepared a 6*H*-dibenzo[*d,f*][1,3]dioxepin by allowing 2,2′-dihydroxybiphenyl to react with dichlorodiphenylmethane in a benzene–pinene mixture at 85–95°C for 8 hr (Eq. 88). Purification of the solid formed on cooling gave 6,6-diphenyl-6*H*-dibenzo[*d*,*f*][1,3]dioxepin (**105**) (mp 169.5–169.6°C) with prominent infrared absorption at 1240s, 1190s, 1210s, 1100m, 1050m, 1020s, 995m, 960m, 765s, 755sh, 725m, 710m, 695m, and 650m cm^{-1}.

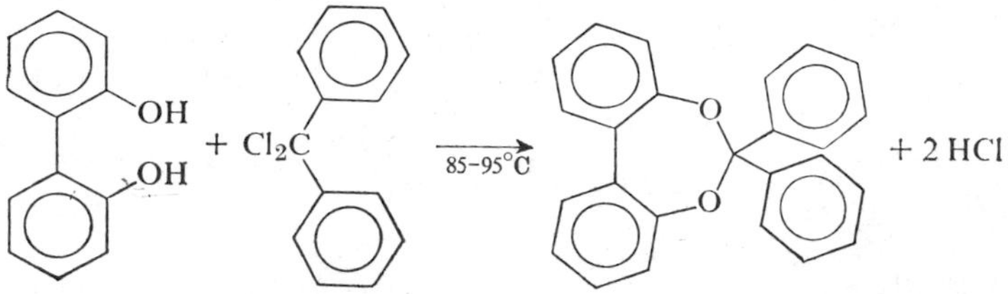

+ Cl_2C → (85–95°C) + 2 HCl (88)

105

6*H*-Dibenzo[*d*,*f*][1,3]dioxepins were also prepared by Hewgill and Hewitt (110) by allowing 2,2′-dihydroxybiphenyl to react with an acetal (Eq. 89). Thus condensation with acetone dimethyl ketal yielded 6,6-dimethyl-6*H*-dibenzo[*d*,*f*][1,3]dioxepin (**100**). Cyclohexanone dimethyl ketal reacted similarly with 2,2′-dihydroxybiphenyl to yield **101**. Other acetals of ketones and aldehydes should react in the same manner, giving 6*H*-dibenzo[*d*,*f*][1,3]-dioxepin derivatives.

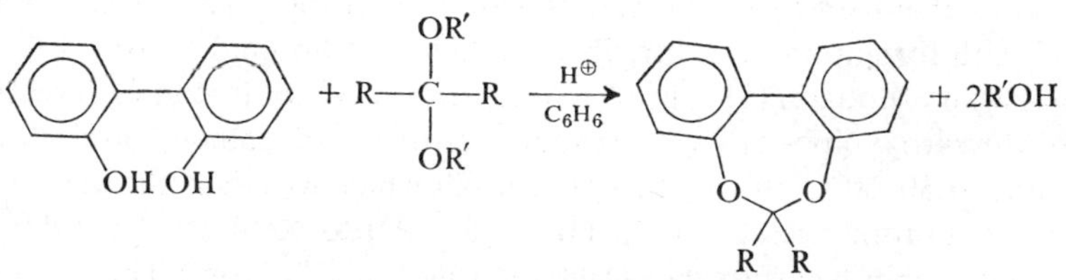

OH OH + R—C(OR′)(OR′)—R → ($H^{\oplus}$, C_6H_6) + 2R′OH (89)

Phenol derivatives of 6*H*-dibenzo[*d*,*f*][1,3]dioxepin can be prepared by allowing a ketone to react with 2,2′,5,5′-tetrahydroxybiphenyl in the presence of an acid catalyst (110) (Eq. 90). Acetone yields 2,10-dihydroxy-6,6-dimethyl-6*H*-dibenzo[*d*,*f*][1,3]dioxepin (**106**), and cyclohexanone yields the 2,10-dihydroxy spiro derivative. Other ketones can be employed similarly.

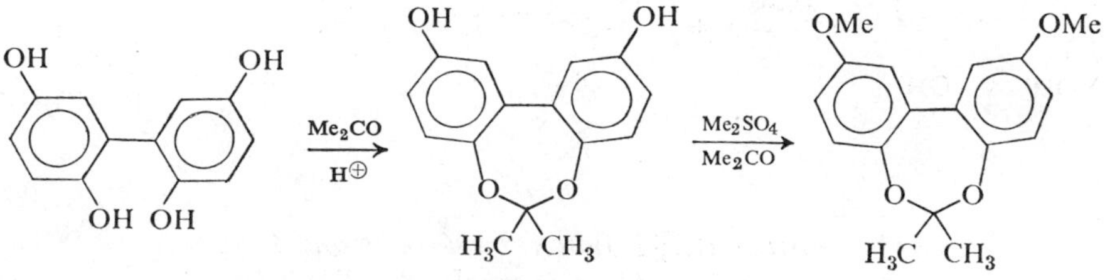

→ (Me_2CO, $H^{\oplus}$) **106** → (Me_2SO_4, Me_2CO) **107**

(90)

The reaction of 2,2′,5,5′-tetrahydroxybiphenyl with an acetal yields a mixture of the desired product and ethers. Thus acetone dimethyl ketal (Eq. 91) yields 2,10-dihydroxy-6,6-dimethyl-6*H*-dibenzo[*d*,*f*][1,3]dioxepin (**106**) and 2,2′,5,5′-tetramethoxybiphenyl.

OH OH OH OH + $CH_3C(OCH_3)_2CH_3$ $\xrightarrow{H^{\oplus}}$ (OH, OH, O, O, H_3C CH_3) **106** + (OMe, OMe, OMe, OMe)

(91)

Ethers of 6*H*-dibenzo[*d*,*f*][1,3]dioxepin can be prepared by condensing 2,2′,5,5′-tetrahydroxybiphenyl with a ketone and tetramethoxysilane in the presence of concentrated sulfuric acid as a catalyst. Cyclohexanone yields 2,10-dimethoxycyclohexanespiro-6′-dibenzo[*d*,*f*][1,3]dioxepin. Dimethoxy derivatives such as **107** can also be prepared by treating a 2,10-dihydroxy-dibenzo[*d*,*f*][1,3]dioxepin with dimethyl sulfate in acetone (110) (Eq. 90).

The 6*H*-dibenzo[*d*,*f*][1,3]dioxepins can be hydrogenated over Pd–C in ethanol with formation of a dihydroxybiphenyl compound (110).

Breslow and Mohacsi (111) used an ester derivative for ir studies. Treatment of 2,2′-dihydroxybiphenyl with sodium hydride in toluene and dimethoxyethane at 50–60°C afforded a sodium salt which was then allowed to react with ethyl dibromoacetate at 50–60°C (Eq. 92) to yield ethyl 6*H*-dibenzo-[*d*,*f*][1,3]dioxepin-6-carboxylate (**108**) (bp 165–170°C/0.7 torr). The ir spectrum showed a strong C═O stretching at 1750 cm^{-1}. Ethyl 2-H^2-6*H*-dibenzo[*d*,*f*]-[1,3]dioxepin-6-carboxylate was prepared via the use of deuterated ethyl dibromoacetate.

(OH OH) $\xrightarrow[\text{PhMe/}\ MeOCH_2CH_2OMe]{\text{NaH}\ 50\text{–}60^\circ C}$ (ONa ONa) $\xrightarrow[50\text{–}60^\circ C]{Br_2CHCO_2Et}$ (O, O, H $CO_2C_2H_5$) **108**

(92)

f. *Dibenzo*[d,f][*1,3*]*dioxepin-6-ones and Ketone Derivatives of* 6H-*Dibenzo*[d,f][*1,3*]*dioxepin*

Prochaska (112) isolated a ketone derivative on treating 2,2′-dihydroxybiphenyl with phosgene and pyridine in methylene chloride (Eq. 93). The product, dibenzo[*d,f*][1,3]dioxepin-6-one (**109**) (mp 101–102°C), polymerized to a polycarbonate in the presence of potassium carbonate. The polymer could be drawn into fibers or used as electric or decorative coatings. Furthermore a polymer formed from 2,2′-dihydroxybiphenyl and diphenyl carbonate at 220°C yielded phenol and dibenzo[*d,f*][1,3]dioxepin-6-one (mp 101–102°C) on being heated to 280°C.

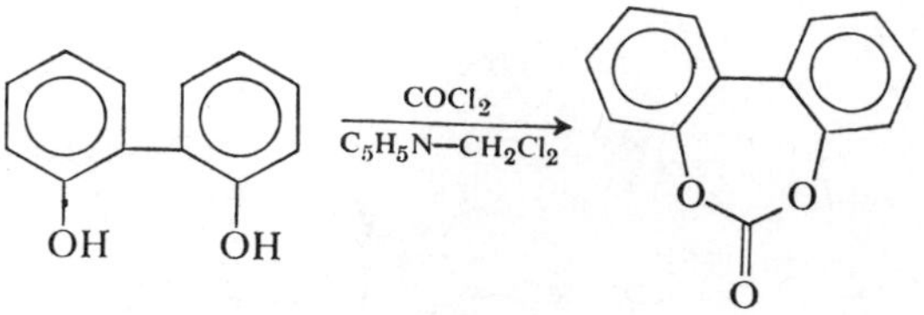

$COCl_2$, $C_5H_5N—CH_2Cl_2$ (93)

OH OH

109

Hewgill (87) prepared a spiro ketone derivative of dibenzo[*d,f*][1,3]dioxepin by oxidation of 4-methoxy-3-*tert*-butylphenol with alkaline ferricyanide solution (Eq. 94). The initial product, 2,5′,10-trimethoxy-3,4′,9-tri-*tert*-butyldibenzo[*d,f*][1,3]dioxepin-6-spiro-2′-cyclohexa-3′,5′-dienone (**110**) (lemon-yellow prisms, mp 209–210°C) reacted with hydroxylamine hydrochloride to yield the oxime (**111**) (mp 195–196°C).

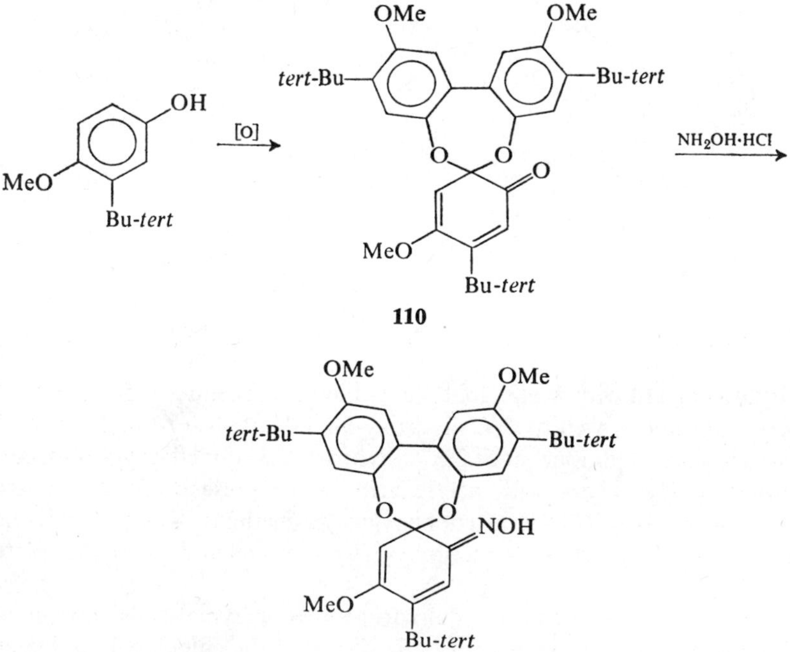

OH, MeO, Bu-*tert* [O] OMe OMe *tert*-Bu Bu-*tert* O O O MeO Bu-*tert* $NH_2OH·HCl$

110

OMe OMe *tert*-Bu Bu-*tert* O O NOH MeO Bu-*tert* (94)

111

Bowman and Hewgill (115) found that treatment of 2-bromo-6-methyl-4-*tert*-butylphenol with solid potassium hydroxide in dimethylformamide or acetone (Eq. 95) gave the spiroketal 3′,4,8-trimethyl-2,5′,10-tri-*tert*-butyldibenzo[*d*,*f*][1,3]dioxepin-6-spiro-2′-cyclohexa-3′,5′-dienone (**112**). When larger groups were substituted for the methyl group (for example, *tert*-butyl) the spiroketal was not formed; instead, a product believed to be a benzoxete was obtained.

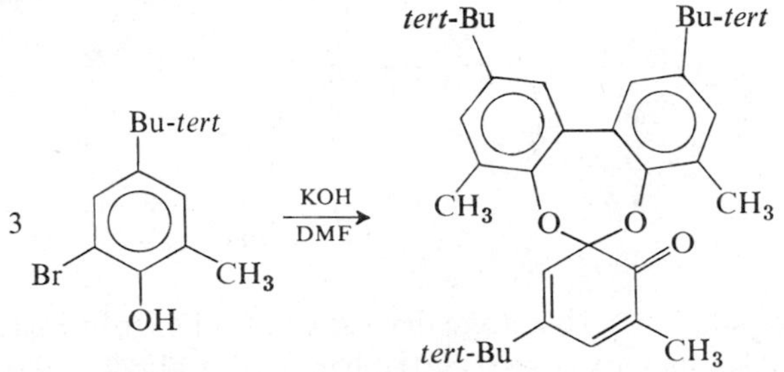

$$3\ \text{2-Br-6-CH}_3\text{-4-}tert\text{-Bu-phenol} \xrightarrow[\text{DMF}]{\text{KOH}} \mathbf{112} \qquad (95)$$

Reduction of spiro ketone **110** with hydrogen in cyclohexane over 10% palladium-charcoal or with lithium hydride in ether gives 2-hydroxy-2′-(2-hydroxy-5-methoxy-4-*tert*-butylphenoxy)-5,5′-dimethoxy-4,4′-di-*tert*-butylbiphenyl as a glass (Eq. 96).

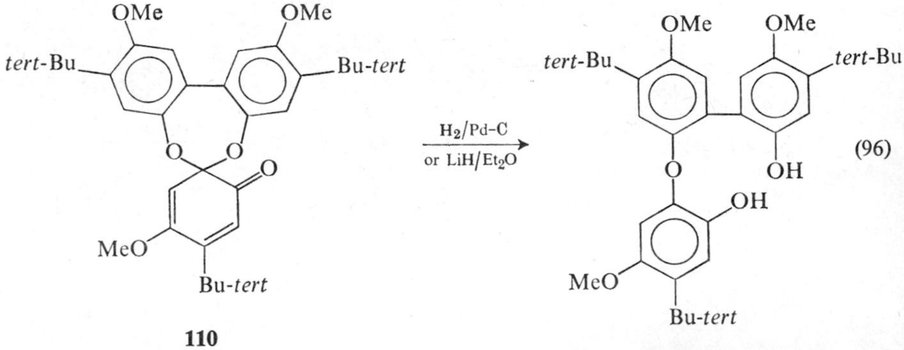

$$\mathbf{110} \xrightarrow[\text{or LiH/Et}_2\text{O}]{\text{H}_2/\text{Pd-C}} \text{product} \qquad (96)$$

Hydrolysis of **110** with acetic acid yields the monoacetate of 5-methoxy-4-*tert*-butyl catechol as well as 2-(2-hydroxy-5-methoxy-4-*tert*-butylphenyl)-5-*tert*-butyl-1,4-benzoquinone and 2,3′-dihydroxy-5,5′-dimethoxy-4,4′-di-*tert*-butylbiphenyl. Hydrolysis with acetic acid in the presence of 2,4-dinitrophenylhydrazine gives 2-(2,4-dinitrophenylazo)-5-methoxy-4-*tert*-butylphenol and 2-(2,4-dinitrophenylazo)-4-methoxy-5-*tert*-butylphenol as brown plates (87).

When α-ethylvanillyl alcohol is dehydrogenated enzymatically in aqueous solution and then treated with as little as 2–3% of the calculated equivalent of peroxide, a spiroketal having the structure **113** is obtained as a major product. Pew and Connors (116) have obtained this product in high yield as a crystalline compound (mp 198–200°C). This spiroketal readily produces colored products when treated with dilute acid or alkali at room temperature.

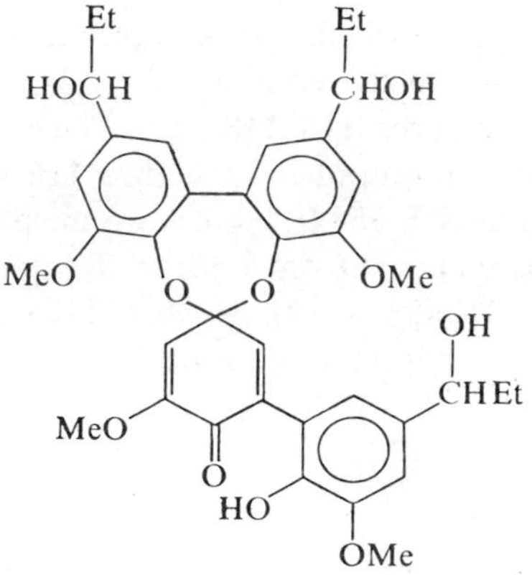

113

g. *6,6′-Spirobi(dibenzo*[d,f][*1,3*]*dioxepin*)

Yoder and Zuckerman (114) found that 2,2′-dihydroxybiphenyl reacted with thiophosgene in aqueous sodium carbonate (Eq. 97) to yield dibenzo-[*d,f*][1,3]dioxepin-6-thione (**114**) (mp 83.5–85°C) with prominent ir absorption at 1450m, 1295s, 1265s, 1245m, 1200sh, 1175sh, 1150s, 1095m, 775sh, and 765s cm^{-1}. This compound, on heating for a prolonged period in the absence of air, yielded 6,6′-spirobi(dibenzo[*d,f*][1,3]dioxepin (**115**) (mp 339.5–340.0°C) as white crystals with prominent ir absorption at 1450m, 1200m, 1160sh, 1120s, 1070s, 1040m, 760s, and 720 cm^{-1}.

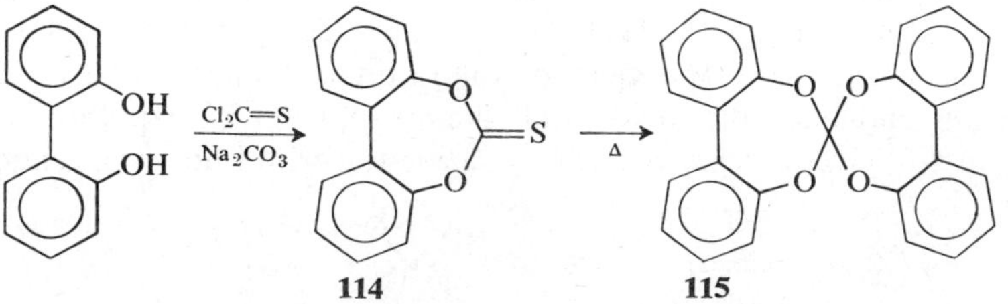

$Cl_2C{=}S$, Na_2CO_3; Δ (97)

114 **115**

h. *Physical Properties of Dibenzodioxepins*

These are listed in Table V-25.

3. Naphthodioxepins

a. *Nomenclature and Structures*

There are three possible types of isomeric naphthodioxepin ring structures. Each type, however, can contain the various 1,2-, 1,3-, 1,4-, 1,5-, 2,3-, and 2,4-dioxa ring systems. Structures **116–118** show the various ring systems and their numbering. The compound of structure **116** type is named 4*H*-naphtho[2,1-*d*][1,3]dioxepin (R.I. 3684). No known members of ring systems **116** or **117** have been prepared apart from a decahydro derivative of structure **116** reported by Stoll and Hinder (117). It should be possible to obtain these derivatives by the procedures employed for the synthesis of simpler benzodioxepin derivatives.

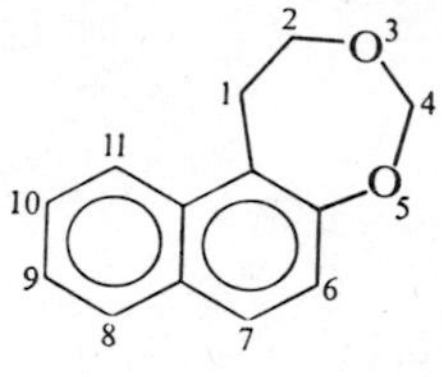

116 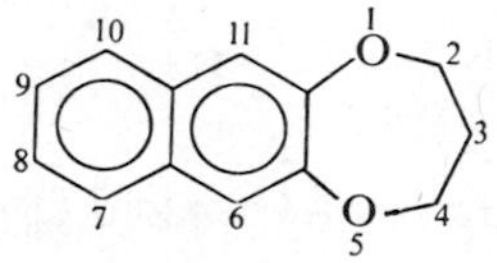 **117** 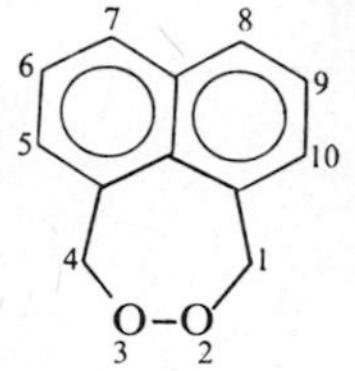 **118**

Ring system **118**, an example of which was prepared by Callighan and coworkers (118), is named 1*H*,4*H*-naphtho[1,8;*d*,*e*][1,2]dioxepin (R.I. 10526).

b. *Preparation*

Oxidation of acenaphthylene (Eq. 98) in methanol at −30°C with 1.06 molar equivalent of ozone gives a 25% yield of 1*H*,4*H*-4-methoxynaphtho-[1,8;*d*,*e*][1,2]dioxepin-1-ol (**119**) as a white powder (mp 115–116°C) (118). Excess ozone decreases the yield of this product. Other products formed are 1,8-naphthaldehydic acid and methyl 8-formyl-1-naphthoate. The dioxepin-1-ol gives a negative hydroperoxide test with lead tetraacetate.

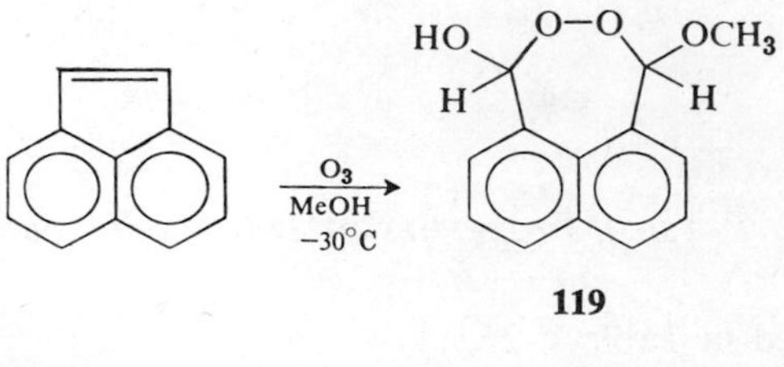

O_3, MeOH, −30°C (98)

119

Another type of naphthodioxepin derivative has been prepared by Criegee and co-workers (119) from 7,8-diphenylacenaphthylene. Ozonolysis of this hydrocarbon at −30°C (Eq. 99) yields 5,6-diphenyl-1,4-epoxynaphtho[1,8;*d*,*e*][1,2]dioxepin (**120**) (mp 156°C) (R.I. 8868).

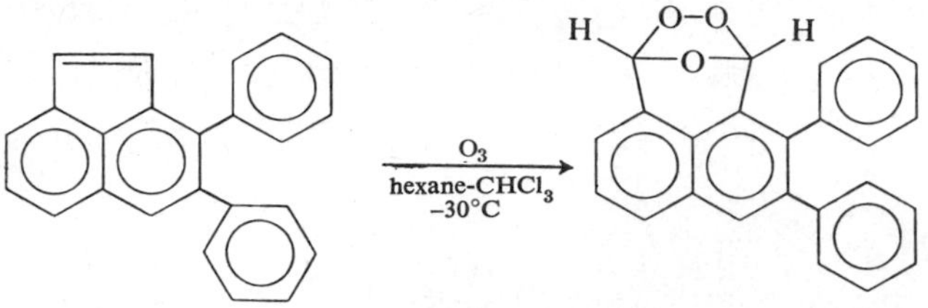

O_3, hexane-$CHCl_3$, −30°C (99)

120

4. Dinaphthodioxepins

a. *Nomenclature and Structures*

The only dinaphthodioxepin ring system presently known is that shown in structure **121**. It is called 4*H*-dinaphtho[2,1-*d*:1′,2′-*f*][1,3]dioxepin (R.I. 6429). A derivative was prepared by Dilthey and co-workers (120). Other derivatives of this type could be prepared, in principle, by the methods used for the synthesis of related dibenzodioxepins. Structure **122** shows another possible type of dinaphthodioxepin. Although no derivatives of this ring system are known at present, it should likewise be possible to obtain them in a manner similar to the preparation of the corresponding dibenzodioxepin analogs.

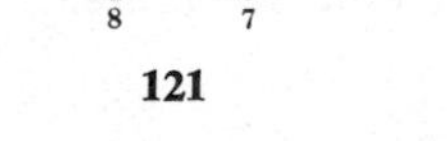

121 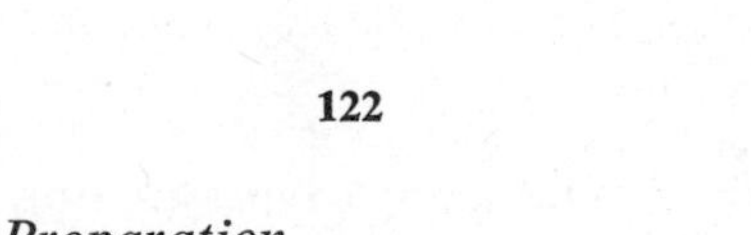 **122**

b. *Preparation*

Dilthey and co-workers (120) prepared 4,4-diphenyl-4*H*-dinaphtho[2,1-*d*:1′,2′-*f*][1,3]dioxepin (**123**) (mp 238°C) by allowing 1,1′-bis(2-naphthol) to react with diphenyldichloromethane at 130°C (Eq. 100).

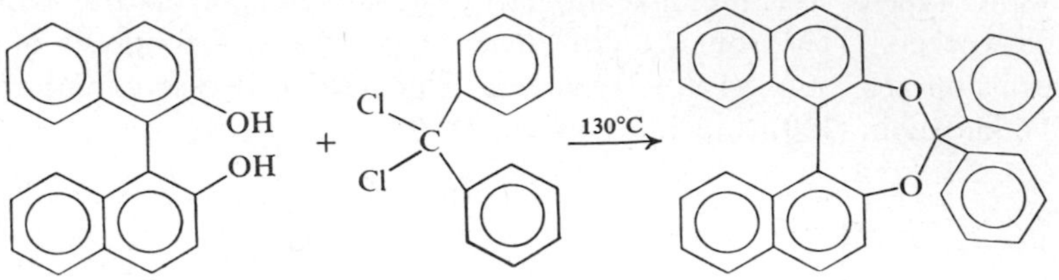

(100)

Structure **124** (R.I. 7713) shows another type of dinaphthodioxepin ring system, a derivative of which was prepared by Max and Schmidt-Nickels (121) by reaction of benzyl-2′-dihydroxydibenzanthrone and 9,9-dichlorofluorene with sodium acetate in nitrobenzene at 200–205°C for 20 hr. The acetic acid formed from the reaction was distilled from the reaction mixture. The product, spiro[dinaphtho[1′,2′,3′:3,4;3″,2″,1″:9,10]perylo[1,12-*def*]-[1,3]dioxepin-17,9′-fluorene-5,10-dione], is useful as a dye.

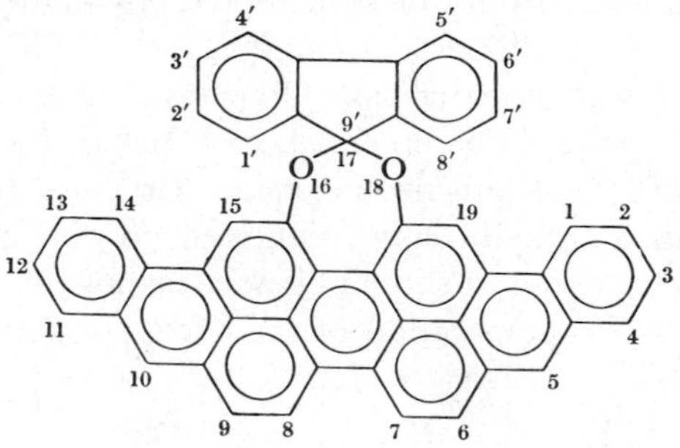

124

5. Other Dioxepin Systems Containing Two or More Rings

a. *History, Types, Nomenclature, and Structures*

There are several known types of ring systems in which a pyridine and dioxepin ring are fused. Eight possible types are shown in structures **125**–**133**. Of these, only ring system **127** is known, in the form of the 1,5-dihydro-3*H*-[1,3]dioxepino[5,6-*c*]pyridine (R.I. 12050) derivatives prepared by Korytnyk (35).

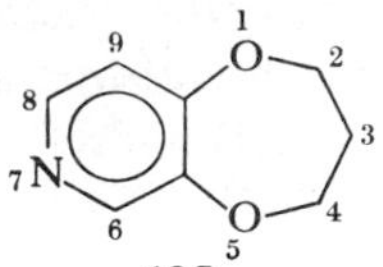

125 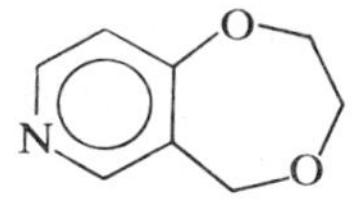 126 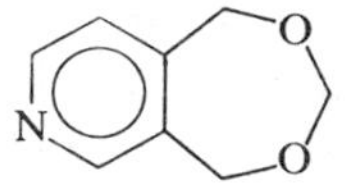 127

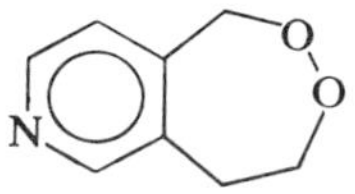

128 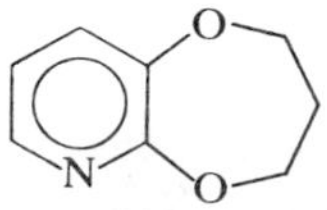 129 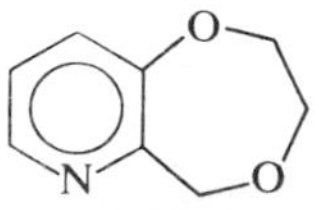 130

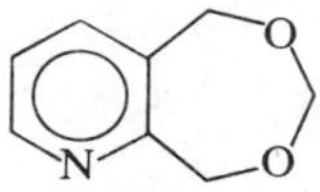

131 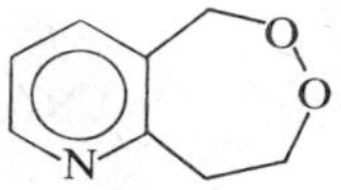 132

Similar seven-membered ring structures can be constructed, in principle, for quinoline and quinoxaline. Structure **133** shows the only type of quinoline derivative known to date, 2*H*-[1,4]dioxepino[2,3-*g*]isoquinoline (R.I. 8616). Members of this ring system were first prepared by Tomita and Takahashi (36).

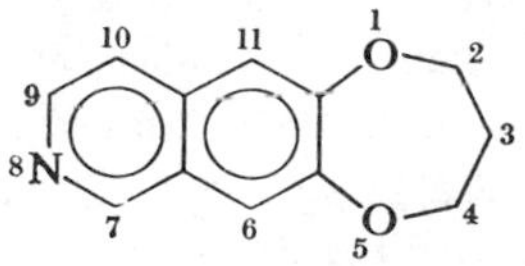

133

The only known quinoxaline ring system, 3*H*-[1,3]dioxepino[5,6-*b*]-quinoxaline (R.I. 12730), is shown as structure **134**. Bird and Jones (37) synthesized the first member of this series in the form of the 1,5-dihydro derivative.

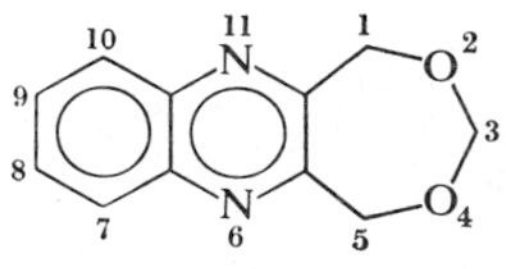

134

b. *1, 5-Dihydro-3*H*-[1,3]dioxepino[5,6-*c*]pyridines*

Korytnyk (35) found that acetone reacts with pyridoxol hydrochloride at −10° to −20°C in the presence of dry hydrogen chloride as a catalyst (Eq. 101), yielding 1,5-dihydro-3,3,8-trimethyl-3*H*-[1,3]dioxepino[5,6-*c*]pyridin-9-ol hydrochloride (**135**·HCl). Other derivatives of this type can be prepared by similar condensation of pyridoxol hydrochloride with other ketones or aldehydes. The free base derivative can be obtained by careful treatment of the hydrochloride salt with potassium carbonate. 1,5-Dihydro-3,3,8-trimethyl-3*H*-[1,3]dioxepino[5,6-*c*]pyridin-9-ol (**135**) (mp 184–185°C) is obtained from its hydrogen chloride salt in this fashion.

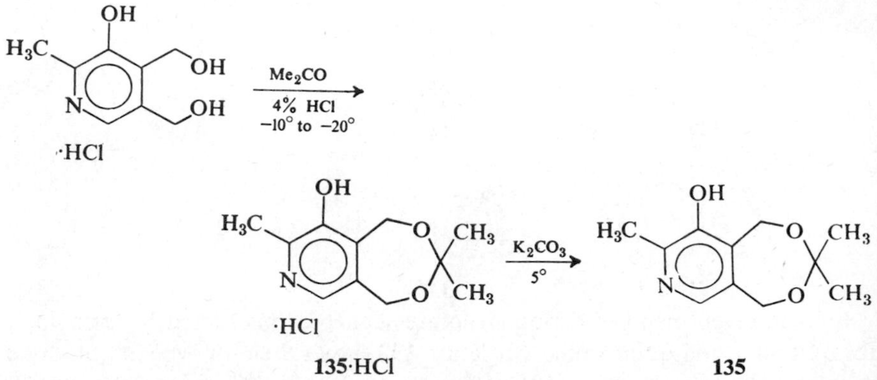

OH, H_3C, OH, OH, N, ·HCl; Me_2CO, 4% HCl, −10° to −20°

OH, H_3C, O, CH_3, N, O, CH_3, ·HCl; K_2CO_3, 5°; OH, H_3C, O, CH_3, N, O, CH_3

135·HCl **135**

(101)

Kimel and Leimgruber (39–42) and Schaeren (43) found that these compounds can also be prepared via the condensation of 1,3-dioxepins with oxazoles (Eq. 102). This reaction is acid-catalyzed and takes place at 80–250°C. It is believed that an unstable intermediate is formed during the reaction. Dioxepins with 2-alkyl, 2-aryl, 2-alkenyl, and 2,2-spiranoalkyl substituents will condense with oxazoles to yield dioxepinopyridinol derivatives; for example, reaction of 4,7-dihydro-1,3-dioxepin with 4-methyloxazole-5-carbonitrile in an autoclave at 150°C in the presence of trichloroacetic acid as a catalyst yields 8-methyl-1,5-dihydro-3*H*-[1,3]dioxepino[5,6-*c*]pyridin-9-ol (**136**) (mp 175–176°C). Other derivatives can be prepared similarly.

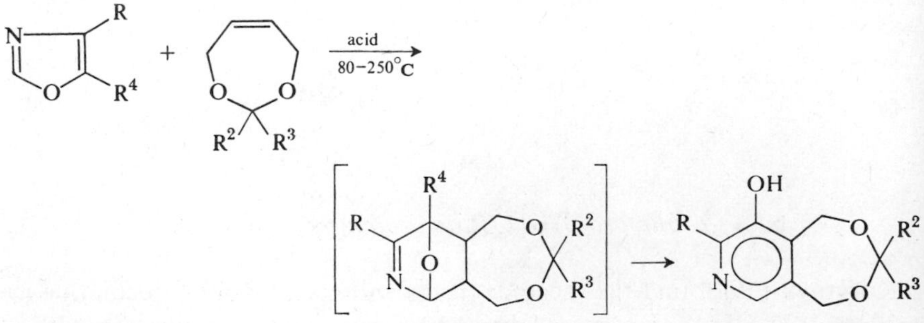

R, N, O, R^4 + O, O, R^2, R^3; acid, 80–250°C

R^4, R, O, R^2, N, O, O, R^3 → OH, R, O, R^2, N, O, R^3

R = alkyl
R^2 = R^3 = H, alkyl, alkenyl, aryl, or polymethylene
R^4 = alkoxy or cyano

(102)

Dioxepinopyridin-9-ols will react with inorganic or organic acids under anhydrous conditions to form acid salts (39, 42). In aqueous hydrogen chloride or methanolic hydrogen chloride at steam bath temperatures they yield the hydrogen chloride salt of pyridoxol (35, 39, 43). Under basic conditions they are stable.

The hydroxy group of dioxepinopyridin-9-ols reacts with acyl halides to produce ester derivatives in high yields, as illustrated in Eq. 103. Structures can be confirmed by comparison of uv spectra with the spectrum of pyridoxol (38).

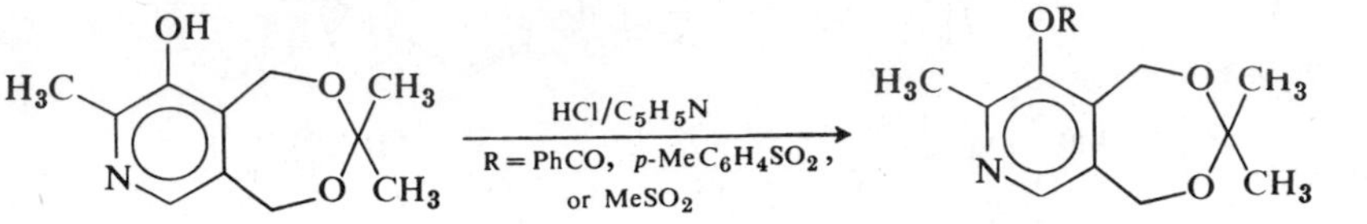
OH
H_3C
O CH_3
N
O CH_3
HCl/C_5H_5N
$R = PhCO$, p-$MeC_6H_4SO_2$, or $MeSO_2$
OR
H_3C
O CH_3
N
O CH_3

135 (103)

Esters can also be prepared by allowing the dioxepinopyridin-9-ols to react with acid anhydrides. As shown in Eq. 104, treatment of 8-methyl-1,5-dihydro-3*H*-[1,3]dioxepino[5,6-*c*]pyridin-9-ol (**136**) with acetic anhydride yields 9-acetoxy-8-methyl-1,5-dihydro-3*H*-[1,3]dioxepino[5,6-*c*]pyridine (**137**). The product (mp 194–195°C) is isolated as a hydrogen chloride salt (**137**·HCl) from ethanol. Other ester derivatives can be obtained similarly (39).

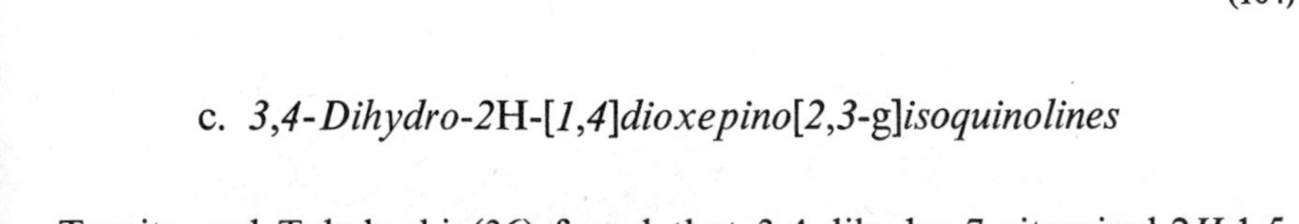
OH
H_3C
O
N
O
Ac_2O
OAc
H_3C
O
N
O
HCl
EtOH
OAc
H_3C
O
N
·HCl
O

136 **137** **137**·HCl (104)

c. *3,4-Dihydro-2*H*-[1,4]dioxepino[2,3-*g*]isoquinolines*

Tomita and Takahashi (36) found that 3,4-dihydro-7-nitrovinyl-2*H*-1,5-benzodioxepin (**138**), on reduction with lithium aluminum hydride in ether and tetrahydrofuran at room temperature (Eq. 105), yielded an amine complex (mp 175–178°C). This complex (**139**), on treatment with formic acid at 120–130°C, gave a N—CHO derivative (**140**); the latter, when allowed to react with phosphorus oxychloride in toluene at 120°C, yielded 3,4,9,10-tetrahydro-2*H*-[1,4]dioxepino[2,3-*g*]isoquinoline (**141**) (picrate derivative, mp 181–185°C dec).

$NO_2CH{=}CH$ — **138** $\xrightarrow[Et_2O-THF,\ 20°C]{LiAlH_4}$ $H_2NCH_2CH_2$ — $\cdot C_2H_4O$ **139** $\xrightarrow[120-130°C]{HCO_2H}$ HCNHCH$_2$CH$_2$ (O=) **140** $\xrightarrow[120°\ C]{POCl_3}$ **141** (105)

Catalytic reduction of 3,4,9,10-tetrahydro-2*H*-[1,4]dioxepino[2,3-*g*]isoquinoline (**141**) with hydrogen in ethanol over Pd–C (Eq. 106) gave 3,4,7,8,9,10-hexahydro-2*H*-[1,4]dioxepino[2,3-*g*]isoquinoline (**142**) (mp 265–268°C). This compound, on treatment with methyl iodide in methanol, yielded a methyl iodide complex (**143**). Treatment of **143** with sodium borohydride furnished 8-methyl-3,4,7,8,9,10-tetrahydro-2*H*-[1,4]dioxepino[2,3-*g*]isoquinoline (**144**) (picrate, mp 181–185°C dec).

141 $\xrightarrow{H_2/Pd-C}$ **142** $\xrightarrow[MeOH]{MeI}$ **143** ($\cdot CH_3I$) $\xrightarrow[MeOH]{NaBH_4}$ **144** (CH_3–N) (106)

d. *Physical Properties*

These are listed in Table V-*26*.

C. Trioxepins

1. Types and Nomenclature

Some seven-membered trioxa unsaturated ring systems are shown in structures **145**, **146**, and **147**. Little is known about these trioxepin compounds.

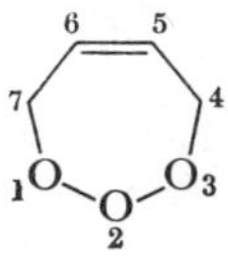

145 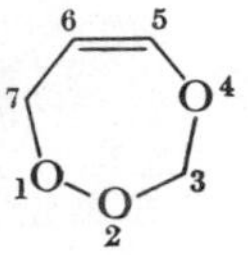 **146** 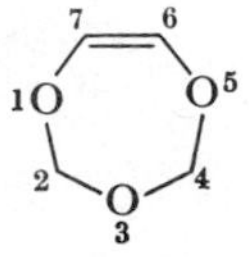 **147** R.I. 335

A number of benzotrioxepin and epoxybenzotrioxepin derivatives are known. Ring systems **148** and **149** are named 2*H*,4*H*-1,3,5-benzotrioxepin (R.I. 1813) and 2,5-epoxy-2*H*,4*H*-1,3,4-benzotrioxepin (R.I. 10305), respectively.

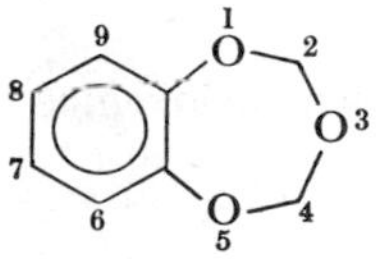

148 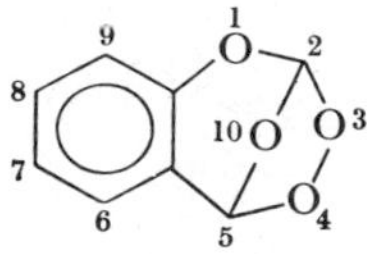 **149**

Other epoxybenzotrioxepin ring systems are shown in structures **150** and **151**. These are called 2,5:8,11-diepoxybenzo[1,2-*e*:4,5-*e*′]bis[1,2,4]trioxepin (R.I. 5525, R.I. 11032) and 2,5-epoxy-5*H*-furo[2,3-*h*]-1,3,4-benzotrioxepin (R.I. 4134), respectively.

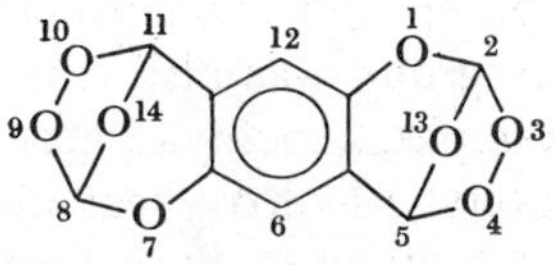

150 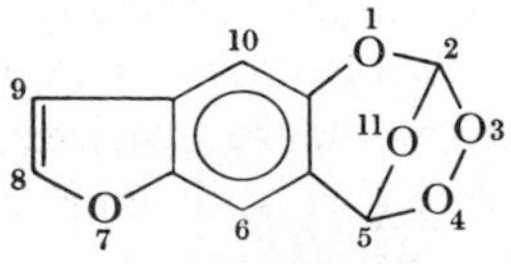 **151**

Other types of trioxepin ring systems are shown in structures **152** and **153**. These are named 9,10-dihydro-9,10-epitrioxyanthracene (R.I. 13376) and

10,15-dihydro-10,15-epoxybenzo[*f*]phenanthro[9,10-*b*][1,4]dioxocin (R.I. 6878).

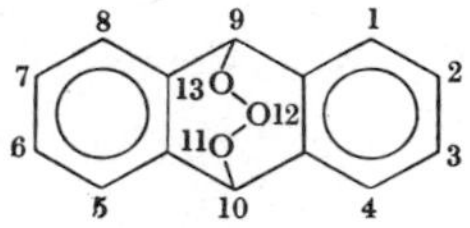

152 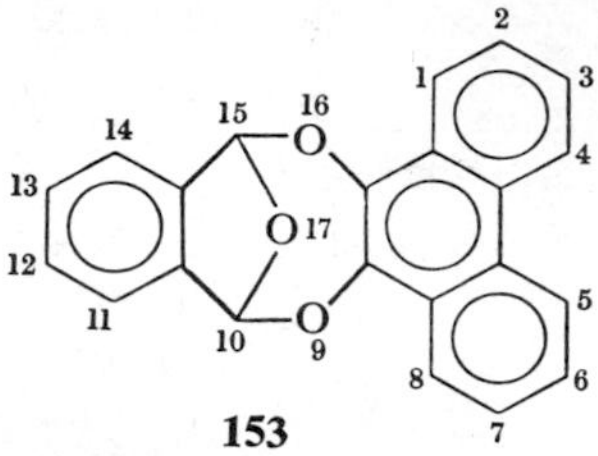 **153**

2. 2*H*,4*H*-1,3,5-Benzotrioxepins

Slooff (122) prepared a derivative of 2*H*,4*H*-1,3,5-benzotrioxepin when he condensed acetaldehyde with 1,2-dihydroxybenzene in the presence of phosphorus pentoxide at $-5°$ to $-10°$C and obtained 2,4-dimethyl-2*H*,4*H*-1,3,5-benzotrioxepin (bp 118°C/20 torrs, mp 34°C). He also prepared 7-nitro-2,4-dimethyl-2*H*,4*H*-1,3,5-benzotrioxepin (mp 111–112°C).

3. 2,5-Epoxy-2*H*,4*H*-1,3,4-benzotrioxepins

A number of 2,5-epoxy-2*H*,4*H*-1,3,4-benzotrioxepin-5-carboxylic acid ester derivatives were prepared by Bernatek and co-workers (123, 124) by treatment of substituted benzofurans in ethyl acetate at 0°C with 4–5% ozone. For example (Eq. 107), 2-methyl-3-ethoxycarbonyl-5-acetoxybenzofuran yielded ethyl 7-acetoxy-2-methyl-2,5-epoxy-5*H*-1,3,4-benzotrioxepin-5-carboxylate (**154**) (mp 105°C). A number of other substituted benzofurans were treated with ozone in a similar manner. These are shown in Table V-27. When only one end of the double bond in the furan ring was substituted, as in methyl benzofuran-2-carboxylate, no stable epoxybenzotrioxepin could be isolated. A free carboxyl group or hydroxy group was also detrimental to the stability of the benzotrioxepin; for example, 2-methyl-5-acetoxybenzofuran-3-carboxylic acid and 2-methyl-3-acetyl-5-hydroxybenzofuran gave unstable products. Ester groups increased stability more than acetyl groups; thus 2-methyl-3-acetyl-5-acetoxybenzofuran yielded 2-methyl-5-acetyl-7-acetoxy-2,5-epoxy-5*H*-1,3,4-benzotrioxepin, which decomposed slowly on standing in the cold, but methyl 2-methyl-5-acetoxybenzofuran-3-carboxylate gave an 81% yield of methyl 7-acetoxy-2-methyl-2,5-epoxy-5*H*-1,3,4-benzotrioxepin-5-carboxylate (mp 88–89°C), which was quite stable.

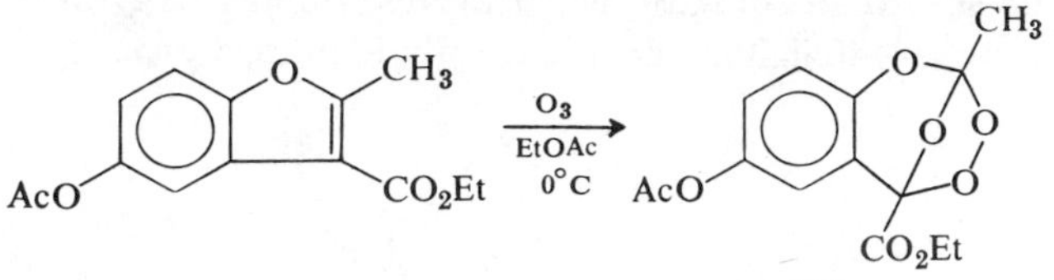

$$\xrightarrow[\text{EtOAc, } 0^\circ\text{C}]{O_3} \quad \textbf{154} \qquad (107)$$

Bernatek and Hvatum (124) calculated from the kinetic data the following activation energies for the formation of ester derivatives of 7-hydroxy-2-methyl-2,5-epoxy-5*H*-1,3,4-benzotrioxepin-5-carboxylic acid:

(*1*) Methyl ester, acetate, 28.2 kcal/mole.
(*2*) Ethyl ester, acetate, 28.7 kcal/mole.
(*3*) Propyl ester, acetate, 28.8 kcal/mole.

Bernatek and Bø (123) found that epoxybenzotrioxepins undergo cleavage on treatment with sodium iodide in glacial acetic acid; for example, ethyl 7-acetoxy-2,9-dimethyl-2,5-epoxy-5*H*-1,3,4-benzotrioxepin-5-carboxylate (**155**) reacted with sodium iodide in glacial acetic acid (Eq. 108) to yield a monoacetate of ethyl 2,5-dihydroxy-3-methylphenylglyoxalate (mp 69°C).

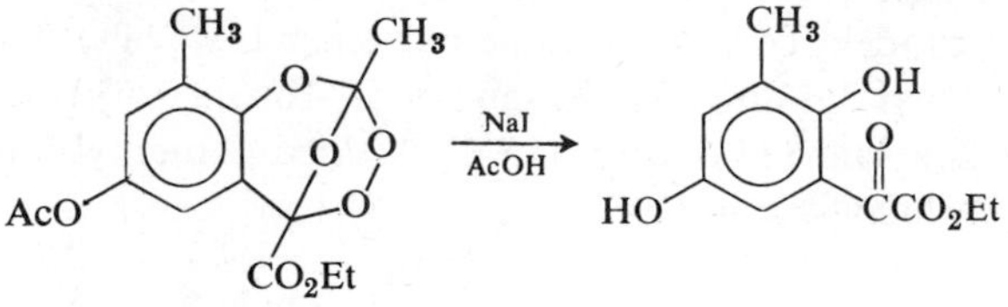

$$\textbf{155} \xrightarrow[\text{AcOH}]{\text{NaI}} \qquad (108)$$

4. 2,5:8,11-Diepoxybenzo[1,2-*e*:4,5-*e*′]bis[1,2,4]trioxepins

Benzotrioxepins of structure **150** were prepared by Bernatek and co-workers (124, 125) by ozonolysis of benzo[1,2-*b*:4,5-*b*′]difuran derivatives and are known as 2,5:8,11-diepoxybenzo[1,2-*e*:4,5-*e*′]bis[1,2,4]trioxepins.

Ozonolysis was carried out in chloroform, ethyl acetate, or glacial acetic acid with 3–4% ozone at room temperature. Products that were somewhat unstable were identified on the basis of their decomposition products. For example, 2,6-dimethylbenzo[1,2-*b*:4,5-*b*′]difuran was treated with ozone (3–4%) in chloroform or acetic acid at room temperature (Eq. 109) to yield

2,8-dimethyl-2,5:8,11-diepoxybenzo[1,2-*e*:4,5-*e*′]bis[1,2,4]trioxepin (**156**). This product was not isolated but was treated with hot water or zinc dust in glacial acetic acid at steam bath temperatures, leading to a 15–20% yield of 2,5-dihydroxyterephthalaldehyde as the only identified product.

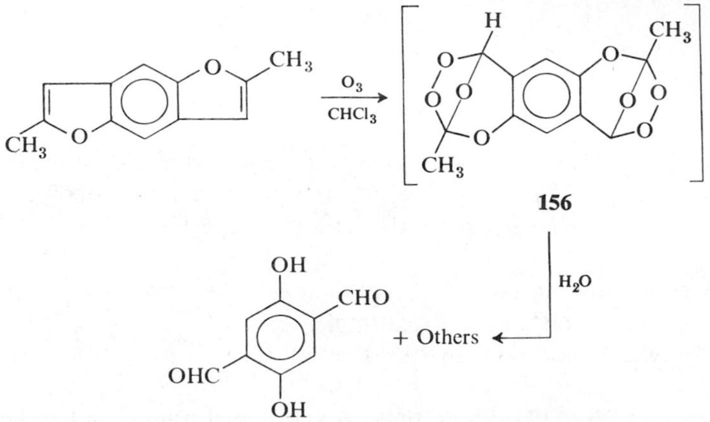

(109)

Ozonolysis of diethyl 2,6-dimethylbenzo[1,2-*b*:4,5-*b*′]difuran-3,7-dicarboxylate in ethyl acetate yielded a mixture of two stable products (Eq. 110). On treatment with hot aqueous sodium hydroxide followed by acidification, diethyl 2,8-dimethyl-2,5:8,11-diepoxybenzo[1,2-*e*:4,5-*e*′]bis[1,2,4]trioxepin-5,11-dicarboxylate (**157**) (mp 132–134°C) yielded 2,5-dihydroxyterephthalic acid which was identified as the dimethyl ether (Eq. 111). Under the same conditions diethyl 2,8-dimethyl-2,5-epoxy-5*H*-furo[2,3-*h*]-1,3,4-benzotrioxepin-5,9-dicarboxylate (**158**) (mp 115°C) yielded 2-methyl-5-hydroxybenzofuran-3,6-dicarboxylic acid (Eq. 112).

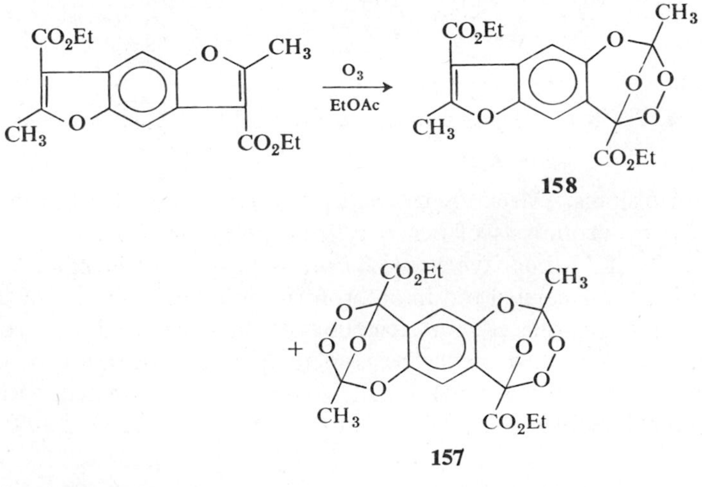

(110)

157 $\xrightarrow{NaOH}$ [2,5-dihydroxyterephthalic acid: OH, CO_2H, HO_2C, OH] (111)

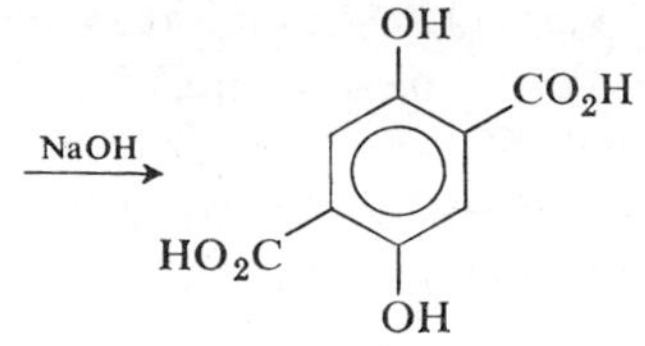

158 $\xrightarrow{NaOH}$ [benzofuran: CO_2H, CH_3, O, OH, CO_2H] (112)

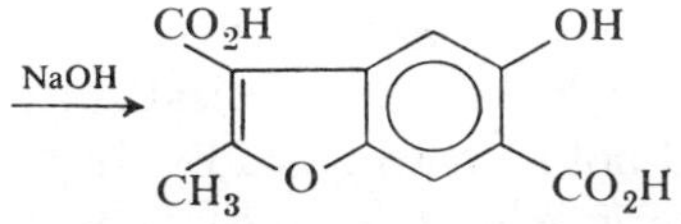

Treatment of 2,6-dimethyl-3,7-diacetylbenzo[1,2-*b*:4,5-*b*′]difuran with ozone in ethyl acetate yielded 2,8-dimethyl-5,11-diacetyl-2,5:8,11-diepoxy-benzo[1,2-*e*:4,5-*e*′]bis[1,2,4]trioxepin which was somewhat unstable and decomposed in hot ethanol. Hydrolysis with aqueous sodium carbonate afforded 2,5-dihydroxyterephthalic acid.

As with the previously mentioned epoxybenzotrioxepins, the more stable derivatives were esters. Derivatives prepared from benzodifurans containing only one substituent on the double bond in the furan ring tended to be unstable. A variety of epoxyfurobenzotrioxepins could be prepared by careful control of the rate of ozonolysis of benzodifurans.

Bernatek and Ledaal (126) found that epoxybenzotrioxepins and diepoxy-benzotrioxepins react with acids (Eqs. 113 and 114) to yield peracids in yields as high as 97%. The peracids can be removed from the reaction mixture by distillation under reduced pressure. Acids such as formic and dichloro-acetic react quite readily without the aid of a catalyst. Reaction with glacial acetic acid and propionic acid was slow, but proceeded quite rapidly when a catalytic amount of perchloric acid was used.

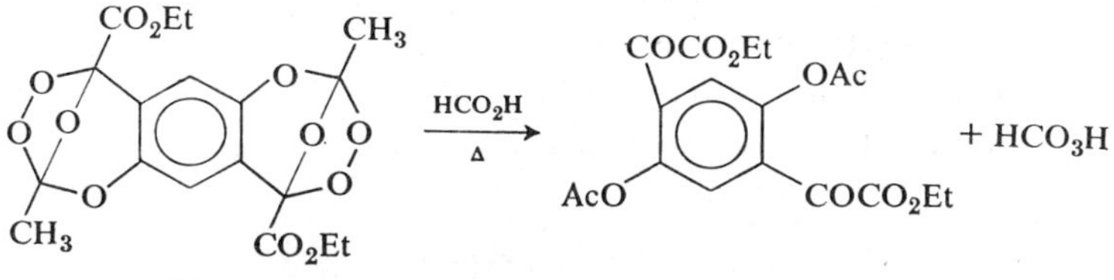

[157: CO_2Et, CH_3, O, O O, O, CH_3, CO_2Et] $\xrightarrow[\Delta]{HCO_2H}$ [$COCO_2Et$, OAc, AcO, $COCO_2Et$] $+ HCO_3H$ (113)

157

[CH_3, O, O O, O, AcO, CO_2R] $\xrightarrow[HClO_4]{AcOH}$ [OAc, AcO, $COCO_2R$] $+ AcO_2H$ (114)

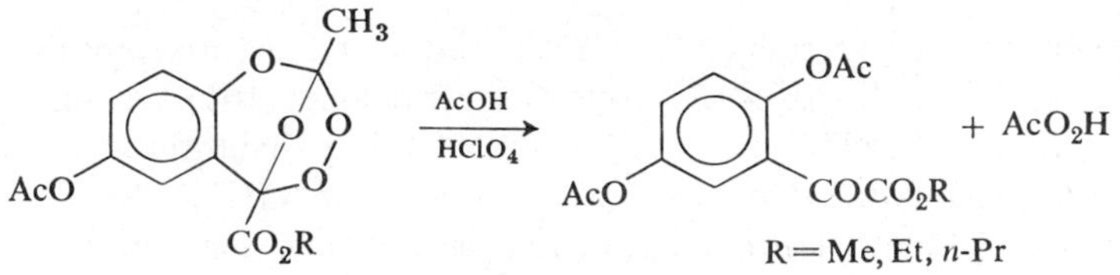

R = Me, Et, *n*-Pr

Bernatek and co-workers (127) found that halogenated diepoxybenzobistrioxepin derivatives could be prepared by ozonolysis of halogenated diethyl 2,6-dimethylbenzo[1,2-*b*:4,5-*b*′]difuran-3,7-dicarboxylates. Thus treatment of diethyl 4,8-dibromo-2,6-dimethylbenzo[1,2-*b*:4,5-*b*′]difuran-3,7-dicarboxylate in ethyl acetate with 4% ozone (Eq. 115) yielded diethyl 6,12-dibromo-2,8-dimethyl-2,5:8,11-diepoxybenzo[1,2-*e*:4,5-*e*′]bis[1,2,4]-trioxepin-5,11-dicarboxylate (**159**) (mp 145°C). Similarly the 6,12-dichloro derivative (**160**) (mp 143°C) was obtained from diethyl 4,8-dichloro-2,6 dimethylbenzo[1,2-*b*:4,5-*b*′]difuran-3,7-dicarboxylate. These products were quite stable and could not be rearranged by treatment with acetic anhydride and concentrated sulfuric acid. They underwent reaction with sodium iodide in glacial acetic acid to yield the product shown in Eq. 115.

O_3 / EtOAc; NaI / AcOH

159; X = Br
160; X = Cl

(115)

Ozonolysis of diethyl 4,8-dibromo-2,6-di(bromomethyl)benzo[1,2-*b*:4,5-*b*′]difuran-3,7-dicarboxylate in ethyl acetate yielded diethyl 6,12-dibromo-2,8-di(bromomethyl)-2,5 : 8,11-diepoxybenzo[1,2-*e*: 4,5-*e*′]bis[1,2,4]trioxepin-5,-11-dicarboxylate (mp 131°C). This compound on reduction with sodium iodide gave the same cleavage product as in the reaction of **159** (Eq. 115).

It may be possible to obtain halogenated diethyl 2,8-dimethyl-2,5-epoxy-5*H*-furo[2,3-*h*]-1,3,4-benzotrioxepin-5,9-carboxylate derivatives by controlled ozonolysis of halogenated benzodifurans.

5. 1,2,3-Trioxepins

Erickson and co-workers (128) proposed a 1,2,3-trioxepin derivative, 9,10-dimethyl-9,10-dihydro-9,10-epitrioxyanthracene (**161**), as the product obtained in 21% yield on oxidation of 9,10-dimethylanthracene with an ozone–nitrogen mixture in methylene chloride at −78°C (Eq. 116). Ozonolysis of 9-methoxy-10-methylanthracene in an acetone–methylene chloride mixture yielded 9-methoxy-10-methyl-9,10-dihydro-9,10-epitrioxyanthracene which

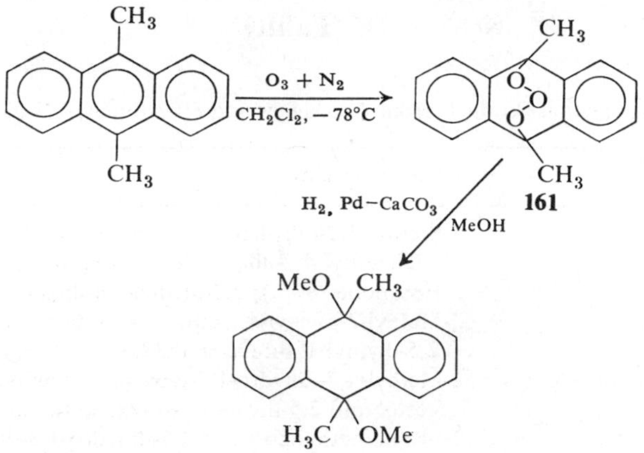

$$\xrightarrow[\text{CH}_2\text{Cl}_2,\ -78^\circ\text{C}]{\text{O}_3 + \text{N}_2}$$ **161** $\xrightarrow{\text{H}_2,\ \text{Pd-CaCO}_3,\ \text{MeOH}}$ (116)

decomposed above $-20°C$ to form 9,10-dimethoxy-9,10-dioxyanthracene and other products. Similar reaction of 9-phenylanthracene failed to give any products of the trioxepin type.

Catalytic hydrogenation of 9,10-dimethyl-9,10-dihydro-9,10-epitrioxyanthracene (**161**) in methanol in the presence of palladium–calcium carbonate (Eq. 116) yields 9,10-dimethoxy-9,10-dimethyl-9,10-dihydroanthracene. Reaction with sodium iodide in methanol leads to the same product after removal of excess iodine with thiosulfate.

6. 1,3,5-Trioxepins

Mustafa (129) found that another type of trioxepin derivative was formed by the action of sunlight for 17 days on phenanthraquinone and 1,3-diphenylisobenzofuran in benzene. Extraction of the crude product obtained after evaporation of the benzene with light petroleum (bp 30–50°C) and recrystallization of the resulting solid from xylene gave 10,15-dihydro-10,15-diphenyl-10,15-epoxybenzo[*f*]phenanthro[9,10-*b*][1,4]dioxocin (mp 220°C) (**162**). This 1,3,5-trioxepin derivative, on being heated with carbon dioxide, regenerated the starting materials as shown in Eq. 117.

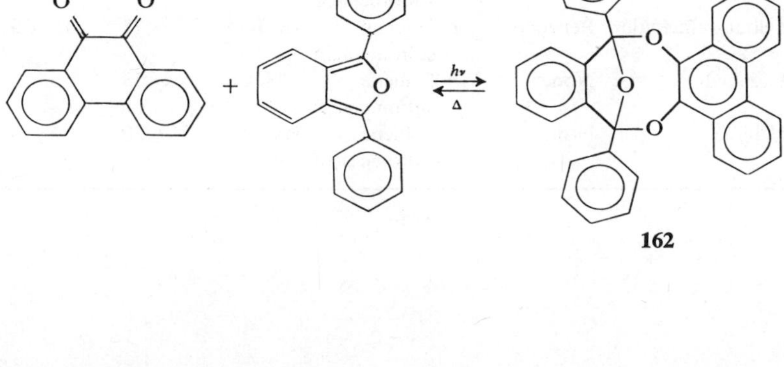

$$\underset{\Delta}{\overset{h\nu}{\rightleftharpoons}}$$ (117)

162

D. Tables

TABLE V-1. Reaction Products from Acetylenes and *cis*-2-Butene-1,4-diol

Acetylene	Products (yield)
Acetylene	2-Methyl-4,7-dihydro-1,3-dioxepin (11%)
Methyl acetylene	2,2-Dimethyl-4,7-dihydro-1,3-dioxepin (35%)
Butyl acetylene	2-Hexanone (87%); 2,5-divinyl-1,4-dioxane (84%)
Phenyl acetylene	2-Methyl-2-phenyl-4,7-dihydro-1,3-dioxepin (6%); 2,5-divinyl-1,4-dioxane (23%)
Propargyl alcohol	2-Methyl-4,7-dihydro-1,3-dioxepin-2-methylol (trace); acetol and 2,5-divinyl-1,4-dioxane (main products)
Propargyl bromide	2-Bromomethyl-2-methyl-4,7-dihydro-1,3-dioxepin (5%); 2,5-divinyl-1,4-dioxane (66%); and bromo acetone (10%)
Propargyl chloroacetate	2,5-Divinyl-1,4-dioxane (57%); and methylaceto chloroacetate (72%)

TABLE V-2. Reaction of *cis*-2-Butene-1,4-diol with Aldehydes

Aldehyde	Solvent	Catalyst	2-R-1,3-Dioxepin	Yield (%)	Refs.
Paraformaldehyde	None	H_2SO_4	H	20	7
Paraformaldehyde	None	*p*-Toluene sulfonic acid	H	69	4
Paraformaldehyde	Benzene	*p*-Toluene sulfonic acid	H	86	3
iso-Butyraldehyde	Benzene	*p*-Toluene sulfonic acid	*i*-Pr	84	3
iso-Butyraldehyde	Hexane	*p*-Toluene sulfonic acid	*i*-Pr	38	15
2-Methylbutyraldehyde	Benzene	*p*-Toluene sulfonic acid	*sec*-Bu	75	3
n-Heptaldehyde	None	*p*-Toluene sulfonic acid	*n*-Hex	53	4
Benzaldehyde	None	*p*-Toluene sulfonic acid	Ph	30	4

TABLE V-3. Reaction of *cis*-2-Butene-1,4-diol with an Acetal or Ketal and Concentrated Sulfuric Acid

Acetal	Solvent	2,2-R,R'-1,3-dioxepin R	R'	Yield (%)	Refs.
Dimethoxymethane	Benzene	H	H	4	9
Acetaldehyde diethyl acetal	None	H	Me	53	9
Acetone dimethyl ketal	None	Me	Me	63, 74	9
Acetone diethyl ketal	None	Me	Me	46	9
Acetone dibutyl ketal	None	Me	Me	50	9
Acetone butyl methyl ketal	None	Me	Me	53	9
2-Butanone dimethyl ketal	None	Et	Me	38	9
Benzaldehyde dimethyl acetal	None	H	Ph	55	9

TABLE V-4. Reaction of *cis*-2-Butene-1,4-diol with Aldehydes and Acetals or Ketals in Benzene with Concentrated Sulfuric Acid

		2,2-R,R′-4,7-dihydro-1,3-dioxepin						
		From aldehyde			From acetal or ketal			
Aldehyde	Acetal or ketal	R	R′	Yield (%)	R	R′	Yield (%)	Refs.
Acetaldehyde	Dimethoxymethane	H	Me	44	H	H	4	10
Acetaldehyde	Acetone dimethyl ketal	H	Me	81	Me	Me	4	10
Paraformaldehyde	Acetone dimethyl ketal	H	H	20	Me	Me	4	10
n-Propionaldehyde	Acetone dimethyl ketal	H	Et	85	None	None		10
n-Butyraldehyde	Acetone dimethyl ketal	H	*n*-Pr	97	None	None		10
Benzaldehyde	Acetone dimethyl ketal	H	Ph	80	None	None		10

TABLE V-5. Reaction of *cis*-2-Butene-1,4-diol with Ketones and Acetals or Ketals in Benzene with Concentrated Sulfuric Acid

		2,2-R,R′-4,7-dihydro-1,3-dioxepin						
		From ketone			From acetal or ketal			
Ketone	Acetal or ketal	R	R′	Yield (%)	R	R′	Yield (%)	Refs.
Acetone	Dimethoxymethane	Me	Me	2	H	H	10	10
2-Butanone	Acetone dimethyl ketal	Et	Me	48	Me	Me	46	10
2-Butanone	Acetone dibutyl ketal	Et	Me	63	Me	Me	8	10
2-Butanone	Benzaldehyde dimethyl acetal	Et	Me	2	H	Ph	51	10
2-Butanone	Cyclohexanone diethyl ketal	None			—$(CH_2)_5$—		71	10
2-Pentanone	Acetone dimethyl ketal	Me	*n*-Pr	38	Me	Me	45	10
2-Pentanone	Acetaldehyde diethyl acetal	Me	*n*-Pr	3	H	Me	78	10
2-Pentanone	Acetone diethyl ketal	Me	*n*-Pr	17	Me	Me	37	10
3-Pentanone	Dimethoxymethane	None			H	H	95	10
3-Pentanone	Acetone dimethyl ketal	Et	Et	40	Me	Me	51	10
Acetophenone	Acetone dimethyl ketal	Me	Ph	24	Me	Me	45	10
Benzophenone	Dimethoxymethane	None			H	H	23	10
Benzophenone	Acetone dimethyl ketal	Ph	Ph	27	Me	Me	58	10

TABLE V-6. Reaction of *cis*-2-Butene-1,4-diol with Ketones in Benzene

Ketone	Catalyst	2,2-R,R′-1,3-dioxepin		Yield (%)	Refs.
		R	R′		
Acetone	H_2SO_4	Me	Me	None	7
2-Butanone	H_2SO_4	Et	Me	None	7
3-Pentanone	H_2SO_4	Et	Et	None	7
4-Methyl-2-pentanone	*p*-Toluene sulfonic acid	Me	2-Methyl-propyl	None	3
Acetophenone	H_2SO_4	Me	Ph	3	7
Benzophenone	H_2SO_4	Ph	Ph	10	7

TABLE V-7. Preparation of 2-Alkenyl and 2-Alkadienyl-4,7-dihydro-1,3-dioxepins by the Reaction of *cis*-2-Butene-1,4-diol with Unsaturated Acetals or Ketals, Unsaturated Aldehydes and Ketones

		2,2-R,R′-4,7-Dihydro-1,3-dioxepin						
		From carbonyl compound			From acetal			
Carbonyl compound	Acetal	R	R′	Yield (%)	R	R′	Yield (%)	Refs.
5-Hexen-2-one	Acetone dimethyl ketal	3-Butenyl	Me	48	Me	Me	36	10
Mesityl oxide	Acetone dimethyl ketal	2-Methylpropenyl	Me	2	Me	Me	72	7
Acrolein	Acetone dimethyl ketal	Polymers			Me	Me	15	7
Tigaldehyde	Acetone dimethyl ketal	Polymers			Me	Me	40	7
Crotonaldehyde	Acetone dimethyl ketal	Propenyl	H	68	None			10
Crotonaldehyde	None	Propenyl	H	25				3
None	Acrolein dimethyl acetal	—	—	—	Polymers			7
2,2-Dimethyl-3,4-octanedienal	Acetone dimethyl ketal	1,1-Dimethyl-2,3-heptadienyl	H	55	None			18
2,2-Dimethyl-3,4-octanedienal	None	1,1-Dimethyl-2,3-heptadienyl	H	81				18
2-Ethyl-2-butyl-5-methyl-3,4-hexadienal	Acetone dimethyl ketal	1-Butyl-1-ethyl-4-methyl-2,3-pentadienyl	H	53	None			18
2-Ethyl-2-butyl-5-methyl-3,4-hexadienal	None	1-Butyl-1-ethyl-4-methyl-2,3-pentadienyl	H	66				18

TABLE V-8. Reaction Products from *cis*-2-Butene-1,4-diol and Cyclic Ketones and Ketals

Carbonyl compound	Ketal	Product	Yield (%)	Refs.
Cyclopentanone	None	6,11-Dioxaspiro[4,6]-undec-8-ene	22	7
Cyclopentanone	Acetone dimethyl ketal	6,11-Dioxaspiro[4,6]-undec-8-ene	78	10
Cyclohexanone	None	7,12-Dioxaspiro[5,6]-dodec-9-ene	73	3
Cyclohexanone	Acetone dimethyl ketal	7,12-Dioxaspiro[5,6]-dodec-9-ene	88	10
None	Cyclohexanone dimethyl ketal	7,12-Dioxaspiro[5,6]-dodec-9-ene	88	9
4-Cyclohexyl cyclohexanone	Acetone dimethyl ketal	3-Cyclohexyl-7,12-dioxaspiro[5,6]dodec-9-ene	80	20
2-Cyclohexyl cyclohexanone	Acetone dimethyl ketal	None, mostly ethers		7
4-*tert*-Butyl cyclohexanone	Acetone dimethyl ketal	3-*tert*-Butyl-7,12-dioxaspiro[5,6]dodec-9-ene	89	20
Camphor	Acetone dimethyl ketal	None		7
Isophorone	Acetone dimethyl ketal	None		7
Tetramethyl-1,3-cyclobutanedione	Acetone dimethyl ketal	None		7
1,4-Benzoquinone	Acetone dimethyl ketal	None		7

TABLE V-9. Reaction Products from *cis*-2-Butene-1,4-diol and Halo Acetals, Halo Aldehydes, or Halo Ketones and Acetals or Ketals

Carbonyl compound	Acetal or ketal	Product	Yield (%)	Refs.
None	Chloroacetaldehyde dimethyl acetal	2-Chloromethyl-4,7-dihydro-1,3-dioxepin	83	7
None	Bromopropanone dimethyl ketal	2-Bromomethyl-2-methyl-4,7-dihydro-1,3-dioxepin	67	9
Bromoacetone	Acetone dimethyl ketal	2-Bromomethyl-2-methyl-4,7-dihydro-1,3-dioxepin	82	10
Chloral	None	Only tars		7
Chloral	Acetone dimethyl ketal	2,2,2-Trichloro-1-methoxyethanol was product		7
Hexachloroacetone	Acetone dimethyl ketal	None		7
o-Chlorobenzaldehyde	Acetone dimethyl ketal	2-(*o*-Chlorophenyl-4,7-dihydro-1,3-dioxepin	70	7
p-Chlorobenzaldehyde	Acetone dimethyl ketal	None		7
p-Chlorobenzaldehyde	None	2-(*p*-Chlorophenyl)-4,7-dihydro-1,3-dioxepin	43	7
p-Chloroacetophenone	None	2-(*p*-Chlorophenyl)-2-methyl-4,7-dihydro-1,3-dioxepin	2	7
p-Chloroacetophenone	Acetone dimethyl ketal	2-(*p*-Chlorophenyl)-2-methyl-4,7-dihydro-1,3-dioxepin	25	10
α-Chloroacetophenone	Acetone dimethyl ketal	2-Chloromethyl-2-phenyl-4,7-dihydro-1,3-dioxepin	27	10

TABLE V-10. Reaction Products from *cis*-2-Butene-1,4-diol and Ring-Substituted Benzaldehydes

Carbonyl compound	Ketal	Product	Yield (%)	Refs.
p-Hydroxybenzaldehyde	None	*p*-(4,7-Dihydro-1,3-dioxepin-2-yl)phenol	10	7
o-Hydroxybenzaldehyde	None	*o*-(4,7-Dihydro-1,3-dioxepin-2-yl)phenol	10	7
o-Hydroxybenzaldehyde	Acetone dimethyl ketal	*o*-(4,7-Dihydro-1,3-dioxepin-2-yl)phenol	57	21
p-Anisaldehyde	Acetone dimethyl ketal	2-(*p*-Methoxyphenyl)-4,7-dihydro-1,3-dioxepin	44	21
m-Nitrobenzaldehyde	None	2-(*m*-Nitrophenyl)-4,7-dihydro-1,3-dioxepin	79	7
Piperonal	None	2-(3,4-Methylenedioxyphenyl)-4,7-dihydro-1,3-dioxepin	52	7

TABLE V-11. Reaction of *cis*-2-Butene-1,4-diol with a Trialkyl Orthoformate and Concentrated Sulfuric Acid

Trialkyl ortho ester	Reaction temperature (°C)	Yield (%)	2,2-R,R′-4,7-Dihydro-1,3-dioxepin		Refs.
			R	R′	
Trimethyl orthoformate	114	10	H	MeO	7
Trimethyl orthoformate	25	65	H	MeO	7
Tri-*n*-butyl orthoformate	25	90	H	n-C_4H_9O	7
Tri-*n*-pentyl orthoformate	25	70	H	n-$C_5H_{11}O$	7
Trimethyl orthoacetate	25	50	Me	MeO	7
Triethylorthopropionate	97	dec	Et	EtO	7

TABLE V-12. Reaction of *cis*-2-Butene-1,4-diol with an Alkoxy Acetal and an Acidic Catalyst

Alkoxy acetal	Solvent	Reaction temperature (°C)	2,2-R,R′-4,7-dihydro-1,3-dioxepin R	R′	Yield (%)	Refs.
Methoxyacetaldehyde dimethyl acetal	Benzene	97[a]	H	Methoxymethyl	67	7
3-Methoxybutyraldehyde dimethyl acetal	Benzene	109[a]	H	2-Methoxypropyl	Mixture	7
3-Methoxypropionaldehyde dimethyl acetal	Benzene	112[b]	H	Ethoxyethyl	80	7
1-*p*-Tolyloxy-2-propanone dimethyl ketal	None	130[b]	Me	(*p*-Tolyloxy)methyl	75	7

[a] *p*-Toluenesulfonic acid.
[b] Concentrated sulfuric acid.

TABLE V-13. 6,9-Methano-2,4-Benzodioxepin Derivatives

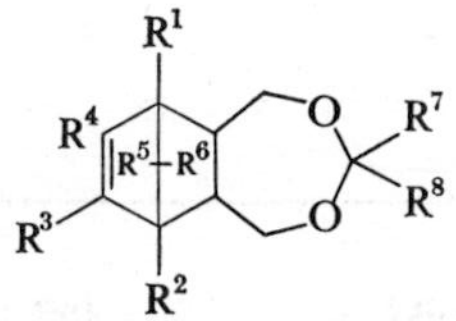

Cyclopentadiene derivative	1,3-Dioxepin derivative	R^1	R^2	R^3	R^4	R^5	R^6	R^7	R^8	Physical properties
Hexahydro	4,7-Dihydro	H	H	H	H	H	H	H	H	bp 117°C/15 mm
Hexahydro	2-Isopropyl	H	H	H	H	H	H	Isopropyl	H	bp 140°C/12 mm
Hexahydro	Spirobutylene	H	H	H	H	H	H	—Butylene—		—
Hexachloro	4,7-Dihydro	Cl	Cl	Cl	Cl	Cl	Cl	H	H	mp 108–109°C
Hexachloro	2-Methyl	Cl	Cl	Cl	Cl	Cl	Cl	Methyl	H	bp 163°C/2 mm
Hexachloro	2-Isopropyl	Cl	Cl	Cl	Cl	Cl	Cl	Isopropyl	H	mp 52–54°C
Hexachloro	2-Hexyl	Cl	Cl	Cl	Cl	Cl	Cl	Hexyl	H	bp 230°C/3 mm
Hexachloro	2-(Methoxyethyl)	Cl	Cl	Cl	Cl	Cl	Cl	$MeOCH_2CH_2$—	H	bp 195–198°C/3 mm
Tetrachloro-5,6-dimethoxy	4,7-Dihydro	Cl	Cl	Cl	Cl	MeO	MeO	H	H	mp 97–98°C
Tetrachloro-5,6-difluoro	4,7-Dihydro	Cl	Cl	Cl	Cl	F	F	H	H	bp 142–144°C/2 mm

TABLE V-14. Physical Properties of 1,3-Dioxepins

R	R^1	Boiling point	Refractive index	Refs.
H	H	126°C/760 mm		1, 2
		127.8–128.2°C/734 mm	1.4570/20°C	3
		124°C/760 mm	1.4508/25°C	9, 10
H	Me	137–138°C/760 mm	1.4478/25°C	9, 10, 14
H	Et	95°C/95 mm	1.4489/25°C, sp. gr. 0.989/25°C	10
H	*n*-Pr	88°C/33 mm	1.4496/25°C, sp. gr. 0.953/25°C	10
H	*iso*-Pr	170–170.6C°/735 mm	1.4484/20°C	3
		105°C/100 mm	1.4472–1.4474/20°C	32
		94–99°C/90 mm	1.4470/25°C	29

TABLE V-14. (*Continued*)

R	R^1	Boiling point	Refractive index	Refs.
H	*sec*-Bu	190–193°C/730 mm	1.4514/20°C	3
H	*n*-Hexyl	93°C/2 mm	1.4527/25°C	4
Me	Me	41°C/6 mm	1.4458/25°C	9
Me	Et	101°C/89 mm	1.4500/25°C	9
Me	*n*-Pr	116°C/41 mm	1.4510/25°C	10
Et	Et	108°C/57 mm	1.4532/25°C	10
H	Propenyl	54–55°C/4.5 mm	1.4739/20°C	3
Me	(3-Butenyl)	122°C/30 mm	1.4652/25°C	9
H	(1-Butyl-1-ethyl-4-methyl-2,3-pentadienyl)	139°C/2.2 mm	1.4890/25°C	18
H	(1,1-Dimethyl-2,3-heptadienyl)	131°C/5.3 mm	1.4875/25°C	18
H	(3-Cyclohexen-1-yl)	83°C/0.4 mm	1.4975/25°C	19
H	(2,4,6-Trimethyl-3-cyclohexen-1-yl)	99°C/1.0 mm	1.4948/25°C	19
H	Ph	114°C/3.5 mm	1.5387/25°C	4
		79°C/0.6 mm	1.5405/25°C, sp. gr. 1.109/25°C	9, 10
Ph	Me	104°C/2.0 mm	1.5282/25°C	10
Ph	Ph	157°C/1.2 mm		10
H	*o*-ClC_6H_4	111°C/0.9 mm	1.5566/25°C	7
H	*p*-ClC_6H_4	121°C/1.3 mm		7
Me	*p*-ClC_6H_4	125°C/1.4 mm	1.5405/25°C	10
Ph	$ClCH_2$	147°C/5.0 mm	1.5542/25°C	7
H	*o*-HOC_6H_4	121°C/0.2 mm		21
H	*p*-$MeOC_6H_4$	136°C/1.9 mm	1.5463/25°C	21

H	3,4-Methylenedioxyphenyl	133°C/0.7 mm	1.5585/25°C	7
H	*m*-$NO_2C_6H_4$	mp 62–64°C		7
———	$(CH_2)_4$ ———	107°C/35 mm	1.4798/25°C, sp. gr. 1.052/25°C	10
———	$(CH_2)_5$ ———	94°C/10 mm	1.4876/20°C	3
		54°C/0.4 mm	1.4850/25°C, sp. gr. 1.047/25°C	9
———	-(3-*tert*-Butylpentamethylene)—	112°C/0.7 mm	1.4825/25°C	20
———	-(3-Cyclohexylpentamethylene)—	152°C/1.0 mm	1.5052/25°C	20
H	$ClCH_2$	100°C/10 mm	1.4795/25°C	9
Me	$BrCH_2$	116°C/10 mm	1.4996/25°C	9
H	(3,4-Dihydro-2*H*-pyran-2-yl)	104°C/1.1 mm	1.4968/25°C	23
H	(2-Furyl)	80°C/1.5 mm	1.5044/25°C	23
H	CH_3COCH_2	82°C/1.6 mm		22
Me	Ac	57°C/0.9 mm	1.4635/25°C	22
Me	EtO_2C	130°C/10 mm	1.4578/25°C	24
H	(2-Methylene-4,7-dihydro-1,3-dioxepin-2-yl)	100°C/1.8 mm mp 58–59°C		16
H	(*p*-4,7-Dihydro-1,3-dioxepin-2-ylphenyl)	mp 134–136°C		16
H	MeO	48°C/4.2 mm	1.4530/25°C	7
H	*n*-BuO	—	1.4435/25°C	7
H	*n*-$C_5H_{11}O$	79°C/110 mm	1.4560/25°C	7
H	$MeOCH_2$	103°C/35 mm	1.4566/25°C	7
H	$EtOCH_2CH_2$	90°C/4.5 mm	—	7
Me	MeO	44°C/2.8 mm	1.4520/25°C	7
Me	(*p*-Tolyloxy)methyl	128°C/0.6 mm	1.4528/25°C	7
H	Styryl	115°C/0.5 mm	1.5708/25°C	7

TABLE V-15. Physical Properties of 5-Substituted 1,3-Dioxepins

Compound	Physical constants	Refs.
5-Methoxy-4,5-dihydro-1,3-dioxepin	bp 48–50°C/7 mm, n_D^{20} 1.4568	3
5-Bromo-4,7-dihydro-1,3-dioxepin	bp 61–61.6°C/7 mm, n_D^{20} 1.5128	3

TABLE V-16. Physical Properties of 5*H*-1,4-Benzodioxepins

Compound	mp (°C)	Refs.
2,3-Dihydro-3,5,7,8-tetramethoxy-2,2-dimethyl-5*H*-1,4-benzodioxepin	120–123.5	45
9-Chloro-2,3-dihydro-5-hydroxy-6,8-dimethoxy-5-methyl-5*H*-1,4-benzodioxepin-3(2*H*)-one	143–144	49
9-Chloro-2,3-dihydro-5,6,8-trimethoxy-5-methyl-5*H*-1,4-benzodioxepin-3(2*H*)-one	97	49
2,3-Dihydro-6,8-dimethoxy-5-(2,4,6-trimethoxy-benzylidene)-5*H*-1,4-benzodioxepin-3(2*H*)-one	240	50
2,3-Dihydro-5*H*-1,4-benzodioxepin-5-one	82	48
2,3-Dihydro-5*H*-1,4-benzodioxepin-2,5-dione	115–115.5	52, 53
2,3-Dihydro-8,9-dimethoxy-5*H*-1,4-benzo-dioxepin-2,5-dione	175	51
3,6-Epoxy-2,3,4,6-tetrahydro-1,5-benzodioxocin	63–64	46
	65	
	bp 113–114/1.0 mm	47

TABLE V-17. 2-Aminoalkyl-3,4-Dihydro-2*H*-1,5-Benzodioxepins (65)

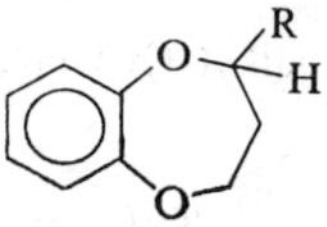

R	Amine	Amine · HCl, mp (°C)
N-Methyl-2-methylamine	bp 110°C/0.1 mm, n_D^{25} 1.5494	311–312
N-Ethyl-2-methylamine		193–195
N-Isopropyl-2-methylamine		188–189
N-*n*-Butyl-2-methylamine		218–219
N-Allyl-2-methylamine		207–208
N-(3-Pyridyl)-2-methylamine		153–154
N-(4-Morpholinyl)-2-methylamine		222–224
N-(1-Pyrrolidinyl)-2-methylamine		230–231
N-(1-Hydroxy-2-propyl)-2-methylamine	bp 150–152°C/0.1 mm	
N-(4-Methylpiperazinyl)-2-methylamine	bp 139°C/0.1 mm	
N-(Hexamethylenimino)-2-methylamine	bp 140–142°C/0.15 mm	
N-(Piperidinyl)-2-methylamine		221–222
N-(*N'*-Ethyl-*N'*-diethylcarbamoylmethyl)-2-mcthylamine	bp 180–182°C/0.3 mm	

TABLE V-18. 2-Guanidinoalkyl-2*H*-3,4-Dihydro-1,5-Benzodioxepin Sulfates (66)

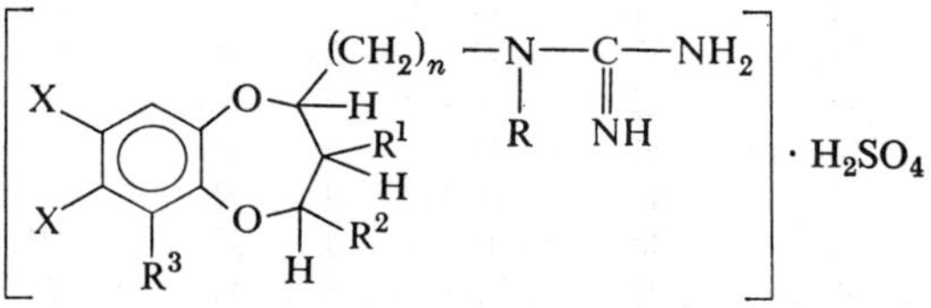

$\cdot H_2SO_4$

R	*n*	R¹	R²	R³	X
H	1	H	H	H	Cl
H	1	Me	H	H	H
H	1	H	H	H	MeO
H	1	H	H	MeO	H
H	2	H	H	H	H
Me	1	H	H	H	H

TABLE V-19. 2-Guanidinoalkyl-3,4-Dihydro-2*H*-1,5-Benzodioxepins (66, 67)

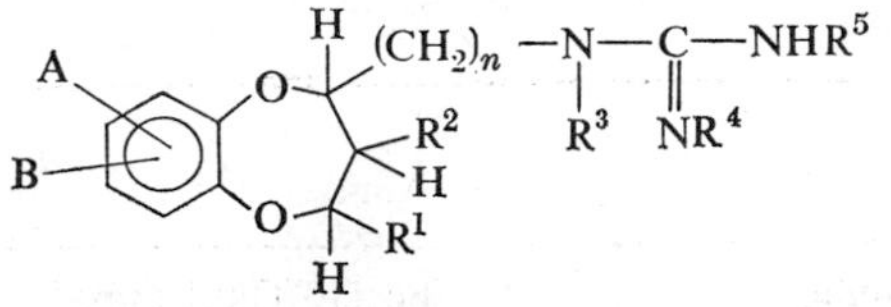

A	B	R^1	R^2	*n*	R^3	R^4	R^5
H	H	Me	H	2	H	H	H
7-EtO	8-EtO	H	H	4	H	H	H
7-Cl	H	H	*n*-Pr	1	H	H	*n*-Bu
7-Br	8-Br	H	H	3	Me	H	H
H	H	H	H	4	H	H	H
6-Me	H	H	H	2	H	Me	Me
7-MeO	H	*n*-Bu	H	4	H	H	H
H	9-Cl	H	Me	1	Et	H	H
6-Cl	H	H	H	2	H	Et	Et
7-Br	H	Me	H	3	Me	H	Me
7-Cl	8-Cl	H	H	4	H	H	H
7-Et	8-Et	H	H	1	H	H	H
6-EtO	H	H	Et	2	H	H	H
H	8-MeO	H	H	4	H	H	H
7-Cl	8-Cl	*n*-Pr	H	1	H	H	Et
6-Br	H	H	Me	2	*n*-Bu	H	H
6-*n*-Bu	H	H	H	3	H	H	H
7-EtO	H	Me	H	4	H	Me	Me
H	9-Br	H	H	2	H	H	*iso*-Pr
6-Cl	H	H	*iso*-Pr	3	Me	H	H
H	9-*n*-Bu	H	H	1	H	H	H
H	H	H	*n*-Bu	4	H	H	H
7-Br	H	H	H	1	H	*n*-Pr	*n*-Pr
7-Me	8-Me	H	Me	4	H	Me	Me
7-MeO	8-MeO	Et	H	3	*iso*-Pr	H	H
H	H	H	Et	4	H	H	Me
H	9-Cl	Me	H	2	Me	H	H
H	H	H	*n*-Pr	4	H	H	H

TABLE V-20. Amide Derivatives of 3,4-Dihydro-2*H*-1,5-Benzodioxepins

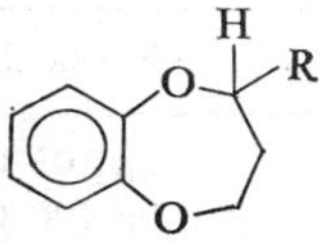

R	mp (°C)	Refs.
2-Carboxamide	169–170	66
N-Ethyl-2-carboxamide	102–103	68
N-Isopropyl-2-carboxamide	113–114	68
N-*n*-Butyl-2-carboxamide	88–89	68
N-Allyl-2-carboxamide	66–67	68
2-Carboxylic acid-, hydrazide	101–103	68
N-(2-Pyridyl)-2-carboxamide	103–105	68
N-(3-Pyridyl)-2-carboxamide	155–157	65, 68
N-(5-Chloro-2-benzoxazolyl)	169–170	68
2-Oxamido	174–176	68
2-(3,4,5-Trimethoxybenzamidomethyl)	169–170	68
2-(Nicotinamidomethyl)	150–151	68
2-(Nicotinamidomethyl)-3-methyl	164–166	68

TABLE V-21. Physical Properties of 3,4-Dihydro-2*H*-1,5-Benzodioxepins

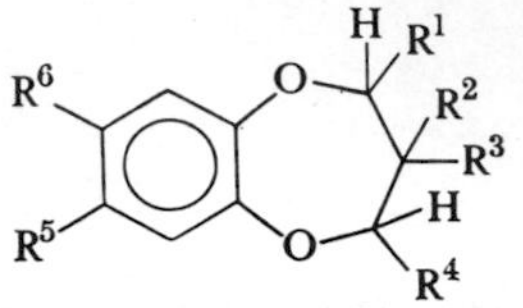

R^n groups are hydrogen unless otherwise noted

R group	Physical properties	Refs.
—	bp 103°C/10 mm	54
	bp 100°C/12 mm	55
	bp 110–114°C/18 mm	56
	bp 100–105°C/11 mm	57
	bp 100–102°C/16 mm	63, 72
	d_4^{20} 1.341	54
R^2R^3 = methylene	bp 113–113.5°C/10 mm, n_D^{20} 1.5553	58
R^2 = Me	bp 111–112°C/12 mm, n_D^{20} 1.5310	58
R^5 = Ac	bp 120°C/0.5 mm, n_D^{20} 1.566	62
R^5 = NH_2	bp 105–110°C/1.0 mm, mp 72–75°C	56
Other amine and amide derivatives listed in Tables V-17, V-18, and V-19		
R^5 = $BrCH_2CO$	mp 75–76°C	62
R^5 = CH_2Cl	bp 128°C/0.11 mm	61
R^1 = CO_2Et	bp 112–121°C/0.5 mm, n_D^{25} 1.5203	68
R^5 = CHO	bp 127–129°C/0.25 mm	61
(semicarbazone, mp 188–190, 183–184°C)		59, 61
R^5 = OH	bp 122–125°C/0.5 mm, mp 94–95°C	56
R^5 = OH; R^6 = *tert*-Bu	mp 133–135°C	56
R^2 = OH and Me	mp 77–79°C	63
R^5 = CH_2OH	bp 146–148°C/0.11 mm	61
(phenylurethane, mp 91–91.5°C)		61
R^5 = $COCH_2NHCHMe_2 \cdot HCl$	mp 204–205°C	62
R^5 = $CH(OH)CH_2NHCHMe_2 \cdot HCl$	mp 163–164°C	62
R^5 = NO_2	mp 106–107°C	56
R^5 = nitrovinyl	mp 92–93°C	59
R^2R^3 = keto	mp 40–42°C	63
(4-nitrophenylhydrazone, mp 210–211°C dec) (2,4-dinitrophenylhydrazone, mp 164°C dec) (semicarbazone, mp 230–233°C)		58, 63
R^2 = MeO and (2,5-dihydroxy-4-methoxybenzyl), R^4 = keto, R^6 = MeO	mp 218°C	71
R^2 = MeO and (2-hydroxy-4,5-dimethoxybenzyl), R^4 = keto, R^6 = MeO	mp 160°C	71
R^1 = keto, R^2 = $PhCH_2$, R^4 = keto	mp 253–255°C	74

TABLE V-22. Physical Properties of 1*H*-2,3-Benzodioxepins

Compound	mp (°C)	Refs.
1*H*-2,3-Benzodioxepin-1-ol: 4,5-dihydro-4-ethoxy	105–108	75
1*H*-2,3-Benzodioxepin: 4,5-dihydro-1,4-diethoxy	80–81	75
1,4-Epoxy-1*H*-2,3-benzodioxepin: 4,5-dihydro		
(a) 1-*tert*-butyl-5-(*tert*-butylphenylvinylidiene)	140	78
(b) 1,4-diphenyl	122–123	76
(c) 6-methoxy-8-(2-methoxyethyl)	95–96	79
(d) 4-phenyl	98	77
(e) 5-phenyl	89	77
1,4-Epoxy-1*H*-2,3-benzodioxepin-5-one: 4,5-dihydro		
(a) 1,4-diphenyl	105 dec	78
(b) 1-ethyl-4-phenyl	50 dec	80
(c) 4-methyl-1-phenyl	90 dec	78
(d) 1-phenyl-4-(*o*-carbomethoxyphenyl)	122–128 dec	81

TABLE V-23. Physical Properties of 3*H*-2,4-Benzodioxepins

Compound	Physical properties	Rcfs.
3*H*-2,4-Benzodioxepin: 1,5-dihydro	mp 37–38°C, bp 97–107°C/0.7 mm	83
(a) 3-methylene	mp 44°C, bp 110°C/0.01 mm	82
(b) 3-ethylene, spiro derivative	mp 78°C, bp 90°C/0.01 mm	82
(c) 3-(bromomethyl)	mp 98°C	82
1,3-Dioxolo[4,5-*h*][2,4]benzodioxepin	mp 148.5–149.5°C	85

TABLE V-24. 11*H*-Dibenzo[*b*,*e*][1,4]dioxepin-11-ones

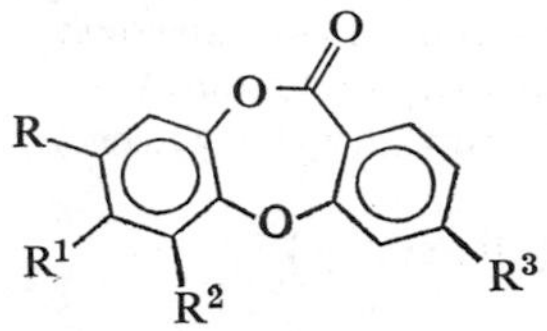

Starting acid	R	R^1	R^2	R^3	mp (°C)	Refs.
2-(2-Hydroxyphenoxy)-benzoic acid	H	H	H	H	65.5–66, rhombs	105
2-(2-Hydroxy-4-methyl-phenoxy)benzoic acid	Me	H	H	H	97–99, needles	107
2-(2-Hydroxy-5-methyl-phenoxy)benzoic acid	H	Me	H	H	48–50, needles	107
2-(2-Hydroxy-6-methyl-phenoxy)benzoic acid	H	H	Me	H	105–106, needles	107
2-(2-Hydroxy-4-formyl-phenoxy)benzoic acid	CHO	H	H	H	192–193, needles	107
2-(2-Hydroxy-5-formyl-phenoxy)benzoic acid	H	CHO	H	H	149–150, prisms	107
Dibromo-2-(2-hydroxy-phenoxy)benzoic acid	H	7,9-Br_2(?)	H	H	222–223, prisms	107
Dibromo-2-(2-hydroxy-4-propylphenoxy)benzoic acid	*n*-Pr	7,9-Br_2(?)	H	H	176–176.5, prisms	107
2-(2,6-Dihydroxyphenoxy)-benzoic acid	H	H	OH	H	142–144	107
2-(2-Hydroxyphenoxy)-4-methoxybenzoic acid	H	H	H	MeO	143.1–143.6, prisms	106
2-(2,4-Dihydroxyphenoxy)-4-methoxybenzoic acid	OH	H	H	MeO	230–231	106

TABLE V-25. Physical Properties of Dibenzodioxepins

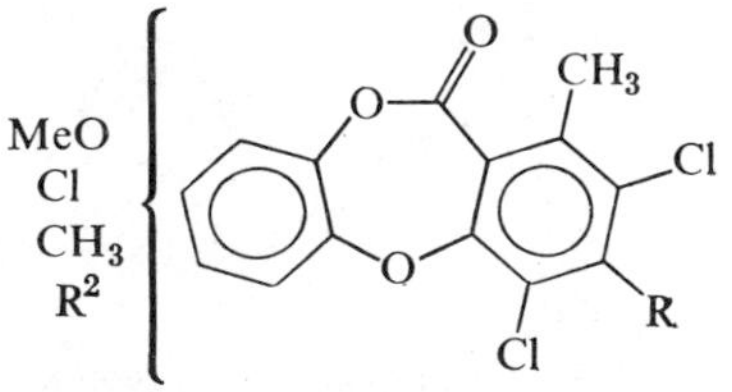

Compound	Physical constants	Refs.
11*H*-Dibenzo[*b*,*e*][1,4]dioxepin	bp 112–114°C/0.2 mm	102
	bp 130–140°C/0.5 mm	103
11*H*-Dibenzo[*b*,*e*][1,4]dioxepin-11-one	mp 58–62°C, needles	104
	mp 62–64°C	105
	mp 65.5–66.0°C, rhombs	105
11*H*-Dibenzo[*b*,*e*][1,4]dioxepin-11-one: 3(R)-2,4-dichloro-8-methoxy-1,6,9-trimethyl	(Vicanicin derivatives)	
(a) R = OH	mp 248–250°C, needles	93
(b) R = AcO	mp 213–214°C, rectangular prisms	93
(c) R = BzO	mp 190–191°C, tablets	93
(d) R = EtO	mp 185–186°C, needles	93
(e) R = MeO	mp 193–194°C, needles	93
11*H*-Dibenzo[*b*,*e*][1,4]dioxepin-11-one: 3-(R^1)-*z*-(R^2)-1,*y*-dimethyl-*x*-methoxy-2 4,*w*-trichloro	(Nidulin derivatives)	
(a) R^1 = OH, R^2 = (1-methylpropenyl)	mp 180–181°C	98
	mp 180°C	97
(b) R^1 = OH, R^2 = *tert*-Bu	mp 147–150°C	98
(c) R^1 = MeO, R^2 = (1-methylpropenyl)	mp 143–145°C	98
	mp 145°C	97
(d) R^1 = MeO, R^2 = *tert*-Bu	mp 172–173°C	98
	mp 168°C	97
(e) R^1 = OAc, R^2 = (1-methylpropenyl)	mp 167–168°C	97

TABLE V-25. (*Continued*)

Compound	Physical constants	Refs.
Dibenzo[*d,f*][1,3]dioxepin, spiro derivatives		
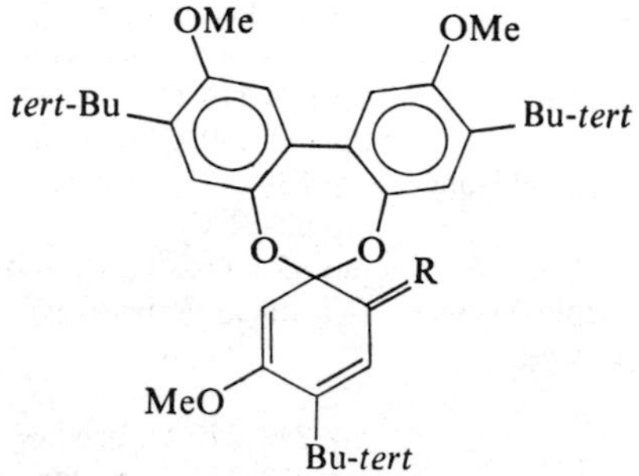		
(a) R=O	mp 209–210°C	113
(b) R=NOH	mp 195–196°C	113
Dibenzo[*d,f*][1,3]dioxepins		
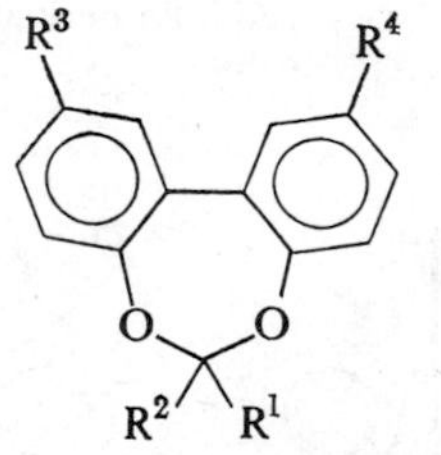		

	R^1	R^2	R^3	R^4	Physical constants	Refs.
(a)	H	Me	H	H	mp 48–49°C, prisms	110
(b)	Me	Me	H	H	mp 77.5–78.5°C, rectangular plates	110
(c)	*n*-Bu	*n*-Bu	H	H	mp 56.8–57.0°C	114
(d)	Ph	Ph	H	H	mp 168–169.5°C needles	110
					mp 169.5–169.6°C	114
(e)	Me	Ph	H	H	mp 100.3–100.5°C	114
(f)	H	EtO_2C	H	H	bp 165–170°C/0.7 mm	111
(g)	Me	Me	OH	OH	mp 193–194°C, prisms	110
(h)	MeO	MeO	MeO	MeO	mp 154–154.5°C, prisms	110
(i)	$—(CH_2)_5—$		OH	OH	mp 175–218°C, prisms	110
(j)	$—(CH_2)_5—$		H	H	mp 67–68°C, prisms	110
(k)	$—(CH_2)_5—$		MeO	MeO	mp 135–136°C	110
(l)	—O—		H	H	mp 101–102°C	111, 112
(m)	—S—		H	H	mp 83.5–85.0°C	114
6,6′-spirobi(dibenzo[*d,f*][1,3,2]-dioxepin)					mp 339.5–340.0°C	114

TABLE V-26. Physical Properties of Dioxepino Derivatives

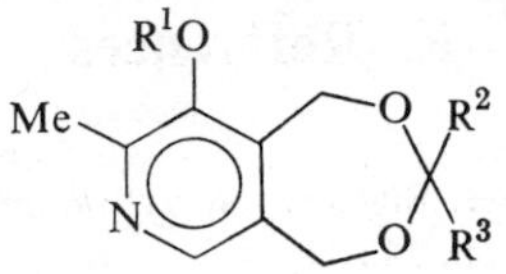

R^1	R^2	R^3	mp (°C) free base	mp (°C) HCl salt	Refs.
H	H	H	175–176	208–208.5 dec	39, 42
H	Me	Me	184–185		35
H	H	*i*-Pr	164–164.5	190–191 dec	39, 42
H	H	Ph	160–160.5		39, 42
H	—$(CH_2)_5$—		167–169		39, 42
Ac	H	H	86.5–87.5	194–195 dec	39, 42
Ac	H	*i*-Pr		173–173.5 dec	39, 42
Bz	Me	Me	107–109		35
Ts	Me	Me	145–146		35
Ms	Me	Me	72–73		35

2*H*-[1,4]Dioxepino[2,3-*g*]isoquinoline

	Refs.
(a) 3,4,9,10-tetrahydro, picrate derivative, mp 181–185°C	36
(b) 3,4,7,8,9,10-hexahydro derivative, mp 265–268°C	36
(c) 8-methyl-3,4,7,8,9,10-hexahydro, picrate derivative, mp 181–185°C dec	36

TABLE V-27. 2,5-Epoxy-2*H*,4*H*-1,3,4-Benzotrioxepins

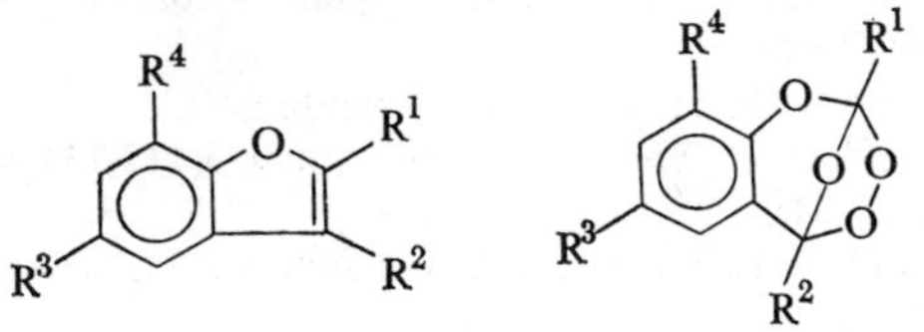

R^1	R^2	R^3	R^4		Refs.
Me	CO_2Et	AcO	H	mp 105°C	123, 124
Me	COMe	HO	H	Unstable	123
Me	COMe	AcO	H	Unstable	123, 124
Me	CO_2Et	HO	Me	Unstable	123
Me	CO_2Et	AcO	Me	mp 112°C	123, 124
Me	COMe	HO	Me	Unstable	123
Me	COMe	AcO	Me	Unstable	123, 124
Me	CO_2H	AcO	H	Unstable	124
Me	CO_2Me	AcO	H	mp 88–89°C	124
Me	CO_2Pr-*n*	AcO	H	mp 68°C	124
CO_2Me	H	H	H	Unstable	124

E. References

1. J. W. Copenhaver and M. H. Bigelow, *Acetylene and Carbon Monoxide Chemistry*, Reinhold, New York, 1949, p. 143.
2. J. W. Reppe, *Justus Liebigs Ann. Chem.*, **596,** 60 (1955).
3. K. C. Brannock and G. R. Lappin, *J. Org. Chem.*, **21,** 1366 (1956).
4. D. B. Pattison, *J. Org. Chem.*, **22,** 662 (1957).
5. G. Rio, *C. R. Acad. Sci., Paris*, **239,** 982 (1954); *Chem. Abstr.*, **49,** 12494 (1955).
6. S. R. Forrester, *Dissertation Abstr.*, **23,** 4352 (1963); *Chem. Abstr.*, **59,** 10054 (1963).
7. C. E. Pawloski, unpublished results.
8. W. Kimel and W. Leimgruber, French Patent 1,384,099 (Jan. 4, 1965); *Chem. Abstr.*, **63,** 4263 (1965); U.S. Patent 3,250,778 (May 10, 1966).
9. G. B. Sterling, E. J. Watson, and C. E. Pawloski, U.S. Patent 3,116,299 (Dec. 31, 1963); *Chem. Abstr.*, **60,** 6856 (1964).
10. G. B. Sterling, E. J. Watson, and C. E. Pawloski, U.S. Patent 3,116,298 (Dec. 31, 1963); *Chem. Abstr.*, **60,** 6856 (1964).
11. G. B. Sterling, E. J. Watson, and C. E. Pawloski, unpublished results.
12. H. Friederick, U.S. Patent 2,911,445 (Nov. 3, 1959); *Chem. Abstr.*, **54,** 3206 (1960).
13. J. W. Copenhaver and M. H. Bigelow, *Acetylene and Carbon Monoxide Chemistry*, Reinhold, New York, 1949, pp. 119–130.
14. A. Seib, German Patent 855,864 (Nov. 17, 1952); through *Chem. Abstr.*, **52,** 9197 (1958).
15. C. S. Rondestvedt, Jr., *J. Org. Chem.*, **26,** 2247 (1961).
16. G. B. Sterling and C. E. Pawloski, U.S. Patent 3,280,148 (Oct. 18, 1966); *Chem. Abstr.*, **66,** 2826 (1967).
17. H. F. Reinhardt, U.S. Patent 3,232,907 (Feb. 1, 1966); *Chem. Abstr.*, **64,** 11234 (1966).
18. C. E. Pawloski and G. B. Sterling, U.S. Patent 3,268,559 (Aug. 23, 1966); *Chem. Abstr.*, **65,** 15409 (1966).
19. C. E. Pawloski and G. B. Sterling, patent pending.
20. G. B. Sterling and C. E. Pawloski, U.S. Patent 3,117,134 (Jan. 7, 1964); *Chem. Abstr.*, **60,** 8050 (1964).
21. G. B. Sterling and C. E. Pawloski, U.S. Patent 3,134,787 (May 26, 1964); *Chem. Abstr.*, **61,** 4385 (1964).
22. G. B. Sterling and C. E. Pawloski, U.S. Patent 3,134,788 (May 26, 1964); *Chem. Abstr.*, **61,** 4384 (1964).
23. G. B. Sterling and C. E. Pawloski, U.S. Patent 3,122,562 (Feb. 25, 1964); *Chem. Abstr.*, **60,** 12036 (1964).
24. C. E. Pawloski and G. B. Sterling, U.S. Patent 3,135,768 (June 2, 1964); *Chem. Abstr.*, **61,** 5655 (1964).
25. A. Hausweiler, K. Schwarzer, H. Wollweber, R. Hiltmann, and G. Unterstenhöfer, German Patent 1,075,373 (Feb. 11, 1960); through *Chem. Abstr.*, **55,** 18003 (1961).
26. M. S. Mulla, L. W. Isaak, and H. Axelrod, *J. Econ. Entomol.*, **56,** 184 (1963); *Chem. Abstr.*, **59,** 1041 (1963).
27. A. Hausweiler, K. Schwarzer, H. Wollweber, and R. Hiltmann, German Patent 1,088,066 (July 12, 1958); *Chem. Abstr.*, **56,** 3498 (1962).
28. Farbwerke Hoechst A.-G., Belgian Patent 631,685 (Nov. 18, 1963); Belgian Patent 631,682 (Nov. 18, 1963); *Chem. Abstr.*, **60,** 14683 (1964); *Chem. Abstr.*, **61,** 4511 (1964).

29. C. S. Rondestvedt, Jr., and G. J. Mantell, *J. Amer. Chem. Soc.*, **84,** 3307 (1962).
30. G. B. Sterling, U.S. Patent 3,219,692 (Nov. 23, 1965); *Chem. Abstr.*, **64,** 5262 (1966).
31. R. F. Monroe, U.S. Patent 3,240,702 (March 15, 1966); *Chem. Abstr.*, **64,** 13995 (1966).
32. A. Sturzenegger and J. J. Zelauskas, U.S. Patent 3,410, 871 (Nov. 12, 1968); *Chem. Abstr.*, **73,** 87950 (1970).
33. Y. Odaira, T. Shimodaira, and S. Tsutsumi, *Chem. Commun.*, **1967,** 757.
34. B. E. Jennings and J. B. Rose, U.S. Patent 3,382,832 (May 28, 1968); *Chem. Abstr.*, **69,** 19996 (1968).
35. W. Korytnyk, *J. Org. Chem.*, **27,** 3724 (1962).
36. M. Tomita and T. Takahashi, *Yakugaku Zasshi*, **77,** 1041 (1957); *Chem. Abstr.*, **52,** 3816 (1958).
37. J. W. Bird and J. K. N. Jones, *Can. J. Chem.*, **41,** 1877 (1963); *Chem. Abstr.*, **59,** 7626 (1963).
38. W. Korytnyk, E. J. Kris, and R. P. Singh, *J. Org. Chem.*, **29,** 574 (1964).
39. W. Kimel and W. Leimgruber, French Patent 1,384,099 (Jan. 4, 1965); *Chem. Abstr.*, **63,** 4263 (1965).
40. F. Hoffman-La Roche & Co., Netherlands Patent Appl. 6,403,004 (Sept. 22, 1964); *Chem. Abstr.*, **62,** 7733 (1965).
41. F. Hoffman-La Roche & Co., Netherlands Patent Appl. 6,404,750 (Nov. 23, 1964); *Chem. Abstr.*, **62,** 11819 (1965).
42. W. Kimel and W. Leimgruber, U.S. Patent 3,250,778 (May 10, 1966).
43. S. F. Schaeren, U.S. Patent 3,296,275 (Jan. 3, 1967).
44. H. E. Zaugg and R. J. Michaels, *J. Org. Chem.*, **28,** 1801 (1963).
45. A. R. Alertsen, *Chem. Met. Ser.*, **13**(10), 1–56 (1961); *Chem. Abstr.*, **57,** 16619 (1962).
46. H. L. Yale, E. J. Pribyl, W. Braker, F. H. Bergeim, and W. A. Lott, *J. Amer. Chem. Soc.*, **72,** 3710 (1950).
47. O. Stephenson, *J. Chem. Soc.*, **1954,** 1571.
48. C. Weizmann, E. D. Bergmann, and M. Sulzbacher, *J. Org. Chem.*, **13,** 796 (1948).
49. A. W. Dawkins and T. P. C. Mulholland, *J. Chem. Soc.*, **1959,** 2211.
50. S. A. N. N. Bokhari and W. B. Whalley, *J. Chem. Soc.*, **1963,** 5322.
51. W. H. Perkins, Jr., and J. Yates, *J. Chem. Soc.*, **81,** 242 (1902).
52. F. Kagan and R. D. Birkenmeyer, *J. Amer. Chem. Soc.*, **81,** 1986 (1959).
53. F. Kagan and R. D. Birkenmeyer, U.S. Patent 2,956,064 (Oct. 11, 1960); *Chem. Abstr.*, **55,** 7450 (1961).
54. K. Ziegler, A. Lüttringhaus, and K. Wohlgemuth, *Justus Liebigs Ann. Chem.*, **528,** 162 (1937); *Chem. Abstr.*, **31,** 6624 (1937).
55. K. Ziegler and A. Lüttringhaus, German Patent 671,840 (Feb. 14, 1939); *Chem. Abstr.*, **33,** 6532 (1939).
56. R. B. Thompson, U.S. Patent 2,698,329 (Dec. 28, 1954); *Cem. Abstr.*, **50,** 1911 (1956).
57. R. C. Cass, S. E. Fletcher, C. T. Mortimer, P. G. Quincey, and H. D. Springall, *J. Chem. Soc.*, **1958,** 2595.
58. R. Hartmut, S. Klaus, and M. Manfred, *Z. Chem.*, **8,** 220 (1968).
59. M. Tomita and T. Takahashi, *Yakugaku Zasshi*, **77,** 1041 (1957); *Chem. Abstr.*, **52,** 3816 (1958).
60. L. Gattermann, *Chem. Ber.*, **31,** 1149 (1894).
61. G. Baddeley and N. H. P. Smith, *J. Chem. Soc.*, **1961,** 2516.
62. Imperial Chemical Industries Ltd., Belgian Patent 633,973 (Dec. 23, 1963); *Chem. Abstr.*, **61,** 672 (1964).
63. V. Rosnati and F. De Marchi, *Tetrahedron*, **18,** 289 (1962).

64. Imperial Chemical Industries Ltd., Netherlands Patent Appl. 6,408,650 (Feb. 1, 1965); *Chem. Abstr.*, **63,** 621 (1965).
65. F. Leonard and J. Koo, Belgian Patent 613,212 (July 30, 1962); *Chem. Abstr.*, **57,** 16639 (1962).
66. Pfizer Corp., Belgian Patent 659,663 (Aug. 12, 1965); *Chem. Abstr.*, **64,** 744 (1966); Netherlands Patent Appl. 6,507,652 (Dec. 16, 1966); through *Chem. Abstr.*, **66,** 115747 (1967).
67. J. Augstein, A. Monro, G. W. H. Patter and T. I. Wrigley, U.S. Patent 3,306,913 (Feb. 28, 1967).
68. F. Leonard and J. Koo, Belgian Patent 613,210 (July 30, 1962); *Chem. Abstr.*, **57,** 16640 (1962).
69. F. Leonard and J. Koo, Belgian Patent 613,215 (July 30, 1962); *Chem. Abstr.*, **57,** 15134 (1962).
70. R. B. Thompson and T. Symon, *J. Amer. Oil Chem. Soc.*, **33,** 414 (1956); *Chem. Abstr.* **50,** 17256 (1956).
71. P. Engles, W. H. Perkins, Jr. and R. Robinson, *J. Chem. Soc.* **93,** 1127 (1908).
72. G. Baddeley, G. Holt, N. H. P. Smith, and F. A. Whittaker, *Nature*, **168,** 386 (1951).
73. W. Gerrard, A. M. A. Mincer, and P. L. Wyvill, *J. Appl. Chem.* (London), **10,** 115 (1960); *Chem. Abstr.*, **54,** 16116 (1960).
74. E. Ziegler, H. Junek, and E. Nolken, *Monatsh. Chem.*, **90,** 206 (1959).
75. J. L. Warnell and R. L. Shriner, *J. Amer. Chem. Soc.*, **79,** 3165 (1957).
76. P. S. Bailey, *Chem. Ber.*, **87,** 993 (1954).
77. R. Criegee, A. Kerckow, and H. Zinke, *Chem. Ber.*, **88,** 1878 (1955).
78. R. Kuhn and B. Schulz, *Chem. Ber.*, **96,** 3200 (1963).
79. T. C. McMorris and M. Anchel, *J. Amer. Chem. Soc.*, **87,** 1594 (1965).
80. R. Criegee, P. Bruyn, and G. Lohaus, *Justus Liebigs Ann. Chem.*, **583,** 19 (1953).
81. J. Rigaudy and P. Auburn, *C. R. Acad. Sci., Paris*, **256,** 3143 (1963).
82. R. Grewe and A. Struve, *Chem. Ber.*, **96,** 2819 (1963).
83. P. A. Small, U.S. Patent 3,252,939 (May 24, 1966).
84. H. Friebolin, R. Meche, S. Kabuss, and A. Lüttringhaus, *Tetrahedron Lett.*, **1964,** 1929.
85. F. Dallacker, K. W. Glombitza, and M. Lipp, *Justus Liebigs Ann. Chem.*, **643,** 91 (1961).
86. V. E. Davidson, J. Keane, and T. J. Nolan, *Sci. Proc. Roy. Dublin Soc.*, **23,** 143 (1943); *Chem. Abstr.*, **38,** 1221 (1944).
87. F. R. Hewgill, *J. Chem. Soc.*, **1962,** 4987.
88. Y. Asahina, H. Kondo, and M. Tasaka, *Chem. Ber.*, **70B,** 810 (1937); *Chem. Abstr.*, **33,** 4661 (1939).
89. H. Hayashi, *J. Pharm. Soc. Jap.*, **57,** 598 (1937); *Chem. Abstr.*, **31,** 6219 (1937).
90. Y. Asahina, and S. Sibata, *Chem. Ber.*, **72B,** 1399 (1939); *Chem. Abstr.*, **33,** 7762 (1939).
91. A. W. Evans, *Rhodora*, **45,** 417 (1943); *Chem. Abstr.*, **38,** 1543 (1944).
92. D. Murphy, J. Keane, and T. J. Nolan, *Sci. Proc. Roy. Dublin Soc.*, **23,** No. 8, 71–82 (1943); *Chem. Abstr.*, **37,** 3748 (1943).
93. S. Neelakantan, T. R. Seshadri, and S. S. Subramanian, *Tetrahedron*, **18,** 597 (1962).
94. A. Schatz, *J. Agr. Food Chem.*, **11,** 112 (1963); *Chem. Abstr.*, **58,** 123883 (1963).
95. A. J. Birch, R. A. Massy-Westropp, and C. J. Moye, *Aust. J. Chem.*, **8,** 539 (1955).
96. M. Tomita and W. Watanabe, *J. Pharm. Soc. Jap.*, **72,** 478 (1952); *Chem. Abstr.*, **47,** 7028 (1953).
97. F. M. Dean, J. C. Roberts, and A. Robertson, *J. Chem. Soc.*, **1954,** 1432.

98. W. F. Beach and J. H. Richards, *J. Org. Chem.*, **26,** 3011 (1961).
99. W. F. Beach and J. H. Richards, *J. Org. Chem.*, **26,** 1339 (1961).
100. B. W. Bycroft, J. A. Knight, and J. C. Roberts, *J. Chem. Soc.*, **1963,** 5148.
101. H. Seigfried and L. Pierre, *Z. Naturforsch*, **23,** 715 (1968).
102. Y. Inubushi, *J. Pharm. Soc. Jap.*, **72,** 1223 (1952); *Chem. Abstr.*, **47,** 12408 (1953).
103. M. Tomita, T. Ujiie, and S. Tanaka, *Yakugaku Zasshi*, **80,** 358 (1960); *Chem. Abstr.*, **54,** 18432 (1960).
104. M. Tomita, Y. Inubuse, and F. Kusuda, *J. Pharm. Soc. Jap.*, **64,** 173 (1944); *Chem. Abstr.*, **45,** 6173 (1951).
105. D. S. Noyce and J. W. Weldon, *J. Amer. Chem. Soc.*, **74,** 401 (1952).
106. D. S. Noyce and J. W. Weldon, *J. Amer. Chem. Soc.*, **74,** 5144 (1952).
107. F. Fujikawa, K. Hirai, N. Ishikawa, and A. Maki, *Yakugaku Zasshi*, **63,** 1172 (1963); *Chem. Abstr.*, **60,** 10685 (1964).
108. H. E. Ungnade and L. Rubin, *J. Org. Chem.*, **16,** 1311 (1951).
109. V. C. Barry and D. Twomey, *Proc. Roy. Irish Acad.*, **53B,** 55 (1950); *Chem. Abstr.*, **45,** 9500 (1951).
110. F. R. Hewgill and D. G. Hewitt, *J. Chem. Soc.*, **1965,** 1536.
111. R. Breslow and E. Mohacsi, *J. Amer. Chem. Soc.*, **85,** 431 (1963).
112. R. J. Prochaska, Belgian Patent 626,345 (June 20, 1963); *Chem. Abstr.*, **60,** 38186 (1964).
113. R. J. Prochaska, French Patent 1,352,580 (Feb. 14, 1964); *Chem. Abstr.*, **61,** 4269 (1964).
114. C. M. S. Yoder and J. J. Zuckerman, *J. Heterocycl. Chem.*, **4,** 166 (1967).
115. D. F. Bowman and F. R. Hewgill, *Chem. Commun.*, **1968,** 524.
116. J. C. Pew and W. J. Connors, *Nature*, **215,** 624 (1967).
117. M. Stoll and M. Hinder, *Helv. Chim. Acta*, **36,** 1984 (1953).
118. R. H. Callighan, M. F. Tarker, Jr., and M. H. Wilt, *J. Org. Chem.*, **26,** 1379 (1961).
119. R. Criegee, A. Kerckow, and H. Zinke, *Chem. Ber.*, **88,** 1878 (1955).
120. W. Dilthey, F. Quint, and J. Heinen, *J. Prakt. Chem.*, **152,** 49 (1939); *Chem. Abstr.*, **33,** 4254 (1939).
121. F. Max and W. Schmidt-Nickels, U.S. Patent 2,515,723 (July 18, 1950); *Chem. Abstr.*, **45,** 1775 (1951).
122. G. Slooff, *Rec. Trav. Chim. Pays-Bas*, **54,** 995 (1935); *Chem. Abstr.*, **30,** 2184 (1936).
123. E. Bernatek and I. Bo, *Acta Chem. Scand.*, **13,** 337 (1959); *Chem. Abstr.*, **54,** 24621 (1960).
124. E. Bernatek and M. Hvatum, *Acta Chem. Scand.*, **14,** 836 (1960); *Chem. Abstr.*, **56,** 9927 (1962).
125. E. Bernatek and F. Thoresen, *Acta Chem. Scand.*, **9,** 743 (1955); *Chem. Abstr.*, **50,** 7099 (1956).
126. E. Bernatek and T. Ledaal, *Tetrahedron Lett.*, **26,** 30 (1960).
127. E. Bernatek, T. Ledaal, and S. Steinsvik, *Acta Chem. Scand.*, **15,** 429 (1961); *Chem. Abstr.*, **55,** 27259 (1961).
128. R. E. Erickson, P. S. Bailey, and J. C. Davis, Jr., *Tetrahedron*, **18,** 389 (1962).
129. A. Mustafa, *J. Chem. Soc.*, **1949,** S83.

CHAPTER VI

Terpene Oxepins

A. N. STARRATT

Research Institute, Canada Department of Agriculture, London, Ontario, Canada

A. Introduction

Most terpene oxepins have been prepared during the course of structural studies and have not been investigated for their own interest. The two ring systems comprising the majority of the representatives of this class are the 2-oxepinones or ε-lactones and the 2,7-oxepindiones or adipic anhydrides. Several ε-lactones (Table VI-1) are known to occur naturally. The largest class consists of limonoid ε-lactones which may be formed by a biological process analogous to the Baeyer-Villiger reaction. In this review an attempt will be made to draw together the widely scattered literature reports of terpene oxepins, including both those of natural origin and those derived from terpenes or obtained during the course of terpene synthesis or degradation.

B. ε-Lactones and ε-Lactols

1. Naturally Occurring ε-Lactones

Recent studies of limonoid bitter principles, especially limonin itself, occurring in members of the family *Rutaceae*, have culminated in the assignment of structures to five related substances, obacunone, nomilin, deacetylnomilin, 7α-obacunol, and zapoterin, which possess a seven-membered lactone ring A. Another recently isolated ε-lactone, methyl ivorensate, is the first A-seco-limonoid to be obtained from plants of the family *Meliaceae*. Apetalactone, a triterpene of the friedelin group, and compound B from

Dammar resin are the only other terpenoid ε-lactones found to date in nature. The structures and some derivatives of these compounds and reactions involving the oxepin ring will be discussed briefly below.

a. *Structure*

Although obacunone (**1**), shown to be identical with the earlier reported casimirolid, has been a subject of investigation for many years (1–6), only recently has it been possible to assign a structure. Nomilin (**2**), co-occurring with obacunone and limonin (**3**) in Florida citrus seed oil (3, 4, 7), undergoes loss of the elements of acetic acid to afford obacunone upon treatment with acetic anhydride and pyridine (1). On the basis of a study by Dean and Geissman (4) of the functional groups of these two compounds and the newly elucidated structure of limonin, structures with a seven-membered oxygen ring were proposed for obacunone and nomilin (8). The same functionality was also suggested independently by Japanese workers (9, 10) for obacunone on the basis of their chemical investigations. Further studies by Barton and co-workers (11) enabled them to propose structure **1** for obacunone, assuming that it is related to limonin both stereochemically and constitutionally. In support, obacunone has been chemically related to limonin by transformation to a common degradation product (**4**) (12–14). This work established the stereochemistry of all positions except that at $C_{(17)}$.

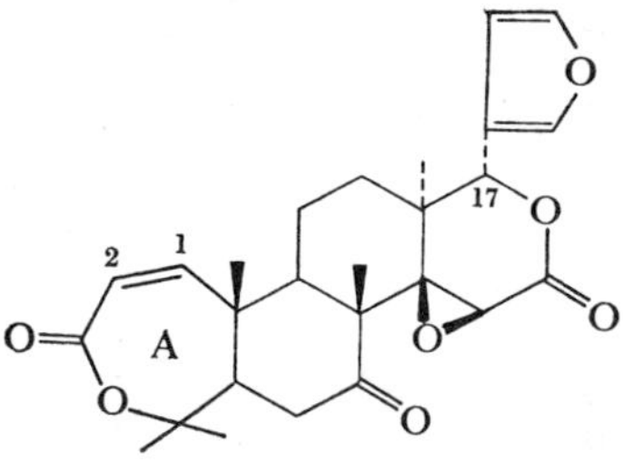

1

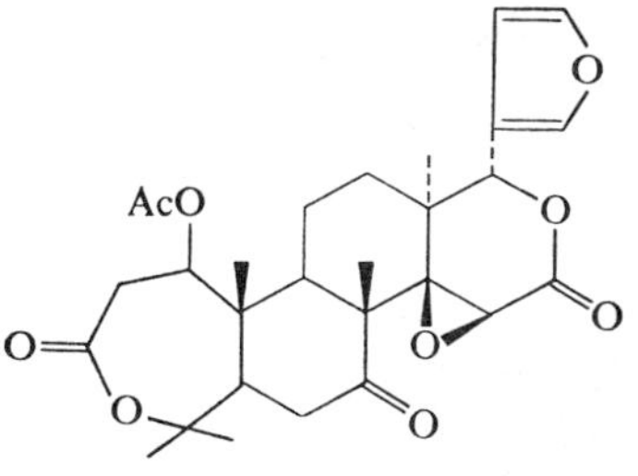

2

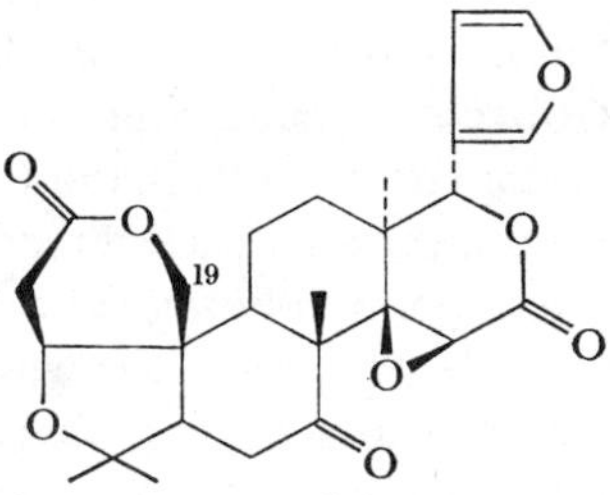

3

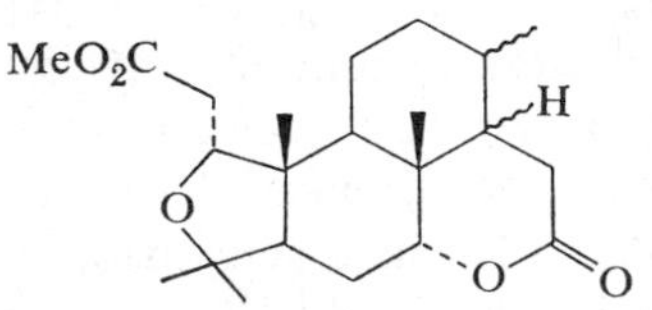

4

The conversion of nomilin to obacunone mentioned above enabled Emerson (1) to propose that nomilin is an acetoxy dihydro derivative with the acetoxy group in the β-position to one of the lactone groups. On this basis Barton (8, 11) proposed **2** as a probable representation of nomilin. Deacetylnomilin (3, 15, 16), a closely related member of this series, has been shown to have structure **5** by spectroscopic studies and by obtention of obacunone (**1**) and nomilin (**2**) on reaction with acetic anhydride and pyridine (15), and is probably identical with the previously reported isolimonin (17). Evidence establishing the orientation of the $C_{(1)}$ substituent has been difficult to obtain, particularly because of uncertainty concerning the A ring conformation (18, 19). The application of Brewster's benzoate rule to this problem was impossible since nmr studies indicated that deacetylnomilin benzoate (**6**) exists in a conformation different from that of nomilin and deacetylnomilin (19). However, consideration of the observed coupling of the $C_{(1)}$ proton in nomilin, deacetylnomilin, deacetylnomilin benzoate (**6**), and phenylglyoxylate (**7**) has led to the proposal that the $C_{(1)}$ hydroxyl is probably in the α-position. Further work will be needed to establish the stereochemistry unequivocally.

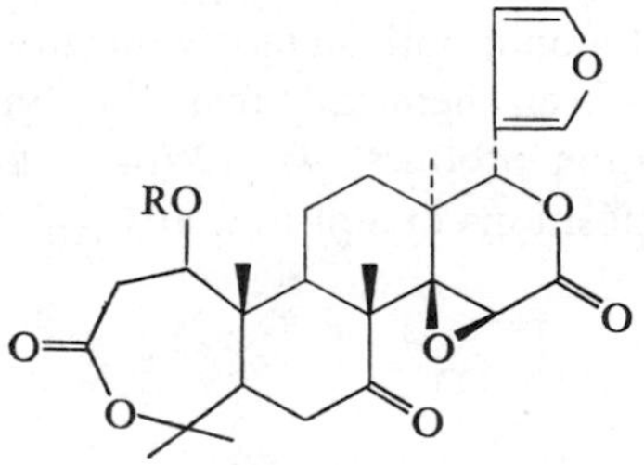

RO

O

O

O

O

O

O

5; R = H
6; R = PhCO
7; R = PhCOCO

An nmr study of limonin, obacunone, and related compounds shows that the band positions for the $C_{(17)}$ furfurylic proton correspond very closely (18). Inasmuch as the position of the band for the proton at $C_{(17)}$ would be expected to be very sensitive to changes of stereochemistry due to the proximity of the epoxide ring, it has been suggested that obacunone and limonin have the same configuration at $C_{(17)}$ (i.e., β-H). The similarity of ORD and CD curves of nomilin, deacetylnomilin, and limonin confirm identical stereochemistry at $C_{(17)}$, as was to be expected in view of their common botanical origin (19).

Dreyer (16) has isolated a new limonoid, 7α-obacunol (**8**), which yields obacunone on chromic acid oxidation and corresponds to β-obacunol, the minor product of borohydride reduction of obacunone (4).

Zapoterin (**9**), first assigned the empirical formula $C_{19}H_{24}O_6$ (20), has been shown by two groups to be a hydroxyobacunone (16, 21). From nmr considerations Dreyer concluded that the hydroxyl was located at either $C_{(11)}$ or $C_{(12)}$ and proposed that the spectra of zapoterin and zapoterin acetate

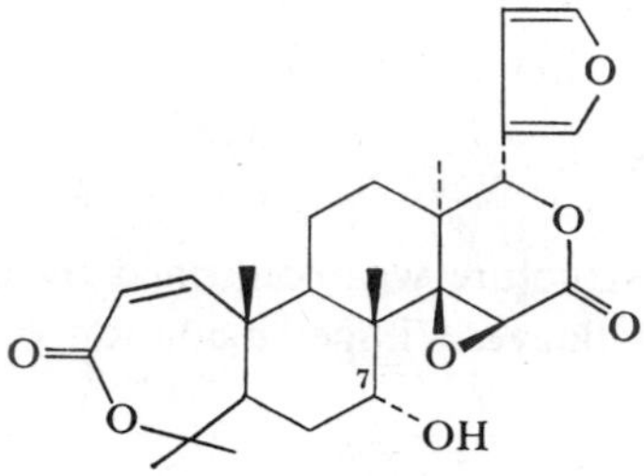

8

are most consistent with a 12α-hydroxy derivative of obacunone (16). Independently, by an analysis of nmr spectra of zapoterin and derivatives, Murphy, Toube, and Cross were led to conclude that zapoterin is 11β-hydroxyobacunone (21). This latter structure has been confirmed by use of the nuclear Overhauser effect (22).

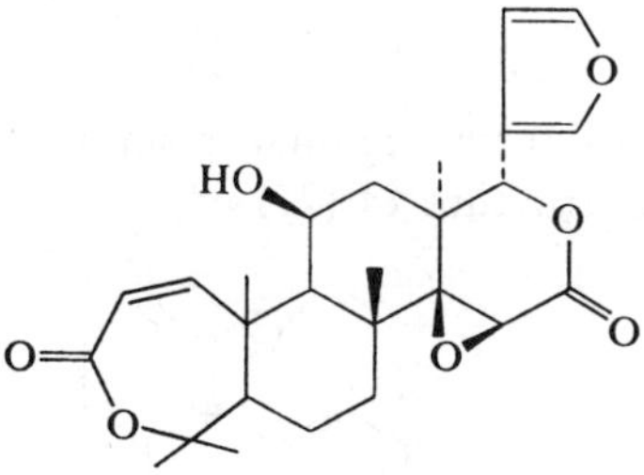

9

Physical techniques are increasingly being used in investigations of limonoids, as is the case, in fact, for all structural studies. Consideration of the ir and nmr spectra of methyl ivorensate (**10**) enabled Adesogan and Taylor (23, 24) to propose a structure which was confirmed by partial synthesis from methyl angolensate. Dreyer has published results of studies concerning the application of nmr (18) and ORD and CD (19) to structural and stereochemical problems in the limonoid field which should prove very useful for the structural elucidation of new members of the series.

Apetalactone, a triterpene ε-lactone biogenetically unrelated to the limonoids, has been shown to be 4,28-dihydroxy-3,4-secofriedelan-3-oic acid

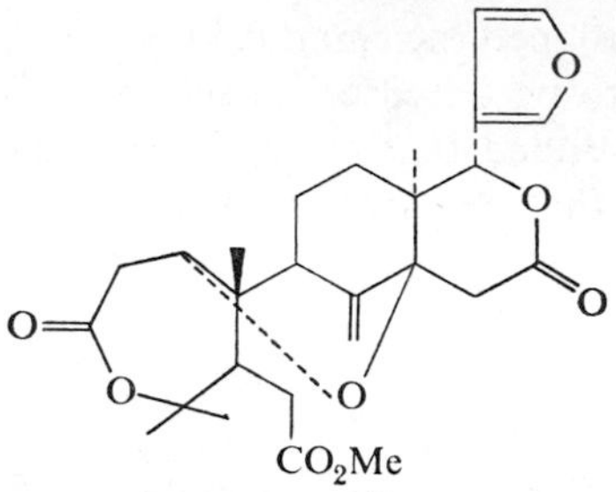

CO$_2$Me

10

lactone (**11**) (25). The structure was confirmed by the obtention of acetylapetalactone (**12**) upon Baeyer-Villiger oxidation of acetylcanophyllol (**13**) (Eq. 1).

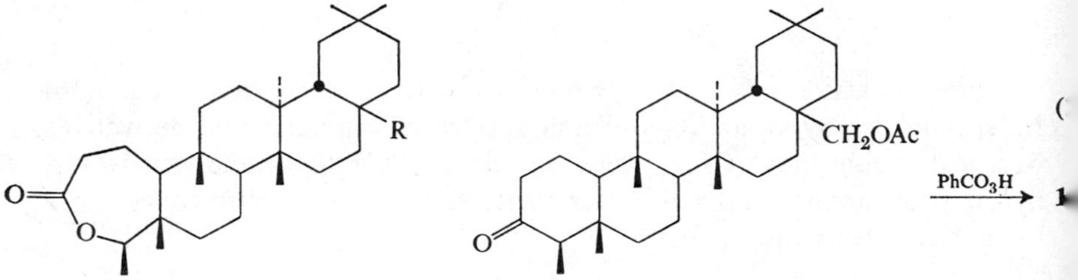

R

11; R = CH$_2$OH
12; R = CH$_2$OAc

CH$_2$OAc

PhCO$_3$H

13

Compound B, isolated from dammar resin (26), was shown by X-ray crystallography to have structure **14** (27).

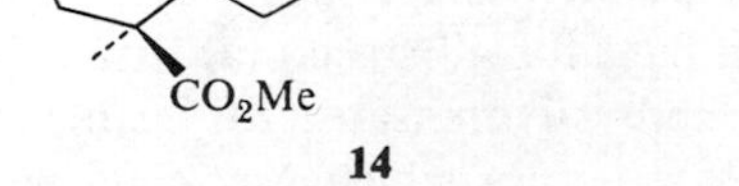

CO$_2$Me

CO$_2$Me

14

b. *Derivatives and Ring A Reactions*

In addition to the already discussed reactions which correlate the structures of the limonoid ε-lactones, some further aspects of the chemistry of ring A in these compounds will be summarized.

Obacunone (**1**) has been reported to afford a crystalline oxime, and yields a hydrochloride which regenerates obacunone upon treatment with pyridine (1, 4). Ozonolysis of obacunone hydrochloride results in the loss of three carbons and formation of etio-obacunoic acid hydrochloride, which can be dehydrochlorinated in pyridine to etio-obacunoic acid (**15**) (28). Oxidation of methyl etio-obacunoate (**16**) with potassium permanganate gives oxalic acid and a dilactone (**17**) (Eq. 2) (9). Permanganate oxidation of obacunone leads to a dilactone (**18**) (convertible to **17** by ozonolysis), a hydroxy dilactone (**19**), and a glycol (**20**) (Eq. 3) (9, 10). The glycol, which forms a diacetate, can be oxidized further to **18** with potassium permanganate.

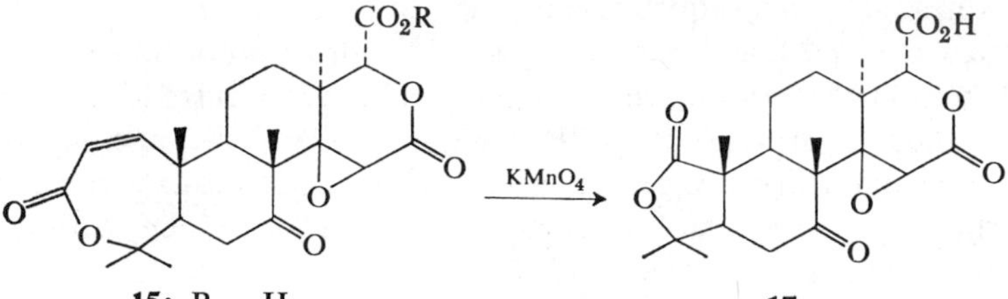

CO_2R CO_2H $KMnO_4$ (2)

15: R = H
16; R = Me

17

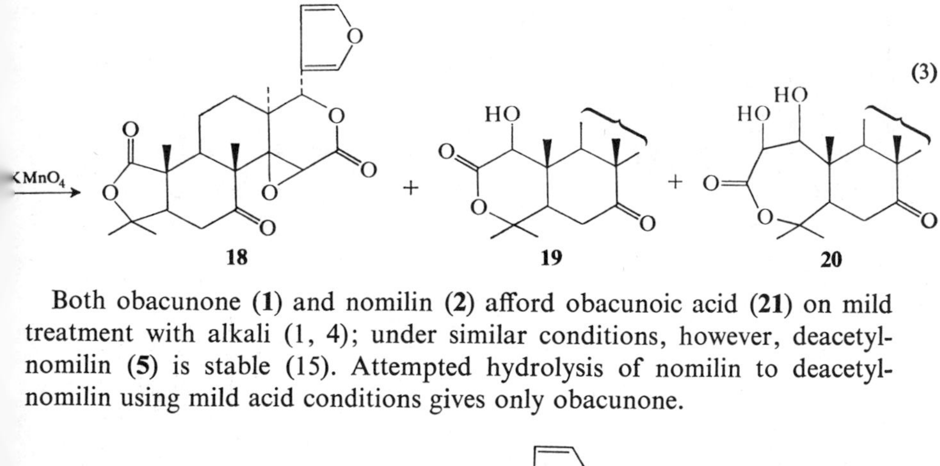

$KMnO_4$ (3)

18 + **19** + **20**

Both obacunone (**1**) and nomilin (**2**) afford obacunoic acid (**21**) on mild treatment with alkali (1, 4); under similar conditions, however, deacetylnomilin (**5**) is stable (15). Attempted hydrolysis of nomilin to deacetylnomilin using mild acid conditions gives only obacunone.

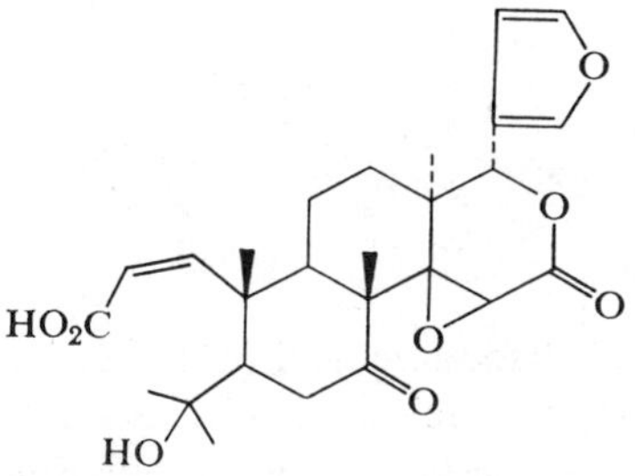

HO_2C HO

21

Although obacunone and other members of the series undergo irreversible opening of the seven-membered lactone ring, glycol **20** can be recovered after alkaline hydrolysis by acidification of the hydrolysate (9, 10). This abnormal behavior was felt to be due to steric effects of the two hydroxyl groups. On the other hand, removal of α,β-unsaturation in certain members of the obacunone and etio-obacunoic acid series by catalytic hydrogenation leads to extremely easy opening of the seven-membered ring. Catalytic reduction of obacunone (**1**), followed by treatment with methanol, gives **22**, in which methanolysis of the seven-membered lactone ring has occurred (29). The product of catalytic reduction of etio-obacunoic acid (**15**), after recrystallization from aqueous acetone, affords the saturated hydroxy dicarboxylic acid **23**. In contrast, methyl dihydroetio-obacunoate (**24**) obtained from methyl etio-obacunoate (**16**) can be recrystallized from acetone–water without hydrolysis of the lactone ring. The facile opening of ring A occurring in the first two examples was attributed to the catalytic action of the carboxyl group attached to carbon bearing an α-oxygen substituent at the opposite end of the molecule (29).

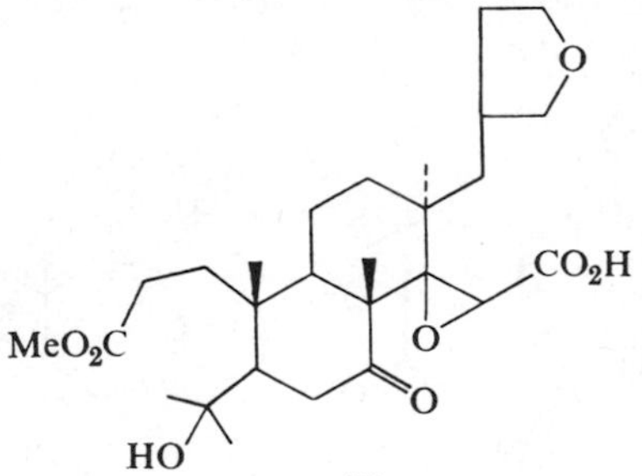

MeO_2C HO CO_2H O O

22

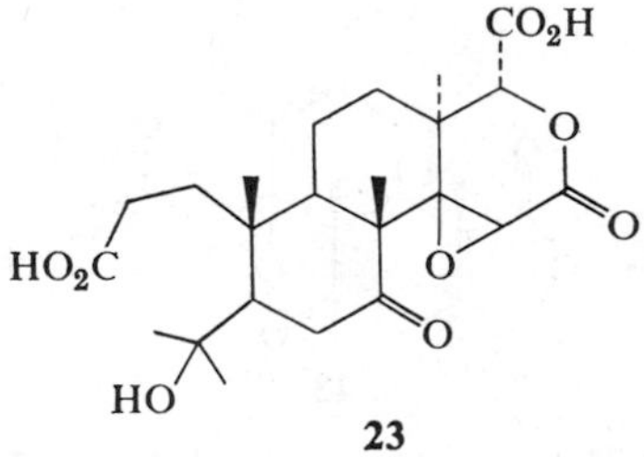

HO_2C HO CO_2H O O O

23

CO_2Me O O O O O

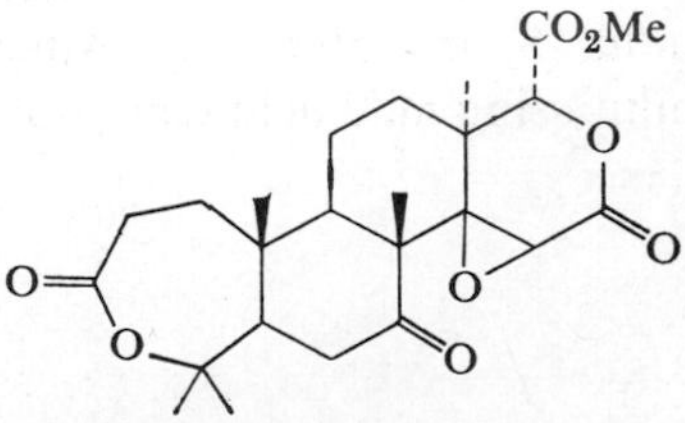

24

Alkaline autoxidation of obacunoic acid (**21**) afforded a diosphenol whose acetate was neutral rather than acidic (11). This new substance was interpreted as being ε-lactone **25**, the formation of which may be represented as shown in Eq. (4).

The interesting observation that obacunone gives a positive iodoform test (4) has been demonstrated to be due to the hydroxyisopropyl group of obacunoic acid separated by two carbon atoms from the ketone group (11).

21 →

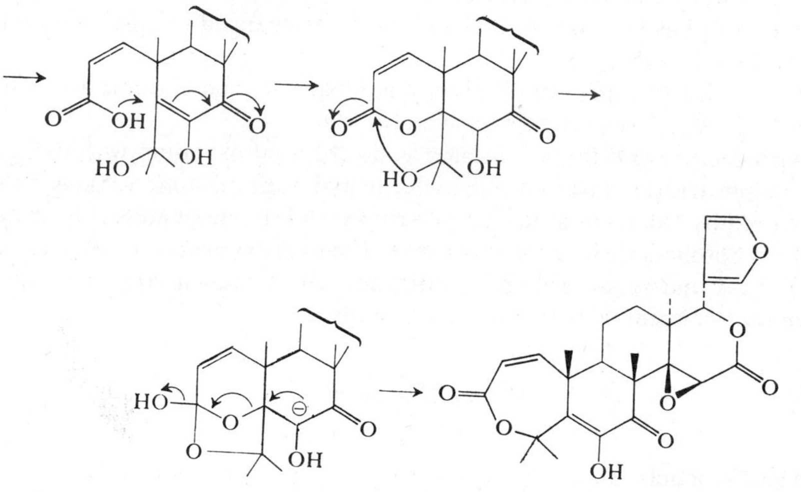

(4)

25

Potassium borohydride reduction of obacunone gives two alcohols, α-and β-obacunol (**26** and **27**) (Eq. 5) (4). From the available data it is impossible to correlate these alcohols with obacunol and isoobacunol (both characterized as monoacetates) obtained by reduction of obacunone with sodium borohydride (13). However, it is probable that α-obacunol, the major product of potassium borohydride reduction, corresponds to obacunol, the

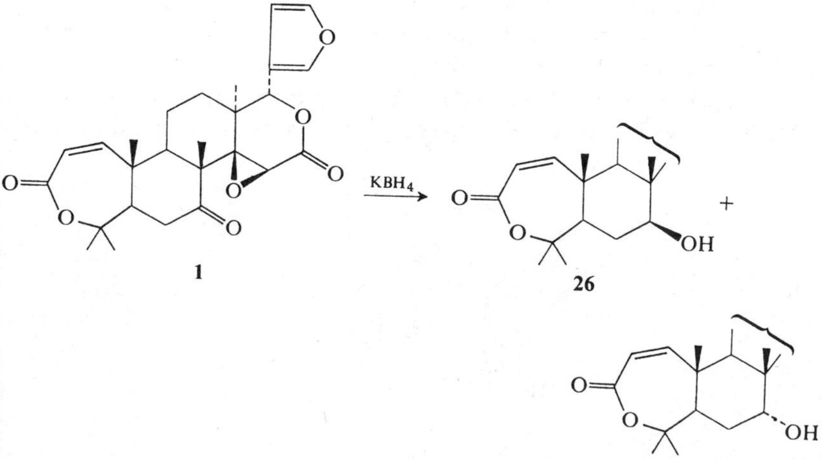

1 $\xrightarrow{KBH_4}$ 26 + 27 (5)

major product of sodium borohydride reduction. Chromic acid oxidation of obacunol gave obacunone, indicating that no other reaction had occurred. Since α-obacunol has the β-configuration and β-obacunol has the α-configuration at $C_{(7)}$, by analogy with the borohydride reduction of limonin (8, 11, 30), Dreyer (16) has proposed that these names be changed to 7β-obacunol and 7α-obacunol, respectively.

Acetylation of zapoterin (**9**) gives a monoacetate and chromic acid oxidation leads to a diketone, zapoterone (16, 21).

Apetalactone (**11**) forms a monoacetate (**12**) and oxidation with pyridine–chromium trioxide gives an aldehyde, dehydroapetalactone (25). A triol is produced by reduction of the lactone ring with lithium aluminum hydride or sodium borohydride in tetrahydrofuran. The hydroxy acid formed by alkaline hydrolysis undergoes facile relactonization on attempted crystallization, or even on treatment with ethereal diazomethane.

c. *Biogenesis*

The limonoids, a rapidly expanding group of compounds related to limonin, may be regarded as degraded tetracyclic triterpenoids (8, 11). The ring D lactone of limonin (**3**) could arise by a Baeyer-Villiger-like oxidation step; a further such oxidation of ring A would afford the seven-membered lactone ring found in obacunone and related compounds. Since deacetylnomilin fails to undergo elimination of water, whereas nomilin easily yields acetic acid but does not undergo hydrolysis of the acetoxy group, Dreyer was led to propose that limonin may be formed from deacetylnomilin (**5**) through nomilin (**2**) and obacunone (**1**) (15). Hydrolysis of obacunone to obacunoic acid (**21**), Michael addition, and oxidation of the $C_{(19)}$ angular methyl group would then afford limonin. The isolation of ichangin (**28**) and its *in vitro* conversion to limonin has led to the suggestion that the alternative biogenetic route, deacetylnomilin to ichangin to limonin, must also be considered (31), However, the co-occurrence of nomilin with obacunone in widely

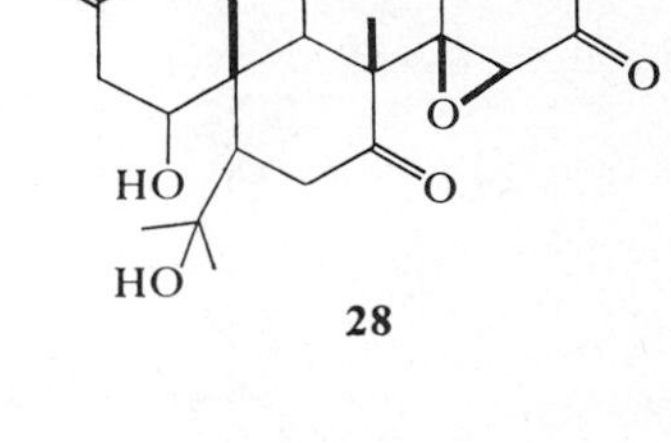

HO

HO

28

distributed species of the *Rutaceae* (3) supports the proposal that nomilin is a biogenetic precursor of obacunone. It may be noted also that obacunoic acid has been reported to co-occur with limonin (32).

ε-Lactone **29** has been proposed to be an intermediate in the biogenetic formation of the limonoid nimbin (**30**), as depicted in (Eq. 6) (33).

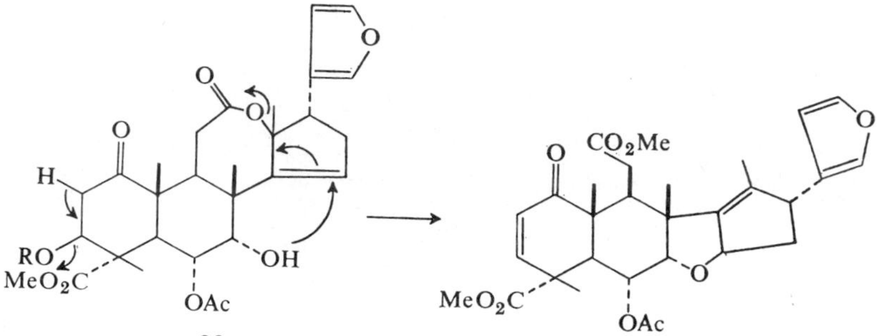

H RO MeO_2C OAc OH O O O O **29** → O CO_2Me O O MeO_2C OAc **30** (6)

The partial synthesis of methyl angolensate (**33**) from 7-oxo-7-deacetoxy-khivorin (**31**) via ε-lactone **32** (Eq. 7) follows a possible biogenetic pathway for the formation of ring B cleaved tetranortriterpenoids (34).

Recently Fried and co-workers (35, 36) postulated that the naturally occurring triterpene seco acids, such as dammarenolic acid (**34**) (37), nyct-anthic acid (**35**) (37, 38), roburic acid (**36**) (39), and canaric acid (**37**) (40), are formed via ε-lactones which arise from an enzymatic Baeyer-Villiger-type reaction. The intermediate ε-lactone could yield the final product either by direct elimination or after hydrolysis by conversion of the tertiary hydroxyl group into a suitable leaving group. This idea was suggested after isolation of the 3,4-seco compound **39** from fermentation of eburicoic acid (**38**) with the fungus *Glomerella fusarioides* (Eq. 8) (35), and was supported by the observation that pyrolysis (200°) of **40** smoothly afforded the 4-methylene acid **41** (Eq. 9) (36).

A related study has shown that peracid oxidation of 4,4-dimethylcholestan-3-one (**42**) in the presence of sulfuric acid leads to the formation of a mono-methyl lactone (**46**) (41, 42). Evidence was obtained which indicated that the reaction proceeds by way of the intermediates shown in Eq. 10. Treatment of the epoxy acid **43** with dilute sulfuric acid was found to give as the major products aldehydo acid **44** and monomethyl ketone **45**. The methyl ester of the former compound yields the monomethyl ketone under similar acid conditions. The removal under relatively mild conditions of a methyl group, with formation of monomethyl ketone **45**, may have biogenetic significance. This product sequence does not rule out variants in the original acid-catalyzed

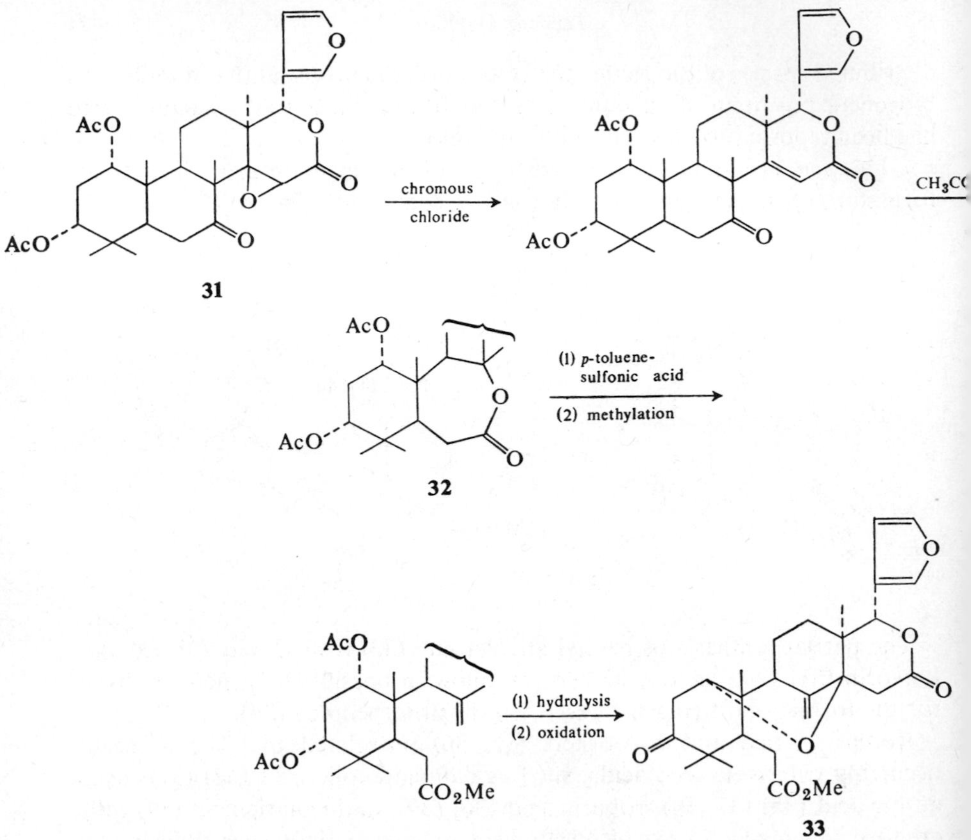

AcO
AcO
chromous chloride
AcO
AcO
CH_3C(
31

AcO
AcO
(1) *p*-toluenesulfonic acid
(2) methylation
32

AcO
AcO
CO_2Me
(1) hydrolysis
(2) oxidation
CO_2Me
33

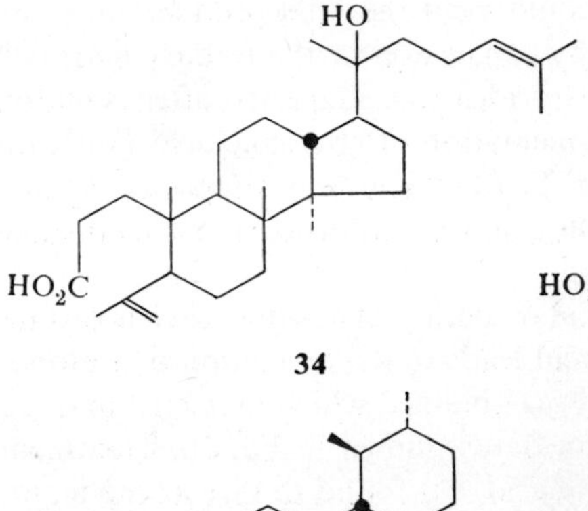

HO
HO_2C
34 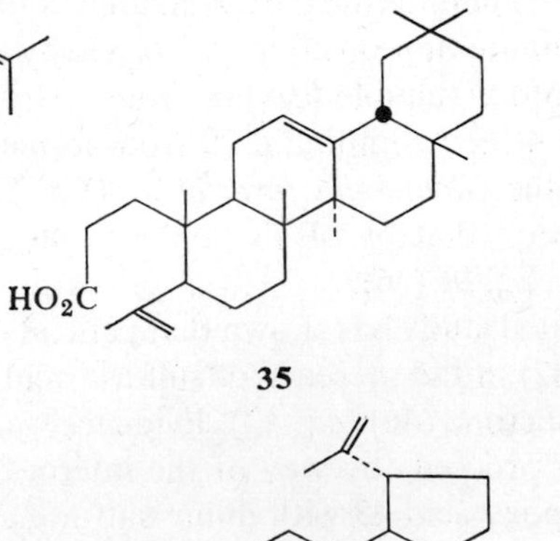
HO_2C
35 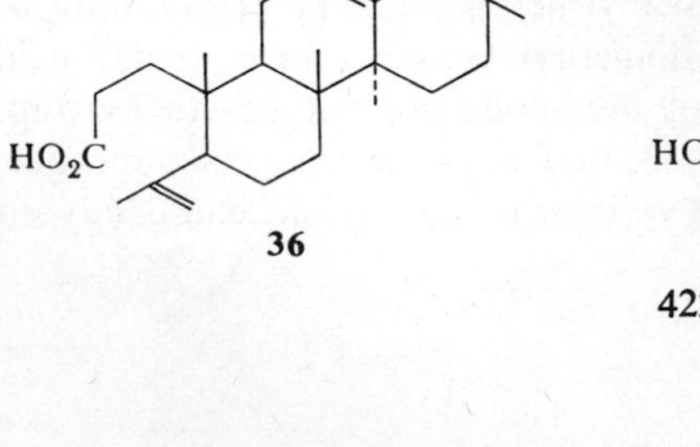
HO_2C
36 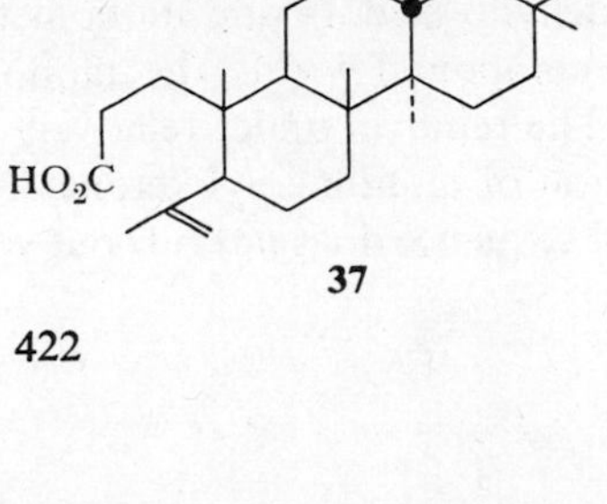
HO_2C
37

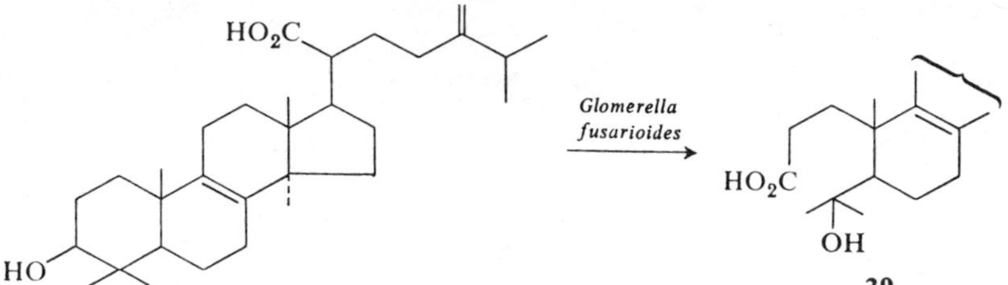

HO_2C ... *Glomerella fusarioides* → HO_2C ... OH (8)

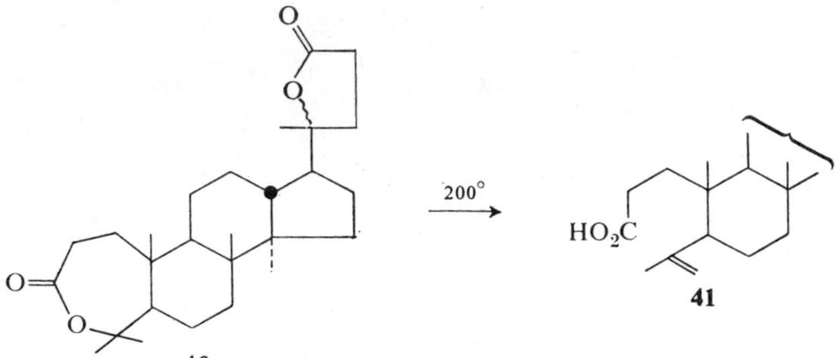

38 39

200° → HO_2C (9)

40 41

Baeyer-Villiger reaction, such as further Baeyer-Villiger oxidation of aldehydo acid **44** to the formate, which would readily lead to monomethyl lactone **46**. Such was the situation when the triterpene unsaturated nitrile **47** was oxidized with performic acid (Eq. 11) (43). The product, formate **48**, yielded upon hydrolysis a hydroxy acid which was lactonized under the acidic conditions. Other pathways which proceed through 3,4-seco compounds have been proposed for the removal of one or both of the $C_{(4)}$ methyl groups from 4,4-dimethyl compounds (44).

It has been proposed that compound B (**14**) from dammar resin may arise from asiatic acid (**49**) with which it occurs (27). Oxidation of asiatic acid to the 2,3-seco acid, lactonization of the $C_{(2)}$ carboxy group with the $C_{(23)}$ hydroxy group, and methylation of the free carboxyl groups could afford compound B. A compound assigned this structure and having a similar melting point was formed from methyl asiatate by periodate oxidation, chromic acid oxidation of the aldehydo hemiacetal, and methylation of the free carboxyl (see Section D) (45). This route represents another possible biogenetic pathway to compound B.

The isolation of triterpene ε-lactones and the known lability of certain compounds having this functionality make attractive the possibility that many other ring-opened, triterpene-derived natural products, of which the limonoids constitute the largest group, have arisen by a biochemical process formally analogous to the Baeyer-Villiger reaction. With this in mind, using

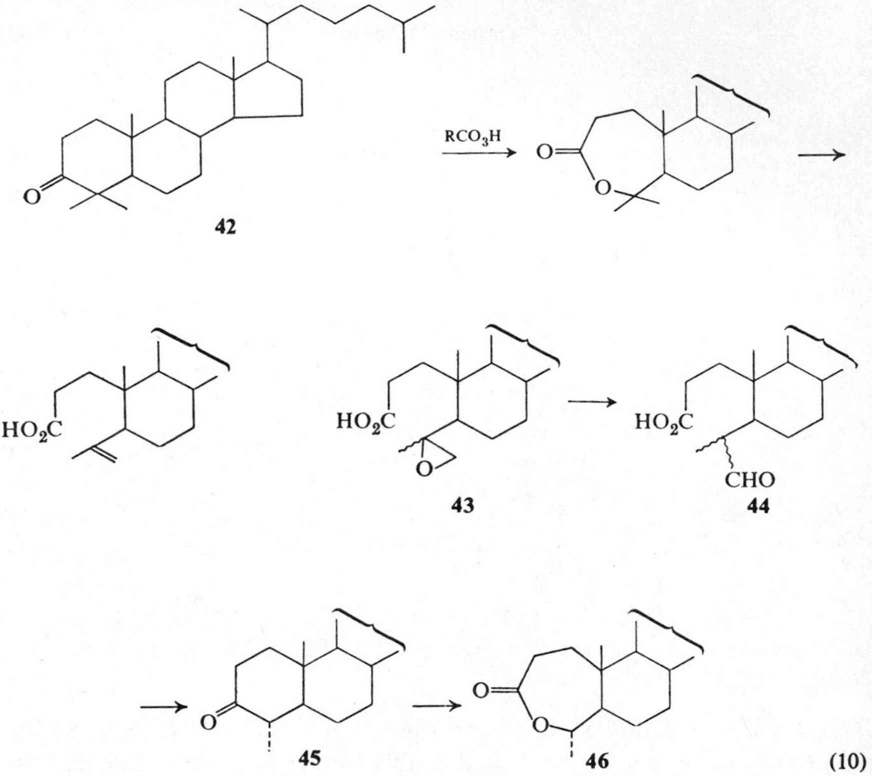

RCO$_3$H

O=

O

42

HO_2C

HO_2C

O

43

HO_2C

CHO

44

O

45

O=

O

46

(10)

HCO$_3$H

N≡C

NaOH

N≡C

HCO

O

47

48

HO_2C

HO

$H^{\oplus}$

O=

O

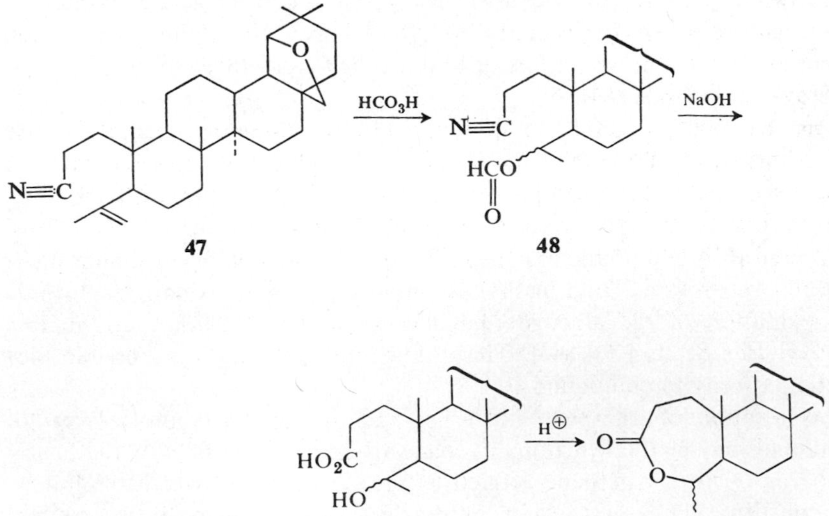

(11)

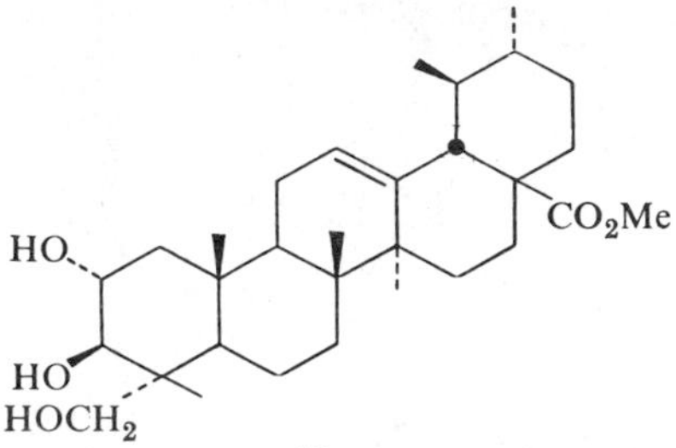

HO, HO, $HOCH_2$, CO_2Me

49

appropriate isolation procedures, it may be possible to isolate ε-lactones from sources now known to afford ring-cleaved compounds.

2. ε-Lactones Originating by Peroxy Acid Oxidation

Baeyer-Villiger oxidation of terpene cyclohexanones to ε-lactones has been the degradative method of choice in many structural investigations because of the mild reaction conditions, good yields, and high degree of selectivity. Terpenes also provided the models for the early exploration of this important reaction. On the other hand, the Baeyer-Villiger reaction has had limited application to the area of terpene synthesis.

a. *Classical Investigations*

In 1896 Baeyer (46) reported the probable formation of an ε-lactone on distillation of 6-hydroxy-3,7-dimethyloctanoic acid derived from menthone (**50**). The lactone (**51**) was later reported to be obtained in two crystalline modifications (mp 8–10° and 47°) (47) probably representing two possible isomers. Subsequently Baeyer and Villiger (48) demonstrated that oxidation of menthone with permonosulfuric acid (Caro's acid) leads to the formation of an ε-lactone (**51**) identical with the higher melting form (Eq. 12). Similarly

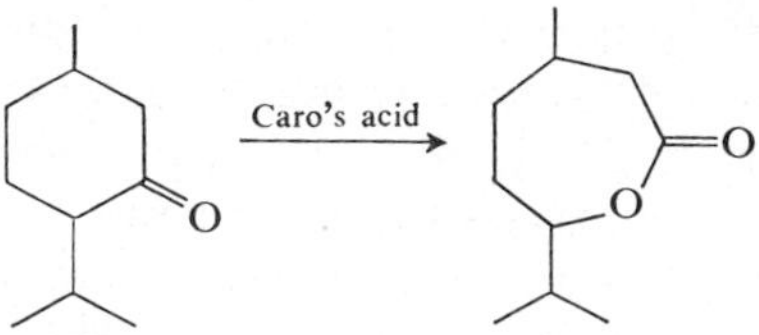

50 —Caro's acid→ **51** (12)

lactone **53** was obtained from tetrahydrocarvone (carvomenthone) **(52)** (Eq. 13).

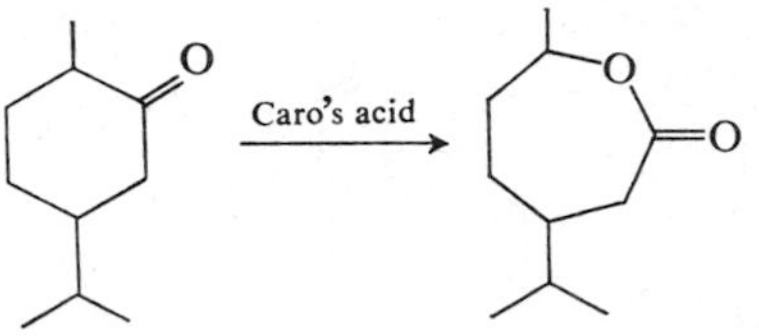

Caro's acid (13)

52 **53**

Wedekind (49) obtained α-tetrahydrosantonilide, which now can be formulated as **55**, by oxidation of α-tetrahydrosantonin (**54**) (50) with Caro's acid (Eq. 14). In the same manner the corresponding ε-lactone was obtained (51) from α-tetrahydroartemisin (**56**) (52).

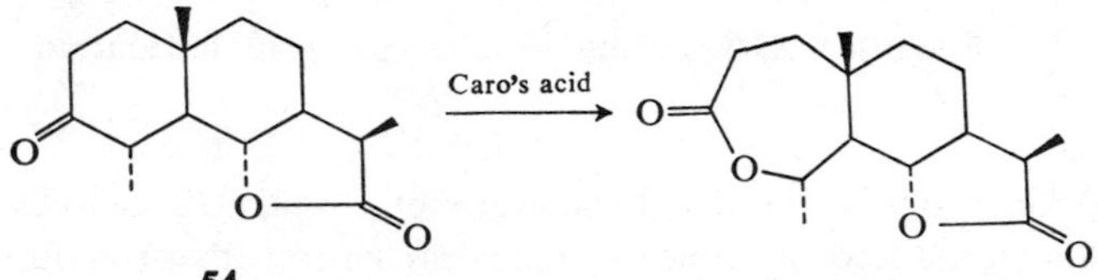

Caro's acid (14)

54 **55**

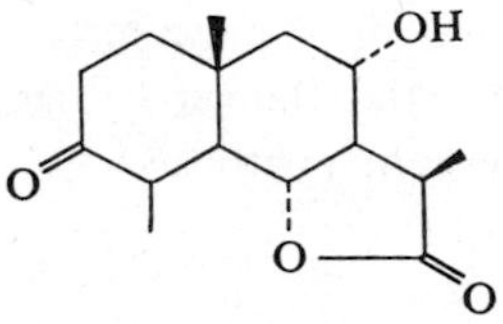

56

b. *Examples*

In recent years a number of organic peroxy acids other than Caro's acid have become the reagents of choice for the oxidation of ketones to esters. ε-Lactones obtained by the oxidation of terpenoid cyclohexanones with these reagents are presented in Table VI-2. References are included to a few uncharacterized ε-lactones which were prepared as intermediates during the course of structural or synthetic studies. In every instance but one the bond undergoing rearrangement is that attached to the most substituted carbon. In the exception, *trans*-(*R*)-(−)-2-bromo-5-methylcyclohexanone derived from the monoterpene (*R*)-(+)-pulegone, it is possible to rationalize the preferential migration of the least substituted bond. Lowering of the electron density by the inductive effect of the bromine atom would allow the more electron-rich bond to migrate (53).

c. *Special Aspects*

One failure of a ketone to undergo the Baeyer-Villiger reaction has been noted; diketolanostane derivative **57** was found to be inert to peracetic acid (90). In certain other instances low yields, attributed to the hindered keto group, have been obtained (81, 82).

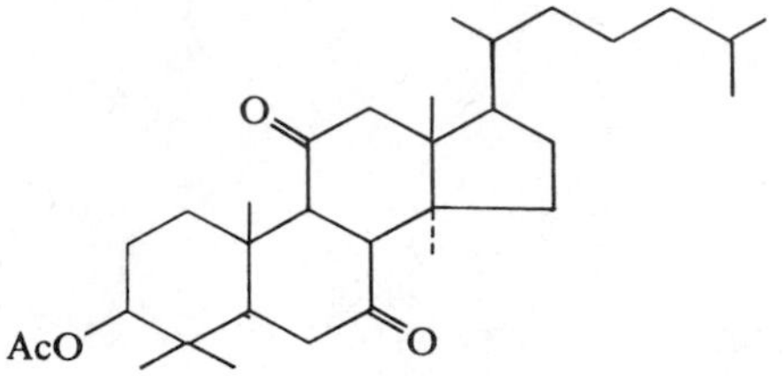

57

In contrast to the oxidation of (—)-dihydrocarvone (**58**) with perbenzoic acid in chloroform which afforded the oxide, Howe, McQuillin, and Temple (55) discovered that lactone **59** was the main product of oxidation with monoperphthalic acid in ether. The acid strength of phthalic acid compared with benzoic acid and the basicity of the solvent were discussed as possible factors responsible for the differential reactivity. A lactone diol, regarded as **60**, was isolated in small yield from oxidation of (—)-dihydrocarvone with perbenzoic acid in chloroform.

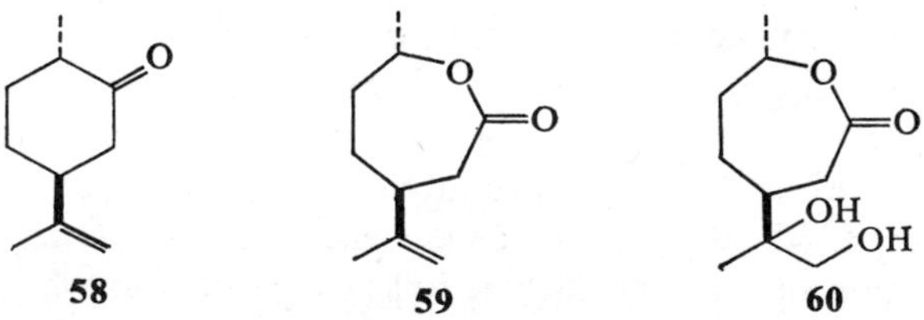

58 **59** **60**

ε-Lactones have played a significant role in the elucidation of the structure of valeranone (**61**) (61–64). An X-ray investigation (91), as well as other work (64), established the relative stereochemistry and indicated structure **61** or **62** for valeranone. Křepinsky and co-workers (64) determined the absolute configuration at $C_{(5)}$ of the ε-lactone (**63**) obtained by peracid oxidation of valeranone to be *S* by applying the extended Hudson-Klyne lactone rule (92). On this basis, since the Baeyer-Villiger oxidation is known to proceed with retention of configuration (93, 94), they proposed structure

62 for valeranone. Independently, Hikino and co-workers (61) in a preliminary communication proposed the enantiomeric structure **61**. Subsequently, further evidence was presented in support of the latter structure for valeranone, proving that the extended lactone rule had failed in this instance (62).

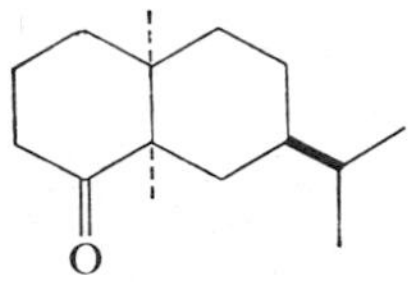

61 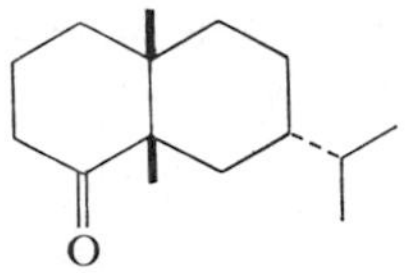 **62**

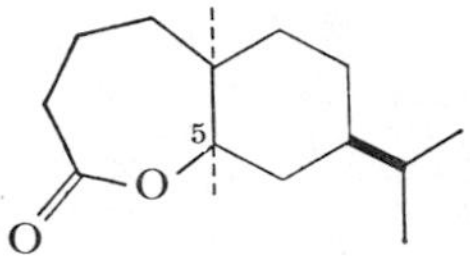

63 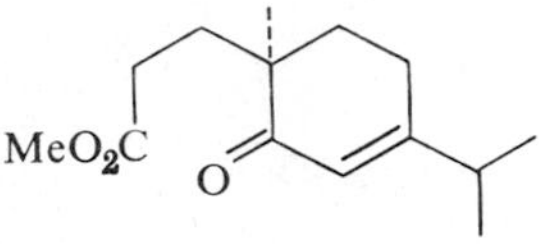 **64**

The enantiomeric keto esters **64** and **67** were prepared from valeranone and β-eudesmol (**65**), respectively, via the corresponding ε-lactones **63** and **66** (Eq. 15) (61, 62).

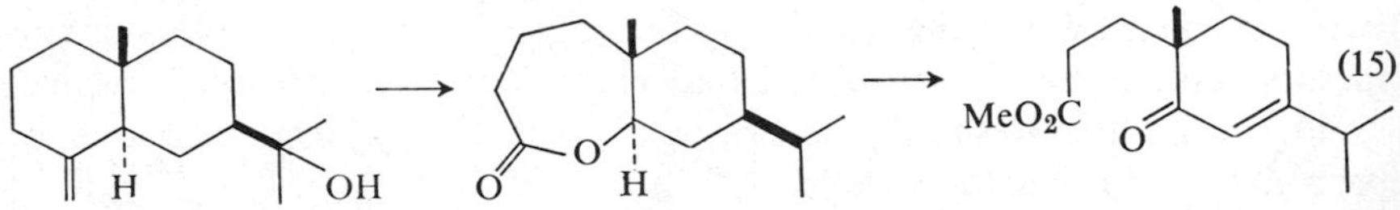

(15)

65 **66** **67**

When the perbenzoic acid oxidation of valeranone was catalyzed by *p*-toluenesulfonic acid, it was observed to take an anomalous course leading to hydroxy lactone **68** (64). The position of the hydroxyl was proved by dehydration to **69** and by the formation of formaldehyde on ozonolysis. The formation of **68** was regarded as evidence that the isopropyl group had the configuration opposite to the methyl groups since it was assumed that spatial vicinity of the keto group and the attached isopropyl group was required for the reaction. ε-Lactone **63** was isomerized to a substance believed to have structure **70** on treatment with boron trifluoride etherate (64).

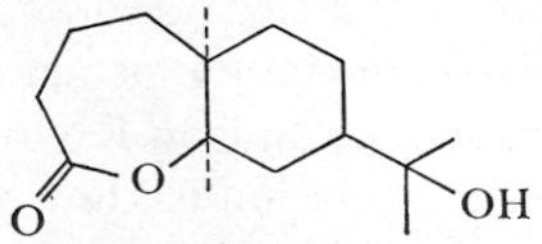

68 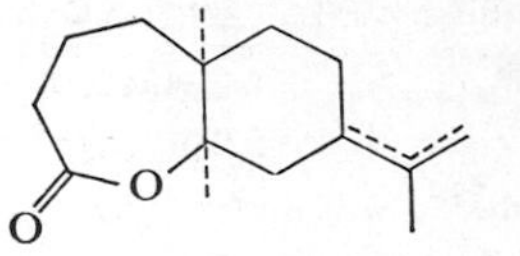 **69** 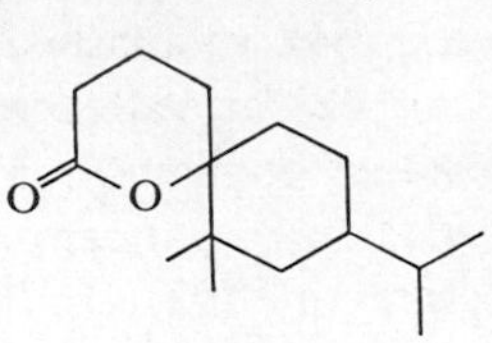 **70**

The cyclohexanone ring of acorone (**71**) (58, 59) and its stereoisomers, isoacorone and cryptoacorone, was oxidized selectively by perbenzoic acid, affording keto ε-lactones (58). By application of the Hudson-Klyne lactone rule the absolute configuration of the methyl group attached to the six-membered ring of the parent compound was established.

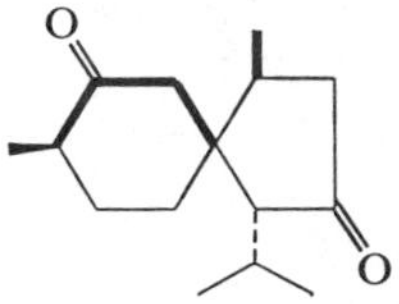

71

Reaction of isolongifolene (**72**) with excess perbenzoic acid in chloroform gave a good yield of lactone **74**, whereas reaction with 1 mole of oxidant led to the isolation of the intermediate ketone **73** (95). On the other hand, oxidation of isologifolene with peracetic acid buffered with sodium acetate gave ketone **73**, lactone **74**, and alcohol **75** in a ratio of 14:1:4, together with a trace of epoxide **76** (Eq. 16) (96). The epoxide, when treated under the same conditions, afforded the above three major products in a similar ratio, suggesting its probable intermediate role.

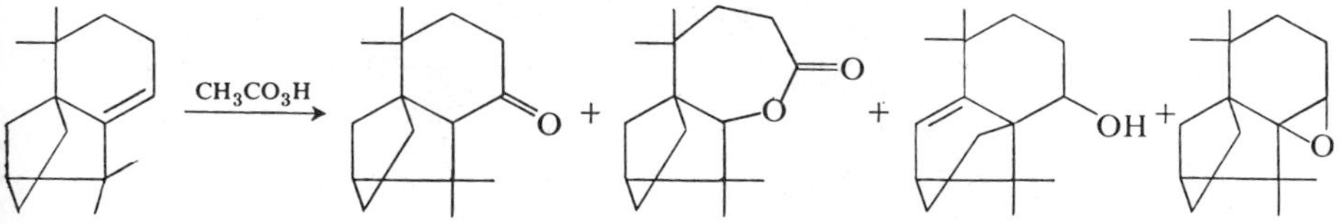

CH_3CO_3H

72 **73** **74** **75** **76**

(16)

Treatment of methyl abietate derivative **77** with *m*-chloroperbenzoic acid (Eq. 17) yields either a mixture of enol ε-lactone **78** and epoxy ε-lactone **79**, or the latter compound exclusively, depending upon the reaction conditions (73, 74). On further oxidation, **78** gives **79**. The fact that epoxyketone **80** could not be detected suggested that **77** is converted into **79** via **78** rather than **80**. Under a variety of conditions **79** undergoes facile rearrangement to the isomeric β-aldehydo δ-lactone **81**. It is believed that this compound arises from the isomer of **79** with the α-epoxy configuration (74).

Although not strictly within the confines of this review, since that portion of the molecule possessing the oxepin grouping is not terpene-derived, two additional examples of ε-lactones may be cited. Compounds **82** and **83** have

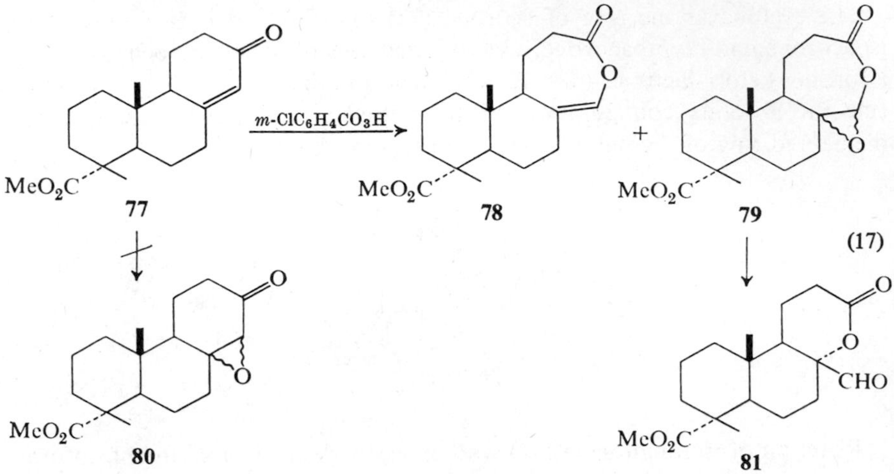

$m\text{-ClC}_6H_4CO_3H$

MeO$_2$C **77** **78** **79** (17) **80** **81** CHO

been prepared by oxidation of the corresponding cyclohexanone derivatives with peracetic acid (97).

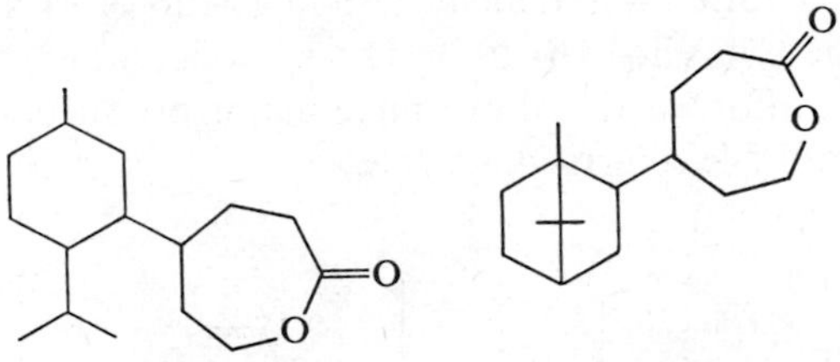

82 **83**

3. Miscellaneously Derived ε-Lactones and ε-Lactols

(*R*)-(+)-Pulegone (**84**) has been the starting material for the synthesis of four optically active methyl-substituted ε-lactones of known absolute configuration (53). (*R*)-(+)-γ-Methyl-ε-caprolactone (**85**) was obtained by acid-catalyzed cyclization of (*R*)-(+)-6-hydroxy-4-methylhexanoic acid with hydrogen chloride in the presence of molecular sieves (Eq. 18). Similarly, (*R*)-(+)-6-hydroxy-3-methylhexanoic acid gave (*R*)-(−)-β-methyl-ε-caprolactone (**86**). Attempts to effect cyclization by several other methods met with less success. (*R*)-(−)-δ-Methyl-ε-caprolactone (**87**) and *trans*-(*R*)-(−)-α-bromo-δ-methyl-ε-caprolactone (**88**) were prepared by Baeyer-Villiger oxidation of the appropriate ketone (Eqs. 19 and 20). Assuming planarity of the lactone group (98), models indicate that only two conformations, a boat and a chair form, are possible for the ε-lactone ring. On the basis of an optical rotatory dispersion study of these compounds it was proposed that the conformation of the ε-lactone ring was a slightly deformed chair (99).

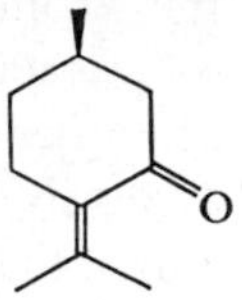

84

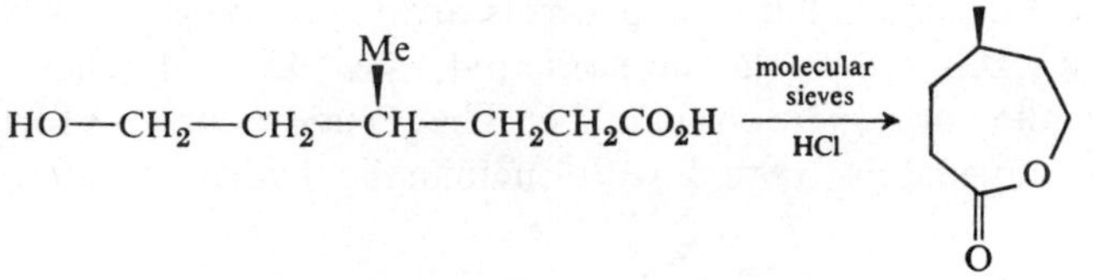

HO—CH_2—CH_2—CH(Me)—$CH_2CH_2CO_2H$ —(molecular sieves, HCl)→ **85** (18)

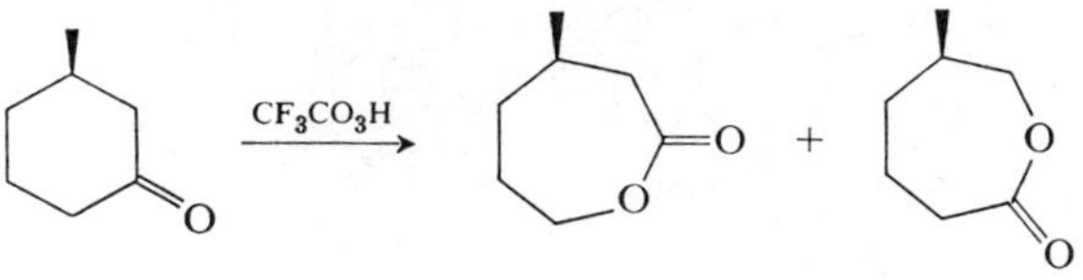

—(CF_3CO_3H)→ **86** + **87** (19)

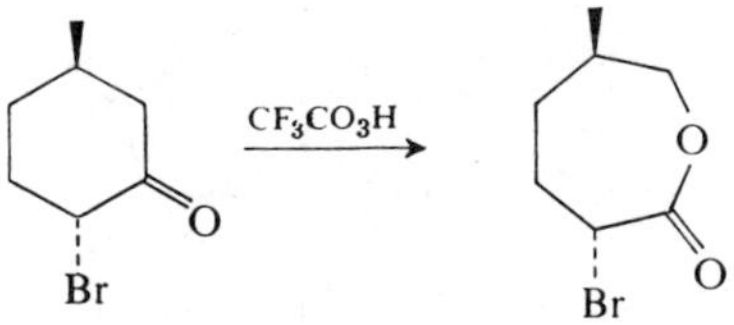

—(CF_3CO_3H)→ **88** (20)

Attempted dehydration of hydroxy acid **89** with phosphorus oxychloride in pyridine gave ε-lactone **90** (Eq. 21); treatment with a catalytic amount of *p*-toluenesulfonic acid gave γ-lactone **91** (100).

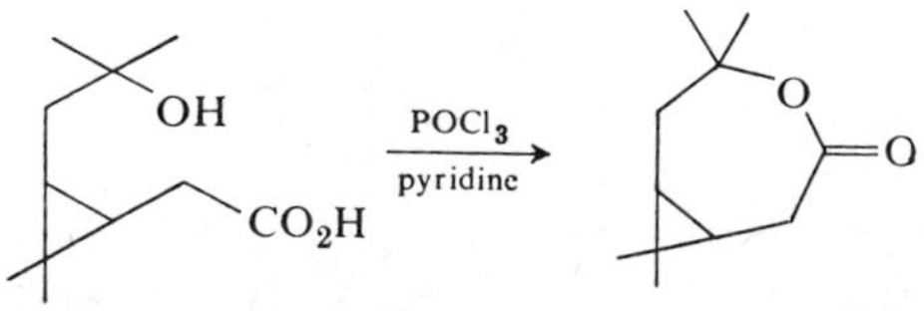

89 —($POCl_3$, pyridine)→ **90** (21)

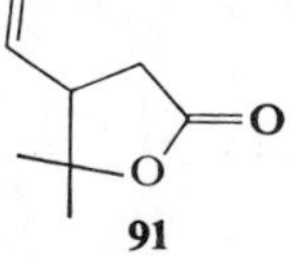

91

Acidification of an aqueous sodium hydroxide solution of lactol **92** yields a new compound believed to have structure **93** (Eq. 22) since it lacks ir absorption for an aldehyde on the aromatic ring but retains that for a hydroxyl group (101). Formation of substances **95** and **96**, having an oxepinone ring, occurs when the diterpene fujenal (**94**) in methanol is heated in a sealed tube (Eq. 23) (102, 103). Hydrolysis of **95** with mineral acid affords the monobasic acid **97** in which hydration of the terminal methylene has occurred. Ozonolysis of **95** gives the pseudo-ester nor-ketone **98** (Eq. 24) which is resistant to further oxidation and forms an oxime. Acid hydrolysis of this compound yields either the monomethyl ester **99** or fujenal nor-ketone (**100**) depending on conditions (Eq. 24). The pseudo-ester **95** is formed in low yield when fujenal is treated with methanolic hydrogen chloride. Other

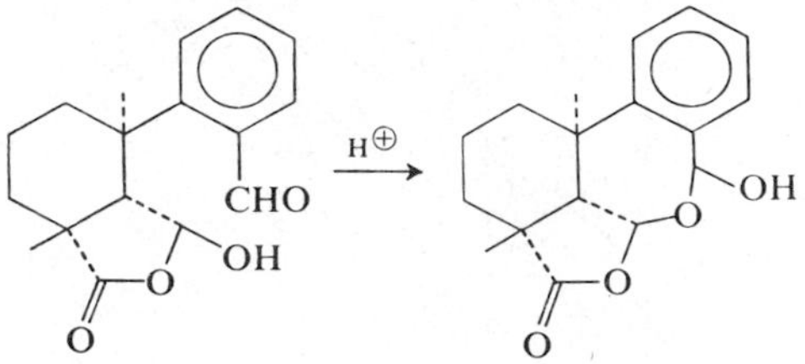

CHO, OH, **92** $\xrightarrow{H^{\oplus}}$ OH, **93** (22)

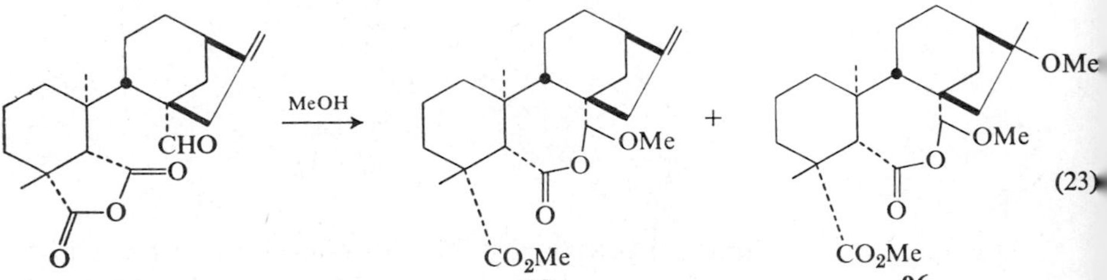

CHO, **94** $\xrightarrow{MeOH}$ OMe, CO_2Me, **95** + OMe, OMe, CO_2Me, **96** (23)

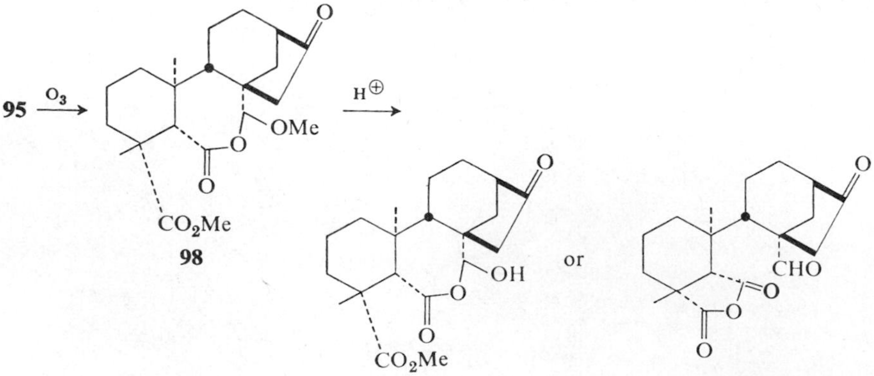

95 $\xrightarrow{O_3}$ OMe, CO_2Me, **98** $\xrightarrow{H^{\oplus}}$ OH, CO_2Me, **99** or CHO, **100** (24)

products are an isomeric dimethyl ester formulated as **101** and a chlorodimethyl ester believed to be **102**.

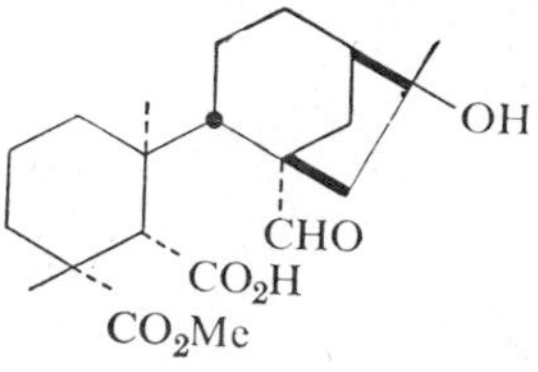

OH CHO CO_2H CO_2Me **97** OMe O O CO_2Me **101** Cl OMe O O CO_2Me **102**

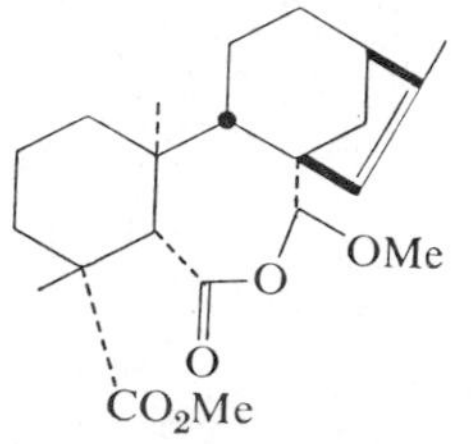

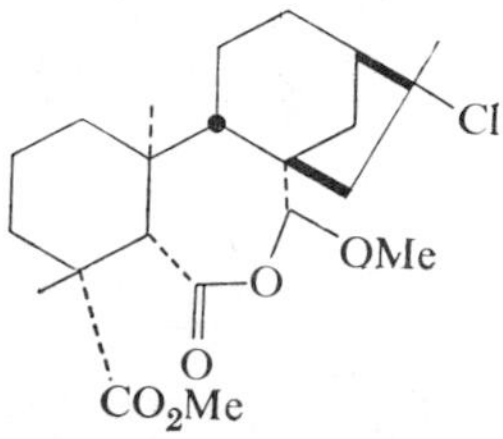

Hydroxydiketone **103** gives the oxepin **104** and carbon dioxide under the conditions of the benzilic acid rearrangement (Eq. 25) (104). It was felt that carbon dioxide was lost from an intermediate β-keto ester. Treatment of the oxepin derivative with phosphoric acid liberates carbon monoxide.

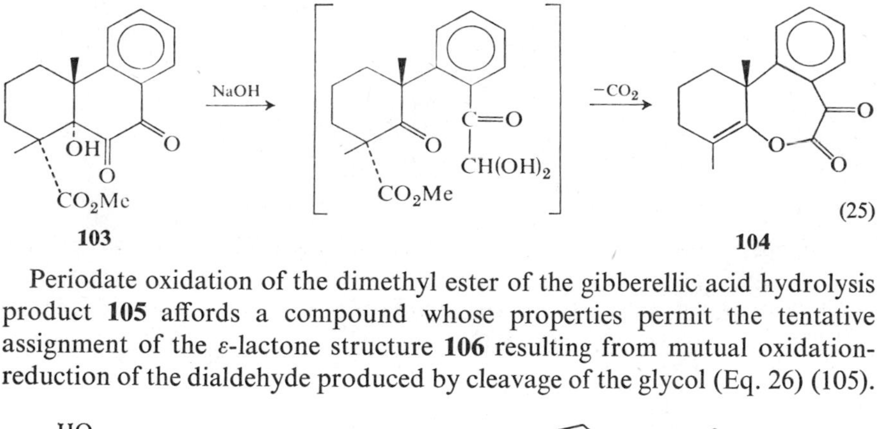

OH O O CO_2Me **103** $\xrightarrow{\text{NaOH}}$ [C=O O $CH(OH)_2$ CO_2Me] $\xrightarrow{-CO_2}$ O O O **104** (25)

Periodate oxidation of the dimethyl ester of the gibberellic acid hydrolysis product **105** affords a compound whose properties permit the tentative assignment of the ε-lactone structure **106** resulting from mutual oxidation-reduction of the dialdehyde produced by cleavage of the glycol (Eq. 26) (105).

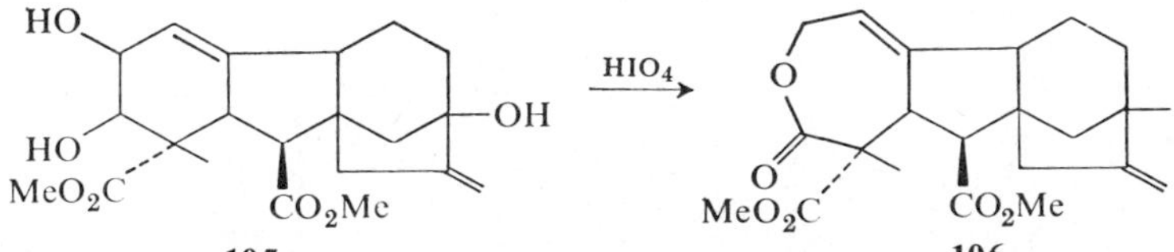

HO HO MeO_2C OH CO_2Me **105** $\xrightarrow{HIO_4}$ O O MeO_2C OH CO_2Me **106** (26)

The borohydride reduction product (106) of rosoic acid (**107**) or of methyl rosoate is identical with the dilactone **108** from Baeyer-Villiger oxidation of dihydroisorosenolactone (Eq. 27) (81). Likewise borohydride reduction of ε-keto acid **109**, followed by acid treatment and sublimation, affords ε-lactone **110** (Eq. 28) (107, 108). Steric factors, in general of great importance in determining the chemistry of terpene oxepins, probably account for the observed ease of cyclization. Van Tamelen and McCormick (109) have reported the isolation of tricyclic lactone **112** on alkaline hydrolysis of bicyclic hydroxyester **111** (Eq. 29).

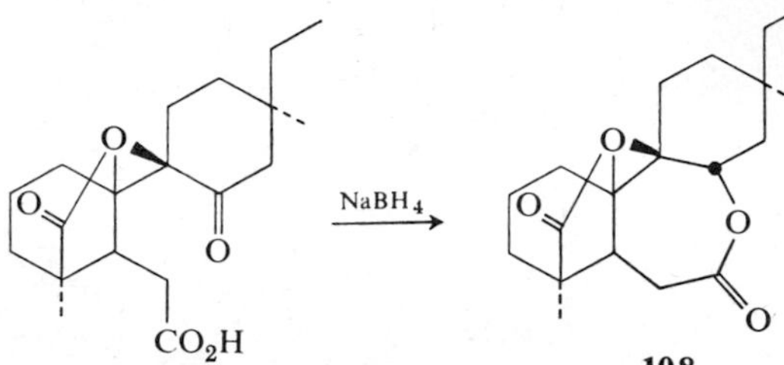

107 $\xrightarrow{NaBH_4}$ **108** (27)

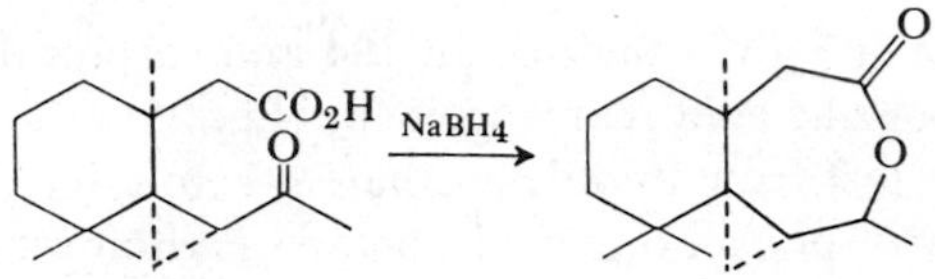

109 $\xrightarrow{NaBH_4}$ **110** (28)

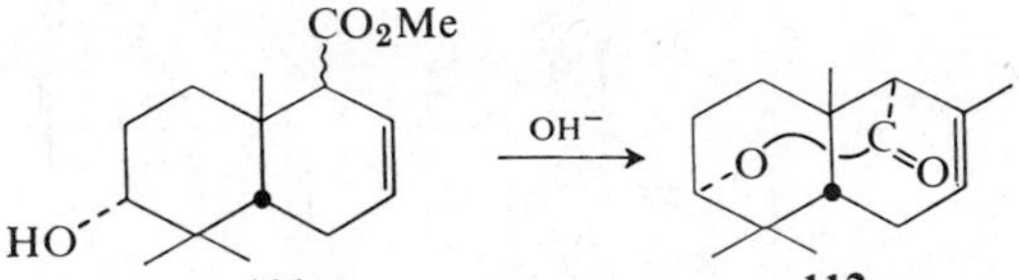

111 $\xrightarrow{OH^-}$ **112** (29)

Hydroperoxide **114**, formed by low-temperature ozonolysis of methyl podocarpate (**113**) in methanol–methylene chloride, was converted by alkaline hydrogen peroxide treatment into ε-lactone **115**, characterized as its methyl ester (**116**) (Eq. 30) (110).

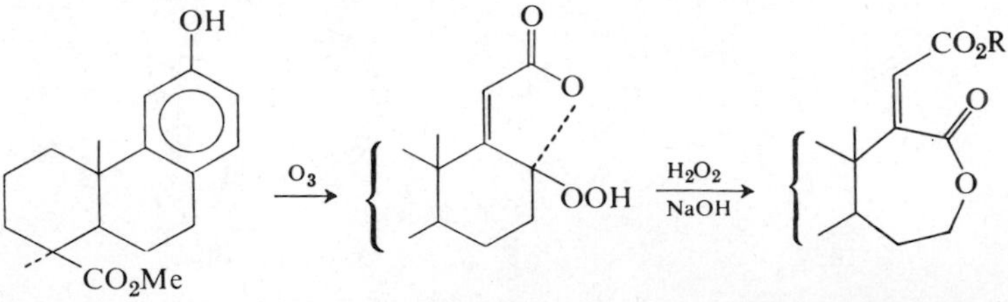

113 $\xrightarrow{O_3}$ **114** $\xrightarrow[NaOH]{H_2O_2}$ **115**; R = H, **116**; R = Me (30)

Ozonolysis of cativic acid (**117**) produces a crystalline product for which structure **118** has been proposed (Eq. 31) (111).

Pyrolysis of the isomeric dimers **119** from (*d*)-camphor-3-carbonyl chloride produces an isomer assigned structure **120**, presumably via a stabilized

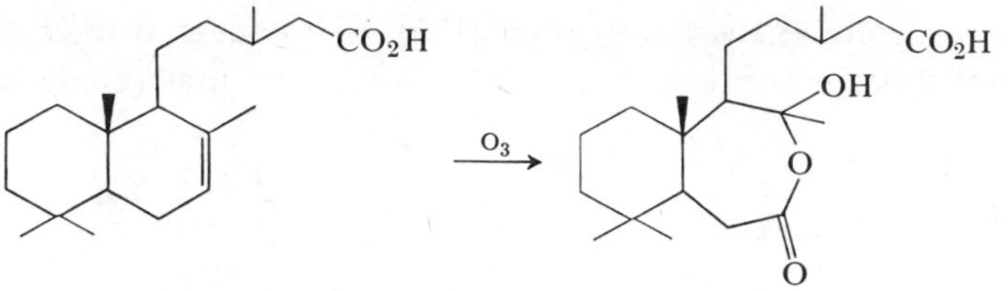

117 118 (31)

zwitterion (Eq. 32) (112). Treatment of this product with concentrated hydrochloric acid leads to a mixture of acids **121** and **122**; the former, on heating *in vacuo*, regenerates the pyrolysis product.

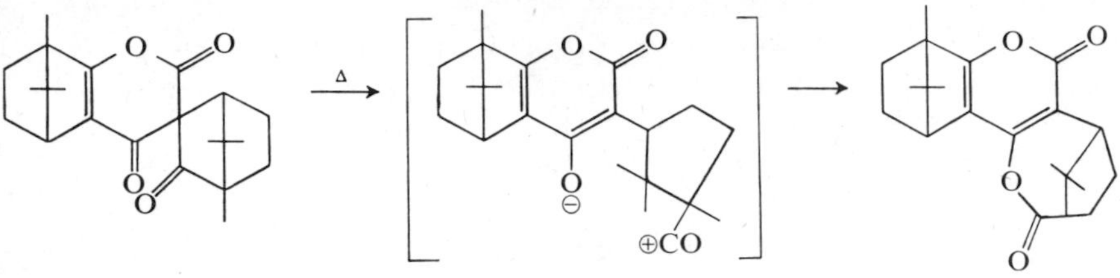

119 120 (32)

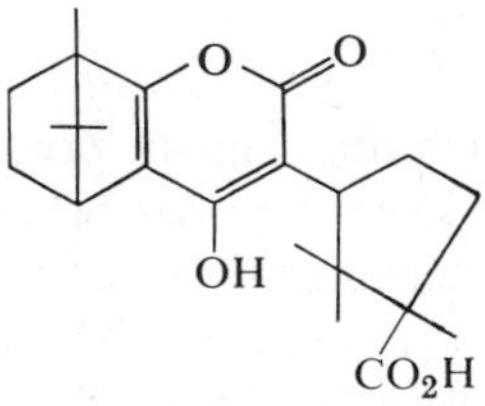

121 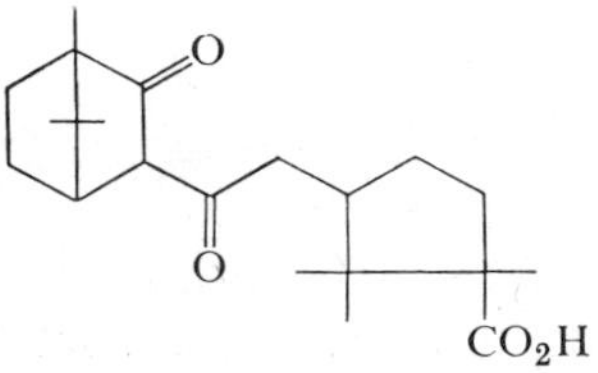 122

Stevenson (87) has reported that nitrosation of friedelolactam **(123)** with dinitrogen tetroxide in carbon tetrachloride at low temperatures gives friedelolactone **(124)** identical with the product of Baeyer-Villiger oxidation of friedelin. Chromic acid oxidation of friedelolactone yields friedonic acid **(125)** (Eq. 33). The hydroxy acid obtained by reduction of friedonic acid cyclizes spontaneously to friedelolactone (113, 114). Friedelolactone has been converted into 2-keto-3-oxa-friedelane **(126)** by the Barbier-Wieland procedure (115).

ε-Lactones have also been obtained by oxidation of the appropriate hemiacetal (see Section D). The literature contains a number of reports of the preparation of ε-lactones whose structures have been shown by subsequent studies to require correction. For example, tricycloekasantalic acid **(127)** and bicycloekasantalic acid **(128)** were reported to furnish the same lactone, for which structure **129** was suggested (116). Subsequent work showed the structure of this lactone to be **130**, the methyl group undergoing migration

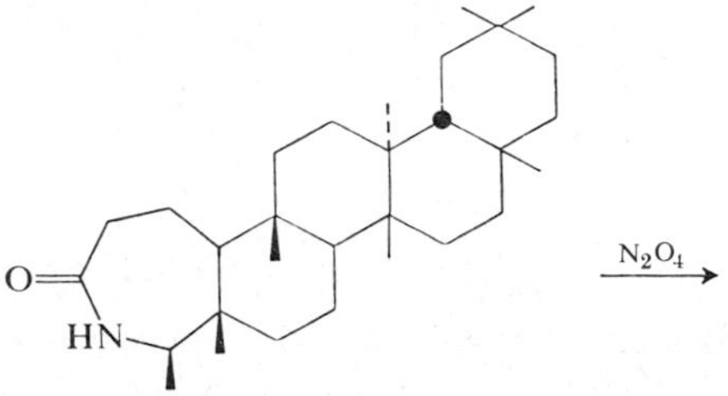

$\xrightarrow{N_2O_4}$ (33)

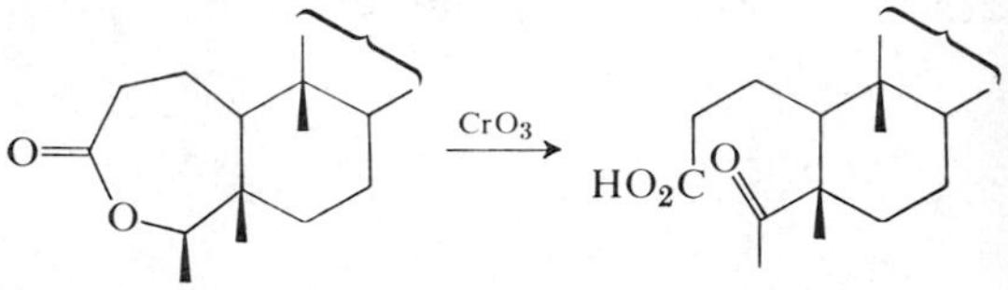

123 **124** $\xrightarrow{CrO_3}$ **125**

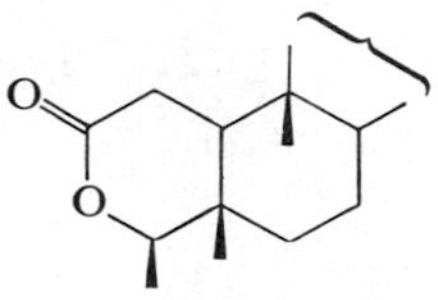

126

as part of the rearrangement (117). Another instance concerns the product of oxidation with Caro's acid of santonic acid, whose chemistry has been interpreted by Woodward, Brutschy, and Baer (118) in terms of tricyclic structure **131**; the product cannot be the ε-lactone **132** proposed by Abkin and Medvedev (119) on the basis of an incorrect structure for santonic acid.

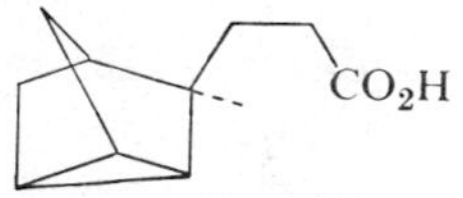

127 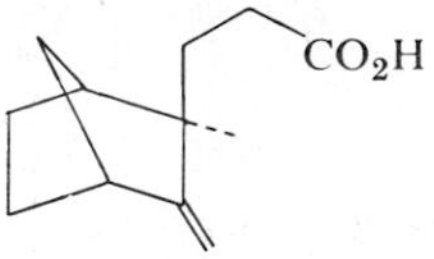 **128** 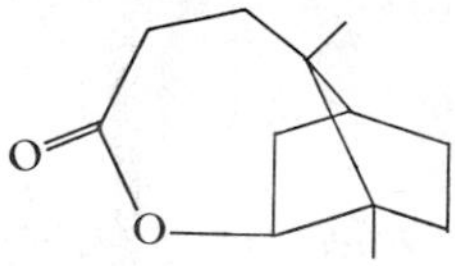 **129**

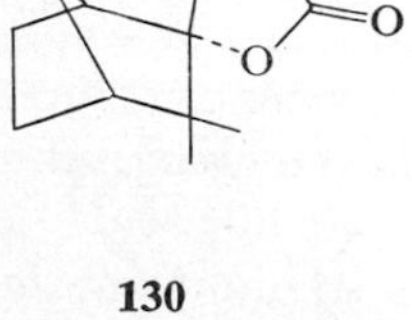

130 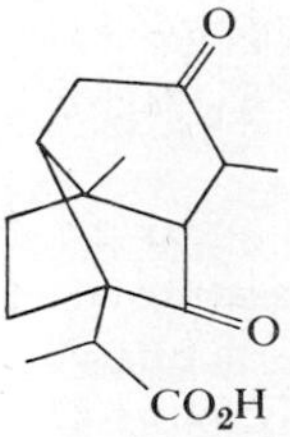 **131** 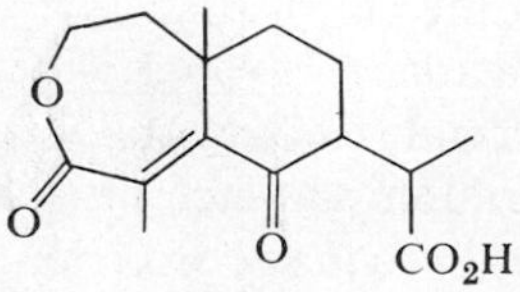 **132**

C. Anhydrides

Examples of 2,7-oxepindiones (adipic anhydrides) derived on treatment of the corresponding diacid with acetic anhydride are presented in Table VI-3. Most of these anhydrides, which are diverse in constitution, have been obtained during structural elucidation studies. A few of them will be discussed briefly below, together with some seven-membered anhydrides obtained by other means.

Wallach and collaborators reported the formation of cineolic acid (**134**) on oxidation of 1,8-cineole (**133**) with potassium permanganate, and observed that **134** was converted into anhydride **135** upon distillation under reduced pressure or heating with acetic anhydride (Eq. 34) (121, 145). The anhydride decomposed on distillation, yielding 2-methylhept-2-en-6-one. Detailed studies of cineolic acid and derivatives were important in the establishment of the structure of 1,8-cineole (146).

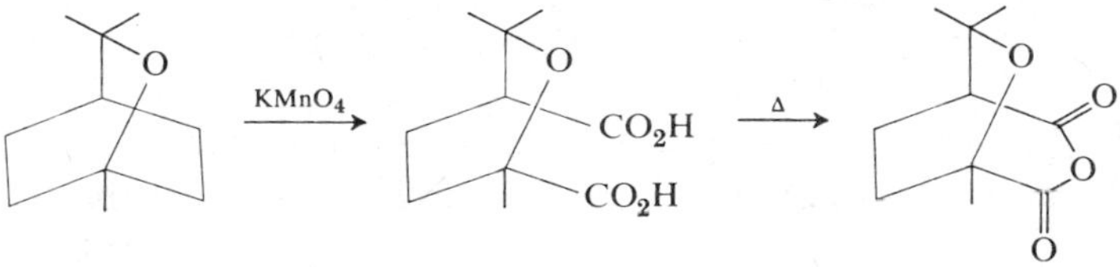

$KMnO_4$; CO_2H ; CO_2H ; Δ (34)

133 **134** **135**

During a study directed towards the synthesis of iridomyrmecin, glycidic acid **136** was heated in the presence of copper powder (147). The desired decarboxylation did not occur and the only product reported was the anhydride **137**, which is believed to have a *trans* ring fusion (Eq. 35).

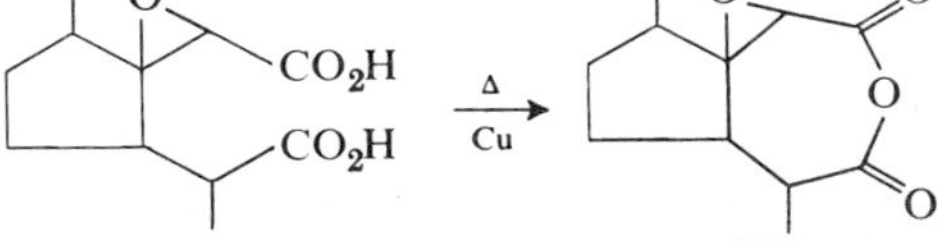

CO_2H ; CO_2H ; Δ, Cu (35)

136 **137**

It has long been known that α-longiforic acid (**138**) affords anhydride **139** (122); the same product has also been reported to be formed from longidione (**140**) and as a minor product from longifolene (**141**) on treatment with perbenzoic acid (Eq. 36) (123). A more recent study of the action of perbenzoic acid on longifolene has failed to reveal the presence of the anhydride among the products although a small amount of longidione was detected spectroscopically (148). It is probable that α-longiforic anhydride was obtained by Ogura by the action of refluxing acetic anhydride on the hydrogen peroxide oxidation product of longidione (149).

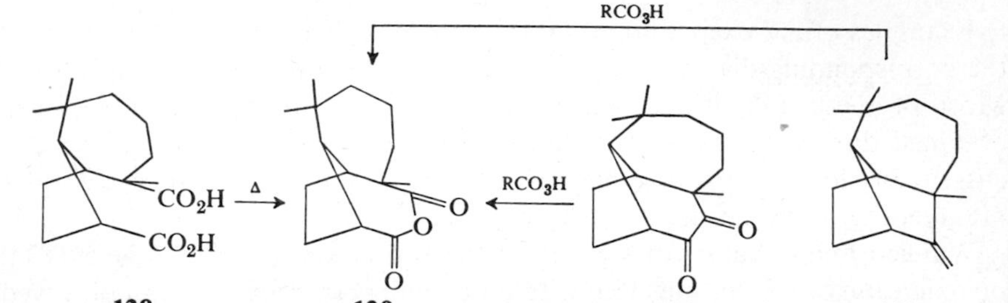

RCO_3H

CO_2H $\xrightarrow{\Delta}$ ... O $\leftarrow$ RCO_3H

CO_2H

138 **139** **140** **141**

(36)

A small amount of anhydride **144** was formed during the preparation of the racemic dicarboxylic acid **143** by oxidative ozonolysis of **142** (Eq. 37) (150). Improved yields of **144** were realized when base treatment was omitted during the isolation of products. Ozonization of **142** in acetic acid–ethyl acetate or in methylene chloride gave a mixture of anhydride and diketone **145** (Eq. 38), indicating that **143** is not a precursor of **144** via dehydration (151). The authors have proposed that anhydride formation is the result of a Baeyer-Villiger-type oxidation of the α-diketone by a peracid generated during ozonolysis, and have obtained support for this suggestion from a study of the ozonization of the 3-hydroxymethylene derivative of camphor. Formation of a small amount of anhydride **148** during autoxidation of euphadien-3-one (**146**) to the diosphenol **147** probably occurs by a similar

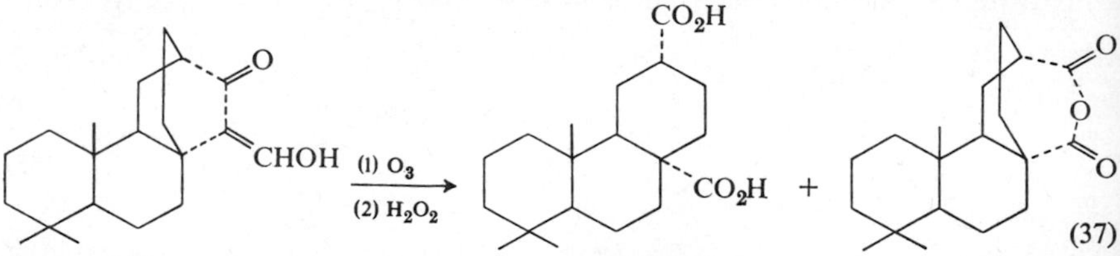

CHOH $\xrightarrow[(2)\ H_2O_2]{(1)\ O_3}$ CO_2H + (37)

142 **143** **144**

142 $\xrightarrow{O_3}$ **145** + **144** (38)

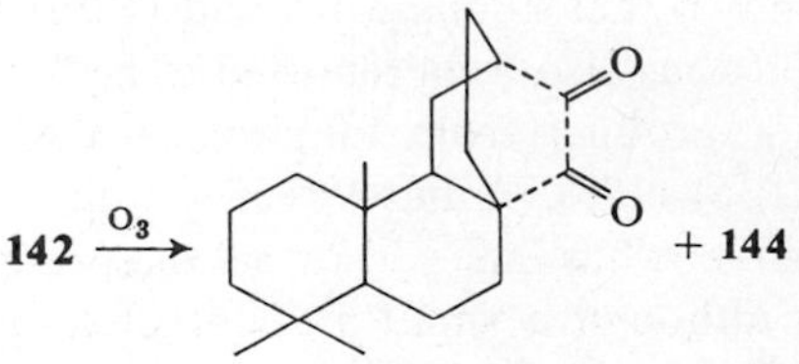

mechanism rather than by the proposed route involving cyclization of the 2,3-secodicarboxylic acid (Eq. 39) (134).

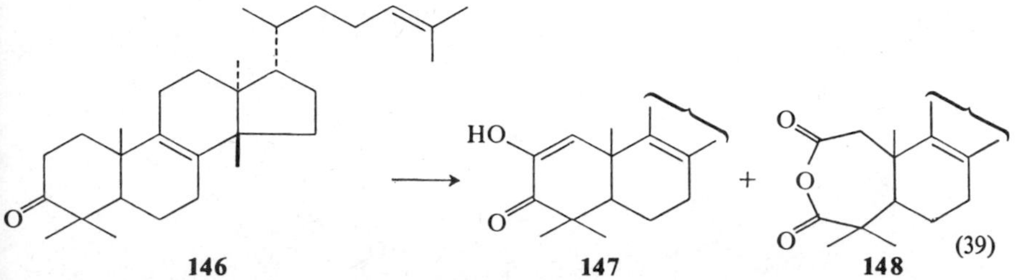

A number of terpene adipic anhydrides have been converted to cyclopentanones. For example, the C-norketone has been obtained in both the α- and β-amyrin cases by pyrolysis of the anhydride (see Table VI-3) derived from the seco diacid on treatment with acetic anhydride (135, 136). Pyrolysis of triacid **149** *in vacuo* yields the A-norketone directly (143, 152) but similar treatment of diacid ester **150** gives anhydride **151** in addition to the cyclopentanone (Eq. 40) (153). In the latter instance it can be assumed that the higher volatility of the ester anhydride allows it to distill before rearrangement to the cyclopentanone can occur.

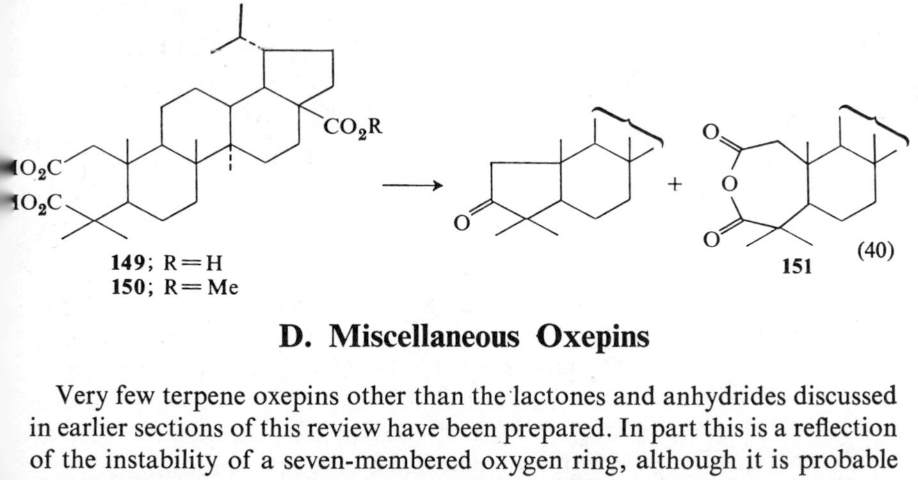

D. Miscellaneous Oxepins

Very few terpene oxepins other than the lactones and anhydrides discussed in earlier sections of this review have been prepared. In part this is a reflection of the instability of a seven-membered oxygen ring, although it is probable that few direct attempts have been made to obtain such compounds. However, three naturally occurring oxepins, which do not belong to the classes already discussed, have been described recently.

Occidenol, identical with occidiol (154), has been shown to have the novel structure **152** (155). Hydrogenation yielded a tetrahydro derivative and acid treatment gave diol **153**, postulated to arise by aldol condensation after hydrolysis (Eq. 41). The absolute configuration was determined by conversion to a substance synthesized from elemol (**154**) (Eq. 42). It was proposed that occidenol may arise in the manner shown in Eq. 43.

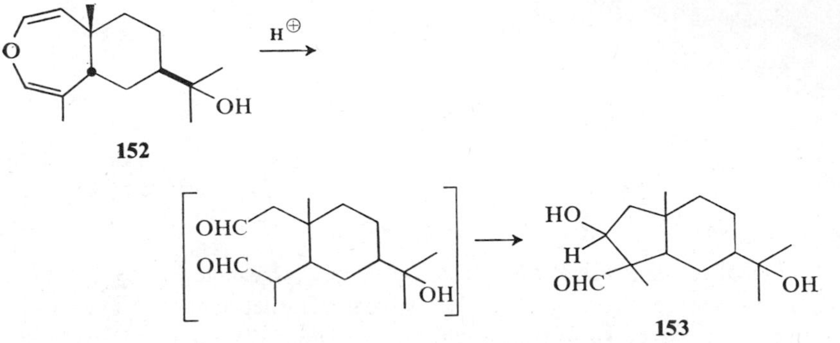

$H^{\oplus}$

152

OHC OHC OH

HO H OHC OH

153

(41)

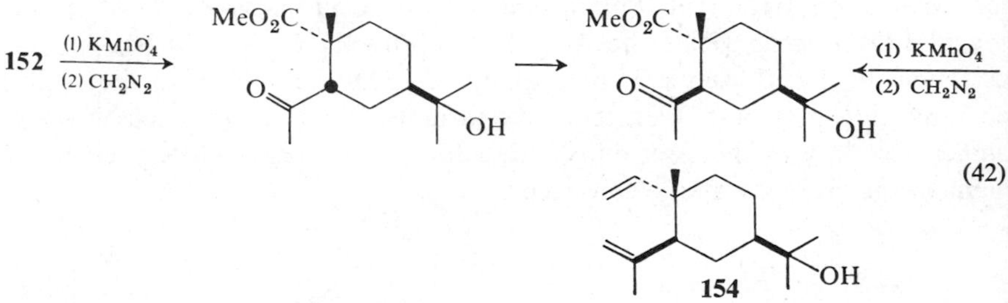

152 (1) $KMnO_4$ (2) CH_2N_2

MeO_2C O OH

MeO_2C O OH

(1) $KMnO_4$ (2) CH_2N_2

(42)

OH

154

OH OH O OH 152

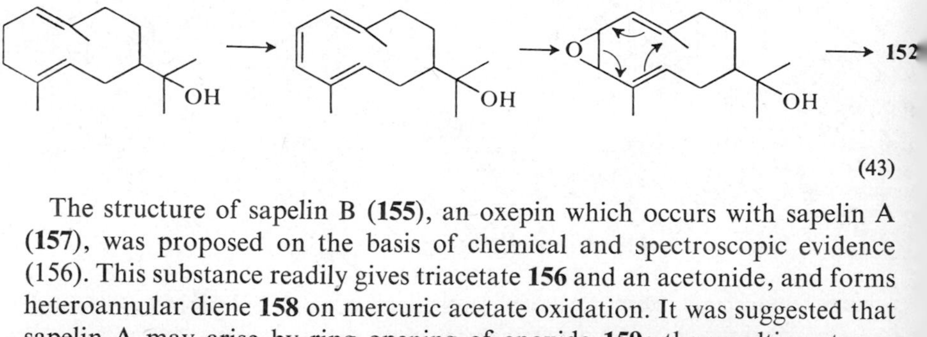

(43)

The structure of sapelin B (**155**), an oxepin which occurs with sapelin A (**157**), was proposed on the basis of chemical and spectroscopic evidence (156). This substance readily gives triacetate **156** and an acetonide, and forms heteroannular diene **158** on mercuric acetate oxidation. It was suggested that sapelin A may arise by ring opening of epoxide **159**; the resulting stereochemistry (Eq. 44) agrees with that indicated by the nmr spectrum.

The third natural product to be discussed in this section is nimbolin B (**160**) (157). Its spectral properties indicated a close relationship to nimbolin A (**161**), with which it occurs. However, differences in the nmr spectrum suggested cleavage of ring C possibly via an intermediate such as **162**. The latter could yield a hydroxy aldehyde which in turn might form nimbolin B as the internal hemiacetal. The hemiacetal hydroxy group could be methylated, and oxidation yielded a lactone.

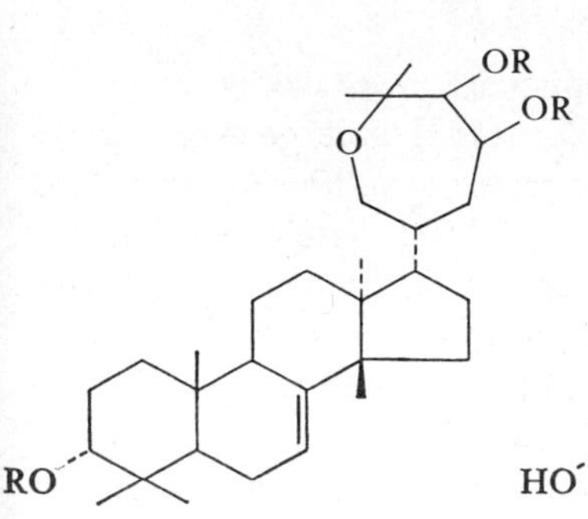

155; R=H
156; R=Ac

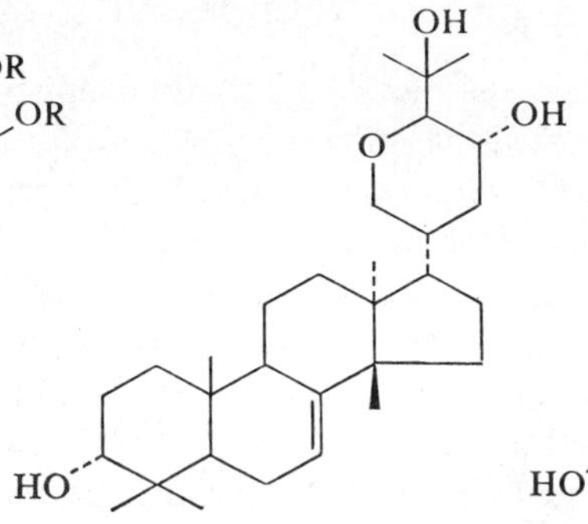

157

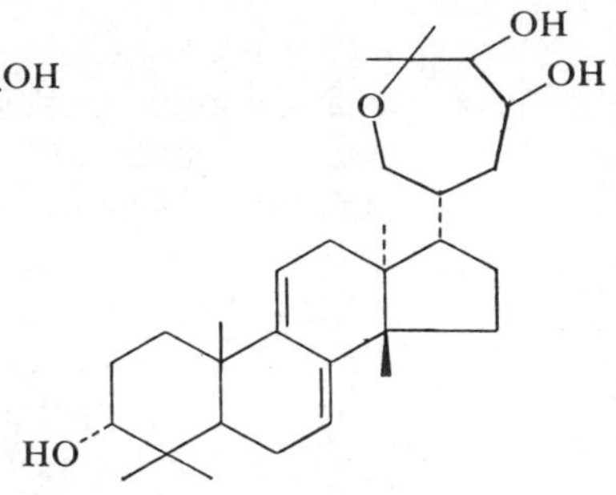

158

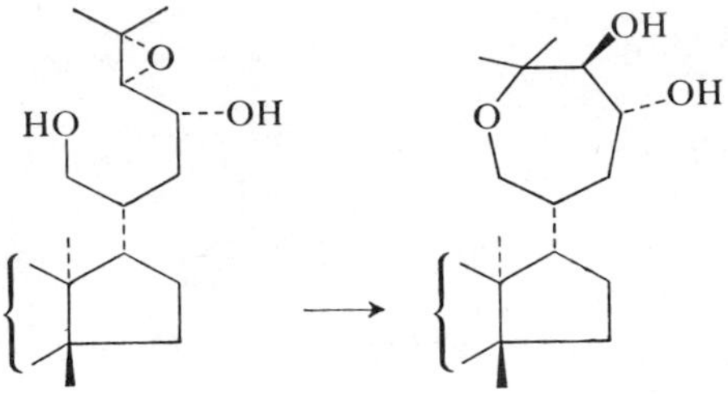

(44)

159

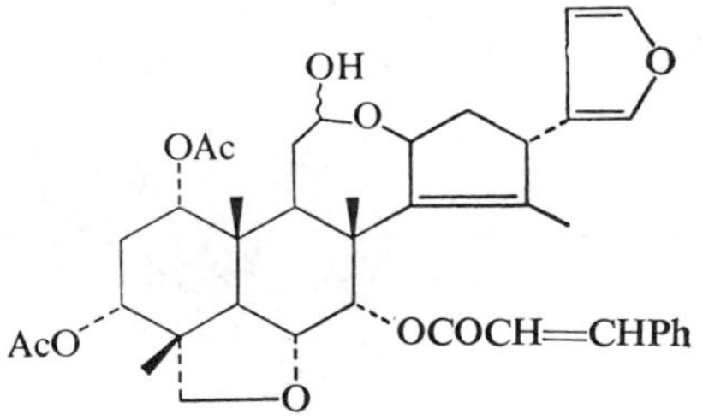

160

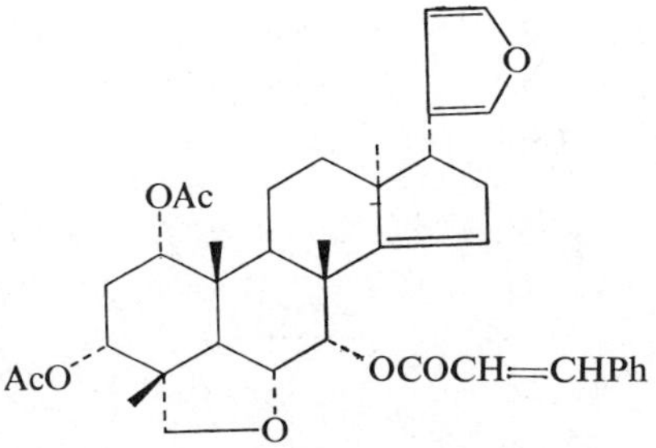

161

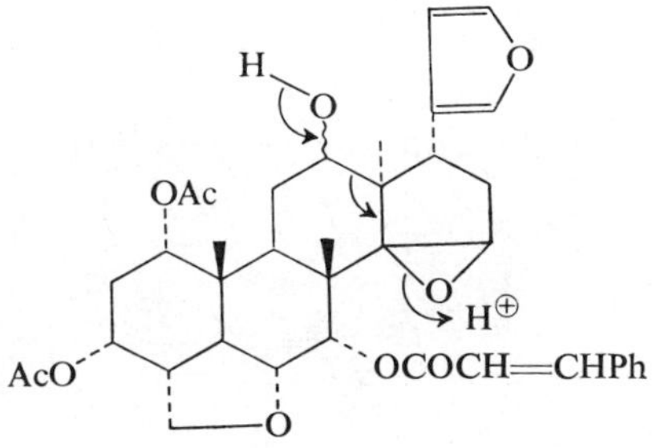

162

Some additional examples of synthetic oxepins are given below.

It has been reported that pinane hydroperoxide (**163**) undergoes rearrangement when treated with acetic anhydride or benzoyl chloride to afford substances assigned structure **164** (R = CH_3 or C_6H_5); these can be hydrolyzed to cyclobutane derivatives (158). The ease of rearrangement in this instance was attributed to the strained ring system. However, later workers (159) have reported that the peresters **165** could be obtained easily but could not be rearranged to the isomeric esters (**164**); instead, decomposition products having structures dependent on the reaction conditions were obtained (160).

OOH **163** OCR(=O) O **164** OOCR(=O) **165**

In the course of studies involving the cyclization of acyclic alcohols, two compounds were obtained from **166** (161) and subsequently assigned the oxepin structures **167** and **168** (Eq. 45) (162). Both compounds afford the same product upon hydrogenation. Further support for the proposed structures was obtained by identification of the products from degradation of **167**.

HO O **166** $\xrightarrow{H^{\oplus}}$ OH O $\xrightarrow{-H_2O}$ O **167** + O **168**

(45)

The acetoxy ketone **169** derived from kessyl glycol yields on permanganate oxidation oxepin **170** (163–165). This can be decarboxylated in refluxing acetic acid to α-hydroxy acid **171** (Eq. 46).

AcO O =O **169** $\xrightarrow{KMnO_4}$ AcO CO_2H O CO_2H **170** $\xrightarrow{AcOH}$ AcO CO_2H OH **171**

(46)

Laurenisol (**172**), an algal metabolite, is reported to be unstable at room temperature; it yields an oily mixture from which two isomeric bromo ethers, **173** and the seven-membered oxide **174**, can be isolated in a 5:1 ratio (Eq. 47) (166).

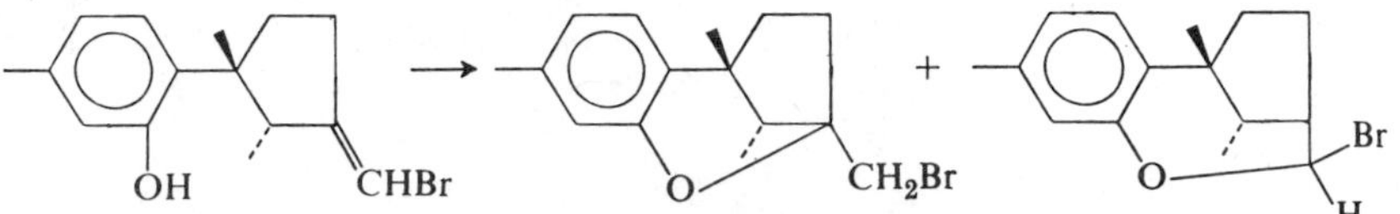

OH CHBr **172** → O CH_2Br **173** + O Br H **174** (47)

Reaction of phenyl magnesium bromide with methyl ester **175**, obtained from isolongifolene (**72**), yielded the expected diphenyl carbinol **176**, together with a minor product which is believed to be hemiacetal **177** (Eq. 48) (95).

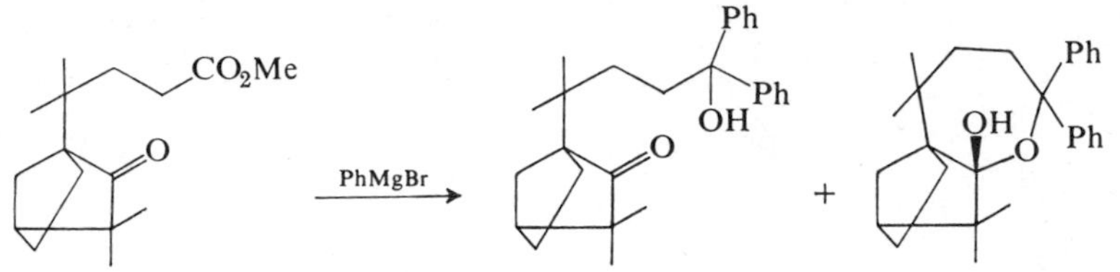

CO_2Me O **175** $\xrightarrow{\text{PhMgBr}}$ Ph Ph OH O **176** + Ph OH O Ph **177** (48)

Recently there have been reports of periodate oxidation of several 2α,3β,23-trihydroxy triterpenes which results in the consumption of 1 mole of reagent and leads to hemiacetal formation between the primary hydroxyl group and the newly generated $C_{(2)}$ aldehyde group. Periodate oxidation of methyl arjunolate (**178**) produces a substance which has been shown to be hemiacetal **179** (Eq. 49) (167). It forms a monoacetate and a mono-2,4-dinitrophenylhydrazone and its nmr spectrum shows only one aldehydic proton. Consideration of coupling constants exhibited by the $C_{(2)}$ proton in the nmr spectrum led to its assignment to the β-configuration. The same product has been obtained by periodate oxidation of bayogenin methyl ester (**180**), the $C_{(2)}$ epimer of methyl arjunolate (168). Likewise methyl brahmate (**181**), brahmol (**182**), and methyl asiatate (**186**) on treatment with sodium periodate yield products indicated by nmr to be hemiacetals **183**, **184**, and **187**, respectively (Eqs. 50 and 51) (45, 169). Chromic acid oxidation of the methyl asiatate product yields monocarboxy ε-lactone **188**, which gives crystalline methyl ester **189** (Eq. 51) (45). A dilactone is obtained by chromic acid oxidation of the methyl brahmate product (169). In the presence of catalytic amounts of piperidine in acetic acid the hemiacetal ring opens and recyclization occurs with formation of an α,β-unsaturated aldehyde; for example (Eq. 50), **185** is formed from the methyl brahmate derived hemiacetal **183**. During another study, it was

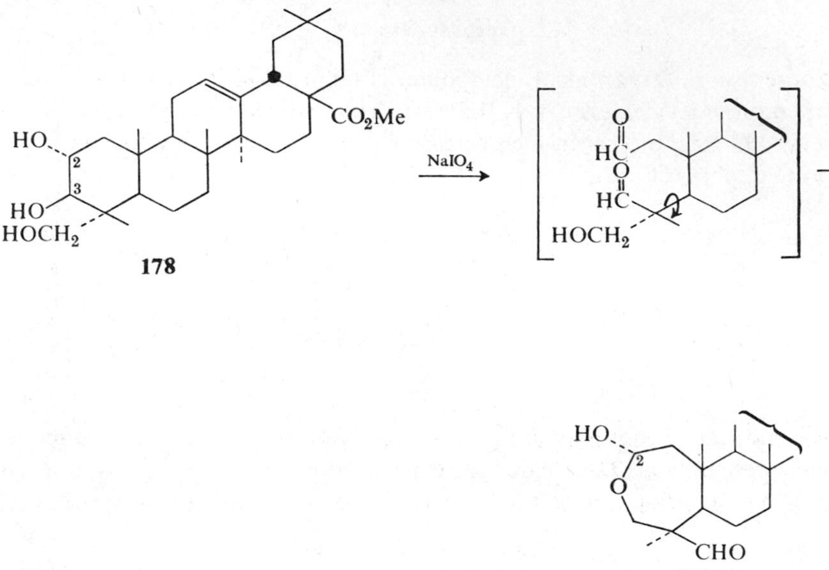

HO
2
3
HO
$HOCH_2$
CO$_2$Me
178

$NaIO_4$ →

O
HC
O
HC
$HOCH_2$
→

HO
2
O
CHO
179

(49)

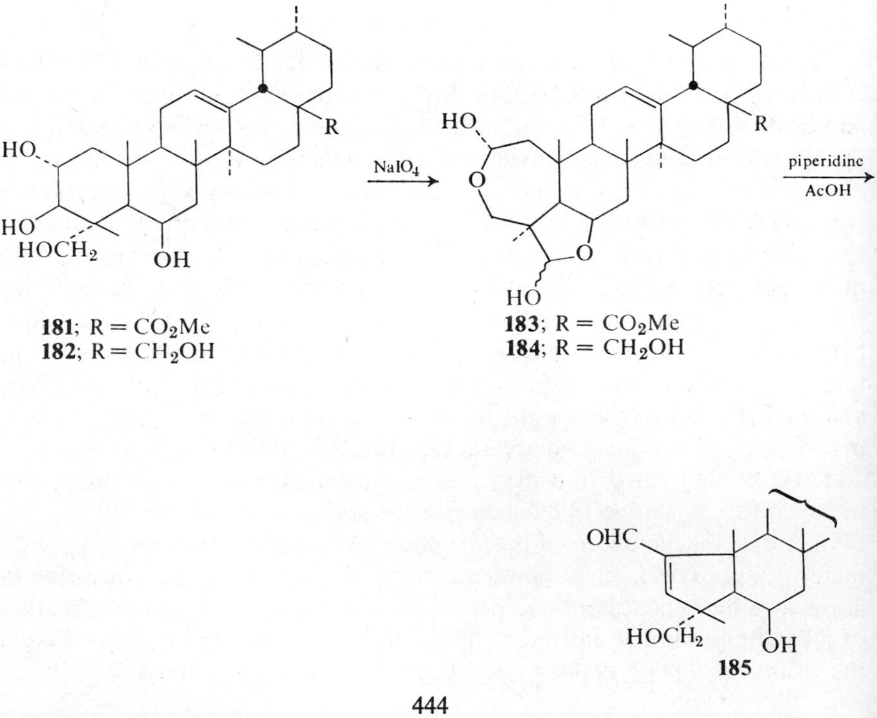

HO
HO
$HOCH_2$
OH
R
181; R = CO_2Me
182; R = CH_2OH

$NaIO_4$ →

HO
O
O
HO
R
183; R = CO_2Me
184; R = CH_2OH

piperidine / AcOH →

OHC
$HOCH_2$
OH
185

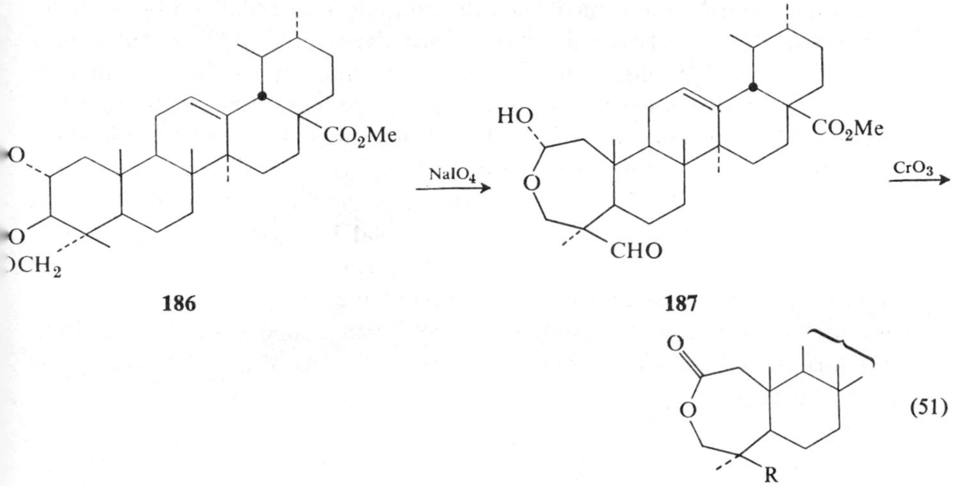

186 $\xrightarrow{NaIO_4}$ 187 $\xrightarrow{CrO_3}$

188; R = CO_2H
189; R = CO_2Me

(51)

observed that periodate oxidation of the tricyclic diterpene virescenol A (**190**) yields an aldehydohemiacetal **191**, which can be oxidized with silver carbonate to ε-lactone **192** (Eq. 52) (170, 171). Similarly it was found that the isomer **193** yielded the corresponding hemiacetal and ε-lactone (171). The coupling constants observed for the $C_{(2)}$ proton of the latter hemiacetal were nearly identical to those for the corresponding proton of **179**, indicating similar stereochemistry.

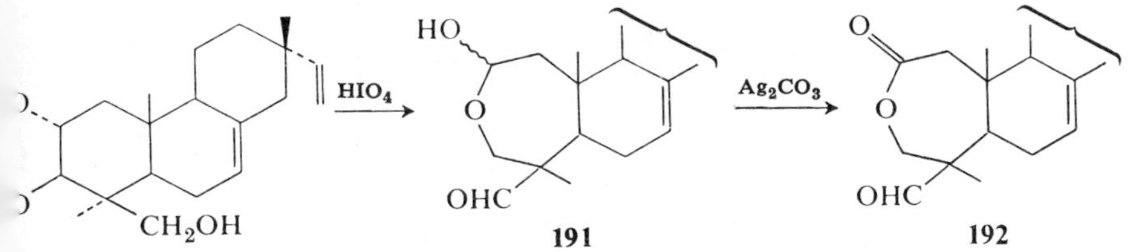

190 $\xrightarrow{HIO_4}$ 191 $\xrightarrow{Ag_2CO_3}$ 192 (52)

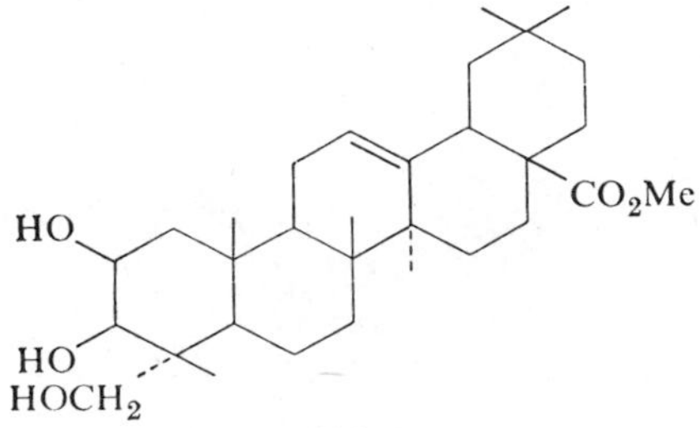

180

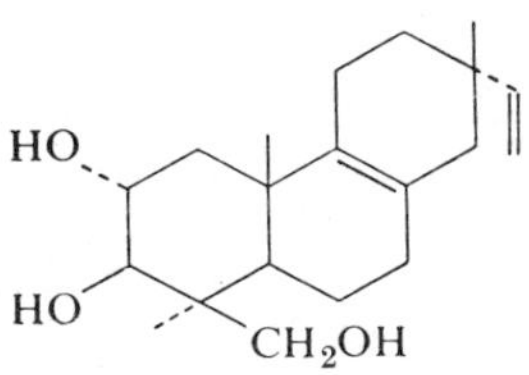

193

Cyclamigenin A[1] and C have been shown to be 30β- and 30α-ethoxy-28,30-epoxyolean-12-en-3β,16α-diol (**194**), respectively (172). These substances form diacetates **195** which can be reconverted into the alcohols by alkaline hydrolysis. Mild acid hydrolysis results in the opening of the oxepin ring and the formation of **196**. A conformational argument was used to assign the stereochemistry of the ethoxy group. The preferred conformation was felt to be as shown in **198**. Since the ratio of cyclamigenin A[1] to C is 30:1, it was proposed that cyclamigenin A[1] is the equatorial 30β-isomer, and that cyclamigenin C is the axial 30α-epimer. Mass spectral evidence is cited in support of these assignments. Both compounds are obviously artifacts, although it has not been possible to prepare them from cyclamiretin D (**197**), which is unchanged when subjected to the conditions of the hydrolysis stage of the extraction.

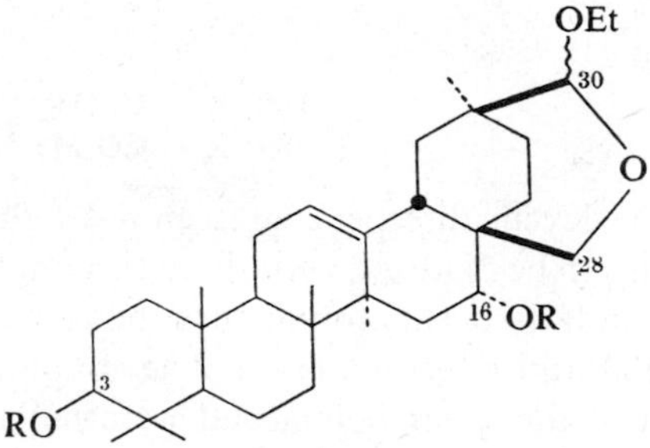

194; R = H
195; R = Ac

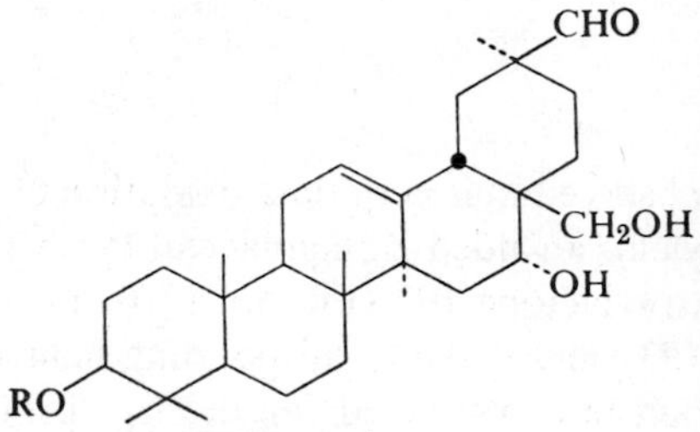

196; R = Ac
197; R = H

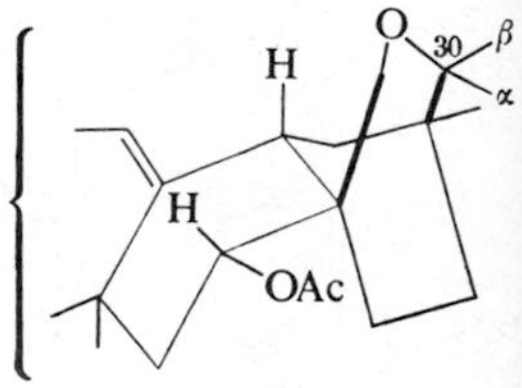

198

E. Tables

TABLE VI-1. Naturally Occurring Terpene Oxepins

Name	Formula	Structure	mp (°C)	$[\alpha]_D$	$\nu_{(C=O)}$ (cm^{-1})	Plant source	Refs.
Obacunone (Casimirolid)	$C_{26}H_{30}O_7$	**1**	229–231	−49.8° ($CHCl_3$)	1730, 1701 (KBr)	*Casimiroa*, *Citrus*, and *Phellodendron* species, *Dictamnus albus*, *Fortunella margeritia*, *Poncirus trifoliata*	1–5
Nomilin	$C_{28}H_{34}O_9$	**2**	278–279	−95.7° (acetone)	1730–1700 (film)	*Casimiroa* and *Citrus* species, *Poncirus trifoliata*	3, 4, 7
Deacetylnomilin	$C_{26}H_{32}O_8$	**5**	263–265	−112° ($[\alpha]_{600}$, dioxane)	1746 (nujol)	*Casimiroa edulis*, *Citrus* species, *Poncirus trifoliata*	3, 15, 16
7α-Obacunol	$C_{26}H_{32}O_7$	**8**	242–245	+72° (acetone)	1750, 1684 (nujol)	*Casimiroa edulis*	4, 16
Zapoterin	$C_{26}H_{30}O_8$	**9**	269–271	−51° ($CHCl_3$)	1758, 1720, 1668 (nujol)	*Casimiroa edulis*	16, 20–22
Methyl ivorensate	$C_{27}H_{34}O_8$	**10**	279–281	−97.5° ($CHCl_3$)	1740	*Khaya ivorensis*	23, 24
Apetalactone	$C_{30}H_{50}O_3$	**11**	335–336 dec	+37.5° ($CHCl_3$)	1720 (nujol)	*Calophyllum apetalum*, *C. tomentosum*	25
Dammar resin compound B	$C_{32}H_{48}O_6$	**14**	234–236	+117° ($CHCl_3$)	1745 (CCl_4)	Dammar resin from trees of *Dipterocarpaceae* family	26, 27
Occidenol	$C_{15}H_{24}O_2$	**152**	42–44	−139°		*Thuja koraiensis*, *T. occidentalis*	155
Sapelin B	$C_{30}H_{50}O_4$	**155**	173–177	−18° ($CHCl_3$)		*Entandrophragma cylindricum* Sprague	156
Nimbolin B	$C_{39}H_{46}O_{10}$	**160**	243–245	−93°		*Azadirachta indica* L., *Melia azedarach* L.	157

TABLE VI-2. Products of Baeyer-Villiger Oxidation of Terpene Cyclohexanones

Ketone	Peracid	Product	Physical data	$\nu_{(C=O)}$ (cm^{-1})	Yield (%)	Refs.
O	Trifluoro-peracetic	O, O +	bp 55°C (0.2 mm) $[\alpha]_D$ −32.18° n_D^{25} 1.4583	1740	64	53
		O, O	mp 32.2–32.4°C $[\alpha]_D$ −36.11°	1740		
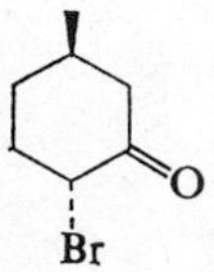 O, Br	Trifluoro-peracetic	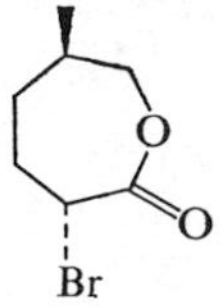 O, O, Br	bp 81–81.5°C (0.09 mm) n_D^{25} 1.5056 $[\alpha]_D$ −15.4°	1745, 1725	87	53
O	Trifluoro-peracetic	O, O	—	1725	—	54
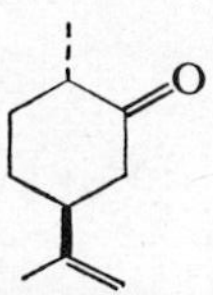 O	Monoper-phthalic	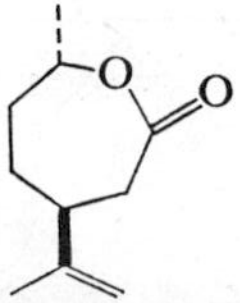 O, O	bp 93–95°C (2 mm) n_D^{20} 1.4735	—	—	55

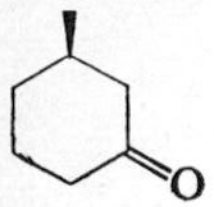

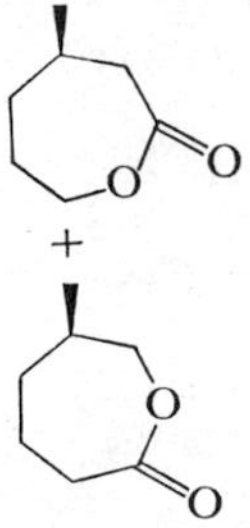

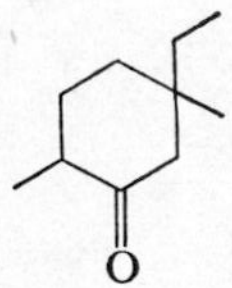

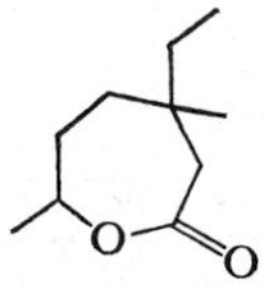

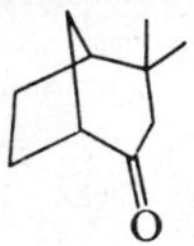	Trifluoro-peracetic	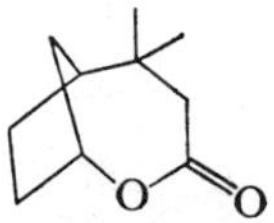	—	—	50 (by nmr)	56
	Peracetic		mp 81.5–82.5°C	—	48	57
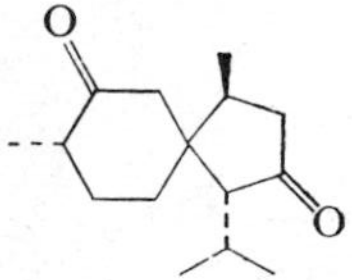	Perbenzoic	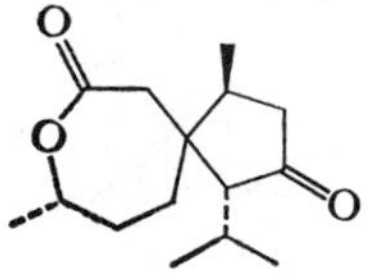	mp 188°C $[\alpha]_D$ +112.4° (EtOH)	—	—	58 (cf.59)
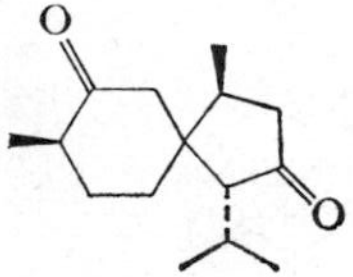	Perbenzoic	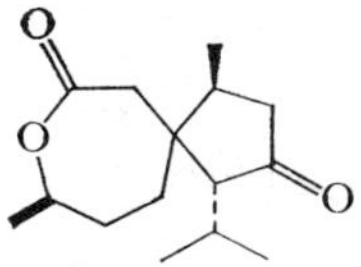	mp 138°C $[\alpha]_D$ −110.5° (EtOH)	—	—	58
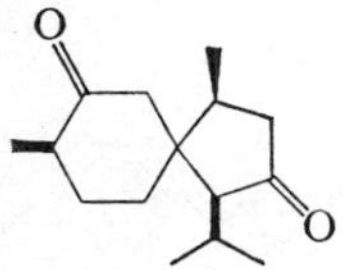	Perbenzoic	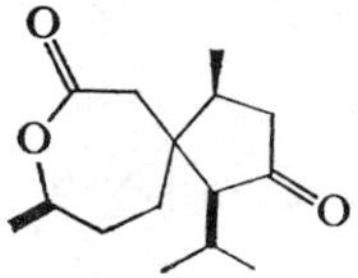	mp 134–135°C $[\alpha]_D$ −75.6° (EtOH)	—	—	58
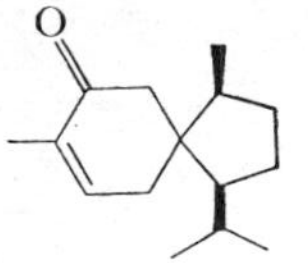	—	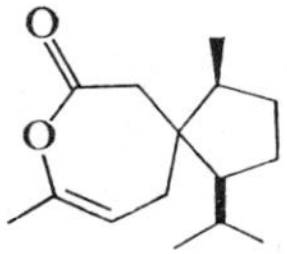	—	—	—	60
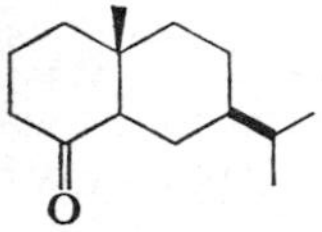	Perbenzoic	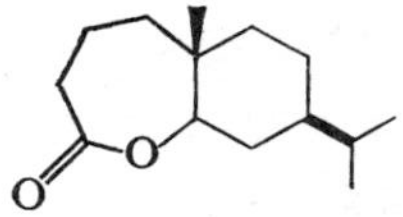	—	1727	—	61, 62

TABLE VI-2.—(*contd.*)

Ketone	Peracid	Product	Physical data	$\nu_{(C=O)}$ (cm^{-1})	Yield (%)	Refs.
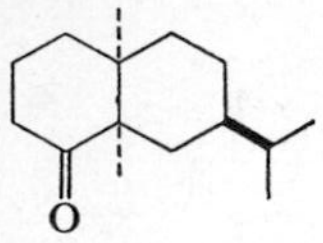	Perbenzoic	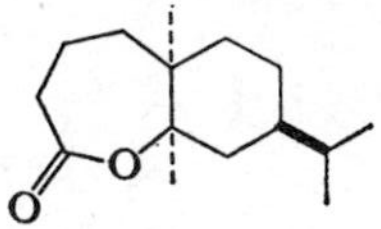	mp 71°C $[\alpha]_D$ −15° (MeOH)	1737	47	61–65
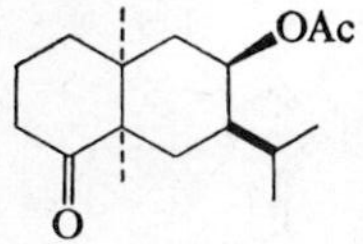 OAc	Perbenzoic	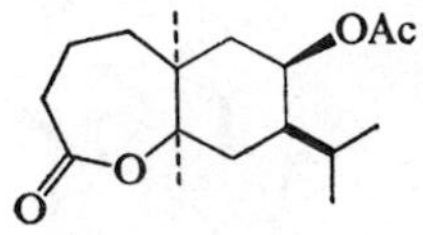 OAc	mp 98.5–100°C $[\alpha]_D$ −58°	1733, 1724 (KBr)	88	66
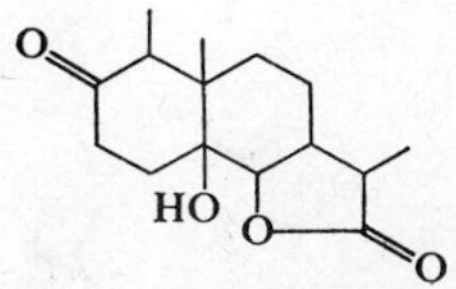 HO	Perbenzoic	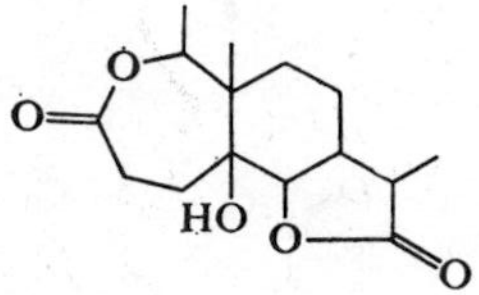 HO	mp 201–203°C	—	27	67
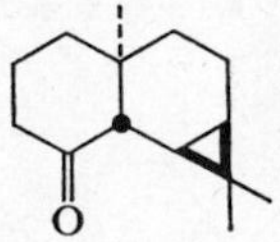	Perbenzoic	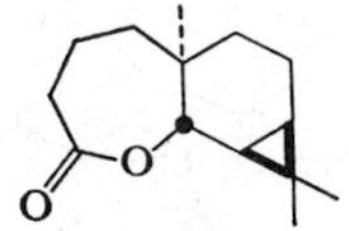	mp 113–114°C	1719	89	68
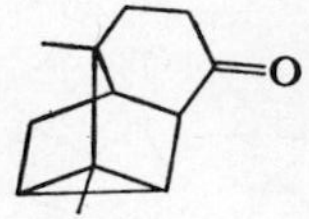	—	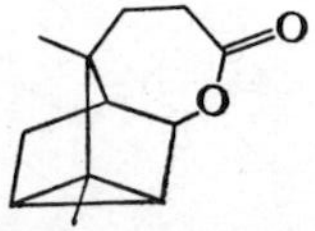	mp 109–111°C	1715 ($CHCl_3$)	—	69

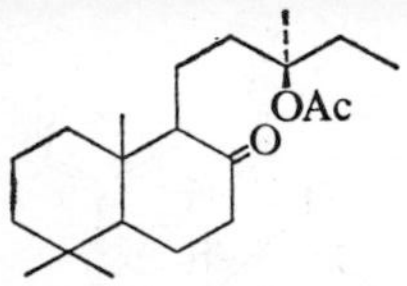	Perbenzoic	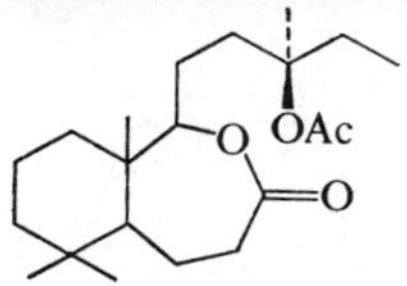	—	—	—	70
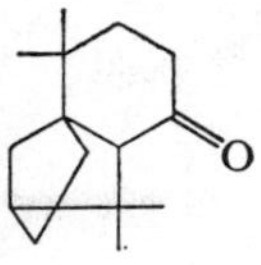	Perbenzoic	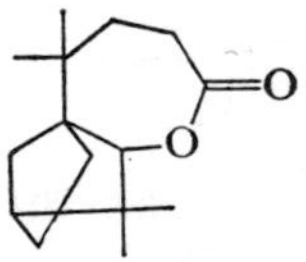	mp 79–80°C	1730	30	71, 72
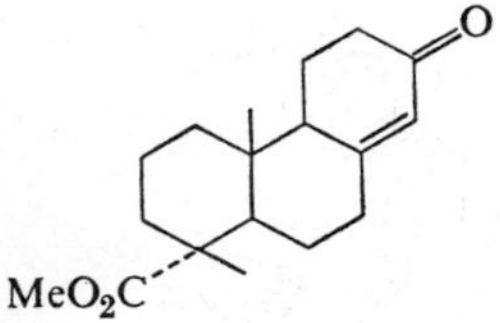	*m*-Chloro-perbenzoic	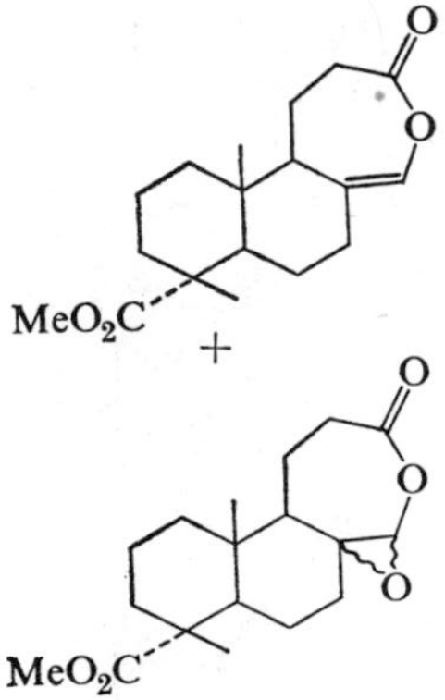	mp 129–130°C mp 151–152°C	1770, 1764, 1739 (CCl_4), 1770, 1739 ($CHCl_3$) 1754, 1709 (Nujol)	Yields dependent on conditions	73, 74
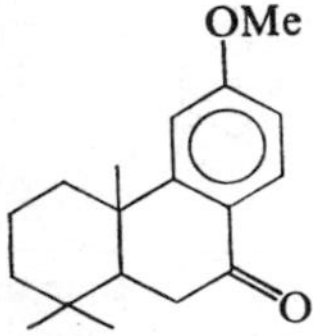	Trifluoro-peracetic	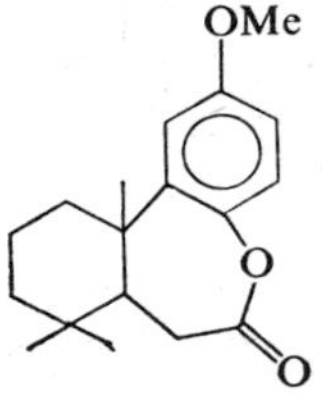	mp 114–115°C $[\alpha]_D$ −254°	1748 ($CHCl_3$)	—	75

TABLE VI-2.—(*contd.*)

Ketone	Peracid	Product	Physical data	$\nu_{(C=O)}$ (cm^{-1})	Yield (%)	Refs.
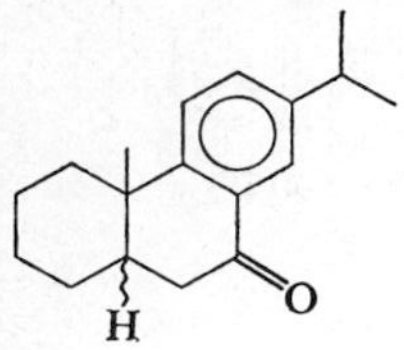 H	Perbenzoic	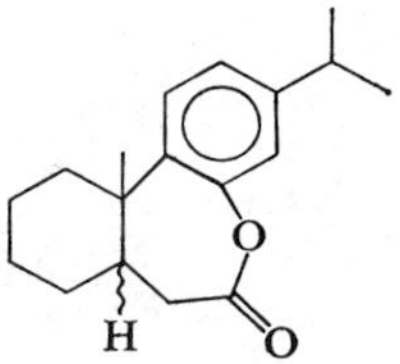 H	—	—	—	76
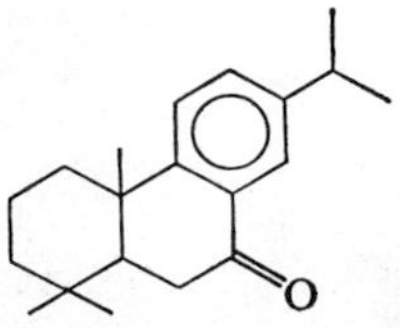	Trifluoro-peracetic	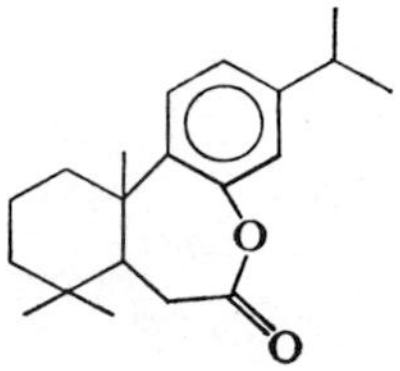	mp 86.5–87.5°C $[\alpha]_D$ −234°	1742 ($CHCl_3$)	51	75
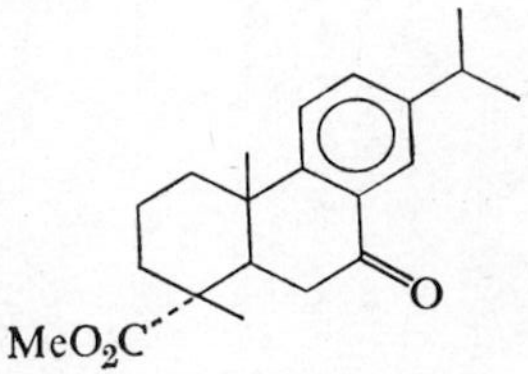 MeO_2C	Trifluoro-peracetic	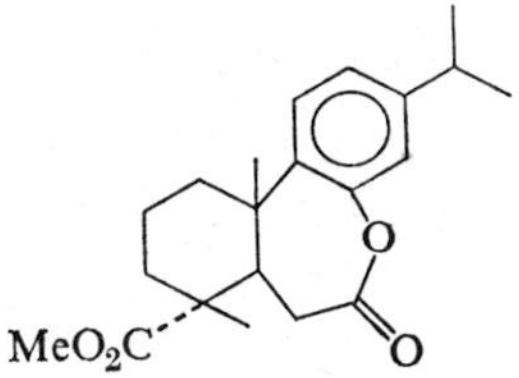 MeO_2C	—		—	77
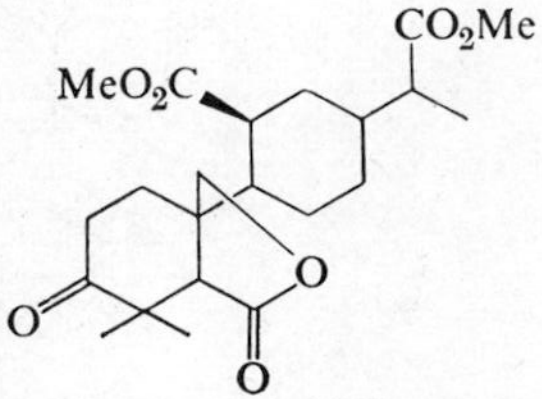 MeO_2C CO_2Me	Perbenzoic	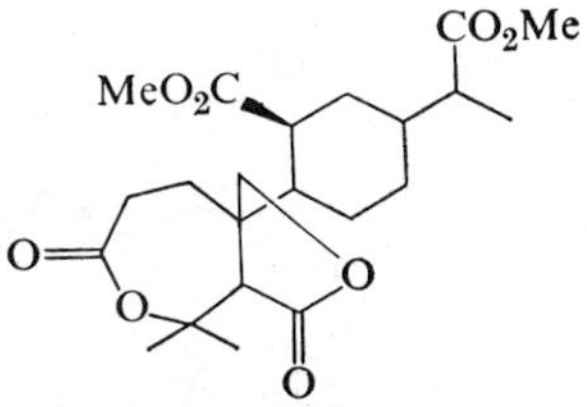 MeO_2C CO_2Me	—	—	—	78, 79

Ketone	Peracid	Lactone	Physical constants	IR (cm^{-1})	Yield (%)	Ref.
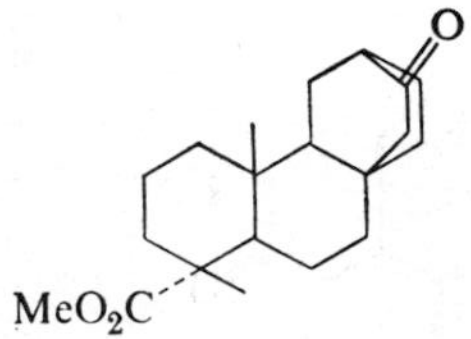 MeO$_2$C	*m*-Chloro- perbenzoic	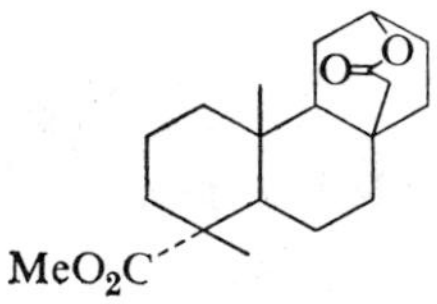 MeO$_2$C	mp 187–188°C	1735–1720 (Nujol)	60	80
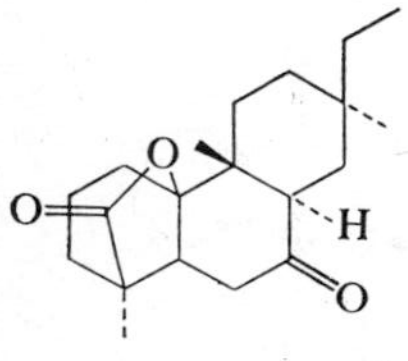	Trifluoro- peracetic	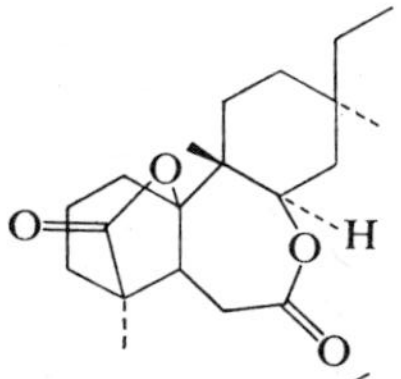	mp 267–270°C $[\alpha]_D$ −19°	1775, 1750 (Nujol)	70	81
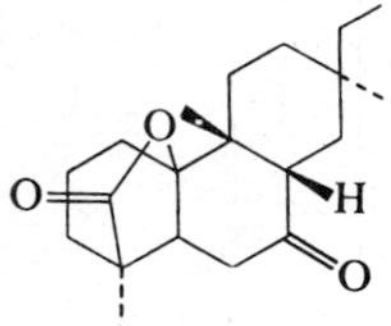	Trifluoro- peracetic	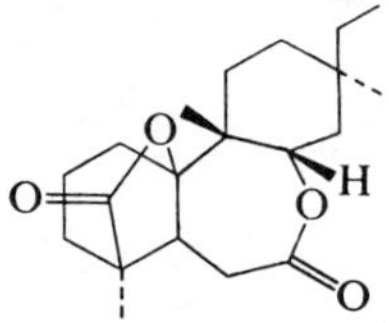	mp 239°C	1775, 1740 (Nujol)	4	81
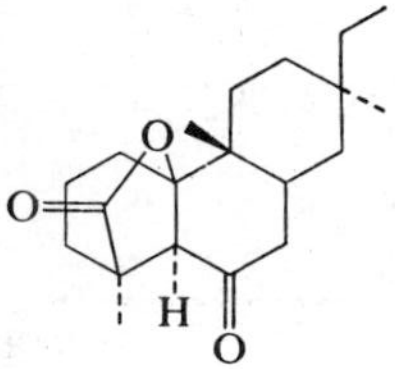	Trifluoro- peracetic	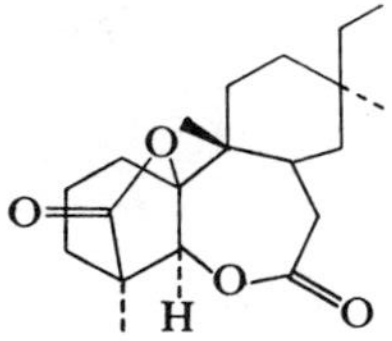	mp 176°C $[\alpha]_D$ −16°	1786, 1736 ($CHCl_3$)	29	82

TABLE VI-2.—(*contd.*)

Ketone	Peracid	Product	Physical data	$\nu_{(C=O)}$ (cm^{-1})	Yield (%)	Refs.
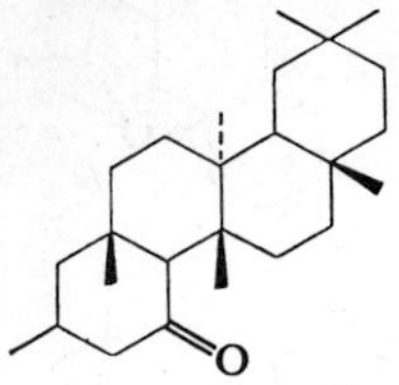	Peracetic	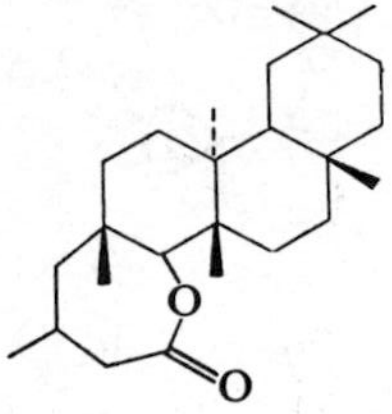	—	1737	—	83
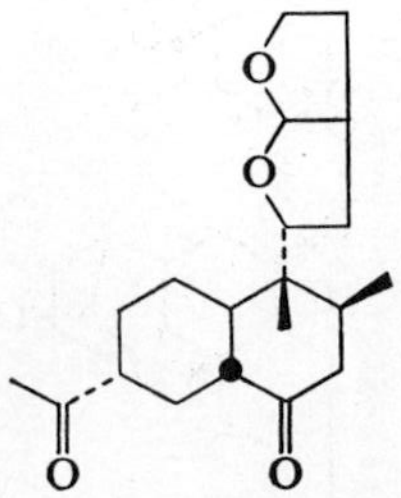	Monoperphthalic	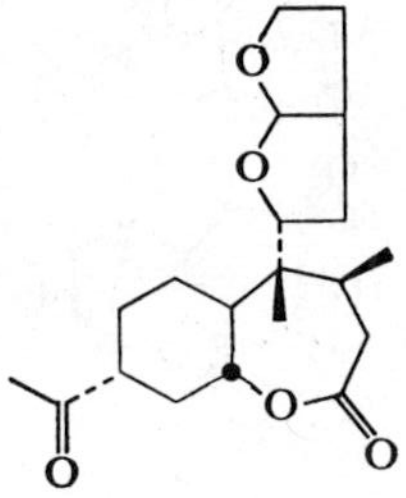	mp 143–145°C $[\alpha]_D$ −6°	1715 (Nujol)	49	84
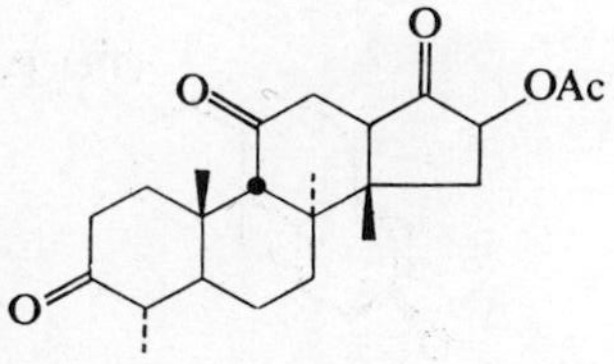 OAc	*m*-Chloroperbenzoic	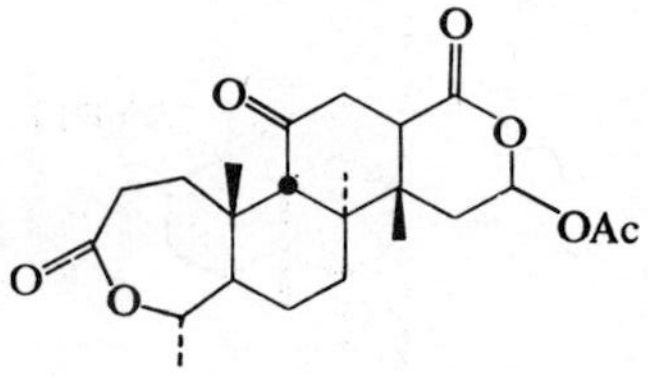 OAc	mp 229–231°C $[\alpha]_D$ +21°	1786, 1724 (KBr)	58	85

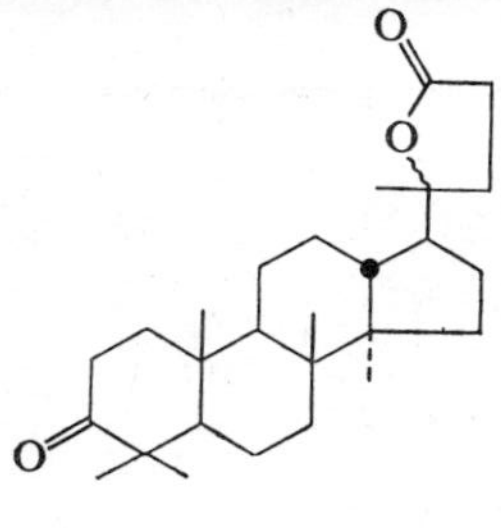	*m*-Chloro-perbenzoic	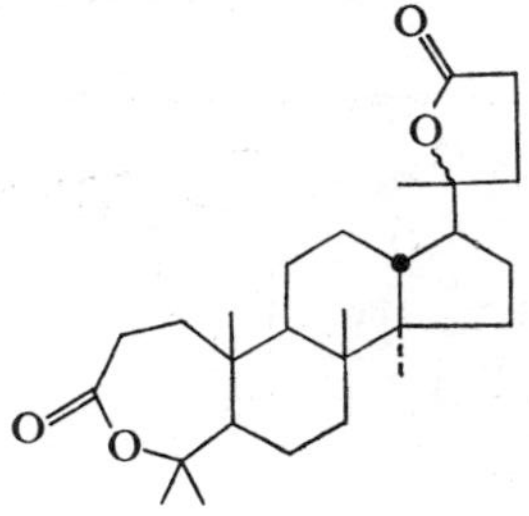	mp 178–180 and 192–194°C $[\alpha]_D$ +104°	1770, 1724 (KBr)	62	36
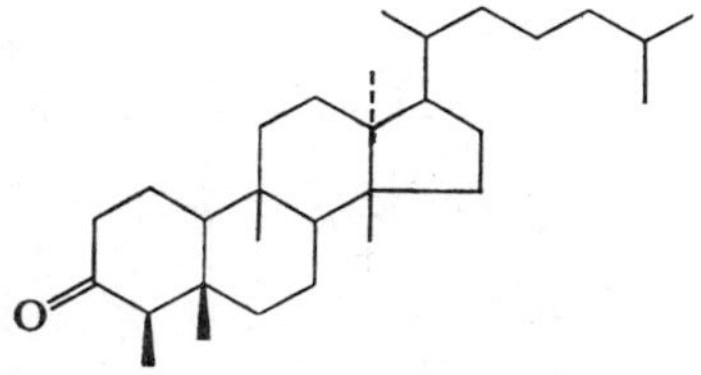	Perbenzoic	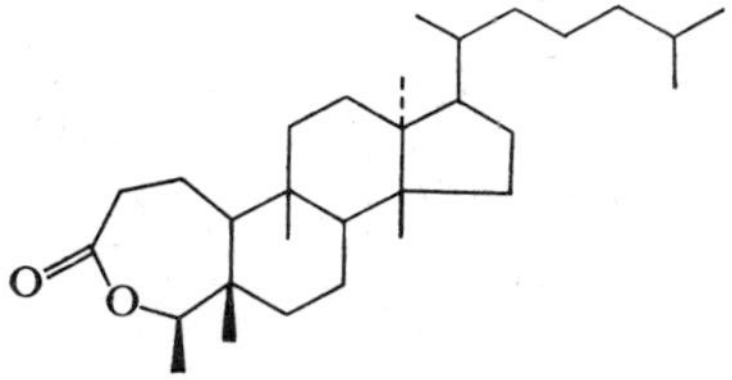	mp 189.5–190.5°C $[\alpha]_D$ +6.7°	1736 (Nujol)	—	86
	Peracetic		mp 303–306°C $[\alpha]_D$ +38°	1727 ($CHCl_3$)	—	83, 87
	Perbenzoic		mp 273–274 and 292–293°C	1734 (Nujol)	62	88
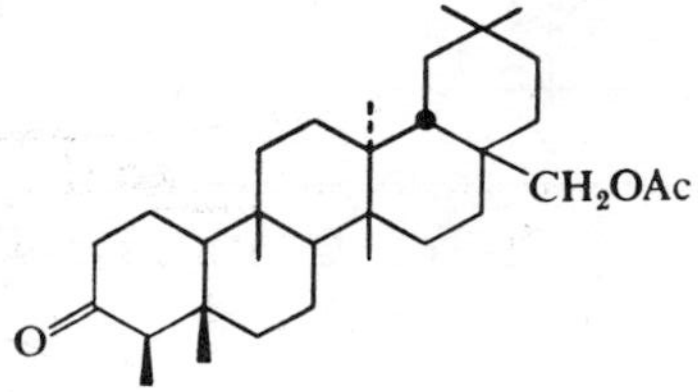 CH_2OAc	Perbenzoic	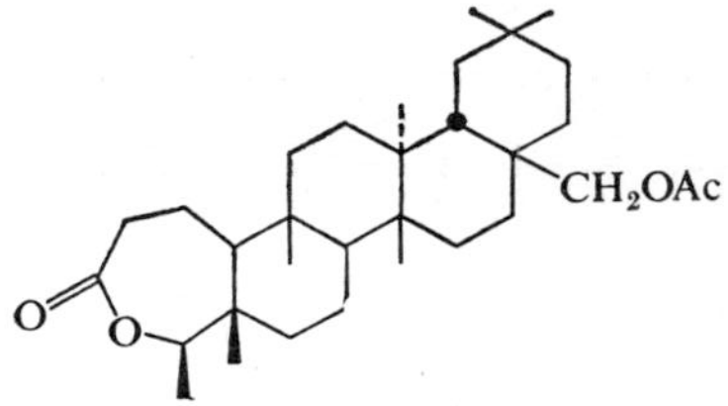 CH_2OAc	mp 230–231°C $[\alpha]_D$ +22.5°	1732 (Nujol)	19	25

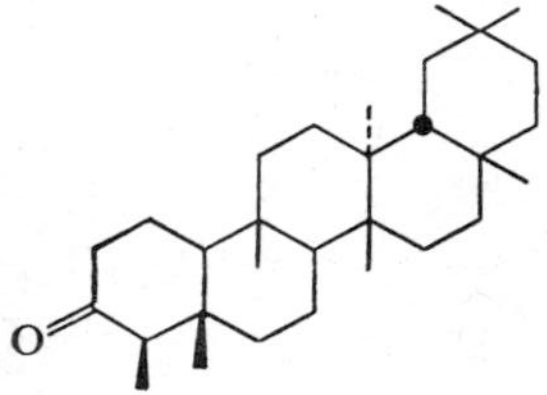

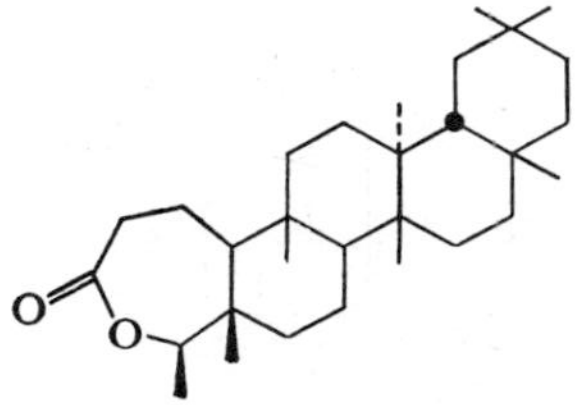

TABLE VI-2.—(*contd.*)

Ketone	Peracid	Product	Physical data	$\nu_{(C=O)}(cm^{-1})$	Yield (%)	Refs.
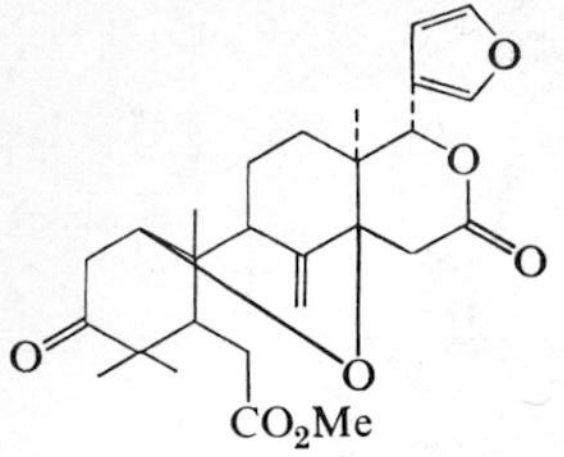	Perbenzoic	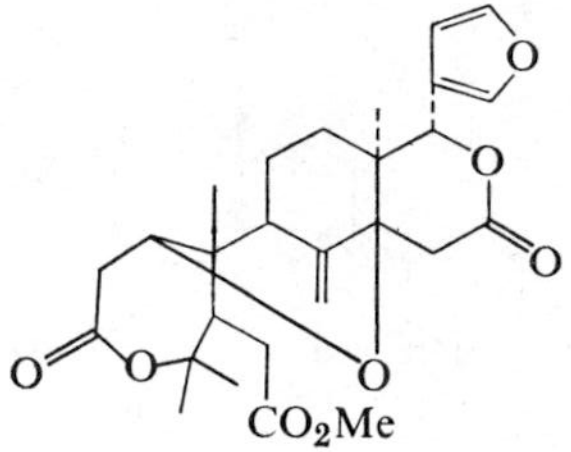	mp 279–281°C $[\alpha]_D$ −97.5°	1740	3	23, 24
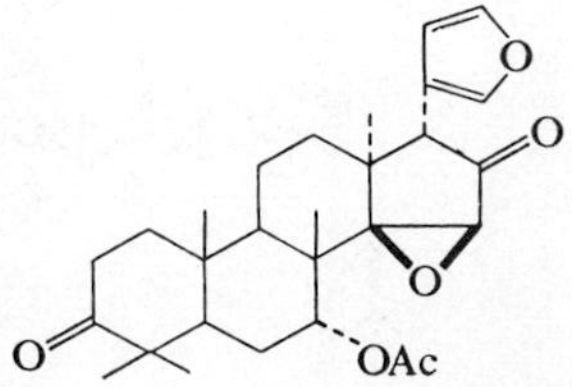	Perbenzoic	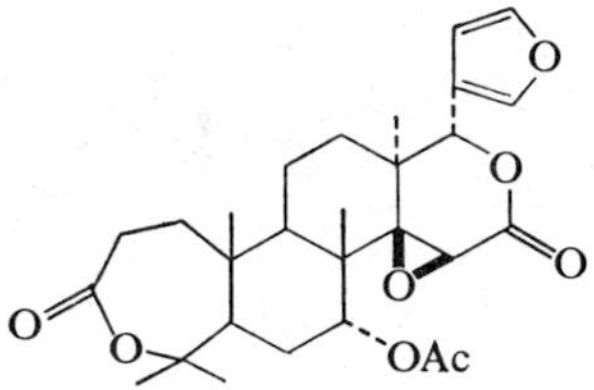	mp 197–199°C $[\alpha]_D$ +2°	1770 (KBr)	—	89
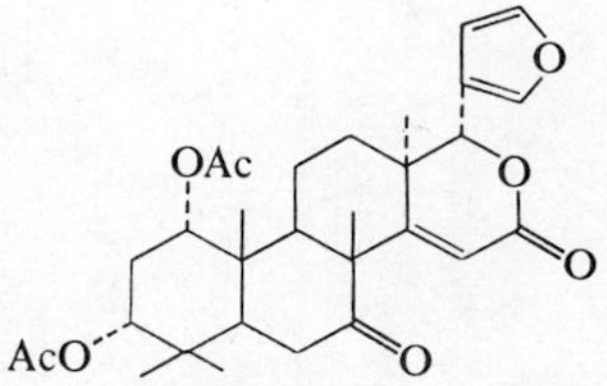	Peracetic	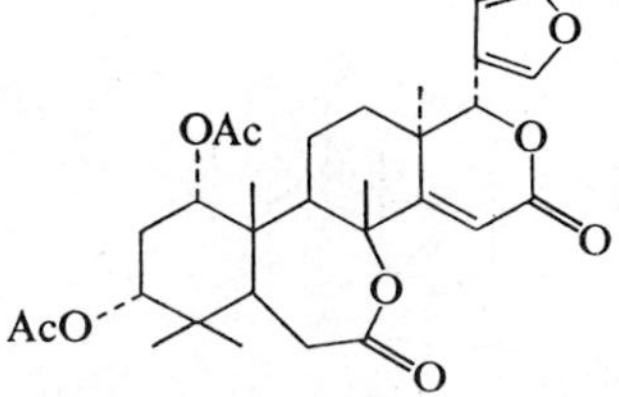	mp 289–291°C	—	—	34

TABLE VI-3. Terpene Related Adipic Ahydrides

Diacid	Anhydride	mp (°C)	$[\alpha]_D$	$\nu_{(\text{anhydride C=O})}$ cm^{-1}	Refs.
CO_2H, CO_2H		—	—	—	120
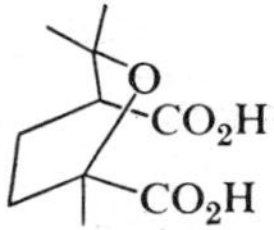 CO_2H, CO_2H	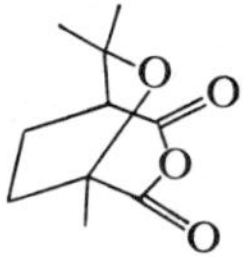	77–78	—	—	121
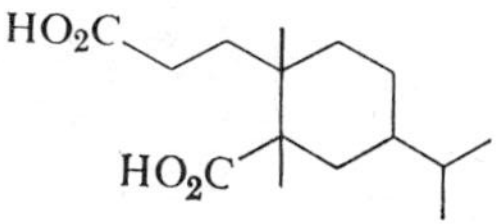 HO_2C, HO_2C	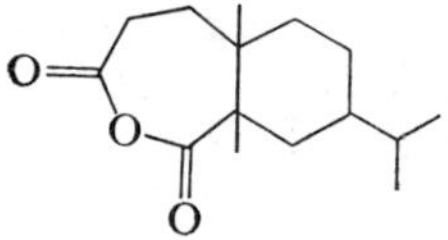	—	—	1796, 1752	63
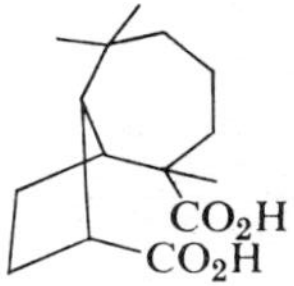 CO_2H, CO_2H	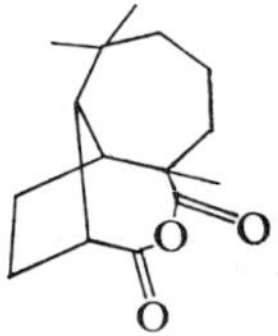	94–95	—	1792, 1750	122, 123
CO_2H, CO_2H		113	+39°	1803, 1762 (CCl_4) 1778, 1740 (KBr)	107, 108, 124
		109–110		1799, 1752 (CCl_4)	125, 126

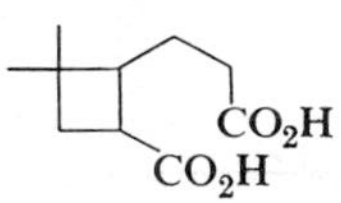

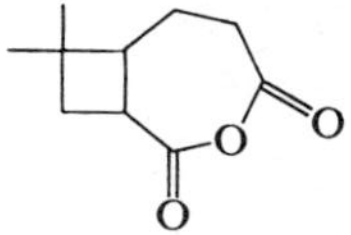

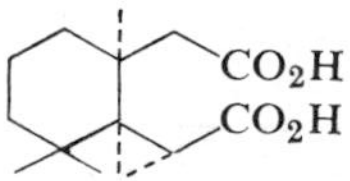

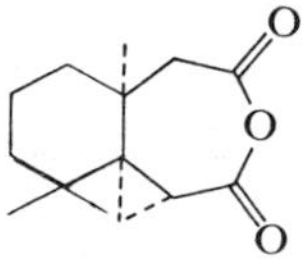

TABLE VI-3—(*contd.*)

Diacid	Anhydride	mp (°C)	$[\alpha]_D$	$\nu_{(\text{anhydride C=O})}$ cm^{-1}	Refs.
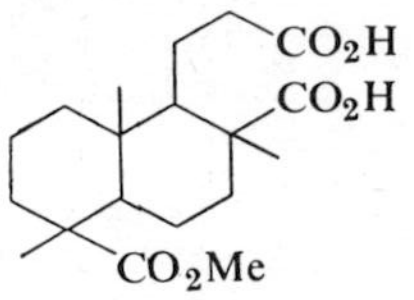 CO_2H, CO_2H, CO_2Me	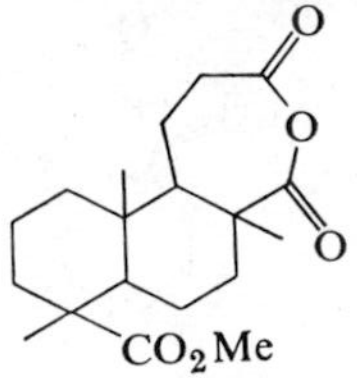 CO_2Me	206–208	+6°	1808, 1764 ($CHCl_3$)	126, 127
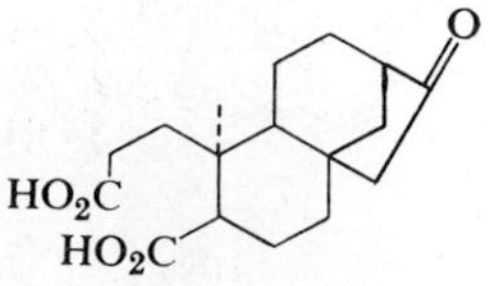 HO_2C, HO_2C	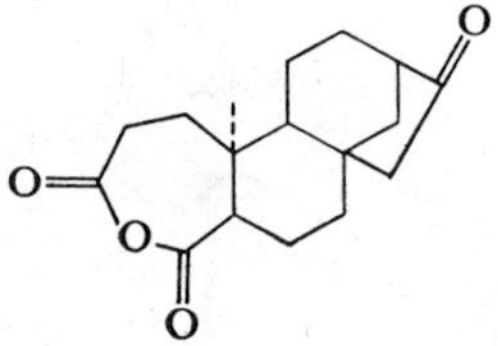	201–202	−137°	1808, 1745	128
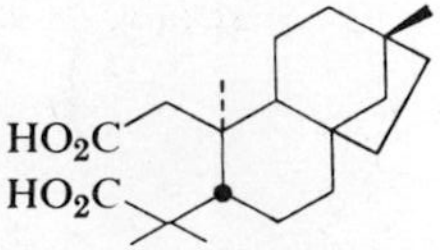 HO_2C, HO_2C	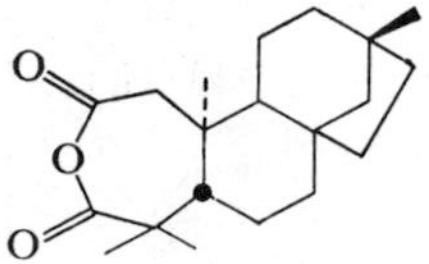	123–124	—	1792, 1749	129
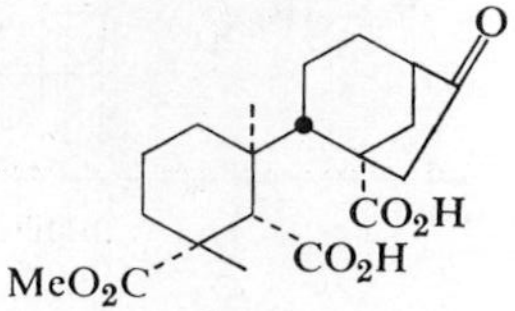 MeO_2C, CO_2H, CO_2H	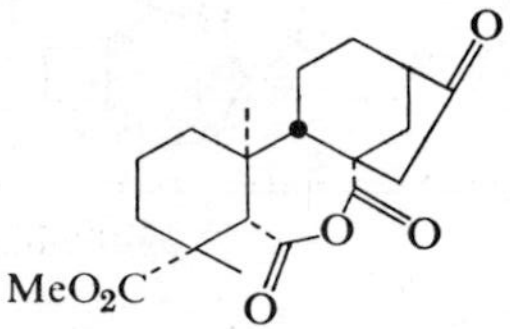 MeO_2C	195–198	—	1794, 1742	130

Starting material	Anhydride	m.p.	$[\alpha]_D$	IR	Ref.
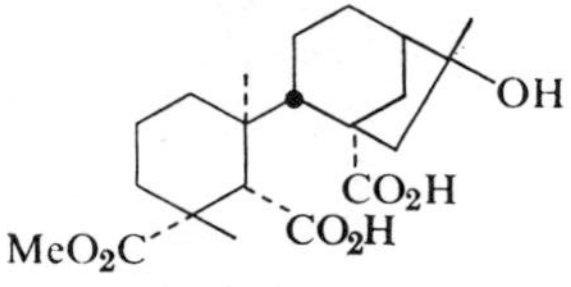 MeO_2C, CO_2H, CO_2H, OH	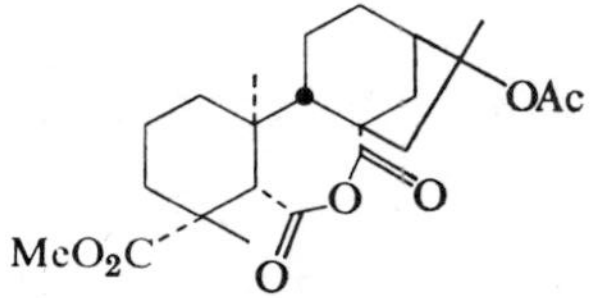 MeO_2C, O, O, O, OAc	200–202	—	1795, 1751	102
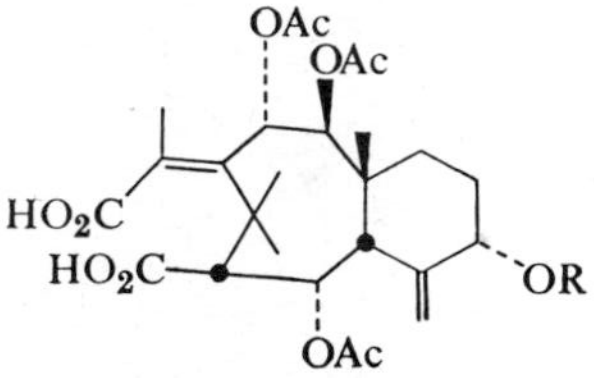 OAc, OAc, HO_2C, HO_2C, OR, OAc R = cinnamoyl or R = dihydrocinnamoyl; methyl rather than terminal methylene	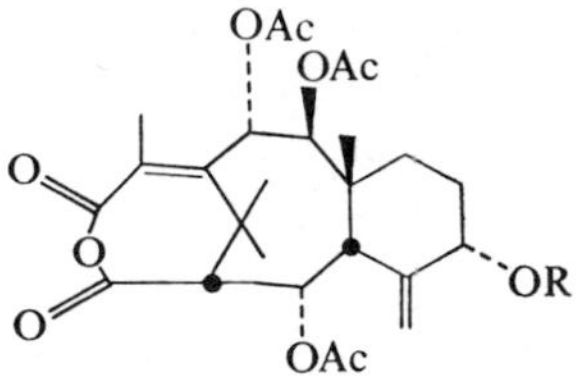 OAc, OAc, O, O, O, OR, OAc	—	—	—	131–133
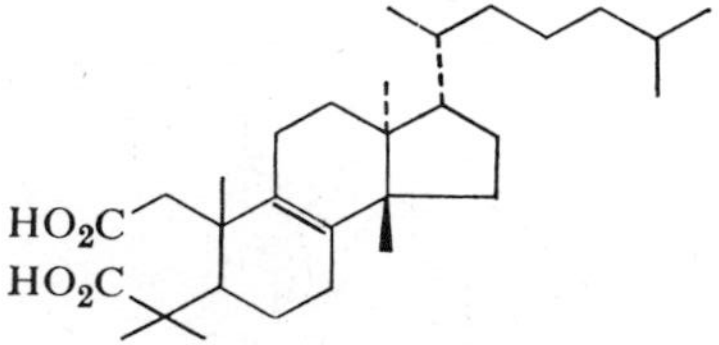 HO_2C, HO_2C	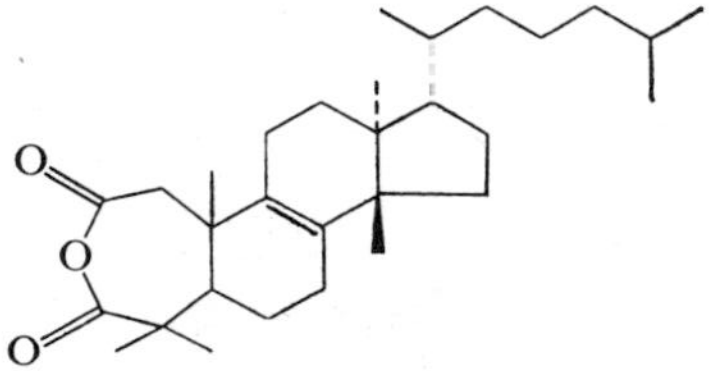 O, O, O	192–193	—	1802, 1759	134
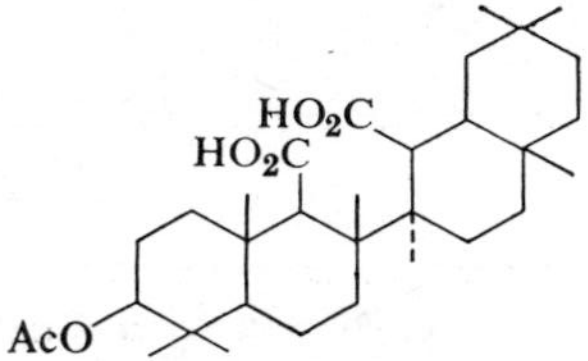 HO_2C, HO_2C, AcO	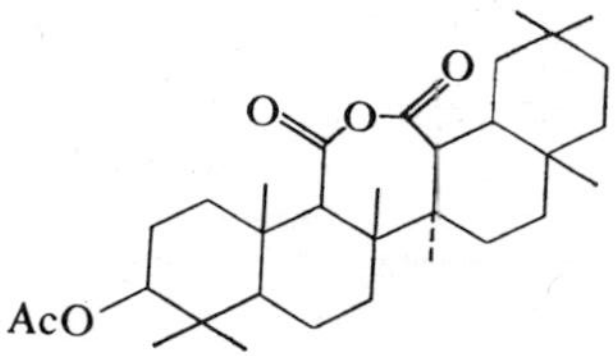 O, O, O, AcO	317–318	−27°	—	135

TABLE VI-3—(*contd.*)

Diacid	Anhydride	mp (°C)	$[\alpha]_D$	$\nu_{(anhydride\ C{=}O)}$ cm^{-1}	Refs.
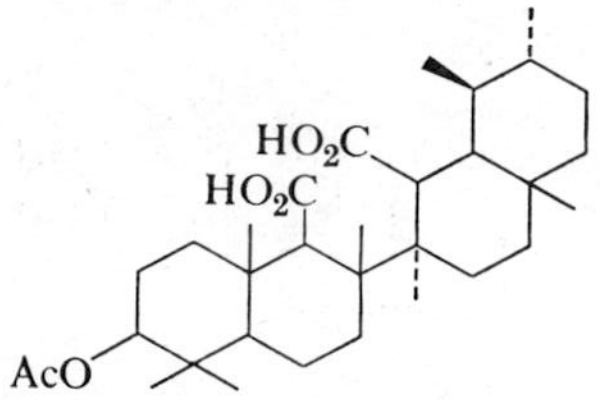 HO_2C, HO_2C, AcO	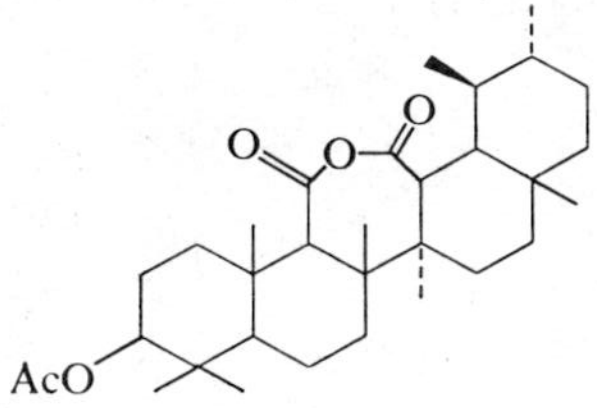 O, O, O, AcO	275–276	+15°	—	136
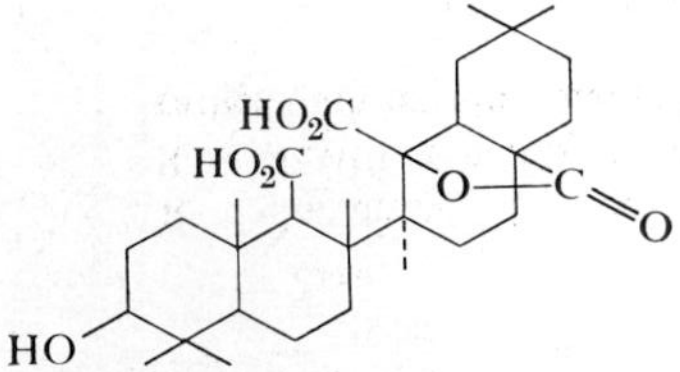 HO_2C, HO_2C, O, C, O, HO	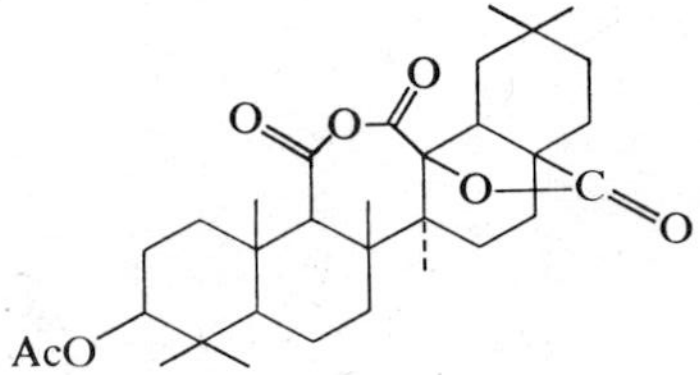 O, O, O, O, C, O, AcO	306–308	—	—	137, 138
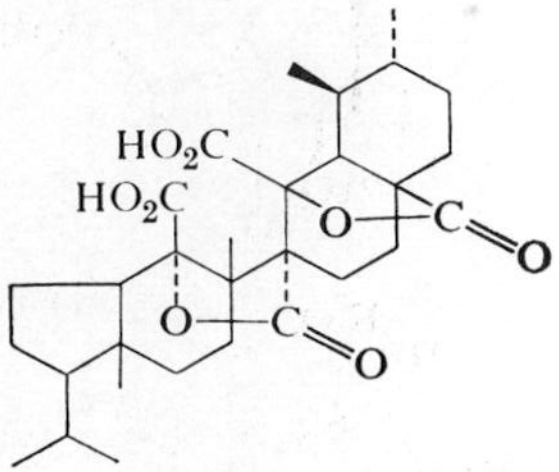 HO_2C, HO_2C, O, C, O, O, C, O	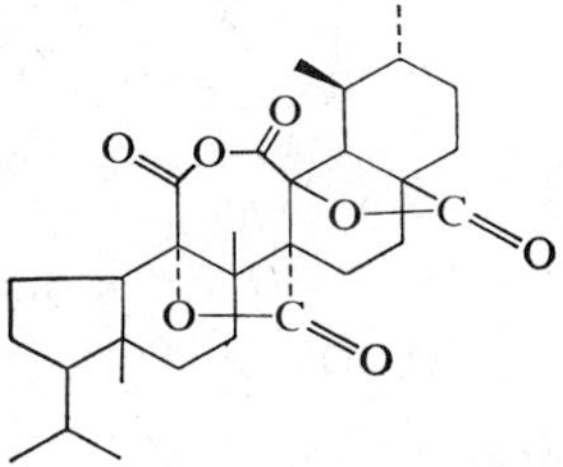 O, O, O, O, C, O, O, C, O	260	—	—	139, 140

Starting material	Product	M.p.	$[\alpha]_D$	IR	Ref.
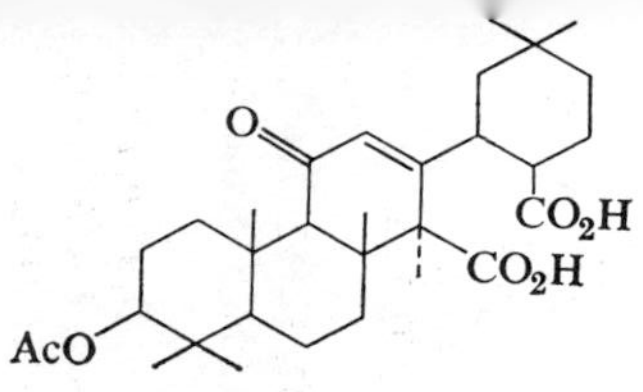 AcO, O, CO_2H, CO_2H	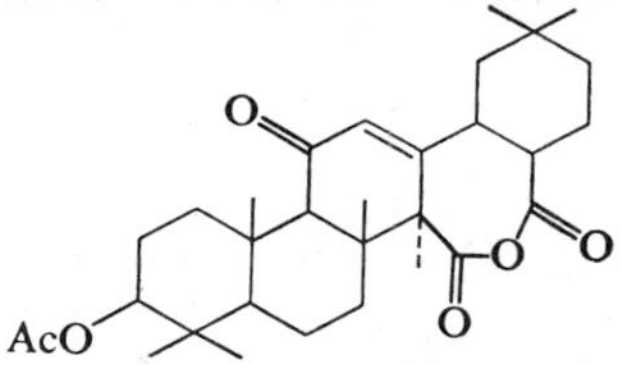 AcO, O, O, O, O	263.5–264.5	+28°	—	141
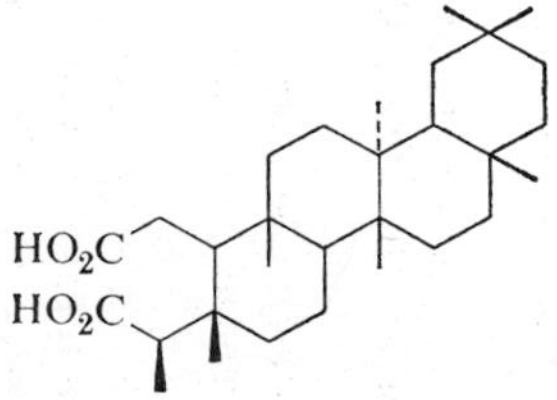 HO_2C, HO_2C	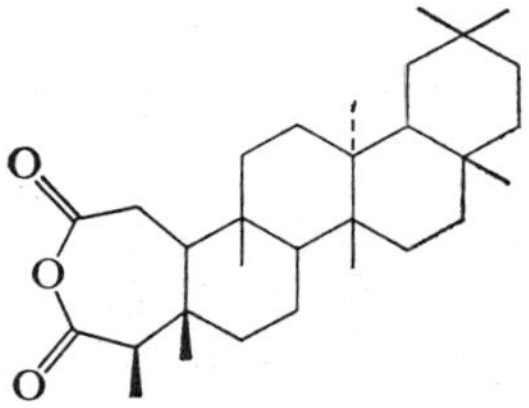 O, O, O	270–272 dec	+75°	1802, 1755	83, 142
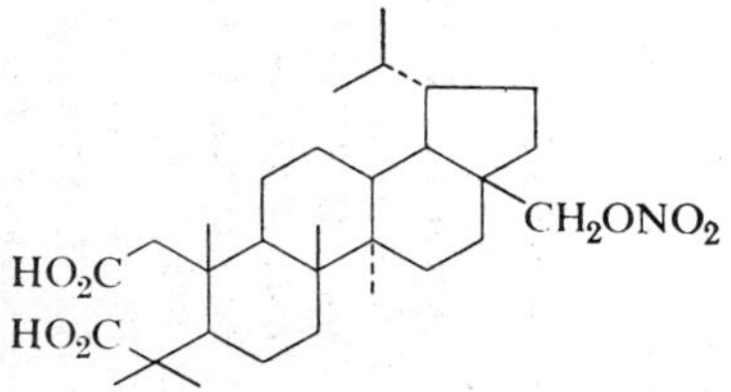 HO_2C, HO_2C, CH_2ONO_2	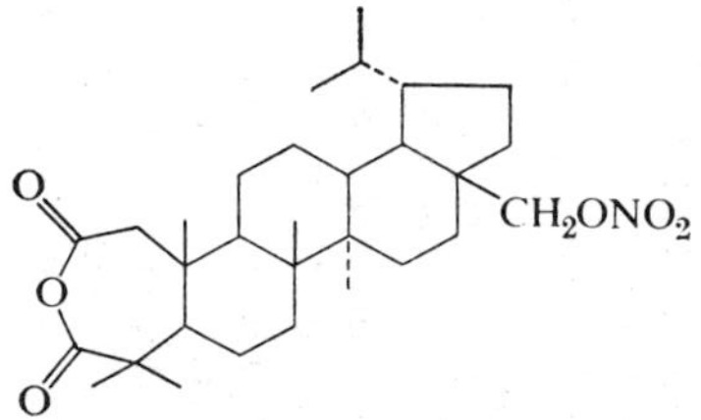 O, O, O, CH_2ONO_2	215	—	—	143
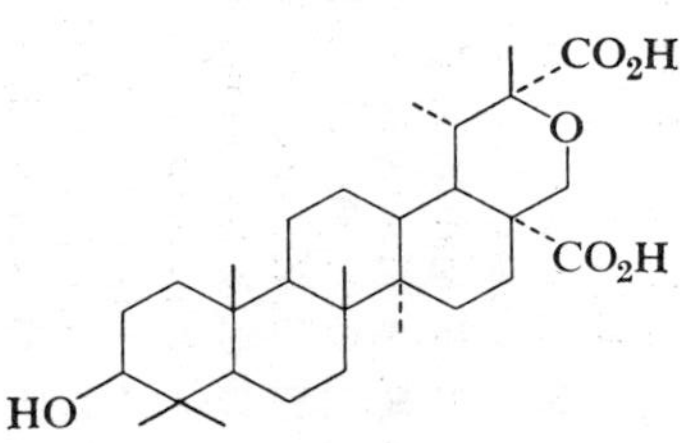 HO, CO_2H, O, CO_2H	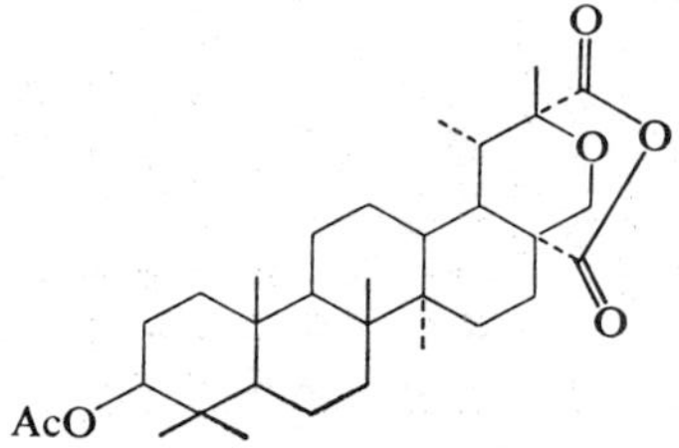 AcO, O, O, O, O	333–336 dec	+82°	1800, 1758 ($CHCl_3$)	144

F. References

1. O. H. Emerson, *J. Amer. Chem. Soc.*, **73,** 2621 (1951).
2. F. Sondheimer, A. Meisels, and F. A. Kincl, *J. Org. Chem.*, **24,** 870 (1959).
3. D. L. Dreyer, *Phytochemistry*, **5,** 367 (1966).
4. F. M. Dean and T. A. Geissman, *J. Org. Chem.*, **23,** 596 (1958).
5. T. Kaku and H. Ri, *J. Pharm. Soc. Jap.*, **55,** 219 (1935); *Chem. Abstr.*, **31,** 6642 (1937).
6. F. B. Power and T. Callan, *J. Chem. Soc.*, **99,** 1993 (1911).
7. O. H. Emerson, *J. Amer. Chem. Soc.*, **70,** 545 (1948).
8. D. Arigoni, D. H. R. Barton, E. J. Corey, O. Jeger, L. Caglioti, S. Dev, P. G. Ferrini, E. R. Glazier, A. Melera, S. K. Pradhan, K. Schaffner, S. Sternhell, J. F. Templeton, and S. Tobinaga, *Experientia*, **16,** 41 (1960).
9. T. Kamikawa and T. Kubota, *Tetrahedron*, **12,** 262 (1961).
10. T. Kubota, T. Kamikawa, T. Tokoroyama, and T. Matsuura, *Tetrahedron Lett.*, No. 8, 1 (1960).
11. D. H. R. Barton, S. K. Pradhan, S. Sternhell, and J. F. Templeton, *J. Chem. Soc.*, **1961,** 255.
12. T. Kamikawa, *Nippon Kagaku Zasshi*, **83,** 625 (1962); *Chem. Abstr.*, **59,** 534 (1963).
13. T. Kubota, T. Matsuura, T. Tokoroyama, T. Kamikawa, and T. Matsumoto, *Tetrahedron Lett.*, **1961,** 325.
14. T. Tokoroyama and T. Matsuura, *Nippon Kagaku Zasshi*, **83,** 630 (1962); *Chem. Abstr.*, **59,** 534 (1963).
15. D. L. Dreyer, *J. Org. Chem.*, **30,** 749 (1965).
16. D. L. Dreyer, *J. Org. Chem.*, **33,** 3577 (1968).
17. R. H. Higby, *J. Amer. Chem. Soc.*, **60,** 3013 (1938).
18. D. L. Dreyer, *Tetrahedron*, **21,** 75 (1965).
19. D. L. Dreyer, *Tetrahedron*, **24,** 3273 (1968).
20. F. A. Kincl, J. Romo, G. Rosenkranz, and F. Sondheimer, *J. Chem. Soc.*, **1956,** 4163.
21. J. W. Murphy, T. Toube, and A. D. Cross, *Tetrahedron Lett.*, **1968,** 5153.
22. G. P. Moss, T. P. Toube, and J. W. Murphy, *J. Chem. Soc., C*, **1970,** 694.
23. E. K. Adesogan and D. A. H. Taylor, *J. Chem. Soc., D*, **1969,** 889.
24. E. K. Adesogan and D. A. H. Taylor, *J. Chem. Soc., C*, **1970,** 1710.
25. T. R. Govindachari, D. Prakash, and N. Viswanathan, *J. Chem. Soc., C*, **1968,** 1323.
26. S. Brewis and T. G. Halsall, *J. Chem. Soc.*, **1961,** 646.
27. S. Brewis, T. G. Halsall, H. R. Harrison, and O. J. R. Hodder, *J. Chem. Soc., D*, **1970,** 891.
28. T. Tokoroyama, T. Kamikawa, and T. Kubota, *Bull. Chem. Soc. Jap.*, **34,** 131 (1961).
29. T. Matsuura, T. Kamikawa, and T. Kubota, *Tetrahedron*, **12,** 269 (1961).
30. A. Melera, K. Schaffner, D. Arigoni, and O. Jeger, *Helv. Chim. Acta*, **40,** 1420 (1957).
31. D. L. Dreyer, *J. Org. Chem.*, **31,** 2279 (1966).
32. G. K. Nikonov, *Med. Prom. SSSR*, **18,** 15 (1964); *Chem. Abstr.*, **62,** 12157 (1965).
33. C. R. Narayanan, R. V. Pachapurkar, S. K. Pradhan, V. R. Shah, and N. S. Narasimhan, *Indian J. Chem.*, **2,** 108 (1964).
34. J. D. Connolly, I. M. S. Thornton, and D. A. H. Taylor, *J. Chem. Soc., D*, **1970,** 1205.
35. A. I. Laskin, P. Grabowich, C. D. Meyers, and J. Fried, *J. Med. Chem.*, **7,** 406 (1964).
36. D. Rosenthal, A. O. Niedermeyer, and J. Fried, *J. Org. Chem.*, **30,** 510 (1965).

37. D. Arigoni, D. H. R. Barton, R. Bernasconi, C. Djerassi, J. S. Mills, and R. E. Wolff, *J. Chem. Soc.*, **1960,** 1900.
38. G. H. Whitham, *J. Chem. Soc.*, **1960,** 2016.
39. L. Mangoni and M. Belardini, *Tetrahedron Lett.*, **1963,** 921; *Gazz. Chim. Ital.*, **94,** 382 (1964).
40. R. M. Carman and D. Cowley, *Aust. J. Chem.*, **18,** 213 (1965).
41. J. S. E. Holker, W. R. Jones, and P. J. Ramm, *J. Chem. Soc., D*, **1965,** 435.
42. J. S. E. Holker, W. R. Jones, and P. J. Ramm, *J. Chem. Soc., C*, **1969,** 357.
43. J. Klinot, E. Úlehlová, and A. Vystrčil, *Collect. Czech. Chem. Commun.*, **32,** 2890 (1967).
44. R. Kazlauskas, J. T. Pinhey, J. J. H. Simes, and T. G. Watson, *J. Chem. Soc., D*, **1969,** 945.
45. B. Singh and R. P. Rastogi, *Phytochemistry*, **8,** 917 (1969).
46. A. Baeyer and E. Oehler, *Chem. Ber.*, **29,** 27 (1896).
47. A. Baeyer and O. Seuffert, *Chem. Ber.*, **32,** 3619 (1899).
48. A. Baeyer and V. Villiger, *Chem. Ber.*, **32,** 3625 (1899).
49. E. Wedekind, *Chem. Ber.*, **47,** 2483 (1914).
50. W. Cocker and T. B. H. McMurry, *J. Chem. Soc.*, **1956,** 4549.
51. K. Tettweiler, O. Engel, and E. Wedekind, *Justus Liebigs Ann. Chem.*, **492,** 105 (1932).
52. M. Sumi, *J. Amer. Chem. Soc.*, **80,** 4869 (1958).
53. C. G. Overberger and H. Kaye, *J. Amer. Chem. Soc.*, **89,** 5640 (1967).
54. J. N. T. Gilbert, A. J. Hannaford, K. Minami, and W. B. Whalley, *J. Chem. Soc., C*, **1966,** 627.
55. R. Howe, F. J. McQuillin, and R. W. Temple, *J. Chem. Soc.*, **1959,** 363.
56. J. Wolinsky, *J. Org. Chem.*, **26,** 704 (1961).
57. Y. Matsubara, *Nippon Kagaku Zasshi*, **78,** 719 (1957); *Chem. Abstr.*, **53,** 21716 (1959).
58. J. Vrkoč, J. Jonáš, V. Herout, and F. Šorm, *Collect. Czech. Chem. Commun.*, **29,** 539 (1964).
59. C. E. McEachan, A. T. McPhail, and G. A. Sim, *J. Chem. Soc., D*, **1965,** 276.
60. R. J. McClure, K. S. Schorno, J. A. Bertrand, and L. H. Zalkow, *J. Chem. Soc., D*, **1968,** 1135.
61. H. Hikino, Y. Hikino, Y. Takeshita, K. Meguro, and T. Takemoto, *Chem. Pharm. Bull.* (Tokyo), **11,** 1207 (1963).
62. H. Hikino, Y. Hikino, Y. Takeshita, K. Meguro, and T. Takemoto, *Chem. Pharm. Bull.* (Tokyo), **13,** 1408 (1965).
63. J. Křepinský, M. Romaňuk, V. Herout, and F. Šorm, *Tetrahedron Lett.*, **1962,** 169; *Collect. Czech. Chem. Commun.*, **27,** 2638 (1962).
64. J. Křepinský, M. Romaňuk, V. Herout, and F. Šorm, *Collect. Czech. Chem. Commun.*, **28,** 3122 (1963).
65. T. R. Govindachari, N. Viswanathan, B. R. Pai, P. S. Santhanam, and M. Srinivasan, *Indian J. Chem.*, **6,** 475 (1968).
66. H. Hikino, Y. Takeshita, Y. Hikino, and T. Takemoto, *Chem. Pharm. Bull.* (Tokyo), **14,** 735 (1966).
67. E. Yoshii, *Yakugaku Zasshi*, **83,** 825 (1963); *Chem. Abstr.*, **60,** 4191 (1964).
68. G. Büchi, M. S. V. Wittenau, and D. M. White, *J. Amer. Chem. Soc.*, **81,** 1968 (1959).
69. F. Kido, R. Sakuma, H. Uda, and A. Yoshikoshi, *Tetrahedron Lett.*, **1969,** 3169.
70. M. Souček and P. Vlad, *Chem. Ind.* (London), **1962,** 1946; *Collect. Czech. Chem. Commun.*, **28,** 1211 (1963).

71. J. R. Prahlad, R. Ranganathan, U. R. Nayak, T. S. Santhanakrishnan, and S. Dev., *Tetrahedron Lett.*, **1964,** 417.
72. R. Ranganathan, U. R. Nayak, T. S. Santhanakrishnan, and S. Dev., *Tetrahedron*, **26,** 621 (1970).
73. C. W. J. Chang and S. W. Pelletier, *Tetrahedron Lett.*, **1966,** 5483.
74. S. W. Pelletier, C. W. J. Chang, and K. N. Iyer, *J. Org. Chem.*, **34,** 3477 (1969).
75. E. Wenkert and D. P. Strike, *J. Amer. Chem. Soc.*, **86,** 2044 (1964).
76. D. Arigoni, J. Kalvoda, H. Heusser, O. Jeger, and L. Ruzicka, *Helv. Chim. Acta*, **38,** 1857 (1955).
77. G. Defaye, *Publ. Sci. Tech. Min. Air (Fr), Notes Tech.*, No. N.T. 165 (1969); *Chem. Abstr.*, **72,** 79254 (1970).
78. T. Kubota, T. Matsuura, T. Tsutsui, S. Uyeo, M. Takahashi, H. Irie, A. Numata, T. Fujita, T. Okamoto, M. Natsume, Y. Kawazoe, K. Sudo, T. Ikeda, M. Tomoeda, S. Kanatomo, T. Kosuge, and K. Adachi, *Tetrahedron Lett.*, **1964,** 1243.
79. T. Kubota, T. Matsuura, T. Tsutsui, S. Uyeo, H. Irie, A. Numata, T. Fujita, and T. Suzuki, *Tetrahedron*, **22,** 1659 (1966).
80. W. Herz, R. N. Mirrington, H. Young, and Y. Y. Lin, *J. Org. Chem.*, **33,** 4210 (1968).
81. G. A. Ellestad, B. Green, A. Harris, W. B. Whalley, and H. Smith, *J. Chem. Soc.*, **1965,** 7246.
82. M. R. Cox, G. A. Ellestad, A. J. Hannaford, I. R. Wallwork, W. B. Whalley, and B. Sjöberg, *J. Chem. Soc.*, **1965,** 7257.
83. E. J. Corey and J. J. Ursprung, *J. Amer. Chem. Soc.*, **78,** 5041 (1956).
84. D. H. R. Barton, H. T. Cheung, A. D. Cross, L. M. Jackman, and M. Martin-Smith, *J. Chem. Soc.*, **1961,** 5061.
85. G. W. Krakower, H. A. Van Dine, P. A. Diassi, and I. Basco, *J. Org. Chem.*, **32,** 184 (1967).
86. Y. Tanahashi, T. Takahashi, F. Patil, and G. Ourisson, *Bull. Soc. Chim. Fr.*, **1964,** 584.
87. R. Stevenson, *J. Org. Chem.*, **28,** 188 (1963).
88. T. Takahashi and G. Ourisson, *Bull. Soc. Chim. Fr.*, **1956,** 353.
89. D. Lavie and E. C. Levy, *Tetrahedron Lett.*, **1970,** 1315.
90. C. S. Barnes, D. H. R. Barton, J. S. Fawcett, and B. R. Thomas, *J. Chem. Soc.*, **1952,** 2339.
91. E. Höhne, *Collect. Czech. Chem. Commun.*, **28,** 3128 (1963).
92. W. Klyne, *Chem. Ind.* (London), **1954,** 1198.
93. R. B. Turner, *J. Amer. Chem. Soc.*, **72,** 878 (1950).
94. T. F. Gallagher and T. H. Kritchevsky, *J. Amer. Chem. Soc.*, **72,** 882 (1950).
95. T. S. Santhanakrishnan, U. R. Nayak, and S. Dev, *Tetrahedron*, **26,** 641 (1970).
96. L. K. Lala and J. B. Hall, *J. Org. Chem.*, **35,** 1172 (1970).
97. V. N. Belov and L. A. Kheifits, *J. Gen. Chem.* (*USSR*), **27,** 1459 (1957).
98. A. M. Mathieson, *Tetrahedron Lett.*, **1963,** 81.
99. C. G. Overberger and H. Kaye, *J. Amer. Chem. Soc.*, **89,** 5646 (1967).
100. T. Sasaki, S. Eguchi, M. Ohno, and T. Oyobe, *Bull. Chem. Soc. Jap.*, **42,** 3582 (1969).
101. A. Tahara, K. Hirao, and Y. Hamazaki, *Chem. Pharm. Bull.* (Tokyo), **12,** 1458 (1964).
102. B. E. Cross, R. H. B. Galt, and J. R. Hanson, *J. Chem. Soc.*, **1963,** 5052.
103. B. E. Cross, R. H. B. Galt, J. R. Hanson, and W. Klyne, *Tetrahedron Lett.*, **1962,** 145.
104. J. F. Grove and B. J. Riley, *J. Chem. Soc.*, **1961,** 1105.
105. B. E. Cross, *J. Chem. Soc.*, **1960,** 3022.
106. A. Harris, A. Robertson, and W. B. Whalley, *J. Chem. Soc.*, **1958,** 1799.

107. T. Norin, *Acta Chem. Scand.*, **15,** 1676 (1961).
108. S. Forsén and T. Norin, *Acta Chem. Scand.*, **15,** 592 (1961).
109. E. E. Van Tamelen and J. P. McCormick, *J. Amer. Chem. Soc.*, **91,** 1847 (1969).
110. R. A. Bell and M. B. Gravestock, *Can. J. Chem.*, **48,** 1105 (1970).
111. H. H. Zeiss and F. W. Grant, *J. Amer. Chem. Soc.*, **79,** 1201 (1957).
112. P. Yates and E. A. Chandross, *Tetrahedron Lett.*, No. 20, 1 (1959).
113. N. L. Drake and W. P. Campbell, *J. Amer. Chem. Soc.*, **58,** 1681 (1936).
114. N. L. Drake and J. K. Wolfe, *J. Amer. Chem. Soc.*, **61,** 3074 (1939).
115. T. Nishihama and T. Takahashi, *Bull. Chem. Soc. Jap.*, **39,** 200 (1966).
116. J. L. Simonsen and D. H. R. Barton, *The Terpenes*, Vol. III, 2nd ed., University Press, Cambridge, 1952, pp. 103–105.
117. A. Bhati, *J. Org. Chem.* **27,** 2135 (1962).
118. R. B. Woodward, F. J. Brutschy, and H. Baer, *J. Amer. Chem. Soc.*, **70,** 4216 (1948).
119. A. Abkin and S. Medvedev, *J. Gen. Chem. (USSR)*, **4,** 1407 (1934); *Chem. Abstr.*, **29,** 3682 (1935).
120. V. Jarolím, M. Streibl, L. Dolejš, and F. Šorm, *Chem. Listy*, **50,** 1289 (1956); *Chem. Abstr.*, **51,** 297 (1957).
121. O. Wallach, *Justus Liebigs Ann. Chem.*, **258,** 319 (1890).
122. J. L. Simonsen, *J. Chem. Soc.*, **123,** 2642 (1923).
123. P. Naffa and G. Ourisson, *Bull. Soc. Chim. Fr.*, **1954,** 1115.
124. H. Erdtman and T. Norin, *Chem. Ind.* (London), **1960,** 622.
125. S. Akiyoshi and S. Nagahama, *Bull. Chem. Soc. Jap.*, **30,** 886 (1957).
126. S. Nagahama, H. Kobayashi, and S. Akiyoshi, *Bull. Chem. Soc. Jap.*, **32,** 366 (1959).
127. L. Ruzicka and E. Bernold, *Helv. Chim. Acta*, **24,** 1167 (1941).
128. C. Djerassi, E. Wilfred, L. Visco, and A. J. Lemin, *J. Org. Chem.*, **18,** 1449 (1953).
129. W. H. Baarschers, D. H. S. Horn, and L. F. Johnson, *J. Chem. Soc.*, **1962,** 4046.
130. R. H. B. Galt and J. R. Hanson, *J. Chem. Soc.*, **1965,** 1565.
131. M. Kurono, Y. Nakadaira, S. Onuma, K. Sasaki, and K. Nakanishi, *Tetrahedron Lett.*, **1963,** 2153.
132. K. Nakanishi, M. Kurono, and N. S. Bhacca, *Tetrahedron Lett.*, **1963,** 2161.
133. M. C. Woods, K. Nakanishi, and N. S. Bhacca, *Tetrahedron*, **22,** 243 (1966).
134. D. Lavie, E. Glotter, and Y. Shvo, *Tetrahedron*, **19,** 1377 (1963).
135. A. Meyer, O. Jeger, V. Prelog, and L. Ruzicka, *Helv. Chim. Acta*, **34,** 747 (1951).
136. A. Meisels, O. Jeger, and L. Ruzicka, *Helv. Chim. Acta*, **32,** 1075 (1949).
137. L. Ruzicka and S. L. Cohen, *Helv. Chim. Acta*, **20,** 1192 (1937).
138. L. Ruzicka and K. Hofmann, *Helv. Chim. Acta*, **19,** 114 (1936).
139. D. H. R. Barton and P. DeMayo, *J. Chem. Soc.*, **1953,** 3111.
140. W. Schmitt and H. Wieland, *Justus Liebigs Ann. Chem.*, **542,** 258 (1939).
141. L. Ruzicka, C. Nisoli, and O. Jeger, *Helv. Chim. Acta*, **29,** 2017 (1946).
142. L. Ruzicka, O. Jeger, and P. Ringnes, *Helv. Chim. Acta*, **27,** 972 (1944).
143. L. Ruzicka and O. Isler, *Helv. Chim. Acta*, **19,** 506 (1936).
144. E. Říhová and A. Vystrčil, *Collect. Czech. Chem. Commun.*, **34,** 240 (1969).
145. O. Wallach and E. Gildemeister, *Justus Liebigs Ann. Chem.*, **246,** 265 (1888).
146. J. L. Simonsen, *The Terpenes*, Vol. I, 2nd ed., University Press, Cambridge, 1947, p. 427.
147. K. Sisido, K. Utimoto, and T. Isida, *J. Org. Chem.*, **29,** 3361 (1964).
148. U. R. Nayak and S. Dev, *Tetrahedron*, **19,** 2269 (1963).
149. I. Ogura, *Proc. Sci. Inst. Kinki Univ.*, **1959,** 17; *Chem. Abstr.*, **54,** 3488 (1960).
150. S. W. Pelletier, D. T. C. Yang, and A. Ogiso, *J. Chem. Soc., D*, **1968,** 830.
151. D. Yang and S. W. Pelletier, *J. Chem. Soc., D*, **1968,** 1055.

152. L. Ruzicka, M. Brenner, and E. Rey, *Helv. Chim. Acta*, **24,** 515 (1941).
153. P. DeMayo and A. N. Starratt, *Can. J. Chem.*, **40,** 788 (1962).
154. E. Von Rudloff and G. V. Nair, *Can. J. Chem.* **42,** 421 (1964).
155. B. Tomita and Y. Hirose, *Tetrahedron Lett.*, **1970,** 235.
156. W. R. Chan, D. R. Taylor, and T. Yee, *J. Chem. Soc.*, *C*, **1970,** 311.
157. D. E. U. Ekong, C. O. Fakunle, A. K. Fasina, and J. I. Okogun, *J. Chem. Soc.*, *D*, **1969,** 1166.
158. V. P. Wystrach and R. K. Madison, U.S. Patent 2,824,138; *Chem. Abstr.*, **52,** 11116 (1958).
159. H. François, R. Lalande, and G. Bex, *Bull. Soc. Chim. Fr.*, **1963,** 833.
160. H. François, G. Bex, and R. Lalande, *Bull. Soc. Chim. Fr.*, **1965,** 3698, 3705.
161. P. Teisseire and B. Corbier, *Recherches* (Paris), **13,** 78 (1963); *Chem. Abstr.*, **61,** 7048 (1964).
162. B. Corbier and P. Teisseire, *Recherches* (Paris), **14,** 93 (1964); *Chem. Abstr.*, **63,** 4334 (1965).
163. S. Itô, M. Kodama, T. Nozoe, H. Hikino, Y. Hikino, Y. Takeshita, and T. Takemoto, *Tetrahedron Lett.*, **1963,** 1787.
164. S. Itô, H. Hikino, M. Kodama, Y. Hikino, and Y. Takeshita, *Kogyo Kagaku Zasshi*, **68,** 804 (1965); *Chem. Abstr.*, **65,** 7220 (1966).
165. S. Itô, M. Kodama, T. Nozoe, H. Hikino, Y. Hikino, Y. Takeshita, and T. Takemoto, *Tetrahedron*, **23,** 553 (1967).
166. T. Irie, A. Fukuzawa, M. Izawa, and E. Kurosawa, *Tetrahedron Lett.*, **1969,** 1343.
167. S. Sasaki, H. C. Chiang, K. Habaguchi, H. Hsü, and K. Nakanishi, *Bull. Chem. Soc. Jap.*, **39,** 1816 (1966).
168. R. A. Eade, K. Hunt, J. J. H. Simes, and W. Stern, *Aust. J. Chem.*, **22,** 2703 (1969).
169. B. Singh and R. P. Rastogi, *Phytochemistry*, **7,** 1385 (1968).
170. J. Polonsky, Z. Baskevitch, N. C. Bellavita, and P. Ceccherelli, *J. Chem. Soc.*, *D*, **1968,** 1404.
171. J. Polonsky, Z. Baskevitch, N. C. Bellavita, and P. Ceccherelli, *Bull. Soc. Chim. Fr.*, **1970,** 1912.
172. R. Ó. Dorchaí, H. E. Rubalcava, J. B. Thomson, and B. Zeeh, *Tetrahedron*, **24,** 5649 (1968).

CHAPTER VII

Steroidal Oxepins

JOHN A. ZDERIC

Syntex International, S.A., Mexico, D.F.

A. Introduction

This review will exclude the examination of steroidal oxepins containing only bridged oxepin rings when the bridges are comprised of single bonds or one carbon atom. In such compounds one logically is not inclined to think of the substances as oxepins, but more correctly as members of simpler ring systems such as oxiranes, oxolanes, and oxanes. Systems thereby omitted from further consideration include, for example, structures **1–3**.

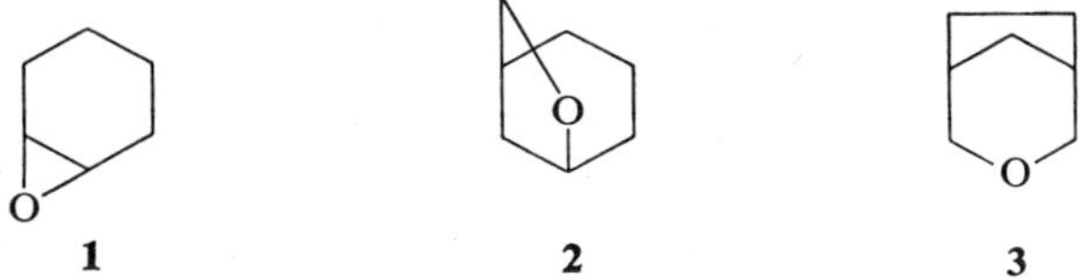

In considering the structural nature of steroidal oxepins, one observes that they can readily be classified as seven-membered cyclic lactones, anhydrides, ethers, or hemiketals occurring in either rings A, B, or C. An arbitrary division on this basis will be a convenient format for reviewing this subject.

B. Nomenclature

The IUPAC-IUB replacement system (1) for naming steroidal oxepins is explicit and simple to use. It is based on the following set of rules: Rule

2S-7.1 provides that ring expansion should be indicated by adding to the basic name of the compound the prefix "Homo," preceded by an italic letter indicating the ring affected. According to Rule 2S-7.3, on ring expansion the small letter a "is added to the highest number in the ring enlarged, exclusive of ring junctions, and this letter and number are assigned to the last peripheral carbon atom in the order of numbering of the ring affected." Rule 2S-10.1 states, "If hetero atoms occur in the ring system of a steroid the replacement ('oxa . . .') system of nomenclature is used with steroid names and numbering." Examples of names derived by the application of these rules are given for structures **4**–**6**.

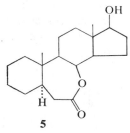

4

4-Oxa-*A*-homo-5α-androstan-17β-ol
3,4,5,5a,5bα,6,7,7a,8α,9,10,10aα,10bβ,11,12,12aα-Hexadecahydro-5aβ,7aβ-dimethyl-1*H*-cyclopenta[5,6]-naphth[2,1-*c*]oxepin-8-ol
3,4-Epoxy-3,4-seco-5α-androstan-17β-ol

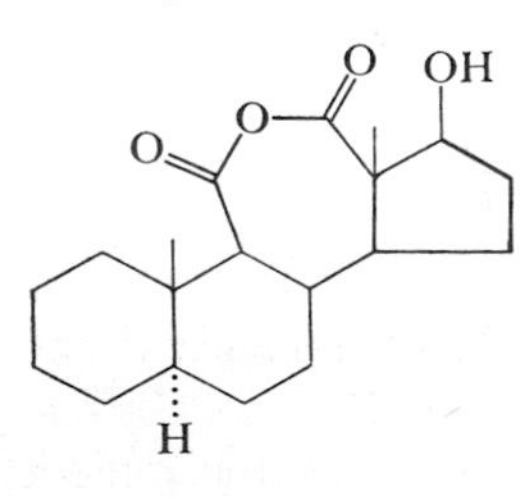

5

17β-Hydroxy-7a-oxa-*B*-homo-5α-androstan-7-one
1,2,3,3aα,3bβ,6,6aα,7,8,9,10,10a,10bα,11,12,12a-Hexadecahydro-1β-hydroxy-10aβ,12aβ-dimethyl-5*H*-benz[*d*]indeno[4,5-*b*]oxepin-5-one
8,17β-Dihydroxy-7,8-seco-5α-androstan-7-oic acid ε-lactone

6

17β-Hydroxy-12-oxa-*C*-homo-5α-androstane-11,12a-dione
3,4,4a,7a,8,9,10,10aα,10bβ,11,12,12aα-Dodecahydro-8β-hydroxy-4aβ,7aβ-dimethyl-1*H*-cyclopenta[*c*]naphth[2,1-*e*]-oxepin-5,7(2*H*,4bβ*H*)-dione
17β-Hydroxy-11,12-seco-5α-androstane-11,12-dioic anhydride

The *Chemical Abstracts* (2) approach to systematic nomenclature employing a fused ring designation is considerably more complex. A familiarity with this system as well as *Chemical Abstracts* preferred names (1, 3) will be desirable for readers wishing to make further independent surveys. Examples of these nomenclature systems are also provided for structures **4**–**6**.

C. Ring A Oxepins

During 1942 Burckhardt and Reichstein (4) described certain room temperature experiments involving the peroxidation of 3-oxo-5β-chol-11-en-24-oic acid methyl ester **7** (Eq. 1). While they had anticipated that this compound would react with one equivalent of perbenzoic acid to yield the keto epoxide **8**, their results clearly indicated that two equivalents were consumed. Moreover, the product of the reaction contained one oxygen more than could be accounted for by the simple epoxide structure originally expected.

CO_2CH_3 **7** $\xrightarrow{PhCO_3H}$ **8** (not formed) / **9** CO_2CH_3 (1)

Further investigation (5) indicated that the product could be formulated as **9**, wherein an extra oxygen had been inserted between C-3 and C-4. During this same investigation it was shown that the reaction was a general one which proceeded with C-3 ketones in both the 5α and 5β-series. Among the substrates investigated were 5α-cholestan-3-one, 5β-cholestan-3-one, 12-acetoxy-3-oxo-5β-cholan-24-oic acid methyl ester, and 3-oxo-5β-cholan-24-oic acid methyl ester. All of these C-3 ketones gave rise to lactones bearing the newly introduced oxygen between C-3 and C-4. As noted by the authors of this paper (5), Gardner and Godden (6) a number of years earlier had probably observed similar lactone formation from 5β-cholestan-3-one by the action of ammonium persulfate.

Proof for the position of oxygen entry was established by hydrolysis of the lactones **10** to the hydroxy acids **11**, which were then subjected to chromium trioxide oxidation (Eq. 2). Under these conditions dicarboxylic acids of the type indicated in partial formula **12** were obtained for each of the products derived from the various starting materials noted above. Since workers in other laboratories had already established the correct structure of the resultant 3,4-seco dibasic acids, it was clear that such acids could only have arisen by oxygen insertion between C-3 and C-4, as shown in partial structure **10.**

O= O. **10** ⟶ HOOC HOH_2C **11** $\xrightarrow{CrO_3}$ HOOC HOOC **12** (2)

Although in Burckhardt and Reichstein's work none of the isomeric C-2, C-3 lactones were observed, it will be seen later that these compounds are indeed formed. At the time their absence was noted as being unusual, since it was known that substitution in 5α-steroids, usually took place at C-2.

In the earlier investigation (4) an attempt was made to treat the 4-ene isomer of **7** with perbenzoic acid. This seemed especially likely to lead to an enol lactone, since it had already been reported in the literature (7) that treatment of benzal acetone with perbenzoic acid led to a styrene ester. While 2 equivalents of perbenzoic acid were in fact consumed by the 4-ene isomer, the only product isolated was a small amount of the corresponding 11,12-epoxide. Presumably any enol lactone produced in the reaction was lost during column chromatography of the reaction mixture.

Finally, Burckhardt and Reichstein (5) reported their attempt to cleave the side chain of 3β-hydroxy-5α-pregnan-20-one acetate with perbenzoic acid at room temperature. While under their usual conditions no conversion was observed, it was found that reaction at reflux temperature in chloroform led to the recovery of low yields of the corresponding 5α-androstane-3β,17β-diol diacetate. Similar studies had been carried out by Marker et al. (8) a few years earlier with potassium persulfate.

One of the first preparative syntheses which made use of the observations of Burckhardt and Reichstein (4, 5) was carried out by Ruzicka's group (9, 10). These investigators were interested in attempting to relate the steroid structures of certain keto-*A*-nor androstanes, that is, **15** and **18**, with musk odors.

The preparation of **15** proceeded by direct chromium trioxide oxidation of 5α-androstan-3-one **13** to the dibasic acid **14**, which was then heated in acetic anhydride (Eq. 3). Pyrolysis of the resulting ring A anhydride yielded the 2-keto-*A*-nor derivative **15**, which incidentally possessed an odor resembling cedar oil.

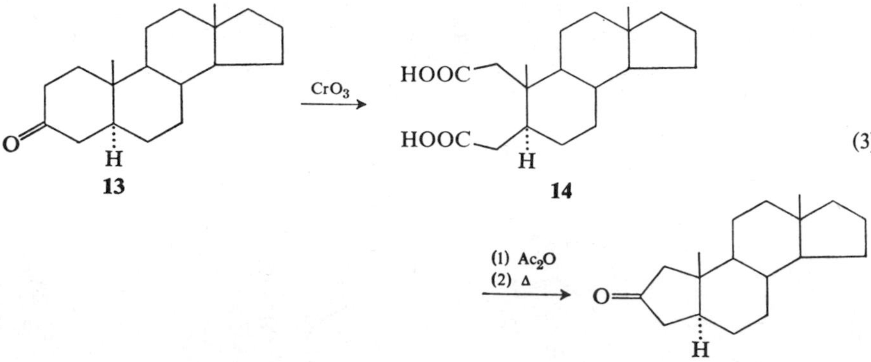

13 $\xrightarrow{CrO_3}$ 14 (HOOC, HOOC) $\xrightarrow{(1)\ Ac_2O,\ (2)\ \Delta}$ 15 (3)

For the preparation of the corresponding *A*-nor-3-ketone, **13** was converted in low yield by means of perbenzoic acid to the lactone **16**, which was then transformed by hydrolysis and oxidation to the dibasic acid **17** (Eq. 4). *In situ* anhydride formation followed by pyrolysis gave the desired product **18**, which proved to be almost odorless.

An alternate synthesis of this same *A*-nor-3-ketone, **18**, was effected (Eq. 4) from **19**, via **20** and **21**. It may be noted that both this synthesis, as well as the previously described one, led to the 3-keto product bearing the β-configuration at C-5.

A synthetic pathway similar to that of the Ruzicka group (10) was followed by Rull and Ourisson (11), who were interested in preparing ring *A*-nor analogs of various natural hormones for purposes of biological evaluation. Instead of using perbenzoic acid for the oxygen insertion, they chose to employ a reagent system developed by Emmons and co-workers (12, 13), namely trifluoroperacetic acid in the presence of finely divided sodium hydrogen phosphate. Indeed, treatment of **22** with this peracid, which is prepared readily from 90% hydrogen peroxide and trifluoroacetic acid, led to a very rapid reaction with formation of both possible lactone isomers (Eq. 5). Thus following chromatography on alumina there was obtained a 6% yield of **23** and a 30% yield of **24**. Continued elution of the column provided an acid fraction which was derived from **23**. The remaining steps involved the conversion of **23** and **24** to their respective *A*-nor hormone analogs by methods previously described.

Another use for this general scheme of oxidative ring enlargement was demonstrated by Heymann and Fieser (14), who were interested in a chemical proof for the configuration of the hydroxyl group at C-7 in dextrorotary cholest-5-ene-3β,7-diol (**25**). By a series of steps (Scheme I), this compound was converted to the 3-keto 7-benzoate **26**, which upon treatment with perbenzoic acid in chloroform yielded an intractable mixture of the lactones

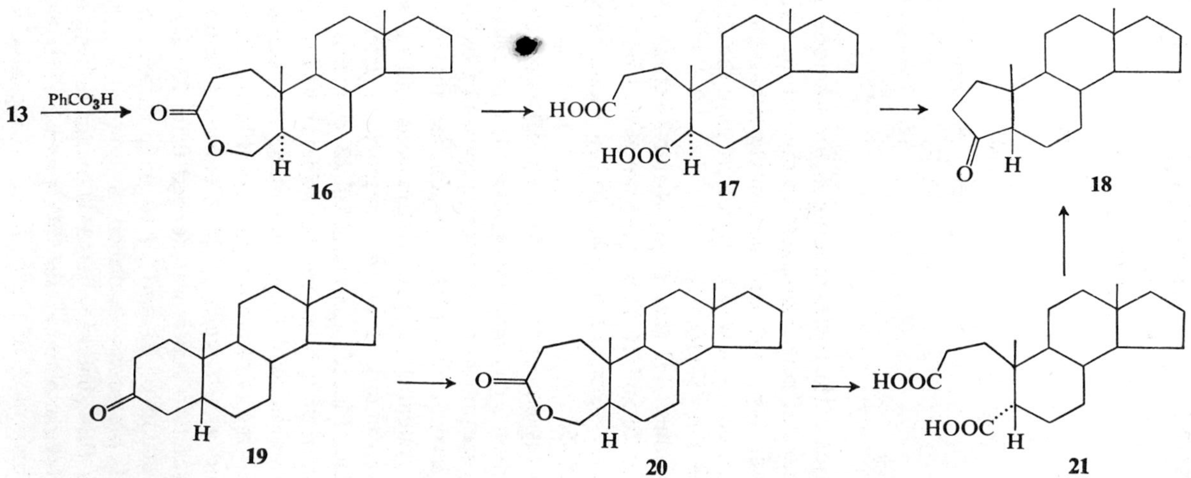

13 $\xrightarrow{PhCO_3H}$ 16 → 17 → 18

19 → 20 → 21 → 18 (4)

HOOC

HOOC H

O H

H

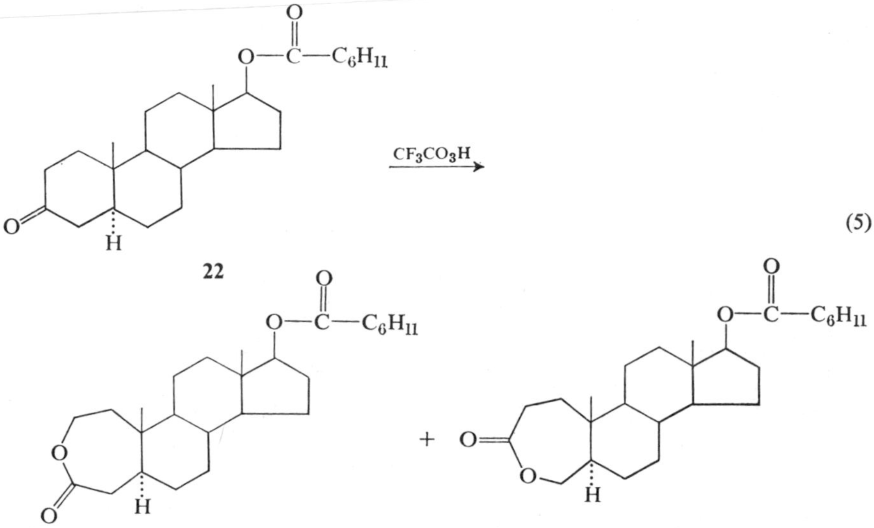

$$\mathbf{22} \xrightarrow{CF_3CO_3H} \mathbf{23} + \mathbf{24} \qquad (5)$$

27 and **28**. Mild hydrolysis of the gross reaction product, followed by diazomethane esterification, oxidation, and further treatment with diazomethane, gave a mixture of the benzoate diesters **29** and **32** from which **29** could be obtained pure by fractional crystallization. The structure of **29** was proven by vigorous hydrolysis to the dicarboxylic acid alcohol **30**, which was then oxidized to the corresponding 7-ketone. After Wolff-Kishner reduction, a previously known 2,3-secodicarboxylic acid **31** was isolated, thereby leaving no doubt that the lactone **27** arising from oxygen insertion between C-2 and C-3 was present in the original perbenzoic acid oxidation product.

In an attempt to pursue the original objective of this study, the mother liquors from the isolation of **29** were first subjected to vigorous saponification, followed by treatment with acetic anhydride and finally diazomethane. Under these conditions the 4-oxa isomer **28** was converted to the γ-lactone methyl ester **33**, which on the basis of its analytical date and ir spectrum could easily be shown to have the structure indicated. Heymann and Fieser therefore concluded that since this lactone could be formed, the C-7 hydroxyl must have had the same configuration as the C-5 carboxylic acid function, thereby establishing the C-7 hydroxyl group as occupying the β-position.

Until the above work of Heymann and Fieser (14) and the previously mentioned observations by Rull and Ourisson (11), the location of oxygen insertion by Baeyer-Villiger reactions on C-3 ketones was believed to be specific for the 4-position. In addition to the above investigators, Hara et al. (15) have recently further emphasized that insertion can occur at the 3-position. Thus by reexamining the work of Burckhardt and Reichstein (5),

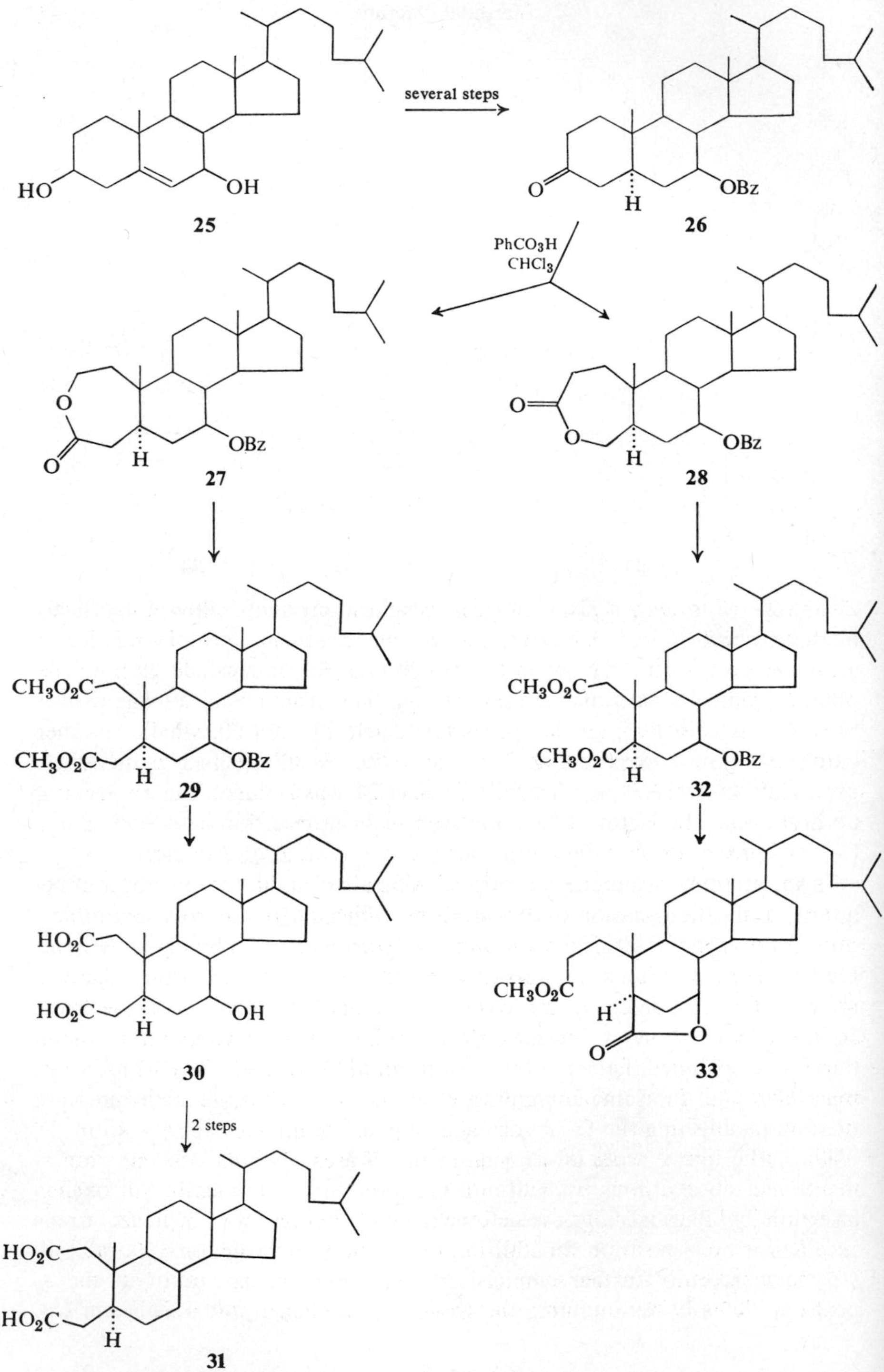

HO OH

25

several steps

O OBz
H

26

$PhCO_3H$
$CHCl_3$

O
O OBz
H

27

O
O OBz
H

28

CH_3O_2C
CH_3O_2C OBz
H

29

CH_3O_2C
CH_3O_2C OBz
H

32

HO_2C
HO_2C OH
H

30

CH_3O_2C
H
O
O

33

2 steps

HO_2C
HO_2C
H

31

Scheme I

these Japanese authors proved that the product of the 5α-cholestan-3-one reaction was indeed a mixture as shown by the following data. Hydrolysis of the reaction product containing both lactones **34** and **37** (Eq. 6), followed by diazomethane esterification and chromium trioxide oxidation, led to a mixture of the mono-esterified dibasic acids **35** and **38**. Mild Fischer esterification then gave a mixture in which only the 2,3-dicarboxylic acid had been converted to its diester **36**. The 3,4 dibasic acid **38**, on the other hand, proved to be too hindered to allow esterification to proceed at C-4, and for this reason alkaline extraction was capable of effecting a quantitative separation of **36** and **38**. The presence of these two different derivatives further demonstrated

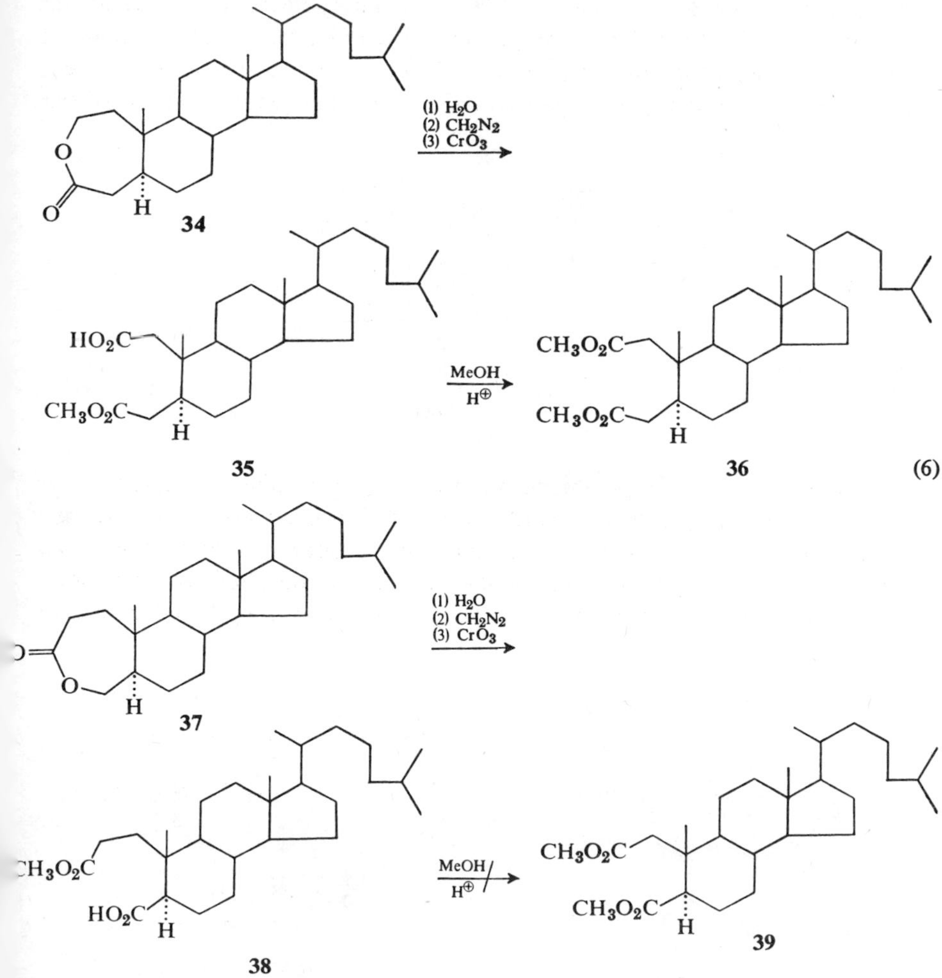

(1) H_2O
(2) CH_2N_2
(3) CrO_3

34

HO_2C CH_3O_2C H **35** $\xrightarrow[H^{\oplus}]{MeOH}$ CH_3O_2C CH_3O_2C H **36** (6)

(1) H_2O
(2) CH_2N_2
(3) CrO_3

37

CH_3O_2C HO_2C H **38** MeOH/$H^{\oplus}$ (no reaction) CH_3O_2C CH_3O_2C H **39**

that Burckhardt and Reichstein's initial reaction mixture was composed of isomers where oxygen insertion had occurred between C-2,3 and C-3,4. In the same paper the authors also noted that reaction times in the 5β-series may be twice as long as in the 5α-series.

During the late 1940s investigations by Wieland and Miescher (16) were aimed at developing new methods for the degradation of pregnanes to androgens. Marker et al. (8) as well as Burckhardt and Reichstein (5) had earlier noted that monopersulfuric acid or perbenzoic acid were capable of effecting such conversions. Sarett (17) had also carried out studies in this area, noting that with progesterone, in addition to side-chain cleavage, some change probably also occurred in ring A. A similar report concerning progesterone was due to Salamon (18), who many years earlier found that treatment of this compound with persulfuric acid led to a product for which he proposed structure **40**. In contrast to the methods of these earlier workers, Wieland and Miescher (16) employed large excesses of peracid and found that

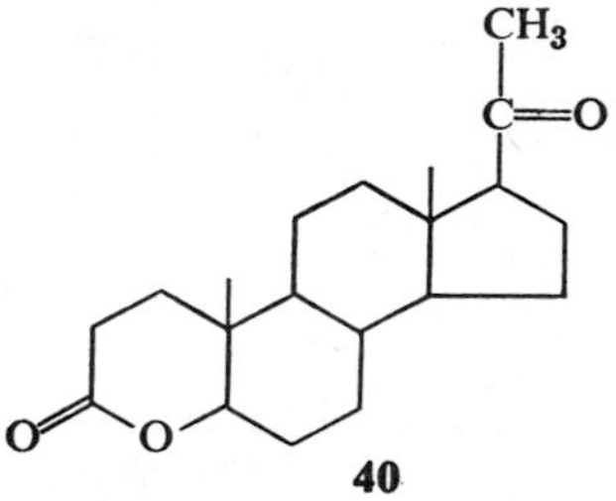

40

either sulfuric acid or *p*-toluenesulfonic acid were effective catalysts for degrading the pregnane side chain. In addition they also reported their results on peracid treatment of **41**, and for the product obtained after debromination (Eq. 7) they assigned structure **42**, noting that no independent proof was available either for the location of the double bond or the position of oxygen insertion.

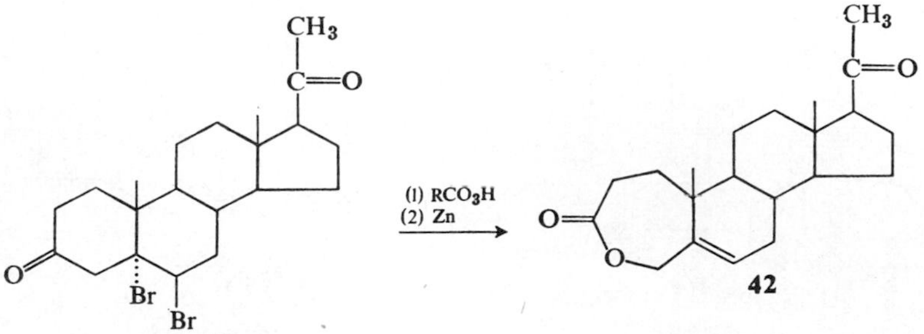

41 —(1) RCO_3H, (2) Zn→ **42** (7)

While Marker (19) had reported that α-brominated ketones were stable to persulfuric acid oxidation, it was found (16) that **43** undergoes peracid attack and produces after pyridine dehydrobromination a mixture of the unsaturated lactones **44** and **45** (Eq. 8). The latter compound was further converted to its 17-hydroxy and 17-ketone derivatives **46** and **47**.

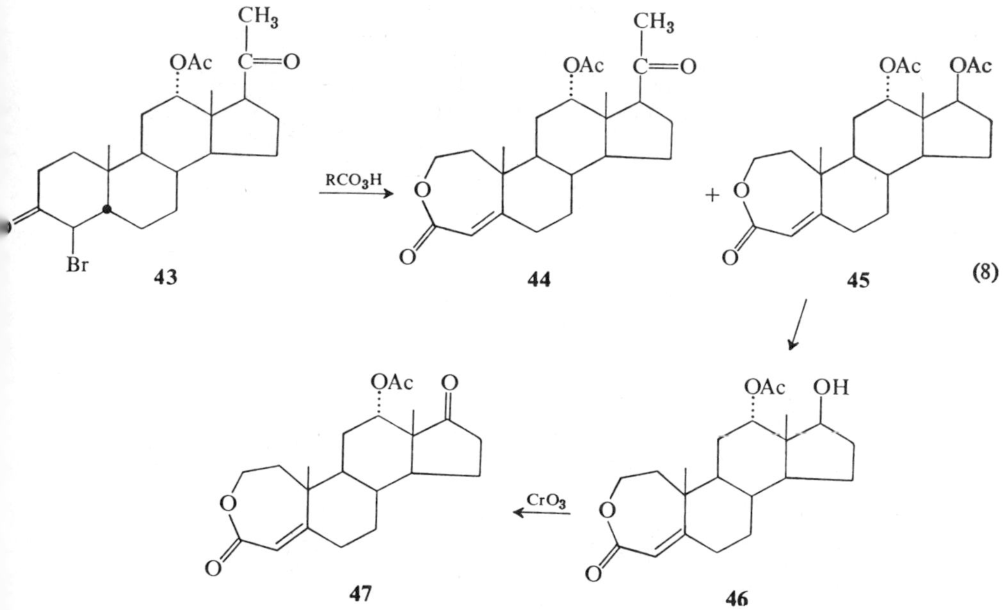

Another example (20) of an α-bromo ketone that undergoes the Baeyer-Villiger reaction is **48**, which upon treatment with trifluoroperacetic acid in the presence of phosphate buffer (Eq. 9) leads to the bromo lactone oxepin **49** in fair yield. The structure of this substance was established by its reduction with chromous chloride, which yielded the known 4-oxa-*A*-homo lactone **50**. It is of interest that when the bromo derivative **49** was warmed in methanol solution a molecular rearrangement took place, presumably via the mechanism shown in structure **51**, thereby quantitatively yielding the methyl ester **52**. While the 2α-chloro ketone corresponding to **48** was also found to undergo Baeyer-Villiger oxidation, the corresponding 2-iodo ketone yielded only 5α-cholest-1-en-3-one.

The preparation of 3-keto-4-oxa enol lactones has also received the attention of numerous other workers. Thus in connection with the need to develop

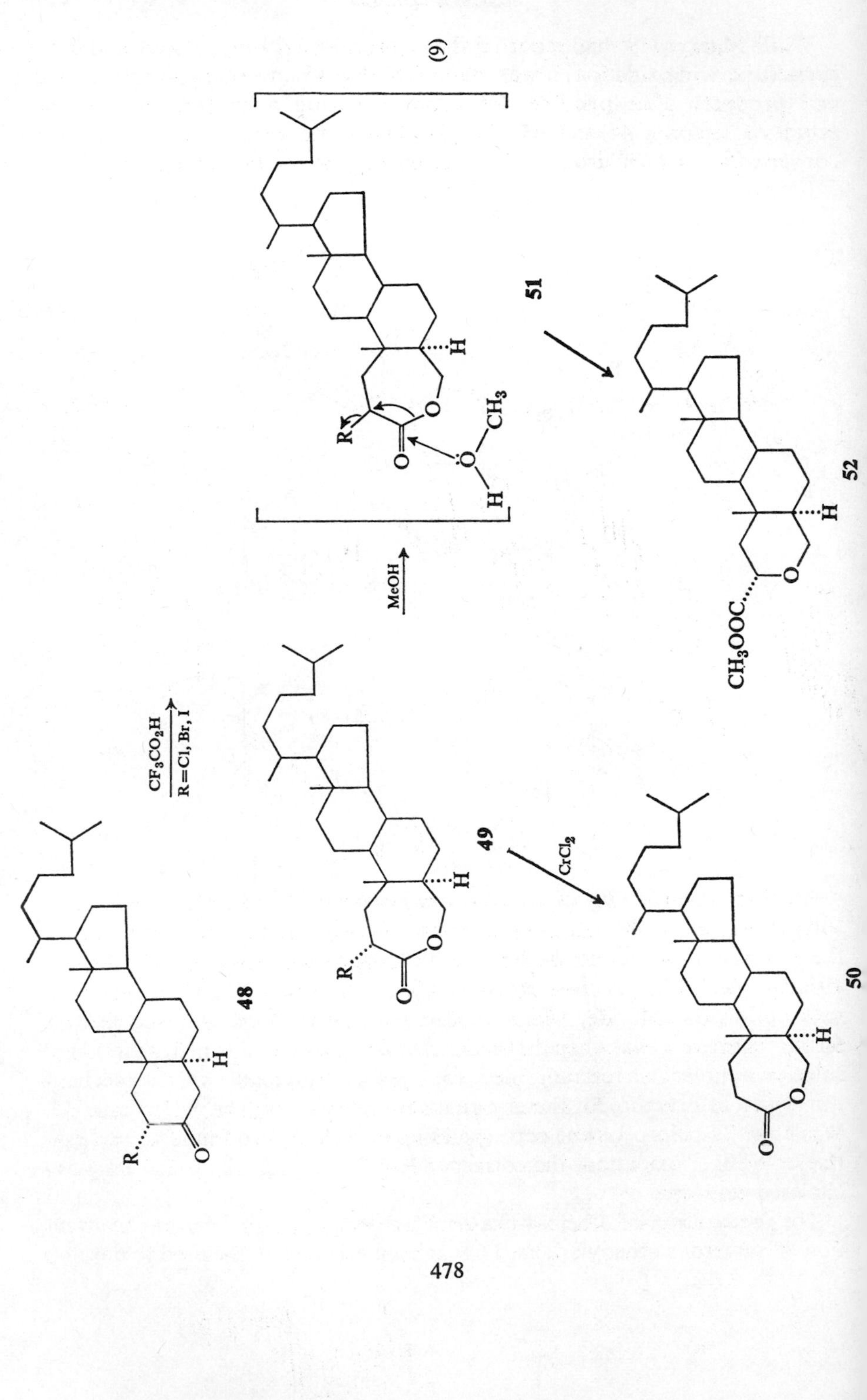

48

$\xrightarrow[R = Cl, Br, I]{CF_3CO_2H}$

49

$\xrightarrow{MeOH}$

51

(9)

$\xrightarrow{CrCl_2}$

50

52

CH_3OOC

precise degradative methods for corticosteroids, one of the Worcester Foundation groups (21) treated **53** with perbenzoic acid in the presence of perchloric acid as a catalyst (Eq. 10). By this means they obtained the enol lactone **54** in reasonably good yield. The assignment of structure followed from the analytical data, the strong end absorption in the uv, and a relatively intense ir band at 1640 cm^{-1} which, considered in conjunction with the carbonyl band, indicated an enol lactone. These observations clearly ruled out the possibility that this product was an epoxide. The final proofs employed to support the proposed structure were the observations that **54** could be catalytically reduced to the lactone **56** or ozonized to yield the nor-acid **57**.

In addition to the enol lactone **54**, trace amounts of a D-homo lactone and 8% of another product were isolated from the original reaction mixture. This latter product was assigned the epoxide structure **55** since it contained one oxygen more than the enol lactone **54** and did not possess any ir or uv characteristics compatible with the presence of an enol. These points were further emphasized by the fact that **55** was stable to catalytic hydrogenation. By way of a more direct proof, **55** was treated with methanolic sulfuric acid and then periodic acid (Eq. 10), whereupon the nor-acid **57** was isolated. On these grounds the structure of **55** seems well established, and it may be surmised that it arises by epoxidation of the main enol lactone product **54**.

Pettit and Kasturi (22), in studies which required the preparative use of 6-membered lactones of the type indicated by structure **61**, speculated on their preparation by the method of Salamon (18), which involved the use of persulfuric acid. Thus when **58** was treated in this fashion (Eq. 11), the reaction was viewed as initially producing the oxepin enol lactone **59**, which under the reaction conditions underwent hydrolysis to the aldehydic intermediate **60**; further Baeyer-Villiger oxidative attack followed by ring closure then led to the main reaction product **61**.

It has already been noted that reaction products arising from the use of different peracids can show considerable structural variation. A further example of this is demonstrated by the work of Pinhey and Schaffner (23, 24), who treated the ene-one **62** with trifluoroperacetic acid (Eq. 12). The product of this reaction was shown to be the aldehyde **63**. Further oxidation led to the carboxylic acid **64**, which had previously been prepared by an alternate route (25) involving the action of alkaline hydrogen peroxide on 4,5-epoxy-3-ketones. Interestingly, when the aldehyde **63** was treated with potassium acetate in acetic anhydride, a facile Claisen condensation took place to yield **65**. In the case of 4-methylcholest-4-en-3-one **66**, this same sequence of reactions (Eq. 13) provided the methyl ketone **67**, which was readily convertible to **68**.

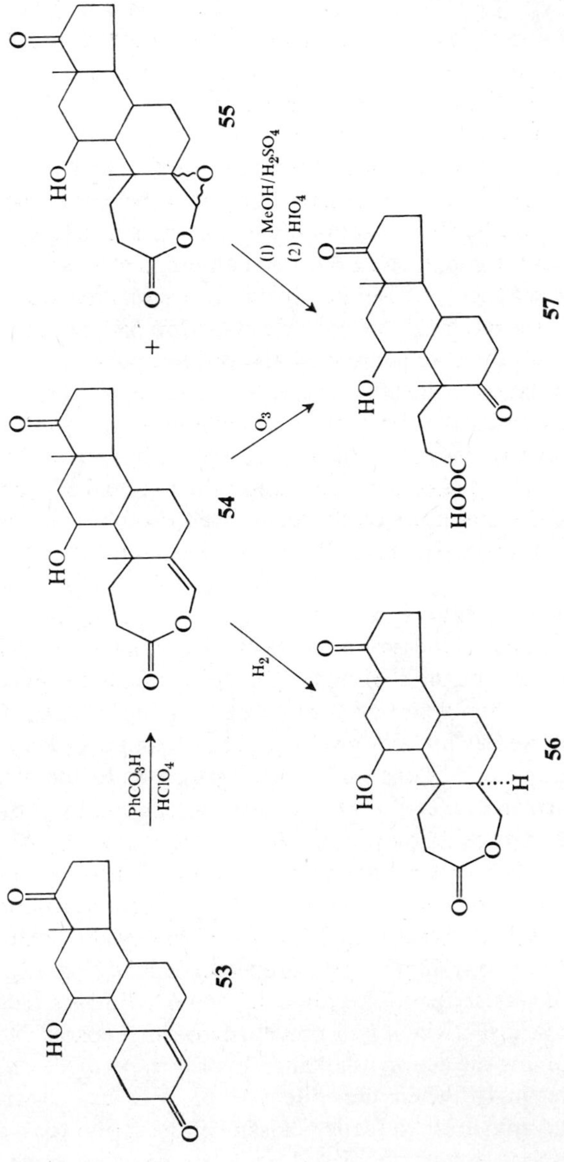

53

$\frac{PhCO_3H}{HClO_4}$

54 + 55

H_2

O_3

(1) $MeOH/H_2SO_4$
(2) HIO_4

56

57

(10)

HO

O

HOOC

H

$OCOC_2H_5$

persulfuric acid

58

$OCOC_2H_5$

59

$OCOC_2H_5$

HOOC

OHC

60

$OCOC_2H_5$

61

(11)

CF_3CO_3H

62

CHO

63

(12)

KOAc
Ac_2O

H

65

COOH

64

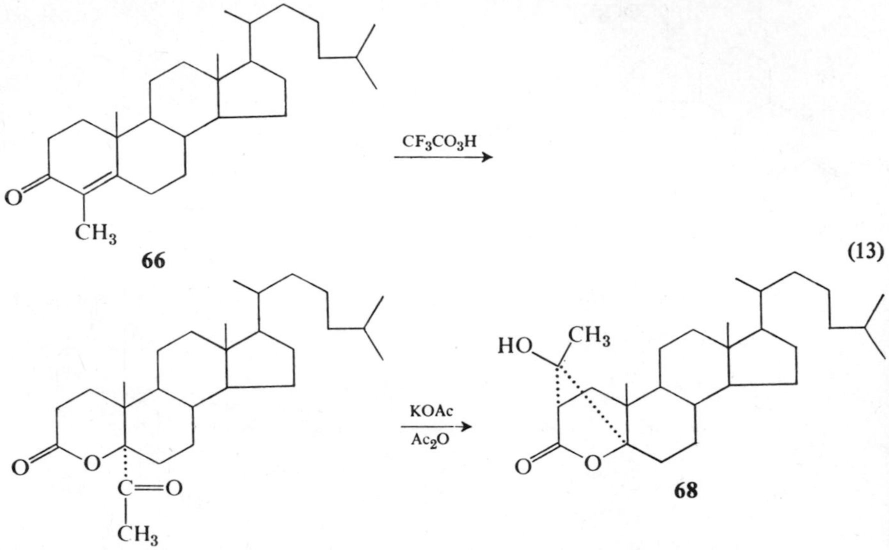

While Pinhey and Schaffner (23) suggested that an epoxy oxepin lactone such as **69** might be an intermediate in the transformation of **62** to **63**, they

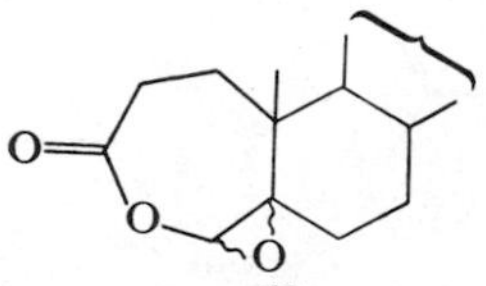

were unsuccessful in finding it in their reaction mixtures. This, of course, contrasts with Caspi's isolation (21) of such a product when perbenzoic acid was employed.

A considerable amount of work has been devoted to studies (26) concerning eburicoic acid and certain hormone analogs derived therefrom. During the course of these studies it was noted (27) that incubation of eburicoic acid **70** with the fungus *Gloremella fusarioides* resulted in ring cleavage (Eq. 14) to give the 3,4-seco acid **71**, which possessed some antibacterial activity.

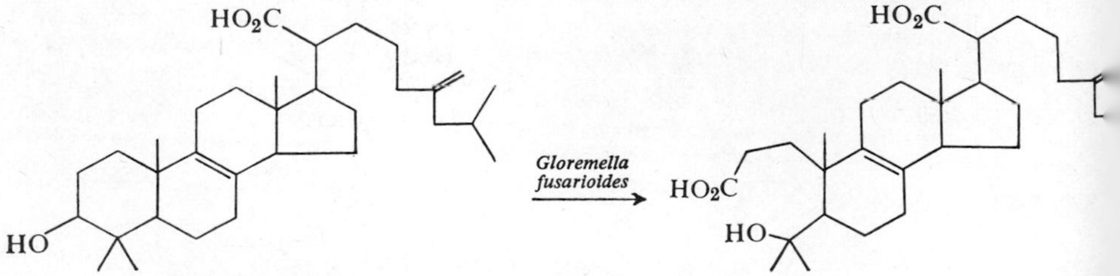

In view of this observation it was decided to examine (26) the antibiotic properties of a number of related 3,4-seco acids and, for this purpose, preparation via a Baeyer-Villiger reaction seemed ideally suited. Indeed, the authors commented that the biochemical transformation in the conversion of **70** to **71** could be considered the microbiological equivalent of a Baeyer-Villiger reaction; numerous examples of such biochemical conversions already appear in the literature (28–30).

Since it was necessary to be absolutely sure that the structure of the Baeyer-Villiger products could be ascertained with ease and complete certainty, the work of Hara and his associates (15) was repeated. It will be recalled that these investigators had been among those who were able to show that Burckhardt and Reichstein's original reaction mixtures (5) were composed of both the C-2,3 and C-3,4 lactones.

Following the lengthy separation procedure of Hara (15), it was found that 5α-cholestan-3-one upon treatment with *m*-chloroperbenzoic acid yielded a product which, after hydrolysis, esterification, oxidation, and further esterification, gave two vapor phase chromatography peaks in the ratio of 7:3. Oddly in contrast to this facile separation, which clearly shows that both isomers are present, the vapor phase chromatography method proved incapable of effecting any fractionation of the original lactones or their derived hydroxy esters.

When these studies were extended to the 4-substituted series, the results were almost completely stereospecific. Thus when the 4α and 4β-methyl-5α-cholestan-3-ones, **72** and **75**, were treated with *m*-chloroperbenzoic acid (Eqs. 15 and 16), 82% and 75% yields of the lactones **73** and **76**, respectively,

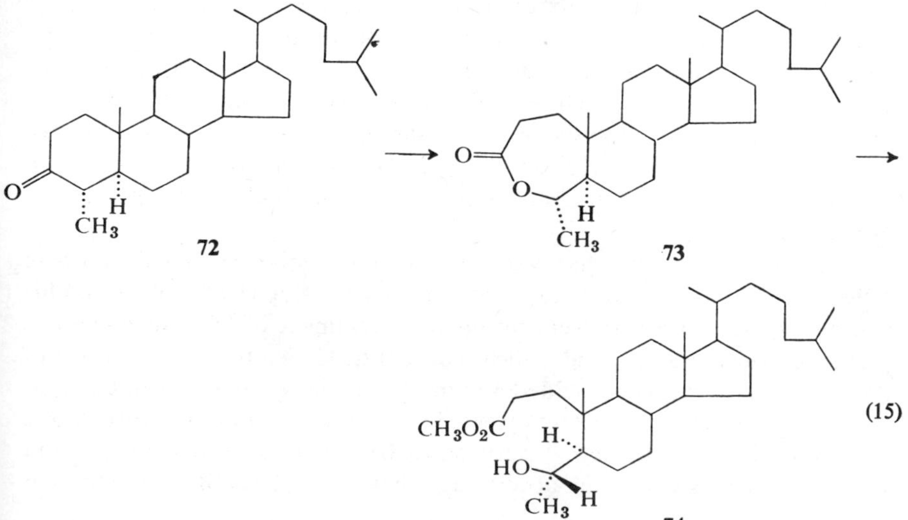

O H CH_3 **72** → O O H CH_3 **73** →

CH_3O_2C H HO CH_3 H **74** (15)

were obtained. Conversion to the hydroxy esters **74** and **77** followed by vapor phase chromatographic analysis showed that in both cases only a single lactone had been produced. On this basis the authors concluded that these reactions were almost completely stereospecific and involved migration of the more highly substituted C-4 carbon atom.

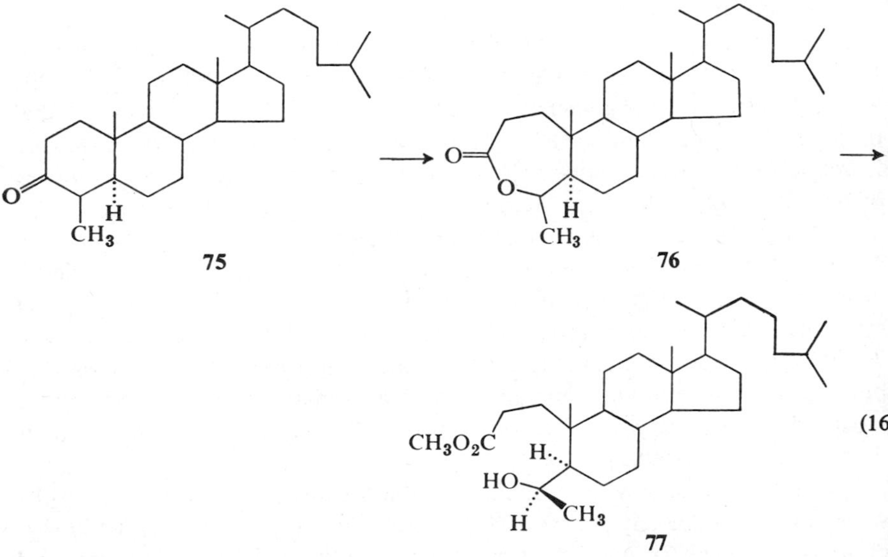

O CH_3 H **75** → O O CH_3 H **76** → CH_3O_2C H HO H CH_3 **77** (16)

In the case of 4,4-dimethyl-5α-cholestan-3-one **78**, the reaction with *m*-chloroperbenzoic acid in a mixture of dichloromethane–chloroform (Eq. 17) was found (26) to be considerably slower, but again yielded only a single lactone **79**. As a proof of structure, this product was converted in two steps to the hydroxy ester **80**, which proved to be resistant to Jones-type oxidation. On the other hand, pyrolysis of **79** at slightly more than 200°C gave the unsaturated acid **81**. Later work showed (31) that the methylene group in this product was derived in almost equal parts from both the 4α-methyl and 4β-methyl groups.

The strong influences that various reaction media may have on these types of oxidative processes are exemplified by the studies of Holker and his colleagues (32). These workers found that treatment of **78** with *m*-chloroperbenzoic acid in acetic acid–sulfuric acid (Eq. 17) led to the loss of one of the C-4 methyl groups and yielded 4*a*α-methyl-4-oxa-*A*-homo-5α-cholestan-3-one **82**. Moreover, when **79** was treated with either sulfuric or hydrochloric acid in acetic acid, the methylene product **81** could be isolated in excellent yield. This fact, as well as the observation that either **79** or **81** upon treatment

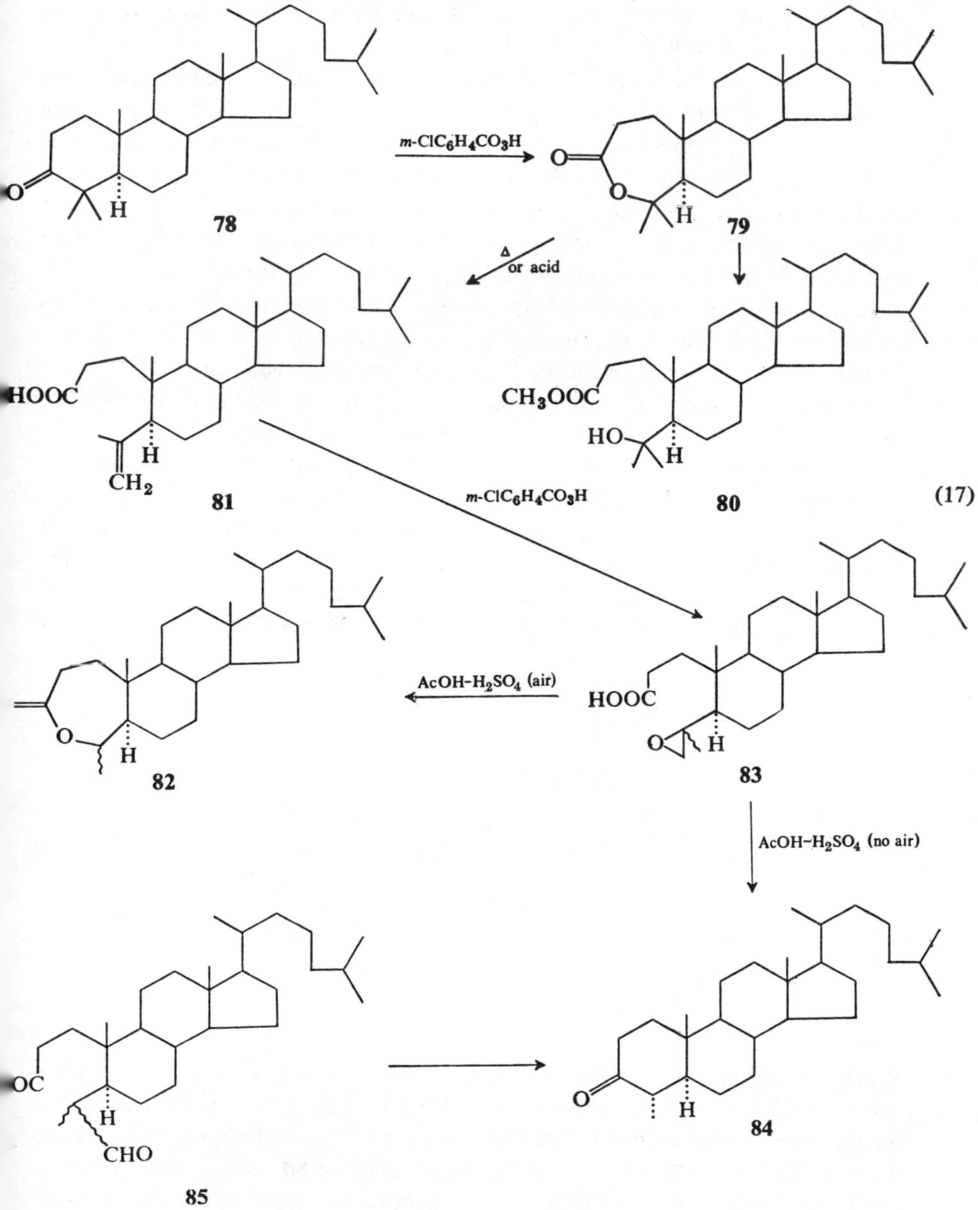

O 78 $m\text{-}ClC_6H_4CO_3H$ O O 79

Δ or acid

HOOC H CH_2 81

CH_3OOC HO H 80

$m\text{-}ClC_6H_4CO_3H$

(17)

O H 82

$AcOH-H_2SO_4$ (air)

HOOC O H 83

$AcOH-H_2SO_4$ (no air)

OC H CHO 85

O H 84

with peracid led to the 4aα-methyl-*A*-homo product **82**, as noted above, made it appear probable that both **79** and **81** were intermediates in the conversion of **78** to **82**.

It was additionally observed (32) that when the epoxy acid **83** derived from **81** was treated with sulfuric acid and acetic acid in the absence of oxygen (Eq. 17), two products could be isolated. One of these was identical with 4α-methyl-5α-cholestan-3-one **84**, whereas the other was indirectly shown to be the formyl derivative **85**. When this same reaction was run in the presence of air, the major product was the 4aα-methyl-*A*-homo derivative **82**. Smaller quantities of the 4α-methyl structure **84**, as well as the formyl derivative **85**, were also isolated. Holker's group has suggested that the formation of **84** arises from **85** by means of an acid-catalyzed Claisen condensation, followed by deformylation of the resultant β-oxoaldehyde intermediate.

Related reactions involving the loss of either or both of the methyl groups present at C-4 in **78** have just recently been reported (33).

Alternate routes to ring-A lactones which have not followed the original procedures employing various peracids are exemplified by the following investigations. In 1957 a general method for preparing hydroperoxide derivatives of steroidal ketones was described (34). Thus, progesterone **86**, when treated with concentrated hydrogen peroxide in an inert solvent (Eq. 18), provided the bishydroperoxide **87**. In view of the observation by

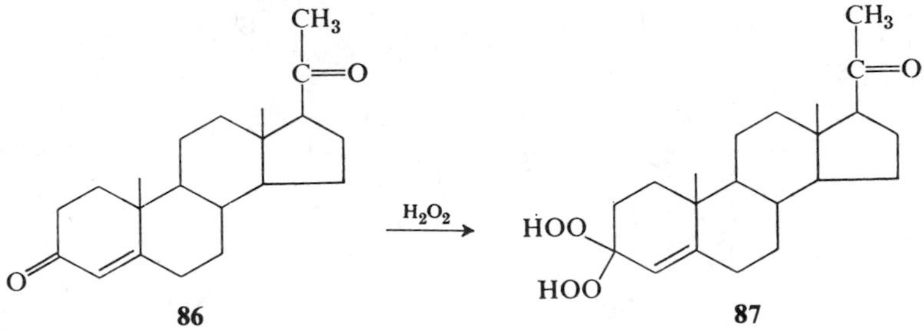

CH$_3$ C=O ... $\xrightarrow{H_2O_2}$ HOO HOO ... CH$_3$ C=O (18)

86 **87**

Wittig and Pieper (35) that fluorene-9,9-bishydroperoxide could be converted to a lactone under acetylation conditions, Velluz et al. (36) investigated the treatment of **87** with succinic anhydride at 0° in pyridine. By these means there was obtained in 85% yield the enolic lactone **88**, whose structure was firmly established (Eq. 19) by its permanganate oxidation to the keto acid **89** and by its hydrolysis to the aldehydic acid **90**. This latter compound was further characterized as the dicarboxylic acid **91**. A similar reaction sequence was reported (36) in the desoxycorticosterone series.

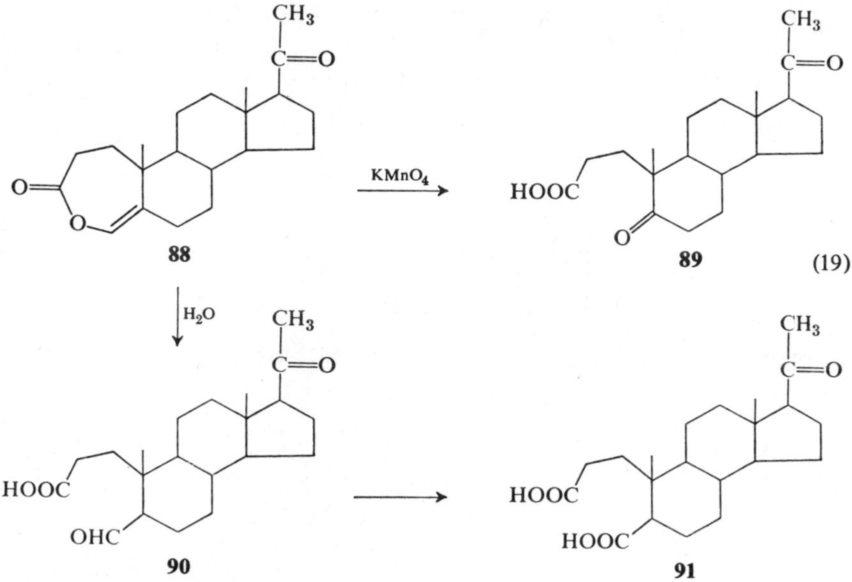

(19)

Another path to oxepin synthesis (Eq. 20) was employed by Shoppee and Sly (37, 38) who, in connection with an investigation of the Beckman rearrangement of 3-oximino steroids, wished to establish the structure of the products in an unequivocal fashion. Thus when the oxime of 5α-cholestan-3-one **92** was treated with thionyl chloride, the 3-aza lactam **93** was obtained as the major reaction product. Acid hydrolysis provided the amino acid salt **94**, which after treatment with nitrous acid yielded the 3-oxa lactone **96** by way of the intermediary hydroxy acid **95**.

The structure proof of **96**, differentiating it from its 4-oxa isomer, was based on indirect evidence. First of all the authors (37) noted that the constants for their lactone **96** were different from those reported by Burckhardt and Reichstein (5) for the authentic 4-oxa lactone. On the other hand their constants were in good agreement with those of Nes and Lettré (39), who had proposed the 3-oxa lactone structure **96** for an oxidation product they had obtained on lithium aluminum hydride reduction of dicarboxylic acid **97** to diol **98** followed by chromium trioxide oxidation (Eq. 21). Under these conditions, in addition to the dibasic acid **97**, a neutral product was isolated and tentatively assigned structure **99**. While one cannot accept this as a complete proof on the basis of the data presented, there seems little question but that Nes and Lettré were correct, thereby allowing the structural claim of Shoppee and Sly (37) in the 5α-series.

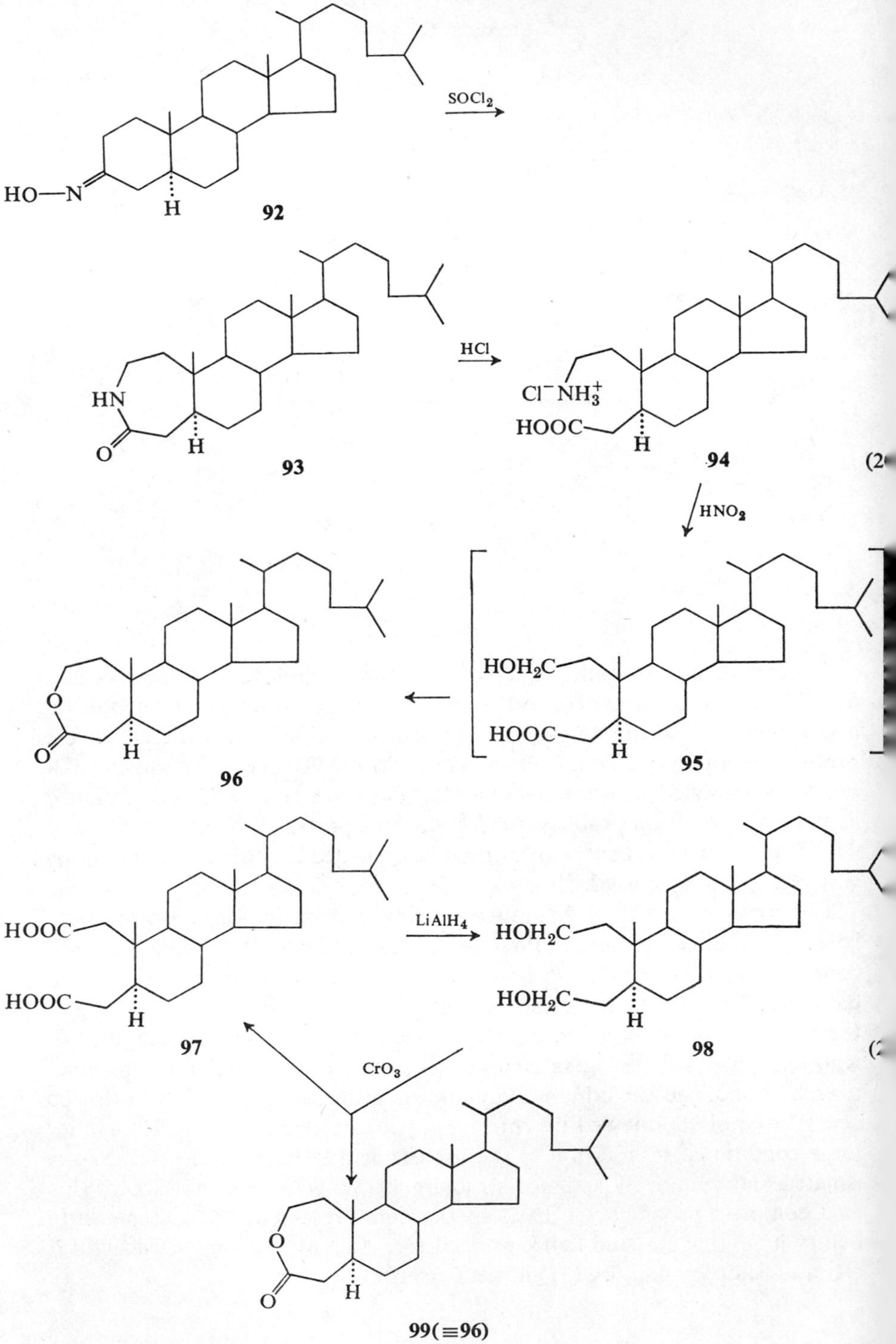

HO—N

H

92

$SOCl_2$

HN

O

H

93

HCl

Cl^- NH_3^+

HOOC

H

94

(2

HNO_2

O

O

H

96

HOH_2C

HOOC

H

95

HOOC

HOOC

H

97

$LiAlH_4$

HOH_2C

HOH_2C

H

98

(2

CrO_3

O

O

H

99(≡96)

Recently Shoppee et al. (40) reviewed the earlier work on the Beckman rearrangement of 5α-cholestan-3-one oxime and concluded that besides the 3-aza lactam, the isomeric 4-aza lactam is also produced. Contrastingly, in the 5α-pregnane series, Nace and Watterson (38) have found that rearrangement leads to the 4-aza lactam in 93% yield. It has been suggested (38) that these differences arise from the fact that the 3-oxime of 5α-pregnane-3,20-dione exists completely in the *anti*-configuration, whereas the 5α-cholestan-3-one oxime employed was a mixture of both the *syn* and *anti*-forms.

In the 5β-series (37), 5β-cholestan-3-one oxime **100** was isomerized to the lactam **101** and then further transformed to the lactone **102** via the nitrous acid reaction sequence (Eq. 22). A rigid proof for the structure of the lactone

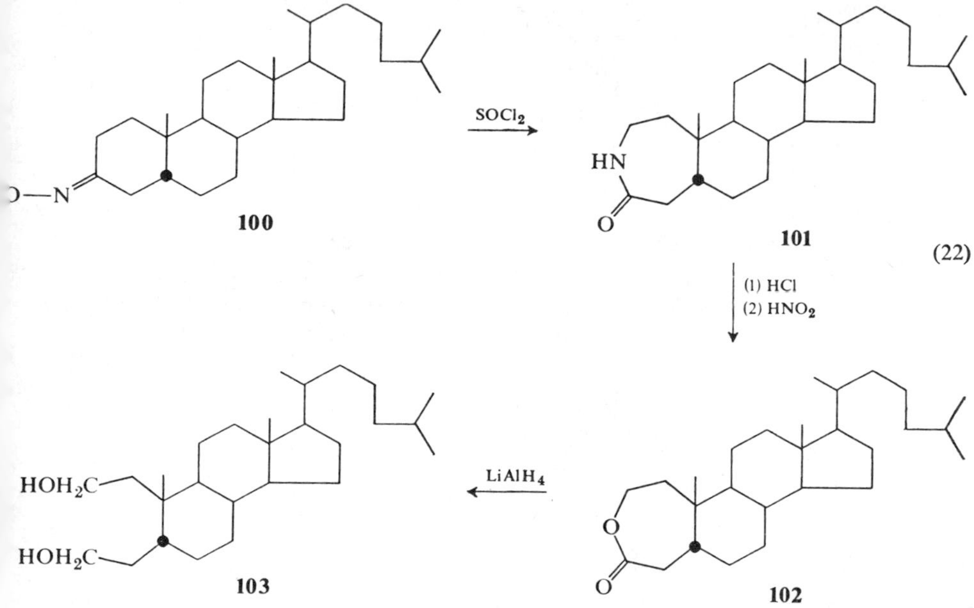

(22)

produced in this instance was obtained by lithium aluminum hydride reduction of **102** to the diol **103**, which was also prepared by an independent and unequivocal route. The fact that this compound was a 2,3-seco derivative left no question but that the inserted oxygen of the lactone **102** was located at the 3-position.

Similar studies to those noted above have also been carried out by Shoppee and Krüger in the androstane series (41).

Caspi et al. (42) have observed ring A oxepin formation during treatment of dihydrotestosterone acetate **104** with hydrogen peroxide and selenium dioxide in *tert*-butanol solvent (Eq. 23). In addition to the lactone **105**, whose yield is minimal by this method, the *A*-nor acid **106** was also formed, along with a series of other closely related compounds.

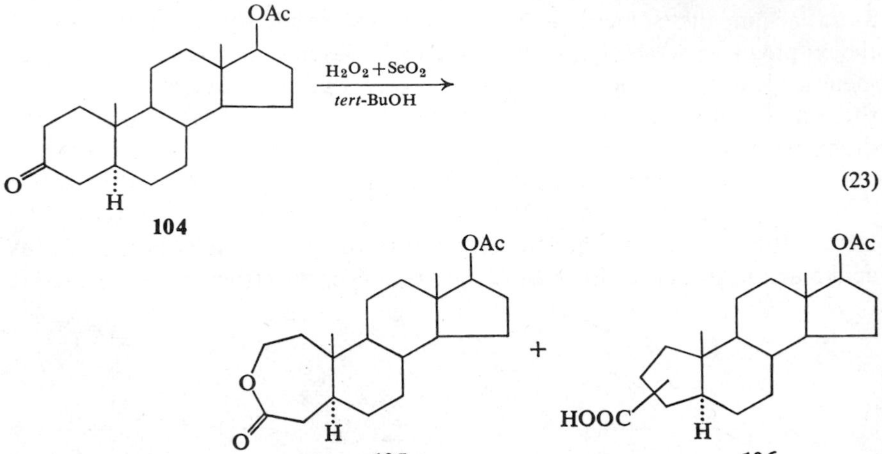

$\xrightarrow[\textit{tert}\text{-BuOH}]{H_2O_2 + SeO_2}$

104 (23)

105 + **106**

The formation of an oxepin involving the C-11 hydroxyl of a steroid has been observed (43, 44) by the sequence shown in Eq. 24. Ozonolysis of the ene-one system in **107** yielded the diketohydroxy acid **108**. Whereas treatment of this compound with sodium acetate and acetic anhydride had been

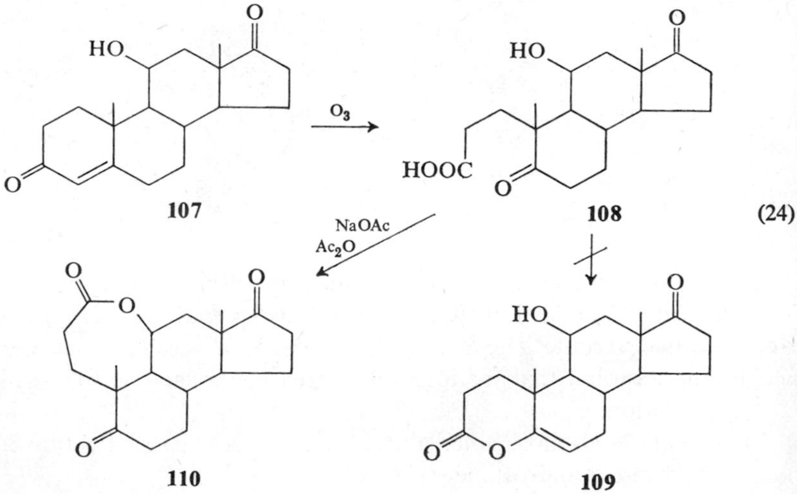

107 $\xrightarrow{O_3}$ **108** (24)

108 $\xrightarrow{NaOAc,\ Ac_2O}$ **110**

108 ↛ **109**

expected to yield the enol lactone **109**, the spectral properties of the resulting product were not in accord with such a proposal. Indeed, the ir spectrum gave no evidence for the presence of hydroxyl or acetate bands; moreover, the product itself was stable to the action of chromium trioxide. On the basis of these data and the fact that the nmr spectrum showed a single equatorial proton attached to an oxygen-bearing carbon atom, the authors proposed the seven-membered lactone structure **110** for their product, noting that it must be thermodynamically more stable than the expected enol lactone **109**.

Another unusual steroidal oxepin system was discovered (45) when the *B*-nor steroid **111** was treated in benzene with lead tetraacetate in the presence of calcium carbonate (Eq. 25). While previous experience led the authors to expect either oxidation of the alcohol function accompanied by bond rupture or ether formation by attack of activated hydroxyl on the adjacent methyl groups, neither event took place. Instead, two products could be isolated in yields of 40 and 25%, respectively, and their structures were determined in the following manner.

The first product, **112**, was a diacetate as shown by its nmr spectrum which possessed 6 hydrogens located at 123 Hz. Hydrolysis provided the diol **114**, and in both this derivative and its precursor, ir bands were present at 1675 and 1670 cm^{-1}, suggesting that the presence of an enol ether was possible. When the hydrolysis product **114** was treated with manganese dioxide (Eq. 25), the compound underwent allylic oxidation to yield the ene-one **115**.

Examination of the second product **113** indicated that upon pyrolysis it underwent acetate elimination, albeit in low yield, to yield the diacetate **112** described above. Supplementary information tending to establish the structure of this material as a triacetate was forthcoming from its nmr spectrum, which exhibited 3 singlets corresponding to acetates at 122, 123, and 126 Hz. Examination of the ir, on the other hand, showed it to be void of hydroxyl absorption, although strong ester carbonyl absorption was present at 1734 cm^{-1}. Treatment of this substance with potassium carbonate (Eq. 25) led to the triol **116**, whose nmr spectrum was consistent with two secondary and one tertiary alcohol functions. Indeed, this spectral observation was confirmed when it was noted that treatment of **116** with acetic anhydride pyridine gave a diacetate corresponding to structure **117**.

Investigations (46) involving 6β-methyl-5α,6α-oxido steroids have also led to oxepin derivatives. Thus treatment of **118** with *p*-toluenesulfonic acid in dry benzene at reflux temperature for 30 min. (Scheme II) yielded the 6-methylene compound **119** as the main product. Epoxidation of this substance led to the oxirane **120**, which upon treatment with dilute sulfuric acid did not give the expected vicinal glycol, but rather a carbonyl bearing product shown to be the *A*-homo derivative **121**. A somewhat similar expansion-contraction of the A,B rings to an *A*-homo-*B*-nor system has been observed by Kirk and Petrow (47). In addition to the new ketone in **121**, there was also present a new primary hydroxyl function which was assigned

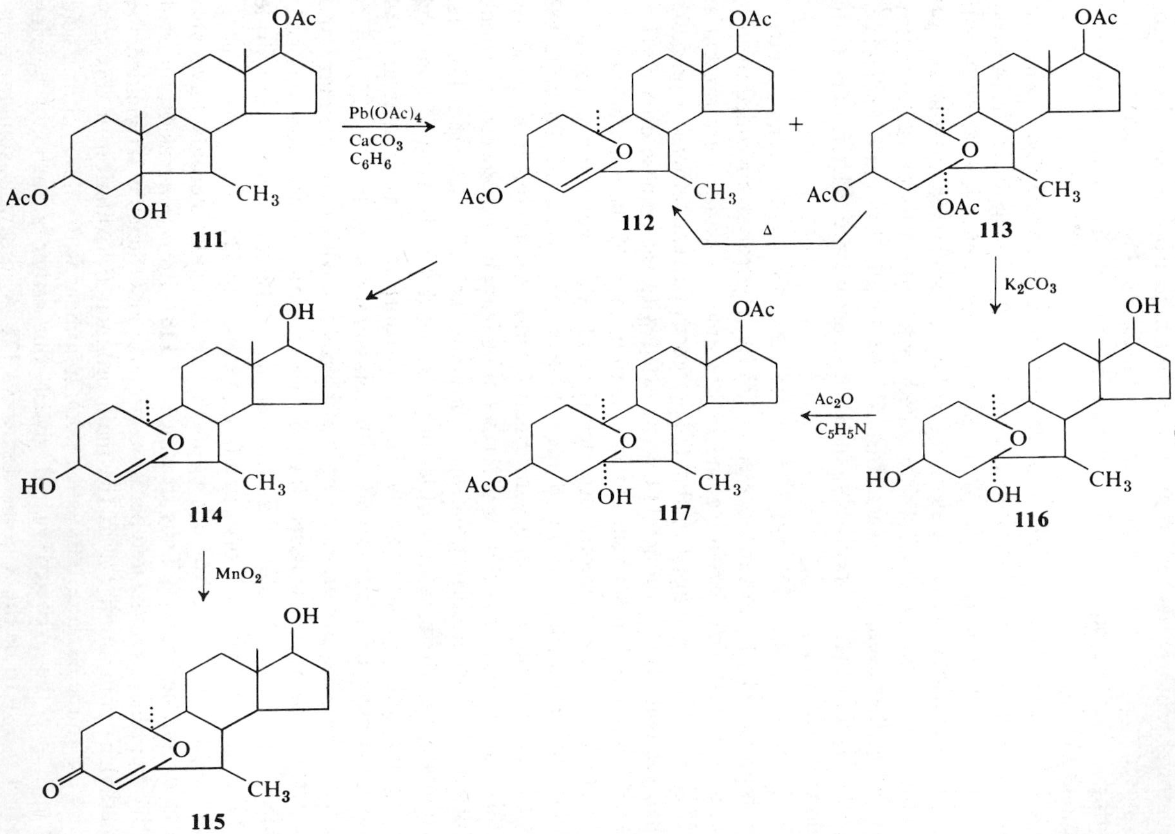

$$\mathbf{111} \xrightarrow[\text{CaCO}_3,\ \text{C}_6\text{H}_6]{\text{Pb(OAc)}_4} \mathbf{112} + \mathbf{113}$$

$$\mathbf{113} \xrightarrow{\Delta} \mathbf{112}$$

$$\mathbf{112} \longrightarrow \mathbf{114} \xrightarrow{\text{MnO}_2} \mathbf{115}$$

$$\mathbf{113} \xrightarrow{\text{K}_2\text{CO}_3} \mathbf{116} \xrightarrow[\text{C}_5\text{H}_5\text{N}]{\text{Ac}_2\text{O}} \mathbf{117} \qquad (25)$$

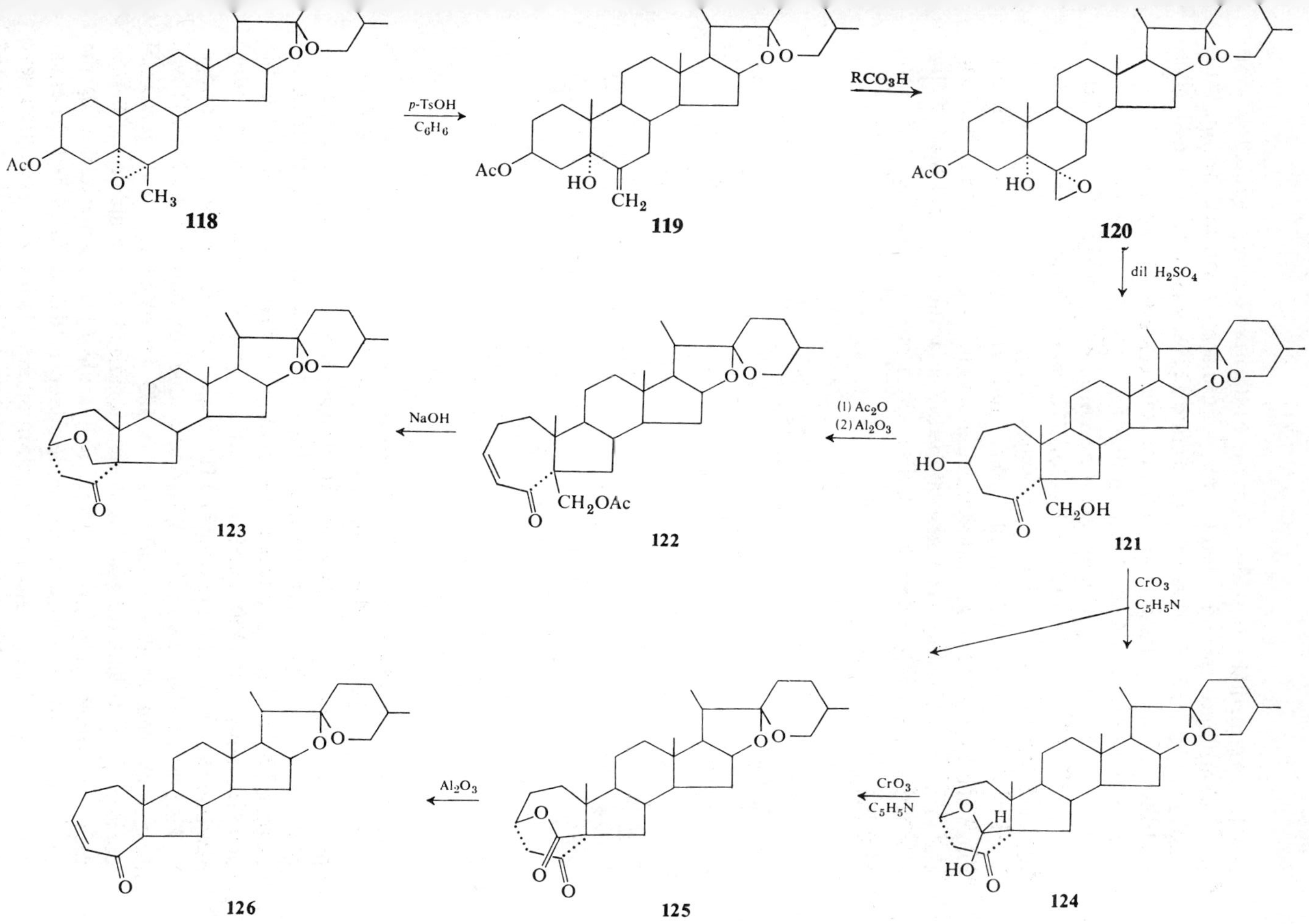

118 $\xrightarrow{p\text{-TsOH},\ C_6H_6}$ 119 $\xrightarrow{RCO_3H}$ 120 $\xrightarrow{\text{dil } H_2SO_4}$ 121

121 $\xrightarrow{(1)\ Ac_2O\ (2)\ Al_2O_3}$ 122 $\xrightarrow{NaOH}$ 123

121 $\xrightarrow{CrO_3,\ C_5H_5N}$ 124 $\xrightarrow{CrO_3,\ C_5H_5N}$ 125 $\xrightarrow{Al_2O_3}$ 126

121 $\xrightarrow{CrO_3,\ C_5H_5N}$ 125

AcO, O, CH_3 — 118

AcO, HO, CH_2 — 119

AcO, HO, O — 120

HO, O, CH_2OH — 121

O, CH_2OAc — 122

O, O — 123

O, H, HO, O — 124

O, O, O — 125

O — 126

Scheme II

as a hydroxymethyl group located at the bridgehead corresponding to C-5 in the unrearranged steroid system. Following diacetylation and heating with basic alumina, the unsaturated keto acetate **122** was isolated. Upon alkaline hydrolysis, this substance then afforded the bridged oxepin **123** by means of an intramolecular Michael-type addition.

When **121** was treated with pyridine–chromium trioxide (Scheme II), a mixture of the hemiacetal **124** and lactone **125** was obtained. The relationship of these substances was readily established by the interconversion of **124** to **125** on further mild pyridine–chromium trioxide oxidation. Finally, when the lactone **125** was passed through a column of alumina in benzene, the unsaturated ketone **126** was obtained. The authors have proposed that this latter reaction is a decarboxylation which proceeds via the mechanism shown.

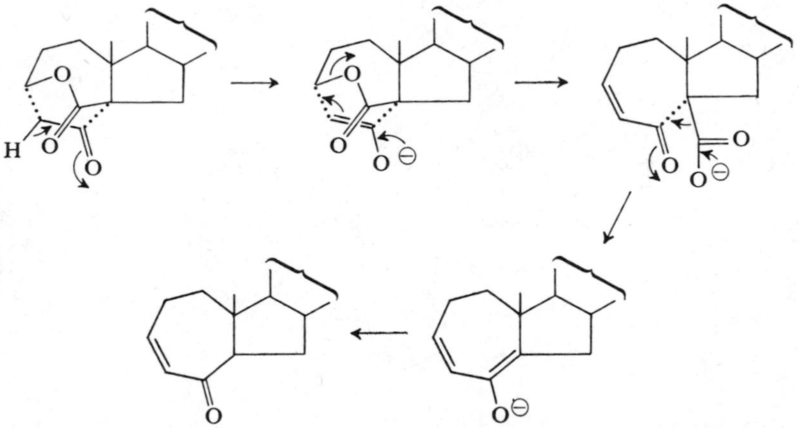

A series of unusual oxepin structures (Eq. 26) have resulted from studies carried out by Swiss workers (48). Thus when **127** was treated with hydrogen peroxide and catalytic quantities of osmium tetroxide, the glycol **128** was obtained in good yield. Irradiation of this substance with uv light led to the lactone **129**, which was reduced with sodium borohydride to a mixture of the epimeric hemiacetals **130**. Treatment with anhydrous copper sulfate in boiling acetone then provided the oxepin **131** which contains a 1α,4α-oxide bridge.

To prepare the isomeric 1β,4β-oxide bridge compound (Eq. 27) **132** was first converted to the methoxy acetal **133** by treatment with methanol containing hydrochloric acid, and then oxidized to the aldehyde **134** by means of dimethyl sulfoxide–acetic anhydride (49). Epimerization of the axial formyl group with sodium methoxide in methanol furnished the equatorial 1α-formyl derivative **135**, which upon acetylation and reduction gave **136**. Treatment of this with aqueous acetic acid then yielded the oxepin **137**

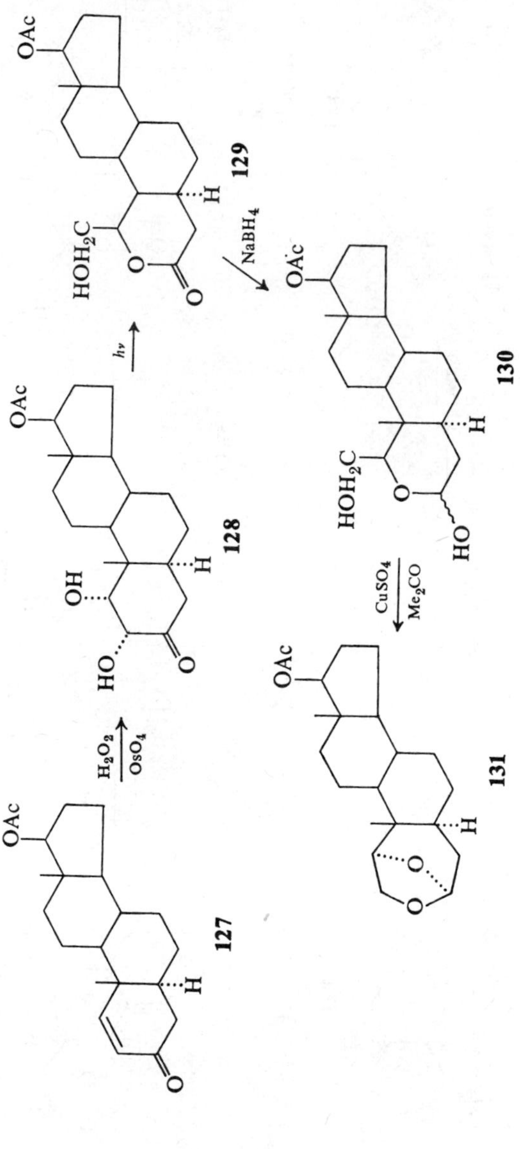

127 $\xrightarrow[OsO_4]{H_2O_2}$ 128 $\xrightarrow{h\nu}$ 129 $\xrightarrow{NaBH_4}$ 130 $\xrightarrow[Me_2CO]{CuSO_4}$ 131 (26)

HOH_2C OAc HO H **132** $\xrightarrow[HCl]{MeOH}$ HOH_2C OAc CH_3O H **133** $\xrightarrow[Ac_2O]{DMSO}$ OHC OAc CH_3O H **134**

NaOMe (27)

OHC OH CH_3O H **135** $\longrightarrow$ HOH_2C OAc CH_3O H **136** $\xrightarrow{AcOH}$ O O OAc H **137**

bearing the 1β,4β-oxide bridge. Similar structures were also prepared (48) in the 5β-series.

D. Ring B Oxepins

The action of perbenzoic acid on the 5α-*C*-6 ketone **138** has been found by Fonken and Miles (50, 51) to provide the 6-oxa-7-keto-*B*-homo steroid lactone **139** in good yield (Eq. 28). Furthermore, as these authors were able to prove, the newly introduced oxygen occupied the 5β position relative to ring A, a fact which accords with the stereospecific retention of configuration (52) ordinarily associated with this reaction.

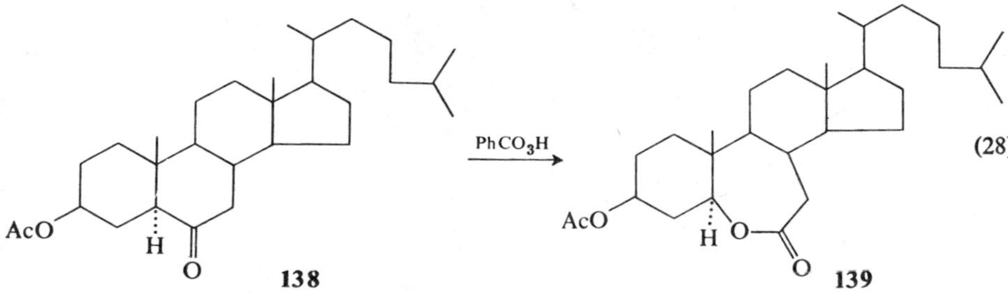

AcO H O **138** $\xrightarrow{PhCO_3H}$ AcO H O O **139** (28)

As previously noted, Heymann and Fieser (14) examined the configuration of C-7 in cholest-5-ene-3β,7-diol by lactone formation. A somewhat related earlier but unsuccessful investigation was due to Heusser et al. (53). In an attempt at a direct proof for the β-configuration of the C-3 hydroxyl in cholesterol, these authors first converted cholesterol by a number of steps to 3β-hydroxy-5α-cholest-7-ene **140**. While oxidation of this compound (Eq. 29)

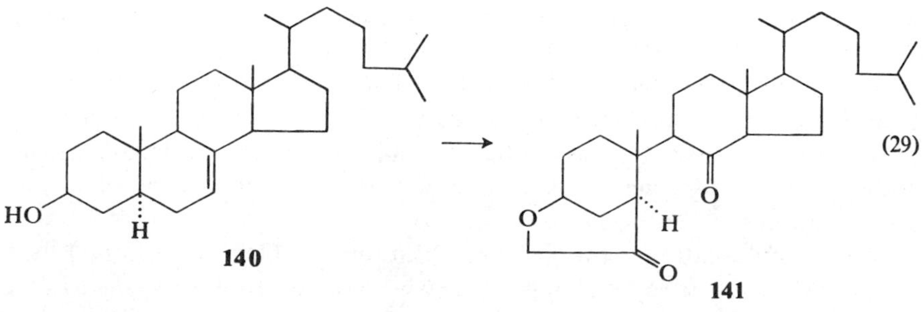

HO H **140** $\longrightarrow$ O H O O **141** (29)

did in fact yield a lactone, presumably with structure **141**, it could not be obtained crystalline; for that reason, an alternate approach was investigated. When the C-7 ketone **142** was treated under Baeyer-Villiger conditions with perbenzoic acid as reported by Burckhardt and Reichstein (5), the corresponding 7a-oxa lactone **143** was obtained in good yield (Eq. 30). The use of 40% peracetic acid has also been reported (54) in this reaction. Proof (53) for the position of the oxygen insertion was forthcoming from hydrolysis,

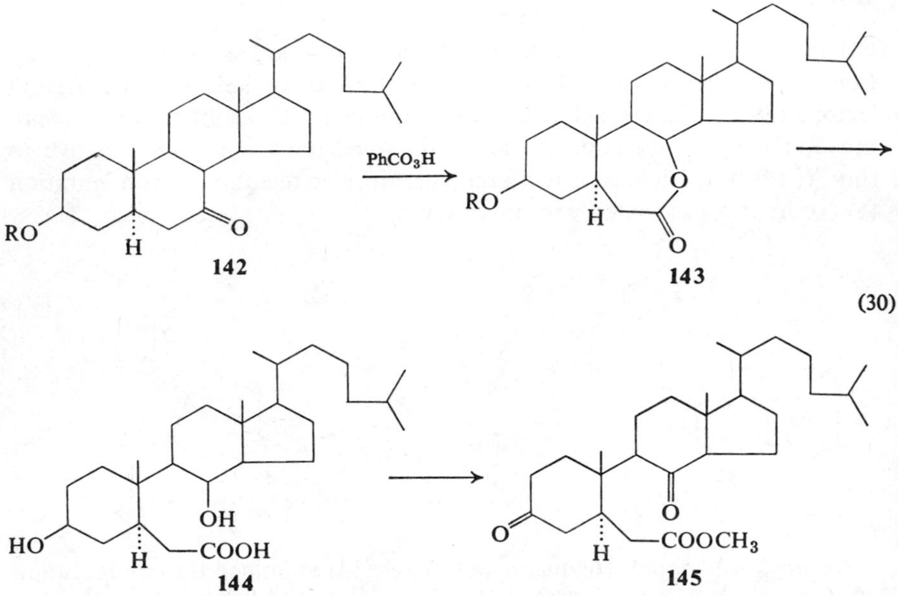

RO H O **142** $\xrightarrow{PhCO_3H}$ RO H O O **143** $\longrightarrow$ (30)

HO H OH COOH **144** $\longrightarrow$ O H O $COOCH_3$ **145**

which gave the dihydroxy acid **144**; since esterification of this compound followed by oxidation yielded the diketo ester **145** rather than a C-6,7 dibasic acid, it was clear that oxygen insertion had occurred between C-7 and C-8.

In view of these results, in a further attempt (53) to prepare the crystalline keto lactone **141**, the above sequence was executed on the corresponding 3-acetate of **142**. When the resulting lactone **143** was subjected to hydrolysis, it was found impossible to effect lactone hydrolysis without simultaneous hydrolysis of the corresponding 3-acetate. Similar failures in obtaining selective lactone hydrolysis were observed in investigations of the 3-benzoate and 3-pivalate; for these reasons this approach for the proof of the β-configuration at C-3 was abandoned.

A successful solution to this C-3 configurational problem was accomplished at about the same time by Shoppee (55) who was able to prepare the 3α and 3β-hydroxy compounds **147** and **150**. Upon treatment of these epimers with acetic anhydride (Eqs. 31 and 32), the 3β isomer **147** was found to give the lactone **148**, whereas the 3α isomers yielded only the unsaturated anhydride **151**.

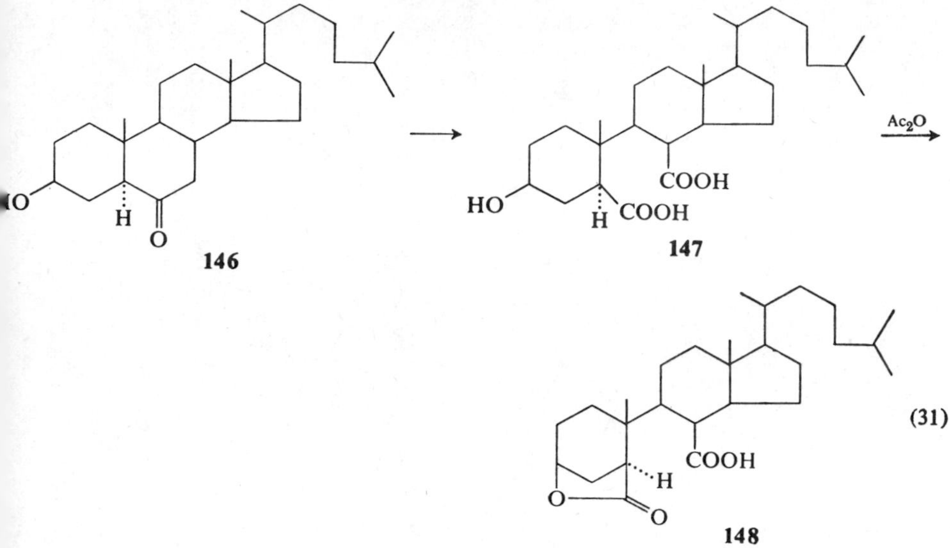

146 → 147 $\xrightarrow{Ac_2O}$ 148 (31)

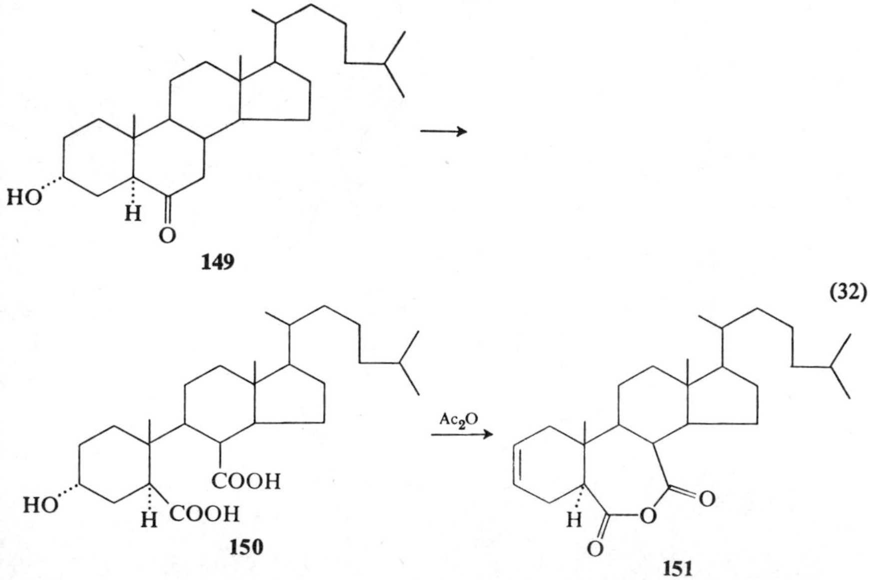

149 → 150 $\xrightarrow{Ac_2O}$ 151 (32)

The preparation of anhydrides in ring B has played a part in Fieser's (56) studies on exhaustive dichromate oxidation of cholestanol and in Gates' and Wallis' proof of structure (57) for iso steroids, but there is no need to review these preparations here. Indeed, no further attention will be devoted to steroid anhydrides unless they have served as intermediates in the preparation of other compounds.

$COOCH_3$ O OAc **152**

several steps

$COOCH_3$ O **153**

two steps

$COOCH_3$ O O **154**

(33)

$COOCH_3$ O O O **155**

$COOCH_3$ NH **156**

Such a use appears in the work of Hara (58) who employed an anhydride intermediate to prepare 7-aza-*B*-homo derivatives in the bile acid series. Wolff-Kishner reduction of the diketo-7α-acetoxy bile acid derivative **152** gave 7α-hydroxycholanic acid, which was then esterified and oxidized (Eq. 33) to the corresponding ring-B ketone **153**. Following perbenzoic acid treatment, it was shown that the main attack on the 7-ketone led primarily to a 7a-oxa rather than a 7-oxa lactone.

On the other hand, nitric acid oxidation of the free acid of **153** was found to proceed between C-6 and C-7, thereby providing the corresponding dicarboxylic acid derivative, which was readily converted to the anhydride **155**. Treatment of this substance with urea, followed by hydride reduction, then gave the 7-aza-*B*-homo azepine derivative **156**. It may be of interest to note that Beckman rearrangement of the oxime derivative of **153** provided a 7a-aza-amide.

An unusual approach to syntheses of steroidal oxepin ethers has been carried out by Lettré and his co-workers (59, 60). As an example (Eq. 34) ozonization of 3-desoxycholesterol **157A** in ethanol–chloroform solution led to a molozonide which upon reduction with lithium aluminum hydride gave the 5,6-seco-dihydroxy compound **158A**. Treatment of this product with mesyl chloride in pyridine for 3 days at 0°C gave directly the ring B oxepin **159A**, presumably by displacement of the intermediary primary mesylate. A similar reaction sequence was carried out on the molozonide of

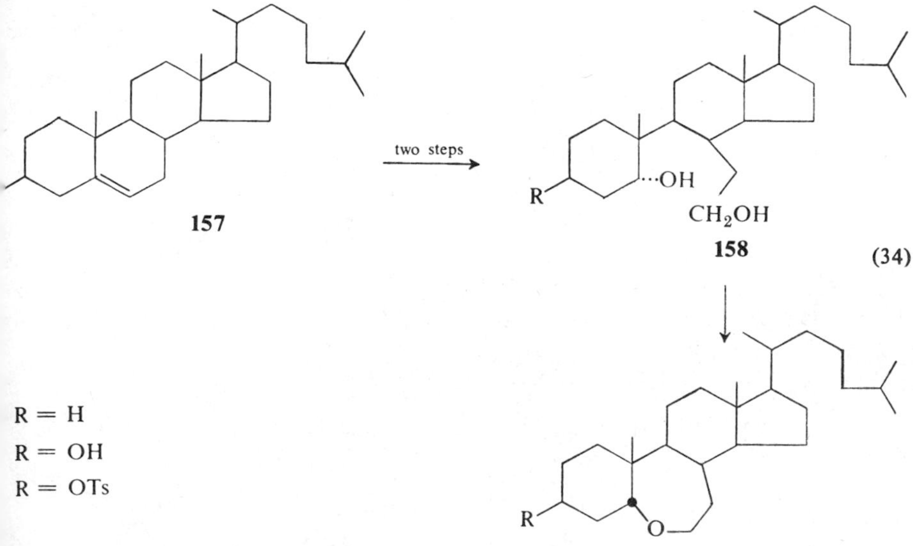

two steps

157

R ···OH CH_2OH

158

(34)

R O

159

R = H

R = OH

R = OTs

cholesterol **157B**; after treatment with lithium aluminum hydride, tosylation led to the 3-tosyl ring B oxepin **159C**.

Still another route to oxepin ethers is due to Pettit and his co-workers (61). By methods already discussed, these workers prepared 7a-oxa-7-keto-*B*-homo-3β-hydroxycholesterol **160** and subjected it to reduction with a reagent prepared from boron trifluoride and lithium aluminum hydride. Following acetylation (Eq. 35), there was obtained the 7a-oxepin ether **161**. The boron trifluoride–lithium aluminum hydride mixture had already been observed to effect reduction of certain lactones and esters to ethers (62).

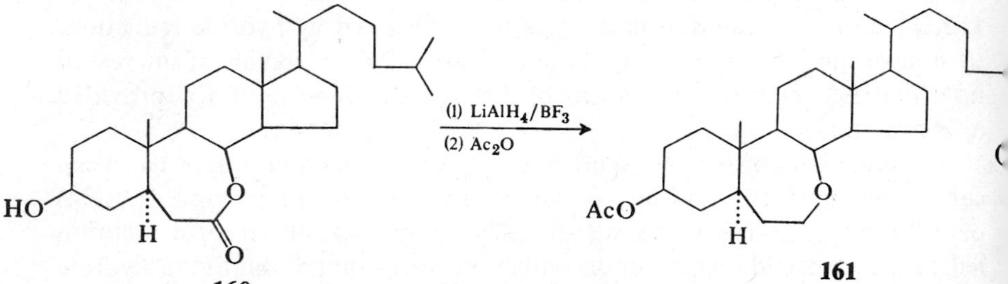

HO H O O **160** (1) $LiAlH_4/BF_3$ (2) Ac_2O AcO H O **161**

It is of interest that when a similar reduction of the ring A lactone **162** was attempted (61) (Eq. 36), there was obtained the seco dialcohol **164** instead of the desired cyclic ether **163**. These different results in rings B and A

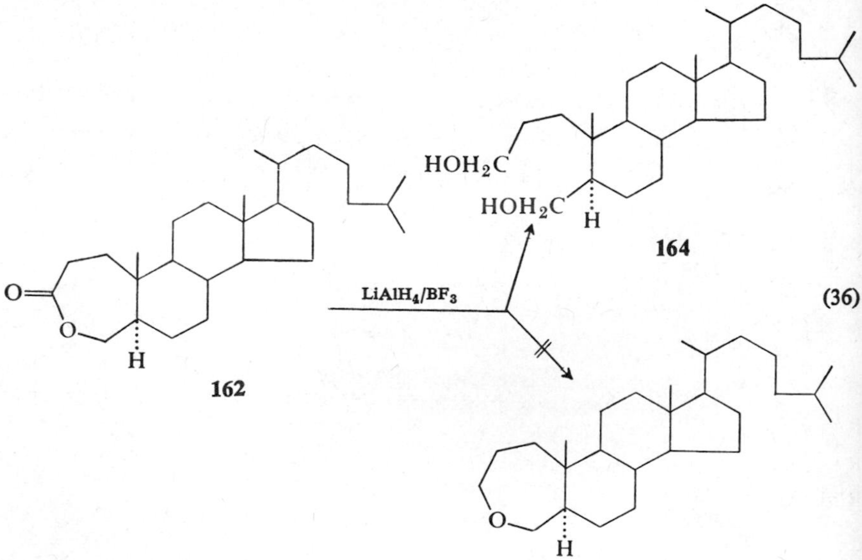

O O H **162** $LiAlH_4/BF_3$ HOH_2C HOH_2C H **164** (36) O H **163**

suggested to the authors that branching adjacent to the alkyl oxygen segment of the ester was required in order to favor ether formation over cleavage.

As noted earlier (42), investigations on the action of hydrogen peroxide in the presence of catalytic amounts of selenium dioxide have been carried out on dihydrotestosterone acetate. When these same conditions were applied (63) to testosterone propionate **165**, the ring B lactone **166** was isolated in good yields (Eq. 37). Alkaline hydrolysis of this product then gave the corresponding 17-alcohol, which was isolated as the five-membered ring A lactone **167**.

An alternate route to **166** involved ozonolysis of testosterone propionate (Eq. 37). After decomposition of the molozonide with water and partitioning of the residue with a weak base, the lactol **168** was obtained. Treatment of this substance with hydrogen peroxide in glacial acetic acid then provided the seven-membered ring lactone **166** described above.

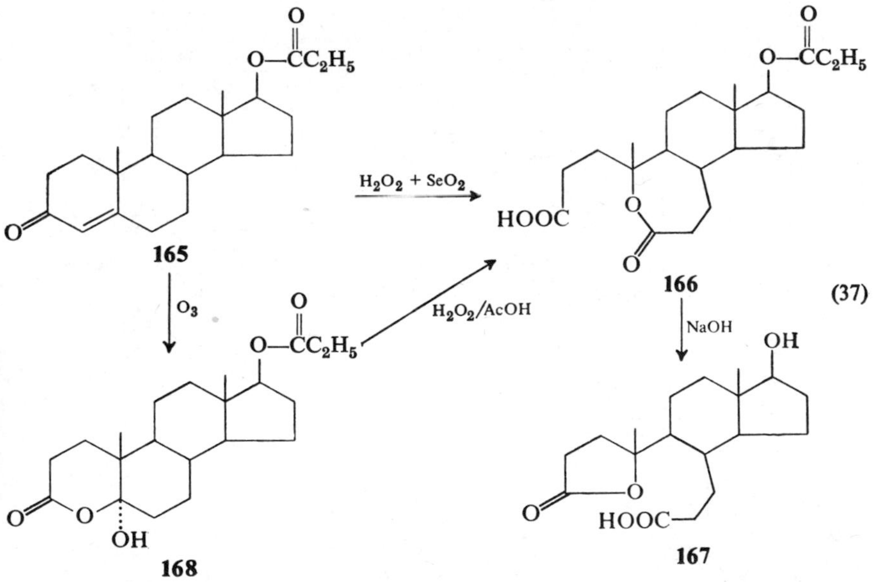

(37)

Starting in 1948, Sorm and co-workers (64) published several papers on *B*-nor steroids prepared as shown in Eq. 38. Chromium trioxide oxidation of cholesterol acetate **169** led in moderate yield to the keto acid **170**. After treatment with benzoyl chloride in pyridine, a product was obtained for which the seven-membered ring enol lactone structure **171** was proposed. At the time this structure seemed reasonable since the substance did not possess a titratable function nor did it react with diazomethane. Moreover, its uv spectrum above 220 mμ resembled that of vinyl acetate. Upon pyrolysis,

AcO 169

CrO_3

AcO O COOH 170

BzCl
C_5H_5N

AcO O O 171

(1) NaOH
(2) CH_2N_2

MeOH

$H_2/Pd\text{-}SrCO_3$

Δ

(38)

HO O $COOCH_3$ 173

COOH 174

AcO 172

carbon dioxide was liberated, thereby providing a new product which was correctly assigned the *B*-nor structure **172**.

Dauben and Fonken (65) also investigated the structures of the acetoxy keto acid **170** and the benzoyl chloride reaction product **171**. With **170**, alkaline hydrolysis and diazomethane treatment led to the hydroxy keto ester **173**, whose ir spectrum clearly showed the presence of hydroxyl and carbomethoxy groups as well as a six-membered ring ketone function. It was also observed that the lactone **171** could be converted by methanolysis to this same product **173**. Aside from these reactions, the authors felt that support for the seven-membered ring enolic lactone structure **171** could be obtained by its transformation to the acid **174**. This conversion seemed especially likely to be significant since it was known that enolic lactones and allylic esters, functional groups that were presumably present in **171**, undergo ready hydrogenolysis. Indeed, hydrogenolysis with palladium-on-strontium carbonate did give rise to the acid **174**, thereby appearing to add further confirmation to the structure assignment in **171** originally proposed by Sorm (64).

Several years later the real nature of the benzoyl chloride reaction product was questioned by Rull and Ourisson (66). These authors, who were working in the androstane series (Eq. 39), noted that the benzoyl chloride reaction product derived from **175** exhibited an ir spectrum band at 1825 cm^{-1}, and that while this was not compatible with a seven-membered ring lactone, it could be considered indicative of a β-lactone structure (Eq. 39). Moreover, they recalled that pyrolysis with loss of carbon dioxide is a general reaction

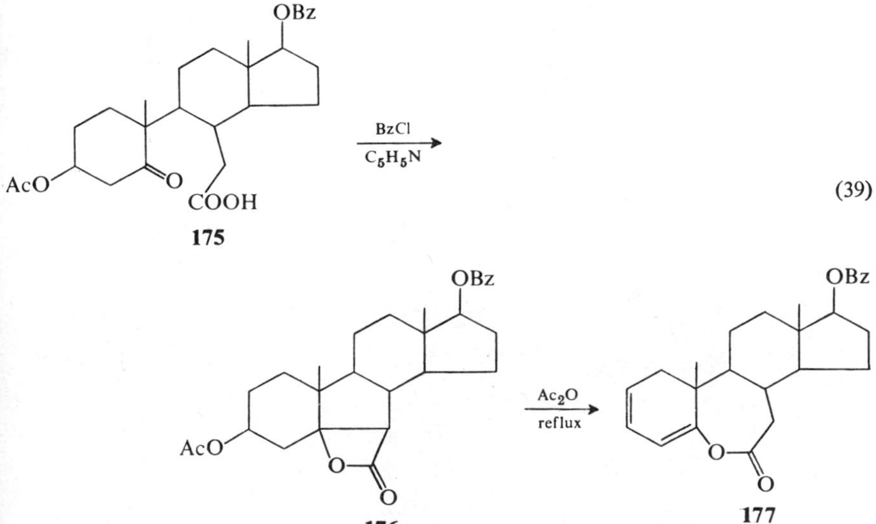

OBz

AcO O COOH

175

$\xrightarrow[C_5H_5N]{BzCl}$ (39)

OBz

AcO O O

176

$\xrightarrow[\text{reflux}]{Ac_2O}$

OBz

O O

177

AcO O O 178

AcO O O $^{\ominus}OCH_3$ 179

HO O OCH_3 $O^{\ominus}$ 180

(40)

HO O $^{\ominus}$ $COOCH_3$ 181

HO O $COOCH_3$ 182

of β-lactones not ordinarily shared by seven-membered ring cyclic lactones. In addition to these observations, the authors noted that when the lactone and acetic anhydride were heated at reflux temperature, the product isolated was the diene lactone **177**, whose structure was clearly evident from its uv absorption bands at 230 and 268 mμ.

A solution to the problem was finally presented by Boswell, Dauben, Ourisson, and Rull (67), who reported that the uv spectrum of the benzoyl chloride reaction product possessed a very weak band at 205 mμ; such a weak absorption was incompatible with the enolic double bond required if the seven-membered lactone was the correct structure. The final and conclusive proof, however, rested on the nmr spectrum of **176**. While this did not exhibit the vinyl proton that must be present if the seven-membered lactone was the correct structure, it was in accord with the β-lactone formulation **176**. This, coupled with the earlier observations, left no question but that Sorm's lactone product (64) was represented by the β-lactone structure **176**.

The methanolysis reaction involved in the conversion of the β-lactone **178** to the hydroxy keto ester **182** may be explained (67) on the basis of the mechanistic pathway shown in Eq. 40.

E. Ring C Oxepins

Relatively little work has been carried out involving the preparation of ring C oxepins. Rothman et al. (68) have observed that treatment of the bile acid ester **183** with perbenzoic acid (Eq. 41) provides a lactone which has a structure corresponding to **184**. Moreover, they were able to carry out a similar reaction (68) by treating hecogenin acetate **185** with perbenzoic acid, or hydrogen peroxide using sulfuric acid catalysis. Under either of these

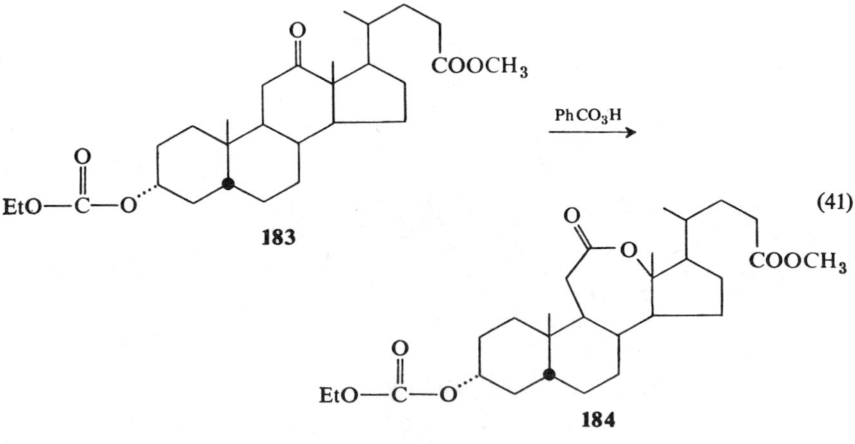

EtO—C(=O)—O ... COOCH$_3$ **183** $\xrightarrow{PhCO_3H}$ EtO—C(=O)—O ... COOCH$_3$ **184** (41)

Scheme III

conditions, after 12-day reaction times, the lactone **186** was obtained in high yield (Scheme III). The position of the inserted oxygen atom was apparent when it was noted that alkaline hydrolysis provided a new compound **187** wherein the extra oxygen was present as a tertiary hydroxyl function. On the other hand, when a similar Baeyer-Villiger conversion was attempted on a C-11 keto steroid, no reaction at all was observed. Undoubtedly this observation is a further indication of the relatively nonreactive nature of C-11 ketones as compared with similar functions located in other positions of the steroid molecule.

Conversion of the ring C lactone **186** to the corresponding 16-ene **188** was carried out (69) by the usual Marker procedure for degradation of the sapogenin side chain. As would be expected, the 16-ene was susceptible to either reduction or epoxidation (Scheme III), thereby yielding **189** and **191**, respectively. It should be noted, however, that this particular epoxidation step could only be effected with hydrogen peroxide and not with perbenzoic acid. In addition to the lactone epoxide **191**, some of the corresponding ring C seco hydroxy acid **190** was also produced under the alkaline conditions. By treatment with acetic anhydride, this latter material was readily recyclized to the lactone epoxide **191**.

In a separate study Mueller et al. (70) also prepared a compound with constants identical to those described (69) for the epoxy lactone **191**. While Mueller's group did not prove the structure of their product, its assignment seems reasonable from the mode of preparation (Scheme III) which involved perbenzoic acid treatment of the 12-keto pregnene derivative **192**, and from the fact that the product contained one more oxygen atom than could be accounted for by epoxide formation alone.

In an additional paper Rothman and Wall (71) mentioned that difficulties were encountered in attempting to open the epoxide ring of the epoxy lactone **191**. For that reason they turned their attention to an alternate approach which involved the lactonization of a more advanced intermediate. This effort was based on perbenzoic acid treatment of **193**, which then provided the ring C lactone **194** bearing a completed cortical side chain (Eq. 42). In

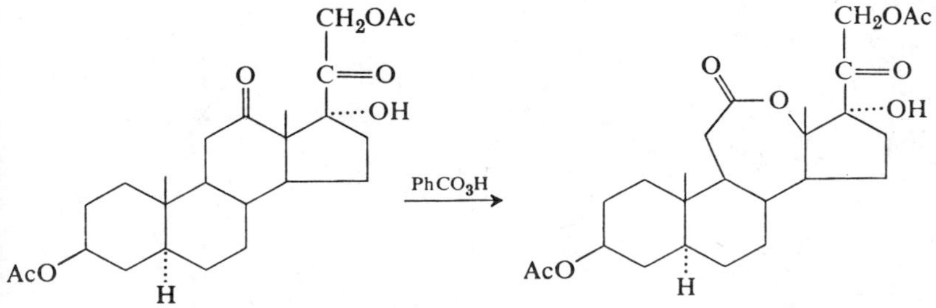

CH_2OAc C=O OH O AcO H **193** $\xrightarrow{PhCO_3H}$ CH_2OAc C=O OH O O AcO H **194** (42)

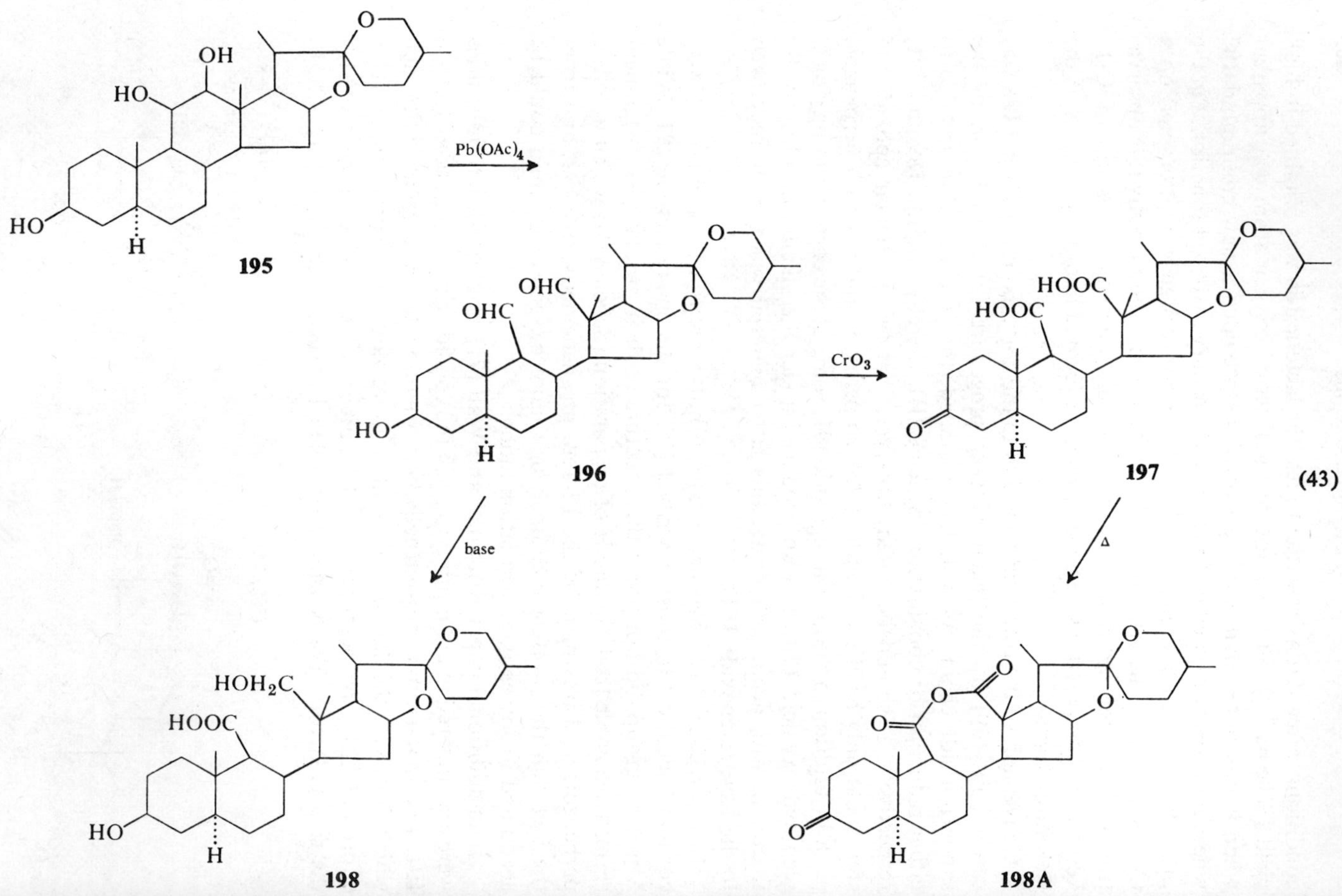

195

$\xrightarrow{Pb(OAc)_4}$

196

$\xrightarrow{CrO_3}$

197

(43)

base → 198

Δ → 198A

198

Ac_2O

199

200 (44)

H_2

201

(1) NaOH
(2) CrO_3

202

DDQ

203

CH_3 C=O O O O AcO H

204

HBr

CH_3 C=O OH Br O O AcO H

205

Ni

CH_3 C=O OH O O AcO H

206

NaOH

CH_3 C=O OH O O HO H

207

Br_2

CH_2Br C=O OH O O HO H

208

AcOK

CH_2OAc C=O OH O O HO H

209

O_3

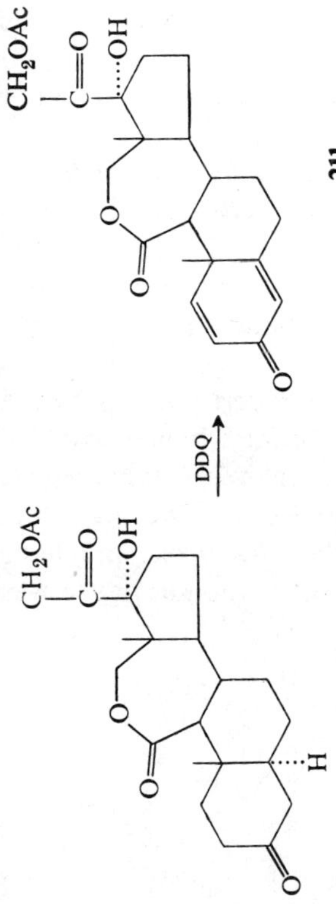

210 $\xrightarrow{\text{DDQ}}$ **211**

Scheme IV

this case no side-chain attack was observed, a fact that stands in contrast to the work of Leeds et al. (72), who observed that treatment of a 17-hydroxy-20-ketone with perbenzoic acid led to the corresponding C-17 ketone.

In 1962 a paper was published (73) reporting the preparation (Eq. 45) of a cortical hormone analog possessing a cyclic seven-membered ring lactone structure in ring C. This preparation started (74) from the 11β,12β-dihydroxy sapogenin **195** which was treated with lead tetraacetate, thereby providing the 11,12-seco-dialdehyde **196**. The structure of this compound was established by its oxidation to the ketodicarboxylic acid **197**, as well as the resultant anhydride **198A** derived therefrom.

In an attempt to effect an aldol condensation of the bisaldehyde **196** with base, even under anhydrous conditions, there was instead observed (Eq. 43) a Canizzaro-type reaction which provided the ring C seco hydroxy acid **198**. Upon acetylation (Eq. 44), this substance was converted in very high yields to the seven-membered ring lactone **199**; indeed, this reaction was so rapid that the preparation of the corresponding carboxylic acid diacetate could not be effected.

Following the Marker procedure, it was found possible to degrade the spiroketal side chain of the ring C lactone **199** to its 16-ene derivative **200**, which was then reduced (Eq. 44) to provide the saturated pregnane **201**. Hydrolysis of the acetate function at C-3 followed by oxidation and dicyano-dichloroquinone introduction of the C-1,4-double bonds (75) gave the dehydroprogesterone analog **203**.

Further work showed that epoxidation of **200** with *tert*-butyl peroxide proceeded smoothly to yield the epoxide **204**. In contrast to the work of Rothman and Wall (71) with a positional ring C isomer of **204**, this epoxide could be opened with hydrogen bromide to give the bromohydrin **205**, which was then converted with Raney nickel to the 17-hydroxy pregnane derivative **206**. Following the steps shown in Scheme IV, this substance was transformed to the prednisone acetate analog **211**.

F. Tables

TABLE VII-1. Melting Points and Rotations for Most of the Compounds Discussed in the Chapter as Well as Certain Additional Selected Products

Compound	Empirical formula	mp (°C)	$[\alpha]_D$	*c*	Solvent	Temperature (°C)	Refs.
9	$C_{25}H_{38}O_5$	122–123	+38.1 ± 4°	0.761	Acetone	17.5	4
a	$C_{25}H_{40}O_4$	126–128	+50.0 ± 4°	0.560	Acetone	18	5
b	$C_{27}H_{42}O_6$	187–189					5
16	$C_{19}H_{30}O_2$	185.5–186	−37.8 ± 3°	0.793	Chloroform	17	9
c	$C_{19}H_{28}O_2$	172.5–173	−26.6 ± 3°	0.677	Chloroform	17	9
d	$C_{26}H_{40}O_4$	195.5–196	−36 ± 3°	0.788	Chloroform	16	9
20	$C_{19}H_{30}O_2$	142–143	+33.4 ± 2°	1.165	Chloroform	20	10
e	$C_{19}H_{28}O_2$	127–128	+43.8 ± 2°	1.098	Chloroform	22	10
f	$C_{26}H_{34}O_4$	208.5–209.5	−20.8 ± 2°	1.103	Chloroform	20	10
23	$C_{26}H_{40}O_4$	230–232	+26°[g]	1%	Chloroform	20	11
24	$C_{26}H_{40}O_4$	187–188	±0°[g]	1%	Chloroform	20	11
37	$C_{27}H_{46}O_2$	186–187	+1.2 ± 2°	0.785	Acetone	15	5, 37
42	$C_{21}H_{30}O_3$	227–228.5	+115 ± 5°	0.139	Chloroform	22	16
44	$C_{23}H_{32}O_5$	178.5–179	+188 ± 4°	1.083	Chloroform	21	16
45	$C_{23}H_{32}O_6$	181.5–182.5	+109 ± 4°	1.025	Chloroform	23	16
46	$C_{21}H_{30}O_5$	211.5–212.5	+55 ± 3°	0.546	Chloroform	22	16
47	$C_{21}H_{28}O_5$	206.5–207.5	+215 ± 4°	0.653	Chloroform	21	16

[a] 4-Oxa-*A*-homo-3-oxo-5β-cholan-24-oic acid methyl ester.
[b] 12-Acetoxy-4-oxa-*A*-homo-3-oxo-5β-cholan-24-oic methyl ester.
[c] 4-Oxa-*A*-homo-3-oxo-5α-androst-16-ene.
[d] 17β-Hydroxy-4-oxa-*A*-homo-5α-androstan-3-one cyclohexanoate.
[e] 4-Oxa-*A*-homo-3-oxo-5β-androst-16-ene.
[f] 17β-Hydroxy-4-oxa-*A*-homo-5β-androstan-3-one benzoate.
[g] Measured at the *J* line of mercury.

TABLE VII-1.—*(contd.)*

Compound	Empirical formula	mp (°C)	$[\alpha]_D$	*c*	Solvent	Temperature (°C)	Refs.
49, R = Br,	$C_{27}H_{45}O_2Br$	192–193	−4.5°	1.0	Chloroform		20
49, R = Cl,	$C_{27}H_{45}O_2Cl$	194–195	−4.3°	1.1	Chloroform		20
h	$C_{27}H_{46}O_2$	157–158	+49.2 ± 2°	1.444	Acetone	20	5, 37
54	$C_{19}H_{26}O_4$	224–226					21
55	$C_{19}H_{26}O_5$	250–251					21
56	$C_{19}H_{28}O_4$	275–276					21
73	$C_{28}H_{48}O_2$	191–191.5	−6°	0.66	Chloroform	25	26
76	$C_{28}H_{48}O_2$	177–179*r*	0 ± 1.6°*r*	—	Chloroform	23	26
79	$C_{29}H_{50}O_2$	123–124	−1.7°	—	Chloroform	20	26
88	$C_{21}H_{30}O_3$	126	+59 ± 2°	1%	Acetone	—	36
i	$C_{23}H_{32}O_5$	162	+72.5°	1%	Acetone	—	36
96	$C_{27}H_{46}O_2$	181–183	+46°	0.7	Chloroform	—	37, 39
102	$C_{27}H_{46}O_2$	182–183	+50°	1.0	Chloroform	—	37
105	$C_{21}H_{32}O_4$	218–223					42
j	$C_{21}H_{32}O_4$	217–224					42
k	$C_{21}H_{32}O_4$	201–204					42
l	$C_{21}H_{32}O_4$	240–242					42
110	$C_{18}H_{24}O_4$	196–198					43
112	$C_{23}H_{34}O_5$	130–131	−255°			25	45
113	$C_{25}H_{38}O_7$	161–171 dec	−38.5°			25	45
114	$C_{19}H_{30}O_3$	169–171	−274°			25	45
115	$C_{19}H_{28}O_3$	154–154.5					45
116	$C_{19}H_{32}O_4$	156.5–158	−25°			25	45
117	$C_{23}H_{36}O_6$	172–172.5	−5°			25	45
123	$C_{28}H_{42}O_4$	225–227	−113.2°	0.68	Chloroform	?	46
124	$C_{28}H_{42}O_5$	254–256	−155.6°	1.015	Chloroform	28	46
125	$C_{28}H_{40}O_5$	280–281	−94.5°	0.73	Chloroform	22	46
131	$C_{21}H_{32}O_4$	163–164	+85°	0.39	Chloroform	—	48
137	$C_{21}H_{32}O_4$	134	−56.5°	0.49	Chloroform	—	48

[m]	$C_{19}H_{30}O_3$	192–193	+35°	0.25	Chloroform	—	48
[n]	$C_{21}H_{32}O_4$	102	−68°	0.25	Chloroform	—	48
139	$C_{29}H_{48}O_4$	162–163	+28°	—	Chloroform	20	50
[o]	$C_{27}H_{46}O_2$	143–144	+28°	—	Chloroform	20	50
[p]	$C_{27}H_{46}O_3$	139–141	+32°	—	Chloroform	20	50
[q]	$C_{19}H_{24}O_2$	115–118	−330.1°	0.42	Dioxane	20	51
143, R = H	$C_{27}H_{46}O_3$	184–185	−11.2°	0.784	Chloroform	18	53
143, R = Ac	$C_{29}H_{48}O_4$	148–149	−14.6°	0.818	Chloroform	—	53
[s]	$C_{34}H_{50}O_4$	160–161	−5.2°	0.727	Chloroform	20	53
[t]	$C_{32}H_{54}O_4$	188–190	−12.1°	1.155	Chloroform	21	53
151	$C_{27}H_{42}O_3$	134					55
159A	$C_{27}H_{48}O$	45–46					59
159C	$C_{34}H_{54}O_4S$	129					60
161	$C_{29}H_{50}O_3$	103–104	+42.2°	1.2	Chloroform	22	61
166	$C_{21}H_{32}O_6$	154–155					63
177	$C_{26}H_{30}O_4$	207–208	−70°[g]	1%	Chloroform		66
184	$C_{28}H_{44}O_7$	105–106.5	+2.95°	—	Chloroform	25	68
186	$C_{29}H_{44}O_6$	292–292.5	−65.1°	—	Chloroform	25	68

[h] 4-Oxa-*A*-homo-5β-cholestan-3-one.
[i] 3,20-Dioxo-21-hydroxy-4-oxa-*A*-homo-pregna-4a-ene acetate.
[j] 3-Oxa-*A*-homo-5β-androstan-17β-ol acetate.
[k] 4-Oxa-*A*-homo-5β-androstan-17β-ol acetate.
[l] 4-Oxa-*A*-homo-5α-androstan-17β-ol acetate.
[m] 3-Oxa-*A*-homo-1α,4α-oxido-5β-androstan-17β-ol.
[n] 3-Oxa-*A*-homo-1β,4β-oxido-5β-androstan-17β-ol acetate.
[o] 6-Oxa-*B*-homo-5α-cholestan-7-one.
[p] 3β-Hydroxy-6-oxa-*B*-homo-5α-cholestan-6-one.
[q] 4-Methyl-6-oxa-*B*-homo-7-oxo-estra-1,3,5(10)-triene.
[r] This material still contains 5% of the 4aα isomer.
[s] 3β-Hydroxy-7a-oxa-*B*-homo-5α-cholestan-7-one benzoate.
[t] 3β-Hydroxy-7a-oxa-*B*-homo-5α-cholestan-7-one pivalate.

TABLE VII-1.—(*contd.*)

Compound	Empirical formula	mp (°C)	$[\alpha]_D$	*c*	Solvent	Temperature (°C)	Refs.
188	$C_{23}H_{32}O_5$	203–205[u]	−26°	—	—	25	69
189	$C_{23}H_{34}O_5$	184–189[v]	+60.4°	—	—	25	69
191	$C_{23}H_{32}O_6$	269–270	+17.8°	—	—	25	69, 70
194	$C_{25}H_{36}O_8$	209–220[w]	+13°	—	Chloroform	25	71
198A	$C_{27}H_{38}O_6$	232–235	−33°	—	Chloroform	—	74
199	$C_{29}H_{44}O_6$	293–295	−92°		Chloroform	—	74
200	$C_{23}H_{32}O_5$	220–222	+38°		Chloroform	—	75
201	$C_{23}H_{34}O_5$	159–160	−7°		Chloroform		78
202	$C_{21}H_{30}O_4$	253–255	+21°		Chloroform		78
203	$C_{21}H_{26}O_4$	255–257	+15°		Chloroform		78
204	$C_{23}H_{32}O_6$	281–282	±0°		Chloroform		78
205	$C_{23}H_{33}O_6Br$	205–206	−47°		Chloroform		78
206	$C_{23}H_{34}O_6$	194–195	−77°		Chloroform		78
207	$C_{21}H_{32}O_5$	281–282	−71°		Chloroform		78
208	$C_{21}H_{31}O_5Br$	248–250	−29°		Pyridine		78
209	$C_{23}H_{34}O_7$	112–117°	−32°		Chloroform		78
210	$C_{23}H_{32}O_7$	247–251	−30°		Chloroform		78
211	$C_{23}H_{28}O_7$	251–253	−13°		Chloroform		78
x	$C_{21}H_{26}O_4$	150–151	−219.1°	0.39	Dioxane	20	51

[u] Double melting point; 173.5°C, then 203–205°C.
[v] Double melting point; 172.5°C, then 184–189°C.
[w] Double melting point; 203–205°C, then 209–220°C.
[x] 17β-Hydroxy-4-methyl-6-oxa-*B*-homo-7-oxo-estra-1,3,5(10)-triene acetate.

G. References

1. "IUPAC–IUB Revised Tentative Rules for Nomenclature of Steroids," *J. Org. Chem.*, **34,** 1517 (1969).
2. A. M. Patterson, L. T. Capell, and D. F. Walker, *The Ring Index*, 2nd ed., American Chemical Society, Washington, D.C., 1960.
3. It is a pleasure to acknowledge the kind efforts of Dr. K. L. Loening, Nomenclature Director for Chemical Abstracts Service, in not only providing the correct systematic and preferred names for structures **4–6** but also for his useful comments relative to replacement nomenclature.
4. V. Burckhardt and T. Reichstein, *Helv. Chim. Acta*, **25,** 821 (1942).
5. V. Burckhardt and T. Reichstein, *Helv. Chim. Acta*, **25,** 1434 (1942).
6. J. A. Gardner and W. Godden, *Biochem. J.*, **7,** 588 (1913).
7. J. Boeseken and A. Kremer, *Rec. Trav. Chim. Pays-Bas*, **50,** 827 (1931).
8. R. E. Marker, E. Rohrmann, E. L. Wittle, H. M. Crooks, Jr., and E. M. Jones, *J. Amer. Chem. Soc.*, **62,** 650 (1940).
9. V. Prelog, L. Ruzicka, P. Meister, and P. Wieland, *Helv. Chim. Acta*, **28,** 618 (1945).
10. L. Ruzicka, V. Prelog, and P. Meister, *Helv. Chim. Acta*, **28,** 1651 (1945).
11. T. Rull and G. Ourisson, *Bull. Soc. Chim. Fr.*, **1958,** 1573.
12. W. D. Emmons and A. S. Pagano, *J. Amer. Chem. Soc.*, **77,** 89 (1955).
13. W. D. Emmons and G. B. Lucas, *J. Amer. Chem. Soc.*, **77,** 2287 (1955).
14. H. Heymann and L. F. Fieser, *Helv. Chim. Acta*, **35,** 631 (1952).
15. S. Hara, N. Matsumoto, and M. Takeuchi, *Chem. Ind.* (London), **1962,** 2086.
16. P. Wieland and K. Miescher, *Helv. Chim. Acta*, **32,** 1768 (1949).
17. L. H. Sarett, *J. Amer. Chem. Soc.*, **69,** 2899 (1947).
18. A. Salamon, *Z. Physiol. Chem.*, **272,** 61 (1941).
19. R. E. Marker, *J. Amer. Chem. Soc.*, **62,** 2543 (1940).
20. J. E. Bolliger and J. L. Courtney, *Aust. J. Chem.*, **17,** 440 (1964).
21. E. Caspi, Y. W. Chang, and R. I. Dorfman, *J. Med. Pharm. Chem.*, **5,** 714 (1962).
22. G. R. Pettit and T. R. Kasturi, *J. Org. Chem.*, **26,** 4557 (1961).
23. J. T. Pinhey and K. Schaffner, *Tetrahedron Lett.*, **1965,** 601.
24. See also M. Gorodetsky, N. Danieli, and Y. Mazur, *J. Org. Chem.*, **32,** 760 (1967).
25. W. Reusch and R. LeMahieu, *J. Amer. Chem. Soc.*, **85,** 1669 (1963).
26. D. Rosenthal, A. O. Niedermeyer, and J. Fried, *J. Org. Chem.*, **30,** 510 (1965) and references therein.
27. A. I. Laskin, P. Grabowich, C. de L. Meyers, and J. Fried, *J. Med. Chem.*, **7,** 406 (1964).
28. G. S. Fonken, H. C. Murray, and L. M. Reineke, *J. Amer. Chem. Soc.*, **82,** 5507 (1960).
29. R. L. Prairie and P. Talalay, *Biochemistry*, **2,** 203 (1963).
30. J. Fried, *J. Amer. Chem. Soc.*, **75,** 5764 (1953).
31. D. Rosenthal, *J. Org. Chem.*, **32,** 4084 (1967).
32. J. S. E. Holker, W. R. Jones, and P. J. Ramm, *J. Chem. Soc., C*, **1969,** 357 and an earlier paper noted therein.
33. R. Kazlauskàs, J. T. Pinhey, J. J. H. Simes, and T. G. Watson, *Chem. Commun.*, **1969,** 945.
34. J. Warnant, R. Joly, J. Mathieu, and L. Velluz, *Bull. Soc. Chim. Fr.*, **1957,** 331.
35. G. Wittig and G. Pieper, *Chem. Ber.*, **73,** 295 (1940).
36. L. Velluz, G. Amiard, J. Martel, and J. Warnant, *Bull. Soc. Chim. Fr.*, **1957,** 1484.
37. C. W. Shoppee and J. C. P. Sly, *J. Chem. Soc.*, **1958,** 3458.

38. See also H. R. Nace and A. C. Watterson, Jr., *J. Org. Chem.*, **31,** 2109 (1966).
39. W. R. Nes and H. Lettré, *Justus Liebigs Ann. Chem.*, **598,** 65 (1956).
40. C. W. Shoppee, G. Krüger, and R. N. Mirrington, *J. Chem. Soc.*, **1962,** 1050.
41. C. W. Shoppee and G. Krüger, *J. Chem. Soc.*, **1961,** 3641.
42. E. Caspi, Y. Shimizu, and S. N. Balasubrahmanyam, *Tetrahedron*, **20,** 1271 (1964).
43. E. Caspi, B. T. Kahn, and W. Schmid, *J. Org. Chem.*, **26,** 3894 (1961).
44. See also M. Akhtar, D. H. R. Barton, J. M. Beaton, and A. G. Hortmann, *J. Amer. Chem. Soc.*, **85,** 1512 (1963); T. Williams, P. Philion, J. Lacobelli, and M. Uskoković, *J. Org. Chem.*, **33,** 509 (1968); W. Koch, M. Carson, and R. W. Kierstead, *J. Org. Chem.*, **33,** 1272 (1968).
45. D. Rosenthal, C. F. Lefler, and M. E. Wall, *Tetrahedron Lett.*, **1965,** 3203.
46. W. Z. Chow, D. C. Huang, and Huang-Minlon, *Tetrahedron*, **22,** 1053 (1966).
47. D. N. Kirk and V. Petrow, *J. Chem. Soc.*, **1960,** 4657.
48. G. Eggart and H. Wehrli, *Helv. Chim. Acta*, **50,** 2362 (1967) and references therein.
49. J. D. Albright and L. Goldman, *J. Amer. Chem. Soc.*, **87,** 4214 (1965).
50. G. J. Fonken and H. M. Miles, *J. Org. Chem.*, **28,** 2432 (1963).
51. See also E. Caspi, D. M. Piatak, and P. K. Grover, *J. Chem. Soc., C*, **1966,** 1034.
52. T. F. Gallagher and T. H. Kritchevsky, *J. Amer. Chem. Soc.*, **72,** 882 (1950) and R. B. Turner, *J. Amer. Chem. Soc.*, **72,** 878 (1950).
53. H. Heusser, A. Segre, and Pl. A. Plattner, *Helv. Chim. Acta*, **31,** 1183 (1948).
54. H. Lettré and D. Werner, *Justus Liebigs Ann. Chem.*, **697,** 217 (1966).
55. C. W. Shoppee, *J. Chem. Soc.*, **1948,** 1032.
56. L. F. Fieser, *J. Amer. Chem. Soc.*, **75,** 4386 (1953).
57. S. Gates and E. S. Wallis, *J. Org. Chem.*, **20,** 610 (1955).
58. S. Hara, *Yakugaku Zasshi*, **78,** 1030 (1958); *Chem. Abst.*, **53,** 3273 (1959).
59. H. Lettré and D. Hotz, *Justus Liebigs Ann. Chem.*, **620,** 63 (1959).
60. See also H. Lettré and A. Jahn, *Justus Liebigs Ann. Chem.*, **608,** 43 (1957); H. Lettré and H. Schelling, *Justus Liebigs Ann. Chem.*, **669,** 160 (1963); H. Lettré and R. Pfirrmann, *Justus Liebigs Ann. Chem.*, **656,** 163 (1962).
61. G. R. Pettit, B. Green, T. R. Kasturi, and U. R. Ghatak, *Tetrahedron*, **18,** 953 (1962).
62. G. R. Pettit, U. R. Ghatak, B. Green, T. R. Kasturi, and D. M. Piatak, *J. Org. Chem.*, **26,** 1685 (1961).
63. E. Caspi and S. N. Balasubrahmanyam, *J. Org. Chem.*, **28,** 3383 (1963).
64. F. Sorm and H. Dyková, *Collect. Czech. Chem. Commun.*, **13,** 407 (1948) and J. Joska and F. Sorm, *Collect. Czech. Chem. Commun.*, **23,** 1377 (1958).
65. W. G. Dauben and G. J. Fonken, *J. Amer. Chem. Soc.*, **78,** 4736 (1956).
66. T. Rull and G. Ourisson, *Bull. Soc. Chim. Fr.*, **1958,** 1581.
67. G. A. Boswell, W. G. Dauben, G. Ourisson, and T. Rull, *Bull. Soc. Chim. Fr.*, **1958,** 1598.
68. E. S. Rothman, M. E. Wall, and C. R. Eddy, *J. Amer. Chem. Soc.*, **76,** 527 (1954).
69. E. S. Rothman and M. E. Wall, *J. Amer. Chem. Soc.*, **77,** 2228 (1955).
70. G. P. Mueller, R. T. Stobaugh, and R. S. Winniford, *J. Amer. Chem. Soc.*, **75,** 4888 (1953).
71. E. S. Rothman and M. E. Wall, *J. Amer. Chem. Soc.*, **77,** 2229 (1955).
72. N. S. Leeds, D. K. Fukushima, and T. F. Gallagher, *J. Amer. Chem. Soc.*, **76,** 2265 (1954).
73. J. A. Zderic, H. Carpio, and D. Chávez Limón, *J. Org. Chem.*, **27,** 1125 (1962).
74. J. A. Zderic, H. Carpio, D. Chávez Limón, and A. Ruiz, *J. Org. Chem.*, **26,** 2842 (1961).
75. D. Burn, D. N. Kirk, and V. Petrow, *Proc. Chem. Soc.*, **1960,** 14.

CHAPTER VIII

Oxepins Derived from Sugars

T. R. HOLLANDS

Food & Drug Directorate, Ottawa, Canada

A. Introduction

As early as the end of the nineteenth century, it was known that acid treatment of certain sugars results in a significant yield of an anhydride. However, not until much later was it proven that an oxepin system is formed under these conditions. More recently, improved techniques have made possible the isolation and characterization of minor components of some of these acid-catalyzed equilibria.

In this chapter, oxepin derivatives are organized according to the work carried out on the parent hexoses and heptoses. A brief review of some dioxepins derived from sugars is also included.

The ring systems, listed in *Chemical Abstracts* and shown here with their Ring Index numbers, will be considered as oxepinoid derivatives for the purpose of this review, with the understanding that their classification as oxepins may be somewhat unorthodox.

2,8-Dioxabicyclo[3.2.1]octanes
R.I. 8004

3,8-Dioxabicyclo[3.2.1]octanes
R.I. 1303

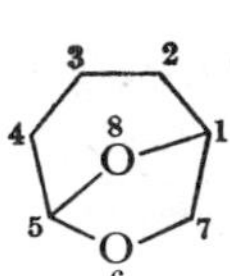

6,8-Dioxabicyclo[3.2.1]octanes
R.I. 1232

3,7,9-Trioxatricyclo[4.2.1.$0^{2,4}$]nonanes
R.I. 2069

3,8,9-Trioxatricyclo[4.2.1.0^{2,4}]nonanes
R.I. 2070

4,7-Epoxy-1,3-dioxolo[4,5-*c*]oxepins
R.I. 8315

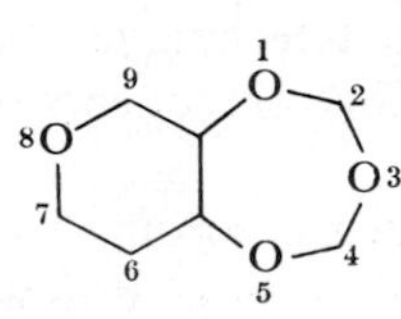

4,7-Epoxy-1,3-dioxolo[4,5-*d*]oxepins
R.I. 2405

m-Dioxino[5,4-*e*][1,4]dioxepins
R.I. 8129

6*H*-Pyrano[3,4-*f*][1,3,5]trioxepins
R.I. 1785

Bis-*m*-dioxino[5,4-*d*:4′,5′-*f*][1,3]dioxepins
R.I. 8615

m-Dioxino[4′,5′:5,6]pyrano[3,4-*f*][1,3,5]trioxepins
R.I. 3641

B. Oxepins Derived from Hexoses

1. 1,6-Anhydro-β-D-allopyranoses

Pratt and Richtmyer (1) succeeded in demonstrating the formation of an anhydride from D-allose, showing that the equilibrium attained in dilute hydrochloric acid solution contains approximately 14% of the 1,6-anhydro-β-D-allopyranose (**1A**). In order to confirm the structure of **1A**, they carried out an oxidation with sodium metaperiodate and found that two molecular equivalents of oxidant lead to the formation of one equivalent of formic acid and one equivalent of a dialdehyde. Upon further oxidation with bromine water, the dialdehyde yielded a dibasic acid identical with the diacid obtained earlier via a similar oxidation of other sugar anhydrides (see Section B-6). The expected tri-*O*-acetate (**1B**) and tri-*O*-*p*-toluenesulfonate (**1C**) were also reported.

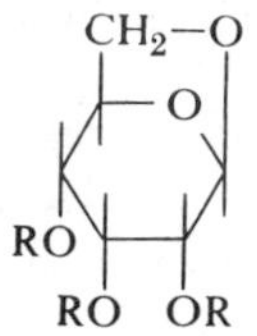

1A; R = H
1B; R = Ac
1C; R = Ts

In the field of antiradiation drugs, Goodman and Christensen (2) used the allose skeleton in connection with the synthesis of nontoxic β-amino-mercaptans. After heating the methyl glycoside (**2**) with 6*N* hydrochloric acid, they obtained a crystalline product which they believed to be the 1,6-anhydride (**3**) on the basis of its large negative rotation. They also found that milder conditions led to a mixture of what was assumed to be the anhydride (**3**) and the free sugar (**4**) (Eq. 1). Efforts to separate **3** and **4** via paper chromatography or fractional crystallization of derivatives met with failure.

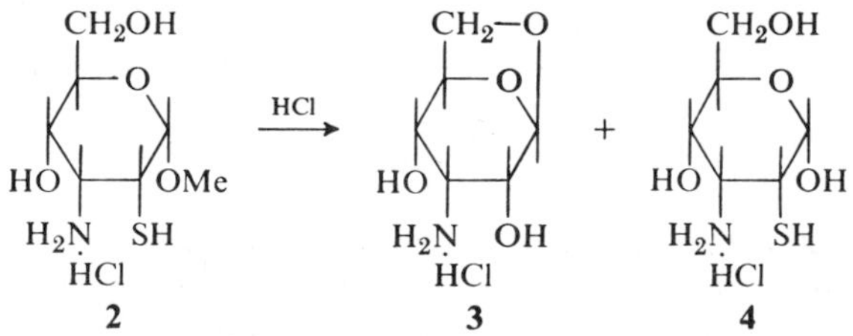

2 $\xrightarrow{HCl}$ **3** + **4** (1)

2. 1,6-Anhydro-β-D-altropyranoses

Richtmyer and Hudson (3) observed the formation of an equilibrium mixture when D-altrose was refluxed in 1*N* hydrochloric acid for $2\frac{1}{2}$ hr. They found that these conditions caused the rotation of the solution to change from $+34.3°$ initially to $-98.2°$, with an accompanying loss of reducing power. After oxidation of the remaining hexose to the more readily removable acid, the nonreducing portion of the equilibrium mixture was isolated. A later communication (4) reported the eventual crystallization of this syrup after standing for 3 years. The rotation of −215° (in water) corroborated previous calculations indicating that only 43% of the D-altrose remained uncyclized when equilibrium was reached.

Newth and Wiggins (5) found that acetylation of such an equilibrium mixture led to the slow formation and crystallization of a tri-*O*-acetyl derivative. The parent anhydride was obtained by allowing the triacetate to stand overnight in methanol containing a trace of sodium methoxide. After evaporation of the solvent, seeding with an authentic specimen yielded the crystalline anhydride.

Although there was initial indecision concerning the structure of the anhydride (6, 7), the problem was resolved readily by means of a periodate oxidation (4). With the uptake of two molar equivalents of periodic acid, one equivalent each of formic acid and a dialdehyde were obtained. The dialdehyde was identical with the compound isolated from a similar oxidation of 1,6-anhydro-β-D-glucopyranose. Further oxidation of the dialdehyde yielded a diacid, which was isolated in the form of a strontium salt and identified as L′-oxy-D-methylenediglycolic acid (see Section B-6). Inasmuch as this finding was inconsistent with the earlier suggestion of an ethylene oxide ring such as in structure **5**, the anhydride had to be 1,6-anhydro-β-D-altropyranose (**6A**).

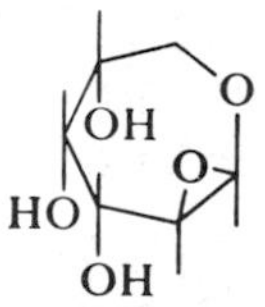

5

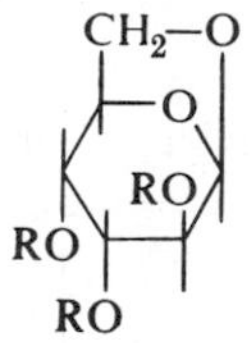

6A; R = H
6B; R = Ac

Richtmyer and Hudson (8) also reported the acetylation of **6A** with acetic anhydride in pyridine. The stability of the rings under these conditions permitted the isolation of a 2,3,4-tri-*O*-acetate (**6B**). On the other hand, the 1,6-anhydride ring in **6B** underwent acetolysis when subjected to acidic conditions (2% sulfuric acid in acetic anhydride), a mixture of 64% penta-*O*-acetyl-α-D-altrose and 36% penta-*O*-acetyl-β-D-altrose being obtained.

Newth and Wiggins (5) found that treatment of **6A** with acetone containing a catalytic amount of sulfuric acid gave 3,4-isopropylidene-1,6-anhydro-β-D-altropyranose (**7A**). Treatment of this material with methyl iodide and silver oxide led to a monomethyl derivative (**7B**) (Eq. 2). After removal of the isopropylidene blocking group, lead tetraacetate oxidation of the vicinal diol system occurred in the expected manner, with the uptake of 1 mole of oxidant.

Newth (9) also employed the 3,4-isopropylidene blocking group in order to synthesize 2-*O*-*p*-toluenesulfonyl-1,6-anhydro-*β*-D-altropyranose (**8**). The latter compound was intended to be an intermediate in the synthesis of the

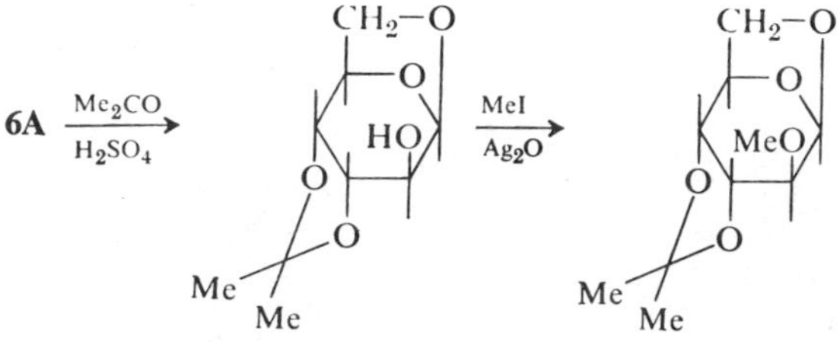

corresponding 2,3-epoxide,1,6:2,3-dianhydro-*β*-D-altropyranose (**9**). Somewhat unexpectedly, however, the *p*-toluenesulfonate derivative proved resistant to both alkoxide and hydroxide treatment, being recovered unchanged even after 5 hr in refluxing 0.5*N* sodium methoxide.

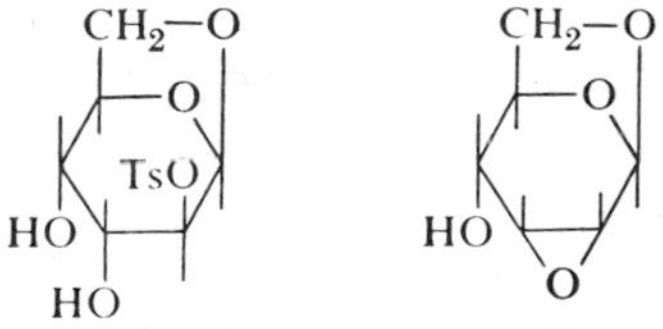

The same report (9) also described the synthesis of the 3-*O*-*p*-toluenesulfonate derivative (**10**), which was accomplished directly from **6A** through the use of only one molecular equivalent of sulfonyl reagent, the more reactive equatorial 3-hydroxy group being esterified preferentially. Although the yield was poor, treatment of **10** with strong alkoxide did give a dianhydride, in contrast to the lack of reaction of **8**. It was assumed that intramolecular displacement of the sulfonate group by the ionized hydroxyl on $C_{(2)}$ would lead to the formation of 1,6:2,3-dianhydro-*β*-D-mannose (**11**) as the primary product. However, a second intramolecular displacement was envisaged (Eq. 3), involving the hydroxy group at $C_{(4)}$ which is *trans* to the epoxide ring. This would result in the formation of 1,6:3,4-dianhydro-*β*-D-altropyranose (**12**). In order to ascertain the structure of the dianhydride actually formed, Newth examined the products generated by hydrolysis of the epoxide ring. Anhydride **12** would be expected to yield D-mannose and D-idose derivatives, while **11** would give those of D-glucose and D-altrose; the first-mentioned derivatives would predominate in each instance because of preferrential diaxial ring opening. Inasmuch as the sulfuric acid hydrolysis of the dianhydride yielded a compound which appeared to be 1,6-anhydro-β-D-mannopyranose rather than 1,6-anhydro-β-D-glucopyranose, the 1,6:3,4-dianhydro-β-D-altropyranose structure (**12**) could be assigned with confidence.

CH_2—O, O, HO, HO, TsO **10** $\xrightarrow{RO^{\ominus}}$ [CH_2—O, O, O, HO **11**] ⟶ CH_2—O, O, HO, O **12** (3)

3. 1,6-Anhydro-3-deoxy-β-D-arabinohexopyranoses

A suspension of methyl-4,6-*O*-benzylidene-3-deoxy-α-D-arabinohexopyranoside in 0.2*N* hydrochloric acid was refluxed for 3 hr (10). After destruction of the unreacted sugar with alkali, the solution yielded 1,6-anhydro-3-deoxy-β-D-arabinohexopyranose (**13A**) as a syrup (Eq. 4). On the basis of the change of rotation, it was calculated that the mixture contained 29% of the anhydride at equilibrium. The diacetate (**13B**) and dibenzoate (**13C**) derivatives were also characterized. As would be expected from the absence of vicinal hydroxy groups in **13A**, no reaction occurred with sodium metaperiodate.

O—CH_2, OMe, O, HO, PhCH—O $\xrightarrow{HCl}$ CH_2—O, O, HO, HO (4)

13A; R = H
13B; R = Ac
13C; R = Bz

4. 1,6-Anhydro-β-D-galactopyranoses

An anhydride derivative of D-galactose was reported first in 1929 by Micheel (11), who obtained it by treating 2,3,4,6-tetra-*O*-acetyl-D-galactopyranosyl trimethylammonium bromide with hot aqueous barium hydroxide (Eq. 5). He also reported obtaining the same material in low yield on pyrolysis of β-D-galactose at about 300°C (*ca.* 3 torrs). On the basis of the limited experimental evidence then available, he suggested structure **14** for this anhydride.

Subsequently, McCreath and Smith (12) obtained as a by-product in the formation of 1,2:3,4-diisopropylidene-D-galactose a compound which they showed to be 3,4-isopropylidene-1,6-anhydro-β-D-galactopyranose. They confirmed the structural assignment made earlier by Micheel.

CH_2OH HO O OH NMe ⊕ Br⊖ OH $\xrightarrow{Ba(OH)_2}$ CH_2—O HO O OH OH **14** (5)

Hann and Hudson (13, 14) described another preparation involving pyrolysis of the inexpensive starting material α-lactose. The yield of 3,4-isopropylidene derivative after appropriate treatment of the pyrolysis product was slightly more than 10% by weight.

Richtmyer (15) also obtained 1,6-anhydro-β-D-galactopyranose (**14**) by treating D-galactose with hot dilute sulfuric acid. The yield of anhydro product recovered from the equilibrium was a mere 1.9% under these conditions. This material was fractionated further into 1,6-anhydro-β-D-galactopyranose and 1,6-anhydro-β-D-galactofuranose (see Section B-5). The pyranose component was separated on the basis of its ability to form a borate complex.

In studying the anhydride from galactose, Hann and Hudson (14) found that periodate oxidation required two molecular equivalents of oxidant. In addition to obtaining one molar equivalent of formic acid, they also isolated one equivalent of L′-oxy-D-methylenediglycolic aldehyde, the same dialdehyde as had been prepared earlier by similar oxidation of the sugar anhydride derivatives of glucose (see Section B-6).

3,4-Isopropylidene-1,6-anhydro-β-D-galactopyranose was obtained from **14** by condensation with acetone in the presence of a trace of acid. Acetylation, benzoylation, and *p*-toluenesulfonylation were found to occur at the remaining 2-hydroxy group. Mild acid treatment of these derivatives sufficed to cleave the isopropylidene ring, with formation of the 2-*O*-acetate, 2-*O*-benzoate, and 2-*O*-*p*-toluenesulfonate, respectively. Free hydroxy groups were thereby generated at $C_{(3)}$ and $C_{(4)}$. Further reaction of the 2-*O*-*p*-toluenesulfonate with alkali afforded a compound which is apparently 1,6:2,3-dianhydro-β-D-galactopyranose.

More recently, Cerny, Gut, and Pacak (16) reported that 2,4-di-*O*-*p*-toluenesulfonyl-1,6-anhydro-β-D-glucopyranose yields 2-*O*-*p*-toluenesulfonyl-1,6:3,4-dianhydro-β-D-galactopyranose (**15**) in alkaline medium. They also found (17) that further treatment with alkali results in the formation of three isomeric epoxy alcohols (Eq. 6). The first, 1,6:3,4-dianhydro-β-D-galactopyranose (**16**), was believed to originate directly from hydrolysis of the 2-*O*-*p*-toluenesulfonate ester. Formation of the second isomer, 1,6:2,3-dianhydro-β-D-gulopyranose (**17**), was suggested to occur by migration of the epoxide ring in the alkaline medium. The third product, 1,6:3,4-dianhydro-β-D-altropyranose (**19**), was postulated to have been formed via two successive epoxide migrations. It was assumed the 1,6:2,3-dianhydro-β-D-mannopyranose (**18**) is the primary product arising from intramolecular displacement of the 2-*O*-*p*-toluenesulfonate group. That double migrations of this type can indeed occur was demonstrated by the isolation of some galactopyranose from a reaction mixture of alkali and 1,6:2,3-dianhydro-β-D-gulopyranose.

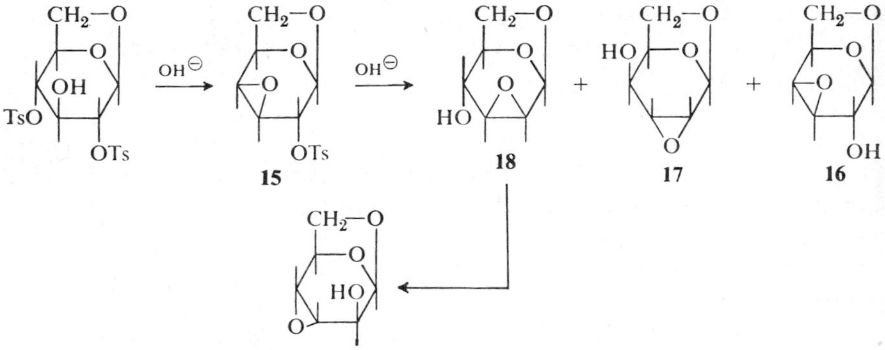

(6)

Heyns and co-workers (18) have found that 1,6-anhydro-β-D-galactopyranose (**20**) may be oxidized "selectively" (up to 40% at one of three free hydroxyl functions) with oxygen and Adams catalyst to 1,6-anhydro-β-D-xylohexopyrano-3-ulose (**21**). In this oxidation the 1,6-anhydride ring serves two distinct purposes. The first, and more obvious, is the blocking of the primary alcohol function at $C_{(6)}$. Second, however, the 1,6-anhydride function is useful in controlling the stereochemistry of the molecule with the hydroxy group at $C_{(3)}$ fixed rigidly in an axial position. This allows for a much more selective oxidation, since axial hydroxyl groups are oxidized in preference to equatorial ones. Reduction of the ketone function in **21** with sodium amalgam yielded 1,6-anhydro-β-D-gulopyranose (**22**), in which the $C_{(3)}$ hydroxyl is equatorial. Treatment of **21** with hydroxylamine afforded the oxime (**23**), which was transformed into 3-amino-1,6-anhydro-3-deoxy-β-D-galactopyranose (**24A**) by catalytic hydrogenation (Eq. 7).

It is of interest to note that in the nmr spectrum of 1,6-anhydro-β-D-galactopyranose (**20**), long-range coupling ($J_{1,3} = 1.4$ Hz) occurs between the two equatorial protons at $C_{(1)}$ and $C_{(3)}$, in addition to the expected larger coupling ($J_{1,2} = 2.0$ Hz) of the adjacent protons at $C_{(1)}$ and $C_{(2)}$. However,

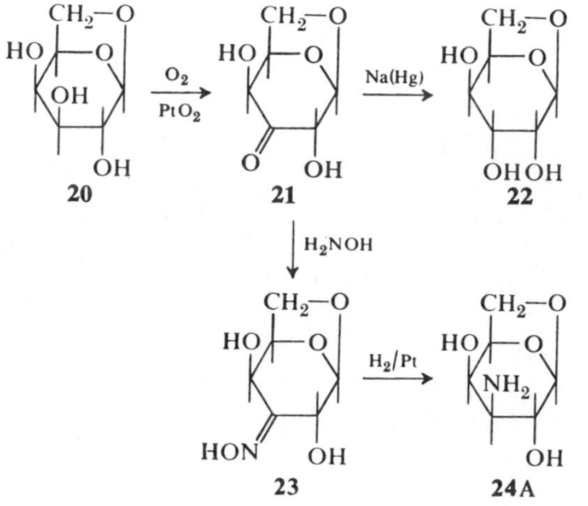

(7)

in 1,6-anhydro-β-D-gulopyranose, in which the five bonds describing a "W" are no longer planar, the signal for the $C_{(1)}$ proton is only split by the adjacent proton at $C_{(2)}$ ($J_{1,2} = 2.2$ Hz).

Cerny and co-workers (19) treated 1,6:3,4-dianhydro-2-*O*-*p*-toluenesulfonyl-β-D-galactopyranose (**15**) with hydrogen and Raney nickel and obtained 1,6-anhydro-4-deoxy-2-*O*-*p*-toluenesulfonyl-β-D-glucopyranose (**24B**). The latter yielded 1,6:2,3-dianhydro-4-deoxy-β-D-mannopyranose (**24C**) on contact with ethoxide ion (Eq. 8).

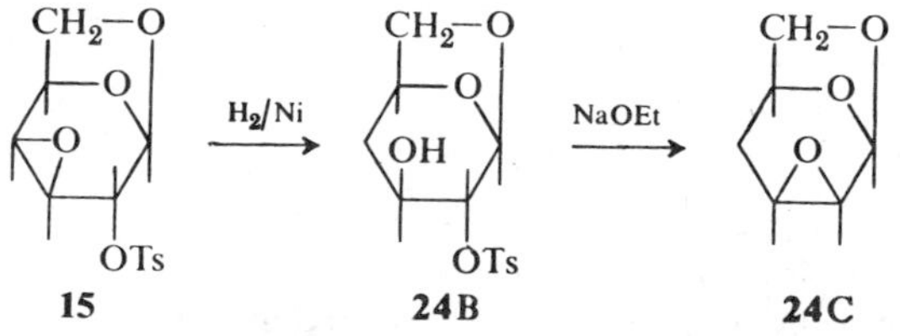

(8)

5. 1,6-Anhydro-β-D-galactofuranoses

Pyrolysis of D-galactopyranose has yielded a mixture of anhydro sugars (20). An anhydride capable of forming an isopropylidene derivative was prepared and identified as 1,6-anhydro-β-D-galactofuranose (**25A**). This compound was identical with the product obtained earlier by Hann and Hudson (21) but erroneously formulated as 1,3-anhydro-β-D-galactopyranose (**25B**).

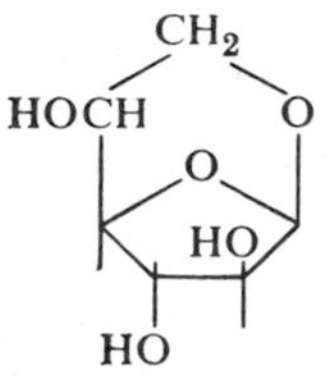

CH_2 HOCH O O HO HO **25A**

CH_2OH HO O O OH **25B**

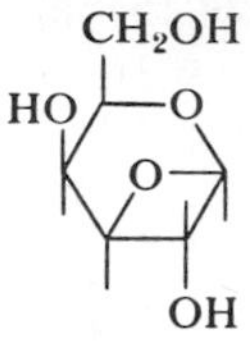

On treatment of D-galactose with hot dilute acid, Richtmyer (15) obtained a mixture of anhydrides and, making use of the fact that a borate complex did not form, separated the 1,6-anhydro-β-D-galactofuranose (**25A**) from the isomeric pyranose (**14**) (see Section B-4).

Alexander and co-workers (20) have reported the acetylation and methylation of the three free hydroxyl groups in **25A**. They also found this compound to be unreactive toward both periodic acid and lead tetraacetate. Reaction of **25** with methanesulfonyl chloride and *p*-toluenesulfonyl chloride yielded the expected tri-*O*-sulfonates (**15**).

6. 1,6-Anhydro-β-D-glucopyranoses

a. *Methods of Preparation*

As early as 1894 Tanret (22) reported that refluxing an aqueous solution of glucose and barium hydroxide yielded a new compound. In fact, his analyses showed this product to be an anhydride of the parent material, and he termed it "levoglucosan" because of its strongly levorotatory effect on plane polarized light. He also isolated the same material after hydrolysis of various naturally-occurring glucosides. Later Pictet and Sarasin (23) reported the recovery of "levoglucosan" from the distillation of cellulose under reduced pressure. This process made the anhydride readily available in large quantities. Karrer and Smirnoff (24) developed another preparation involving the reaction of tetra-*O*-acetylglucosyl bromide with trimethylamine, followed by treatment of the quaternary ammonium salt with barium hydroxide. Dry distillation of cellulose and starch under reduced pressure likewise yielded "levoglucosan" according to Irvine and Oldham (25).

Hann and Hudson (14) showed that destructive distillation of either α- or β-D-glucose gave the anhydride with equal ease. On the other hand, Montgomery and co-workers (26) found that while phenyl β-D-glucoside could be converted readily into the anhydride by refluxing in aqueous alkaline solution, the α-glucoside remained unaffected under similar conditions. In fact they were able to recover 85% of the latter anomer by crystallization even after alkaline treatment lasting for a period of 2 weeks. Other workers (27) obtained the same material by alkaline treatment of tri-*O*-acetylglucose 6-nitrate. Dilute sulfuric acid treatment of glucose has also been found to lead to some anhydride, together with disaccharide (28).

Wood and co-workers (29, 30) have reported that alkaline hydrolysis of the naturally-occurring glucoside stevioside yields the anhydride of D-glucose. This indicates that D-glucose must be the sugar residue cleaved from stevioside. Akagi and co-workers (31) reported a synthesis of the same anhydride by reaction of cold sodium methoxide with 6-*O*-*p*-toluenesulfonyl-1,2,3,4-tetra-*O*-acetyl-β-D-glucopyranose, the product being isolated in approximately 60% yield as a triacetate.

More recently cationic exchange resins have been used to cause D-glucose to undergo dehydration, usually as a secondary reaction during polymerization studies. O'Colla and co-workers (32, 33) have isolated the anhydride, as well as polymeric materials, after contact with several different ion exchange media. However, due to the numerous side products, this process cannot be classed as immediately valuable from a synthetic point of view.

The commercial-scale production of the anhydride of D-glucose has been carried out by pyrolysis of "hardwoods and agricultural wastes" with steam under reduced pressure (34).

b. *Proof of Structure*

It was recognized early (22, 23) that three readily acylable hydroxy groups were present in the anhydride of D-glucose. Pictet and Sarasin (23) realized that since the compound failed to reduce Fehling's solution, there was no free aldehyde function in the molecule. Accordingly they suggested several possible structures, all of which incorporated bicyclic systems.

Methylation of "levoglucosan" gave Irvine and Oldham (25) a trimethylated material, from which they obtained 2,3,4-tri-*O*-methylglucose (35). Since the methylated oxygens had to represent the three free hydroxy groups, those oxygens, involved in anhydride formation were placed on $C_{(1)}$ and $C_{(6)}$. Accordingly "levoglucosan" was formulated as 1,6-anhydro-β-D-glucopyranose (**26A**).

Further evidence supporting structure **26A** came from the work of Jackson and Hudson (36) who carried out periodic acid (and sodium metaperiodate) oxidation and obtained a dialdehyde which they assumed to be L′-oxy-D-methylenediglycolic aldehyde (**27**). The oxidation required two molecular equivalents of oxidant and yielded on equivalent of formic acid in addition to the dialdehyde. The latter was characterized by further oxidation with

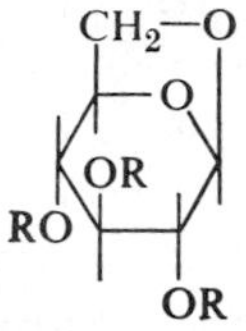

26A; R = H
26B; R = Ac
26C; R = Bz
26D; R = $PhCH_2$
26E; R = Me

bromine water to a diacid (**28**) easily separated as a strontium salt. This material was found to be unusually stable under acidic conditions and could not be hydrolyzed to its two components, glyoxylic and D-glyceric acids.

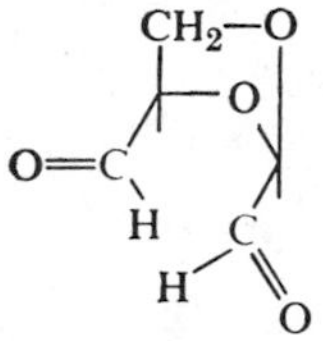

27 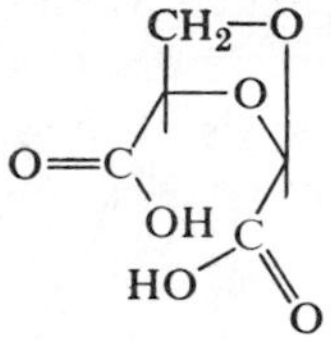 **28**

The foregoing oxidation has been repeated on most of the sugar anhydrides and has yielded the same products whenever it has been performed with 1,6-anhydrohexopyranoses. Consequently it has become a classic tool for correlation and proof of structure in the sugar field.

c. *Reactions and Derivatives*

Various authors have reported the preparation of the tri-*O*-acetate (**26B**) (22, 23, 31, 37, 38) and tri-*O*-benzoate (**26C**) (22, 23, 31) of **26A**. Zemplen and co-workers (37) treated **26B** with alkaline benzyl chloride to obtain the tri-*O*-benzyl derivative (**26D**) from which they made 1,6-di-O-acetyl-2,3,4-tri-*O*-benzylglucopyranose by acetolysis of the 1,6-anhydride function. Irvine and Oldham (35) described the formation of the 2,3,4-tri-*O*-methyl derivative (**26E**). Treatment of this compound with acid yielded 2,3,4-tri-*O*-methylglucose (39), the anhydride being used, in effect, to block the oxygen functions on $C_{(1)}$ and $C_{(6)}$.

Jeanloz and co-workers (40) have reported success in connection with attempted partial esterification of 1,6-anhydro-β-D-glucopyranose. They verified experimentally their initial assumption that the hydroxy groups on $C_{(2)}$ and $C_{(4)}$ would possess at least equal, and probably greater, reactivity than the one on $C_{(3)}$. By the use of two molecular equivalents of *p*-toluenesulfonyl chloride or benzoyl chloride in pyridine, the major products were 2,4-diesters. Cerny and co-workers (41) likewise prepared the 2,4-di-*O*-*p*-toluenesulfonate which was converted into 2-*O*-*p*-toluenesulfonyl-1,6:3,4-dianhydro-β-D-galactopyranose (**29A**) on treatment with sodium methoxide in chloroform. Jeanloz and Rapin (42) have found that the 2,4-di-*O*-*p*-toluenesulfonate undergoes ammonolysis to yield 2,4-diamino-2,4-dideoxy-1,6-anhydro-β-D-glucopyranose (**29B**) on heating with methanolic ammonia (Eq. 9).

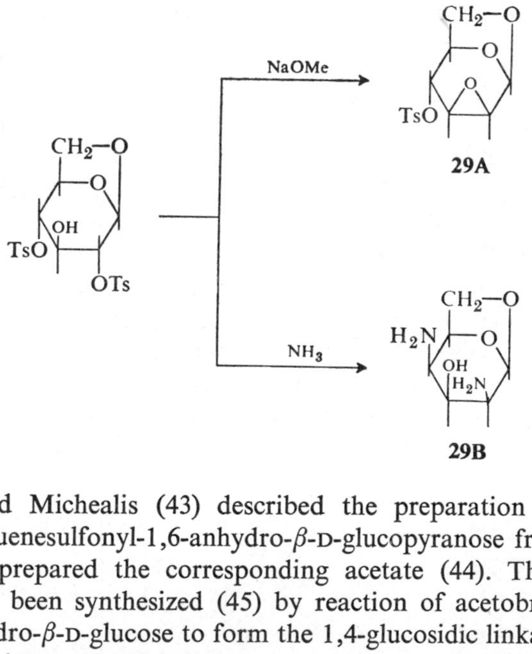

(9)

Micheel and Michealis (43) described the preparation of 2-amino-2-deoxy-*N*-*p*-toluenesulfonyl-1,6-anhydro-β-D-glucopyranose from which they subsequently prepared the corresponding acetate (44). The disaccharide cellobiose has been synthesized (45) by reaction of acetobromo-D-glucose with 1,6-anhydro-β-D-glucose to form the 1,4-glucosidic linkage.

It has also been reported (46) that hydrogenation of 1,6-anhydro β-D-glucopyranose yields an anhydrohexitol with physical and chemical properties similar to glycerol. Another publication (47) lists R_f values for both paper and thin-layer chromatography of diverse sugar derivatives, including the 1,6-anhydride of glucose described here.

d. *Mechanism of Formation*

Two simultaneous papers in 1945 (48, 49) clearly outlined a proof for a plausible mechanistic pathway involving the formation of a 1,6-anhydride under alkaline conditions. This was expanded subsequently (50) to include participation by an acetate group on $C_{(2)}$ which, under acidic conditions, could provide anchimeric assistance with the β-anomer, but not with the α-form. Other authors (51) pointed out that no mechanism so far devised could be operative for all known instances of anhydride formation, citing the example of 1-fluoro-2,3-di-*O*-methyl-α-D-glucose which in alkaline medium forms 2,3-di-*O*-methyl-1,6-anhydro-β-D-glucopyranose (Eq. 10).

CH2OH, F, O, OMe, HO, OMe —NaOH→ CH2—O, O, OMe, HO, OMe (10)

Later work (30) involving the reaction of 6-*O*-*p*-toluenesulfonyl-1,2,3,4-tetra-*O*-acetyl-β-D-glucopyranose with sodium methoxide likewise yielded the 1,6-anhydride (**26A**) (Eq. 11). It was proposed that the ionized hydroxy group on $C_{(1)}$ can attack $C_{(6)}$ with loss of the *p*-toluenesulfonate ion. To account for the fact that no 3,6-anhydride was formed, the suggestion was advanced that the hemiacetal oxygen at $C_{(1)}$ is more nucleophilic than the hydroxyl at $C_{(3)}$, and that the latter is hindered sterically by substitution at $C_{(4)}$. A more likely explanation is that interaction between the $C_{(2)}$ and $C_{(4)}$ substituents tends to deform the ring in a manner that moves the $C_{(3)}$ substituent away from the primary *p*-toluenesulfonate at $C_{(6)}$. It is of interest to note that the β-anomer containing an amino instead of a hydroxy group at $C_{(2)}$ behaves similarly under alkaline conditions, giving a 1,6-anhydride.

CH2OTs, O, OAc, OAc, AcO, OAc —NaOMe→ [TsO—CH2, O⊖, O, OH, HO, OH] → **26A** (11)

It should be pointed out that, because of the diverse conditions employed by various workers in this field, generalizations involving D-glucopyranose and other sugars must always be viewed with caution.

e. *Polymerization*

Wolfrom and co-workers (53) reported that thermal polymerization of 1,6-anhydro-β-D-glucopyranose leads to the formation of glucosidic linkages of the type found in pyrodextrin. Further studies indicated that high temperature dextrination of starch may lead to depolymerization with the generation of terminal anhydro groups that undergo repolymerization to give new glucosidic linkages. Subsequent findings (54) suggested that the ethanol-soluble portion of this material consisted of disaccharides and trisaccharides containing 1,6-anhydro-β-D-glucopyranose end groups.

Upon being heated in the presence of zinc chloride (55) or zinc dust (56), 1,6-anhydro-β-D-glucopyranose yields a mixture of several polymers having anhydro end groups. Polymerization in the presence of boron trifluoride and water has also been reported (57) to give a material with high rotation. This indicates high stereoselectivity, resulting in a preponderance of α-linkages in the polymer. A slower reaction rate coincided with a greater degree of stereospecificity, while catalysis with boron trifluoride etherate in toluene had the opposite effect. Other work (32) has shown that heating glucose with cationic exchange resins yields, in addition to the 1,6-anhydride, some anhydropolysaccharides. "Trimethyllevoglucosan polymers" have been prepared (58) via cationic catalysis in methylene chloride solution in the cold. Although the pure polymer is too brittle to be practical, dibutyl phthalate has been found to be a useful plasticizer.

Results of the pyrolysis of cellulose (59) have led workers to believe that depolymerization proceeds by scission of the 1,4-glucosidic linkages, whereupon an intramolecular rearrangement forms 1,6-anhydro-β-D-glucopyranose. This material either breaks down to give volatile products or repolymerizes. The technique of cellulose pyrolysis has been refined recently in order to expand its usefulness (60).

7. 1,6-Anhydro-β-D-glucofuranoses

After distillation of the pyrolyzate of starch, Dimler and co-workers (61) discovered that, in addition to the well-characterized 1,6-anhydro-β-D-glucopyranose, they could isolate a second anhydride of glucose which they formulated as 1,6-anhydro-β-D-glucofuranose (**30A**). This material has also been isolated after treatment of glucose with dilute sulfuric acid solution (28).

1,6-Anhydro-β-D-glucofuranose has been found to resist cleavage by periodate and lead tetraacetate in acetic acid. It has been suggested that this lack of reactivity is due to the rigidity of the *trans* diol system imposed by the

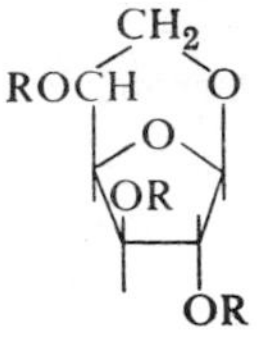

CH_2
ROCH O
O
OR
OR

30A; R = H
30B; R = Ac
30C; R = Me
30D; R = Ts

bicyclic nature of the ring system. The corresponding tri-*O*-acetate (**30B**), tri-*O*-methyl ether (**30C**), and tri-*O*-*p*-toluenesulfonate (**30D**) have been characterized.

8. 1,6-Anhydro-β-D-gulopyranoses

Stewart and Richtmyer (62) reported the treatment of D-gulose with dilute acid. After equilibrium was attained the unchanged material was destroyed by heating with alkali and the new anhydride was isolated in approximately 43% yield. The same compound was also obtained on reduction with sodium amalgam (18) of 1,6-anhydro-*β*-D-xylohexopyranose-3-ulose (see Section B-4).

The anhydride was characterized as 1,6-anhydro-*β*-D-gulopyranose (**31A**) on the basis of sodium metaperiodate oxidation and subsequent formation of the diacid that had been obtained from 1,6-anhydro-*β*-D-glucopyranose via a similar sequence. The tri-*O*-acetate (**31B**), tri-*O*-benzoate (**31C**), and tri-*O*-*p*-toluenesulfonate (**31D**) derivatives of **31A** were also prepared and characterized.

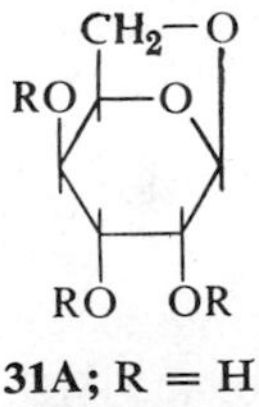

CH_2—O
RO —O
RO OR

31A; R = H
31B; R = Ac
31C; R = Bz
31D; R = Ts

During the hydrolytic degradation of some *Streptomyces* antibiotics, van Tamelen, Carter, and co-workers (63, 64) obtained 2-amino-1,6-anhydro-2-deoxy-*β*-D-gulopyranose (**32**). This substance has thus been useful in confirming the presence of a gulose residue as the sugar moiety in a number of natural products.

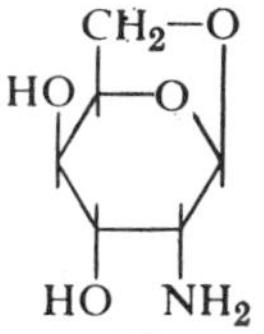

CH_2—O
HO —O
HO NH_2
32

9. 1,6-Anhydro-β-D-idopyranoses

The preparation of an anhydride derived from D-idose was first described by Sorkin and Reichstein (65). They obtained an equilibrium mixture upon slight warming of the free sugar in 5% aqueous sulfuric acid followed by neutralization with barium carbonate. Oxidation of the equilibrium mixture with bromine water yielded the more readily separable mixture of 20% idonic acid and 75% 1,6-anhydro-*β*-D-idopyranose (**33A**). Reeves (66) reported the use of a similar method for the preparation of 1,6-anhydro-3-*O*-methyl-*β*-D-idopyranoside (**33B**).

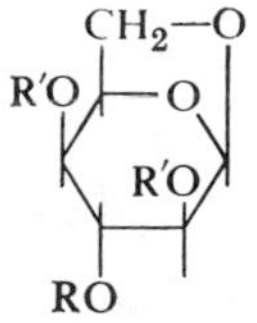

CH_2—O
R′O —O
R′O
RO
33A; R = R′ = H
33B; R = Me, R′ = H
33C; R = R′ = Ac

In the course of characterizing 1,6-anhydro-*β*-D-idopyranose, Sorkin and Reichstein (65) obtained a triacetate (**33C**) with acetic anhydride and pyridine, and prepared D-idose pentaacetate by acetolysis of **33C** with acetic anhydride and sulfuric acid. In addition 6-*O*-*p*-toluenesulfonyl-*D*-idose was obtained on treatment of 1,6-anhydro-*β*-D-idopyranose with *p*-toluenesulfonyl chloride in pyridine.

Two deoxy amino derivatives of 1,6-anhydro-*β*-D-idopyranose have also been reported. Kuhn and Bister (67) and Jeanloz and co-workers (68) independently obtained the crystalline 2-acetamido-3,4-di-*O*-acetyl-1,6-anhydro-2-deoxy-*β*-D-idopyranose (**34A**) by refluxing aqueous hydrochloric acid solutions of 2-amino-2-deoxy-*β*-D-idopyranose (or the corresponding tetra-*O*-acetate), and then acetylating the amino group. Jeanloz (69) reported that milder treatment of the 3-amino derivative yielded 3-acetamido-1,6-anhydro-3-deoxy-2,4-di-*O*-acetyl-*β*-D-idopyranose (**34B**), the same substance that James and co-workers (70) had synthesized earlier by ammonolysis of 1,6:2,3-dianhydro-β-D-talopyranose (see Section B-10).

Baggett and Jeanloz (71) have described the synthesis of 1,6-anhydro-3-*O*-methyl-β-L-idopyranose (**35**) and 1,6-anhydro-2,4-di-*O*-methyl-β-L-idopyranose (**36**) as shown in Eq. 12. Treatment of 1,2-isopropylidene-L-idofuranose with acetone yielded the 1,2:5,6-diisopropylidene derivative, which was subjected to methylation followed by hydrolysis in refluxing 1.3*N* sulfuric acid, to give **35**. Benzoylation of the diisopropylidene derivative at the free 3-hydroxy group furnished 1,6-anhydro-3-*O*-benzoyl-β-L-idopyranose, and methylation and reductive debenzoylation of this led to **36**.

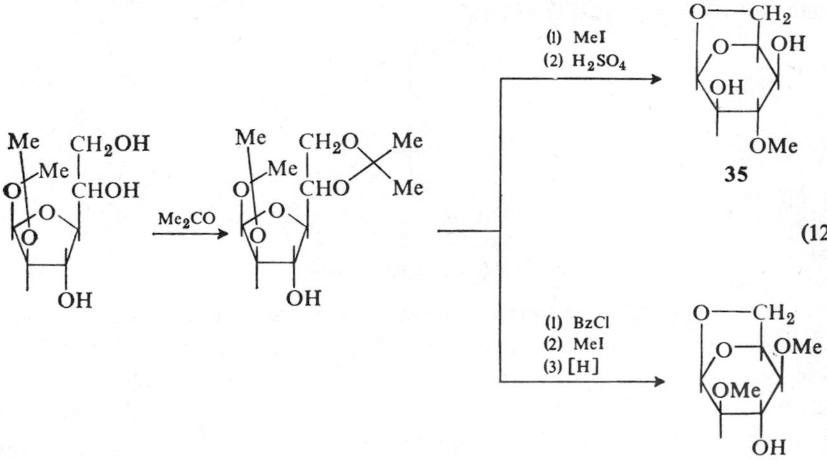

(12)

10. 1,6-Anhydro-β-D-mannopyranoses

A 1,6-anhydro derivative of D-mannose was reported first by Zemplen and co-workers (72) who prepared it by pyrolyzing "vegetable ivory," the endosperm of ivory-nut palm seeds (*Phytelephas macrocarpa*), which contains a large concentration of the polysaccharide D-mannosan.

Although Hudson and co-workers (73) initially employed the same method to obtain the anhydride in approximately 8% yield by weight, they later reported another route (74) which originated from phenyl β-D-mannoside. On being refluxed with aqueous potassium hydroxide for 5 days and being subjected to a suitable extraction procedure, this material yielded a syrup which could be acetylated with acetic anhydride in pyridine to give 2,3,4-tri-*O*-acetyl-1,6-anhydro-β-D-mannopyranose (**37A**). Catalytic deacetylation with barium methoxide led to 1,6-anhydro-β-D-mannopyranose (**37B**) in approximately 55% yield.

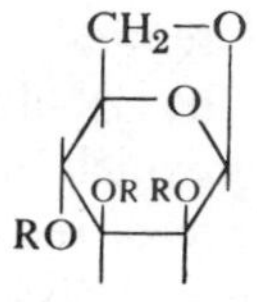

37A; R = Ac
37B; R = H

Richtmyer and co-workers (75) found that they could obtain up to 0.7% of the same anhydride (actually isolating approximately 0.5%, however) by allowing mannose to equilibrate in 0.2*N* sulfuric acid at 83°C.

In their study of this anhydride, Hudson and co-workers (73) found that allowing the 1,6-anhydride of D-mannose to stand overnight in acetone solution sufficed to cause precipitation of a crystalline isopropylidene derivative (**38**). As expected, this condensation was promoted by the addition of anhydrous copper sulfate. Hydrolysis of **38** with 0.1*N* sulfuric acid at room temperature of 1 day furnished a 91% yield of 1,6-anhydro-β-D-mannopyranose. However, stronger treatment with 1*N* hydrochloric acid on a steam bath for 2 hr generated D-mannose almost quantitatively.

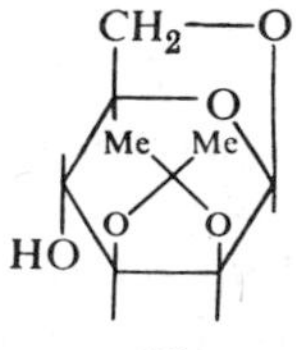

38

Periodic acid (or sodium metaperiodate) oxidation of 1,6-anhydro-β-D-mannopyranose (**37B**) required two molecular equivalents of oxidant and liberated one equivalent of formic acid together with one equivalent of a dialdehyde. Further oxidation of the dialdehyde yielded a diacid identical with that obtained from similar treatment of the 1,6-anhydride of D-glucose (see Section B-6). Zemplen and co-workers (72) reported that the three free hydroxy groups of **37B** could be acetylated readily to **37A**.

The isopropylidene derivative (**38**) has only one free hydroxyl function as shown by the formation of the customary monoesters. The dioxolane and 1,6-anhydride rings of the mono-*O*-methyl ether were hydrolyzed selectively on treatment with hot 1*N* hydrochloric acid for 4 hr. Since the product yielded the 4-*O*-methyl-D-glucose phenylosazone synthesized earlier by other workers, the free hydroxy group was clearly situated at $C_{(4)}$.

Aspinall and Zweifel (76) obtained other *O*-methyl derivatives of D-mannose by making use of the stereochemistry of 1,6-anhydro-β-D-mannopyranose. As with cyclohexanols, esterification of the free hydroxy groups of sugars was shown to occur preferentially at the equatorial positions. In 1,6-anhydro-β-D-mannopyranose the 1,6-anhydride ring serves to fix the pyranose ring in a conformation in which only the hydroxyl function at $C_{(2)}$ can be equatorial. Taking advantage of the selective reactivity of equatorial hydroxy groups in order to form a 2-*O*-*p*-toluenesulfonate, these authors synthesized 3,4-di-*O*-methyl-D-mannose readily by methylation of the remaining free hydroxy groups on $C_{(3)}$ and $C_{(4)}$ with methyl iodide and calcium sulfate, followed by hydrolysis of the sulfonate ester and 1,6-anhydride ring. The blocking action of the 1,6-anhydride ring with respect to the primary alcohol function is of no consequence in this instance because ester hydrolysis causes recyclization after the methylation step. This type of procedure illustrates the usefulness of the 1,6-anhydride ring in fixing the conformation of the pyranose ring in sugars.

1,6-Anhydro-β-D-mannopyranose derivatives have been used in combination with other sugars in disaccharide syntheses of cellobiose (77, 78) and lactose epimers (79).

11. 1,6-Anhydro-3-deoxy-β-D-ribohexopyranoses

The parent anhydride (**39A**) was prepared by refluxing phenyl 3-deoxy-β-D-ribohexopyranoside with 2.6*N* sodium hydroxide for 19 hr (10). Neutralization and extraction yielded the syrupy anhydride which was characterized as the crystalline di-O-acetate (**39B**). Equilibration in acid led to a mixture containing 10% anhydride according to rotational measurements. As might

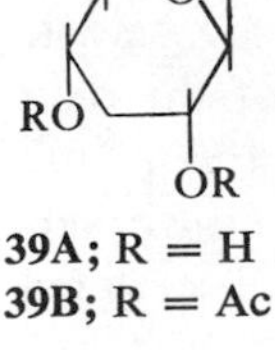

CH_2—O

O

RO

OR

39A; R = H
39B; R = Ac

be anticipated in view of the absence of a 1,2-diol system in **39A**, the extent of reaction with sodium metaperiodate in a 24-hr period was negligible.

12. 1,6-Anhydro-β-D-talopyranoses

Although the anhydride itself has not been reported, two epoxide derivatives have been prepared from 1,6-anhydrides of other sugars.

Hann and Hudson (14, 80) reported the synthesis of 1,6:3,4-dianhydro-*β*-D-talopyranose from 2,3-isopropylidene-1,6-anhydro-*β*-D-mannopyranose (**40A**) (Eq. 13). The free hydroxy group at $C_{(4)}$ of **40A** was esterified to **40B** with *p*-toluenesulfonyl chloride and the isopropylidene function was cleaved selectively by refluxing in aqueous acetic acid. The resulting 4-*O*-*p*-toluenesulfonyl-1,6-anhydro-*β*-D-mannopyranose (**40C**) was then treated with sodium methoxide in methanol to give 1,6:3,4-dianhydro-*β*-D-talopyranose (**41**).

CH₂—O, Me, Me, O, O, O, RO → AcOH → CH₂—O, O, OH HO, TsO → NaOMe → CH₂—O, O, O, HO (13)

40A; R = H
40B; R = Ts
40C
41

James and co-workers (70) have described the synthesis of a second dianhydride derivative of D-talose, 1,6:2,3-dianhydro-*β*-D-talopyranose (**42**), as shown in Eq. 14. They converted 3,4-isopropylidene-1,6-anhydro-*β*-D-galactopyranose (see Section B-4) into 3,4-isopropylidene-2-*O*-methanesulfonyl-1,6-anhydro-*β*-D-galactopyranose and hydrolyzed the isopropylidene blocking group by mild acid treatment. Further reaction of the sulfonate ester with sodium methoxide brought about loss of the methanesulfonate anion with concurrent formation of **42**.

CH_2—O, O, O, O, Me, Me, OH → (1) $MsCl/C_5H_5N$ (2) $H^{\oplus}$ → CH_2—O, HO, O, OH, OMs → NaOMe → CH_2—O, HO, O, O (14)

42

Hudson (80) treated 1,6:3,4-dianhydro-*β*-D-talopyranose (**41**) with sodium methoxide in order to effect cleavage of the epoxide ring. The resulting mixture of products was characterized via further reduction to D-idose and D-mannose, the isomers expected from *trans*-diaxial opening of the epoxide ring.

Further work in the same connection was carried out by James and coworkers (70) who utilized both ammonia and sodium methoxide as alkaline ring-cleaving agents. Heating 1,6:2,3-dianhydro-β-D-talopyranose (**42**) and 1,4:3,4-dianhydro-β-D-talopyranose (**41**) with ammonia under pressure (Eq. 15) afforded two products in each instance, apparently in nearly equivalent yields. As one might predict, one of these compounds, which was discovered to be common to both reaction mixtures, was 3-amino-3-deoxy-1,6-anhydro-β-D-idopyranose. However, when 1,6:2,3-dianhydro-β-D-talopyranose was treated with aqueous ammonia, the predominant product (56% yield) was 2-amino-2-deoxy-1,6-anhydro-β-D-galactopyranose. This outcome was exactly the reverse of what had been anticipated on the basis of the corresponding hydrolysis of methyl 2,3-anhydro-β-D-taloside.

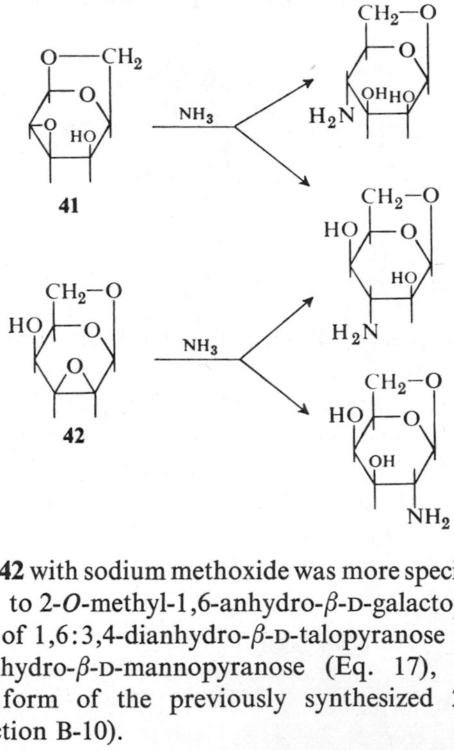

(15)

Ring scission of **42** with sodium methoxide was more specific than ammonolysis, leading solely to 2-*O*-methyl-1,6-anhydro-β-D-galactopyranose (Eq. 16). Similar treatment of 1,6:3,4-dianhydro-β-D-talopyranose (**41**) afforded only 4-*O*-methyl-1,6-anhydro-β-D-mannopyranose (Eq. 17), which was characterized in the form of the previously synthesized 2,3-isopropylidene derivative (see Section B-10).

The above facts are explained most readily in terms of the steric influence of the substrate on the attacking nucleophile. The larger the incoming species, the less likely it would be, in each instance, to approach the most highly hindered terminal of the epoxide ring.

The work on 1,6-anhydro-β-D-talopyranose derivatives and the attendant theoretical considerations outlined above have proved extremely useful

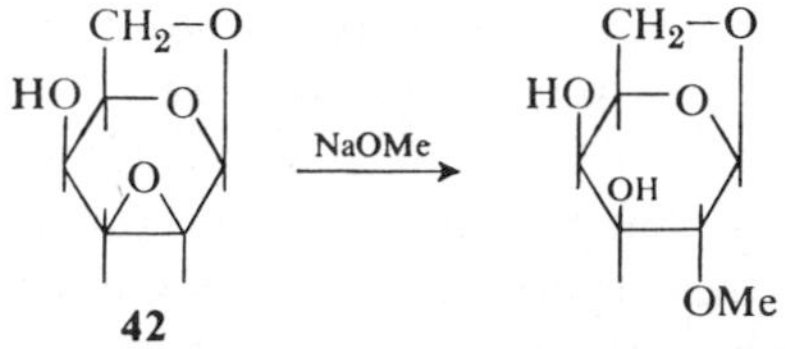

CH_2—O HO O O NaOMe → CH_2—O HO O OH OMe (16)

42

CH_2—O O O HO NaOMe → CH_2—O O OH HO MeO (17)

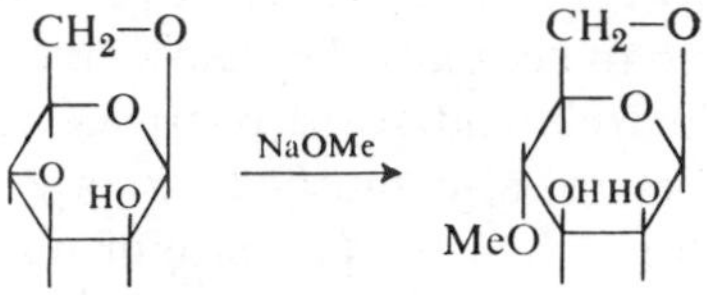

41

in the complete characterization of the naturally-occurring sugar chondrosamine through synthesis.

C. Oxepins Derived from Heptoses

1. 2,7-Anhydro-β-D-altroheptulopyranoses

LaForge and Hudson (81) were the first to conclude that either the naturally-occurring sugar "sedoheptulose" or its anhydride, called "sedoheptulosan," could be transformed into an equilibrium mixture of the two compounds by heating in dilute acid media. The equilibrium was thought initially to contain 20% of the sugar and 80% of the anhydride.

Zill and co-workers (82), using ion exchange chromatography of borate complexes, first obtained evidence indicating that a third minor component was present. Later Zill and Tolbert (83) reported the isolation of this new material and assigned to it the structure 2,7-anhydro-β-D-altroheptulofuranose. It was found to comprise approximately 2% of the equilibrium mixture.

Zissis and Richtmyer (84) have discussed the formation of this anhydride during alkaline treatment of the α- and β-anomers of phenyl D-altroheptulopyranoside, especially with reference to the intermediacy of a 2,3-epoxide ring.

Since the solution of the structure of 2,7-anhydro-β-D-altroheptulopyranose has involved so much time, and since it has been applied generally in this group of anhydro sugars, the arguments used in its derivation will be outlined in detail.

After some previous inconclusive work on the crystalline sugar anhydride called "sedoheptulosan," Hudson (85) suggested structure **43** in 1938. In fact he used only experimental evidence already reported by other workers in order to prove that this structure, with its novel combination of both an epoxide and an oxepin, must be the correct one.

For 13 years structure **43** remained unchallenged. In 1951, however, Hudson and co-workers (86) published new experimental evidence pertaining to this problem. First they reported, on the basis of periodate oxidation data, that three consecutive hydroxy groups must be present. Since "sedoheptulose" had already been proven to be D-altroheptulose, the number of structures for the corresponding anhydride had to be restricted to three possibilities (**43**, **45**, and **47**). These structures also satisfied earlier evidence involving tritylation experiments which required the presence of a single primary alcohol.

Hudson and co-workers also discovered that hydrogenation of the dialdehyde resulting from periodate oxidation led to a trihydroxy diether (**44**, **46**, or **48**). Acid hydrolysis of this compound, followed by treatment with *p*-nitrobenzoyl chloride, led to a single tri-*O*-*p*-nitrobenzoate ester of glycerol. This new observation eliminated the 2,3:2,7-ring system from consideration as a possibility for "sedoheptulosan." In addition, the tetra-*O*-*p*-toluenesulfonate derivative was found to be unreactive toward sodium iodide, an observation made previously in connection with other sugars bearing primary *p*-toluenesulfonate functions adjacent to a carbonyl group. This evidence led Hudson and co-workers to favor the 2,6:2,7-ring system of structure **47** over the 1,2:2,7-ring system of **45**. In accord with this choice, "sedoheptulosan" was found to be stable to strong alkaline treatment, suggesting that no epoxide function is present.

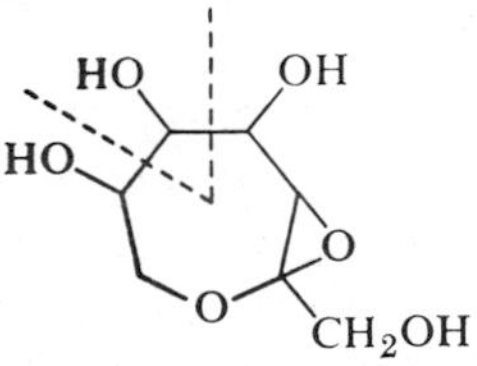

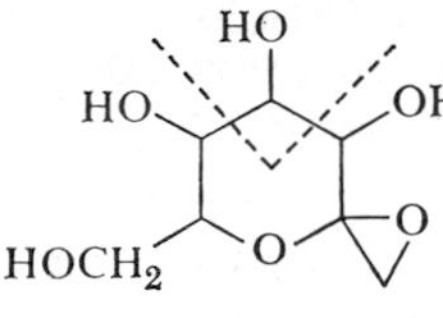

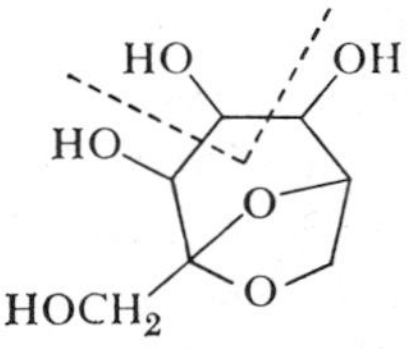

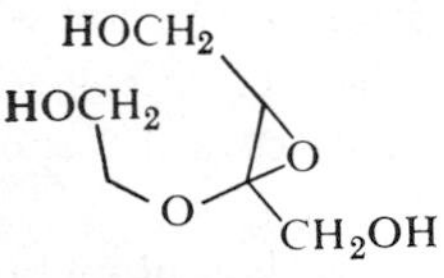

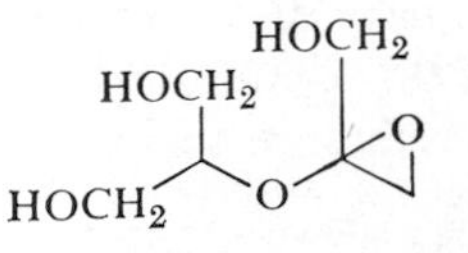

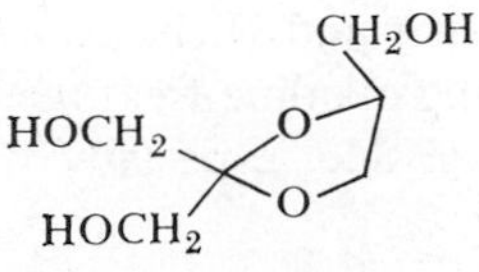

Hudson and co-workers (88) cited some observations made in connection with the oxidation of "sedoheptulosan" with sodium metaperiodate as an interesting and more definitive proof of structure **47**. They based their conclusions on rules elucidated earlier by other authors for the oxidation of 1,2-diols in the presence of lead tetraacetate, the most important assumption being that α-hydroxy aldehydes are oxidized much more slowly than vicinal diols. Two molecular equivalents of oxidant were consumed in 1 hr, but the single molecular equivalent of formic acid liberated as a result of oxidation was recovered in its entirety only after 12 days. Moreover, the rotation of the reaction mixture did not attain a constant value until after a similar period of time. On this basis Hudson and co-workers stated with confidence that "sedoheptulosan" was indeed 2,7-anhydro-β-D-altroheptulopyranose. It was assumed that a molecule having structure **47** would undergo initial oxidation to dialdehyde **49**, which could then recyclize to the five-membered ring hemiacetal (**50**). This intermediate could immediately undergo further oxidation to the formate ester (**51**), which would be hydrolyzed slowly with formation of monohydroxy dialdehyde **52** and formic acid (Eq. 18).

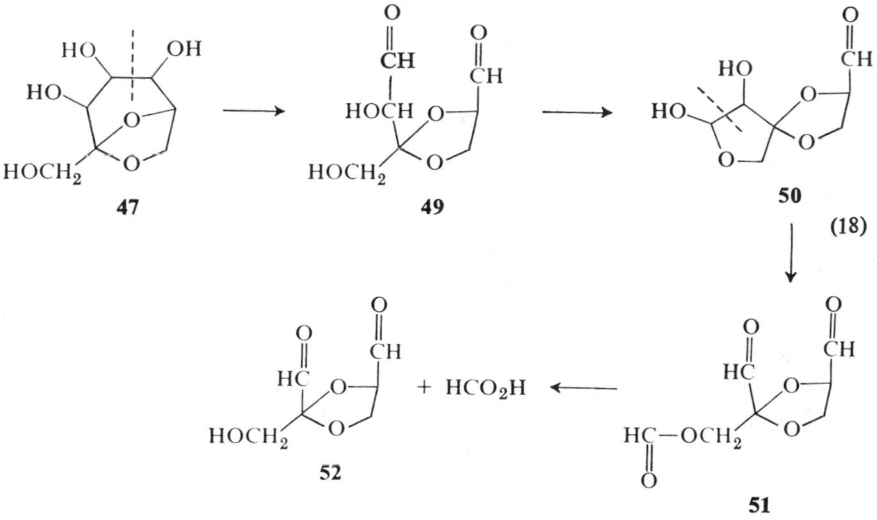

47 → **49** → **50** → **51** → **52** + HCO_2H (18)

Although there can be little doubt as to the correctness of structure **47**, it is interesting that the possibility, at least, of a similar sequence of intermediates was also discussed for structure **45**. In this instance the first oxidation would lead stepwise to dialdehyde **53** and seven-membered hemiacetal **54**, as indicated in Eq. 19. The hemiacetal, containing a 1,2-diol system, could be oxidized further to a formate ester (**55**) that would be hydrolyzed slowly to **56**. While the plausibility of such a seven-membered hemiacetal might be expected to be less than that of the five-membered ring derivative arising from **47**, its existence in low concentration cannot be ruled out completely. In fact, a solution containing traces of this intermediate in equilibrium with the open form is all that would be required to account for the oxidation results.

45 **53** **54** (19)

56 + HCO_2H **55**

As mentioned above, in the course of proving the structure of "sedoheptulosan" Hudson (89) obtained a dialdehyde with six carbons on oxidation with sodium periodate. Further oxidation with bromine water furnished the corresponding dibasic acid. Isopropylidene and dimethylene derivatives, as well as a tetra-*O-p*-toluenesulfonate and tetra-*O*-benzoate, were likewise characterized.

Formation of a syrupy tetra-*O*-acetate with acetic anhydride and pyridine was also reported, further treatment with acetic anhydride containing a small amount of perchloric acid giving a crystalline "sedoheptulose hexa-*O*-acetate" (90).

2. 2,7-Anhydro-β-D-altroheptulofuranoses

Zill and co-workers (82, 83) were the first to show the existence of a second anhydride in equilibrium with D-altroheptulose. Isolation of this material was effected via ion-exchange chromatography of the borate derivatives.

However, Richtmyer and Pratt (90) found that they could isolate this scarce anhydride chemically from the acid-catalyzed equilibrium mixture in the following manner. Neutralization and concentration of the acidic medium caused crystallization of a major portion of 2,7-anhydro-β-D-altroheptulopyranose. Oxidation of the remaining pyranose anhydride with sodium metaperiodate, followed by treatment with bromine water and calcium carbonate and removal of the solid by filtration, gave 2,7-anhydro-β-D-altroheptulofuranose (**57**) in about 2% yield.

The structure of this anhydride was shown to be **57** for the following reasons. No reaction occurred on treatment with sodium metaperiodate. Only a mono-*O*-trityl derivative was produced even in the presence of excess tritylating reagent, and the tetra-*O*-*p*-toluenesulfonate failed to react when heated vigorously with sodium iodide. The single primary hydroxy group therefore had to be situated next to a potential carbonyl function.

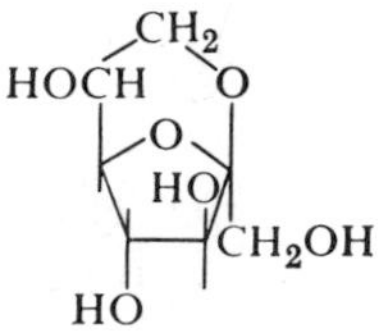

57

3. 2,7-Anhydro-α-L-galactoheptulopyranoses

Richtmyer and co-workers (91) found that heating L-galactoheptulose with 0.2*N* sulfuric acid for 3 hr at 80°C gave approximately a 5% yield of nonreducing material which could be separated into two fractions by elution from an ion-exchange column with sodium tetraborate. The fraction forming the borate complex was 2,7-anhydro-α-L-galactoheptulopyranose (**58**) whereas the other material was found to be the furanoid isomer.

In agreement with its assigned structure this anhydride consumed two molecular equivalents of sodium metaperiodate while slowly liberating one equivalent of formic acid. The resulting dialdehyde had a rotation equal but opposite in sign to that derived from 2,7-anhydro-β-D-altroheptulopyranose (see Section C-1).

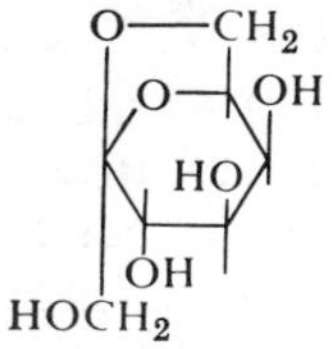

58

4. 2,7-Anhydro-α-L-galactoheptulofuranoses

In addition to 2,7-anhydro-α-L-galactoheptulopyranose (**58**) (see Section C-3), acid treatment of L-galactoheptulose generated 2,7-anhydro-α-L-galactoheptulofuranose (**59**), which could be separated from **58** on the basis of its inability to form a borate complex readily (91).

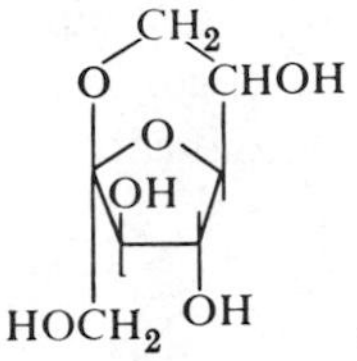

59

In order to prove structure **59**, the tetra-*O*-*p*-toluenesulfonate ester was prepared and determined to be unreactive toward sodium iodide treatment. Only a mono-*O*-tritylated product could be obtained, moreover, in accord with the presence of only one primary alcohol. Furthermore, oxidation with lead tetraacetate in pyridine occurred readily, in agreement with the fact that this reagent will cleave even sterically hindered 1,2-diols.

5. 2,7-Anhydro-β-D-glucoheptulopyranoses

Richtmyer and co-workers (92) found that free D-glucoheptulose could be transformed into the anhydride (**60**) in 2% yield by heating with 0.5*N* sulfuric acid. They also obtained the same material in 34% yield by treating phenyl α-D-glucoheptulopyranoside with hot aqueous 2*N* potassium hydroxide for 2½ hr.

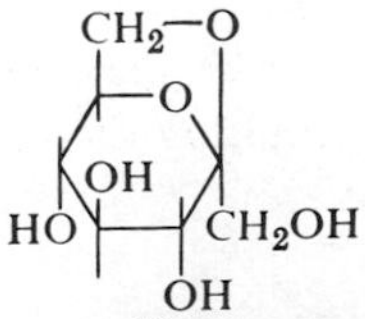

60

Metaperiodic acid oxidation of **60** afforded the expected product which was identical with the dioxane derivative obtained from 2,7-anhydro-β-D-altroheptulopyranose under similar conditions. Normal tetrasubstitution was found to occur during acetylation, benzoylation, and *p*-toluenesulfonylation.

6. 1,6-Anhydro-D-gluco-β-D-idoheptopyranose

Hudson and co-workers (93) reported that acid treatment of D-gluco-D-idoheptose generated an equilibrium mixture containing 43% 1,6-anhydro-D-gluco-β-D-idoheptopyranose (**61**).

HOCH$_2$CH—O
HO —O
HO
OH
61

7. 1,6-Anhydro-D-glycero-β-D-guloheptopyranoses

Hudson and co-workers (94) succeeded in obtaining an anhydro sugar on alkaline treatment of phenyl D-glycero-β-D-guloheptopyranoside and carried out experiments indicating that this anhydride could be represented by structure **62**.

HOCH$_2$CH—O
HO —O
OH OH
62

Stewart and Richtmyer (95) subsequently obtained an equilibrium mixture containing about 7% of this compound by treatment of D-glycero-D-guloheptose with 0.5*N* sulfuric acid. It might be noted further that the non-reducing portion of this equilibrium was also found to contain approximately 12% of a new anhydride which was formulated as 1,7-anhydro-D-glycero-D-guloheptopyranose.

According to Hudson and co-workers (94), both a tetra-*O*-benzoate and a tetra-*O*-*p*-nitrobenzoate could be prepared from this anhydride. Moreover, under normal conditions of *p*-toluenesulfonylation, only a trisubstituted derivative was obtained. That a fourth hydroxyl function was still present was shown on the basis of the fact that acetylation yielded the mono-*O*-acetyl-tri-*O*-*p*-toluenesulfonyl derivative, and that further treatment of this with sodium iodide yielded an iodide. Finally, sodium metaperiodate oxidation required two molecular equivalents of oxidant and gave one equivalent of formic acid.

The foregoing data required the presence in the molecule of three contiguous secondary hydroxy groups, and a fourth which had to be primary. On the basis of the starting material's structure, and with the knowledge that the conditions of synthesis would not affect the pyranose ring, the authors assigned to the anhydride the 1,6-anhydro-D-glycero-β-D-gulo-heptopyranose structure (**62**) with certainty.

The tri-*O*-*p*-toluenesulfonate was assumed to be the 2,3,7-substituted derivative (**63**) in which the hydroxy group at $C_{(4)}$ is sterically hindered by the sulfonylated primary hydroxyl function on $C_{(1)}$. It was also discovered that heating **63** with sodium iodide caused elimination of one *p*-toluene-sulfonate group with formation of an iodine-free derivative. Since this new compound did not contain a double bond and could not be acetylated further, the hydroxyl function at $C_{(4)}$ had apparently become masked in some fashion. Inasmuch as the same material was obtained on deacetylation of the mono-*O*-acetyl-di-*O*-*p*-toluenesulfonyl iodide, they assigned to this new compound structure **64** in which a new five-membered had obviously been created (Eq. 20). The tricyclic compound, 2,3-di-*O*-*p*-toluenesulfonyl-1,6:4,7-dianhydro-D-glycero-β-D-guloheptopyranose, was also synthesized.

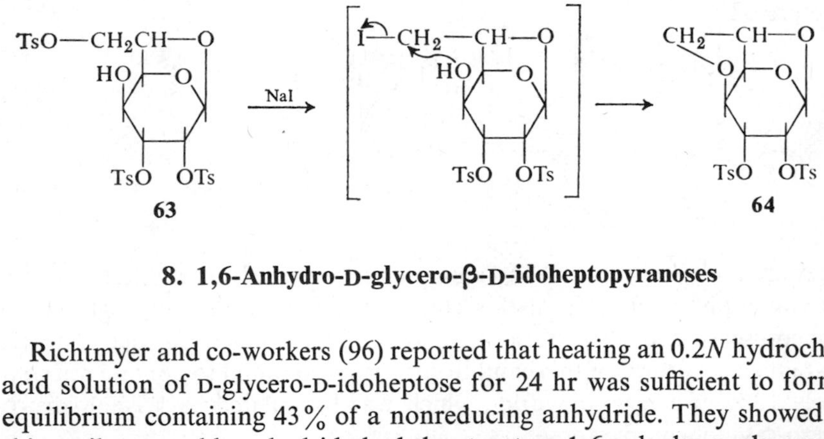

TsO—CH₂CH—O, HO, O, TsO, OTs — **63** —NaI→ [I—CH₂—CH—O, HO, O, TsO, OTs] → CH₂—CH—O, O, O, TsO, OTs — **64** (20)

8. 1,6-Anhydro-D-glycero-β-D-idoheptopyranoses

Richtmyer and co-workers (96) reported that heating an 0.2*N* hydrochloric acid solution of D-glycero-D-idoheptose for 24 hr was sufficient to form an equilibrium containing 43% of a nonreducing anhydride. They showed that this easily separable anhydride had the structure 1,6-anhydro-D-glycero-β-D-idoheptopyranose (**65**).

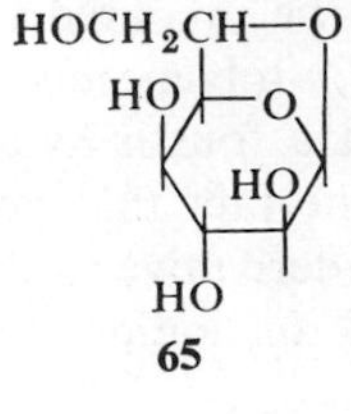

HOCH₂CH—O, HO, O, HO, HO — **65**

In the same communication the above authors reported that the usual sodium metaperiodate treatment gave results similar to those obtained with 1,6-anhydro-D-glycero-β-D-guloheptopyranose, the $C_{(2)}$-epimer (see Section C-7). The primary *p*-toluenesulfonate group could be replaced with sodium iodide, subsequent hydrogenation of the intermediate iodo derivative leading to a C-methyl group.

When *p*-toluenesulfonylation was attempted, two products were obtained. The first, formed in about 70% yield, was a tetra-*O*-*p*-toluenesulfonate; the second, formed in about 20% yield, was a tri-*O*-*p*-toluenesulfonate. Recalling the similar tri-substituted derivative isolated from 1,6-anhydro-D-glycero-β-D-guloheptopyranose (see Section C-7), Richtmyer and co-workers demonstrated that deacetylation of the mono-*O*-acetyl-di-*O*-*p*-toluenesulfonyl iodide led to a new anhydro compound. They assumed that they had accomplished the conversion of **66** into 1,6:4,7-dianhydro-2,3-di-*O*-*p*-toluenesulfonyl-D-idoheptopyranose (**67**) as indicated in Eq. 21.

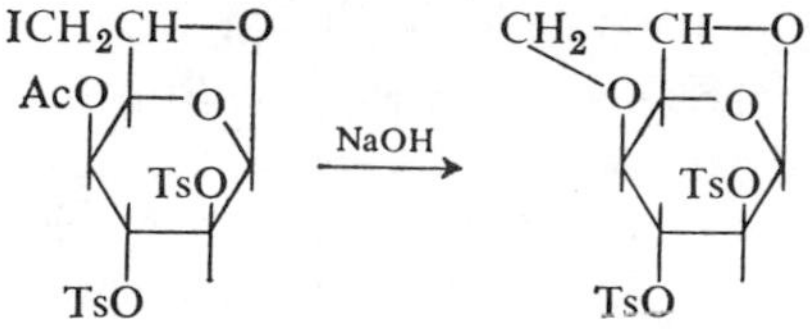

66 **67** (21)

9. 2,7-Anhydro-β-L-guloheptulopyranoses

Hudson and co-workers (97) have reported the formation of a nonreducing anhydride on refluxing L-guloheptulose for 1 hr in 0.2*N* hydrochloric acid. In fact, much to their surprise, the equilibrium mixture contained 80% of the 2,7-anhydro-β-L-guloheptulopyranose (**68**) which they were able to isolate readily in crystalline form.

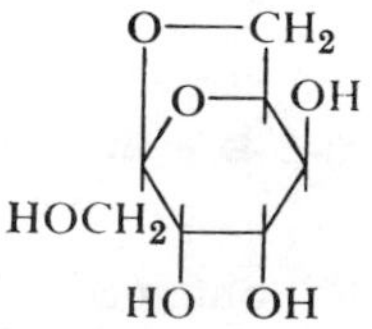

68

The proof of structure for this new anhydride involved the usual sodium metaperiodate oxidation, which yielded formic acid. The accompanying dialdehyde underwent hypobromite oxidation to diacid **69** which was shown to be the enantiomorph of the diacid derived from 2,7-anhydro-β-D-altroheptulopyranose. This conclusion was reached on the basis of rotations and melting points as well as X-ray powder photographs.

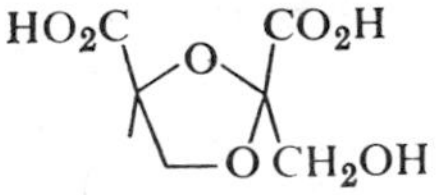

HO_2C CO_2H O O CH_2OH

69

10. 2,7-Anhydro-β-D-idoheptulopyranose

It was observed by Hudson and co-workers (93) that heating D-idoheptulose for 1 hr with 0.2N hydrochloric acid was sufficient to generate an equilibrium mixture containing 85% of the nonreducing anhydride. This new material was isolated with little difficulty by crystallization after destruction of the unchanged ketose by alkaline treatment.

The anhydride was shown to have three contiguous hydroxy groups by sodium metaperiodate oxidation which consumed two molecular equivalents of oxidant and released one equivalent of formic acid. The dialdehyde obtained in this manner was shown to be the same as that derived from 2,7-anhydro-β-D-altroheptulopyranose. The tetra-*O*-*p*-toluenesulfonate was also prepared and proved to be unaffected by sodium iodide treatment. By analogy with the preparation of other anhydrides, 2,7-anhydro-β-D-idoheptulopyranose (**70**) was selected as the structure.

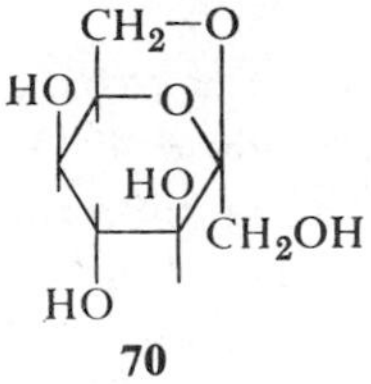

CH_2—O HO O HO CH_2OH HO

70

11. 2,7-Anhydro-β-D-mannoheptulopyranoses

Richtmyer and co-workers (75) obtained 2,7-anhydro-β-D-mannoheptulopyranose (**71**) from D-mannoheptulose by heating in 0.2N sulfuric acid solution for 3 hr at 90°C. Appropriate isolation procedures afforded the pure anhydride in approximately 8% yield. Further search of the equilibrium mixture failed to reveal any trace of other anhydride derivatives such as the furanose isomer.

The same anhydride was also obtained in approximately 8% yield on treatment of phenyl α-D-mannoheptulopyranoside with 1*N* aqueous potassium hydroxide under reflux for 3 hr.

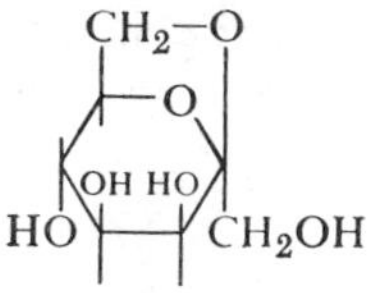

71

2,7-Anhydro-β-D-mannoheptulopyranose (**71**) was found to undergo the usual periodate cleavage with liberation of one molecular equivalent of formic acid. Tetra-*O*-benzoate and tetra-*O*-*p*-toluenesulfonate derivatives were also characterized. In addition, 2,7-anhydro-3,4-isopropylidene-β-D-mannoheptulopyranose was synthesized by reaction of **71** with acetone in the presence of copper sulfate.

D. Oxepins Derived from Sugars and Aldehydes

1. Formaldehyde Derivatives

Head (98) has discussed the formation of *trans*-hexahydro-1,3,5-trioxepin from paraldehyde and *trans*-1,2-dihydroxycyclohexane, and has shown that with equimolar aldehyde concentrations the yield of five-membered ring from the *cis*-diol is higher than the yield of seven-membered ring from the *trans*-diol. Nonetheless, the literature does contain some examples of dioxepins derived by condensation of sugars with aldehydes.

Hudson and co-workers (99) found that allowing D-mannitol to stand in acidic formaldehyde solution resulted in the slow crystallization of trimethylene-D-mannitol. After acetolysis with sulfuric acid and acetic anhydride, and subsequent hydrolysis with sodium methoxide, they isolated a compound to which they assigned the structure 2,5-methylene-D-mannitol (**72**). Thus the trimethylene derivative had to be 1,3:2,5:4,6-trimethylene-D-mannitol. Later papers (100, 101) suggested several possible factors leading to dioxepin formation instead of 1,3-dioxane formation, a plausible alternative.

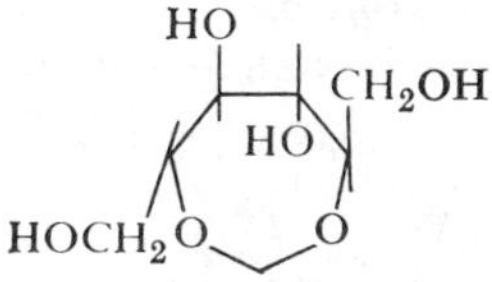

72

Baker and Kohanyi (102) have reported that treatment of 2,5-methylene-2,4-di-*O*-*p*-toluenesulfonyl-1,6-di-*O*-trityl-D-mannitol with either sodium methoxide in methanolic chloroform or potassium hydroxide in water leads to 3,4-anhydro-2,5-methylene-1,6-di-*O*-trityl-D-talitol (**73**). While this derivative proved stable in alkali, hydrolysis in refluxing 10% hydrochloric acid for 2 hr removed the trityl groups, opened the epoxide ring, and partially cleaved the methylene ring. Stronger acid conditions had a similar effect.

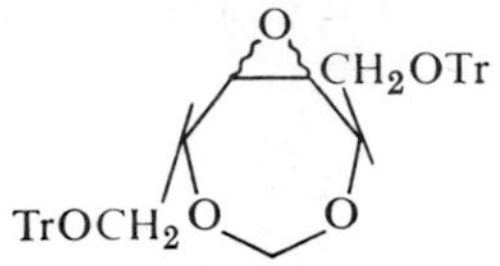

73

The major product of the acid-catalyzed reaction of L-rhamnose and formaldehyde has been shown to be 3,4-dimethyleneoxy-L-rhamnose (**74**) (103). A minor product has been found to be the isomeric 2,3-dimethyleneoxy-L-rhamnose (**75**), as depicted in Eq. 22.

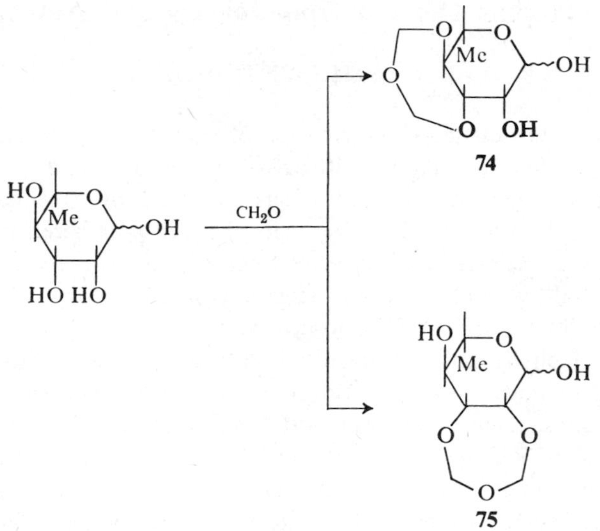

74

75

(22)

Zissis and Richtmyer (104) observed that acid-catalyzed condensation of aqueous formaldehyde with volemitol leads to a trimethylene derivative formulated as 1,3:2,5:4,6-trimethylene-D-glycero-D-mannoheptitol (**76**).

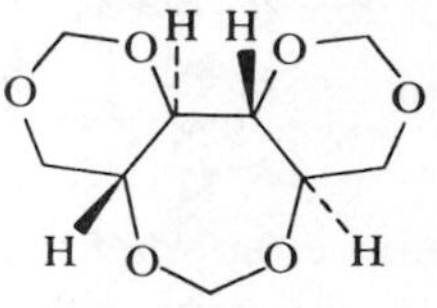

76

2. Acetaldehyde and Higher Aldehyde Derivatives

Appel and co-workers (105) condensed both α- and β-methyl glucoside with acetaldehyde and sulfuric acid to form the corresponding 2,3-oxido-diethylidene-4,6-ethylidenemethyl glucosides (77). As expected, the seven-membered ring in **77** proved to be acid-labile. Helferich and Porck (106) made the same acetal from benzyl β-D-glucoside. Upon reductive debenzylation they isolated a compound with alkali-stable substituents blocking the hydroxyl functions at $C_{(2)}$, $C_{(3)}$, $C_{(4)}$, and $C_{(6)}$.

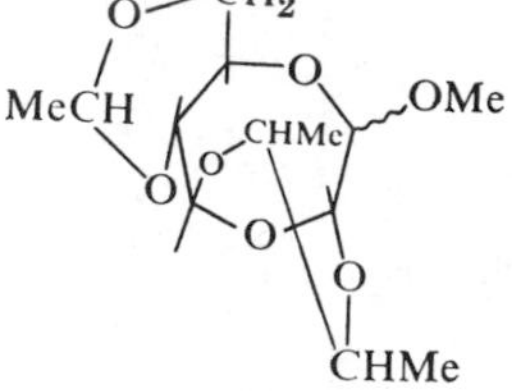

77

Mellies and co-workers (107) have condensed several different aldehydes with D-glucose in the preparation of trioxepin derivatives.

E. Oxepins Related to Sugars

The scission of vicinal diols leads to dialdehydes which are believed, in some instances, to undergo cyclization to hemiacetal hydrates. An example of such a reaction given by Goldschmid and Perlin (108) involves the oxidation of methyl 4,6-benzylidene-α-D-glucopyranoside to compound **78**, as indicated in Eq. 23.

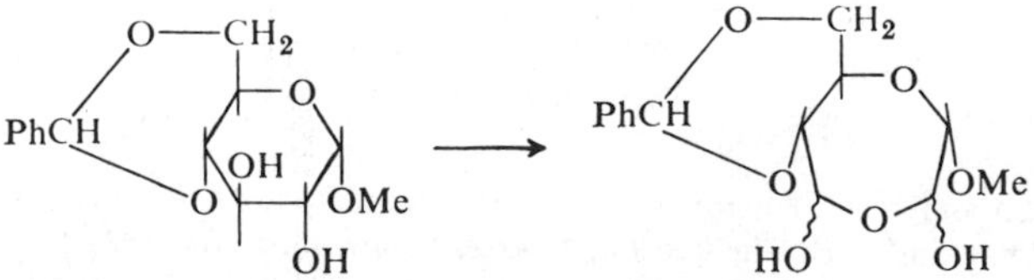

(23)

78

In the course of work on diethylsulfonyl derivatives of some heptoses (109), a hemiacetal was reported to form on periodate oxidation of compound **79**.

Derivatives of the two similar acids quinic and shikimic have been shown to lactonize internally to give compounds **80** and **81**, respectively (110, 111).

In the epiquinicol series, Gorin (110) found that acetic acid treatment of 5-*O*-*p*-toluenesulfonylepiquinicol yield 2′,5-anhydroquinicol (**82**). Likewise

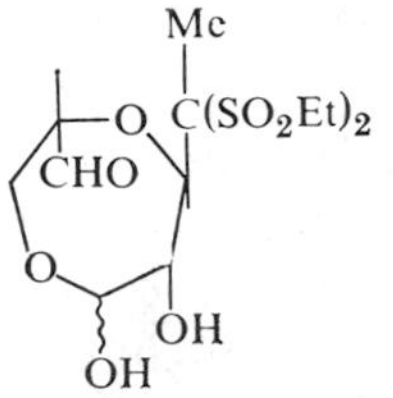

Me

$C(SO_2Et)_2$

CHO

O

O

OH

OH

79

the 3-*O*-*p*-toluenesulfonate isomer yields 2′,3-anhydroquinicol (**83**) under alkaline conditions.

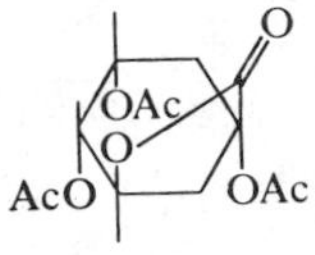

Me Me

O O

O

OAc

O

Br

AcO

OAc

Br

O

O

80

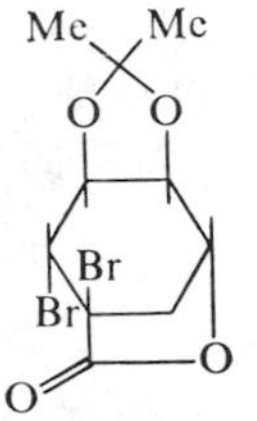

81

O CH_2

HO

OH

OH

CH_2

OH

O

HO

OH

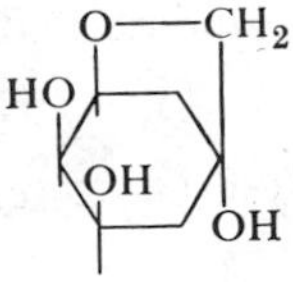

82

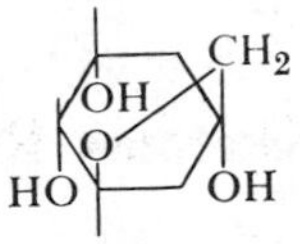

83

Some work has been reported on hexitol anhydrides in which ir data were used for the correlation of ring size (112). A growing interest in branched-chain sugars has led to some synthetic work that has yielded some novel dioxepin sugar derivatives (113).

F. References

1. J. W. Pratt and N. K. Richtmyer, *J. Amer. Chem. Soc.*, **77,** 1906 (1955).
2. L. Goodman and J. E. Christensen, *J. Amer. Chem. Soc.*, **83,** 3823 (1961).
3. N. K. Richtmyer and C. S. Hudson, *J. Amer. Chem. Soc.*, **57,** 1716 (1935).
4. N. K. Richtmyer and C. S. Hudson, *J. Amer. Chem. Soc.*, **61,** 214 (1939).
5. F. H. Newth and L. F. Wiggins, *J. Chem. Soc.*, **1950,** 1734.
6. G. J. Robertson and H. G. Dunlop, *J. Chem. Soc.*, **1938,** 472.
7. N. K. Richtmyer and C. S. Hudson, *J. Amer. Chem. Soc.*, **62,** 961 (1940).
8. N. K. Richtmyer and C. S. Hudson, *J. Amer. Chem. Soc.*, **63,** 1727 (1941).
9. F. H. Newth, *J. Chem. Soc.*, **1956,** 441.

10. J. W. Pratt and N. K. Richtmyer, *J. Amer. Chem. Soc.*, **79,** 2597 (1957).
11. F. Micheel, *Chem. Ber.*, **62,** 687 (1929).
12. D. McCreath and F. Smith, *J. Chem. Soc.*, **1939,** 387.
13. R. M. Hann and C. S. Hudson, *J. Amer. Chem. Soc.*, **63,** 1484 (1941).
14. R. M. Hann and C. S. Hudson, *J. Amer. Chem. Soc.*, **64,** 2435 (1942).
15. N. K. Richtmyer, *Arch. Biochem. Biophys.*, **78,** 376 (1958).
16. M. Cerny, V. Gut, and J. Pacak, *Collect. Czech. Chem. Commun.*, **26,** 2542 (1961).
17. M. Cerny, I. Buben, and J. Pacak, *Collect. Czech. Chem. Commun.*, **28,** 1569 (1963).
18. K. Heyns, J. Weyer, and H. Paulsen, *Chem. Ber.*, **98,** 327 (1965).
19. M. Cerny, J. Pacak, and J. Stanek, *Chem. Ind.* (London), **1961,** 945.
20. B. H. Alexander, R. J. Dimler, and C. L. Mehltretter, *J. Amer. Chem. Soc.*, **73,** 4658 (1951).
21. R. M. Hann and C. S. Hudson, *J. Amer. Chem. Soc.*, **63,** 2241 (1941).
22. M. Tanret, *Bull. Soc. Chim. Fr.*, **11,** 949 (1894).
23. A. Pictet and J. Sarasin, *Helv. Chem. Acta*, **1,** 87 (1918).
24. P. Karrer and A. P. Smirnoff, *Helv. Chem. Acta*, **4,** 817 (1921).
25. J. C. Irvine and J. W. H. Oldham, *J. Chem. Soc.*, **119,** 1744 (1921).
26. E. M. Montgomery, N. K. Richtmyer, and C. S. Hudson, *J. Amer. Chem. Soc.*, **65,** 3 (1943).
27. E. K. Gladding and C. B. Purves, *J. Amer. Chem. Soc.*, **66,** 76 (1944).
28. S. Peat, W. J. Whelan, T. E. Edwards, and O. Owen, *J. Chem. Soc.*, **1958,** 586.
29. H. B. Wood Jr., R. Allerton, H. W. Diehl, and H. G. Fletcher, *J. Org. Chem.*, **20,** 875 (1955).
30. H. B. Wood and H. G. Fletcher, *J. Amer. Chem. Soc.*, **78,** 207 (1956).
31. M. Akagi, S. Tejima, and M. Haga, *Chem. Pharm. Bull.* (Tokyo), **10,** 905 (1962).
32. P. S. O'Colla and E. Lee, *Chem. Ind.* (London), **1956,** 522.
33. P. S. O'Colla, E. E. Lee, and D. McGrath, *J. Chem. Soc.*, **1962,** 2730.
34. O. P. Golova, Ya. V. Epshtein, N. S. Maksimenko, V. N. Sergeeva, A. I. Kalnin'sh, P. N. Odintsov, and V. G. Panasyuk, *Gidrolizn. i Lesokhim. Prom.*, **14,** 4 (1961); *Chem. Abstr.*, **59,** 15485e (1963).
35. J. C. Irvine and J. W. H. Oldham, *J. Chem. Soc.*, **127,** 2729 (1925).
36. E. L. Jackson and C. S. Hudson, *J. Amer. Chem. Soc.*, **62,** 958 (1940).
37. G. Zemplen, A. Csuros, and S. Angyal, *Chem. Ber.*, **70,** 1848 (1937).
38. A. Thompson, K. Anno, M. L. Wolfrom, and M. Inatome, *J. Amer. Chem. Soc.*, **76,** 1309 (1954).
39. Wm. Charlton, W. N. Haworth, and R. Wm. Herbert, *J. Chem. Soc.*, **1931,** 2855.
40. R. W. Jeanloz, A. M. C. Rapin, and S-I. Hakomori, *J. Org. Chem.*, **26,** 3939 (1961).
41. M. Cerny, V. Gut, and J. Pacak, *Collect. Czech. Chem. Commun.*, **26,** 2542 (1961).
42. R. W. Jeanloz and A. M. C. Rapin, *J. Org. Chem.*, **28,** 2978 (1963).
43. F. Micheel and E. Michaelis, *Chem. Ber.*, **91,** 188 (1958).
44. F. Micheel and E. Michaelis, *Chem. Ber.*, **96,** 1959 (1963).
45. K. Freudenberg and W. Nagai, *Chem. Ber.*, **66,** 27 (1933).
46. G. S. Barysheva, N. A. Vasyunina, and S. V. Chepigo, *Sb. Tr. Gos. Nauchn. Issled. Inst. Gidrolizn. i Sul'fitno-Spirt. Prom.*, **11,** 94 (1963); *Chem. Abstr.*, **61,** 3285h (1964).
47. F. Micheel and O. Berendes, *Mikrochim. Ichnoanal. Acta*, **1963,** 519; *Chem. Abstr.*, **59,** 9294e (1963).
48. C. M. McCloskey and G. H. Coleman, *J. Org. Chem.*, **10,** 184 (1945).
49. E. M. Montgomery, N. K. Richtmyer, and C. S. Hudson, *J. Org. Chem.*, **10,** 194 (1945).

50. R. U. Lemieux and C. Brice, *Can. J. Chem.*, **30,** 295 (1952).
51. F. Micheel and A. Klemer, *Chem. Ber.*, **91,** 194 (1958).
52. M. Akagi, S. Tejima, and M. Haga, *Chem. Pharm. Bull.* (Tokyo), **10,** 1039 (1962).
53. M. L. Wolfrom, A. Thompson, and R. B. Ward, *J. Amer. Chem. Soc.*. **81,** 4623 (1959).
54. M. L. Wolfrom, A. Thompson, R. B. Ward, D. Horton, and R. H. Moore, *J. Org. Chem.*, **26,** 4617 (1961).
55. H. Prugsheim and K. Schmalz, *Chem. Ber.*, **55,** 3001 (1922).
56. J. C. Irvine and J. W. H. Oldham, *J. Chem. Soc.*, **127,** 2903 (1925).
57. C-C. Tu and C. Schuerch, *J. Polym. Sci.*, *Part B*, **1,** 163 (1963).
58. V. V. Korshak, V. A. Sergeev, J. Surna, and R. Pernikis, *Vysokomol. Soedin.*, **5,** 1593 (1963); *Chem. Abstr.*, **61,** 3285g (1964).
59. R. F. Schwenker Jr. and L. R. Beck Jr., *J. Polym. Sci.*, *Part C*, **2,** 331 (1963).
60. S. Glassner and A. R. Pierce, *Anal. Chem.*, **37,** 525 (1965).
61. R. J. Dimler, H. A. Davis, and G. E. Hilbert, *J Amer. Chem. Soc.*, **68,** 1377 (1946).
62. L. C. Stewart and N. K. Richtmyer, *J. Amer. Chem. Soc.*, **77,** 1021 (1955).
63. E. E. van Tamelen, J. R. Dyer, H. E. Carter, J. V. Pierce, and E. E. Daniels, *J. Amer. Chem. Soc.*, **78,** 4817 (1956).
64. H. E. Carter, J. V. Pierce, G. B. Whitfield Jr., J. E. McNary, E. E. van Tamelen, J. R. Dyer, and H. A. Whaley, *J. Amer. Chem. Soc.*, **83,** 4289 (1961).
65. E. Sorkin and T. Reichstein, *Helv. Chem. Acta*, **28,** 1 (1945).
66. R. E. Reeves, *J. Amer. Chem. Soc.*, **71,** 2116 (1949).
67. R. Kuhn and W. Bister, *Justus Liebigs Ann. Chem.*, **617,** 92 (1958).
68. R. W. Jeanloz, Z. T. Hlazer, and D. A. Jeanloz, *J. Org. Chem.*, **26,** 532 (1961).
69. R. W. Jeanloz and D. A. Jeanloz, *J. Org. Chem.*, **26,** 537 (1961).
70. S. P. James, F. Smith, M. Stacey, and L. F. Wiggins, *J. Chem. Soc.*, **1946,** 625.
71. N. Baggett and R. W. Jeanloz, *J. Org. Chem.*, **28,** 1845 (1963).
72. G. Zemplen, A. Gerecs, and T. Valatin, *Chem. Ber.*, **73,** 575 (1940).
73. A. E. Knauf, R. M. Hann, and C. S. Hudson, *J. Amer. Chem. Soc.*, **63,** 1447 (1941).
74. E. M. Montgomery, N. K. Richtmyer, and C. S. Hudson, *J. Amer. Chem. Soc.*, **64,** 1483 (1942).
75. E. Zissis, L. C. Stewart, and N. K. Richtmyer, *J. Amer. Chem. Soc.*, **79,** 2593 (1957).
76. G. O. Aspinall and G. Zweifel, *J. Chem. Soc.*, **1957,** 2271.
77. W. T. Haskins, R. M. Hann, and C. S. Hudson, *J. Amer. Chem. Soc.*, **64,** 1289 (1942).
78. W. T. Haskins, R. M. Hann, and C. S. Hudson, *J. Amer. Chem. Soc.*, **63,** 1725 (1941).
79. W. T. Haskins, R. M. Hann, and C. S. Hudson, *J. Amer. Chem. Soc.*, **64,** 1852 (1942).
80. R. M. Hann and C. S. Hudson, *J. Amer. Chem. Soc.*, **64,** 925 (1942).
81. F. B. LaForge and C. S. Hudson, *J. Biol. Chem.*, **30,** 61 (1917).
82. L. P. Zill, J. X. Khym, and G. M. Cheniae, *J. Amer. Chem. Soc.*, **75,** 1339 (1953).
83. L. P. Zill and N. E. Tolbert, *J. Amer. Chem. Soc.*, **76,** 2929 (1954).
84. E. Zissis and N. K. Richtmyer, *J. Org. Chem.*, **30,** 462 (1965).
85. C. S. Hudson, *J. Amer. Chem. Soc.*, **60,** 1241 (1938).
86. J. W. Pratt, N. K. Richtmyer, and C. S. Hudson, *J. Amer. Chem. Soc.*, **73,** 1876 (1951).
87. H. Hibbert and C. G. Anderson, *Can. J. Res.*, **3,** 306 (1930); *Chem. Abstr.*, **25,** 498 (1931).
88. J. W. Pratt, N. K. Richtmyer, and C. S. Hudson, *J. Amer. Chem. Soc.*, **74,** 2200 (1952).
89. W. T. Haskins, R. M. Hann, and C. S. Hudson, *J. Amer. Chem. Soc.*, **74,** 2198 (1952).
90. N. K. Richtmyer and J. W. Pratt, *J. Amer. Chem. Soc.*, **78,** 4717 (1956).
91. L. C. Stewart, E. Zissis, and N. K. Richtmyer, *J. Org. Chem.*, **28,** 1842 (1963).

92. L. C. Stewart, E. Zissis, and N. K. Richtmyer, *Chem. Ber.*, **89,** 535 (1956).
93. J. W. Pratt, N. K. Richtmyer, and C. S. Hudson, *J. Amer. Chem. Soc.*, **74,** 2210 (1952).
94. E. M. Montogomery, N. K. Richtmyer, and C. S. Hudson, *J. Amer. Chem. Soc.*, **65,** 1848 (1943).
95. L. C. Stewart and N. K. Richtmyer, *J. Amer. Chem. Soc.*, **77,** 424 (1955).
96. J. W. Pratt, N. K. Richtmyer, and C. S. Hudson, *J. Amer. Chem. Soc.*, **75,** 4503 (1953).
97. L. C. Stewart, N. K. Richtmyer, and C. S. Hudson, *J. Amer. Chem. Soc.*, **74,** 2206 (1952).
98. F. S. H. Head, *J. Chem. Soc.*, **1960,** 1778.
99. A. T. Ness, R. M. Hann, and C. S. Hudson, *J. Amer. Chem. Soc.*, **65,** 2215 (1943).
100. R. M. Hann and C. S. Hudson, *J. Amer. Chem. Soc.*, **66,** 1909 (1944).
101. A. T. Ness, R. M. Hann, and C. S. Hudson, *J. Amer. Chem. Soc.*, **70,** 765 (1948).
102. S. B. Baker and G. Kohanyi, *J. Amer. Chem. Soc.*, **75,** 2140 (1953).
103. P. Andrews, L. Hough, and J. K. N. Jones, *J. Amer. Chem. Soc.*, **77,** 125 (1955).
104. E. Zissis and N. K. Richtmyer, *J. Org. Chem.*, **22,** 1528 (1957).
105. H. Appel, W. N. Haworth, E. G. Cox, and F. J. Llewellyn, *J. Chem. Soc.*, **1938,** 793.
106. B. Helferich and A. Porck, *Justus Liebigs Ann. Chem.*, **582,** 225 (1953).
107. R. L. Mellies, C. L. Mehltretter, and C. E. Rist, *J. Amer. Chem. Soc.*, **73,** 294 (1951).
108. H. R. Goldschmid and A. S. Perlin, *Can. J. Chem.*, **38,** 2280 (1960).
109. L. D. Hall, L. Hough, S. H. Shute, and T. J. Taylor, *J. Chem. Soc.*, **1965,** 1154.
110. P. A. J. Gorin, *Can. J. Chem.*, **41,** 2417 (1963).
111. B. A. Bohm, *Chem. Rev.*, **65,** 435 (1965).
112. P. Sohar, L. Vargha, and E. Kasztreimer, *Tetrahedron*, **20,** 647 (1964).
113. J. S. Burton, W. G. Overend, and N. R. Williams, *Chem. Ind.* (London), **1961,** 175.

CHAPTER IX

Alkaloids Containing a Seven-Membered Oxygen Ring

PAUL J. SCHEUER

Department of Chemistry, University of Hawaii, Honolulu, Hawaii

A. Introduction

Basic nitrogen, the characteristic heteroatom of alkaloids, commonly occurs in this class of compounds as a member of a ring. A majority of alkaloids also contain oxygen, but with this hetero element the reverse situation obtains: cyclic oxygen in alkaloids, if one regards the dioxymethylene grouping as a variant of a methoxyl rather than as a true cycle, is uncommon and oxygen in a seven-membered ring occurs vary rarely. In fact, there would be little justification for a separate treatment of this group of compounds were it not for the circumstance that, among others, the strychnine alkaloids, which are the subject of an extensive literature, contain this rare structural feature. Because of this situation, however, it might be advantageous to examine this wealth of information and to focus on only one part of the molecule, the seven-membered oxygen ring. Thereby it might be possible to see the properties of this structural moiety from a new angle. When viewed in this manner, the implications of a given reaction of the oxepin system for the total structural problem, while historically important, will no longer be dominant. Whether this is a fruitful approach may to some extent become apparent in the course of this review, since the strychnine bases are not the sole representatives of alkaloids containing an oxepin system. The cularine and insularine groups, though little known, have this structural feature in common with the strychnine alkaloids and may thus serve as a convenient internal standard for the present treatment.

The organization of this chapter is aligned with the approach just outlined. The properties of the oxepin system in strychnine and its congeners will be discussed first, followed by a treatment of oxepin chemistry as exhibited in the cularine and insularine alkaloids.

B. The Strychnine Group

The most recent comprehensive review of strychnine chemistry is that of Smith (1). It includes a historical survey of the field and a discussion of some of the transformations of strychnine and its congeners. Boit's treatment (2) is briefer than Smith's; it emphasizes recent developments and offers useful tabular summaries. A set of tables, which includes references to spectral data, was published by Hesse (3). Selected aspects of strychnine chemistry are discussed by Battersby and Hodson in Volume XI of the Manske series (4).

It is not the intent of the present chapter to duplicate these excellent sources. Our focal point will be the oxepin portion of the strychnine alkaloids. But for those readers who are unfamiliar with strychnine chemistry, some background material will be presented briefly so that the discussion of oxepin chemistry which follows may be placed in its proper setting.

1. General Background

The correct structure of strychnine (**1**) has been amply confirmed by X-ray crystallography in Robertson's (5) and Bijvoet's (6) laboratories and by Woodward's synthesis (7). Chemical degradative work, which led to structure **1** (exclusive of some stereochemical aspects) had extended over many years. The central portion of the molecule comprising $C_{(14)}$, $C_{(15)}$, and $C_{(16)}$ yielded direct evidence only with difficulty, and long after the sequence of the peripheral chain of atoms was established. The structural problem was fully solved by Woodward (8) in 1948 after he had proferred the correct structure a year earlier (9). Although Robinson claimed to have considered the correct structure as early as 1942 (10), he regarded it as one of several possible structures; in his 1947 paper Robinson (11) favored the correct structure over his other choices, but did not consider the structural problem

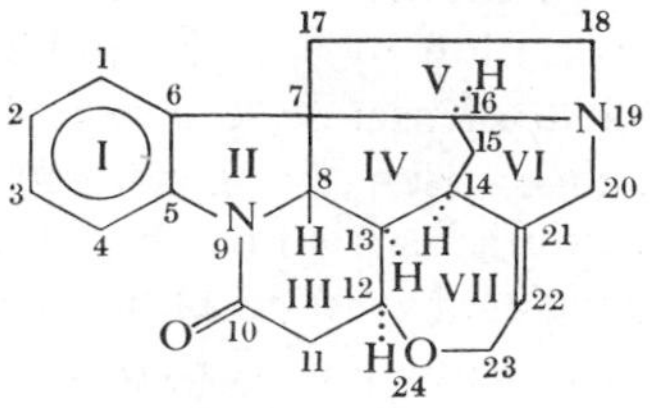

1

solved. The respective sizes of rings IV, V, and VI were the last aspects of the strychnine problem to be solved, while remarkably enough the size of ring VII, the rare oxepin cycle, had been suggested by Robinson (12) as early as 1932 and was retained in all subsequent formulations of the alkaloid. Our discussion will deal with the properties of this oxepin system as it is found in a majority of the known strychnine alkaloids.

2. Opening of the Oxepin Ring

Ring VII can be opened with water at 180°C (13) or with methanolic ammonia at 150°C (14) and leads to isostrychnine I (**2**). The ease of this transformation is without precedent in an isolated oxepin system, but is clearly a consequence of the position of the ether oxygen atom β to the lactam carbonyl in ring III. The geometry of ring III apparently suppresses normal lactam resonance and places a high degree of carbanion activity at $C_{(11)}$, as evidenced by the ease of formation of a benzal derivative. Removal of a proton at $C_{(11)}$ clearly triggers the opening of ring VII, and in the absence of a $C_{(10)}$ carbonyl, as in strychnidine derivatives, the ring opening fails to take place. Again the geometry of the six intact rings forces the newly created double bond of ring III out of conjugation. The ring-opening reaction is stereospecifically reversible. When isostrychnine I is treated with alcoholic potassium hydroxide, strychnine is regenerated, as demonstrated by Prelog and co-workers (15). This facile ring closure constituted the final step in Woodward's total synthesis (7). Although two asymmetric carbon atoms are reformed in this reaction, only one of the four possible isomers is strain-free.

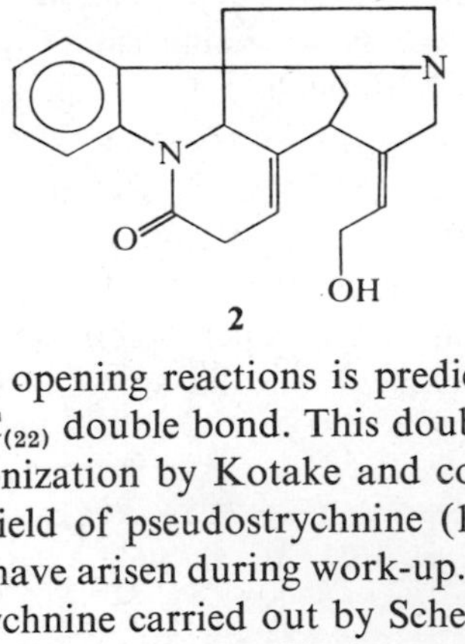

2

Another series of ring opening reactions is predicated upon the presence in ring VII of the $C_{(21)}$-$C_{(22)}$ double bond. This double bond of strychnine is not readily cleaved. Ozonization by Kotake and collaborators (16) has led only to an unspecified yield of pseudostrychnine (16-hydroxystrychnine), a product which may well have arisen during work-up. In contrast, ozonization of the neutral 18-oxostrychnine carried out by Scheuer (17) under relatively drastic conditions, in glacial acid at 10–15°C, has produced the 21,22-epoxide and a cleavage product with subsequent loss of $C_{(22)}$ and $C_{(23)}$. Formation of an epoxide has only occasionally been observed with hindered olefins (18).

Other oxidizing agents which likewise fail to attack ring VII are chromic acid in aqueous sulfuric acid (19, 20) and nitric acid (21). Perbenzoic acid does not attack the olefinic linkage in strychnine. Kotake and co-workers (22) showed that it furnishes the *N*-oxide. The neutral 18-oxo compound, however, yields a $C_{(21)}$-$C_{(22)}$ epoxide, which is resistant to ring opening under drastic conditions (17).

Attack on the $C_{(21)}$-$C_{(22)}$ double bond can be achieved with permanganate. This reaction, under a variety of conditions, has been among the most revealing degradative pathways in strychnine chemistry. The 21,22-diol is only a minor product when strychnine is treated with permanganate in acetone or for reasons of solubility in acetone–chloroform (23). The compound was also obtained readily in Kotake's laboratory by reaction of strychnine with osmium tetroxide in chloroform (24). In contrast, 18-oxostrychnine furnishes the corresponding diol as the sole product (17). Solubility may well be the chief reason for this contrasting behavior. In dilute aqueous acid the product of oxidation with permanganate is ketol **3**, as demonstrated by Prelog and Kathriner (25). Catalytic hydrogenation of **3** leads to the 21,22-diol, long presumed to have identical (*cis*) configuration, although no direct comparison was made (25) until the recent studies of Wheeler and co-workers (26).

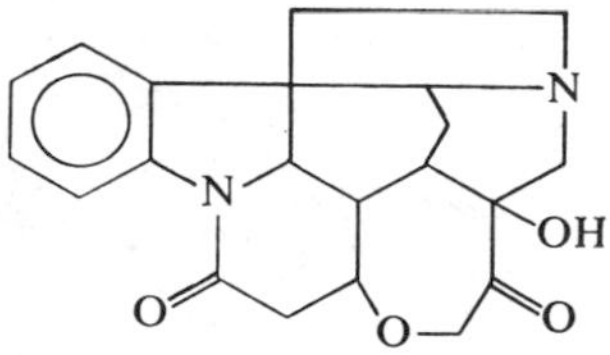

N N OH O O O

3

Wheeler and co-workers showed that the major product of catalytic hydrogenation was indeed the *cis* diol, and that the *trans* isomer was formed as a by-product. From this result they conclude that ketol **3** exists in both chair and boat conformations and that the chair form predominates. Chemical reduction of **3** with zinc and acid, on the other hand, leads to opening of ring VII. Prelog and Kathriner (25) assigned structure **4** to this compound. The acids, strychninonic (**5**) and a lesser amount of dihydrostrychninonic

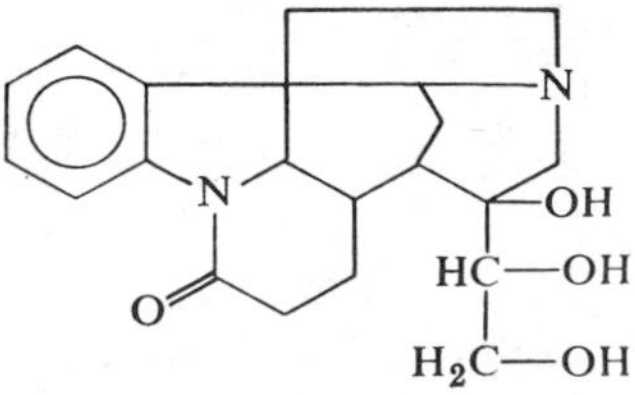

N N OH HC—OH H_2C—OH O

4

(**6**), in which the oxepin ring is cleaved, are the major products of permanganate oxidation in acetone and were in fact the only compounds isolated when the reaction was first carried out by Leuchs (27, 28). The $C_{(21)}$ epimer of **6**, strychninolic acid, was obtained from **5** by sodium amalgam reduction by Leuchs and Schneider (29) or catalytically by Prelog and Szpilfogel (30). A rationale for the stereochemical course of these transformations was provided by Edward (31).

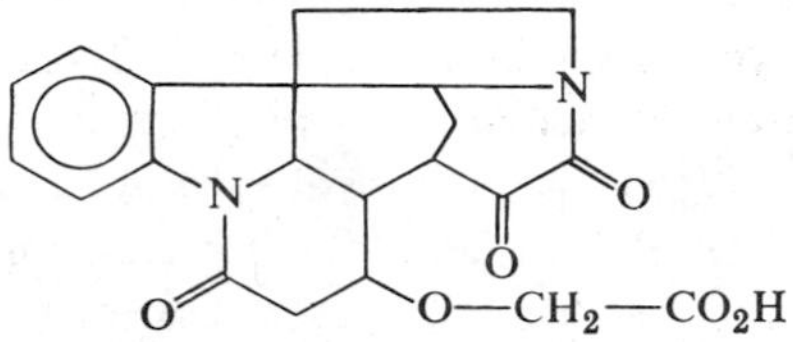

5 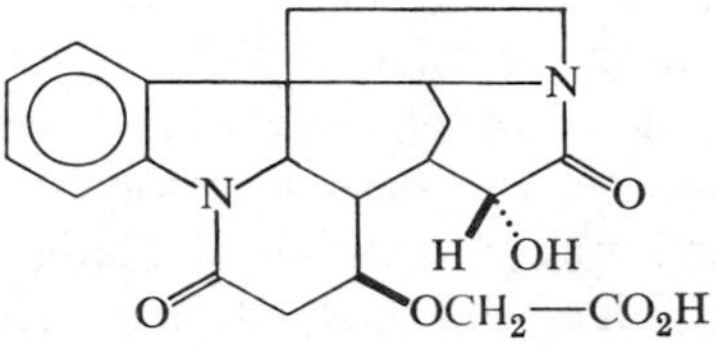 **6**

Ring opening of the 21,22-diol or of ketol **3** with lead tetraacetate to the keto aldehyde **7** was shown by Kotake and co-workers (32) to be unexceptional, and was studied in detail on 16-hydroxy- and 16-methoxystrychnine (ψ-strychnine and its methyl ether) by Prelog and Kocor (33).

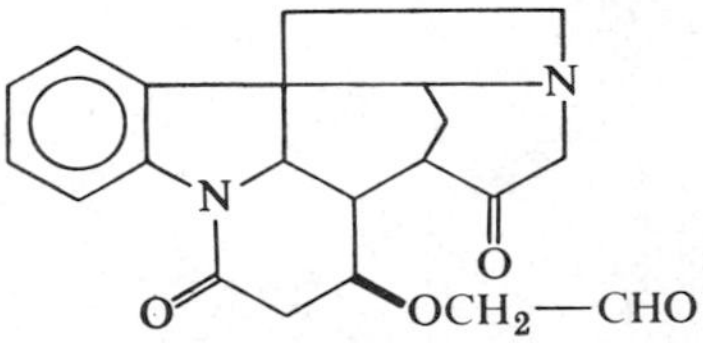

7

Cleavage of ring VII under reductive conditions has also been reported. According to Leuchs and co-workers (34, 35), when strychnidine (**8**) is hydrogenated in dilute hydrochloric acid at elevated temperature in the presence of a platinum catalyst, the resulting dodecahydrostrychnine (**9**) is a compound with a reduced benzene ring, a cleaved allylic ether in ring VII, and a saturated $C_{(21)}$-$C_{(22)}$ double bond.

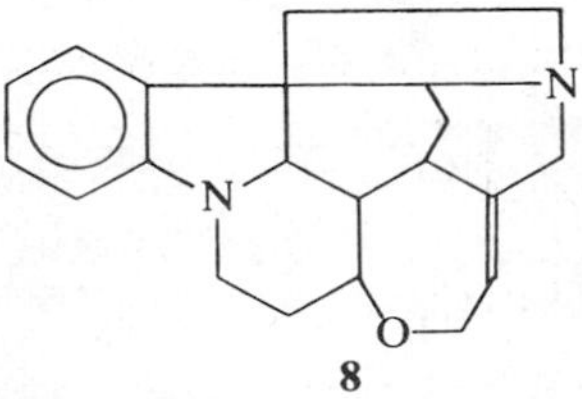

8 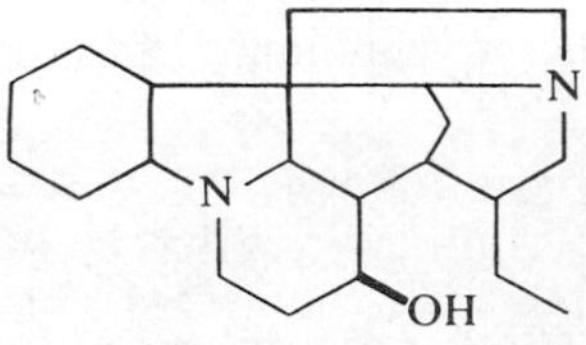 **9**

The Emde degradation (sodium amalgam) of strychnine methosulfate not only leads to opening of ring VI as expected, but also involves hydrogenolysis of the allylic ether in ring VII (36). Structure **10** has been proposed for this compound and it rests on reasonably good evidence.

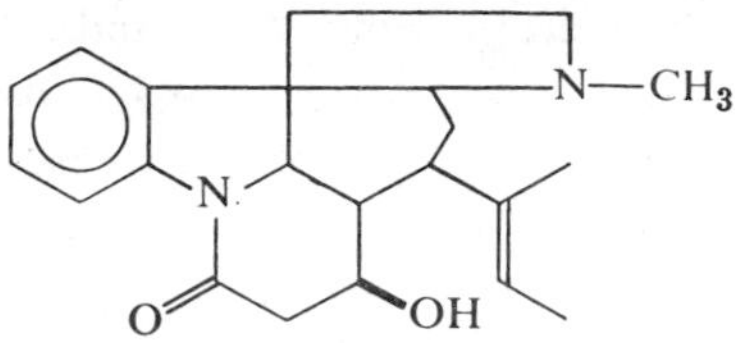

10

An interesting ring contraction of ring VII occurred when the compound methoxymethylchanodihydrostrychnone, subsequently shown by Woodward (8) to have structure **11**, was subjected to Clemmensen reduction conditions. Reynolds and Robinson (37), as well as Woodward (8), demonstrated that **12** is the correct structure of the product. Although the $C_{(21)}$ carbonyl disappears during the reaction, it does so almost certainly not as a result of a normal Clemmensen reduction. Instead, an elimination reaction of the ring VII oxygen atom β to the carbonyl function is most likely the initial step; this is followed by ring closure through hemiketal formation and eventual reduction of the $C_{(21)}$ oxygen function.

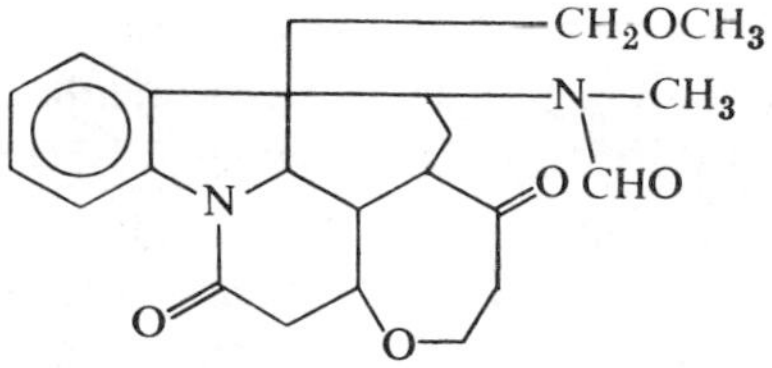

11 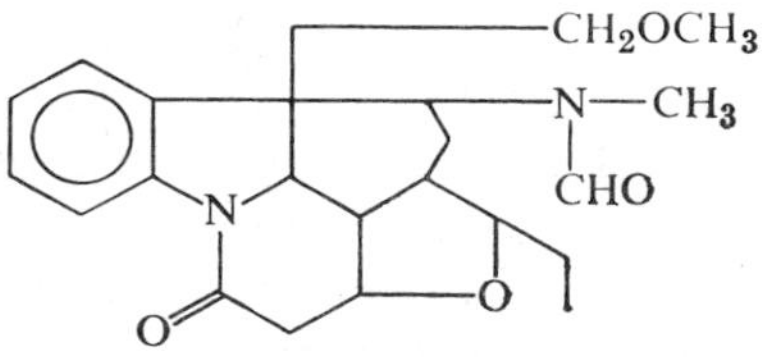 **12**

3. Congeners of Strychnine

Among the most interesting members of the strychnine group of alkaloids *per se* are those having a modified ring VII. The parent compound is the Wieland-Gumlich aldehyde (**13**) which was first obtained by Wieland and Kaziro (38) following a Beckman rearrangement of 11-isonitrosostrychnine prepared previously by Wieland and Gumlich (39). Its tautomeric nature was recognized much later by Anet and Robinson (40), and its identity with the naturally-occurring caracurine VII, which was first described by Asmis, Schmid, and Karrer (41), was established by Bernauer and co-workers (42). It is thus apparent that removal of ring III of the strychnine molecule allows ready entry into the normally stable ring VII. A number of such pentacyclic strychnine bases have been isolated from natural sources. They are properly considered derivatives of the Wieland-Gumlich aldehyde. The alkaloid diaboline, isolated from *Strychnos diaboli*, is N-acetyl Wieland-Gumlich aldehyde. Acetylation with acetic anhydride and pyridine furnishes two isomeric O-acetyl derivatives, N,O-diacetyl Wieland-Gumlich aldehyde A and B. These hemiacetal acetates are epimeric with respect to C-17 (numbering as in formula **13**). The B epimer occurs naturally. It is the alkaloid jobertine (O-acetyldiaboline B), which was recently isolated from *Strychnos jobertiana* (43). These C-17 epimers differ in molecular rotation by 323°, which is closely parallel to the molecular rotation shift observed in the anomeric six-membered acetylated D-aldopyranoses. Apparently the geometry of the oxepin hemiacetals does not differ greatly from that of the pyran hemiacetals.

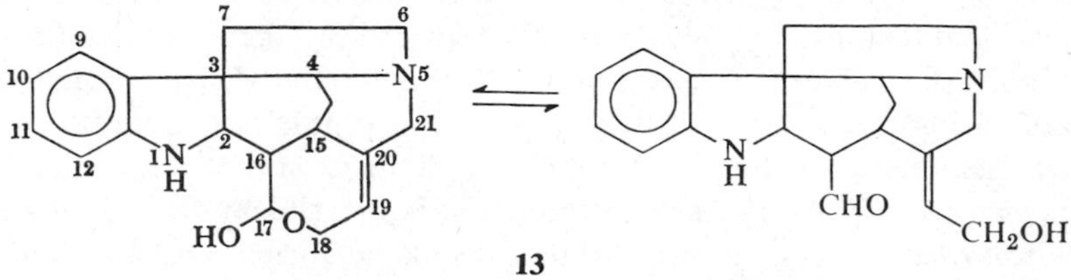

13

With the recognition of the biogenetic significance of the Wieland-Gumlich aldehyde occurrence of alkaloids with a six-membered ring VII becomes apparent. Reduction of the aldehyde function of **13** leads to an alcohol, as exemplified by the alkaloid retuline (**14**), whose structure was recently determined (44). Derivatives of this type of compound, where ring VII is reformed as a six-membered ring are also known, for example, in strychnosplendine (**15**), whose structure was determined by LeMen and co-workers (45).

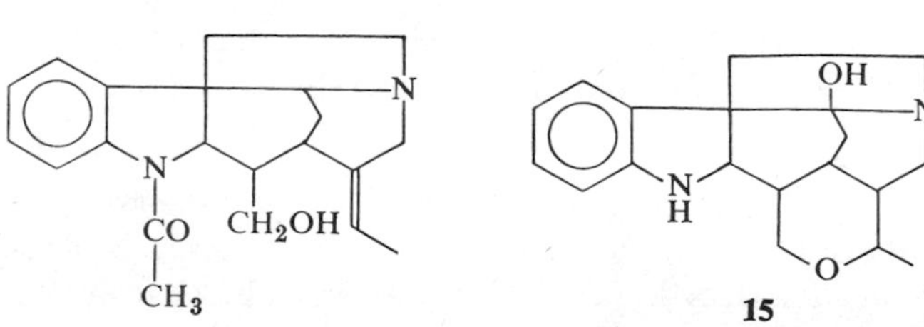

14

15

In view of the large number of strychnine derivatives which have been described in the literature it would be interesting to examine the influence of various structural modifications on ring VII. It would be worthwhile, for example, to know whether the reactivity of ring VII in vomicine (**16**) differs markedly from that in strychnine (**1**). At first sight this seems to be a fruitful line of approach, since permanganate oxidation of vomicine, for example, leads to entirely different results from the same reaction on strychnine. The reason for this behavior may well be the interference of the sensitive phenolic function rather than the positive influence of opened ring VI. However, an unequivocal answer to this question is not possible because of a lack of data. Virtually all degradations in the strychnine series have been carried out with strychnine (**1**) or brucine. Brucine, it may be recalled, is 2,3-dimethoxystrychnine, and exhibits fully analogous behavior in its ring VII reactions.

One aspect of vomicine chemistry is of interest in connection with ring VII transformations. When vomicine (**16**) is treated with hydriodic acid and red phophorus in glacial acetic acid, so-called yellow deoxyvomicine is formed. Structure **17** has been proposed for it by Huisgen and co-workers (46), but not all reported data are in agreement with this structure. Alkali in pyridine, among other reagents, readily converts yellow deoxyvomicine into colorless deoxyvomicine for which structure **18** has been proposed by Huisgen (47). The point of interest rests in the greatly enhanced basicity of these two compounds as compared with vomicine itself. Huisgen's explanation (47) of increased transannular interaction in the nine-membered ring is certainly reasonable and there seems to be little doubt that this interaction is made possible by the removal of ring VII. Unfortunately, data which might corroborate this view, such as ir spectra or the pK value of isovomicine (**19**) are lacking.

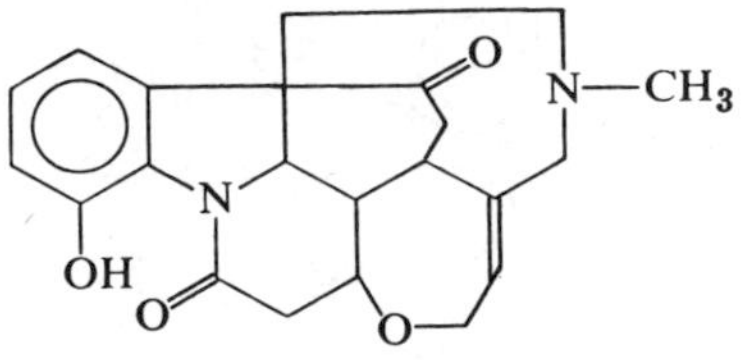

16 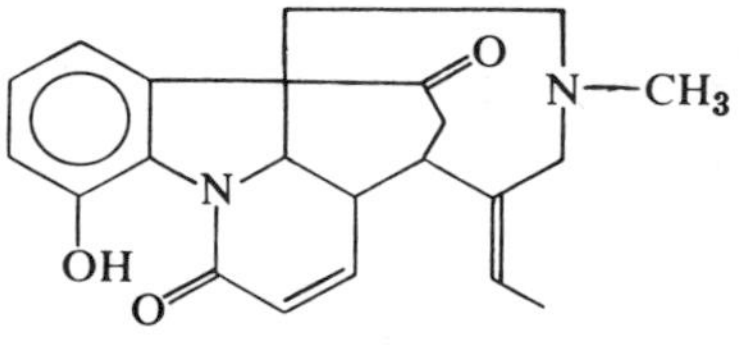 **17**

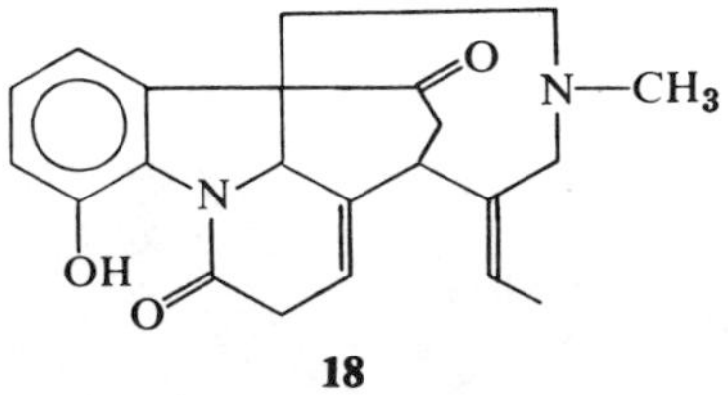

18 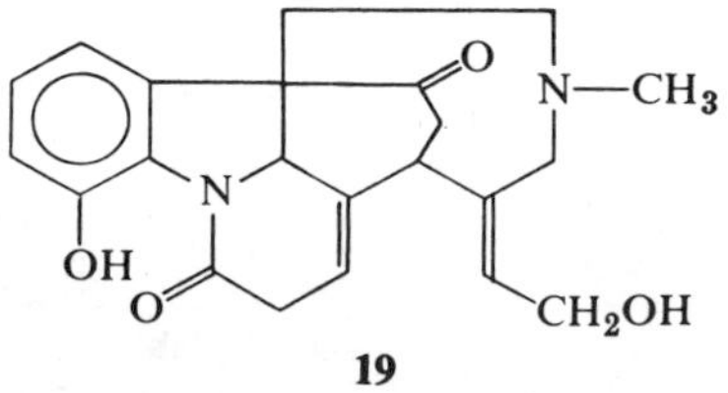 **19**

C. The Cularine Group

This small group of alkaloids has been isolated from only two genera of the Papaveraceae. It belongs to the class of benzylisoquinoline bases but possesses an oxepin ring which bridges the phenyl and isoquinoline moieties of the molecules. The chemistry of these compounds was reviewed by Manske (48), Boit (49), and Bentley (50). Manske himself carried out the pioneering research by establishing the structure of cularine in 1950 (51). Cularine may be considered a derivative of dibenz[*b*,*f*]oxepin (**20**) (R.I. 3696) and possesses structure **21**. The nature of the oxepin ring was unambiguously established by Manske (51) through ether cleavage with sodium in liquid ammonia as previously described by Sartoretto and Sowa (50). This reaction cleanly led to a single product, the phenol **22**, without a trace of the alternate compound. Further degradation of phenol **22** confirmed its structure. The ether cleavage reaction is of some intrinsic interest for dibenzoxepins. The isolation of a single product from cularine raises the question whether this course might be predicted on the basis of the known mechanism of the reaction. Eargle (53) followed the reaction by esr techniques and demonstrated that the initial ion radical is transformed into a dianion which in turn undergoes cleavage to a phenoxide ion and a carbanion. Eargle's aryl ethers were all symmetrical and therefore offer no immediate clue to the present question. One would expect, however, that an initial ion radical involving $C_{(5)}$ of cularine is formed more readily than one involving $C_{(5')}$ which bears a methoxyl. Instances involving unsymmetrically substituted dibenzoxepins without distinct bias toward one product, as is true of cularine, have apparently not been studied.

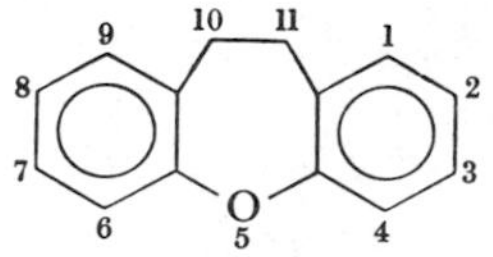

20

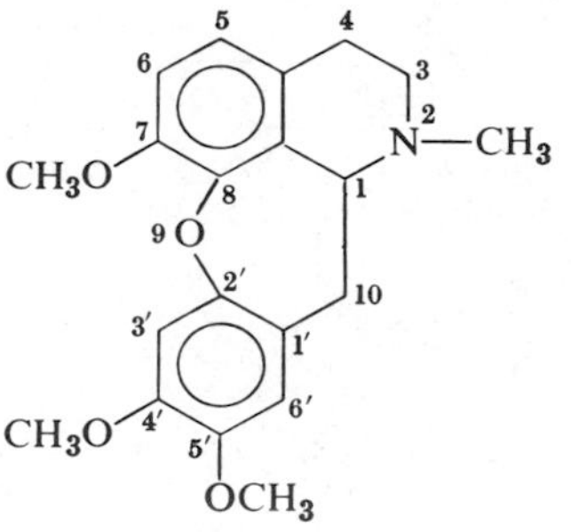

21

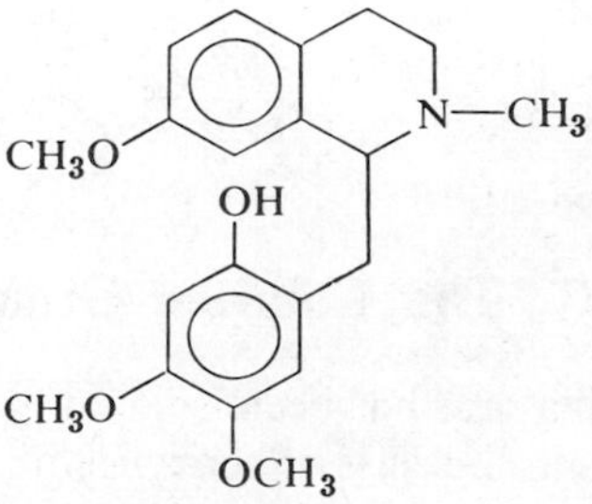

22

No other reactions of the dibenzoxepin system in cularine have been reported. Attempts to synthesize cularine and its congeners date back to 1955. Kulka and Manske (54) carried out an Ullmann reaction with two appropriately substituted benzene derivatives to the expected diphenyl ether, which could be cyclodehydrated to the oxepin only with difficulty. Kametani's (55, 56) reports of his successful cularine synthesis did not elaborate on his mode of preparation of the oxepin system. Judging from related publications by Kametani and collaborators (57), one may assume that they prepared appropriate diphenyl ethers by the Ullmann reaction and then cyclized the piperidine and oxepin rings. Oxepin ring closure followed by construction of the piperidine ring was apparently difficult. In further synthetic approaches toward cularine alkaloids in Kametani's laboratory the resolution of the key intermediate (**23**) has been reported (58). An entirely different synthetic approach toward alkaloids of the cularine type was recently outlined by Kametani and co-workers (59). This new route proceeds from a phenolic benzyltetrahydroisoquinoline via oxidative coupling to a spiro-dienone and hence, by dienone-phenol rearrangement, to an oxepin. Compound (**24**) was oxidized with potassium ferricyanide in 3% yield to a single dienone (**25**). This dienone yielded upon treatment with hydrochloric acid in glacial acetic acid the cularine derivative (**26**) in unspecified yield. Structural assignment of compound (**26**) was based on spectral evidence.

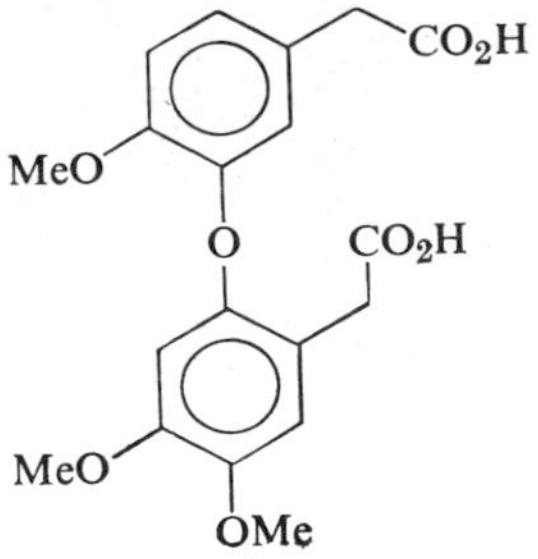

23

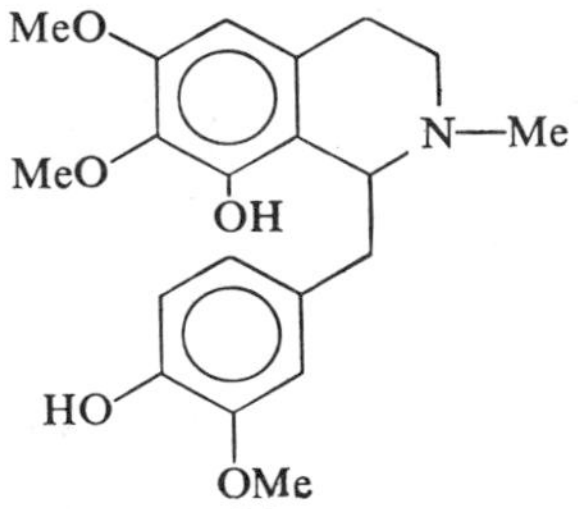

24

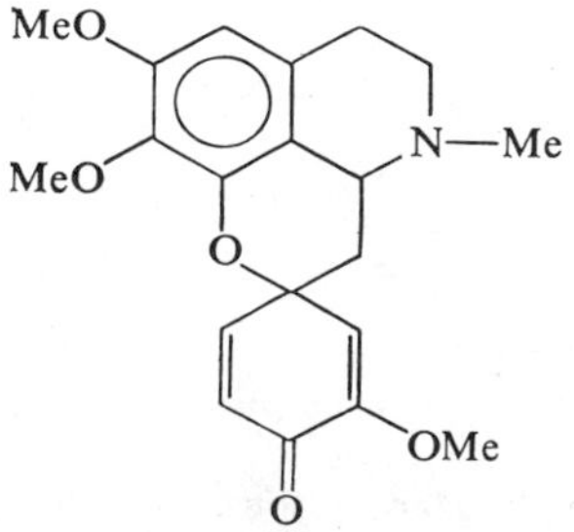

25

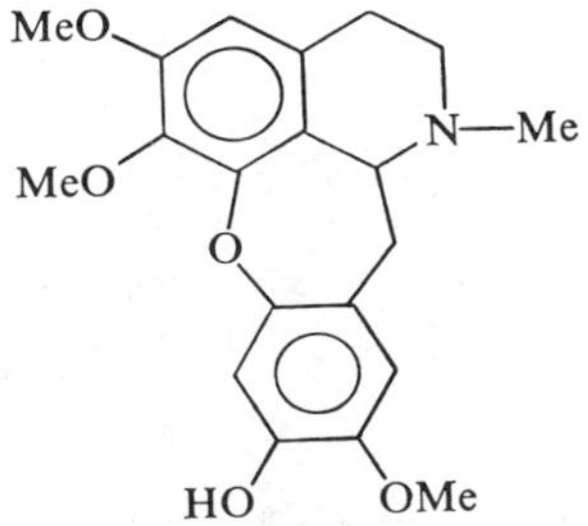

26

The few reported congeners of cularine differ from the parent alkaloid only by lack of the *N*-methyl group (cularimine) (51), by the occurrence of a free phenol (cularidine) (60), or by a free phenol and a dioxymethylene (cularicine) (61). These compounds therefore offer no further insight into oxepin chemistry.

D. Insularine and Insulanoline

The cularine alkaloids, as we have seen, are benzylisoquinoline derivatives in which an oxepin ring is formed by phenol coupling. A similar origin may be envisaged for the two alkaloids, insularine and insulanoline, which are derivatives of bisbenzylisoquinoline. Their structures (**27A** and **27B**), which contain the rare 1,4-dioxepin ring, were established by Kikuchi and coworkers (62, 63) through ether cleavage with sodium in liquid ammonia. Interestingly, in the reaction product **28** two of the ether linkages are cleaved in the expected sense, but one remains intact. Since the remaining ether link exists in the ring which originally contains two such linkages, the result is not surprising.

No additional reaction or syntheses of these two compounds have been reported.

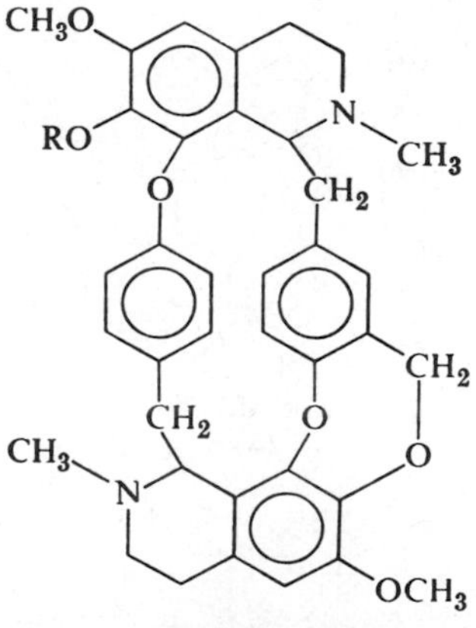

27A; R = CH_3
27B; R = H

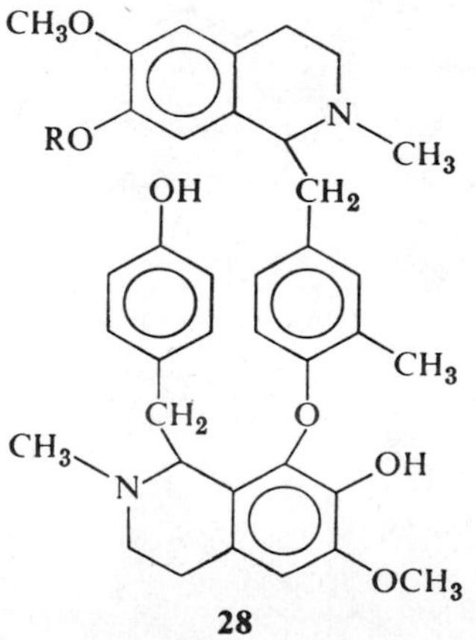

28

E. References

1. G. F. Smith, "Strychnos Alkaloids," in R. H. F. Manske, Ed., *The Alkaloids*, Vol. VIII, Academic Press, New York, 1965, pp. 591–671.

2. H. G. Boit, *Ergebnisse der Alkaloid-Chemie bis 1960*, Akedemie-Verlag, Berlin, 1961, pp. 594–630.
3. M. Hesse, *Indolalkaloide in Tabellen*, Springer-Verlag, Berlin, 1964, pp. 44–53, and 1968 supplement.
4. A. R. Battersby and H. F. Hodson, "Alkaloids of Calabash Curare and *Strychnos* Species," in R. F. Manske, Ed., *The Alkaloids*, Vol. XI, Academic Press, New York, 1968, pp. 189–204.
5. J. H. Robertson and C. A. Beevers, *Nature*, **165,** 690 (1950); *Acta Crystallogr.*, **4,** 270 (1951).
6. C. Bokhoven, J. C. Schoone, and J. M. Bijvoet, *Acta Crystallogr.*, **4,** 275 (1951).
7. R. B. Woodward, M. P. Cava, W. D. Ollis, A. Hunger, H. V. Daeniker, and K. Schenker, *J. Amer. Chem. Soc.*, **76,** 4749 (1954); *Tetrahedron*, **19,** 247 (1963).
8. R. B. Woodward and W. J. Brehm, *J. Amer. Chem. Soc.*, **70,** 2107 (1948).
9. R. B. Woodward, W. J. Brehm, and A. L. Nelson, *J. Amer. Chem. Soc.*, **69,** 2250 (1947).
10. L. H. Briggs, H. T. Openshaw, and R. Robinson, *J. Chem. Soc.*, **1946,** 903.
11. R. N. Chakravarti and R. Robinson, *Nature*, **160,** 18 (1947).
12. K. N. Menon and R. Robinson, *J. Chem. Soc.*, **1932,** 780.
13. A. Bacovescu and A. Pictet, *Chem. Ber.*, **38,** 2787 (1905).
14. H. Leuchs and R. Nitschke, *Chem. Ber.*, **55,** 3171 (1922).
15. V. Prelog, J. Battegay, and W. I. Taylor, *Helv. Chim. Acta*, **31,** 2244 (1948).
16. M. Kotake, T. Sakan, and S. Kusumoto, *Sci. Pap. Inst. Phys. Chem. Res.* (Tokyo), **35,** 415 (1939); *Chem. Abstr.*, **33,** 5858 (1939).
17. P. J. Scheuer, *J. Amer. Chem. Soc.*, **82,** 193 (1960).
18. P. D. Bartlett and M. Stiles, *J. Amer. Chem. Soc.*, **77,** 2806 (1955).
19. H. Hansen, *Chem. Ber.*, **18,** 1917 (1885).
20. F. Cortese, *Justus Liebigs Ann. Chem.*, **476,** 283 (1929).
21. H. Leuchs, F. Osterburg, and H. Kaehrn, *Chem. Ber.*, **55,** 564 (1922).
22. M. Kotake and T. Mitsuwa, *J. Chem. Soc. Jap.*, **57,** 236 (1936); *Chem. Abstr.*, **30,** 5229 (1936).
23. M. Kotake and T. Mitsuwa, *Bull. Chem. Soc., Jap.*, **11,** 231 (1936); *Chem. Abstr.*, **30,** 5228 (1936).
24. A. Kogure and M. Kotake, *J. Inst. Polytech. Osaka City Univ., Ser. C.*, **2,** 39 (1951); *Chem. Abstr.*, **46,** 6131 (1952).
25. V. Prelog and A. Kathriner, *Helv. Chim. Acta*, **31,** 505 (1948).
26. D. M. S. Wheeler, M. C. Smith, A. M. Kneece, and J. B. McMaster, *Can. J. Chem.*, **43,** 985 (1965).
27. H. Leuchs, *Chem. Ber.*, **41,** 1711 (1908).
28. H. Leuchs and G. Schwaebel. *Chem. Ber.*, **46,** 3693 (1913).
29. H. Leuchs and W. Schneider, *Chem. Ber.*, **42,** 2494 (1909).
30. V. Prelog and S. Szpilfogel, *Helv. Chim. Acta*, **28,** 1669 (1945).
31. J. T. Edward, *Tetrahedron*, **2,** 356 (1958).
32. A. Kogure, T. Sakan, and M. Kotake, *J. Chem. Soc. Jap.*, **70,** 182 (1949); *Chem. Abstr.*, **45,** 7129 (1951).
33. V. Prelog and M. Kocor, *Helv. Chim. Acta*, **30,** 360 (1947).
34. H. Leuchs and H. S. Overberg, *Chem. Ber.*, **66,** 951 (1933).
35. H. Leuchs, J. Beyer, and H. S. Overberg, *Chem. Ber.*, **66,** 1378 (1933).
36. W. H. Perkin, Jr., R. Robinson, and J. C. Smith, *J. Chem. Soc.*, **135,** 1239 (1932).
37. T. M. Reynolds and R. Robinson, *J. Chem. Soc.*, **1934,** 592.
38. H. Wieland and K. Kaziro, *Justus Liebigs Ann. Chem.*, **506,** 60 (1933).

39. H. Wieland and W. Gumlich, *Justus Liebigs Ann. Chem.*, **494,** 191 (1932).
40. F. A. L. Anet and R. Robinson, *J. Chem. Soc.*, **1955,** 2253.
41. H. Asmis, H. Schmid, and P. Karrer, *Helv. Chim. Acta*, **37,** 1983 (1954).
42. K. Bernauer, S. K. Pavanaram, W. V. Philipsborn, H. Schmid, and P. Karrer, *Helv. Chim. Acta*, **41,** 1405 (1958).
43. F. Delle Monache, E. Corio, and G. B. Marini-Bettolo, *Ann. Ist. Super. Sanita*, **3,** 564 (1967); *Chem. Abstr.*, **68,** 87440u (1968).
44. J. L. Occolowitz, K. Biemann, and J. Bosly, *Farmaco, Ed. Sci.*, **20,** 751 (1965).
45. M. Koch, M. Plat, B. C. Das, and J. LeMen, *Tetrahedron Lett.*, **1966,** 2353.
46. R. Huisgen, H. Eder, L. Blatzejewicz, and E. Mergenthaler, *Justus Liebigs Ann. Chem.*, **573,** 121 (1951).
47. R. Huisgen, H. Wieland, and H. Eder, *Justus Liebigs Ann. Chem.*, **561,** 193 (1949).
48. R. H. F. Manske, "The Cularine Alkaloids," in R. H. F. Manske and H. L. Holmes, Eds., *The Alkaloids*, Vol. IV, Academic Press, New York, 1954, pp. 249–252, 258–260; and Vol. X, 1968, pp. 463–465.
49. H. G. Boit, *Ergebnisse der Alkaloid-Chemie bis* 1960, Akademie-Verlag, Berlin, 1961, pp. 258–260.
50. K. W. Bentley, *The Isoquinoline Alkaloids*, Pergamon Press, London, 1965, pp. 60–64.
51. R. H. F. Manske, *J. Amer. Chem. Soc.*, **72,** 55 (1950).
52. P. A. Sartoretto and F. J. Sowa, *J. Amer. Chem. Soc.*, **59,** 603 (1937).
53. D. H. Eargle, Jr., *J. Org. Chem.*, **28,** 1703 (1963).
54. M. Kulka and R. H. F. Manske, *J. Amer. Chem. Soc.*, **75,** 1322 (1955).
55. T. Kametani and K. Fukumoto, *Chem. Ind.* (London), 291 (1963).
56. T. Kametani and K. Fukumoto, *J. Chem. Soc.*, **1963,** 4289.
57. T. Kametani and K. Ogasawara, *J. Chem. Soc.*, **1964,** 4142.
58. T. Kametani and S. Shibuya, *Yakugaku Zasshi*, **87,** 196 (1967); *Chem. Abstr.*, **67,** 54313f (1967).
59. T. Kametani, T. Kikuchi, and K. Fukumoto, *Chem. Commun.*, **1967,** 546.
60. T. Kametani, S. Shibuya, and C. Kibayashi, *Tetrahedron Lett.*, **1966,** 3215.
61. R. H. F. Manske, *Can. J. Chem.*, **43,** 989 (1956).
62. M. Tomita and T. Kikuchi, *Yakugaki Zasshi*, **77,** 997 (1957); *Chem. Abstr.*, **52,** 3832 (1958).
63. T. Kikuchi and K. Bessho, *Yakugaki Zasshi*, **78,** 1413 (1958); *Chem. Abstr.*, **53,** 7220 (1959).

CHAPTER X

Monocyclic Seven-Membered Rings Containing Sulfur

LAMAR FIELD AND DAVID L. TULEEN

Department of Chemistry, Vanderbilt University, Nashville, Tennessee

A. Introduction

This chapter deals with systems consisting of one seven-membered ring in which only atoms of sulfur and carbon are present. Spiran systems are included, but fused-ring systems or rings containing other elements are to be treated elsewhere in these volumes.

Specific nomenclature will be illustrated as each class is discussed. The general approach of *Chemical Abstracts* is to name unsaturated derivatives as thiepins, "ep" signifying a seven-membered ring. An example is the name 1,2,5-trithiepin-3,4,6,7-tetracarbonitrile for structure **1**. Sulfur atoms are

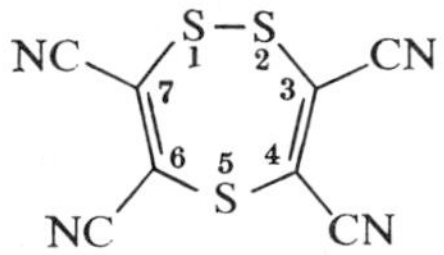

1

given the lowest possible numbers, and numbering then usually proceeds clockwise. Saturated thiepins are named as tetrahydrothiepins or thiepanes.

Numbers which have been assigned to structures in *The Ring Index* (158) are shown wherever possible. The chapter deviates from the practice of *Chemical Abstracts* in that rings usually are shown with the sulfur atom down instead of up for conformity with the style of these volumes and frequent general practice.

The chapter itself and the tabulation of compounds in Section F are organized according to the number of sulfur atoms, their numerical position in the ring, and the degree of saturation for a particular type of ring system. Thus thiepins are treated before dithiepins, 1,2-dithiepins before 1,3-dithiepins, and thiepins before dihydrothiepins.

Section F lists all compounds within the scope of the chapter which were located in a systematic search of the literature to 1969, with partial coverage later. It summarizes common physical properties, but only calls attention to the availability in the citation of more involved types of data (spectral, crystallographic, etc.) since to abstract such data in part without interpretation might be more misleading than helpful. Conventional abbreviations for terms such as nuclear magnetic resonance (nmr), infrared (ir), and ultraviolet (uv) spectra are used in the summary table of Section F and in the text. In the absence of comment, elemental analyses for compounds discussed ordinarily were satisfactory.

Prospects for research will be seen to abound. Many of the theoretically possible ring systems are unknown. Although some of the early work was well done and is obviously valid, many of the structures reported should be confirmed by modern methods. Molecular weights have not been determined for some of these compounds, and the reader should be wary on this point particularly. Only slight attention has been given, thus far, to the reactions of many of these ring systems.

B. Rings Containing One Sulfur Atom (Thiepins and Derivatives)

1. Thiepins and Derivatives

No monocyclic thiepins involving divalent sulfur (**2**) are known (157). Much of the interest in the parent ring system, as well as in the analogous heterocyclic compounds oxepin (**3**) and azepine (**4**), has resulted from a

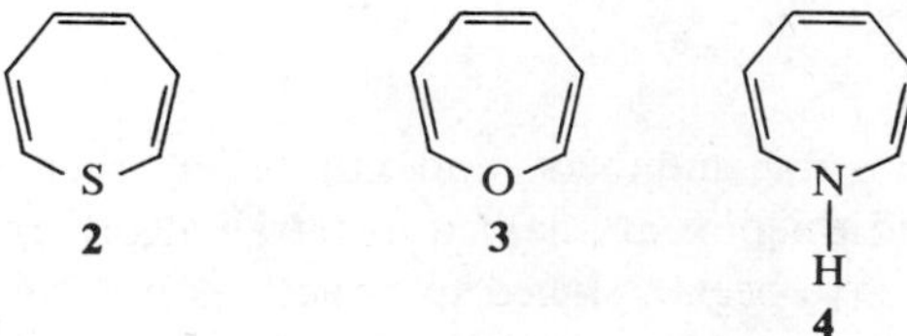

consideration of the possibility of aromatic stabilization in ring systems containing other than $(4n + 2)\pi$ electrons in systems which may be planar (99).

The three heterotropilidines mentioned above are isoelectronic with the cycloheptatrienyl anion (91) and all have been predicted by molecular orbital methods to be unstable (184, 206). Calculations of thiepin indicate it to be slightly antiaromatic, with carbon-carbon bond lengths close to those expected for polyene single and double bonds (64). Its stability has been predicted to be less than expected from a consideration of canonical forms involving carbon to sulfur double bonds, such as **5**.

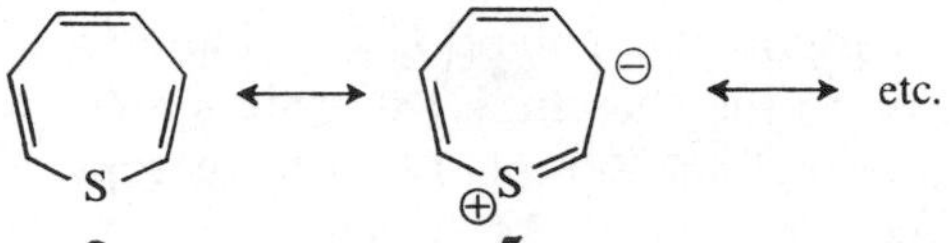

S ⟷ ⊕S ⊖ ⟷ etc.

2 **5**

The dithiol obtained from the reaction of thiothiophthene **6** with hot methanolic potassium hydroxide has occasionally been considered to be the substituted thiepin **7**. The structure favored by Arndt and co-workers for this compound generally has been the substituted thiophene **8** (7–9), although at one time Arndt preferred the thiepin **7** (5). Bothner-By and Traverso

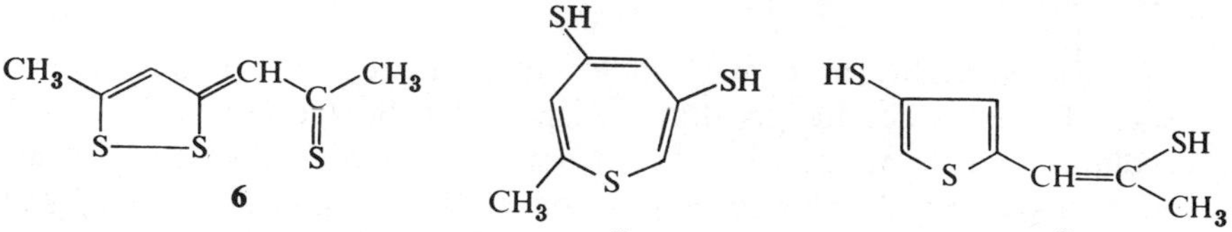

CH_3 CH CH_3 C S—S S **6**; SH SH S CH_3 **7**; HS S CH=C SH CH_3 **8**

examined the nmr spectrum (40 MHz) of the dimethyl thioether and assigned the thiepin structure to this derivative of the dithiol on the basis of the chemical shifts of the vinylic protons (27). Bothner-By later communicated privately to Arndt the possibility of overlapping vinylic and aromatic regions in the nmr spectrum (9). Breslow and Skolnik (36) have reviewed the data pertaining to the dithiol and have reasonably concluded that additional structural evidence would be desirable. The weight of the evidence seems to indicate, however, that the product is not a substituted thiepin.

The synthesis of thiepin 1,1-dioxide (**9**), which eluded Maerov (117), has been achieved recently by Mock (125, 126). The synthesis is shown by Eq. 1.

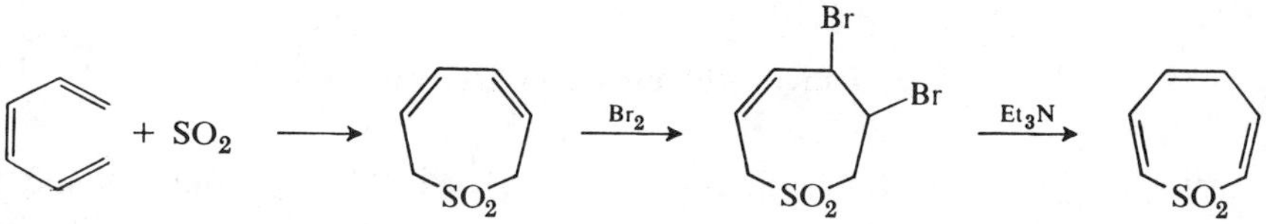

$$+ SO_2 \longrightarrow \underset{\mathbf{10}}{SO_2} \xrightarrow{Br_2} \underset{\mathbf{11}}{SO_2\ (Br,\ Br)} \xrightarrow{Et_3N} \underset{\mathbf{9}}{SO_2} \quad (1)$$

The vinylic protons of **9** appear in the nmr spectrum in the region $\tau 2.8$–3.5; this chemical shift is sufficiently close to the chemical shift of the vinyl protons of **10** to indicate that there is no appreciable diamagnetic ring current. The sulfur-to-oxygen stretching frequencies in the ir spectrum occur at 1300 and 1120 cm^{-1}, values which are not atypical from normal sulfones. The uv spectrum of thiepin 1,1-dioxide is quite similar to that of cycloheptatriene. An X-ray crystallographic study of **9** has shown that thiepin 1,1-dioxide exists as a boat (3). Some electron delocalization is reflected in a torsional angle (i.e., the dihedral angle of the S-$C_{(2)}$-$C_{(3)}$ and $C_{(2)}$-$C_{(3)}$-$C_{(4)}$ planes) of 8.4°. Non-planarity of the thiepin 1,1-dioxide system renders the methyl groups of the isopropyl moiety of **12** diastereotopic. An nmr study at varied frequency (60, 100, and 250 MHz) and temperature (room temperature to −150°C) has shown a free energy barrier to inversion for **12** of 6.4 kcal/mole (4).

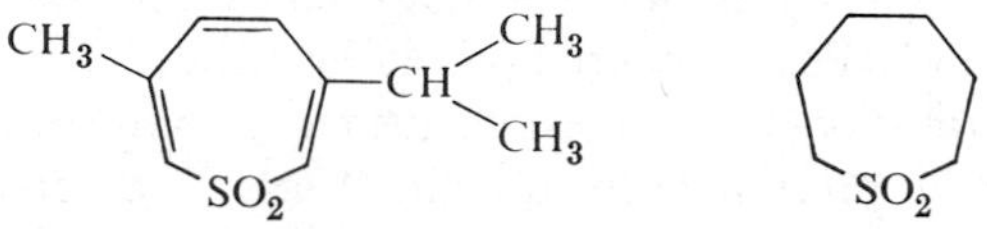

CH_3 — $CH(CH_3)_2$ — SO_2

12

SO_2

13

Sulfone **9** readily absorbs 3 moles of hydrogen (with palladium-on-charcoal in ethyl acetate) to afford thiepane 1,1-dioxide (**13**). It undergoes decomposition when heated above its melting point to give benzene and sulfur dioxide (Eq. 2). This decomposition is reasonably accounted for by the postulated intermediacy of the episulfone (**14**).

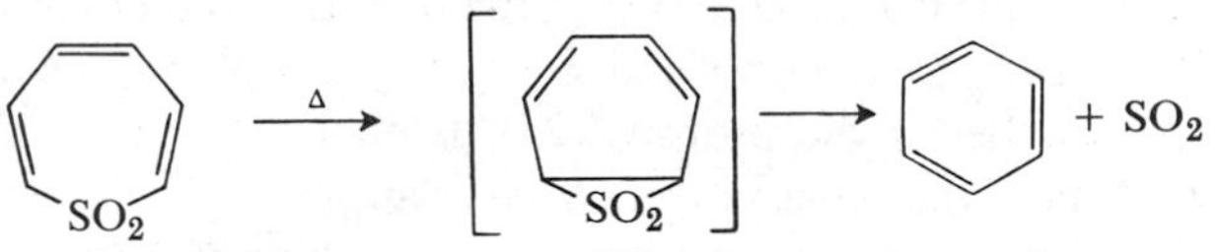

SO_2 $\xrightarrow{\Delta}$ [SO_2] ⟶ + SO_2 (2)

9 **14**

Although no monocyclic thiepins have been prepared, many compounds are known which contain the thiepin ring as part of a fused aromatic system. The assumption has often been made that such fused thiepins are more readily obtained because they are stabilized by benzenoid moieties (91). A review of these fused systems will appear elsewhere in these volumes.

2. Dihydrothiepins and Derivatives

There are four possible dihydrothiepins, **15**–**18**. Of these, only **15** and **16** have been reported. There are no known monocyclic 2,3- or 2,5-dihydrothiepin derivatives (**17**, **18**).

Stogryn and Brois have reported that 4,5-dihydrothiepin (**15**) results from the thermally induced Cope rearrangement of *cis*-1,2-divinylthiirane (Eq. 3) (182); experimental details have not yet appeared. The authors noted that a

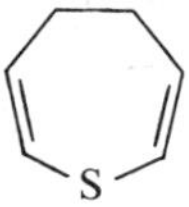

15 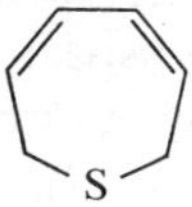 **16** 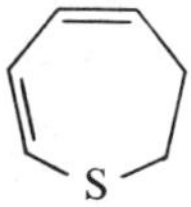 **17** 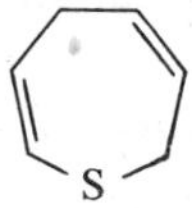 **18**

temperature of 100°C is required for this isomerization, somewhat higher than the 60°C necessary to effect the analogous rearrangement of *cis*-1,2-divinyloxirane to 4,5-dihydrooxepin (34, 183), and much higher than is

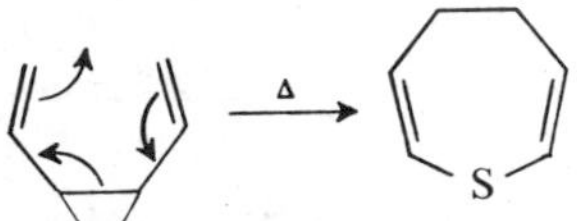

$$\xrightarrow{\Delta} \quad \mathbf{15} \qquad (3)$$

required for the isomerization of *cis*-1,2-divinylaziridines (181) or *cis*-1,2-divinylcyclopropane (66, 194).

The product of the hydrogen peroxide oxidation of 2,5-diphenyl-1,4-dithiadiene (**19**) was originally thought to be the 4,5-dihydrothiepin derivative **20** on the basis of molecular weight and elemental analysis (187). In subsequent work Szmant and Alfonso (186) reported that the oxidation product

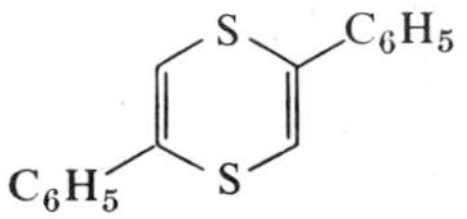

19 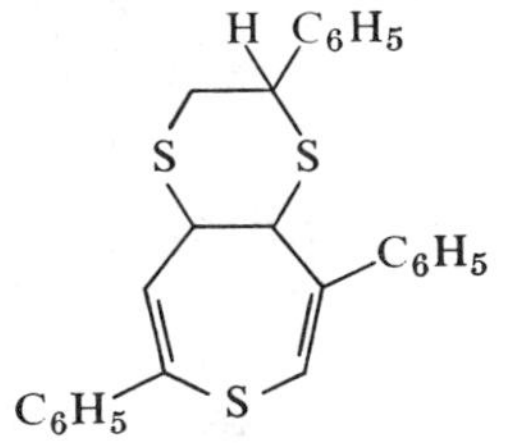 **20** 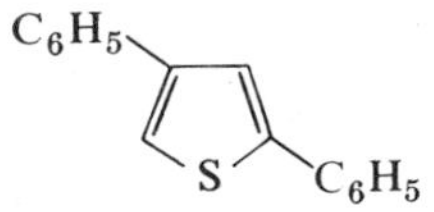 **21**

was instead a solid complex of the starting dithiadiene together with 2,4-diphenylthiophene (**21**).

The synthesis of 2,7-dihydrothiepin 1,1-dioxide (**10**) was mentioned above in connection with the synthesis of thiepin 1,1-dioxide (Eq. 1). This cyclic sulfone (mp 107–108°C) is formed in the reaction of *cis*-hexatriene and sulfur dioxide in ether at room temperature (Eq. 4) (126). The cyclization is

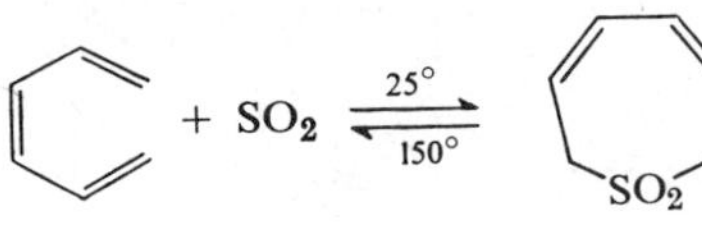

$$+ SO_2 \underset{150^\circ}{\overset{25^\circ}{\rightleftharpoons}} \mathbf{10} \qquad (4)$$

reversible; *cis*-hexatriene and sulfur dioxide are regenerated from the liquid sulfone at 150–160°C. Treatment of **10** with hydrogen (over palladium-on-carbon) resulted in the uptake of 2 moles of hydrogen and formation of the saturated sulfone, **13**. The structure of the dihydrothiepin **10** was substantiated by uv and nmr spectra, as well as by mass-spectral molecular weight (126).

Mock has also prepared the *trans*- and *cis*-isomers of 2,7-dimethyl-2,7-dihydrothiepin 1,1-dioxide (**22A**, **22B**) (127). Reduction of **23A** with sodium borohydride, followed by dehydration, affords **22A**. A similar sequence applied to a mixture of **23A** and the nonisolable **23B** gives some **22B** (Eqs. 5 and 6). Thermal decomposition of **22A** and **22B** affords the

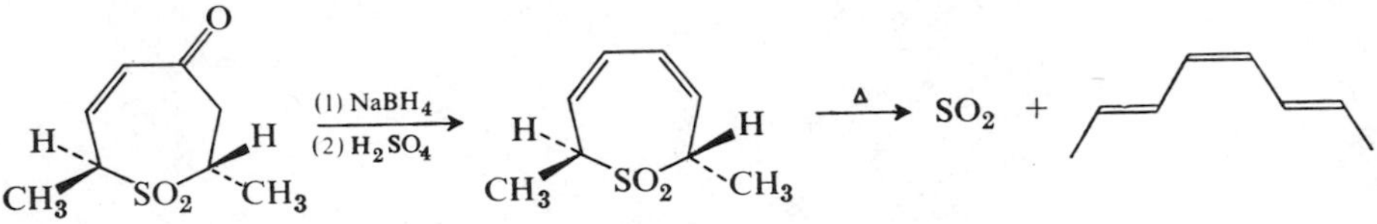

O; H; H; CH_3; SO_2; CH_3 — **23A** —(1) $NaBH_4$ (2) H_2SO_4→ H; H; CH_3; SO_2; CH_3 — **22A** —Δ→ SO_2 +

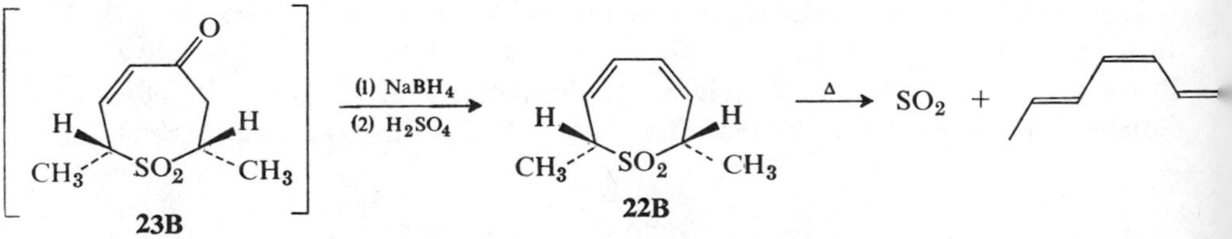

O; H; H; CH_3; SO_2; CH_3 — [**23B**] —(1) $NaBH_4$ (2) H_2SO_4→ H; H; CH_3; SO_2; CH_3 — **22B** —Δ→ SO_2 +

indicated octatrienes. The stereochemistry of these processes demands interpretation in terms of a *trans* concerted elimination which is antarafacial (127).

Dodson and Nelson have recently prepared 4,5-diphenyl-2,7-dihydrothiepin 1-oxide (**24**) by the reaction of 3,4-diphenyl-1,3,5-hexatriene with sulfur monoxide generated *in situ* by the thermal decomposition of thiirane 1-oxide (65). This sulfoxide is oxidized cleanly by *m*-chloroperbenzoic acid to afford the sulfone (**25**); reduction by lithium aluminum hydride gives 4,5-diphenyl-2,7-dihydrothiepin (**26**) (65).

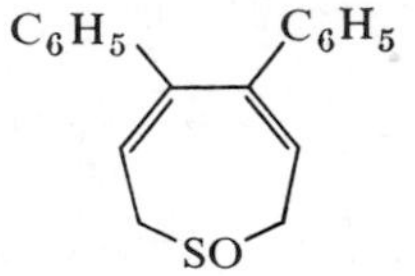

C_6H_5 C_6H_5 SO **24** C_6H_5 C_6H_5 SO_2 **25** C_6H_5 C_6H_5 S **26**

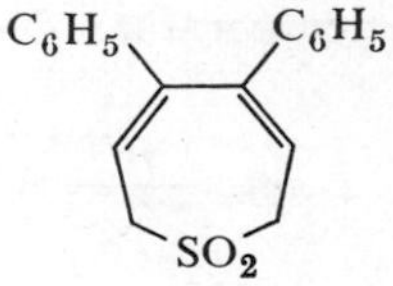

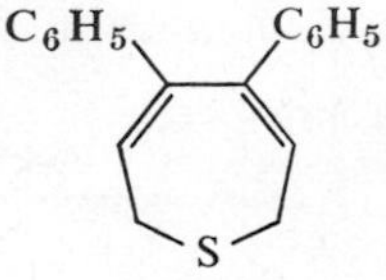

3. Tetrahydrothiepins and Derivatives

Few examples of monocyclic tetrahydrothiepin derivatives have been reported. Overberger and Katchman observed that pyrolysis of 4-acetoxythiepane 1,1-dioxide (**27**) affords an olefinic product (151). The synthesis of **27** is described in Section B-5-c. The olefinic product was believed to be a mixture of 2,3,4,7-tetrahydrothiepin 1,1-dioxide (**28**) and 2,3,6,7-tetrahydrothiepin 1,1-dioxide (**29**) on the basis of combustion analysis and a rather wide melting point range (83.5–88.5°C). Attempted separation of the mixture by recrystallization and chromatography, as well as attempted isomerization to a single olefin, were unsuccessful. Dehydrogenation of the mixture of olefins with selenium dioxide resulted in a very small yield of a monoolefin, mp 114–115°C, which was converted by *N*-bromosuccinimide to an allylic bromide, mp 92–93°C. Structures for these two compounds have not been proposed.

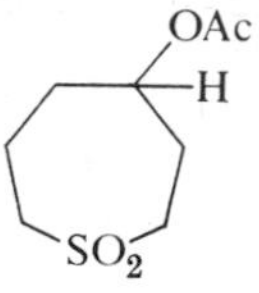

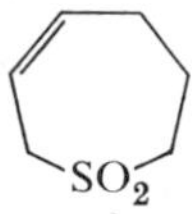

Addition of bromine to the mixture of olefins (**28**, **29**) afforded an inseparable mixture of dibrominated sulfones. Dehydrohalogenation of this mixture, followed by repeated recrystallization afforded a vinylic bromide, mp 142–144°C, of unknown structure (151).

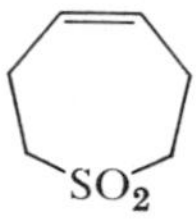

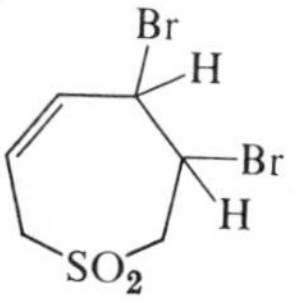

Treatment of 2,7-dihydrothiepin 1,1-dioxide (**10**) with bromine in chloroform affords a dibromide, mp 128–129°C. The structure of this compound has been provisionally formulated by Mock as *cis*- or *trans*-3,4-dibromo-2,3,4,7-tetrahydrothiepin 1,1-dioxide (**11**) on the basis of its nmr spectrum (125, 126).

A novel entry into the 2,3,4,7-tetrahydrothiepin ring system has been reported recently by Mock (127). Reduction of a mixture of epimers of **30** with lithium aluminum hydride followed by treatment with triethylamine in benzene results in the formation of *trans*- and *cis*-2,7-dimethyl-4-oxo-2,3,4,7-tetrahydrothiepin 1,1-dioxides (**23A**, **23B**). The stereochemical integrity of the methyl groups is destroyed during these reactions, presumably because of the incursion of enolate anion **31** under basic conditions. Compound **23B** was not isolated; its presence, however, seems well established by further reactions already described (Section B-2).

The products of the attempted Wolff rearrangement of diazoketone **32** include a 30% yield of an α,β-unsaturated ketone (83). Spectral evidence

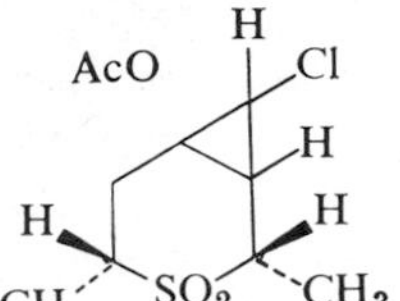

30 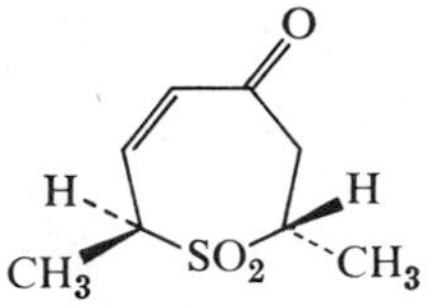 **23A** 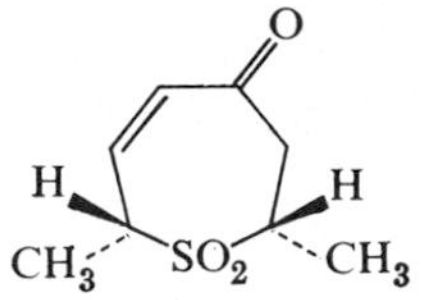 **23B** 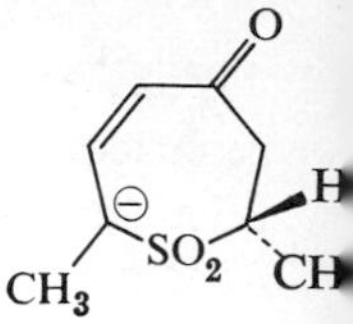 **31**

is not sufficient to distinguish between **33** and **34** for the structure of this product.

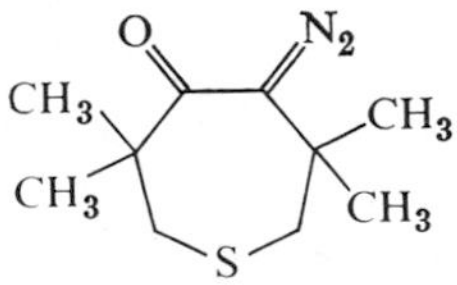

32 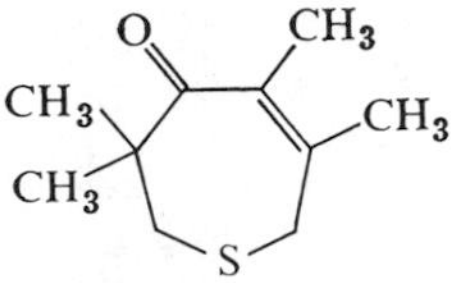 **33** 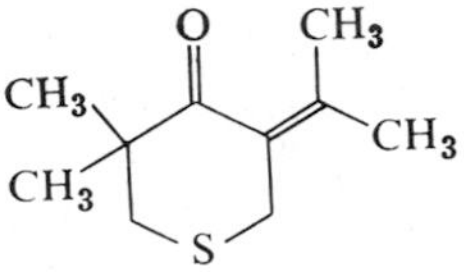 **34**

4. Thiepanes

The name preferred by *Chemical Abstracts* for the saturated monocyclic sulfide **35** is thiepane. This name will be used henceforth in preference to others which often appear in the literature, such as hexahydrothiepin, hexamethylene sulfide, and thiacycloheptane.

a. *Occurrence*

In 1906 Mabery and Quale (115) reported isolation of a substance of formula $C_6H_{12}S$ from Canadian petroleum. They tentatively identified this material as either thiepane (**35**) or 2-methylthiacyclohexane (**36**). More recent studies

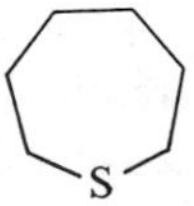

35 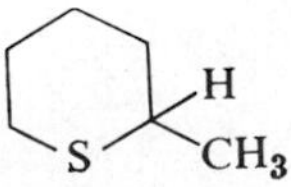 **36**

(23, 47) indicate, however, that although many five- and six-membered rings are found, the presence of seven-membered cyclic sulfides in petroleum is unlikely (92).

b. *Preparation and Properties*

Thiepane was originally synthesized (6% yield) in 1910 by von Braun from potassium sulfide and 1,6-diiodohexane (Eq. 7) (33). Grishkevitch-Trokhimovsky performed a similar cyclization using sodium sulfide and 1,6-dibromohexane (82). The latter method usually has been employed for

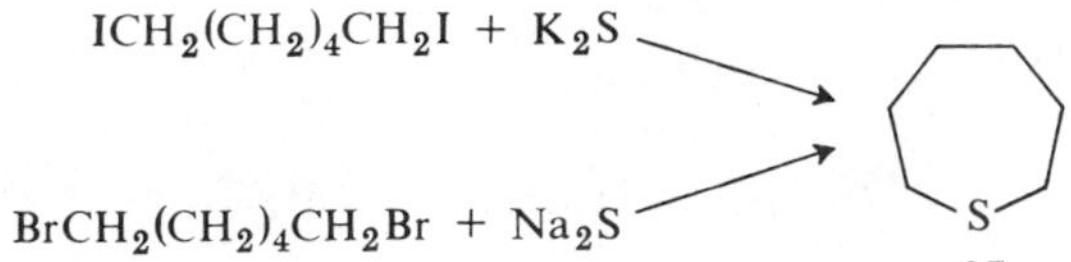

$$ICH_2(CH_2)_4CH_2I + K_2S \rightarrow \mathbf{35} \leftarrow BrCH_2(CH_2)_4CH_2Br + Na_2S \quad (7)$$

the synthesis of thiepane (58, 108, 133, 147, 150); yields as high as 52% have been reported (58, 133).

Thiepane also has been produced in low yield by the reduction of thiepane 1,1-dioxide (**13**) with lithium aluminum hydride (145), and by thermal degradation of the polythiolcarbamate **37** (68). Thermal degradation of

$$\left[-\overset{O}{\overset{\|}{C}}-NH-C_6H_4-CH_2-C_6H_4-NH-\overset{O}{\overset{\|}{C}}-S-CH_2(CH_2)_4CH_2S-\right]_n$$

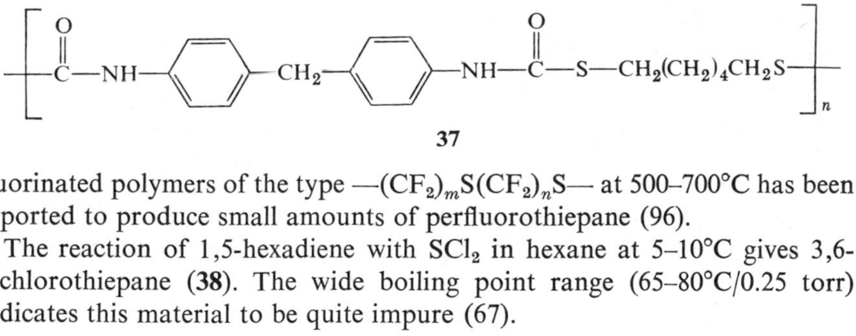

37

fluorinated polymers of the type —$(CF_2)_mS(CF_2)_nS$— at 500–700°C has been reported to produce small amounts of perfluorothiepane (96).

The reaction of 1,5-hexadiene with SCl_2 in hexane at 5–10°C gives 3,6-dichlorothiepane (**38**). The wide boiling point range (65–80°C/0.25 torr) indicates this material to be quite impure (67).

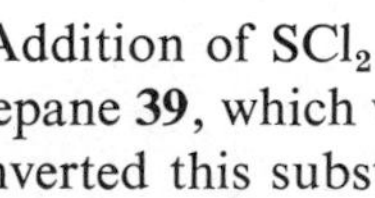

38 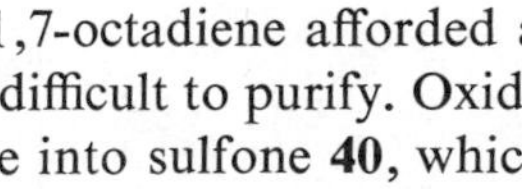 **39** 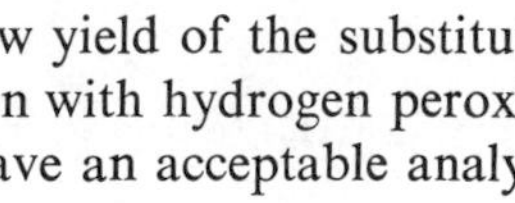 **40**

Addition of SCl_2 to 1,7-octadiene afforded a low yield of the substituted thiepane **39**, which was difficult to purify. Oxidation with hydrogen peroxide converted this substance into sulfone **40**, which gave an acceptable analysis (106).

An interesting attempted cyclization to a thiepane derivative was reported recently by Truce and coworkers (Eq. 8) (190). Benzyl ε-chloropentyl sulfone reacts with amide ion to afford benzyl cyclopentyl sulfone (**41**) as the only cyclic product. The mechanism presumably involves abstraction of a proton α to the sulfone moiety followed by internal displacement of chloride ion. The alternative mode of cyclization, which would give 2-phenylthiepane

$$C_6H_5CH_2SO_2CH_2(CH_2)_3CH_2Cl \underset{}{\overset{NH_2^{\ominus}}{\rightleftarrows}} C_6H_5CH_2SO_2\overset{\ominus}{C}H(CH_2)_3CH_2Cl \longrightarrow \text{cyclopentyl}(H)\text{-}SO_2CH_2C_6H_5\ (\mathbf{41})$$

$$C_6H_5CH_2SO_2CH_2(CH_2)_3CH_2Cl \underset{}{\overset{NH_2^{\ominus}}{\rightleftarrows}} C_6H_5\overset{\ominus}{C}HSO_2CH_2(CH_2)_3CH_2Cl \not\longrightarrow \text{2-}C_6H_5\text{-2-H-thiepane-}SO_2\ (\mathbf{42})$$

1,1-dioxide (**42**), did not occur. Since the benzylic hydrogens would be expected to be the more acidic α-protons, the observed product may be explained in terms of the entropy of activation for the cyclization.

This reticence to form the seven-membered ring also is seen in the results of Bennett and co-workers (20, 21) who studied the thermally induced cyclization of several ω-halo sulfides (Eq. 9). Cyclic sulfonium halides with

$$C_6H_5S(CH_2)_nCH_2Cl \longrightarrow C_6H_5\overset{\oplus}{S}\begin{matrix}(CH_2)_n\\ |\\ CH_2\end{matrix}\ Cl^{\ominus} \tag{9}$$

five- and six-membered rings ($n = 3, 4$) formed much more readily than the thiepane derivative ($n = 5$). In fact, 6-chlorohexyl phenyl sulfide was not cyclized under conditions much more stringent than those which sufficed to close the other two rings. Formation of the thiepane ring in this instance should be adversely affected not only by entropy considerations but also by the reduced basicity of the sulfur atom adjacent to an electron-withdrawing phenyl ring. Cyclization to a seven-membered ring was accomplished with ethyl 6-chlorohexyl sulfide (21) (Eq. 10); the initially formed sulfonium chloride **43** was converted to the yellow chloroplatinate **44**.

$$C_2H_5S(CH_2)_5CH_2Cl \xrightarrow{\Delta} \text{1-}C_2H_5\text{-thiepanium}\ Cl^{\ominus}\ (\mathbf{43}) \longrightarrow [\text{1-}C_2H_5\text{-thiepanium}]_2\ PtCl_6^{\ominus\!\ominus}\ (\mathbf{44}) \tag{10}$$

Thiepane is a liquid of strong (133) but not particularly disagreeable odor. Reported physical properties are included in Table X-3. The nmr spectrum of thiepane (in carbon tetrachloride) consists of a multiplet at τ 7.2–7.5 (4H) for the hydrogens on carbons α to sulfur, and a multiplet at τ 7.9–8.5 (8H) attributed to the remainder of the protons (58). The uv spectrum of thiepane displays only end absorption below 280 mμ, with points of inflection at 230–250 mμ (149). The heat of formation of thiepane has been reported to be −27 kcal/mole (116, 185).

c. *Reactions*

Thiepane is oxidized by an equimolar amount of hydrogen peroxide to thiepane 1-oxide (**45**) (54). Oxidation with potassium permanganate (82) or a peracid (126) has been employed for the synthesis of thiepane 1,1-dioxide (**13**).

SO
45

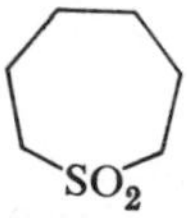

SO_2
13

Thiepane reacts in a normal manner with methyl iodide to form sulfonium salt **46** [m.p. 147°C (33), 137–138°C (143), 141.5–142°C (108), 161–162°C (68)]. With a large excess of methyl iodide in the presence of acetone at 133–149°C for 17 hr the ring is cleaved with the formation of 1,6-diiodohexane and trimethylsulfonium iodide (133). This ring cleavage presumably involves the reaction path depicted in Eq. 11.

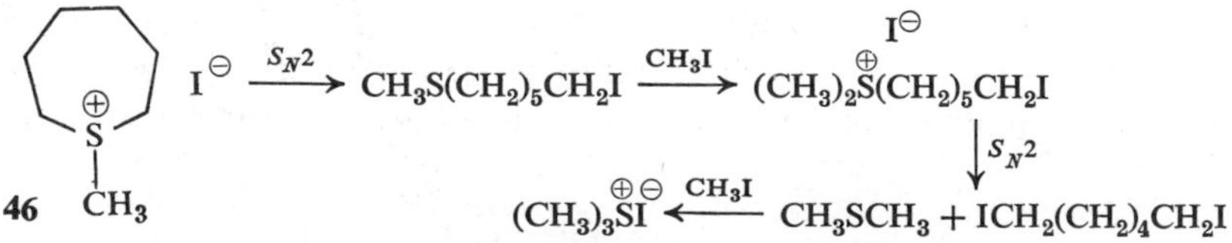

$$\textbf{46}\ (\overset{\oplus}{S}\text{–}CH_3)\ I^{\ominus} \xrightarrow{S_N2} CH_3S(CH_2)_5CH_2I \xrightarrow{CH_3I} (CH_3)_2\overset{\oplus}{S}(CH_2)_5CH_2I\ I^{\ominus}$$
$$\downarrow S_N2$$
$$(CH_3)_3\overset{\oplus\ominus}{SI} \xleftarrow{CH_3I} CH_3SCH_3 + ICH_2(CH_2)_4CH_2I \qquad (11)$$

Thiepane forms a 1:1 complex with $HgCl_2$, mp 152–153°C (144). One of the methods used for the purification of thiepane involves the formation of this complex and the regeneration of thiepane from it (54).

Nickel catalysts in the presence of thiepane reportedly produce a new catalyst which is effective for the reduction of diolefins to monoolefins (29). This application apparently has not been used widely.

Johnson and Jones have shown that the oxidation of thiepane with either *tert*-butyl hypochlorite or isopropyl hypochlorite in methylene chloride at −78C° affords the alkoxysulfonium chlorides **47** (Eq. 12, R = *tert*-butyl or isopropyl) (97). These salts were too unstable for isolation, but were trans-

$$\underset{\mathbf{35}}{(CH_2)_6S} \xrightarrow{ROCl} \underset{\mathbf{47}}{(CH_2)_6\overset{\oplus}{S}OR\ Cl^{\ominus}} \xrightarrow{SbCl_5} \underset{\mathbf{48}}{(CH_2)_6\overset{\oplus}{S}OR\ SbCl_6^{\ominus}} \tag{12}$$

formed into the readily isolable hexachloroantimonates **48** by addition of antimony pentachloride. The hexachloroantimonates were characterized by elemental analysis and nmr spectra.

Cox and Owen (58) have shown recently that thiepane is converted in good yield to 2-acetoxythiepane (**49**) by the method of Sosnovsky (180) (Eq. 13).

$$\text{thiepane} \xrightarrow[Cu_2Cl_2]{CH_3COOBu\text{-}tert} \left[\underset{\mathbf{50}}{\text{thiepane-}\overset{\oplus}{S}\text{-OAc}} \quad {}^{\ominus}OBu\text{-}tert \right] \longrightarrow \underset{\mathbf{49}}{\text{2-(H)(OAc)-thiepane}} \tag{13}$$

This reaction of a sulfide with *tert*-butyl peracetate in the presence of cuprous halide seems analogous to the Pummerer reaction and presumably involves the intermediacy of the acetoxysulfonium salt **50**. The structure of acetate **49** was confirmed by its ir and nmr spectra.

Hydrolysis of 2-acetoxythiepane gave an excellent yield of 2-hydroxythiepane (**51**) (58). The ir spectrum of **51** in the solid state displays not only a strong O—H stretching band at 3350 cm^{-1}, but also a very weak carbonyl band at 1720 cm^{-1}. In chloroform solution the carbonyl band in the ir spectrum is quite strong and the OH band at 3600 cm^{-1} is weak. In addition, the solution spectrum contains a band at 2720 cm^{-1} attributed to aldehydic C—H stretching. These data indicate that in solution the cyclic hemithioacetal **51** undergoes rapid ring-chain tautomerization to form 6-mercaptohexanal (**51A**). In agreement with this interpretation, the nmr spectrum in

$$\underset{\mathbf{51}}{\text{2-(H)(OH)-thiepane}} \rightleftharpoons \underset{\mathbf{51A}}{HS(CH_2)_5CHO} \qquad \underset{\mathbf{52}}{\text{2-(H)(OH)-thiolane}} \qquad \underset{\mathbf{53}}{\text{2-(H)(OH)-thiane}}$$

$CDCl_3$ displays a single proton triplet at τ 0.23 and a singlet at τ 7.54 which is removed by exchange with D_2O. In methanol solution the compound readily reacts with one equivalent of iodine, indicating a free —SH moiety.

The ring opening of hemithioacetal **51** is much more rapid than the analogous reactions of the lower ring homologs (58), possibly reflecting entropy considerations similar to those mentioned in Section B.4b. The nmr spectra of **52** and **53** indicate them to be cyclic structures. Unlike the five- and six-membered hemithioacetals, however, 2-hydroxythiepane does not appear to form a mercaptal on standing (Eq. 14) (58). No experiments involving the attempted catalysis of mercaptal formation from **51** have been attempted.

H OH S ⟶(not) S H $S(CH_2)_5CHO$ (14)

Thiepane reacts with *N*-chlorosuccinimide in carbon tetrachloride or benzene to form 2-chlorothiepane (**54**) (192). The chloro sulfide was not

35 —(N-chlorosuccinimide, NCl)→ **54** (2-chlorothiepane; H, Cl)

isolated, but was characterized by its nmr spectrum, by peracetic acid oxidation to 2-chlorothiepane 1,1-dioxide (**55**), and by reaction with methyl and phenyl Grignard reagents to form 2-methylthiepane (**56**) and 2-phenylthiepane (**57**). The over-all yields of the two steps were 66 and 56% for the methyl and phenyl derivatives (192). Chlorosulfone **55** is converted by hot aqueous sodium hydroxide into cyclohexene (Eq. 15). When the reaction medium is

35 —(NCl)→ **54** —(CH_3MgBr)→ **56** (H, CH_3)
54 —(C_6H_5MgBr)→ **57** (H, C_6H_5)
54 —(CH_3CO_3H)→ **55** (SO_2, H, Cl) —(NaOH, H_2O)→ cyclohexene (15)

NaOD–D_2O, this Ramberg-Backlund reaction affords an excellent yield of 1,2-dideuteriocyclohexene (193).

Thiepane has been reported to be decomposed by an aluminum silicate bead catalyst at 400°C; hydrogen sulfide is the principal product (146–148).

5. Thiepanones and Derivatives

All of the three possible monocyclic thiepanones **58–60** are known.

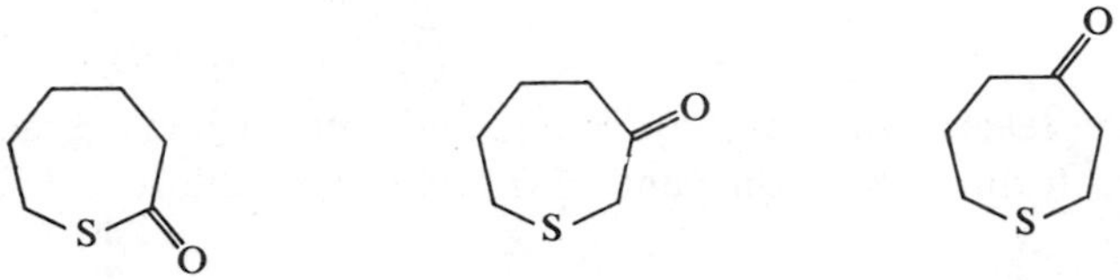

58 **59** **60**

a. *2-Thiepanones* (*ε-Thiocaprolactones*)

Overberger and Weise have prepared several 2-thiepanones by the cyclization of the corresponding substituted 6-mercaptohexanoic acids. Cyclizations to **58** and **61–63** were performed best by distillation of the thiolactones from phosphorus pentoxide (153–155). The optically active (*R*)-(−)-4-methyl-2-thiepanone (**61**) and (*R*)-(+)-5-methyl-2-thiepanone (**62**) were synthesized elegantly from (*R*)-(+)-pulegone (**64**) (154).

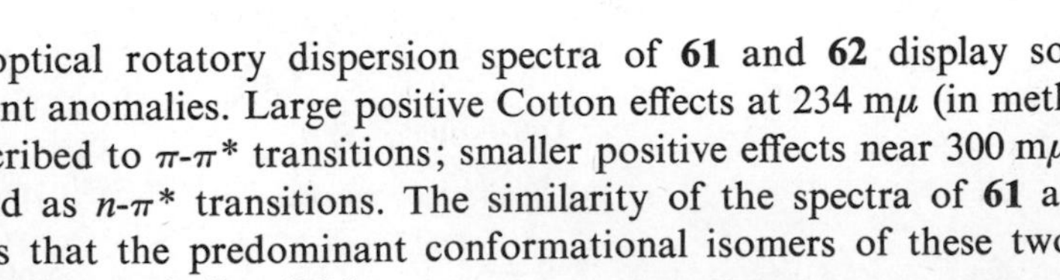

61 **62** **63** **64**

The optical rotatory dispersion spectra of **61** and **62** display solvent-dependent anomalies. Large positive Cotton effects at 234 mμ (in methanol) were ascribed to π-π^* transitions; smaller positive effects near 300 mμ were explained as n-π^* transitions. The similarity of the spectra of **61** and **62** indicates that the predominant conformational isomers of these two (*R*)-thiolactones are similar (154).

Treatment of the 2-thiepanones with strong bases, such as potassium *tert*-butoxide at elevated temperatures or *n*-butyllithium at room temperature, resulted in the formation of polythioesters **65** of high molecular weight.

Initiation of the reaction with polymerization catalysts which are normally cationic, such as aluminum chloride–water, produces colored polymers of low molecular weight. The base-catalyzed reactions are consistent with an ionic ring-opening polymerization initiated by nucleophilic attack at the carbonyl carbon atom and propagated by the resultant thiolate anion (153, 155, 156).

$$\left[\overset{\overset{\displaystyle O}{\|}}{C}-CH_2-CHR-CHR-CH_2CH_2S\right]_n$$

65

b. *3-Thiepanones*

Dieckmann cyclization of diester **66** affords a liquid which presumably is 2-carbethoxy-3-thiepanone (**67**), 4-carbethoxy-3-thiepanone (**68**), or a mixture of these two β-keto esters (108). The product appeared to decompose

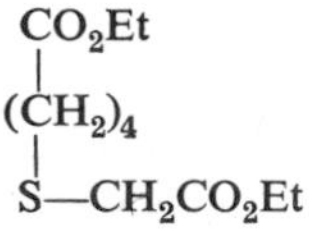

CO_2Et–$(CH_2)_4$–$S-CH_2CO_2Et$

66

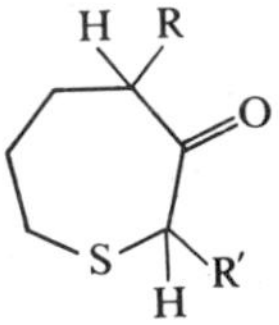

67; $R = H$, $R' = CO_2Et$
68; $R = CO_2Et$, $R' = H$

under the conditions used for distillation. Acid-catalyzed hydrolysis and decarboxylation gave 3-thiepanone (**59**), which was characterized by its ir spectrum and as the oxime and benzoylated oxime derivatives (108).

Addition of diazoalkanes to thiacyclohexan-3-one produces mixtures of substituted 3-thiepanones and 4-thiepanones which can be separated by distillation (Eq. 16) (26). Reduction of the substituted 3-thiepanones with

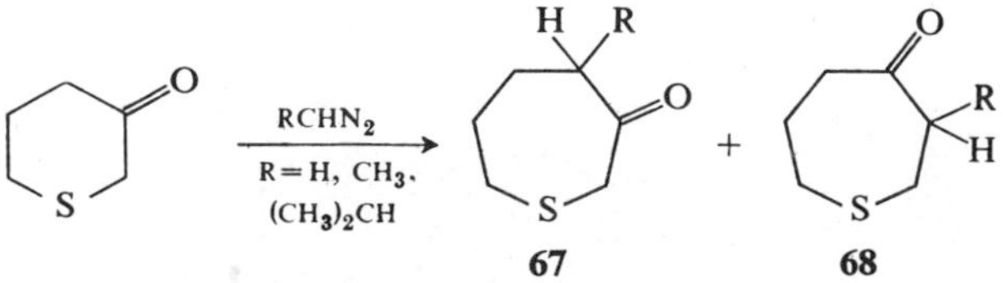

$RCHN_2$; $R=H, CH_3, (CH_3)_2CH$ → **67** + **68** (16)

lithium aluminum hydride affords mixtures of the *cis* and *trans* alcohols (Eq. 17). Pure *cis* alcohols **69** were isolated by chromatography (26). Infrared spectra of the *cis* alcohols **69** exhibit two O—H stretching bands near 3620 and 3490 cm^{-1}. The band at lower frequency is ascribed to transannular association. The proportion of the transannularly associated conformation is greater when R is methyl or isopropyl than when it is hydrogen.

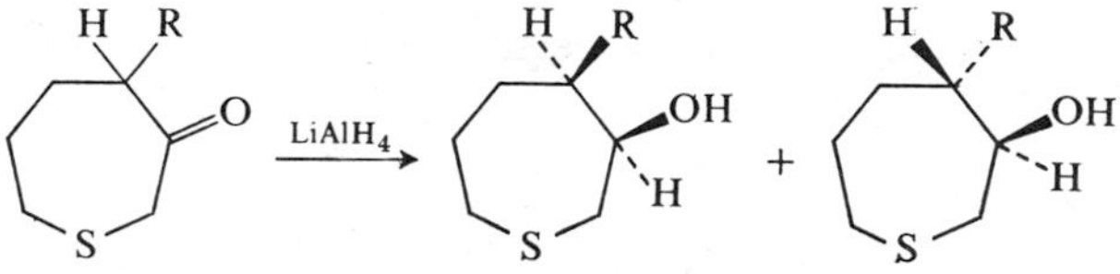

$$\mathbf{67} \xrightarrow{LiAlH_4} \mathbf{69} + \mathbf{70} \qquad (17)$$

Additional steric bulk at the 4-position thus seems to increase the proportion of molecules containing a pseudoaxial OH (necessary for transannular interaction) (26).

An interesting ring contraction of 3-thiepanone occurs under the conditions of the Clemmensen reduction (Eq. 18). The ketone is converted to 2-methylthiacyclohexane **36** (108). A similar rearrangement, resulting in the formation

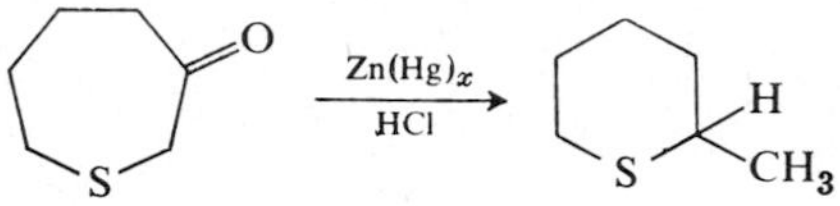

$$\mathbf{59} \xrightarrow[HCl]{Zn(Hg)_x} \mathbf{36} \qquad (18)$$

of five-membered rings, has been observed with the six-membered β-keto sulfides 3-ketothiacyclohexane (108) and 4-ketoisothiochroman (32). A reasonable mechanistic proposal, which involves initial hydrogenolysis of the S—C_2 bond followed by reduction of the carbonyl group and internal dehydration, was advanced by von Braun and Weissbach (32). Another plausible pathway is indicated by Eq. 19.

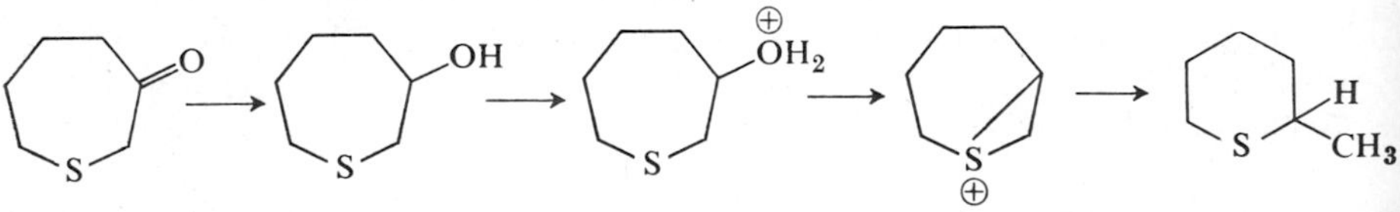

(19)

c. *4-Thiepanones*

(1) PREPARATION. Overberger and Katchman (151) synthesized 4-thiepanone (**60**) by ring enlargment of tetrahydro-1,4-thiapyrone (**71**) as shown by Eq. 20. The diazomethane was generated *in situ* from nitrosomethylurethan, using barium oxide as the base. The desired ketone **60** (42%)

was formed together with epoxide **72** (40%); the two liquids were separated by distillation. Ketone **60** was characterized by its strong ir absorption at 1710 cm^{-1} and as its 2,4-dinitrophenylhydrazone.

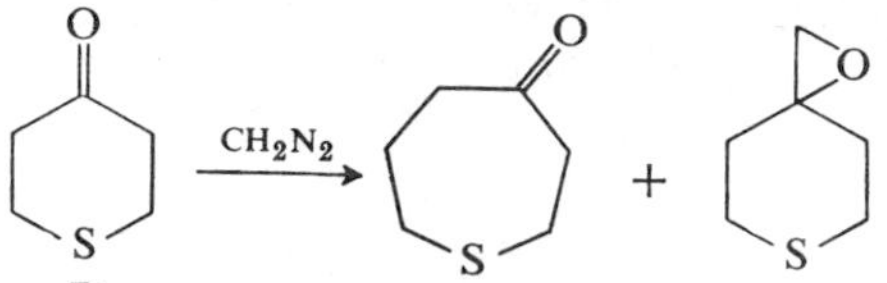

CH_2N_2 (20)

71 **60** **72**

The dipole moment of 4-thiepanone (3.04 D) as well as its single, unsplit carbonyl band in the ir led to the conclusion that there is no appreciable transannular S—C_{co} interaction in this ketone (109). This behavior may be contrasted with that of 5-thiocanone **(73)** which is believed to exist as a mixture of "non-interacted" and "interacted" forms **73A** and **73B** on the basis of a large dipole moment (3.8 D) and a split carbonyl band in the ir spectrum (107, 109, 152).

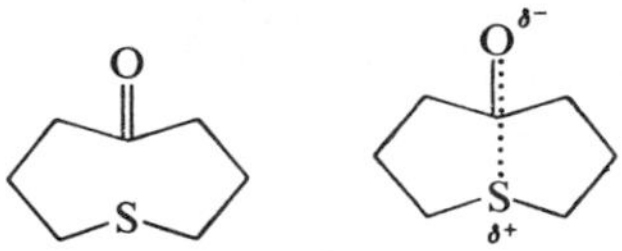

73A **73B**

The acyloin condensation of the tetramethylthiodipropionate **74** affords 3,3,6,6-tetramethylthiepan-4-on-5-ol **(75)** in 75% yield (88, 202). This acyloin **(75)** was oxidized by lead tetraacetate in pyridine to 3,3,6,6-tetramethyl-4,5-thiepandione **(76)** (88). Wolff-Kishner reduction of the mono-

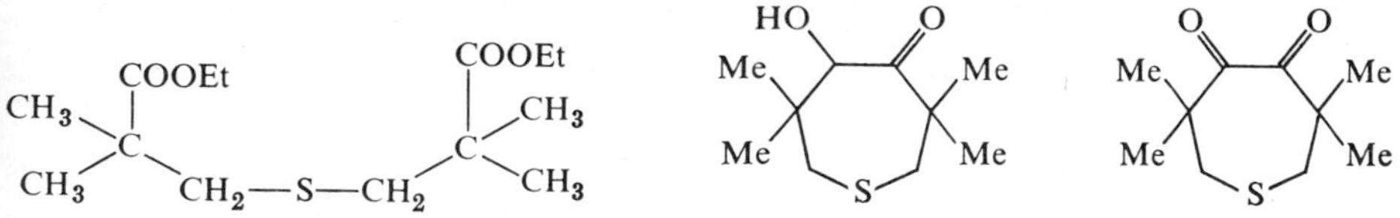

74 **75** **76**

hydrazone **(77)** affords a good yield of 3,3,6,6-tetramethyl-4-thiepanone **(78)**. A poor yield of the same monoketone resulted from treatment of the hydrazone with potassium *tert*-butoxide in xylene (88).

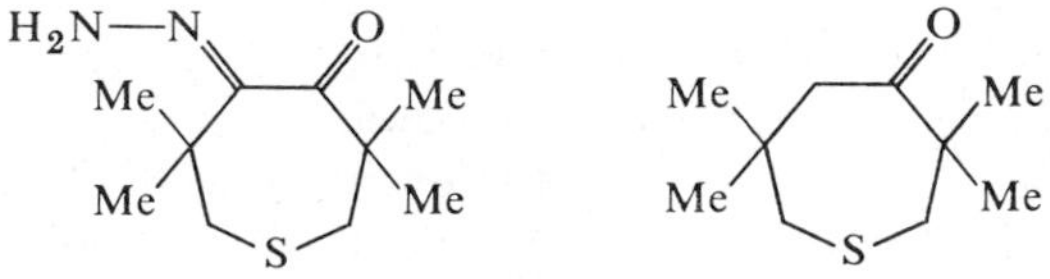

77 **78**

Diketone **76** reacts with 3 equivalents of methylenetriphenylphosphorane in dimethyl sulfoxide to form the 1,3-diene **79** (Eq. 21) (87). The uv spectrum of this unusual diene displays no maxima above 185 mμ, indicating that there is no appreciable conjugation between the two olefinic linkages. The authors believe that the angle of twist between the π orbitals of the double bonds is about 90°. This hypothesis seems substantiated by the fact that **79** did not

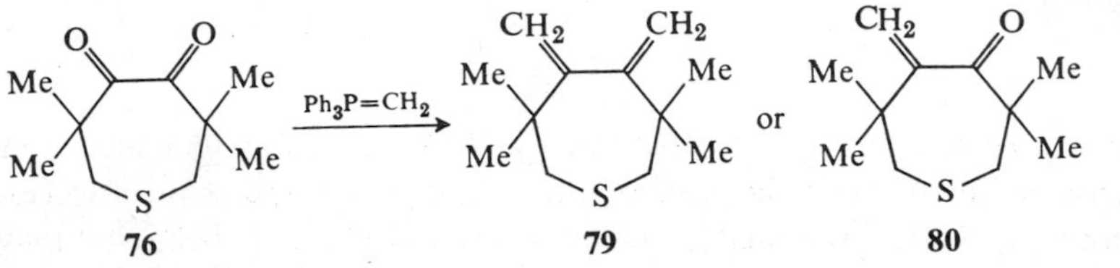

$Ph_3P{=}CH_2$ (21)

76 **79** or **80**

form a Diels-Alder adduct with diethyl acetylenedicarboxylate, tetracyanoethylene, benzyne, or fluorenethione. Rigidity in the cyclic system also was adduced from nmr spectra of **79** at various temperatures. The peaks for the nonequivalent methyl groups, which are displayed as separated singlets at room temperature, coalesce at 92°C; similarly, the AB pattern for the $C_{(2)}$ and $C_{(7)}$ methylene protons coalesces at 112°C. The barrier to inversion for **79** calculated from these spectra is 8.3 kcal/mole (87).

Reaction of **76** with 1.5 equivalents of methylenetriphenylphosphorane gives **80**. The uv spectrum of this methylene ketone (λ_{max} = 300 mμ) also indicates a lack of 1,3-conjugation in this sterically hindered ring (87).

Reaction of ethyl oxalate with thiodipropionitrile has been observed to give 3,6-dicyanocatechol (Eq. 22) (112). Loudon and Steel remarked that the reaction probably proceeds through the intermediate 3,6-dicyano-4,5-thiepandione (**81**), which undergoes loss of hydrogen sulfide and cyclization to the aromatic nucleus.

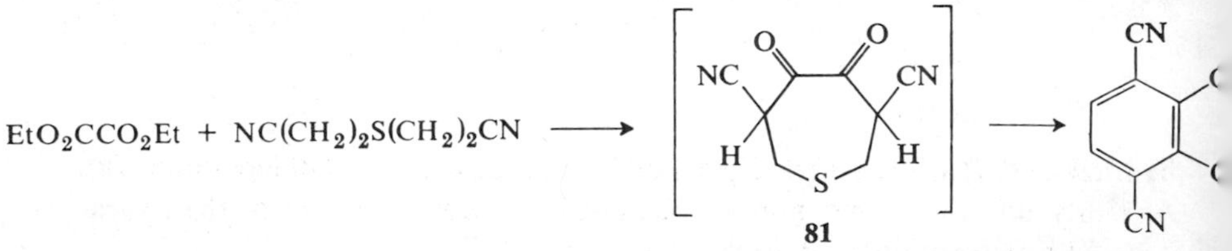

$EtO_2CCO_2Et + NC(CH_2)_2S(CH_2)_2CN \longrightarrow$ [**81**] $\longrightarrow$

Irradiation of 3-thiacyclohexanone in Freon 113 has recently been shown to produce 4-thiepanone in poor yield (120).

(2) Reactions. Oxidation of 4-thiepanone with hydrogen peroxide affords the expected sulfone **82** (151); oxidation with sodium metaperiodate produces the hygroscopic sulfoxide **83** (110). Treatment of this sulfoxide

with perchloric or fluoboric acid reportedly produces the interesting bicyclic sulfoxonium salts **84** and **85**. Transannular bonding between the carbon atom of a protonated carbonyl group and the electron-donating sulfoxide

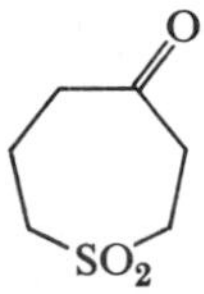

82 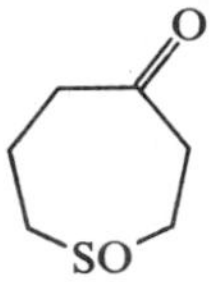 **83** 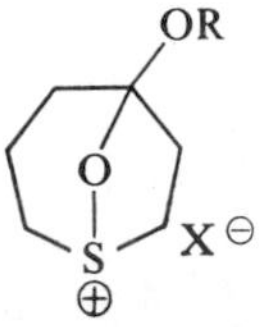 **84**; R = H, X = ClO_4
85; R = H, X = BF_4
86; R = CH_3, X = BF_4

moiety accounts for the formation of these salts. The bicyclic salts exhibit strong absorption in the ir at 3320 cm^{-1} (indicative of O—H stretching), and an absence of absorption at 1245 cm^{-1} (S—O stretching). Titration of either **84** or **85** with base results in the regeneration of thiepane 1-oxide. Methylation of **85** with 2,2-dimethoxypropane gives 5-methoxy-8-oxo-1-thioniabicyclo[3.2.1]octane tetrafluoborate (**86**).

The carbonyl group of 4-thiepanone reacts normally with nucleophiles. Addition of *n*-butyl Grignard reagent gives the tertiary alcohol **87** in 33% yield (204). Normal carbonyl derivatives, such as the 2,4-dinitrophenylhydrazone (151), phenylhydrazone, *p*-tolylhydrazone, and *p*-carbethoxyphenylhydrazone, have been prepared and characterized (2).

The arylhydrazones of 4-thiepanone are converted via the Fischer indole synthesis into tetrahydrothiepino[4,5-*b*] indoles **88** (2). Analogous indoles **89** are formed from the substituted hydrazones of 4-thiepanone 1,1-dioxide (**82**).

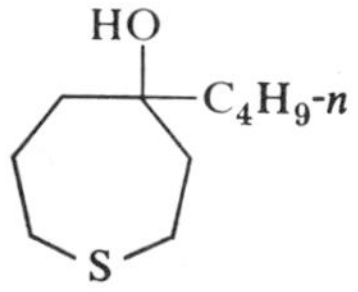

87 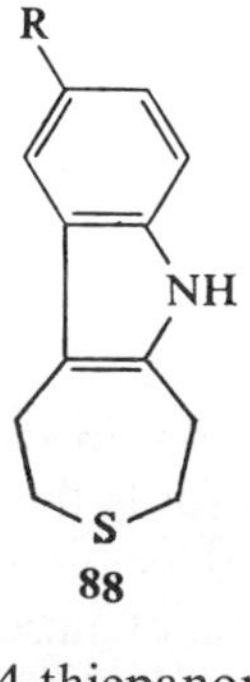 **88** 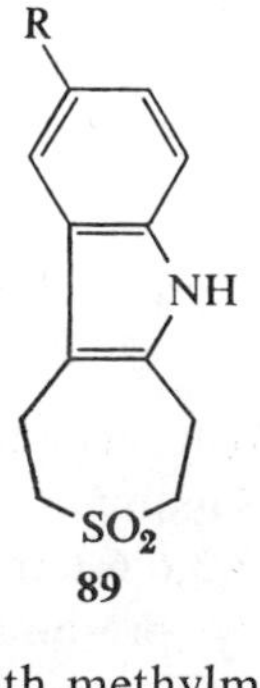 **89**

Reaction of 3,3,6,6-tetramethyl-4-thiepanone (**78**) with methylmagnesium bromide, isopropyllithium, or *tert*-butyllithium proceeds to give the expected substituted 4-thiepanols **90** in good yield (83, 86). The ir spectra of alcohols **90** where R is either a methyl or an isopropyl substituent display evidence of only a small amount of transannular interaction of the OH group with sulfur. Only when the substituent at $C_{(4)}$ of **90** is the *tert*-butyl group is there clear evidence of such an interaction (86). These results are in marked

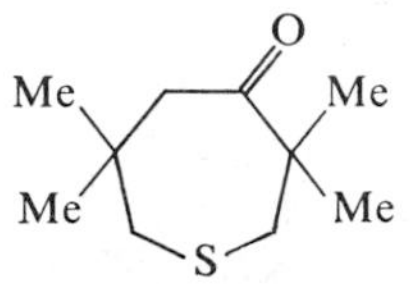

78

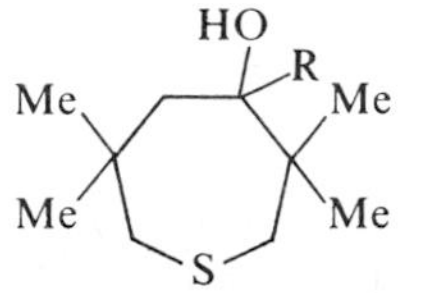

R=Me, *i*-Pr, *tert*-Bu
90

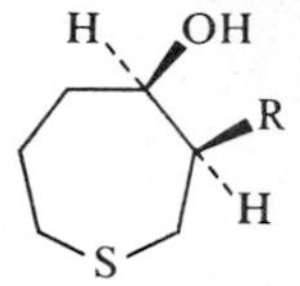

R=Me, *i*-Pr
91

contrast to those of Borsdorf, Kasper, and Repp (26) who report substantial amounts of the transannularly associated conformer of **91** where R is either a methyl or an isopropyl group.

Reaction of 3,3,6,6-tetramethyl-4,5-thiepandione (**76**) with nucleophiles generally involves only a single carbonyl group. The oxime, hydrazone, *p*-toluenesulfonylhydrazone, and semicarbazone have been prepared and characterized (88, 202).

Reaction of **76** with methyllithium or *tert*-butyllithium also results in addition at only one carbonyl group to afford alcohols **92** in good yield. Allyllithium, however, adds to both carbonyl groups to form an 83% yield of the vicinal diol **93** (83).

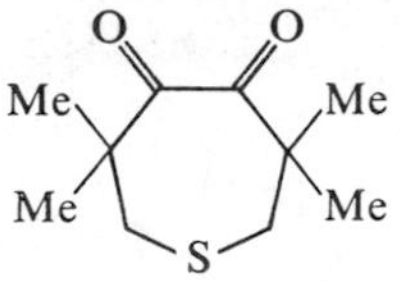

76

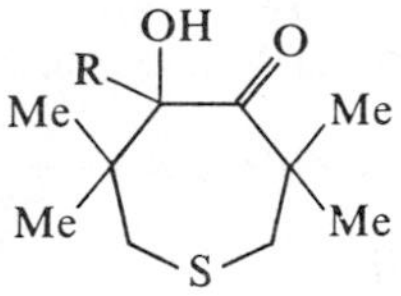

92

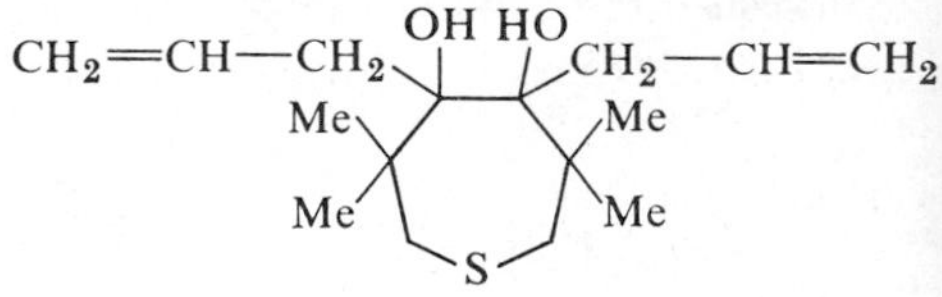

93

R=Me, *tert*-Bu

Wittig reagents add normally to **76** to produce **94** and **95** in good yield (83, 89). Saponification of **94** affords the acid **96** (83). Hydrolysis of **95** gives 5-formyl-3,3,6,6-tetramethyl-4-thiepanone (**97**) which is not enolized appreciably (presumably because of steric factors) (89).

The benzilic acid rearrangement of 3,3,6,6-tetramethyl-4,5-thiepandione (**76**) proceeds quite cleanly (Eq. 23) (83). Oxidation of monohydrazone **98** with manganese dioxide presumably affords the α-diazoketone **99** which undergoes spontaneous decomposition to afford a 49% total yield of

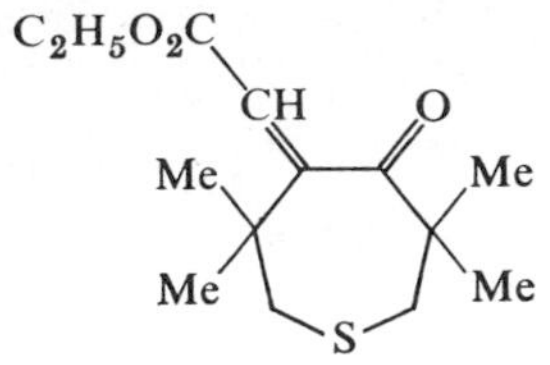

94

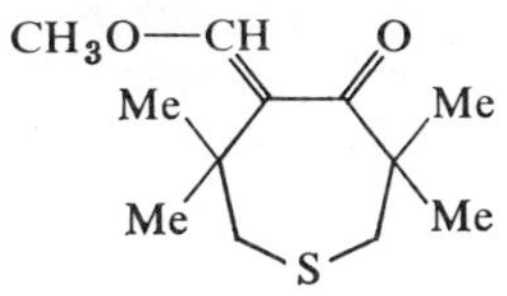

95

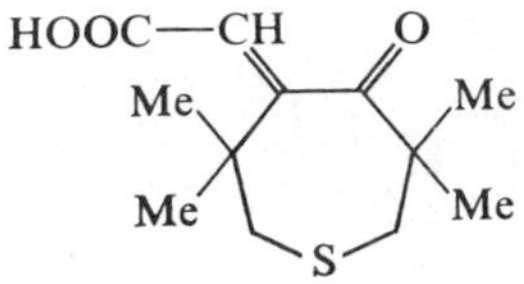

96

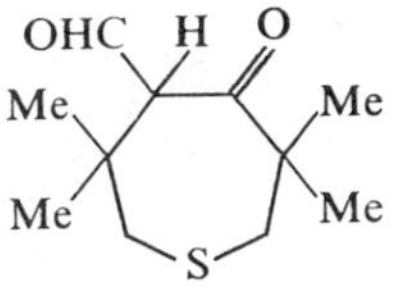

97

Wolff rearrangement products (Eq. 24) (83). An acyclic model, pivalil (**100**), is quite resistant to the action of hot caustic and fails to undergo the

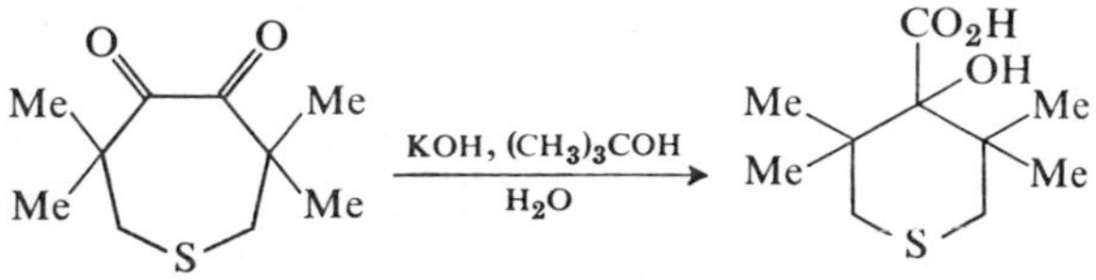

KOH, $(CH_3)_3COH$, H_2O (23)

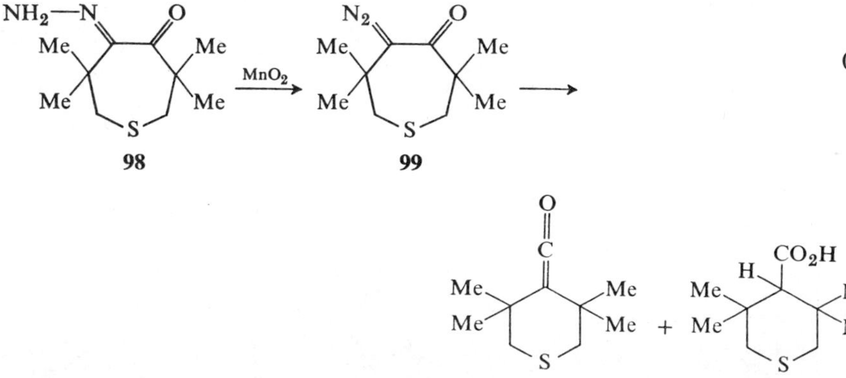

76

MnO_2 (24)

98

99

benzilic acid rearrangement (162). Diazoketone **101** gives only a minute amount of the ketene expected from Wolff rearrangement (134). The greater propensity for rearrangement of **99** (compared to **101**) is in accord with the reported (101) *cis* relationship of carbonyl and diazo groups in the transition state for the Wolff rearrangement. No explanation has been offered for the differences of behavior of **76** and **100** under the conditions of the benzilic acid rearrangement.

$$\underset{\mathbf{100}}{(CH_3)_3C\overset{O}{\overset{\|}{C}}\overset{O}{\overset{\|}{C}}C(CH_3)_3} \qquad \underset{\mathbf{101}}{(CH_3)_3C\overset{N_2}{\overset{\|}{C}}-\overset{O}{\overset{\|}{C}}C(CH_3)_3}$$

Reduction of 4-thiepanone with lithium aluminum hydride (98) or by the Meerwein-Pondorff-Verley method (151) affords 4-thiepanol (**102**) which has been characterized by its ir and mass spectra. Reaction of the alcohol with acetyl chloride gives acetate **103**. Oxidation of **102** by hydrogen peroxide in

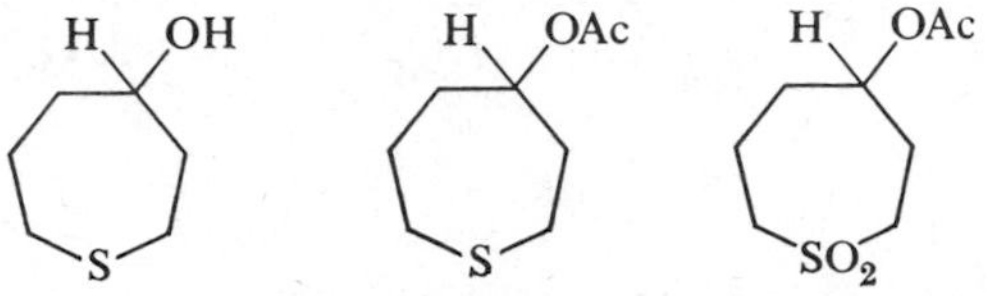

H OH / S — **102**   H OAc / S — **103**   H OAc / SO_2 — **27**

acetic acid followed (without isolation of the intermediate impure 4-hydroxythiepane 1,1-dioxide) by acetylation gives 4-acetoxythiepane 1,1-dioxide (**27**); this substance also has been prepared by a more difficult method involving hydrogenation of 4-thiepanone 1,1-dioxide (**82**) over Raney nickel followed by acetylation (151). The pyrolysis of **27** has been discussed in Section B-3 of this chapter.

Reduction of acyloin **75** with lithium aluminum hydride provides vicinal glycol **104** (84, 86). This diol undergoes an interesting "abnormal pinacol rearrangement" in a mixture of acetic acid and sulfuric acid to give, as the only observed product (66% yield), the bicyclic compound **105** (Eq. 25) (84).

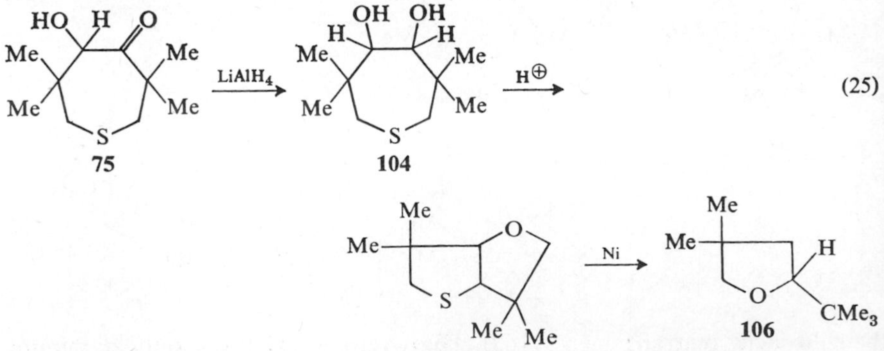

$$\mathbf{75}\ \xrightarrow{LiAlH_4}\ \mathbf{104}\ \xrightarrow{H^{\oplus}}\ \mathbf{105}\ \xrightarrow{Ni}\ \mathbf{106} \tag{25}$$

The structure of **105** seems unequivocally assigned on the basis of ir, nmr, and mass spectra, together with Raney nickel desulfurization, which affords 2-*tert*-butyl-4,4-dimethyltetrahydrofuran (**106**).

The abnormal dehydration of **104** would appear necessarily to involve the transannular formation of a transient sulfonium salt. De Groot, Boerma, and Wynberg reasonably suggest a mechanism in which sulfur displaces water from the protonated glycol **107** to generate sulfonium ion **108** (84).

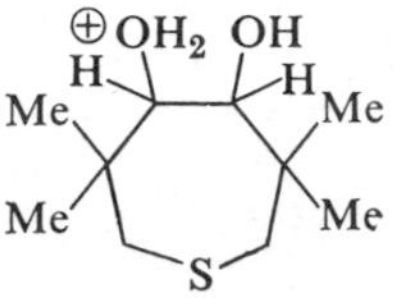

107 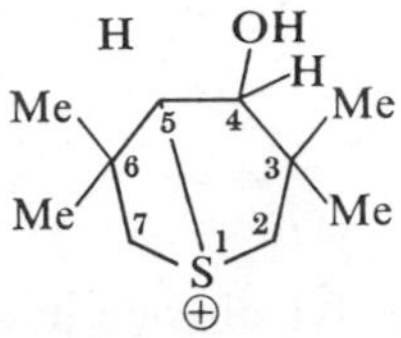 **108** 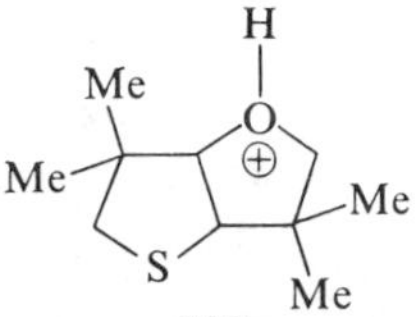 **109**

Cleavage of the four-membered ring at the S—$C_{(7)}$ linkage to generate a short-lived primary carbonium ion which is attacked by the alcoholic oxygen, or else displacement of sulfide via internal S_N2 attack by the alcoholic oxygen on $C_{(7)}$, would readily lead to the protonated product **109**.

A similar abnormal rearrangement has been observed for the treatment of acyloin **75** with triphenylphosphine dibromide in boiling dimethyl formamide (Eq. 26) (86). Transannular displacement of triphenylphosphine

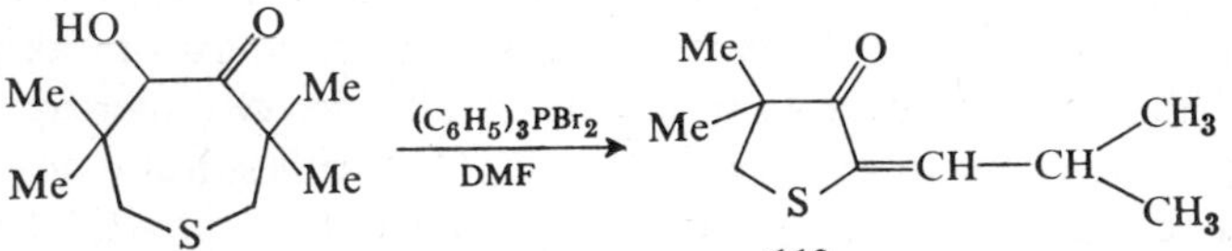

$(C_6H_5)_3PBr_2$, DMF (26)

75 **110**

oxide from an initially formed salt **111** is believed to result in the formation of sulfonium ion **112** which suffers cleavage of the four-membered ring at the S-$C_{(7)}$ linkage. Typical carbonium-ion processes then lead to **110** (86).

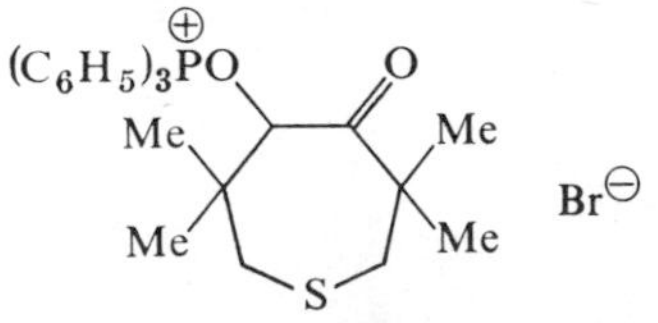

111 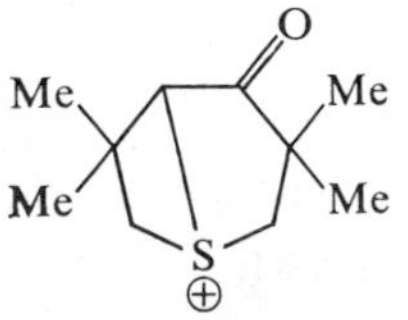 **112**

The reaction of acyloin **75** with thionyl chloride has been studied by de Groot, Boerma, and Wynberg (85). The expected 3,3,6,6-tetramethyl-5-chloro-4-thiepanone (**113**) does not form (Eq. 27). Surprisingly, the stable enediol sulfite **114** is produced. This novel structure is believed by the authors to be stabilized by the steric protection of the four methyl groups. The enediol sulfite is converted to thiepandione **76** by heat (85).

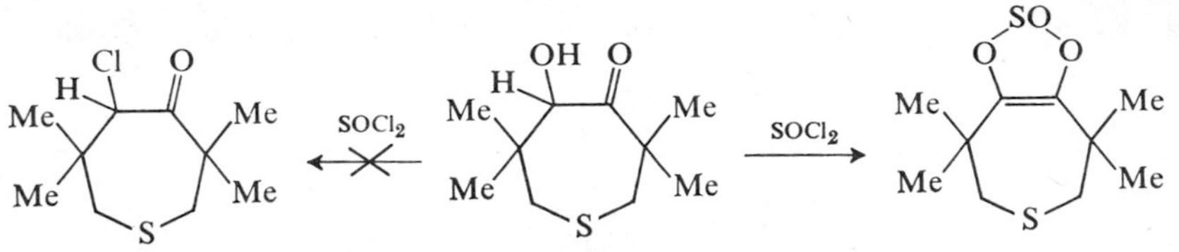

(27)

113 **75** **114**

Bromination of 4-thiepanone 1,1-dioxide in acetic acid gives a dibrominated product, mp 156.7–157.7°C, in excellent yield. This compound was presumed to be 3,5-dibromo-4-thiepanone 1,1-dioxide (151), but the structure has not been proved.

6. Thioseptanoses

The first example of a monocyclic seven-membered glycoside containing sulfur as the heteroatom was reported in 1965 by Cox and Owen (56) as part of a larger study designed to compare the formation of rings containing five, six, and seven members (57, 58). For simplicity, these glycosides will not be named as thiepanes, but will be regarded as thioseptanoses in keeping with accepted practice. Acid-catalyzed hydrolysis of 6-deoxy-6-mercapto-2,3,4,5-tetra-*O*-methyl-D-galactose ethylene acetal (**115**) with sulfuric acid in acetic acid gave a mixture of anomers of 6-deoxy-6-mercapto-2,3,4,5-tetra-*O*-methyl-D-galactoseptanose (**116**).

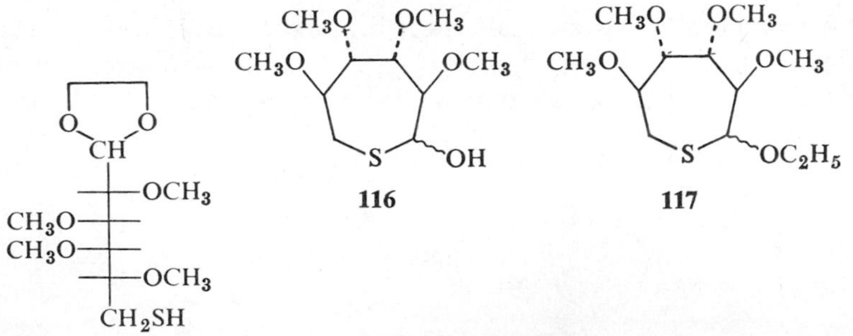

115 **116** **117**

Examination of the areas of the peaks corresponding to the anomeric protons in the nmr spectrum showed that the α and β forms were present in essentially equal amounts. The ir spectrum of **116** exhibited no band attributable to S—H stretching; a strong O—H band was present. Furthermore a nitroprusside test for the mercaptan group was negative. Thus the thio-septanose ring is more stable than the open-chain form, both as a pure liquid and in deuteriochloroform solution. This equilibrium may be displaced under certain conditions; thus the compound reacts as a free thiol with iodine within 5 min. This behavior is similar to that of thiofuranose rings but suggests considerably less stability than the thiopyranose ring system.

Ethanolysis of **116** in the presence of hydrochloric acid produced the ethyl glycoside **117** (56). Again, examination of the nmr spectrum indicated that essentially equal amounts of the two anomers were present. This ethyl glycoside also was formed by treatment of **116** with hydrochloric acid in ether, followed (without isolation of the intermediate chloride) by treatment with ethanol and silver carbonate. Material obtained in this manner was identical with the earlier sample except that analysis of the nmr spectrum indicated the ratio of anomers to be 1:3 (α:β). The result is analogous to ring closures of pyranose rings where the glycoside prepared by way of the chloride generally contains a higher proportion of β anomer.

Synthesis of the thioseptanose ring from acyclic precursors also has been achieved by Whistler and Campbell (200). Treatment of 6-deoxy-6-mercapto-2,3,4,5-tetra-*O*-acetyl-D-galactose (**118**) with pyridine effects mutarotation and ring formation. Acetylation affords a mixture of the two anomers of **119** which were separated by chromatography. Treatment of a mixture of the anomers

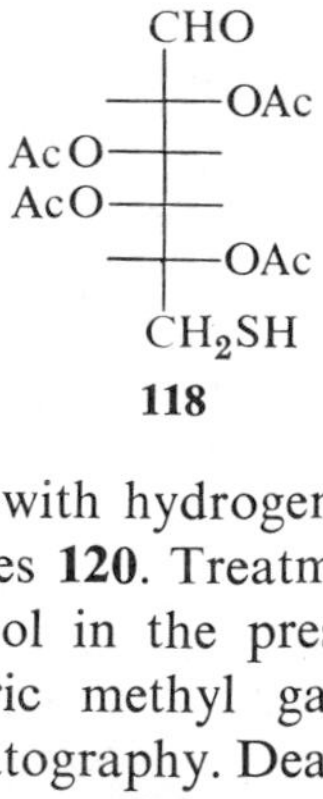

118

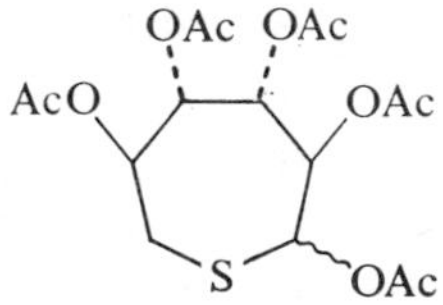

119

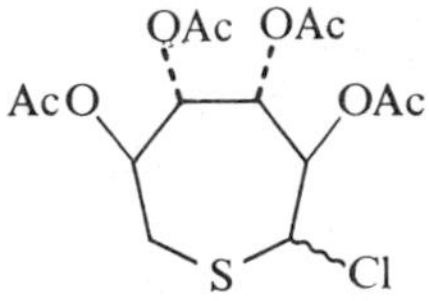

120

of **119** with hydrogen chloride in ether afforded a mixture of the anomeric chlorides **120**. Treatment of this mixture of anomeric α-chloro sulfides with methanol in the presence of silver carbonate effected conversion to the anomeric methyl galactothioseptanosides **121** which were separated by chromatography. Deacetylation of the β anomer with methanol and ammonia gave methyl β-D-galactothioseptanoside (**122**) (200).

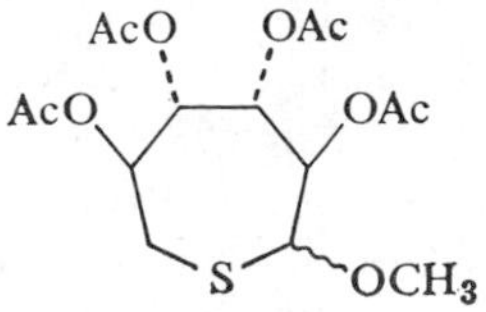

121

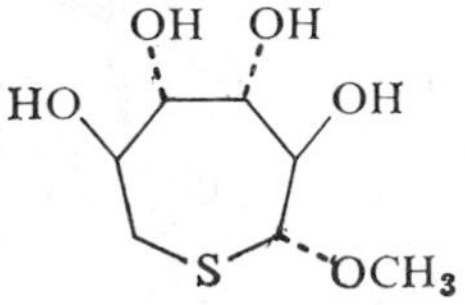

122

C. Rings Containing Two Sulfur Atoms (Dithiepins and Derivatives)

All of the three possible arrangements of two sulfur atoms in a seven-membered ring (1,2; 1,3; 1,4) have been reported. In the sense that sulfur is isosteric with a carbon-carbon double bond (30) and can participate in resonance delocalization (as, for example, in thiophene compared with benzene) (170), anions with two double bonds in a seven-membered ring are analogs of the cyclononatetraenide anion, which is a stable delocalized system (45). Zahradník and Párkányi, interested in this analogy, performed theoretical calculations on structures **123**, **124**, and **125** using the Hückel approximation of the MO LCAO method (208). They concluded that although

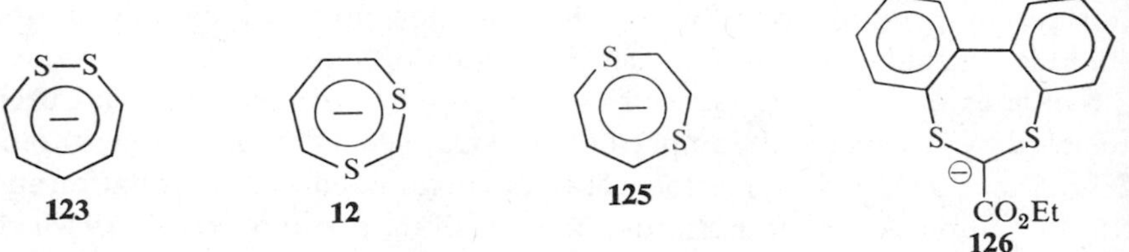

all three should be unstable and hard to prepare, the most promising for synthetic attack was structure **123**. Breslow and Mohacsi, in considering aromatic systems with ten electrons in a seven-membered ring, became interested in the possible aromatic character of the 1,3-dithiepinyl anion **124** but could detect no aromaticity in structure **126** (46); the chemistry of **126** will be dealt with in Chapter XI of this volume.

1. 1,2-Dithiepins and Derivatives

5*H*-1,2-Dithiepin (**127**) has been referred to also as 1,2-dithia-3,6-cyclo-

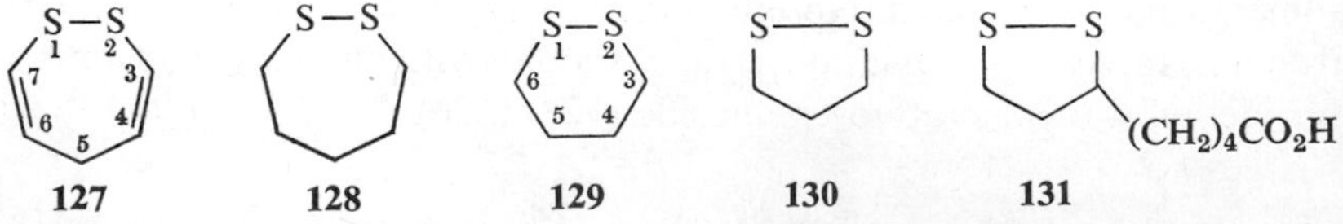

heptadiene, although it has not been prepared; the literature does not seem to mention its isomer, 3*H*-1,2-dithiepin. The fully reduced form, 1,2-dithiepane (**128**), has been called 1,2-dithiacycloheptane and pentamethylene disulfide. Most studies on 1,2-dithiepin systems have been done with the reduced forms, as have many of those on 1,2-dithianes (**129**) (41), in efforts to understand more fully the chemistry of 1,2-dithiolanes (**130**) (38). The 1,2-dithiolanes typically are the most reactive of the three classes, and are a particularly significant class because of the biological importance of lipoic acid (**131**).

a. *Physical Properties*

Schöberl and Gräfje described 1,2-dithiepane (**128**) as a colorless, water-insoluble liquid stable for several months (169); the ir spectrum did not change during storage in a Pyrex vessel for 9 months (71). It can be distilled under reduced pressure without significant polymerization (169). This behavior contrasts markedly with that of the 1,2-dithiolanes which typically are yellow and often polymerize readily (169). The odors of cyclic disulfides with 6–15 members in the ring are more like those of reduced aromatic compounds than of low aliphatic disulfides (169).

1,2-Dithiepane 1,1-dioxide (**132**), mp *ca.* 25°C, is stable for more than 10 months; the 1,1,2,2-tetraoxide **133**, mp 159–160°C, is stable for at least 5

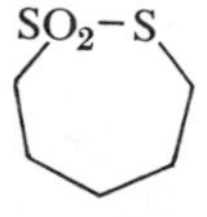

SO_2–S

132

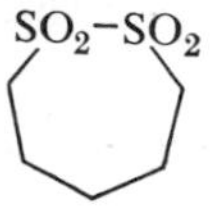

SO_2–SO_2

133

months (71). Oxidized forms of 1,2-dithiolane are likewise more stable than the disulfide; the mp (22–22.5°C) of the dioxide remains unchanged after more than a month, and that of the tetraoxide (166–167°C) after 9 months (71).

Foss and Schotte reported X-ray crystallographic parameters for several cyclic disulfides (76). (±)-1,2-Dithiepane-3,7-dicarboxylic acid (**134**) forms

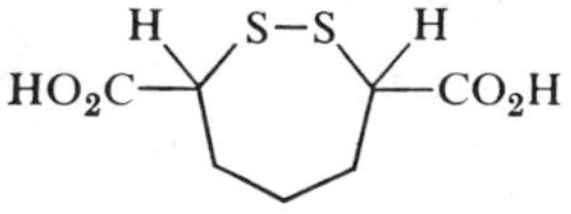

H S–S H

HO_2C– –CO_2H

134

monoclinic prisms, which are mostly poorly developed, having four molecules per unit cell and a density of 1.57 g/cm³. The *meso* form of **134**, also monoclinic, has eight molecules per unit cell and a density of 1.49 g/cm³ (76). Considering the —CSSC— moiety of the dithiepane ring to be almost certainly nonplanar, Foss and Schotte stated that two enantiomorphous forms of the *meso* acid, which are interconverted in solution, must be present in the crystal (76). This behavior resembles that of *meso*-1,2-dithiane-3,6-dicarboxylic acid (76).

Barltrop, Hayes, and Calvin plotted the uv spectra of "4-thioctic acid" (**135**), *n*-propyl disulfide, and the five- and six-membered isomers of **135** (18).

S—S
$(CH_2)_2CO_2H$
H

135

The absorption maximum of **135** occurred at somewhat longer wavelength than that of *n*-propyl disulfide. The six-membered disulfide showed a considerably larger bathochromic shift, and the five-membered ring displayed an even greater one (18). The bathochromic shift with decreasing ring size was attributed to increasing ring strain (18); reasons for the unusually high strain which seems to be characteristic of 1,2-dithiolanes are discussed in an earlier volume (39). The effect of ring size in 1,2-dithiepanes thus is much less significant than for 1,2-dithianes, and particularly 1,2-dithiolanes. A similar trend is seen in the uv spectra of the dioxides of **128**, **129**, and **130**, but the differences are rather slight (71).

Wood found that the hydrochlorides of 4-amino-1,2-dithiepane (**136**) and 5-amino-1,2-dithiepane (**137**) showed only a very slight blue shift with

S—S
$NH_3^{\oplus}Cl^{\ominus}$
H
136

S—S
H
$NH_3^{\oplus}Cl^{\ominus}$
137

increasing solvent polarity, comparable to that shown by ethyl disulfide. The data for the longest absorption bands are given in Table X-1 (201). A shift toward the red in comparison with an open-chain linear polymeric disulfide (λ_{max} 249 mμ, ε 393) was attributed to a small amount of ring strain (201).

TABLE X-1. Uv Spectra of Amino-1,2-dithiepane Hydrochlorides

	EtOH		H_2O	
Substituent	λ_{max} (mμ)	ε	λ_{max} (mμ)	ε
4-$NH^{\oplus}Cl^{\ominus}$ (**136**)	256	412	255	394
5-$NH_3^{\oplus}Cl^{\ominus}$ (**137**)	258	418	257	404

Ramakrishnan, Thompson, and McGlynn reported somewhat later that 1,2-dithiepane itself (**128**) had λ_{max} 256 mμ. They associated this absorption with the removal of an S_A-S_B nonbonding electron and its placement in an S_A-S_B antibonding orbital. 1,2-Dithiepane also exhibits two broad bands at 206 and 200 mμ, as well as a diffuse band at 191.5 mμ (159).

In a paper summarizing most of his earlier work on 1,2-dithiepanes, Schotte recognized that special properties must be conceded to five-membered (and to some extent to the six-membered) cyclic disulfides (177). He concluded that larger rings require no appreciable deviation from the normal value of the dihedral angle of the —CSSC— bonds, and therefore more closely resemble linear disulfides (177). He found the uv spectra of (±)- and *meso*-**134** (and of their salts) to be continuous (174, 177).

No assignment of ir stretching vibrations was possible for **134** because of its complicated spectrum; the spectrum approximately resembled those obtained with polymers containing —SS— linkages (174, 177).

Partial resolution of (±)-**134**, not completed because of lack of material (176), indicated to Schotte that the optical rotation of the enantiomers would be higher than for the 1,2-dithiane analog (177). This high value might be a consequence either of alternation in the homologous series, or of greater symmetry in the dithiane structure (177). Mixtures of the *meso* and racemic forms of **134** give a melting point diagram of the common eutectic type; the eutectic-point composition is 43% *meso*–57% (±) (176).

Reports indicating that lipoic acid (**131**) produces a dithiyl radical (**138**) during photosynthesis, and perhaps in other enzymatic systems, led Smissman and Sorenson to study the esr spectra of **131** and of six- and seven-membered analogs (179). Lipoic acid (λ_{max} 334 mμ) did indeed give rise to an esr spectrum when subjected to monochromatic light (334 mμ), filtered light (> 300 mμ), or unfiltered uv light (Eq. 28). 1,2-Dithiane (**129**),

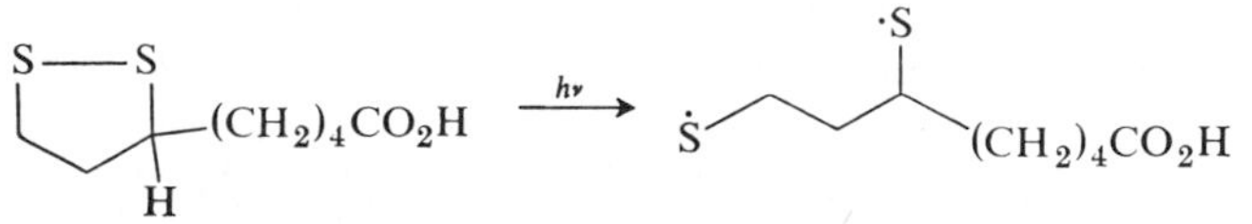

S—S, H, $(CH_2)_4CO_2H$ $\xrightarrow{h\nu}$ ·S, Ṡ, $(CH_2)_4CO_2H$ (28)

131 **138**

1,2-dithiepane (**128**), and methyl disulfide also gave thiyl radicals (e.g., **139**) when photolyzed either monochromatically at their uv maxima (stated as 259 mμ for **128**) or with unfiltered uv light (Eq. 29). The spectra indicated that

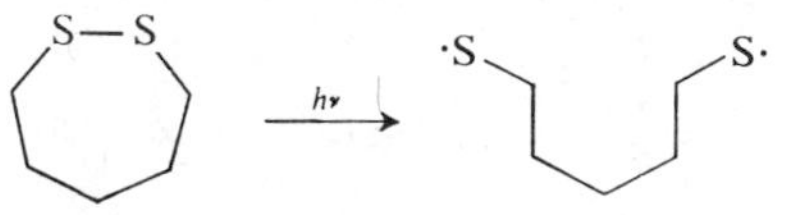

S—S $\xrightarrow{h\nu}$ ·S S· (29)

128 **139**

only S—S homolytic cleavage had occurred. The hyperfine structure expected from the products of C—S fission was not seen, and there was no evidence for hydrogen abstraction or for a triplet state (179).

b. *Preparation*

In 1925 Arndt, Nachtwey, and Pusch reported that reaction of diacetylacetone with phosphorus pentasulfide (Eq. 30) gave an orange solid which they formulated tentatively as the 1,2-dithiepin **140** (6). However, the weight of subsequent evidence is that their product actually was the resonance-stabilized 1,2-dithiole **6**, now commonly called thiothiophthene (see Section B-1). This initial work and a large amount of later research on thiothiophthene

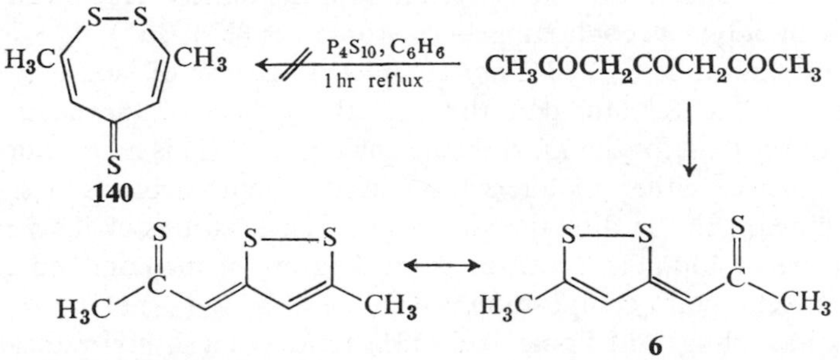

$$\mathbf{140} \xleftarrow[\text{1 hr reflux}]{P_4S_{10},\ C_6H_6} CH_3COCH_2COCH_2COCH_3 \longrightarrow \mathbf{6} \qquad (30)$$

and related substances reported originally as dithiepins was reviewed in detail earlier in these volumes (40).

Affleck and Dougherty described the first general study of 1,2-dithiacycloalkanes in 1950 (1). Prior to their work the only synthesis of cyclic disulfides seems to have been one in which 1,4,5-oxadithiepane was slowly obtained by heating its polymer (63). Although this was the only specific example given, essentially the same method was later patented (under extremely broad claims) as a preparation of 1,2-dithiepane and its 5-substituted derivatives from polysulfide polymer (61, 62). Reconversion to useful polymers also was claimed (61, 62).

Affleck and Dougherty tried unsuccessfully to cyclize polymethylene Bunte salts (e.g., **141**) by oxidizing them with iodine or hydrogen peroxide. Polymerization invariably occurred, even when low concentrations of the Bunte salts were used. Bunte salt **141** reacted with aqueous cupric chloride,

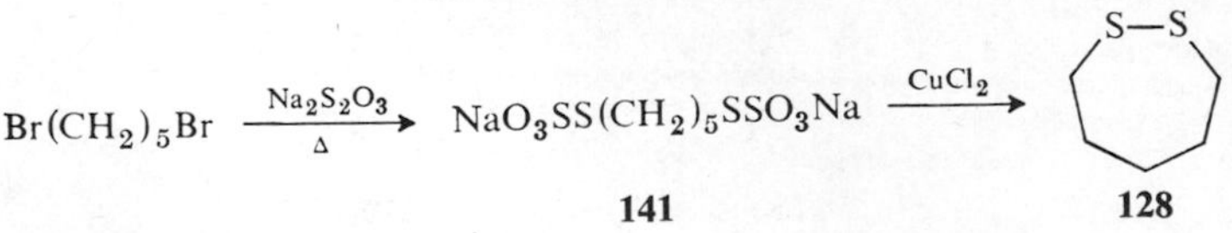

$$Br(CH_2)_5Br \xrightarrow[\Delta]{Na_2S_2O_3} \underset{\mathbf{141}}{NaO_3SS(CH_2)_5SSO_3Na} \xrightarrow{CuCl_2} \mathbf{128} \qquad (31)$$

however, to produce a molecular complex from which 1,2-dithiepane could be obtained in low yield by steam distillation (Eq. 31). The constitution of an intermediate complex was not determined, nor was the nature of the reaction understood. Later Schöberl and Gräfje speculated that partial hydrolysis of the bis-Bunte salt **141** gave a monothiol which then displaced sulfite ion to close the ring (169). Still later Rosenthal patented a method for preparing cyclic disulfides from α,ω-bis-Bunte salts or bisthiocyanates by adding sodium sulfide and concomitantly distilling the product to minimize polymerization; **128** was included in the patent but no yield was given (165).

Affleck and Dougherty characterized 1,2-dithiepane (**128**) by determining its molecular weight and sulfur content as well as by showing the absence of thiol groups and the presence of a disulfide moiety (using Grote's solution) (1). Unfortunately, since only boiling points (and usually molecular weights) were determined, it is difficult to assess the purity of their products. The fact that Schöberl and Gräfje later obtained three of the reported liquids as crystalline solids (169), along with differences in relative reactivities, indicates that at least some of the disulfides were impure (169).

In 1952, shortly after the work of Affleck and Dougherty, 4-thioctic acid (1,2-dithiepane-3-propionic acid, **135**) was reported as a minor by-product in a synthesis which simultaneously gave 5-thioctic acid (**142**) and 6-thioctic acid (i.e., lipoic acid, **131**) (49, 50); some of this work on 4-thioctic acid was reviewed earlier in these volumes in relation to lipoic acid (44).

$(CH_2)_3CO_2H$ (H) $\xrightarrow{\text{(1) KI, } H_3PO_4;\ \text{(2) HBr, } (H_2N)_2CS,\ \Delta;\ \text{(3) NaOH};\ \text{(4) Aq } KI_3}$ S–S $(CH_2)_2CO_2H$ (H) + (32)

135 1%

S–S $(CH_2)_3CO_2H$ (H) + S—S $(CH_2)_4CO_2H$ (H)

142 22% **131** 5%

The three isomers **135**, **142**, and **131** were isolated later in the yields specified in Eq. 32 (50). The 1,2-dithiepane **135** was the only compound isolated when 4-hydroxy-8-bromooctanoic acid, or its lactone, was treated with hydrobromic acid and thiourea, followed by hydrolysis and iodine oxidation (7% yield) (50). 4-Thioctic acid had only slight biological activity, rather less than that of the dithiane **142**, and both **135** and **142** were far less active than lipoic acid (**131**) (50).

Optically active derivatives of 1,2-dithiepane were first considered in several papers by Schotte (1954–1956); one of these presents a summary of the work (177). Schotte synthesized the racemic and *meso* diastereoisomers of 1,2-dithiepane-3,7-dicarboxylic acid (**134**) by reaction of sodium disulfide with the appropriate dibromo diacids (both diastereoisomers of which had been characterized) as shown in Eq. 33, and by oxidation of the dimercapto diacids with hydrogen peroxide, oxygen, or iodine (Eq. 34) (176, 177).

$$HO_2CCHBr(CH_2)_3CHBrCO_2H + Na_2S_2 \longrightarrow$$

$$\mathbf{134} + \mathbf{143} \qquad (33)$$

$$HO_2CCH(SH)(CH_2)_3CH(SH)CO_2H \underset{Zn}{\overset{(O)}{\rightleftharpoons}} \mathbf{134} \qquad (34)$$

Iodine worked best for the pimelic acid series leading to **134** (Eq. 34), although no differences among oxidants were found for the oxidation of (±)-α,α′-dimercaptoglutaric acid (176, 177). Elemental analyses, molecular weights, and neutralization equivalents were satisfactory for *meso* and racemic **134** (176). Reduction of **134** to the dithiol (Eq. 34) was said to rule out the alternative structure of a six-membered cyclic sulfide with a branching sulfur atom, that is —S(=S)— (173), but this type of structure seems unlikely anyway. Racemic **134** was not fully resolved, but preliminary work pointed to rather high rotation for the enantiomers (176, 177). The partial resolution, together with the fact that the other diastereoisomer of **134** could not be resolved at all, established the configuration of both the racemic and *meso* forms (176); (±)-**134** was best obtained (albeit in low yield) by heating *meso*-**134** briefly at about 200°C.

As shown in Eq. 33, reaction with sodium disulfide gave monosulfide **143** (and polymers) as well as **134**. These competitive reactions differ from those with dibromoglutaric acid where lactonization was the chief competitor (177). Both *meso*-**134** and *meso*-**143** resulted from the appropriate disulfide displacement, and Schotte thought the same kind of situation would be likely for the racemic series (177). He considered that the reactions of the dibromoadipic and dibromopimelic acids proceed by an S_N2 mechanism.

Since polymers were always by-products (though in lesser amount at high dilution), and since racemic ⇌ *meso* conversions did not occur, Schotte concluded that after the first S_N2 replacement at an asymmetric center, one of three reactions ensued: (*1*) ring closure to form **134**, (*2*) ring closure followed by loss of sulfur to form **143**, or (*3*) formation of polymer (177).

Possibly, however, sulfur may have been lost from the RSS⁻ species *before* **134** formed, so that the residual RS⁻ species then may have either cyclized to the monosulfide (**143**) or reacted intermolecularly to form polymer.

Wood also has synthesized optically active 1,2-dithiepanes (1962). She prepared *S*-(−)-4-amino-1,2-dithiepane hydrochloride (**136**) and its sulfonamide (**144**) from *S*-(+)-glutamic acid, as well as (±)-5-amino-1,2-dithiepane hydrochloride (**137**) from bis(2-chloroethyl)ether (201).

The last step for *S*-(−)-4-amino-1,2-dithiepane hydrochloride (**136**) proceeded from the dithiol as shown by Eq. 35. This oxidation was best

$$\text{HS}(CH_2)_3CH(SH\text{-}CH_2)NH_3^{\oplus}Cl^{\ominus} \xrightarrow[(72\%)]{\text{Aq } KI_3} \mathbf{136} \quad (35)$$

136

achieved by simultaneous addition of solutions of the dithiol and iodine to water; nearly complete conversion to polymer occurred when the dithiol was maintained in excess. After the oxidation the mixture was neutralized and extracted with benzene for 48 hr to recover the free base of **136**, which then was converted to the salt **136**. The free base of **136** thus seems to be much more stable than the free bases of amino disulfides such as 2-aminoethyl aryl disulfides, which decompose rapidly (74).

The racemic 5-amino counterpart (**137**) of **136** was obtained similarly from the dithiol (82% yield; 7% from the chloro ether); oxidation with air was unsuccessful, as were several other approaches via Bunte or isothiuronium salts. Potentiometric titration with silver nitrate afforded an equivalent weight consistent with **137** (201) although apparently molecular weights were not determined for any of the compounds in the 4- or 5-substituted series.

Wood achieved the preparation of *S*-(−)-4-(methylsulfonamido)-1,2-dithiepane (**144**) as shown in Eq. 36. *S*-(−)-4-(Benzenesulfonamido)-1,2-dithiepane was obtained in like manner from the appropriate dithiocyanate (43% yield) (201).

In 1957 Schöberl and Gräfje reported a systematic study, similar to that of Affleck and Dougherty, of disulfides having eight ring sizes ranging from five to fifteen members; the study focused attention on yields, characterization, and effects of ring size. Their preliminary report (168) was soon followed by full details (169). Schöberl and Gräfje considered oxidative cyclization of the dithiols (Eq. 37) to afford the best and most reliable synthesis of cyclic disulfides, although ferric chloride was not always used as the oxidant.

Oxidation with ordinary concentrations of the dithiol **145** gave 1,2-dithiepane (**128**) in 30% yield along with polymers (169). In contrast when an ether solution of the dithiol was added during 2 days to a boiling solution of ferric chloride in ether–acetic acid by means of a high-dilution technique, the yield was 80% (even higher dilutions were used for dithiols with chains

$$\text{(CH}_2)_3\text{CH(OSO}_2\text{CH}_3)\text{CH(OSO}_2\text{CH}_3)... \xrightarrow{\text{NaSCN}} \text{SCN, SCN, NHSO}_2\text{CH}_3 \xrightarrow[(24\%)]{\text{KOH, 50°C}} \mathbf{144}$$

$$\xrightarrow{\text{Na}_2\text{S}_2\text{O}_3} \text{SSO}_3\text{Na, SSO}_3\text{Na, NHSO}_2\text{CH}_3 \xrightarrow[\text{(37\% overall)}]{\text{KOH, 85°C}} \mathbf{144} \quad (36)$$

longer than that of **145**). Schöberl and Gräfje usually distilled their cyclic disulfides in the dark under reduced pressure in a quartz apparatus; although the absence of light and the use of quartz are not essential, these precautions tend to improve the yield of product (54% reported in ref. 71).

$$\underset{\mathbf{145}}{HS(CH_2)_5SH} \xrightarrow{FeCl_3 \cdot 6H_2O} \underset{\mathbf{128}}{\overline{(CH_2)_5SS}} \quad (37)$$

Field and Barbee concluded, after a study of the synthesis of cyclic disulfides, that the method of Schöberl and Gräfje (Eq. 37) is the best route to 1,2-dithiepane (**128**) (71). Oxidation of 1,5-pentanedithiol with hydrogen peroxide in acetic acid containing potassium iodide, as well as the methods indicated by Eqs. 38 and 39, afforded only low yields of **128**, although such reactions worked well for 1,2-dithiane (**129**) or 1,2-dithiolane (**130**) (71).

$$HS(CH_2)_5SH + ArSO_2Cl \xrightarrow{2OH^{\ominus}} [ArSO_2S(CH_2)_5S^{\ominus}] \longrightarrow \mathbf{128} \quad (38)$$

$$[(CH_2)_5S_2]Pb + S \longrightarrow \mathbf{128} + PbS \quad (39)$$

Other preparative work on 1,2-dithiepanes has been less extensive. Benning patented the 1,5-addition of sulfur to perfluoro-1,4-pentadiene under pressure, which gives a 49% yield of perfluoro-3,4-dithiabicyclo[4.1.0]heptane (**146**) as shown in Eq. 40 (22). Vacuum pyrolysis of fluorinated copolymers having the structure —$(CF_2)_mS(CF_2)_nS$— at 500–700°C, which gave small amounts of perfluorothiepane (Section B-4-b), also gave small amounts of perfluoro-1,2-dithiepane; the structure of the perfluorodithiepane was based on nmr and mass spectra (strong molecular ion) (96).

$$F_2C{=}CF{-}CF_2{-}CF{=}CF_2 \xrightarrow{\text{S, 300°C, 18 hr}} \textbf{146} \qquad (40)$$

146
R.I. 9860

In a study of mercapto analogs of lysine, Yuan and Shchukina reported the isolation of a significant amount of 3-carbethoxy-1,2-dithiepane from the reaction of potassium hydrosulfide with ethyl 2,6-dibromohexanoate (Eq. 41) (203). The relatively high yield of the undesired dithiepane is noteworthy;

$$\text{KSH} + Br(CH_2)_4CHBrCO_2Et \xrightarrow[\text{reflux}]{\text{EtOH, 18 hr}} \text{3-carbethoxy-1,2-dithiepane (42\%)} + HS(CH_2)_4CH(SH)CO_2Et\ (25\%) \qquad (41)$$

perhaps it was a consequence of slow formation of the dithiol, subsequent oxidation being able to proceed at high dilution.

All of the foregoing preparations have involved ring-closure reactions. The only substitutions of the 1,2-system which have been reported have been oxidations involving the sulfur atoms. Thus 1,2-dithiepane (**128**) was oxidized to the 1,1-dioxide **132** with hydrogen peroxide in acetic acid; the yield was only 17%, in marked contrast to 66% for the 1,1-dioxide of 1,2-dithiane **129** (71). The ir spectrum of **132** was consistent with the assigned structure, as was a test for the $—SO_2S—$ moiety in which a thiol causes cleavage to a sulfinic acid (71). Oxidation of **128** with ozone in nitromethane gave the dioxide **132** but in only 3% yield; the main product (65%) was the sulfonic anhydride (71). 1,2-Dithiepane 1,1,2,2-tetraoxide (**133**) was obtained in 10% yield by oxidation of the dioxide **132** with hydrogen peroxide in acetic acid containing tungstic acid; ir and nmr spectra were consistent with structure **133** (71). The tetraoxide could not be obtained by oxidation of 1,2-dithiepane (71).

c. *Reactions*

Affleck and Dougherty estimated relative reactivities of cyclic disulfides qualitatively by the ease of ring opening to form linear polymers (1). All of their disulfides polymerized in a few hours in the presence of a trace of aluminum chloride. On a reactivity scale of 1 to 6, where 1 was least reactive, 1,2-dithiepane (**128**) was assigned a value of 1. This may be compared with values of 5 to 5+ for the five-, six-, and eight-membered rings. Affleck and Dougherty concluded that the seven-, nine-, and eleven-membered rings were stable indefinitely, but that the others polymerized rather easily; they associated relative reactivities with varying amounts of restriction to rotation about the sulfur-sulfur bond. This attribute has since been invoked by others in explaining the considerable differences in reactivity of cyclic disulfides.

Schöberl and Gräfje criticized the foregoing comparisons on the basis of their own comparisons with purer substances (169). For example Affleck and Dougherty had ranked 1,2-dithiane (**129**; isolated only as an oil) as highly reactive, while Schöberl and Gräfje considered it particularly stable (isolated as a solid, mp 30.8–31.5°C). They noted the approximate nature of their conclusions and suggested that rigorous comparisons based on methods such as heat of combustion would be of much interest (169). They used three methods to obtain their approximations:

I. The tendency of cyanide ion to effect cleavage (Eq. 42) as assessed by a color test with sodium nitroprusside. Relative reactivities

$$\mathrm{RSSR + KCN \rightleftharpoons RSK + RSCN} \qquad (42)$$

on this basis were assigned in four categories, with the known substances in parentheses providing a comparison: (*1*) no cleavage; (*2*) slow, possibly minor cleavage (butyl disulfide); (*3*) slow, possibly considerable cleavage (dithiodihydracrylic acid); (*4*) fast, possibly complete cleavage (cystine).

II. The tendency toward polymerization, adjudged as follows: (*1*) no polymerization; (*2*) polymerization only with catalysis, for example, HCl; (*3*) monomers isolable, but onset of slow polymerization in a few hours; (*4*) isolation difficult or impossible.

III. Estimates of ring strain based on allocation of (*1*) 1 kcal for each close approach of a hydrogen atom to another hydrogen atom or to sulfur, and (*2*) an amount judged appropriate (using the cosine squared law) for the energy needed to distort the two C—S bonds from the preferred angle of 80–90° to the actual angles shown by models (e.g., 10–14 kcal for **130**; see Ref. 39 for a discussion of this topic).

Table X-2 shows these three criteria for 1,2-dithiepane in context with other unsubstituted cyclic disulfides studied by Schöberl and Gräfje (169); it includes the colors reported, since these may be related to ring strain.

Notable features of Table X-2, according to Schöberl and Gräfje, were the marked reactivity of 1,2-dithiolane (**130**) and the marked stability of 1,2-dithiane (**129**). 1,2-Dithiepane (**128**) showed quite noticeable, albeit minor strain, which they attributed to hydrogen-atom interactions (169).

TABLE X-2. Reactivity of Cyclic Disulfides (169)

Total number of ring members (including S)	Compound number	Cyanide cleavage	Polymerization	Estimated ring strain (kcal/mole)	Color
5	**130**	4	4	14–18	Deep gold
6	**129**	1	1	0–2	Colorless
7	**128**	2	2	4–5	Colorless
8		3	3	8–10	Colorless
9		3	3	8	Very pale yellow
10		4	3–4	10–14	Faint gold
12		3	3	10–14	Pale yellow
15		2	2	0–5	Colorless

An effort to obtain precise data bearing on ring strain was reported subsequently by Dainton, Ivin, and Walmsley (60). They determined the heats for addition polymerization of 19 monomers, including ethylenic compounds and cyclic ethers as well as cyclic disulfides. The heat of polymerization to an open-chain polymer should be a direct measure of the strain energy of a cyclic molecule if the polymer is unstrained (60). Because of difficulties in obtaining the amounts needed, they studied only disulfides having six-, seven-, and eight-membered rings; 1,2-dithiolane was omitted because removal of solvent led to polymerization, in common with general experience (see Refs. 18 and 169). It should be reiterated that since the method of Affleck and Dougherty was used for synthesis, the disulfides compared in this study (particularly 1,2-dithiane) may have contained impurities. Polymerization was initiated with boron trifluoride etherate. Values of 0.5 (six-membered), 2.5 (seven-membered), and 3.8 (eight-membered) kcal/mole were obtained for $-\Delta H$. These values were compared with one of 6.3 kcal/mole derived indirectly for 1,2-dithiolane (**130**), which had been considered to reflect ring strain (18); reasons for the greater strain in 1,2-dithiolane are discussed earlier in this series (39). Dainton and his associates also were interested in the lower values of 2.5–3.8 kcal/mole shown by the seven- and eight-membered disulfides, as compared with those of 5.2–8.3 kcal/mole which they calculated to be theoretically (but not practicably) possible for cycloheptane and cyclooctane (59); they felt the lower values of the disulfides accorded with their earlier views that replacement of —CH_2CH_2— by —SS— should reduce steric strain in a cyclic monomer of greater than six-members because of (*1*) easier angular deformation of sulfur *p* orbitals than of carbon sp^3 orbitals, and (*2*) reduced interhydrogen repulsions in the larger rings (60).

Some reference was made to the photolysis of 1,2-dithiepane (**128**) in relation to its esr spectrum (Section C-1-a). The chemistry of the photolytic reaction has been explored by Ramakrishnan, Thompson, and McGlynn who photolyzed liquid **128** and explained the results quantum-mechanically (159). Irradiation of **128** under nitrogen in the region of its long-wavelength absorption band (256 mμ; see Section C-1-a) led to dithiol **145** in the absence of alkali. Irradiation in the presence of alkali apparently gave rise to monosulfide **147** with at least a threefold rate increase (Eq. 43); extensive polymerization also occurred. Products were identified only by the uv absorption spectrum of vapor in equilibrium with the liquid substrate, but the approach was said to offer several advantages: pure substrate can be photolyzed readily; irradiation time can be reduced considerably, thus minimizing secondary products; and product identification can be immediate. The explanation for these results was that, in the absence of alkali, removal of a nonbonding electron from sulfur and its placement in an antibonding orbital by excitation at 256 mμ resulted in the lowest energy excitation to a singlet state and effected S—S scission, probably through the initial formation of a diradical; in the presence of alkali, on the other hand, a contributing structure was invoked with a double bond between carbon and sulfur, which led to a free-radical type of C—S scission between carbon and sulfur (159).

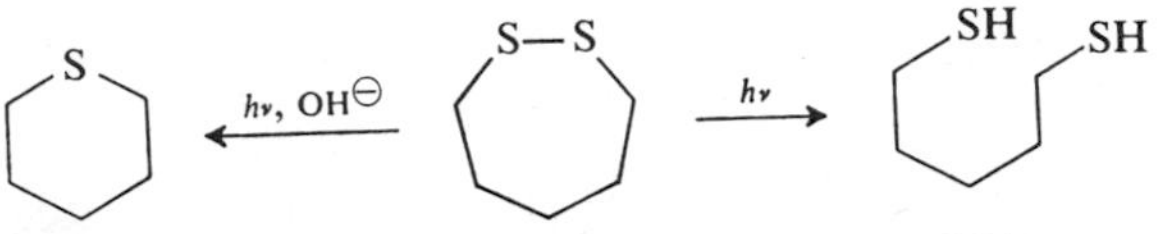

S S—S SH SH

$h\nu$, $OH^{\ominus}$ $h\nu$ (43)

147 **128** **145**

Polarographic study of the acids **148**, **149**, and **134** was undertaken by Nygård and Schotte in an attempt to find a method for comparing ring stability of cyclic carboxydisulfides. The results appeared in two nearly simultaneous papers covering much the same ground (137, 177). The studies later were extended by Nygård (136).

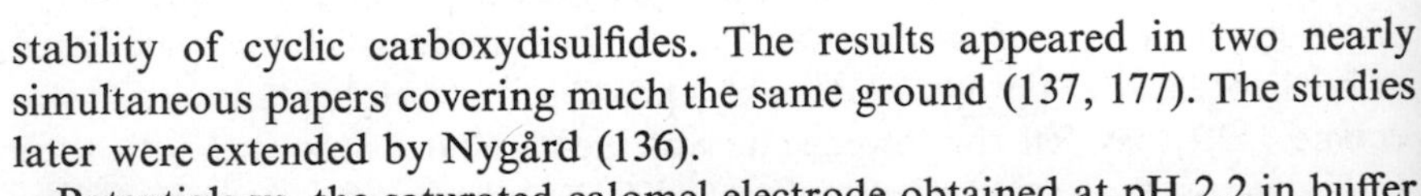

S—S S—S S—S

HO_2C CO_2H HO_2C CO_2H HO_2C CO_2H

H H H H H H

148 **149** **134**

(±), −0.33 V (±), −0.76 V (±), −0.92 V

meso, −0.75 V *meso*, −0.93 V

Potentials vs. the saturated calomel electrode obtained at pH 2.2 in buffer by means of a dropping mercury electrode are shown under the structures (177). The dithiolane **148** gave very good waves in a reversible two-electron process. The six- and seven-membered disulfides **149** and **134** gave polarographic waves only in acidic solution, and the waves were rather drawn out, indicating irreversible electroreduction (177). Since the polarographic effect seemed to be based on an intermediate mercury complex, it was felt that direct comparison was vitiated for establishing differences in ring stability, but that the five-membered ring (**148**) reacted more readily, with the seven-membered ring (**134**) giving waves resembling those of dithiodiacetic acid (177).

The polarographic behavior of 1,2-dithiepane (**128**) has been compared with that of 1,2-dithiane (**129**) and 1,2-dithiolane (**130**) in diglyme (72). The apparent increasing ease of electrochemical reduction for the three ring sizes was 7 (−2.01 V) < 6(−1.80 V) < 5(−1.36 V) (72), but the irreversibility of the reduction mitigated against a firm conclusion (72). The three 1,1-dioxides and the three 1,1,2,2-tetraoxides were studied as well, reduction being irreversible once again. The half-wave potentials (in parentheses) for the five-, six-, and seven-membered dioxides suggested increasingly easy reduction in the order 6(−1.05 V) < 7(−0.85 V) < 5(−0.83 V), and those for the tetraoxides suggested the same order 6(−1.23 V) < 7(−0.93 V) < 5(−0.88 V) (72). The apparent order of increasing ease of functional group reduction might be interpreted as SS < SO_2SO_2 < SO_2S from the data, but a cautionary note was issued about this conclusion owing to possible differences in the mechanisms of reduction (72).

Field and Barbee compared the hydrolysis of the seven-, six-, and five-membered cyclic disulfides **128**, **129**, and **130**, and also of the 1,1-dioxides and 1,1,2,2-tetraoxides, in refluxing aqueous dioxane (72). The disulfides were quite resistant; 1,2-dithiepane (**128**) showed no apparent change after 52 hr and only little change after 120 hr. The dioxides underwent hydrolysis only somewhat more readily; 62% of the dithiepane dioxide **132** survived after 46 hr and the other two were roughly comparable to it. The three tetraoxides were hydrolyzed far more readily; thus, with the dithiepane tetraoxide **133**, alkali consumption after about 7 hr was consistent with 90% conversion to the acid shown in Eq. 44, and the reactivity of the three rings seemed to increase in the order 6 < 7 < 5 (72).

$$\underset{\textbf{133}}{\overline{(CH_2)_5SO_2SO_2}} + H_2O \rightarrow HO_3S(CH_2)_5SO_2H \qquad (44)$$

When alkali was added to maintain aqueous acetonitrile solutions of the three tetraoxides at pH 8 (*ca.* 25°C), the reactivities for the three ring sizes were (with times for cessation of alkali uptake in parentheses): 6 (30 min) < 5 (20 min) < 7 (15 min) (72).

With thiophenol, the tetraoxide **133** reacted only to the extent of 10% after 20 hr in ethanol, about like 1,2-dithiane tetraoxide (9%, acetonitrile) but less than 1,2-dithiolane tetraoxide (50%). Thiophenoxide *ion*, on the other hand, reacted quantitatively with the dithiane tetraoxide used as a model substance (Eq. 45) and presumably would do so with **133** (72).

In the hope of developing synthetic methods useful with cyclic disulfides in general, Field and Barbee used 1,4-dithiane 1,1-dioxide and 1,1,2,2-tetraoxide as models for the preparation of acylic disulfides containing sulfinate,

$$\overline{(CH_2)_4SO_2SO_2} \xrightarrow{C_6H_5SNa} [NaO_2S(CH_2)_4SO_2SC_6H_5] \xrightarrow{C_6H_5SNa} NaO_2S(CH_2)_4SO_2Na + (C_6H_5S)_2 \qquad (45)$$

sulfone, or sulfonate moieties, and also of sulfinate-sulfonates and sulfonate-disulfide dioxides (72). Illustrations of this type of "oxodisulfide cleavage" are shown in Eq. 46. Although these reactions were intended also to be applicable for 1,2-dithiepanes ($n = 5$), conversion of the dithiepane tetraoxide **133** to pentamethylenesulfide 1,1-dioxide at 280°C failed; this breakdown

$$\begin{array}{ccccc}
\overline{(CH_2)_nSO_2} & & [R'SO_2(CH_2)_nS]_2 & \xleftarrow{100°C,H_2O} & R'SO_2(CH_2)_nSSR \\
\uparrow{\scriptstyle 280°C} & & & & \uparrow{\scriptstyle R'X} \\
\overline{(CH_2)_nSO_2SO_2} & \xleftarrow{H_2O_2} & \overline{(CH_2)_nSO_2S} & \xrightarrow{RSNa} & NaO_2S(CH_2)_nSSR \\
\downarrow{\scriptstyle NaOH} & & & & \downarrow{\scriptstyle H_2O_2} \\
NaO_3S(CH_2)_nSO_2Na & \xrightarrow[RSH]{H^+,n\text{-}BuONO} & NaO_3S(CH_2)_nSO_2SR & & NaO_3S(CH_2)_nSSR
\end{array} \qquad (46)$$

in analogy seems insignificant since the 1,2-dithiane tetraoxide prototype itself gave a maximum yield of only 12%.

Field and Barbee concluded from the various studies mentioned above that generalizations concerning the effect of ring size and oxidation level on reactivity of five-, six-, and seven-membered disulfides and their oxides were difficult (72). Usually, however, generalities seemed to be, in ring size, for easier cleavage of the five-membered systems than of the six (with the seven variable) and in oxidation level, for greater resistance to self-polymerization or to attack of a thiol by the more oxidized forms, but for less resistance to hydrolysis or electrochemical reduction (72).

2. 1,3-Dithiepins and Derivatives

a. *1,3-Dithiepin*

No monocyclic compounds containing the 1,3-dithiepin ring **(150)** are known. The anion derived from **150** is of theoretical interest as a potentially aromatic substance, as mentioned at the beginning of Section C. To the extent that a sulfur atom can resemble a carbon-carbon double bond capable of participating in electron delocalization through resonance (30, 170), anion **124** is analogous to the cyclononatetraenide anion (208). As a cyclic system containing ten π-electrons, the 1,3-dithiepinide anion **124** fulfills Hückel's $(4n + 2)$ requirement for aromaticity.

Valence-bond forms of the 1,3-dithiepinide anion which involve only *p*-orbitals, such as **124B**, might be expected to be unimportant in the description of the anion, because of charge separation. Structures such as **124C** in which sulfur utilizes both an electron-donating *p*-orbital and an electron-accepting *d*-orbital may also be considered; stabilization by significant *d*-orbital involvement of this type might render the anion aromatic if it could achieve planarity without appreciable steric strain (46). Canonical form **124C** is analogous to **151**, which has been invoked in order to rationalize

150 R.I. 7768 **124**

124A **124B** **124C** **151**

the aromaticity of thiophene (111, 170). As mentioned earlier (Section C), the 1,3-dithiepinide anion **124** has been predicted to be unstable and difficult to prepare (207, 208).

b. *Dihydro-1,3-dithiepins*

The 4,5-dihydro-1,3-dithiepin ring **152** is unknown. Only two examples of monocyclic 4,7-dihydro-1,3-dithiepins have been reported. Friebolin, Mecke, Kabuss, and Lüttringhaus have examined the nmr spectra of 4,7-dihydro-1,3-dithiepin (**153**) and 2,2-dimethyl-4,7-dihydro-1,3-dithiepin (**154**) as part of a continuing conformational analysis of rings containing sulfur. Other physical constants, as well as the method of synthesis for these dihydrodithiepins, have not yet appeared (78).

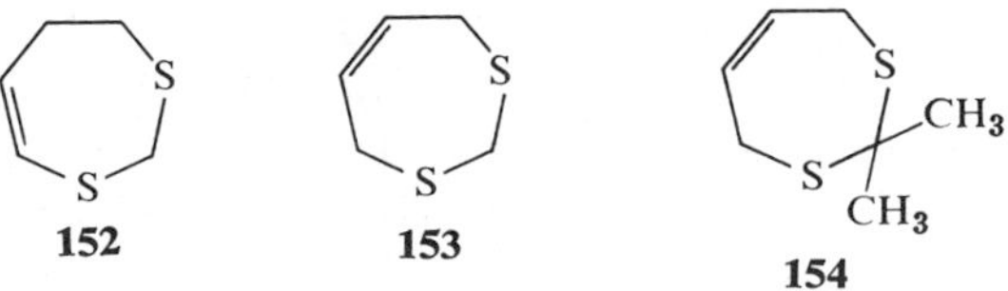

The nmr spectra of **153** and **154** at low temperature show that the preferred conformation of the 4,7-dihydro-1,3-dithiepin ring is that of the chair (**155**) rather than of the twisted tub (**156**). The two $C_{(2)}$ positions are equivalent in **156**, and are quasi-axial and quasi-equatorial in **155**. From the separation of chemical shifts for the AB doublet of doublets of the $C_{(2)}$ protons of **153** and the singlets corresponding to the nonequivalent methyl groups of **154** at the coalescence temperature, the enthalpies of activation for chair interconversion of these compounds were determined to be 8.5 and 8.2 kcal/mole, respectively (78).

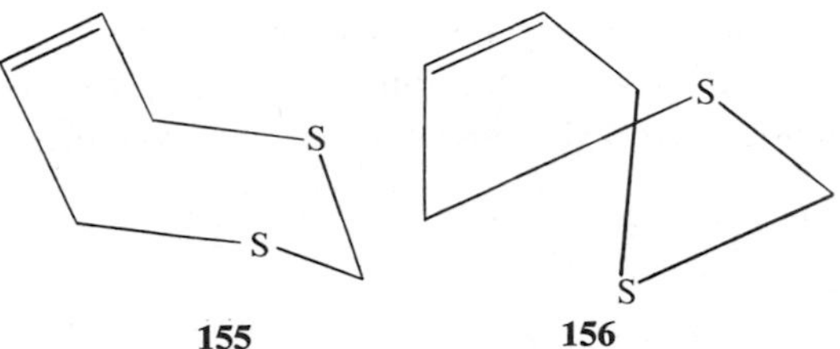

155 **156**

Introduction of a fused benzene ring into the 4,7-dihydrodithiepin, as in **157**, increases the barrier to inversion to 10.2 kcal/mole (78). Replacement of $C_{(2)}$ by sulfur also results in an increase in the energy of activation for ring

157 **158** **159**

interconversion; the value of 8.9 kcal/mole for 4,7-dihydro-1,2,3-trithiepin (**158**) is attributed to "stiffening" of the ring by the third sulfur atom (100).

It is interesting that an unusually high value of the energy of activation for interconversion (17.4 kcal/mole) is observed for the fused ring dihydrotrithiepin **159** (100).

c. *1,3-Dithiepanes*

(1) Physical Properties. The first-order rate constants for the base-catalyzed T-H exchange of a variety of tritiated mercaptals and orthothioformates have been determined by Oae and co-workers (139, 140). The rate of exchange of 2-ethyl-2-tritio-1,3-dithiepane **160** is 5.37 times that of the tritiated acyclic mercaptal **161**. The observed facility for exchange in cyclic mercaptals is attributed to stabilization of the incipient anion by nonclassical 2*p*-3*d* overlap between carbon and the two sulfur atoms, together with 3*p*-3*d* sulfur-sulfur overlap, as in **162**. This mode of stabilization of the anion is presumably much more important for the relatively rigid cyclic structures

than for acyclic mercaptals. The effects of alkyl substitution at $C_{(2)}$ in cyclic mercaptals have been examined rather thoroughly for the 1,3-dithiolane and 1,3-dithiane systems. The significant rate retardations engendered by alkyl substitution in these 5- and 6-membered cyclic systems have been discussed in terms of the geometry of the rings. Unfortunately, rate data for 2-tritio-1,3-dithiepane (**163**) have not been measured (139, 140).

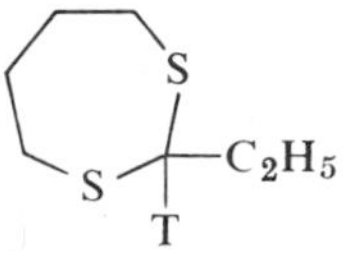

$(C_2H_5S)_2CTC_2H_5$

160 **161** **162** 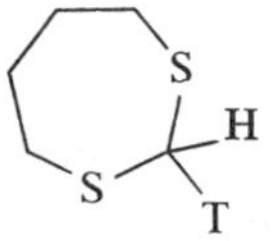 **163**

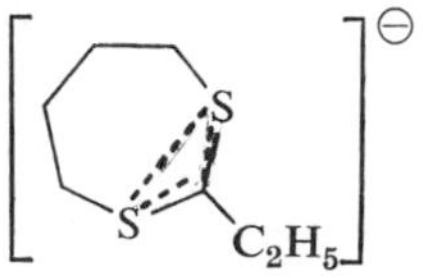

The uv spectra of acyclic mercaptals exhibit maxima near 235 mμ. A distinct shift to longer wavelength is seen in the spectra of 1,3-dithiolanes (near 246 mμ) and 1,3-dithianes (near 250 mμ). This red-shift has been interpreted in terms of mesomeric stabilization involving expansion of the valence of the sulfur octet to a singlet diradical (**164**). Such interaction is believed to be important only for relatively rigid cyclic systems. The uv spectra of 1,3-dithiepanes **165–167** display shoulders between 235 and 250 mμ; however, it is difficult to conclude with certainty whether this represents a bathochromic shift (relative to acyclic mercaptals) requiring explanation in terms of some special stabilization such as in **164** (138, 141).

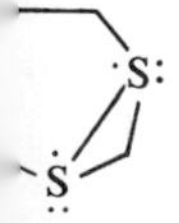

164 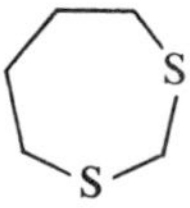 **165** 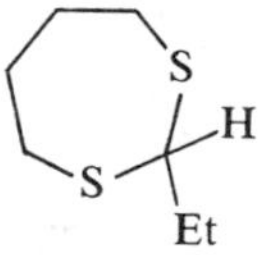 **166** 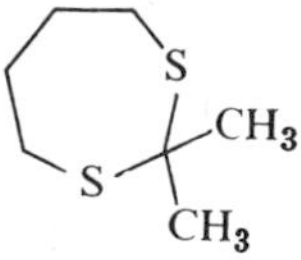 **167**

In a study of the stereochemistry of α-sulfonyl carbanions Corey, Koenig, and Lowry examined the acidity of the $C_{(2)}$ hydrogen of a series of cyclic 1,3-disulfones. In agreement with their hypothesis that an asymmetric α-sulfonyl carbanion is preferentially oriented as in **168**, with restricted rotation about the C_α—SO_2 bond, and that this conformation is more stable than that of **169**, the pK_a of cyclic 1,3-disulfones **170** decreases from 13.9

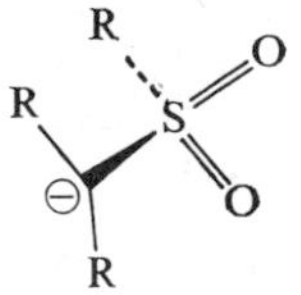

168 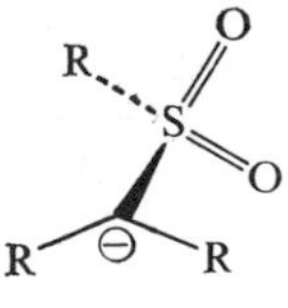 **169** 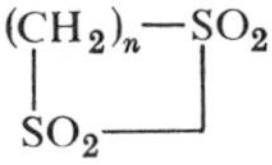 **170**

($n = 2$) to 11.0 ($n = 5$) as the flexibility of the ring increases (55). Experimental details for the synthesis of 1,3-dithiepane 1,1,3,3-tetraoxide **170** ($n = 4$) have not yet appeared.

(2) PREPARATION. Oae and co-workers recently reported that 1,3-dithiepane (**165**) and 2-ethyl-1,3-dithiepane (**166**) result from the Lewis acid-catalyzed reaction of 1,4-butanedithiol with the appropriate diethyl acetal (Eq. 47). Reaction of 1,4-butanedithiol with acetone in the presence of

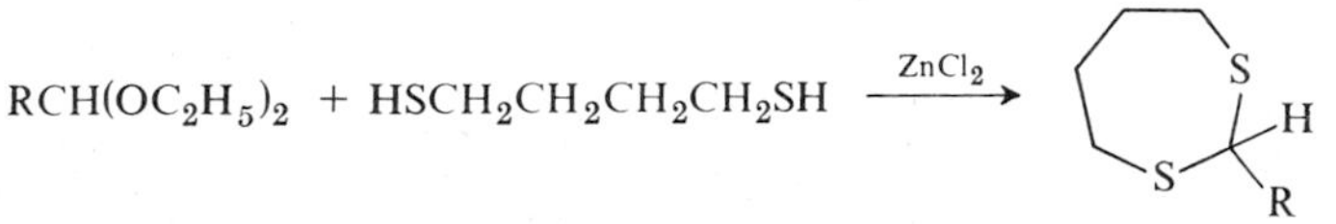

$$RCH(OC_2H_5)_2 + HSCH_2CH_2CH_2CH_2SH \xrightarrow{ZnCl_2} \text{2-R-2-H-1,3-dithiepane} \quad (47)$$

hydrogen chloride afforded 2,2-dimethyl-1,3-dithiepane (**167**). These condensations afford the 1,3-dithiepanes in low yield together with high boiling solids. The authors reported that "special care" was necessary for purification; **166** formed in 25% yield. Physical constants for **165** and **167**, as well as details for these syntheses, have not yet appeared (140, 141).

A similar condensation (Eq. 47) in which R = $CO_2CH_2CH_3$ and boron trifluoride etherate is employed as a catalyst affords 2-carbethoxy-1,3-dithiepane (**171**). Hydrolysis readily gives acid **172** (124). Decarboxyla-

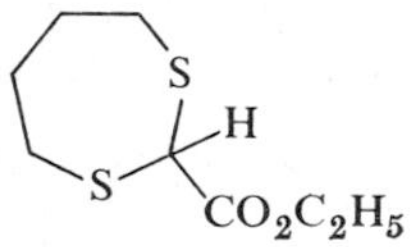

S, H, $CO_2C_2H_5$ — **171**; S, H, CO_2H — **172**; S, C=O, ⊖O, S — **173**

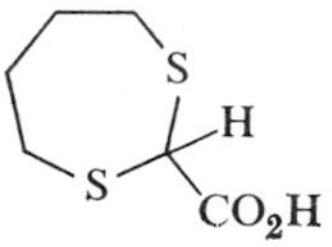

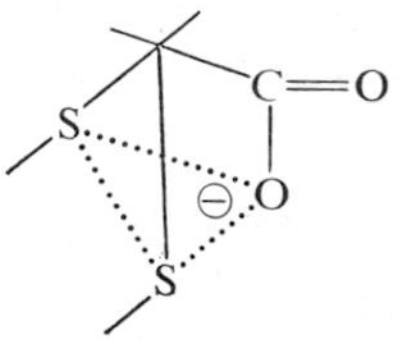

tion of **172** has been studied by Oae's group; small differences in the rates of decarboxylation of 2-carboxy-1,3-dithiacycloalkanes are interpreted in terms of the possible incursion of anions such as **173** (124, 142).

Reaction of 1,4-dichlorobutane with alkali trithiocarbonates (Eq. 48) affords 1,3-dithiepan-2-thione (**174**). The thione, an unstable oil, was characterized by means of the mercuric chloride complex (166).

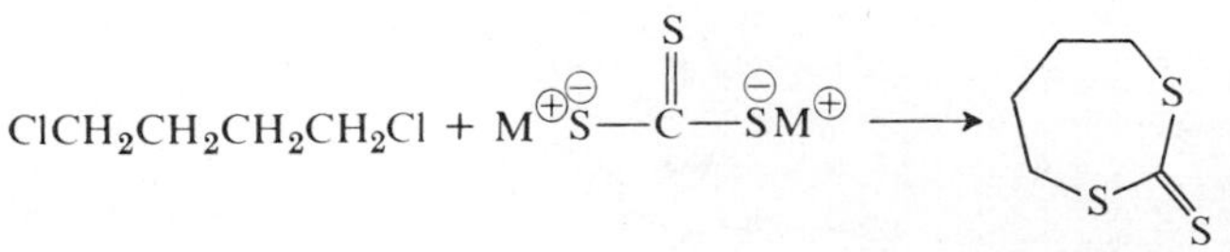

$$ClCH_2CH_2CH_2CH_2Cl + M^{\oplus}\overset{\ominus}{S}-\overset{S}{\overset{\|}{C}}-\overset{\ominus}{S}M^{\oplus} \longrightarrow \textbf{174} \quad (48)$$

Cyclization of 4-hydroxybutyl-*N*,*N*-diethyldithiocarbamate (**175**) in the presence of *p*-toluenesulfonyl chloride produces the dithiepane **176**, which was isolated as the perchlorate salt. A reasonable mechanism for this interconversion (Eq. 49) involves intramolecular displacement of the tosylate anion from an intermediate tosylate ester (**177**) (102).

$$\mathrm{HOCH_2CH_2CH_2CH_2S\overset{S}{\overset{\|}{C}}N(Et)_2}\ (\mathbf{175}) \xrightarrow{\mathrm{TsCl}} [\mathbf{177}] \longrightarrow \mathbf{176}\ (\overset{\oplus}{N}Et_2\ \overset{\ominus}{O}Ts) \qquad (49)$$

3. 1,4-Dithiepins and Derivatives

a. *1,4-Dithiepin*

The *Chemical Abstracts* name for the parent compound of this series is 5*H*-1,4-dithiepin although the designation of the position of the hydrogen seems superfluous in this instance. No monocyclic structures containing the 1,4-dithiepin ring **178** are known. Molecular orbital calculations similar to those discussed in Sections C and C-2-1 for the 1,3-dithiepinide anion indicate that the 1,4-dithiepinide anion **125** is unstable and would be difficult to prepare (207, 208).

178
R.I. 353

b. *Dihydro-1,4-dithiepins*

There are two possible dihydro-1,4-dithiepins, 2,3-dihydro-5*H*-1,4-dithiepin (**179**) and 6,7-dihydro-5*H*-1,4-dithiepin (**180**). Neither of these parent compounds is known.

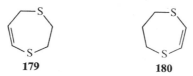

179 **180**

Fuson and Speziale synthesized the only known monocyclic 2,3-dihydro derivative by dehydration of 6-hydroxy-1,4-dithiepane 1,1,4,4-tetraoxide

(**181**). The product (Eq. 50), 2,3-dihydro-5*H*-1,4-dithiepin 1,1,4,4-tetraoxide (**182**), was formed in 74% yield; its structure was confirmed by reaction with aqueous potassium permanganate and by hydrogenation to the known saturated disulfone **183** (80).

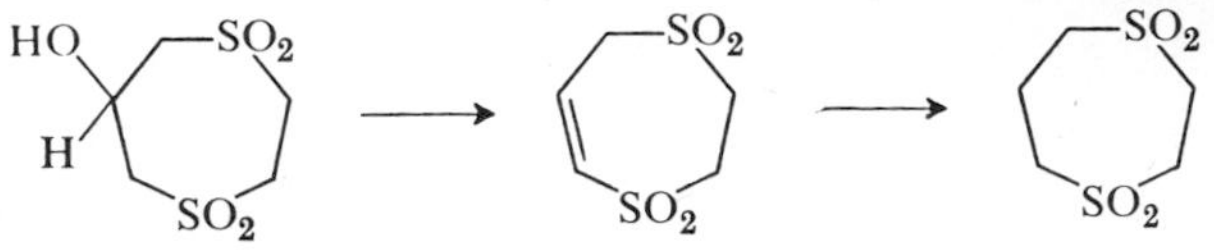

(50)

181 **182** **183**

An interesting entry into the 6,7-dihydro-5*H*-1,4-dithiepin system has recently been reported by Bottini and Böttner. Allylic sulfide **184** undergoes cyclization in the presence of potassium *tert*-butoxide (Eq. 51) to form a small amount of 2-methyl-6,7-dihydro-5*H*-1,4-dithiepin (**185**). The structure of the product was confirmed by consistent ir, uv, and nmr spectra (28).

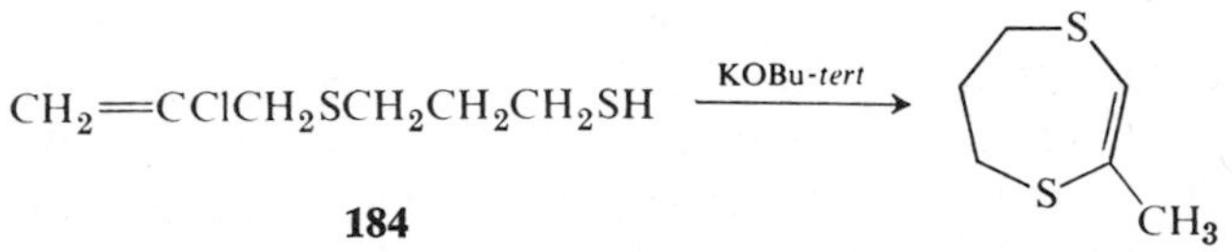

$$CH_2{=}CClCH_2SCH_2CH_2CH_2SH \xrightarrow{\text{KOBu-}tert} \quad (51)$$

184 **185**

A double displacement of chloride ion from chloromethylallyl chloride is reported to give a 15% yield of dihydrodithiepin derivative **186** (Eq. 52) (163).

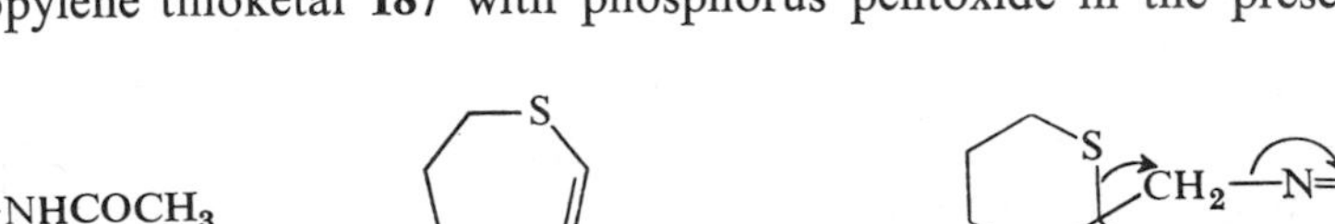

$$ClCH_2{-}C({=}CH_2){-}CH_2Cl + NaSCH{=}CHSNa \xrightarrow[NH_3]{H_2O} \quad (52)$$

186

The 6,7-dihydro-5*H*-1,4-dithiepin system is also generated from treatment of propylene thioketal **187** with phosphorus pentoxide in the presence of

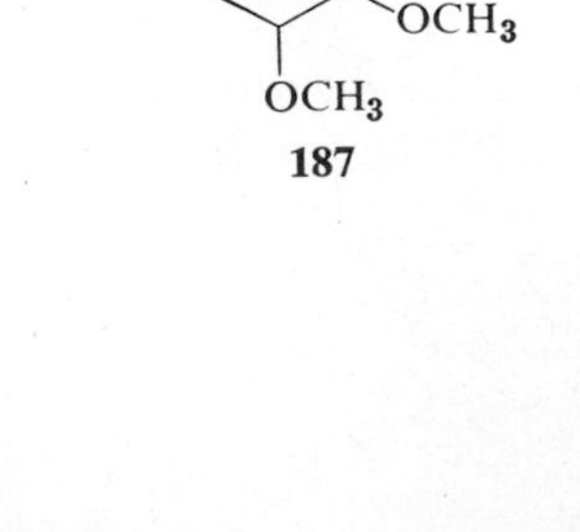

$CH_2{-}NHCOCH_3$ OCH_3 OCH_3

187

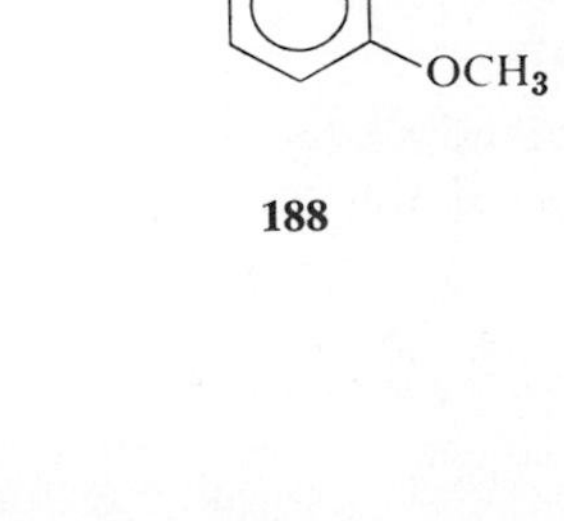

OCH_3 OCH_3

188

CH_3 $CH_2{-}N{=}C$ OP(OCH_3 OCH_3

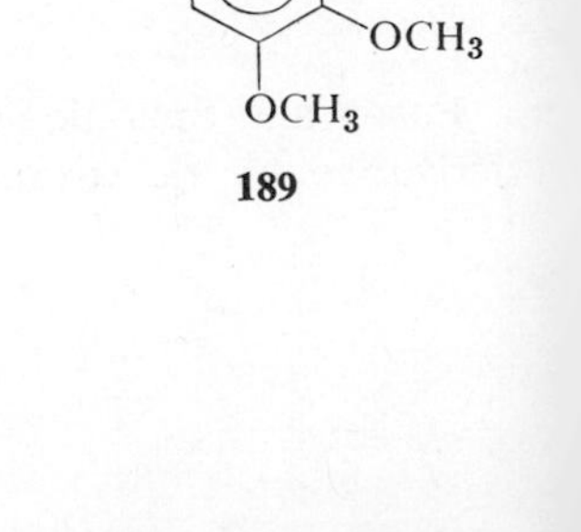

189

pyridine. The product, **188**, is believed to arise from the rearrangement which accompanies the expulsion of acetonitrile from the Vilsmeir-Haack adduct **189**, followed by the loss of a proton (121).

c. *1,4-Dithiepanes*

(1) 1,4-DITHIEPANE AND DERIVATIVES. In 1899 Autenrieth and Wolff reported the preparation of 1,4-dithiepane (**190**) by reaction of 1,3-propanedithiol with ethylene bromide (Eq. 53) (16). They did not isolate this 1,4-disulfide but converted it by oxidation with potassium permanganate to disulfone **183**, mp 282°C. After an interval of 35 years, Reid and co-workers

$HSCH_2CH_2CH_2SH + BrCH_2CH_2Br$

$HSCH_2CH_2SH + BrCH_2CH_2CH_2Br$

$\xrightarrow{NaOEt}$ **190** $\xrightarrow{[O]}$ **183** (53)

isolated 1,4-dithiepane in 8% yield from the reaction of 1,2-ethanedithiol with 1,3-dibromopropane (Eq. 53) (122, 191). The seven-membered ring was isolable only after the removal of a considerable amount of material of higher molecular weight, which also forms in the reaction. The molecular weight of the monomeric 1,4-disulfide was determined to be 140 (compared to the theoretical value of 134) from the melting point depression of borneol. Oxidation with potassium permanganate afforded 1,4-dithiepane 1,1,4,4-tetraoxide (**183**) of mp 281–282°C ("in agreement with Autenrieth and Wolff") (191) or of 287–288°C (122). Fuson and Speziale have repeated and verified the results of Tucker and Reed (191). They found that disulfone **183** melts with decomposition at 279–280°C (80).

(2) 1,4-DITHIEPAN-6-OL AND DERIVATIVES. Fuson and Speziale observed that 1,4-dithiepan-6-ol (**191**) results from the condensation of 1,2-ethanedithiol with 1,3-dibromo-2-propanol in ethanolic sodium ethoxide (Eq. 54) (80). A similar reaction with 2,3-dibromo-1-propanol, however, did not produce the expected 2-hydroxymethyl-1,4-dithiane (**192**). Instead, 1,4-dithiepan-6-ol (**191**) was obtained once again (80). The course of the latter reaction may be conveniently rationalized by assuming that 2,3-dibromo-1-propanol is converted to epibromohydrin under the reaction conditions

employed. The structure of 1,4-dithiepan-6-ol (**191**) was established unequivocally by transformation to 1,4-dithiepane-1,1,4,4-tetraoxide (**183**) via **181** and **182** (80).

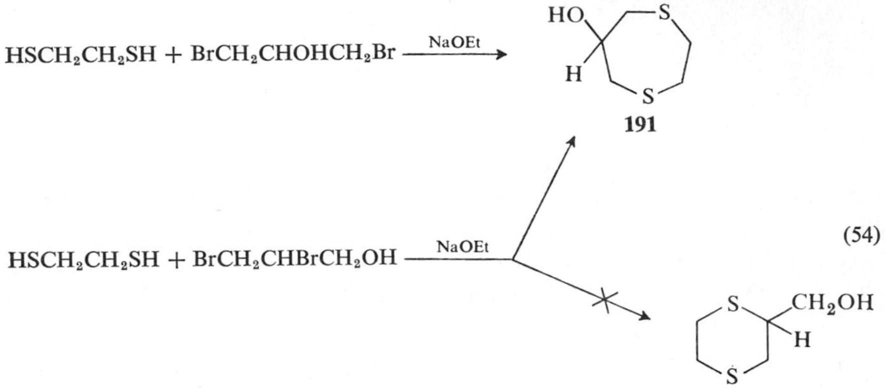

$HSCH_2CH_2SH + BrCH_2CHOHCH_2Br \xrightarrow{NaOEt}$ **191**

$HSCH_2CH_2SH + BrCH_2CHBrCH_2OH \xrightarrow{NaOEt}$ **191** (not **192**) (54)

1,4-Dithiepan-6-ol reacts normally with acyl halides to form the benzoate and *p*-nitrobenzoate esters **193**. Oxidation with hydrogen peroxide gives 1,4-dithiepan-6-ol 1,1,4,4-tetraoxide (**181**) (80). An interesting ring contraction occurs when the alcohol is allowed to react with thionyl chloride (Eq.

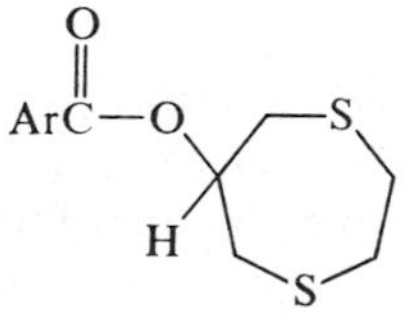

193 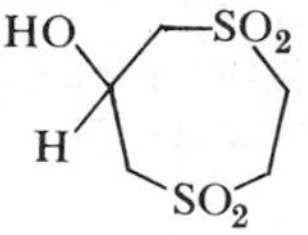 **194**

55). The six-membered cyclic product **194** presumably results from ring-opening of the intermediate bicyclic sulfonium chloride **195** (81, 90). The immediate precursor to the bicyclic structure **195** has been postulated to be

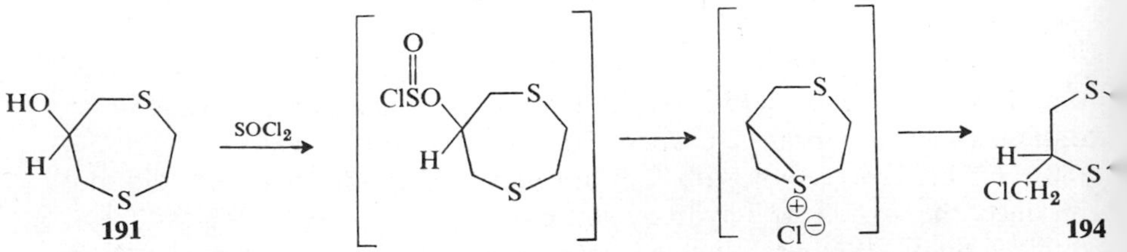

191 $\xrightarrow{SOCl_2}$ [**196**] $\longrightarrow$ [**195**] $\longrightarrow$ **194**

either 6-chloro-1,4-dithiepane (81) or the corresponding monocyclic secondary carbonium ion (90). Displacement of chlorosulfite anion from chlorosulfite **196** would also seem to represent a reasonable mechanistic pathway.

(3) 1,4-DITHIEPAN-6-ONE AND DERIVATIVES. (a) *Preparation.* Lüttringhaus and Prinzbach observed that Dieckmann cyclization of diester **197**, using potassium *tert*-butoxide in benzene, gave a 45% yield of ketoester **198**; hydrolysis and decarboxylation afforded 1,4-dithiepan-6-one (**199**) in good yield (Eq. 56) (114). This ketone also has been prepared in 70% yield by

$$(EtO_2CH_2SCH_2\text{-})_2 \xrightarrow{\text{KOBu-}tert} \mathbf{198} \xrightarrow{H^{\oplus}} \mathbf{199} \quad (56)$$

197 **198** **199**

condensation of 1,2-ethanedithiol with 1,3-dichloroacetone under alkaline conditions (Eq. 57) (93, 94).

$$HSCH_2CH_2SH + ClCH_2C(=O)CH_2Cl \xrightarrow{OH^{\ominus} \text{ or } OCH_3^{\ominus}} \mathbf{199} \quad (57)$$

199

(*b*) *Reactions.* Oxidation of 1,4-dithiepan-6-one (**199**) with 2 moles of hydrogen peroxide gives a dioxide, identified by Howard and Lindsey as the 1,4-disulfoxide **200** on the basis of an ir spectrum which displays sulfoxide bands at 1015 and 1025 cm^{-1} and a transparent sulfone region (93, 94)

200 **201**

Oxidation with 4 moles of hydrogen peroxide affords 1,4-dithiepan-6-one-1,1,4,4-tetraoxide (**201**), mp 279–282°C. This melting point is somewhat higher than the decomposition point of 252–253°C reported earlier for the disulfone obtained by Fuson and Speziale by dichromate oxidation of **181** (71), but it seems likely that both groups of investigators had the same product.

The carbonyl group of 1,4-dithiepan-6-one readily forms a variety of interesting derivatives. Ethylene ketal **202** and mercaptole **203**, as well as hydantoin **204**, are formed by standard methods (93, 94). In addition Howard and Lindsey found that reaction with malononitrile (in the absence of any catalyst) affords 6-dicyanomethylene-1,4-dithiepane (**205**) in quantitative yield. Reaction with phenol in the presence of hydrogen chloride gave the 6,6-bis(*p*-hydroxyphenyl) derivative **206** (93, 94).

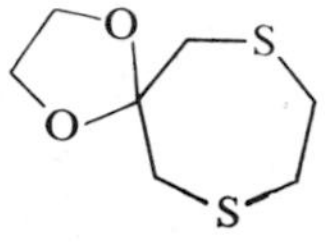

202

R.I. 9992

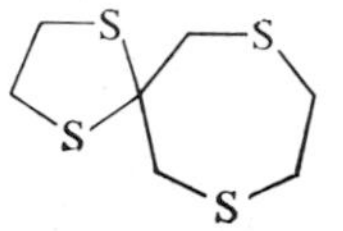

203

R.I. 9993

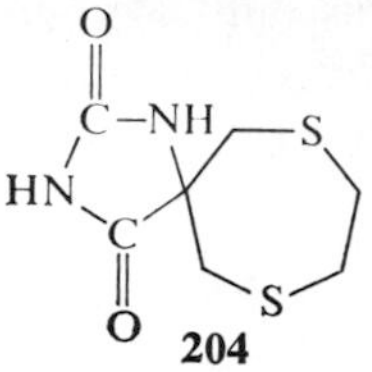

204

R.I. 9986

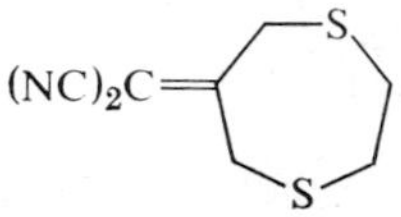

$(NC)_2C$

205

p-HOC_6H_4

p-HOC_6H_4

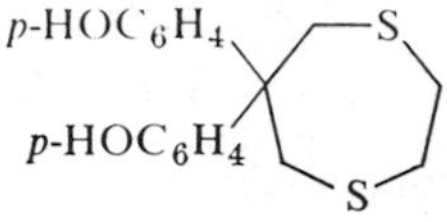

206

Condensation of 1,4-dithiepan-6-one with sulfur in the presence of ammonia has been reported to give the spiro compound **207** in 53% yield; reaction with 2-mercaptocyclooctanone in the presence of ammonia reportedly affords **208** in 93% yield (12).

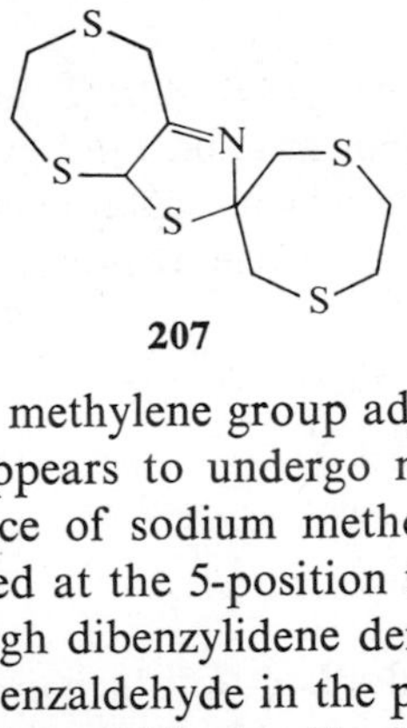

207

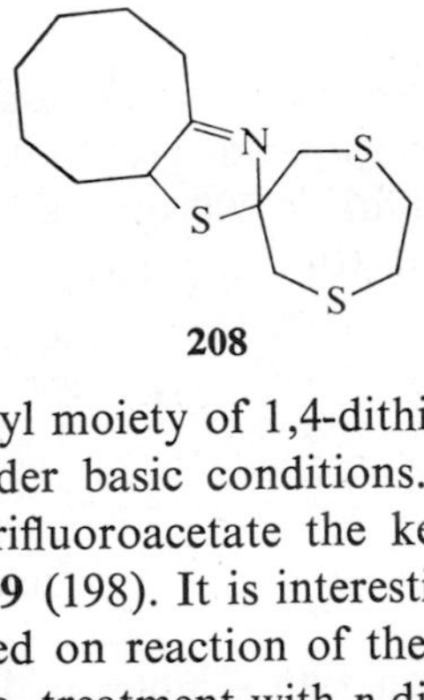

208

The methylene group adjacent to the carbonyl moiety of 1,4-dithiepan-6-one appears to undergo normal reactions under basic conditions. In the presence of sodium methoxide and methyl trifluoroacetate the ketone is acylated at the 5-position to form ketoester **209** (198). It is interesting that although dibenzylidene derivative **210** is formed on reaction of the ketone with benzaldehyde in the presence of piperidine, treatment with *p*-dimethylaminobenzaldehyde affords only monobenzylidene derivative **211** (93, 94).

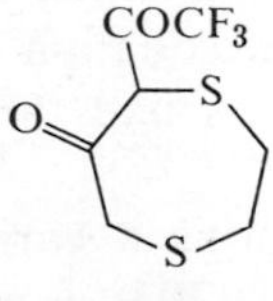

$COCF_3$

209

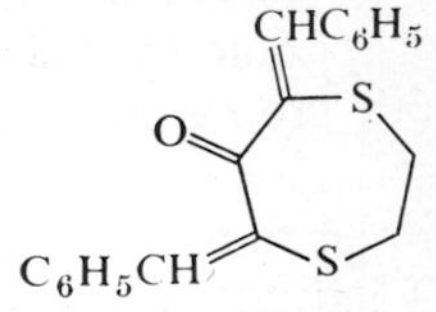

CHC_6H_5

C_6H_5CH

210

$CHC_6H_4N(CH_3)_2$

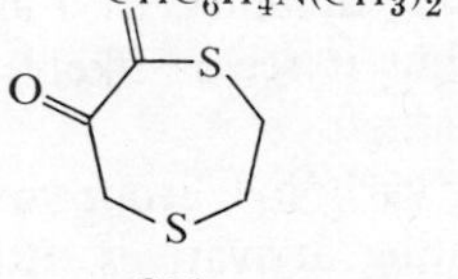

211

D. Rings Containing Three Sulfur Atoms (Trithiepins and Derivatives)

Four seven-membered ring systems containing three sulfur atoms are possible. These would have the sulfur atoms in the following positions: 1,2,3; 1,2,4; 1,2,5; and 1,3,5. Of these, only the 1,2,4 system remains unclaimed. According to Zahradník and Párkányi, molecular orbital calculations for structure **212** suggest that this particular 1,2,4-type of structure, an analog of the cyclononatetraenide anion, should be unstable and difficult to prepare (208).

Zahradník has reviewed theoretical aspects relating to the possible preparation of the fully-unsaturated seven-membered systems 1,2,3-trithiepin (**213**) and 1,2,5-trithiepin (**214**) (206). One can regard either **213** or **214** as an analog of cyclodecapentaene to the extent that a sulfur atom can simulate a

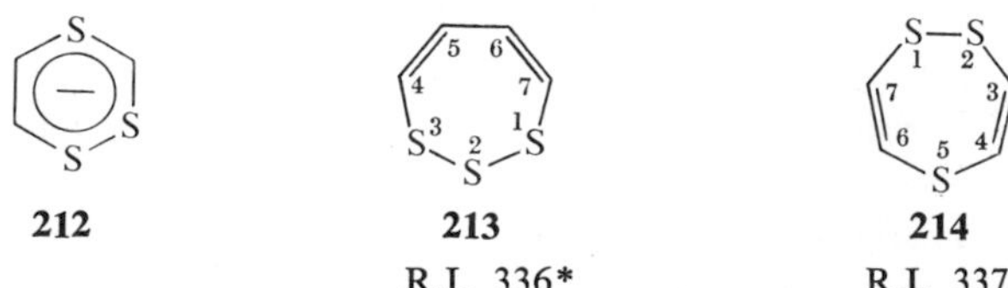

212 **213** R.I. 336* **214** R.I. 337

carbon-carbon double bond as it does in thiophene, an isostere of benzene (30, 170). Zahradník states that the prospect for synthesis of such compounds is unattractive because of the predicted low specific delocalization energy, the presence of positions with high π-electron density, and the high energy of certain molecular orbitals. The Hückel molecular orbital treatment suggested that structure **214** might be prepared, but that **213** probably could not (206).

1. 1,2,3-Trithiepins and Derivatives

The nomenclature of *Chemical Abstracts* and the *Ring Index* (158) is illustrated by the name 1,2,3-trithiepin for **213** and of 1,2,3-trithiepane for

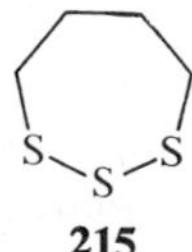

215

its tetrahydro form **215**. The nuclear name trithiacycloheptane has been used for **215**; it is less preferable, and without numbering is ambiguous.

* The *Ring Index* citation under this number, erroneously to *J. Chem. Soc.*, should have been to *J. Amer. Chem. Soc.*

The 1,2,3-trithiepin system was reported first in reduced form as an incidental result from studies of the synthesis of *gem*-dithiols (51). The reaction of acetonylacetone with hydrogen sulfide at high pressures afforded only a trace of *gem*-dithiols. Instead, 4,7-dimethyl-1,2,3-trithiepane (**216**) was isolated in 25% yield (Eq. 58) along with 3,6-dimethyl-1,2-dithiane (9%). Only a suggestive elemental analysis could be obtained, and no other evidence was given for the structure of **216**. Formulation in terms of the unbranched sulfur chain was preferred, despite the seven-membered ring, in view of previous work cited on the general occurrence of sulfur atoms in unbranched instead of branched chains.

$CH_3COCH_2CH_2COCH_3 + H_2S \xrightarrow[7500-8500 \text{ atm}]{80°C, 15 \text{ hr}}$ 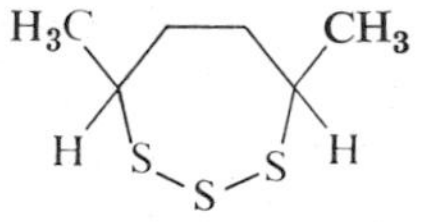 (58)

216

A well-characterized fused-ring treated in detail in Chapter XI of this volume is 1,2,3-trithiepin **159** (1,5-dihydro-2,3,4-benzotrithiepin according to *Chemical Abstracts*, or 2,3,4-benzotrithiepin as named by Milligan and Swan) (123). This substance was prepared by treating the appropriate bis-Bunte salt with sodium sulfide and formaldehyde (123).

Milligan and Swan did not mention the use of the analogous tetramethylene bis-Bunte salt **217** (Eq. 59), but the fact that the analogous disodium penta-

$$\underset{\mathbf{217}}{NaO_3SS(CH_2)_4SSO_3Na} + Na_2S \rightarrow \underset{\mathbf{215}}{\underbrace{S(CH_2)_4SS}} + 2Na_2SO_3 \qquad (59)$$

methylene and hexamethylenebis(thiosulfate), which could have given eight- and nine-membered cyclic trisulfides, led only to polymer might suggest dim prospects for success by this method. Indeed the Bunte salt **217** could not be prepared from 1,4-dibromobutane; the bis-Bunte salt was obtained from 1,4-dichloro-2-butene in 58% yield, but no mention was made of its use in preparing 4,7-dihydro-1,2,3-trithiepin (**158**) (123).

Despite these ill omens, Kabuss and co-workers succeeded in making both 1,2,3-trithiepane (**215**) and 4,7-dihydro-1,2,3-trithiepin (**158**), presumably by the approach of Milligan and Swan (Eq. 59) (100). Experimental details have not yet appeared, but in mentioning these compounds relative to a study of nmr spectra, the bis-Bunte salt method was stated to be well suited for the preparation of cyclic compounds even though the yield of 1,2,3-trithiepane (**215**) was less than 5%.

Conclusions of Kabuss and co-workers regarding the conformations of the dihydrotrithiepin **158** were referred to earlier (see Section C-2-b). Rate constants for conformational inversions determined by variation of nmr spectra with temperature were 120 sec^{-1} at $-90°C$ for the dihydrotrithiepin **158** and 152 sec^{-1} at $+83°C$ for 1,5-dihydro-2,3,4-benzotrithiepin (**159**). No rate constant was given for 1,2,3-trithiepane itself (**215**) because the incipient splitting of signals reflecting freezing of inversion occurred at $-130°C$, only 3°C above the lowest temperature reached in this experiment; the free energy of activation was estimated to be 6–7 kcal/mole, which may be compared with values of 8.9 kcal/mole for the dihydrotrithiepin **158** and 17.4 kcal/mole for the dihydrobenzotrithiepin **159**. Seemingly, stiffening of the seven-membered ring by the third sulfur atom occurs in **215** since inversions of other saturated seven-membered rings studied could not be frozen at all. The energy values are higher than those of the corresponding disulfides but still are not high enough for the preparative separation of conformational isomers; Kabuss and co-workers suggested the possibility of eventually obtaining separable isomers by suitable substitution of the ring (the energy barrier required would be at least 23 kcal/mole).

2. 1,2,5-Trithiepins and Derivatives

a. *Preparation and Structure*

1,2,5-Trithiepin (**214**) in its tetrahydro form **218** is properly named 1,2,5-trithiepane, but has also been called 1,2,5-trithiacycloheptane, diethylene trisulfide-1,2,5, or (ambiguously) diethylene trisulfide. Those who have reported on the chemistry of the 1,2,5- system have rarely summarized prior work, and cross references are sparse.

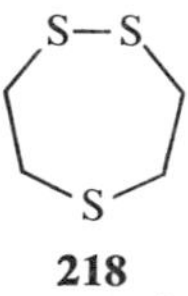

218
R.I. 337

Syntheses of the 1,2,5-trithiepane system **218** first were reported by Rây (160) and Fromm and Jörg (79). Westlake and co-workers also commented more recently on its possible formation (199).

Rây boiled 1,2-ethanedithiol ("dithioethylene glycol") with benzylidene chloride for 30 hr (160). After washing the product with alkali, probably to remove thiols (although such washing may have failed if they were polymeric), he isolated as one of several products an unstated amount of "diethylene trisulfide," mp 96°C. From Rây's statement that this solid was formed by intermolecular condensation of the thiol, it may be presumed that he considered it to be 1,2,5-trithiepane (**218**). Rây also isolated "diethylene trisulfide," again as a by-product, after heating another of his fractions with ethyl iodide (160).

Fromm and Jörg obtained what they believed to be **218** by the procedures shown in Eq. 60 (79). In each instance, the product had a melting point

$$[Cl(CH_2)_2S]_2 + Na_2S \rightarrow \mathbf{218} \leftarrow Na_2S_2 + [Cl(CH_2)_2]S \qquad (60)$$

of 74–75°C and could be reduced to β-mercaptoethyl sulfide. Although the structures of the intermediates seemed well corroborated, it is now known that the key step of reduction to the dithiol would be just as characteristic of a dimer or of polymers as it would be of **218**.

The melting points of 96°C and of 74–75°C for the two compounds believed to be **218** suggest that they were probably oligomers or polymers. Although both products had satisfactory analyses for sulfur, no molecular weights were reported. These analyses presumably rule out the polymeric unit —$S(CH_2)_2S$—, but are consistent with one such as —$S(CH_2)_2S(CH_2)_2S$—, which could have been produced by partial loss of sulfur from [—$S(CH_2)_2S$—]$_n$ or could have taken this form in a linear polymer. The high melting points seem inconsistent with the fact that 1,2-dithiepane (**128**) (Table X-3) and 1,4,5-oxadithiepane (**219**) (63) are liquids. These substances could be poor models, however; the trithiepane **218** conceivably might have a symmetry more like that of 1,4-dithiane for which melting points range from 108°C to 113°C (161). Furthermore the products may have had greater solubilities than anticipated for polymer. Thus Rây's product was recrystallized from benzene; Fromm's and Jörg's seems to have been less soluble in organic solvents but was recrystallized from nitrobenzene. Davis and Fettes felt that polymer rather than **218** was formed in the Fromm-Jörg reactions (Eq. 60) and claimed to have prepared polymers by essentially the same method (63).

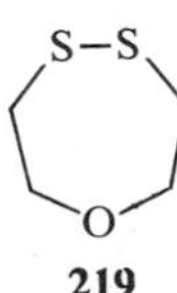

219

According to Westlake and co-workers, ethylene reacted under pressure with molten sulfur to give not only polymers but also a colored volatile oil (of unspecified boiling point) which was sparingly soluble in alkali but miscible with organic solvents (199). Catalysts had no effect on the reaction. The oil evidently was impure but gave an elemental analysis and molecular weight which suggested the formulation of $(C_2H_4)_2S_3$. Oxidation with chlorine–water gave 2-chloroethanesulfonyl chloride and a little ethanedisulfonyl chloride (*sic*). A six-membered 1,4-dithiane structure, with the third sulfur branched from one of those in the ring, was preferred for the oil. However, this formulation is unlikely since sulfur atoms tend to form linear unbranched chains (75). It was felt that structure **218** could not be excluded for the oil. The report of "no reaction" with several reducing agents is inconsistent with structure **218** (199) unless the failure of reduction can be explained by inadequate contact of the oil with the aqueous reducing media. The refractive index of the oil of 1.5746 is about what might be expected for **218** since 1,2-dithiepane (**128**) has a value of 1.570 (see Table X-3) and 1,4,5-oxadithiepane (**219**) one of 1.5823 (63).

Davis has patented a procedure embodying, among very broad and general claims, the preparation of **218** by distillation of polysulfide polymers (61, 62). However, an earlier paper of Davis suggests the major concern of the patents to be oxadithiepane **219**; in any event, even the yield of **219** was very low (63).

Recently, an unequivocal synthesis for 1,2,5-trithiepane (**218**) has made it possible to establish its properties. The trithiepane **218** was obtained in 55% yield as shown by Eq. 61; other methods of cyclization failed (73). Good

$$HS(CH_2)_2S(CH_2)_2SH \xrightarrow{FeCl_3} \underset{\mathbf{218}}{\overline{S(CH_2)_2S(CH_2)_2S}} \tag{61}$$

evidence for the structure of **218** included chemical reactions, nmr, Raman, uv, and mass spectra, and molecular weight determination in solution (73). The trithiepane **218** was a colorless liquid with properties quite different from any reported earlier. The refractive index (n_D^{25} 1.6424) was sufficiently above that of Westlake and co-workers (n_D^{25} 1.5746) to suggest that the earlier workers probably dealt with a different substance. The trithiepane **218** resembled 1,2-dithiepane (**128**) in being quite resistant to polymerization, could be converted to a solid methiodide (72% yield), and could be oxidized to the 5-sulfoxide with sodium metaperiodate (66% yield). The latter results carry the general implication that a sulfide moiety can undergo such reactions in the presence of a 1,2-disulfide moiety in the same ring. Comparison of the reactivities of **218**, the methiodide, the sulfoxide, and 1,2-dithiepane (**128**) in polymerization and cleavage with cyanide ion suggested that the sulfide function or its derivatives have no marked effect on the disulfide linkage; the uv spectra also seem to reflect little interaction.

Krespan and Brasen were the first to report a carbon-substituted 1,2,5-trithiepane (105). They found that sulfur reacted with tetrafluoroethylene at 445°C to give a trithiolane, a dithiane, and a tetrathiane. In studying the reaction intermediates they also used the less vigorous conditions shown in Eq. 62. Along with the dithiane **220**, they isolated octafluoro-1,2,5-trithiepane (**221**) as a colorless liquid. Elemental analyses agreed well for **221**, but a determination of the molecular weight was not reported.

Despite its seeming complexity, Krespan and Brasen stated that this reaction, in contrast to that of sulfur with hydrocarbons, yields few byproducts, so that individual polysulfides are easily obtained where tetrafluoroethylene is used (105). The trithiepane **221** was thought to result from interaction of intermediate polysulfides with tetrafluoroethylene. However, higher

$$F_2C{=}CF_2 + S \xrightarrow[\text{pressure, 3 hr}]{CS_2,\ 300°C} \mathbf{221}\ (24\%) + \mathbf{220}\ (37\%) \qquad (62)$$

proportions of sulfur than one equivalent gave less trithiepane (**221**) and dithiane (**220**) and more of other products. The possibility that **221** could be formed by direct addition of tetrafluoroethylene to tetrafluoro-1,2,3-trithiolane was thought likely but was not demonstrated.

Pyrolysis of a disulfide polymer containing the repetitive unit —CF_2CF_2SS— under pressure also gave trithiepane **221** (105), a reaction reminiscent of that mentioned above as patented by Davis (61, 62). The method of Krespan and Brasen, with somewhat varied conditions, was likewise patented (103, 104).

The first report of a trithiepin seems to have been made in 1936 by Brass, Pfluger, and Honsberg (31). They treated a dibromonaphthazarin with sodium sulfide and assumed the product to be a dinaphtho-1,2,5-trithiepin although they offered no evidence other than elemental analysis. This work is discussed in Chapter XI of this volume.

A monocyclic trithiepin, 1,2,5-trithiepin-3,4,6,7-tetracarbonitrile (**1**, Eq. 63), was claimed in two patents as a product from the oxidation of di-

$$\mathbf{1} \xleftarrow[-20°\,C]{Br_2,\text{etc.}}\!\!\!\!\!\!/\ \ \mathbf{222} \longrightarrow \mathbf{224} \longrightarrow \mathbf{223} \qquad (63)$$

sodium dimercaptomaleonitrile (**222**) or of sodium cyanodithioformate (which readily gives **222**) (113, 172). Simmons and co-workers, after thorough study of the oxidation product, concluded that in reality it was the isothiazole **223** (178). They proposed a mechanism to account for the conversion of the proved intermediate dithiin **224** to isothiazole **223** by sulfur formed during the oxidation (178). Had an elimination at one stage of this mechanism occurred to give a *cis* rather than *trans* arrangement of cyano groups, they felt that trithiepin **1** would have been a likely product. That the oxidation product was in fact isothiazole **223** was shown convincingly by the consistency of ir and uv spectra, by hydrolysis of **223** to a nitrile-free amide of the proper analysis (to give a proper analysis **1** would have had to retain a cyano group), and by agreement of the ^{13}C nmr spectrum of product synthesized using ^{13}C cyano groups. Furthermore dithiin **224** was converted by sulfur to a sample of isothiazole **223** identical with one synthesized according to the Merck group (113, 172). Evidently these workers also recognized **223** to be the correct structure, since another patent application prior to the appearance of the paper of Simmons and co-workers claimed essentially the original process but with isothiazole **223** instead of trithiepin **1** as the product (128). It seems likely that the chemistry of trithiepin **1** would be closely related to that of the substances shown in Eq. 63.

Boberg believed in 1965 that he was the first to report the 1,2,5-trithiepin system (25). Accepting the conclusions of Simmons and co-workers, he discounted the claims of the Merck group. His certainly is the most extensive study of the 1,2,5-ring system (25). In view of Boberg's careful work, in which, for example, he determined the molecular weight of each of ten products, one can scarcely doubt that he did indeed have 1,2,5-trithiepins, or 1,2,5-trithiacycloheptadienes as he named them.

Boberg was interested in the alkaline cleavage of 5-amino-4-chloro-1,2-dithiacyclopenten-3-ones of general structure **225** (Eq. 64). He found that the cleavage at temperatures up to 55°C gave 1,2,5-trithiepins **226**. Table X-3 summarizes the 1,2,5-trithiepins prepared in this manner.

Evidence for structure **226** was extensive (25). In addition to analysis for all five elements present, for methoxyl, and for the molecular weight (ebullioscopic and Rast), it included comparison of ir spectra with those of model *p*-dithiins and thiophenes. It also was based significantly on desulfurization of several products with Raney nickel to two types of amines (Eq. 64); for the amines, in turn, typical structures were confirmed appropriately, for example, by isolation of several saturated amino esters as picrates and by conversion of an unsaturated amino ester to derivatives of ethyl formylacetate. Boberg considered six-membered alternatives to the trithiepin structure **226** with a third sulfur atom in a transannular bridge, but ruled them out. The chemical behavior of the products represented by the trithiepin structure **226** seems quite consistent with expectation; it is described in Section D-2-c.

$2R''OH(KOH)$; $-S, -2HCl$; Ni

225 → **226** (yields, 30–83%) → $CHNRR' = CHCO_2R''$ + $CH_2NRR'–CH_2CO_2R''$

$R''O^{\ominus}$; R = Aryl; R′ = Alkyl; R″ = CH_3 or C_2H_5

227 —(−S)→ **228**; **227** → **226**

Boberg explained the transformation of **225** into **226** in terms of a three-step mechanism (Eq. 65) (25). A principal point of interest had been whether nucleophilic cleavage of **225** would occur between $S_{(2)}$ and $C_{(3)}$ or $S_{(1)}$ and $S_{(2)}$; both of these modes had been reported for other types of sulfur compounds. Boberg concluded from the presence of the —SS— linkage in thiepin **226** that —SS—CO— cleavage did in fact occur between $S_{(2)}$ and $C_{(3)}$. He thus invoked such an attack by alkoxide anion for the cleavage of **225** to **227**. He proposed that **227** lost sulfur to give **228**, which reacted with the initial fragment **227** to produce trithiepin **226**. The latter reaction, which forms the trithiepin, has the net effect of a displacement of two vinylic chlorine atoms by two thiolate-type ions; perhaps it occurs by addition of the thiolates to the double bonds, followed by elimination of two chloride ions as the double bonds are re-established.

Boberg provided additional support for his mechanism (Eq. 66) by carrying out a similar reaction with the model compound **229** in the presence of dimethyl sulfate (25). The same type of intermediate as **227** (i.e., **230**) similarly lost sulfur and then was trapped as the methyl sulfide (i.e., **231**). In the absence of dimethyl sulfate, *p*-dithiin **232** resulted (Eq. 67). In an earlier study with hydrogen atoms in place of both the phenyl and halogen of **229**, no primary products had been found (25).

Boberg considered the possible intermediacy of a four-membered ring containing a disulfide linkage in the formation of **226**; such a ring could arise from intramolecular displacement of chloride ion in **227** by the thiolate-type anion. He thought this unlikely because he never observed isomerism of the substituents in the dithiin **232**; dimerization of four-membered rings, with loss of sulfur, he felt, should have led to two different isomers. He also envisioned but rejected several other pathways for conversion of the dithiacyclopentenone **225** to the trithiepin **226** (25).

C_6H_5, Br, S, S, C=O — $\xrightarrow{CH_3O^{\ominus}}$ [C_6H_5, Br, S, $S^{\ominus}$, CO_2CH_3] $\xrightarrow[2)\ (CH_3O)_2SO_2]{1)\ -S}$ C_6H_5, Br, H_3CS, CO_2CH_3 (66)

229 **230** **231**

–S

H_3CO_2C, S, C_6H_5, C_6H_5, S, CO_2CH_3 (67)

232

b. *Stereochemistry*

In characterizing the perfluoro-1,2,5-trithiepane **221** referred to in Section D-2-a, Krespan and Brasen determined the ^{19}F nmr spectrum (105). The perfluoro-1,4-dithiane **220** had shown an nmr spectrum consisting of a single resonance at lower than usual field for CF_2. Since the ring of the dithiane **220** must be puckered, this result was interpreted to mean rapid interconversion of the various possible conformations to give an average planar form. For the trithiepane **221**, however, the spectrum showed two peaks in a 1:1 ratio. Relative to 1,1,2,2-tetrachloro-1,2-difluoroethane, a single peak occurred at +700 Hz, and an AB doublet of doublets, of equal area, was seen at higher field (850, 1077, 1135, and 1366 Hz; w-s-s-w). The splitting, a result of nonequivalence of fluorine atoms on one of the two types of difluoromethylene groups, was attributed to the existence of trithiepane **221** in "one preferred puckered conformation as a mixture of optical isomers." Since hindrance to intramolecular rotation occurs about sulfur-sulfur bonds and since dithiane **220** inverted rapidly, the spin-spin coupling responsible for the AB pattern was attributed to nonequivalent fluorine atoms on the carbon atoms adjoining the disulfide bond. The single peak was assigned to the remaining fluorine atoms, which must be too nearly magnetically equivalent to show observable coupling.

c. *Reactions*

Section D-2-a described reduction of what Fromm and Jörg thought was 1,2,5-trithiepane (**218**) (79), as well as oxidation but failure of reduction with what Westlake and co-workers thought might be the trithiepane (199).

Krespan and Brasen reported that octafluoro-1,2,5-trithiepane (**221**), when heated at 300°C for 10 hr, gave octafluoro-1,4-dithiane (**220**) in 42% yield; 10% of **221** was recovered (105). They pointed out that this transformation is unusual in light of conversions of acyclic fluoroalkyl polysulfides at high temperature to give disulfides with little monosulfide. They explained the unexpected reaction in terms of intramolecular formation of a sterically favored six-membered ring, noting that if initial dissociation of the disulfide bond into thiyl radicals is assumed, the conversion of **221** to **220** must involve a rare type of attack by a thiyl radical on a saturated carbon atom with displacement of sulfur; the slowness of the reaction of trithiepane **221** might result from the difficulty of such a process relative to recombination or reaction with sulfur to form higher polysulfides.

Octafluoro-1,2,5-trithiepane (**221**) did not give insoluble polymer in acetonitrile (used as a base) even with dimethyl sulfoxide as a catalyst, but catalytic amounts of triethylamine in acetonitrile at −40°C did cause precipitation of poly(octafluoro-1,2,5-trithiepane) in 80% yield (105). The tendency to polymerize on heating reflected the same influence of ring size upon reactivity as has been noted for nonfluorinated cyclic disulfides. The order of increasing stability found in this work was:

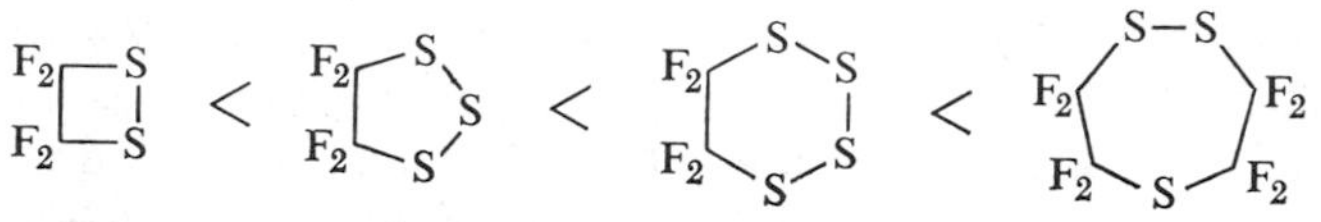

233 < **234** < **235** < **221**

The four-membered heterocycle **233** could not be isolated as the monomer, in contrast to **221** which required amine catalysis to promote ionic ring opening. The relation of ring size to stability was attributed not only to compression of normal bond angles but also to strong repulsions between *p*-electrons on adjoining sulfur atoms, which result in energy minima when the dihedral angle for —S—S— is about 90–100°. The ^{19}F nmr spectrum for **234** shows that its ring is nearly planar; unlike **221**, trithiolane **234** shows only one singlet (105). Its ease of cleavage relative to **221** thus is understandable. Octafluoro-1,2,5-trithiepane (**221**) did not react with tetrafluoroethylene at 300°C, even in the presence of sulfur (105). In contrast, other fluorinated higher cyclic polysulfides such as **235** absorb the olefin to give a disulfide polymer (105).

The trithiepane **221** behaved with chlorine as disulfides generally do, giving the sulfenyl chloride **236** (Eq. 68) (104).

$$\text{(}F_2C)_4S_3 \text{ ring: } S{-}S,\ F_2,\ F_2,\ F_2,\ F_2,\ S \xrightarrow{Cl_2} ClS(CF_2)_2S(CF_2)_2SCl \qquad (68)$$

236

Boberg investigated a few reactions of 1,2,5-trithiepanes **226** mainly to support his assignment of structure (25). These amines dissolved in concentrated hydrochloric or hydrobromic acid and could be recovered after 10 days by addition of water; destruction began at the boiling point and only the amine moiety then could be isolated.

The trithiepin ring system of **226** was destroyed by alkali, even at room temperature. Consequently conversion of the esters to acids could not be achieved and only the amine moiety could be isolated. Despite the great alkali sensitivity of compounds of structure **226**, Boberg had obtained them in yields of up to 80% by treating dithiacyclopentenones **225** with alcoholic alkali. He explained the paradox by stating that the products precipitated rapidly after formation and thus were protected against further attack by alkali, and also that the alkali was quickly neutralized by the HCl formed in the reaction. As would be expected, he found that excess alkali in the synthesis (Eq. 64) caused decreases in yields of trithiepins with increasing time of reaction.

d. *Uses*

The group at E. Merck patented the alleged tetracyano-1,2,5-trithiepin (**1**) as a fungicide and nematocide (113, 172); as stated above, however, this product actually was the isothiazole **223** (Section D-2-a). Bremer reported that "tetranitrilotrithiacycloheptadiene" is effective as a fungicide when used in a propellant; the result, called Deftan-Fog, was recommended for use in greenhouses (35). From the name used by Bremer, this compound *could* be 1,2,3-trithiepin-4,5,6,7-tetracarbonitrile, but this author's association with E. Merck A.-G. suggests that the name was meant to represent the alleged 1,2,5-trithiepin **1**, which was really an isothiazole.

Section D-2-c referred to the polymerization of octafluoro-1,2,5-trithiepane (**221**). These polymers had relatively low molecular weights even when prepared at low temperatures, as gauged by the brittleness of film pressed from them (105). Films pressed at 100°C were weak. The polymers were not soluble in toluene or chloroform (105). On the other hand the bis-sulfenyl chloride **236** available from **221** by reaction with chlorine (Section D-2-c) was patented for making polymers, as a cross-linking agent for crepe rubber, and as an agent to confer lubricity or water repellency to cellulose products (104).

In Section D-2-a reference was made to inclusion of claims for the preparation of 1,2,5-trithiepane in two extremely broad patents (61, 62). These patents also claimed polymerization of 1,2,5-trithiepane (**218**) to afford protective coatings, adhesives, plastic compositions, and other uses.

3. 1,3,5-Trithiepins and Derivatives

A group at E. Merck A.-G. patented the preparation of 1,3,5-trithiepin-6,7-dicarbonitrile ("2,3-dicyano-1,4,6-trithia-2-cycloheptene") (**237**) by the reaction of disodium dimercaptomaleonitrile with bis(chloromethyl) sulfide (Eq. 69) (171).

$$\text{NC}(\text{S}^{\ominus}\text{Na}^{\oplus})\text{C}{=}\text{C}(\text{S}^{\ominus}\text{Na}^{\oplus})\text{CN} + (\text{ClCH}_2)_2\text{S} \longrightarrow \mathbf{237} \quad (69)$$

237 (R.I.9772)

Compound **237** was claimed as one of several pesticides having in common the unit RSC(CN)=C(CN)SR, where the R groups might be linked together. A subsequent patent reiterated the method and claimed that **237** repels houseflies (95).

E. Rings Containing Four or More Sulfur Atoms (Tetrathiepins, Pentathiepins, Hexathiepins, S_7, and Derivatives)

1. Tetrathiepins and Derivatives

Four seven-membered ring systems containing four sulfur atoms each are possible. These would have the four sulfur atoms in the following arrangements: 1,2,3,4; 1,2,3,5; 1,2,4,5; and 1,2,4,6. Of these possibilities, the sole example characterized so far has been the 1,2,4,6-ring system.

Some basis for optimism that 1,2,3,4-tetrathiepanes might be isolable, or at least that a chain of four sulfur atoms in a cyclic environment might be shown not to be inherently unstable, may be inferred from the formation in 60% yield of the aforementioned 1,2,3,4-tetrathiane **235** from tetrafluoroethylene and sulfur (105); this appears to be the first known cyclic tetrasulfide (105). Milligan and Swan felt that an impurity in 1,2,3-trithiane, which was prepared from disodium trimethylene di(thiosulfate) and sodium sulfide, might be the corresponding tetrasulfide, 1,2,3,4-tetrathiepane. Their suspicion was based solely on the retention volume of an impurity separated from the trithiane by gas chromatography (123).

Backer and Evenhuis carried out studies which deserve attention and further experimental scrutiny because of the possibility that an unstable 1,2,3,4-tetrathiepane might have been an intermediate in the formation of a more stable six-membered 1,2,3-trithiane (17). They prepared "sodium tetrasulfide" (Eq. 70) and observed that treatment of this "tetrasulfide" with

$$2NaSH + 3S \rightarrow Na_2S_4 + H_2S \qquad (70)$$

$$C(CH_2Br)_4 + 2Na_2S_4 \rightarrow C_5H_8S_6 + 4NaBr + 2S \qquad (71)$$

pentaerithrityl tetrabromide (Eq. 71) led to a hexasulfide (**238**) in 54% yield (mp 182–184°C). The analysis was good, and Schotte later stated that the molecular weight proves **238** to be a monomer (175). Backer and Evenhuis considered formulating the hexasulfide **238** as **238A**, perhaps envisioning

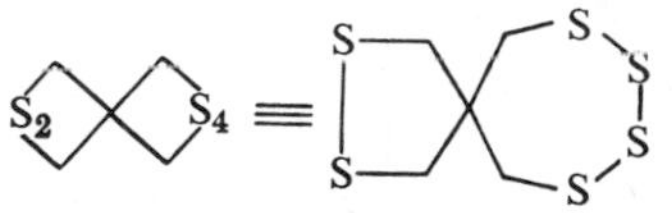

238A **238B**  **238C** 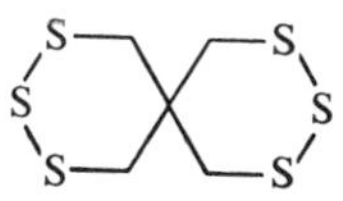 **238D**

238A as the 1,2,3,4-tetrathiepane **238B**; they rejected **238A** in favor of **238C**, however, because hexasulfide **238** formed no derivative with mercuric chloride, a property which they had found characteristic of dithiolanes. On the other hand they suggested that the *initial* reaction product might nevertheless have had a tetrasulfide function in each ring, and that its instability could have resulted in the loss of an atom of sulfur from each ring to give hexasulfide **238**. If this view were correct, the unstable tetrasulfide reasonably could be formulated as **239**. Backer and Evenhuis cited supporting literature

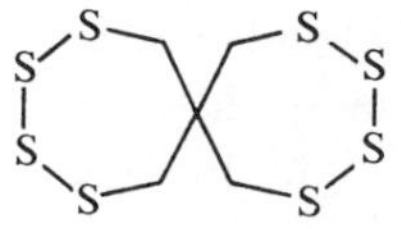

239

references for the decomposition of tetrasulfides to trisulfides (17). From spectrochemical studies Schotte later assigned structure **238D** to the hexasulfide but did not comment on the possible intermediacy of a tetrathiepane such as **239** (175). Additional related work of the Backer group with 1,3-dihalides and sodium tetrasulfide is summarized by Breslow and Skolnik earlier in these volumes (43). The work of Backer and associates is intriguing in its implications for the 1,2,3,4-tetrathiepane system, since even though "sodium tetrasulfide" may be more of a stoichiometric notation than a chemical entity, the isolation of trithianes under stoichiometric conditions favorable for tetrathiepanes suggests that 1,2,3,4-tetrathiepanes either form with difficulty or decompose readily. More substantial experimental evaluation of the stability of the 1,2,3,4-system would be helpful.

The only known tetrathiepane, 1,2,4,6-tetrathiepane (**240**), was first prepared in 1890 by Baumann (19). For brevity, he called it trimethylene tetrasulfide; it has also been called 1,2,4,6-tetrathiacycloheptane. As shown by Eq. 72. Baumann saturated a solution of formaldehyde with hydrogen

$$H_2CO + H_2S \xrightarrow{HCl} HS\text{-}CH_2\text{-}SH \xrightarrow[H_2S,\ HCl]{H_2CO} HS\text{-}CH_2\text{-}S\text{-}CH_2\text{-}SH\ (\mathbf{242}) \xrightarrow{H_2CO,\ H_2S,\ HCl} HS\text{-}CH_2\text{-}S\text{-}CH_2\text{-}S\text{-}CH_2\text{-}SH\ (\mathbf{241}) \xrightarrow{I_2} \mathbf{240} \tag{72}$$

$HS\text{-}CH_2\text{-}SH \rightarrow CH_2(SO_2CH_3)_2$ **244**; **242** $\rightarrow SO_2(CH_2SO_2CH_3)_2$ **243**

240

R.I. 314

sulfide at low temperature, added an equal volume of 5% hydrochloric acid, and then washed the resulting precipitate. Treatment of the precipitate with iodine in ethanol gave a sparingly soluble resin from which 1,2,4,6-tetrathiepane (**240**) was extracted in unstated yield with hot ethanol (19). The elementary analysis and molecular weight agreed with expectation for 1,2,4,6-tetrathiepane (**240**) (19).

At first glance, one is inclined to regard the structure of a product from such a potentially complex reaction mixture with scepticism. However, with acceptance of the molecular formula $C_3H_6S_4$ and of the reasonable assumption that no carbon-carbon bonds are formed, one can agree that three methylene groups and two sulfur atoms were present, in the form of the unit

$$—CH_2SCH_2SCH_2—.$$

The analysis and molecular weight establish the presence of two more sulfur atoms in a monomeric structure; these can scarcely be accommodated other than by visualizing them as closing the ring to give the tetrathiepane structure **240**. Demonstration of the absence of a thiol group in the product would have been a key point, but this absence apparently follows from Baumann's observation that an ethanol solution gave no precipitate with lead acetate (silver nitrate did give a precipitate, but this can be attributed to rupture of the disulfide linkage).

1,2,4,6-Tetrathiepane (**240**) formed colorless crystals (mp 83–84°C) which could not be distilled without decomposition. It was insoluble in water and soluble in organic solvents. Hot aqueous alkali caused loss of sulfur and formation of insoluble, probably polymeric amorphous material. Reduction with zinc and acid gave hydrogen sulfide and what must have been thiols (leeklike odor). Oxidation with nitric acid led to an explosion, although sulfuric acid was said to form; permanganate also led to the sulfate stage, but no sulfone could be obtained.

Baumann's interest in the tetrathiepane **240** apparently was more or less incidental to his study of the products which result from the reaction of formaldehyde with hydrogen sulfide. His interest in the isolation of **240** seems to have been prompted mainly by the fact that it provided evidence for the structure of the dithiol **241**. By extracting an uncatalyzed mixture of formaldehyde and hydrogen sulfide, he was able to isolate the earlier intermediary dithiols, methanedithiol, and **242** (Eq. 72); these were identified by dimethylation, oxidation, and separation of the sulfones **243** and **244**. The overall sequence (Eq. 72) thus seems unequivocal, both in itself and as a buttress for the structure of **240**. It should be pointed out, however, that reactions of formaldehyde with hydrogen sulfide can produce other products. A discussion of such reactions, with leading references, can be found in Reid's treatise (162).

1,2,4,6-Tetrathiepane (**240**), evidently identical with Baumann's product, recently has been isolated in minor amounts from a species of mushroom, along with a pentathiepane and hexathiepane (*vide infra*) (131). It could be purified readily by sublimation as well as by recrystallization. The identity was confirmed by nmr and mass spectra. It could be obtained, together with a trithiolane, after the reaction of $Na_2S_{2.5}$ with methylene chloride at a pH of about 12 (131). Compound **245** also could be isolated from the mushroom. Since **245** gradually decomposed in organic solvents to give **240**, as well as the aforementioned pentathiepane, evidently it is a precursor of both. Compound **245** must be formed from still another precursor, however, since it could not be isolated until the mushroom was treated with water (131).

$$CH_3S(O_2)SCH_2OCH_2SCH_2S(O)CH_3$$

245

The nmr spectrum of **240** has been studied in detail (129). Two sharp peaks are seen at τ 5.86 and 5.82 with an intensity ratio of 2:1 at room temperature; these correspond to the structurally nonequivalent methylene groups. The spectrum did not change significantly at $-90°C$. The energy barrier for equilibration is lower than in lenthionine (see Section E-2). The polarographic behavior of **240** also has been studied (10).

By inclusion in broad formulations covering a wide array of thiepanes, 1,2,4,6-tetrathiepane (**240**) was claimed in two patents to result from distillation of polysulfide polymers in the presence of alkali (61, 62). However, these patents were centered on 1,4,5-oxadithiepane (**219**), and this route seems unlikely to be feasible, at least to the pure tetrathiepane, particularly in light of Baumann's report that the tetrathiepane decomposes on distillation or treatment with alkali (19). Again only as a part of very broad claims, 1,2,4,6-tetrathiepane (**240**) was stated to be polymerizable to materials useful as protective coatings and in other applications (61, 62).

2. Pentathiepins and Derivatives

Of the three possible pentathiepane ring systems (the 1,2,3,4,5; the 1,2,3,4,6; and the 1,2,3,5,6), only that having the 1,2,3,5,6 arrangement of the sulfur atoms is known (**246**). Structure **246** is named 1,2,3,5,6-pentathiepane. It is perhaps significant for the relative stability of the 1,2,3,4,5 system

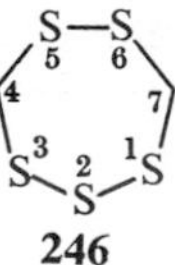

246

R.I. 9769

that reaction of tetrafluoroethylene with sulfur has given the four-membered ring with two adjacent sulfur atoms (**233**; highly unstable), the five-membered ring with three adjacent sulfur atoms (**234**; 10%), and the six-membered ring with four adjacent sulfur atoms (**235**; 60%), but evidently not the seven-membered ring with five adjacent sulfur atoms (tetrafluoro-1,2,3,4,5-pentathiepane) (105). As remarked in relation to 1,2,3,4-tetrathiepanes (Section E-1), however, the existence of the tetrathiane **235** with its internal chain of four sulfur atoms lends a measure of hope for the possible stability of the seven-membered counterpart, 1,2,3,4,6-pentathiepane.

Much of the work with 1,2,3,5,6-pentathiepanes thus far has been concerned with the highly complicated but useful reactions which carbonyl compounds undergo with ammonium sulfides and related substances. To illustrate this complexity, depending on the circumstances, reactions of sulfur, hydrogen sulfide, and ammonia or amines with ketones can lead to α-mercaptoketones, dithiazines, Δ^3-thiazolines, 1,2,4-trithiolanes, and 1,2,4,5-tetrathianes as well as to thiepanes (15). Although yields sometimes are either low or unstated, many are astonishingly good, and the reactions seem to have the virtue of providing a synthetic "grab bag" from which otherwise very difficultly obtainable compounds may be retrieved. Earlier volumes of this series have reviewed applications of such reactions in the synthesis of trithiolanes (37) and tetrathianes (42).

A pentathiepane was first reported only as late as 1958 by Fredga (77). In connection with a study of disulfides, Fredga and Magnusson investigated compounds obtained from reactants of the type named above, which are like the components of the Willgerodt reaction. They intended to use milder conditions than those usual in Willgerodt reactions. In various experiments both cyclopentanone and cyclohexanone gave only 1,2,4,5-tetrathianes. However, when cycloheptanone was allowed to stand for a week with a concentrated solution of ammonium disulfide (Eq. 73), a solid unexpectedly resulted (in unstated yield) which had an analysis and molecular weight consistent with thc formula $C_{14}H_{24}S_5$. In view of the improbability of branched chains of sulfur (75), structure **247** was assigned (77). Ir absorption at

$$+ (NH_4)_2S_2 \longrightarrow \quad (73)$$

247

R.I. 8719

about 671 cm^{-1} (C—S) and 490 cm^{-1} (S—S) showed the ring system to be different from that obtained using cyclohexanone. The index name for **247** used by *Chemical Abstracts* is 8,9,17,18,19-pentathiadispiro[6.2.6.3]nonadecane; Fredga used the name bis-4,4,7,7-hexamethylene-1,2,3,5,6-pentathiepane. As would be expected, the product was colorless and easily soluble in organic solvents. Preliminary crystallographic studies indicated that the pentathiepane **247** has a very complicated crystal structure with a rather large unit cell, but they led to no conformational details (77).

Fredga was unable to explain the atypical reaction of cycloheptanone but suggested that the crude reaction mixture might have represented a dynamic equilibrium of various cyclic and polymeric species from which **247** separated on the basis of its ease of crystallization (77). In 1959 Magnusson was able to obtain a pentathiepane from cyclohexanone by using ammonium disulfide

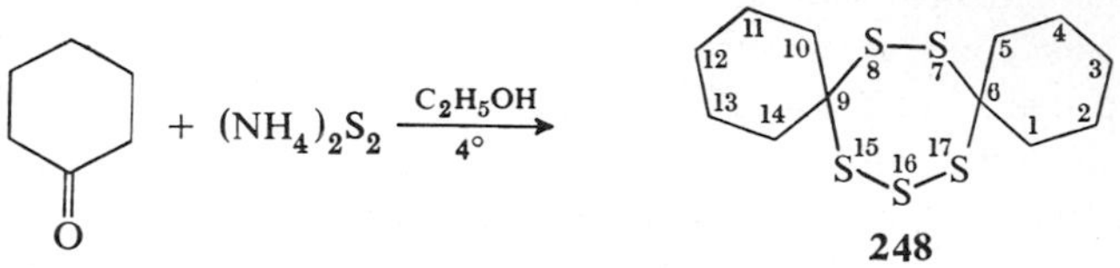

$$\text{cyclohexanone} + (NH_4)_2S_2 \xrightarrow[4°]{C_2H_5OH} \mathbf{248} \qquad (74)$$

248
R.I. 10520

in ethanol at 4°C (Eq. 74) (118). The conclusion thus is reinforced that the ring size of the product is highly dependent on the reaction conditions as well as the choice of starting materials. Certain other ketones, however, still gave only tetrathianes or trithiolanes (118). The colorless solid product 7,8,15,16,17-pentathiadispiro[5.2.5.3]heptadecane (**248**) had the correct molecular weight, but neither the melting point of the crude compound nor the yield of pure **248** was given.

Also in 1959 in a paper dealing mainly with tetrathianes and trithiolanes, Asinger and co-workers showed by obtaining a pentathiepane from diethyl ketone that formation of pentathiepanes may not be restricted to cyclic

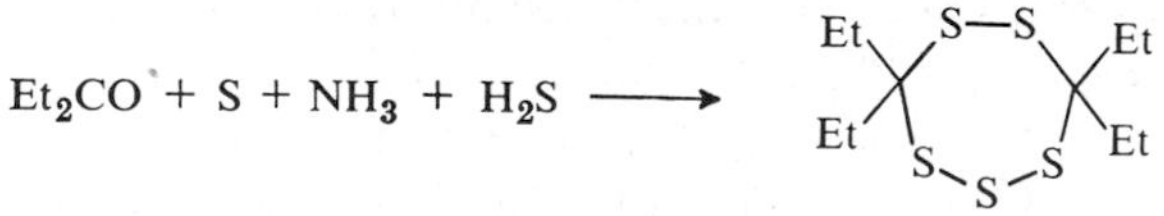

$$Et_2CO + S + NH_3 + H_2S \longrightarrow \mathbf{249} \qquad (75)$$

249

ketones. When diethyl ketone and sulfur were treated with ammonia and hydrogen sulfide for 8 hr (Eq. 75), subsequent fractionation of the products gave the presumed pentathiepane **249** in unstated yield (15). Elemental analyses were satisfactory, but no molecular weight was given; the report of **249** as a liquid is disturbing since all five of the other known 1,2,3,5,6-pentathiepanes are solids with melting points of at least 54°C (see Table X-3). Marked sensitivity to conditions was noted in this work as well. Thus when the flow of hydrogen sulfide was decreased from 7 to 4 lit/hr, the variety of products evidently no longer included **249**.

Another synthetic approach has been reported by Magnusson (119). The only application thus far of this method is shown in Eq. 76. It probably

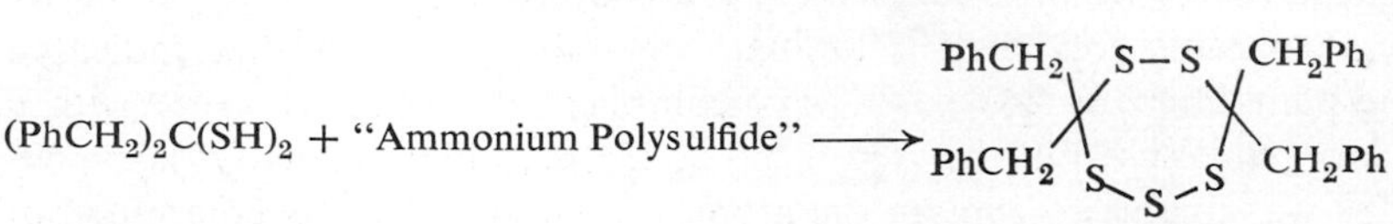

$$(PhCH_2)_2C(SH)_2 + \text{"Ammonium Polysulfide"} \longrightarrow \mathbf{250} \qquad (76)$$

250

involves reactions much like those involving ketones.

In still another approach not involving a carbonyl compound, the reaction shown in Eq. 77 has been reported to give a pentathiepane; similar reactions could be used also to give tetrathianes and trithiolanes (53).

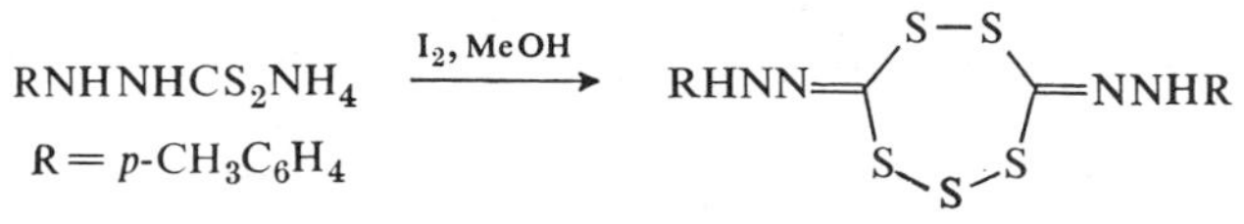

$$\text{RNHNHCS}_2\text{NH}_4 \xrightarrow{I_2,\ \text{MeOH}} \text{RHNN}=\text{(1,2,3,5,6-pentathiepane ring)}=\text{NNHR} \qquad (77)$$

$R = p\text{-}CH_3C_6H_4$

251

One pentathiepane occurs naturally, the prototype ring itself, 1,2,3,5,6-pentathiepane (**246**). It was isolated by Morita and Kobayashi from *Shiitake*, a mushroom highly prized for its edible qualities in Asian countries (130). This pentathiepane, which Morita and Kobayashi named lenthionine, is the characteristic odorous material of the mushroom. For its isolation, 5 kg of dried mushrooms was immersed in water; evolution of the odor at this point suggests intriguingly that the pentathiepane **246** is formed by enzymatic action. The resulting mass was extracted with methylene chloride, and the oil thus removed was chromatographed on silica gel using chloroform and hexane to give as the initial fraction 0.4 g of crystals (mp 60–61°C). Analysis and molecular weight (osmotic pressure) agreed excellently with the formulation of $C_2H_4S_5$. The high-resolution mass spectrum gave a parent ion and fragmentation pattern consistent with structure **246** (e.g., S_3 and S_2, but evidently not S_4 or S_5 as might be expected for the isomeric pentathiepanes); isotopic abundance ratios also were consistent. Virtual absence of a peak for $—CH_2SCH_2—$, clearly seen in trithioformaldehyde, excluded the $—CH_2SCH_2—$ grouping of 1,2,3,4,6-pentathiepane. The nmr spectrum afforded further evidence for the arrangement of the methylene groups in showing a singlet at τ 5.67 ppm.

Two total syntheses, although in low yield, were used to confirm the structure of **246** (Eqs. 78 and 79) (130). The first of these (Eq. 78) is reminiscent of reactions described above; presumably "$Na_2S_{2.5}$" is intended to

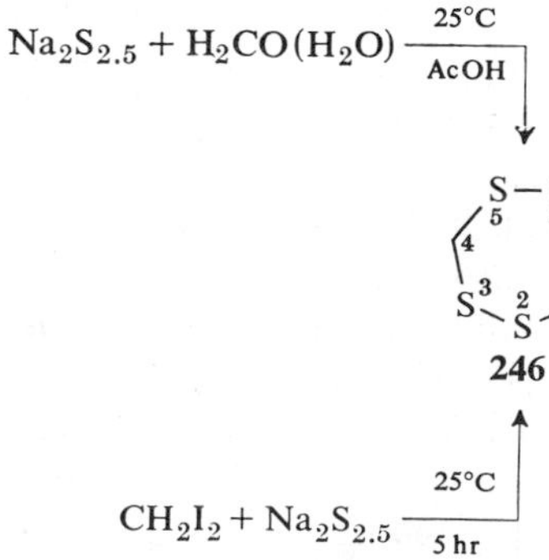

$$Na_2S_{2.5} + H_2CO(H_2O) \xrightarrow[\text{AcOH}]{25°\text{C}} \mathbf{246} \qquad (78)$$

246

$$CH_2I_2 + Na_2S_{2.5} \xrightarrow[5\ \text{hr}]{25°\text{C}} \mathbf{246} \qquad (79)$$

afford H_2S_2 and H_2S_3 as in the approach of Baumann (see Eq. 72). The second route (Eq. 79) embodies a novel approach, particularly apt for this instance because it excludes the possibility of a carbon-carbon linkage. Identity of the synthetic and natural compounds was assured by ir spectra, thin-layer chromatography, and mixture melting point, as well as by the characteristic odor.

Further information soon appeared. A simple synthesis for lenthionine (**246**) was developed from methylene chloride and $Na_2S_{2.5}$ at pH 8; hexathiepane also resulted (131). The reaction depends importantly on the pH since at pH 12 the products were a trithiolane and a tetrathiepane. A crystalline precursor of lenthionine was isolated in the form of **245**, the same compound referred to in Section E-1 as a precursor for the tetrathiepane **240**; as mentioned, **245** is itself formed only when the dried mushroom is treated with water (131). Favorable activities for lenthionine were reported against bacteria and fungi (131). The chemical structure of lenthionine was established by the X-ray method; all atomic positions, except these of hydrogen, were found unambiguously (135). The X-ray results also revealed that the conformation resembled a chair form with two nonequivalent methylene groups (135).

Considerable attention has been given to the practical applications of lenthionine (**246**). Essentially the aforementioned methods (Eqs. 78 and 79) were patented, and **246** was claimed as a fungicide; the LD_{50} was 500–1000 mg/kg (*per os* in mice) (188). The extraction procedure for lenthionine (**246**) was patented (132), as was a product for flavor enhancement (0.1–10 ppm) (189). It was nearly insoluble in water (0.002%) but somewhat soluble in nonpolar solvents (0.3–2%); among various stability studies was one which showed that half of **246** decomposed in 6 days in water (195). Studies were made of the amounts of **246** necessary for a shiitakelike flavor (197). Much of the foregoing information has been summarized (196).

The nmr spectrum of **246** has recently been studied more extensively (129). The sharp singlet at $\tau 5.67$ seen at 30°C broadens at lower temperatures (coalescence temperature, −60°C). At −90°C, two sharp singlets were seen. This behavior was attributed to ring inversion between two conformational isomers with nonequivalent sets of methylene groups. Failure to observe a geminal coupling constant was attributed to pseudorotation which rapidly equilibrates the geminal protons. The Arrhenius activation energy for ring inversion was 12.9 ± 0.4 kcal/mole. The energy barrier for equilibration of the nonequivalent methylene groups thus is higher than for 1,2,4,6-tetrathiepane (**240**; see Section E-1). The very low barrier for pseudorotation in cycloheptane (2–3 kcal/mole) is not increased significantly in **240** or **246**, although **246** does have a fairly high barrier for ring inversion.

The polarographic behavior of **246** has been studied (10).

For much of the pentathiepane work the multiplicity of products possible from the complex reactions leads to some uncertainty concerning assignments of structures among the three possible pentathiepane ring systems. The answer can be given that for all products the improbability of carbon-carbon bond formation seems to rule out 1,2,3,4,5-pentathiepane structures. For the tetrabenzyl derivative **250** (Eq. 76), a 1,2,3,4,6-structure seems unlikely because of the improbability that a —CSC— linkage would be formed, even though formation of such a linkage is not impossible (e.g., via a thioketone). Only with lenthionine (**246**) does it seem certain that the structure must be that of 1,2,3,5,6-pentathiepane and not that of the 1,2,3,4,5 or 1,2,3,4,6 isomer.

The only reaction explored thus far with pentathiepanes has been reduction. Although use of sodium in liquid ammonia was unsuccessful, Bobbio and Bobbio were able to reduce the bis(pentamethylene)pentathiepane (**248**) with zinc amalgam under acidic conditions to cyclohexane-1,1-dithiol (Eq. 80) and with lithium aluminum hydride to cyclohexanethiol (Eq. 81)

248 $\xrightarrow[\text{CHCl}_3,\ 6\text{hr}]{\text{Zn-Hg, HCl}}$ $C_6H_{10}(SH)_2$ 34% (80)

$\xrightarrow[\text{THF-Et}_2\text{O, 6 hr reflux}]{\text{LiAlH}_4,}$ $C_6H_{11}SH$ 92% (81)

(24). Since dithiols are stable to strong acids but apparently are converted to thiocarbonyl compounds by bases (52), the difference in products probably reflects nothing more profound mechanistically than reduction in both instances to a dithiol which is stable under acidic conditions (Eq. 80) but is converted under basic conditions (Eq. 81) to cyclohexanethione, which then is reduced to cyclohexanethiol. These reactions may have useful synthetic applications for the conversion of ketones into mono or dithiols via the pentathiepanes (24).

3. Hexathiepanes and S_7

The only structure possible, of course, for a seven-membered ring containing six sulfur atoms is 1,2,3,4,5,6-hexathiepane. One claim to the synthesis of such a structure was later withdrawn, but the chemistry warrants brief attention here. Complex reactions related to the Willgerodt-Kindler reaction, such as those discussed under the pentathiepins (Section E-2), gave a compound at first believed to be a hexathiepane (14). Depending on the circumstances, other products were thiazolines, imidazolinethiones, disulfides, and thioamides (14). Thus, methyl phenyl *N*-butylketimine reacted with *n*-butylamine and sulfur at 20°C (Eq. 82) to give a compound which had the empirical formula $C_{12}H_{15}NS_7$ (14). On the basis of its solubility in basic solution and of ir absorption at 1530 cm^{-1} (13), this product at first was formulated as structure **251** (14). In a brief sequel, however, structure **251** was rejected in favor of the isomeric thiocane structure **252** (13). The methyl analog was assigned structure **253** because its nmr spectrum showed

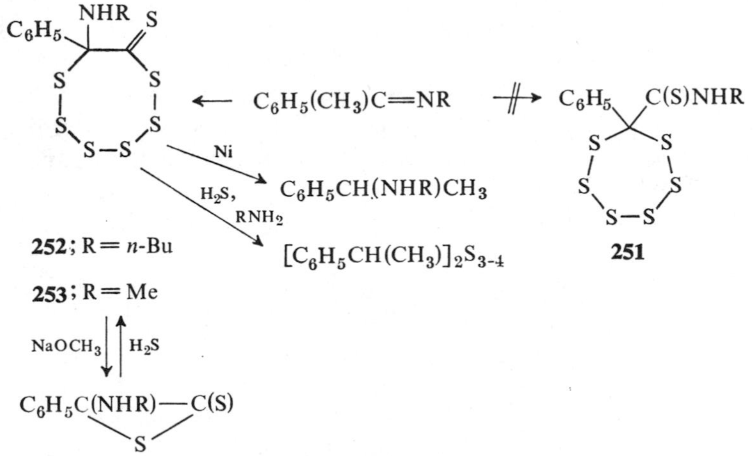

$C_6H_5(CH_3)C{=}NR$ → **252**; R = *n*-Bu; **253**; R = Me (not → **251**, C_6H_5, C(S)NHR hexathiepane)

252 —Ni→ $C_6H_5CH(NHR)CH_3$; **252** —H_2S, RNH_2→ $[C_6H_5CH(CH_3)]_2S_{3\text{-}4}$ (82)

253 ⇌ (NaOCH$_3$ / H_2S) $C_6H_5C(NHR)$—C(S) (three-membered ring with S)

only signals attributable to a monosubstituted phenyl nucleus, a secondary amine hydrogen, and a methyl doublet. Evidence was adduced for structure **252** by desulfurization with Raney nickel, by reduction with an amine plus hydrogen sulfide, and by cleavage with an alkoxide (13). Full details later appeared (11). When these interesting and unusual structures and reactions, and especially the possibilities for rearrangement, elimination, and reduction are considered, it is tempting to speculate as to whether thiepanes may eventually re-enter the picture. For its bearing of analogy on hexathiepane, it is worth adding that the eight-membered heptathiocane $(CH_2)S_7$ has been reported to be formed in low yield by reaction of methanedithiol and dichloropentasulfane; it underwent slow polymerization at room temperature (70).

1,2,3,4,5,6-Hexathiepane (**254**) itself, however, has recently been isolated from a species of mushroom, although only in very small quantities, along

254

with the tetrathiepane and pentathiepane mentioned already (131). Compound **254**, obtained as colorless prisms from dioxane, sintered about 79°C but showed no distinct melting point. The identity of **254** was established by elemental analysis, by the mass spectrum which showed the molecular ion, and by a single nmr peak at τ 5.46 (131). The reaction of $Na_2S_{2.5}$ and methylene chloride at pH 8 also gave **254**, along with the pentathiepane **246** (see Section E-2) (131). The polarographic behavior of **254** has been studied (10).

Cycloheptasulfur can be regarded as heptathiepane. Although cyclohexasulfur (S_ρ) has been known for some years, as has the long familiar cyclooctasulfur (S_λ), of the numerous allotropes of sulfur only S_8 seemed likely to be very stable (69). Mass spectrometric studies did not seem to have demonstrated the existence of S_7 in the vapor state very clearly, but the suggestion has been made that S_7 should be a puckered ring with two possible isomers, a chair and a boat form (69). These matters were reviewed in the report of a symposium (69). Recently work of Schmidt and his associates on thermodynamically unstable element modifications has made available several of the cyclic sulfur molecules: S_6, S_7, S_{10}, and S_{12} (167).

Cycloheptasulfur was prepared from bis(π-cyclopentadienyl)titanium(IV) pentasulfide (**255**) by using an equimolar amount of S_2Cl_2 (Eq. 83) (167).

$$\underset{\textbf{255}}{(C_5H_5)_2TiS_5} + S_2Cl_2 \xrightarrow{0^\circ C,\ \text{dark},\ CS_2} (C_5H_5)_2TiCl_2 + S_7 \tag{83}$$

Chilling of a toluene extract of the product gave S_7 in 23% yield as long, intensely yellow needles. Good cryoscopic and osmometric molecular weights were obtained for S_7 (167). Mass spectrometric studies, published simultaneously, also showed the expected molecular weight; the molecular-ion peak was much enhanced at energies below the usual 70 eV; peaks for the ions of S, S_2, S_3, S_4, S_5, and S_6 were seen (205).

The S_7 ring thus becomes the first odd-numbered sulfur ring to be reported (167). Cycloheptasulfur melts reversibly at 39°C. It can be kept for weeks in the cold and dark, but it changes readily in light or at room temperature into S_8, by way of a polymer. X-Rays rapidly decompose S_7 at room temperature, but good crystal photographs were obtained at -80°C, and the structure is being calculated (167).

F. Tables

TABLE X-3. Monocyclic Seven-Membered Rings Containing Sulfur

Substituents	Yield (%)	mp, °C, or (bp, °C/torr)	Other data	Refs.
Thiepins and Derivatives				
Thiepin 				
1,1-Dioxide		117–118	ir, uv, nmr, mass spectrum	125, 126
3-*i*-Pr-6-Me-1,1-dioxide				4
2,7-Dihydrothiepin 				
1,1-Dioxide		107–108	uv, nmr, mass spectrum	125, 126
2,7-Me_2-1,1-dioxide (*cis*)		53		127
2,7-Me_2-1,1-dioxide (*trans*)		83		127
4,5-Ph_2		142–143	nmr	65
4,5-Ph_2-1-oxide		162	ir, nmr	65
4,5-Ph_2-1,1-dioxide		190	ir, uv, nmr	65
4,5-Dihydrothiepin 				
None				182
2,3,6,7-Tetrahydrothiepin 				
1,1-Dioxide			not isolated	151
3,3,6,6-Me_4; 4,5-enediol sulfite		86–88	ir, uv, nmr, mass spectra	85

	2,3,4,7-Tetrahydrothiepin 		
1,1-Dioxide		not isolated	151
3,4-Br_2-1,1-dioxide	128–129		126
2,7-Me_2-4-oxo-1,1-dioxide (*trans*)			127
3,3,5,6-Me_4-4-oxo ?		ir, uv	83
	Thiepane 		
None	(171–172)	n_D^{20} 1.5138, d_4^{20} 0.9876	150
	(96–97/139)	n_D^{20} 1.5110	68
	(171.5–172.5)	n_D^{20} 1.5150, d_4^{20} 0.9874	147
	(173–174)	n_D^{20} 1.5137, d_4^{20} 0.9883, ir	133
	(174/750)	n_D^{20} 1.5125, ir	108
	(167/755)	n_D^{19} 1.5134, nmr	58
	(169–171/747)	n_D^{18} 1.5044, d_4^{18} 0.9743	82
	(170–171)	n_D^{20} 1.5134	54
		uv	149
Methiodide salt	161–162		68
	147		33
	137.5–138.5		82, 143
	141.5–142		108
$HgCl_2$ complex	152–153		144
	149		82
Me chloroplatinate salt	193		33
Et chloroplatinate salt			21
1,1-Dioxide	70.5–71, 70–71		82, 126
1-Oxide	(144–145/14)		54

TABLE X-3. (*contd.*)

Substituents	Yield (%)	mp, °C, or (bp, °C/torr)	Other data	Refs.
tert-Bu hexachloro-antimonate salt		80–81		97
i-Pr hexachloro-antimonate salt		139–141		97
2-AcO		(65/0.08)	n_D^{24} 1.4974, n_D^{15} 1.5006, ir, nmr	58
2-Cl			not isolated	192
2-Cl-1,1-dioxide		57.5–58		193
2-OH		Variable, 100–118	ir, nmr	58
2-Me		(94–96/52)	n_D^{26} 1.5007	192
2-Ph		(100–101/0.35)	n_D^{25} 1.5748	192
2,7-$(ClCH_2)_2$		(92–94/0.1)	n_D^{20} 1.5432, nmr	106
2,7-$(ClCH_2)_2$-1,1-dioxide		127–129	nmr	106
2,2,3,3,4,4,5,5,6,6,7,7-F_{12}			mass spectrum	96
3-OH			ir	26
3-OH-4-Me (*cis* and *trans*)				26
3-OH-4-isopropyl (*cis* and *trans*)				26
3,6-Cl_2		(65–80/0.25)	n_D^{20} 1.5504	67
4-AcO		(80–82/2.1)	n_D^{25} 1.4998	151
4-AcO-1,1-dioxide		75–76 (164–168/0.4)		151
4,5-$(OH)_2$-3,3,6,6-Me_4		179–180	ir, nmr	84, 86
4,5-$(OH)_2$-4,5-diallyl-3,3,6,6-Me_4		82.5–83	ir	83
4,5-$(OH)_2$-3,3,4,6,6-Me_5		157–158		83
4,5-Dimethylene-3,3,6,6-Me_4		(78/1.9)	n_D^{20} 1.5135, ir, uv, nmr	87

4-OH	(75–77/1)	n_D^{24} 1.5379, d_4^{20} 1.1306, ir, mass spectrum	26, 98, 151
4-OH-1,1-dioxide	85–86		151
4-OH-3-Me (*cis* and *trans*)			26
4-OH-3-isopropyl (*cis* and *trans*)			26
4-OH-3,3,6,6-Me_4	(74–76/0.8)	ir	86
4-OH-3,3,4,6,6-Me_5	47–48.5	ir	86
4-OH-4-*n*-butyl	(128–130/3)	n_D^{20} 1.5130	204
4-OH-4-isopropyl-3,3,6,6-Me_4	60.5–62	ir	86
4-OH-4-*tert*-butyl-3,3,6,6-Me_4	(164–166/14)	ir, nmr	83, 86
	2-Thiepanone (ring positions 1 S, 2–7; C=O at 2)		
None	(74–75/1)	n_D^{26} 1.5257, ir, nmr	153
(*R*)-(−)-4-Me	(76/1.1)	n_D^{25} 1.5153, ir, uv, nmr	154
(*R*)-(+)-5-Me	(74/1.0)	n_D^{25} 1.5135, ir, uv, nmr	154
racemic-5-Me	(75/1.0)	n_D^{25} 1.5129, ir, uv, nmr	154
	3-Thiepanone (ring positions 1 S, 2–7; C=O at 3)		
None	20.5–21 (54/0.55)	n_D^{20} 1.5285	108
Oxime	96–97		108
Benzoylated oxime	91–92		108
Semicarbazone	189.5–190		108
2-? or 4-? Carbethoxy	(102/0.9)	n_D^{20} 1.5177	108

TABLE X-3.—(*contd.*)

Substituents	Yield (%)	mp, °C, or (bp, °C/torr)	Other data	Refs.
		4-Thiepanone 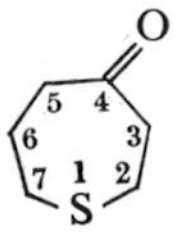		
None		(72–75/1.5)	n_D^{23} 1.5299, d_4^{20} 1.1351, ir, mass spectrum	98, 109, 151
p-Carbethoxyphenylhydrazone		139–140.5		2
2,4-Dinitrophenylhydrazone		189.5–191		151
Phenylhydrazone		101–103		2
p-Tolylhydrazone		123–125		2
1,1-Dioxide		121.5–123		151
p-Carbethoxyphenylhydrazone		165–166		2
2,4-Dinitrophenylhydrazone		218–220		151
1-Oxide		75.5–76	ir	110
3,5-Dibromo-1,1-dioxide		156.7–157.7		151
3,3,6,6-Me_4		47.5–49	nmr	88
5-AcO-3,3,6,6-Me_4		127–128	ir, nmr	88
5-Carbethoxymethylene-3,3,6,6-Me_4		(98–128/0.1)	ir, nmr	83
5-Formyl-3,3,6,6-Me_4		97–99	ir, uv, nmr	89
5-Carboxymethylene-3,3,6,6-Me_4		184–186	ir	83
5-Diazo-3,3,6,6-Me_4			not isolated	83

5-Hydroxy-3,3,6,6-Me_4	80–82	ir, nmr	88
p-Toluenesulfonate	125–126	ir, nmr	86
5-Hydroxy-5-*tert*-butyl-3,3,6,6-Me_4	155–156	ir, nmr	83
5-Hydroxy-3,3,5,6,6-Me_5	80–81	ir, nmr	83
5-Keto-3,3,6,6-Me_4	(68/0.35)	n_D^{20} 1.4954, uv, nmr	88, 202
Monooxime	124–125	nmr	88
Monohydrazone	45.5–48		88
Mono-*p*-toluenesulfonylhydrazone	150–151	nmr	88
Monosemicarbazone	234–235		88
5-Methoxymethylene-3,3,6,6-Me_4	109–110°	ir, uv, nmr	89
5-Methylene-3,3,6,6-Me_4	40.5–41(60/0.5)	ir, uv, nmr	88
6-Deoxy-6-mercapto-D-galactoseptanose			
1-Hydroxy-2,3,4,5-tetra-*O*-methyl	($120–130/10^{-4}$)	$[\alpha]_D^{22}$ −112° (c 0.3, H_2O); n_D^{19} 1.4996, ir, nmr; mixture of anomers	56, 57
1-Ethoxy-2,3,4,5-tetra-*O*-methyl	($92–93/10^{-4}$)	$[\alpha]_D^{24}$ −117° (c 0.8, $CHCl_3$), n_D^{20} 1.4820, nmr; mixture of anomers	56, 57
1,2,3,4,5-Penta-*O*-acetyl			
α-Anomer	154–155	$[\alpha]_D^{25}$ −2.8° (c 1.52, $CHCl_3$)	200
β-Anomer	148–149	$[\alpha]_D^{25}$ −233° (c 1.54, $CHCl_3$)	200
1-Chloro-2,3,4,5-tetra-*O*-acetyl α-Anomer	143–144	$[\alpha]_D^{25}$ −48.9° (c 1.63, $CHCl_3$)	200

TABLE X-3.—(*contd.*)

Substituents	Yield (%)	mp, °C, or (bp, °C/torr)	Other data	Refs.
1-Methoxy-2,3,4,5-tetra-*O*-acetyl				
α-Anomer		Glass	$[\alpha]_D^{25}$ +45° (c 1.46, $CHCl_3$)	200
β-Anomer		99–100	$[\alpha]_D^{25}$ −211° (c 1.49, $CHCl_3$)	200
1-Methoxy-2,3,4,5-tetrahydroxy		90–91	$[\alpha]_D^{25}$ −235° (*c* 1.09, CH_3OH)	200
1-Methoxy-2,3,4,5-tetra-*O*-*p*-nitrobenzoyl		236–237	$[\alpha]_D^{25}$ −120° (c 1.08, $CHCl_3$)	200

Dithiepins and Derivatives

1,2-Dithiepanes (S—S ring, positions 1–7)

Substituents	Yield (%)	mp, °C, or (bp, °C/torr)	Other data	Refs.
None		(70/10)	nearly colorless, uv	159
		(57–59/2.5)		165
	5	(40–41/1.5)	uv, esr	179
	13	(57–60/5)	mol wt 137	1
	80	(82/14; 41/2)	colorless, n_D^{25} 1.570 mol wt 133.8	169
			−Δ*H*, polym, 2.5 kcal/mole	60
	60		n_D^{25} 1.5700, uv	73
	6–54	(55–60/1.7)	n_D^{25} 1.5690–1.5710, uv, ir	71
			$E_{1/2}$ −2.01 V	72

1,1-Dioxide	17	~25	uv, ir, +test for $-SO_2S-$	71
			$E_{1/2}$ −0.85 V	72
1,1,2,2-Tetraoxide	10	159–160	ir, nmr	71
			$E_{1/2}$ −0.93 V	72
3-CO_2Et	42	(187.5–189/0.9)		203
3-$(CH_2)_2CO_2H$	7	79–81, 81–86	mol wt 200	48
			uv	18
3,7-$(CO_2H)_2$ (+ form)		162–166	impure, $[\alpha]_D^{25}$ +220° (C_2H_5OH)	176
racemic	15–34	171–173	cryst data; uv, ir	76, 176, 177
			$E_{1/2}$ (pH 2.2) −0.92 V	136, 137, 177
			pK_a 3.73, 4.73	
meso	20–50	191–193	cryst data; uv, ir	76, 137, 174, 176, 177
		193–194	$E_{1/2}$ (pH 2.2) −0.93 V equiv mol wt	
S-4-$NH_3^+Cl^-$	72	229–230, sint 226	$[\alpha]_D^{26}$ −35° (c 0.7444, H_2O), ir, uv	201
S-4-$NHSO_2CH_3$	24–37	122.5–123.5	$[\alpha]_D^{25}$ −31° (c 3.989, $CHCl_3$), ir, uv	201
S-4-$NHSO_2C_6H_5$	43	105–105.5, sint 98	$[\alpha]_D^{25}$ −5° (c 3.3576, $CHCl_3$), ir, uv	201
5-$NH_3^+Cl^-$	82	228, sint 224	ir, uv, equiv wt 185	201
F_2 S–S F_2 F F F_2	49	91.5 (107.5/739)		22
3,3,4,4,5,5,6,6,7,7-Decafluoro			nmr, mass spectrum	96

TABLE X-3.—(*contd.*)

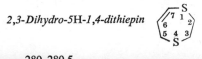

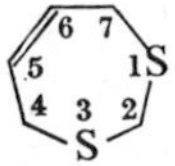

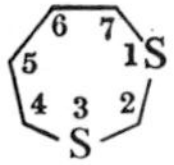

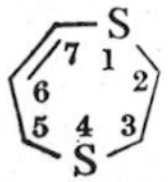

Substituents	Yield (%)	mp, °C, or (bp °C/torr)	Other data	Refs.
1,3-Dithiepins and Derivatives				
4,7-Dihydro-1,3-dithiepins				
None			variable temp nmr	78
$2,2\text{-}Me_2$			variable temp nmr	78
1,3-Dithiepanes				
None		Oil	uv	138, 141
1,1,3,3-Tetraoxide			pK_a 11.75	55
$2\text{-}CO_2Et$		(115–117/1.8)		124
$2\text{-}CO_2H$		85–86		124
$2,2\text{-}Me_2$			uv	141
2-Et		(94/7)	uv	138, 140, 141
2-Thiono		Oil		166
$HgCl_2$ complex				166
2-(Diethylimino), perchlorate salt		99–100	uv	102
1,4-Dithiepins and Derivatives				
2,3-Dihydro-5H-1,4-dithiepin				
1,1,4,4-Tetraoxide		280–280.5		80

6,7-Dihydro-5H-1,4-dithiepin			
2-Me	(73–75/0.5)	n_D^{26} 1.5881, ir, uv, nmr	28
2-(3,4-Dimethoxy)phenyl	101–103	nmr	121
6-Methylene	(76.5–77/0.06)	n_D^{20} 1.6260	163
1,4-Dithiepane			
None	47	mol wt 140	16, 80, 122, 191
	(221–222)		
1,1,4,4-Tetraoxide	281–282		191
	287–288		122
	282		16
	279–280 dec		80
6-Benzoate	75–76		80
1,1,4,4-Tetraoxide	229–230		80
6,6-Bis-(*p*-hydroxyphenyl)	186–187		94
6-Dicyanomethylene	167–168		94
6-Hydroxy	64.5–65.5		80
1,1,4,4-Tetraoxide	233–234		80
6-*p*-Nitrobenzoate	134–134.5		81
6-Keto	(85–90/0.3) 13.5–14	n_D^{25} 1.5925, ir, nmr	93, 94, 114
	(75/0.1)		
Semicarbazone	217–218		114
Ethylene ketal	101		93, 94
Ethylene thioketal	140–142		93, 94
Hydantoin	294–294.5		94
6-Keto-1,4-dioxide	177–181 dec	ir	93, 94
6-Keto-1,1,4,4-	279–282 dec	ir	93, 94
Tetraoxide	252–253 dec		80
Phenylhydrazone	235–236 dec		80

TABLE X-3.—(*contd.*)

Substituents	Yield (%)	mp, °C, or (bp, °C/torr)	Other data	Refs.
6-Keto-5-carbethoxy		(149–150/12)		114
6-Keto-5-trifluoroacetyl		66–67		198
6-Keto-5,7-dibenzylidene		118–119	uv	93, 94
6-Keto-5-*p*-dimethyl-aminobenzylidene		129.5–130.5	uv	93, 94
Spiro-1,4-dithiepanes				
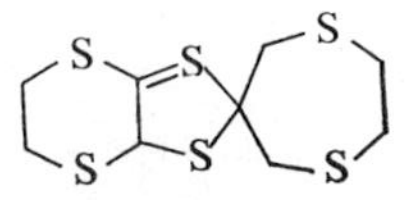		182–183°		12
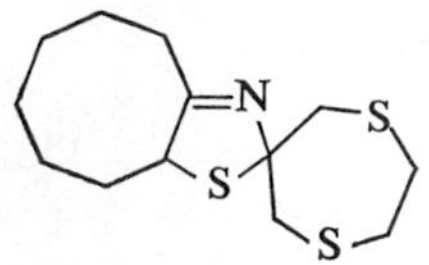		100		12
Picrate		152–153		12
Trithiepins and Derivatives				
1,2,3-Trithiepins		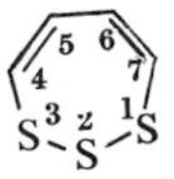		
4,7-H_2			nmr	100
4,5,6,7-H_4	<5		nmr	100
4,7-Me_2-4,5,6,7-H_4	25	(54/0.5)	n_D^{25} 1.5639, impure	51

Substituents	Yield, %	mp, °C (bp, °C/mm)	Properties	Ref.
3,6-$(C_6H_5NCH_3)_2$-4,7-$(CO_2CH_3)_2$	83	142	mol wt 504, slightly to moderately sol in ROH, ir	25
3,6-$(C_6H_5NC_2H_5)_2$-4,7-$(CO_2CH_3)_2$	78	98	mol wt 517	25
3,6-$(C_6H_5NCH_3)_2$-4,7-$(CO_2C_2H_5)_2$	60	122	mol wt 532	25
3,6-$(o\text{-}CH_3C_6H_4NCH_3)_2$-4,7-$(CO_2CH_3)_2$	69	167	mol wt 499	25
3,6-$(m\text{-}CH_3C_6H_4NCH_3)_2$-4,7-$(CO_2CH_3)_2$	60	132–133	mol wt 504	25
3,6-$(p\text{-}CH_3C_6H_4NCH_3)_2$-4,7-$(CO_2CH_3)_2$	73	147	mol wt 510	25
3,6-$(o\text{-}CH_3OC_6H_4NCH_3)_2$-4,7-$(CO_2CH_3)_2$	43	149	mol wt 493–615	25
3,6-$(p\text{-}CH_3OC_6H_4NCH_3)_2$-4,7-$(CO_2CH_3)_2$	62	134	mol wt 550	25
3,6-$(o\text{-}C_2H_5OC_6H_4NCH_3)_2$-4,7-$(CO_2CH_3)_2$	51	142	mol wt 566	25
3,6-$(p\text{-}C_2H_5OC_6H_4NCH_3)_2$-4,7-$(CO_2CH_3)_2$	30	132	mol wt 579	25
3,4,6,7-H_4- (?)		96	sol benzene	160
(see text for		74–75	insol alc, H_2O, sol $C_6H_5NO_2$	79
discussion)			n_D^{25} 1.5746, mol wt 144	199
	55	(61–63/0.2)	nmr, Raman, ir, uv, mass spectrum, n_D^{25} 1.6424; mol wt	73

TABLE X-3.—(*contd.*)

Substituents	Yield (%)	mp, °C, or (bp, °C/torr)	Other data	Refs.
Methiodide	72	131–132°	mass spectrum, uv, Raman, ir, nmr	73
3,4,6,7-H_4-5-oxide	66	95–96	nmr, ir, Raman, uv, mass spectrum	73
Perfluoro-3,4,6,7-H_4	24	(40–42/20) (71–72/100)	^{19}F nmr	105
		1,3,5-Trithiepins		
6,7-$(CN)_2$		112–114, sint 109		95, 171

Tetrathiepins and Derivatives

Substituents	Yield (%)	mp, °C, or (bp, °C/torr)	Other data	Refs.
		1,2,4,6-Tetrathiepanes		
None		83–84 (bp, dec)	insol H_2O, moderately to very sol in org solv, mol wt	19
			low temp nmr	129
		79	ir, nmr, mass spectrum	131
			polarography	10

Pentathiepins and Derivatives

Substituents	Yield (%)	mp (bp/mm) °C	Properties	Ref.
1,2,3,5,6-Pentathiepanes			H_2C_4 (S5–S6) $_7CH_2$, S3–S2–S1	
None	Naturally occurring, synthetic	60–61, 54–56	mass spectrum, nmr, ir, mol wt 188	130, 131, 188
			low temp nmr	129
			nearly insol H_2O, sparingly in org solv	195
			X-ray cryst data	135
			polarography	10
4,7-(=$NNHC_6H_4Me$-*p*)$_2$		103–104		53
4,4,7,7-Et_4		(157–160/0.5)	n_D^{21} 1.610	15
4,4-$(CH_2)_5$-7,7-$(CH_2)_5$-		85–87	very sol in org solv, mol wt 316	118
	66	84–85		24
4,4-$(CH_2)_6$-7,7-$(CH_2)_6$-		127–128	sol in org solv, mol wt 336–363, ir, cryst properties	77
4,4,7,7-$(PhCH_2)_4$	49	130–131.5	mol wt 542–550	119

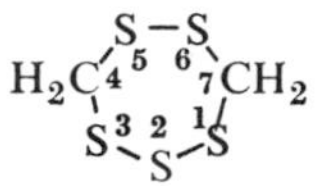

TABLE X-3.—(*contd.*)

Substituents	Yield (%)	mp, °C, or (bp, °C/torr)	Other data	Refs.
Hexathiepins and Derivatives				
1,2,3,4,5,6-Hexathiepane 				
None	Naturally occurring, synthetic	79 (sinter)	ir, nmr, mass spectrum polarography	131 10
1,2,3,4,5,6,7-Heptathiepane				
(*Cycloheptasulfur*) 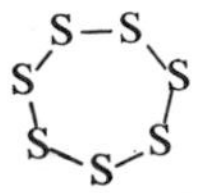				
None	23	39	mol wt 218–224, X-ray mass spectrum	167 205

Acknowledgments

The authors wish to record appreciation for generous support of research in organic sulfur chemistry related to that reviewed here to the U.S. Army Medical Research and Development Command, Department of the Army (Research Contracts DA-49-193-MD-2030 and DADA-17-69-C-9128; L. F.), to the U.S. Public Health Service (Research Grant No. AM11685 from the National Institute for Arthritis and Metabolic Diseases; L. F.), and to the Petroleum Research Fund administered by the American Chemical Society (D. L. T.).

G. References

1. J. G. Affleck and G. Dougherty, *J. Org. Chem.*, **15,** 865 (1950).
2. L. A. Aksanova, N. F. Kucherova, and V. A. Zagorevskii, *Zh. Org. Khim.*, **1,** 2215 (1965); *Chem. Abstr.*, **64,** 11189 (1966).
3. H. L. Ammon, P. H. Watts, Jr., J. M. Stewart, and W. L. Mock, *J. Amer. Chem. Soc.*, **90,** 4501 (1968).
4. F. A. L. Anet, C. H. Bradley, M. A. Brown, W. L. Mock, and J. H. McCausland, *J. Amer. Chem. Soc.*, **91,** 7782 (1969).
5. F. Arndt, *Chem. Ber.*, **89,** 730 (1956).
6. F. Arndt, P. Nachtwey, and J. Pusch, *Chem. Ber.*, **58B,** 1633 (1925).
7. F. Arndt, R. Schwarz, C. Martius, and E. Aron, *Rev. Fac. Sci. Univ. Istanbul*, **A13,** 57 (1948); *Chem. Abstr.*, **42,** 4176 (1948).
8. F. Arndt and G. Traverso, *Chem. Ber.*, **89,** 124 (1956).
9. F. Arndt and W. Walter, *Chem. Ber.*, **94,** 1757 (1961).
10. Y. Asahi, K. Terada, and M. Ishio, *Rev. Polarogr.*, **14,** 382 (1967); *Chem. Abstr.*, **71,** 21598 (1969).
11. F. Asinger, H.-W. Becker, W. Schäfer, and A. Saus, *Monatsh. Chem.*, **97,** 301 (1966).
12. F. Asinger, W. Schäfer, M. Baumann, and H. Römgens, *Justus Liebigs Ann. Chem.*, **672,** 103 (1964).
13. F. Asinger, W. Schäfer, and H.-W. Becker, *Angew. Chem.*, **77,** 41 (1965).
14. F. Asinger, W. Schäfer, K. Halcour, A. Saus, and H. Triem, *Angew. Chem.*, **75,** 1050 (1963).
15. F. Asinger, M. Thiel, and G. Lipfert, *Justus Liebigs Ann. Chem.*, **627,** 195 (1959).
16. W. Autenrieth and K. Wolff, *Chem. Ber.*, **32,** 1375 (1899).
17. H. J. Backer and N. Evenhuis, *Rec. Trav. Chim. Pays-Bas*, **56,** 174 (1937).
18. J. A. Barltrop, P. M. Hayes, and M. Calvin, *J. Amer. Chem. Soc.*, **76,** 4348 (1954).
19. E. Baumann, *Chem. Ber.*, **23,** 1869 (1890).
20. G. M. Bennett, F. Heathcoat, and A. N. Mosses, *J. Chem. Soc.*, **1929,** 2567.
21. G. M. Bennett and E. G. Turner, *J. Chem. Soc.*, **1938,** 813.
22. C. J. Benning, U.S. Patent 2,968,659 (1961); *Chem. Abstr.*, **55,** 13460 (1961).
23. S. F. Birch, T. V. Cullum, R. A. Dean, and R. L. Denyer, *Ind. Eng. Chem.*, **47,** 240 (1955).
24. F. O. Bobbio and P. A. Bobbio, *Chem. Ber.*, **98,** 998 (1965).
25. F. Boberg, *Justus Liebigs Ann. Chem.*, **683,** 132 (1965).
26. R. Borsdorf, H. Kasper, and H.-D. Repp, *Angew. Chem., Int. Ed. Engl.*, **6,** 872 (1967).
27. A. A. Bothner-By and G. Traverso, *Chem. Ber.*, **90,** 453 (1957).

28. A. T. Bottini and E. F. Böttner, *J. Org. Chem.*, **31,** 586 (1966).
29. K. H. Bourne, P. D. Holmes, and R. C. Pitkethly, *Proc. Int. Congr. Catalysis 3rd, Amsterdam*, 1964, **2,** 1400 (Pub. 1965); *Chem. Abstr.*, **63,** 15609 (1965).
30. H. L. Bradlow, C. A. Vanderwerf, and J. Kleinberg, *J. Chem. Educ.*, **24,** 433 (1947).
31. K. Brass, R. Pfluger, and K. Honsberg, *Ber.*, **69B,** 80 (1936).
32. J. v. Braun and K. Weissbach, *Chem. Ber.*, **62,** 2416 (1929).
33. J. v. Braun, *Chem. Ber.*, **43,** 3220 (1910).
34. R. A. Braun, *J. Org. Chem.*, **28,** 1383 (1963).
35. H. G. Bremer, *Pflanzenkrankh. Pflanzenschutz*, **70,** 321 (1963); *Chem. Abstr.*, **59,** 12101 (1963).
36. D. S. Breslow and H. Skolnik, in *The Chemistry of Heterocyclic Compounds*, Vol. 21, Part 1, A. Weissberger, Ed., Wiley-Interscience, New York, 1966, pp. 416–417.
37. Ref. 36, pp. 68–75.
38. Ref. 36, pp. 313–347.
39. Ref. 36, pp. 323 ff.
40. Ref. 36, pp. 410 ff.
41. D. S. Breslow and H. Skolnik, in *The Chemistry of Heterocyclic Compounds*, Vol. 21, Part 2, A. Weissberger, Ed., Wiley-Interscience, New York, 1966, pp. 952–967.
42. Ref. 41, pp. 626–632.
43. Ref. 41, pp. 689–692.
44. Ref. 41, p. 956.
45. R. Breslow, *Organic Reaction Mechanisms*, Benjamin, New York, 1965, p. 29.
46. R. Breslow and E. Mohacsi, *J. Amer. Chem. Soc.*, **85,** 431 (1963).
47. R. H. Brown and S. Meyerson, *Ind. Eng. Chem.*, **44,** 2620 (1952).
48. M. W. Bullock, J. A. Brockman, Jr., E. L. Patterson, J. V. Pierce, M. H. Von Saltza, F. Sanders, and E. L. R. Stokstad, *J. Amer. Chem. Soc.*, **76,** 1828 (1954).
49. M. W. Bullock, J. A. Brockman, Jr., E. L. Patterson, J. V. Pierce, and E. L. R. Stokstad, *J. Amer. Chem. Soc.*, **74,** 1868 (1952).
50. M. W. Bullock, J. A. Brockman, Jr., E. L. Patterson, J. V. Pierce, and E. L. R. Stokstad, *J. Amer. Chem. Soc.*, **74,** 3455 (1952).
51. T. L. Cairns, G. L. Evans, A. W. Larchar, and B. C. McKusick, *J. Amer. Chem. Soc.*, **74,** 3982 (1952).
52. T. L. Cairns, G. L. Evans, A. W. Larchar, and B. C. McKusick, *J. Amer. Chem. Soc.*, **74,** 3984 (1952).
53. L. Cambi, G. Bargigia, L. Colombo, and E. Paglia Dubini, *Gazz. Chim. Ital.*, **99,** 780 (1969); *Chem. Abstr.*, **71,** 124394 (1969).
54. A. Cerniani, G. Modena, and P. E. Todesco, *Gazz. Chim. Ital.*, **90,** 382 (1960); *Chem. Abstr.*, **55,** 12421 (1961).
55. E. J. Corey, H. König, and T. H. Lowry, *Tetrahedron Lett.*, **1962,** 515.
56. J. M. Cox and L. N. Owen, *Chem. Commun.*, **1965,** 513.
57. J. M. Cox and L. N. Owen, *J. Chem. Soc., C*, **1967,** 1121.
58. J. M. Cox and L. N. Owen, *J. Chem. Soc., C*, **1967,** 1130.
59. F. S. Dainton, T. R. E. Devlin, and P. A. Small, *Trans. Faraday Soc.*, **51,** 1710 (1955).
60. F. S. Dainton, K. J. Ivin, and D. A. G. Walmsley, *Trans. Faraday Soc.*, **56,** 1784 (1960).
61. F. O. Davis, U.S. Patent 2,657,198 (1953); *Chem. Abstr.*, **48,** 4247 (1954).
62. F. O. Davis, U.S. Patent 2,715,635 (1955); *Chem. Abstr.*, **50,** 1353 (1956).
63. F. O. Davis and E. M. Fettes, *J. Amer. Chem. Soc.*, **70,** 2611 (1948).
64. M. J. S. Dewar and N. Trinajstic, *J. Amer. Chem. Soc.*, **92,** 1453 (1970).
65. R. M. Dodson and J. P. Nelson, *J. Chem. Soc., D*, **1969,** 1159.

66. W. v. E. Doering and W. R. Roth, *Tetrahedron*, **19,** 715 (1963).
67. Dunlop Rubber Co., Ltd., French Patent 1,427,429; *Chem. Abstr.*, **65,** 12180 (1966).
68. E. Dyer and D. W. Osborne, *J. Polym. Sci.*, **47,** 349 (1960).
69. B. Meyer, Ed., *Elemental Sulfur—Chemistry and Physics*, Wiley-Interscience, New York, 1965; see especially pp. 79, 81, 89, 126–159, 327.
70. F. Fehér and W. Becher, *Z. Naturforsch*, **206,** 1125 (1965).
71. L. Field and R. B. Barbee, *J. Org. Chem.*, **34,** 36 (1969).
72. L. Field and R. B. Barbee, *J. Org. Chem.*, **34,** 1792 (1969).
73. L. Field and C. H. Foster, *J. Org. Chem.*, **35,** 749 (1970).
74. L. Field, T. F. Parsons, and D. E. Pearson, *J. Org. Chem.*, **31,** 3550 (1966).
75. O. Foss, in *Organic Sulfur Compounds*, Vol. 1, N. Kharasch, Ed., Pergamon Press, New York, 1961, pp. 75–77.
76. O. Foss and L. Schotte, *Acta Chem. Scand.*, **11,** 1424 (1957).
77. A. Fredga, *Acta Chem. Scand.*, **12,** 891 (1958).
78. H. Friebolin, R. Mecke, S. Kabuss, and A. Lüttringhaus, *Tetrahedron Lett.*, **1964,** 1929.
79. E. Fromm and H. Jörg, *Chem. Ber.*, **58,** 304 (1925).
80. R. C. Fuson and A. J. Speziale, *J. Amer. Chem. Soc.*, **71,** 823 (1949).
81. R. C. Fuson and A. J. Speziale, *J. Amer. Chem. Soc.*, **71,** 1582 (1949).
82. E. Grishkevich-Trokhimovskii, *J. Russ. Phys. Chem. Soc.*, **48,** 944 (1916); *Chem. Abstr.*, **11,** 786 (1917).
83. A. de Groot, J. A. Boerma, J. de Valk, and H. Wynberg, *J. Org. Chem.*, **33,** 4025 (1968).
84. A. de Groot, J. A. Boerma, and H. Wynberg, *Tetrahedron Lett.*, **1968,** 2365.
85. A. de Groot, J. A. Boerma, and H. Wynberg, *Chem. Commun.*, **1968,** 347.
86. A. de Groot, J. A. Boerma, and H. Wynberg, *Rec. Trav. Chim. Pays-Bas*, **88,** 994 (1969).
87. A. de Groot, B. Evenhuis, and H. Wynberg, *J. Org. Chem.*, **33,** 2214 (1968).
88. A. de Groot and H. Wynberg, *J. Org. Chem.*, **31,** 3954 (1966).
89. A. de Groot and H. Wynberg, *J. Org. Chem.*, **33,** 3337 (1968).
90. K. D. Gundermann, *Angew. Chem., Int. Ed. Engl.*, **2,** 674 (1963).
91. K. Hafner, *Angew. Chem., Int. Ed. Engl.*, **3,** 165 (1964).
92. H. D. Hartough, in *The Chemistry of Heterocyclic Compounds*, Vol. 3, A. Weissberger, Ed., Wiley-Interscience, New York, 1952, p. 78.
93. E. G. Howard, Jr., U.S. Patent 2,965,650; *Chem. Abstr.*, **55,** 14492 (1961).
94. E. G. Howard and R. V. Lindsey, Jr., *J. Amer. Chem. Soc.*, **82,** 158 (1960).
95. E. Jacobi, D. Erdmann, S. Lust, and W. Wirtz, German Patent 1,134,552 (1962); *Chem. Abstr.*, **57,** 12961 (1962).
96. R. James and D. G. Rowsell, *Chem. Commun.*, **1969,** 1274.
97. C. R. Johnson and M. P. Jones, *J. Org. Chem.*, **32,** 2014 (1967).
98. P. Y. Johnson and G. A. Berchtold, *J. Org. Chem.*, **35,** 584 (1970).
99. M. J. Jorgenson, *J. Org. Chem.*, **27,** 3224 (1962).
100. S. Kabuss, A. Lüttringhaus, H. Friebolin, and R. Mecke, *Z. Naturforsch*, **21b,** 320 (1966). See also H. Friebolin and S. Kabuss, *Nucl. Magnetic Resonance Chem. Proc., Symp. Cagliari, Italy*, 125 (1964); *Chem. Abstr.*, **66,** 3530 (1967).
101. F. Kaplan and G. K. Meloy, *J. Amer. Chem. Soc.*, **88,** 950 (1966).
102. K. C. Kennard and J. A. Van Allan, *J. Org. Chem.*, **24,** 470 (1959).
103. C. G. Krespan, U.S. Patent 3,088,935 (1963); *Chem. Abstr.*, **59,** 10096 (1963).
104. C. G. Krespan, U.S. Patent 3,099,688 (1963); *Chem. Abstr.*, **60,** 1597 (1964).
105. C. G. Krespan and W. R Brasen, *J. Org. Chem.*, **27,** 3995 (1962).
106. F. Lautenschlaeger, *J. Org. Chem.*, **33,** 2620 (1968).

107. N. J. Leonard, T. L. Brown, and T. W. Milligan, *J. Amer. Chem. Soc.*, **81**, 504 (1959).
108. N. J. Leonard and J. Figueras, Jr., *J. Amer. Chem. Soc.*, 74, 917 (1952).
109. N. J. Leonard, T. W. Milligan, and T. L. Brown, *J. Amer. Chem. Soc.*, **82**, 4075 (1960).
110. N. J. Leonard and W. L. Rippie, *J. Org. Chem.*, **28**, 1957 (1963).
111. H. C. Longuet-Higgins, *Trans. Faraday Soc.*, **45**, 173 (1949).
112. J. D. Loudon and D. K. V. Steel, *J. Chem. Soc.*, **1954**, 1163.
113. S. Lust, O. W. Mueller, A. v. Schoor, and E. Jacobi, German Patent 1,095,582 (1960); *Chem. Abstr.*, **56**, 12,038 (1962).
114. A. Lüttringhaus and H. Prinzbach, *Justus Liebigs Ann. Chem.*, **624**, 79 (1959).
115. C. F. Mabery and W. O. Quale, *Amer. Chem. J.*, **35**, 404 (1906).
116. H. Mackle and P. A. G. O'Hare, *Tetrahedron*, **19**, 961 (1963).
117. S. B. Maerov, *Dissertation Abstr.*, **14**, 765 (1954).
118. B. Magnusson, *Acta Chem. Scand.*, **13**, 1031 (1959).
119 B. Magnusson, *Acta Chem. Scand.*, **13**, 1715 (1959).
120. K. K. Maheshwari and G. A. Berchtold, *Chem. Commun.*, **1969**, 13.
121. J. L. Massingill, Jr., M. G. Reinecke, and J. E. Hodgkins, *J. Org. Chem.*, **35**, 823 (1970).
122. J. R. Meadow and E. E. Reid. *J. Amer. Chem. Soc.*, **56**, 2177 (1934).
123. B. Milligan and J. M. Swan, *J. Chem. Soc.*, **1965**, 2901.
124. I. Minamida, Y. Ikeda, K. Uneyama, W. Tagaki, and S. Oae, *Tetrahedron*, **24**, 5293 (1968).
125. W. L. Mock, Abstracts of the 153rd Meeting of the American Chemical Society, Spring, 1967, Paper 157.
126. W. L. Mock, *J. Amer. Chem. Soc.*, **89**, 1281 (1967).
127. W. L. Mock, *J. Amer. Chem. Soc.*, **91**, 5682 (1969).
128. G. Mohr and K. G. Schmidt, German Patent 1,159,452 (1963); *Chem. Abstr.*, **60**, 9287 (1964).
129. R. M. Moriarty, N. Ishibe, M. Kayser, K. C. Ramey, and H. J. Gisler, Jr., *Tetrahedron Lett.*, **1969**, 4883.
130. K. Morita and S. Kobayashi, *Tetrahedron Lett.*, **1966**, 573.
131. K. Morita and S. Kobayashi, *Chem. Pharm. Bull.*, **15**, 988 (1967).
132. K. Morita and S. Kobayashi, Japanese Pat. 20,319 (1967); *Chem. Abstr.*, **69**, 3354 (1968).
133. A. Muller, E. Funder-Fritzsche, W. Konar, and E. Rintersbacher-Wlasak, *Monatsh. Chem.*, **84**, 1206 (1953).
134. M. S. Newman and A. Arkell, *J. Org. Chem.*, **24**, 385 (1959).
135. M. Nishikawa, K. Kamiya, S. Kobayashi, K. Morita, and Y. Tomiie, *Chem. Pharm. Bull.* **15**, 756 (1967).
136. B. Nygård, *Ark. Kemi*, **28**, 75 (1968).
137. B. Nygård and L. Schotte, *Acta Chem. Scand.*, **10**, 469 (1956).
138. S. Oae, W. Tagaki, and A. Ohno, *Proc. Int. Symp. Mol. Struct. Spectry., Tokyo* 1962; *Chem. Abstr.*, **61**, 9056 (1964).
139. S. Oae, W. Tagaki, and A. Ohno, *Tetrahedron*, **20**, 417 (1964).
140. S. Oae, W. Tagaki, and A. Ohno, *Tetrahedron*, **20**, 427 (1964).
141. S. Oae, W. Tagaki, and A. Ohno, *Tetrahedron*, **20**, 437 (1964).
142. S. Oae, W. Tagaki, K. Uneyama, and I. Minamida, *Tetrahedron*, **24**, 5283 (1968).
143. R. D. Obolentsev, V. G. Bukharov, and N. K. Faizullina, *Khim. Sera-i Azotorgan. Soedin. Soderzhasch. v Neft. i Nefteprod., Akad. Nauk SSSR, Bashkirsk. Filial*, **3**, 67 (1960); *Chem. Abstr.*, **57**, 7217 (1962).

144. R. D. Obolentsev, V. G. Bukharov, and N. K. Fairzullina, *Khim. Sera-i Azotorgan. Soedin. Soderzhashch. v Neft. i Nefteprod., Akad. Nauk SSSR, Bashkirsk. Filial*, **3**, 51 (1960); *Chem. Abstr.*, **57**, 7218 (1962).
145. R. D. Obolentsev, V. G. Bukharov, and M. M. Gerasimov, *Khim. Sera-i Azotorgan. Soedin. Soderzhashch. v Neft. i Nefteprod., Akad. Nauk SSSR, Bashkirsk. Filial*, **3**, 43 (1960); *Chem. Abstr.*, **57**, 7231 (1962).
146. R. D. Obolentsev and V. I. Dronov, *Dokl. Akad. Nauk SSSR*, **130**, 98 (1960); *Chem. Abstr.*, **54**, 9468 (1960).
147. R. D. Obolentsev and V. I. Dronov, *Khim. Sera-i Azotorgan. Soedin. Soderzhashch. v Neft. i Nefteprod., Akad. Nauk SSSR, Bashkirsk. Filial*, **3**, 271 (1960); *Chem. Abstr.*, **56**, 5933 (1962).
148. R. D. Obolentsev and V. I. Dronov, *Khim. Sera-i Azotorgan. Soedin. Soderzhasch. v Neft. i. Nefteprod., Akad. Nauk SSSR, Bashkirsk. Filial*, **4**, 151 (1961); *Chem. Abstr.*, **57**, 9817 (1962).
149. R. D. Obolentsev, N. S. Lyubopytova, and E. A. Makova, *Khim. Sera-i Azotorgan. Soedin. Soderzhashch. v Neft. i Nefteprod., Akad. Nauk SSSR, Bashkirsk. Filial*, **3**, 93 (1960); *Chem. Abstr.*, **56**, 4254 (1962).
150. R. D. Obolentsev, S. V. Netupskaya, L. K. Gladkova, V. G. Bukharov, and A. V. Mashkina, *Khim. Sera-i Azotorgan. Soedin. Soderzhashch. v Neft. i Nefteprod, Akad. Nauk SSSR, Bashkirsk. Filial, Materialy Vtoroi Sessii*, 87 (1956); *Chem. Abstr.*, **54**, 249 (1960).
151. C. G. Overberger and A. Katchman, *J. Amer. Chem. Soc.*, **78**, 1965 (1956).
152. C. G. Overberger and A. Lusi, *J. Amer. Chem. Soc.*, **81**, 506 (1959).
153. C. G. Overberger and J. Weise, *J. Polym. Sci., Part B.*, **2**, 329 (1964).
154. C. G. Overberger and J. K. Weise, *J. Amer. Chem. Soc.*, **90**, 3525 (1968).
155. C. G. Overberger and J. K. Weise, *J. Amer. Chem. Soc.*, **90**, 3533 (1968).
156. C. G. Overberger and J. K. Weise, *J. Amer. Chem. Soc.*, **90**, 3538 (1968).
157. L. A. Paquette and J. H. Barrett, *J. Amer. Chem. Soc.*, **88**, 1718 (1966).
158. A. M. Patterson, L. T. Capell, and D. F. Walker, *The Ring Index*, 2nd ed., American Chemical Society, Washington, D.C. (1960).
159. V. Ramakrishnan, S. D. Thompson, and S. P. McGlynn, *Photochem. Photobiol.*, **4**, 907 (1965).
160. P. C. Rây, *J. Chem. Soc.*, **125**, 1141 (1924).
161. E. E. Reid, *Organic Chemistry of Bivalent Sulfur*, Vol. III, Chemical Publishing Co., New York, 1960, p. 104.
162. Ref. 161, pp. 150–153.
163. H. Richter, K. Schulze, and M. Mühlstädt, *Z. Chem.*, **8**, 220 (1968).
164. T. G. Roberts and P. C. Teague, *J. Amer. Chem. Soc.*, **77**, 6258 (1955).
165. N. A. Rosenthal, U.S. Patent 3,284,466 (1966); *Chem. Abstr.*, **68**, 1254 (1968).
166. F. Runge, Z. El-Hewehi, H.-J. Renner, and E. Taeger, *J. Prakt. Chem.*, **11**, 284 (1960).
167. M. Schmidt, B. Block, H. D. Block, H. Köpf, and E. Wilhelm, *Angew. Chem., Int. Ed. Engl.*, **7**, 632 (1968).
168. A. Schöberl and H. Gräfje, *Angew. Chem.*, **69**, 713 (1957).
169. A. Schöberl and H. Gräfje, *Justus Liebigs Ann. Chem.*, **614**, 66 (1958).
170. V. Schomaker and L. Pauling, *J. Amer. Chem. Soc.*, **61**, 1769 (1939).
171. A. v. Schoor, E. Jacobi, S. Lust, and H. Flemming, German Patent 1,060,655 (1959); *Chem. Abstr.*, **55**, 7748 (1961).
172. A. v. Schoor, E. Jacobi, S. Lust, H. Flemming, and O. W. Mueller, U.S. Patent 3,000,780; *Chem. Abstr.*, **56**, 5162 (1962).
173. L. Schotte, *Acta Chem. Scand.*, **8**, 131 (1954).

174. L. Schotte, *Ark. Kemi*, **8**, 579 (1955–56).
175. L. Schotte, *Ark. Kemi*, **9**, 361 (1956).
176. L. Schotte, *Ark. Kemi*, **9**, 413 (1956).
177. L. Schotte, *Ark. Kemi*, **9**, 441 (1956).
178. H. E. Simmons, R. D. Vest, D. C. Blomstrom, J. R. Roland, and T. L. Cairns, *J. Amer. Chem. Soc.*, **84**, 4746 (1962).
179. E. E. Smissman and J. R. J. Sorenson, *J. Org. Chem.*, **30**, 4008 (1965).
180. G. Sosnovsky, *Tetrahedron*, **18**, 15 (1962).
181. E. L. Stogryn and S. J. Brois, *J. Org. Chem.*, **30**, 88 (1965).
182. E. L. Stogryn and S. J. Brois, *J. Amer. Chem. Soc.*, **89**, 605 (1967).
183. E. L. Stogryn, M. H. Gianni, and A. J. Passannante, *J. Org. Chem.*, **29**, 1275 (1964).
184. A. Streitwieser, Jr., *Molecular Orbital Theory for Organic Chemists*, John Wiley and Sons, New York, 1961, p. 280.
185. S. Sunner, *Acta Chem. Scand.*, **17**, 728 (1963).
186. H. H. Szmant and L. M. Alfonso, *J. Amer. Chem. Soc.*, **78**, 1064 (1956).
187. H. H. Szmant and J. Dixon, *J. Amer. Chem. Soc.*, **75**, 4354 (1953).
188. Takeda Chemical Industries, Ltd., French Patent 1,502,924 (1967); *Chem. Abstr.*, **69**, 9069 (1968).
189. Takeda Chemical Industries, Ltd., French Patent 1,498,739; *Chem. Abstr.*, **69**, 7233 (1968).
190. W. E. Truce, K. R. Hollister, L. B. Lindy, and J. E. Parr, *J. Org. Chem.*, **33**, 43 (1968).
191. N. B. Tucker and E. E. Reid, *J. Amer. Chem. Soc.*, **55**, 775 (1933).
192. D. L. Tuleen and R. H. Bennett, *J. Heterocycl. Chem.*, **6**, 115 (1969).
193. D. L. Tuleen and R. H. Bennett, METV, *J. Pure and Appl. Sci.*, **4**, 121 (1971).
194. E. Vogel, K.-H. Oh, and K. Gajek, *Justus Liebigs Ann. Chem.*, **644**, 172 (1961).
195. S. Wada, H. Nakatani, S. Fujinawa, H. Kimura, and M. Hagaya, *Eiyo To Shokuryo*, **20**, 355 (1968); *Chem. Abstr.*, **69**, 172 (1968).
196. S. Wada, H. Nakatani, and K. Morita, *J. Food Sci.*, **32**, 559 (1967); *Chem. Abstr.*, **68**, 183 (1968).
197. S. Wada, H. Nakatani, J. Toda, and M. Hagaya, *Eiyo To Shokuryo*, **20**, 360 (1968); *Chem. Abstr.*, **69**, 172 (1968).
198. H. A. Wagner, U.S. Patent 3, 158,619; *Chem. Abstr.*, **62**, 7784 (1965).
199. H. E. Westlake, Jr., M. G. Mayberry, M. H. Whitlock, J. R. West, and G. J. Haddad, *J. Amer. Chem. Soc.*, **68**, 748 (1946).
200. R. L. Whistler and C. S. Campbell, *J. Org. Chem.*, **31**, 816 (1966).
201. R. H. Wood, *The Synthesis of Amino-1,2-dithiepanes*, Ph.D. thesis, Rensselaer Polytechnic Institute, Troy, N.Y., December, 1962; available from University Microfilms, Ann Arbor, Michigan 48103, under Order No. 64-3721. This work has recently been published: H. F. Herbrandson and R. H. Wood, *J. Med. Chem.*, **12**, 617, 620 (1969).
202. H. Wynberg and A. de Groot, *Chem. Commun.*, **1965**, 171.
203. C-E. Yuan and M. N. Shchukina, *Zh. Obshch. Khim.*, **27**, 1103 (1957); *Chem. Abstr.*, **52**, 3681 (1958).
204. V. A. Zagorevskii and K. I. Lopatina, *Zh. Org. Khim.*, **1**, 366 (1965); *Chem. Abstr.*, **62**, 16190 (1965).
205. U.-I. Záhorszky, *Angew. Chem., Int. Ed. Engl.*, **7**, 633 (1968).
206. R. Zahradník, in *Advances in Heterocyclic Chemistry*, Vol. 5, A. R. Katritzky, Ed., Academic Press, New York, 1965, pp. 21–23.
207. Ref. 206, p. 31.
208. R. Zahradník and C. Párkányi, *Collect. Czech. Chem. Commun.*, **30**, 3016 (1965)

CHAPTER XI

Condensed Thiepins

VINCENT J. TRAYNELIS

Department of Chemistry, West Virginia University, Morgantown, West Virginia

A. Introduction

The chemistry of condensed thiepins and derivatives containing fused carbocyclic rings and/or selected fused heterocyclic rings as well as bridged thiepin compounds is reviewed in this chapter. The literature covered extends through 1968 although numerous more recent references are cited. Nomenclature follows the systematic approach (64) practiced in *Chemical Abstracts.*

The systems containing one sulfur atom are considered first and are organized in increasing number and complexity of condensed rings, that is, cyclopentathiepin, 1-benzothiepins, 2-benzothiepins, 3-benzothiepins, naphthothiepins, dibenzothiepins, etc. Next appear the furothiepins and thienothiepins which serve as a transition to compounds with two or more sulfur atoms in the seven-membered ring. Condensed dithiepins are followed by trithiepins, etc., again ordered by increasing number and complexity of condensed rings. Finally, the bridged systems, bicyclo and tricyclo, which contain sulfur as part of a seven-membered ring complete the chapter.

Initial synthetic interest in the fused thiepins arose from consideration of the manner in which these compounds fit into the concept of aromaticity. Molecular orbital calculations have been made on several model thiepin derivatives which are yet to be synthesized (65, 330). Most recently Dewar and Trinajstic (66) concluded on the basis of theoretical calculations that thiepin itself was antiaromatic and that 3-benzothiepin and thieno[3,4-*d*]-thiepin have resonance energies close to those of benzene and thiophene, respectively. These theoretical predictions are supported by experimental observations on a few thiepin derivatives and lead to the position that the thiepin ring should not be aromatic.

Another factor which has promoted interest in condensed thiepins is their applicability as model compounds in stereochemical and mechanistic studies. A more recent stimulus for synthetic work in this area can be traced to the broad physiological effects of certain fused thiepins on the central nervous system.

B. 2*H*-Cyclopenta[*b*]thiepins

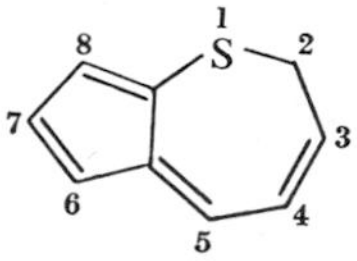

(R.I. 8049)

One brief report has appeared in the literature (158) which described the conversion of 2-(4-bromobutyl)cyclopentanone (**1**) on treatment with alcoholic potassium hydroxide saturated with hydrogen sulfide into 3,4,5,6,7,8-hexahydro-2*H*-cyclopenta[*b*]thiepin (**2**) (Eq. 1).

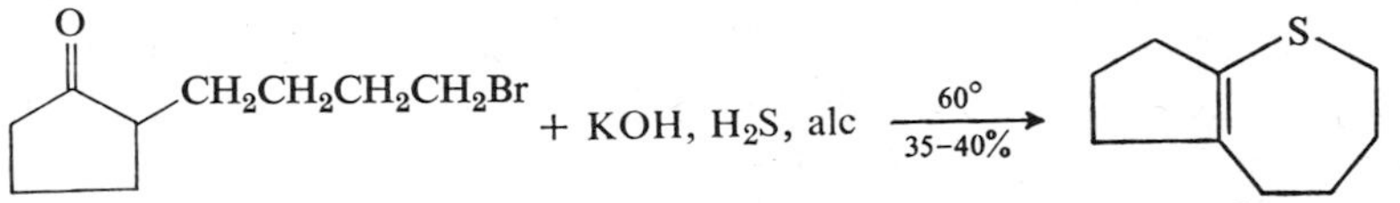

$$\mathbf{1}\ (\text{2-}CH_2CH_2CH_2CH_2Br\text{-cyclopentanone}) + KOH,\ H_2S,\ alc \xrightarrow[35\text{–}40\%]{60°} \mathbf{2} \qquad (1)$$

C. 1-Benzothiepins

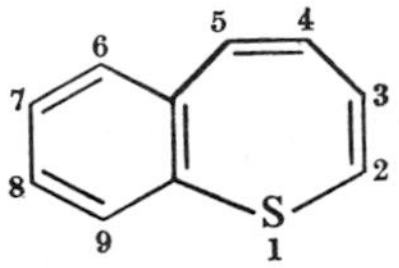

(R.I. 1851)

1. 1-Benzothiepin and Derivatives

The parent compound, 1-benzothiepin, has not been isolated or characterized; however, it has been suggested as a reaction intermediate and stable substituted 1-benzothiepins have been reported recently. The synthetic approaches employed in these investigations were basically twofold: (*1*) fusion of the saturated seven-membered ring to the benzene ring, followed by functional group manipulation on the sulfur ring to introduce the 4,5-double bond and then the 2,3-double bond, or vice versa; and (*2*) ring-expansion reactions utilizing compounds containing a cyclopropane or epoxide ring joined to the six-membered sulfur ring. The latter approach produced the key intermediate which was subsequently converted to a stable, isolable substituted 1-benzothiepin.

One application of synthetic approach *1* (Eq. 2) utilized 2,3-dihydro-1-benzothiepin (**4**) which was prepared by dehydration of 5-hydroxy-2,3,4,5-tetrahydro-1-benzothiepin (**3**) in dimethyl sulfoxide (295). Attempts to introduce the remaining 2,3-double bond in **4** by dehydrogenation or via bromination of **4** with *N*-bromosuccinimide followed by reaction of the allylic bromide with base were unsuccessful (152). However, reaction of **4** with sulfuryl chloride (35, 187, 302) produced low yields of sulfur and naphthalene (*ca.* 5–10%) along with an unidentified dichloro compound (296). The isolation of sulfur and naphthalene in comparable amounts was offered as evidence for the conversion of the α-chloro compound **5** to 1-benzothiepin

OH DMSO SO_2Cl_2 S S **3** **4**

(2)

H S Cl S S + **5** **6**

(**6**) which then suffered sulfur extrusion. Such extrusion reactions (149, 284) where sulfur or sulfur dioxide is eliminated and a more stable aromatic system is produced are common in the thiepin series and are useful in supporting structural assignments.

4-Chloro-1-benzothiepin (**8**) proved to be more stable than 1-benzothiepin and was recently isolated from the reaction of 2,4-dichloro-2,3-dihydro-1-benzothiepin (**7**) with potassium *tert*-butoxide at room temperature (Eq. 3) (300). When the above reaction was carried out at 70°C or when **7** was heated in LiCl/DMF, the only product observed was β-chloronaphthalene (**9**), presumably formed by the extrusion of sulfur from **8**. Purification of 4-chloro-1-benzothiepin by column chromatography gave a yellow oil which showed a molecule ion peak (194) in the mass spectrum and had the following nmr peaks: an AB quartet ($J_{AB} = 9$ Hz) centered at τ 3.80 ($C_{(2)}$ and $C_{(3)}$ hydrogens) and a complex multiplet (6 hydrogens) at τ 2.60–2.80 which included the aromatic protons and the $C_{(5)}$ proton. 4-Chloro-1-benzothiepin slowly underwent sulfur extrusion even at room temperature but required 5 days for complete conversion to β-chloronaphthalene; however, when **8** was heated under reflux in benzene, β-chloronaphthalene formed rapidly (300). Additional support for structure **8** was obtained by oxidation to 4-chloro-1-benzothiepin-1,1-dioxide (**10**) (Eq. 3) (300). β-Chloronaphthalene

Cl H Cl S **7** —KOBu-*tert*→ Cl S **8** —Δ→ Cl **9** +

8 —m-$ClC_6H_4CO_3H$, 10°C→ Cl SO_2 **10** +

was also isolated in the oxidation reaction but does not appear to arise from the extrusion of sulfur from **8**.

The alternative application of synthetic approach *1*, where the 2,3-double bond is introduced first, utilized 5-hydroxy-2-chloro-4,5-dihydro-1-benzothiepin (**14**) as the key intermediate. A convenient preparation of **14** via compounds **11**, **12**, and **13** is outlined in Eq. 4. When the alcohol **14** was exposed to *p*-toluenesulfonic acid in anhydrous benzene, ring contraction with elimination of hydrogen chloride occurred with formation of compound **15** as the only identified product (296). Discussion of this carbonium ion rearrangement will be presented in a subsequent section of this chapter.

In order to avoid carbonium ion intermediates, 5-acetoxy-2-chloro-4,5-dihydro-1-benzothiepin (**16**) was pyrolyzed at 300–320°C. The isolation of 1-chloronaphthalene and 1,1′-naphthyl disulfide (**18**) from this reaction again suggests the intermediacy of a 1-benzothiepin (Eq. 5) (296). The sulfur extrusion reaction with 2-chloro-1-benzothiepin (**17**) provided 1-chloronaphthalene and sulfur; the extruded sulfur, which was described as an active form, presumably reacted with 1-chloronaphthalene to generate 1,1′-naphthyl disulfide (**18**). Support for this explanation included the lack of any reaction with elemental sulfur in refluxing 1-chloronaphthalene and the formation of 1,1′-naphthyl disulfide when 2,5-diphenyl-1,4-dithiin (**19**)

3 $\xrightarrow[C_5H_5N\ (94\%)]{p\text{-}NO_2C_6H_4COCl}$ 11 $\xrightarrow[87\%]{2SO_2Cl_2}$ 12 $\xrightarrow{\Delta\ (92\%)}$ 13 $\xrightarrow{81\%}$ 14 $\xrightarrow[30\%]{p\text{-}MeC_6H_4SO_3H}$ 15 (4)

$$14 \longrightarrow \mathbf{16} \xrightarrow{\Delta} [\mathbf{17}] \longrightarrow S + \text{1-chloronaphthalene} \longrightarrow \mathbf{18} \qquad (5)$$

was added to refluxing 1-chloronaphthalene (Eq. 6). Sulfur extrusion of **19** occurred at 200°C (191), and successful reaction of the extruded sulfur from **19** with 1-chloronaphthalene was offered in support of an activated form of extruded sulfur (296).

$$\mathbf{19} \xrightarrow[\text{reflux}]{\text{1-chloronaphthalene}} \mathbf{18} + S + \text{2,4-diphenylthiophene} \qquad (6)$$

The reaction of alcohol **14** with thionyl chloride (146) or calcium chloride and hydrochloric acid (294) led to 2,5-dichloro-4,5-dihydro-1-benzothiepin (**20**) along with an unidentified compound. Treatment of **20** with potassium *tert*-butoxide gave a yellow oil which decomposed at room temperature to sulfur and 1-chloronaphthalene (Eq. 7) (146). A tentative structure assigned to the yellow oil, whose ir spectrum differs from that of 1-chloronaphthalene, was 2-chloro-1-benzothiepin (**17**).

$$14 \xrightarrow[\text{CaCl}_2,\ \text{HCl}]{\text{SOCl}_2 \text{ or}} \mathbf{20} \xrightarrow{\text{KOBu-}tert} \mathbf{17} \xrightarrow{\text{room temp}} S + \text{1-chloronaphthalene} \qquad (7)$$

Parham and Schweizer (260) described an elegant method for the ring expansion of dihydropyran to 6-chloro-2,3-dihydrooxepin via a dichlorocyclopropyl adduct. Extension of this reaction to 4*H*-1-benzothiopyran (**21**) produced the cyclopropyl adduct **22**; the latter, upon being heated in quinoline at 210°C, gave 2-chloronaphthalene as indicated in Eq. 8 (190).

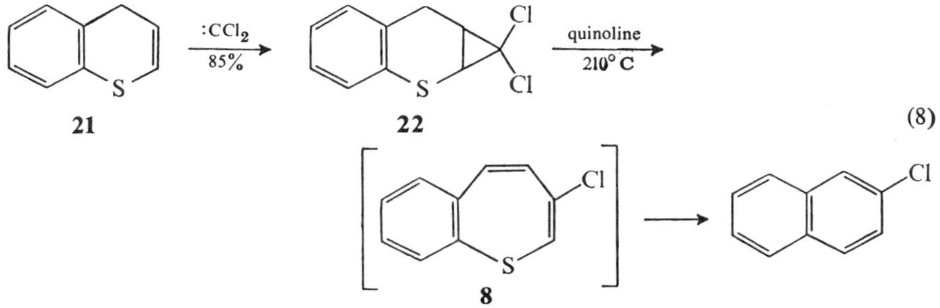

:CCl_2 85% quinoline 210° C Cl Cl S S Cl Cl S

21 **22** **8** (8)

The explanation of the foregoing results given by Parham and Schweizer involved formation of 3-chloro-1-benzothiepin (**8**) which suffered sulfur extrusion under their experimental conditions. Attempts to apply the above scheme to 2*H*-1-benzothiopyran (**23**) failed to form the cyclopropyl adduct; instead, insertion occurred at the $C_{(2)}$ position, producing 2-dichloromethyl-2*H*-1-benzothiopyran (**24**) (Eq. 9). In contrast to the behavior of **23**, 4-ethoxy-2*H*-1-benzothiopyran (**25**) readily formed adduct **26**; however, when **26** was heated in quinoline, rearrangement to 3-chloro-2-methylthiochromone (**27**) took place (188). A rationalization of the rearrangement offered by Parham and Bhavsar (188) is outlined in Eq. 10. Most recently, Parham and

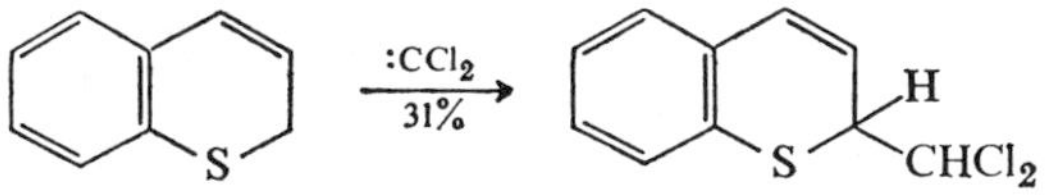

:CCl_2 31% S S H $CHCl_2$ (9)

23 **24**

OEt :CCl_2 33% EtO Cl Cl quinoline 210° C EtO Cl S S S

25 **26** (10)

O Cl S Me Et O Cl S Me OEt Cl S

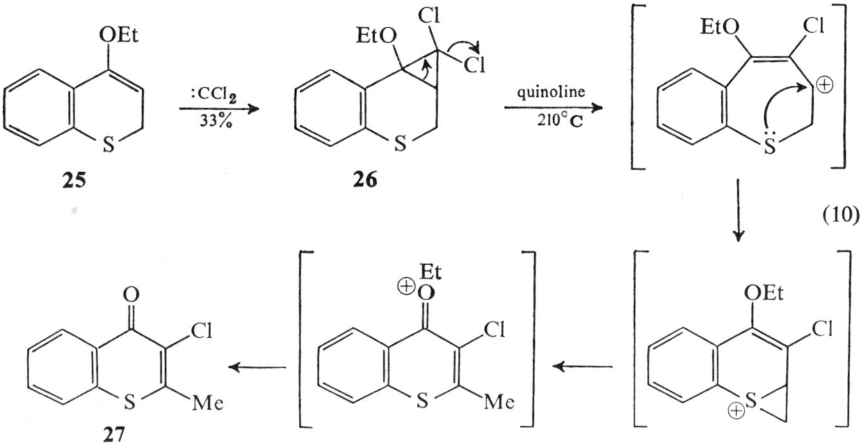

27

28 $\xrightarrow[71\%]{:CCl_2}$ 29 $\xrightarrow[DMSO]{NaOMe}$ [30] + 31 + (11)

30 ⟶ 32 + 33 + 34

Weetman (192) reported the results of a comparable study with 4-ethoxy-2-methyl-2*H*-1-benzothiopyran (**28**). The corresponding cyclopropyl adduct **29**, when treated with sodium methoxide or ethoxide in dimethyl sulfoxide at 20–25°C, gave a variety of 1-benzothiepin derivatives (Eq. 11). Compounds **33**, **34**, and particularly 2-methoxy-4-ethoxy-1-methylnaphthalene (**32**) appeared to arise from the intermediate 4-methoxy-5-ethoxy-2-methyl-1-benzothiepin (**30**), while the origin of **31** was somewhat obscure. Again, sulfur extrusion of **30** was proposed for the origin of **32**. When sodium ethoxide was used as the base, 2,4-diethoxy-1-methyl-naphthalene was the sulfur extrusion product.

Wynberg and co-workers (181) have proposed the intermediacy of the 1-benzothiepin **36** in the thermal (> 200°C) conversion of **35** to dimethyl naphthalene-1,2-dicarboxylate (Eq. 12).

+ $MeO_2CC{\equiv}CCO_2Me$ $\xrightarrow{h\nu}$ H, S, CO_2Me, CO_2Me **35** $\xrightarrow{\Delta}$

(12)

[S, CO_2Me, CO_2Me] **36** ⟶ CO_2Me, CO_2Me

Recently Hofmann and Westernacher (104, 106) reported the isolation of the first stable 1-benzothiepin derivative. The synthetic sequence which resulted in the preparation of 3,5-diacetoxy-4-phenyl-1-benzothiepin (**38**) is outlined in Eq. 13; it involves formation of the thiepin ring via a ring

O, O, H, Ph, S $\xrightarrow[\text{BF}_3\cdot\text{Et}_2\text{O},\ 70\%]{\text{H}_2\text{SO}_4\ \text{or}}$ O, H Ph, =O, S **37** $\xrightarrow[\text{C}_5\text{H}_5\text{N},\ 98\%]{\text{Ac}_2\text{O}}$

(13)

OAc, Ph, OAc, S **38**

enlargement rearrangement. The resulting diketone **37** did not exhibit any measurable concentration of its enolic or dienolic tautomer; however, reaction of **37** with acetic anhydride in pyridine at room temperature trapped the dienol form and gave the benzothiepin **38**.

The dienol acetate **38** is a colorless, crystalline compound with mp 86–87°C, carbonyl absorption at 1770 cm^{-1}, and an nmr spectrum which shows the $C_{(2)}$ hydrogen singlet at τ 4.0 and two different methyl singlets at τ 8.15 and τ 8.35. According to a preliminary account (105) of the chemical properties of **38**, thermal sulfur extrusion in pyridine at 60–70°C produced 1,3-diacetoxy-2-phenylnaphthalene (**40**). When the reaction of **38** was performed in refluxing pyridine with sodium acetate, on the other hand, a ring contraction resulted, forming 1,3-diacetoxy-4-acetylthio-2-phenylnaphthalene (**41**). As indicated

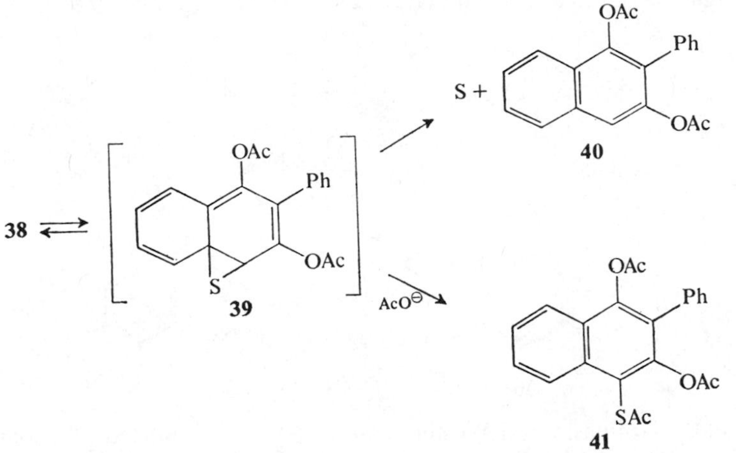

(14)

in Eq. 14, an episulfide **39** which is a valence bond tautomer of **38** seems to be a reasonable precursor to both naphthalene derivatives (105).

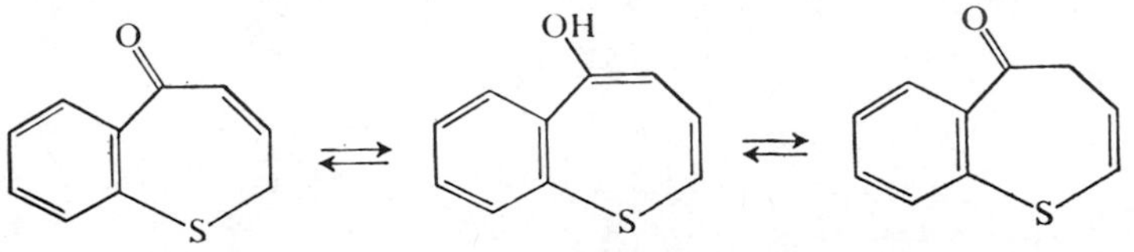

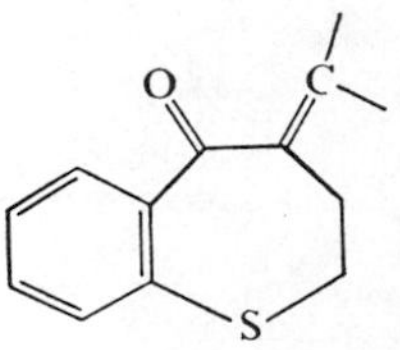

(15)

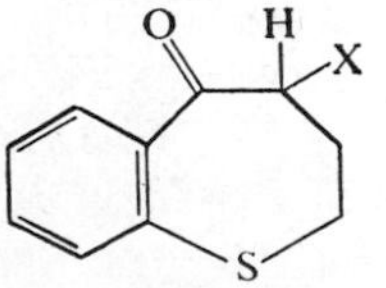

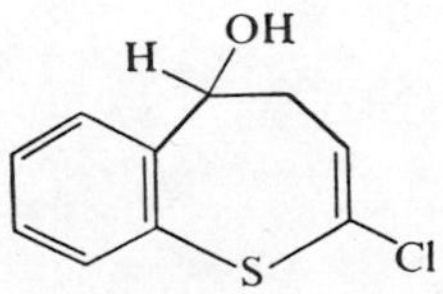

The chemical and physical properties of 3,5-diacetoxy-4-phenyl-1-benzothiepin (**38**) are consistent with an 8π electron model with no significant aromatic stabilization in the sulfur ring. The facile sulfur extrusion to a more stable naphthalene ring and the position of the $C_{(2)}$ hydrogen in the nmr, which shows olefinic character, both support the above conclusion.

Other attempts to prepare unsaturated 5-oxo-1-benzothiepins with the intent of capturing the tautomeric 5-hydroxy-1-benzothiepin (Eq. 15) in the form of stable derivatives were unsuccessful. These approaches included dehydrohalogenation of **42** (X = Br, I) (152, 294), isomerization of structure **43** (36), and oxidation of compound **14** (146).

2. 1-Benzothiepin-1-oxide and Derivatives

Essentially no work has been reported in the literature on the synthesis or attempted synthesis of 1-benzothiepin-1-oxide. However, one reaction sequence has been investigated which suggested formation of the intermediate 2-chloro-1-benzothiepin-1-oxide (**45**) and subsequent SO extrusion (294). Oxidation of compound **14** with perbenzoic acid provided the corresponding sulfoxide **44**; the latter, when subjected to acid-catalyzed dehydration conditions, gave 1-chloronaphthalene. This dehydration pathway of **44** (Eq. 16) is in contrast to the acid-catalyzed ring contraction of **14** (see Eq. 36).

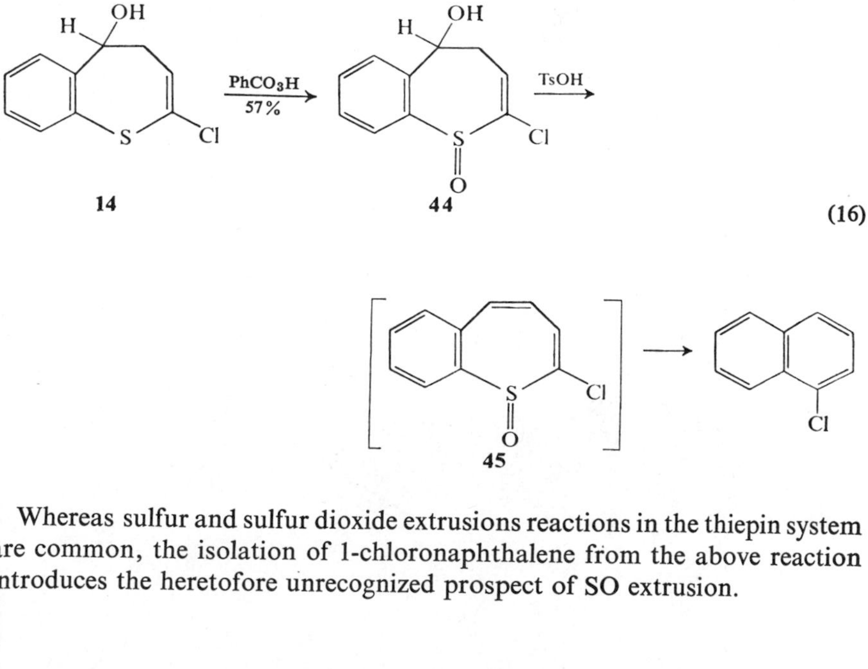

H OH S Cl **14** $\xrightarrow[57\%]{PhCO_3H}$ H OH S O Cl **44** $\xrightarrow{TsOH}$ (16)

[S O Cl **45**] → Cl

Whereas sulfur and sulfur dioxide extrusions reactions in the thiepin system are common, the isolation of 1-chloronaphthalene from the above reaction introduces the heretofore unrecognized prospect of SO extrusion.

46 $\xrightarrow{NaBH_4,\ 96\%}$ 3 (OH, H)

46 $\xrightarrow{H_2O_2,\ 88\%}$ 47 (SO_2)

46 $\xrightarrow{Zn(Hg),\ HCl}$ 53 (S)

3 $\xrightarrow{H_2O_2,\ 91\%}$ 48 (OH, H; SO_2)

47 $\xrightarrow{NaBH_4,\ 88\%}$ 48

48 $\xrightarrow{H_3PO_4,\ 81\%}$ 49 (SO_2)

49 $\xrightarrow{NBS,\ Bz_2O_2,\ 40\%}$ 50 (Br, H; SO_2)

50 $\xrightarrow{Et_3N,\ 88\%}$ 51 (SO_2)

51 $\xrightarrow{2H_2,\ Pt,\ 92\%}$ 52 (SO_2)

53 $\xrightarrow{H_2O_2,\ 61\%}$ 52

(17)

3. 1-Benzothiepin-1,1-dioxide Derivatives

The synthesis of 1-benzothiepin-1,1-dioxide (**51**) via compounds **46–50** (Eq. 17) was reported initially in 1958 (297, 298). Confirmation of the structural assignments was achieved by the quantitative absorption of 2 moles of hydrogen by **51** to produce the known 2,3,4,5-tetrahydro-1-benzothiepin-1,1-dioxide (**52**). In contrast to other thiepin dioxides, compound **51**, a colorless crystalline solid, could be distilled with only slight decomposition at 250°C and showed no tendency to form naphthalene via SO_2 extrusion; however, prolonged exposure of **51** at 250°C led to resinification.

2-Chloro-1-benzothiepin-1,1-dioxide (**55**) has been generated by dehydration of 5-hydroxy-2-chloro-4,5-dihydro-1-benzothiepin-1,1-dioxide (**54**) (Eq. 18) (294). In addition Hofmann and Westernacher (104, 106) have reported the preparation of 1-benzothiepin-1,1-dioxide derivatives **58** and **59** via intermediates **56** and **57** (Eq. 19).

56; R = H
57; R = Me

58; R = H
59; R = Me

A uv and ir spectral study of 1-benzothiepin-1,1-dioxide (**51**) and related compounds (298) revealed no appreciable conjugative involvement of the sulfur function with the unsaturation of the heterocyclic ring. The uv spectrum of **51** was essentially the same in either polar (95% ethanol) or nonpolar (cyclohexane) solvents, and the symmetric and asymmetric ir stretching absorption bands of the sulfone in **51** were essentially at the same frequency as in the tetrahydro derivative **52**. The nmr spectrum (36) of 2-chloro-1-benzothiepin-1,1-dioxide (**55**) showed olefinic protons on the heterocyclic ring in the range τ 2.5 to τ 3.5 while the single $C_{(2)}$ proton on the heterocyclic ring in compound **58** appeared at τ 3.25 (106). These physical properties do not support consideration of appreciable aromatic character in the thiepin ring.

A study of the chemical properties of 1-benzothiepin-1,1-dioxide (299) revealed that nucleophilic addition fails to take place with **51** and piperidine or hydrobromic acid. Normal addition of bromine to **51** likewise failed; however, in the presence of sunlight or uv light (Eq. 20), 1 or 2 moles of

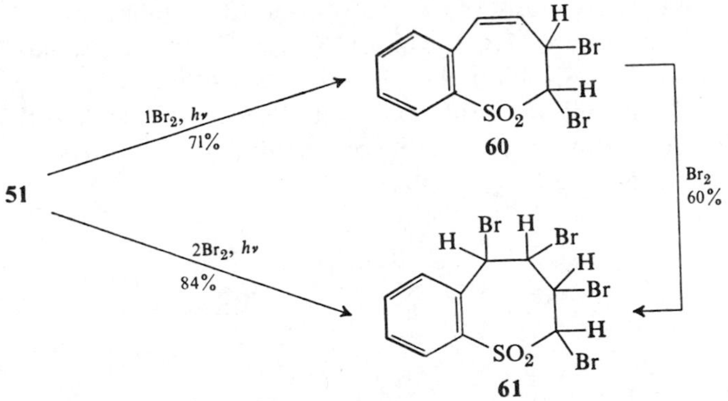

51 → ($1Br_2$, $h\nu$, 71%) **60**; **51** → ($2Br_2$, $h\nu$, 84%) **61**; **60** → (Br_2, 60%) **61** (20)

bromine were taken up, giving **60** or **61**, respectively (299). Structural assignments for compounds **60** and **61** were based on a comparison of their uv spectra with those of model compounds. An electrophilic nitration of 1-benzothiepin-1,1-dioxide with red fuming nitric acid produced two isomeric

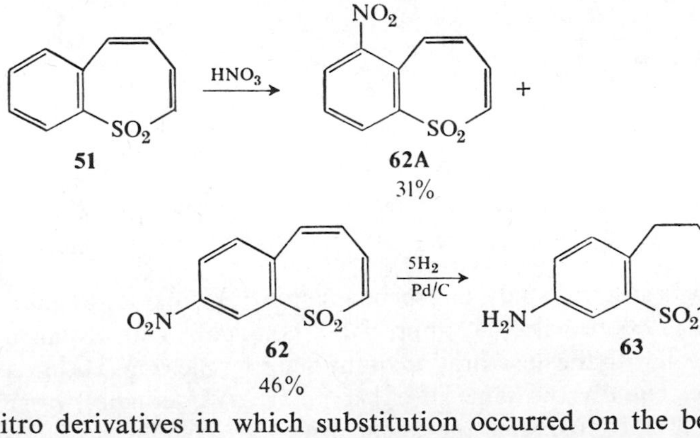

51 → (HNO_3) **62A** (31%) + **62** (46%) (21)

62 → ($5H_2$, Pd/C) **63**

mononitro derivatives in which substitution occurred on the benzene ring (Eq. 21). The structure of one of the isomers, 8-nitro-1-benzothiepin-1,1-dioxide (**62**), was established conclusively by reduction to 8-amino-2,3,4,5-tetrahydro-1-benzothiepin-1,1-dioxide (**63**) which was identical to an authentic sample prepared by an alternate route. Although the structure of the second nitro isomer was not established, directive effects of the sulfone function and the vinyl group lead one to expect substitution into the 6-position on the benzene ring, as in **62A**.

In a preliminary report Hofmann and Westernacher (105) described an unusual reaction of **58** with sodium acetate in acetic anhydride under reflux (Eq. 22). Again no SO_2 was lost to produce a naphthalene derivative; instead rearrangement occurred to form 2a-acetyl-2,2a-dihydro-4-phenyl-cyclobutadieno[1,2-*b*]thionaphthene-2-one-3,3-dioxide (**64**). The ir spectrum of **64** contained bands at 1733 cm^{-1} ($\nu_{C=O}$), 1652 cm^{-1} ($\nu_{C=C}$), 1313 cm^{-1} and

OAc Ph OAc SO$_2$ —NaOAc / Ac$_2$O→ Ph SO$_2$ C=O Me O —several steps→

58 **64** (22)

O SO$_2$

65

1160 cm^{-1} ($\nu_{S=O}$); the nmr spectrum showed a signal in the aromatic region (9 hydrogens) and at τ 7.3 (3 hydrogens). Chemical evidence for **64** involved degradation to 2,3-dihydrothionaphthene-3-one-1,1-dioxide (**65**).

4. Dihydro-1-benzothiepin and Derivatives

There are three possible isomeric dihydro-1-benzothiepins with an endocyclic double bond; the parent compound and/or derivatives for each of these systems are known. In addition, location of the double bond in the exo

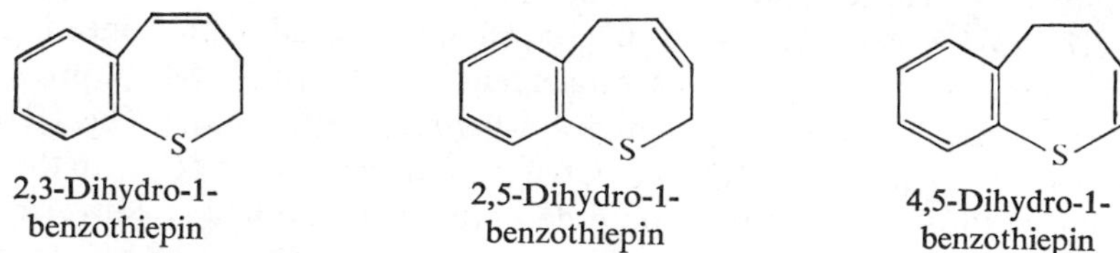

2,3-Dihydro-1-benzothiepin

2,5-Dihydro-1-benzothiepin

4,5-Dihydro-1-benzothiepin

position relative to the heterocyclic ring leads to isomers which are also named as dihydro-1-benzothiepins. Of the four possible types, derivatives of three systems are known. The partial name listed under the compounds can be completed by adding the name of the residue attached to the sulfur ring. Examples appear later in the text.

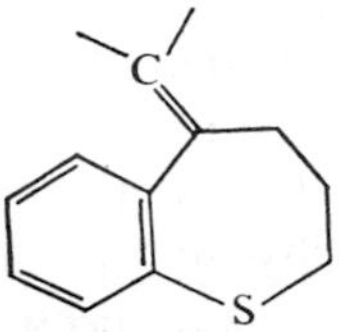

3,4-Dihydro-1-benzothiepin-$\Delta^{5(2H)}$

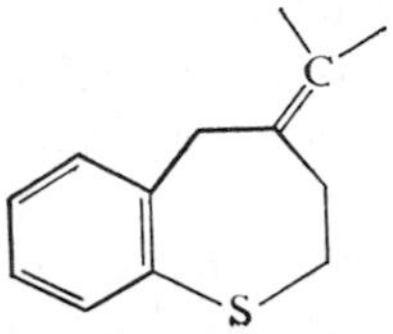

2,5-Dihydro-1-benzothiepin-$\Delta^{4(3H)}$

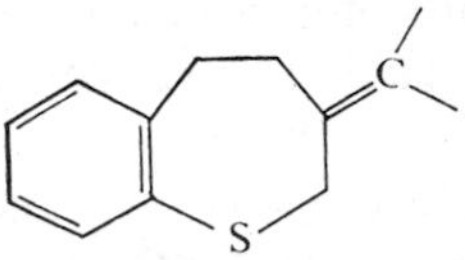

2,5-Dihydro-1-benzothiepin-$\Delta^{3(4H)}$

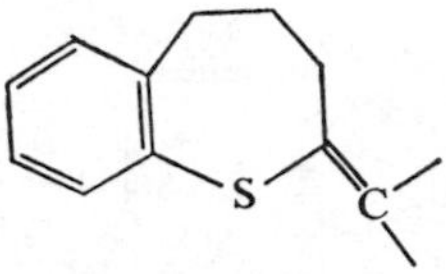

3,4-Dihydro-1-benzothiepin-$\Delta^{2(5H)}$

The preparation of 2,3-dihydro-1-benzothiepin (**4**) was accomplished by dehydration of 5-hydroxy-2,3,4,5-tetrahydro-1-benzothiepin (**3**) with phosphorus pentoxide (152) (64% yield) or dimethyl sulfoxide (295) (81% yield), or by conversion of alcohol **3** to 5-chloro-2,3,4,5-tetrahydro-1-benzothiepin (**66**) which was dehydrochlorinated with *sym*-collidine (Eq. 23) (269). Other

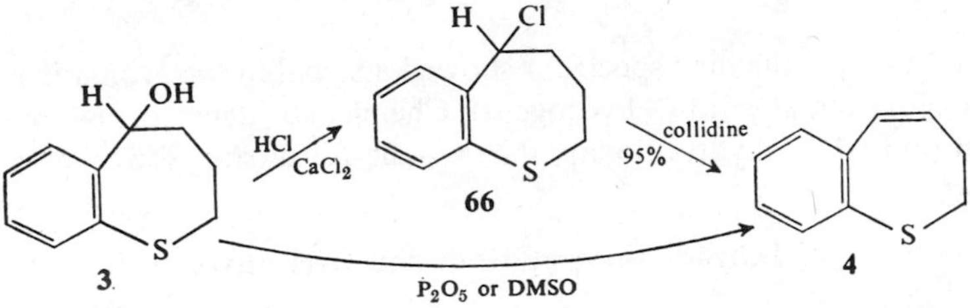

HCl, $CaCl_2$ → **66** → collidine, 95% → **4**; **3** → P_2O_5 or DMSO → **4** (23)

alcohols have been dehydrated to yield substituted 2,3-dihydro-1-benzothiepin; for examples see Table XI-2.

A second source of 5-substituted-2,3-dihydro-1-benzothiepins involves an interesting rearrangement of a substituted dihydrodibenzo[*b*,*e*]thiocin (**67**) (18, 19). When compound **67** or **68** was heated with concentrated hydrobromic acid (Eq. 24), rearrangement of the ring system provided 5-[2-(2-bromoethyl)phenyl]-2,3-dihydro-1-benzothiepin (**69**). The corresponding chloro derivative **70** is available from the reaction of **67** with thionyl chloride. Both the chloro and bromo derivatives **70** and **69** are converted to the dimethylamino or monomethylamino derivative **71** or **72** by reaction with the appropriate amine. Bickelhaupt, Stach, and Thiel (18) confirmed the structure of the rearranged products by the independent synthesis of compound **71** outlined in Eq. 25. The rationalization offered for the above rearrangement (Eq. 26) involves conversion of **67** or **68** to intermediate **73**

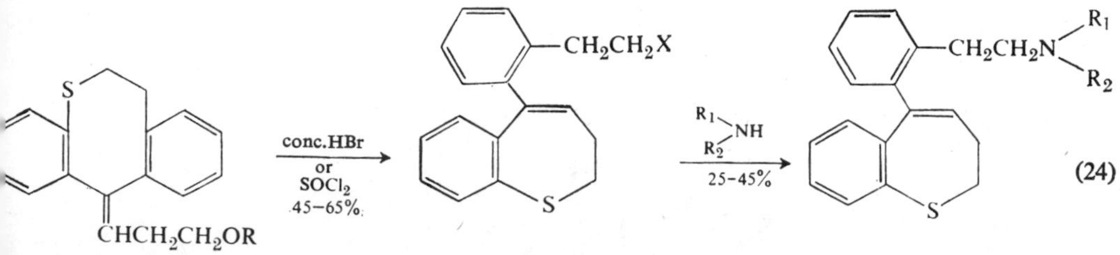

$CHCH_2CH_2OR$ —(conc.HBr or $SOCl_2$, 45–65%)→ $-CH_2CH_2X$ —(R_1R_2NH, 25–45%)→ $-CH_2CH_2NR_1R_2$ (24)

67; R = H
68; R = $PhCH_2$
69; X = Br
70; X = Cl
71; $R_1 = R_2 =$ Me
72; R_1 = Me, R_2 = H

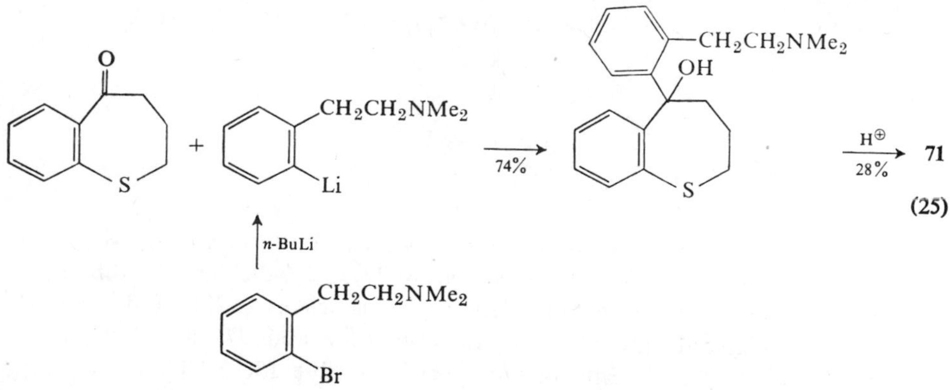

$CH_2CH_2NMe_2$, Br —(n-BuLi)→ $CH_2CH_2NMe_2$, Li; + ketone —(74%)→ $-CH_2CH_2NMe_2$, OH —($H^{\oplus}$, 28%)→ **71** (25)

which undergoes intramolecular sulfonium salt formation to give **74**. Bromide ion attack with ring opening of the sulfonium salt leads to the isolated product **69**. A bromine free by-product was also isolated in the reaction of **67**

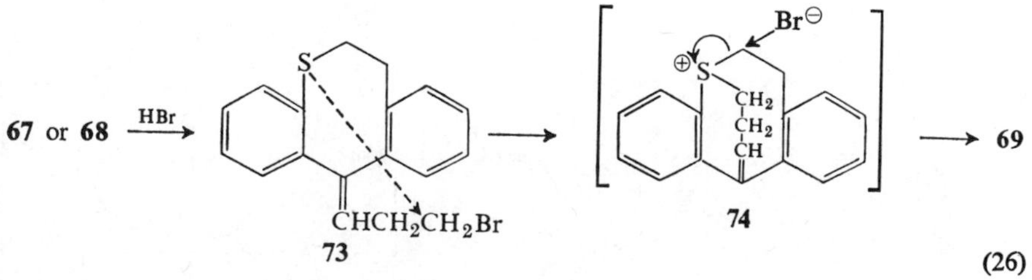

67 or **68** —(HBr)→ $CHCH_2CH_2Br$ **73** → [$Br^{\ominus}$, $\oplus$S, CH_2, CH_2, CH **74**] → **69** (26)

with concentrated hydrobromic acid and assigned structure **75** on the basis of its uv and nmr spectra (18). The formation of **75** is explained (Eq. 27) by an electrophilic substitution at the $C_{(4)}$ olefinic carbon in **69**.

The third approach leading to 2,3-dihydro-1-benzothiepin derivatives employs the 3,4-dihydro-1-benzothiepin-5(2*H*)-one system (**46**). Conversion of these substituted ketones to enol acetates was accomplished by reaction

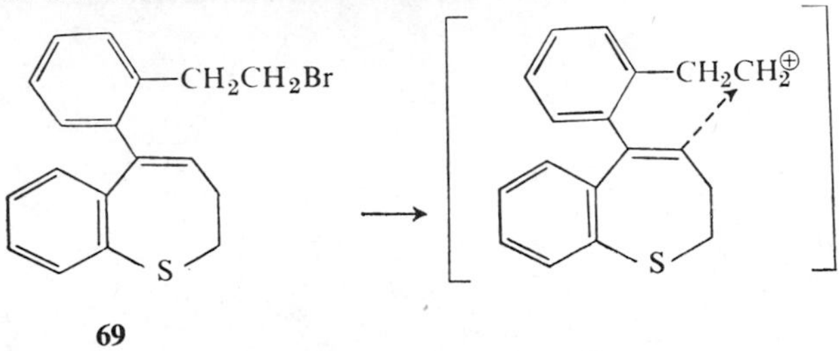

$-CH_2CH_2Br$ $-CH_2CH_2^{\oplus}$

69 **75**

(27)

with acetic anhydride in pyridine and will be discussed with other ketone reactions (Section C-5). Another route involving the conversion of a ketone to a dihydrobenzothiepin is illustrated by the application of the Vilsmeier reaction on ketone **46** which yielded compound **76** (Eq. 28) (312).

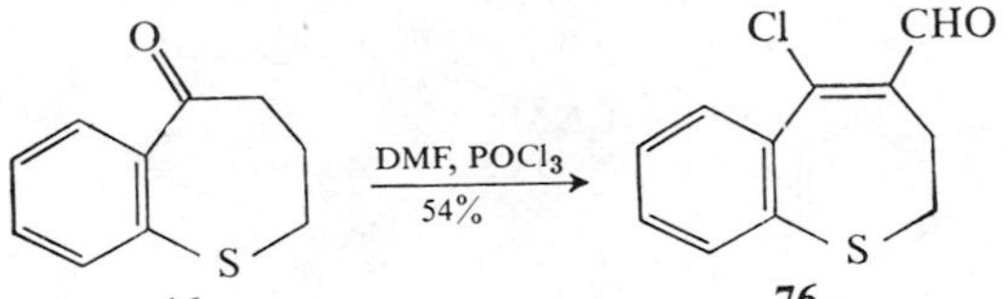

O Cl CHO

DMF, $POCl_3$ 54%

S S

46 **76**

(28)

The fourth synthetic method is illustrated by the conversion of 7-hydroxy-7a-chlorocyclopropa[*b*][1]benzothiapyran (**77**) to 2,4-dichloro-2,3-dihydro-1-benzothiepin (**7**) by the action of hydrochloric acid (Eq. 29) (300). Similarly an acid-catalyzed reaction of acetic anhydride and **77** gave 2-acetoxy-4-chloro-2,3-dihydro-1-benzothiepin (**78**) (Eq. 29) (300). These reactions presumably proceed via the homoallylic cation **79**.

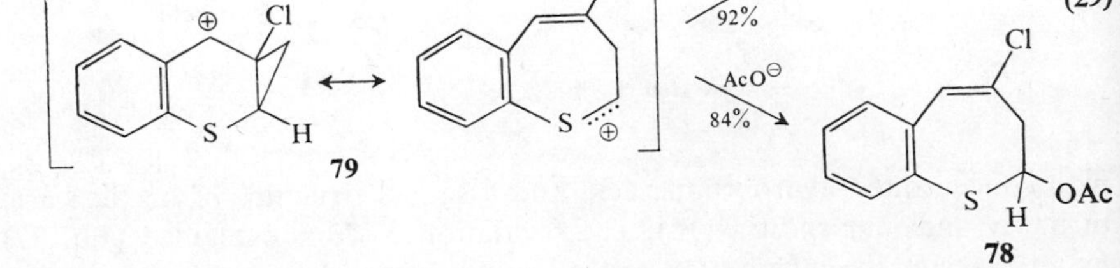

OH H Cl $H^{\oplus}$ S H

77

Cl S H Cl

7

$\oplus$ Cl S H ⟷ Cl S $\oplus$

79

$Cl^{\ominus}$ 92%

$AcO^{\ominus}$ 84%

(29)

Cl S H OAc

78

The physical properties of **4** are described in Table XI-2 and its chemical properties are as expected. Additions of hydrogen and bromine occur in high yield to produce the tetrahydro derivative **53** and 4,5-dibromo-2,3,4,5-tetrahydro-1-benzothiepin, respectively (152). The reaction of **4** with red fuming nitric acid (Eq. 30) forms a low yield of 4-nitro-2,3-dihydro-1-benzothiepin (**80**) (146) which was identified by oxidation to the sulfone

$$\mathbf{4} \xrightarrow[\text{d. } 1.5]{HNO_3} \mathbf{80}\ (NO_2) \xrightarrow{H_2O_2} \mathbf{81}\ (NO_2, SO_2) \qquad (30)$$

derivative **81** (see structural evidence for sulfone **81** below).

Alkylation on sulfur results when **4** is treated with methyl iodide or ethyl iodide in the presence of silver fluoroborate (Eq. 31) (300). Oxidation of **4** with sodium metaperiodate (117, 144) in acetic acid (Eq. 32) provides 2,3-dihydro-1-benzothiepin 1-oxide (**82**) as a viscous oil (300) whose ir spectrum shows an absorption band at 1030 cm^{-1} ($\nu_{S=O}$); its nmr spectrum reveals olefinic protons at τ 3.44 (doublet) and τ 3.99 (two triplets), and methylene protons as a complex multiplet between τ 6.5 and τ 8.0. Microanalytical data were obtained on the solid derivative 1-methoxy-2,3-dihydro-1-benzothiepinium fluoroborate (**83**) which was prepared by reaction of **82** with methyl iodide and silver fluoroborate (300). The nmr spectrum of **83** shows olefinic protons at τ 3.4 (doublet) and τ 3.79 (two triplets) as well as methylene protons in the expected regions. Attempts to prepare **82** by dehydration

$$\mathbf{4} \xrightarrow[AgBF_4]{RI} \text{S}^{\oplus}\text{–R}\ BF_4^{\ominus} \qquad R = \text{Me, Et} \qquad (31)$$

$$\mathbf{4} \xrightarrow[AcOH]{NaIO_4} \mathbf{82}\ (S{=}O) \xrightarrow[AgBF_4]{MeI} \mathbf{83}\ (\text{S}^{\oplus}\text{–OMe}\ BF_4^{\ominus}) \qquad (32)$$

of 5-hydroxy-2,3,4,5-tetrahydro-1-benzothiepin-1-oxide or dehydrohalogenation of 5-chloro-2,3,4,5-tetrahydro-1-benzothiepin-1-oxide were unsuccessful (146). In the latter reaction the only product isolated and identified was 2,3-dihydro-1-benzothiepin (**4**).

With a more vigorous oxidizing agent such as hydrogen peroxide (Eq. 33), oxidation of **4** leads to 2,3-dihydro-1-benzothiepin-1,1-dioxide (**49**) in fair yield (152); however, dehydration of 5-hydroxy-2,3,4,5-tetrahydro-1-benzothiepin-1,1-dioxide (**48**) with phosphoric acid (Eq. 33) is the superior method for the preparation of **49**. A second approach leading to substituted

$$\mathbf{4} \xrightarrow[\text{AcOH}]{H_2O_2} \mathbf{49} \xleftarrow[81\%]{H_3PO_4} \mathbf{48} \qquad (33)$$

derivatives of 5-acetoxy-2,3-dihydro-1-benzothiepin-1,1-dioxide involves the preparation of enol acetates mainly from 4-halo-3,4-dihydro-1-benzothiepin-5(2*H*)-one-1,1-dioxides. Discussion of these reactions will appear under reactions of ketones (Section C-5).

The reaction of **49** with bromine gives 4,5-dibromo-2,3,4,5-tetrahydro-1-benzothiepin-1,1-dioxide; treatment of **49** with red fuming nitric acid (d = 1.5) produces 4-nitro-2,3-dihydro-1-benzothiepin-1,1-dioxide (**81**) in 60% yield (Eq. 34) (299). Compound **81** had absorption maxima in the uv (95% ethanol) at 225 mμ (log ε 3.00) and 311 mμ (log ε 3.10) which correspond to the β-nitrostyrene chromophore (38). Partial reduction of **81** forms 2,5-dihydro-1-benzothiepin-4(3*H*)-one-1,1-dioxide oxime (**84**). Other examples of 2,3-dihydro-1-benzothiepin-1,1-dioxide derivatives, particularly brominated compounds, were presented earlier in Section C-3.

$$\mathbf{49} \xrightarrow[\text{d. 1.5, } 60\%]{HNO_3} \mathbf{81} \xrightarrow[68\%]{\text{Pd/C}, H_2} \mathbf{84} \qquad (34)$$

A recent preliminary report by Morin, Spry, and Mueller (176) cites the formation of 3-methyl-2,5-dihydro-1-benzothiepin (**86**) and 2,5-dihydro-1-benzothiepin-$\Delta^{3(4H)}$-methane (**87**) as rearrangement products in a Pummerer reaction (Eq. 35). The reaction of sulfoxide **85** with *p*-toluenesulfonic acid produced a higher yield of olefins. Structural assignments for **86** and **87** are based primarily on the nmr spectra of these compounds.

The third class of dihydro derivatives, comprising the 4,5-dihydro-1-benzothiepin system, have already been described previously in the form of the sulfoxide and sulfone (Sections C-1, 2, and 3). It is of interest to compare, in this series, the acid-catalyzed reaction of **14**, **44**, and **54** which all lead to different end products. The carbonium ion **88** generated in the acid-catalyzed reaction of **14** can find stabilization as a homoallylic cation (44,

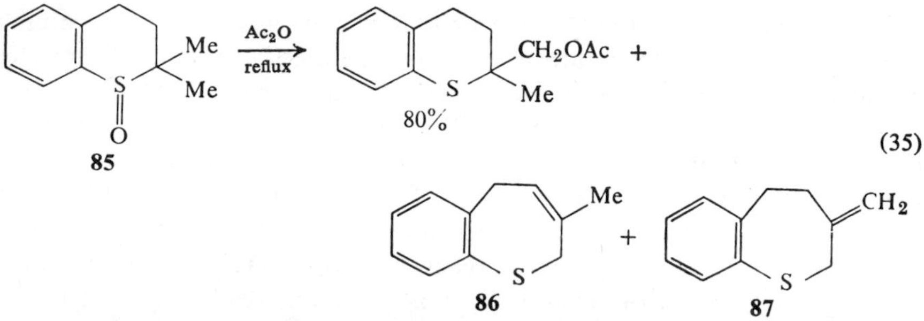

(35)

85   80%   86   87

268, 325) (Eq. 36). The electron-donating character of the sulfide and chloro groups aid this stabilization and direct the reaction toward the formation of **15**. In contrast the electron-withdrawing nature of the sulfoxide and sulfone

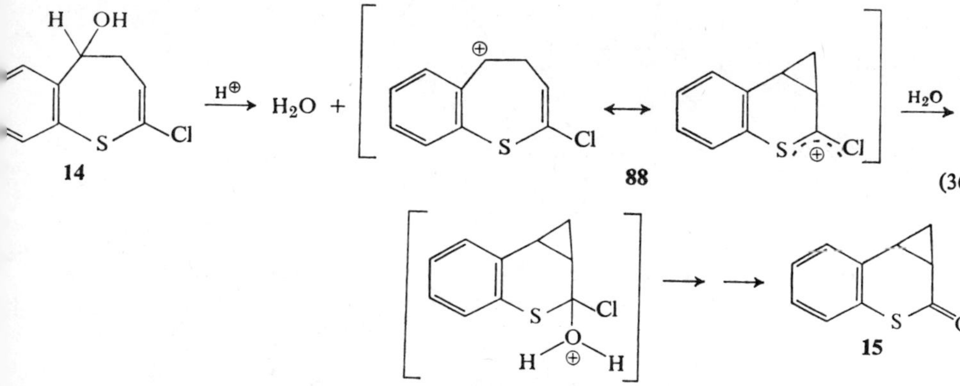

(36)

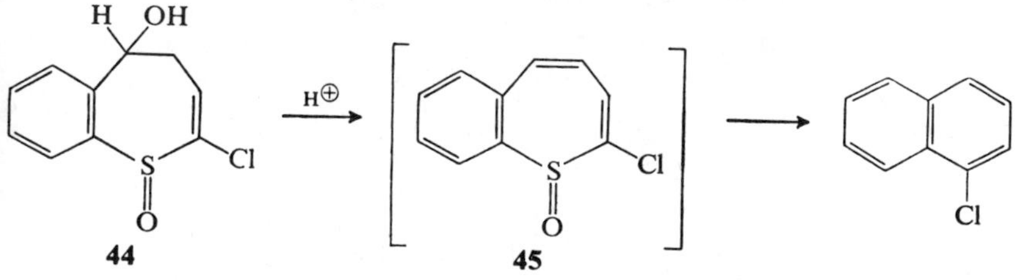

14   88   15

(37)

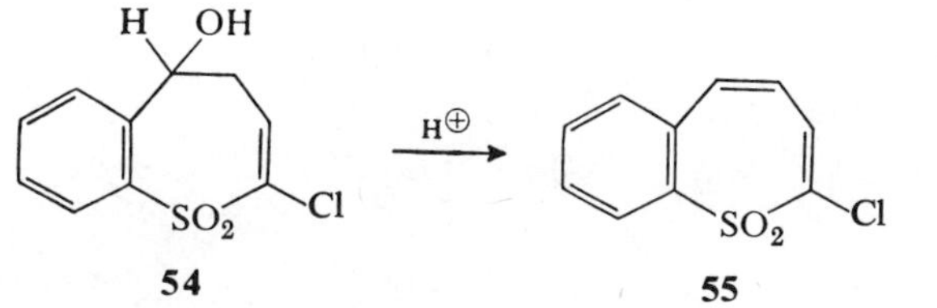

44   45

(38)

54   55

groups in **44** and **54**, respectively, destabilize homoallylic resonance and thus favor normal dehydration to the corresponding 1-benzothiepins (Eqs. 37 and 38). The instability of **45** leads to subsequent SO extrusion and formation of α-chloronaphthalene.

In the category of dihydro-1-benzothiepins with an exocyclic double bond, 2,5-dihydro-1-benzothiepin-$\Delta^{3(4H)}$-methane (**87**) is the only reported example of the $\Delta^{3(4H)}$ system and has already been described. Several members of the 2,5-dihydro-1-benzothiepin-$\Delta^{4(3H)}$ series are known; however, since they also contain the keto function, they will be reviewed under reactions of ketones (Section C-5). A large group of derivatives in the 3,4-dihydro-1-benzothiepin-$\Delta^{5(2H)}$ system have been prepared by Mohrbacher (170). One approach (Eq. 39) entailed a Wittig reaction with ketone **46** for the synthesis of 3,4-dihydro-1-benzothiepin-$\Delta^{5(2H)}$-α-acetonitrile (**89**) and ethyl 3,4-dihydro-1-benzothiepin-$\Delta^{5(2H)}$-α-acetate (**90**). An alternate procedure for the preparation of compound **90** involved a Reformatsky reaction to form the

$(EtO)_2\overset{\oplus}{P}(\rightarrow O)-\overset{\ominus}{C}HCN$ → **89** (CHCN)

$(EtO)_2\overset{\oplus}{P}(\rightarrow O)-\overset{\ominus}{C}HCO_2Et$ → **90** ($CHCO_2Et$) (39)

46 — Zn, $BrCH_2CO_2Et$ → **91** (HO, CH_2CO_2Et) — H_2SO_4 → **90**

46 + $Me_2N(CH_2)_3MgCl$ → **92** (HO, $(CH_2)_3NMe_2$) — H_2SO_4 → **93** ($CHCH_2CH_2NMe_2$) (40)

hydroxy ester **91** and subsequent acid-catalyzed dehydration. The third synthetic sequence (Eq. 40) utilized a Grignard reaction on ketone **46** followed by acid-catalyzed dehydration of the resulting alcohol (**92**) to produce *N*,*N*-dimethyl-3,4-dihydro-1-benzothiepin-$\Delta^{5(2H)}$-γ-propylamine (**93**), isolated as the hydrochloride salt. It is interesting to note that all of the dehydration reactions in Mohrbacher's work lead to an exocyclic double bond. Since these results are revealed in the patent literature, evidence for structural assignments is not reported. Extension of the Grignard reaction-dehydration procedure provided compounds **94** and **95**; **96** on the other hand, was prepared from **89** by lithium aluminum hydride reduction of the

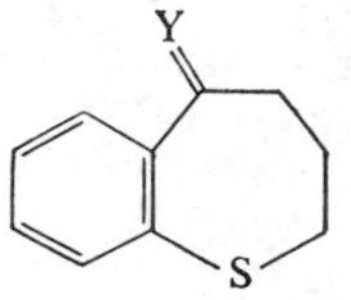

Y

S

94; Y = Et_2NCH_2CH
95; Y = $Et_2NCH_2CHCH(Me)$
96; Y = Me_2NCH_2CH

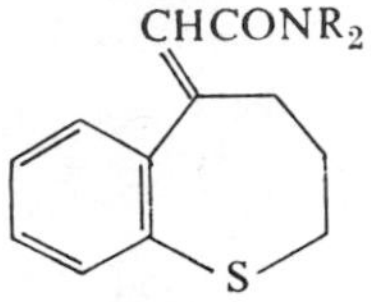

$CHCONR_2$

S

97; R = Me
98; R = Et
99; R = *n*-Pr
100; R = *i*-Bu

cyano function to the amine, followed by a Clarke-Eschweiler methylation. Saponification of ester **90** gave two isomeric acids (no structural assignments were offcred) and the lower-melting acid was converted to a series of amides **97–100**. Interest in these compounds was stimulated by their hypotensive activity.

5. 3,4-Dihydro-1-benzothiepin-5(2*H*)-one and Derivatives

The most important and most studied ketone in the 1-benzothiepin system is 3,4-dihydro-1-benzothiepin-5(2*H*)-one (**46**) (homothiochromanone) which was first prepared in 66% yield by Cagniant and Deluzarche (53) via Friedel-Crafts ring closure of γ-thiophenoxybutyryl chloride. An improved procedure

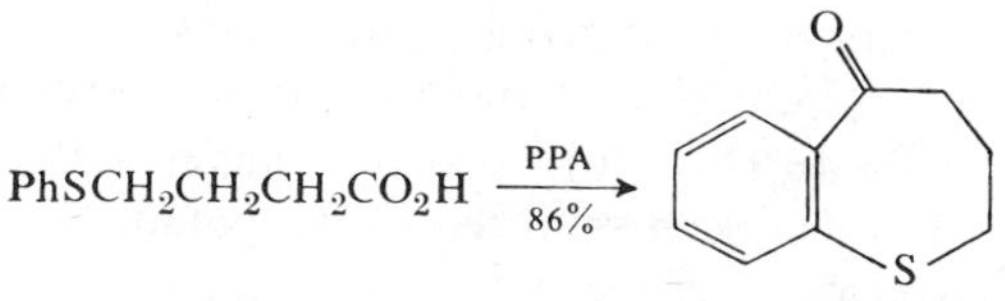

$$PhSCH_2CH_2CH_2CO_2H \xrightarrow[86\%]{PPA} \textbf{46} \qquad (41)$$

O

S

46

employs a polyphosphoric acid (PPA) cyclization (298) of γ-thiophenoxybutyric acid (Eq. 41) [prepared (241) by reaction of sodium thiophenoxide and γ-butyrolactone]. The Friedel-Crafts acid chloride ring closure reaction

(Eq. 42) appears to be more general and was extended to the preparation of ketones **102–106** (51, 177). A complication observed with the PPA ring closure (51) is illustrated in the conversion of **107** to **108** which proceeds with loss of

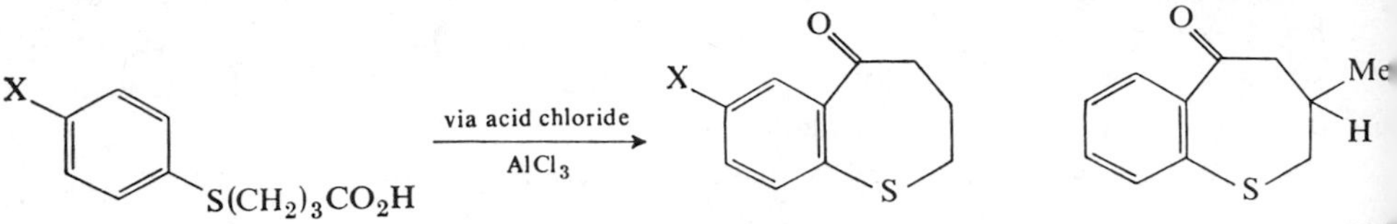

X S(CH$_2$)$_3$CO$_2$H **101** —via acid chloride, AlCl$_3$→ **102**; X = F; **103**; X = Cl; **104**; X = Br (20% yield); **105**; X = *tert*-Bu (59% yield); **106** (42)

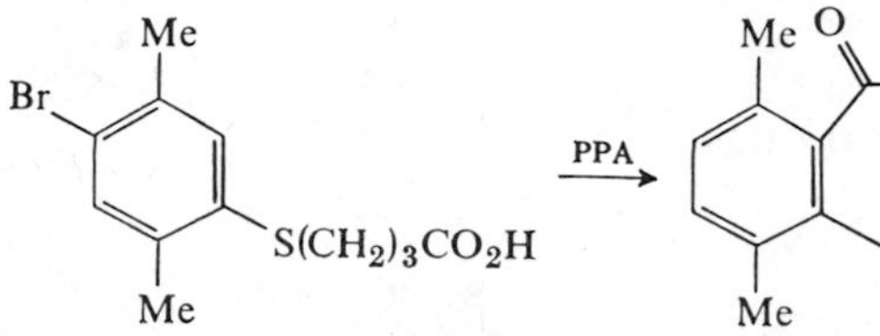

107 —PPA→ **108** (43)

the bromine (Eq. 43). Additional examples involving loss of substituents *para* to the thio function are reported in reactions of PPA with **101** (X = Br and X = *tert*-Bu) which in both instances lead to the unsubstituted ketone **46** (51).

Physical properties for 3,4-dihydro-1-benzothiepin-5(2*H*)-one and its ring substituted derivatives are listed in Table XI-2 together with their corresponding carbonyl derivatives such as oximes, semicarbazones, 2,4-dinitrophenylhydrazones, and azines. A number of these carbonyl derivatives, some not isolated or characterized, are key intermediates in the synthesis of other heterocyclic compounds (Eq. 44). Aksanova, Kucherova, and Zagorevskii (4) have prepared a large number of indole derivatives by application of the Fischer indole synthesis with ketone **46** (see Table XI-2 for the new indoles and their physical constants). Cagniant and Deluzarche (53) initially prepared ketone **46** as a component of the Pfitzinger quinoline synthesis illustrated in Eq. 45, and additional quinoline derivatives (126) were obtained by a Friedlander synthesis on **46** (Eq. 46). Also the conversion of ketone **46** to the spiro hydantoin **109** has been reported in the patent literature (6), and the formation of a spiro dioxolane **110** can be achieved upon prolonged reaction of **46** and ethylene glycol in refluxing benzene (Eq. 47) (294).

The oxime derivative **111** has received particular attention in its conversion to a variety of *O*-acyl and *O*-sulfonyl derivatives (see Table XI-2) which appear to be central nervous system depressants (171). In addition the oxime

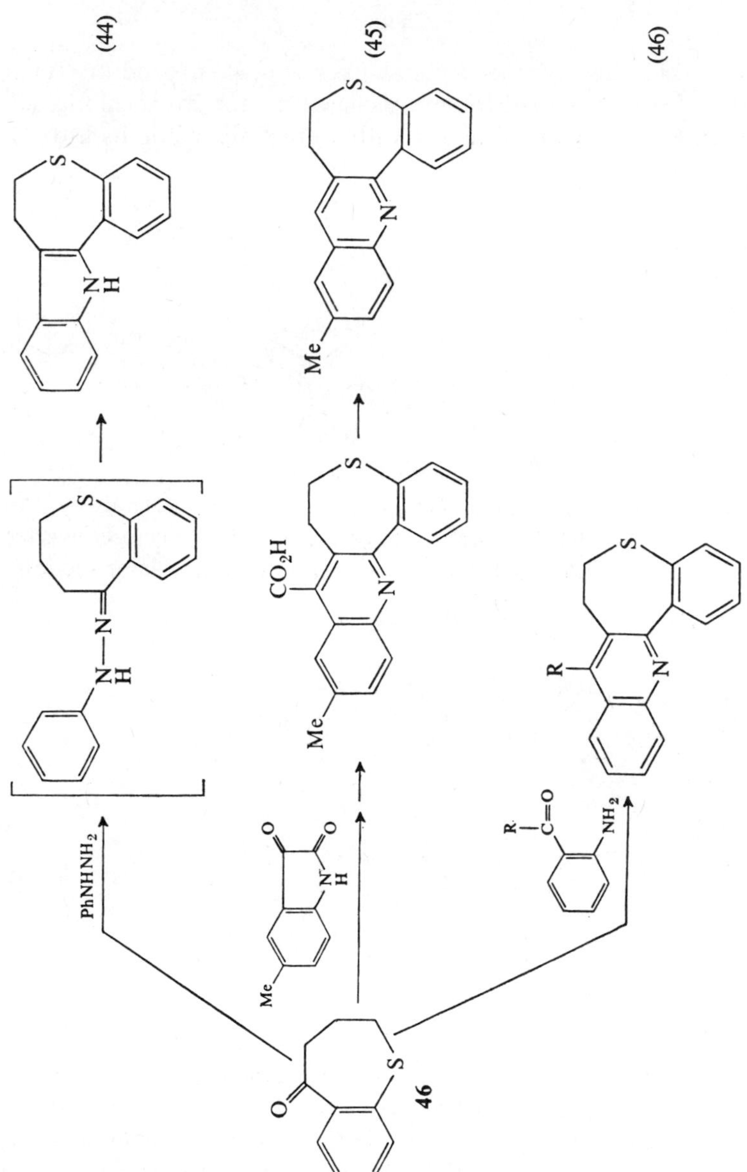

(44)
(45)
(46)
PhNHNH2
46

tosylate **112** can undergo the Neber reaction (Eq. 48) to produce 4-amino-3,4-dihydro-1-benzothiepin-5(2*H*)-one, isolated as the hydrochloric salt (**113**) (76, 173). Reduction of oxime **111** with lithium aluminum hydride (74, 174)

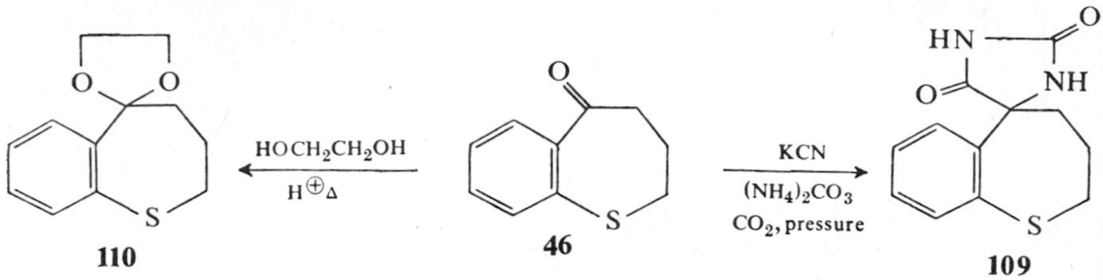

110 ← $HOCH_2CH_2OH$, $H^{\oplus}\Delta$ — **46** — KCN, $(NH_4)_2CO_3$, CO_2, pressure → **109**

(47)

or sodium and alcohol (269) (11% yield) proceeds predictably to 5-amino-2,3,4,5-tetrahydro-1-benzothiepin (**114**) which has been converted to a variety of *N*-substituted derivatives to be described in a later section.

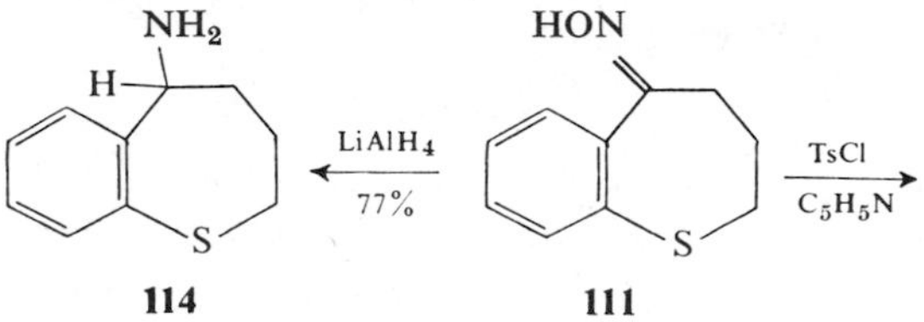

114 ← $LiAlH_4$, 77% — **111** — TsCl, C_5H_5N →

(48)

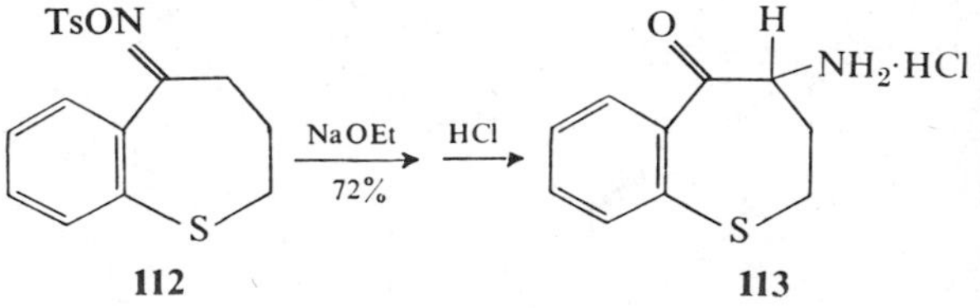

112 — NaOEt, 72% → HCl → **113**

Reduction of ketone **46** to 2,3,4,5-tetrahydro-1-benzothiepin (**53**) via the Clemmensen-Martin method was reported by Cagniant (53) and by Truce (306); however, varying amounts of 2,3-dihydro-1-benzothiepin are also formed in this reaction (152, 294). Alternate conversions of ketone **46** to **53** were achieved by the Wolff-Kishner reduction (294) and by preparation of the tosylhydrazone **115** followed by a lithium aluminum hydride reduction. The reduction of ketone **46** to alcohol **3** is accomplished in high yield by reaction with sodium borohydride (298). Other simple addition reactions at the carbonyl function have been described using Grignard reagents (18, 170), phenyl and pyridyl lithiums (176, 269), and the Reformatsky reaction (170) (Eq. 49).

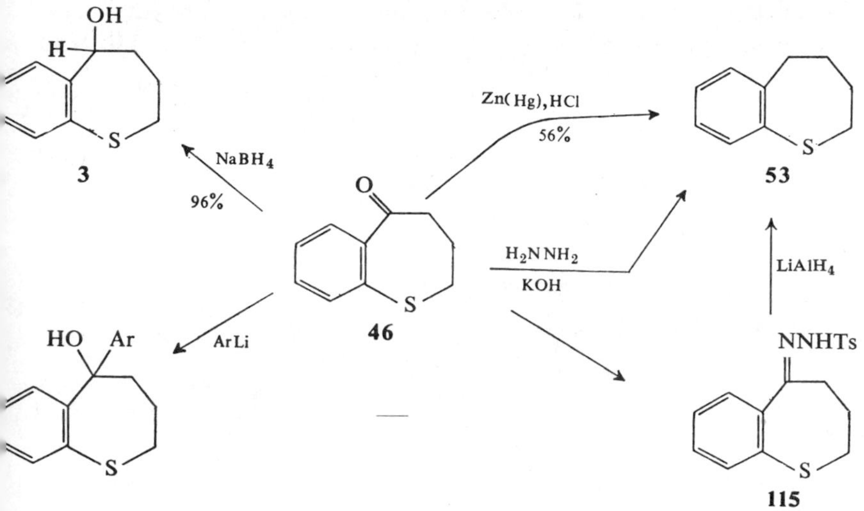

(49)

Derivatives of the sulfide function are also easily prepared (Eq. 50); they include the sulfonium salt **116** (300), the sulfoxide **117** (146), and the sulfone **47** (298). Attempts to effect the Pummerer reaction with **117** were unsuccessful (146).

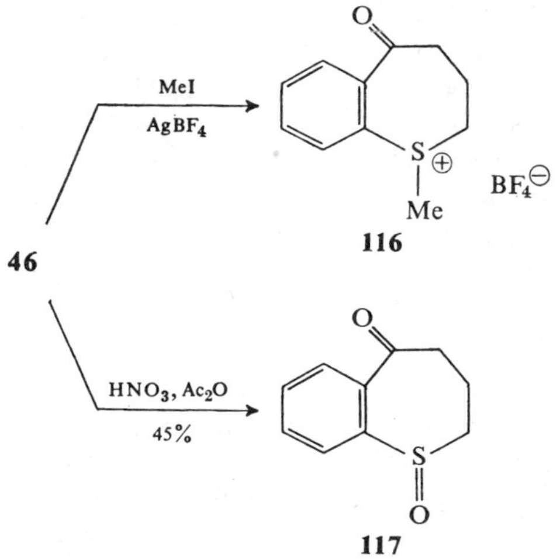

(50)

A variety of halogenated derivatives of ketone **46** have been prepared (Eq. 51). Bromination occurs readily in acetic acid, yielding 4-bromo-3,4-dihydro-1-benzothiepin-5(2*H*)-one (**118**) (152, 269). Evidence for attachment of bromine in the 4-position involves reduction of **118** to the bromohydrin **119** which when treated with base suffers HBr elimination to regenerate ketone **46** (300). Although bromoketone **118** is essentially inert to silver nitrate, reaction of **118** with sodium iodide proceeds readily to form 4-iodo-3,4-dihydro-1-benzothiepin-5(2*H*)-one (**120**) (294). Also, reaction of **118** with

46 —(Br_2, AcOH, 71%)→ **118** —($NaBH_4$)→ **119** —(NaOH)→ 4

46 —(NCS 30% or SO_2Cl_2 50%)→ *cis*-**123**

118 —(NCS 25–30% or SO_2Cl_2 63%)→ *cis*-**121**, mp 134–135°; *trans*-**122**, mp 98–99°

118 —(NaI)→ **120**

(5

N-chlorosuccinimide or sulfuryl chloride results in the introduction of chlorine at $C_{(2)}$ and the formation of two isomers, *cis* and *trans* (300). Sulfuryl chloride provides the *trans* isomer (**122**) as the major component with only a trace of *cis* (**121**); *N*-chlorosuccinimide yields about a 2.5 to 1 ratio of the *trans* to *cis* isomer (300). Attempts to introduce one chlorine into ketone **46** by reaction with 1 mole of sulfuryl chloride led to a mixture of products from which some *cis*-2,4-dichloro-3,4-dihydro-1-benzothiepin-5(2*H*)-one (**123**) was isolated (300). Good yields of **123** are obtainable on treatment of **46** with 2 moles of sulfuryl chloride.

Attempts to use a variety of bases in order to eliminate HBr or HI from ketones **118** and **120**, respectively, were unsuccessful in producing an unsaturated product (152, 294). Either starting material was recovered or under more vigorous conditions resinification occurred. However, Sindelar and Protiva (269) recently reported the conversion of **118**, by action of diethylamine or *N*-methylpiperazine, into the 4-amino ketones **124** and **125** (Eq. 52).

118 + Et_2NH or MeN(piperazine)NH ⟶ **124** (NEt_2) or **125** (N-methylpiperazinyl) (5

A plausible explanation for difficulty in effecting elimination in **118** or **120** involves an unfavorable conformation for *trans* coplanar elimination. An ir spectral study of the influence of the α-bromo group on the carbonyl frequency in **46** and other model compounds revealed a carbonyl frequency shift in **46** of 17 cm^{-1}. This observation suggests that the 4-bromo group exists in a quasiequatorial conformation which is unfavorable for facile elimination (152). A similar ir spectral study with the dichloro ketone **123** shows a carbonyl frequency shift of 25 cm^{-1} likewise indicating the existence of a pseudoequatorial 4-chloro function (300). Supporting evidence for this assignment is found in the solvent effect of benzene on the $C_{(4)}$ proton shift of ketone **123** in the nmr spectrum (300) as generalized by the "carbonyl plane rule" (17, 323, 324).

The 2,4-dichloro ketone **123**, in sharp contrast to bromo ketone **118**, reacts exceedingly rapidly with triethylamine in chloroform (Eq. 53), giving cyclopropyl ketone **126** via 1,3-elimination (300). Other bases such as alcoholic KOH or KOBu-*tert* are equally effective in promoting the 1,3-elimination. The structural assignment of ketone **126** was based primarily on its nmr spectrum and conversion to sulfone **127**. The nmr spectrum of **126** showed the cyclopropyl resonances at τ 6.80 (quartet, $C_{(2)}$—H) and τ 7.76, τ 8.24 (triplet and quartet, respectively, methylene bridge). Further support for the direction of 1,3-elimination was found in the rapid conversion of each chlorobromo ketone **121** and **122** on treatment with trimethylamine in chloroform (Eq. 54) into the same bromocyclopropyl ketone **128** which was also oxidized to its solid sulfone derivative **129** (300).

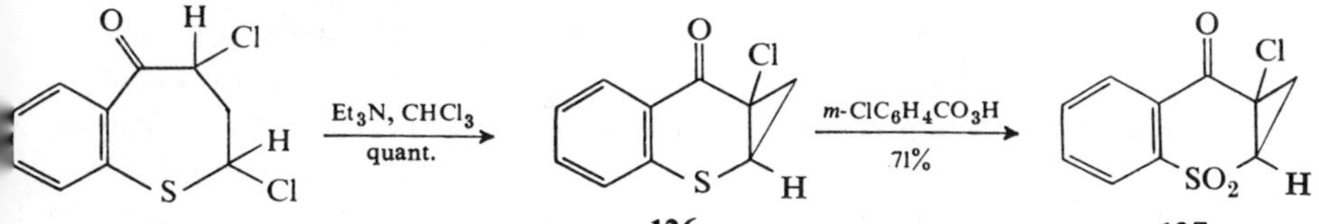

123 —(Et_3N, $CHCl_3$; quant.)→ **126** —(m-$ClC_6H_4CO_3H$; 71%)→ **127** (53)

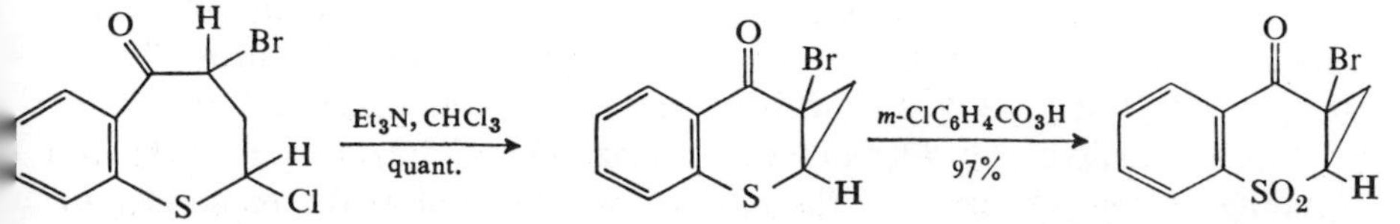

121 or **122** —(Et_3N, $CHCl_3$; quant.)→ **128** —(m-$ClC_6H_4CO_3H$; 97%)→ **129** (54)

Cyclopropyl ketone **126** undergoes a rapid stereospecific ring opening reaction with hydrogen chloride giving *trans*-2,4-dichloro-3,4-dihydro-1-benzothiepin-5(2*H*)-one **(130)** (Eq. 55) (300). The *cis*-dichloroketone **123** was stable under the experimental conditions of the ring-opening reaction and thus should have been detected if formed. Reaction of *trans*-**130** with triethylamine in chloroform resulted in rapid ring closure to regenerate **126** (300).

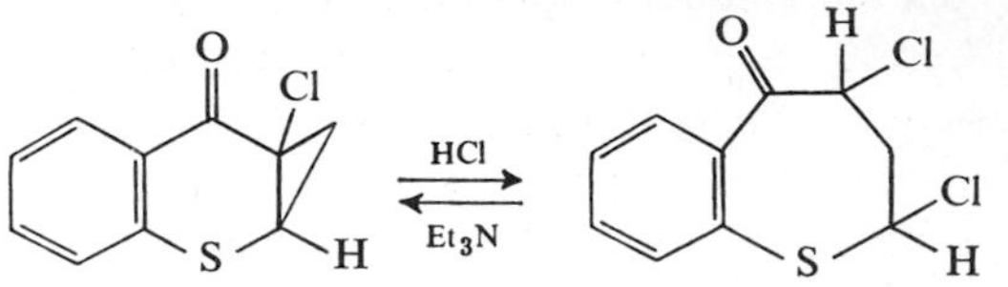

(55)

126 $\underset{Et_3N}{\overset{HCl}{\rightleftharpoons}}$ *trans*-**130**

Although bromo ketone **118** is inert to exposure to pyridine and acetic anhydride, the *cis*- and *trans*-dichloro ketones **123** and **130** both react readily (Eq. 56) to give enol acetate **131** (300). A similar enol acetate **132** is available from cyclopropyl ketone **126** upon treatment with acetic anhydride and pyridine (Eq. 57). The facile conversion of enol acetate **131** to the cyclopropyl ketone **126** by reaction with alcoholic KOH is noteworthy.

Another broad group of reactions of ketone **46** involves condensations at the activated $C_{(4)}$ position. When **46** was subjected to the Mannich reaction using dimethylamine hydrochloride and paraformaldehyde in isoamyl alcohol and a small amount of concentrated hydrochloric acid (Eq. 58), two products were isolated, 4-[(dimethylamino)methyl]-3,4-dihydro-1-benzothiepin-5(2*H*)-one hydrochloride (**133** · HCl) (23%) and 2,3,3′,4′,5′,6′-hexahydrospiro(1-benzothiepin-4(5*H*)-2′(2*H*)-1-benzothiepino[5,4-*b*]pyran)-5-one (**134**) (25%) (172, 294). The use of piperidine hydrochloride in place of dimethylamine hydrochloride in the preceding reaction gave the Mannich base **135** as well as **134** (269). A modification of the above procedure employing ketone **46**, formalin, dimethylamine, ethanol, and acetic acid leads to an 86% yield of compound **134** (294). Compound **134** is also available by heating the free base of **133** (172). The structural assignment (269, 294) for **134** was based on elemental analysis, molecular weight, ir spectrum, and analogy to similar Diels-Alder dimerizations of 2-methylenecyclohexanone and 2-methylenecycloheptanone (245, 246).

The base-catalyzed condensation reaction between ketone **46** and benzaldehyde (Eq. 59) can lead to 4-(α-hydroxybenzyl)-3,4-dihydro-1-benzothiepin-5(2*H*)-one (**136**), 4-benzylidene-1-benzothiepin-5(2*H*,3*H*)-one (**137**), or 4,4′-benzylidene bis(3,4-dihydro-1-benzothiepin-5(2*H*)-one) (**141**) depending on temperature and the alcohol solvent used in the reaction (36). The hydroxy ketone **136** was also converted into **137** by acid-catalyzed dehydration. An alternate procedure used in the preparation of **137** involved an extension (36) of the scheme devised by Ireland (108) for the synthesis of an α-benzylidene ketone. Reaction of **46** with ethyl formate and sodium methoxide resulted in the formation of hydroxymethylene derivative **138**

123 or 130 $\xrightarrow[63\%]{Ac_2O,\ C_5H_5N}$ 131 $\xrightarrow[ROH]{KOH}$ 126 (56)

126 $\xrightarrow{Ac_2O}$ [⟷] ⟶ 132 (57)

46 → 133·HCl + 134 (CH_2O, $Me_2NH \cdot HCl$, *i*-AmOH)

133·HCl → 133 → (Δ) → [ ] → 134

135

46 → 141 (NaOH, PhCHO, 60% EtOH reflux, 36%)

46 → 136 (NaOH, PhCHO, 80% *tert*-BuOH, room temp, 58%)

136 → 137 (TsOH, C_6H_6, 29%)

46 → 137 (NaOH, PhCHO, 60% EtOH, room temp, 52%)

46 → 138 (HCO_2Et, MeONa)

138 → 139 (morpholine, 71% from 46)

139 → 137 (PhMgBr, 57%)

139 → 140 (MeMgI)

137 → 142

137 → 143 (H_2O_2, AcOH, 57%)

47 + PhCHO → 143 (NaOH, EtOH, 50%)

(59)

which had the characteristic ir peaks at 1629 and 1585 cm^{-1}. Conversion of **138** to enamine **139** was effected by reaction with morpholine; further reaction of phenyl magnesium bromide with **139** gave the benzylidene derivative **137**. The enamine synthesis was also extended to the preparation of 4-ethylidene-1-benzothiepin-5(2*H*,3*H*)-one (**140**).

The sulfoxide **142** was prepared by mild hydrogen peroxide oxidation of **137**; more vigorous oxidative conditions produced the sulfone **143** (36). An alternate synthesis of **143** was accomplished by base-catalyzed condensation of **47** and benzaldehyde. Structural assignments were based on (*1*) the ozonolysis of **143** to benzaldehyde and (*2*) the uv and nmr spectra of the benzylidene derivatives which showed conjugation of the benzylidene group with the carbonyl function (36). The nmr spectrum of **140**, which contained a doublet at τ 8.11 (Me), a multiplet at τ 7.25 (ring methylene groups), a quartet at τ 3.06 (the olefinic H), and a multiplet at τ 2.61 (phenyl ring hydrogens), provided strong evidence for the assigned structure.

Interest in 4-benzylidene-1-benzothiepin-5(2*H*,3*H*)-one (**137**) and the corresponding ethylidene derivative **140** centered on attempts to isomerize the exocyclic double bond into the ring (36). The procedure described by Leonard (142, 145) utilizing 10% Pd/C was employed; however, only unreacted starting materials were recovered (75–85%) in these reactions (36).

The preparation of 3,4-dihydro-1-benzothiepin-5(2*H*)-one-1,1-dioxide (**47**) is accomplished conveniently (Eq. 60) by oxidation of ketone **46** with hydrogen peroxide in acetic acid (298). Physical properties of **47** and its carbonyl derivatives are listed in Table XI-2. Some of the derivatives of the keto sulfone **47** undergo reactions similar to those described earlier for the corresponding sulfide series (Eq. 60). Oxime **144** is reduced readily by lithium aluminum hydride to the corresponding amine **145** (75); numerous

46 $\xrightarrow[\text{AcOH}]{H_2O_2}$ **47** → **144** (NOH, SO_2) $\xrightarrow{\text{LiAlH}_4}$ **145** (H NH_2, SO_2)

47 → **146** (NNHPh, SO_2) $\xrightarrow{H^{\oplus}}$ (indole-fused product, N–H, SO_2)

(60)

phenylhydrazone derivatives (**146**) are converted by acid into the corresponding indoles (4) (see Table XI-2 for a list of new indoles).

Direct bromination of **47** leads to 4-bromo-3,4-dihydro-1-benzothiepin-5-(2*H*)-one-1,1-dioxide (**147**) in 92% yield (Eq. 61); alternatively, oxidation of **118** with hydrogen peroxide affords **147** in 39% yield (152, 269). The introduction of chlorine into the 4-position to generate **148** can be accomplished by reaction of **47** with sulfuryl chloride (300). Oxidation of *cis*-2,4-dichloro-3,4-dihydro-1-benzothiepin-5(2*H*)-one (**123**) with *m*-chloroperbenzoic acid (Eq. 62) yields the corresponding *cis*-dichlorosulfone **149**. On treatment of **149** with triethylamine in benzene, instead of a 1,3-elimination reaction as observed with **123** to produce the chlorocyclopropyl sulfone **127**, isomerization to *trans*-2,4-dichloro-3,4-dihydro-1-benzothiepin-5(2*H*)-one-1,1-dioxide (**150**) is

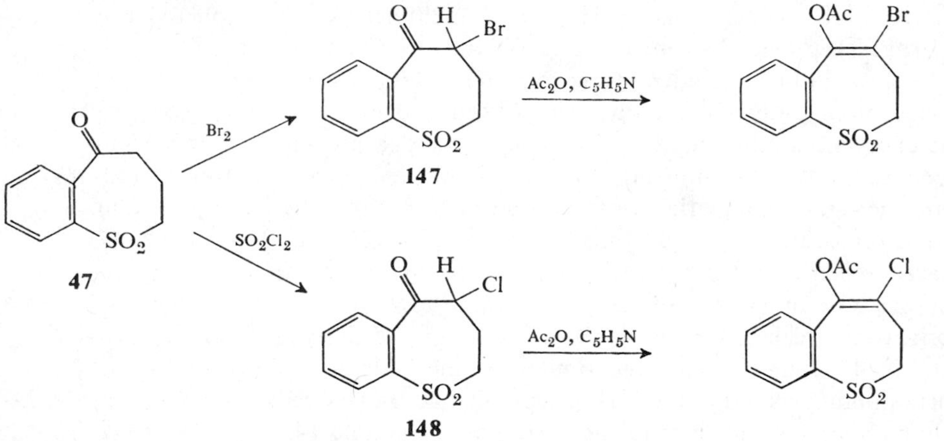

O H Br; SO_2; **147**; Ac_2O, C_5H_5N; OAc Br; SO_2; O; SO_2; **47**; Br_2; SO_2Cl_2; O H Cl; SO_2; **148**; Ac_2O, C_5H_5N; OAc Cl; SO_2

(61)

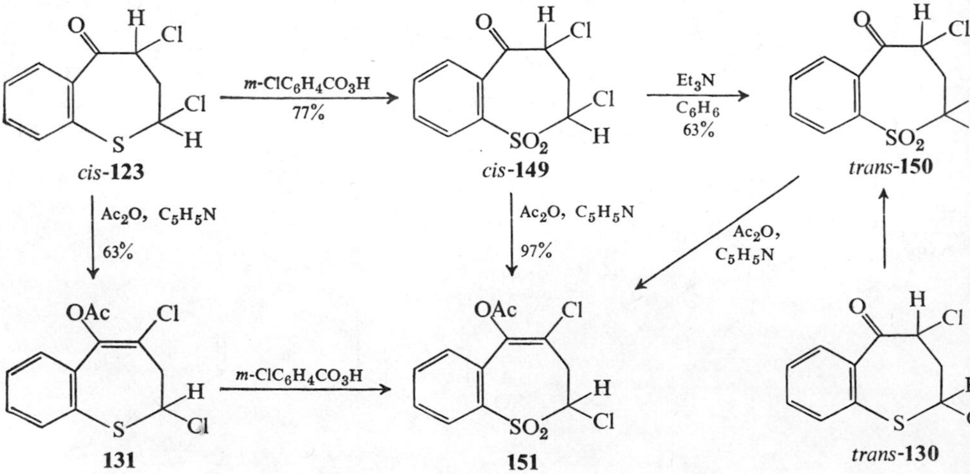

O H Cl; Cl; S H; *cis*-**123**; *m*-$ClC_6H_4CO_3H$ 77%; O H Cl; Cl; SO_2 H; *cis*-**149**; Et_3N C_6H_6 63%; O H Cl; H; SO_2 Cl; *trans*-**150**; Ac_2O, C_5H_5N 63%; Ac_2O, C_5H_5N 97%; Ac_2O, C_5H_5N; OAc Cl; H; S Cl; **131**; *m*-$ClC_6H_4CO_3H$; OAc Cl; H; SO_2 Cl; **151**; O H Cl; H; S Cl; *trans*-**130**

(62)

obtained (300). The *trans*-sulfone **150** was also obtained by oxidation of the *trans*-dichloroketone **130** with *m*-chloroperbenzoic acid. A plausible explanation for the fact that isomerization is observed instead of 1,3-elimination may be that in **149** abstraction of the $C_{(2)}$ hydrogen occurs in preference to removal of the $C_{(4)}$ hydrogen (300). However, the reaction of either **149** or **150** with acetic anhydride and pyridine (Eq. 62) leads to an enol acetate **151** which is the same as that obtained from oxidation of enol acetate **131**. The formation of enol acetate **151** raises the possibility of generating an enolate ion at $C_{(4)}$. Thus failure to obtain the cyclopropyl ring may be a reflection of an unfavorable conformation for 1,3-elimination (300). These matters are still under investigation. Enol acetates can also be prepared from the bromo ketone **147** and chloro ketone **148** (Eq. 61).

6. 1-Benzothiepin-3,5(2*H*,4*H*)-dione and Derivatives

Examples of diketones in this system were described earlier (Sections C-1 and C-2). One synthetic sequence (Eq. 63) involved ring expansion of an α,β-epoxyketone to produce 4-phenyl derivatives **37** and **56** (184). Subsequently this approach was extended to other examples (106); however, rearrangements in the sulfoxide series were unsuccessful. Compound **37**

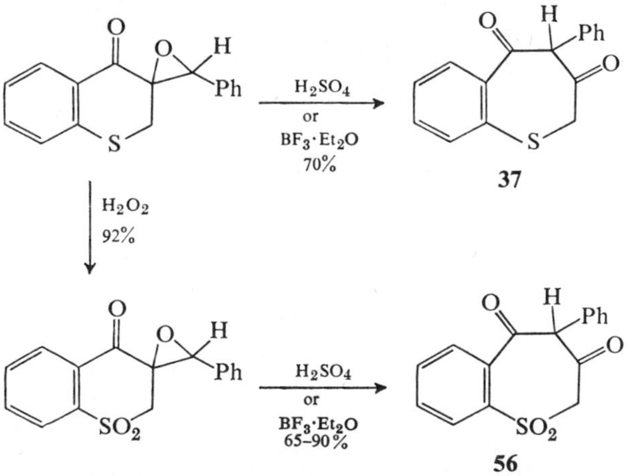

H_2SO_4 or $BF_3 \cdot Et_2O$ 70% → **37**

H_2O_2 92%

H_2SO_4 or $BF_3 \cdot Et_2O$ 65–90% → **56**

(63)

exhibited two carbonyl frequencies in the ir region at 1732 and 1673 cm^{-1} and showed no enolic hydroxyl group. The nmr spectrum of **37** exhibited a singlet at τ 6.5 for the $C_{(2)}$ methylene group and a singlet at τ 3.7 for the $C_{(4)}$ proton.

Reaction of **37** with acetic anhydride and pyridine (Eq. 64) for 1 hr produced a monoenol acetate **152**; however, upon prolonged reaction

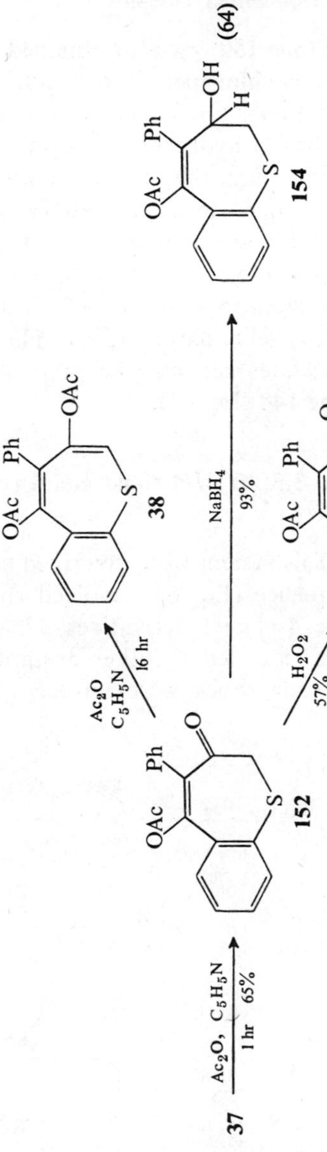

37 $\xrightarrow[\text{1 hr} \quad 65\%]{Ac_2O,\ C_5H_5N}$ 152

152 $\xrightarrow[\text{16 hr}]{Ac_2O,\ C_5H_5N}$ 38

152 $\xrightarrow[93\%]{NaBH_4}$ 154 (64)

152 $\xrightarrow[57\%]{H_2O_2}$ 153

OAc Ph O S **152**

OAc Ph OAc S **38**

OAc Ph O SO_2 **153**

OAc Ph OH H S **154**

(16 hr) the dienol acetate **38** was isolated. When **56** was treated with acetic anhydride and pyridine, only the dienol acetate **58** was isolated even after 15 min reaction time. For the preparation of monoenol acetate **153**, hydrogen peroxide oxidation of **152** was required. The structural assignment for **152** was based primarily on its nmr spectrum, which showed a singlet at τ 6.52 for the $C_{(2)}$ methylene hydrogens, and on the sodium borohydride reduction to the allyl alcohol **154**, which showed the $C_{(2)}$ hydrogens as a doublet (106).

Parham and Weetman (192) reported the isolation of 2-methyl-1-benzothiepin-3,5(2*H*,4*H*)-dione (**34**) which apparently arose from hydrolysis of 2-methyl-3-methoxy-5-ethoxy-1-benzothiepin (**30**) (see Section C-1). Also, hydrolysis of enol ether **33** formed simultaneous in this reaction led to dione **34**. Structural assignment for **34** was supported by nmr, ir, uv, and mass

(65)

spectral data, and by its oxidation with sodium hypochlorite (Eq. 65) to compound **155**, identified by independent synthesis.

7. 2,3,4,5-Tetrahydro-1-benzothiepin and Derivatives

The parent compound in this series has been prepared from 3,4-dihydro-1-benzothiepin-5(2*H*)-one (**46**) by a variety of reductive procedures described earlier (see Section C-5). One interesting aspect of this system relates to the study of electron release by sulfur as manifest in effects on the uv spectrum. Studies of the effect of ring size and conformation including 2,3,4,5-tetrahydro-1-benzothiepin have been reported (83, 133). In addition, systems in which the ring sulfur is conjugated to keto functions as in 7-acetyl-2,3,4,5-tetrahydro-1-benzothiepin (**156**) (83) and compound **46** (133) were also studied. Preparation of ketone **156** was achieved (Eq. 66) via a Friedel-Crafts acylation of **53** (58, 83). Initial interest in the preparation of **156** was for its conversion into quinoline derivatives by the Pfitzinger procedure (53).

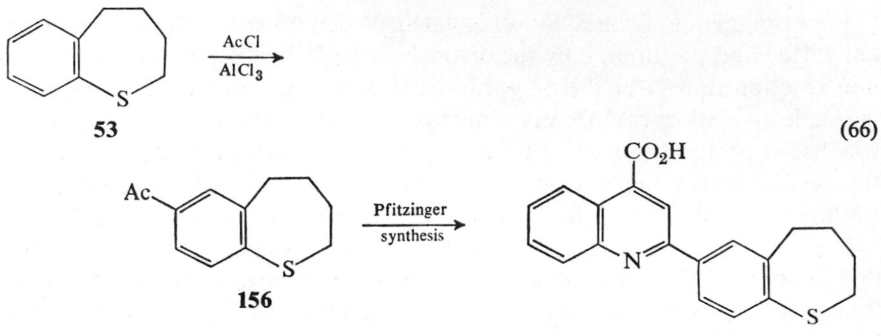

(66)

Preparations of 2,3,4,5-tetrahydro-1-benzothiepin-5-ol (**3**) and derivatives which have been described earlier (see Section C-5) entail either hydride reduction or organometallic addition reactions with ketone **46**. Reduction of amino ketone **113** with sodium borohydride leads to amino alcohol **157**, which served as a starting point for the synthesis (Eq. 67) of derivatives containing an isoxazole ring fused to the 1-benzothiepin ring (173).

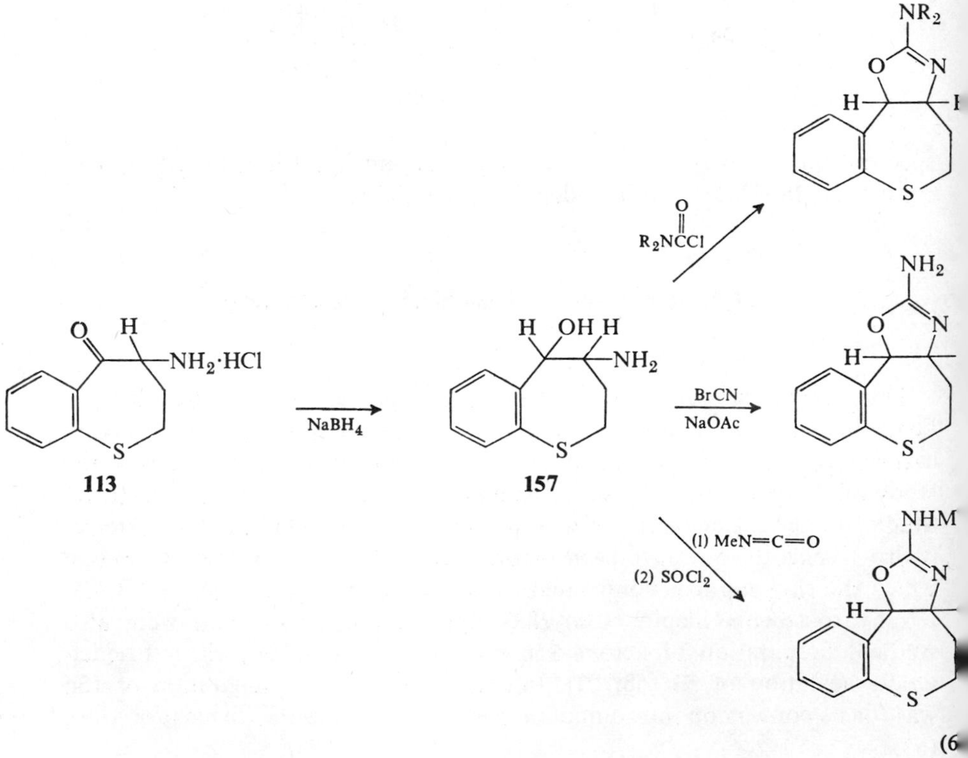

(6

The oxidation of alcohol **3** to the corresponding sulfone **48** is accomplished by reaction with hydrogen peroxide in acetic acid (298); however, when **3** is treated with red fuming nitric acid in acetic anhydride, the sulfoxide **158** is isolated in 64% yield (Eq. 68). Also, oxidation of **3** with sodium metaperiodate leads to sulfoxide **158** (294). Characterization of the sulfoxide involved conversion to the known sulfone **48** and to the methoxysulfonium fluoroborate salt **159** (300).

Typical reactions at the hydroxy function of **3** have already been described and include dehydration and esterification procedures. Conversion of **3** to an ether has been accomplished by the acid-catalyzed reaction with methanol (152) and by application of the Williamson ether synthesis (174). The latter

H OH S **3** $\xrightarrow[\text{or } NaIO_4]{HNO_3,\ Ac_2O\ 67\%}$ H OH S=O **158** $\xrightarrow[69\%]{H_2O_2}$ H OH SO_2 **48**; **158** $\xrightarrow[AgBF_4]{MeI}$ H OH $S^{\oplus}$–O–Me $BF_4^{\ominus}$ **159** (68)

method has been used in the synthesis of a variety of alkylamino ethers (Eq. 69). Interest in these amino ethers arises from their hypotensive activity (174). Reaction of **3** with concentrated hydrochloric acid easily formed 5-chloro-2,3,4,5-tetrahydro-1-benzothiepin (**66**) (146, 269). Since **66** appeared to be unstable, it was characterized by conversion to the sulfoxide and sulfone derivatives **160** and **161**, respectively (146).

The preparation of 5-amino-2,3,4,5-tetrahydro-1-benzothiepin (**114**) (74, 174) or its 1,1-dioxide (**145**) (75) has been mentioned previously (Section C-5). In addition to the parent compound **114**, various *N*-alkylated derivatives were prepared (Eq. 70) by formation of the appropriate amide followed by lithium aluminum hydride reduction to the amine (77, 174). Two other reported methods of synthesis of amines entail reaction of 5-chloro-2,3,4,5-tetrahydro-1-benzothiepin (**66**) with various cyclic amines (Eq. 71) (269) or the application of the Leuckart reaction (Eq. 72) on ketone **46** (269). The secondary amines are capable of further alkylation reactions (Eq. 72).

8. 2,3,4,5-Tetrahydro-1-benzothiepin-1,1-dioxide and Derivatives

Several of the sulfone derivatives belonging to this group have been described in the preceding section. The parent compound **52** is readily available by oxidation of the corresponding sulfide **53** or reduction of 1-benzothiepin-1,1-dioxide (**51**) (see Section C-3). A second method for preparing fused thiepin sulfones involves Friedel-Crafts ring closure with sulfonyl chlorides as described by Truce (306). The reaction of 4-phenyl-1-butanesulfonyl chloride (**162**) and aluminum chloride in nitrobenzene (Eq. 73) provided **52** in about 30% yield. Similar ring closure reactions to form the five- or six-membered sulfone were more facile; in a competitive

114

ClC(O)C(O)Cl, 80%

H NH—C(=O)—C(=O)—NH H

$LiAlH_4$, 70%

H HN—CH_2—CH_2—NH H

(70)

(1) EtC(O)Cl (2) $LiAlH_4$

H NHPr-*n*

(1) AcCl (2) $LiAlH_4$

H N(Et)—*n*-Pr

H Cl

66 + N H → N H (71)

O

46 + MeNHCH(=O) $\xrightarrow[70\%]{HCO_2H}$ H NHMe $\xrightarrow{HC \equiv CCH_2Br}$ Me H N—$CH_2C \equiv CH$ (72)

experiment between formation of a five- or seven-membered sulfone (Eq. 74), the five-membered ring was generated (304).

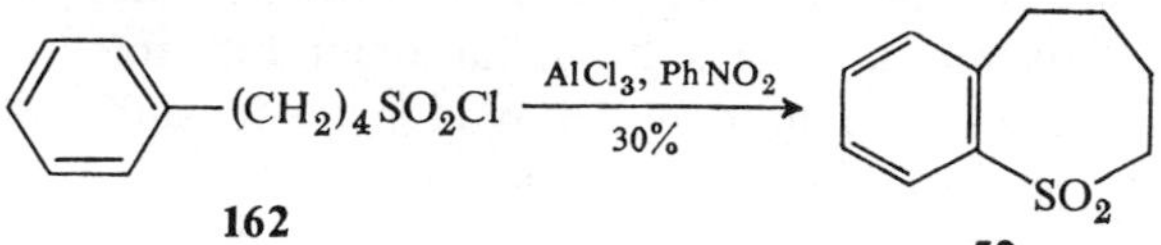

(73)

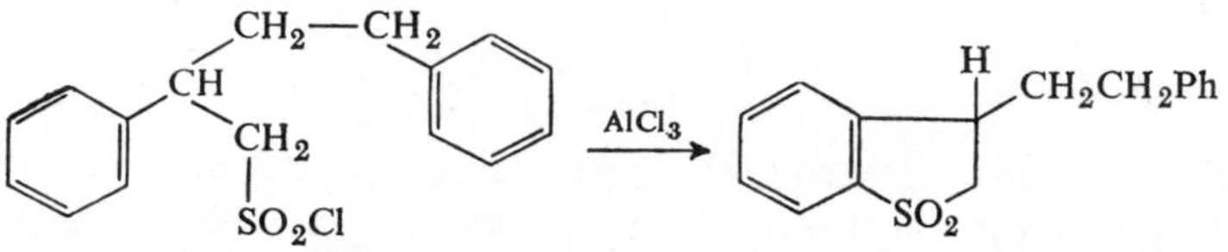

(74)

Nitration of **52** yields 8-nitro-2,3,4,5-tetrahydro-1-benzothiepin-1,1-dioxide (**163**) (299) which can be reduced catalytically to the amine **63** (Eq. 75). This serves as an independent synthesis of **63** which has also been obtained by reduction of 8-nitro-1-benzothiepin-1,1-dioxide (**62**). The location of the nitro function in **163** was established by oxidation to 4-nitro-2-sulfobenzoic acid.

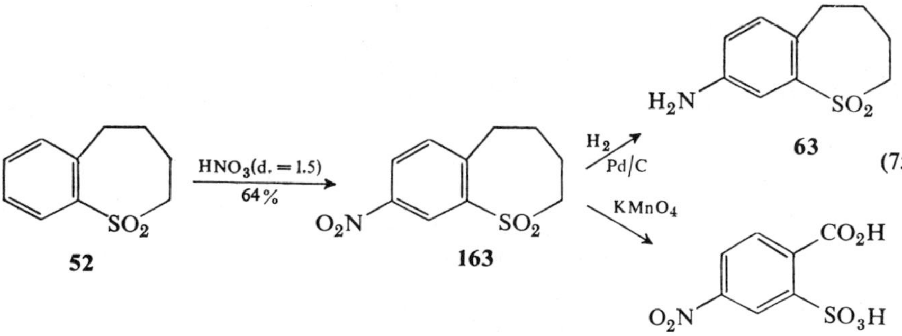

(75)

D. 2-Benzothiepins

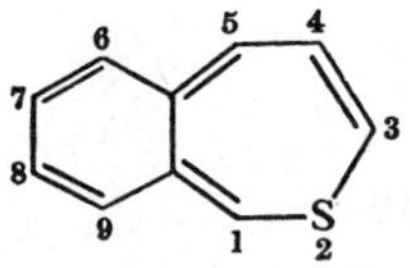

The parent member (**164**) of this system is unknown and no attempted syntheses have been reported. One would anticipate that **164**, with its quinoid structure, would be less stable than the isomeric 1-benzothiepin or 3-benzothiepin and that the sulfur extrusion reaction would lead to naphthalene (Eq. 76). One interesting consideration in the extrusion reaction is that if **164** is converted into its valence bond tautomer **165**, the latter contains a more favorable benzenoid unit which may provide sufficient stability to allow the presence of this species to be detected.

164 ⇄ **165** → naphthalene + S (76)

The first compound reported in this series (Eq. 77) was 1,3,4,5-tetrahydro-2-benzothiepin (**166**), which was initially prepared by von Braun in 1925 by

167 (CH_2Br, CH_2Br) $\xrightarrow[40\%]{Na_2S}$ **166** $\xleftarrow[KOH]{H_2NNH_2}$ **168**

$Ph(CH_2)_3SH + RCHO + HCl \xrightarrow{92\%} Ph(CH_2)_3SCHCl(R)$ **169** (R = H, Me, Et) $\xrightarrow{AlCl_3, CS_2;\ 44\%, R=H}$ **166** (77)

166 $\xrightarrow[h\nu]{Cl_2}$ **170** (Cl, H)

170 $\xrightarrow[20\%]{RMgX}$ (R, H) R = Me, Ph

170 $\xrightarrow[22\%]{EtOH,\ C_5H_5N}$ (H, OEt)

170 $\xrightarrow[26\%]{Hg(CN)_2}$ **171** (H, CN) → (H, CO…)

$Ph(CH_2)_3SCH_2CN \xrightarrow[49\%]{Cl_2} Ph(CH_2)_3SCHCN(Cl) \xrightarrow[20\%]{AlCl_3}$ **171**

reaction of sodium sulfide with **167** (39). Subsequently, two additional preparations were reported by Cagniant (49) and by Böhme (33). Cagniant's procedure entailed a Wolff-Kishner reduction of ketone **168**. In his account Cagniant did not assign structures to ketone **168** or the reduction product **166**; however, in view of the similarity of his physical properties for **166** and those reported for the same compound by Böhme, the proposed assignments appear to be correct. Böhme developed a general ring-closure method involving intramolecular Friedel-Crafts reaction of α-chloro sulfides (**169**). In addition to the parent compound **166**, this method also provided a route to the 1-methyl and 1-ethyl derivatives (33). An alternate pathway permitting the introduction of substituents into the $C_{(1)}$ position (Eq. 78) involved chlorination of **166** to form 1-chloro-1,3,4,5-tetrahydro-2-benzothiepin (**170**) followed by alkylation with Grignard reagents or nucleophilic substitution with alcohol and cyanide. The nitrile **171** was also prepared by Friedel-Crafts ring closure (Eq. 78). Most of the above-mentioned sulfides were oxidized readily with 30% hydrogen peroxide to the corresponding sulfones (see Table XI-3 for examples).

Another method for preparing 1-methyl-1,3,4,5-tetrahydro-2-benzothiepin (**173**) involved a ring contraction of ketone **172** under Clemmensen conditions

O S Zn·Hg HCl S + Me H S (79)

172 **173**

(Eq. 79). The yields were poor and mixtures of sulfides were obtained. Ring contraction reactions of β-keto sulfides during Clemmensen reduction have been observed previously by von Braun and Weissbach (41) with 4-keto-isothiochroman, and by Leonard and Figueras (143) with 3-thiepanone

O Me Cl—C—CH S CH_2 CH_2CH_2 $AlCl_3$ H Me S (80)

174 **173**

O Me Cl—C—CH S CH_2CH_2 $AlCl_3$ H Me S (81)

175 **176**

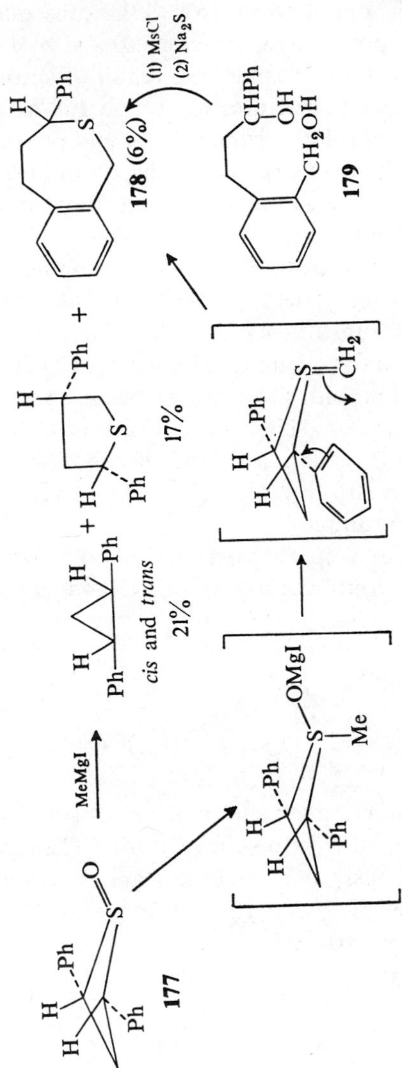
MeMgI
177
cis and trans
21%
17%
178 (6%)
(1) MsCl
(2) Na$_2$S
179
CHPh
OH
CH$_2$OH
OMgI
Me
CH$_2$
(82)

(see Chapter X in this Volume). A third process which led to **173** involved loss of CO during the Friedel-Crafts ring closure of **174** (Eq. 80) (49). Similar ring-closure reactions where CO is lost from α-amino acid chlorides had been reported earlier by von Braun (40, 42). Cagniant (49) also added an example (Eq. 81) wherein a six-membered ring product **176** resulted from acid chloride **175**.

A recent communication by Dodson and Hammen (72) disclosed an interesting reaction of *cis*-2,4-diphenylthietane-1-oxide (**177**) and methylmagnesium iodide which gave as a minor product 3-phenyl-1,3,4,5-tetrahydro-2-benzothiepin (**178**) (Eq. 82). Evidence for the assignment of structure **178** entailed oxidation to its sulfone and an independent synthesis of **178** from **179**. An explanation for the formation of **178** involves a rearrangement of a sulfonium ylide (Eq. 82).

Introduction of substituents into the $C_{(1)}$ position can be effected by alkylation of **171** or 1,3,4,5-tetrahydro-2-benzothiepin-2,2-dioxide (**180**) as shown in Eq. 83 (34).

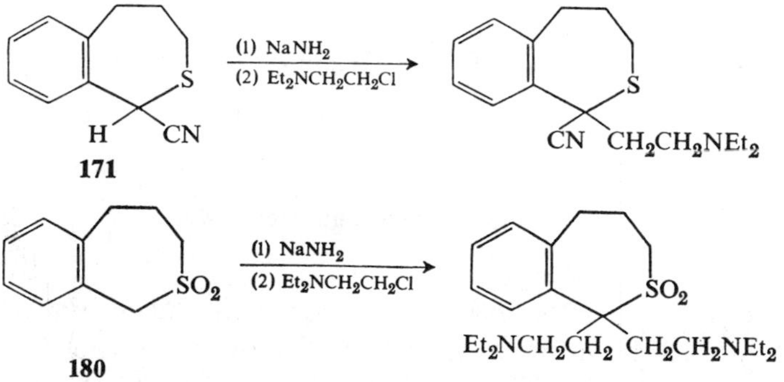

(1) $NaNH_2$ (2) $Et_2NCH_2CH_2Cl$

H CN **171** → CN $CH_2CH_2NEt_2$ (83)

(1) $NaNH_2$ (2) $Et_2NCH_2CH_2Cl$

SO_2 **180** → SO_2 $Et_2NCH_2CH_2$ $CH_2CH_2NEt_2$

The reaction of 3-benzylthiopropionyl chloride (**181**) with aluminum chloride (Eq. 84) produced a ketone which appears to be 3,4-dihydro-2-benzothiepin-5(1*H*)-one (**168**) (49). Although Cagniant (49) did not make a structural assignment for this ketone, its characterization and reduction to a sulfide (assigned structure **166** above) support the proposed structure.

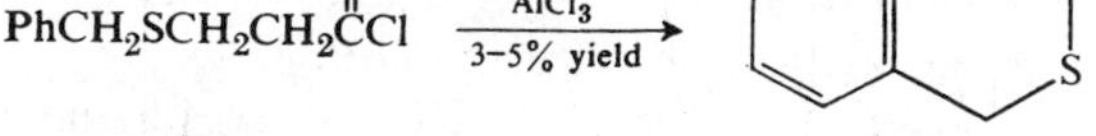

$PhCH_2SCH_2CH_2CCl$ (C=O) **181** $\xrightarrow[3\text{–}5\%\ \text{yield}]{AlCl_3}$ **168** (84)

The synthesis of ketone **168** was recently achieved in good yield by an interesting ring-expansion sequence outlined in Eq. 85 (154).

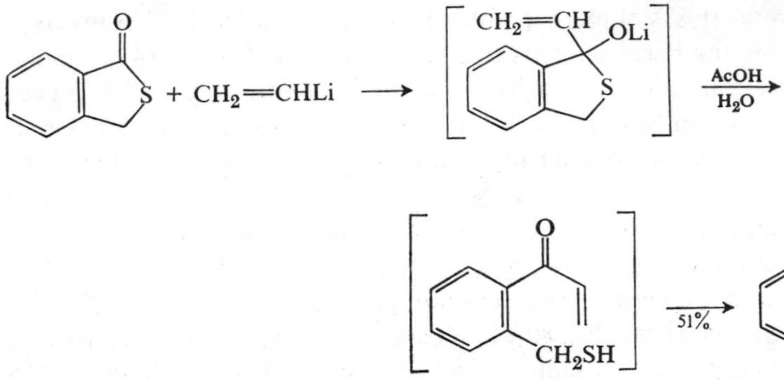

$$\text{[1(3H)-benzothiophenone...]} + CH_2{=}CHLi \longrightarrow \left[CH_2{=}CH,\ OLi \right] \xrightarrow[H_2O]{AcOH}$$

(85)

$$\left[CH_2SH \right] \xrightarrow{51\%} \mathbf{168}$$

E. 3-Benzothiepins

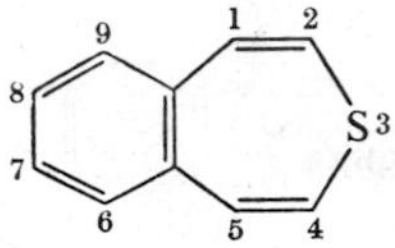

R.I. 1853

1. 3-Benzothiepins and Derivatives

The first example of a fully unsaturated benzothiepin derivative was reported in this system when Scott (261) described the preparation of 3-benzothiepin-2,4-dicarboxylic acid (**182**) in 1953. The acid **182** was isolated from the base-catalyzed (NaOMe) condensation reaction of *o*-phthalaldehyde and ethyl thiodiacetate (Eq. 86) but appeared to be unstable, losing sulfur upon standing, or more rapidly when heated in ethanol, to form naphthalene-2,3-dicarboxylic acid. Dimroth and Lenke (70, 71) observed that conversion of **182** into its methyl ester **183** (via diazomethane) or ethyl ester **184** (via esterification of the diacid chloride of **182**) led to enhanced stability and provided compounds suitable for further study. By application of these reactions Dallacker, Glombitza, and Lipp (61) also prepared 7,8-methylenedioxy-3-benzothiepin-2,4-dicarboxylic acid and its methyl ester **185** which readily undergoes thermal sulfur extrusion (Eq. 87).

The structure for acid **182** and its ester derivatives assigned by Scott and Dimroth was questioned by Schonberg and Fayez (259) who favored the quinoid structure **186**. Their argument was based on the color of the acid and the ease with which **182** was reduced with metal and mineral acid to

CHO CHO + CH_2CO_2Et —S— CH_2CO_2Et $\xrightarrow{NaOMe}$ $\xrightarrow[22\%]{H^{\oplus}}$ **182** (CO_2H, S, CO_2H)

182 $\xrightarrow[\Delta \; 92\%]{(1)\ EtOH}$ CO_2H, CO_2H + S

182 $\xrightarrow[60\%]{CH_2N_2}$ **183** (CO_2Me, S, CO_2Me) (86)

182 $\xrightarrow{(1)\ SOCl_2 \; (2)\ EtOH}$ **184** (CO_2Et, S, CO_2Et)

185 (CO_2Me, S, CO_2Me; O, O) $\xrightarrow[5\ min \; 90\%]{140°}$ CO_2Me, CO_2Me (O, O) + S (87)

R.I. 10,400

naphthalene-2,3-dicarboxylic acid (Eq. 88). However, Jorgenson (118) subsequently presented spectral evidence supporting structure **182**. On the

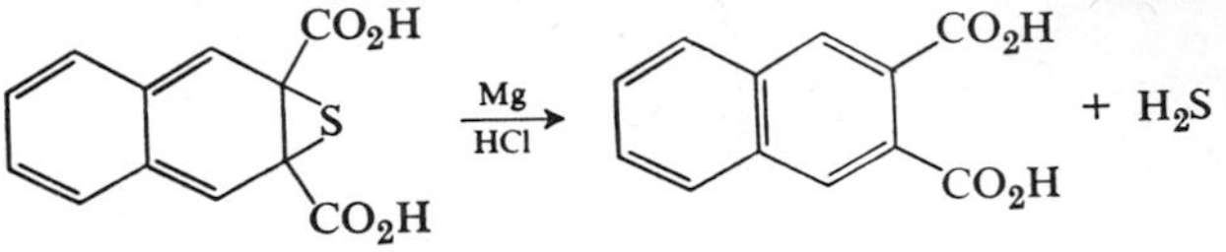
CO2H
S
CO2H
Mg
HCl
CO2H
CO2H
+ H2S

186

(88)

basis of a detailed analysis of the nmr and other spectral properties for the analogous 3-benzoxepin-2,4-dicarboxylic acid and some of its reduced derivatives, the benzothiepin structure was confirmed. A similar application of the nmr argument to the dimethyl ester **183** (τ 2.70m, aromatic H; τ 2.06s, $C_{(1)}$ and $C_{(5)}$ hydrogens) provided support for the structural assignment made by Scott and Dimroth.

The most striking feature in this system is the ease with which the acid **182** undergoes sulfur extrusion in contrast to its ester derivatives. On the basis of this instability Scott (261) concluded that the unsaturated seven-membered ring possesses no aromatic properties; however, Dimroth (70, 71) implied aromatic character by using the trivial name "4,5-benzothiatropilidene dicarboxylic acid (2,7)." Dimroth and Lenke (71) showed that ester **183**

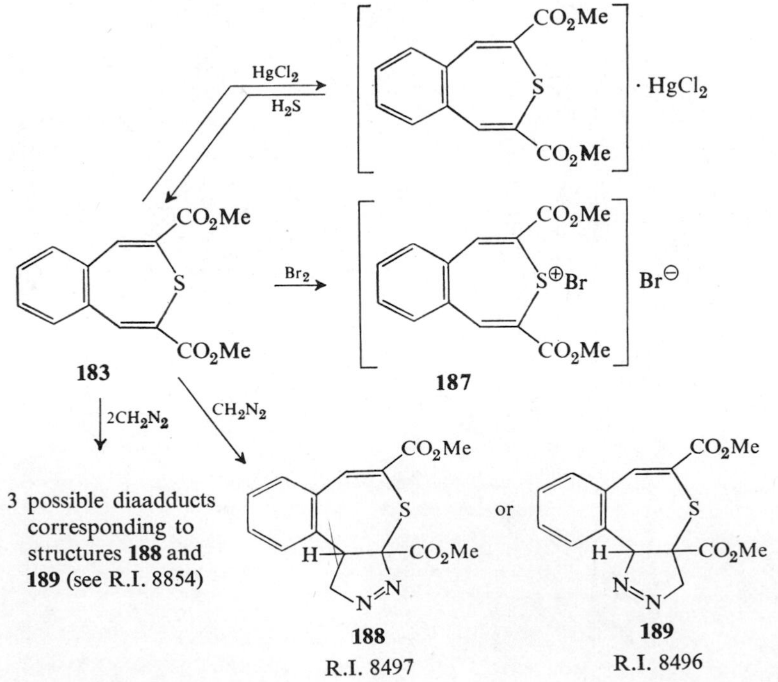
HgCl2
H2S
CO2Me
S
CO2Me
· HgCl2
CO2Me
S
CO2Me
183
Br2
CO2Me
S⊕Br
CO2Me
Br⊖
187
2CH2N2
CH2N2
3 possible diaadducts corresponding to structures 188 and 189 (see R.I. 8854)
CO2Me
S
H
CO2Me
N=N
188
R.I. 8497
or
CO2Me
S
H
CO2Me
N=N
189
R.I. 8496

(89)

can be dissolved in concentrated sulfuric acid and can be recovered unchanged upon dilution. They suggested protonation of the heterocyclic ring. Similar stability of 3-benzoxepin-2,4-dicarboxylic acid toward sulfuric acid has been reported (67, 68); however, Jorgensen (118) established that protonation occurred on the oxygen and that the stability of the benzoxepin system is not a manifestation of aromatic character. Instead stability is attributed to the divinyl ether structure and the presence of electron withdrawing ester functions on the $C_{(2)}$ and $C_{(4)}$ ring carbons. Similar considerations may be in order in methyl 3-benzothiepin-2,4-dicarboxylate.

Other chemical properties for ester **183** reported by Dimroth (71) include conversion to a chloromercuri derivative which upon treatment with H_2S regenerates ester **183**, and formation of adduct **187** by reaction with 1 mole of bromine (Eq. 89). The latter adduct did not add any more bromine, contained bromide ion, and decomposed slowly; however, further work appears necessary to confirm the assignment of structure **187**. Although ethyl diazoacetate does not react with **183**, diazomethane adds readily to either ester **183** or **184** to form a monoadduct or diadduct with a pyrazoline structure. The final choice between structures **188** and **189** remains to be settled.

2. 3-Benzothiepin-3,3-dioxide and Derivatives

Attempts to convert 3-benzothiepin-2,4-dicarboxylic acid to 3-benzothiepin-3,3-dioxide (**195**) by oxidation and decarboxylation proved unsuccessful, as were attempts to condense *o*-phthalaldehyde with bis(carbethoxymethyl)sulfone (305); however, Truce and Lotspeich (305) succeeded in preparing 3-benzothiepin-3,3-dioxide via intermediates **190–194** as indicated in Eq. 90. All the reactions proceeded in high yield except the bromination of **193** which appeared to undergo a competing brominolysis of the sulfone linkage. Sulfur dioxide was given off and apparently α,β,β-tribromo-*o*-diethylbenzene was one of the products (although not isolated and proven). The structural assignment for **195** was established by catalytic hydrogenation to 1,2,4,5-tetrahydro-3-benzothiepin-3,3-dioxide (**196**) which was prepared independently by an alternate procedure from α,α'-dibromo-*o*-xylene (Eq. 90).

3-Benzothiepin-3,3-dioxide is a crystalline solid which melts without decomposition; however, at 250°C sulfur dioxide is eliminated and naphthalene is formed (305). This observation is in contrast to the ease of sulfur extrusion in 3-benzothiepin-2,4-dicarboxylic acid and its derivatives, and in contrast to the stability of the isomeric 1-benzothiepin-1,1-dioxide (see Section C-3). The uv spectrum of **195** exhibits absorption maxima at λ_{max} 232 mμ (log ε 4.5) and 268 mμ (log ε 3.9) (305). It is interesting to compare the uv spectrum of

$(CH_2)_2SCH_2CCl$ (O) $\xrightarrow[\text{Cl}_2\text{CHCHCl}_2,\ 80\%]{\text{AlCl}_3}$ **190** $\xrightarrow[70\%]{\text{H}_2\text{O}_2}$ **191** $\xrightarrow[99\%]{\text{H}_2/\text{Ni}}$ **192** (H, OH, SO_2)

192 $\xrightarrow[80\%]{\text{PBr}_3}$ **193** (H, Br, SO_2) $\xrightarrow[18\%]{\text{Br}_2,\ h\nu}$ **194** (H, Br; H, Br; SO_2) $\xrightarrow[99\%]{\text{Et}_2\text{N}}$ **195** (SO_2) $\xrightarrow[98\%]{\text{H}_2/\text{Ni}}$ **196** (SO_2)

o-$C_6H_4(CH_2CH_2Br)_2$ $\xrightarrow{\text{(1) Na}_2\text{S};\ \text{(2) H}_2\text{O}_2}$ **196** $\qquad$ (90)

1-benzothiepin-1,1-dioxide [λ_{max} 233 mμ (log ε 4.10) and 288 mμ (log ε 3.98)] (298) with that for **195**. Both compounds appear to have a similar absorption band at 232 mμ which may reflect the benzothiepin dioxide ring system.

A study of the palladium-catalyzed hydrogenation of β-styryl methyl sulfone, 1,2-dihydro-3-benzothiepin-3,3-dioxide (**197**), and 3-benzothiepin-3,3-dioxide led Truce and Lotspeich to conclude that **195** possesses no more resonance stabilization than β-styryl methyl sulfone. Furthermore sulfone **195** decolorized potassium permanganate and bromine in the presence of light. Attempts to dehydrogenate **197** with selenium dioxide were unsuccessful. These results and conclusions were reported without experimental detail (305). The reaction of 3-benzothiepin-3,3-dioxide with fuming nitric acid introduced a nitro function into the benzene ring (Eq. 91). Reduction of the nitro compound (absorption of 5 moles H_2) gave an aromatic amine which underwent diazotization and coupling with β-naphthol to give an orange

NO_2 SO_2 $\xrightarrow{5H_2}$ NH_2 SO_2

HNO_3

SO_2 (91)

195 $\xrightarrow{\Delta}$ + SO_2

dye (90% yield). In view of the chemical and physical properties of 3-benzothiepin-3,3-dioxide, one can conclude there is little resonance stability in this compound.

3. 1,2-Dihydro-3-benzothiepin-3,3-dioxide

One report is found in the literature of a dihydro-3-benzothiepin derivative. As an outgrowth of the synthetic work on 3-benzothiepin-3,3-dioxide, Lotspeich (147) prepared 1,2-dihydro-3-benzothiepin-3,3-dioxide (**197**) by dehydrohalogenation of 1-bromo-1,2,4,5-tetrahydro-3-benzothiepin-3,3-dioxide (**198**) with triethylamine (Eq. 92). Attempts to brominate **197** at the

H Br SO_2 $\xrightarrow[85\%]{Et_3N}$ SO_2 $\xrightarrow[\text{24 hr, 60°C, 8\%}]{\text{NBS, }CCl_4}$ H Br H Br SO_2 (92)

198 **197** **199**

Br_2(dark), $CHCl_3$ 70%

$C_{(5)}$ position with *N*-bromosuccinimide led only to bromine addition at the double bond and formation of dibromide **199**.

4. 4,5-Dihydro-3-benzothiepin-1(2*H*)-one

The major method for generating 4,5-dihydro-3-benzothiepin-1(2*H*)-one (**190**) and alkylated derivatives involves an intramolecular Friedel-Crafts ring closure of 5-phenyl-3-thiapentanoyl chloride (Eq. 93). This reaction was initially described by von Braun and Weissbach (41) and later investigated in detail by Cagniant and co-workers (49, 55) and also by Truce and Lotspeich (305), who reported an 80% yield of ketone **190**. A variety of methylated 4,5-dihydro-3-benzothiepin-1(2*H*)-ones were described by Cagniant (55),

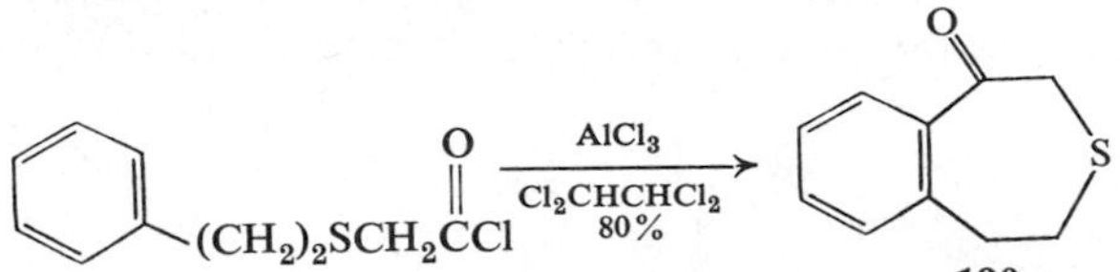

$$C_6H_5(CH_2)_2SCH_2\overset{O}{\overset{\|}{C}}Cl \xrightarrow[\text{80\%}]{AlCl_3,\ Cl_2CHCHCl_2} \mathbf{190} \qquad (93)$$

and a detailed discussion of their ir spectra was given (see Table XI-4 for the list of examples). In ring closure reactions with acid chlorides **200** and **201**, mixtures of two ketones were observed; however, reaction at the least hindered *para* position predominated, giving **202** and **203** (Eq. 94).

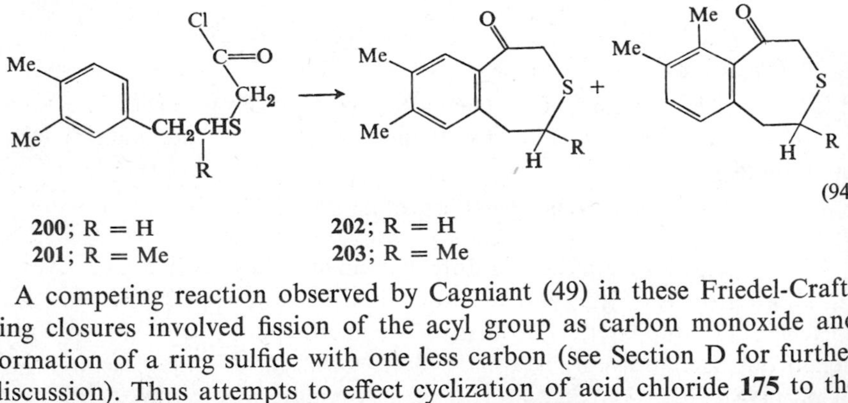

$$\text{(94)}$$

200; R = H **202**; R = H
201; R = Me **203**; R = Me

A competing reaction observed by Cagniant (49) in these Friedel-Crafts ring closures involved fission of the acyl group as carbon monoxide and formation of a ring sulfide with one less carbon (see Section D for further discussion). Thus attempts to effect cyclization of acid chloride **175** to the corresponding seven-membered ring ketone resulted in formation of the isothiochroman **176**.

5. 1,2,4,5-Tetrahydro-3-benzothiepin

Reduction of the ketones described in Section E-4 by the Wolff-Kishner method has led to the corresponding 1,2,4,5-tetrahydro-3-benzothiepins. The problems of ring contraction encountered in the Clemmensen reduction of cyclic keto sulfides (see Section D) are avoided by the Wolff-Kishner procedure. Cagniant and co-workers (49, 55) have reported the preparation (62–90% yield) of a large variety of methyl-substituted 1,2,4,5-tetrahydro-3-benzothiepins and have provided a detailed interpretation of the ir spectra for these compounds. An alternate procedure for the preparation of 1,2,4,5-tetrahydro-3-benzothiepin (**204**), as described by Truce (305), involves the reaction of *o*-bis(β-bromoethyl)benzene and sodium sulfide (Eq. 95).

O

S $\xrightarrow[\text{KOH, 90\%}]{H_2NNH_2}$ S $\xleftarrow{Na_2S}$ $(CH_2)_2Br$ $(CH_2)_2Br$ (95)

190 **204**

The parent sulfide **204** is readily oxidized with 30% hydrogen peroxide to the corresponding sulfone **196** (305). Friedel-Crafts acylation of **204** leads to 7-acyl derivatives (56). 7-Acetyl-1,2,4,5-tetrahydro-3-benzothiepin (**205**) has been converted into quinoline derivatives via the Pfitzinger reaction (56) (Eq. 96). On Wolff-Kishner reduction, **205** and other 7-acyl derivatives of **204** gave the corresponding 7-alkyl-1,2,4,5-tetrahydro-3-benzothiepin derivatives (56) (77–95% yield, see Table XI-4 for examples).

Me O N H O $\xrightarrow{}$ $\xrightarrow[-CO_2]{\Delta}$ Me N S

$\xrightarrow[\text{AlCl}_3\text{, 60\%}]{\text{AcCl}}$ Ac S (96)

205

$\xrightarrow[\text{KOH, 77\%}]{H_2NNH_2}$ Et S

Derivatives of 1,2,4,5-tetrahydro-3-benzothiepin-3,3-dioxide **196** have been described earlier (Sections E-2 and E-3).

F. Naphthothiepins

1. Naphtho[*m*,*n*-*b*]thiepins

Three possible isomeric ring systems of this type are naphtho[1,2-*b*]thiepin (**206**), naphtho[2,1-*b*]thiepin (**207**), and naphtho[2,3-*b*]thiepin (**208**).

Although none of the fully unsaturated compounds are known, partially reduced derivatives in each system have been reported.

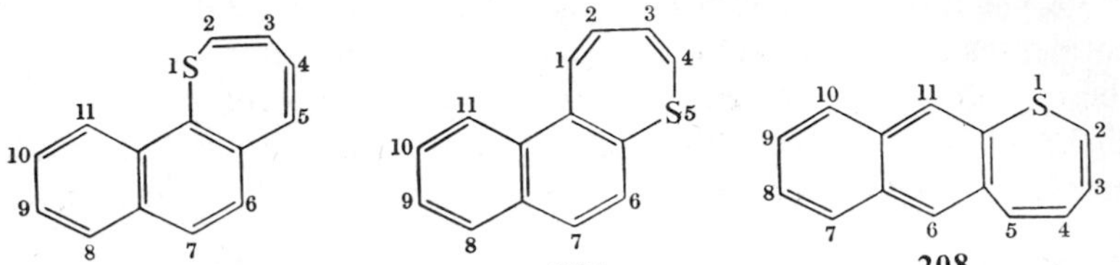

206 **207** **208**

R.I. 3706

The preparations of 3,4-dihydronaphtho[1,2-*b*]thiepin-5(2*H*)-one (**210**) and its 7-methyl derivative have been accomplished (Eq. 97) by cyclodehydration of 4-(1-naphthylthio)butyric acid (**209**) with polyphosphoric acid or via a Friedel-Crafts ring closure of the acid chloride of **209** (50). Each

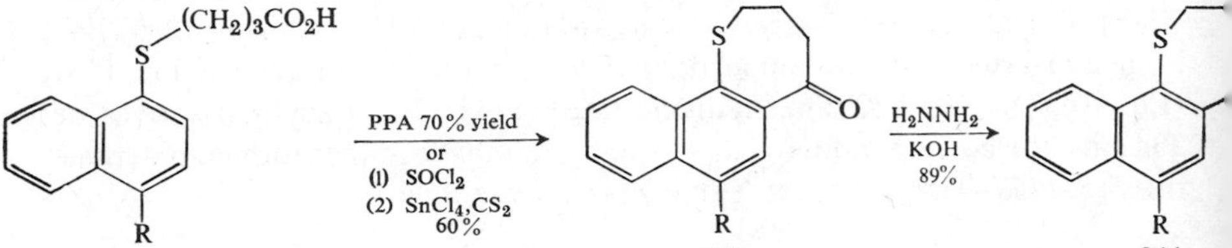

$(CH_2)_3CO_2H$

PPA 70% yield or (1) $SOCl_2$ (2) $SnCl_4,CS_2$ 60%

H_2NNH_2 KOH 89%

209 R = H,Me

210 R = H,Me

211 R = H,Me

(9

reaction proceeded in comparable yield. Interest in the ketones has been as intermediates in the synthesis, via a Wolff-Kishner reduction (Eq. 97), of 2,3,4,5-tetrahydronaphtho[1,2-*b*]thiepins (**211**). Preparation of 7 ethyl-2,3,4,5-tetrahydronaphtho[1,2-*b*]thiepin has also been achieved via a Friedel-Crafts acetylation of **211** (R = H) followed by a Wolff-Kishner reduction (50). The structural assignment for the 7-ethyl derivative involved comparison of its spectral properties with those of **211** and the corresponding 7-methyl derivative. Cagniant (46) has also presented an extensive discussion of the spectral characteristics (uv, ir, and nmr) of all the naphtho[*m*,*n*-*b*]-thiepins.

When one examines the ring closure reaction of 4-(2-naphthylthio)butyric acid (**212**), two sites for intramolecular acylation are possible, one at the α-carbon and the other at the β-carbon of the naphthalene nucleus. Cagniant (50) found that both modes of ring closure occur and are dependent on the experimental method employed (Eq. 98). When the Friedel-Crafts procedure is applied with the acid chloride of **212**, 2,3-dihydronaphtho[2,1-*b*]thiepin-1(4*H*)-one (**213**) is obtained as the exclusive product. The initial report of this observation was made by Truce and Toren (307) and latter confirmed by

$S(CH_2)_3CO_2H$

212

(1) PCl_5
(2) $SnCl_4$
65–72%

213

H_2NNH_2, KOH
90%

215

PPA
50%

214

H_2NNH_2, KOH
40%

216

(98)

Cagniant (50). On the other hand the action of polyphosphoric acid on **212** leads predominantly to ring closure at the β-position (50) with formation of 3,4-dihydronaphtho[2,3-*b*]thiepin-5(2*H*)-one (**214**) and only small amounts of **213**. Wolff-Kishner reduction of **213** and **214** afforded **215** and **216**, respectively (50).

Truce and Toren (307) have extended their studies of the Friedel-Crafts ring-closure reaction of a sulfonyl chloride to the naphthalene series with the reported conversion of **217** to 1,2,3,4-tetrahydronaphtho[2,1-*b*]thiepin-5,5-dioxide (**218**). This compound was also prepared (Eq. 99) by Wolff-Kishner reduction of keto sulfone **219**.

$(CH_2)_4SO_2Cl$ **217** $\xrightarrow[PhNO_2,\ 16\%]{AlCl_3}$ **218** (SO_2) $\xleftarrow[KOH]{H_2NNH_2}$ **219** (SO_2) $\xleftarrow{H_2O_2}$ **213** (99)

2. Naphtho[*m*,*n*-*d*]thiepins

The literature contains one report dealing with the naphtho[1,2-*d*]thiepin system, namely a description of two synthetic pathways to 1,2,4,5-tetrahydronaphtho[1,2-*d*]thiepin (**220**). Cagniant and co-workers (57) carried out Friedel-Crafts ring closure reactions with 2-(α-naphthyl)ethylthioglycolic and 2-(β-naphthyl)ethylthioglycolic acid to generate the seven-membered sulfur ring (Eq. 100). Wolff-Kishner reduction of the resulting ketones, 4,5-dihydronaphtho[1,2-*d*]thiepin-1(2*H*)-one (**221**) and 1,2-dihydronaphtho-[1,2-*d*]thiepin-5(4*H*)-one (**222**), provided the sulfide **220**. The uv and ir spectral properties of these compounds were discussed (57).

$(CH_2)_2SCH_2CO_2H$ (α-naphthyl) $\xrightarrow[(2)\ AlCl_3,\ CS_2,\ 37\%]{(1)\ SOCl_2}$ → **222** $\xrightarrow{H_2NNH_2,\ KOH,\ 88\%}$ **220**

$(CH_2)_2SCH_2CO_2H$ (β-naphthyl) $\xrightarrow[(2)\ AlCl_3,\ CS_2,\ 60\%]{(1)\ SOCl_2}$ → **221** → **220** (100)

The other system possible under this grouping is naphtho[2,3-*d*]thiepin. Although the parent compound is unknown, a number of 2,4-disubstituted derivatives have been prepared (150). The method of synthesis involves base catalyzed condensation of naphthalene-2,3-dialdehyde with ethyl thiodiacetate or phenacyl sulfide (Eq. 101), which leads to ethyl naphtho[2,3-*d*]-thiepin-2,4-dicarboxylate (**223**) or 2,4-dibenzoylnaphtho[2,3-*d*]thiepin (**224**), respectively. In the former instance some of the monoethyl ester (**225**) is also

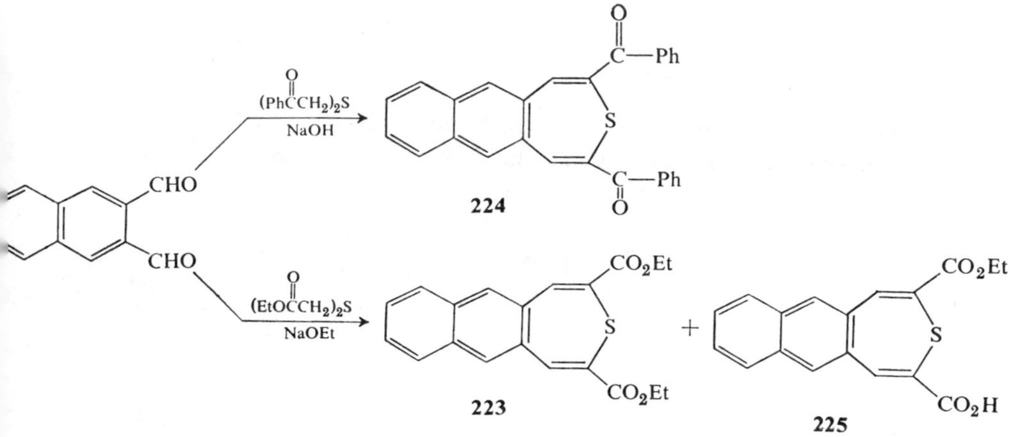

$(PhCOCH_2)_2S$, NaOH → **224** (2,4-dibenzoyl)
$(EtOCOCH_2)_2S$, NaOEt → **223** (CO_2Et, CO_2Et) + **225** (CO_2Et, CO_2H)

(101)

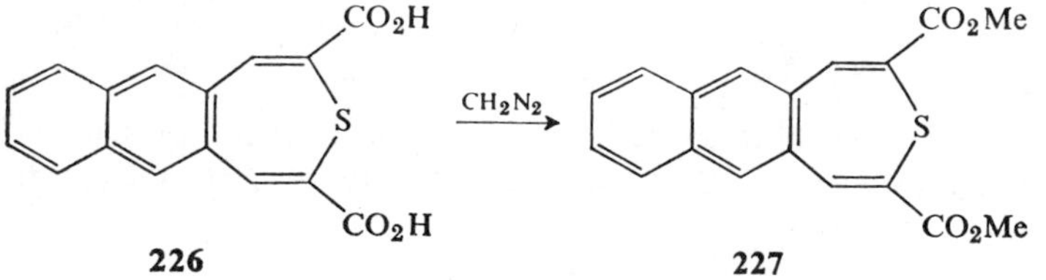

226 (CO_2H, CO_2H) $\xrightarrow{CH_2N_2}$ **227** (CO_2Me, CO_2Me) (102)

obtained; when a longer reaction time is employed, naphtho[2,3-*d*]thiepin-2,3-dicarboxylic acid (**226**) is the product. The diacid **226** can be readily esterified with diazomethane (Eq. 102) to its dimethyl ester **227**.

The naphtho[2,3-*d*]thiepin system appears to be more stable toward the sulfur extrusion reaction than the corresponding 3-benzothiepin derivatives. Whereas compound **224** was readily isolated in the above reaction and required temperatures of 180–250°C to promote sulfur extrusion, the reaction of *o*-phthalaldehyde with phenacyl sulfide in the presence of base produced 2,3-dibenzoylnaphthalene with no isolation of the intermediate 2,4-dibenzoyl-3-benzothiepin. Other naphtho[2,3-*d*]thiepins from which sulfur has been thermally extruded to give anthracene derivatives (Eq. 103) are the esters **225** and **227**. In each instance temperatures over 200°C were used.

Although the naphtho[2,3-*d*]thiepin system appears to be more stable than 3-benzothiepin, Cagniant and co-workers (56) attempted to dehydrogenate 1,2,4,5,7,8,9,10-octahydronaphtho[2,3-*d*]thiepin (**228**) with selenium and

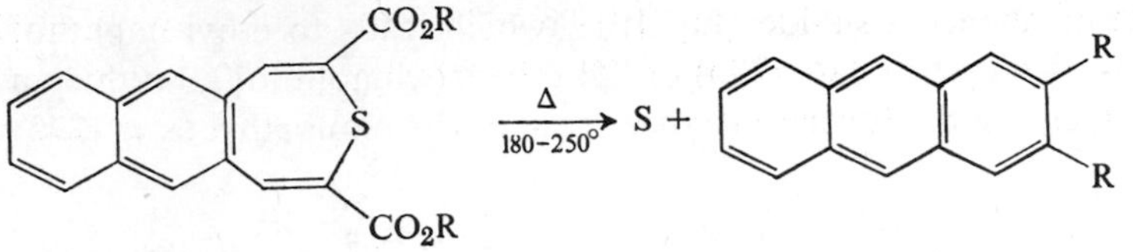

CO_2R S CO_2R $\xrightarrow[180-250°]{\Delta}$ S + R R (103)

225; R = Et
227; R = Me

obtained resinous material. The synthesis of **228** (Eq. 104) involved Wolff-Kishner reduction of ketone **229** which was prepared via Friedel-Crafts ring closure of **230**.

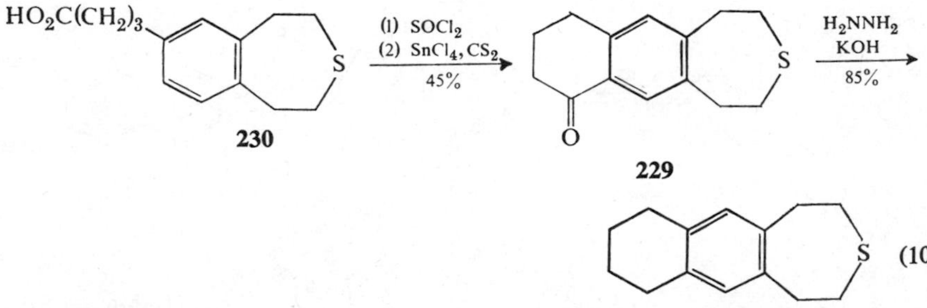

$HO_2C(CH_2)_3$ S $\xrightarrow[45\%]{(1)\ SOCl_2\ (2)\ SnCl_4, CS_2}$ O S $\xrightarrow[85\%]{H_2NNH_2,\ KOH}$

230 **229**

S (104)

228

A recent report (132) described the formation of a complex naphthothiepin derivative **232** via a Diels-Alder reaction between **231** and 1,3-diphenylbenzo[*c*]furan (Eq. 105).

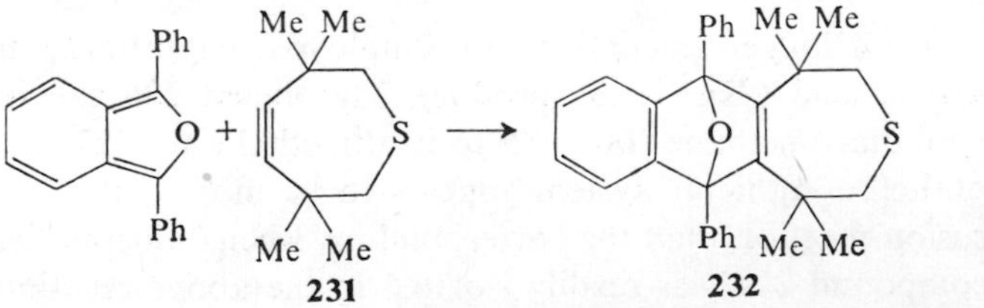

Ph O Ph + Me Me S Me Me $\longrightarrow$ Ph Me Me O S Ph Me Me (105)

231 **232**

3. Naphtho[1,8-*c,d*]thiepins

In the study of the Friedel-Crafts cyclization of 4-(1-naphthyl)-3-thiobutanoic acid (**233**) via its acid chloride, Cagniant and co-workers (58) reported that ring closure with aluminum chloride proceeded predominantly to the β-position (Eq. 106) but also gave a small amount of the naphthothiepin **234**. When stannic chloride was used as the catalyst, ring closure went exclusively to the β-position. The mixture of ketones was not separated but reduced directly by the Wolff-Kishner procedure to a mixture of **235** and **236**. No evidence was offered in support of compounds **234** and **236**.

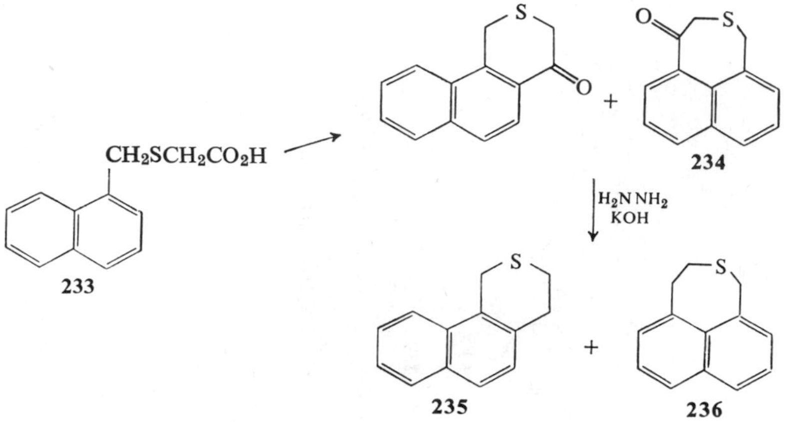

$CH_2SCH_2CO_2H$

233

234

H_2NNH_2
KOH

(106)

235

236

G. Dibenzo[*b*,*d*]thiepins

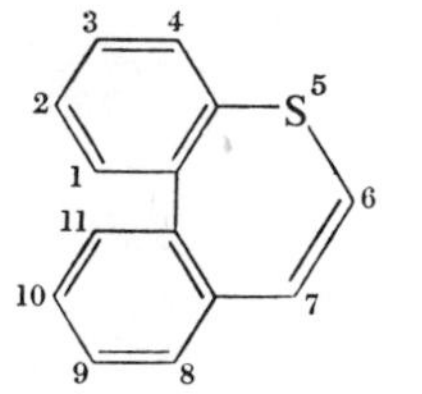

R.I. 10,530

In the course of synthetic studies directed toward structures related to colchicine, Lotspeich and Karickhoff (148) described the preparation of *cis*- and *trans*-1,2,3,4,4a,11b-hexahydro-9,10,11-trimethoxydibenzo[*b*,*d*]thiepin-7(6*H*)-one (**237**) and (**238**), respectively. These ketones resulted from a Friedel-Crafts cyclization of the *cis* acid **239** or *trans* acid **240** via the corresponding acid chlorides. Although ring closure involved attachment to the activated trimethoxyphenyl ring, stannic chloride was ineffective as a catalyst; nonetheless the reaction was promoted by aluminum chloride. Attempts to cyclodehydrate acid **239** with polyphosphoric acid or hydrogen fluoride were unsuccessful. The stereochemical assignments for the ketones relied upon

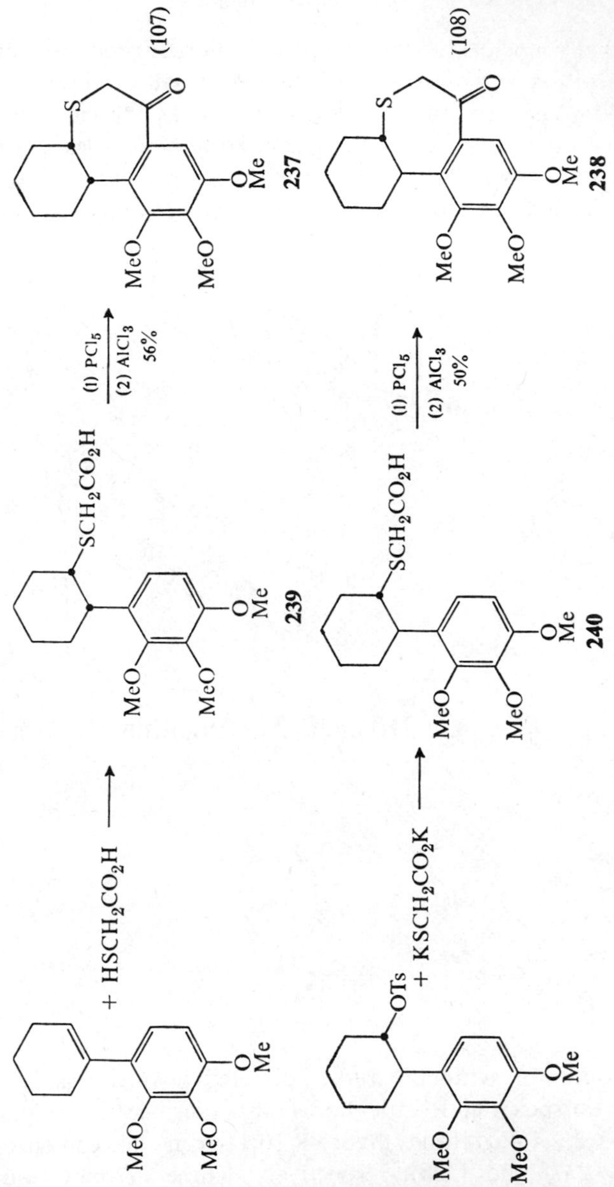

$$\text{+ HSCH}_2\text{CO}_2\text{H} \longrightarrow \mathbf{239} \xrightarrow[\text{(2) AlCl}_3,\ 56\%]{\text{(1) PCl}_5} \mathbf{237} \tag{107}$$

$$\text{+ KSCH}_2\text{CO}_2\text{K} \longrightarrow \mathbf{240} \xrightarrow[\text{(2) AlCl}_3,\ 50\%]{\text{(1) PCl}_5} \mathbf{238} \tag{108}$$

those made for the precursor acids and were corroborated by nmr spectral data. Preparation of the *cis* acid **239** involved peroxide-catalyzed addition of thioglycolic acid to 1-(2′,3′,4′-trimethoxyphenyl)cyclohexene (Eq. 107); the *trans* acid **240** resulted from a nucleophilic displacement reaction, with retention of configuration, between *trans*-2-(2′,3′,4′-trimethoxyphenyl)-cyclohexyl tosylate and dipotassium mercaptoacetate (Eq. 108).

The synthesis of dibenzo[*b*,*d*]thiepin-7(6*H*)-one-5,5-dioxide (**241**), as reported by Emrick and Truce (80), was achieved by Dieckmann ring closure of ethyl 2′-methylsulfonylbiphenyl-2-carboxylate (Eq. 109). Among the chemical properties reported for **241** were that it was readily soluble in ammonia or sodium carbonate but not in sodium bicarbonate; that it did not form stable salts with alkaloids having an ionization constant of 10^{-6} or less; that it gave no color with ferric chloride or ceric sulfate; and that it formed a phenylhydrazone, underwent coupling with diazonium salts, and yielded a colorless bromide. However, experimental details of these reactions were not presented. Interpretation of the chemical and spectral (ir and uv) properties of **241** led Emrick and Truce to conclude that the sulfur-containing ring was nonplanar. Moreover the data excluded the likelihood of any appreciable quantities of enol form **242** in equilibrium with **241**.

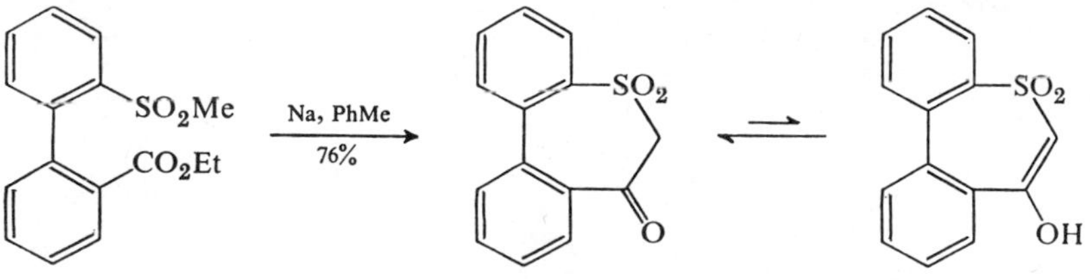

SO_2Me CO_2Et —(Na, PhMe; 76%)→ **241** ⇌ **242** (109)

H. Dibenzo[*b*,*e*]thiepins

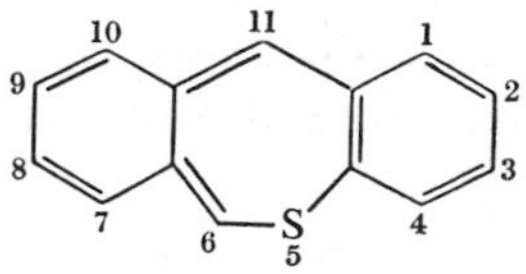

1. Dibenzo[*b*,*e*]thiepin and Derivatives

Although dibenzo[*b*,*e*]thiepin (**243**) or its derivatives have not been described in the literature, an isomeric unsaturated system **244** was reported in 1962 by four independent groups of workers (91, 226, 281, 327). The psychotropic activity and other physiological properties of dibenzo[*b*,*e*]-thiepin-$\Delta^{11,6H}$-γ-propyldimethylamine (**245**) (Prothiaden) and related compounds have stimulated a high degree of synthetic activity in this heterocyclic system and have resulted in extensive patent literature.

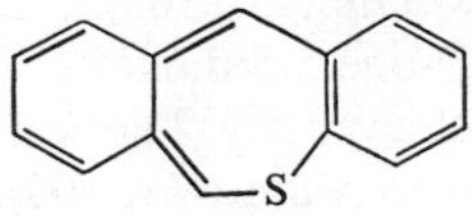

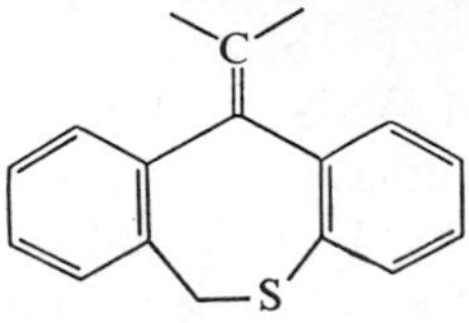

243 **244**

The primary synthetic scheme (24, 224, 228, 280) leading to **245** involved ring closure of acid **246** to ketone **247** which upon reaction with 3-dimethylaminopropylmagnesium chloride (Eq. 110) gave alcohol **248**. Dehydration of **248**, which produced **245**, was readily achieved under mild conditions and in high yield with HCl or HBr in ethanol or acetic acid or with 3*N* H_2SO_4. This basic synthetic pathway has been employed to produce derivatives of **245** and related compounds with a modified aminoalkyl side chain (see Table XI-7 for examples). The sulfone **249** was prepared from the keto sulfone **250** by the scheme (327) of Eq. 111 while the sulfoxide **251** was obtained by hydrogen

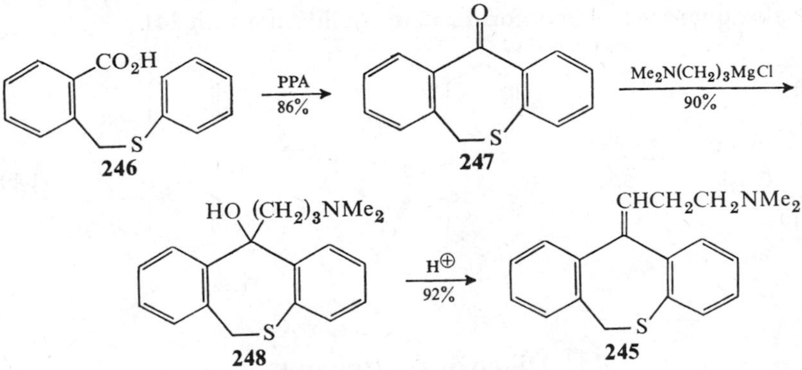

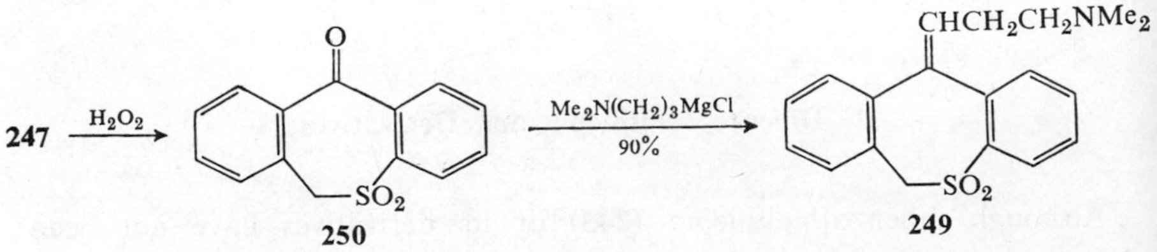

CO_2H; S; **246** —PPA, 86%→ O; S; **247** —$Me_2N(CH_2)_3MgCl$, 90%→

HO $(CH_2)_3NMe_2$; S; **248** —$H^{\oplus}$, 92%→ $CHCH_2CH_2NMe_2$; S; **245** (110)

247 —H_2O_2→ O; SO_2; **250** —$Me_2N(CH_2)_3MgCl$, 90%→ $CHCH_2CH_2NMe_2$; SO_2; **249**

(111)

peroxide oxidation of **245** (Eq. 112) at room temperature (224). Application of the Pummerer reaction (Eq. 112) to sulfoxide **251** led to a nearly quantitative conversion to 6-acetoxydibenzo[*b*,*e*]thiepin-$\Delta^{11,6H}$-γ-propyldimethylamine (**252**) which appeared unstable even when isolated as the fumarate salt (270).

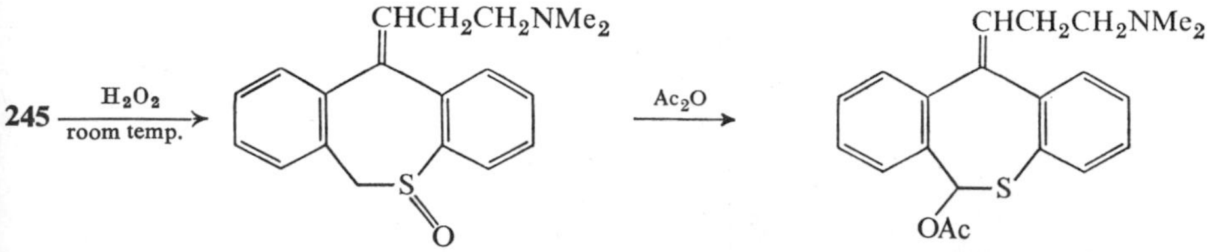

(112)

The synthesis of *N*,*N*-dimethyldibenzo[*b*,*e*]thiepin-$\Delta^{11,6H}$-acetamide (**253**) via alcohol **254** (Eq. 113), which illustrates another method for preparing the precursor alcohol, entails base-catalyzed addition of *N*,*N*-dimethylacetamide to ketone **247** (30).

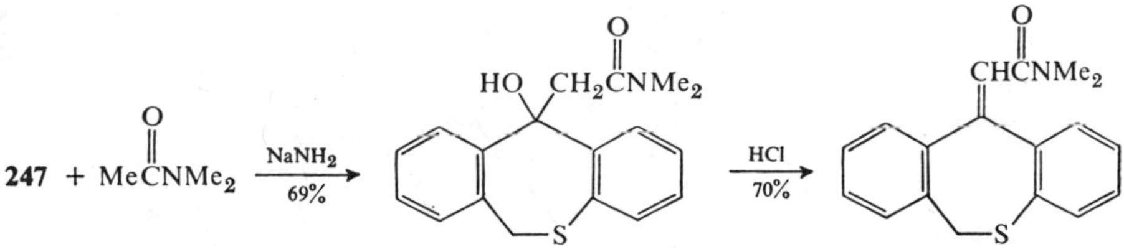

(113)

Similar base-catalyzed addition of acetonitrile and other nitriles (Eq. 114) to **247** were reported (29) and were utilized for the conversion of **247** into unsaturated nitriles **255** or amines **256**.

The few reactions reported to date in this system either involve the exocyclic double bond, which can be hydrogenated in the customary manner (281) or reduced chemically with hydrogen iodide and phosphorus (102) or aluminum powder (30), or involve transformations of functional groups in the side chain. The former reaction is illustrated by the preparation of **257** and **258** (Eq. 115).

An example of the latter reaction is the alternate synthesis of **245** shown in Eq. 116 (32). The three-carbon side chain was introduced in the standard manner via a Grignard reaction and the resulting alcohol **259** was dehydrated to **260** which was subjected to further transformations leading to **245**. The reaction of methylamine with **261** produced compound **262** (32) which was also available from **245** by the demethylation sequence outlined in Eq. 117

247 + MeCN → HO CH_2CN (S) —$LiAlH_4$→ HO $CH_2CH_2NH_2$ (S) —$H^{\oplus}$→ $CHCH_2NH_2$ (S) **256** (114)

HO CH_2CN (S) —$H^{\oplus}$→ CHCN (S) **255**

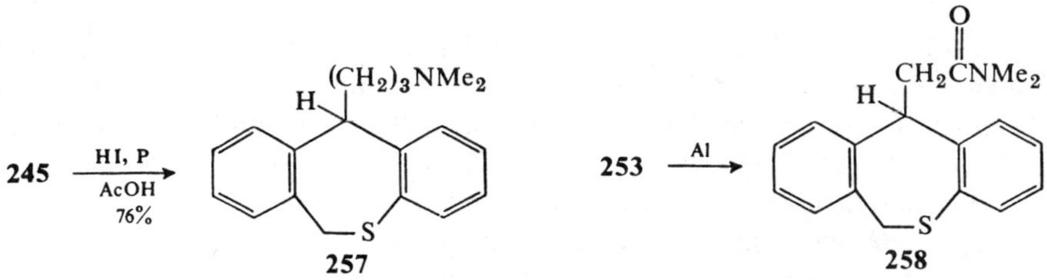

245 $\xrightarrow[\text{AcOH, 76\%}]{\text{HI, P}}$ 257 [H, $(CH_2)_3NMe_2$; dibenzothiepin, S]

253 $\xrightarrow{\text{Al}}$ 258 [H, $CH_2C(=O)NMe_2$; dibenzothiepin, S]

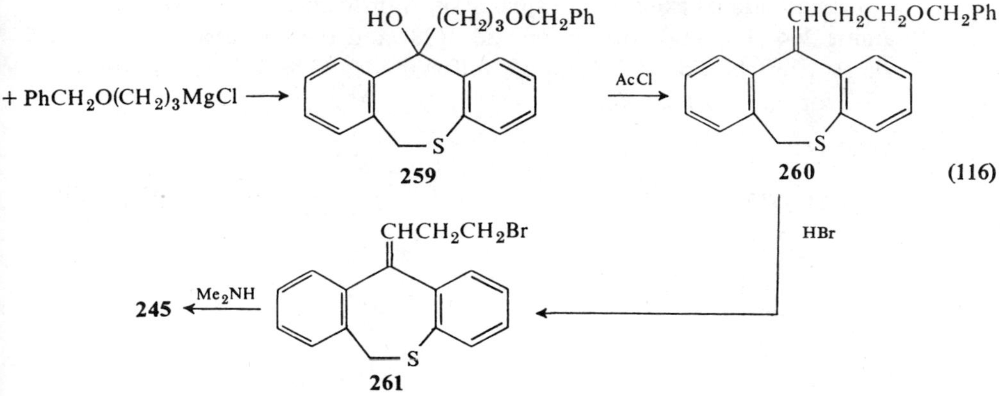

(115)

$+ PhCH_2O(CH_2)_3MgCl \longrightarrow$ 259 [HO, $(CH_2)_3OCH_2Ph$; S] $\xrightarrow{\text{AcCl}}$ 260 [$=CHCH_2CH_2OCH_2Ph$; S] (116)

260 $\xrightarrow{\text{HBr}}$ 261 [$=CHCH_2CH_2Br$; S] $\xrightarrow{\text{Me}_2\text{NH}}$ 245

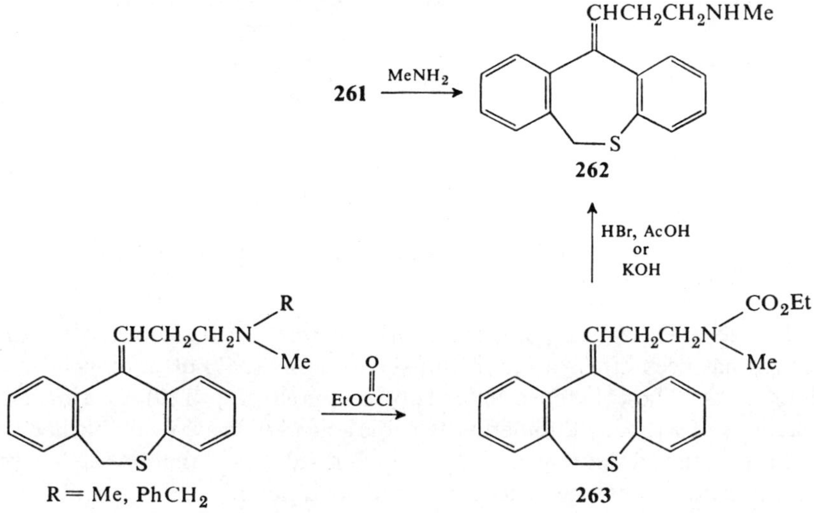

261 $\xrightarrow{\text{MeNH}_2}$ 262 [$=CHCH_2CH_2NHMe$; S]

[$=CHCH_2CH_2N(R)Me$; S], R = Me, $PhCH_2$ $\xrightarrow{\text{EtOC(=O)Cl}}$ 263 [$=CHCH_2CH_2N(CO_2Et)Me$; S] $\xrightarrow[\text{or KOH}]{\text{HBr, AcOH}}$ 262

(117)

(27, 121, 224, 239, 280). When the carbamate **263** was hydrolyzed under alkaline conditions (224, 239), two isomeric (*cis* and *trans*) monomethyl amines **262** were isolated as hydrochloride salts. Although the unsymmetrically substituted system **244** is capable of geometric isomerism, the only three examples thus far reported have been that cited above, the 2-methyl derivatives of **262**, and 2-fluorodibenzo[*b*,*e*]thiepin-$\Delta^{11,6H}$-γ-propyldimethylamine. The latter compound was generated by acid-catalyzed dehydration of the corresponding carbinol (226); however, all other examples of acid-catalyzed dehydration of alcohols as a method of introduction of the exocyclic double bond appear to proceed stereoselectively to one isomer.

Two additional examples of side-chain transformation involve reaction of amine **264** (Eq. 118) with an epoxide (28) and the conversion of acid **265** via its acid chloride into complex sulfonamides (326) (Eq. 119) (see Table XI-7 for other examples).

$CHCH_2NH_2$ (**264**) + $PhOCH_2CH(O)CH_2$ → $CHCH_2NHCH_2CH(OH)CH_2OPh$ (1

$CHCO_2H$ (**265**) → $CHC(=O)NHCH_2CH_2$–C_6H_4–$SO_2NHC(=O)NHN$ (piperidinyl)–M (1

2. Dibenzo[*b*,*e*]thiepin-11(6*H*)-one and Derivatives

The general starting point for synthetic work in the dibenzo[*b*,*e*]thiepin system has been dibenzo[*b*,*e*]thiepin-11(6*H*)-one (**247**) or its ring-substituted derivatives. These ketones are easily prepared (Eq. 120) by ring closure reactions of *o*-(phenylthiomethyl)benzoic acids which are available in 80–95% yield from the addition of sodium thiophenoxides to phthalide (226). Cyclization of acid **246** has been accomplished with polyphosphoric acid (91, 226, 281, 283), concentrated sulfuric acid (121), boron trifluoride etherate (121), phosphorus pentoxide (327), and zinc chloride and phosphorus oxychloride in nitrobenzene (234); however, the most frequently used reagent was polyphosphoric acid which provided ketones in yields of 60–95%, (see Table XI-7) (226, 281). In addition, acid **246** was converted into its acid chloride

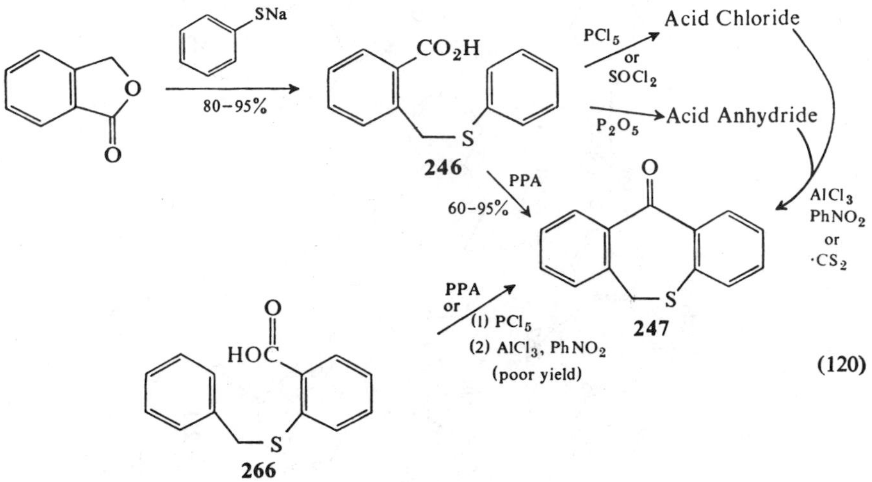

(120)

and anhydride and each of these derivatives was cyclized to ketone **247** via the Friedel-Crafts reaction (91, 226, 237). The alternate acid **266** was unaffected by polyphosphoric acid or liquid hydrogen fluoride (327) but afforded ketone **247** in poor yield on Friedel-Crafts cyclization of the acid chloride or anhydride.

The spectral properties of ketone **247** were consistent with the assigned structure and showed a carbonyl frequency (1651 cm^{-1}) identical with that of benzophenone. In the uv spectrum the hypsochromic shift of the conjugation band in **247** relative to dibenzo[*b,e*]oxepin-11(6*H*)-one suggested that neither benzene ring in **247** is coplanar with the carbonyl group (238). Although an initial report claimed that ketone **247** did not react with carbonyl reagents (281), Protiva (102) found that prolonged heating of **247** with hydroxylamine in pyridine produced the oxime **267** in good yield (Eq. 121). The oxime was *O*-alkylated by reaction of its sodium salt with dimethylaminoalkyl chlorides (3). Oxidation of **247** with 30% hydrogen peroxide proceeded readily and in good yield to produce the sulfoxide **271** (226, 237) (room temperature) or sulfone (226, 237, 327) (refluxing acetic acid). A Pummerer reaction (Eq. 121) with **271** resulted in a quantitative conversion into the acetate **272** which was characterized by conversion into its sulfone (270). Reduction of **247** to 6,11-dihydrodibenzo[*b,e*]thiepin-11-ol (**268**) was effected with sodium and alcohol (102), sodium borohydride (102, 226, 237), or lithium aluminum hydride (62, 102, 327). In the sodium borohydride

H OH

268

$NaBH_4$
or
$LiAlH_4$
70%

269

247

NH_2OH
75%

NOH

267

(1) Na
(2) $R_2N(CH_2)_xC$

$NO(CH_2)_xNR_2$

Zn, AcOH
80%

Zn, AcOH
93%

270

H_2O_2, AcOH
room temp.

271

Ac_2O
140°C

OAc

272

(121)

reduction, ether **269** was also observed as a by-product; however, with lithium aluminum hydride and an acid work-up, ether **269** became the chief product (102). When ketone **247** was treated with zinc and acetic acid, reduction proceeded to 6,11-dihydrobenzo[*b*,*e*]thiepin (**270**) which was also the product from the reaction of oxime **267** with zinc and acetic acid (102).

3. 6,11-Dihydrodibenzo[*b*,*e*]thiepin and Derivatives

Preparation of the parent compound **270** is best achieved as described in the preceding paragraph; however, reduction of alcohol **268** with hydrogen iodide and phosphorus also gives **270**, albeit in poor yield (224). Similar reduction of tertiary alcohol **248** with hydrogen iodide and phosphorus (81, 208) or hydrogen iodide and calcium hypophosphite (224, 272) (Eq. 122) produced 11-(3-dimethylaminopropyl)-6,11-dihydrodibenzo[*b*,*e*]thiepin (**257**) which was reported earlier (Section H-1) from a comparable reduction of olefin **245**. Another pathway (102) to **257** employed a coupling reaction between dimethylaminopropylmagnesium chloride and 11-chloro-6,11-dihydrodibenzo[*b*,*e*]thiepin (**273**). The reduction reaction with other tertiary alcohols has been reported (208, 272).

HO $(CH_2)_3NMe_2$ … S **248** $\xrightarrow[\text{P or } CaHPO_2,\ 70\%]{HI}$

H $(CH_2)_3NMe_2$ … S **257** $\xleftarrow[55\%]{Me_2N(CH_2)_3MgI}$ H Cl … S **273** (112)

Alcohol **268** was converted into ether derivatives (Eq. 123) by reaction of its sodium salt with dialkylaminoalkyl halides (204, 224), or by conversion via the tosylate into the chloro compound **273** followed by reaction with a dialkylaminoalkanol (204, 224, 252). These reactions can be illustrated by the preparation of **274** (204, 224), **275** (204, 224), and the more exotic ether **276** (252). 11-Chloro-6,11-dihydrodibenzo[*b*,*e*]thiepin also undergoes displacement reactions with silver cyanide to give **277** (poor yield) (102) and with amines to form **278** and analogous 11-amino derivatives (102, 230, 329) (see Table XI-7). Although reduction of oxime **267** did not yield an amine, the

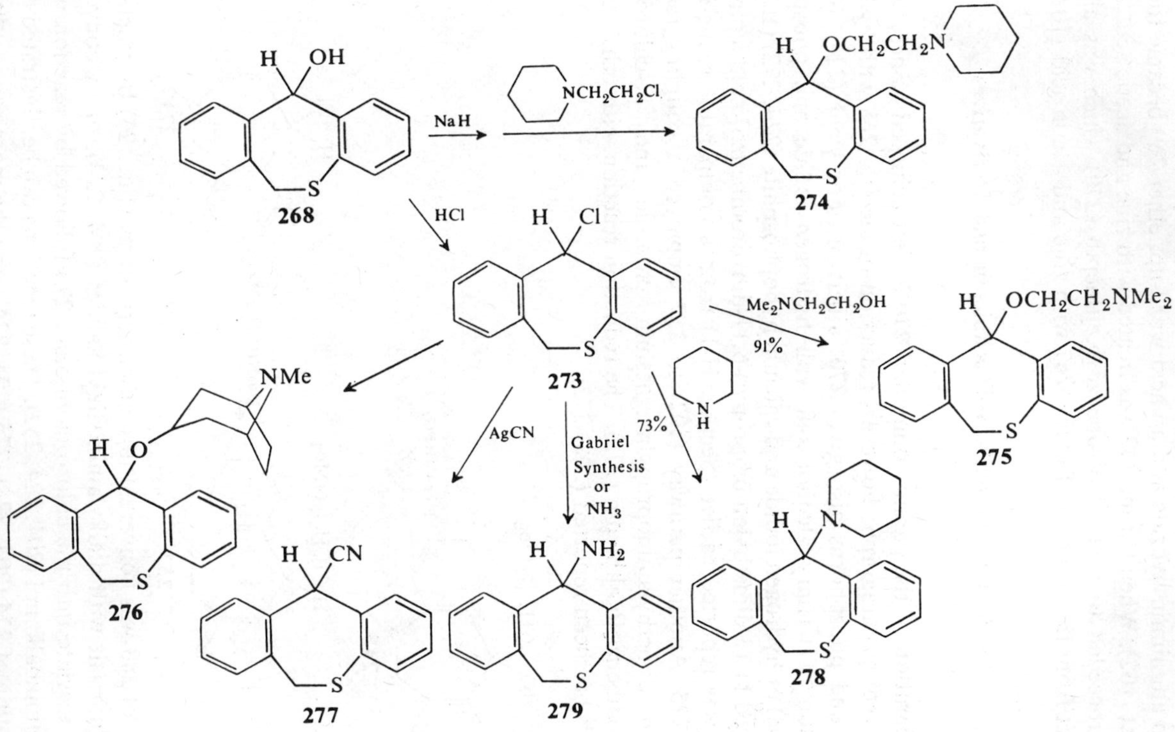

H OH

268

$\xrightarrow{NaH}$ $\xrightarrow{\text{(piperidino)}NCH_2CH_2Cl}$

H OCH_2CH_2N(piperidino)

274

HCl

H Cl

273

$Me_2NCH_2CH_2OH$ 91%

H $OCH_2CH_2NMe_2$

275

(123)

NMe

H O

276

AgCN

H CN

277

Gabriel Synthesis or NH_3

H NH_2

279

73% piperidine (N–H)

H N(piperidino)

278

primary amino compound **279** was prepared by a Gabriel synthesis (102) or by reaction of **273** with ammonia in an autoclave (230). The Pummerer reaction with the sulfoxide of **268** was complicated and did not lead to identifiable products; however, reaction of **280** with acetic anhydride yielded the acetoxy sulfide **281** (Eq. 124) which upon hydrolysis gave alcohol **282** in equilibrium with the ring opened form **283** (270). Evidence for this equilibration was supported by nmr and the isolation of disulfide **284**.

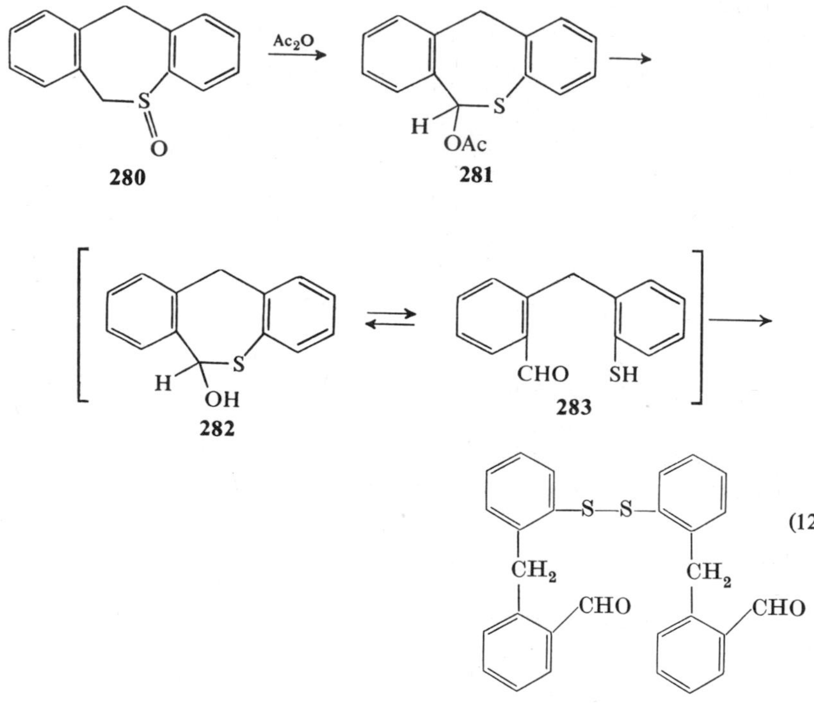

Ac_2O

S, O, **280**; H, S, OAc, **281**; H, S, OH, **282**; CHO, SH, **283**; S—S, CH_2, CHO, CH_2, CHO

(124)

I. Dibenzo[*b*,*f*]thiepins

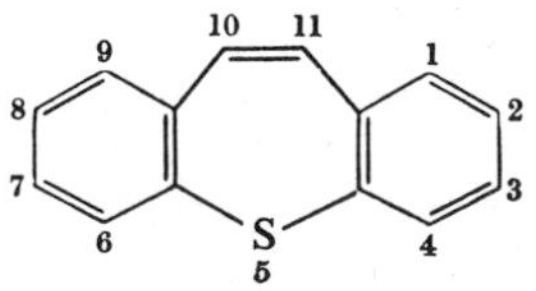

10, 11, 1, 2, 3, 4, S 5, 6, 7, 8, 9

R.I. 8637

The initial work in the dibenzo[*b*,*f*]thiepin system, reported by Loudon, Sloan, and Summers (151) in 1957, was stimulated by a study of the sulfur extrusion reaction. A few years later Bergmann and Rabinovitz (16) prepared the parent compound as part of their continuing interest in the correlation of structure with aromatic properties. But the greatest impetus in recent years, which has produced an abundant literature on this system, stems from the pharmacodynamic activity of several amine derivatives. Zirkle (331) claimed utility for 10-aminoalkyldibenzo[*b*,*f*]thiepins (**285**) as central nervous system stimulants, antihistaminics, and antispasmodics, whereas Protiva and co-workers (115) reported neurotropic and psychotropic activity for 10-(2-dimethylaminoethoxy)-10,11-dihydrodibenzo[*b*,*f*]thiepin (**286**) and 10-(4-methylpiperazino)-10,11-dihydrodibenzo[*b*,*f*]thiepin (**287**) ("perathiepin") (116).

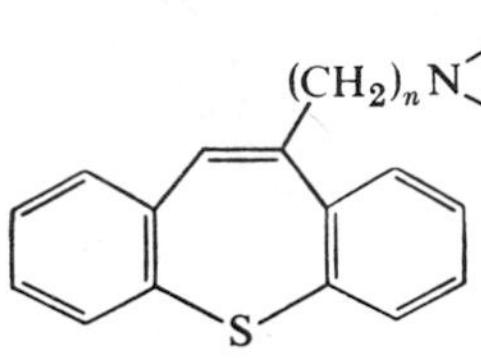

$(CH_2)_nNRR'$ **285** H $O(CH_2)_2NMe_2$ **286** H N NMe **287**

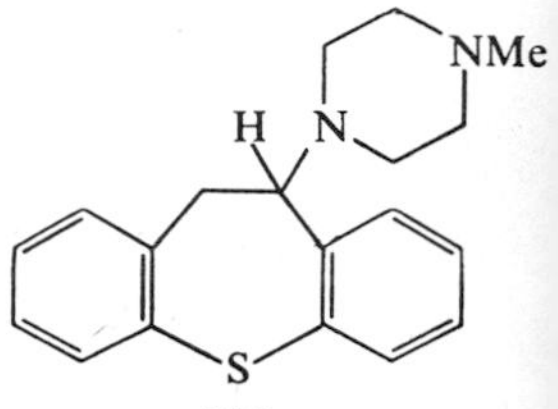

1. Dibenzo[*b*,*f*]thiepin and Derivatives

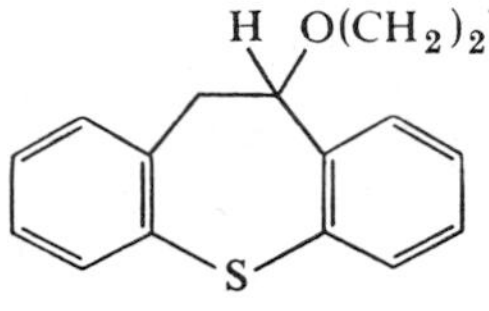

The first approach described in the literature for generating the dibenzo-[*b*,*f*]thiepin ring system entailed a ring closure reaction of 2-arylthiophenylpyruvic acid derivatives as illustrated (Eq. 125) by the synthesis of 2-nitro-8-methyldibenzo[*b*,*f*]thiepin-10-carboxylic acid (**288**) (151). These pyruvic acid

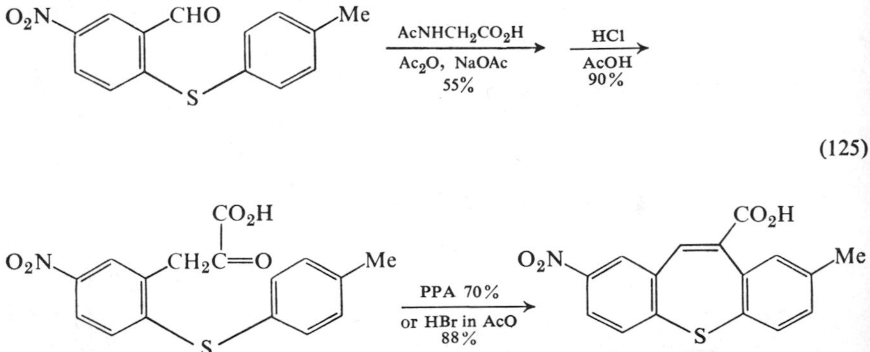

O_2N CHO S Me $\xrightarrow[\text{Ac}_2\text{O, NaOAc}\ 55\%]{\text{AcNHCH}_2\text{CO}_2\text{H}}$ $\xrightarrow[\text{AcOH}\ 90\%]{\text{HCl}}$

(125)

O_2N $CH_2C(=O)CO_2H$ S Me $\xrightarrow[\text{or HBr in AcO}\ 88\%]{\text{PPA}\ 70\%}$ O_2N CO_2H S Me **288**

derivatives were prepared from 2-arylthiobenzaldehydes via the azlactone intermediate, and ring closure to dibenzo[*b*,*f*]thiepin-10-carboxylic acids was effected easily with polyphosphoric acid; however, the use of concentrated HBr (48%) in acetic acid gave better yields. A second example of a similar ring closure method (Eq. 126) is the conversion of 2-phenylthiophenylacetic acid by the action of phosphorus oxychloride and zinc chloride into 10-chlorodibenzo[*b*,*f*]thiepin (**289**) (116). The possibility that ring closure proceeds first to ketone **290** which then undergoes subsequent reaction to form **289** cannot be excluded because **290** leads to **289** under the experimental

CH_2CO_2H — $POCl_3$, $ZnCl_2$ / $PhNO_2$ / 80% → **289** (Cl); **290** (O) — $POCl_3$, $ZnCl_2$ / 62% → **289** (126)

conditions used for the ring closure reaction (194). These methods for producing dibenzo[*b*,*f*]thiepins have not been used very extensively.

A second general pathway to dibenzo[*b*,*f*]thiepins is illustrated by the ring expansion reaction reported by Bergmann and Rabinovitz (16) for the synthesis of the parent compound **291** via thioxanthene intermediates **292**–**294** (Eq. 127). The structure assignment of **291** was based on its nonidentity with 9-methylenethioxanthene (63), the similarity between its uv spectrum and that of *cis*-stilbene, and its oxidation by potassium permanganate in acetone to diphenylsulfide-2,2′-dicarboxylic acid. Further support for the occurrence of ring expansion was found in the rearrangement of 9-(α-hydroxybenzyl)-thioxanthene (**295**) to 10-phenyldibenzo[*b*,*f*]thiepin (**296**) (Eq. 128) (15). 9-Benzylidenethioxanthene was prepared independently and proved to be different from **296**.

More recently several groups of workers have extended this synthetic approach to the preparation of 10-substituted dibenzo[*b*,*f*]thiepins (15, 331) as well as various ring-methylated derivatives (see Table XI-8 for examples) (308). Starting alcohol **293** was readily available from the metalation of thioxanthene (**292**) with *n*-butyllithium followed by treatment with formaldehyde (16); use of benzaldehyde in place of formaldehyde (15), and the reaction of acetaldehyde with 9-methylthioxanth-9-yl lithium provided **295** and other precursor alcohols. An alternate approach to the synthesis of **293** and its ring-methylated derivatives entailed lithium aluminum hydride reduction of thioxanthene-9-carboxylic acid (**297**) (308) or its ethyl ester (Eq. 129). The ethyl ester **298** may first be alkylated in the $C_{(9)}$ position

292 —(1) n-BuLi; (2) CH_2O; 75%→ 293 (H, CH_2OH) —$TsCl/C_5H_5N$; 78%→ 294 (H, CH_2OTs) —HCO_2H; 77% crude, 44% pure→ 291 (12[?])

295 (H, CH(Ph)OH) —TsOH, HCO_2H→ 296 (Ph) (1[?])

292 —(1) n-BuLi; (2) CO_2→ 297 (H, CO_2H) —$LiAlH_4$→ 293 (129)

(Eq. 130). Reduction and subsequent rearrangement can then provide 10-substituted dibenzo[*b*,*f*]thiepins as illustrated by the preparation of **299** (331). Rearrangement of alcohol **293** was effected with phosphorus pentoxide (48%) (15), polyphosphoric acid (44%) (15), or *p*-toluenesulfonic acid (80–91%) (308), or by solvolysis of the tosylate in formic acid; derivatives of **293** were converted into the corresponding dibenzo[*b*,*f*]thiepins by the action of *p*-toluenesulfonic acid (308) or phosphorus pentoxide (331).

298 (H, CO_2Et) —(1) K; (2) $Me_2NCH_2CH_2Cl$→ ($Me_2NCH_2CH_2$, CO_2Et) —$LiAlH_4$→ ($Me_2NCH_2CH_2$, CH_2OH) —P_2O_5→ 299 ($CH_2CH_2NMe_2$) (1[?])

A reasonable mechanism for ring expansion (Eq. 131) involves rearrangement of cation **300** to the more stable 10,11-dihydrodibenzo[*b*,*f*]thiepin-10-yl cation (**301**) which upon deprotonation generates **291**. Whitlock (318) used the rearrangement of **300** to **301** to detect the degree of reaction between thioxanthylium perchlorate (**302**) and diazomethane and reported the formation of dibenzo[*b*,*f*]thiepin (**291**) in 22% yield.

293 $\xrightarrow{H^{\oplus}}$ 300 $\longrightarrow$ 301 $\xrightarrow{-H^{\oplus}}$ 291

302 + CH_2N_2 $\longrightarrow$ [$CH_2N_2^{\oplus}$ adduct] $\xrightarrow{-N_2}$ 301

(131)

In an exhaustive study of Whitlock's reaction, Seidl and Biemann (262) found five by-products, four of which were oxygen-containing dibenzo[*b*,*f*]-thiepin derivatives (Eq. 132). Formation of 10,11-dihydrodibenzo[*b*,*f*]-thiepin-10-ol (**303**), 10-methoxy-10,11-dihydrodibenzo[*b*,*f*]thiepin (**304**), and bis(10,11-dihydrodibenzo[*b*,*f*]thiepin-10-yl) ether (**305**) was attributed to reaction of cation **301** with water, methanol, or **303**. The other by-product, dibenzo[*b*,*f*]thiepin-10(11*H*)-one (**290**), was explained by a subsequent oxidation of **303**. Further substantiation of the origin of **304**, **303**, and **305** and the intermediacy of cation **301** was found in the isolation of appreciable quantities (as high as 33% yield) of 10-isopropoxy-10,11-dihydrodibenzo-[*b*,*f*]thiepin (**306**) when the reaction of **302** and diazomethane was quenched with isopropyl alcohol. Seidl and Biemann also reported the conversion of alcohol **303** or ethers **304**, **305**, and **306** into dibenzo[*b*,*f*]thiepin (**291**) in 35–98% yield by reaction of each oxygen derivative with boron trifluoride etherate in acetic anhydride. These observations of acid-catalyzed elimination of water or alcohol led the authors (262) to a procedural modification of the reaction of **302** and diazomethane which produced dibenzo[*b*,*f*]thiepin in 55% yield.

The elimination reactions cited above illustrate the third general approach for generating dibenzo[*b*,*f*]thiepins by introduction of the 10,11-double bond. Additional examples of acid-catalyzed eliminations leading to the

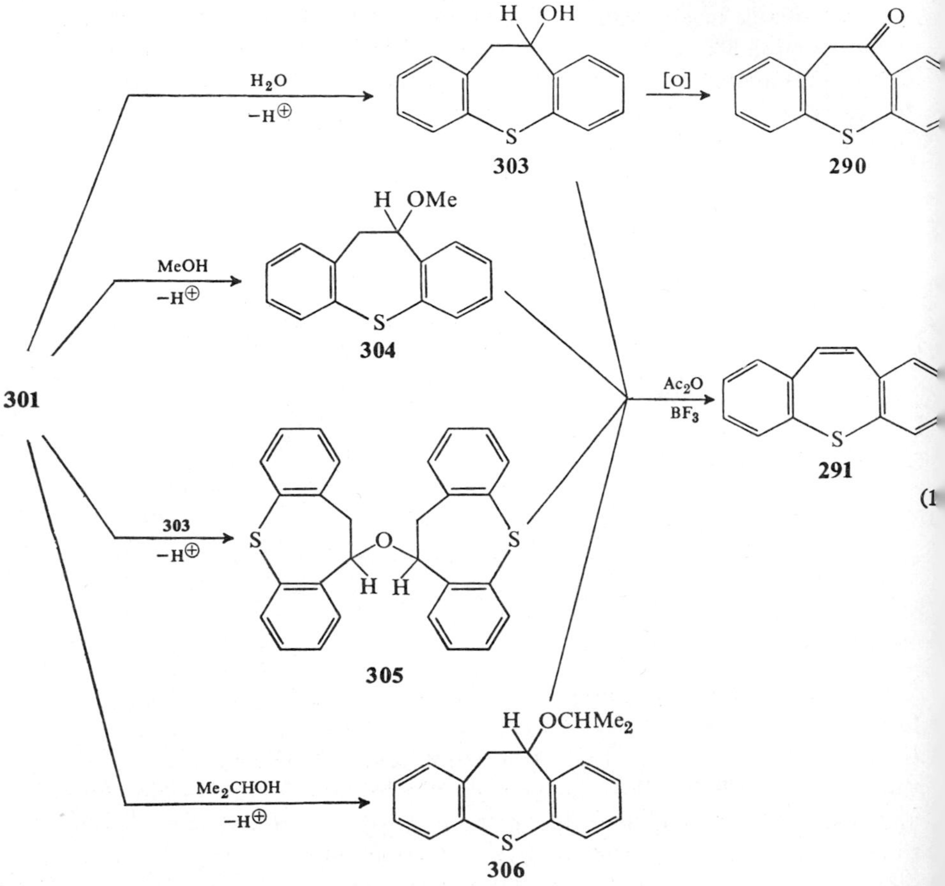

301 → (H_2O, $-H^{\oplus}$) 303 → ([O]) 290
301 → (MeOH, $-H^{\oplus}$) 304
301 → (303, $-H^{\oplus}$) 305
301 → (Me_2CHOH, $-H^{\oplus}$) 306
303, 304, 305, 306 → (Ac_2O, BF_3) 291

(1

10,11-double bond (Eqs. 133 and 134) include the dehydration of **303** with *p*-toluenesulfonic acid (70%) (115), the dehydration of **307** with acetic anhydride (85), and elimination reactions of amines such as **308** (116). The mechanism of these amine reactions is not clear, particularly the origin of **310** from **309** or **311**.

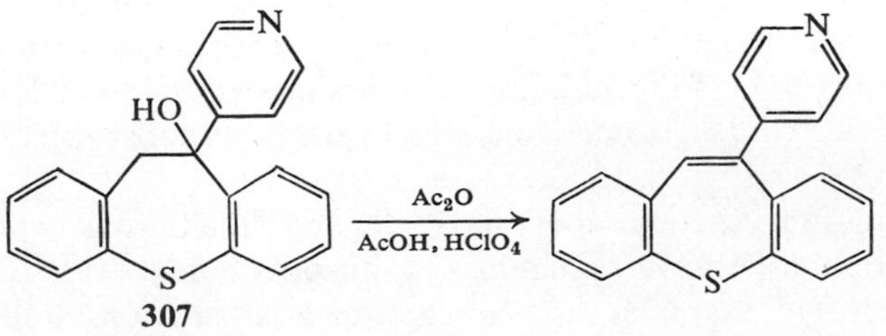

307 → (Ac_2O; AcOH, $HClO_4$) (133)

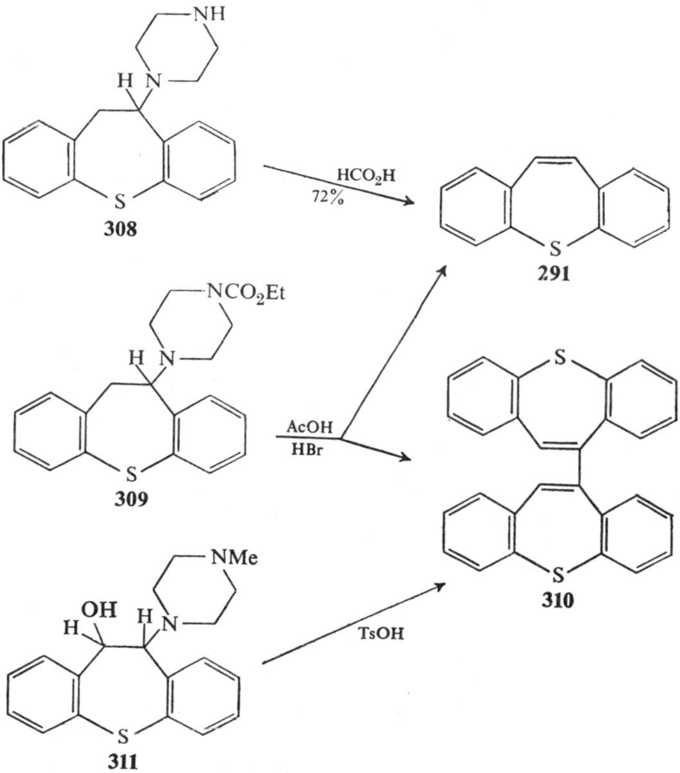

(134)

Base-catalyzed elimination reactions with 10-halo-10,11-dihydrodibenzo-[*b*,*f*]thiepins have also been used for generating dibenzo[*b*,*f*]thiepins. The reaction of **312** with potassium *tert*-butoxide (95) (Eq. 135) formed the 10-methyl derivative **313**; treatment of **314** with dimethylamine or other secondary amines (Eq. 136), on the other hand, resulted in concurrent elimination

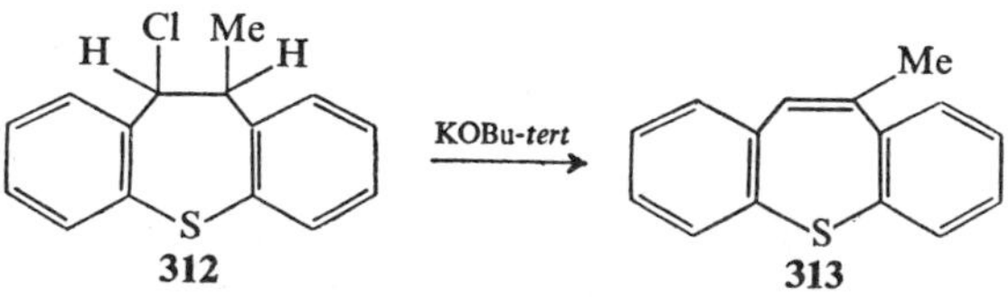

(135)

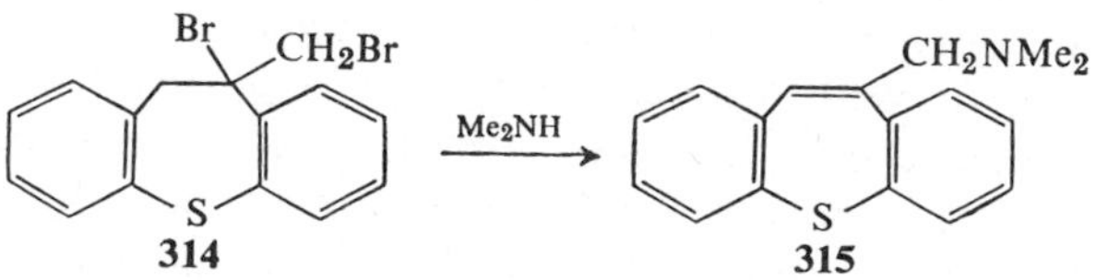

(136)

and substitution to give **315** (97). Although the reaction of 10-chloro-10,11-dihydrodibenzo[*b*,*f*]thiepin (**316**) and 4-methylpiperazine (Eq. 137) has been employed for the synthesis (115) of 10-(4-methylpiperazino)-10,11-dihydrodibenzo[*b*,*f*]thiepin (**287**) ("perathiepin"), dehydrochlorination to dibenzo-[*b*,*f*]thiepin (**291**) was a serious competing reaction (yields were 40–65% for several chloro derivatives (194).

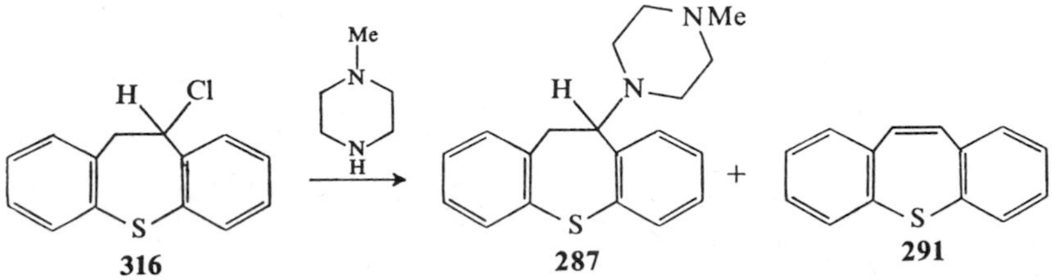

(137)

316 **287** **291**

The tricyclic dibenzo[*b*,*f*]thiepin ring system appears to be nonplanar, with a uv spectrum resembling that of *cis*-stilbene. After a study of the properties of dibenzo[*b*,*f*]thiepin (**291**), Bergmann and Rabinovitz (16) concluded that the double bond in **291** is less olefinic than the corresponding double bond in 1,2,5,6-dibenzo-1,3,5,7-cyclooctatetraene (**317**). Urberg and Kaiser (308) reported that the esr spectrum of the dibenzo[*b*,*f*]thiepin anion radical (stable below −40°C) showed a spin distribution resembling the stilbene anion radical more closely than the anion radical of **317**. They also concluded that conjugation through the sulfide bridge was much weaker than conjugation through the vinyl residue. From a study of the free energies of activation for conformational inversion of 10-(2-hydroxyprop-2-yl)dibenzo[*b*,*f*]thiepin (**318**) and related compounds, Ollis and co-workers (182) suggested an

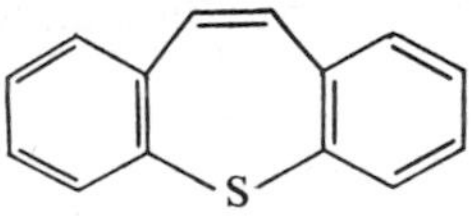

291 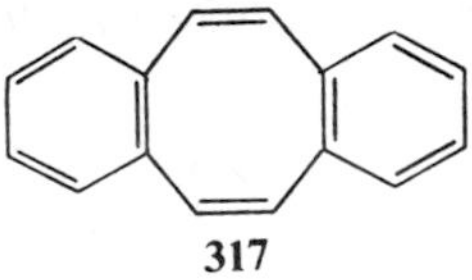 **317** 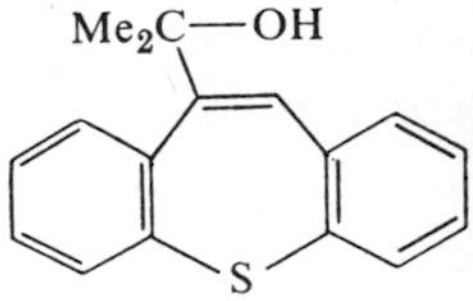 **318**

increase in conjugation energy arising from the interaction of the sulfide bridge with the benzene rings; however, there appears to be no significant increase or decrease in delocalization energy of the type which would arise from cyclic conjugation in the planar seven-membered ring. The experimental technique used to determine the conformational inversion activation energies involved application of nmr to measure the coalescence temperature of the methyl groups in tertiary alcohol **318**, whose nonplanar ring conformation is chiral. Ring inversion is attributed to a planar achiral transition state.

In contrast to the benzothiepins, dibenzo[*b*,*f*]thiepin (**291**) is a stable yellow crystalline solid which melts without decomposition. Sulfur extrusion with subsequent formation of phenanthrene derivatives required prolonged exposure of the dibenzo[*b*,*f*]thiepin derivatives to elevated temperatures in the presence of copper bronze. Even under these conditions, carboxylic acid derivatives such as **319** underwent decarboxylation without loss of sulfur to form **320** (Eq. 138). Other examples of these reactions (Eqs. 139 and 140) may provide useful synthetic pathways to substituted phenanthrene derivatives (151).

CO_2H, O_2N, S — **319** →(Cu bronze; quinoline, reflux 4 hr, 45–50%) O_2N, S — **320** →(Cu bronze; diethyl phthalate, 250°C, 25%) O_2N (138)

CO_2H, O_2N, Me, S →(Cu bronze; quinoline, reflux 4 hr) Me, O_2N (139)

CO_2Me, O_2N, R, S, R=H, Me →(Cu bronze; diethyl phthalate, 30–50%) R, O_2N, CO_2Me, R=H, Me (140)

Oxidation of dibenzo[*b*,*f*]thiepin (**291**) can involve reaction at either the sulfur or olefinic site depending on the oxidant and the experimental conditions. Earlier mention was made of the potassium permanganate oxidation of **291** in acetone which led to the formation of diphenylsulfide-2,2′-dicarboxylic acid (**321**). However, treatment of **291** with potassium permanganate in water containing sodium carbonate followed by acidification with sulfuric acid (Eq. 141) gave thioxanthone (**322**) (16). The explanation for the course of this reaction is analogous to one given for a similar reaction

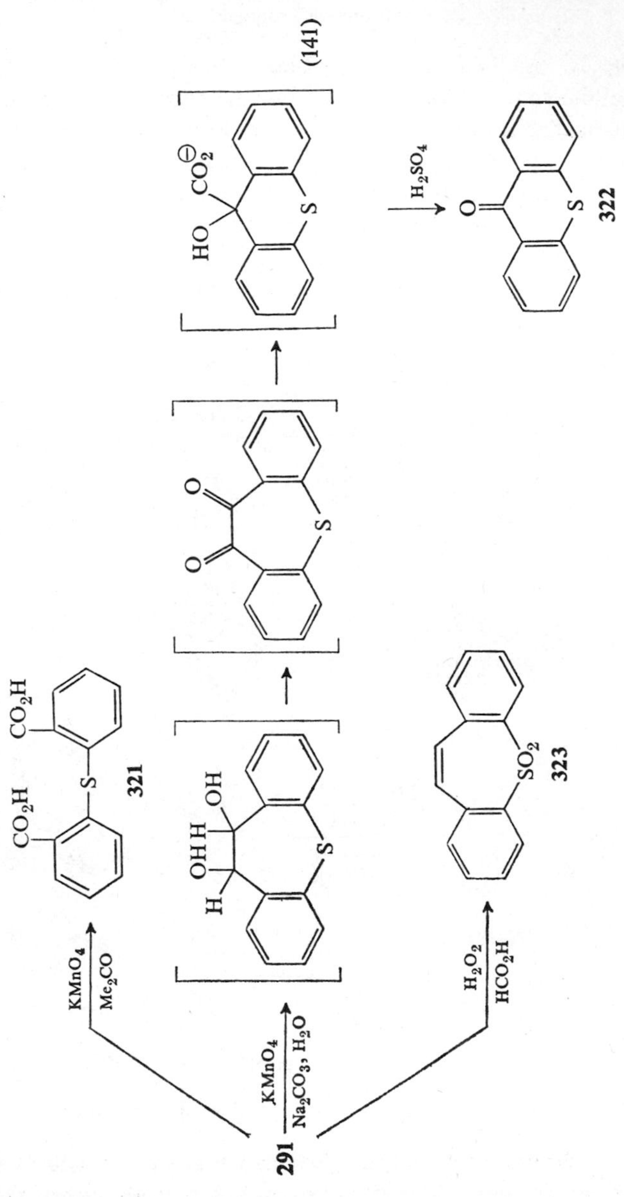
291
KMnO4
Me2CO
CO2H
S
CO2H
321
KMnO4
Na2CO3, H2O
H
OH H
OH
S
O
O
S
HO
CO2⊖
S
(141)
H2SO4
O
S
322
H2O2
HCO2H
SO2
323

of dibenzo[*b*,*f*]oxepins (156, 157) involving a benzylic acid rearrangement as the ring contraction step. When **291** was heated with hydrogen peroxide in formic acid, oxidation occurred at sulfur with easy conversion to the sulfone **323** (16).

Catalytic hydrogenation of 2-nitrodibenzo[*b*,*f*]thiepin (**320**) can be controlled to reduce only the nitro group (Eq. 142) with formation of the amine (**324**) (1), whereas hydrogenation of **325** reduces both the pyridine ring and the 10,11-double bond (Eq. 143) to give **326** (85). Chemical reduction

NO_2 ; H_2/Pt, 48% ; NH_2 ; S ; **320** → **324** (142)

$\oplus$NMe $I^{\ominus}$; S ; **325** ; H_2/PtO_2 → **326** (H, NMe, S) (143)

of dibenzo[*b*,*f*]thiepin (**291**) to the 10,11-dihydro derivative **327** was effected with sodium and alcohol (Eq. 144), whereas reaction with hydrogen iodide, phosphorus, and calcium hypophosphite resulted in an unusual rearrangement and reduction to **328** (116).

291 —Na, EtOH→ **327** (S); **291** —HI,P, $CaHPO_2$→ **328** (H, Me, S) (144)

The 10-methyl group of **313** is readily brominated with *N*-bromosuccinimide to produce 10-bromomethyldibenzo[*b*,*f*]thiepin (**329**), a valuable intermediate in the synthesis of alkylamino dibenzo[*b*,*f*]thiepins such as **299** (95). The preparation of this compound is outlined in Eq. 145.

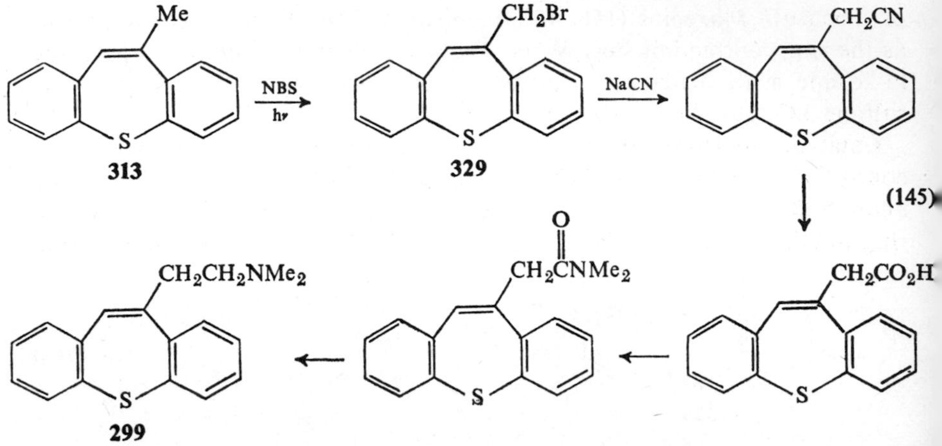

Me NBS hν 313 CH_2Br NaCN 329 CH_2CN (145) CH_2CO_2H $CH_2\overset{O}{\overset{\|}{C}}NMe_2$ $CH_2CH_2NMe_2$ 299

Relocation of the 10,11-double bond of the dibenzo[*b*,*f*]thiepin ring into an *exo* position at $C_{(10)}$ provides an isomeric system which has been described in the literature. Dehydration of **330** with acetyl chloride (Eq. 146) gave an olefinic product whose spectral properties were consistent with its formulation as dibenzo[*b*,*f*]thiepin-$\Delta^{10(11H)}$-γ-propyldimethylamine (**331**). Other examples

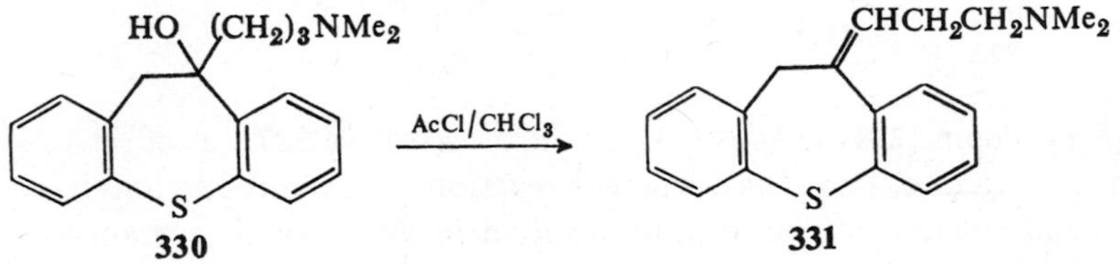

HO $(CH_2)_3NMe_2$ AcCl/$CHCl_3$ **330** $CHCH_2CH_2NMe_2$ **331** (1

HO CH_2R 0.2 *N* HCl CHR Br_2 Br R CH_2Br Me_2

332; R = H
334; R = Me

333; R = H
335; R = Me

314; R = H

R $CHNMe_2$ (

315; R = H

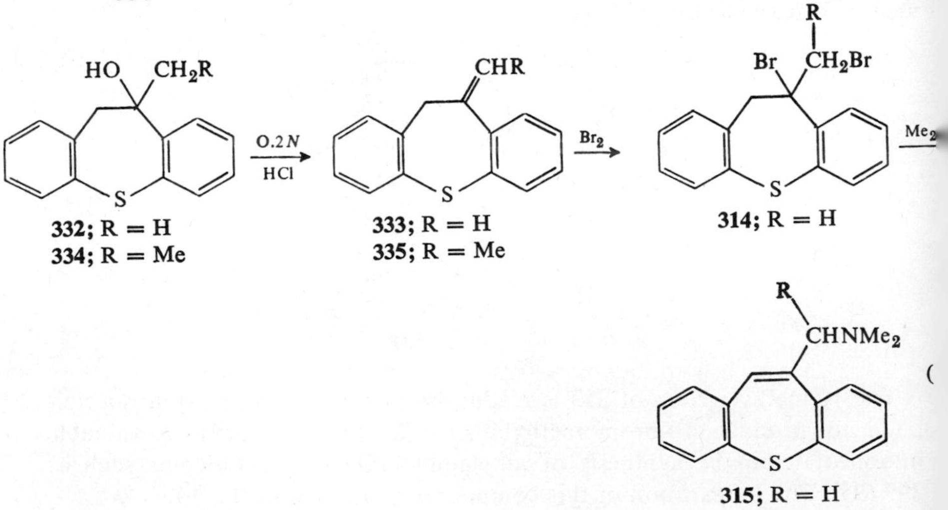

leading to formation of the exocyclic double bond include the acid-catalyzed dehydration of **332** and **334** to **333** and **335**, respectively (Eq. 147) (97). Although evidence for the structural assignments of **333** and **335** was not reported, addition of bromine and subsequent reaction with a variety of secondary amines provided chemical support for the exocyclic nature of the double bond.

2. Dibenzo[*b*,*f*]thiepin-5-oxide and Derivatives

The preparation of dibenzo[*b*,*f*]thiepin-5-oxide and its derivatives was accomplished by oxidation of the corresponding dibenzo[*b*,*f*]thiepin with iodosobenzene diacetate (301). The sulfoxides were converted into anion radicals by the action of potassium metal in 1,2-dimethoxyethane. From a study of the esr spectra of these dibenzo[*b*,*f*]thiepin-5-oxide anion radicals, Trifunac and Kaiser (301) concluded that the sulfoxide group can withdraw electron spin density from the benzo rings more effectively than the sulfide moiety in the dibenzo[*b*,*f*]thiepin anion radical, and that conjugation through the sulfoxide bridge is much weaker than conjugation through the vinyl residue.

3. Dibenzo[*b*,*f*]thiepin-5,5-dioxide and Derivatives

Dibenzo[*b*,*f*]thiepin-5,5-dioxide (**323**) (16) and its 2-chloro derivative (111) were prepared by oxidation of the corresponding sulfide with 30% hydrogen peroxide in formic and acetic acid. An alternate synthesis in this system involves the preparation of 10-bromodibenzo[*b*,*f*]thiepin-5,5-dioxide (**336**) (293) by an addition–elimination sequence (Eq. 148).

SO_2 **323** $\xrightarrow[56\%]{Br_2}$ H, Br, H, Br, SO_2 **337** $\xrightarrow[THF\ 88\%]{Et_3N}$ Br, SO_2 **336**

(148)

This class of compounds has not been studied very extensively; however, chemical interest in **336** stemmed from Tochtermann's work on triple bond (arynelike) intermediates in unsaturated seven-membered rings. When the dibromo adduct **337** was treated with potassium hydroxide in methanol (Eq. 149), Tochtermann (293) obtained the expected product **336** along with some 10-methoxydibenzo[*b*,*f*]thiepin-5,5-dioxide (**339**). A rationalization for the formation of **339** involved addition of methanol to the intermediate aryne **338**. Evidence supporting the generation of **338** was found in the isolation of the Diels-Alder adduct **340** when the 10-bromo derivative **336** was treated with potassium *tert*-butoxide in the presence of furan (293).

337 $\xrightarrow[\text{MeOH}]{\text{KOH}}$ **336** + **339** (OMe, SO_2)

336 $\xrightarrow[\textit{tert}\text{-BuOH}]{\textit{tert}\text{-BuOK}}$ [**338** (SO_2)] — KOH; MeOH → **339**; piperidine (N–H), 67% → **342** (SO_2); NaN_3, 94% → **343** (N=N, NH, SO_2); furan, 50% → **340** (O, SO_2); → **344** (O_2S, O, SO_2) (149)

340 $\xrightarrow{H_2,\ Pd/C}$ $\xrightarrow[140°C]{PPA}$ **341** (SO_2) 60% overall

Hydrogenation of **340**, followed by treatment with polyphosphoric acid, produced tribenzo[*b*,*d*,*f*]thiepin-9,9-dioxide (**341**). Reactions of **336** with potassium *tert*-butoxide and piperidine or sodium azide likewise appeared to proceed via intermediate **338**, yielding adducts **342** and **343**, respectively. When **336** was allowed to react with potassium *tert*-butoxide in *tert*-butyl alcohol, a dimeric product assigned structure **344** was formed along with a substance tentatively formulated as 10-*tert*-butoxydibenzo[*b*,*f*]thiepin-5,5-dioxide. The origin of dimer **344** most probably entails reaction of **338** with the enol form of dibenzo[*b*,*f*]thiepin-10(11*H*)-one-5,5-dioxide.

4. Dibenzo[*b*,*f*]thiepin-10(11*H*)-one and Derivatives

The key intermediate for synthetic work in the dibenzo[*b*,*f*]thiepin ring system has been dibenzo[*b*,*f*]thiepin-10(11*H*)-one (**290**) and its derivatives. The earliest report in the literature (Eq. 150) described the preparation of 2-nitro-8-methyldibenzo[*b*,*f*]thiepin-10(11*H*)-one (**346**) via polyphosphoric acid ring closure of **345** (151). Application of this ring closure technique was extended by Protiva and co-workers to the synthesis of the parent compound **290** (95% yield) (115) and numerous derivatives (75–95% yield) (1, 111, 128, 194) (see Table XI-8 for examples). It is of interest that cyclization of 2-(4

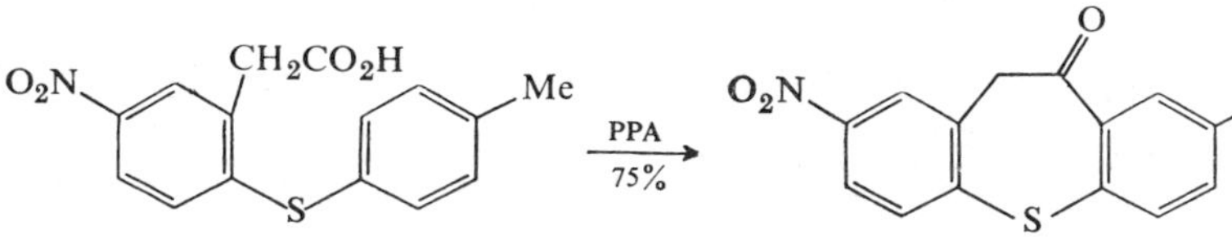

O_2N, CH_2CO_2H, S, Me — **345** $\xrightarrow[75\%]{PPA}$ O_2N, O, S, Me — **346** (150)

trifluoromethylphenylthio)phenylacetic acid to the corresponding ketone **347** required anhydrous hydrogen fluoride (197), and that in the reaction of 2-(4-bromophenylthio)phenylacetic acid with polyphosphoric acid some of the debrominated ketone **290** accompanied the expected product **348** (112).

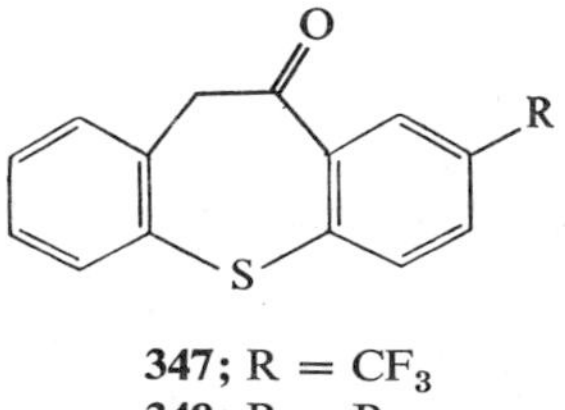

347; R = CF_3
348; R = Br

The major method for preparing the precursor 2-arylthiophenylacetic acid (115) is illustrated in Eq. 151 with the synthesis of 2-phenylthiophenylacetic acid (**349**).

Dibenzo[*b*,*f*]thiepin-10(11*H*)-one (**290**) contains a puckered heterocyclic ring which can assume two half-chair conformations; however, ring inversion at room temperature is rapid and the nmr spectrum shows only one peak for the methylene protons (78). On the other hand, dibenzo[*b*,*f*]thiepin-10(11*H*)-one-5-oxide (**350**) exhibits an AB pattern for the methylene protons. Duerr (78) concluded from a variable temperature nmr study that the free energy of activation for ring inversion in **290** is small ($\Delta G^{\ddagger} = 9.3$ kcal/mole), whereas

(151)

the value for the sulfoxide **350** ($\Delta G^{\ddagger} = 23.3$ kcal/mole) is sufficiently large to allow possible separation into optical antipodes.

In Duerr's nmr study of **290** no mention was made of the presence of enol **351**; however, Protiva and co-workers (115) had suggested earlier that **290**

undergoes extensive enolization. Their suggestion was based on the reaction of **290** with 3-dimethylaminopropylmagnesium chloride which gave only 3% of the expected tertiary alcohol and a 76% recovery of starting ketone. Also, Seidl and Biemann (262) reported that the ir spectrum of **290** contains a broad OH peak at 3600 cm^{-1} along with a strong carbonyl peak at 1675 cm^{-1}. Another pertinent observation is the facile conversion of **290** into 10-chlorodibenzo[*b*,*f*]thiepin by the action of phosphorus oxychloride and zinc chloride (116). Although the foregoing evidence provides ample support for the existence of enol **351**, the precise extent of enolization remains to be settled.

Very few of the dibenzo[*b*,*f*]thiepin-10(11*H*)-ones have been characterized by the preparation of classical derivatives (see Table XI-8); however, the oxime **352** undergoes some interesting reductive transformations (Eq. 152). Reduction of **352** with sodium amalgam and methanol (128) or reductive acetylation followed by saponification (115) gave the expected 10-amino-10,11-dihydrodibenzo[*b*,*f*]thiepin (**353**), also available by the Leuckart

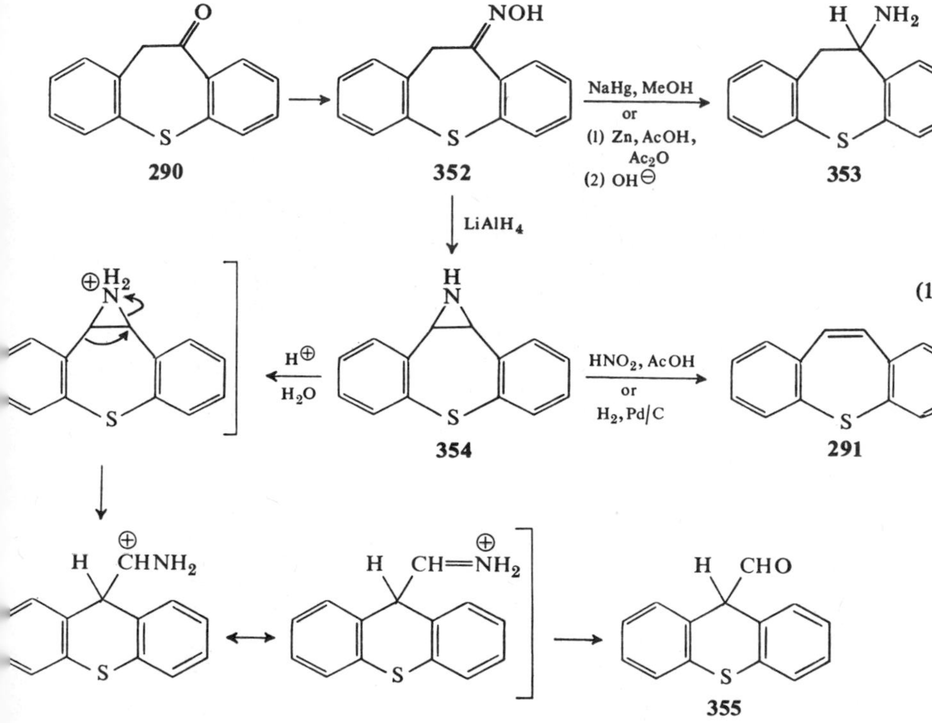

(152)

reaction of **290** with formamide (115). However, reaction of **352** with lithium aluminum hydride gave the aziridine **354** (84) which was converted to dibenzo-[*b*,*f*]thiepin (**291**) by the action of nitrous acid or by hydrogenolysis with palladium on charcoal. Treatment of aziridine **354** with acid led to a rearrangement and ring contraction to 9-formylthioxanthene (**355**) (84). A suggested pathway for the rearrangement is outlined in Eq. 152.

The chemical properties of the dibenzo[*b*,*f*]thiepin-10(11*H*)-ones are typical for carbonyl compounds with an α-methylene group. Reduction of the carbonyl function with sodium borohydride (115, 194) or lithium aluminum hydride (95) leads to the corresponding alcohol; application of the Wolff-Kishner reduction (95, 115) has been used extensively to remove the carbonyl function. These reactions are illustrated (Eq. 153) with the preparation of **303** and **327**. The Grignard reaction of **290** with 3-dimethylaminopropyl-magnesium chloride has already been discussed; other Grignard reagents used successfully to prepare tertiary alcohols have been methyl- and ethylmagnesium iodide (97).

Reactions at the $C_{(11)}$ methylene group include alkylation (95, 96, 255), cyanoethylation (95), Mannich condensation (115), bromination (116), and

nitrosation (116, 128). Examples of these reactions appear in Eq. 154. In addition to the expected Mannich product from the reaction of **290**, formaldehyde, and piperidine, a neutral substance **356** was also isolated. Condensation of bromoketone **357** with 4-methylpiperazine gave the substitution product **358** and an unexpected diketone **359**. Exposure of **358** to Wolff-Kishner conditions led to a rearrangement and formation of thioxanthone.

Oxidation of **290** with hydrogen peroxide at room temperature gave sulfoxide **350** in 50% yield; however, under more vigorous oxidative conditions, but with 30% hydrogen peroxide (Eq. 155), **290** gave the diketo sulfone **360** (116). Structural assignment for **360** was aided by its facile benzilic acid rearrangement and decarboxylation to thioxanthone-5,5-dioxide (**361**). The diketo sulfide **359** was also formed as a by-product (116). This diketone likewise undergoes benzilic acid rearrangement and has been proposed as an oxidative intermediate in the reaction of dibenzo[*b*,*f*]thiepin with alkaline potassium permanganate in acetone (see Section I-1). Protiva and co-workers (116) observed that nitrosation of **290** formed the mono oxime **362**. Subsequently Lüttringhaus and Creutzburg (153) reported that acid-catalyzed

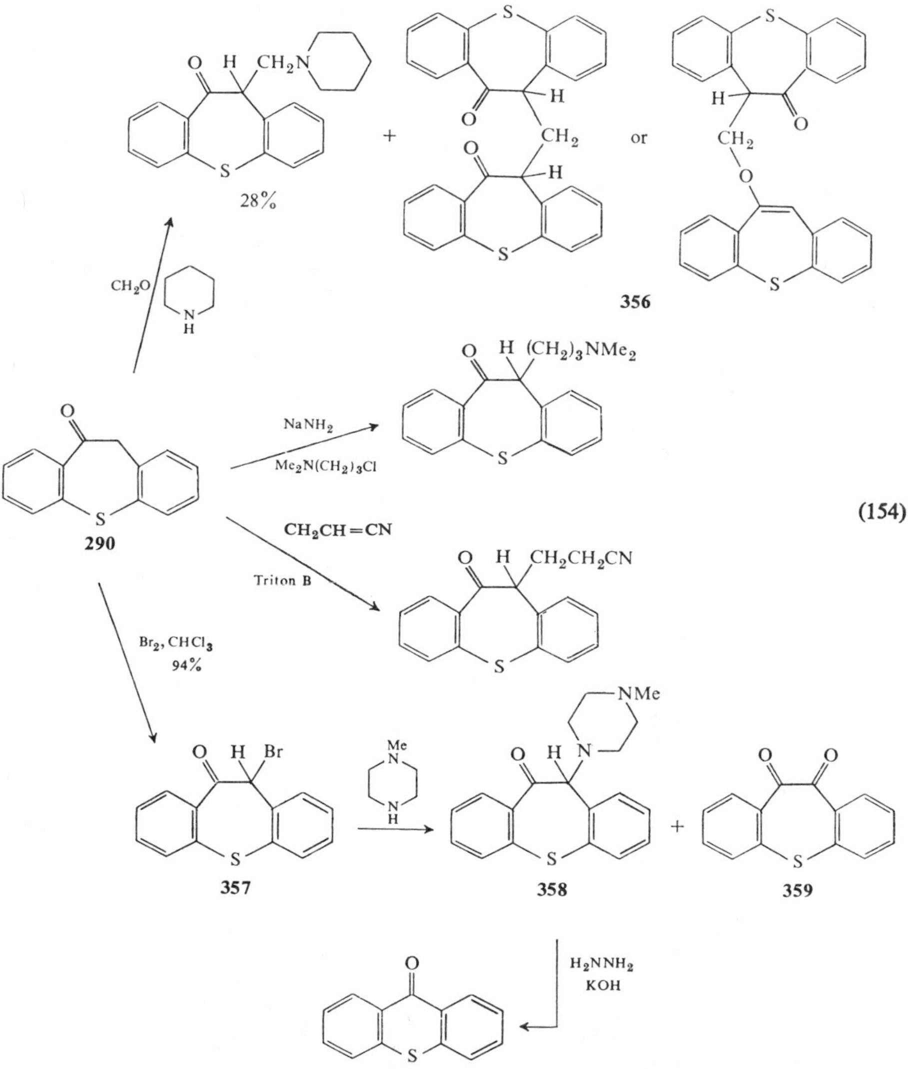

(154)

rearrangement of **362** and **359** in the presence of ethanol led to the formation of ethyl thioxanthene-9-carboxylate (**363**). In a separate experiment they found that the conversion of **364** into **363** was faster than the above reaction of **359** or **362**. This observation, along with the identification of acetaldehyde

290 $\xrightarrow{H_2O_2}$ 360 $\xrightarrow{NaOH}$ $\xrightarrow[\Delta]{H^{\oplus}}$ 361

290 $\xrightarrow[NaOEt]{n\text{-}BuONO}$ (53%)

$\xrightarrow[H^{\oplus}]{EtOH}$ 363

359 $\xrightarrow[H^{\oplus}]{EtOH}$ 363

364 $\xrightarrow[H^{\oplus}]{EtOH}$ 363

(155)

in the reaction of **362** with ethanol and acid, suggests preliminary rearrangement to **364** followed by reduction and esterification as the mechanism of formation of **363**.

5. 10,11-Dihydrodibenzo[*b*,*f*]thiepin and Derivatives

The preparation of 10,11-dihydrodibenzo[*b*,*f*]thiepin (**327**) by reduction of dibenzo[*b*,*f*]thiepin (**291**) or dibenzo[*b*,*f*]thiepin-10(11*H*)-one (**290**) has been described earlier (see Sections I-1 and I-4); the latter intermediate (via Wolff-Kishner reduction) provides the best synthetic procedure. The Wolff-Kishner reduction has been applied extensively in the preparation of 10-aminoalkyl-10,11-dihydrodibenzo[*b*,*f*]thiepins as illustrated (Eq. 156) by the synthesis of **365** (95).

O H CH_2CH_2CN — $\xrightarrow[\text{KOH}]{H_2NNH_2}$ — H $CH_2CH_2CO_2H$ — several steps, via N,N-dimethyl amide → H $(CH_2)_3NMe_2$ (156)

365

A large variety of 10,11-dihydrodibenzo[*b*,*f*]thiepin-10-ols (see Table XI-8) have been obtained by sodium borohydride reduction of the corresponding ketones (111, 115, 194). These alcohols have been converted into 10-dialkylaminoalkoxy-10,11-dihydrodibenzo[*b*,*f*]thiepins (115, 194) as in the synthesis of **286** shown in Eq. 157. By reaction with thionyl chloride (115, 194), or with hydrogen chloride and calcium chloride in benzene (111) (the preferred procedure), the alcohols form the corresponding chloro derivatives (see Table XI-8). Interest in 10-chloro-10,11-dihydrodibenzo-[*b*,*f*]thiepins revolved about their reaction with amines to produce the 10-amino derivatives (111, 115, 116, 194). An example of this reaction which was cited earlier was the synthesis of 10-(4-methylpiperazino)-10,11-dihydrodibenzo[*b*,*f*]thiepin (**287**); the same type of reaction has been employed in preparing numerous other amine derivatives (see Table XI-8). An alternative method of synthesis of simple 10-amino derivatives involving a

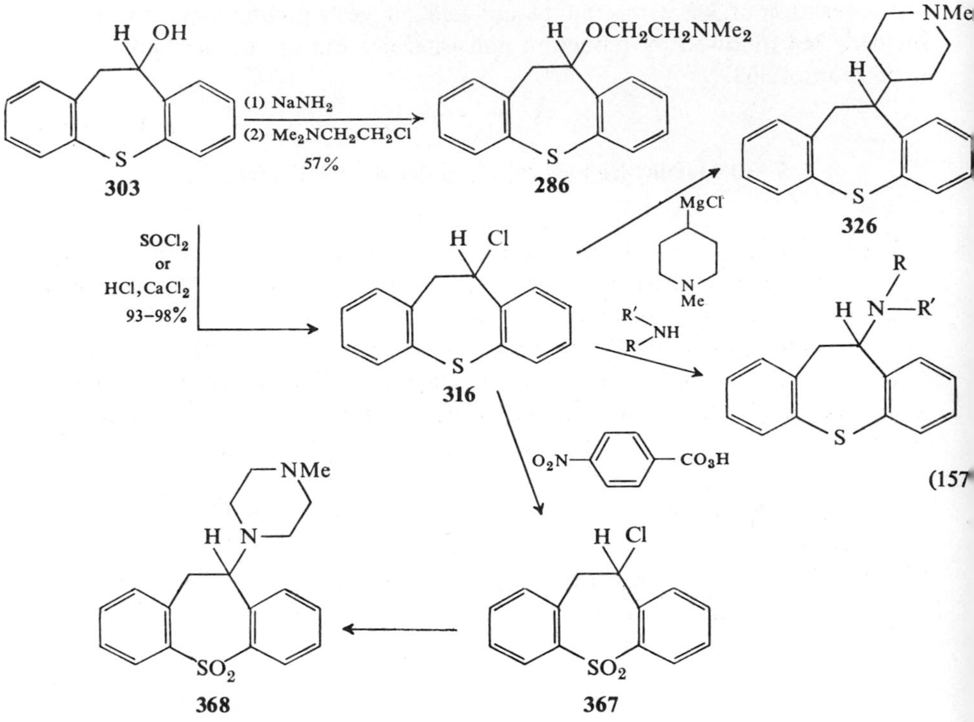

(1) $NaNH_2$
(2) $Me_2NCH_2CH_2Cl$
57%

303 → 286

$SOCl_2$ or HCl, $CaCl_2$ 93–98%

316

MgCl (N-methylpiperidin-4-yl) → 326

R′R NH →

O_2N–C_6H_4–CO_3H

367 → 368

(157)

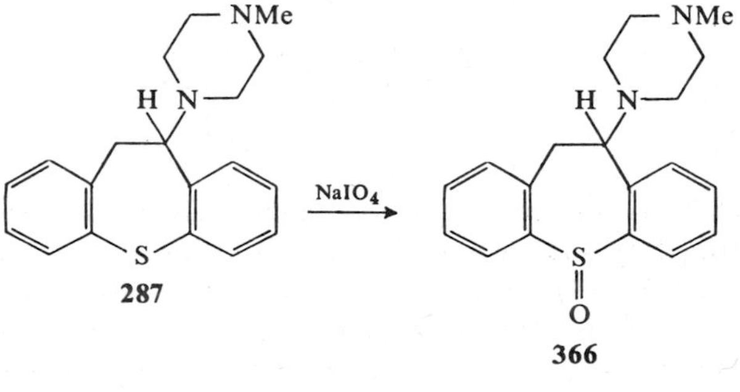

$NaIO_4$

287 → 366

(158)

Leuckart-type reaction has also been discussed in the preceding Section (I-4).

The sulfoxide **366** was available from oxidation of **287** with sodium metaperiodate (Eq. 158); however, gentle treatment of **287** with hydrogen peroxide led to a complex of dibenzo[*b*,*f*]thiepin-5-oxide and dibenzo[*b*,*f*]thiepin-5,5-dioxide. Under more vigorous conditions **287** and hydrogen peroxide gave dibenzo[*b*,*d*]thiepin-5,5-dioxide exclusively (116). The synthesis of sulfone **368** required oxidation of the chloro compound **316** to the chloro sulfone

$$\mathbf{362} \xrightarrow[H_2]{PtO_2} \mathbf{369} \xrightarrow[R = Me, Ph]{(RC(O))_2O} \text{RC(O)NH-ketone} \xrightarrow{NaBH_4} \text{RC(O)NH-alcohol} \tag{159}$$

362

369

R = Me, Ph

$NH_2 \cdot HCl$

OAc

367 with *p*-nitroperbenzoic acid, followed by reaction of **367** with 4-methylpiperazine (Eq. 157) (243).

10-Chloro-10,11-dihydrodibenzo[*b*,*f*]thiepin (**316**) undergoes a coupling reaction with Grignard reagents as illustrated (Eq. 157) by the preparation of **326** (116).

Derivatives of 11-amino-10,11-dihydrodibenzo[*b*,*f*]thiepin-10-ol have been obtained by reduction of 11-aminodibenzo[*b*,*f*]thiepin-10(11*H*)-one (**369**) which was prepared by reduction of the mono oxime **362** (128). Further reactions of **369** are shown in Eq. 159.

6. Hexahydro and Octahydrodibenzo[*b*,*f*]thiepins

A synthetic sequence (Eq. 160) was recently described (73) leading to the synthesis of *cis*- and *trans*-8-chloro-1,2,3,4,4a,11a-hexahydrodibenzo[*b*,*f*]-thiepin-10(11*H*)-ones **374** and **375**, respectively. The reactions parallel

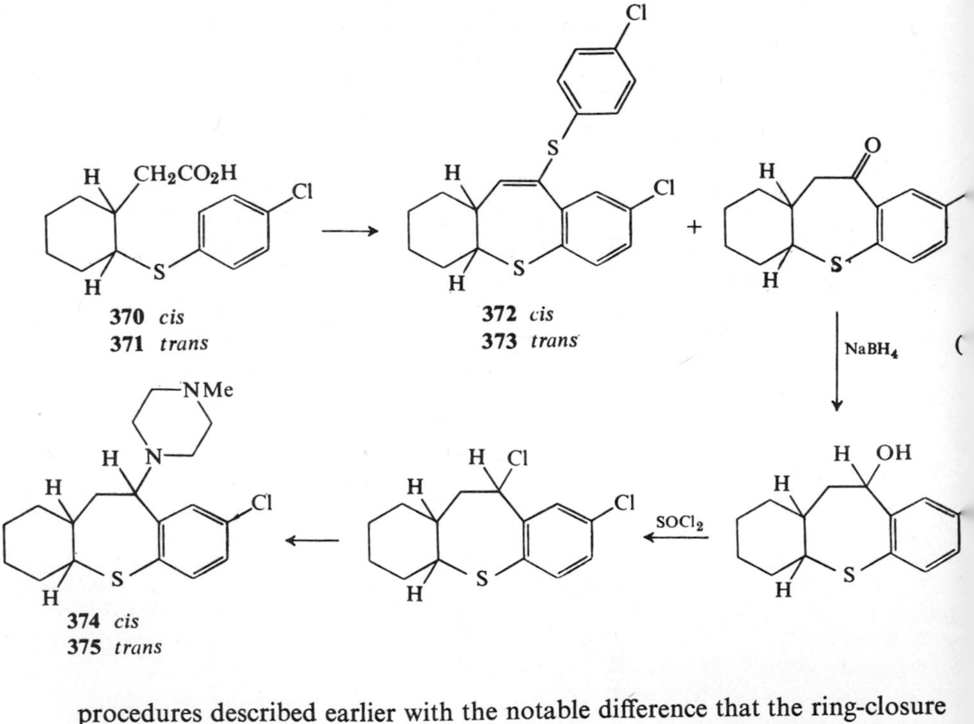

370 *cis*
371 *trans*

372 *cis*
373 *trans*

$NaBH_4$

$SOCl_2$

374 *cis*
375 *trans*

procedures described earlier with the notable difference that the ring-closure reactions of **370** and **371** led to formation of by-products **372** and **373**, respectively.

J. Dibenzo[*c*,*e*]thiepins

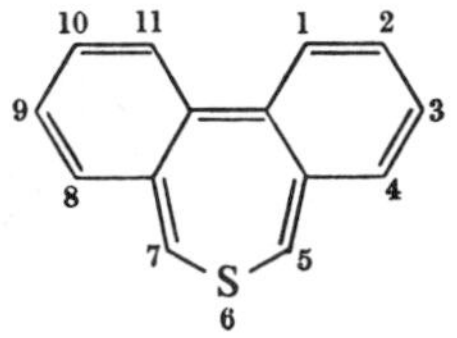

R.I. 8638

The parent compound (**376**) of this system has not been reported. If **376** were available, one would expect it to undergo a facile sulfur extrusion reaction with the formation of phenanthrene (Eq. 161). This system, as in 2-benzothiepin, provides the potential for observing the influence of valence bond tautomer **377**, with its stabilizing benzenoid units, upon the sulfur extrusion reaction.

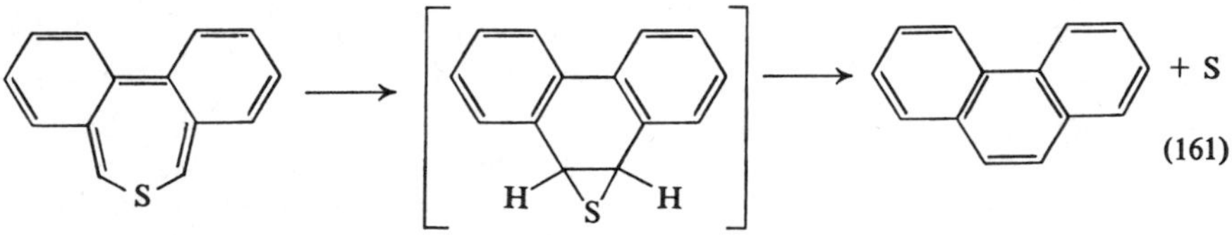

(161)

All the work reported in this area deals with 5,7-dihydrodibenzo[*c*,*e*]-thiepin (**378**), substituted derivatives of **378**, and their corresponding sulfoxides and sulfones. On numerous occasions these compounds have been utilized as models for testing reaction mechanisms and various facets of structural theory. The preparation of 5,7-dihydrodibenzo[*c*,*e*]thiepin (**378**) was reported initially by Truce and Emrick (303) who employed the reaction of 2,2′-bis(bromomethyl)biphenyl and sodium sulfide (Eq. 162). This

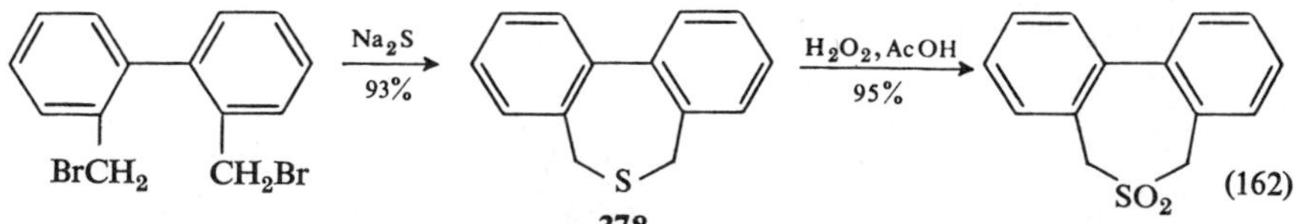

(162)

procedure has been applied in the synthesis of substituted derivatives of **378** (167, 185, 288) (see Table XI-9) and is the major synthetic avenue for entry into this system. Oxidation of **378** with hydrogen peroxide leads to sulfone **379** in excellent yield (303); Paquette (185, 186) and more recently Fraser and Schuber (88) were successful in preparing substituted sulfones by oxidation of the corresponding sulfides with *m*-chloroperbenzoic acid. Sulfoxide **380** was available from the oxidation of **381** with sodium metaperiodate (Eq. 163) (88).

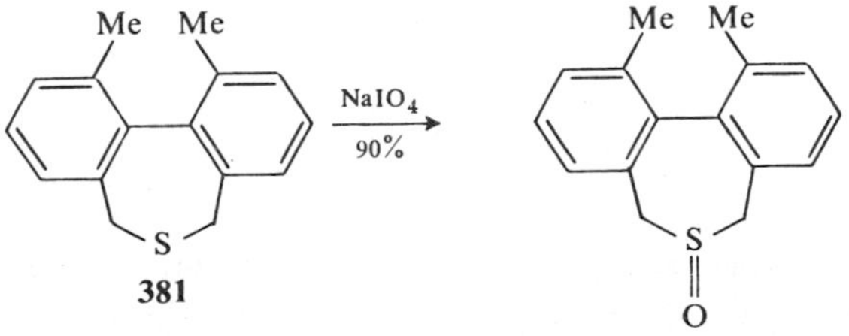

Me Me → ($NaIO_4$, 90%) Me Me (163)

381 **380**

Examination of a molecular model of **378** suggested that the rings are dissymmetrically twisted, with the phenyl rings in a noncoplanar position (167). Mislow calculated the angle of torsion about the biphenyl pivotal 1,1′-bond for **378** and found a value of 56.6°, as compared to the 50° angle calculated for biphenyl in heptane (287). One consequence of twisting the two phenyl rings out of a planar position is the decrease in biphenyl conjugation. The conjugation band of **378** reported by Truce (303) (λ_{max} 245 mμ, ε 10,000) and by Mislow (167) (λ_{max} 244 mμ, ε 10,200), when compared to that of biphenyl ($\lambda_{max}^{neat,\ crystalline}$ 275 mμ, $\lambda_{max}^{heptane}$ 247.5 mμ, ε 19,000), shows reduced biphenyl conjugation. 1,11-Dimethyl-5,7-dihydrodibenzo[*c*,*e*]thiepin (**381**) (conjugation band at λ_{max} 236 mμ, ε 10,300) represents even greater twisting of the phenyl rings out of plane (167).

A second consequence of the dissymmetrically twisted ring system in **378** is the prospect of obtaining optically active enantiomers. Truce and Emrich (303) succeeded in resolving 5,7-dihydrodibenzo[*c*,*e*]thiepin-6,6-dioxide-3,9-dicarboxylic acid (**382**) into one of its optical antipodes via the acid cinchonine salt; however, their attempts to obtain an optically active isomer of 6-methyl-5,7-dihydrodibenzo[*c*,*e*]thiepinium salt were unsuccessful. On the other hand, placement of substituents into the 1- and 11-positions enhanced

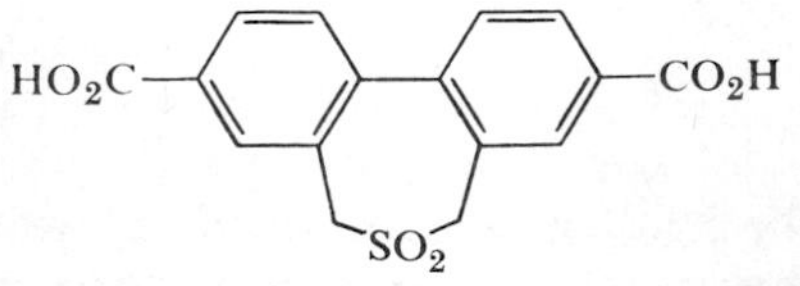

HO_2C CO_2H SO_2 R R S

382

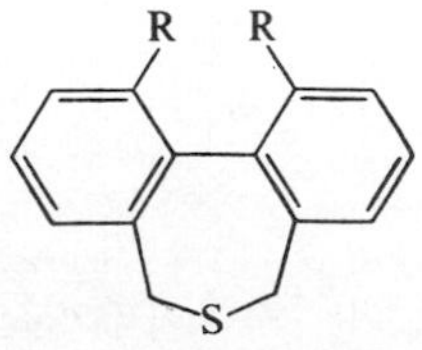

383; R = CO_2Me
384; R = CH_2OH
385; R = CH_2Br

optical stability, allowing Mislow and co-workers (167) to prepare several optically active compounds **381** and **383–385**.

The third consequence of the dissymmetrically twisted ring system in **378** is the diastereomeric relationship of the methylene protons at the $C_{(5)}$ and $C_{(7)}$ positions. This prediction is supported by the observation of an AB pattern for these protons in the 100 MHz nmr spectrum of **378** in deuteriochloroform (167). Since the two protons H_A and H_B can exchange environments on ring inversion (see Fig. XI-1), the appearance of the AB pattern

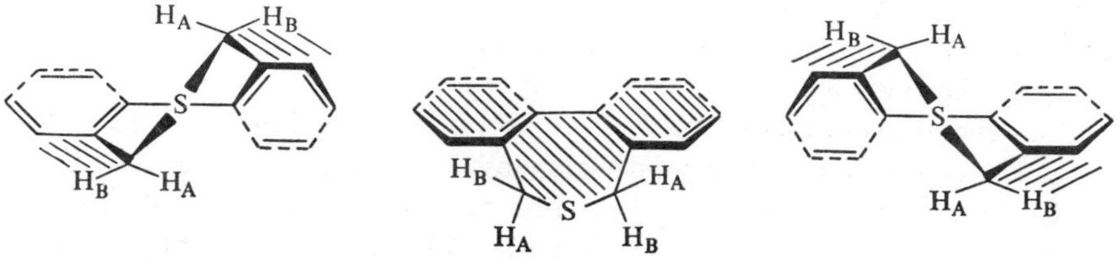

Figure XI-1.

in the nmr spectrum at room temperature reflects some conformational stability for **378**. Kurland (136) and Sutherland and Ramsay (286) independently conducted a variable temperature nmr study of **378** in deuteriochloroform and obtained a value of 16 kcal/mole for the free energy of activation for ring inversion, whereas Mislow and co-workers (167) reported a calculated value of 17 kcal/mole. The $\Delta G^{\ddagger}$ for ring inversion in 5,7-dihydrodibenzo[*c*,*e*]thiepin-6,6-dioxide (**379**) was 18.2 kcal/mole (286). The optical stability of 1,11-dimethyl-5,7-dihydrodibenzo[*c*,*e*]thiepin (**381**) suggests a conformationally rigid ring system, yet the nmr spectrum for **381** in deuteriochloroform shows a singlet for the $C_{(5)}$ and $C_{(7)}$ methylene protons. A singlet peak was also observed for these protons in a variety of solvents, but in benzene the expected AB pattern appeared (167).

Fraser and Schuber (86, 88) have described a detailed study of the nmr spectrum of 1,11-dimethyl-5,7-dihydrodibenzo[*c*,*e*]thiepin (**381**), its sulfoxide (**380**), its sulfone (**386**), and its *S*-methyl perchlorate salt (**387**). They employed various deuterated derivatives and applied the nuclear Overhauser effect in making configurational assignments for the $C_{(5)}$ and $C_{(7)}$ methylene

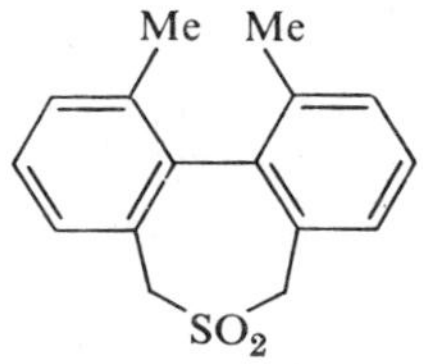

386

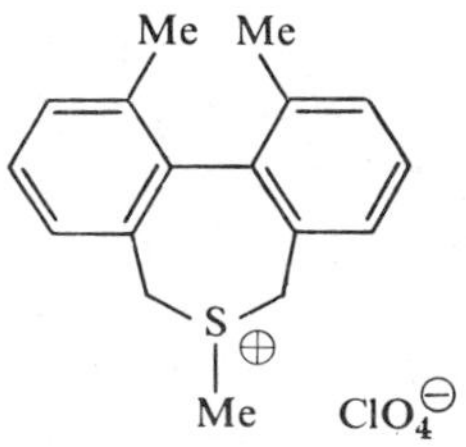

387

(benzylic) protons. In sulfoxide **380** there were four magnetically nonequivalent protons, two nonequivalent methyl groups, and six nonequivalent aromatic protons. The sulfonium perchlorate **387** nmr spectrum was quite similar to that of the sulfoxide **380**, suggesting to Fraser and Schuber (88) that the effect of the sulfoxide group on the shift of adjacent protons should be attributed to the unshared electron pair on sulfur. Fraser and Schuber also reported exchange rates for the diastereotopic protons in sulfoxide **380** (86) and sulfone **386** (87). In the latter study evidence was presented that the proton located within the O—S—O bond angle is the one more readily removed, in accord with the stabler carbanion conformation predicted by MO calculations (328).

The chemical properties of 5,7-dihydrodibenzo[*c*,*e*]thiepins are characteristic of the sulfide group and biphenyl ring system. Oxidation to the sulfoxide or sulfone has been described above. Electrophilic substitution reactions, such as the nitration of sulfone **379**, lead to the 3-nitro or 3,9-dinitro derivatives (303) characteristic of the 4,4′-substitution of biphenyl. 6-Methyl-5,7-dihydrodibenzo[*c*,*e*]thiepinium iodide (303) or the 1,11-dimethyl derivative **387** (88) are prepared in high yield by standard reaction of the appropriate sulfide and methyl iodide. Raney nickel desulfurization studies with 5,7-dihydrodibenzo[*c*,*e*]thiepin (**378**) have been reported by Badger and co-workers (Eq. 164) (9). Raney nickel W-7 gave an excellent

S

378

RaNi →

$\dot{C}H_2$ $\dot{C}H_2$

388

H-rich RaNi →

Me Me

389

H-poor RaNi →

(164)

yield of 2,2′-dimethylbiphenyl (**389**), whereas the use of Raney nickel W-7J (hydrogen poor) gave lower yields of **389** and some phenanthrene. The proposed origin of the phenanthrene involves cyclization of the initial diradical **388** to 9,10-dihydrophenanthrene which is then converted into phenanthrene.

Another reaction characteristic of sulfides is α-chlorination with sulfuryl chloride (Eq. 165) (35, 187, 302). Paquette has described the monochlorination (185) of **378**, **381**, and 3,9-dimethyl-5,7-dihydrodibenzo[*c*,*e*]thiepin

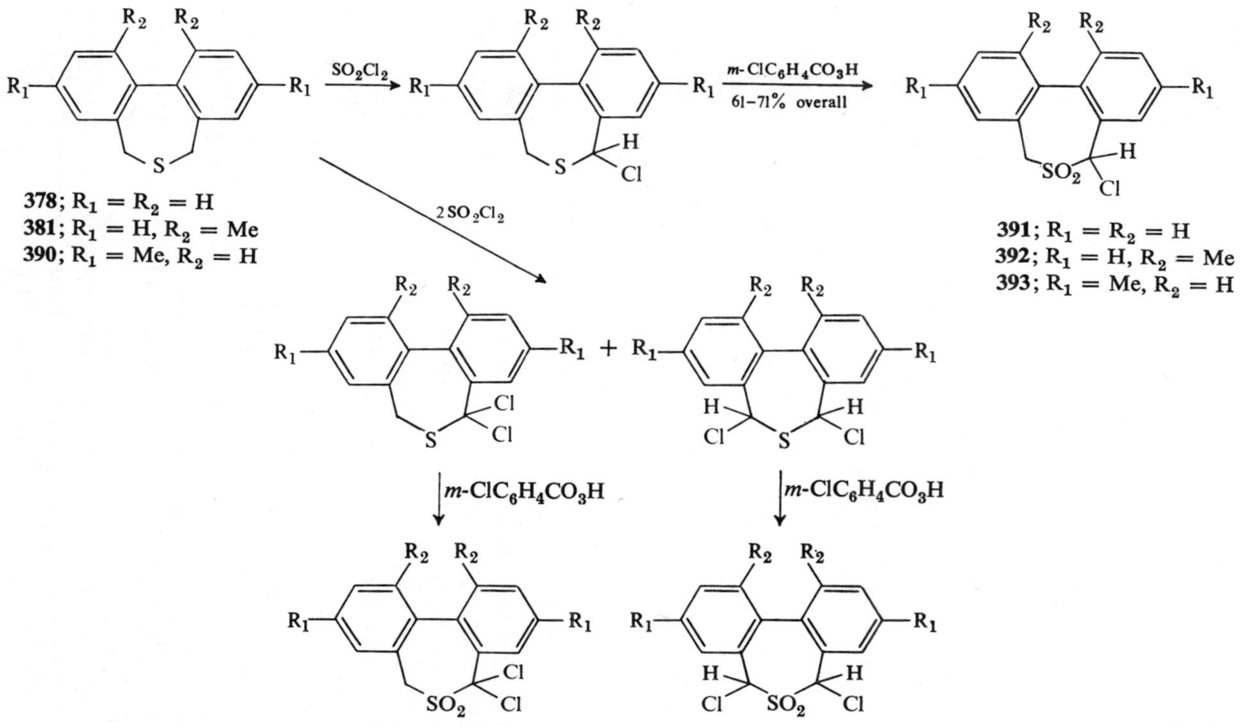

378; $R_1 = R_2 = H$
381; $R_1 = H, R_2 = Me$
390; $R_1 = Me, R_2 = H$

SO_2Cl_2

$m\text{-}ClC_6H_4CO_3H$
61–71% overall

391; $R_1 = R_2 = H$
392; $R_1 = H, R_2 = Me$
393; $R_1 = Me, R_2 = H$

$2\,SO_2Cl_2$

$m\text{-}ClC_6H_4CO_3H$

$m\text{-}ClC_6H_4CO_3H$

(165)

394; $R_1 = R_2 = H$
395; $R_1 = H, R_2 = Me$

396; $R_1 = R_2 = H$
397; $R_1 = H, R_2 = Me$

(**390**), as well as the dichlorination (186) of **378** and **381** using sulfuryl chloride. In all instances the products were oxidized with *m*-chloroperbenzoic acid and isolated as the corresponding sulfones. The dichlorination reactions gave predominately the α,α-dichloro compound but also formed some of the α,α′-dichloro product, not previously observed in sulfuryl chloride reactions.

Interest in the chloro sulfones centered in their application toward mechanistic studies of the Ramberg-Backlund reaction. The geometry of the

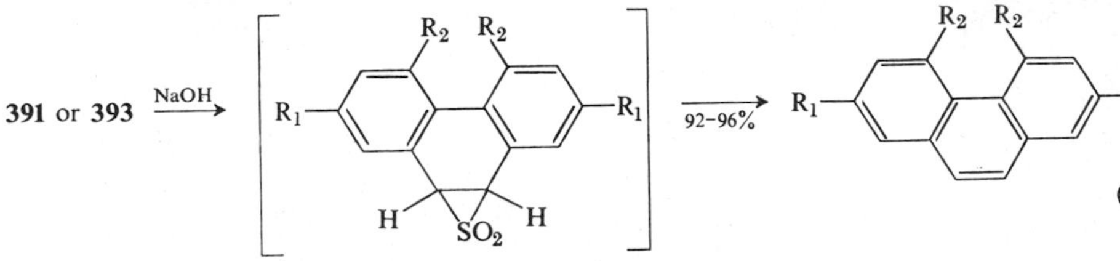

391 or **393** $\xrightarrow{\text{NaOH}}$ [**398**] $\xrightarrow{92\text{–}96\%}$ (166)

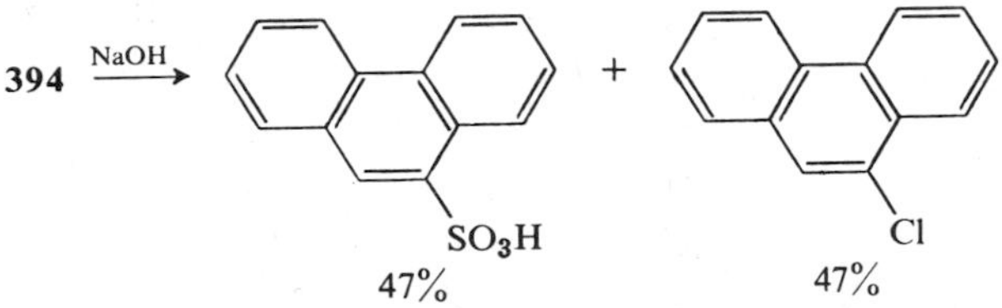

394 $\xrightarrow{\text{NaOH}}$ (SO_3H) 47% + (Cl) 47% (167)

center ring is such as to favor the formation of intermediate **398** with the three six-membered rings coplanar. Thus conversion of **391** and **393** by action of base into the corresponding phenanthrene (Eq. 166) can be achieved readily and in high yield (185). With methyl groups in the 1,11-positions, as in **392**, the intermediate **398** is less favored due to steric crowding of the methyl groups and reaction fails to occur. The conversion of the dichloro sulfone **394** into phenanthrene-9-sulfonic acid and 9-chlorophenanthrene (Eq. 167), and the concomitant failure of **395** to react with base, is also explained by an intermediate analogous to **398** (186). In this instance SO_2 extrusion leads to 9-chlorophenanthrene, and attack of base on the sulfone group with ring opening and elimination of chloride ion provides phenanthrene-9-sulfonic acid. Base catalyzed reactions with **396** and **397** have not been described.

Mislow and co-workers (166, 167) reported some interesting ring-closure reactions of 1,11-disubstituted-5,7-dihydrodibenzo[*c*,*e*]thiepins (Eq. 168). With the 1,11-bis(hydroxymethyl) derivative **399** (either racemic or optically active), reaction with *p*-toluenesulfonic acid produced the tetracyclic compound (either racemic or optically active) **400**. The reaction of racemic **399** with phosphorus tribromide formed the crystalline dibromide **401** which on treatment with sodium sulfide gave racemic 4*H*,6*H*-[2]benzothiepino[6,5,4-*d,e,f*][2]benzothiepin (**402**). The latter compound can also be prepared by reaction of 2,2′,6,6′-tetrakis(bromomethyl)biphenyl (**403**) with two equivalents of sodium sulfide. When racemic dibromide **401** was heated above its melting point a disproportionation occurred with formation of equal amounts of tetrabromide **403** and the tetracyclic compound **402**. The reaction

(168)

of optically active (+)-diol **399** with phosphorous tribromide gave the corresponding (+)-dibromide **401** as an oil which upon standing disproportionated to **403** and racemic **402**; however, if (+)-**401** was treated with sodium sulfide shortly after isolation, (+)-**402** was obtained in high yield. The facile disproportionation of (+)-**401** and (±)-**401** as a melt was rationalized via sulfonium salt intermediates as pictured in Eq. 169. Configurational assignments and a full discussion of the stereochemistry of these systems has been presented (167).

K. Benzonaphthothiepins

Of the 21 possible isomeric benzonaphthothiepin ring systems, only three have been described in the literature. The synthetic work reported in connection with the benzo[*e*]naphtho[1,2-*b*]thiepin derivative parallels the work on

BrCH$_2$ BrCH$_2$ S

BrCH$_2$ BrCH$_2$ S

BrCH$_2$ BrCH$_2$ $Br^{\ominus}$ CH_2 $S^{\oplus}$ CH_2 Br S

Br CH_2 BrCH$_2$ BrCH$_2$ S CH_2 Br S

(169)

Br CH_2 BrCH$_2$ BrCH$_2$ $S^{\oplus}$ $Br^{\ominus}$ S

BrCH$_2$ BrCH$_2$ CH_2Br CH_2Br + **402**

dibenzo[*b*,*e*]thiepins (see Section H). Entry into this ring system involved polyphosphoric acid ring closure of acid **404** to benzo[*e*]naphtho[1,2-*b*]-thiepin-7(12*H*)-one (**405**) which upon treatment with 3-dimethylamino-propylmagnesium chloride gave alcohol **406** (Eq. 170). Acid-catalyzed dehydration of **406** led to **407** which showed an antiserotonin effect (224).

CO_2H **404** —PPA, 80%→ **405**; **405** —$Me_2N(CH_2)_3MgCl$, 43%→ **406** (HO, $(CH_2)_3NMe_2$); **406** —H⊕, 71%→ **407** ($CHCH_2CH_2NMe_2$) (170)

The second isomeric system described in the literature is represented by 10-nitrobenzo[*f*]naphtho[1,2-*b*]thiepin-7-carboxylic acid (**409**) which resulted from polyphosphoric acid ring closure (Eq. 171) of the pyruvic acid **408**; however, attempts to cyclize 2-(2-naphthylthio)-5-nitrophenylpyruvic acid were unsuccessful and gave only starting material (151). No reactions with **409** were reported.

408 (O_2N, $CH_2C(=O)CO_2H$, S) —PPA→ **409** (O_2N, CO_2H, S) (171)

R.I. 9103

Knapp (129) studied the cyclization of 2-(2-methoxy-1-naphthylthio)-benzoic acid (**410**) with polyphosphoric acid (Eq. 172) and observed the formation of 1-methoxybenzo[*f*]naphtho[1,8-*bc*]thiepin-12-one (**411**). When the β-position in the naphthalene ring is unsubstituted, as in 2-(naphthylthio)benzoic acid (**412**), ring closure furnishes **413** rather than the thiepin system. Little evidence was offered for the structural assignment of **411**.

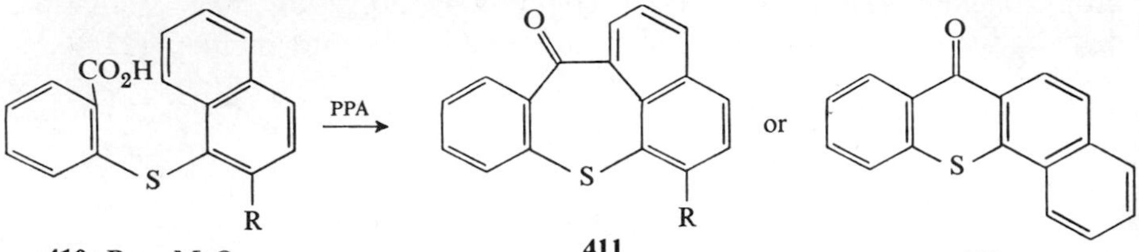

CO_2H S R PPA → O S R or O S

410; R = MeO
412; R = H

411

413

R.I. 5308 (172)

L. Tribenzo[*b*,*d*,*f*]thiepins

The synthesis of tribenzo[*b*,*d*,*f*]thiepin (**415**) was achieved (Eq. 173) by thermal decomposition of 2-chlorosulfonyl-2′-phenyldiphenyl sulfide (**414**) (160, 180). Reaction occurred at 250–260°C with cuprous chloride added to the melt of **414** or by heating **414** in octachloronaphthalene; interestingly,

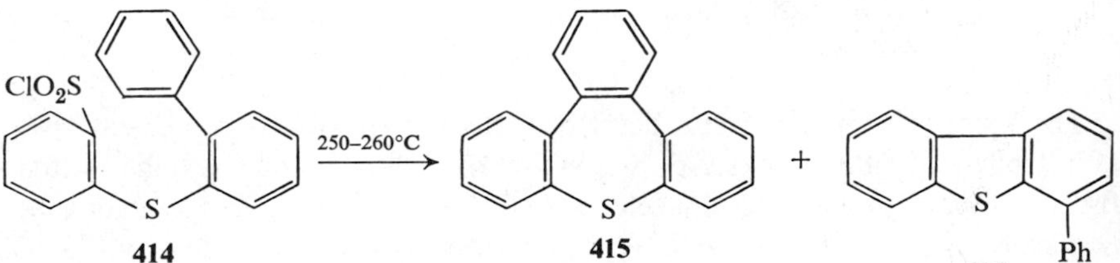

ClO_2S S 250–260°C → S + S Ph

414

415 40%

416 14%

(173)

ring closure favored formation of the seven-membered sulfur ring rather than **416**. The reaction mechanism has been compared to the Gomberg-Bachmann arylation reaction and the Pschorr ring closure.

Tribenzo[*b*,*d*,*f*]thiepin is a colorless crystalline solid with a uv spectrum resembling that of *o*-terphenyl. Although **415** is stable indefinitely up to 400°C in the absence of air, exposure of a bisphenoxyterphenyl solution of **415** to the action of copper bronze at 380°C for 5 hr (Eq. 174) results in sulfur extrusion and formation of triphenylene (**417**). Thiepin **415** also undergoes desulfurization with Raney nickel (W-7) to form *o*-terphenyl (**418**) and is oxidized with hydrogen peroxide to tribenzo[*b*,*d*,*f*]thiepin-9,9-dioxide (**341**) (180).

A unique alternate synthesis of **341** described by Tochtermann and coworkers (293) is outlined in Eq. 175. The key first step involves trapping of

Raney Ni
(W-7)

418

415

Cu bronze 380° C

417

(174)

H_2O_2
HOAc

SO_2
341

Br

SO_2
336

tert-BuOK

O

50%

O

SO_2

340

H_2

O

SO_2

419

PP A
140°C

341

Ph
Ph
O
Ph
Ph

O

SO_2 +

Ph
Ph
Ph
Ph

+ CO

Δ
xylene

Ph
Ph
CO
O
SO_2
Ph
Ph

(175)

421

420

CH_2=$CHCO_2Et$

O

SO_2

O_2C
H

422

H_2SO_4
Ac_2O

SO_2

EtO_2C

423

NaOH

SO_2

HO_2C

424

an arynelike intermediate with furan to produce **340**. Subsequent conversion of **340** via **419** to the sulfone **341** proceeds in 60% over-all yield via routine reactions. An interesting extension of this method entails a Diels-Alder reaction between **340** and tetraphenylcyclopentadienone to form **420** which undergoes thermal elimination of CO and fragmentation into 1,2,3,4-tetraphenylbenzene and **421** (291, 292). Reaction of **421** with ethyl acrylate yields **422** which undergoes dehydration to the tribenzothiepin **423** (292). One can in principle substitute other dienophiles for ethyl acrylate in the above reaction with **421**, thereby gaining access to a variety of substituted tribenzo[*b*,*d*,*f*]thiepin-9,9-dioxides.

Saponification of **423** produced tribenzo[*b*,*d*,*f*]thiepin-9,9-dioxide-2-carboxylic acid (**424**) which was resolved via its brucine salt into both enantiomers (292). The enantiomers had appreciable optical stability but suffered racemization when heated at 80°C for an extended period. These observations supported the assignment of a nonplanar structure for the tribenzo[*b*,*d*,*f*]thiepin-9,9-dioxide ring system with the central thiepin dioxide ring in a boat conformation. Racemization occurred by ring inversion and was found to have the following reaction parameters: $\Delta G^{\ddagger} = 29.5$ kcal/mole, $E_A = 32$ kcal/mole, $\Delta S^{\ddagger} = +5\ (\pm\ 5)$ cal/(mole)(degree) (292).

When **340** was exposed to light of $\lambda > 330$ mμ, a rapid and extensive photochemical cycloaddition process gave **425** which was photostable under the experimental conditions (200). However, thermal isomerization of

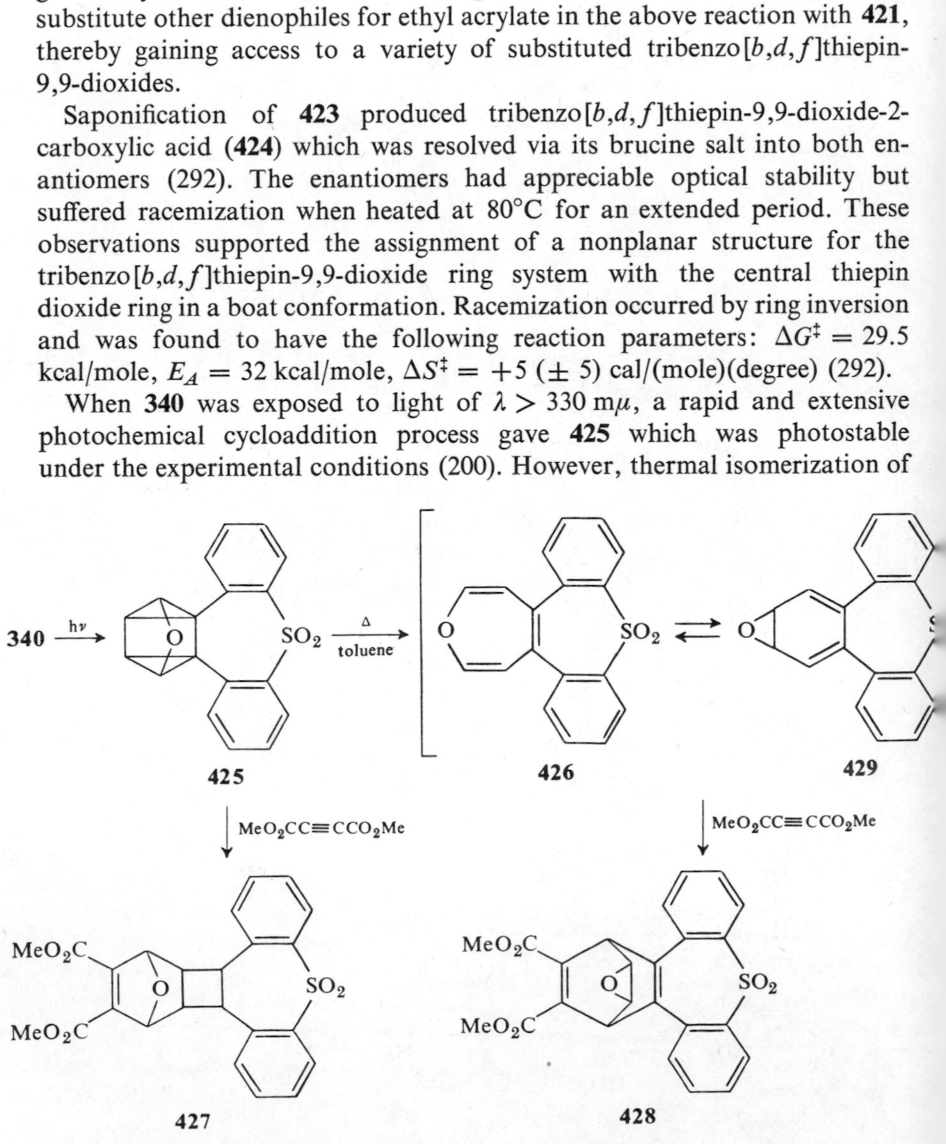

340 —hν→ **425** —Δ, toluene→ [**426** ⇄ **429**]

425 —$MeO_2CC{\equiv}CCO_2Me$→ **427**

429 —$MeO_2CC{\equiv}CCO_2Me$→ **428**

425 to the oxepin **426** was accomplished on a preparative scale by heating **425** in toluene at 110°C for a few minutes (Eq. 176). The highly strained ring system in **425** undergoes reaction with dimethyl acetylenedicarboxylate to form an adduct assigned structure **427** on the basis of spectral data (200). The oxepin derivative **426** also forms an adduct with dimethyl acetylenedicarboxylate; however, the proposed structure **428** suggests reaction of the acetylene derivative with the valence bond tautomer **429**. The structures of the adducts have been formulated tentatively on the basis of spectral evidence, including particularly nmr data.

M. Thienothiepins and Benzothienothiepins

1. Thieno[3,2-*c*]thiepins and Derivatives

Cagniant and co-workers (47, 48, 52) have described an extensive study of the uv, ir, and nmr spectral properties of 6,7-dihydro-4*H*,8*H*-thieno[3,2-*c*]-thiepin-8-one (**431**), its 6-methyl derivative, and 6-methyl-6,7-dihydro-4*H*,8*H*-thieno[3,2-*c*]thiepin. Synthetic entry into this ring system (Eq. 177) entailed Friedel-Crafts ring closure of acid chloride **430** with stannic chloride (52).

$CH_2SCH_2CH_2\overset{O}{\overset{\|}{C}}Cl$ — $\xrightarrow[10\%]{SnCl_4}$ (177)

430 **431**

2. Thieno [2,3-*d*] thiepins and Derivatives

Friedel-Crafts cyclization of acid chloride **432** to 4,5-dihydrothieno[2,3-*d*]-thiepin-8(7*H*)-one (**433**) (ring closure to α-carbon) and of acid chloride **434** to 7,8-dihydrothieno[2,3-*d*]thiepin-4(5*H*)-one (**435**) (ring closure to β-carbon) both proceeded in comparable yield with stannic chloride catalyst (Eq. 178) (45). When aluminum chloride was used, resinification occurred. Wolff-Kishner reduction of either **433** (R = H) or **435** (R = H) led to the same tetrahydro derivative **436**. Reduction of the four ketones with sodium borohydride produced the corresponding alcohols **437** and **438** which were easily dehydrated with polyphosphoric acid to the respective olefins **439** and **440**

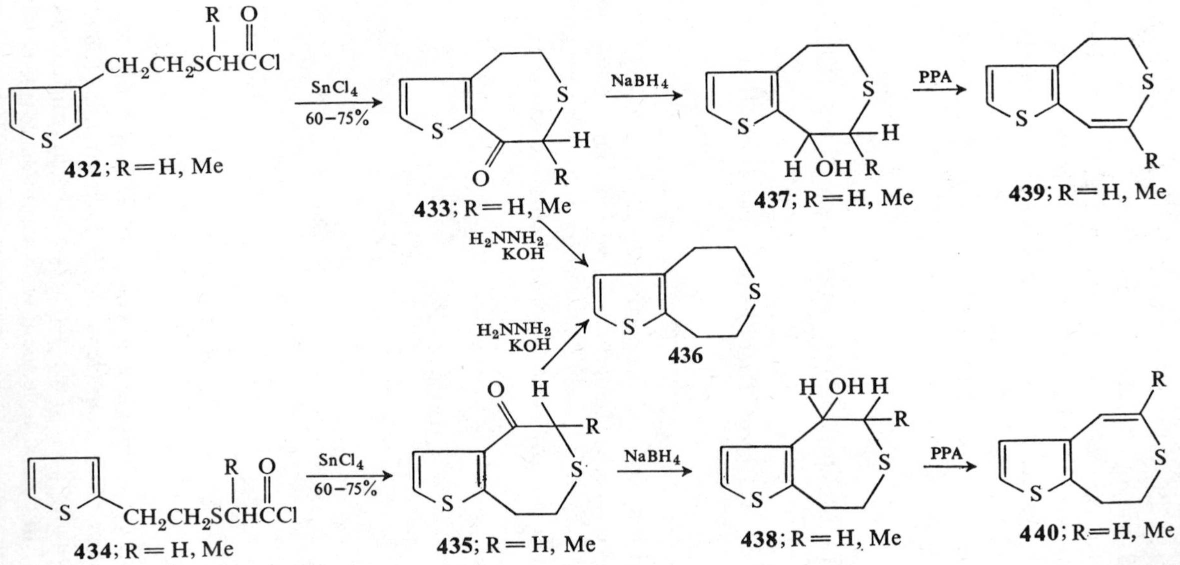

$SnCl_4$, 60–75% (432 → 433; 434 → 435)

$NaBH_4$ (433 → 437; 435 → 438)

PPA (437 → 439; 438 → 440)

H_2NNH_2, KOH (433 → 436; 435 → 436)

432; R = H, Me — $CH_2CH_2SCH(R)C(O)Cl$

433; R = H, Me

437; R = H, Me

439; R = H, Me

436

434; R = H, Me — $CH_2CH_2SCH(R)C(O)Cl$

435; R = H, Me

438; R = H, Me

440; R = H, Me

(178)

(Eq. 178) (45). Cagniant and co-workers published the results of an extensive ir, uv, and nmr study of ketones **433** and **435** (47), and also of 4,5,7,8-tetrahydrothieno[2,3-*d*]thiepin (**436**) (48).

3. Thieno[3,4-*c*]thiepins and Derivatives

A detailed interpretation of the ir, nmr, and uv spectra of 1,3,6-trimethyl-6,7-dihydro-4*H*,8*H*-thieno[3,4-*c*]thiepin-8-one has been made by Cagniant and co-workers (47).

4. Thieno[3,4-*d*]thiepins and Derivatives

Schlessinger and Ponticello have published a preliminary description of the synthesis of thieno[3,4-*d*]thiepin (**443**) (256), thieno[3,4-*d*]thiepin-6-oxide (**444**) (257), and thieno[3,4-*d*]thiepin-6,6-dioxide (**445**) (257) (Eq. 179). Periodate oxidation of sulfide **441** gave a 97% yield of the corresponding sulfoxide **442** which upon treatment with acetic anhydride gave rise to thieno-[3,4-*d*]thiepin (**443**). Controlled oxidation of **443** with *m*-chloroperbenzoic acid led to sulfoxide **444** or sulfone **445**. Structural assignments for these compounds were supported by mass spectral, nmr, and uv data.

441 $\xrightarrow[97\%]{NaIO_4}$ **442** $\xrightarrow[50\%]{Ac_2O}$ [**443** ⟷ **446**]

443 $\xrightarrow[0°C]{m\text{-}ClC_6H_4CO_3H}$ **444** (90%)

443 $\xrightarrow{m\text{-}ClC_6H_4CO_3H}$ **445**

(179)

Interest in the thieno[3,4-*d*]thiepins arises from a consideration of how these 4*n* π-electron compounds fit into the theory of aromaticity. A study of spectral and chemical properties (particularly the Diels-Alder reaction discussed below) of **443**, **444**, and **445** led Schlessinger and Ponticello (256, 257) to conclude that sulfide **443** possesses more aromatic character than its sulfoxide or sulfone. This view was further supported by the X-ray crystallographic analysis of **443** which showed considerable similarity to azulene (247) and was consistent with the suggestion of Schlessinger and Ponticello that structure **446** contributes significantly to the ground state of thieno[3,4-*d*]thiepin. However, a ring-puckered structure for sulfone **445** precluded significant π-electron delocalization in its seven-membered ring (247). Dewar and Trinajstic (66) challenged the amount of resonance contribution from structure **446** to compound **443**. On the basis of their calculations Dewar and Trinajstic view **443** as essentially a thiophene ring with a nonaromatic or weakly aromatic thiepin ring attached to it.

Thieno[3,4-*d*]thiepin (**443**) is a stable crystalline solid exhibiting the following nmr spectrum: τ 3.36 (two protons, singlet), τ 3.94 and τ 4.78 (four protons, quartet). Its uv spectrum indicates extended conjugation (256). The mass spectra of **443**, **444**, and **445** all showed a base peak at *m*/*e* 134 which appeared to correspond to benzo[*c*]thiophene formed by S, SO, or SO_2 extrusion. The respective molecule ion peaks were also observed in **443**, **444**, and **445** (257). When **443** was heated at 150°C with *N*-phenylmaleimide, quantitative conversion to *exo* adduct **447** was observed (Eq. 180) (256).

443 + N-phenylmaleimide $\xrightarrow{\Delta}$ **447**

(180)

The possibility that **443** undergoes sulfur extrusion from the seven-membered ring to form benzo[*c*]thiophene (**448**) which then adds *N*-phenylmaleimide to give **447** could be excluded because reaction of **448** with *N*-phenylmaleimide gave equal amounts of *exo*- and *endo*-adducts. Treatments of **443** with dimethyl acetylenedicarboxylate at elevated temperatures led to dimethyl naphthalene-2,3-dicarboxylate (Eq. 181) (257). A study of the Diels-Alder reaction of sulfoxide **444** and sulfone **445** with *N*-phenylmaleimide (Eq. 182) revealed that these reactions proceed at a lower temperature and more rapidly than the comparable reaction with sulfide **443** (257). In addition approximately equal amounts of *exo*- and *endo*-adducts **447** and **448** were formed. At higher reaction temperature, generation of the *exo*-adduct was favored. Again the authors suggest direct addition of *N*-phenylmaleimide to **444** and **445** with subsequent SO or SO_2 extrusion.

448 + N-phenylmaleimide $\xrightarrow{\Delta}$ **447** + **448**

443 + $MeO_2CC{\equiv}CCO_2Me$ ⟶ (naphthalene-2,3-$(CO_2Me)_2$) (181)

444 or **445** + N-phenylmaleimide ⟶ **447** + **448** (182)

The synthesis of 4,5-dihydrothieno[3,4-*d*]thiepin (**441**) entails an interesting rearrangement-transannular reaction (Eq. 183) (79) of 1,6-dithiacyclodeca-3,8-diyne (**449**) under the influence of potassium *tert*-butoxide. Dihydrothiepin **441** is a stable crystalline solid which has been characterized by its uv,

449 $\xrightarrow[tert\text{-BuOH}]{tert\text{-BuOK}}$ [intermediate; ⊖OBu-*tert*, H, HO-*tert*-Bu] ⟶ **441** (183)

ir, nmr, and mass spectra. The structural assignment of **441** was confirmed by X-ray crystallographic analysis (79).

4,5-Dihydrothieno[3,4-*d*]thiepin-6-oxide (**442**) and the corresponding sulfone **450** have been described by Schlessinger and Ponticello (256, 257). Catalytic hydrogenation of **441** or **442** over palladium/charcoal catalyst

441 or **442** $\xrightarrow{H_2, Pd/C}$ **451** (184)

450 $\xrightarrow{H_2, Pd/C}$ **452**

provided 4,5,7,8-tetrahydrothieno[3,4-*d*]thiepin (**451**); catalytic hydrogenation of sulfone **450** produced **452** (Eq. 184).

5. Benzo[*e*]thieno[2,3-*b*]thiepins and Derivatives

Interest in this system revolves about the potential physiological properties of benzo[*e*]thiepino[2,3-*b*]thiepin-$\Delta^{4(9H)}$-γ-propyldimethylamine (**453**) and related componds (273), which are thiophene isosteres of dibenzo[*b,e*]thiepin-$\Delta^{11(6H)}$-γ-propyldimethylamine (**245**). The synthetic pathway leading to **453** parallels the synthesis of **245** (Eq. 185) (224,273). Difficulty was encountered in the ring closure step which proceeded with polyphosphoric acid in yields of less than 25%.

CO_2H $\xrightarrow[<25\%]{PPA}$ O $\xrightarrow[70\%]{}$ (185)

HO $(CH_2)_3NMe_2$ $\xrightarrow[78\%]{H_2SO_4}$ $CH(CH_2)_2NMe_2$

6. Benzo[*f*]thieno[2,3-*b*]thiepins and Derivatives

The few synthetic reports in this area have been stimulated by the potential physiological effect of replacing a benzo ring by a thiophene ring in 10-(4-methylpiperazino)-10,11-dihydrobenzo[*b,f*]thiepin ("perathiepin") (**287**) and related derivatives. 4-(Methylpiperazino)-4,5-dihydrobenzo[*b*]thieno[3,2-*f*]

CH_2CO_2H — $\xrightarrow[PhCH_3\ 77\%]{P_2O_5}$ — $\xrightarrow[87\%]{NaBH_4}$ H OH **457**

457 $\xrightarrow{HCl}$ H Cl **456**; R = H, Cl

457 $\xrightarrow{(1)\ NaNH_2;\ (2)\ Me_2NCH_2CH_2Cl}$ H $OCH_2CH_2NMe_2$ **458**; R = H (186)

456 $\xrightarrow{\text{N-methylpiperazine}}$ H NMe **454**; R = H (40%) **455**; R = Cl (60%) + **459**; R = H **460**; R = Cl thiepin (**454**) ("peradithiepin") showed a high central depressant activity comparable to **287** (236); the 2-chloro derivative **455**, on the other hand, had a weaker central effect than "octoclothepin," its benzo isostere.

The reaction sequence employed to synthesize **454** or **455** was similar to that used to produce the benzo analogs and is outlined in Eq. 186 (235, 236). The reaction of the 4-chloro compounds **456** not only formed the desired piperazino derivatives **454** or **455**, but also gave benzo[*f*]thieno[2,3-*b*]-thiepin **459** or its 2-chloro derivative **460** by hydrogen chloride elimination. Other amines similar to **454** have been prepared (279) and alcohol **457** has been converted into its dimethylaminoethyl ether **458** (236).

N. Furothiepins

Schlessinger and Ponticello (258) have published a preliminary report of the synthesis of furo[3,4-*d*]thiepin (**461**), its sulfoxide (**462**), and its sulfone (**463**) as shown by Eq. 187. The spectroscopic properties of **461**, **462**, and **463**

CH_2CH_2Cl CH_2CH_2Cl Na_2S 90% S SO Ac_2O S $m\text{-}ClC_6H_4CO_3H$ 80% SO **462** $m\text{-}ClC_6H_4CO_3H$ S Ac_2O 150°C 22% S $m\text{-}ClC_6H_4CO_3H$ **461** SO_2 **463** (18

were comparable to those of the corresponding thieno analogs. Compound **461** showed extended conjugation in the uv and visible spectrum; in the nmr spectrum the vinylic protons in the fully unsaturated thiepin ring absorbed at a higher field than in the corresponding dihydro derivative. Similar spectral effects were present in **462** and **463**.

A second synthetic entry into the furo[3,4-*d*]thiepin system involved the condensation of dimethyl thiodiglycolate with furan dialdehyde **464** (Eq. 188) (103). The resulting diacid **465** was converted into dimethyl 1,3-dimethyl-furo[3,4-*d*]thiepin-5,7-dicarboxylate (**466**), a stable golden yellow crystalline

464 + $S(CH_2CO_2Me)_2$ $\xrightarrow[\text{(2) } H^{\oplus}]{\text{(1) NaH}}$ 465 $\xrightarrow[\text{60\% overall}]{}$ 466 $\xrightarrow{\text{McMgBr}}$ 467 (188)

Me, CHO, CHO, Me, O — 464

CH_2CO_2Me, S, CH_2CO_2Me

Me, O, Me, CO_2H, S, CO_2H — 465

Me, O, Me, CO_2Me, S, CO_2Me — 466

Me, O, Me, OH, CMe_2, S, CMe_2, OH — 467

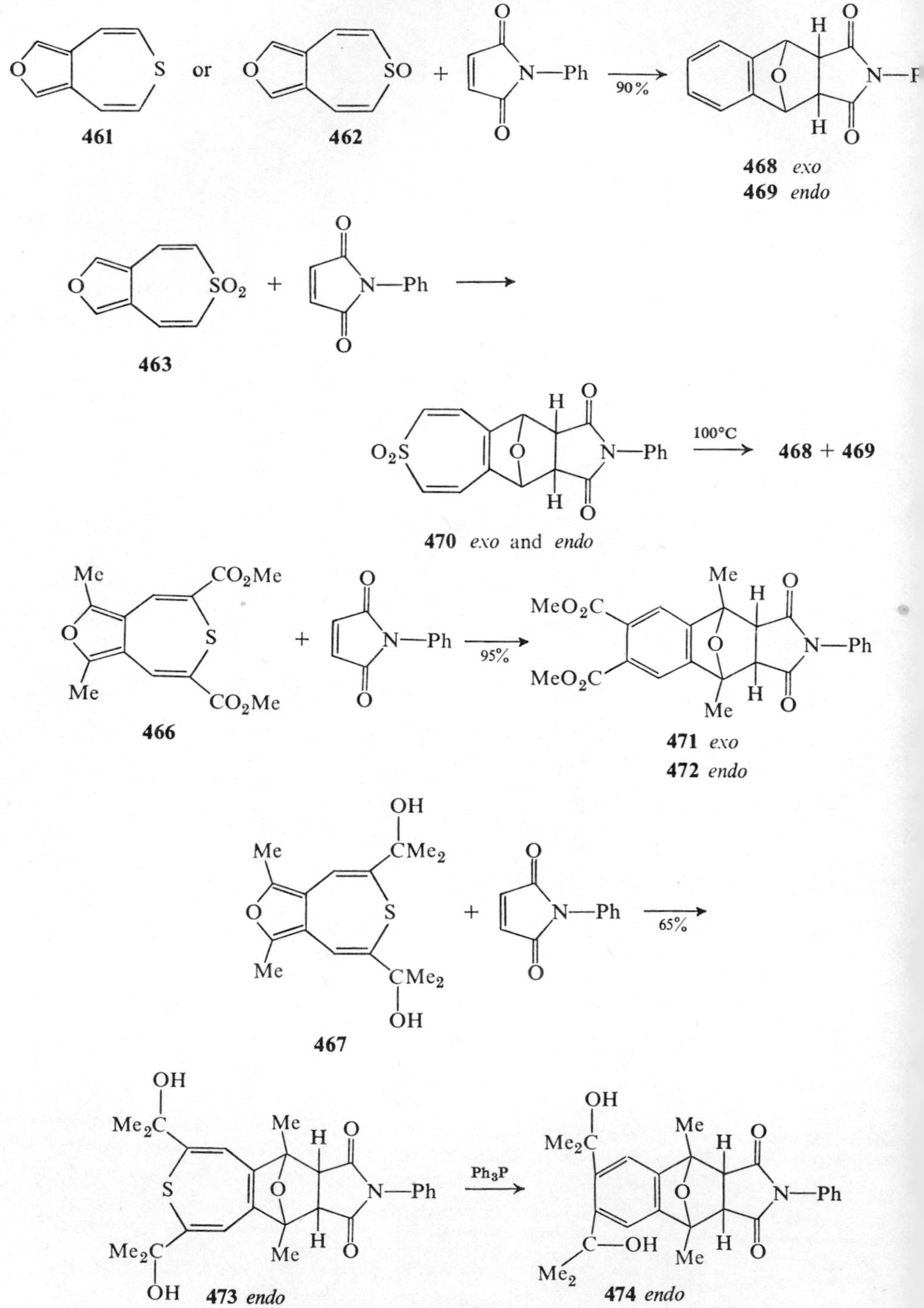

or + N—Ph $\xrightarrow{90\%}$

461 **462**

468 *exo*
469 *endo*

SO_2 + N—Ph →

463

O_2S N—Ph $\xrightarrow{100°C}$ **468** + **469**

470 *exo* and *endo*

Me CO_2Me + N—Ph $\xrightarrow{95\%}$ MeO_2C Me H N—Ph

Me CO_2Me MeO_2C H Me

466

471 *exo*
472 *endo*

OH
Me CMe_2 + N—Ph $\xrightarrow{65\%}$
Me CMe_2
OH

467

OH Me_2C Me H N—Ph $\xrightarrow{Ph_3P}$ OH Me_2C Me H N—Ph

Me_2C OH H Me C—OH Me_2 H Me

473 *endo* **474** *endo*

(189)

solid which could be converted into the highly reactive furothiepin diol **467** (103). Structural assignments for these products were supported by spectral properties.

Furothiepins are more reactive than the corresponding thienothiepins in the Diels-Alder reaction with *N*-phenylmaleimide and in contrast to the thienothiepins give both *exo*- and *endo*-adducts. Reaction of either **461** or **462** with *N*-phenylmaleimide at room temperature or −65°C resulted in the rapid formation of a mixture of **468** and **469** in a ratio of 3:2 (Eq. 189) (258). This value differs from the product ratio observed in the reaction of benzo-[*c*]furan and *N*-phenylmaleimide; however, no intermediates with the thiepin ring intact were detected. In contrast to the behavior of sulfide **461** and sulfoxide **462**, condensation of sulfone **463** with *N*-phenylmaleimide at 25°C gaves adduct **470**, which underwent SO_2 extrusion and yielded **468** and **469** on being heated to 100°C (258). Although the reaction of thiepin **466** and *N*-phenylmaleimide at 120°C produced a mixture of adducts **471** and **472** (Eq. 189) (103), the reaction of *N*-phenylmaleimide and diol **467** gave a bright yellow, highly crystalline adduct **473** in which the thiepin ring was intact. It was thermally stable, melting at 180°C without decomposition; however, treatment of **473** with triphenylphosphine led to removal of sulfur from the thiepin ring and formation of **474**. Thiepin **473** represents the most thermally stable annulated thiepin derivative reported to date. Structural assignments for the above-mentioned adducts were consistent with the observed spectral data. An X-ray crystallographic structure determination of **473** is in process.

O. Cycloalkadithiepins

1. Cyclobutadithiepins. (3,5-Dithiabicyclo[5.2.0]nonanes)

One report has appeared in the literature (159) describing the formation of cyclobutadithiepins **475** and **476** by an interesting ring closure reaction (Eq. 190).

$Me_2C(SO_2CH_2CH_2CH_2Cl)_2$ —KOH, 64%→ **475**

(190)

Cyclohexane-1,1-$(SO_2CH_2CH_2CH_2Cl)_2$ —KOH, 85%→ **476**

2. Cyclopentadithiepins. (2,6-Dithiabicyclo[5.3.0]decane)

A study of the photolysis of 1,5-dithiaspiro[5.4]decane **477** led to the isolation of *cis*- and *trans*-2,6-dithiabicyclo[5.3.0]decanes **478** (Eq. 191) which were also prepared independently by alkylation of *cis*- or *trans*-cyclopentane-1,2-dithiol (**479**) with 1,3-diiodopropane (322). The explanation suggested for the photochemical conversion of **477** to **478** entailed cleavage to diradical **480**, formation of olefin intermediate **481**, and intramolecular radical addition of the thiol group across the double bond.

$h\nu$; 25–51% ; $+ I(CH_2)_3I$

477 ; *cis* and *trans* **478** ; *cis* or *trans* **479** ; **480** ; **481** (191)

3. Cyclohexadithiepins. (2,6-Dithiabicyclo[5.4.0]undecane)

The photochemical study described in the preceding section (O-2) also included the photolysis of 1,5-dithiaspiro[5.5]undecane (**482**) which gave *cis* and *trans* 2,6-dithiabicyclo[5.4.0]undecanes (**483**) (Eq. 192) (322).

$h\nu$; $+ I(CH_2)_3I$ (192)

482 ; *cis* and *trans* **483** ; *cis* or *trans*

P. Benzodithiepins

There are six possible isomeric benzodithiepin systems as shown. Although none of the fully unsaturated compounds or derivatives have been reported, dihydro derivatives of the four systems **484**, **485**, **486**, and **487**, are described briefly in the literature.

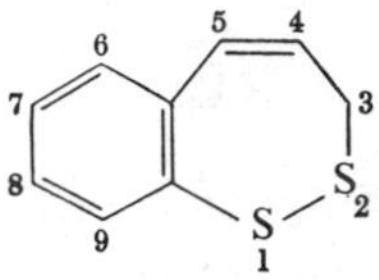

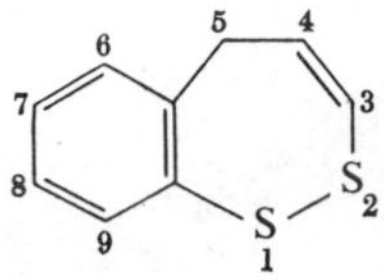

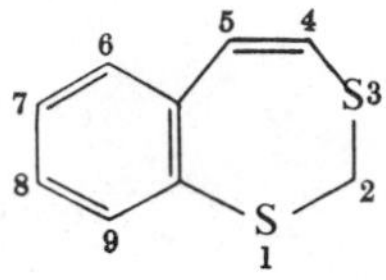

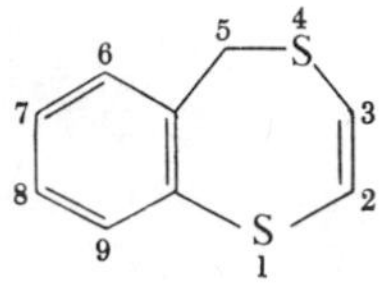

484
R.I. 8147

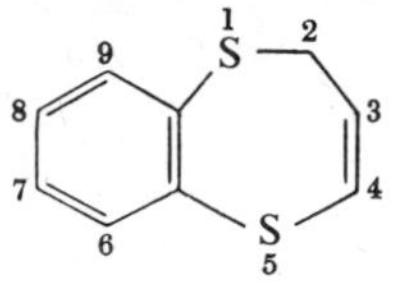

485

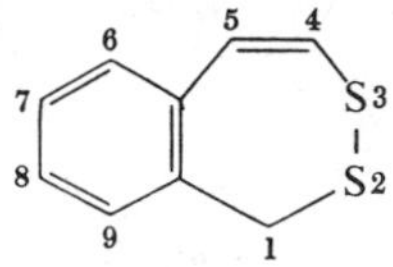

486

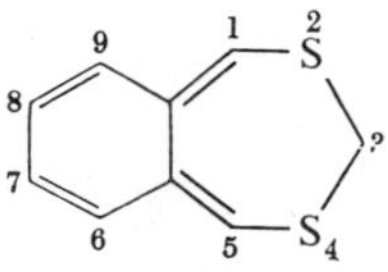

487
R.I. 1838

Kulka (135) reported the isolation of 7-chloro-2,3-dihydro-5*H*-1,4-benzodithiepin (**488**) from the action of base on 2-(2-chloroethylthio)-5-chlorobenzyl isothiouronium chloride (Eq. 193). In a study of 1,3-dichloro-2-methylenepropane, Richter, Schulze, and Muehlstädt (244) reported that reaction with 4,5-dimethyl-1,2-benzenedithiol (Eq. 194) produced 7,8-dimethyl-1,5-benzodithiepin-$\Delta^{3(2H)}$-methane (**489**).

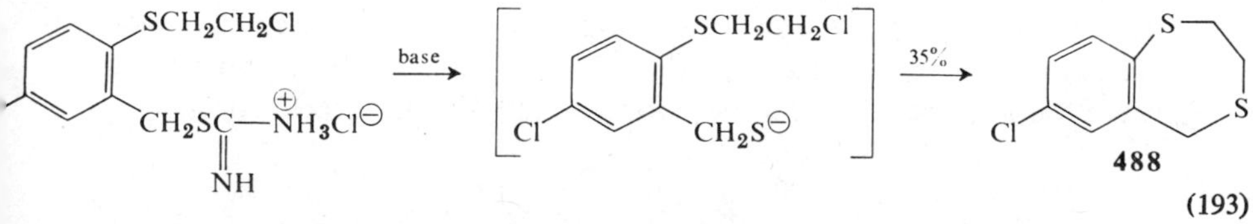

SCH_2CH_2Cl; $CH_2SC(=NH)—\overset{\oplus}{N}H_3Cl^{\ominus}$ —base→ [Cl; SCH_2CH_2Cl; $CH_2S^{\ominus}$] —35%→ **488** (193)

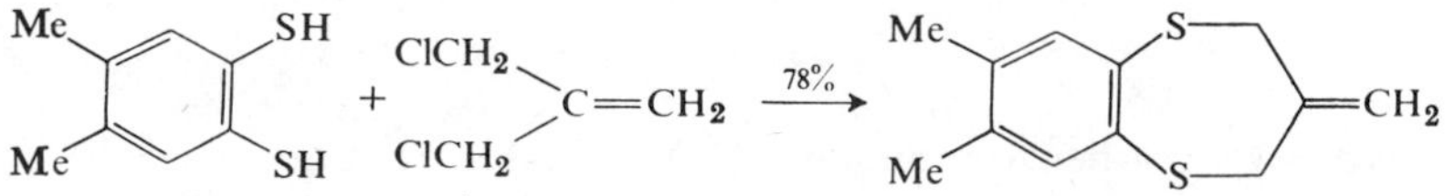

Me, Me, SH, SH + $ClCH_2$, $ClCH_2$ >C=CH_2 —78%→ Me, Me, S, S, =CH_2 **489** (194)

4,5-Dihydro-1*H*-2,3-benzodithiepin was cited in a preliminary report dealing with the determination of the conformational preference in seven-membered disulfides and trisulfides (125). A discussion of these data will be presented in the 2,3,4-benzotrithiepin section.

The system which has received the greatest attention is 1,5-dihydro-3*H*-2,4-benzodithiepin (**490**), isolated initially by Kotz (130) in 1900 from the reaction of **491** and methylene iodide. A general method of preparation of this system, developed by Autenrieth and Hennings (7, 8), involved the reaction of **491** with carbonyl compounds in the presence of dry HCl (Eq. 195).

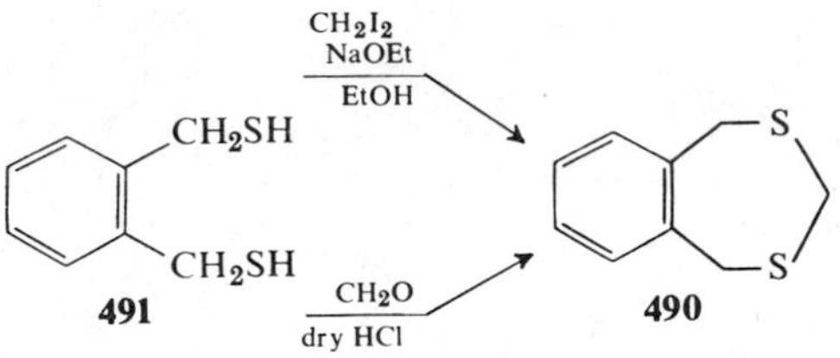

CH_2SH, CH_2SH **491** — CH_2I_2, NaOEt / EtOH → **490**; CH_2O / dry HCl → **490**

(195)

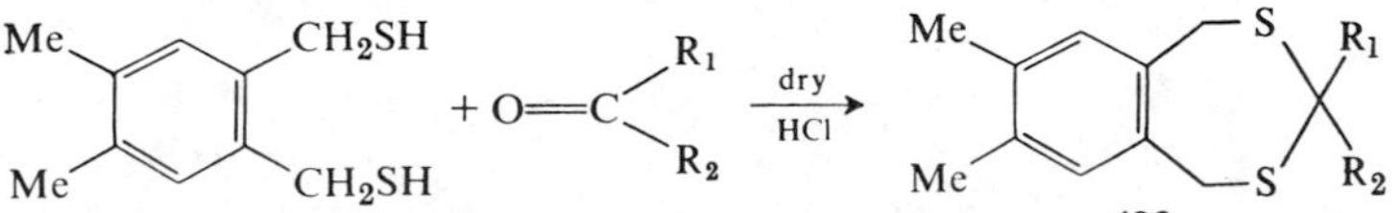

492 (Me, Me, CH_2SH, CH_2SH) + $O{=}C(R_1)(R_2)$ —dry HCl→ **493** (Me, Me, S, S, R_1, R_2)

This reaction has been extended to steroidal ketones (123), and the reaction of **492** with a variety of aldehydes and ketones has been reported (266). Specific examples are presented in Table XI-15. The condensation of **492** with aldehydes is so rapid that Shahak and Bergmann (266) suggested the use of this technique to distinguish and separate aldehydes and ketones.

A variable temperature nmr study provided evidence for a strong preference for the chair conformation in **490** and its 3,3-dimethyl derivative **494** (90). The free activation enthalpies for ring inversion in these two compounds were found to be 10.9 kcal/mole for **490** and 12.1 kcal/mole for **494**.

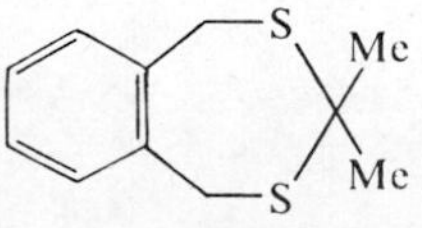

494 (S, S, Me, Me)

1,5-Dihydro-3*H*-2,4-benzodithiepin (**490**) is stable to aqueous acid or alkali; however, treatment with a mercuric salt (266) in a hydroxylic solvent regenerates the carbonyl compound from which the dithiepin was prepared.

$$\mathbf{492} + \text{HOCH}{=}\text{C(R)}{-}\text{C(=O)}{-}\text{R}' \longrightarrow \text{[dithiepane adduct: 2-(CH(R)–C(=O)R')]}$$

$$\xrightarrow{\text{LiAlH}_4} \text{[dithiepane: 2-(CH(R)–CHR'(OH))]} \xrightarrow[\text{EtOH}]{\text{(1) HCl, (2) HgCl}_2} \text{HC(=O)}{-}\text{C(R)}{=}\text{CHR}'$$

$$\xrightarrow{\text{R}''\text{MgX}} \text{[dithiepane: 2-(CH(R)–C(R')(R'')OH)]} \xrightarrow[\text{EtOH}]{\text{(1) HCl, (2) HgCl}_2} \text{HC(=O)}{-}\text{C(R)}{=}\text{CR}'\text{R}'' \quad (196)$$

The ring system appears to be stable to Grignard reagents, organolithium compounds, sodium borohydride, and lithium aluminum hydride. These observations permitted Shahak and Bergmann (266) to develop synthetic applications utilizing these reactions for the preparation of substituted α,β-unsaturated aldehydes (Eq. 196). Reaction of the benzodithiepin derivatives **493** with Raney nickel led to desulfurization with the net effect of reducing the carbonyl group from which the benzodithiepin was prepared to a methylene group (266) (see also Ref. 123 for application to steroidal ketones). When "degassed" Raney nickel was used with **493** (R_1 = H, R_2 = aromatic), the reaction products were stilbenes (266).

Q. Dibenzodithiepins and Dinaphthodithiepins

Only the dibenzo[*b*,*d*][1,3]dithiepin system has been reported in the literature. Two general methods of preparation which generate the 1,3-dithiepin ring involve condensation of 2,2′-biphenyldithiol (**495**) with carbonyl compounds (10) and 1,1-dihalo compounds (43). These reactions are illustrated in Eq. 197 with the synthesis of 6-phenyldibenzo[*b*,*d*][1,3]dithiepin (**496**) (10) and ethyl dibenzo[*b*,*d*][1,3]dithiepin-6-carboxylate (**497**) (43).

SH
SH
495
PhCHO
dry HCl
S H
S Ph
496
R.I. 3685
(197)
NaH
Cl_2CHCO_2Et
S H
S CO_2Et
497

Interest in compound **497** has been directed toward comparison of its properties with those of open-chain analog **498**. Thus deuterium exchange occurred somewhat more rapidly with **498** than with **497**, and acidity measurements showed **498** to be a stronger acid than **497**. These observations led

Breslow and Mohacsi (43) to conclude that no aromaticity can be associated with the cyclic ten π-electron system of **497**.

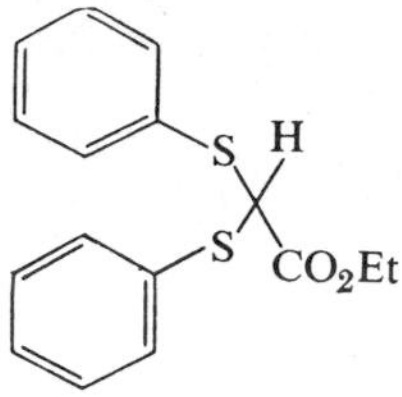

498

Armarego and Turner (5) described the synthesis of optically active 8,8-dimethyldinaphtho[2,1-*d*:1′2′-*f*][1,3]dithiepin (**499**) via the reaction sequence shown in Eq. 198.

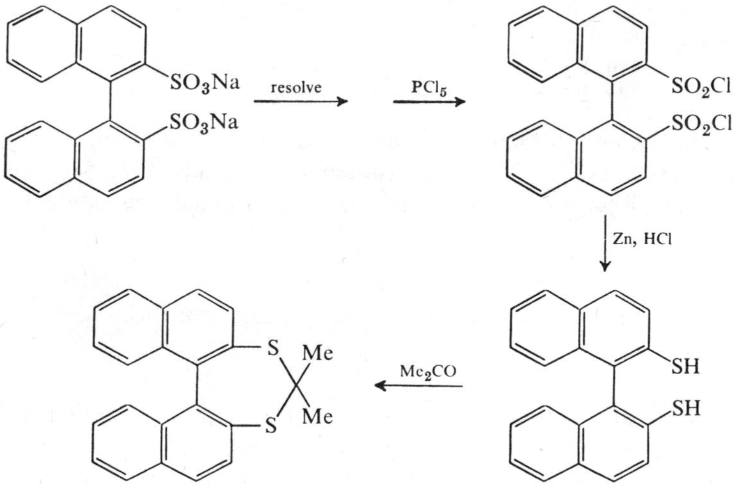

(198)

499
R.I. 9392

R. Benzotrithiepins

1,5-Dihydro-2,3,4-benzotrithiepin (**500**) was first prepared by Milligan and Swan (165) by reaction of the *bis*-Bunte salt **501** with sodium sulfide (Eq. 199). This procedure was repeated by Lüttringhaus and co-workers (124, 125) and extended to the preparation of 6,9-dimethyl-1,5-dihydro-2,3,4-benzotrithiepin (**502**). Oxidation of **500** with monoperphthalic acid produced a monosulfoxide which was assigned structure **503** (165).

Of major interest in this system has been a conformational study conducted with the aid of variable temperature nmr spectrometry. The mechanism for ring inversion of a chair conformation entails two equilibria,

R

CH_2SSO_3Na

$+ Na_2S \xrightarrow{48\%}$

CH_2SSO_3Na

R

501; R=H, Me

R

S

S

S

R

500; R=H

502; R=Me

CO_3H

CO_2H

R=H

S

S

S

O

503

(199)

conversion of chair to boat and boat to twist boat (Eq. 200). Lüttringhaus and co-workers (124, 125) have studied compounds **502**, **500**, and 4,5-dihydro-1*H*-2,3-benzodithiepin, and have determined the conformational

Twist Boat

Chair ⇄ Boat   Boat ⇄ Chair   (200)

Twist Boat

preference at various temperatures and the free activation enthalpy for the two conversions. Compound **502** showed the appearance of a mixture of two components at 114°C; at room temperature, the mixture was analyzed to contain 55% of the twist boat and 45% of the chair conformation. At −45°C the equilibria between the twist boat conformations are retarded and a further resolution of proton coupling in the twist boat is observed. The free activation enthalpy for conversion of chair to twist boat is 19.8 kcal/mole; the value for the interconversion of twist boat forms is 11.5 kcal/mole (125). In the absence of 6,9-dimethyl groups, as in **500**, two components were observed at +84°C, with a mixture composition at room temperature of 85% chair and 15% twist boat. The free activation enthalpy for this process was 17.4 kcal/mole (124, 125). Conditions permitting the observation of further splitting in the twist boat were not achieved; however, line broadening was observed and the temperature at which splitting would occur was estimated to be −80°C, with a free activation enthalpy of approximately 10 kcal/mole.

Line splitting occurs at 0°C, in 4,5-dihydro-1*H*-2,3-benzodithiepin, with an equilibrium composition at −35°C of 70% chair and 30% twist boat and a free activation enthalpy of 13.5 kcal/mole. Lowering the temperature to −60°C inhibits the interconversion of the twist boat forms, the free activation enthalpy being reduced to 10.4 kcal/mole.

From the foregoing data and similar studies on 1,5-dihydro-3*H*-2,4-benzodithiepin (**490**), Lüttringhaus and co-workers conclude that bond rotation in a disulfide bond is more hindered than in a sulfide bond. Thus rotation in a trisulfide (two disulfide bonds) is even further hindered in accord with the higher activation enthalpies observed in **500** and **502**.

S. Dinaphthotrithiepins

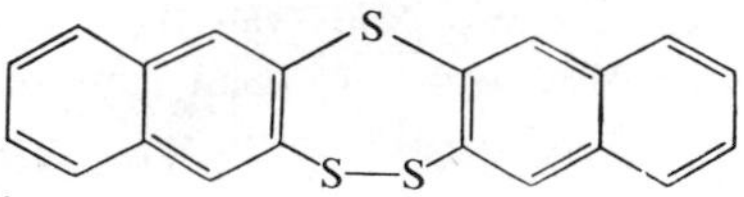

S
S—S

R.I. 6428

One report has appeared in the literature proposing dinaphthotrithiepin **504** as the structure of a secondary product from the reaction of dibromonaphthazarin (**505**) and sodium sulfide (Eq. 201) (37). No physical constants were given for **504** and no evidence supporting the structural assignment was offered.

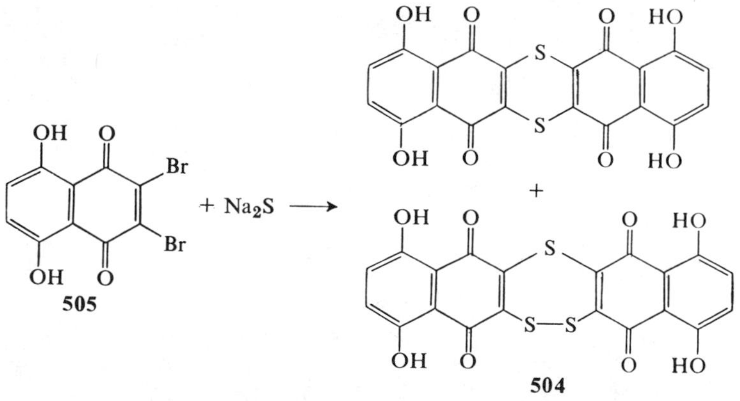

OH O Br Br OH O **505** + Na_2S ⟶ OH O S S O HO OH O O HO + OH O S S—S O HO OH O O HO **504** (201)

T. Benzopentathiepins

Hexahydrobenzopentathiepin (**506**) was described in a preliminary communication (82) as a product from the reaction of *trans*-1,2-cyclohexanedithiol and dichlorotrisulfane (Eq. 202). Attempts to extend this reaction

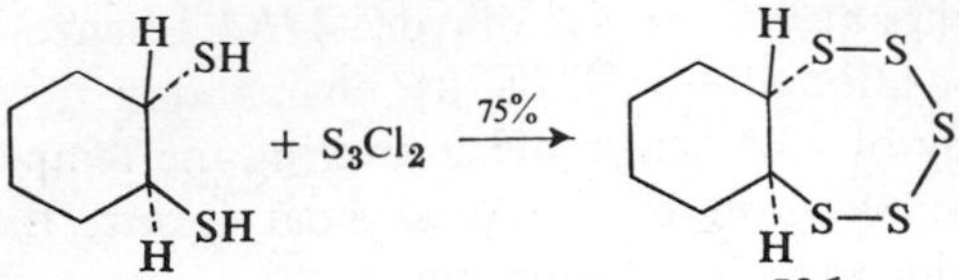

(202)

506

to 1-substituted 1,2-cyclohexanedithiols or *o*-benzenedithiol were unsuccessful. Evidence for the structural assignment **506** involves ir, nmr, and mass spectral data.

U. Physiological Activity

Three ring systems which have yielded compounds of potential pharmacological interest are the 1-benzothiepins, dibenzo[*b*,*e*]thiepins, and dibenzo-[*b*,*f*]thiepins. Derivatives of these ring systems are described in numerous patents. Major interest in this area has been centered on compounds with potential neurotropic and psychotropic activity.

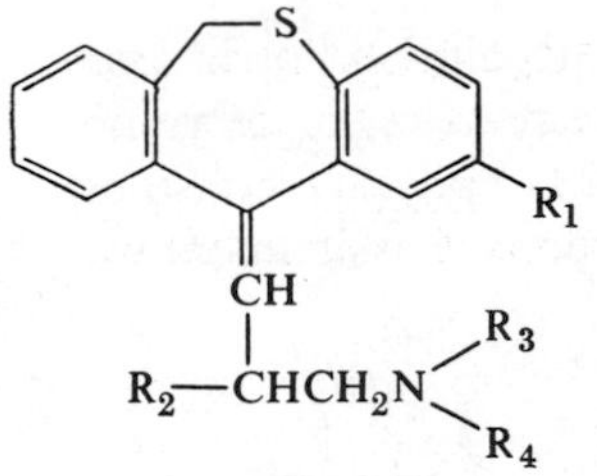

507 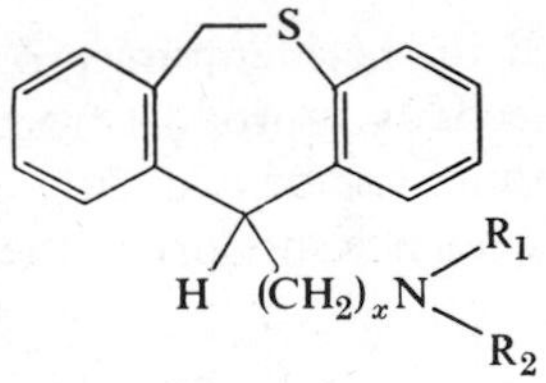 **508** 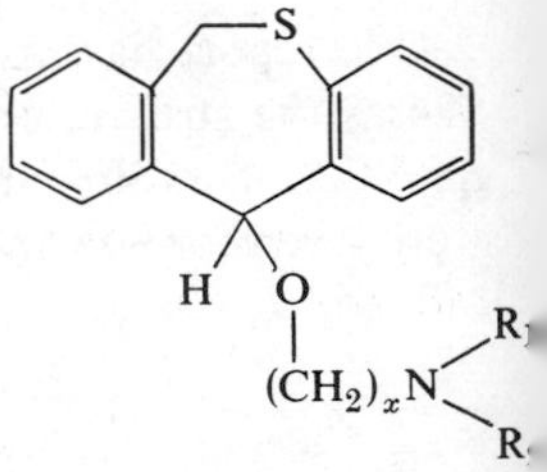 **509**

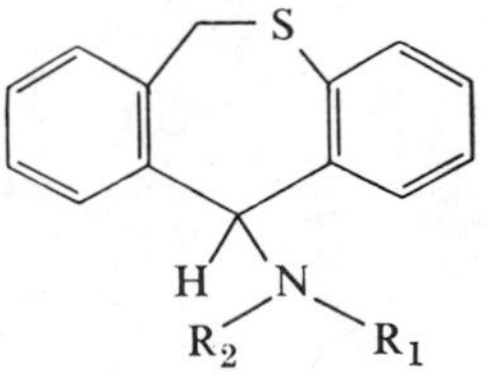

510

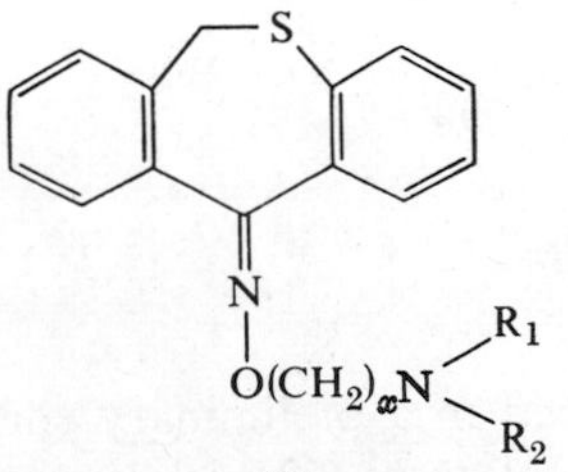

511 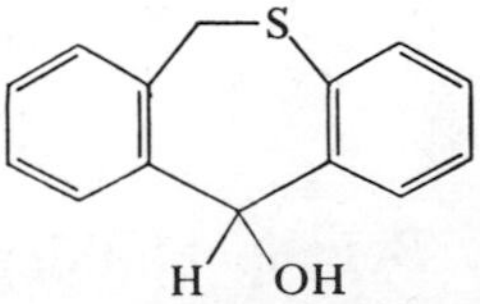 **268** 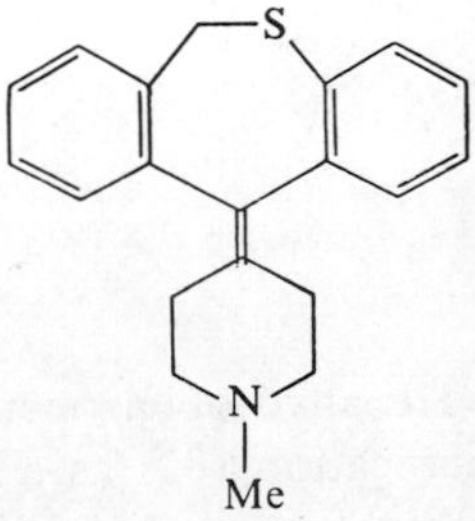 **512**

Several systems in the dibenzo[*b*,*e*]thiepin series have exhibited significant physiological properties and have been studied extensively. Dibenzo[*b*,*e*]-thiepin-$\Delta^{11,6H}$-γ-propyldimethylamine hydrochloride (prothiaden) (**507**, $R_1 = R_2 = H$, $R_3 = R_2 = Me$) (91, 201, 226, 281, 327) shows central depressant activity (11–13, 163, 164, 168) with a tranquilizing, antidepressive effect, and has been suggested for possible treatment of schizophrenia and psychosomatic disturbance (251). In addition, prothiaden has antihistaminic, anti-serotonin, and other types of activity (109, 119, 161, 164, 193, 199, 228, 309). A large number of compounds with the basic structure of **507** have been prepared and evaluated (24, 119, 162, 205, 227). Northiaden (**507**, $R_1 = R_2 = R_3 = H$, $R_4 = CH_3$) shows significant antireserpine activity (224), perithiadene (**512**) is a more potent antihistamine and possesses a more powerful central depressant action than prothiaden (164); oxidation of prothiaden to the sulfoxide reduces the sedative effect but leaves moderately strong antihistaminic activity (224).

Reduction of the exocyclic double bond of prothiaden produces hydrothiaden (**508**, $x = 3$) which still retains the antireserpine effect but has weaker antihistamine and antiserotonin activity than prothiaden (224). Compounds of structure **508**

$$(x = 2,\ R_1 = H,\ R_2 = CH_2\overset{\displaystyle OH}{\overset{|}{C}}HCH_2OAr)$$

possess cardiac and circulatory action (28).

Amethobenzepin (**509**, $x = 3$, $R_1 = R_2 = Me$) has a weak central depressive effect but is the most potent antihistamine of the dibenzo[*b*,*e*]thiepin type thus far tested (162, 224). The parent alcohol **268** from which these ethers were prepared showed activity as an anticonvulsant, especially against convulsions caused by electroshock (62).

9-Aminodibenzo[*b*,*e*]thiepins of general structure **510** display a weak central depressant activity (102) but have neurotropic activities (antihistamine, local anesthetic, analgesic, and antispasmodic) (230). Attachment of

—N(piperazine)N—N=CH—(4-pyridyl)

as the amine moiety in **510** provides a compound useful as an anticonvulsant which antagonizes electroshock seizures (329). The oxime ethers **511** may be useful as antidepressants (3).

In the dibenzo[*b*,*f*]thiepin system the 10-amino-9,10-dihydro derivatives have been studied most extensively and compared to the carbocyclic analog (263). Simple amino compounds such as methylamino, dimethylamino, and diethylaminoethylamino have weak central depressant activity but significant antihistamine activity (115); however, the most potent member of the series is

10-(4-methylpiperazino)-10,11-dihydrodibenzo[*b*,*f*]thiepin (**287**) (perathiepin), which shows central depressant activity two to three times greater than

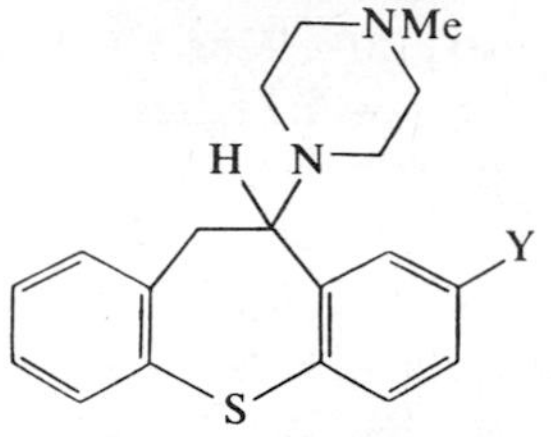

287; Y = H, perathiepin
513; Y = Cl, octoclothiepin
514; Y = CH_3, meperathiepin
515; Y = OCH_3, octometothepin
516; Y = SCH_3, methiothepin
517; Y = CN, cyanothepin
518; Y = CF_3, trifluthepin

chlorpromazine (216). In addition, perathiepin has been found to exhibit antihistamine, antiserotonin, and analgesic activity (215, 216). It has been tested for antileukemic effects and the ability to enhance the effect of fluoxydine (233), and also for bacteriostatic activity (267), antimicrobial potency (267), and tuberculostatic activity (267). Variation of substituents on the 4-nitrogen of piperazine produced no improved activity, although some compounds occasionally showed anticonvulsant properties (216). The 8-chloro derivative octoclothepin (**513**) (216) had central depressant activity three times greater than perathiepin, whereas chloro derivatives in other locations (194, 216) were less effective than perathiepin. Among other examples of 8-substituted perathiepins which have been prepared, meperathiepin (**514**), octometothepin (**515**), and methiothepin (**516**) were stronger central depressants than perathiepin (196). Cyanothepin (**517**) (112) was even more active than octochlothepin but also more toxic; trifluthepin (**518**) (197) showed weaker central depressant activity but surpassed all compounds in this series in catalepsy tests. Oxidation of perathiepin to its sulfoxide decreased activity and increased toxicity (116).

10-(2-Dimethylaminoethoxy)-10,11-dihydrodibenzo[*b*,*f*]thiepin (**286**) as well as its 2-chloro derivative has intensive and long-lasting central depressant and antihistaminic activity (111, 194, 216). Nerve blocking and anticonvulsive activity has been claimed for 11-(4-methylpiperazino)dibenzo[*b*,*f*]thiepin-10(11*H*)-one (**358**) and related compounds (99). When the amino group is separated from the ring ketone by two or three carbons as in **519** (x = 2, 3), these compounds are antagonists to reserpine and tetrabenazine (96, 98, 290).

A variety of aminomethyldibenzo[*b*,*f*]thiepins of general structure **520** and their acid addition salts have adrenolytic, sedative, and narcotic activity

(97). Separation of the amino group and the dibenzo[*b*,*f*]thiepin ring by two or three carbons as in **521** ($x = 2, 3$) leads to compounds which show activity

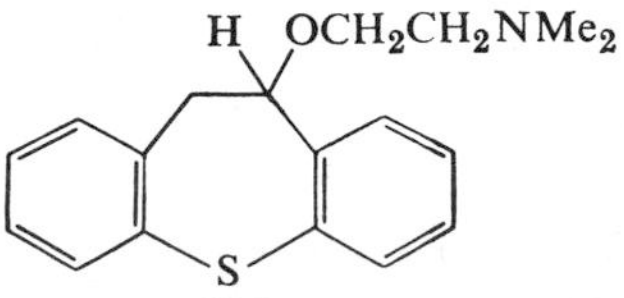

286 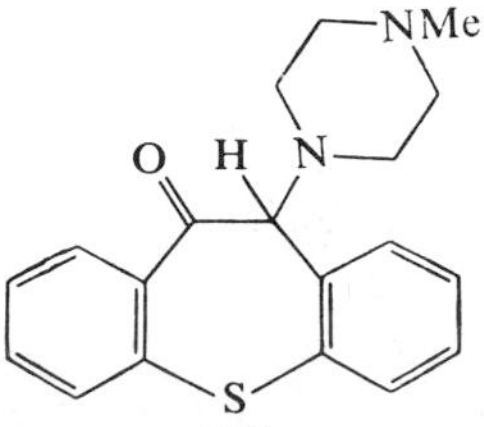 **358** 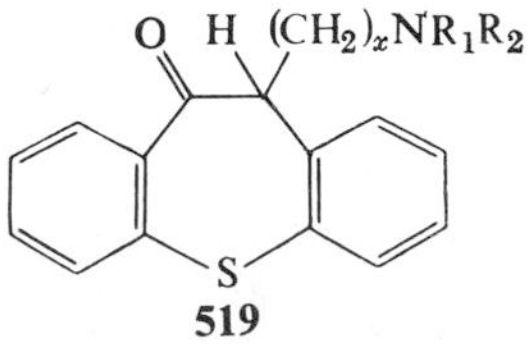 **519**

as central nervous system stimulants, antihistamines, and antispasmodics (331). With the double bond exocyclic, compound **331** exhibits central depressant activity (114).

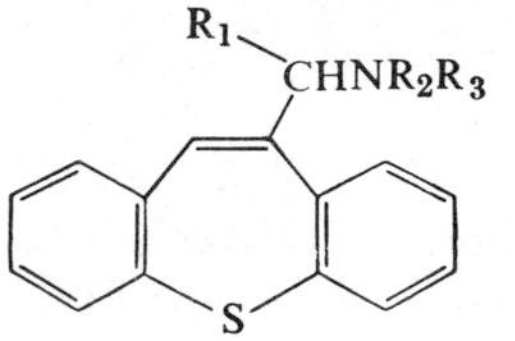

520 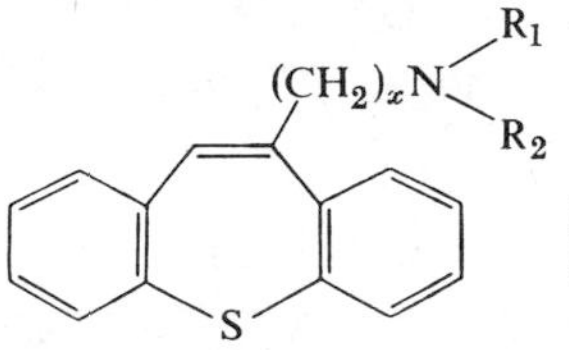 **521** 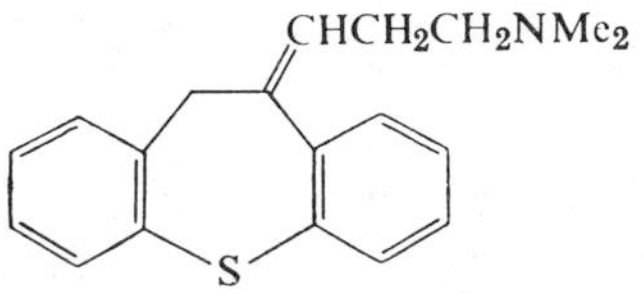 **331**

The derivatives of the 1-benzothiepin system which have potentially interesting physiological properties parallel in many instances structural types in the dibenzo[*b*,*f*]thiepin series. The following compounds have been reported as potential hypotensive agents: **522** ($x = 1, 2$) (106), **523** ($x = 2, 3$) (174), **524** (269), **525** (R_1 and R_2 = H, simple alkyl) (174), **526** (Ar = phenyl or

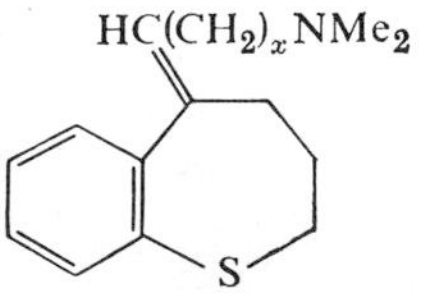

522 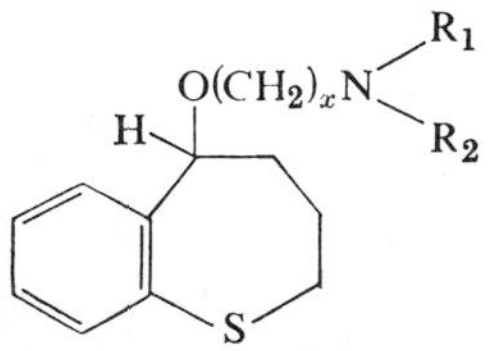 **523** 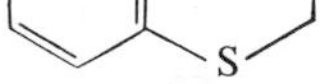 **524** 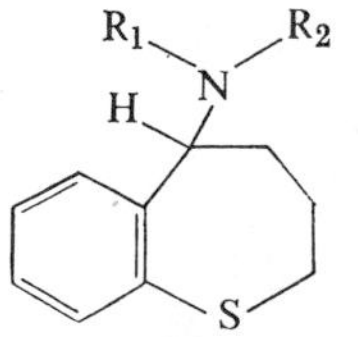 **525**

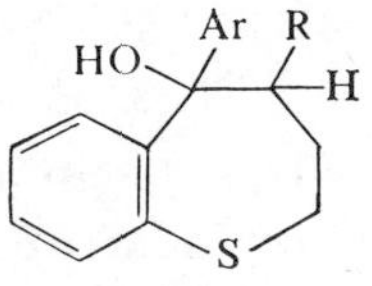

526 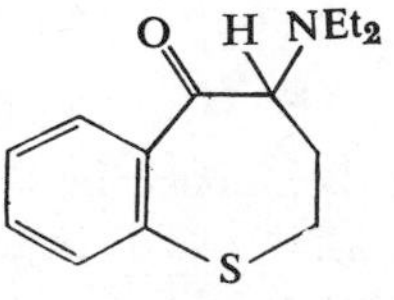 **527**

pyridyl and R = CH_2NMe_2) (175), and **527** (269). In addition, compounds **528** (173), **529** (171), and **530** (171) are central nervous system depressants. The hydantoin **109** has anticonvulsant activity (6).

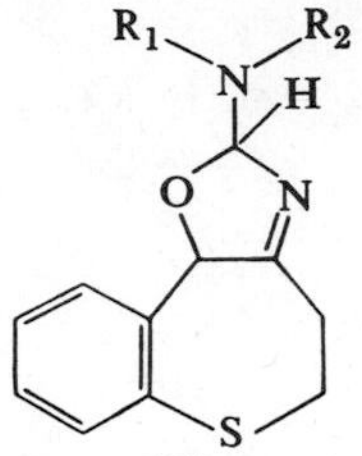

528 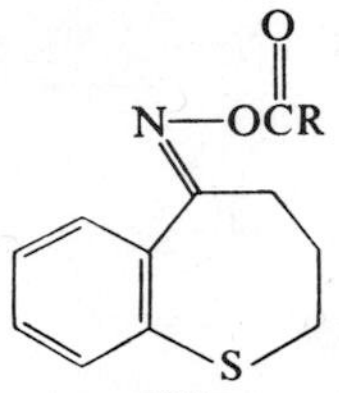 **529** 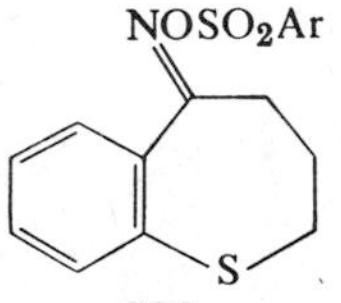 **530** 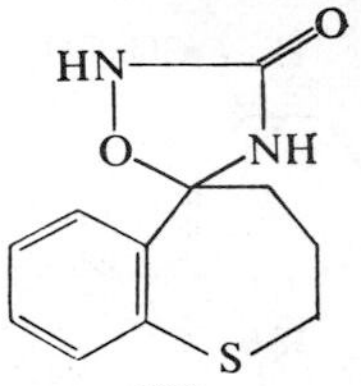 **109**

V. Bridged Systems

1. *x*-Thiabicyclo[4.1.0]heptane Systems

Three isomeric structures are possible in this system and derivatives of each type have been reported in the literature. 7-Thiabicyclo[4.1.0]heptane (**531**), commonly named cyclohexene sulfide, has been reviewed in the chapter on thiiranes in Volume 19 of this series (242). Parham and co-workers (189) described the cycloaddition of dichlorocarbene to Δ^2-dihydrothiapyran (**532**) which produced 7,7-dichloro-2-thiabicyclo[4.1.0]heptane (**533**) (Eq. 203);

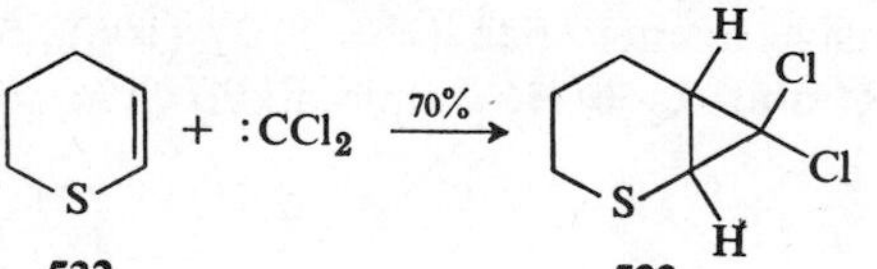

$$\mathbf{532} + :CCl_2 \xrightarrow{70\%} \mathbf{533} \qquad (203)$$

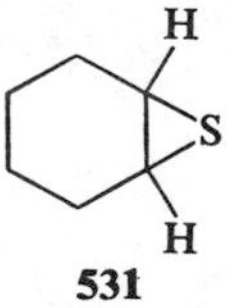

531

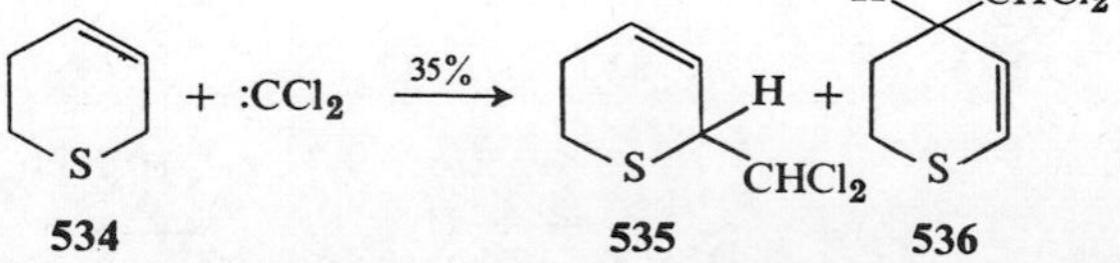

$$\mathbf{534} + :CCl_2 \xrightarrow{35\%} \mathbf{535} + \mathbf{536} \qquad (204)$$

however, reaction of dichlorocarbene with Δ^3-dihydrothiapyran (**534**) led to approximately equal amounts of insertion products **535** and **536** (Eq. 204). Adduct **533** was stable to distillation under vacuum and upon treatment with Raney nickel formed *n*-hexane and 2-methylpentane in a ratio of 3:1. Dimroth and co-workers (69) applied the cycloaddition reaction of dichlorocarbene under neutral conditions to 4*H*-thiapyran (**537**) to obtain a mixture of 7,7-dichloro-2-thiabicyclo[4.1.0]hept-3-ene (**538**) (contaminated with insertion compound **539**) and the diadduct **540** (Eq. 205). Purification of **538** entailed destruction of the more reactive **539** by heating with alcoholic silver nitrate for a short time. Adducts **538** and **540** were readily converted to their sulfones **541** and **542**, respectively, by oxidation with 30% hydrogen peroxide.

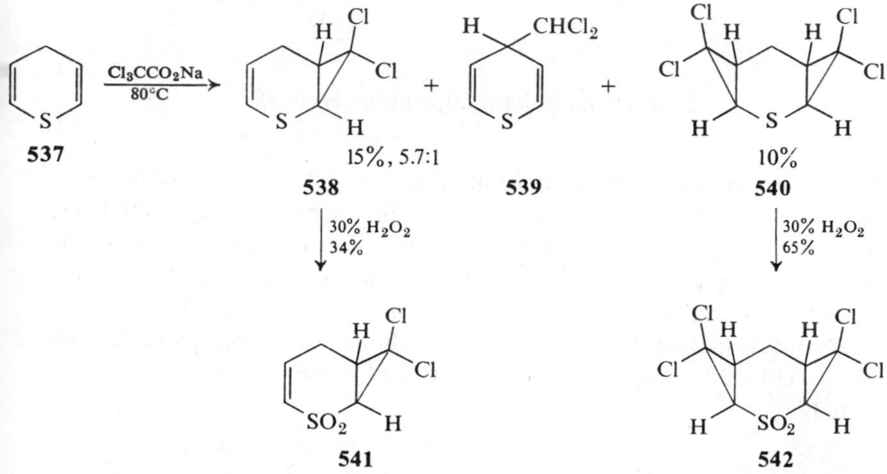

Cl_3CCO_2Na, 80°C: **537** → **538** (15%, 5.7:1) + **539** + **540** (10%)

538 → ($30\%\ H_2O_2$, 34%) → **541**; **540** → ($30\%\ H_2O_2$, 65%) → **542** (205)

The structures assigned to these compounds were confirmed by ir and nmr spectral data. Attempts to convert **538** or **541** into thiepin derivatives by the elimination and ring expansion procedure of Parham (260) were unsuccessful.

Benzo fused derivatives of the thiabicyclo[4.1.0]heptane system shown and others have been discussed in earlier sections of this chapter (see Section C-1 and C-5).

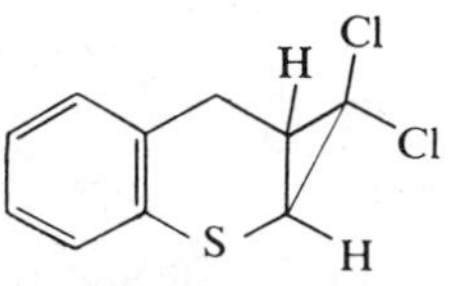

22 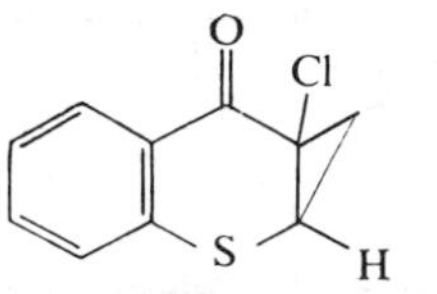 **126** 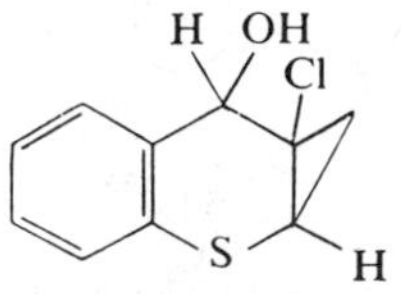 **77**

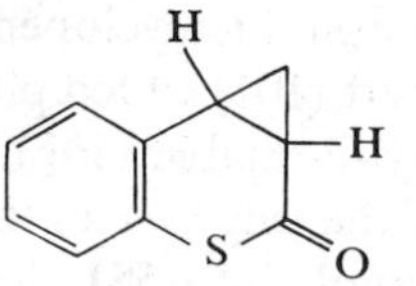

15 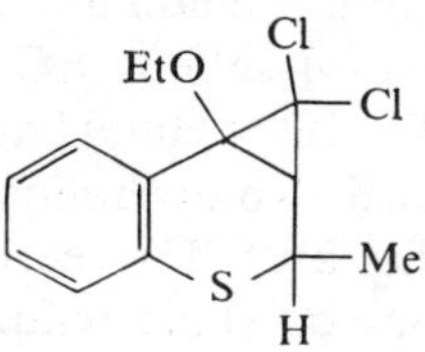 **29**

One example appears in the patent literature (14) describing the formation of 3,4-dithiabicyclo[4.1.0]perfluoroheptane (**543**) by 1,5-addition of sulfur to 1,5-perfluoropentadiene (Eq. 206).

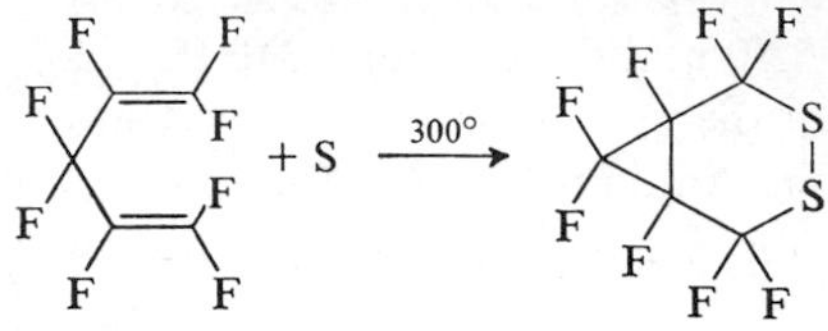

$$\text{C}_5\text{F}_8 + \text{S} \xrightarrow{300°} \mathbf{543} \qquad (206)$$

2. *x*-Thiabicyclo[3.2.0]heptane Systems

The photosensitized cycloaddition reaction of dimethylmaleic anhydride and thiophene in the presence of benzophenone (Eq. 207) led to 6,7-dimethyl-2-thiabicyclo[3.2.0]hept-3-ene-6,7-dicarboxylic acid anhydride (**544**) in 80% yield (254). A comparable photocatalyzed cycloaddition reaction of dichloromaleimide and **545** (Eq. 208) gave two isomeric adducts of 6,7-dichloro-3-thiabicyclo[3.2.0]heptane-3,3-dioxide-6,7-dicarboximide (**546**) in 54% yield (253). The chemical properties of these systems have not been reported.

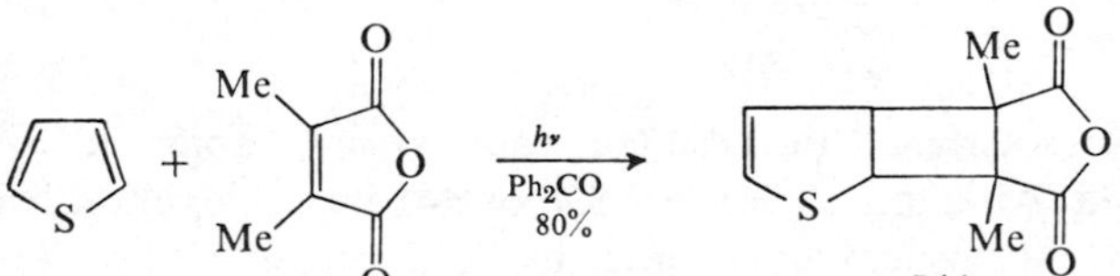

$$\xrightarrow[\text{Ph}_2\text{CO},\ 80\%]{h\nu} \mathbf{544} \qquad (207)$$

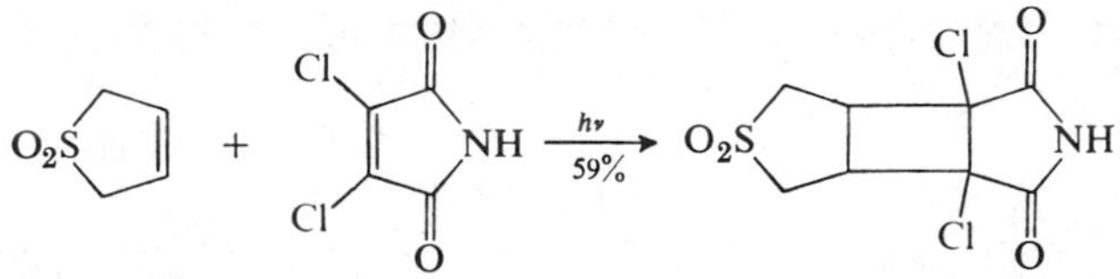

$$\mathbf{545} + \xrightarrow[59\%]{h\nu} \mathbf{546} \qquad (208)$$

Entry into the 6-thiabicyclo[3.2.0]heptane system is accomplished by the addition of sulfenes to appropriate enamines. Opitz and Rieth (183) reported that cycloaddition of cyclohexyl sulfene **547** to 1-morpholinocyclopentene **548** gave **549** in 53% yield (Eq. 209); Hamid and Trippett (101) added phenyl sulfene to **548** and obtained 1-morpholino-7-phenyl-6-thiabicyclo[3.2.0]-heptane-6,6-dioxide (**550**) (Eq. 210). The sensitivity of the ring system in **550** was shown by its conversion, on short reflux in ethanol, into **551**; longer reflux in ethanol led to 2-benzylsulfonylcyclopentanone (**552**) (101).

3. *x*-Thiabicyclo[3.2.1]octane Systems

The three isomeric parent compounds in this system, 2-thiabicyclo[3.2.1]-octane (**553**), 3-thiabicyclo[3.2.1]octane (**554**), and 6-thiabicyclo[3.2.1]octane (**555**), have been isolated from Iranian kerosene (20). Structural assignments for these bicyclic sulfides were confirmed by Birch and co-workers (21, 22) by comparison with synthetic samples prepared by reaction of sodium sulfide with the appropriate dibromide or ditosylate (Eqs. 211–213). The bicyclic sulfides were waxlike solids with an odor resembling camphor.

Strom and co-workers (285) have studied the nmr spectra of iodine–alkyl sulfide complexes and used these data to determine the equilibrium constant for complex formation. Mention is made of the use of the iodine complex of 1,8,8-trimethyl-3-thiabicyclo[3.2.1]octane (**556**) in aiding the interpretation of its nmr spectral assignment. Dodson and co-workers (179) described the oxidation of **556** (synthesized from (+)-camphor) to a mixture of the *exo* and *endo*-1,8,8-trimethyl-3-thiabicyclo[3.2.1]octane-3-oxides **557** and **558**,

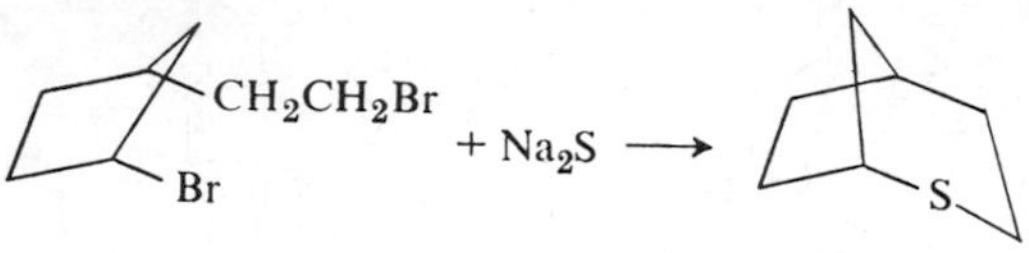

CH_2CH_2Br, Br + Na_2S ⟶ **553** (211)

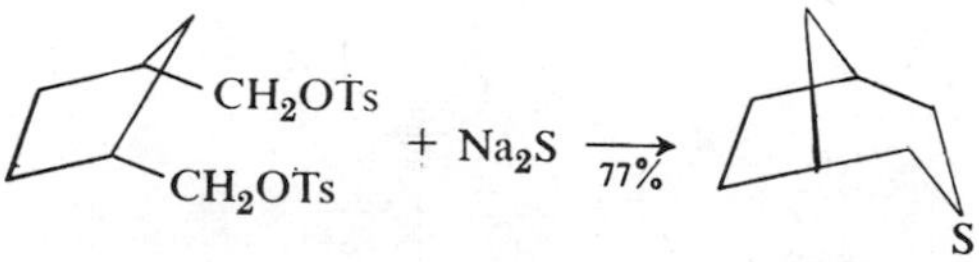

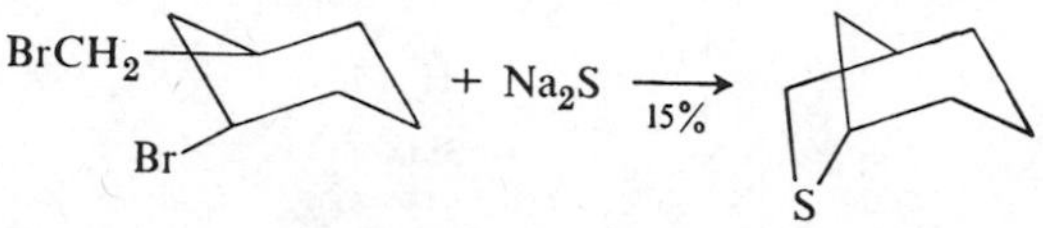

CH_2OTs, CH_2OTs + Na_2S $\xrightarrow{77\%}$ **554** (212)

$BrCH_2$, Br + Na_2S $\xrightarrow{15\%}$ **555** (213)

respectively (Eq. 214). Oxidation of **556** with *m*-chloroperbenzoic acid produced predominantly the *exo* isomer **557**, whereas oxidation with *tert*-butyl hypochlorite gave predominantly the *endo* isomer **558**. Determination of the ORD curves of **557** and **558**, as well as other considerations, led Dodson to assign the absolute configuration *R* for **557** and *S* for **558**.

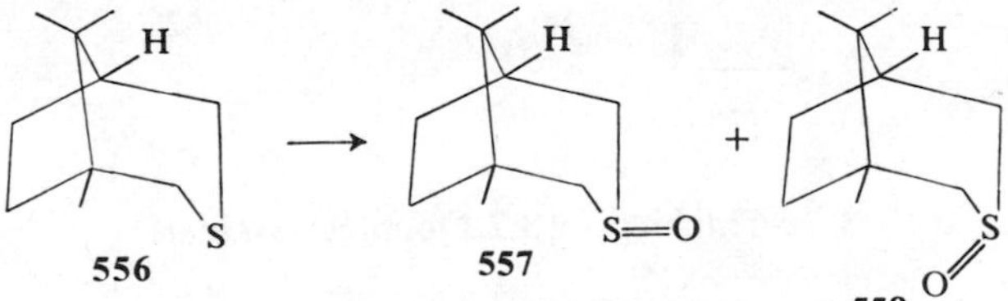

556 ⟶ **557** + **558** (214)

Oxidant	Yields (%)	
m-$ClC_6H_4CO_3H$	95 ± 2	5 ± 2
tert-BuOCl	4 ± 2	96 ± 2

As part of the complex mixture of products from the reaction of limonene (**559**) and sulfur, Weitkamp (313–317) isolated *trans*-4,7,7-trimethyl-6-thiabicyclo[3.2.1]octane (**560**) and 2,6,6-trimethyl-7-thiabicyclo[3.2.1]oct-2-ene (**561**) (Eq. 215). Treatment of **561** with Raney nickel under mild conditions led to racemization of **561**; however, under more vigorous conditions desulfurization occurred, producing limonene, 1-*p*-menthene, *p*-cymene, and *trans*- and *cis*-*p*-menthane (315). Catalytic hydrogenation of **561** gave both **560** and *cis*-4,7,7-trimethyl-6-thiabicyclo[3.2.1]octane **562** (316). Raney nickel desulfurization of **560** or its sulfone **563** led primarily to *trans-p*-menthane (94–96% pure) together with a small amount of *cis-p*-menthane (315). Mechanistic explanations of the Raney nickel desulfurizations and the

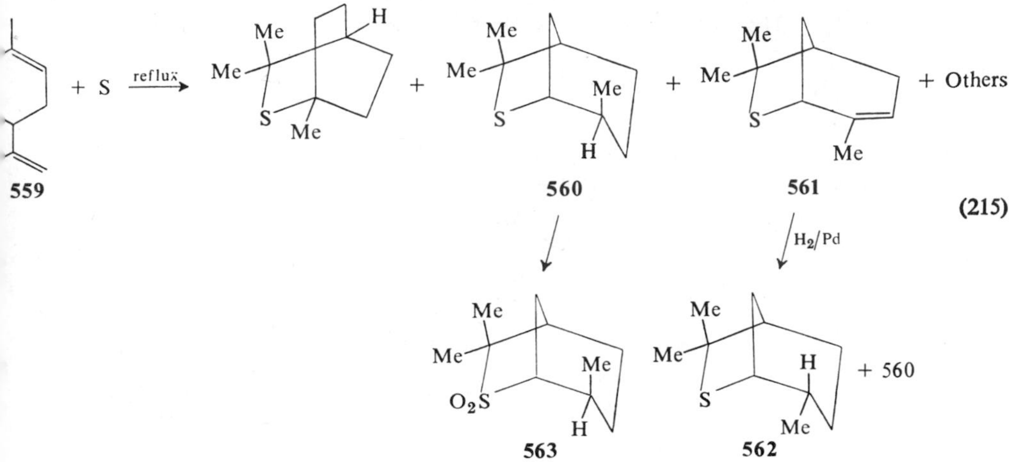

+ S —reflux→ ... + ... + ... + Others (215)

559 560 561

560 → 563; 561 —H_2/Pd→ 562 + 560

563 562

formation of the bicyclic sulfides from limonene and sulfur have been offered (315, 316).

Van Tamelen and Grant (289) obtained 7-phenyl-6-thiabicyclo[3.2.1]-octane-6,6-dioxide (**565**) by the base-catalyzed ring closure of **564** (Eq. 216). Bochwie and Frankowski (23) observed the formation of 4-acetoxy-6-thiabicyclo[3.2.1]octan-7-one (**567**) from the reaction of **566** with sodium acetate in acetic acid (Eq. 217).

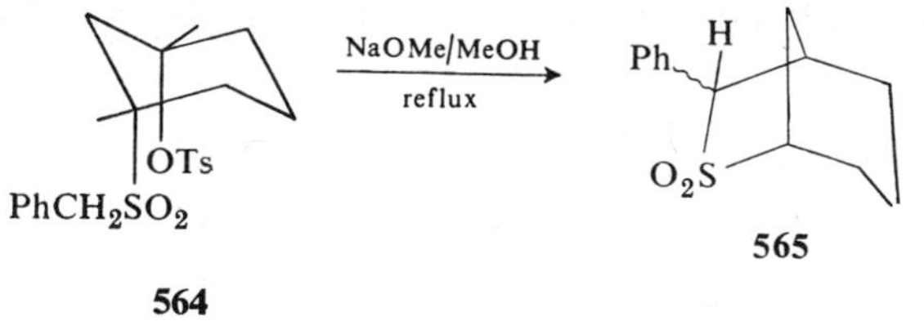

564 —NaOMe/MeOH, reflux→ **565** (216)

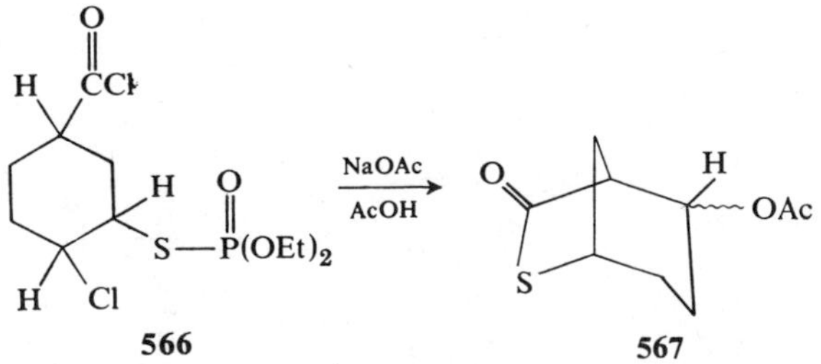

566 —NaOAc, AcOH→ **567** (217)

An example of a dithiabicyclo[3.2.1]octane has been described by Cox and Owen (60) in the sugar series. Acid hydrolysis of methyl 4,6-dideoxy-4,6-dimercapto-2,3-di-*O*-methyl-*β*-D-glucopyranoside (**568**) led in 41% yield to 1,6-anhydro-4,6-dideoxy-4,6-dimercapto-2,3-di-*O*-methyl-*β*-D-glucofuranose (**569**) (Eq. 218) which was characterized as its 3,5-dinitrobenzoate **570**.

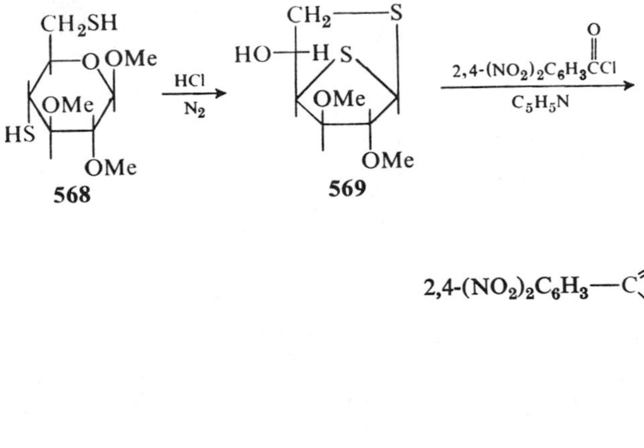

568 **569** (218)

570

4. *x*-Thiabicyclo[4.2.1]nonane Systems

1,4-Addition of sulfur dioxide to 1,3,5-cyclooctatriene has been found to give 9-thiabicyclo[4.2.1]nona-2,7-diene 9,9-dioxide (**571**) (Eq. 219) (169). Conversion of **571** into **576** can be effected by protection of the $C_{(2,3)}$ double bond via bromine addition, reduction of the $C_{(7,8)}$ double bond in **572** with

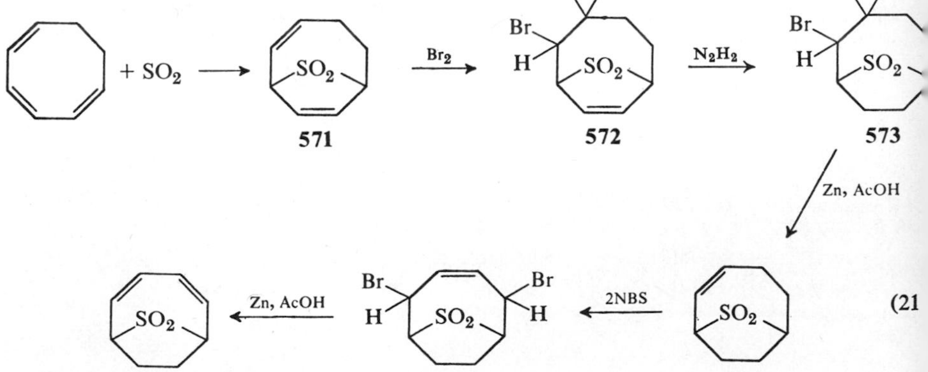

571 **572** **573**

576 **575** **574** (21

diimide to give **573**, regeneration of the $C_{(2,3)}$ double bond to give **574**, double allylic bromination of **574** with rearrangement to **575**, and finally debromination with zinc and acetic acid.

Interest in **571** and **576** centered on the thermal elimination of SO_2 and on orbital symmetry considerations in the mechanism of this elimination. Compound **571** decomposed readily at temperatures of 100–120°C with the following activation parameters: $\Delta H^{\ddagger} = 31.2 \pm 0.9$ kcal/mole, $\Delta S^{\ddagger} = +7.0 \pm 3$ eu, $\Delta G^{\ddagger} = 28.5 \pm 0.9$ kcal/mole. Compound **576** was stable at 110°C and decomposed at a convenient rate at 240–275°C giving the following activation parameters: $\Delta H^{\ddagger} = 30.0 \pm 1.5$ kcal/mole, $\Delta S^{\ddagger} = -18 \pm 5$ eu, $\Delta G^{\ddagger} = 39.5 \pm 1.5$ kcal/mole. Mock (169) interpreted the differences of the activation free energies as a minimal measure of the magnitude of the effect of orbital symmetry on these elimination reactions. He proposed that the synchronous bond rupture in **576** via a symmetrical transition state is less favored by at least 10 kcal/mole with respect to **571** and other sulfolenes.

Weil and co-workers (311) reported the synthesis of 7,8-dichloro-9-thiabicyclo[4.2.1]nonane (**578**) via transannular addition of sulfur dichloride to 1,3-cyclooctadiene (**577**) (Eq. 220). Attempts to prepare the parent compound **582** by direct lithium aluminum hydride reduction of **578** were unsuccessful; however, compound **582** was obtained by a more circuitous scheme involving oxidation to the sulfone **579**, zinc dehalogenation to **580**, and finally two reduction steps via **581** (311). 9-Thiabicyclo[4.2.1]nonane was distinguished from 9-thiabicyclo[3.3.1]nonane by characteristic bands in the ir spectrum. The structural assignment for **578** is supported by its nmr spectrum and the known *trans*-stereospecific nature of sulfenyl chlorides additions to double bonds (127).

The addition of sulfur dichloride to 1,5-cyclooctadiene (**583**) (59, 138, 311) or cyclooctatetraene (**584**) (141) gave 9-thiabicyclo[3.3.1]nonane derivatives

+ SCl_2 → ≡

577 **578** **578**

H_2O_2 84% (220)

LiAlH$_4$ Raney Ni 52% Zn 75%

582 **581** **580** **579**

585 and **586**, respectively, and no 9-thiabicyclo[4.2.1]nonane derivatives (Eq. 221). However, Mueller (178) reported that addition of methanesulfenyl chloride to 1,5-cyclooctatetraene (Eq. 222) produced predominantly a mixture of the 2:1 adducts **589** and **590** (Eq. 223), and some 2-chloro-5-methylthio-9-methyl-9-thiabicyclo[4.2.1]nonanesulfonium chloride (**588**) as well. Slow formation of crystalline **588** from the equilibrium mixture of **589** and **590** was observed and attributed to regeneration of the sulfonium intermediate **587**. A similar transannular participation occurred in the addition of methanesulfenyl chloride and phenylsulfenyl chloride to 5-methylthiacyclooctene (**591**) with the formation of 9-thiabicyclo[4.2.1]nonane derivatives **592** (Eq. 224). However, addition of hydrogen chloride or

583 + SCl_2 —96%→ **585**; **584** + SCl_2 —31%→ **586** (22

583 —MeSCl→ —MeSCl→ [**587**] $Cl^\ominus$ → **588** $Cl^\ominus$ (22

[**589** ⇄ **590**] → [587] → 588 (22

593 $X^\ominus$ + **594** $X^\ominus$ ←HX, X=Cl, I— **591** —RSCl, R=Me,Ph→ **592** $Cl^\ominus$ (22

hydrogen iodide to **591** gave both **593** and **594**, with the latter compound being the minor (20–25%) component (178).

Another entry into the 9-thiabicyclo [4.2.1]nonane system was reported by Ganter and Moser (93) in the course of their chemical studies of a photoproduct **595** obtained in previous work (92). When **595** was exposed to concentrated hydrochloric acid in chloroform (Eq. 225), cycloaddition of the mercaptan occurred across the double bond with formation of 6-hydroxy-7-oxa-2-thiatricyclo[4.3.1.0^{3,8}]decane (**596**). Reaction of **596** with acetic anhydride/pyridine produced 7-acetoxy-9-thiabicyclo[4.2.1]nonan-3-one (**597**). Wolff-Kishner reduction of **597** gave **598** which was transformed in three steps into 9-thiabicyclo[4.2.1]nonane (**582**). Stereochemical assignments placing the 7-substituent *trans* to the sulfur bridge in **597**, **598**, **599**, and **600** are supported by nmr data. The action of Raney nickel on **596** and **597** led to the expected desulfurized products.

(225)

5. 2-Thiatricyclo[3.2.1.0^{3,7}]octane and Derivatives

Cycloaddition of sulfur dichloride to norbornadiene (Eq. 226) led to *exo,exo*-4,8-dichloro-2-thiatricyclo[3.2.1.0^{3,7}]octane (**601**) which could be oxidized with hydrogen peroxide to its sulfone **602** or to a mixture of isomeric sulfoxides **603** and **604** (139). Sulfoxide **604** was thermally stable and could be converted into sulfoxide **603** via the ethoxysulfonium salt **605**. Both sulfoxides were oxidized to the same sulfone **602**. Structural assignments were based on the nature of the oxidation products and an extensive nmr analysis of these compounds. A similar addition (Eq. 227) of sulfur dichloride to 2-methylene-5-norbornene gave *exo,exo*-8-chloro-3-chloromethyl-2-thiatricyclo[3.2.1.0^{3,7}]octane (**606**) (140). Compound **602** has broad insecticidal activity (155).

+ SCl_2 →(81%) **601**; **601** →(H_2O_2, 99%) **602** (SO_2); **601** →(H_2O_2, 95%) **603** + **604**; **603** →(H_2O_2, 75%) **602**; **604** → **602**; **604** →($Et_3O^{\oplus}BF_4^{\ominus}$) **605** (EtO–$S^{\oplus}$ $BF_4^{\ominus}$); **605** →(NaOH) **603** (226)

+ SCl_2 → **606** (CH_2Cl) (227)

6. 6-Thiatricyclo[3.2.1.1^{3,8}]nonane and Derivatives

Wilder and Feliu-Otero (319, 320) have reported reactions of 4-thiatricyclo-[5.2.1.0^{2,6}]dec-8-ene (**607**) with hydrogen bromide and bromine which produced tetracyclothionium salts **608** and **609** (Eq. 228). The sulfonium salt formation entails transannular bridging of the sulfide via intermediate **610** and subsequent *exo* attack of bromide to give a *trans* addition to the $C_{(8,9)}$ double bond.

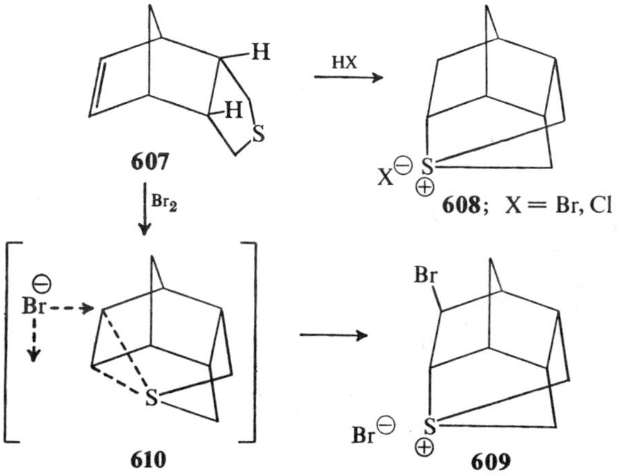

607 —HX→ **608**; X = Br, Cl (228)

607 —Br_2→ [**610**] → **609**

When **608** was treated with lithium aluminum hydride (Eq. 229), carbon sulfur bond cleavage led to a 47% yield of 2-methyl-6-thiatricyclo-[3.2.1.1^{3,8}]-nonane (**611**) and a 53% yield of 4-thiatricyclo[5.2.1.0^{2,6}]decane (**612**) (321).

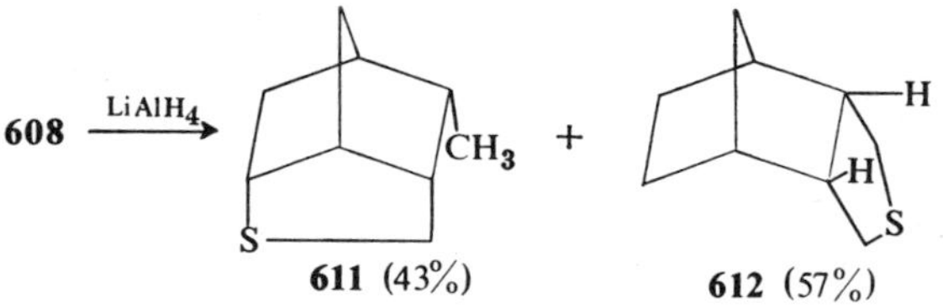

608 —$LiAlH_4$→ **611** (43%) + **612** (57%) (229)

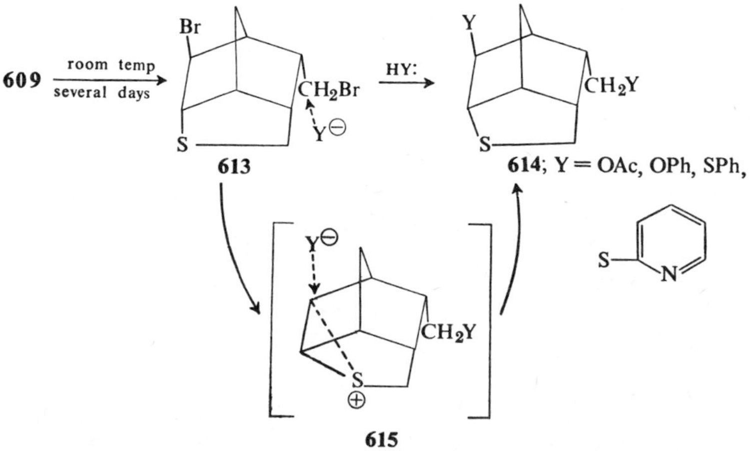

609 —room temp, several days→ **613** —HY:→ **614**; Y = OAc, OPh, SPh, S-(2-pyridyl) (230)

613 → [**615**] → **614**

Ring opening also occurred (Eq. 230) with **609**, simply on standing, to form 2-bromomethyl-4-bromo-6-thiatricyclo[3.2.1.$1^{3,8}$]nonane (**613**). The structural assignment of **613** was confirmed by an X-ray crystal structure determination of its picrate. A variety of nucleophilic substitution reactions with **613** were also reported (321) to give **614** which contains the same stereochemistry as the starting dibromo compound **613**. Retention of the stereochemistry at $C_{(4)}$ is attributed to sulfur participation via intermediate **615**.

TABLE XI-1. Cyclopenta[*b*]thiepins (R.I. 8147)

Substituents	mp, °C or (bp, °C/torr)	Other data	Refs.
Octahydro-2H-Cyclopenta[b]thiepins			
None	(85–86/4)	n_D^{20} 1.5473, d_{20}^{20} 1.0437	158

TABLE XI-2. *1-Benzothiepins* (R.I. 1851)

Substituents	mp, °C or (bp, °C/torr)	Other data	Refs.
1-Benzothiepins			
-Cl		ir	146
-Cl		ir, nmr	300
,5-$(AcO)_2$-4-Ph	86–87	ir, nmr	104, 106
1-Benzothiepin-1,1-dioxides			
one	140–141	ir, uv	152, 297, 298
Cl	137–137.5	ir, nmr, uv	36
Cl	155–157 (dec)	ir, nmr	300
NO_2 (?)	185–186	ir	152, 299
NO_2	180–181	ir	152, 299
5-$(AcO)_2$-4-Ph	140–141	ir, nmr	104, 106
5-$(AcO)_2$-2-Me-4-Ph	212–213	ir, nmr	106
2,3-Dihydro-1-benzothiepins			
one	(86/0.18)		300
	(95–96/0.4)	n_D^{20} 1.6470, ir, nmr, uv	295, 296
	(133–135/10)	ir, uv	269
Me (BF_4^- salt)	87–88	nmr	300
Et (BF_4^- salt)	83.5–85	nmr	300

TABLE XI-2. —(*contd.*)

Substituents	mp, °C or (bp, °C/torr)	Other data	Refs.
1-oxide	Oil	ir, nmr	300
1-MeO (BF_4^- salt)	103–104	ir, nmr	300
4-NO_2	116.5–117.5	ir	146
5-(2-CH_2CH_2Cl-phenyl)	(205–210/0.2)	uv	18
5-(2-CH_2CH_2Br-phenyl)	(195–196/0.05)		18
	(197–211/0.1)		
5-(2-CH_2CH_2NHMe-phenyl)	(178–182/0.1)		19
	(180–185/0.2)		18
hydrochloride	248–251	uv	18, 19
5-(2-$CH_2CH_2NMe_2$-phenyl)	(184–186/0.01)		19
	(196–201/0.2)		18
hydrochloride	228–230	uv	18, 19
2,4-Cl_2	55–57	ir, nmr	300
2-AcO-4-Cl		ir, nmr	300
4-CHO-5-Cl	73–74		312
4-(piperidino)NCH$_2$-5-Ph	124–124.5	ir, uv	269
4-(piperidino)NCH$_2$-5-Ph hydrochloride	195 (dec)		269
2-Me-3-oxo-4-Cl	(100–180/0.05)	ir, nmr, uv	192
2-Me-3-oxo-5-EtO	101–102	ir, nmr, uv	192
2,4-DNP	199.5–201	nmr	192
2,4-Cl_2-5-AcO		ir, nmr	300
2,5-$(AcO)_2$-4-Cl	104–105	nmr	300
3-OH-4-Ph-5-AcO	140–141	nmr	106
3-Oxo-4-Ph-5-AcO	124–125	ir, nmr	104 106,

ABLE XI-2.—(*contd.*)

2,3-Dihydro-1-benzothiepin-1,1-dioxides

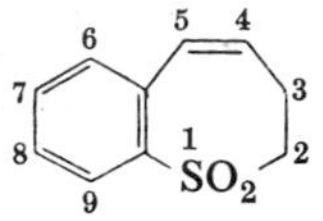

ubstituents	mp, °C or (bp, °C/torr)	Other data	Refs.
one	106–107	ir, uv	297, 298, 299
·Br	144–145	ir, uv	297, 298, 299
$\cdot NO_2$	197–198	ir, uv	299
$,3\text{-}Br_2$	147–148	ir, uv	299
$,4\text{-}Cl_2$	187–188.5		300
·AcO-4-Cl	163–165	ir, nmr	300
·Cl-5-AcO	147.5–149	ir, nmr	300
Br-5-AcO	161–162	ir, nmr	300
$4\text{-}Cl_2\text{-}5\text{-}AcO$	147.5–148.5	ir, nmr	300
Oxo-4-Ph-5-AcO	161–162	ir	106
2,5-Dihydro-1-benzothiepins			
Me		nmr	176
4,5-Dihydro-1-benzothiepins			
$5\text{-}Cl_2$		ir, nmr	152
Cl-5-OH	98–99	ir, nmr, uv	296
acetate	111–112	nmr, uv	296
p-nitrobenzoate	201–202	ir	296
Cl-5-OH-1-oxide	141–142 (dec)	ir, nmr, uv	36
4,5-Dihydro-1-benzothiepin-1,1-dioxides			
Cl-5-OH	116–117	ir, nmr, uv	36

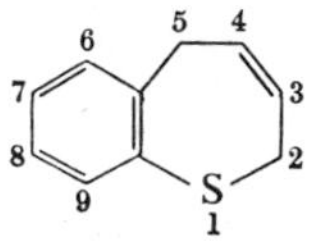

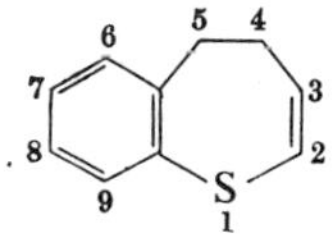

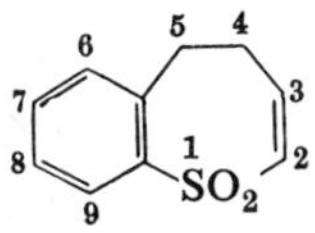

3,4-Dihydro-1-benzothiepin-Δ$^{5(2H)}$-

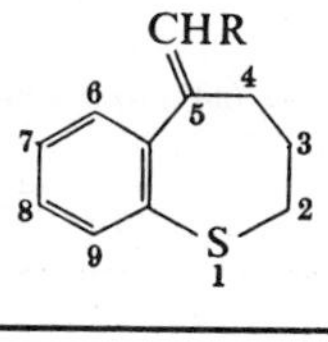

TABLE XI-2.—(*contd.*)

Substituents	mp, °C or (bp, °C/torr)	Other data	Refs.
R = CN	78–80		170
R = CH_2NH_2	Oil		170
R = CH_2NH_2 hydrochloride	245–246		170
R = Me_2NCH_2-			170
R = $Me_2NCH_2CH_2$-	(distil 128–134/0.05)		170
R = $Me_2NCH_2CH_2$- hydrochloride	155–156		170
R = CO_2H	163–165 197–199	*cis* and *trans*	170
R = CO_2Et	(122–140/0.075)		170
	(123/0.175)		170
R = $CONMe_2$	(172–178/0.45)		170
R = Et_2NCH_2-			170
R = $CH(Me)CH_2NEt_2$			170
R = $CONEt_2$			170
R = CON(*i*-Bu)$_2$			170
4,5-Dihydro-1-benzothiepin-Δ$^{3(2H)}$- 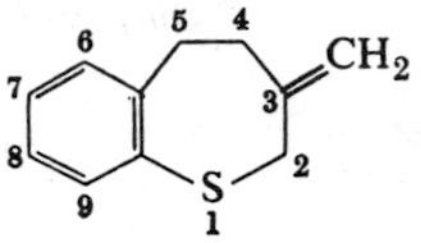			
None		nmr	176
4-Alkylidene-1-benzothiepin-5(2H,3H)*-ones* 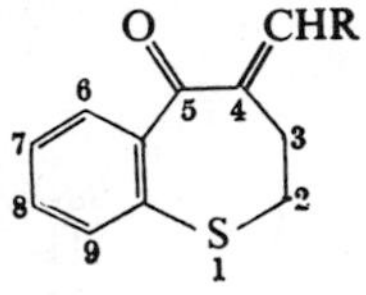			
R = Me	79–79.5	ir, nmr, uv	36
R = Ph	88.5–89	ir, nmr, uv	36
R 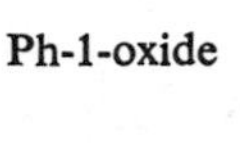 = morpholino (O N—)	166–167	ir, uv	36
R = Ph-1-oxide	137–140	ir, uv	36
4-Alkylidene-1-benzothiepin-5(2H,3H)*-one-1,1-dioxides* 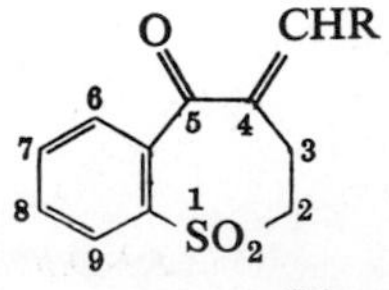			
R = Ph	156–157	ir, nmr, uv	36

TABLE XI-2.—(*contd.*)

Substituents	mp, °C or (bp, °C/torr)	Other data	Refs.
3,4-Dihydro-1-benzothiepin-5(2H)-one			
None	(119.5–120/1.5)	n_D^{20} 1.6228, n_D^{20} 1.6232, ir, nmr, uv	18, 133, 269, 298, 306
	(175–176/22)		53
oxime	97–98, 98–100		171, 173, 174
O-acetate	111–112.5		171
O-butyrate			171
O-benzoate			171
O-benzenesulfonate			171
O-tosylate	96–97		76, 171
O-brosylate			171
semicarbazone	213, 220–222, 222		53, 269, 298, 306
2,4-DNP	231–232		152
p-carboxyphenylhydrazone	166.5–167.5		4
1-Me (BF_4^-)	127–129		300
1-oxide	72.5–73.5	ir	146
2,4-DNP	176.5		146
3-Me	(107–108/0.001), 61–62		177
4-Br	89–90	ir, uv	152, 300
	86.5–87	ir, uv	269
4-I	98–99	ir	36
4-Et_2N hydrochloride	191–192	ir, uv	269
4-CH_3-N N-	(170–175/0.4)		269
hydrogen maleate	158–160		269
4-Me_2NCH_2- hydrochloride	176–179, 182–184	ir	36, 175
4- N—CH_2-			
hydrochloride	188–190	ir, uv	269
4- C_6H_5–CH(OH)-	168.5–169	ir, uv	36
7-Br	(202/15), 86		51
oxime	144.5		51
2,4-DNP	143		51

TABLE XI-2.—(*contd.*)

Substituents	mp, °C or (bp, °C/torr)	Other data	Refs.
7-Cl	(107–108/0.001), 61–62		177
7-F	(73–75/0.001), 59–60		177
7-*tert*-Bu	(207–208/15)	n_D^{23} 1.5800	51
oxime	111		51
2,4-DNP	249–250		51
ketazine	216.5		51
2,4-Cl_2 *cis*	108–109	ir, nmr	300
trans	108–109	ir, nmr	300
2-Br-2-Cl *cis*	134–135	ir, nmr	300
trans	98–99	ir, nmr	300
6,9-Me_2	(183.5/12)		51
3,4-Dihydro-1-benzothiepin-5(2H)-one-1,1-dioxides 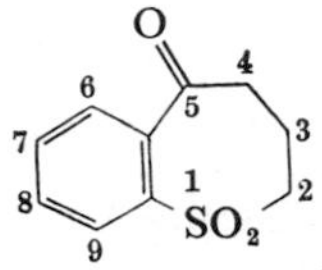			
None	155–156	ir, uv	134, 298
oxime	199–200	nmr	75, 299
phenylhydrazone	235–236 dec		4
p-tolylhydrazone	213–215		4
p-carbethoxyphenyl hydrazone	183–185		4
p-acetylaminophenyl hydrazone	262–263		4
4-Cl	139–143	ir, nmr	300
4-Br	156–157	ir, uv, nmr	152
	143.5–144		269
2,4-Cl_2 *cis*	172–174	ir, nmr	300
trans	178.5–180	ir, nmr	300
2-Cl-4-Br *cis*	167–169	ir, nmr	300
trans	161–163	ir, nmr	300
2,3-Dihydro-1-benzothiepin-4(5H)-one-1,1-dioxides 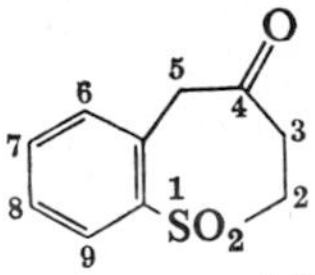			
None	85–87	ir	152
oxime	186–187	ir, nmr	299
1-Benzothiepin-3,5(2H,4H)-diones 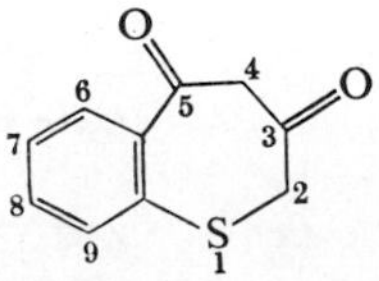			
2-Me	(85/0.05)	ir, nmr, uv	192
4-Ph	93–94	ir, nmr	104, 106

TABLE XI-2.—(*contd.*)

Substituents	mp, °C or (bp, °C/torr)	Other data	Refs.
1-Benzothiepin-3,5-(2H,4H)-dione-1,1-dioxides			
-Ph	160–161	ir	104, 106
-*p*-FC_6H_4-	158–159	ir	106
-*p*-ClC_6H_4-	198–199	ir	106
2,3,4,5-Tetrahydro-1-benzothiepins			
None	(140–141/21)	$n_D^{29.5}$ 1.5996	53
	(103–104/3)		306
	(103–105/3)	n_D^{20} 1.6052, ir, nmr, uv	46, 133, 298
	(142–144/23)	n_D^{18} 1.5939	83
-OH	70–71	ir, nmr, uv	146, 296, 298
acetate	71–72		298
p-nitrobenzoate	149–150	ir	298
-OH-1-oxide	110–113	ir	146
-OH-1-MeO(BF_4^-)	110–112 (soft 104)		300
-OH-4-NH_2	269–270		173
-OH-4-$NHC(=O)NH_2$			173
-OH-4-Br	92–93	ir, nmr	300
-OH-5-CH_2CO_2Et	44.5–46		170
-OH-5-Ph			175
-OH-5-(2-$CH_2CH_2NMe_2$-phenyl)	153–155	uv	18
-OH-5-(2-pyridyl)	79–80.5		175
-OH-5-(3-pyridyl)			175
-OH-5-(4-pyridyl)			175

TABLE XI-2.—(*contd.*)

Substituents	mp, °C or (bp, °C/torr)	Other data	Refs.
5-OH-2,2-Cl_2			
acetate	75–77		296
p-nitrobenzoate	112	ir	296
5-OH-5-Ph-4-Me_2NCH_2-	135–140		175
5-OH-5-Ph-4-(piperidino)N—CH_2-	166–167	ir, uv	269
5-OH-5-(2-pyridyl)-4-Me_2NCH_2-			175
5-MeO	(86/0.4)	n_D^{20} 1.5860, ir	152
5-(phthalimido)NCH_2CH_2O			174
5-$H_2NCH_2CH_2O$- hydrochloride			174
5-EtC(=O)$NHCH_2CH_2O$-			174
5-(phthalimido)$NCH_2CH_2CH_2O$-			174
5-$H_2NCH_2CH_2CH_2O$- hydrochloride			174
5-*n*-$PrNHCH_2CH_2O$-			174
5-$Me_2NCH_2CH_2O$- hydrochloride			174
5-$Et_2NCH_2CH_2O$-	(136–139/0.28)		174
citrate	115–117		174
5-(*n*-Pr)$_2NCH_2CH_2O$- hydrochloride			174
5-$Me_2NCH_2CH_2CH_2O$- hydrochloride·H_2O	144–146		174
5-$Et_2NCH_2CH_2CH_2O$-			174
5-NH_2	(115–117/0.1)	n_D^{20} 1.6135	74
	(113–115/0.9)		269
5-NH_2 hydrochloride	289 (dec)		74
	300		174

ABLE XI-2.—(*contd.*)

ubstituents	mp, °C or (bp, °C/torr)	Other data	Refs.
NH_2 hydrogen maleate	148–150.5		269
HC(=O)NH-	120–122		174
EtC(=O)NH-			174
MeNH	(111/0.8)		269
fumarate	149–150.5		174, 269
HC(=O)NMe-	93–96		174
AcNMe-			174
Me_2N hydrochloride	209.8–210.5		174
Et(*n*-Pr)N hydrochloride			174
$HC{\equiv}CHCH_2NMe$	(128/0.6)		269
hydrogen maleate	91–94		269
(pyrrolidino)N—	(140/1)		269
picrate	172–175 (dec)		269
(piperidino)N—	(142/0.45)		269
ıydrogen maleate	137–138/5		269
) N— (morpholino)	(146/0.4)		269
ıydrogen maleate	124–125.5		269
CH_3N (piperazino) N-			
›is-hydrogen maleate	163–164.5		269
›hN (piperazino) N—	143–144		269
naleate	187–188		269
nethiodide	103–105		269
›hCH_2N (piperazino) N-			
ihydrochloride	177–180		269
l	(79–81/0.05–0.08)	n_D^{20} 1.6159, ir	146
	(105–110/0.4)		269
l-1-oxide	141–142	ir	146
r	(136–137/0.4)	d_4^{18} 1.457, $n_D^{18.6}$ 1.6447	51

TABLE XI-2.—(*contd.*)

Substituents	mp, °C or (bp, °C/torr)	Other data	Refs.
7-*tert*-Bu	(167/15)	$n_D^{21.6}$ 1.5657	51
7-Ac	(217/19)		53
	(200–204/20)	n_D^{17} 1.6121, uv	83
semicarbazone	217–219		53
2,4-DNP	194		83
7-(quinolin-2-yl) [structure: N]	98		53
picrate	193 (soft 187)		53
7-(4-CO_2H-quinolin-2-yl) [structure: CO_2H, N]	170–171		53
4,5-Br_2	107–108	ir	152

2,3,4,5-Tetrahydro-1-benzothiepin-1,1-dioxides [structure with ring positions 1–9; 1 = SO_2]

Substituents	mp, °C or (bp, °C/torr)	Other data	Refs.
None	77–78	ir, uv	134, 298, 299, 3
5-OH	141–142	ir	297, 298
acetate	144.5–145.5		298
5-NH_2	152		75
hydrochloride	313		75
5-Cl	97–98.2	ir	146
7-*tert*-Bu	153		51
8-NO_2	197–198	ir	299
8-NH_2	169–170	ir	299
8-AcNH-	224–225		299
4,5-Br_2	195–196	ir, uv	298, 299
2,3,4,5-Br_4	222	uv	299

Miscellaneous 1-Benzothiepin

Compound	mp, °C	Other data	Refs.
[structure: two 2,3,4,5-tetrahydro-1-benzothiepin-5-one units (O, S) joined at C-4 through Ph–CH]	207–208		36
[structure: 4-(piperidinomethyl) unit (CH_2N, O, S) joined through CH_2 to a second 1-benzothiepin-5-one unit (H, O, S)]	166–168	ir, uv	269

ABLE XI-2.—(*contd.*)

ubstituents	mp, °C or (bp, °C/torr)	Other data	Refs.
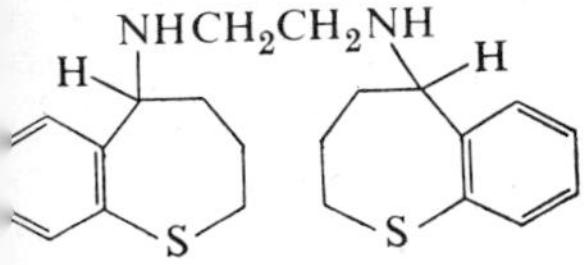 $NHCH_2CH_2NH$ (bis structure)	232–233		77
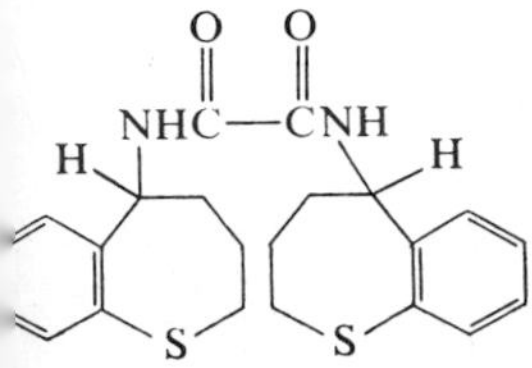 NHC(O)—C(O)NH (bis structure)	320		77
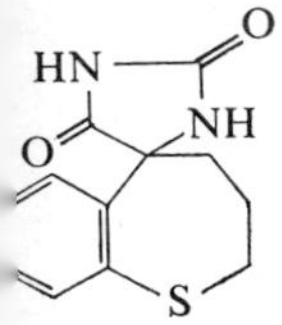 (spiro hydantoin structure)	288–290		6
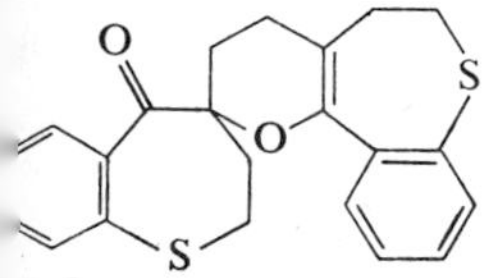 (spiro structure)	155	ir, nmr	36, 172
	152–153		269
oxime	214–217		269
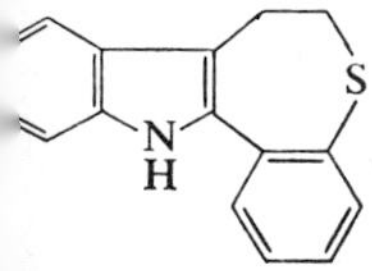 (indole-fused structure)	158.5–159		4
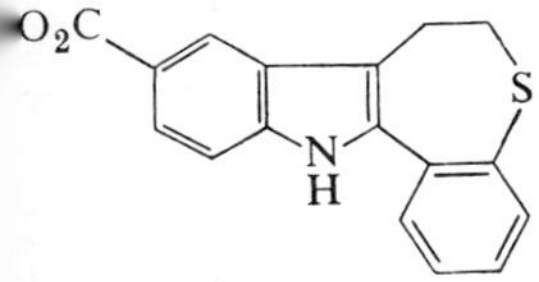 O_2C (indole-fused structure)	265–266 (dec)		5
ethyl ester	198.5–199		4
$Et_2NCH_2CH_2$ ester	168–169		4
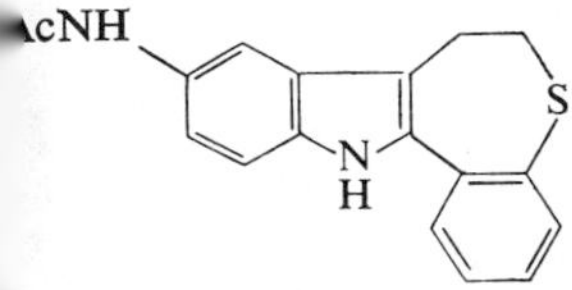 cNH (indole-fused structure)	255–256		4

TABLE XI-2.—(*contd.*)

Substituents	mp, °C or (bp, °C/torr)	Other data	Refs.
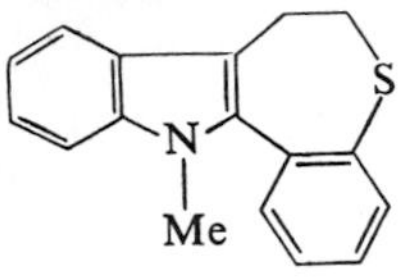 S N Me	168–169		4
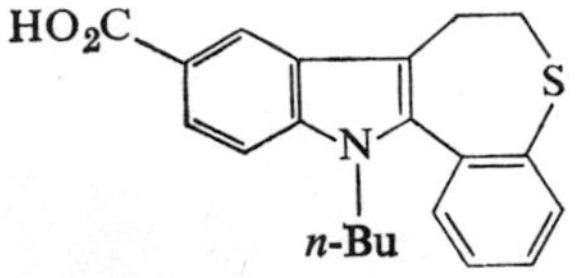 HO_2C S N *n*-Bu	259–261		4
ethyl ester	148–148.5		4
$Me_2NCH_2CH_2$ ester	127–127.5		4
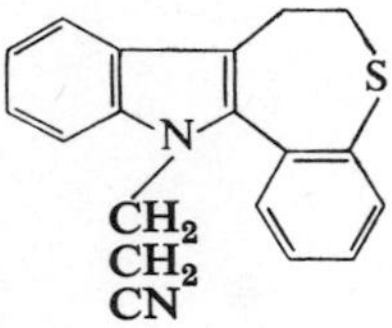 S N CH_2 CH_2 CN	149–150		4
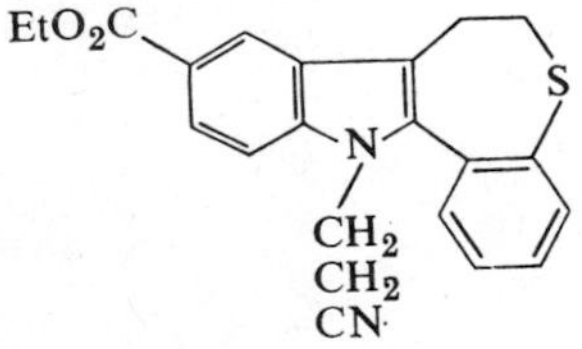 EtO_2C S N CH_2 CH_2 CN	145–146		4
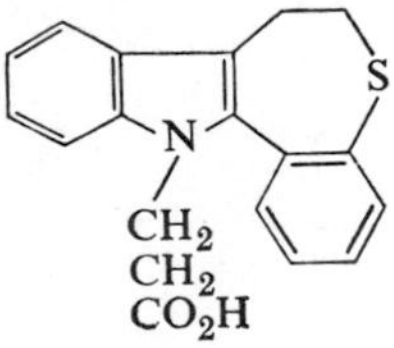 S N CH_2 CH_2 CO_2H	196–197		4
ethyl ester	91–92		4
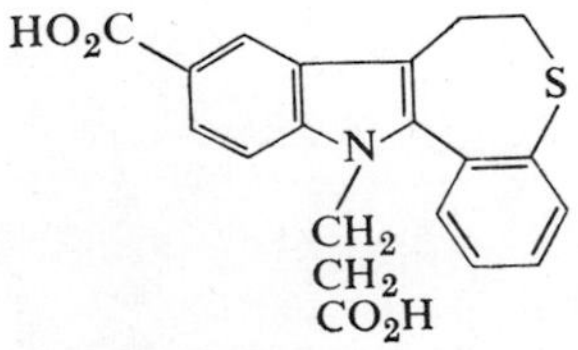 HO_2C S N CH_2 CH_2 CO_2H	268–269		4
diethyl ester	112–113		4

ABLE XI-2.—(*contd.*)

ubstituents	mp, °C or (bp, °C/torr)	Other data	Refs.
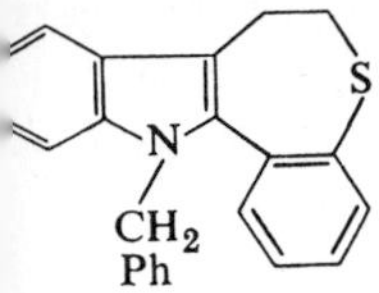 N	145–146		4
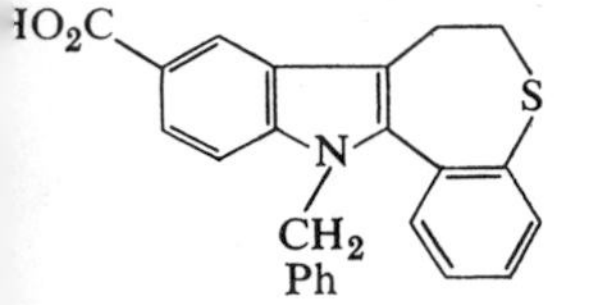 IO₂C	249–250 (dec)		4
ethyl ester	143–144		4
$Me_2NCH_2CH_2$ ester	160–161		4
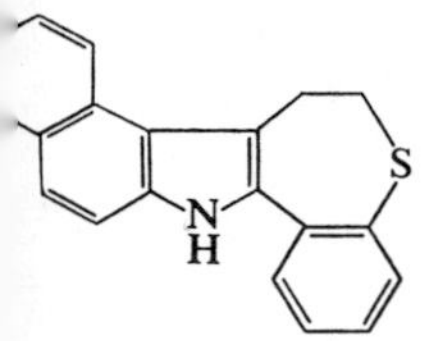 S	218–219		4
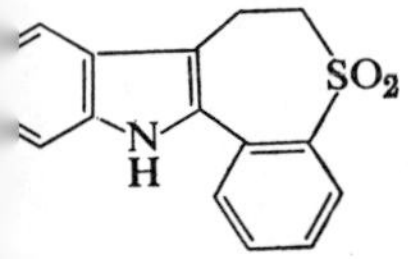 SO₂	295–297		4
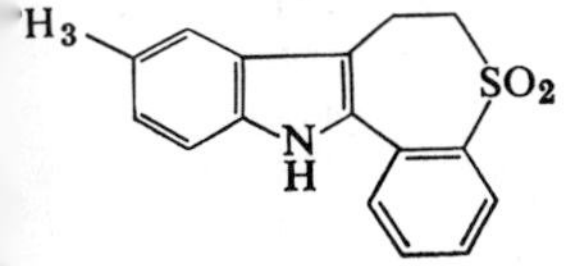 H₃	>350		4
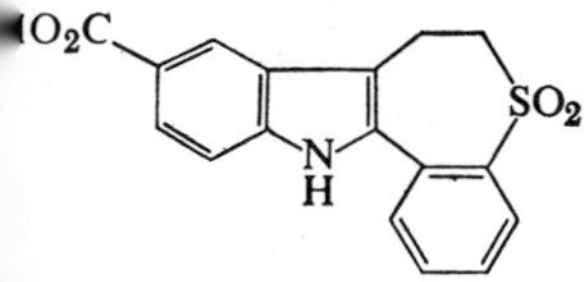 IO₂C	320 (dec)		4
ethyl ester	241–242		4
$Me_2NCH_2CH_2CH_2$ ester			4
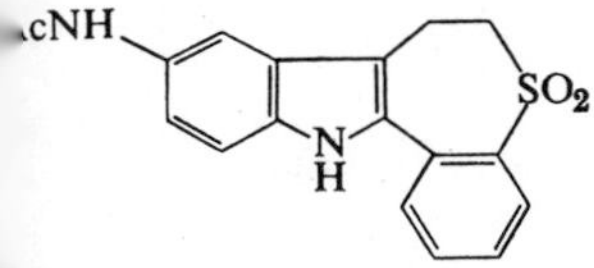 cNH	>320		4

TABLE XI-2.—(*contd.*)

Substituents	mp, °C or (bp, °C/torr)	Other data	Refs.
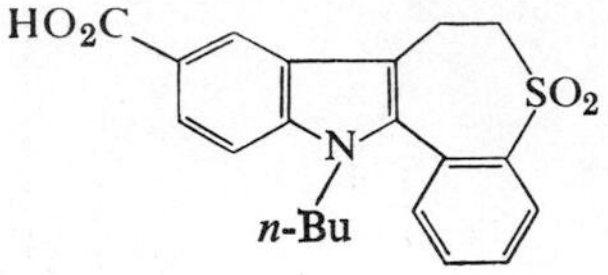 HO_2C, SO_2, N, *n*-Bu	263–265		4
ethyl ester	196.5–197		4
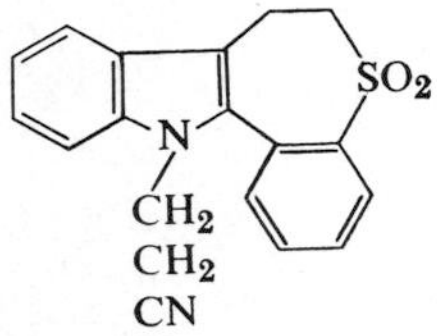 SO_2, N, CH_2, CH_2, CN	211.5–212		4
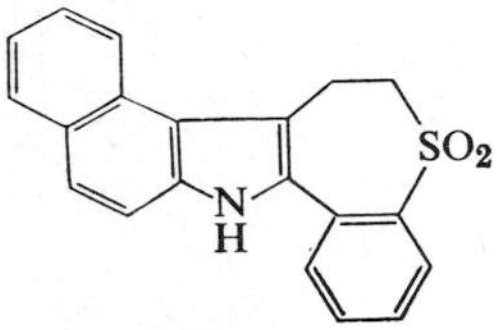 SO_2, N, H	216–218		4
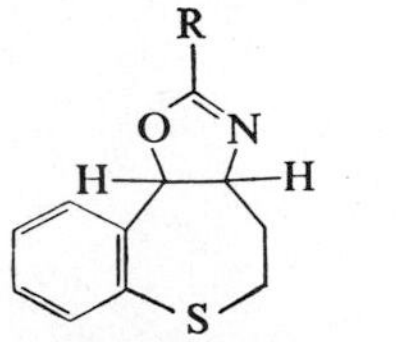 R, O, N, H, H, S			
R = NH_2 hydrochloride	246–247		173
R = MeNH-			173
R = Me_2N-			173
R = EtNH-			173
R = Et_2N-			173
R = *n*-BuNH-			173
R = *n*-Bu_2N-			173
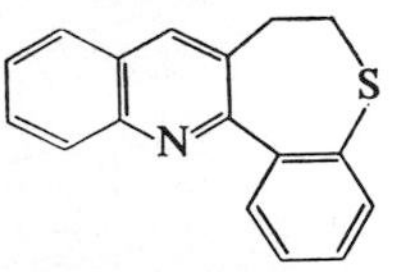 S, N	(240/5), 145		54
picrate	188–190		54
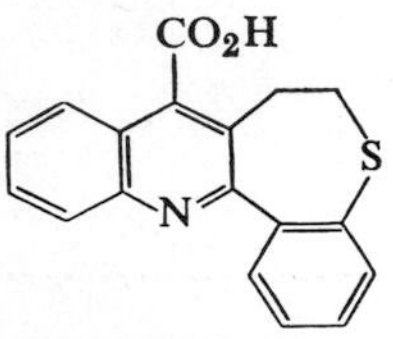 CO_2H, S, N			54

TABLE XI-2.—(*contd.*)

Substituents	mp, °C or (bp, °C/torr)	Other data	Refs.
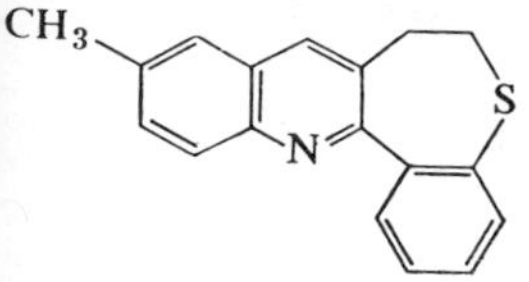 [structure: CH_3-substituted quinoline-fused benzothiepin]	(250/5), 169		54
picrate	202–204		54
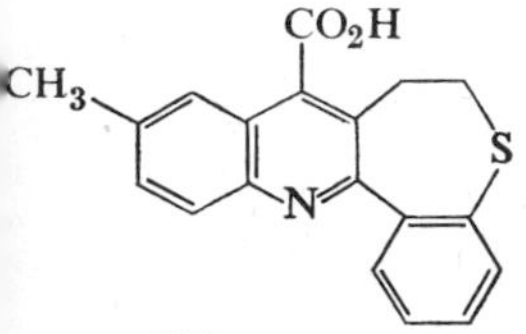 [structure: CH_3, CO_2H-substituted quinoline-fused benzothiepin]			54
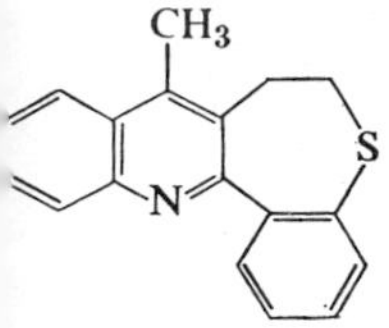 [structure: CH_3-substituted quinoline-fused benzothiepin]	163–165		77
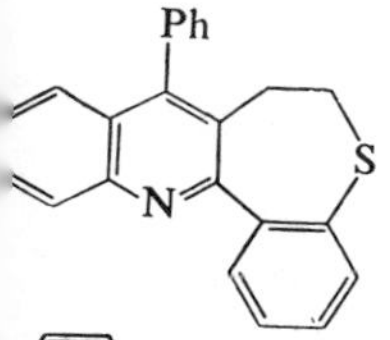 [structure: Ph-substituted quinoline-fused benzothiepin]	236–237		77
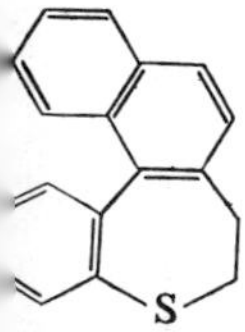 [structure: naphtho-fused benzothiepin]	(173–183/0.2), 99–101	uv	18

ABLE XI-3. 2-Benzothiepins (R.I. 1852)

ıbstituents	mp, °C or (bp, °C/torr)	Other data	Refs.
1,3,4,5-Tetrahydro-2-benzothiepins 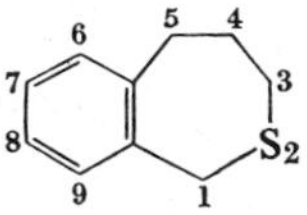			
one	(141–145/14), 95–96		39
	(143/15), 48.5–49	n_D^{24} 1.5810, d_4^{25} 1.044	49
	(134–136/12), 44–45		33
chloromercuri deriv.	117		49

TABLE XI-3.—(*contd.*)

Substituents	mp, °C or (bp, °C/torr)	Other data	Refs.
1-Me	(82–86/0.1)	n_D^{20} 1.5888	33
	(136.5–137/12; 140/13.5)	d_4^{24} 1.083, $n_D^{22.6}$ 1.5885, ir	49
1-Et	(95–97/0.2)	n_D^{20} 1.5783	33
1-Ph	55–56		33
1-Cl			33
1-EtO	(110–130/0.2)	n_D^{20} 1.5803	33
1-CN	79–81		33
1-CO_2H	121–123		33
1-CO_2Me	(100–120/0.1)	n_D^{20} 1.5793	33
2-Me (I^-)	154		39
3-Ph	143–144	ir, mass spectrum	72
6-OH			100
6-SO_3H			100
1-CN-1-$Et_2NCH_2CH_2$-	(140–160/0.1)	n_D^{20} 1.5512	34
perchlorate	106–108		34
methobromide	218–222		34
1,3,4,5-Tetrahydro-2-benzothiepin-2,2-dioxides			
None	176, 176–178		33, 39
1-Me	116–118		33
1-Et	105–107		33
1-Ph	150–151		33
1-CO_2Me	153		33
3-Ph	270–271	ir, mass spectrum	72
1,1-$(Et_2NCH_2CH_2)_2$	(190/210/0.01)	n_D^{20} 1.5367	34
bisperchlorate	238–240		34
bismethobromide	242–244		34
3,4-Dihydro-2-benzothiepin-5(1H)-ones			
None	(183/14)	n_D^{20} 1.6037, d_4^{20} 1.190	49
	(100–130/0.01), 50–51	ir, nmr	154
semicarbazone	280		49
2,4-DNP	207		49
oxime	168–169		154
Octahydro-2-benzothiepins			
None			100

TABLE XI-3.—(*contd.*)

Substituents	mp, °C or (bp, °C/torr)	Other data	Refs.
	Octahydro-2-benzothiepin-6(1H)-ones 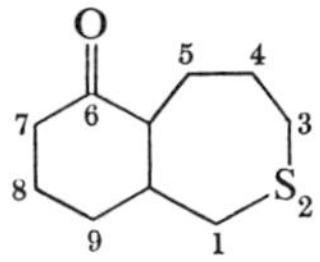		
None			100
2-oxide			100
2,2-dioxide			100
	Miscellaneous 2-Benzothiepins		
Compound			
[structure: S]			100
[structure: S=O]			100
[structure: SO_2]			100

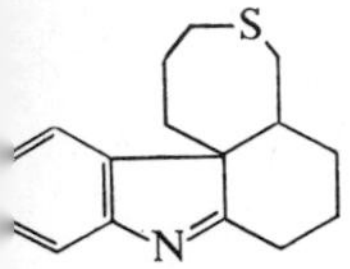

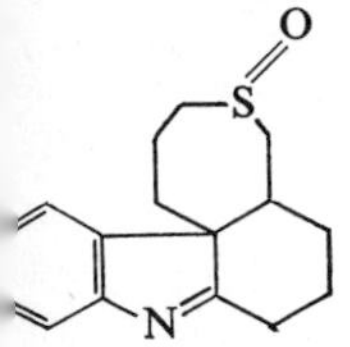

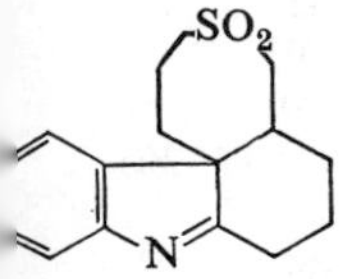

TABLE XI-4. 3-Benzothiepins (R.I. 1853)

Substituents	mp, °C or (bp, °C/torr)	Other data	Refs.
	3-Benzothiepins 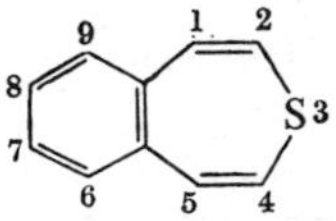		
2,4-$(CO_2H)_2$	dec	uv	261
2,4-$(CO_2Me)_2$	95–97	ir, uv	70, 71
	89–90	ir, nmr, uv	118
chloromercuri deriv	134–135		71

TABLE XI-4.—(*contd.*)

Substituents	mp, °C or (bp, °C/torr)	Other data	Refs.
2,4-$(CO_2Et)_2$	75–77		71
2,4-$(CO_2Me)_2$-3-Br(Br^-)	130–131		71
2,4-$(CO_2H)_2$-7,8-methylenedioxy	120 sudden decolor, 150 dec.		61
2,4-$(CO_2Me)_2$-7,8-methylenedioxy	124.5 sudden decolor, 172.5 (dec.)		61
3-Benzothiepin-3,3-dioxides			
None	161.5–163.5	ir, uv	305
X-NO_2 (most probable 7-sub)	173–176		305
1,2-Dihydro-3-benzothiepin-3,3-dioxides			
None	135–137		147
4,5-Dihydro-3-benzothiepin-1(2H)-ones			
None	(181–183/15)		41
	51–53		305
	(183/13), 50	d_4^{23} 1.200, $n_D^{18.8}$ 1.6140, ir	49
oxime	149, 151–152		41, 49, 305
semicarbazone	244		41
	279		49
2,4-DNP	210		49, 56
6-Me	(218/37), 103		55
oxime	170		55
semicarbazone	273		55
2,4-DNP	247		55
8-Me	(193/12.5), 64–64.5		55
oxime	202		55
2,4-DNP	241		55
6,8-Me_2	(203–205/15), 96		55
oxime	208		55
2,4-DNP	243		55

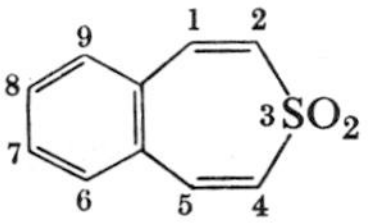

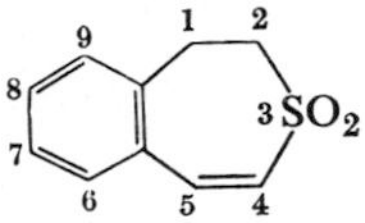

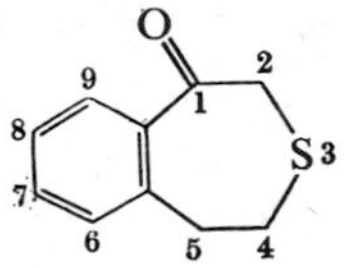

TABLE XI-4.—(*contd.*)

Substituents	mp, °C or (bp, °C/torr)	Other data	Refs.
,9-Me_2	(192/14)(141/2.5) (201/19), 95.5		55
oxime	89.5		55
2,4-DNP	200		55
,8-Me_2	(203.5–204/12.5), 98		55
oxime	192		55
semicarbazone	252		55
2,4-DNP	224		55
,9-Me_2	impure		
oxime	179		55
,7,8-Me_3	(222–223/20.7), 114		55
oxime	195		55
semicarbazone	232		55
2,4-DNP	206.5		55
4,5-Dihydro-3-benzothiepin-1(2H)-one-3,3-dioxides			
None	133–134.5		305
1,2,4,5-Tetrahydro-3-benzothiepins			
None	(136.5/15.5)	$d_4^{20.7}$ 1.070, $n_D^{20.2}$ 1.5907, ir	49
	(140/14.5)	$d_4^{20.5}$ 1.106, $n_D^{19.5}$ 1.6008	56
chloromercuri deriv.	185		56
-Me	(160/15)	$n_D^{22.3}$ 1.5920, ir	55
chloromercuri deriv.	142		55
-Me	(145–146/15)	$d_4^{22.4}$ 1.044, $n_D^{20.2}$ 1.5792, ir	55
chloromercuri deriv.	161–162		55
-Et	(159/12.7)	d_4^{21} 1.056, $n_D^{18.3}$ 1.5750, ir	56
chloromercuri deriv.	148		56
-Ac	(203/13.5), 72.5		56
oxime	139.5		56
semicarbazone	227		56
2,4-DNP	247		56
-*n*-Bu	(200/2.8)	$n_D^{19.5}$ 1.5590, ir	56
chloromercuri deriv.	143		56
-*n*-PrCO	(145–156/2.5)	n_D^{20} 1.5748	56
2,4-DNP	202–202.5		56

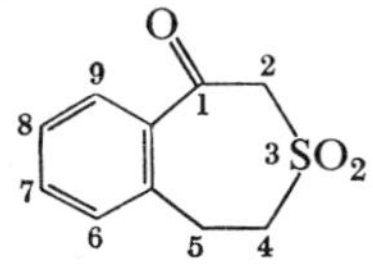

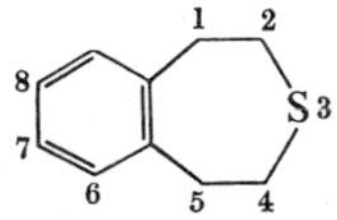

TABLE XI-4.—(*contd.*)

Substituents	mp, °C or (bp, °C/torr)	Other data	Refs.
8-$HO_2CCH_2CH_2CH_2$-	(236–238/12)		
	(212/3), 108		56
amide	166		56
8-$HO_2CCH_2CH_2CO$-	142		56
8-$MeO_2CCH_2CH_2CO$-	(210–215/3), 94	$n_D^{13.5}$ 1.5713	56
2,4-DNP	187		56
8-(6-methyl-2-quinolyl)	135		56
picrate	160		56
8-(4-carboxy-6-methyl-2-quinolyl)	222		56
9-SO_3H			100
9-OH			100
6,8-Me_2	(163–165/15)	$d_4^{20.7}$ 1.043, n_D^{19} 1.5782, ir	55
chloromercuri deriv.	188		55
6,9-Me_2	(152.5–153/14.3)	d_4^{17} 1.045, $n_D^{15.4}$ 1.5805, ir	55
chloromercuri deriv.	129		55
7,8-Me_2	(158–160/15)	d_4^{17} 1.034, $n_D^{14.6}$ 1.5768, ir	55
chloromercuri deriv.	135		55
2,7,8-Me_3	(170–172/18), 65	d_4^{20} 1.033, $n_D^{18.8}$ 1.5702, ir	55
chloromercuri deriv.	134–135		55
6,9-Me_2-8-Et	128		56
6,9-Me_2-8-Ac	(159–160/2), 70		56
2,4-DNP	145		56

1,2,4,5-Tetrahydro-3-benzothiepin-3,3-dioxides

Substituents	mp, °C or (bp, °C/torr)	Other data	Refs.
None	154–156	ir	305
1-OH	185.5–187		305
1-Br	175–177		305
1,2-Br_2	195–199		147
1,5-Br_2	188–194 (impure)		305

Octahydro-3-benzothiepin-6(7H)-ones

Substituents	mp, °C or (bp, °C/torr)	Other data	Refs.
None			100
3-oxide			100
3,3-dioxide			100

ABLE XI-4.—(*contd.*)

ubstituents	mp, °C or (bp, °C/torr)	Other data	Refs.
	Decahydro-3-benzothiepins 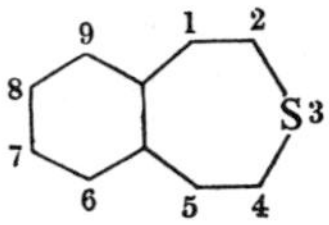		
OH			100
	Miscellaneous 3-Benzothiepins		
ompound 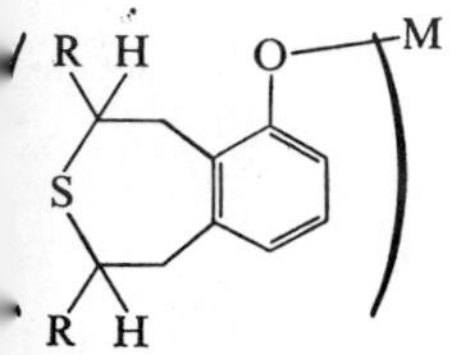 = C_6 to C_{10} = Ba or Ca			107
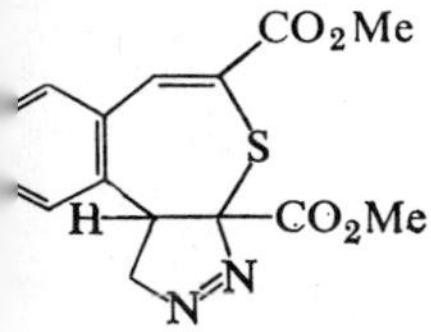 or 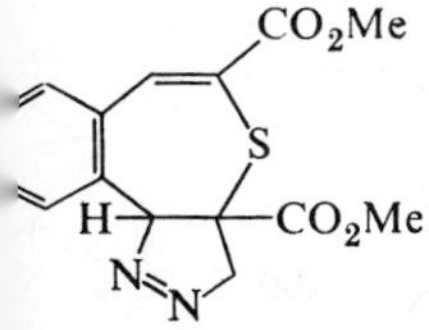	117–119		71
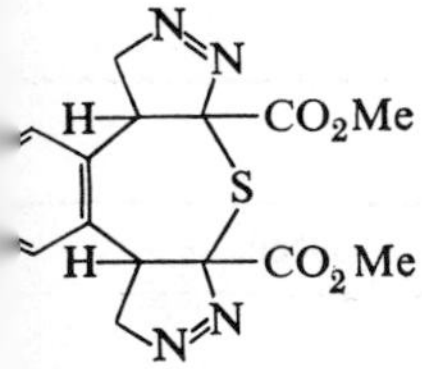 or 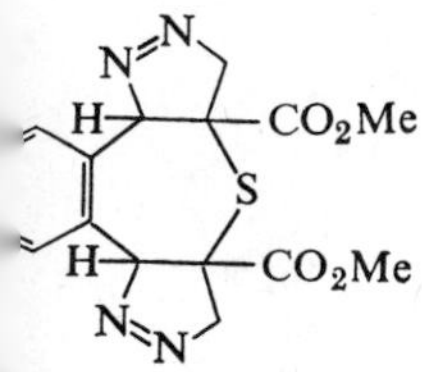	98–99		71

TABLE XI-4.—(*contd.*)

Substituents	mp, °C or (bp, °C/torr)	Other data	Refs.
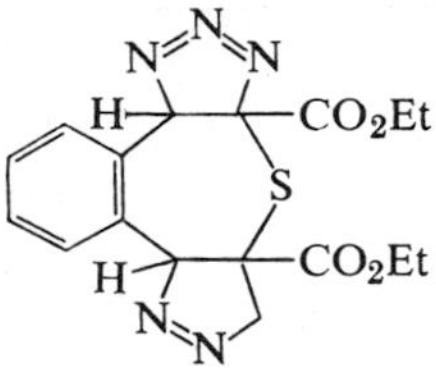 or 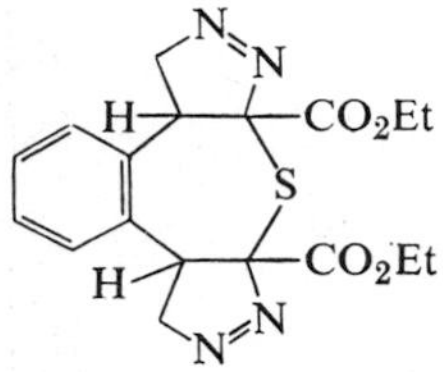	97–99		71
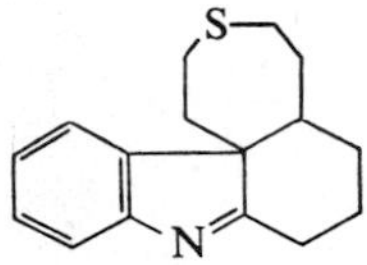			100
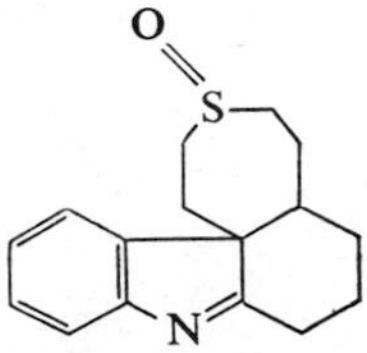			100
SO_2			100

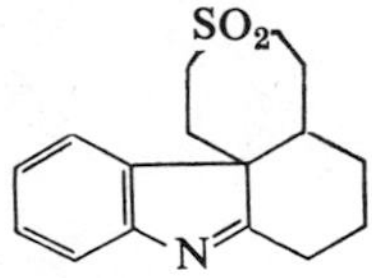

TABLE XI-5. Naphthothiepins

Substituents	mp, °C or (bp, °C/torr)	Other data	Refs.
2,3,4,5-Tetrahydronaphtho[1,2-b]thiepins 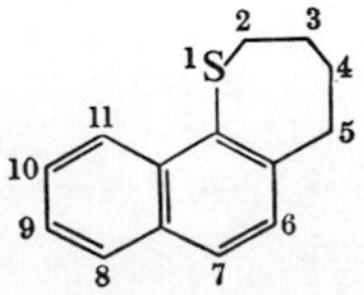			
None	(143/0.4), 65	ir, nmr, uv	46, 50
picrate	80		50
TNF	137		50

ABLE XI-5.—*(contd.)*

ubstituents	mp, °C or (bp, °C/torr)	Other data	Refs.
-Me	(161–162/0.4), 95	ir, nmr, uv	46, 50
picrate	83		50
TNF	119		50
-Et	(163–164/0.4), 65	ir, nmr, uv	46, 50
picrate	70		50
TNF	131		50
-Ac	(203/0.4), 119		50
oxime	116		50
2,4-DNP	216		50
*3,4-Dihydronaphtho[1,2-*b*]thiepin-5(2H)-ones*			
one	(184/1), 82		50
oxime	191		50
2,4-DNP	300		50
-Me	(182/0.3), 75		50
oxime	155		50
2,4-DNP	317		50
*1,2,3,4-Tetrahydronaphtho[2,1-*b*]thiepins* (R.I. 3706)			
one	(164/1)	d_4^{20} 1.159, n_D^{20} 1.6705, ir, nmr, uv	46, 50
picrate	74		50
TNF	163		50
5-dioxide	134–135		307
*2,3-Dihydronaphtho[2,1-*b*]thiepin-1(4H)-ones*			
one	(184/0.8), 78–79		50, 307
oxime	191		50
2,4-DNP	269–270		50
5-dioxide	176–177		307

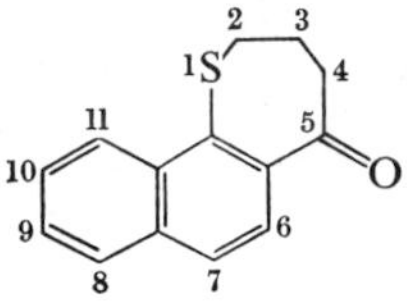

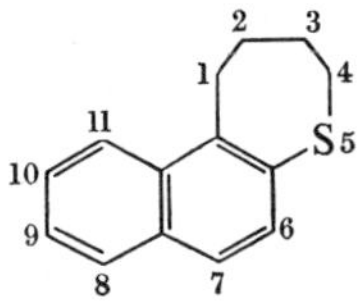

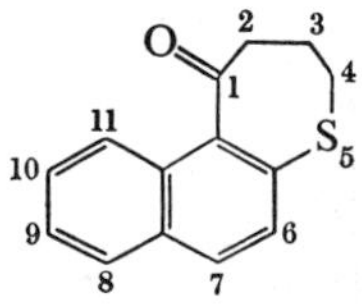

TABLE XI-5—(*contd.*)

Substituents	mp, °C or (bp, °C/torr)	Other data	Refs.
2,3,4,5-Tetrahydronaphtho[*2,3*-b]*thiepins* 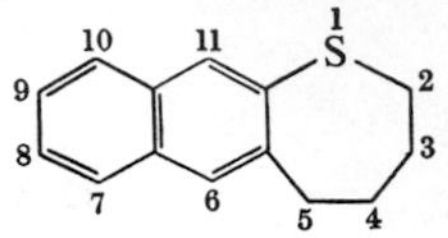			
None	(186/3), 121	ir, nmr, uv	46, 50
1,1-dioxide	151		50
3,4-Dihydronaphtho[*2,3*-b]*thiepin-5*(*2*H)*-ones* 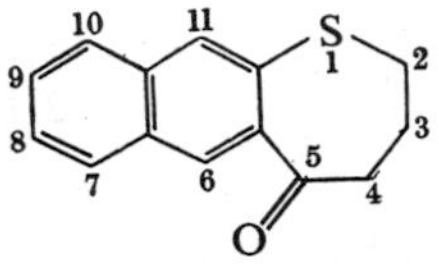			
None	(178–180/1.2), 55		50
4,5-Dihydronaphtho[*1,2*-d]*thiepin-1*(*2*H)*-ones* 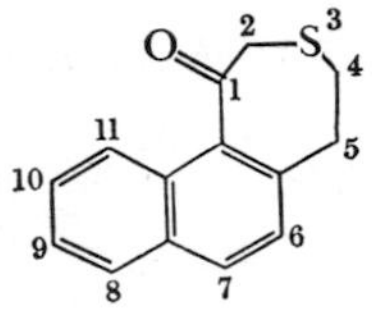			
None	(203/2.5), 78		57
oxime	191		57
1,2,4,5-Tetrahydronaphtho[*1,2*-d]*thiepins* 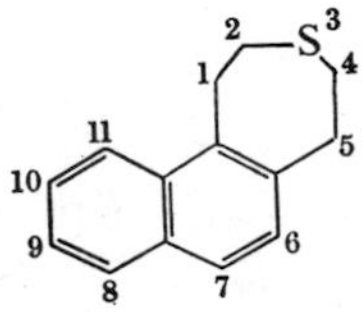			
None	(155/3), 67		57
picrate	105		57
styphnate	121		57
TNF	117		57
chloromericuri deriv.	170		57
1,2-Dihydronaphtho[*1,2*-d]*thiepin-5*(*4*H)*-ones* 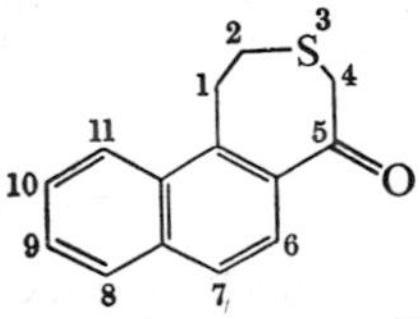			
None	(217/15), 128		57
2,4-DNP	265		57
Naphtho[*2,3*-d]*thiepin* 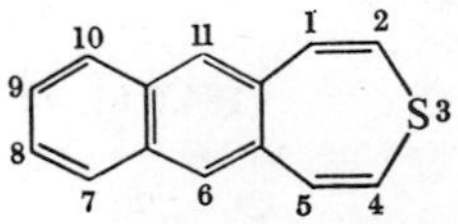			
2,4-$(CO_2H)_2$			150
2,4-$(CO_2Me)_2$	178 (dec)		150

TABLE XI-5.—(*contd.*)

Substituents	mp, °C or (bp, °C/torr)	Other data	Refs.
2,4-$(CO_2Et)_2$	167–170	uv	150
2,4-$(PhCO)_2$	174 (dec)		150
1,2,4,5,8,9-Hexahydronaphtho[2,3-d]*thiepin-7(10*H)-*ones*			
None	(230/11), 108.5–109		56
2,4-DNP	285		56
1,2,4,5,7,8,9,10-Octahydronaphtho[2,3-d]*thiepins*			
None	(200/10), 110.5		56
chloromercuri deriv.	198		56
*1*H-*Naphtho*[*1,8*-c,d]*thiepin-4*(*3*H)-*ones*			
None			58
*3,4-Dihydro-1*H-*naphtho*[*1,8*-c,d]*thiepins*			
None			58

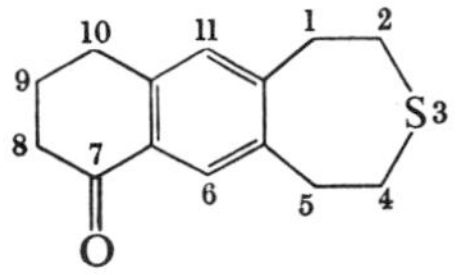

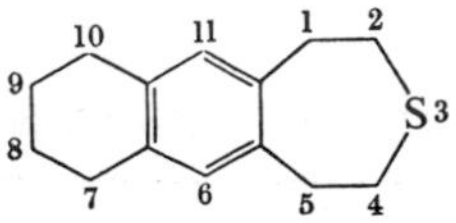

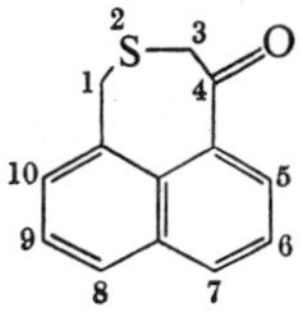

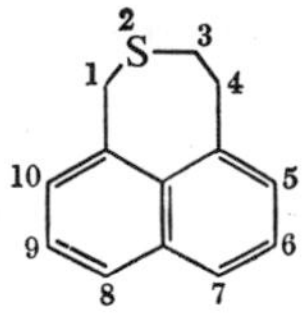

TABLE XI-6. Dibenzo[*b,d*]thiepins (R.I. 10,530)

Substituents	mp, °C or (bp, °C/torr)	Other data	Refs.
1,2,3,4,4a,11b-Hexahydrodibenzo[b,d]*thiepin-7*(*6*H)-*ones*			
-A/B-9,10,11-$(MeO)_3$	84–86	nmr	148
ns-A/B-9,10,11-$(MeO)_3$	119–124	nmr	148

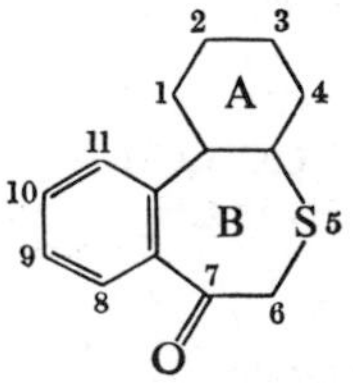

TABLE XI-6.—(*contd.*)

Substituents	mp, °C or (bp, °C/torr)	Other data	Refs.
Dibenzo[b,d]*thiepin-7(6H)-one-5,5-dioxides* 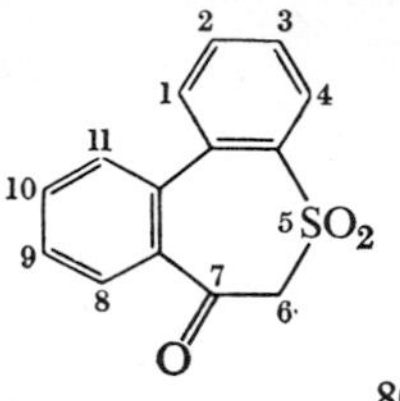			
None	167–168	ir	80

TABLE XI-7. Dibenzo[*b*,*e*]thiepins

Substituents	mp, °C or (bp, °C/torr)	Other data	Refs.
Dibenzo[b,e]*thiepins*-$\Delta^{11(6H)}$- 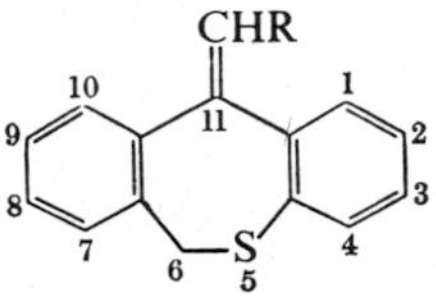			
R = CN	176–177		29
11-Et(CN)C	112–113		29
R = Me—⟨cyclohexyl⟩NNHCONHSO$_2$—⟨phenylene⟩—CH_2CH_2NHCO-	222–223		326
R = Me_2NCO-	137–138		30
R = H_2NCH_2- hydrochloride	217–218		29
11-H_2NCH_2(Et)C hydrochloride	253		29
R = $PhOCH_2CH(OH)CH_2NHCH_2$- succinate			28
R = Me_2NCH_2-	(160–162/0.05)		31
hydrochloride	232–233		31
R = $BrCH_2CH_2$-	142–143		32
R = $PhCH_2OCH_2CH_2$-	(245–250/0.1)		32
R = $MeNHCH_2CH_2$-	(177–181/0.1), 70–72		25, 26, 27, 28
	(183–187/0.1)		32
	(173/0.1)		225
hydrochloride	234–235		25, 26, 27, 32 280

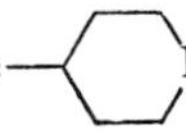

'ABLE XI-7.—(*contd.*)

ubstituents	mp, °C or (bp, °C/torr)	Other data	Refs.
Isomer A	79–80		239
hydrochloride	242–244, 244–246		224, 225, 239
Isomer B			239
hydrochloride	158–160		224, 225, 239
= $MeNHCH_2CH_2$-, 2-Me			
Isomer A hydrochloride	263–265		239
Isomer B hydrochloride	199–200		239
= $Me_2NCH_2CH_2$-	(162–164/0.2)	ir, uv	110, 228, 229, 237
	(170/0.1)		251
	(186–190/0.4), 53–54		224
	(190–192/0.15), (170–172/0.05), 55–57		280
hydrochloride	215–217		24, 228, 229
	218–220		32, 226, 237, 280
oxalate	167–169		91, 210, 251
= $Me_2NCH_2CH_2$-, 5-oxide	100–101	ir	222, 224, 270
succinate	139–140		222, 224
= $Me_2NCH_2CH_2$-, 5,5-dioxide			
hydrochloride	249–252	uv	327
= $Me_2NCH_2CH_2$-, 2-Me	(176–180/0.15)		280
hydrochloride	220		227, 229, 240
	206–208		24, 280
oxalate	189–192		91, 119
= $Me_2NCH_2CH_2$-, 4-Me			
hydrochloride	195–197		226, 227, 229, 240
= $Me_2NCH_2CH_2$-, 2-Et			
hydrochloride	200–201		226, 227, 229, 240
= $Me_2NCH_2CH_2$-, 2-*i*-Pr			
hydrochloride	198–200		226, 227, 229, 240
= $Me_2NCH_2CH_2$-, 2-*n*-Bu			
hydrochloride	98–101		225, 227, 229, 240
= $Me_2NCH_2CH_2$-, 2-$PhCH_2$-			
hydrochloride			229
= $Me_2NCH_2CH_2$-, 2-F			
hydrochloride	200–202		229
Isomer A hydrochloride	229–231		226, 227, 240
somer B hydrochloride	190–194		226, 227, 240
= $Me_2NCH_2CH_2$-, 2-Cl	(178–185/0.1)		280
ydrochloride	244–247		226, 227, 229, 240
ydrochloride·$\frac{1}{2}H_2O$	234–236		24, 280
xalate	215–216		91, 119
= $Me_2NCH_2CH_2$-, 3-Cl			
uccinate	124–126		224
= $Me_2NCH_2CH_2$-, 9-Cl			
ydrochloride	184–185		226, 227, 229, 240
	197–198		224

TABLE XI-7.—(*contd.*)

Substituents	mp, °C or (bp, °C/torr)	Other data	Refs.
R = $Me_2NCH_2CH_2$-, 2-Br			
hydrochloride	260–261, 260–263		226, 227, 229, 240
R = $Me_2NCH_2CH_2$-, 3-Br			
succinate	130–131		224
R = $Me_2NCH_2CH_2$-, 2-MeO			
oxalate	187–189		91, 119
R = $Me_2NCH_2CH_2$-, 6-AcO-			270
R = $Me_2NCH_2CH_2$-, 2-MeS			
oxalate	180–185		91, 119
R = $Me_2NCH_2CH_2$-, 3,8-F_2			
fumarate	218		238
R = $Me_2NCH_2CH_2$-, 2,9-Cl_2			
hydrochloride	233–236		226, 227, 229, 240
R = $PhCH_2N(Me)CH_2CH_2$-	(220–230/0.1)		280
	(210–225/0.15)		24
R = $EtO_2CN(Me)CH_2CH_2$-	(190–205/0.1), 82–84		102, 280
	80–82		26, 27
R = $EtO_2CN(Me)CH_2CH_2$-, 2-Me	(220–225/0.3)		239
R = $Et_2NCH_2CH_2$-	(200–230/0.05)		248
oxalate	174–176		91, 120, 248
R = (pyrrolidine ring, N–Me) CH_2-			
oxalate	150–153		91, 120, 248
R = (pyrrolidine ring, N–Me) CH_2-, 2-Me			
oxalate	170–175		119
R = (pyrrolidine ring, N–Me) CH_2-, 2-Cl			
oxalate	185–187		119
R = (pyrrolidine ring, N–Me)			
fumarate	213–215		91, 120, 248

TABLE XI-7.—(*contd.*)

Substituents	mp, °C or (bp, °C/torr)	Other data	Refs.
R = (piperidino)NCH_2CH_2-	(196–198/0.1)		280
hydrochloride	270–275, 260–262		2, 120, 205, 226 229, 248
hydrochloride·½H_2O	250–251		24, 280
fumarate	193–197		91, 120, 248
R = Ph–(4-piperidino)NCH_2CH_2-			
hydrochloride	230–232		198, 221
R = $HOCH_2$-			
-CH_2–(4-piperidino)NCH_2CH_2-	(235–255/0.01), 50–52		32
R = (piperidino)NCH_2CH_2-, 2-Cl			
fumarate	240–245		91, 119
R = (piperidino)$NCH_2CH(Me)$-			
oxalate	187–189		91, 120
R = (piperidino)$NCH_2CH(Me)$-, 2-Me			
oxalate	170–175		119
R = (piperidino)$NCH_2CH(Me)$-, 2-Cl			
oxalate	211–215		119
R = MeN(piperazino)NCH_2CH_2-	(210–215/0.07)		280
dihydrochloride	255–263		205, 226, 229
	255–257		24, 280
	263		2
R = $HOCH_2$-			
-CH_2N(piperazino)NCH_2CH_2-	(245–255/0.02)		32

TABLE XI-7.—(*contd.*)

Substituents	mp, °C or (bp, °C/torr)	Other data	Refs.
R = O(CH₂CH₂)₂NCH₂CH₂- (morpholino)			
fumarate	165–168		91, 120, 248
R = O(CH₂CH₂)₂N—CH₂CH(Me)- (morpholino)			
fumarate	182–185		91, 120, 248
R = O(CH₂CH₂)₂N—CH₂CHMe-, 2-MeS (morpholino)			
fumarate	156–160		119
R = (1-methyl-2-piperidyl)CH₂-			
hydrochloride	198–201		205, 226, 229
	204–206		2
hydrobromide	210–217		91, 120, 248
R = (1-methyl-2-piperidyl)CH₂-, 2-Me			
hydrobromide	219–225		119
R = (1-methyl-2-piperidyl)CH₂-, 2-Cl			
hydrobromide	245–260		91, 119
R = (1-methyl-2-piperidyl)CH₂-, 2-MeO			
hydrochloride	204–211		91, 119
R = (1-methyl-2-piperidyl)CH₂-, 2-MeS			
hydrochloride	260–261		119

TABLE XI-7.—(*contd.*)

Substituents	mp, °C or (bp, °C/torr)	Other data	Refs.
R = 1-methyl-3-piperidyl (structure: N–Me)			
hydrochloride	191–194		2, 205, 226, 229
fumarate	240–242		91, 120, 248
R = 1-methyl-3-piperidyl (structure: N–Me), 2-Me			
fumarate	215–220		119
R = 1-methyl-3-piperidyl (structure: N–Me), 2-Cl			
fumarate	130–135		119
R = 1-methyl-3-piperidyl (structure: N–Me), 2-MeO			
fumarate	200–207		119
R = 1-methyl-3-piperidyl with CH_2CH_2- (structure: N–Me)			
oxalate	187–189		248
MeN(piperidylidene)=C_{11}	109–110.5		224
hydrochloride	267–272		2, 229
hydrobromide	265–270		91, 120
MeN(piperidylidene)=C_{11}, 2-Me			
hydrobromide	294–297		91, 119
MeN(piperidylidene)=C_{11}, 2-Cl	161–164		91, 119
MeN(piperidylidene)=C_{11}, 2-MeO	120–121		91, 119

TABLE XI-7.—(*contd.*)

Substituents	mp, °C or (bp, °C/torr)	Other data	Refs.
 MeN=C_{11}, 2-MeS	154–155		91, 119
Dibenzo[b,e]*thiepin-11*(*6*H)*-ones* 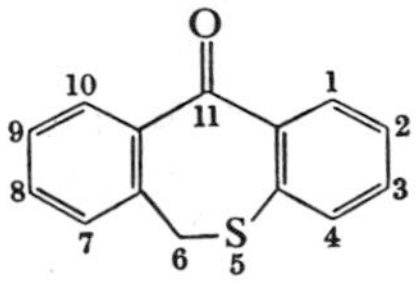			
None	(162–165/0.1), 85–86		228
	(162–165/0.3), 86–88	ir, nmr, uv	281, 283
	(175–180/1), 86–87		226, 237
	84–86		91, 120, 121, 250
	85–86		81
	86–88		234, 282
	88–89		24, 327
oxime	223–225 (dec)		3
$=NOCH_2CH_2NMe_2$	(190–195/0.2)		3
hydrochloride	208–209 (dec)		3
$=NOCH_2CH_2NEt_2$	(198–204/0.1)		3
hydrochloride	136–139		3
$=NOCH_2CH_2CH_2NMe_2$	(196–200/0.2)		3
hydrochloride	199–204		3
5-oxide	97–100		226, 237, 270
5,5-dioxide	127–128		226, 237, 327
2-Me	(167–175/0.2), 119–120	ir, uv	281, 283
	119–121		24, 226, 227, 22 240
	121–122		91, 119, 122, 23
2-Me			
oxime			
$=NOCH_2CH_2NMe_2$			
hydrochloride	174–179		3
$=NOCH_2CH_2NEt_2$	(219–221/0.25)		3
4-Me	109–111		226, 227, 229, 2
2-Et	52–53		226, 227, 229, 2
2-*n*-Pr	86–87		226, 240
2-*i*-Pr	94–95		226, 227, 240
2-*n*-Bu	58–60		226, 227, 229, 2
2-$PhCH_2$	155–156		224, 226, 227, 2
2-CF_3	116–119		91, 122
2-F	101–105		226, 227, 240
2-Cl	(167–168/0.5), 136	ir	240
	(175–181/0.2), 133–134	ir, uv	281, 283
	130–132		24
	134–136		91, 119, 122
	136		226, 227
oxime			
$=NOCH_2CH_2NMe_2$			
hydrochloride	251–252 (dec)		3

TABLE XI-7.—(*contd.*)

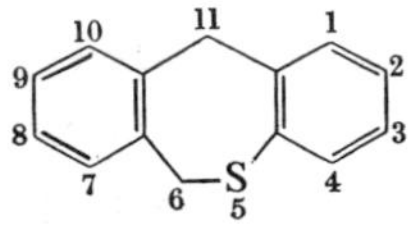

Substituents	mp, °C or (bp, °C/torr)	Other data	Refs.
$=NOCH_2CH_2NEt_2$ A	118–119		3
B	70–71		3
3-Cl	106–108		224
9-Cl	89–90		226, 227, 240
3-Br	116–117		224
2-MeO	89–90		24
	94–96		91, 122
6-AcO-	129–130	ir, uv	270
6-AcO-5,5-dioxide	135–136	ir, nmr, uv	270
2-MeS	92–94		91, 119, 122
3,8-F_2	128–130		238
2,9-Cl_2	135–136		226, 227, 240
6,11-Dihydrodibenzo[b,e]*thiepins*			
None	101–102	ir, uv	224
	103–104		102
	103–105		264
5-Oxide	98–100	nmr	270
11-CH_2CN	124–126		29
1-Me— NNHCONHSO$_2$—			
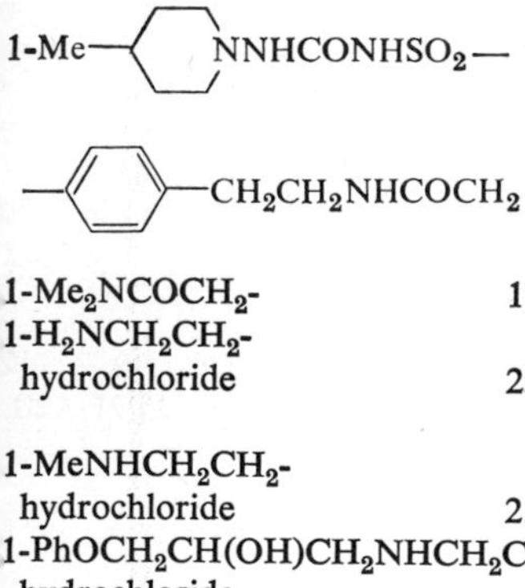 — —$CH_2CH_2NHCOCH_2$	222–223		326
1-Me_2NCOCH_2-	169–171		30
1-$H_2NCH_2CH_2$- hydrochloride	251–252		29
1-$MeNHCH_2CH_2$- hydrochloride	215–216		31
1-$PhOCH_2CH(OH)CH_2NHCH_2CH_2$- hydrochloride			28
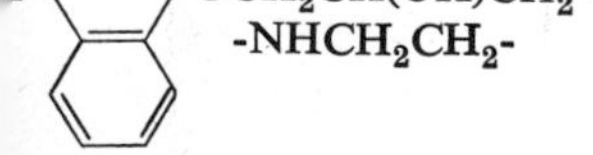 1- -$OCH_2CH(OH)CH_2$- -$NHCH_2CH_2$-	143–144		28
maleate	169–171		28
1-$Me_2NCH_2CH_2$- hydrochloride	201–202		31
1-$MeNHCH_2CH_2CH_2$-	(170–176/0.2)		102, 265
hydrochloride	183–185		102, 265
	173–175		81

TABLE XI-7.—(*contd.*)

Substituents	mp, °C or (bp, °C/torr)	Other data	Refs.
11-$Me_2NCH_2CH_2CH_2$-	(170–175/0.1) (188/1.5)		102
	(175–180/0.2)		231
hydrochloride	198–200		102
hydrochloride·$\frac{1}{2}H_2O$	174		208, 224, 231, 27[illegible]
hydrobromide	75–80		81
hydroiodide	205–206		102, 231
picrate	176–177		102, 231
11-$EtO_2CN(Me)CH_2CH_2CH_2$-	(215–220/0.5)		102, 265
2-F-11-$Me_2NCH_2CH_2CH_2$- hydrochloride	166	ir, uv	208, 224, 272
2-Cl-11-$Me_2NCH_2CH_2CH_2$- hydrochloride	210–213		208, 224, 272
2-Br-11-$Me_2NCH_2CH_2CH_2$- hydrochloride	213–215		208, 272
	216	ir, uv	224
2,9-Cl_2-11-$Me_2NCH_2CH_2CH_2$- hydrochloride·$\frac{1}{2}H_2O$·$\frac{1}{2}$EtOH	121–123		208, 224, 272
6-OH		nmr	270
11-OH	107–108		62, 102, 226, 23[illegible], 327
11-TsO	178–184		252
11-OH-5-oxide	201–202	ir	270
11-OH-5,5-dioxide	157		62, 327
11-AcO-5-oxide	172–174	ir, nmr	270
11-AcO-5,5-dioxide	199–200	ir	270
11-OH-11-Me_2NCOCH_2-	129–130		30
11-OH-11-CH_2CN	119–120		29
11-OH-11-$H_2NCH_2CH_2$	118–119		29
hydrochloride	113–115		29
11-OH-11-$Me_2NCH_2CH_2$ hydrochloride	245		31
11-OH-11-$Me_2NCH_2CH_2CH_2$-	130–131	ir, uv	226, 228, 229, 2[illegible]
	130–132		24, 81, 91, 120, 249
	130–134		137
	131–132		280
11-OH-11-$Me_2NCH_2CH_2CH_2$- 5,5-dioxide	142–144		327
11-OH-11-$Et_2NCH_2CH_2CH_2$-	105–107		120, 249
11-OH-11- $PhCH_2N(Me)CH_2CH_2CH_2$-	108–109		24, 280
11-OH-11-$CH_3CH_2CH(CN)$			29
11-OH-11-$H_2NCH_2CH(Et)$ hydrochloride	253		29

'ABLE XI-7.—(*contd.*)

ubstituents	mp, °C or (bp, °C/torr)	Other data	Refs.
1-OH-11-$Me_2NCH_2CH_2CH_2$-, 2-Me	133–137		24
	139–142		91, 119
	142–143		226, 227, 229, 240
	148–150		280
1-OH-11-$Me_2NCH_2CH_2CH_2$-, 4-Me	164–166		226, 227, 229, 240
-OH-11-$Me_2NCH_2CH_2CH_2$-, 2-Et	138–139		226, 227, 229, 240
-OH-11-$Me_2NCH_2CH_2CH_2$-, 2-*i*-Pr	169–170		226, 227, 229, 240
-OH-11-$Me_2NCH_2CH_2CH_2$-, 2-*n*-Bu	122		226, 227, 229, 240
-OH-11-$Me_2NCH_2CH_2CH_2$-, 2-$PhCH_2$	122–123		226, 227, 229, 240
-OH-11-$Me_2NCH_2CH_2CH_2$-, 2-F	155–156		226, 227, 229, 240
-OH-11-$Me_2NCH_2CH_2CH_2$-, 2-Cl	152–153		226, 227, 229, 240
	154–155		91, 119
	158–160		280
-OH-11-$MeNCH_2CH_2CH_2$-, 3-Cl	155–156		224
-OH-11-$Me_2NCH_2CH_2CH_2$-, 9-Cl	144–145		226, 227, 229, 240
-OH-11-$Me_2NCH_2CH_2CH_2$-, 2-Br	164–165		226, 227, 229, 240
-OH-11-$Me_2NCH_2CH_2CH_2$-, 3-Br	160–161		224
-OH-11-$Me_2NCH_2CH_2CH_2$-, 2-MeO	123–125		91, 119
-OH-11-$Me_2NCH_2CH_2CH_2$-, 2-MeS	137–138		91, 119
OH-11-$Me_2NCH_2CH_2CH_2$-, 3,8-F_2	161–162		238
OH-11-$Me_2NCH_2CH_2CH_2$-, 2,9-Cl_2	166–167		226, 227, 229, 240
OH-11-(pyrrolidine ring, N-Me)CH_2CH_2-			
somer A	192–200		91, 120, 248, 249
somer B	116–120		91, 120, 248, 249
OH-11-(pyrrolidine ring, N-Me)CH_2CH_2-, 2-Me	138–145		119

TABLE XI-7.—(*contd.*)

Substituents	mp, °C or (bp, °C/torr)	Other data	Refs.
11-OH-11-(1-methyl-2-pyrrolidinyl)CH_2CH_2-, 2-Cl	130–132		119
11-OH-11-(1-methyl-3-pyrrolidinyl)CH_2-	(200/0.15)		91, 120, 249
11-OH-11-(piperidino)$NCH_2CH_2CH_2$-	181–183		24
	184–186		91, 280
11-OH-11-Ph-(piperidino)NCH_2CH_2-CH_2-	178–179		198–221
11-OH-11-(piperidino)NCH_2CH_2-CH_2-, 2-Cl	195–197		91, 119
11-OH-11-MeN(piperazino)NCH_2CH_2-CH_2-	204–206		2, 205, 226
11-OH-11-O(morpholino)$NCH_2CH_2CH_2$-	175–177		91, 120, 249
11-OH-11-(piperidino)$NCH_2CH(Me)$-CH_2-	187–190		91, 120, 249
11-OH-11-(piperidino)$NCH_2CH(Me)$-CH_2-, 2-Cl	163–165		119
11-OH-11-O(morpholino)$NCH_2CH(Me)$-CH_2-	163–165		91, 120, 249
11-OH-11-O(morpholino)$NCH_2CH(Me)$-CH_2-, 2-MeS	184–190		119
11-OH-11-(1-methyl-2-piperidyl)CH_2CH_2-	175–180		91, 120, 249
	188–189		2, 205, 226
11-OH-11-(1-methyl-2-piperidyl)CH_2CH_2-, 2-Me	188–189		119

TABLE XI-7.—(*contd.*)

Substituents	mp, °C or (bp, °C/torr)	other data	Refs.
1-OH-11- (N-Me ring) CH_2CH_2-, 2-Cl	Oil		91
1-OH-11 (N-Me ring) CH_2CH_2-, 2-MeO	141–142		91, 119
1-OH-11- (N-Me ring) CH_2CH_2-, 2-MeS	132–133		119
1-OH-11- (N-Me ring) CH_2-	167–170		2, 205, 226
	170–175		91, 120, 249
1-OH-11- (N-Me ring) CH_2-, 2-Me	157–162		119
1-OH-11- (N-Me ring) CH_2-, 2-MeO	175–180		119
-OH-11-MeN (ring) —	184–187		91, 120, 249
	186–186.5		2
-OH-11-MeN (ring) —, 2-Me	181–183		91, 119
-OH-11-MeN (ring) —, 2-Cl	182–184		91, 119
-OH-11-MeN (ring) —, 2-MeO	182–185		91, 119
-OH-11-MeN (ring) —, 2-MeS	178–180		91, 119
-OH-11-$PhCH_2OCH_2CH_2CH_2$-	76–77		32

TABLE XI-7.—(*contd.*)

Substituents	mp, °C or (bp, °C/torr)	Other data	Refs.
11-$Me_2NCH_2CH_2O$-	60–61.5		203, 204, 224
oxalate	188–190 (dec)		203, 204, 224
fumarate	123–125		203, 204, 224
11-$Me_2NCH_2CH_2CH_2O$-			
fumarate·H_2O	148–149		203, 204, 224
11-$Et_2NCH_2CH_2O$-,fumarate	159–160		203, 204, 224
11-(piperidino)NCH_2CH_2O,-fumarate	159.5–161.5		203, 204, 224
11-Cl	81–82		102, 231
11-NH_2	149–150		102, 230
11-(phthalimido)N—	133–136		102, 230
11-$Et_2NCH_2CH_2NH$- dihydrochloride	208–213 (dec)		102, 230
11-(piperidino)N- hydrochloride	198–200		102, 230
11-MeN(piperazino)N-			
maleate·$CH_3C(=O)CH_3$	182–184		102, 230
11-N(pyridinium)—CH=N—N(piperazino)N-			329
11-CN	130–132	ir	102

Miscellaneous Dibenzo[b,e]*thiepins*

Compound	mp, °C	Other data	Refs.
bis(6,11-dihydrodibenzo[b,e]thiepin-11-yl) ether (structure)	204–206	ir	102

TABLE XI-7.—(*contd.*)

Substituents	mp, °C or (bp, °C/torr)	Other data	Refs.
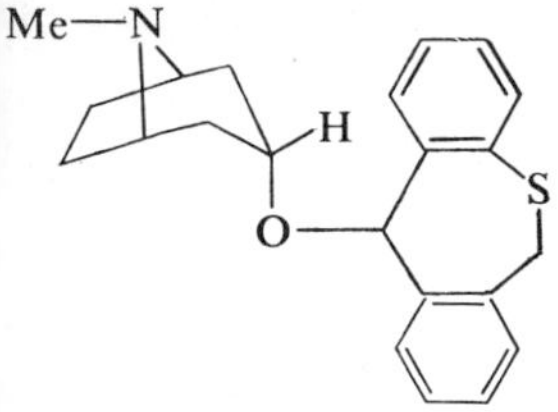 [structure: Me—N tropane ether, O-linked, with H]	(190–200/0.02)		252
oxalate	207–210		252
succinate	188–190		252
maleate	180–182		252
fumarate	234–236		252
methanesulfonate	207–209		252
citrate	168–169		252
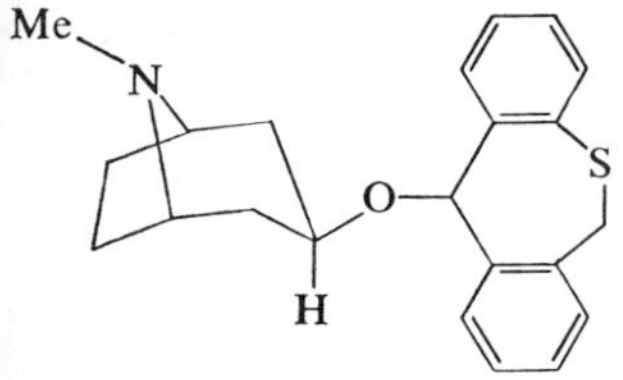 [structure: Me—N tropane ether, O-linked, with H] (pseudotropine)			
hydrochloride	273–278		252
maleate	175–177		252
succinate	162–165		252
fumarate	207–209		252

TABLE XI-8. Dibenzo[*b*,*f*]thiepins (R.I. 8637)

Substituents	mp, °C or (bp, °C/torr)	Other data	Refs.
	Dibenzo[b,f]*thiepins* [structure, positions numbered 1–11, S at 5]	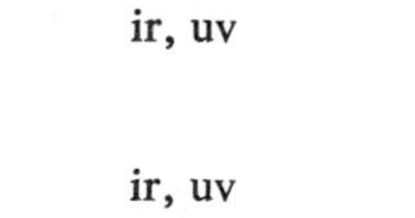	
None	86–87		84, 318
	87–88	ir, esr rad. anion, mass spectrum, uv	115, 262, 308
	88		116
	89–90	ir, uv	15, 16
-F	82–83		111
-Cl	95–96		194
-Cl	78–79	ir, uv	111, 194
-Cl	95–97	ir, uv	194
-Cl	43–45		194
)-Cl	91–93	ir, uv	116
-Br	93–95		111

TABLE XI-8.—(*contd.*)

Substituents	mp, °C or (bp, °C/torr)	Other data	Refs.
10-Br	83–84	ir, uv	116
2-MeO	105–106.5		196
2-MeS	89–91	ir, uv	196
2-$MeSO_2$	138–140	ir, uv	196
2-NO_2	110		151
	111–112	ir, uv	1
2-NH_2	127–128	ir, uv	1
2-CN	160–161	ir, uv	112
2-CO_2H	233–235	ir, uv	112
2-CF_3	(137/0.3)	ir, uv	197
10-Ph	150	ir, uv	15
2-Me	76.5–77.5	esr rad. anion	308
10-Me	(114/0.001), 51–52		94, 95
10-$BrCH_2$-	83–85		94
10-H_2NCH_2-			
hydrochloride	228–231		97
10-$MeNHCH_2$-			
hydrochloride·H_2O	234–237		97
	228–231		94
10-$EtNHCH_2$-			
hydrochloride	234–237		94
10-Me_2NCH_2-	112		97
	114–116		94
10-(pyrrolidino)NCH_2-	37		94
10-MeN(piperazino)NCH_2-			
dihydrochloride	225–228		94, 97
10-$HOCH_2CH_2N$(piperazino)NCH_2-			
dihydrochloride	235–242		94, 97
10-*tert*-BuC(=O)OCH_2CH_2N(piperazino)NCH_2-			
dihydrochloride	199–200		94, 97
10-Et	(131/0.015)		94
10-MeNHCH(Me)-			
hydrochloride	251–252		94, 97
10-Me_2NCH(Me)-	(160/0.012)		94, 97
10-(pyrrolidino)NCH(Me)-	84–85		97

TABLE XI-8.—(*contd.*)

Substituents	mp, °C or (bp, °C/torr)	Other data	Refs.
0-MeN NCH(Me)- 			
dihydrochloride	245–249		97
0-CH_2CN	(165–172/0.01)		95
0-CH_2CO_2H	169–171		95
0-Me_2NCOCH_2-	134–135		95
0-$Me_2NCH_2CH_2$-	(140/0.004)		95
hydrochloride	224		95
0-$Me_2NCH_2CH_2CH_2$-	(172–177/0.22)		331
0-HN $NCH_2CH_2CH_2$- 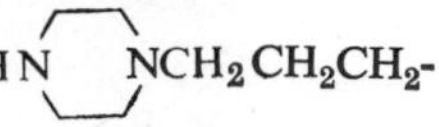			331
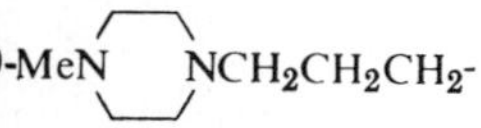 0-MeN $NCH_2CH_2CH_2$-			331
hydrochloride			331
0-HC(=O)N $NCH_2CH_2CH_2$- 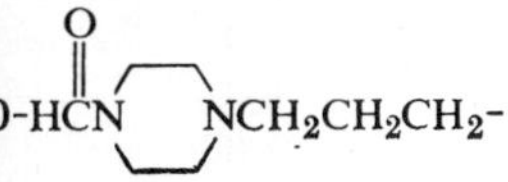			331
0-$HOCH_2CH_2N$ N-			331
0-$AcOCH_2CH_2N$ N- $CH_2CH_2CH_2$- hydrochloride			331
-$HOCH_2CH_2OCH_2CH_2N$ N- $CH_2CH_2CH_2$-			331
- CH_2- N Me			331
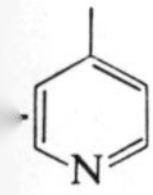 - N	147		85
methobromide			85
-$CH_2CH(Me)CO_2H$			96

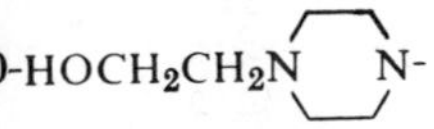

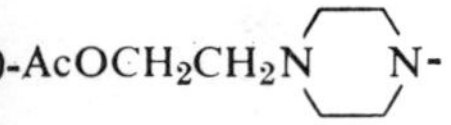

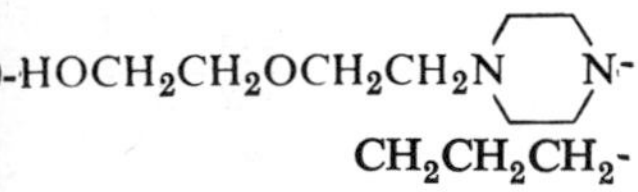

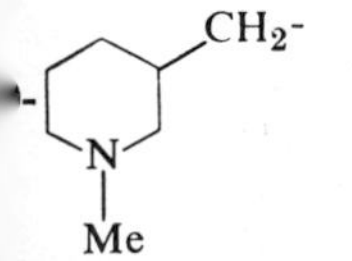

TABLE XI-8.—(*contd.*)

Substituents	mp, °C or (bp, °C/torr)	Other data	Refs.
10-$Me_2NCH_2CH(Me)CH_2$-			96, 331
hydrobromide			331
10-$(EtO_2C)_2C(Me)CH_2$			95
10-$Me_2C(OH)$		nmr	182
1,4-Me_2	81–81.5	esr, rad. anion	308
2,4-Me_2	39–40	esr, rad. anion	308
2,8-Cl_2	164–166	ir, uv	194
2-NO_2-8-Me	117		151
2-Cl-10-Me	68		94
2-Cl-10-$BrCH_2$-	111–112		94
2-Cl-10-Me_2NCH_2-	76–77		94, 97
2-Cl-10-(piperidino)NCH_2-	118		94, 97
2-Cl-10-$Me_2NCH_2CH_2CH_2$-			331
hydrochloride			331
2-NO_2-10-CO_2H	248		151
2-NO_2-10-CO_2Me	156		151
2-MeO-10-$Me_2NCH_2CH_2CH_2$-			331
citrate			331
2-CF_3-10-$Me_2NCH_2CH_2CH_2$-			331
hydrochloride			331
2,10-Me_2	(125–128/0.01)		94
2-Me-10-Me_2NCH_2- salt	245–248		94, 97
2-Cl-11-Me	78		94
2-Cl-11-$BrCH_2$-	114		94
2-Cl-11-Me_2NCH_2-	116		97
2-Cl-11-MeN(piperazino)NCH_2-	111–112		94, 97
2-Cl-11-$Me_2NCH_2CH_2CH_2$-			331
hydrochloride			331
2-MeO-11-Me	(153–155/0.01)		94
2-MeO-11-Me_2NCH_2-	(151/0.01), 92.5		94, 97
hydrochloride	230–231		94
2-MeO-11-$Me_2NCH_2CH_2CH_2$-			331
citrate			331
2-CF_3-11-$Me_2NCH_2CH_2CH_2$-			331
hydrochloride			331
3-Cl-11-Me	(140/0.001)		94
3-Cl-11-$BrCH_2$-	118		94
3-Cl-11-Me_2NCH_2-	70		94, 97

TABLE XI-8.—(*contd.*)

Substituents	mp, °C or (bp, °C/torr)	Other data	Refs.
3-Cl-11- 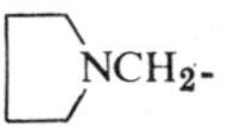 NCH$_2$-	(172–175/0.001)		94, 97
hydrochloride	265–268 (dec)		94, 97
10,11-Me$_2$	94–95	esr rad. anion	308
2-NO$_2$-8-Me-10-CO$_2$H	276		151
2-NO$_2$-8-Me-10-CO$_2$Me	183		151
2-Cl-8-Me-10-Me$_2$NCH$_2$CH$_2$CH$_2$-			331
2-Me-8-Cl-10-Me$_2$NCH$_2$CH$_2$CH$_2$-			331
Dibenzo[b,f]*thiepin-5-oxides* 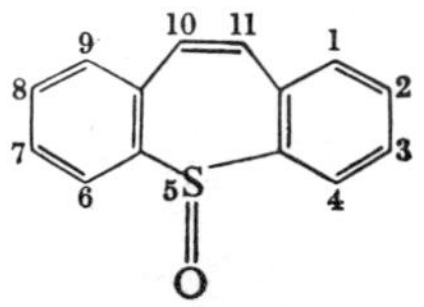			
None		esr rad. anion	301
,4-Me$_2$		esr rad. anion	301
,4-Me$_2$		esr rad. anion	301
10,11-Me$_2$		esr rad. anion	301
Dibenzo[b,f]*thiepin-5,5-dioxides* 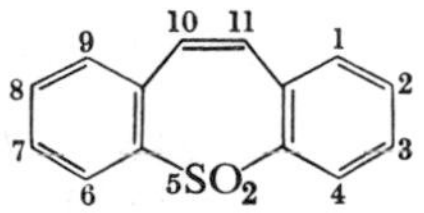			
None	168–175	ir, uv	116
	171–172	ir, uv	16
-Cl	131–132	ir, uv	111
10-Br	179–180	ir, nmr	293
 10- N-	210–212		293
10-MeO			293
Dibenzo[b,f]*thiepin-*$\Delta^{11(10H)}$- 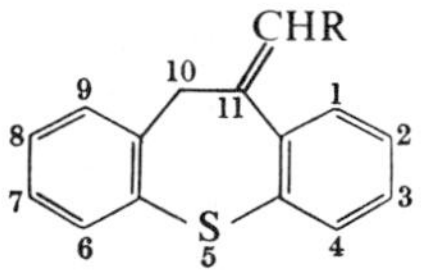			
= H			97
= Me			97
= Me$_2$NCH$_2$CH$_2$-			
hydrochloride	220–225		114
	225	ir, uv	115

TABLE XI-8.—(*contd.*)

Substituents	mp, °C or (bp, °C/torr)	Other data	Refs.
	Dibenzo[b,f]*thiepin-10(11*H)-*one*	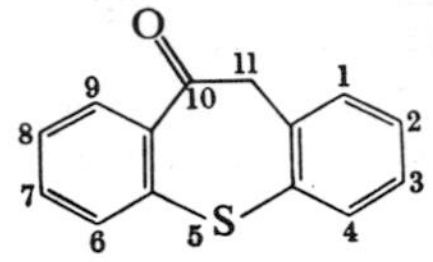	
None	68		94, 95, 97
	68–70	ir, mass spectrum	262
	72–73		115, 128, 219
	73–74		112
		nmr	78
oxime	143	ir	115
6-oxide	184–187	ir, uv	116
		nmr	78
1-Cl	129–130		194, 215, 227
2-Cl	141–143		94, 95, 215, 227
	Sublimes 135°/1 mm		194
2-MeO	131.5		94
3-Cl	143–145	ir, uv	94, 194
	148–150		215, 227
4-Cl	160–162		215, 227
	163–164	ir, uv	194
6-Cl	124–125		226, 227
	125.5–127	ir	194
7-Cl	132–133.5		194, 215, 227
7-CO_2H	215–219	ir, uv	197
7-CF_3	(165–170/1), 89–91	ir, uv	197, 278
8-F	104–106	ir, uv	111, 215, 227
8-Cl	119–120	ir, uv	94, 95
	125–126		111, 215, 220,
8-Br	113	ir	111, 215, 227
	110		112
8-MeO	97.5–98		196
	100–101		128
8-MeS	88–89	ir, uv	196
8-$MeSO_2$	190–192	ir, uv	196
8-NO_2	174–175	ir, uv	1, 278
8-NH_2	186.5–188	ir, uv	1, 207
8-NHAc	215–217		1, 207
8-CN	192–194	ir, uv	1
8-CO_2H	265–267	ir, uv	112
8-CO_2Et	130–131	ir, uv	112
8-Me	65–68		94
	68–69	ir	196, 278
8-*tert*-Bu	(184–186/3)		196, 278
8-CF_3	65–67	ir, uv	197
11-Br	105–106		99
	109–110	ir, uv	116, 213
	68–70	ir, uv	116, 213

ABLE XI-8.—(*contd.*)

bstituents	mp, °C or (bp, °C/torr)	Other data	Refs.
 -MeN⟨piperazine⟩N-	101.5		99
maleate	151–154		116, 213
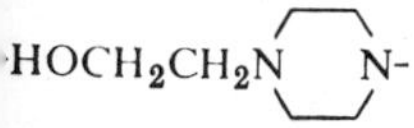 -$HOCH_2CH_2N$⟨piperazine⟩N-			
dihydrochloride	174.5		99
-Me	(148–150/0.002), 62–64		94, 95
	(160–165/0.1)		218
 ⟨piperidine⟩NCH_2-			
maleate	138–140		232
Et	(152–154/0.005)		94
CH_2CH_2CN	96–98		232
$Me_2NCH_2CH_2$-	(160–161/0.005)		96
hydrochloride	206–208		96
$MeNHCH_2CH_2CH_2$-			
hydrochloride	201–203		98
$MeN(CN)CH_2CH_2CH_2$-	104–106		98
$Me_2NCH_2CH_2CH_2$-	(165–168/0.005)		96, 98
hydrochloride	122–124		96
$Et_2NCH_2CH_2CH_2$-	(187–189/0.01)		96
hydrochloride	197–200		96
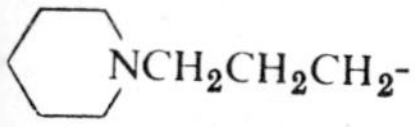 ⟨piperidine⟩$NCH_2CH_2CH_2$-	(193–196/0.01)		96
hydrochloride	203–206		96
$Me_2NCH_2CH(Me)CH_2$-	(181–182/0.005), 85–86		96
O_2, 8-Me	161		151
xime	191		151
Cl_2	164–165		215
	165–167	ir, uv	194
l-11-MeN⟨piperazine⟩N- 	167–178		99
l-11-$HOCH_2CH_2N$⟨piperazine⟩N- 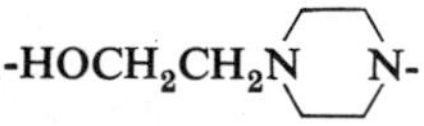			
ihydrochloride	195–197		99
l-11-Me	125–128		94, 95
l-11-CH_2CH_2CN	121		95
eO-11-Me	(116–119/0.02)		94
eO-11-MeN⟨piperazine⟩N- 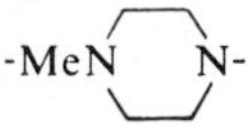	122–124.5		99

TABLE XI-8.—(*contd.*)

Substituents	mp, °C or (bp, °C/torr)	Other data	Refs.
2-MeO-11-HOCH$_2$CH$_2$N(piperazine)N-	140–143		99
2-MeO-11-MeNHCH$_2$CH$_2$CH$_2$-			
hydrochloride	209–210		98
2-MeO-11-Me$_2$NCH$_2$CH$_2$CH$_2$-	(185–189/0.005)		98
hydrochloride	186–188		98
3-Cl-11-MeN(piperazine)N-	136–139		99
3-Cl-11-HOCH$_2$CH$_2$N(piperazine)N-	147–151		99
3-Cl-11-Me	94		94
3-Cl-11-Me$_2$NCH$_2$CH$_2$-	91–92		96
3-Cl-11-Me$_2$NCH$_2$CH$_2$CH$_2$-	83–84		96
8-Cl-11-MeN(piperazine)N-	153–157		99
8-Cl-11-HOCH$_2$CH$_2$N(piperazine)N-	144–148		99
8-Cl-11-Me	113		94
8-Cl-11-Me$_2$NCH$_2$CH$_2$-	108–109		96
8-Cl-11-CH$_2$CH$_2$CN	164		95
8-Cl-11-Me$_2$NCH$_2$CH$_2$CH$_2$-	(180–183/0.01), 62–63		96
8,11-Me$_2$	(145–151/0.06)		94
11-Me-11-Me$_2$NCH$_2$CH$_2$-	(160–163/0.01)		255
hydrochloride	170–174		255
11-Me-11-Me$_2$NCH$_2$CH$_2$CH$_2$-	(165–170/0.002)		255
hydrochloride	185–190		255
2,8-Cl$_2$-11-Me$_2$NCH$_2$CH$_2$CH$_2$-	(197–200/0.002)		96
hydrochloride	168–172		96
2-Cl-8-Me-11-MeNHCH$_2$CH$_2$CH$_2$-			
hydrochloride	252–255		98
2-Cl-8-Me-11-MeN(CN)CH$_2$CH$_2$CH$_2$-	84–86		98
2-Cl-8-Me-11-Me$_2$NCH$_2$CH$_2$CH$_2$-			
hydrochloride	168–172		98
2-Cl-11-Me-11-CH$_2$CH$_2$CN			95
2-Cl-11-Me-11-HO$_2$CCH$_2$CH$_2$-	198–199		95

TABLE XI-8.—(*contd.*)

ubstituents	mp, °C or (bp, °C/torr)	Other data	Refs.
10,11-Dihydrodibenzo[b,f]*thiepin-10,11-diones* 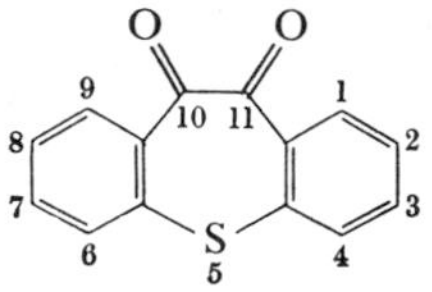			
Jone	135–136	ir, uv	116
mono oxime	222–224	ir, uv	116
,6-Dioxide	278–270	ir	116
10,11-Dihydrodibenzo[b,f]*thiepins* 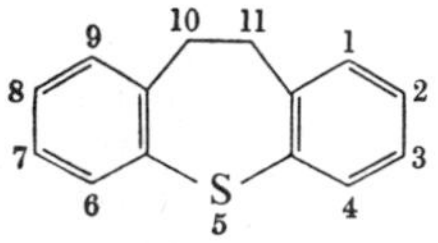			
Tone	(125–130/0.1)		116
	143–144		115
)-$HO_2CCH_2CH_2$-	109–111		95
)-$Me_2NCOCH_2CH_2$-	108–109		95
)-$Me_2NCH_2CH_2CH_2$-	(169/0.01)		95
)-$Et_2NCH_2CH_2CH_2$-	(175/0.05)		95
)-MeN 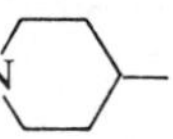	75		85
hydrogen maleate	159.5–160.5		116, 206
Cl-10-$HO_2CCH_2CH_2$-	178		95
Cl-10-$Me_2NCOCH_2CH_2$-	101		95
Cl-10-$Me_2NCH_2CH_2CH_2$-	(182–184/0.02)		95
Cl-10-$MeNHCH_2CH_2CH_2$-	(165/0.001)		95
Cl-10-$Me_2NCH_2CH_2CH_2$-	(149–150/0.001)		95
Cl-10-$HO_2CCH_2CH_2$-	184		95
Cl-10-$Me_2NCH_2CH_2CH_2$-	(175/0.01)		95
Cl-10-MeN 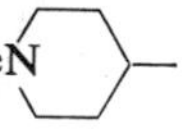			
maleate	195		85
MeO-10-MeN 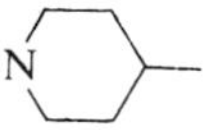			
hydrogen maleate	176–177.5		196, 206
-Me-10-$HO_2CCH_2CH_2$-			95
-Me-10-$Me_2NCH_2CH_2CH_2$-	(155–157/0.001)		95
,11-Br_2	144–146	ir, uv	116
Cl-10-Me-10-$Me_2NCH_2CH_2CH_2$-	(156–157/0.003)		95
-OH	99–100	mass spectrum, nmr	115, 219, 262
Cl-10-OH	118–121		194, 2 15

TABLE XI-8.—(*contd.*)

Substituents	mp, °C or (bp, °C/torr)	Other data	Refs.
2-Cl-10-OH	106.5–107.5		194, 215
3-Cl-10-OH	105–107		194, 215
4-Cl-10-OH	118–120		194, 215
6-Cl-10-OH	87–88°		194, 215
7-Cl-10-OH	95.5–96.5		194, 215
8-Me-10-OH	70–72		196, 278
8-CF_3-10-OH	113–114		197
8-*tert*-Bu-10-OH	(180–182/2 mm)		196, 278
8-CN-10-OH	95–97	ir, uv	112
10-OH-8-CO_2H	214–216		112
10-OH-8-CO_2Et	100–101	ir, uv	112
8-F-10-OH	77–78		111, 215
8-Cl-10-OH	84–85		111, 215, 220, [illegible]
8-Br-10-OH	107		111, 215
8,10-$(OH)_2$	139–140		1
8-MeO-10-OH	93–94		196
8-MeS-10-OH	117–118		196
8-$MeSO_2$-10-OH	127–128		196
8-NO_2-10-OH	191–192		1, 278
8-NH_2-10-OH	123–123.5		1, 207
hydrochloride	196–198		1
8-NHAc-10-OH	227–229		1, 207
8-CF_3CONH-10-OH·½H_2O	185–187		1, 207
10-Me-10-OH	109–111		97
10-Et-10-OH			97
10-OH-10-$Me_2NCH_2CH_2CH_2$-	115	ir, uv	114, 115
10-OH-10-(4-pyridyl)	248		85
10-OH-11-Me	(142–144/0.05) (178–185/0.8)		94, 95 116, 218
10-OH-11-(piperidino)CH_2-	143–144	ir	115, 232
10-OH-11-Et	(163–165/0.005)		94
10-OH-11-NH_2	195–196	ir, uv	116
hydrochloride·H_2O	238–240		116
10-OH-11-(4-methylpiperazin-1-yl)—	148–152, 152–154		116, 212
maleate	171		116, 212
dihydrochloride·H_2O	185–190		116, 212
2,8-Cl_2-10-OH	120–121		194, 215
2-MeO-10-OH	167.5		94

BLE XI-8.—(*contd.*)

bstituents	mp, °C or (bp, °C/torr)	Other data	Refs.
-MeO	(170/0.6)	ir, mass spectrum, nmr	262
-*i*-PrO	(180–185/0.6)	ir, mass spectrum, nmr	262
-$Me_2NCH_2CH_2O$-	(188/0.4)		115, 219
hydrochloride	70–71		115, 219
picrate	163–164		115, 219
maleate	112–113		115, 219
Cl-10-$Me_2NCH_2CH_2O$-	(212–215/1.5)		194, 195
hydrogen maleate	162–165		194, 195
Cl-10-$Me_2NCH_2CH_2O$-			
hydrogen maleate	123–124		194, 195
Cl-10-$Me_2NCH_2CH_2O$-			
hydrogen maleate	161–163		194, 195
Cl-10-$Me_2NCH_2CH_2O$-	(220–223/1.5)		194, 220
hydrogen maleate	123.5–125		194, 216, 220
Cl-10-$Me_2NCH_2CH_2O$-	(214–218/1.2)		194, 220
hydrogen maleate	114–116		194, 216, 220
Cl-10-$Me_2NCH_2CH_2O$-	(194–198/0.4)	ir, uv	111, 220
naleate	108–110		111, 216, 220
Cl-10-$Me_2NCH_2CH_2CH_2O$-	(198–200/0.5)		111, 220
naleate	102–103		111, 220
MeO-10-$Me_2NCH_2CH_2O$-			
ıydrogen maleate	182–184.5		196
Cl	84.5		115, 116, 276
Cl-5,5-dioxide	160		271
0-Cl_2	103.5–104		194, 215, 227
0-Cl_2	115–117		227
	124–124.5		194, 215, 216
0-Cl_2	106–108		194, 215, 227
0-Cl_2	112–114		194, 215, 227
0-Cl_2	104–105		194, 215, 227
0-Cl_2	104–105		194, 215, 227
Ie-10-Cl	127–128		196, 278
F_3-10-Cl	79–80		197
N-10-Cl	133–135		112
Cl-8-CO_2H	203–205	ir, uv	112
-10-Cl	98		111
0-Cl_2	105–106		111, 215, 227
r-10-Cl	105–106		111, 215, 227
IeO-10-Cl	116–118		196
IeS-10-Cl	106–108		196
$IeSO_2$-10-Cl	120–121		196
O_2-10-Cl	137–138		1, 278
=S=O-10-Cl	107–111	ir, uv	1
O ‖ $HCCF_3$-10-Cl	158–160	ir, uv	1, 207

TABLE XI-8.—(*contd.*)

Substituents	mp, °C or (bp, °C/torr)	Other data	Refs.
10-Br-10-BrCH$_2$-			97
10-Br-10-MeCHBr-			97
10-Cl-11-Me	(140/0.05), *trans*-118–119		94, 95
	trans-121–122		116, 218
10-Cl-11-Me	68		94
10,11-Br$_2$-5,5-dioxide	195–197		293
2-MeO-10-Cl-11-Me	80		94
8,11-Me$_2$-10-Cl	(138–141/0.05)		94
10-NH$_2$	56–57		128
	(162–164/0.8)		113, 115, 276
hydrochloride	256–258		113, 115, 276
	249–250		128
10-HC(=O)NH	153–155		113, 115
10-AcNH-	198		113, 115
10-MeNH-,hydrochloride	205–207		113, 115, 276
10-Et$_2$NCH$_2$CH$_2$NH-	(180–185/0.4)		115, 276
maleate	140–141		115, 276
10-Me$_2$N	98		115, 276
hydrochloride	207–208		115, 276
10-(piperidine ring)N-	98–99	ir	115, 276
hydrochloride	225–230		115, 276
10-HO–(piperidine ring)N-	145–146		116
hydrogen maleate	173–174		116
10-(4-Ph-4-EtO$_2$C-piperidine ring)N-	197–198		115, 276
hydrochloride	198–200		115, 276
10-HN(piperazine ring)N-	103–105, 104		116, 216, 217,
maleate	188–190, 190–191		116, 216, 217,
10-EtO$_2$C-N(piperazine ring)N-	112–114		116, 209, 216
maleate	192–193		116, 209, 216
methanesulfonate	211–212		116, 209

.BLE XI-8.—(*contd.*)

bstituents	mp, °C or (bp, °C/torr)	Other data	Refs.
O ‖ HCN N-	135–136		116, 209, 210, 216
ıydrogen maleate	162–164		116, 209, 210, 216
AcN N-	129–131		116, 209, 210, 216
MeN N-	135–136, 134–135		115, 116, 214, 216
lextro base	106–107	$[\alpha]_D^{20} + 28°$	116
evo base	105–108	$[\alpha]_D^{20} - 10°$	116
lihydrochloride	208–212		115, 214, 276
ıaleate	157–158		116, 216
extro maleate	160–164	$[\alpha]_D^{20} + 30°$	116
umarate	199–201		116
ıono methiodide	220–222		116
ydrogen dibenzoyl tartrate·H_2O	149–152	$[\alpha]_D^{20} - 30°$	116
i-*p*-tolyltartrate	180–181	$[\alpha]_D^{20} + 82°$	116
MeN N-5-oxide	161		243
ıaleate	165–167		116
ıaleate·EtOH	135–140		116
MeN N-5,5-dioxide	140		243, 271
ːtN N-	85–86		116, 211, 216, 276
ıaleate	150–151		116, 211, 216, 276
IOCH$_2$CH$_2$N N-	108–110		116, 216, 276
aleate	129–130		116, 216, 276
IOCH$_2$CH$_2$CH$_2$N N-	138		116
aleate	156		116
ιcOCH$_2$CH$_2$CH$_2$N N-			
aleate	171–173		194
s-hydrogen maleate	137–138		116

TABLE XI-8.—(*contd.*)

Substituents	mp, °C or (bp, °C/torr)	Other data	Refs.
10-$PhCH_2N$ 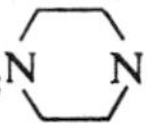 ⟨piperazine⟩ N-	148–149		116
maleate	210–211		116
10-$PhCH_2N$  ⟨piperazine⟩ N-,5,5-dioxide	160		243
10-PhN  ⟨piperazine⟩ N-	185–186		116
mono methanesulfonate	213–214		116
10-ONN  ⟨piperazine⟩ N-	128–129		116
10-H_2NN  ⟨piperazine⟩ N-	157–158		116
maleate	167–168		116
10-MeN 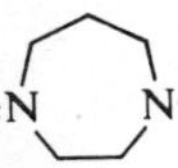 ⟨homopiperazine⟩ N-	82		116, 277
maleate	142		216, 217
3-Cl-10-$Et_2NCH_2CH_2NH$- *bis*-hydrogen maleate	103–104		194
8-Cl-10-HO—⟨piperidine⟩ N- 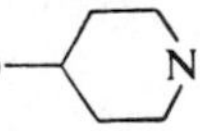	97–99		111
hydrogen maleate	188–189		111
1-Cl-10-MeN  ⟨piperazine⟩ N-	113–114	ir, uv	194, 215, 277
maleate	145, resolidifies, then 173		194, 214, 277
2-Cl-10-MeN  ⟨piperazine⟩ N-,maleate	170.5–171		194, 215, 216,
3-Cl-10-MeN  ⟨piperazine⟩ N-,maleate	156–158		194, 215, 216,
4-Cl-10-MeN 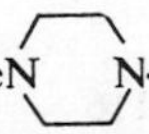 ⟨piperazine⟩ N-	113–115		194, 215, 277
maleate	186–188		194, 215, 277
6-Cl-MeN 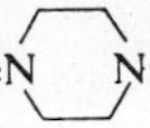 ⟨piperazine⟩ N-	115–117.5		194, 215, 277
maleate	163–163.5		194, 215, 216,

BLE XI-8.—(*contd.*)

bstituents	mp, °C or (bp, °C/torr)	Other data	Refs.
CF_3-10-MeN⟨piperazine⟩N-			278
Cl-10-MeN⟨piperazine⟩N-	138–139		194, 215, 277
naleate	183.5–185		194, 215, 216, 277
Cl-10-			
$HOCH_2CH_2CH_2N$⟨piperazine⟩N-	108–111		194
naleate	139–140		194
Me-10-MeN⟨piperazine⟩N-			
is-hydrogen maleate	160–162		196, 278
CF_3-10-MeN⟨piperazine⟩N-			
naleate	137–138		197
CF_3-10-EtO_2CN⟨piperazine⟩N-			
ydrogen maleate	191–192		197
rt-Bu-10-MeN⟨piperazine⟩N-			
is-hydrogen maleate·H_2O	92–94		196, 278
N-10-MeN⟨piperazine⟩N-	171–173	ir, uv	112
aleate	187–189		112
O_2H-10-MeN⟨piperazine⟩N-	225–235	ir, uv	112
a salt·H_2O	223–228		112
hydrochloride·H_2O	186–190		112
aleate·H_2O	135–140		112
$_2NCH_2$-10-MeN⟨piperazine⟩N-			
is-hydrogen maleate·H_2O	103–107		112

TABLE XI-8.—(*contd.*)

Substituents	mp, °C or (bp, °C/torr)	Other data	Refs.
8-F-10-MeN(piperazine)N-	144		111
maleate	175–177		111
8-Cl-10-MeN(piperazine)N-	99–100	ir, uv	111, 215, 277
hydrochloride·H_2O	130–135		111, 215, 277
dihydrochloride	230		215, 277
	190		111, 216
maleate	206–207		111
8-Cl-5-oxide maleate	186–188		111
8-Br-10-MeN(piperazine)N-	118–119		111, 215, 277
maleate	203–204		111, 215, 277
8-MeO-10-NH_2	60–62		128
hydrochloride	250		128
8-MeO-10-MeN(piperazine)N-	80–82		196, 278
maleate	207–209		196, 278
8-MeS-10-MeN(piperazine)N-			
maleate	160–161		196, 278
8-$MeSO_2$-10-MeN(piperazine)N-			
maleate	141–143		196, 278
8-Me_2NSO_2-10-MeN(piperazine)N-			278
8-NO_2-10-MeN(piperazine)N-	160–163	ir, uv	1
maleate	187.5–189.5		1
8-NO_2-10-MeN(piperazine)N-5-oxide·$\frac{1}{2}H_2O$	211–212.5		1
maleate·$\frac{1}{2}H_2O$	195–198		1
8-NH_2-10-MeN(piperazine)N-	154.5–156	ir, uv	1, 202, 207
tris-hydrochloride	202–205		1, 202, 207

TABLE XI-8.—(contd.)

Substituents		mp, °C or (bp, °C/torr)	Other data	Refs.
-Cl-10-$PhCH_2N$ ⟨piperazine⟩ N- 		150–151		111
maleate		204–205		111
·Cl-10-Ph  N ⟨piperazine⟩ N-		135–136		111
maleate		182–184		111
-Me-10-MeN ⟨piperazine⟩ N- 		119		116, 218
maleate		145–147, 146		116, 218
8-Cl_2-10-MeN ⟨piperazine⟩ N- 				
maleate		180–181		194, 216, 277

1,2,3,4,4a,11a-Hexahydrodibenzo[b,f]*thiepins*

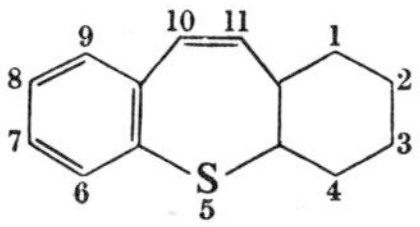

Substituents		mp, °C or (bp, °C/torr)	Other data	Refs.
Cl-10-*p*-ClC_6H_4S-	*cis*	143–145		73
	trans	134–136		73

1,2,3,4,4a,11a-Hexahydrodibenzo[b,f]*thiepin-10(11*H*)ones*

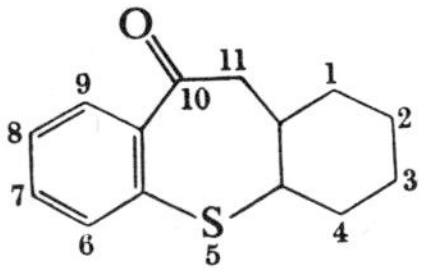

Substituents		mp, °C or (bp, °C/torr)	Other data	Refs.
Cl	*cis*	117–119		73
	trans	90–92		73

1,2,3,4,4a,10,11,11a-Octahydrodibenzo[b,f]*thiepins*

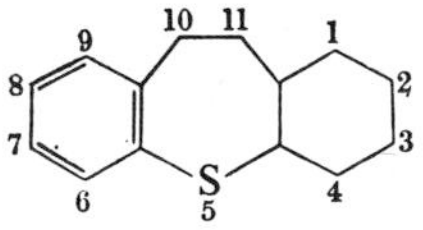

Substituents		mp, °C or (bp, °C/torr)	Other data	Refs.
l-10-OH	*cis*	121–123	ir	73
	trans	123–125	ir	73
-Cl_2	*cis*			73
	trans			73
-10 MeN ⟨piperazine⟩ N- 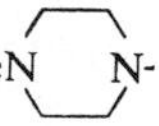	*cis*	140–142	ir, mass spectrum, nmr	73
ydrochloride		262–264		73
	trans		ir, mass spectrum, nmr	73
ihydrochloride		250–252		73

TABLE XI-8.—(*contd.*)

Substituents	mp, °C or (bp, °C/torr)	Other data	Refs.
	Miscellaneous Dibenzo[b,f]*thiepins*		
Compound			
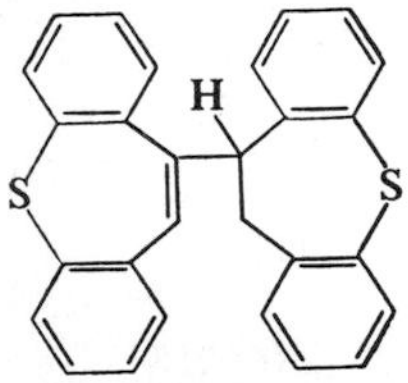 [structure]	227	ir, mass spectrum uv,	116
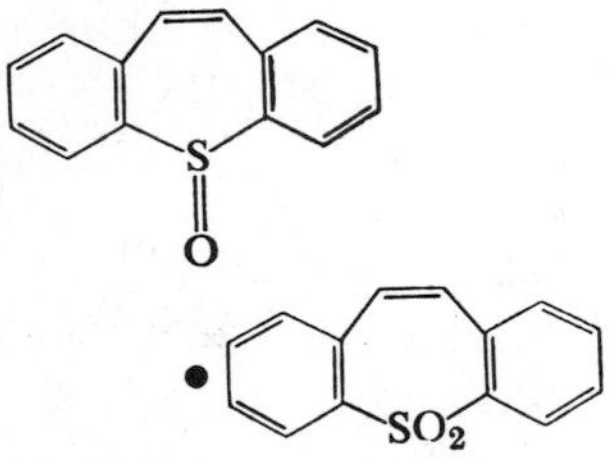 [structure] • [structure] (S, SO_2)	127–128 (1:1 complex)	ir, mass spectrum, uv	116
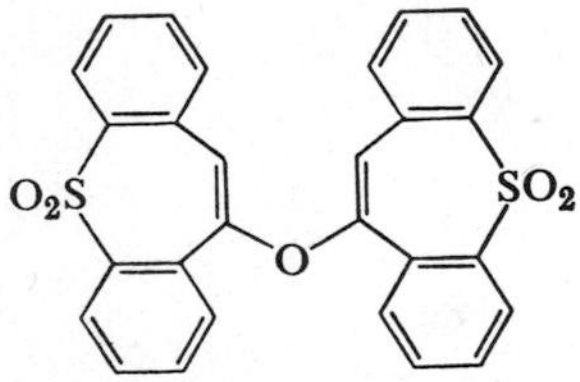 [structure] (O_2S, O, SO_2)	314–316	ir, nmr	293
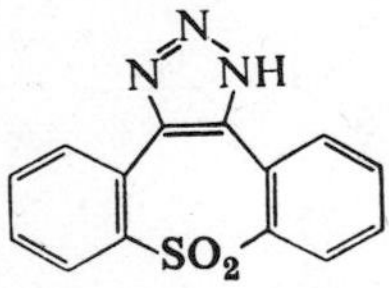 [structure] (N, N, NH, SO_2)	273–277		293
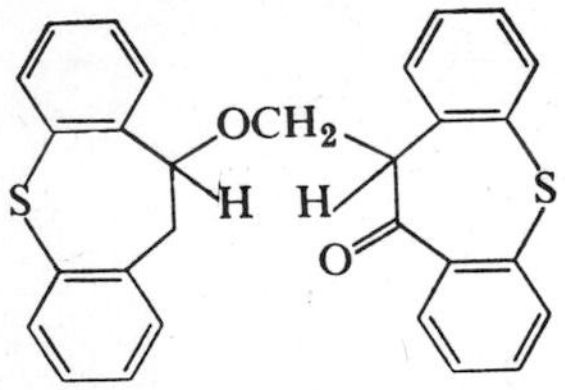 [structure] (OCH_2, H, H, S, O) or [structure] (H, CH_2, H, S, O, O) 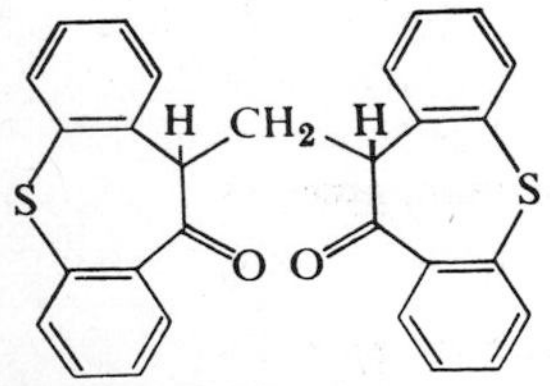	207–208	ir	115

ABLE XI-8.—(*contd.*)

ıbstituents	mp, °C or (bp, °C/torr)	Other data	Refs.
ompound			
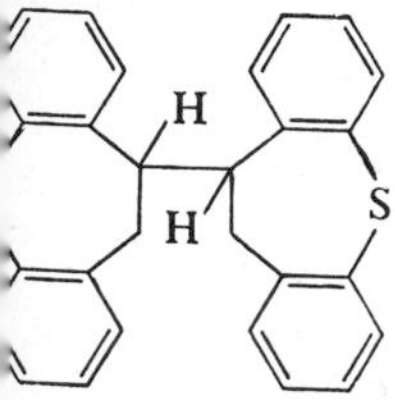	(155–160/0.9), 50	ir	115
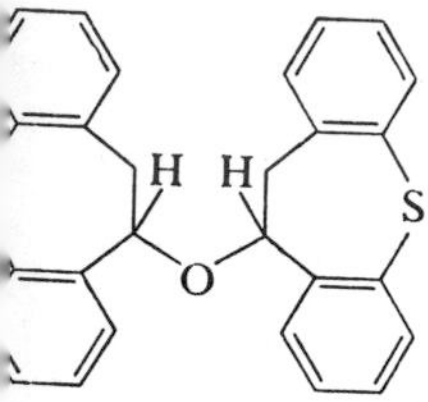	175–177	ir, mass spectrum	262
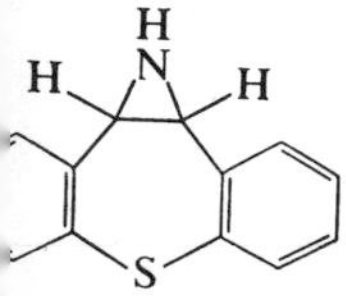	170–172	ir, mass spectrum, nmr	84

BLE XI-9. Dibenzo[*c*,*e*]thiepins (R.I. 8638)

stituents	mp, °C or (bp, °C/torr)	Other data	Refs.
	5,7-Dihydrodibenzo[c,e]*thiepins*	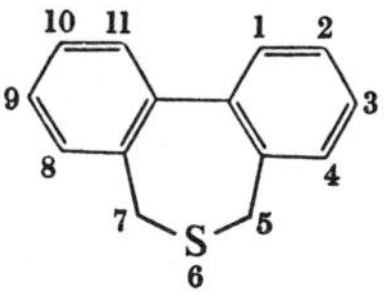 10 11 1 2 9 3 8 4 7 5 S 6	
ıe	89–90	uv	9, 303
		uv, mass spectrum, nmr	136, 167, 286, 310
ıloromercuri deriv.	182.5–183		303
ıethiodide	139.5–140.5	uv	303
-Me_2 (±)	98.5–99		185
	102–103	nmr, uv	167
		mass spectrum	310
(+)	85–85.5	$[\alpha]_D^{26}$ +142° (c 1.3, benzene)	167
ethiodide	189–191	nmr	88
oxide	137–138	nmr	86, 88

TABLE XI-9.—(*contd.*)

Substituents		mp, °C or (bp, °C/torr)	Other data	Refs.
1,11-$(BrCH_2)_2$	(±)	150–151, partial resolidification and remelting 210	ir	167
	(−)	Oil	$[\alpha]_D^{22}$ −258 ± 5° (c 1.3, benzene)	167
1,11-$(HOCH_2)_2$	(±)	154–155	uv	167
	(+)	123–124	$[\alpha]_D^{23}$ +57° (c 1.5, $CHCl_3$), ir, ORD	167
1,11-$(CO_2Me)_2$	(±)	172–173	uv	167
	(−)	95–96	$[\alpha]_D^{20}$ −251° (c 1.6, benzene), ORD, uv	107
3,9-Br_2		191–192	uv	303
3,9-Me_2		128.5–129.5		185
3,9-Ph_2		273.5–274.5	fluorescence, uv	288
1,3,9,11-Me_4			mass spectrum	310

5,7-Dihydrodibenzo[c,e]*thiepin-6,6-dioxides*

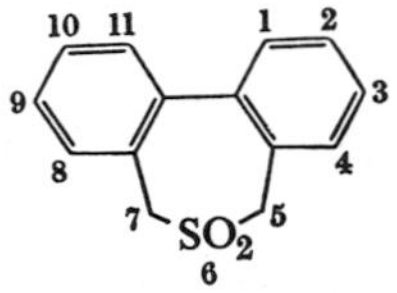

Substituents	mp, °C or (bp, °C/torr)	Other data	Refs.
None	209–210	uv	303
		nmr	286
3-NO_2	230.5	uv	303
3-NH_2		uv	303
5-Cl	204.5–205.5		185
1,11-Me_2	221–222	nmr	87, 88
3,9-Br_2	Did not melt	uv	303
3,9-$(NO_2)_2$	287–288	uv	303
3,9-$(NH_2)_2$	310 (dec)		303
3,9-$(CN)_2$			303
3,9-$(CO_2H)_2$	Did not appear to melt		303
diammonium salt (+)		$[\alpha]_D^{25}$ +3.1° (4 dm, c 1.828 g, 0.5*N* NH_3 in 5% aqueous urea)	303
3,9-$(CO_2Et)_2$	200–201	uv	303
5,5-Cl_2	198.5–199	uv	186
5,7-Cl_2	234–235	uv	186
1,11-Me_2-5-Cl	192–195		185
3,9-Me_2-5-Cl	189–190		185
1,11-Me_2-5,5-Cl_2	208–210		186
1,11-Me_2-5,7-Cl_2	173–175		186

TABLE XI-9.—(*contd.*)

Substituents		mp, °C or (bp, °C/torr)	Other data	Refs.
		Miscellaneous Dibenzo[c,e]*thiepins*		
Compound				
[structure: O-bridged/S-bridged dibenzothiepin]	(±)	161–162	nmr, uv	167
	(+)	161–162	$[\alpha]_D^{27}$ +333°, $[\alpha]_{435}^{26}$ +1060° (c 0.2, *o*-xylene)	167
[structure: S-bridged/S-bridged dibenzothiepin]	(±)	266–267	nmr, uv	167
	(+)	213–215	$[\alpha]_D^{24}$ +440° (c 1.2, benzene) $[\alpha]_{435}^{28}$ +1250° (c 0.17, *o*-xylene), ir, ORD	167

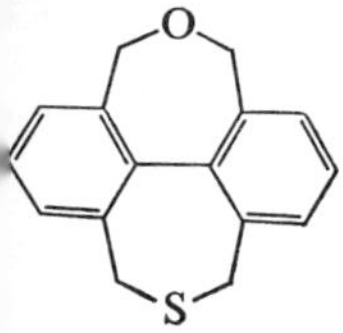

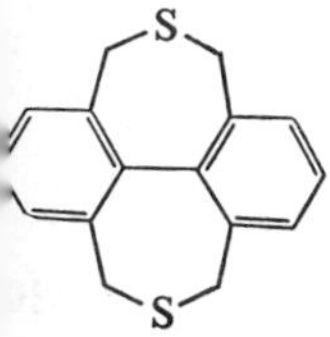

TABLE XI-10. Benzonaphthothiepins

Substituents	mp, °C or (bp, °C/torr)	Other data	Refs.
Benzo[e]*naphtho*[*1*,*2*-b]*thiepin*-$\Delta^{7(12H)}$-		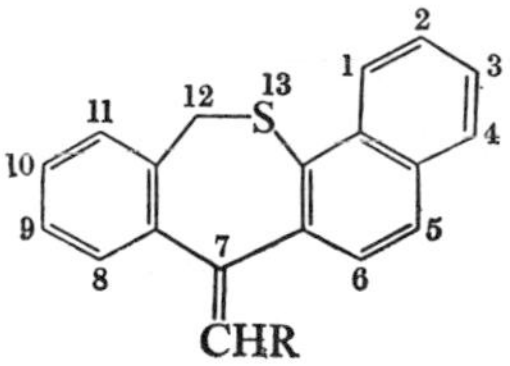 [structure: 7-CHR; positions 1–13]	
= $Me_2NCH_2CH_2$-			224
= $Me_2NCH_2CH_2$-, 9-Cl			224
Benzo[e]*naphtho*[*1*,*2*-b]*thiepin*-*7*(*12*H)-*ones*		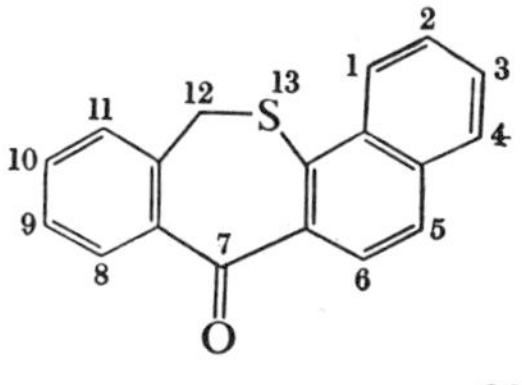 [structure: 7-one; positions 1–13]	
ne			224
l			224
7,*12*-*Dihydrobenzo*[e]*naphtho*[1,2-b]*thiepins*		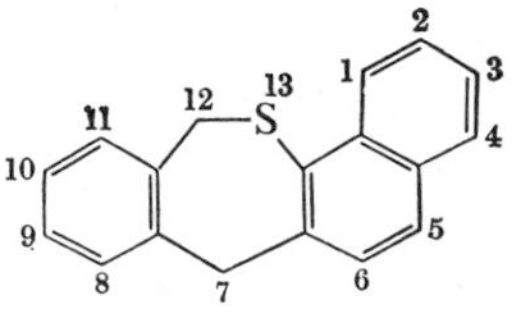 [structure: positions 1–13]	
H-7-$Me_2NCH_2CH_2CH_2$-			224
H-7-$Me_2NCH_2CH_2CH_2$-9-Cl			224

TABLE XI-10.—(*contd.*)

Substituents	mp, °C or (bp, °C/torr)	Other data	Refs.
	Benzo[f]*naphtho*[*1*,*2*-b]*thiepins*	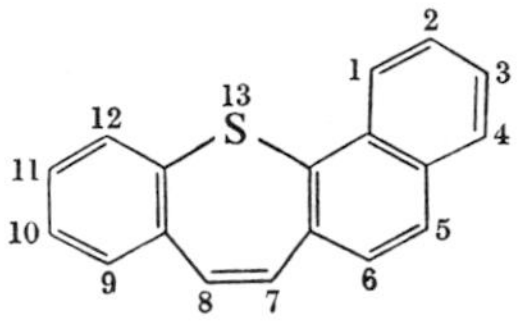	
10-NO_2-7-CO_2H	274 (dec)		151
	Benzo[f]*naphtho*[*1*,*8*-b,c]*thiepin-12-ones*	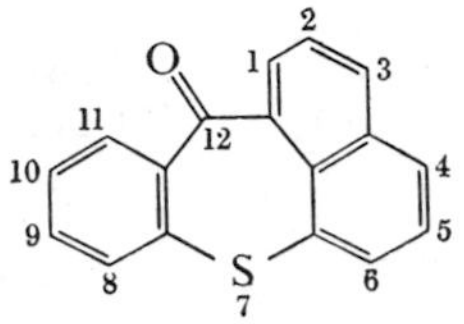	(R.I. 53
1-MeO	184–185		129

TABLE XI-11. Tribenzo[*b*,*d*,*f*]thiepins

Substituents	mp, °C or (bp, °C/torr)	Other data	Refs.
	Tribenzo[b,d,f]*thiepins*	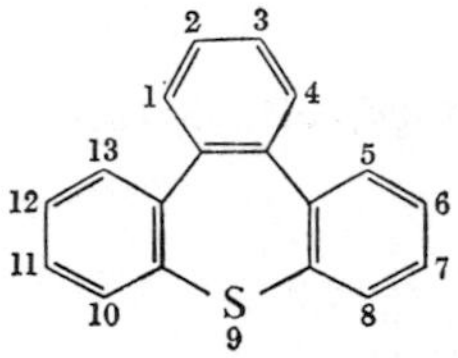	
None	(163–166/0.05), 117–118	mass spectrum, uv	160, 180
	Tribenzo[b,d,f]*thiepin-9,9-dioxides*	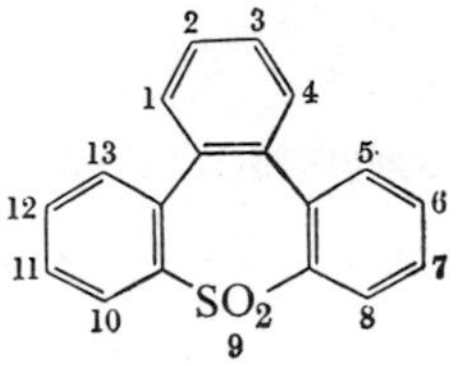	
None	200–201, 192–194	uv	180, 292, 293
2-CO_2H	335–337	ir, mass spectrum, uv, $[\alpha]^{24}_{436}$ −250° (c 0.01 g/cc, diglyme)	291, 292
		$[\alpha]^{24}_{436}$ +200° (c 0.01 g/ml, diglyme), $[\alpha]^{20}_{436}$ +250° (c 1.01 g/cc, diglyme)	292

TABLE XI-11.—(*contd.*)

Substituents	mp, °C or (bp, °C/torr)	Other data	Refs.
2-CO_2Et	197–199		291, 292
	Miscellaneous Tribenzo[b,d,f]*thiepins*		
Compound			
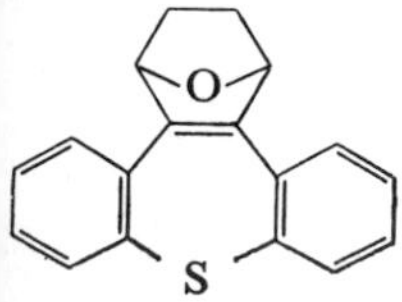	221–223		293
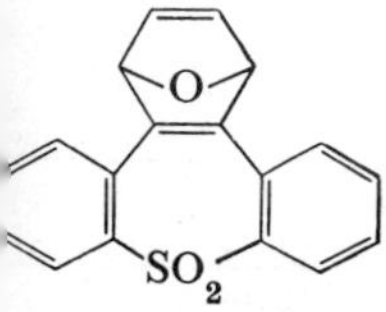	233–235	nmr, uv	200, 293
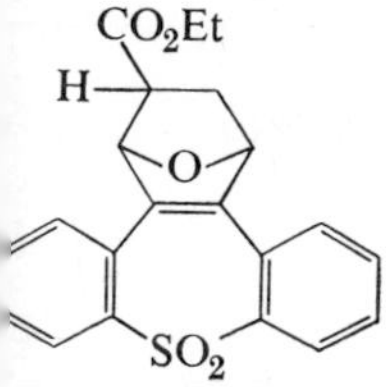	172–173		292
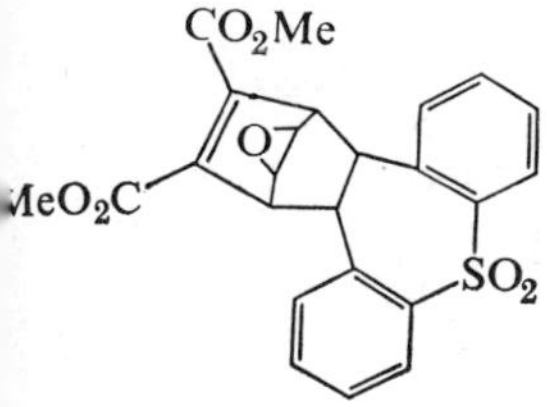	224–226	nmr, uv	200
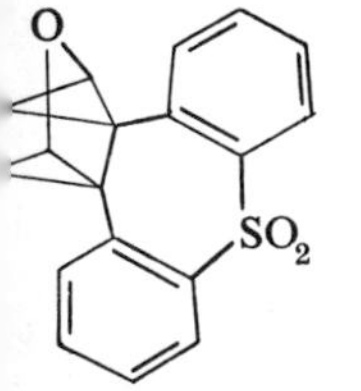	224–226	nmr, uv	200
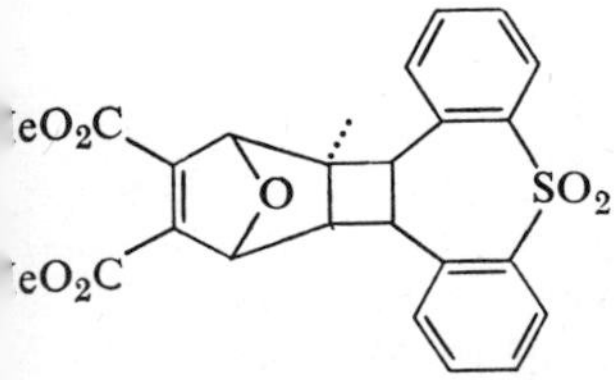	183 (dec)	nmr, uv	200

TABLE XI-11.—(*contd.*)

Substituents	mp, °C or (bp, °C/torr)	Other data	Refs.
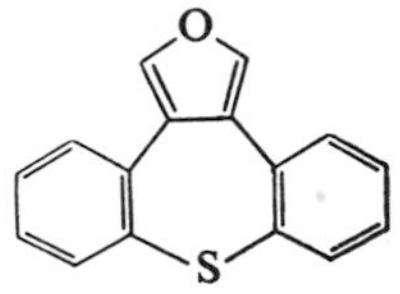 [structure: furan-fused dibenzothiepin]	203–204	nmr	291, 292
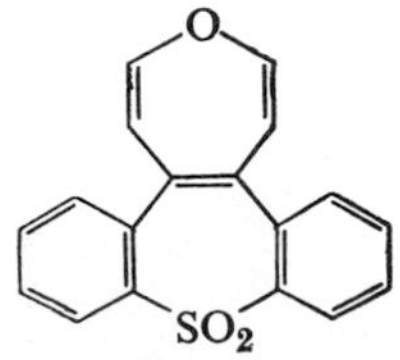 [structure: oxepin-fused dibenzothiepin, SO_2]	190	uv, nmr	200

TABLE XI-12. Thienothiepins

Substituents	mp, °C or (bp, °C/torr)	Other data	Refs.
*6,7-Dihydro-4*H*,8*H*-thieno[3,2-*c*]thiepin-8-ones* 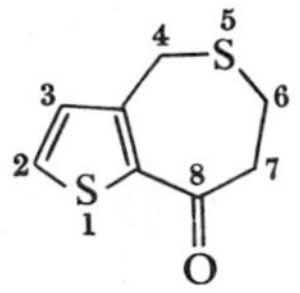			
None	(175–185/13)	ir, nmr, uv	47, 52
2,4-DNP	200		52
6-Me		ir, nmr, uv	47
*6,7-Dihydro-4*H*,8*H*-thieno[3,2-*c*]thiepins* 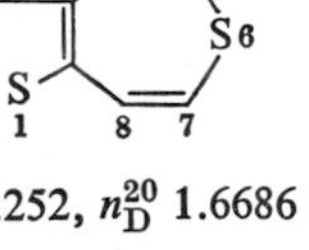			
6-Me		nmr, uv	48
*4,5-Dihydrothieno[2,3-*d*]thiepins* 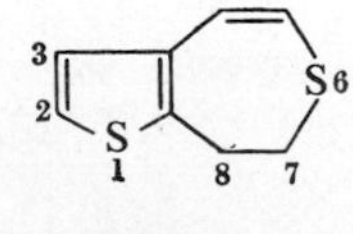			
None	(117.5/1)	d_4^{20} 1.252, n_D^{20} 1.6686	45
7-Me	(115/0.5), 61.5		45
*7,8-Dihydrothieno[2,3-*d*]thiepins*			
None	(105/0.2)	n_D^{20} 1.6629	45
5-Me	(117/0.7)	d_4^{20} 1.211, n_D^{20} 1.6433	45

'ABLE XI-12.—(*contd.*)

ubstituents	mp, °C or (bp, °C/torr)	Other data	Refs.
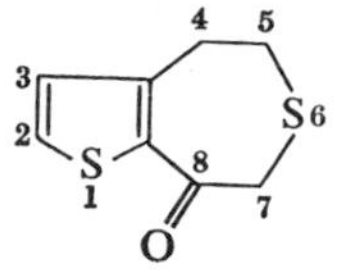	*4,5-Dihydrothieno*[*2,3*-d]*thiepin-8*(*7*H)-*ones*		
one	(203/17), 89	ir, nmr, uv	45, 47
oxime	165		45
2,4-DNP	238		45
Me	(205/15), 127	ir, nmr, uv	45, 47
oxime	131		45
2,4-DNP	110		45
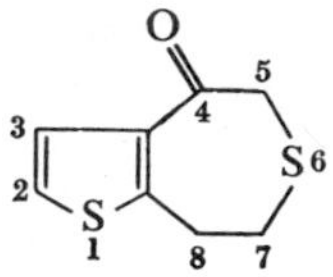	*7,8-Dihydrothieno*[*2,3*-d]*thiepin-4*(*5*H)-*ones*		
ɔne	(195/15), 78	ir, nmr, uv	45, 47
ɔxime	132		45
?,4-DNP	246		45
ʌle	(200/15), 95.5	ir, nmr, uv	45, 47
?,4-DNP	188		45
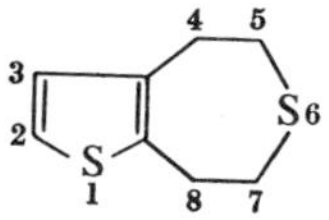	*4,5,7,8-Tetrahydrothieno*[*2,3*-d]*thiepins*		
ne	(142/15)	d_4^{20} 1.167, n_D^{20} 1.6086	45
		nmr, uv	48
·H	111		45
·henylurethane	125		45
·H	121		45
henylurethane	160		45
·H-5-Me	Pale yellow oil	n_D^{20} 1.6036	45
H-7-Me	Pale yellow oil		45
henylurethane	155.5		45
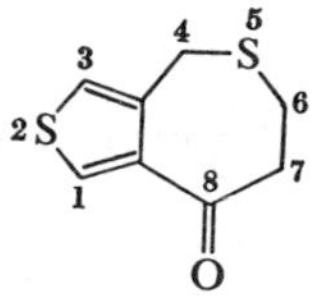	*6,7-Dihydro-4*H,*8*H-*thieno*[*3,4*-c]*thiepin-8-ones*		
5-Me_3		ir, nmr, uv	47

TABLE XI-12.—(*contd.*)

Substituents	mp, °C or (bp, °C/torr)	Other data	Refs.
	Thieno[*3,4*-d]*thiepins* 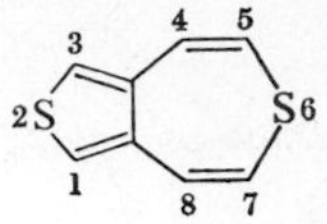		
None	149–151	mass spectrum, nmr, uv	256
		X-ray crystal data	247
6-oxide	75–95 (dec)	ir, mass spectrum, uv	257
6,6-dioxide	183	mass spectrum, nmr, uv	257
		X-ray crystal data	247
	4,5-Dihydrothieno[*3,4*-d]*thiepins* 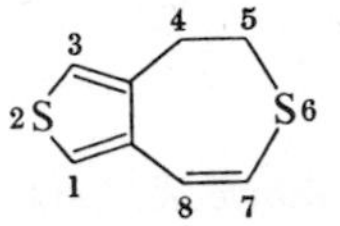		
None	54–56	ir, mass spectrum, nmr, uv	79
6-oxide	78		256, 257
6,6-dioxide			257
	4,5,7,8-Tetrahydrothieno[*3,4*-d]*thiepins* 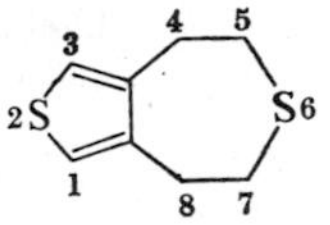		
None	50–52	ir, nmr, uv	79, 257
6,6-dioxide			257
	Benzo[e]*thieno*[*2,3*-b]*thiepin*-$\Delta^{4(9H)}$- 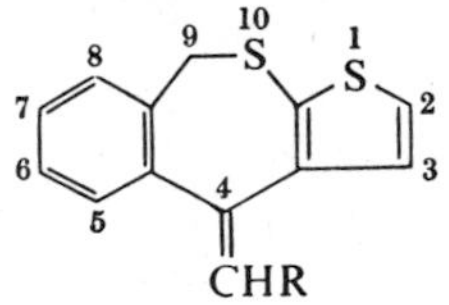		
R = $Me_2NCH_2CH_2$- hydrochloride	243–244		223, 224, 274, 2
	Benzo[e]*thieno*[*2,3*-b]*thiepin-4*(*9*H)*-ones* 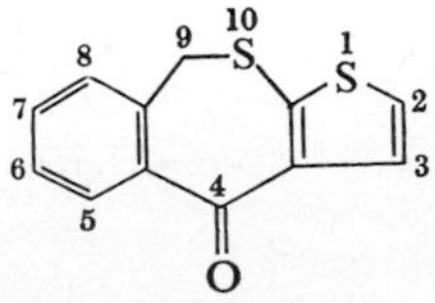		
None	(178/0.25), 53–55		223, 224, 274,

TABLE XI-12.—(contd.)

ubstituents	mp, °C or (bp, °C/torr)	Other data	Refs.
4,9-Dihydrobenzo[e]*thieno*[*2,3*-b]*thiepins*		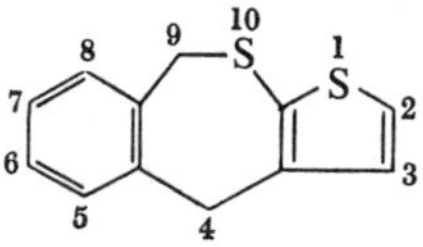	
-OH-4-$Me_2NCH_2CH_2CH_2$-	118–119		223, 224, 274, 275
Benzo[f]*thieno*[*2,3*-b]*thiepins*		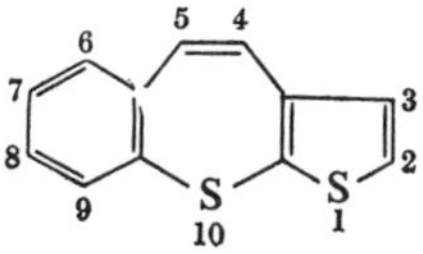	
None	89–90		236
-Cl	70–71	ir, uv	235
Benzo[f]*thieno*[*2,3*-b]*thiepin-4*(*5*H)*-ones*		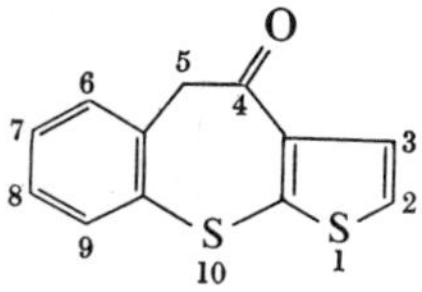	
None	122–124		236, 279
-Cl	121–123	ir, uv	235
4,5-Dihydrobenzo[f]*thieno*[*2,3*-b]*thiepins*		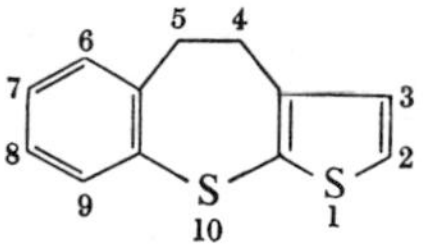	
-OH	101–103		236, 279
-$Me_2NCH_2CH_2O$-	118–119		236
-Cl	106–108		236, 279
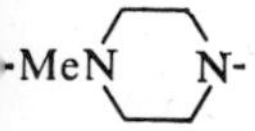 -MeN(piperazine)N-	141–142		236, 279
maleate	164–166		236, 279
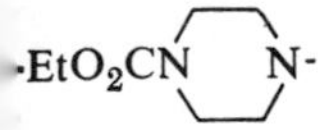 -EtO_2CN(piperazine)N-			279
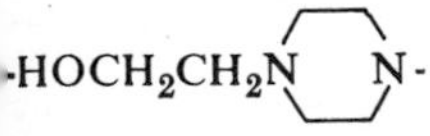 -$HOCH_2CH_2N$(piperazine)N-			279
-Cl-4-OH	128–130	ir, uv	235
,4-Cl_2			235
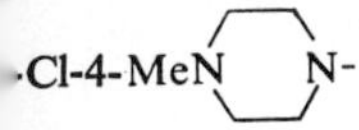 -Cl-4-MeN(piperazine)N-	144–146	nmr, uv	235
maleate	181–183		235

TABLE XI-13. Furothiepins

Substituents	mp, °C or (bp, °C/torr)	Other data	Refs.
Furo[3,4-d]*thiepins*	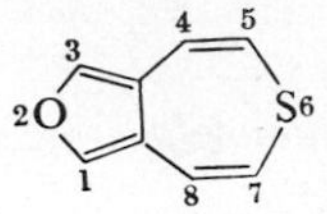		
None	116	nmr, uv	258
6-Oxide		nmr, uv	258
6,6-Dioxide		nmr, uv	258
1,3-Me_2-5,7-$(CO_2H)_2$			103
1,3-Me_2-5,7-$(CO_2Me)_2$	188	mass spectrum, nmr, uv	103
1,3-Me_2-5,7-$(Me_2C(OH)\text{-})_2$		nmr	103
4,5-Dihydrofuro[3,4-d]*thiepins*	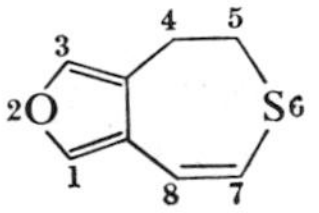		
None		nmr	258
6-Oxide		nmr	258
6,6-Dioxide		nmr	258
4,5,7,8-Tetrahydrofuro[3,4-d]*thiepins*	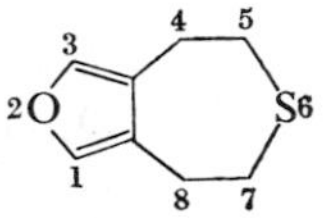		
None			258
6-Oxide			258
Miscellaneous Furothiepins			
Compound			
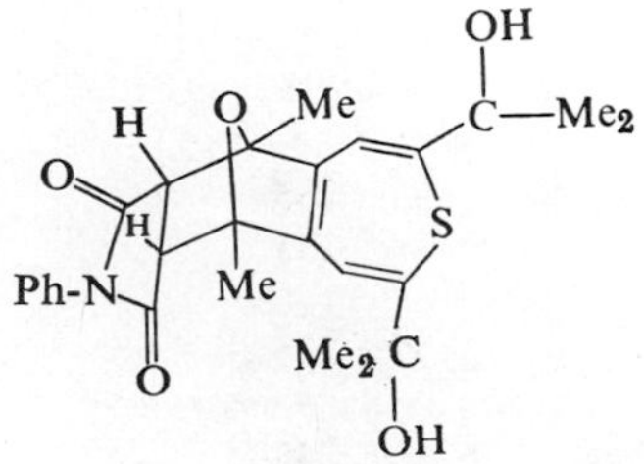 [structure]	180 (dec)	mass spectrum, nmr, uv	103

ABLE XI-14. Cyclopentadithiepins

ıbstituents		mp, °C or (bp, °C/torr)	Other data	Refs.
		*Hexahydro-2*H,*6*H-*cyclopenta*[b]-*1,4-dithiepins*	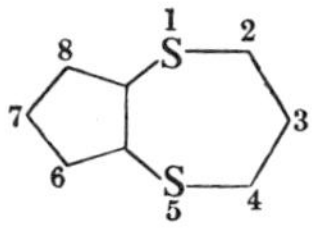	
one	*cis*	(91/25)	ir, mass spectrum, nmr uv	322
	trans		ir, mass spectrum, nmr	322

ABLE XI-15. Benzodithiepins

bstituents		mp, °C or (bp, °C/torr)	Other data	Refs.
		*2,3-Dihydro-5*H-*1,4-benzodithiepins*	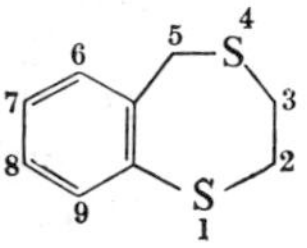	(R.I. 8147)
Cl		(180–182/12), 91–92		135
		2H-*1,5-Benzodithiepin*-$\Delta^{3(4H)}$-	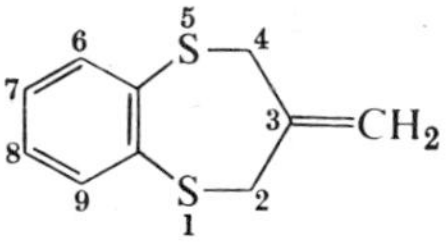	
-Me_2				244
		*Octahydro-2*H-*1,5-benzodithiepins*	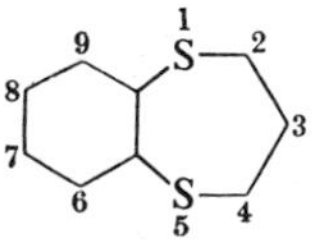	
ne	*cis*	44–45	ir, mass spectrum, nmr, uv	322
	trans	57–58	ir, mass spectrum, nmr, uv	322
		*4,5-Dihydro-1*H-*2,3-benzodithiepins*	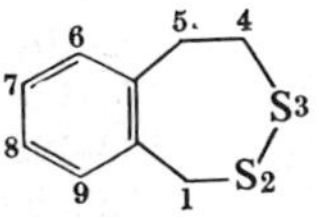	
ıe				125

TABLE XI-15.—(*contd.*)

Substituents	mp, °C or (bp, °C/torr)	Other data	Refs.
*1,5-Dihydro-3*H*-2,4-benzodithiepins* 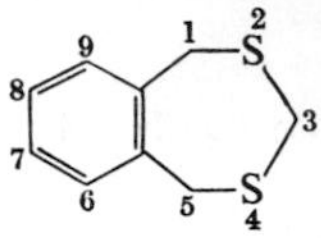 (ring positions 1–9; S_2, S_4)			(R.I. 183
None	152–153	nmr	9, 90, 130, 131
2,2,4,4-Tetraoxide	>300		7
3,3-Br_2-2,2,4,4-tetraoxide	250		7
3,3-Me_2		nmr	89, 90
3-Me-3-Ph	126		7
3-Me-3-Ph-2,2,4,4-tetraoxide	202		7
3,7,8-Me_3	161–163		266
3-*n*-Pr-7,8-Me_2	142–143		266
3-*n*-Hexyl-7,8-Me_2	128–130		266
3-Ph-7,8-Me_2	200		266
3-*p*-Tolyl-7,8-Me_2	221–223		266
3-(*p*-BrC_6H_4)-7,8-Me_2	234–235		266
7,8-Me_2-3-CO_2H	197–199		266
7,8-Me_2-3-CO_2Et	71–72		266
7,8-Me_2-3-$CO_2C_4H_9$-*n*	60		266
7,8-Me_2-3-(2-COOH-phenyl) 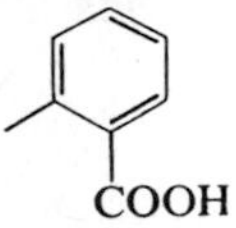	260–270		266
7,8-Me_2-3-$CH_2\overset{O}{\overset{\|}{C}}CH_3$	180–181		266
7,8-Me_2-3-(2-oxocyclohexyl) 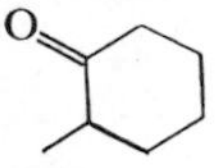	140–141		266
7,8-Me_2-3-$Ph\overset{O}{\overset{\|}{C}}$-	140–145		266
3,3,7,8-Me_4	129–131		266
3,7,8-Me_3-3-Et	110–112		266
3,7,8-Me_3-3-Ph	124–126		266
3,7,8-Me_3-3-CO_2H	70–75		266
3,7,8-Me_3-3-CO_2Me	85–86		266
3,7,8-Me_3-3-EtO_2CCH_2-	102–104		266
3,7,8-Me_3-3-$HO_2CCH_2CH_2$-	174–176		266
3,7,8-Me_3-3-$Me_2C(OH)CH_2$-	100–102 (stable modification)		266
	78–80 (unstable modification)		266

TABLE XI-15.—(*contd.*)

Substituents	mp, °C or (bp, °C/torr)	Other data	Refs.
3,7,8-Me_3-3-$Ph_2C(OH)CH_2$-	225–226		266
3,7,8-Me_3-3-$Ph_2C(OH)CH_2CH_2$-	157–158		266
7,8-Me_2-3-Ph-3-$PhCH_2$-	148–149		266
7,8-Me_2-3,3-$(PhCH_2)_2$-	141–142		266
7,8-Me_2-3,3-Ph_2-	230		266
7,8-Me_2-3-Ph-3-EtO_2CCH_2-	114–115		266
7,8-Me_2-3,3-$(CO_2H)_2$	85–87		266
7,8-Me_2-3,3-$(HO_2CCH_2CH_2$-$)_2$	215–218		266

*6*H-*Dibenzo*[b,f]-*1,3-dithiepins*

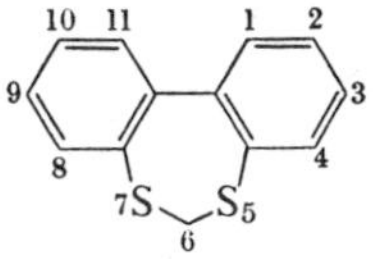

Substituents	mp, °C or (bp, °C/torr)	Other data	Refs.
-Ph	105–106		10
-CO_2Et	(177–1780/.3), 73–74	nmr	43
-*D*-6-CO_2Et			43
-Oxo	101.5		10
,6-Me_2	95		10
-Ph-6-PhCO	198		10

*8*H-*Dinaphtho*[*2,1*-d:*1′,2′*-f]-*1,3-dithiepins*

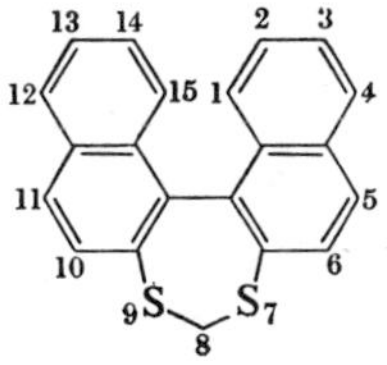

Substituents	mp, °C or (bp, °C/torr)	Other data	Refs.
,8-Me_2		$[\alpha]^{21}_{5461}$ $-368 \pm 1°$	5

*Miscellaneous 3*H-*2,4-Benzodithiepins*

Compound	mp, °C		Refs.
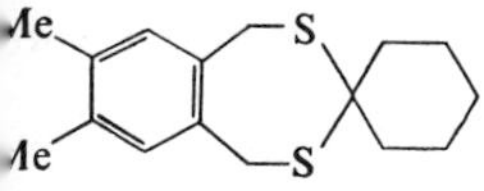	122–123		266
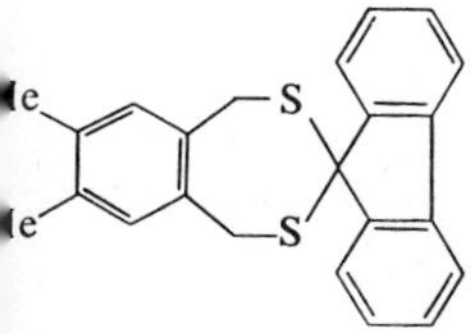	215–216		266

TABLE XI-15.—(*contd.*)

Substituents	mp, °C or (bp, °C/torr)	Other data	Refs.
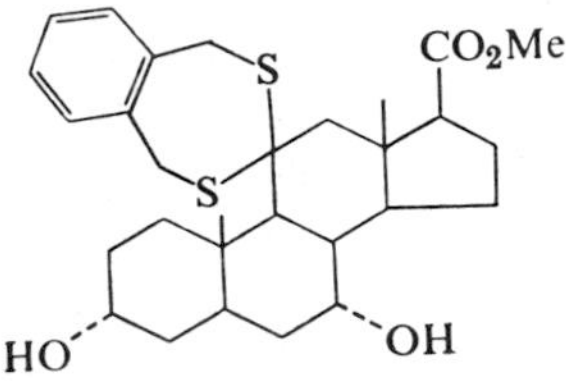 (R.I. 9515)	132–139	$[\alpha]_D^{24}$ −6.3 ± 2°	123
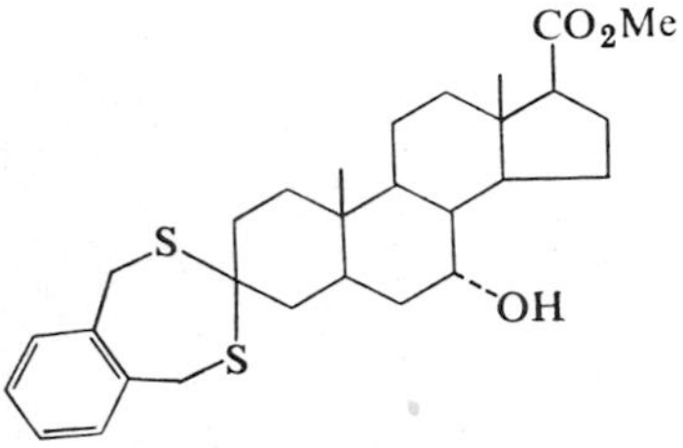 (R.I. 9514)			123
	Miscellaneous Benzodithiepins		
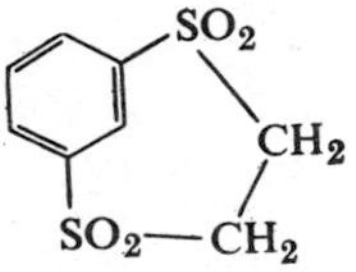 SO_2, CH_2, SO_2—CH_2			184

TABLE XI-16. Benzotrithiepins

Substituents	mp, °C or (bp, °C/torr)	Other data	Refs.
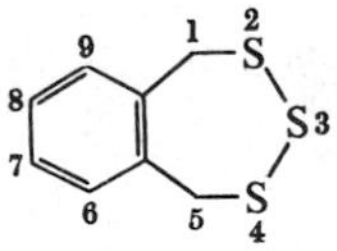	*1,5-Dihydro-2,3,4-benzotrithiepins*		
None	101–102	nmr	89, 124, 125, 16
2-Oxide	133	nmr	165
6,9-Me_2		nmr	125
	Miscellaneous Benzotrithiepins		
Compound			
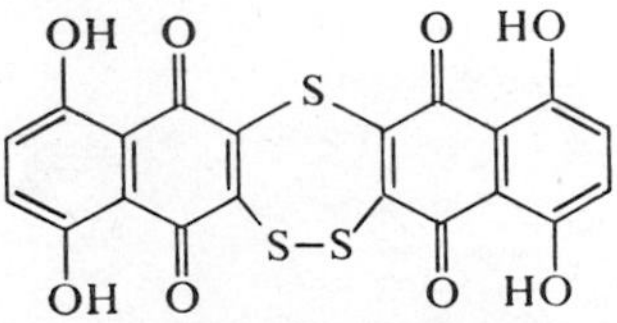			37

TABLE XI-17. Benzopentathiepins

Substituents	mp, °C or (bp, °C/torr)	Other data	Refs.
	1,2,3,4,5-Benzopentathiepins		
None	56.5–70		82

TABLE XI-18. Bridged Systems

Substituents	mp, °C or (bp, °C/torr)	Other data	Refs.
	2-Thiabicyclo[4.1.0]heptanes		
7,7-Cl_2	(71/0.75)	n_D^{25} 1.5604	189
7,7-Cl_2-3-ene	(76–80/0.2)	ir	69
7,7-Cl_2-3-ene-2,2-dioxide	91.5	ir	69
	Miscellaneous Compounds		
Compound			
Cl, Cl, Cl, Cl (S)	171	ir	69
Cl, Cl, Cl, Cl (SO_2)	229	ir	69
	3,4-Dithiabicyclo[4.1.0]heptanes		
Perfluoro	(107.5/739), 91.5		14
	Cyclopropa[b][*1*]*benzothiapyrans*		
1,1-Cl_2	(101/0.2), (94–96/0.07)	n_D^{25} 1.6234, ir, nmr, uv	190

TABLE XI-18.—(*contd.*)

Substituents	mp, °C or (bp, °C/torr)	Other data	Refs.
7a-Cl-7-oxo	61–62	ir, nmr	300
7a-Br-7-oxo	39–42	ir, nmr	300
7a-Cl-7-oxo-2,2-dioxide	176.5–178	ir, nmr	300
7a-Br-7-oxo-2,2-dioxide	187.5–189.5	ir	300
7-OH-7a-Cl		ir, nmr	300
7-OH-7a-Cl-2,2-dioxide	122–123	ir	300
Cyclopropa[c][*1*]*benzothiapyrans* (ring numbering: 1, 1a, 2, S 3, 4, 5, 6, 7, 7a, 7b)			
2-Oxo	80.5–81	ir, nmr, uv	146, 296
1,1-Cl_2-7b-EtO	68	nmr, uv	188
1,1-Cl_2-2-Me-7b-EtO	(120–124/0.025), 44	n_D^{25} 1.5740, nmr, uv	192
Bicyclo[*3.2.0*] *System*			
6-Thiabicyclo[*3.2.0*]*heptanes* (ring numbering: 1, 2, 3, 4, 5, S 6, 7)			
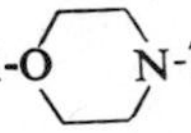 1-(morpholino-N)-7-Ph-6,6-dioxide	134–135	mass spectrum, nmr, uv	101
Miscellaneous Compounds			
Compound			
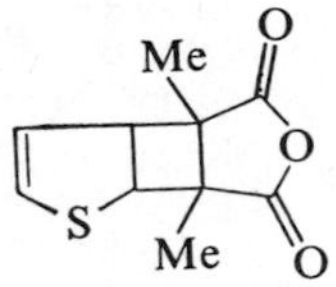 [structure: thieno-fused cyclobutane dicarboxylic anhydride, Me, Me]	158–159	Sublimed 130°/0.2	254
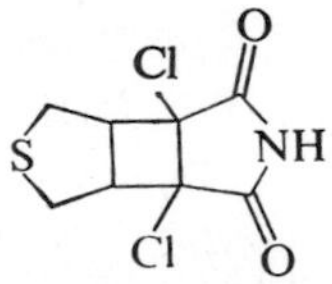 [structure: thiolane-fused cyclobutane dicarboximide (NH), Cl, Cl]	isomer A 245		253
	isomer B 286		253
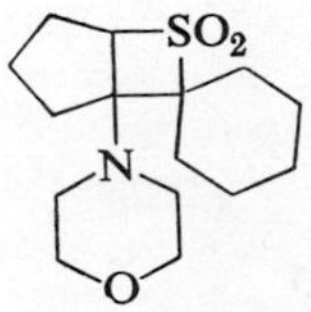 [structure: cyclopentane-fused thietane (SO_2) spiro-cyclohexane, N-morpholino]	117–118		183

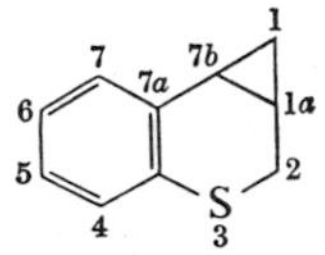

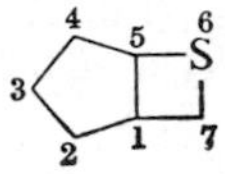

TABLE XI-18.—(*contd.*)

Substituents		mp, °C or (bp, °C/torr)	Other data	Refs.
Bicyclo[3.2.1] System				
2-Thiabicyclo[3.2.1]octanes				
None		(197/744), 165–166		20, 22
chloromercuri deriv.		193–194 (dec)		22
methiodide		210–211 (dec)		22
2,2-Dioxide		257 (dec)		22
3-Thiabicyclo[3.2.1]octanes				
None		174–175		21
chloromercuri deriv.		189.5–190.5		21
3,3-Dioxide		229–230		21
,8,8-Me_3			nmr	285
,8,8-Me_3-3-oxide	(*R*)		nmr, ORD, uv	179
	(*S*)		nmr, ORD, uv	179
6-Thiabicyclo[3.2.1]octanes				
None		(197/769), 172.5–174		20, 22
chloromercuri deriv.		157–157.5 (dec)		22
methiodide		162.5–163.5 (dec)		22
,6-Dioxide		236–237		22
-Ph-6,6-dioxide		90–91		289
-AcO-7-oxo		(74–76/0.005), *ca.* 20	ir	23
,7,7-Me_3 (4-Me-axial) (2,8-*trans-p*-menthylene sulfide)		(84/5), −3.88	n_D^{20} 1.5180, $[\alpha]_D^{25}$ 69.1°, d_4^{20} 1.0016, ir	313, 314
6,6-dioxide	(*dl*)	65.5		314
	(opt act)	106		314

Substituents	mp, °C or (bp, °C/torr)	Other data	Refs.
4,7,7-Me_3 (4-Me-equatorial) (2,8-*cis-p*-menthylene sulfide)	(79–80/5)	n_D^{20} 1.509, $[\alpha]_D^{25}$ −54°, ir	314
4,7,7-Me_3-3-ene	(79.5/5)	n_D^{20} 1.5250, d_4^{20} 1.0057, $[\alpha]_D^{25}$ +79.9°, ir	313, 314
Miscellaneous Compounds			
Compound			
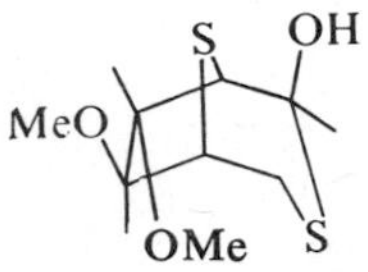 [structure: MeO, OMe, S, OH, S bicyclic compound]	(120–125/0.0001)	n_D^{23} 1.5660 $[\alpha]_D^{24}$ −34° (c 1.9, $CHCl_3$), ir, nmr	60
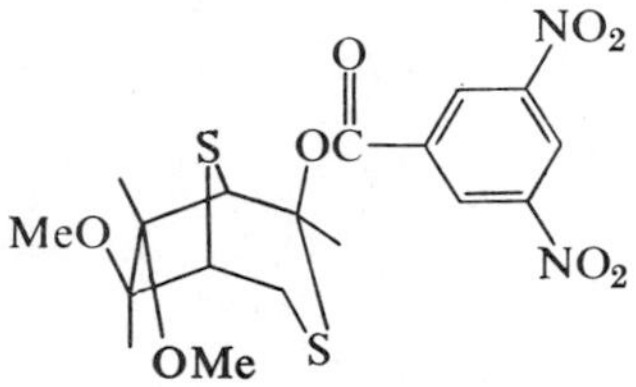 [structure: MeO, OMe, S, S bicyclic compound with 3,5-dinitrobenzoate ester OC(O)–$C_6H_3(NO_2)_2$]	167–168	$[\alpha]_D^{24}$ −42° (c 1.1, $CHCl_3$), ir, nmr	60

Bicyclo[4.2.1] System

9-Thiabicyclo[4.2.1]nonanes [ring-numbering structure: 1–8, 9S]

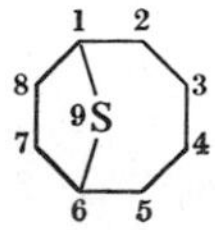

Substituents	mp, °C or (bp, °C/torr)	Other data	Refs.
None	127–128	ir	311
7-OH			93
7-AcO	Oil	ir, mass spectrum, nmr	93
7-Cl	56–57	Sublimed 50°/0.05, ir, mass spectrum, nmr	93
9-Me (Cl^-)		nmr	178
9-Me (I^-)		nmr	178
9,9-Dioxide	235–237	Sublimed 110°/0.5, ir	311
7-AcO-3-oxo	61–62	Sublimed 55/0.02, ir, nmr, uv	93
2-MeS-9-Me (Cl^-)	162–164 (dec)	nmr	178
2-MeS-9-Me (BF_4^-)	103–105 (dec)	nmr	178
2-PhS-9-Me (Cl^-)	151–152 (dec)	nmr	178

TABLE XI-18.—(*contd.*)

Substituents	mp, °C or (bp, °C/torr)	Other data	Refs.
,8-Cl_2	185.5–186.5	ir	311
-Ene-9,9-dioxide			169
-Ene-9,9-dioxide	190–191	ir, nmr	311
,3-Br_2-9,9-dioxide			169
,8-Cl_2-9,9-dioxide	264	ir	311
,3-Br_2-7-ene-9,9-dioxide			311
,5-Br_2-3-ene-9,9-dioxide			311
,4-Diene-9,9-dioxide			311
,7-Diene-9,9-dioxide			311
-Cl-2-MeS-9-Me (Cl^-)	175–177	nmr	178

Bicyclo[5.2.0] System

3,5-Dithiabicyclo[5.2.0]nonane 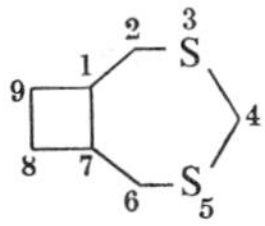 (R.I. 835)

4-Me_2-3,3,5,5-tetraoxide	139–140		159

Miscellaneous Compounds

ompound			
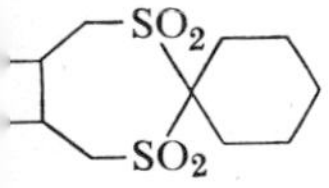 (structure: SO_2, SO_2 spiro compound) (R.I. 2198)	154.5		159

Tricyclo[3.2.1.0] System

2-Thiatricyclo[3.2.1.0^{3,7}]octanes

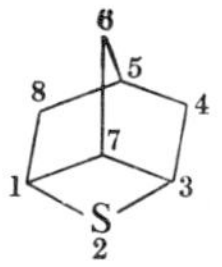

$ClCH_2$-8-Cl	(87/0.09), 44–46.5		140
-Cl_2	(89–91/0.5), 43.5–45	ir	139
-Cl_2-2-oxide (A)	130–131	ir	139
(B)	131–134	ir	139
-Cl_2-2,2-dioxide	139–141.5	ir	139

TABLE XI-18.—(*contd.*)

Substituents	mp, °C or (bp, °C/torr)	Other data	Refs.
	Tricyclo[3.2.1.1] System		
	6-Thiatricyclo[*3.2.1.1*^{3,8}]*nonanes* 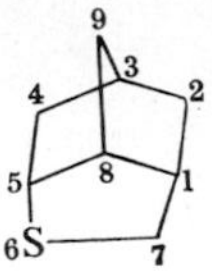		
2-Me (*endo* to C_9)	143–144	nmr	321
methiodide	158		321
2-$BrCH_2$ (*endo* to C_9)-4-Br (*exo* to C_9)	(137–142/0.4)	mass spectrum, nmr	321
picrate	172–173		321
2-$AcOCH_2$ (*endo* to C_9)-4-AcO (*exo* to C_9)		mass spectrum, nmr	321
2-$PhOCH_2$ (*endo* to C_9)-4-PhO (*exo* to C_9)	69.5–71	mass spectrum, nmr	321
2-$PhSCH_2$ (*endo* to C_9)-4-PhS (*exo* to C_9)	80–80.5	mass spectrum, nmr	321
2-(2-pyridyl)—SCH_2 (*endo* to C_9)-4-(2-pyridyl)S (*exo* to C_9) 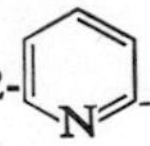 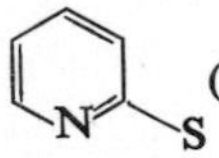	86.5–88	mass spectrum, nmr	321
dipicrate	173.5–175		321
	Miscellaneous Compounds		
Compound			
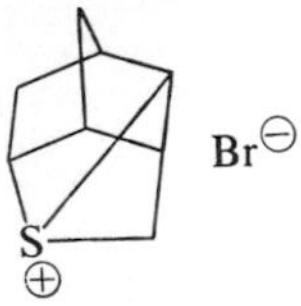 [sulfonium salt] $Br^{\ominus}$	245–247		319
picrate	228–230		319
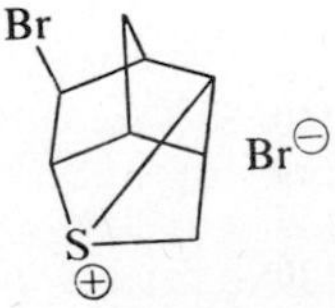 [Br-substituted sulfonium salt] $Br^{\ominus}$	117–118		320

X. References

1. E. Adlerova, I. Ernest, J. Metysova, and M. Protiva, *Collect. Czech. Chem. Commun.*, **33,** 2666 (1968).
2. E. Adlerova, V. Seidlova, and M. Protiva, *Cesk. Farm.*, **12,** 122 (1963); *Chem. Abstr.*, **59,** 8702 (1963).
3. G. Aichinger, S. Scheutz, and F. Hoffmeister, British Patent 1,123,527 (Aug. 14, 1968); *Chem. Abstr.*, **69,** 106577 (1968).
4. L. A. Aksanova, N. F. Kucherova, and V. A. Zagorevskii, *J. Gen. Chem. USSR*, **34,** 3417 (1964).
5. W. L. F. Armarego and E. E. Turner, *J. Chem. Soc.*, **1957,** 13.
6. H. Arnold, E. Kuehas, and N. Brock, German Patent 1,135,915 (Sept. 6, 1962); *Chem. Abstr.*, **58,** 3439 (1963).
7. W. Autenrieth and R. Hennings, *Chem. Ber.*, **34,** 1772 (1901).
8. W. Autenrieth and R. Hennings, *Chem. Ber.*, **35,** 1388 (1902).
9. G. M. Badger, P. Cheuychit, and W. H. F. Sasse, *J. Chem. Soc.*, **1962,** 3241.
10. H. J. Barber and S. Smiles, *J. Chem. Soc.*, **1928,** 1141.
11. O. Benesova, Z. Bohdanecky, and I. Grofova, *Int. J. Neuropharmacol.*, **3,** 479 (1964); *Chem. Abstr.*, **62,** 3287g (1965).
12. O. Benesova, Z. Bohdanecky, Z. Votava, and I. Trinerova, *Arzneim.-Forsch.*, **5,** 100 (1964); *Chem. Abstr.*, **61,** 1139 (1964).
13. O. Benesova and I. Trinerova, *Int. J. Neuropharmacol.*, **3,** 473 (1964); *Chem. Abstr.*, **62,** 3287 (1965).
14. C. J. Benning, U.S. Patent 2,968,659 (Jan. 17, 1961); *Chem. Abstr.*, **55,** 13460 (1961).
15. E. D. Bergmann and M. Rabinovitz, *Israel J. Chem.*, **1,** 125 (1963).
16. E. D. Bergmann and M. Rabinovitz, *J. Org. Chem.*, **25,** 828 (1960).
17. N. S. Bhacca and D. H. Williams, *Tetrahedron Lett.*, **1964,** 3127.
18. F. Bickelhaupt, K. Stach, and M. Thiel, *Chem. Ber.*, **98,** 685 (1965).
19. F. Bickelhaupt K. Stach, and M. Thiel, German Patent 1,204,686 (Nov. 11, 1965); *Chem. Abstr.*, **64,** 3506 (1966).
20. S. F. Birch, T. V. Cullum, and R. A. Dean, *Chem. Data Ser.*, **3,** 359 (1958); *Chem. Abstr.*, **53,** 17486 (1959).
21. S. F. Birch and R. A. Dean, *Justus Liebig Ann. Chem.*, **585,** 234 (1954).
22. S. F. Birch, R. A. Dean, N. J. Hunter, and E. V. Whitehead, *J. Org. Chem.*, **22,** 1590 (1957).
23. B. Bochwie and A. Frankowski, *Tetrahedron*, **24,** 6653 (1968).
24. C. F. Boehringer and G.m.b.H. Soehne, Belgium Patent 623,259 (April 5, 1963); *Chem. Abstr.*, **60,** 10659 (1964).
25. C. F. Boehringer and G.m.b.H. Soehne, Belgium Patent 631,009 (Nov. 4, 1963); *Chem. Abstr.*, **61,** 1843 (1964).
26. C. F. Boehringer and G.m.b.H. Soehne, Belgium Patent 645,877 (Sept. 28, 1964); *Chem. Abstr.*, **63,** 14830 (1965)
27. C. F. Boehringer and G.m.b.H. Soehne, British Patent 996,255 (June 23, 1965); *Chem. Abstr.*, **63**, 8330 (1965).
28. C. F. Boehringer and G.m.b.H. Soehne, British Patent 1,128,938 (Oct. 2, 1963); *Chem. Abstr.*, **70,** 47324 (1969).
29. C. F. Boehringer and G.m.b.H. Soehne, British Patent 1,129,029 (Oct. 2, 1969); *Chem. Abstr.*, **70,** 37664 (1969).

30. C. F. Boehringer and G.m.b.H. Soehne, British Patent 1,129,209 (Oct. 2, 1968); *Chem. Abstr.*, **70,** 19951 (1969).
31. C. F. Boehringer and G.m.b.H. Soehne, British Patent 1,129,210 (Oct. 2, 1968); *Chem. Abstr.*, **70,** 28839 (1969).
32. C. F. Boehringer and G.m.b.H. Soehne, Netherlands Patent Appl. 6,407,758 (Jan. 11, 1965); *Chem. Abstr.*, **62,** 16216 (1965).
33. H. Böhme and B. Haack, *Chem. Ber:*, **101,** 2971 (1968).
34. H. Böhme and B. Haack, *Arch. Pharm.*, **302,** 72 (1969).
35. F. G. Bordwell and B. M. Pitt, *J. Amer. Chem. Soc.*, **77,** 572 (1955).
36. D. M. Borgnaes, Ph.D. Dissertation, Univ. of Notre Dame, August 1968.
37. K. Brass, R. Pfluger, and K. Honsberg, *Chem. Ber.*, **69,** 80 (1936).
38. E. A. Braude, E. R. H. Jones, and G. G. Rose, *J. Chem. Soc.*, **1947,** 1104.
39. J. von Braun, *Chem. Ber.*, **58B,** 2165 (1925).
40. J. von Braun, G. Blessing, and R. S. Cahn, *Chem. Ber.*, **57,** 908 (1924).
41. J. von Braun and K. Weissbach, *Chem. Ber.*, **62B,** 2416 (1929).
42. J. von Braun and K. Wirz, *Chem. Ber.*, **60,** 102 (1927).
43. R. Breslow and E. Mohacsi, *J. Amer. Chem. Soc.*, **85,** 431 (1963).
44. H. C. Brown, K. J. Morgan, and F. J. Chloupek, *J. Amer. Chem. Soc.*, **87,** 2137 (1965).
45. P. Cagniant, *C.R. Acad. Sci., Paris*, **271,** 375 (1970).
46. D. Cagniant and P. Cagniant, *Bull. Soc. Chim. Fr.*, **1966,** 228.
47. D. Cagniant, P. Cagniant, and G. Merle, *Bull. Soc. Chim. Fr.*, **1968,** 3816.
48. D. Cagniant, P. Cagniant, and G. Merle, *Bull. Soc. Chim. Fr.*, **1968,** 3828.
49. P. Cagniant and D. Cagniant, *Bull. Soc. Chim. Fr.*, **1959,** 1998.
50. P. Cagniant and D. Cagniant, *Bull. Soc. Chim. Fr.*, **1966,** 2037.
51. P. Cagniant and D. Cagniant, *Bull. Soc. Chim. Fr.*, **1966,** 3674.
52. P. Cagniant and D. Cagniant, *Bull. Soc. Chim. Fr.*, **1967,** 2597.
53. P. Cagniant and A. Deluzarche, *C.R. Acad. Sci., Paris*, **223,** 677 (1946).
54. P. Cagniant and A. Deluzarche, *C.R. Acad. Sci., Paris*, **223,** 808 (1946).
55. P. Cagniant, G. Jecko, and D. Cagniant, *Bull. Soc. Chim. Fr.*, **1960,** 1798.
56. P. Cagniant, G. Jecko, and D. Cagniant, *Bull. Soc. Chim. Fr.*, **1961,** 1931.
57. P. Cagniant, G. Jecko, and D. Cagniant, *Bull. Soc. Chim. Fr.*, **1961,** 1934.
58. P. Cagniant, G. Jecko, and M. Cagniant, *Bull. Soc., Chim. Fr.*, **1966,** 236.
59. E. J. Corey and E. Block, *J. Org. Chem.*, **31,** 1663 (1966).
60. J. M. Cox and L. N. Owen, *J. Chem. Soc., C*, **1967,** 1121.
61. F. Dallacker, K. W. Glombitza, and M. Lipp, *Justus Liebig Ann. Chem.*, **643,** 82 (1961).
62. M. A. Davis and S. O. Winthrop, U.S. Patent 3,234,235 (Feb. 8, 1966); *Chem. Abstr.*, **64,** 12654 (1966).
63. H. Decker, *Chem. Ber.*, **38,** 2511 (1905).
64. "Definitive Rules for Nomenclature of Organic Chemistry," *J Amer. Chem. Soc.*, **82,** 5566 (1960).
65. M. J. S. Dewar, G. J. Gleicker, and C. C. Thompson, Jr., *J. Amer. Chem. Soc.*, **88,** 1349 (1966).
66. M. J. S. Dewar and N. Trinajstic, *J. Amer. Chem. Soc.*, **92,** 1453 (1970).
67. K. Dimroth and H. Freyschlag, *Chem. Ber.*, **89,** 2602 (1956).
68. K. Dimroth and H. Freyschlag, *Chem. Ber.*, **90,** 1628 (1957).
69. K. Dimroth, W. Kinzebach, and M. Soyka, *Chem. Ber.*, **99,** 2351 (1966).
70. K. Dimroth and G. Lenke, *Angew. Chem.*, **68,** 519 (1956).
71. K. Dimroth and G. Lenke, *Chem. Ber.*, **89,** 2608 (1956).

72. R. M. Dodson and P. D. Hammen, *Chem. Commun.*, **1968,** 1294.
73. P. Dostert and E. Kyburz, *Helv. Chim. Acta*, **53,** 1813 (1970).
74. N. V. Dudykina and V. A. Zagorevskii, *Sintez Prirodn. Soedin., ikh Analogov i Fragmentov, Akad. Nauk SSSR, Otd. Obshch. i Tekhn. Khim.*, **1965,** 134; *Chem. Abstr.*, **65,** 683 (1966).
75. N. V. Dudykina and V. A. Zagorevskii, *Sintez Prirodn. Soedin., ikh Analogov i Fragmentov, Akad. Nauk SSSR, Otd. Obshch. i Tekhn. Khim.*, **1965,** 139; *Chem. Abstr.*, **65,** 8866 (1966).
76. N. V. Dudykina and V. A. Zagorevskii, *Zh. Org. Khim.*, **2,** 2222 (1966); *Chem. Abstr.*, 75878 (1967).
77. N. V. Dudykina and V. A. Zagorevskii, *Khim. Geterotsikl. Soedin.*, **1967,** 250; *Chem. Abstr.*, **68,** 114579 (1968).
78. H. Duerr, *Z. Naturforsch., B*, **22,** 786 (1967).
79. G. Eglinton, J. A. Lardy, R. A. Raphael, and G. A. Sim, *J. Chem. Soc.*, **1964,** 1154.
80. D. D. Emrick and W. E. Truce, *J. Org. Chem.*, **25,** 1103 (1960).
81. E. L. Engelhardt and M. E. Christy, Belgium Patent 641,407 (June 17, 1964); *Chem. Abstr.*, **63,** 5613 (1965).
82. F. Feher and B. Degen, *Angew. Chem. Int. Ed. Engl.*, **6,** 703 (1967).
83. M. J. Y. Foley and N. H. P. Smith, *J. Chem. Soc.*, **1963,** 1899.
84. J. Fouche, *Bull. Soc. Chim. Fr.*, **1970,** 1376.
85. J. C. L. Fouche, R. Gaumont, and C. G. A. Gueremy, French Patent 1,516,783 (March 15, 1968); *Chem. Abstr.*, **70,** 106407 (1969).
86. R. R. Fraser and F. J. Schuber, *Chem. Commun.*, **1969,** 397.
87. R. R. Fraser and F. J. Schuber, *Chem. Commun.*, **1969,** 1474.
88. R. R. Fraser and F. J. Schuber, *Can. J. Chem.*, **48,** 633 (1970).
89. H. Friebolin and S. Kabuss, *Nucl. Magnetic Resonance Chem. Proc. Symp.* **1964,** 125 (1965); *Chem. Abstr.*, **66,** 37265 (1967).
90. H. Friebolin, R. Mecke, S. Kabuss, and A. Luettringhaus, *Tetrahedron Lett.*, **1964,** 1929.
91. F. Gadient, E. Jucker, A. Lindenmann, and M. Taeschler, *Helv. Chim. Acta*, **45,** 1860 (1962).
92. C. Ganter and J. F. Moser, *Helv. Chim. Acta*, **51,** 300 (1968).
93. C. Ganter and J. F. Moser, *Helv. Chim. Acta*, **52,** 725 (1969).
94. J. R. Geigy, A.G., Netherlands Patent Appl. 6,404,862 (Nov. 5, 1964); *Chem. Abstr.*, **62,** 1625 (1965).
95. J. R. Geigy, A.G., Netherlands Patent Appl. 6,407,588 (Jan. 6, 1965); *Chem. Abstr.*, **63,** 1777 (1965).
96. J. R. Geigy, A.G., Netherlands Patent Appl. 6,414,131 (June 8, 1965); *Chem. Abstr.*, **63,** 18130 (1965).
97. J. R. Geigy, A.G., Netherlands Patent Appl. 6,514,187 (May 4, 1966); *Chem. Abstr.*, **65,** 12183 (1966).
98. J. R. Geigy, A.G., Netherlands Patent Appl. 6,515,702 (June 6, 1966); *Chem. Abstr.*, **65,** 13672 (1966).
99. J. R. Geigy, A.G., Netherlands Patent Appl. 6,605,741 (Oct. 31, 1966); *Chem. Abstr.*, **66,** 65507 (1967).
100. V. Georgian, U.S. Patent 2,858,314 (Oct. 28, 1958).
101. A. M. Hamid and S. Trippett, *J. Chem. Soc., C*, **1968,** 1612.
102. V. Hnevsova-Seidlova, M. Rajsner, E. Adlerova, and M. Protiva, *Monatsh. Chem.*, **96,** 650 (1965).
103. J. Hoffman, Jr. and R. Schlessinger, *J. Amer. Chem. Soc.*, **92,** 5263 (1970).

104. H. Hofmann and H. Westernacher, *Angew. Chem. Int. Ed. Engl.*, **5,** 958 (1966).
105. H. Hofmann and H. Westernacher, *Angew. Chem. Int. Ed. Engl.*, **6,** 255 (1967).
106. H. Hofmann and H. Westernacher, *Chem. Ber.*, **102,** 205 (1969).
107. Gh. Iordache, *Petrol. Gaze* (Bucharest), **15,** 30 (1964); *Chem. Abstr.*, **61,** 6834 (1964).
108. R. Ireland and P. Shiess, *J. Org. Chem.*, **28,** 6 (1963).
109. F. Irmis, L. Pickenhaim, and F. Klingberg, *Activitas Nervesa Super.*, **9,** 70 (1967); *Chem. Abstr.*, **67,** 20259 (1967).
110. E. Jancik and B. Kakac, *Cesk. Farm.*, **13,** 3 (1964); *Chem. Abstr.*, **61,** 9358 (1964).
111. J. O. Jilek, J. Metysova, J. Pomykacek, and M. Protiva, *Collect. Czech. Chem. Commun.*, **33,** 1831 (1968).
112. J. O. Jilek, J. Pomykacek, J. Metysova, and M. Protiva, *Collect. Czech. Chem. Commun.*, **35,** 276 (1970).
113. J. O. Jilek and M. Protiva, Czechoslovakian Patent 121,632 (Jan. 15, 1967); *Chem. Abstr.*, **68,** 87212 (1968).
114. J. O. Jilek, M. Protiva, and J. Pomykacek, Czechoslovakian Patent 120,194 (Oct. 15, 1966); *Chem. Abstr.*, **68,** 78159 (1968).
115. J. O. Jilek, V. Seidlova, E. Svatek, M. Protiva, J. Pomykacek, and Z. Sedivy, *Monatsh. Chem.*, **96,** 182 (1965).
116. J. O. Jilek, E. Svatek, J. Metysova, J. Pomykacek, and M. Protiva, *Collect. Czech. Chem. Commun.*, **32,** 3186 (1967).
117. C. R. Johnson and D. McCants, Jr., *J. Amer. Chem. Soc.*, **87,** 1109 (1965).
118. M. J. Jorgenson, *J. Org. Chem.*, **27,** 3224 (1962).
119. E. Jucker, A. J. Lindenmann, and F. Gadient, Belgium Patent 626,301 (June 17, 1963); *Chem. Abstr.*, **60,** 10658 (1964).
120. E. Jucker, A. J. Lindenmann, and F. Gadient, Swiss Patent 384,589 (Feb. 15, 1965); *Chem. Abstr.*, **63,** 17762 (1965).
121. E. Jucker, A. J. Lindenmann, and F. Gadient, Swiss Patent 385,886 (March 31, 1965); *Chem. Abstr.*, **63,** 2962 (1965).
122. E. Jucker, A. J. Lindenmann, and F. Gadient, Swiss Patent 396,940 (Jan. 31, 1966); *Chem. Abstr.*, **64,** 15858 (1966).
123. R. Jungmann, H. P. Sigg, O. Schindler, and T. Reichstein, *Helv. Chim. Acta*, **41,** 1206 (1958).
124. S. Kabuss, A. Lüttringhaus, H. Friebolin, and R. Mecke, *Z. Naturforsch.*, **21,** 320 (1966).
125. S. Kabuss, A. Lüttringhaus, H. Friebolin, H. Schmid, and R. Mecke, *Tetrahedron Lett.*, **1966,** 719.
126. G. Kempter, P. Zaenker, and H. D. Zuerner, *Arch. Pharm.*, **300,** 829 (1967).
127. N. Kharasch, *Organic Sulfur Compounds*, Vol. I, Pergamon Press, New York, 1961, p. 383.
128. S. Kimoto, M. Okamoto, K. Yabe, T. Uchida, and Y. Matsutaka, *Yakugaku Zasshi*, **88,** 1323 (1968); *Chem. Abstr.*, **70,** 47273 (1969).
129. W. Knapp, *Monatsh Chem.*, **71,** 440 (1938).
130. A. Kotz, *Chem. Ber.*, **33,** 729 (1900).
131. A. Kotz and O. Sevin, *J. Prakt. Chem.*, [2] **64,** 526 (1901).
132. A. Krebs and H. Kimling, *Tetrahedron Lett.*, **1970,** 761.
133. G. Kresze and W. Amann, *Spectrochim. Acta, Part A*, **24,** 1283 (1968).
134. G. Kresze and W. Amann, *Spectrochim. Acta, Part A*, **25,** 393 (1969).
135. M. Kulka, *Can. J. Chem.*, **36,** 750 (1958).
136. R. J. Kurland, M. B. Rubin, and W. B. Wise, *J. Chem. Phys.*, **40,** 2426 (1964).

137. F. Kvis and M. Borovicka, Czechoslovakian Patent 126,832 (March 15, 1968); *Chem. Abstr.*, **70,** 57691 (1969).
138. F. Lautenschlaeger, *Can. J. Chem.*, **44,** 2813 (1966).
139. F. Lautenschlaeger, *J. Org. Chem.*, **31,** 1679 (1966).
140. F. Lautenschlaeger, *J. Org. Chem.*, **33,** 2620 (1968).
141. F. Lautenschlaeger, *J. Org. Chem.*, **33,** 2627 (1968).
142. N. J. Leonard and D. Choudhury, *J. Amer. Chem. Soc.*, **79,** 156 (1957).
143. N. J. Leonard and J. Figueras, *J. Amer. Chem. Soc.*, **74,** 917 (1952).
144. N. J. Leonard and C. R. Johnson, *J. Org. Chem.*, **27,** 282 (1962).
145. N. J. Leonard, L. A. Miller, and J. W. Berry, Jr., *J. Amer. Chem. Soc.*, **79,** 1482 (1957).
146. J. R. Livingston, Jr., Ph.D. Dissertation, Univ. of Notre Dame, March 1962.
147. F. J. Lotspeich, *J. Org. Chem.*, **30,** 2068 (1965).
148. F. J. Lotspeich and S. Karickhoff, *J. Org. Chem.*, **31,** 2183 (1966).
149. J. D. Loudon, "The Extrusion of Sulfur," in *Organic Sulfur Compounds*, Vol. 1, N. Kharasch, Ed., Pergamon Press, New York, 1961, p. 299.
150. J. D. Loudon and A. B. D. Sloan, *J. Chem. Soc.*, **1962,** 3262.
151. J. D. Loudon, A. B. D. Sloan, and L. A. Summers, *J. Chem. Soc.*, **1957,** 3814.
152. R. F. Love, Ph.D. Dissertation, Univ. of Notre Dame, May 1960.
153. A. Lüttringhaus and G. Creutzburg, *Angew. Chem. Intern. Ed. Engl.*, **7,** 128 (1968).
154. W. C. Lumma, Jr., G. A. Dutra, and C. A. Voeker, *J. Org. Chem.*, **35,** 3442 (1970).
155. T. A. Magee, U.S. Patent 3,396,225 (Aug. 6, 1968); *Chem. Abstr.*, **69,** 66481 (1968).
156. R. H. F. Manske and A. E. Ledingham, *J. Amer. Chem. Soc.*, **72,** 4797 (1950).
157. F. Mathys, V. Prelog, and R. B. Woodward, *Helv. Chim. Acta.*, **39,** 1095 (1956).
158. R. Mayer and I. Liebster, *Angew. Chem.*, **70,** 105 (1958).
159. Ya. G. Mazover, *Zh. Obshch. Khim.*, **19,** 849 (1949); *Chem. Abstr.*, **44,** 3436 (1950).
160. E. B. McCall and T. J. Rawlings, British Patent 1,016,373 (Jan. 2, 1966); *Chem. Abstr.*, **64,** 9697 (1966).
161. J. Metysova and J. Metys, *Int. J. Neuropharmacol.*, **4,** 111 (1965); *Chem. Abstr.*, **63,** 10536 (1965).
162. J. Metysova, J. Metys, and Z. Votava, *Arzneim.-Forsch.*, **15,** 524 (1965); *Chem. Abstr.*, **67,** 20296 (1967).
163. J. Metysova, J. Metys, and Z. Votava, *Int. J. Neuropharmacol.*, **3,** 361 (1964); *Chem. Abstr.*, **61,** 16662 (1964).
164. J. Metysova-Stramkova, J. Metys, and Z. Votava, *Arzneim.-Forsch.*, **13,** 1039 (1963); *Chem. Abstr.*, **60,** 11251 (1964).
165. B. Milligan and J. M. Swan, *J. Chem. Soc.*, **1965,** 2901.
166. K. Mislow and M. A. W. Glass, *J. Amer. Chem. Soc.*, **83,** 2780 (1961).
167. K. Mislow, M. A. W. Glass, H. B. Hopps, E. Simon, and G. H. Wahl, Jr., *J. Amer. Chem. Soc.*, **86,** 1710 (1964).
168. J. Misurec and K. Nahumek. *Activitas Nervosa Super.*, **7,** 231 (1965); *Chem. Abstr.*, **64,** 4139 (1966).
169. W. L. Mock, *J. Amer. Chem. Soc.*, **92,** 3807 (1970).
170. R. J. Mohrbacher, U.S. Patent 3,287,367 (Nov. 22, 1966); *Chem. Abstr.*, **67,** 11439 (1967).
171. R. J. Mohrbacher, U.S. Patent 3,287,368 (Nov. 22, 1966); *Chem. Abstr.*, **66,** 37788 (1967).
172. R. J. Mohrbacher, U.S. Patent 3,287,369 (Nov. 22, 1966); *Chem. Abstr.*, **66,** 28754 (1967).

173. R. J. Mohrbacher and E. L. Carson, U.S. Patent 3,243,439 (March 29, 1966); *Chem. Abstr.*, **64,** 19623 (1966).
174. R. J. Mohrbacher and V. Paragamian, U.S. Patent 3,287,370 (Nov. 22, 1966); *Chem. Abstr.*, **66,** 28691 (1967).
175. R. J. Mohrbacher and V. Paragamian, U.S. Patent 3,389,144 (June 18, 1968); *Chem. Abstr.*, **69**, 67250 (1968).
176. R. B. Morin, D. O. Spry, and R. A. Mueller, *Tetrahedron Lett.*, **1969,** 849.
177. H. Morren, British Patent 1,112,681 (May 8, 1968); *Chem. Abstr.*, **69,** 35979 (1968).
178. W. H. Mueller, *J. Amer. Chem. Soc.*, **91,** 1223 (1969).
179. R. Nagarajan, B. H. Chollar, and R. M. Dodson, *Chem. Commun.*, **1967,** 550.
180. A. J. Neale, T. J. Rawlings, and E. B. McCall, *Tetrahedron*, **21,** 1299 (1965).
181. D. Neckers, J. Dopper, and H. Wynberg, *Tetrahedron Lett.*, **34,** 2913 (1969).
182. M. Nogradi, W. D. Ollis, and I. O. Sutherland, *Chem. Commun.*, **1970,** 158.
183. G. Opitz and K. Rieth, *Tetrahedron Lett.*, **1965,** 3977.
184. R. Otto, *J. Prakt. Chem.*, [2] **36,** 450 (1887).
185. L. A. Paquette, *J. Amer. Chem. Soc.*, **86,** 4085 (1964).
186. L. A. Paquette, *J. Amer. Chem. Soc.*, **86,** 4089 (1964).
187. L. A. Paquette, L. S. Wittenbrook, and K. Schreiber, *J. Org. Chem.*, **33,** 1080 (1968).
188. W. E. Parham and M. D. Bhavsar, *J. Org. Chem.*, **29,** 1575 (1964).
189. W. E. Parham, L. Christensen, S. H. Groen, and R. M. Dodson, *J. Org. Chem.*, **29,** 2211 (1964).
190. W. E. Parham and R. Koncos, *J. Amer. Chem. Soc.*, **83,** 4034 (1961).
191. W. E. Parham and V. J. Traynelis, *J. Amer. Chem. Soc.*, **76,** 4960 (1954).
192. W. E. Parham and D. G. Weetman, *J. Org. Chem.*, **34,** 56 (1969).
193. I. Pechan, *Biochem. Pharmacol.*, **14,** 1651 (1965).
194. K. Pelz, I. Ernest, E. Adlerova, J. Metysova, and M. Protiva, *Collect. Czech. Chem. Commun.*, **33,** 1852 (1968).
195. K. Pelz, I. Ernest, and M. Protiva, Czechoslovakian Patent 123,691 (July 15, 1967); *Chem. Abstr.*, **69,** 43820 (1968).
196. K. Pelz, I. Jirkovsky, E. Adlerova, J. Metysova, and M. Protiva, *Collect. Czech. Chem. Commun.*, **33,** 1895 (1968).
197. K. Pelz, I. Jirkovsky, J. Metysova, and M. Protiva, *Collect. Czech. Chem. Commun.*, **34,** 3936 (1969).
198. K. Pelz and M. Protiva, *Collect. Czech. Chem. Commun.*, **32,** 2840 (1967).
199. B. Pillat, *Helv. Physiol. Pharmacol. Acta*, **25,** 32 (1967); *Chem. Abstr.*, **67,** 42293 (1967).
200. H. Prinzbach, P. Wuersch, P. Vogel, W. Tochtermann, and C. Franke, *Helv. Chim. Acta*, **51,** 911 (1968).
201. M. Protiva, CNS Drugs Symp., 13 (1966); *Chem. Abstr.*, **70,** 113542 (1969).
202. M. Protiva and E. Adlerova, Czechoslovakian Patent 129,201 (Sept. 15, 1968); *Chem. Abstr.*, **72,** 12608 (1970).
203. M. Protiva, E. Adlerova, and J. Metys, Belgium Patent 646,051 (July 31, 1964); *Chem. Abstr.*, **63,** 13226 (1965).
204. M. Protiva, E. Adlerova, and J. Metys, Czechoslovakian Patent 113,624 (Feb. 15, 1965); *Chem. Abstr.*, **64,** 17560 (1966).
205. M. Protiva, E. Adlerova V. Seidlova, and J. Metysova, Czechoslovakian Patent 107,631 (June 15, 1963); *Chem. Abstr.*, **60,** 1718 (1964).
206. M. Protiva, I. Ernest, E. Adlerova, J. Metysova, Czechoslovakian Patent 123,188 (June 15, 1967); *Chem. Abstr.*, **69,** 43819 (1968).

207. M. Protiva I. Ernest, and J. Metysova, Czechoslovakian Patent 129, 216 (Sept. 15, 1968); *Chem. Abstr.*, **72,** 12607 (1970).
208. M. Protiva, V. Hnevsova-Seidlova, and J. Metysova, Czechoslovakian Patent 111,416 (July 15, 1964); *Chem. Abstr.*, **62,** 14642 (1965).
209. M. Protiva and J. Jilek, Czechoslovakian Patent 120,744 (Dec. 15, 1966); *Chem. Abstr.*, **68,** 105245 (1968).
210. M. Protiva and J. Jilek, Czechoslovakian Patent 121,092 (Dec. 15, 1966); *Chem. Abstr.*, **68,** 105244 (1968).
211. M. Protiva and J. Jilek, Czechoslovakian Patent 121,093 (Dec. 15, 1966); *Chem. Abstr.*, **68,** 105238 (1968).
212. M. Protiva and J. Jilek, Czechoslovakian Patent 124,220 (Sept. 15, 1967); *Chem. Abstr.*, **69,** 67428 (1968).
213. M. Protiva, J. Jilek, and J. Metysova, Czechoslovakian Patent 124,218 (Sept. 15, 1967); *Chem. Abstr.*, **69,** 67425 (1968).
214. M. Protiva, J. Jilek, and J. Metysova, Czechoslovakian Patent 124,533 (Sept. 15, 1967); *Chem. Abstr.*, **69,** 67420 (1968).
215. M. Protiva, J. Jilek, J. Metysova, I. Ernest, K. Pelz, E. Adlerova, Czechoslovakian Patent 121,337 (Dec. 15, 1966); *Chem. Abstr.*, **68,** 105247 (1968).
216. M. Protiva, J. O. Jilek, J. Metysova, V. Seidlova, I. Jirkovsky, J. Metys, E. Adlerova, I. Ernst, K. Pelz, and J. Pomykacek, *Farmaco, Ed. Sci.*, **20,** 721 (1965); *Chem. Abstr.*, **64,** 5090 (1966).
217. M. Protiva, J. Jilek, and J. Pomykacek, Czechoslovakian Patent 121,091 (Dec. 15, 1966); *Chem. Abstr.*, **68,** 105242 (1968).
218. M. Protiva, J. Jilek, J. Pomykacek, and J. Metysova, Czechoslovakian Patent 124,219 (Sept. 15, 1967); *Chem. Abstr.*, **69,** 67426 (1968).
219. M. Protiva, J. Jilek, V. Seidlova, and J. Metys, Czechoslovakian Patent 120,933 (Dec. 15, 1966); *Chem. Abstr.*, **68,** 87214 (1968).
220. M. Protiva, J. Jilek, V. Seidlova, and J. Metysova, Czechoslovakian Patent 124,340 (Sept. 15, 1967); *Chem. Abstr.*, **69,** 86838 (1968).
221. M. Protiva, K. Pelz, F. Hradil, Czechoslovakian Patent 128,004 (June 15, 1968); *Chem. Abstr.*, **72,** 3387 (1970).
222. M. Protiva and M. Rajsner, Czechoslovakian Patent 114,291 (April 15, 1965); *Chem. Abstr.*, **64,** 17560 (1966).
223. M. Protiva and M. Rajsner, Czechoslovakian Patent 115,241 (July 15, 1965); *Chem. Abstr.*, **64,** 17608 (1966).
224. M. Protiva, M. Rajsner, E. Adlerova, V. Seidlova, and Z. J. Vejdelek, *Collect. Czech. Chem. Commun.*, **29,** 2161 (1964).
225. M. Protiva, M. Rajsner, and J. Metysova, Czechoslovakian Patent 115,607 (July 5, 1965); *Chem. Abstr.*, **65,** 2236 (1966).
226. M. Protiva, M. Rajsner, V. Seidlova, E. Adlerova, and Z. J. Vejdelek, *Experientia*, **18,** 326 (1962).
227. M. Protiva, M. Rajsner, V. Seidlova, Z. Vejdelek, and J. Metysova, Czechoslovakian Patent 107,632 (June 15, 1963); *Chem. Abstr.*, **60,** 1718 (1964).
228. M. Protiva, M. Rajsner, Z. Votava, and J. Metysova, Czechoslovakian Patent 105,590 (Nov. 15, 1962); *Chem. Abstr*, **59,** 10010 (1963).
229. M. Protiva, M. Rajsner, Z. Votava, J. Metysova, E. Adlerova, V. Seidlova, and Z. Vejdelek, French Patent 1,332,145 (July 12, 1963); *Chem. Abstr.*, **60,** 2916 (1964).
230. M. Protiva and V. Seidlova, Czechoslovakian Patent 120,573 (Nov. 15, 1966); *Chem. Abstr.*, **68,** 21859 (1968).

231. M. Protiva and V. Seidlova, Czechoslovakian Patent 121,977 (Feb. 15, 1967); *Chem. Abstr.*, **68**, 87213 (1968).
232. M. Protiva, V. Seidlova, and J. Jilek, Czechoslovakian Patent 122,951 (May 15, 1967); *Chem. Abstr.*, **69**, 43817 (1968).
233. V. Pujman, *Int. Congr. Chemother. Proc. 5th*, **2**, 793 (1967); *Chem. Abstr.*, **70**, 27477 (1969).
234. M. Rajsner, M. Borovicka, and M. Protiva, Czechoslovakian Patent 127,965 (June 15, 1968); *Chem. Abstr.*, **72**, 34034 (1970).
235. M. Rajsner, J. Metysova, and M. Protiva, *Collect. Czech. Chem. Commun.*, **35**, 378 (1970).
236. M. Rajsner, J. Metysova, and M. Protiva, *Farmaco, Ed. Sci.*, **23**, 140 (1968); *Chem. Abstr.*, **69**, 19046 (1968).
237. M. Rajsner and M. Protiva, *Cesk. Farm.*, **11**, 404 (1962); *Chem. Abstr.*, **59**, 2772 (1963).
238. M. Rajsner and M. Protiva, *Collect. Czech. Chem. Commun.*, **32**, 2013 (1967).
239. M. Rajsner, Z. Sedivy, M. Borovicka, M. Protiva, Czechoslovakian Patent 128,359 (July 15, 1968); *Chem. Abstr.*, **72**, 3402 (1970).
240. M. Rajsner, V. Seidlova, and M. Protiva, *Cesk. Farm.*, **11**, 451 (1962); *Chem. Abstr.*, **59**, 2773 (1963).
241. W. Reppe, *Justus Liebigs Ann. Chem.*, **596**, 194 (1955).
242. D. D. Reynolds and D. L. Field, "Ethylene Sulfides," in *The Chemistry of Heterocyclic Compounds*, Vol. 19, Part I, A. Weissberger, Ed., Wiley-Interscience, New York, 1964, p. 576.
243. Rhone-Poulenc, S.A., Netherlands Patent Appl. 6,611,458 (Feb. 24, 1967); *Chem. Abst.*, **67**, 100151 (1967).
244. H. Richter, K. Schulze, and M. Muehlstädt, *Z. Chem.*, **8**, 220 (1968).
245. H. Roth and G. Dvorak, *Arch. Pharm.*, **296**, 510 (1963).
246. H. Roth, G. Dvorak, and C. Schwenke, *Arch. Pharm.*, **297**, 298 (1964).
247. T. D. Sakore, R. H. Schlessinger, and H. M. Sobell, *J. Amer. Chem. Soc.*, **91**, 3995 (1969).
248. Sandoz Patents Ltd., British Patent 1,001,822 (Aug. 18, 1965); *Chem. Abstr.*, **64**, 718 (1966).
249. Sandoz Patents Ltd., British Patent 1,001,823 (Aug. 18, 1965); *Chem. Abstr.*, **64**, 718 (1965).
250. Sandoz Patents Ltd., British Patent 1,001,824 (Aug. 18, 1965); *Chem. Abstr.*, **64**, 717 (1966).
251. Sandoz Patents Ltd., British Patent 1,001,825 (Aug. 18, 1965); *Chem. Abstr.*, **64**, 718 (1966).
252. Sandoz Patents Ltd., Netherlands Patent Appl. 6,608,716 (Dec. 29, 1966); *Chem. Abstr.*, **68**, 49472 (1968).
253. H. D. Scharf and F. Korte, *Angew. Chem. Int. Ed. Engl.*, **4**, 429 (1965).
254. G. O. Schenk, W. Hartmann, and R. Steinmetz, *Chem. Ber.*, **96**, 498 (1963).
255. W. Schindler and H. Blattner, Swiss Patent 443,344 (Feb. 15, 1968); *Chem. Abstr.*, **69**, 52035 (1968).
256. R. H. Schlessinger and G. S. Ponticello, *J. Amer. Chem, Soc.*, **89**, 7138 (1967).
257. R. H. Schlessinger and G. S. Ponticello, *Tetrahedron Lett.*, **1968**, 3017.
258. R. H. Schlessinger and G. S. Ponticello, *Tetrahedron Lett.*, **1969**, 4361.
259. A. Schonberg and M. B. E. Fayez, *J. Org. Chem.*, **23**, 104 (1958).
260. E. E. Schweizer and W. E. Parham, *J. Amer. Chem. Soc.*, **82**, 4085 (1960).
261. G. P. Scott, *J. Amer. Chem. Soc.*, **75**, 6332 (1953).

262. H. Seidl and K. Biemann, *J. Heterocycl. Chem.*, **4,** 209 (1967).
263. V. Seidlöva and M. Protiva, *Collect. Czech. Chem. Commun.*, **32,** 1747 (1967).
264. V. Seidlova and M. Protiva, Czechoslovakian Patent 122,942 (May 15, 1967); *Chem. Abstr.*, **69,** 43818 (1968).
265. V. Seidlova, M. Protiva, and J. Metysova, Czechoslovakian Patent 118,520 (May 15, 1966); *Chem. Abstr.*, **67,** 82125 (1967).
266. I. Shahak and E. D. Bergmann, *J. Chem. Soc., C,* **1966,** 1005.
267. A. Simek, A. Copek, J. Turinova, and J. Tuma, *Folia Microbiol.*, **13,** 134 (1968); *Chem. Abstr.*, **68,** 94348 (1968).
268. M. Simonetta and S. Winstein, *J. Amer. Chem. Soc.*, **76,** 18 (1954).
269. K. Sindelar and M. Protiva, *Collect. Czech. Chem. Commun.*, **33,** 4315 (1968).
270. K. Sindelar and M. Protiva, *Collect. Czech. Chem. Commun.*, **35,** 3328 (1970).
271. Societe des Usines Chimiques Rhone-Poulenc, French Patent M. 4,888 (April 10, 1967); *Chem. Abstr.*, **69,** 67427 (1968).
272. Spofa Sdruzeni Podniku pro Zdravotnicku Vyrobu, British Patent 1,002,234 (Aug. 25, 1965); *Chem. Abstr.*, **63,** 16318 (1965).
273. Spofa Sdruzeni Podniku pro Zdravotnicku Vyrobu, Netherlands Patent Appl. 6,410,097 (March 1, 1965); *Chem. Abstr.*, **63,** 2976 (1965).
274. SPOFA United Pharmaceutical Works, French Patent 1,413,978 (Oct. 15, 1965); *Chem. Abstr.*, **64,** 5098 (1966).
275. SPOFA United Pharmaceutical Works, Netherlands Patent Appl. 6,410,097 (March 1, 1965); *Chem. Abstr.*, **63,** 2976 (1965).
276. SPOFA United Pharmaceutical Works, Netherlands Patent Appl. 6,512,186 (March 18, 1966); *Chem. Abstr.*, **65,** 7157 (1966).
277. SPOFA United Pharmaceutical Works, Netherlands Patent Appl. 6,517,282 (July 1, 1966); *Chem. Abstr.*, **66,** 2591 (1967).
278. SPOFA United Pharmaceutical Works, Netherlands Patent Appl. 6,608,618 (Dec. 23, 1966); *Chem. Abstr.*, **67,** 43821 (1967).
279. SPOFA United Pharmaceutical Works, Netherlands Patent Appl. 6,611,420 (Feb. 27, 1967); *Chem. Abstr.*, **67,** 100151 (1967).
280. K. Stach and F. Bickelhaupt, *Monatsh. Chem.*, **93,** 896 (1962).
281. K. Stach and H. Springer, *Angew. Chem. Int. Ed. Engl.* **1,** 50 (1962).
282. K. Stach and H. Springer, German Patent 1,279,682 (Oct. 10, 1968); *Chem. Abstr.*, **70,** 37663 (1968).
283. K. Stach and H. Springer, *Monatsh. Chem.*, **93,** 889 (1962).
284. B. Stark and A. Duke, *Extrusion Reactions*, Pergamon Press, New York, 1967, pp. 72–108.
285. E. T. Strom, W. L. Orr, B. S. Snowden, Jr., and D. E. Woessner, *J. Phys. Chem.*, **71,** 4017 (1967).
286. I. O. Sutherland and M. V. J. Ramsay, *Tetrahedron*, **21,** 3401 (1965).
287. H. Suzuki, *Bull. Chem. Soc. Jap.*, **27,** 597 (1954).
288. R. L. Taber, G. H. Daub, F. N. Hayes, and D. G. Ott, *J. Heterocycl. Chem.*, **2,** 181 (1965).
289. E. E. van Tamelen and E. A. Grant, *J. Amer. Chem. Soc.*, **81,** 2160 (1959).
290. W. Theobald, O. Buech, Cl. Morpurgo, and W. Schindler, *Antidepressant Drugs, Proc. Int. Symp., 1st Milan* **1966,** 205 (1967); *Chem. Abstr.*, **67,** 115430 (1967).
291. W. Tochtermann and C. Franke, *Angew. Chem. Int. Ed. Engl.*, **6,** 370 (1967).
292. W. Tochtermann, C. Franke, and D. Schaefer, *Chem. Ber.*, **101,** 3122 (1968).
293. W. Tochtermann, K. Oppenländer, and M. N-D. Hoang, *Justus Liebig Ann. Chem.*, **701,** 117 (1967).

294. V. J. Traynelis and D. M. Borgnaes, unpublished observations.
295. V. J. Traynelis, W. L. Hergenrother, J. R. Livingston, Jr., and J. A. Valicenti, *J. Org. Chem.*, **27**, 2377 (1962).
296. V. J. Traynelis and J. R. Livingston, Jr., *J. Org. Chem.*, **29**, 1092 (1964).
297. V. J. Traynelis and R. F. Love, *Chem. Ind.* (London), **1958**, 439.
298. V. J. Traynelis and R. F. Love, *J. Org. Chem.*, **26**, 2728 (1961).
299. V. J. Traynelis and R. F. Love, *J. Org. Chem.*, **29**, 366 (1964).
300. V. J. Traynelis and J. C. Sih, unpublished observations.
301. A. Trifunac and E. T. Kaiser, *J. Phys. Chem.*, **74**, 2236 (1970).
302. W. E. Truce, G. H. Birum, and E. T. McBee, *J. Amer. Chem. Soc.*, **74**, 3594 (1952).
303. W. E. Truce and D. D. Emrick, *J. Amer. Chem. Soc.*, **78**, 6130 (1956).
304. W. E. Truce, D. D. Emrick, and R. E. Miller, *J. Amer. Chem. Soc.*, **75**, 3359 (1953).
305. W. E. Truce and F. J. Lotspeich, *J. Amer. Chem. Soc.*, **78**, 848 (1956).
306. W. E. Truce and J. P. Miloinis, *J. Amer. Chem. Soc.*, **74**, 974 (1952).
307. W. E. Truce and G. A. Toren, *J. Amer. Chem. Soc.*, **76**, 695 (1954).
308. M. M. Urberg and E. T. Kaiser, *J. Amer. Chem. Soc.*, **89**, 5931 (1967).
309. V. Vitek, B. Mosinger, and V. Kujalova, *Nature*, **205**, 90 (1965).
310. G. H. Wahl, Jr., *Org. Mass. Spectrosc.*, **3**, 1349 (1970).
311. E. D. Weil, K. J. Smith, and R. J. Gruber, *J. Org. Chem.*, **31**, 1669 (1966).
312. M. Weissenfels, H. Schurig, and G. Huehsam, *Z. Chem.*, **6**, 471 (1966).
313. A. W. Weitkamp, *Flavor Res. Food Acceptance*, **1958**, 331 (1958); *Chem. Abstr.*, **53**, 10276 (1959).
314. A. W. Weitkamp, *J. Amer. Chem. Soc.*, **81**, 3430 (1959).
315. A. W. Weitkamp, *J. Amer. Chem. Soc.*, **81**, 3434 (1959).
316. A. W. Weitkamp, *J. Amer. Chem. Soc.*, **81**, 3437 (1959).
317. A. W. Weitkamp, *Perfum. Essent. Oil Rec.*, **49**, 803 (1958); *Chem. Abstr.*, **53**, 12590 (1959).
318. H. W. Whitlcok, *Tetrahedron Lett.*, **1961**, 593.
319. P. Wilder, Jr. and L. A. Feliu-Otero, *J. Org. Chem.*, **30**, 2560 (1965).
320. P. Wilder, Jr. and L. A. Feliu-Otero, *J. Org. Chem.*, **31**, 4264 (1966).
321. P. Wilder, Jr. and R. F. Gratz, *J. Org. Chem.*, **35**, 3295 (1970).
322. J. D. Willett, J. R. Grunwell, and G. A. Berchtold, *J. Org. Chem.*, **33**, 2297 (1968).
323. D. Williams and N. Bhacca, *Tetrahedron*, **21**, 1641 (1965).
324. D. Williams and N. Bhacca, *Tetrahedron*, **21**, 2021 (1965).
325. S. Winstein, M. Shatavsky, C. Norton, and R. B. Woodward, *J. Amer. Chem. Soc.*, **77**, 4183 (1955).
326. W. Winter, E. Fauland, K. Stach, F. H. Schmidt, and W. Aumueller, South African Patent 6,801,636 (Aug. 12, 1968); *Chem. Abstr.*, **70**, 68204 (1969).
327. S. O. Winthrop, M. A. Davis, F. Herr, J. Stewart, and R. Gaudry, *J. Med. Pharm. Chem.*, **5**, 1207 (1962).
328. S. Wolfe, A. Rauk, and I. G. Csizmadia, *J. Amer. Chem. Soc.*, **91**, 1567 (1969).
329. P. Yonan, U.S. Patent 3,401,165 (Sept. 10, 1968); *Chem. Abstr.*, **70**, 4144 (1969).
330. R. Zahradnik and C. Parkanyi, *Collect. Czech. Chem. Commun.*, **30**, 3016 (1965).
331. C. L. Zirkle, U.S. Patent 3,100,207 (Aug. 6, 1963); *Chem. Abstr.*, **60**, 1719 (1964).

Author Index

Numbers in parentheses are reference numbers and show that an author's work is referred to although his name is not mentioned in the text. Numbers in italics indicate the pages on which the full references appear.

Subject Index